Delmar's Standard Textbook of
Electricity

Eighth Edition

Stephen L. Herman
Derek Vigstol

Cengage

Australia • Brazil • Canada • Mexico • Singapore • United Kingdom • United States

Cengage

Delmar's Standard Textbook of Electricity, 8E
Stephen L. Herman, Derek Vigstol

SVP, Product: Cheryl Costantini

VP, Product: Thais Alencar

Senior Product Director: Mark Santee

Associate Portfolio Product Manager: Carolyn Mako

Product Assistant: Jannell Whitted

Learning Designer: Vanessa Leahr

Senior Content Manager: Sharon Chambliss

Senior Digital Project Manager: Elizabeth Cranston

Senior Director, Product Marketing: Neena Bali

Product Marketing Manager: Morgan Perin

Content Acquisition Analyst: Erin McCullough

Production Service: MPS Limited

Designer: Tim Biddick

Cover Image Source: skrotov/Shutterstock.com

Interior image Source: skrotov/Shutterstock.com

Library of Congress Control Number: 2024914195

ISBN: 978-0-357-93385-5

Cengage
5191 Natorp Boulevard
Mason, OH 45040
USA

Cengage is a leading provider of customized learning solutions. Our employees reside in nearly 40 different countries and serve digital learners in 165 countries around the world. Find your local representative at **www.cengage.com**.

To learn more about Cengage platforms and services, register or access your online learning solution, or purchase materials for your course, visit **www.cengage.com**.

Notice to the Reader
Publisher does not warrant or guarantee any of the products described herein or perform any independent analysis in connection with any of the product information contained herein. Publisher does not assume, and expressly disclaims, any obligation to obtain and include information other than that provided to it by the manufacturer. The reader is expressly warned to consider and adopt all safety precautions that might be indicated by the activities described herein and to avoid all potential hazards. By following the instructions contained herein, the reader willingly assumes all risks in connection with such instructions. The publisher makes no representations or warranties of any kind, including but not limited to, the warranties of fitness for particular purpose or merchantability, nor are any such representations implied with respect to the material set forth herein, and the publisher takes no responsibility with respect to such material. The publisher shall not be liable for any special, consequential, or exemplary damages resulting, in whole or part, from the readers' use of, or reliance upon, this material.

Printed in the United States of America
Print Number: 01 Print Year: 2024

Contents

Section 4

Preface

Intended Use

Delmar's Standard Textbook of Electricity, 8th edition, is intended for students in electrical trade programs at high schools and community colleges as well as those in industry training. It assumes that the reader has had no prior knowledge of electricity but also provides enough comprehensive coverage to be used as a reference tool for experienced electricians.

Subject and Approach

The content itself is presented as a blend of the practical and theoretical. It not only explains the different concepts relating to electrical theory but also provides many practical examples of how to do many of the common tasks the industrial electrician must perform. An extensive art program containing full color photographs and line drawings, as well as the inclusion of practical exercises for the student, also serve to further clarify theoretical concepts.

Design of Text

The subject matter has been divided into 35 separate units—each designed to "stand alone." The "stand-alone" concept permits the information to be presented in almost any sequence the instructor desires, as teaching techniques vary from one instructor to another. The information is also presented in this manner to allow students and instructors quick reference on a particular subject.

Math Level

The math level has been kept to basic algebra and trigonometry, and Appendix B contains a section of electrical formulas—all divided into groups that are related to a particular application. Unit 14 of the text provides an introduction to basic trigonometry and vectors for those students who need practice with the subject.

A Note about Calculations

Delmar's Standard Textbook of Electricity, 8th edition, like all other scientific texts, contains numerous mathematical equations and calculations. Students often become concerned if their answers to problems are not exactly the same as the solutions given in the text. The primary reason for a discrepancy is the rounding off of values. Different scientific calculators carry out numbers to different places depending on the manufacturer and model. Some calculators carry numbers to 8 places, some to 10 places, and some to 12 places. There may also be times when numbers that are reentered into the calculator are carried to only 2 or 3 decimal places of accuracy. For example, the numbers shown below will be multiplied with a calculator that carries numbers out to 8 places of accuracy:

$$3.21 \times 34.6 \times 4.32 \times 0.021 \times 3.098 \times 0.467$$

The answer is 14.577480.

The same problem will again be multiplied, but this time, each answer will be reentered before it is multiplied by the next number. Each time the answer is reentered, it will be rounded off to 3 places after the decimal. If the fourth number after the decimal is 5 or greater, the third decimal place will be rounded up. If the fourth number is less than 5, it will be rounded down. The answer is 14.577405.

The same set of numbers will again be multiplied, but this time, each answer will be reentered after rounding off the number to one place after the decimal. The answer is 14.617100.

Notice that all three answers are different, but all three are essentially correct. The most accurate answer is 14.577480, and the least accurate answer is 14.617100. Although these answers may look substantially different, they are within approximately 1% of each other.

Another consideration is problems that contain multiple steps. The more steps it takes to solve a problem, the more chance there is for inaccuracy. In most instances in this text, the answers were left in the display of the calculator, which permits the greatest degree of accuracy. When numbers had to be reentered, they were taken to 3 places of accuracy. When you work a problem in this text and your answer is different, consider the degree of difference before concluding that your answer is incorrect.

New to This Edition

The 8th edition of *Delmar's Standard Textbook of Electricity* continues to remain true to the comprehensive nature and visually appealing style that are its trademark features but will now offer more emphasis on the practical approach to electrical theory. New to this edition:

- Updated learning objectives to better reflect the concepts and skills students are asked to achieve.
- Review questions have been revised, converted to multiple choice, and made available as auto-graded assignments in MindTap.
- Updated graphics.
- Extended coverage of batteries to include information on lithium-ion batteries.
- Extensive revisions have been made to better align content with modern industry practices.
- Several updates were made to improve the electrical safety concepts included in the text.
- Several units were revised to remove outdated and obsolete material.

Features of the Text

- **"Safety Overview"**

  At the beginning of Section I, the Safety Overview provides information on general safety rules, personal protective equipment, potential job hazards, lock-out/tag-out procedures, GFCI, grounding, and more! Students are acquainted with the all-important safety concerns applicable to working in a lab and on the work site.

- **"Cautions"**

  The author highlights text where students should be aware of potential risks in working with various types of electrical equipment.

Caution

The ammeter, unlike the voltmeter, is a very low-impedance device. The ammeter is used to measure current and must be connected in series with the load to permit the load to limit the current flow (Figure 9–11).

- **Math Presentation**

  Key math concepts that will be used in later units to solve complex electrical circuits are presented earlier in the text in Unit 14, *Basic Trigonometry,* providing a foundation for students as they work through math equations.

Example 14–3

Using the same triangle (*Figure 14–7*), determine the number of degrees in angle Y.

Solution

In this example, the lengths of the hypotenuse and the adjacent side are known. The cosine function can be used to find the angle:

$$\text{cosine} \angle Y = \frac{\text{adjacent}}{\text{hypotenuse}}$$

$$\text{cosine} \angle Y = \frac{9}{14}$$

$$\text{cosine} \angle Y = 0.643$$

To find what angle corresponds to the cosine of 0.643, use the trigonometric tables in Appendices A and B or the COS function of a scientific calculator:

$$\text{COS}^{-1} \text{ or ARC COS } 0.643 = 50^\circ$$

14–4 Formulas

Some formulas that can be used to find the angles and lengths of different sides follow:

$$\sin \angle\theta = \frac{O}{H} \qquad \cos \angle\theta = \frac{A}{H} \qquad \tan \angle\theta = \frac{O}{A}$$

$$\text{Adj.} = \cos \angle\theta \times \text{Hyp.} \qquad \text{Adj.} = \frac{O}{\tan \angle\theta}$$

$$\text{Opp} = \sin \angle\theta \times \text{Hyp.} \qquad \text{Opp.} = \text{Adj.} \times \tan \angle\theta$$

$$\text{Hyp.} = \frac{O}{\sin \angle\theta} \qquad \text{Hyp.} = \frac{A}{\cos \angle\theta}$$

14–5 Practical Application

Although the purpose of this unit is to provide preparation for the study of AC circuits, basic trigonometry can provide answers to other problems that may be encountered on the job. Assume that it is necessary to know the height of a tall building (*Figure 14–8*). Now assume that the only tools available to make this measurement are a 1-foot ruler, a tape measure, and a scientific calculator. To make the measurement, find a relatively flat area in the open sunlight. Hold the ruler upright and measure the shadow cast by the sun (*Figure 14–9*). Assume the length of the shadow to be 7.5 inches. Using the length of the shadow as one side of a right triangle and the ruler as the other side, the angle of the sun can now be determined.

- **"Why You Need to Know"**

  This element at the beginning of each unit explains to students the importance of learning the material presented in each unit and how it may apply to actual job situations.

Unit 33 Single-Phase Motors

Key Terms

Centrifugal switch
Compensating winding
Conductive compensation
Consequent-pole motor
Encoder
Holtz motor
Inductive compensation
Multispeed motors

Why You Need to Know

Single-phase motors are used mainly in residential applications and rarely in industrial and commercial locations with the exception of smaller fractional horsepower applications. Unlike three-phase motors, there are many different types of single-phase motors, and they do not all operate on the same principle. There are some that operate on the principle of a rotating magnetic field, but others do not. Some single-phase motors are designed to operate at more than one speed, while others have more constant speed applications. This unit

- presents several different types of single-phase motors and explains how they operate.
- explains how to determine the appropriate motor to be used in a given situation by evaluating the operating principles of each.

- **"Practical Applications"**

  Word problems step the students through potential situations on the job and encourage them to develop critical thinking skills.

Practical Applications

Industry often employs the use of an electromagnet to pick up metal objects. These large magnets are called round horseshoe magnets. They contain a center core surrounded by an outer core. A direct current winding of magnetic wire is wound around the center core. An illustration of a round horseshoe magnet that has been cut in half is shown in *Figure 4–21*. When direct current flows through the winding, the magnetic field produces the polarities as shown. The dimensions of the magnet are shown in *Figure 4–22*. It is assumed that the magnetic field developed has a flux density of 2 webers per square inch. Determine the maximum pulling force of the magnet.

- **Dedication to Technical Accuracy and Consistency**
  The text was thoroughly reviewed for technical accuracy and consistency, ensuring existing errors were corrected, and enabling students to readily grasp more difficult concepts.

Course Solutions

Online Learning Platform: MindTap

Today's leading online learning platform, MindTap for Herman/Vigstol, *Delmar's Standard Textbook of Electricity*, 8th edition, gives you complete control of your course to craft a personalized, engaging learning experience that challenges students, builds confidence, and elevates performance.

MindTap introduces students to core concepts from the beginning of your course using a simplified learning path that progresses from understanding to application and delivers access to e-textbooks, study tools, interactive media including electrical training simulations, auto-graded assessments, and performance analytics.

Use MindTap for Herman/Vigstol, *Delmar's Standard Textbook of Electricity*, 8th edition as-is or personalize it to meet your specific course needs. You can also easily integrate MindTap into your Learning Management System (LMS). Learn more at https://www.cengage.com/mindtap.

Ancillary Package

Additional instructor resources for this product are available online. Instructor assets include an Instructor's Manual, Educator's Guide, PowerPoint® slides, and a test bank powered by Cognero®. Sign up or sign in at www.cengage.com to search for and access this product and its online resources.

The teaching package for instructors includes:

- ***Instructor's Manual*** outlines activities and resources that can be incorporated into instruction for each unit.
- ***PowerPoint® slides*** support lectures with key terms, concepts, examples, and activities.
- ***Test Bank powered by Cognero®*** provides a flexible, online system that allows you to access, customize, and deliver a test bank from your chosen text to your students through your LMS or another channel outside of MindTap.
- ***Transition Guide*** outlines changes between the 7th and 8th editions.
- ***Solution & Answer Guide*** that provides solutions and answers to the end-of-unit questions.
- ***The Complete Laboratory Manual for Electricity***, by Steve Herman. This manual is designed to be utilized with common lab equipment. (Order #: 978-1-133-67382-8).
- ***Correlation Guide*** to the *Complete Laboratory Manual for Electricity.*
- ***Image Gallery*** with updated images from this revision.
- ***Educator's Guide*** that details activities found in the accompanying MindTap.
- ***Guide to Teaching Online***

Go to https://faculty.cengage.com/, sign in or create an account, then search for this title. You can save the title for easy access and then download the resources that you need.

A Note about the Lab Manual

A laboratory manual is available for *The Standard Textbook of Electricity*. *The Complete Laboratory Manual for Electricity, 4th edition,* provides extensive opportunities for students to apply what they have learned. The manual contains multiple hands-on experiments for selected units of the textbook and has been extensively field-tested to ensure that all the experiments will work as planned. *The Complete Laboratory Manual for Electricity* was field-tested at the Shreveport-Bossier Regional Technical School under the direction of Richard Cameron.

About the Authors

Derek Vigstol is an electrical safety and maintenance consultant and works for e-Hazard Management, LLC, out of Louisville, Kentucky. He is a Class A Master Electrician in the state of Minnesota and has been involved in the electrical teaching industry since 2013. Derek began his teaching career at the Minneapolis Electrical JATC, where he taught apprentices and pre-apprentices in both electrical theory and installation practices. He also spent several years working for the National Fire Protection Association (NFPA) prior to taking the leap into the consulting world. Derek served in multiple roles at NFPA and was involved in everything from the development of standards to the development of NFPA training and educational material. He joined e-Hazard in 2021 and has been delivering the e-Hazard brand of excellence in practical application technical information ever since.

Derek currently resides in Massachusetts with his wife, Katie, and their children, Dylan and Kendra. He is active in contributing to several trade organizations such as IEEE and IAEI and is a member of the NFPA. He also is the cohost of e-Hazard's electrical safety-focused podcast, "Plugged into Safety," contributes regular content to several industry publications, and is an author for Cengage's Electrical Wiring Residential textbook.

Stephen L. Herman has been both a teacher of industrial electricity and an industrial electrician for many years. His formal training was obtained at Catawba Valley Technical College in Hickory, North Carolina. Mr. Herman has worked as a maintenance electrician for Superior Cable Corp. and as a class "A" electrician for National Liberty Pipe and Tube Co. During those years of experience, Mr. Herman learned to combine his theoretical knowledge of electricity with practical application. The books he has authored reflect his strong belief that a working electrician must have a practical knowledge of both theory and experience to be successful.

Mr. Herman was the Electrical Installation and Maintenance instructor at Randolph Technical College in Asheboro, North Carolina, for 9 years. After a return to industry, he became the lead instructor of the Electrical Technology Curriculum at Lee College in Baytown, Texas. He retired from Lee College after 20 years of service and, at present, resides in Pittsburg, Texas, with his wife. He continues to stay active in the industry, write, and update his books.

Acknowledgments

The author would like to express a special thank you to his wife of 18 years, Katie. She was the one who held down the fort as Derek spent several late nights and weekends wrapped up in working on the book. Without her help and encouragement, the process of authoring and editing books like this would not be possible! Derek would also like to personally thank Derrick Atkins from the Minneapolis Electrical JATC. Not only has Mr. Atkins been an incredible mentor and teacher to so many in the electrical industry, including the author, but he has also dedicated countless hours towards making the electrical industry a safer place as he serves on code making panel 5 of NFPA 70: National Electrical Code(NEC) and the NEC Correlating Committee. He has also provided technical review of the manuscript for this book and continues to challenge the author to strive for excellence. As one of the author's most influential instructors and colleagues, Mr. Atkins has played a critical role in the author's career, and Derek is proud to be able to call him a friend.

The author and publisher would also like to express thanks to those reviewers who provided insightful feedback throughout the development of this text:

James Blackett, Thomas Nelson Community College, Hampton, VA

Oscar Buschinelli, Centre for Skills Development and Training, Burlington, ON, Canada

James Cipollone, Antelope Valley Community College, Lancaster, CA

Eduardo Del Toro, MacArthur High School/Independent Electrical Contractors, San Antonio, TX

Anne Koering, Clark College, Vancouver, WA
Paul Lingo, Training Director IECRM/Journeyman Electrician
Randy Ludington, Guilford Community College, Greensboro, NC
Robert B. Meyers, Jr., Harrisburg Area Community College, Harrisburg, PA
J.C. Morrow, Hopkinsville Community College, Hopkinsville, KY
Quinton Phillips, Athens Technical College, Athens, GA
Larry Pogoler, LA Trade Tech College, Los Angeles, CA
Rick Reardon, Eastern Maine Community College, Bangor, ME
Dean Senter, Pratt Community College, Pratt, KS
Justin Shores, Antelope Valley Community College, Lancaster, CA
Elmer Tepper, Gateway Community College, Phoenix, AZ
Raul Vasquez, Independent Electrical Contractors, San Antonio, TX
Kevin Weigman, Northeast Wisconsin Technical College, Green Bay, WI

Electrical Occupations

Organization of the Industry

The electrical industry is one of the largest in the United States and Canada. This industry includes all aspects of construction, maintenance, and servicing of electrical equipment and systems. The opportunity for employment and advancement as an electrician or electrical technician is one of the highest of any industry as the entire country runs on electricity. Industry, commercial locations, and homes all employ electricity as the main source of power. A recent interview with electrical industry leaders indicated that the industry is facing a serious shortage of workers. In fact, one statistic mentioned is that the construction industry is nearly 100,000 workers short of the number of electricians needed to complete the jobs that are already on the books! The future for anyone entering the electrical field is bright because this is one of the best career choices a person could make in both the United States and Canada.

Electrical Personnel

Electrical workers can generally be divided into several categories depending on their specific area of employment. Each of these categories may require special skills.

- **Construction Electricians**

  Electricians working in the construction industry generally require a basic knowledge of electrical theory and an extensive knowledge of *National Electrical Code*® and *NFPA 70E*® requirements and wiring practices. Electricians in the construction area can generally be divided into helpers, apprentices, journeypersons, and masters. Several states require extensive tests for journey and master level licenses.

- **Industrial Electricians**

  Industrial electricians, sometimes referred to as *electrical technicians*, are generally concerned with maintaining equipment that has already been installed. Electricians in an industrial environment require an extensive knowledge of electrical theory and *National Electrical Code*® requirements for installation of motors, capacitor banks, and transformers. In addition to the *NEC*, industrial electrical workers must be well versed in *NFPA 70E*® and *NFPA 70B*® for safe work practices and maintenance in an industrial setting. Industrial electricians should also possess a basic knowledge of electronics and electronic devices. Modern industry employs many electronic devices, such as variable frequency drives, solid state controls for direct current motors, and programmable logic controllers. Another area of concern for most industrial electricians is motor controls. Motor control systems are generally either relay logic or electronic in the form of programmable logic controllers or distributive control systems.

- **Instrumentation Technicians**

  Instrumentation technicians calibrate and maintain devices that sense such quantities as temperature, pressure, liquid level, flow rate, and others. These people should have an extensive knowledge of electrical theory, especially as it pertains to low-voltage and closed-loop systems.

- **Multi-Craft Maintenance**

  A recent trend in the industry is the concept of a *multi-craft maintenance technician*. This role is similar to the industrial electrician. However, these individuals do not only focus on the electrical equipment and systems. These individuals must have a clear understanding of electrical theory, installation, maintenance, and servicing procedures. In addition to all of the electrical specifics that they must know, they must also possess the knowledge necessary to perform work in other disciplines.

- **Related Industries**

  The fields related to the electrical industry are too numerous to mention but include air conditioning and refrigeration, aircraft electronics, automotive, cable TV, broadcast media, energy and utilities, and home appliance and repair, as well as many, many others. The opportunity for employment in the electrical field is almost unlimited.

Union and Nonunion Employees

The largest percentage of electricians are nonunion employees. Many construction electricians receive training at various trade and technical schools. Some employers also sponsor apprenticeship programs. Apprenticeship-type programs generally require the electrician to work on the job as well as attend classes. The advantage to apprenticeship training is that it permits a person to earn money while they attend class. The disadvantage is that it can create an extremely busy schedule. Most industrial electricians, and those in related fields, require special training at a trade or technical school.

The largest electricians' union is the International Brotherhood of Electrical Workers (IBEW). The construction electricians who belong to the IBEW generally receive apprenticeship-type training for a 4- or 5-year timeframe depending on local jurisdiction requirements. Union electricians who work in specific related fields might belong to unions organized for their particular industry, such as United Auto Workers or United Steel Workers.

Apprentices, whether union or nonunion, attend classes several hours a week and work on the job under the supervision of a journeyperson. Most journey and master level electricians have completed their apprenticeship training and a set number of hours of practical work and are required to pass an examination. In several states, journey level electricians must work under the supervision of a master electrician. The master is ultimately responsible for the work performed and is answerable to the architect or owner.

Building Codes

Many cities, counties, and states have their own building codes that supersede the *National Electrical Code*®. The *National Electrical Code*® is law only if the local authority has adopted it as law. Always check local codes before beginning a construction project. Local codes often specify the manner in which wiring is to be installed and the size or type of wire that must be used for a particular application.

Green Building

"Green building" basically means making buildings more energy efficient. This can encompass many areas of the construction such as using "low E" energy-efficient windows, adding extra insulation, adding solar collectors to assist the water heater, and installing solar panels and/or wind generators to assist the electrical service. For the electrician, it may be installing larger wire than necessary to help overcome voltage drop or installing energy-efficient appliances such as heat pump-type water heaters. These water heaters use about half the amount of power of a standard electric water heater. Energy-efficient appliances are generally identified by an Energy Star label. Energy Star is a government-backed symbol awarded to products that are considered energy efficient. Energy Star was established to reduce greenhouse gas emissions and other pollutants caused by inefficient use of energy and to aid consumers in identifying and purchasing energy-efficient products that will save money without sacrificing performance, features, or comfort.

Before a product can receive an Energy Star label, it must meet certain requirements set forth in Energy Star product specifications:

- Product categories must produce significant energy savings nationwide.
- Qualified products must deliver the features and performance demanded by customers as well as increase energy efficiency.
- If the qualified product costs more than a conventional, less-efficient counterpart, purchasers must be able to recover their investment in increased energy efficiency through utility bill savings within a reasonable period of time.
- Energy efficiency must be achievable through broadly available, nonproprietary technologies offered by more than one manufacturer.
- Product energy consumption and performance must be measurable and verified with testing.
- Labeling should effectively differentiate products and be clearly visible to purchasers.

Solar Energy

One of the primary sources of green energy is solar power. Solar energy is the primary source of heating water in many countries and can be as simple as a dark colored container mounted on the roof of a structure (*Figure Occupations 1).* Other types of solar water heaters involve a solar collector, a special tank that contains a heat exchanger, and related equipment (*Figure Occupations 2).* Most of these types of water heaters contain backup electric heating elements for cloudy weather when the solar collector cannot supply enough energy to heat the water.

Figure occupations 1 Solar water heaters mounted on a roof.

Solar energy can do more than heat water, it can even generate electricity with what is generally called PV (photovoltaic) systems. In these types of systems, solar panels are mounted on the roof of a dwelling or in an open area on the ground (*Figure Occupations 3).* Photovoltaic cells generate direct current, which must be changed into alternating current by an inverter (*Figure Occupations 4).* The home typically remains connected to the utility company, but there are several houses that exist completely off the grid. The solar (PV) panels help reduce the energy supplied by the utility company. These systems can even earn a payback as they sell power back to the utility, depending on the utility company and state or local laws. Some systems cause the electric meter to run backward during times that the solar panels are producing more energy than is being supplied by the utility company. Other systems require the use of two separate meters, (*Figure Occupations 5).* One records the amount of power

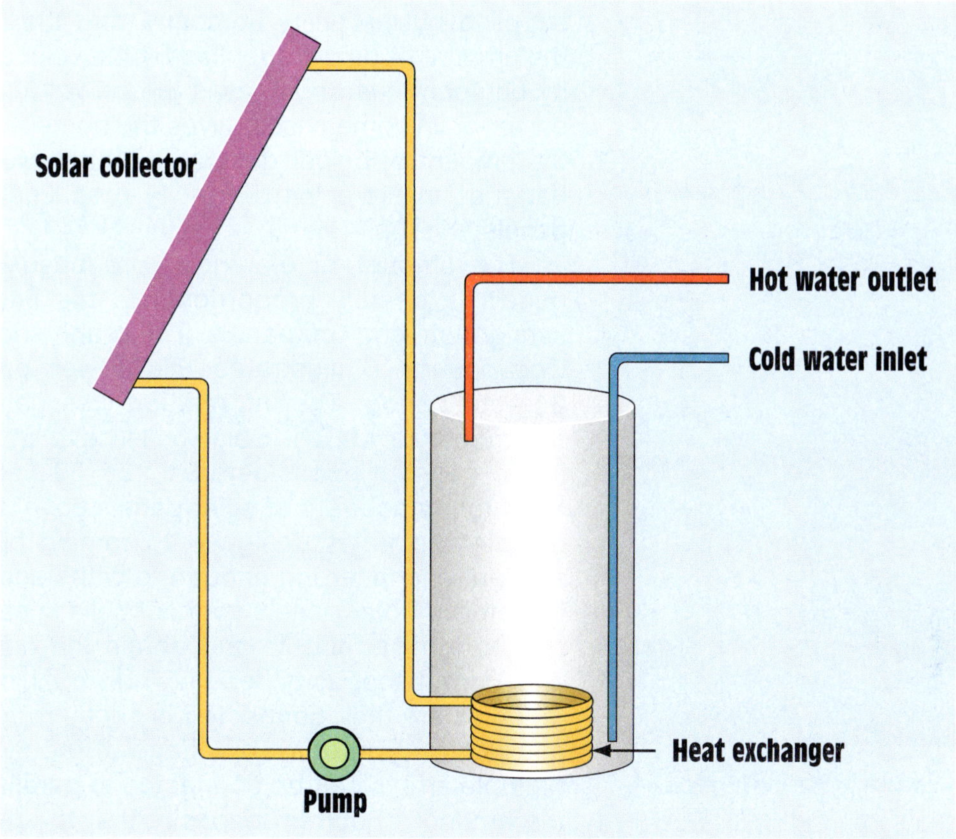

Figure occupations 2 Some solar water heaters use a solar panel and special tank with a heat exchanger.

Figure occupations 3 Solar panels are often mounted on the roof or in an open area.

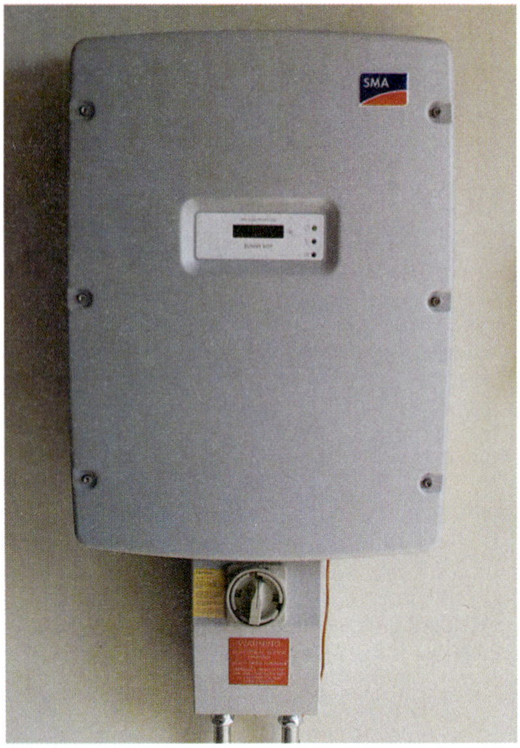

Figure occupations 4 Inverters change the direct current produced by the solar cells into alternating current.

Figure occupations 5 One meter records the power supplied by the utility company, and another records the amount of power supplied by the solar panels.

supplied by the utility company, and the other records the amount of power supplied by the solar cells. The utility company then purchases the power from the homeowner or, in some cases, gives the homeowner credit for the amount of power generated. Other systems employ batteries to store the electricity produced by the solar panels.

The amount of electricity produced by the solar panels is directly proportional to the intensity of the sunlight striking the panels. The graph shown in *Figure Occupations 6* illustrates the power output over a 24-hour period. The information was gathered during the month of March. Solar cells have a very long life span, generally considered to be 50 years or more. Most manufacturers of solar panels cover the cells with a material that is designed to remain clear in direct sunlight and is strong enough to withstand the average hail storm. Solar panels connect cells in series and parallel to form a module and obtain the desired voltage and current capacity for the individual modules. The modules are then connected in series to form the array and achieve the desired voltage of the system output. Multiple arrays can be connected in parallel to increase the amount of current capacity that the overall system can supply.

Regardless of the type of system, there are generally specific procedures that must be followed during the installation of photovoltaic systems. Special circuit breakers designed for direct current and high amperage interrupt capability are often required. Manufacturers' recommendations as well as national and local electrical codes should be followed.

Wind Power

Another widely used form of "Green" energy is wind. Wind is actually a product of solar energy. The Sun heats different areas of the Earth's surface at different rates. Hot air rises at a faster rate than cool air. As the hot air rises, cool air rushes in to replace the void left by the rising hot air, and wind is created. Air has mass, and moving air can contain a lot of energy. Wind generators convert the kinetic energy of moving air into electricity. Wind energy increases by the cube of the speed, which means that each time the wind speed doubles, the amount of energy increases eight times. This is the reason that the shape of many automobiles is designed to move through the air with less friction. The wind resistance of an automobile traveling at 60 miles per hour will be eight times greater than when traveling at 30 miles per hour.

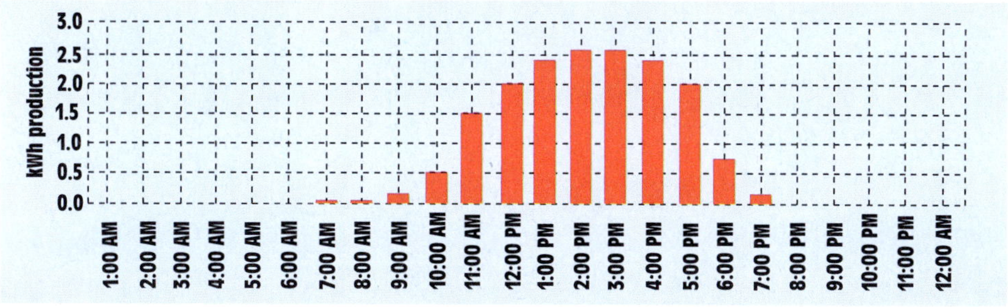

Figure occupations 6 Generation of electric power over a 24-hour period.

Wind generators are often referred to as wind turbines. There are two basic designs of wind turbines, the horizontal axis and the vertical axis (*Figure Occupations 7).* Vertical axis turbines are often called "egg beaters." The main advantage of vertical axis turbines is that they are omnidirectional, meaning that they will operate regardless of wind direction. Although horizontal axis turbines must turn to face the wind, they are the most used type for producing electricity. The size of wind turbines can vary greatly depending on the amount of electricity they are intended to produce. Utility scale turbines used in land-based wind farms (*Figure Occupations 8),* can have rotor diameters that range from 50 meters (164 feet) to 90 meters (295 feet). The tower height is generally the same as the rotor diameter. Utility wind turbines generally feed the electricity they produce directly into the power grid to aid other electricity-generating plants that use fossil fuels.

Wind turbines intended for residential or small commercial use are much smaller. Most have rotor diameters of 8 meters (26 feet) or smaller and are mounted on towers of 40 meters (131 feet) or less. As with photovoltaic installations, wind-powered systems can be installed to connect directly to the power grid or to charge a bank of batteries. An inverter is used to convert the direct current of the batteries into alternating current to supply the home. Inverters used to couple the wind turbine to the power line must be able to maintain a steady power flow with varying wind speeds and varying voltages. They must

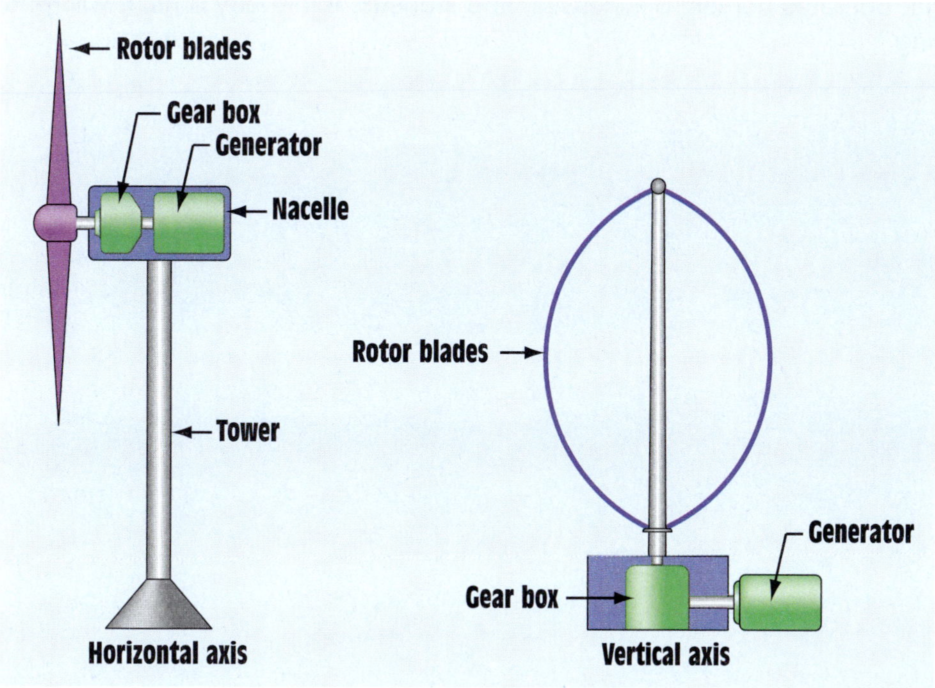

Figure occupations 7 Examples of different wind turbine types.

Figure occupations 8 Wind farms are large installations of several individual turbines grouped together.

also be able to shape the waveform to that of a sine wave. Similar to photovoltaic installations, some wind-powered systems cause the electric meter to run backward when it is producing more power than is required by the home. Some utility companies will give credit for the amount of power generated, and some will purchase the power from the customer. Other utilities require the use of two separate meters to determine the amount of wind power produced.

As with photovoltaic systems, when installing a wind-powered system, manufacturers' instructions and utility requirements should be followed. Before installing a wind-powered system, check to make certain that the area has a high enough average wind speed to justify the cost of the system.

Summary

The electrical field offers many avenues that can lead to success. Most electricians work in the construction industry, but many are employed as maintenance technicians in industry and other related fields. The demand for qualified electricians is expected to increase at a rate of over 8000 new jobs per year over the next 10 years. The lay-off rate for electricians is one of the lowest in the country. Electricity is the power that operates homes, businesses, and industry. If industry runs, it will require electricians to keep it running.

Section 1 Safety, Basic Electricity, and Ohm's Law

Luisa Leal Photography/Shutterstock.com

Safety Overview

Key Terms

Arc-rated clothing

Cardiopulmonary resuscitation (CPR)

Confined spaces

De-energized

Energized circuit

Fibrillation

Horseplay

Idiot proofing

Lockout and tagout

Milliamperes (mA)

Occupational Safety and Health Administration (OSHA)

Safety Data Sheet (SDS)

Scaffolds

Why You Need to Know

Safety is everyone's shared responsibility! You should be concerned not only with your own safety but also with the safety of others around you. This is especially true for people employed in the electrical field. Some simple rules must be followed when working with electric equipment or circuits to ensure both you and everyone around you get to go home at the end of the day the same way they came to work, in one piece and uninjured!

Outline

S–1 General Safety Rules

S–2 Effects of Electric Current on the Body

S–3 On the Job

S–4 Personal Protective Equipment

S–5 Ladders and Scaffolds

S–6 Fires

S–7 Ground-Fault Circuit Interrupters

S–8 Arc-Fault Circuit Interrupters

S–9 Grounding and Bonding

Objectives

After studying this unit, you should be able to

- identify basic safety principles to minimize risk of injury.
- analyze the effects of electric current on the body.
- discuss the origin and responsibilities of OSHA.
- examine what types of protective clothing are required based on the hazards.
- describe the operation of ground-fault circuit interrupters and arc-fault circuit interrupters.
- illustrate the importance of grounding.

S–1 General Safety Rules

Always establish an electrically safe work condition when working with electrical conductors and circuit parts

Regardless of your industry, the Occupational Safety and Health Act of 1970 (OSHA) is going to mandate that conductors and circuit parts be **de-energized** before an employee works on or near them. *NFPA 70E®: Standard for Electrical Safety in the Workplace®* is going to take this requirement a step further and require a state that is referred to as an electrically safe work condition (ESWC). An ESWC is a state in which the conductors and circuit parts have been disconnected from a source of power, have been locked out in accordance with established standards meeting OSHA's lockout/tagout requirements, and have been verified to be at a zero-voltage state. In certain situations, we will also need to temporarily ground the conductors to ensure that the employee is not injured by residual or induced voltages and, in the ultimate case of a safety breakdown, become accidentally re-energized prior to work being complete. To put it simply, an ESWC is a state in which the circuit has been disconnected from a voltage source, steps have been taken to ensure it cannot be re-energized without the knowledge of the person working on the circuit, and that person has verified de-energization prior to performing any work.

Verifying de-energization also requires the use of the following three-step check:

1. Test the meter on a known voltage source to make sure the meter is operating properly.
2. Test the conductors or circuit parts that are being verified as de-energized.
3. Test the meter on the known voltage source again to make certain the meter is still operating properly.

Think

Of all the rules concerning safety, this one is probably the most important. No amount of safeguarding or **idiot proofing** a piece of equipment can protect a person as well as taking time to think before acting. Many technicians have been killed by supposedly "dead" circuits. Do not depend on circuit breakers, switches, or someone else to open a circuit. Test it yourself before you touch it.

Avoid Horseplay

Jokes and **horseplay** have a time and place, but not when someone is working on an electric circuit or a piece of moving machinery. Do not be the cause of someone being injured or killed, and do not let someone else be the cause of you being injured or killed.

Do Not Work Alone

This is especially true when working in a dangerous situation, such as on a live circuit. Have someone with you who can provide first aid and/or **cardiopulmonary resuscitation (CPR)**. A critical component of responding to electrical injuries is the application of an automated external defibrillator (AED). Severe electric shocks can cause breathing difficulties and can cause the heart to go into ventricular fibrillation.

Learn First Aid

Anyone working on electric equipment, especially those working with voltages greater than 50 volts, should make an effort to learn first aid. A knowledge of first aid, especially CPR and AEDs, may save your own or someone else's life.

Avoid Alcohol and Drugs

The use of alcohol and drugs has no place on a worksite. Alcohol and drugs are not only dangerous to users and those who work around them, but they also cost the industry millions of dollars each year. Alcohol and drug abusers kill thousands of people on the highways each year and are just as dangerous on a worksite as they are behind the wheel of a vehicle. Many industries have instituted testing policies to screen for alcohol and drugs. A person who tests positive may or may not receive a warning the first time, depending on the stringency of the company's drug and alcohol policy.

S–2 Effects of Electric Current on the Body

Most people have heard the old saying that it is not the voltage that kills you, but rather it is the current. This is true to an extent, but do not be fooled into thinking that voltage doesn't play its own role. Voltage is the force that causes current to flow in the first place. Without an applied voltage, there would be no current flowing to cause injury or death. Voltage can be compared to the pressure that pushes water through a pipe. When all other parameters remain unchanged, the more pressure available in the system, the greater the volume of water will flow through the pipe. Students often ask how much current will flow through the body at a particular voltage. There is no easy answer to this question. The amount of current that can flow at a particular voltage is determined by the resistance of the current path. Different people have different resistances. A body has less resistance on a hot day when sweating because saltwater is a very good conductor. What one eats and drinks for lunch can have an effect on the body's resistance, as can the length of the current path. Is the current path between two hands or from one hand to one foot? All these factors affect body resistance. However, OSHA bases their shock threshold on the assumption that a shock can encounter the worst-case scenario of an open wound. OSHA uses an internal resistance of the human body of around 500 ohms to justify that any voltage above 50 volts can produce a fatal shock. Unfortunately, they also have real life data to back up this assertion.

Figure S–1 illustrates the effects of different amounts of current on the body. This chart is general—some people may have less tolerance to electricity and others may have a greater tolerance. However, you don't want to find out the hard way what level of tolerance you have!

A current of 1 to 3 **milliamperes (mA)** (0.001 to 0.003 amperes) usually causes a slight tingling sensation, which increases as current increases and becomes very noticeable and painful at about 10 milliamperes (0.010 amperes). Ground-fault circuit interrupter-type devices used for shock protection of personnel are set to interrupt the circuit at current levels-between 4 and 6 milliamperes. The tingling sensation becomes very painful at about 20 milliamperes. Currents above 10 milliamperes can cause a person to seize the line and be unable to let go of the circuit. Currents above 30 milliamperes begin to cause muscular paralysis and breathing difficulty. When the current increases to about 75 to 100 milliamperes, breathing is extremely difficult and the heart can go into ventricular **fibrillation**, a condition in which the heart begins to "quiver" and the pumping action stops. Left untreated, this condition will be fatal.

Currents above 200 milliamperes can cause severe tissue damage, and currents around 400 milliamperes can cause the heart to squeeze shut and stop beating. Large amounts of current can cause severe

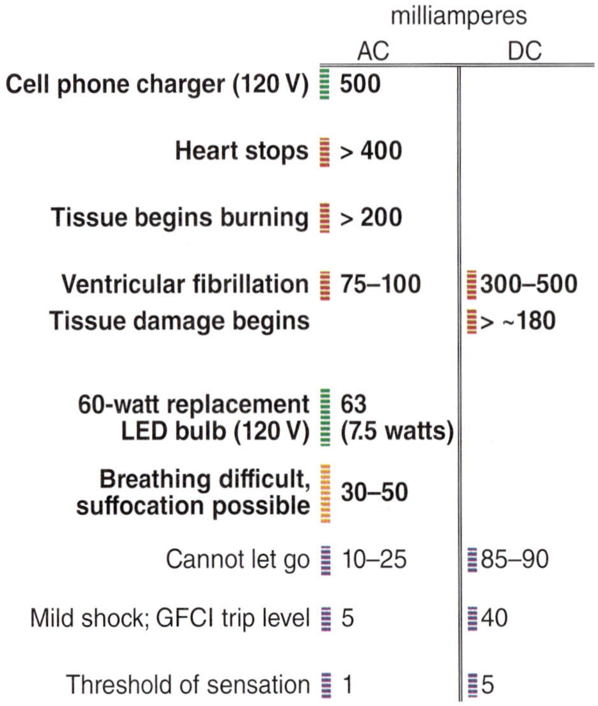

Figure S–1 The effects of electric current on the body.

electric burns that are often very serious because they occur on the inside of the body. The exterior of the body may not look seriously burned, but the inside may be severely burned. Also, keep in mind that it is not always the higher voltage systems that cause the most damage; a system operating at 120 volts can cause a current of around 120 milliamperes to flow hand-to-hand if we use the assumed bodily resistance between the hands of 1000 ohms. This is more than enough current to cause an individual to become "hung up" and unable to let go of the circuit, and it is more than enough current to put the heart into ventricular fibrillation. Never assume that 120 volts cannot hurt you!

S–3 On the Job

OSHA

OSHA is an acronym for the Occupational Safety and Health Act of 1970. However, today the acronym is mostly used to refer to the **Occupational Safety and Health Administration**, a division of the U.S. Department of Labor. Created by Congress in 1971, its mission is to ensure safe and healthful workplaces in the United States. Since its creation, workplace fatalities have been cut in half, and occupational injury and illness rates have declined by 40%. Enforcement of OSHA regulations is the responsibility of the Secretary of Labor.

OSHA standards cover many areas, such as the handling of hazardous materials, fall protection, protective clothing, and hearing and eye protection. There are several parts of the OSHA regulations that you will need to become familiar with, such as Part 1910 for general industry and Part 1926 for construction. Part 1910, Subpart S, deals mainly with the regulations concerning electrical safety in general industry. These regulations are available in books and can be accessed at the OSHA website www.osha.gov. You can also access the regulations in the Electronic Code of Federal Regulations at www.ecfr.gov under Title 29: Labor.

Hazardous Materials

It may become necessary to deal with some type of hazardous material. A hazardous material or substance is anything to which exposure may result in adverse effects on the health or safety of employees. Hazardous materials may be chemical, biological, or nuclear. OSHA sets standards for dealing with many types of hazardous materials. The required response is determined by the type of hazard associated with the material. Hazardous materials are required to be listed as such. Much information concerning hazardous materials is generally found on **safety data sheets (SDS)**. If you are working in an area that contains hazardous substances, always read any information concerning the handling of the material and any safety precautions that should be observed. After a problem exists is not the time to start looking for information on what to do.

Some hazardous materials require a hazardous materials (HAZMAT) response team to handle any problems. A HAZMAT team is any group of employees designated by the employer that is expected to handle and control an actual or potential leak or spill of a hazardous material. They are expected to work in close proximity to the material. A HAZMAT team is not always a fire brigade, and a fire brigade may not necessarily have a HAZMAT team. On the other hand, a HAZMAT team may be part of a fire brigade or fire department.

Employer Responsibilities

Section 5(a)(1) of the Occupational Safety and Health Act basically states that employers must furnish each of their employees a place of employment that is free of recognized hazards that are likely to cause death or serious injury. This places the responsibility for compliance on employers. Employers must identify hazards or potential hazards within the worksite and eliminate them, control them, or provide employees with suitable protection from them. The employer is responsible for obtaining, maintaining, and posting (or making available) SDS for all hazardous materials that may be used in the workplace. It is the employees' responsibility to read and understand the SDS for all chemicals and materials that they may be exposed to in their working environment and to follow the safety procedures set up by the employer in the event of an accident. Employees should ask questions about any information or procedures they do not understand.

To help facilitate these safety standards and procedures, OSHA requires that an employer have a competent person to oversee implementation and enforcement of these standards and procedures. This person must be able to recognize unsafe or dangerous conditions and have the authority to correct or eliminate them. This person also has the authority to stop work or shut down a worksite until safety regulations are met.

Safety Data Sheets

These sheets were formerly referred to as an MSDS or material safety data sheet, and the name change was part of a larger global standard harmonization effort. A sample SDS can be seen in *Table S–1*. Safety data sheet requirements can be found in OSHA 1910.1200 for Hazard Communications. They generally warn users of any hazards associated with the product. They outline the physical and chemical properties of the product; list precautions that should be taken when using the product; and list any potential health hazards, storage considerations, flammability, reactivity, and, in some instances, radioactivity. They sometimes list the name, address, and telephone number of the manufacturer; the SDS date and emergency telephone numbers; and, usually, information on first aid procedures to use if the product is swallowed or comes in contact with the skin. Safety data sheets can be found on many home products, such as cleaning products, insecticides, and flammable liquids.

Section 1	**Identity of Material**
Trade Name	OATEY HEAVY DUTY CLEAR LO-VOC PVC CEMENT
Product Numbers	31850, 31851, 31853, 31854
Formula	PVC Resin in Solvent Solution
Synonyms	PVC Plastic Pipe Cement
Firm Name & Mailing Address	OATEY CO., 4700 West 160th Street, P.O. Box 35906 Cleveland, Ohio 44135, U.S.A. http://www.oatey.com
Oatey Phone Number	1-216-267-7100
Emergency Phone Numbers	For Emergency First Aid call 1-303-623-5716 COLLECT. For chemical transportation emergencies ONLY, call Chemtrec at 1-800-424-9300
Prepared By	Charles N. Bush, Ph.D.

Section 2	**Hazardous Ingredients**		
Ingredients	***%***	***Cas Number***	***Sec 313***
Acetone	0–5%	67-64-1	No
Amorphous Fumed Silica (Nonhazardous)	1–3%	112945-52-5	No
Proprietary (Nonhazardous)	5–15%	N/A	No
PVC Resin (Nonhazardous)	10–16%	9002-86-2	No
Cyclohexanone	5–15%	108-94-1	No
Tetrahydrofuran (See SECTION 11)	30–50%	109-99-9	No
Methyl Ethyl Ketone	20–35%	78-93-3	Yes

Section 3	**Known Hazards Under U.S. 29 CFR 1910.1200**				
Hazards	***Yes***	***No***	***Hazards***	***Yes***	***No***
Combustible Liquid		x	Skin Hazard	x	
Flammable Liquid	x		Eye Hazard	x	
Pyrophoric Material		x	Toxic Agent	x	
Explosive Material		x	Highly Toxic Agent		x
Unstable Material		x	Sensitizer		x
Water Reactive Material		x	Kidney Toxin	x	
Oxidizer		x	Reproductive Toxin	x	
Organic Peroxide		x	Blood Toxin		x
Corrosive Material		x	Nervous System Toxin	x	
Compressed Gas		x	Lung Toxin	x	
Irritant	x		Liver Toxin	x	
Carcinogen NTP/IARC/ OSHA (see SECTION 11)		x			

Section 4	**Emergency and First Aid Procedures—Call 1-303-623-5716 Collect**
Skin	If irritation arises, wash thoroughly with soap and water. Seek medical attention if irritation persists. Remove dried cement with Oatey Plumber's Hand Cleaner or baby oil.
Eyes	If material gets into eyes or if fumes cause irritation, immediately flush eyes with water for 15 minutes. If irritation persists, seek medical attention.

Table S–1 SDS for Heavy-Duty Clear LO-VOC PVC Cement

Inhalation	Move to fresh air. If breathing is difficult, give oxygen. If not breathing, give artificial respiration. Keep victim quiet and warm. Call a poison control center or physician immediately. If respiratory irritation occurs and does not go away, seek medical attention.
Ingestion	**Do Not Induce Vomiting.** This product may be aspirated into the lungs and cause chemical pneumonitis, a potentially fatal condition. Drink water and call a poison control center or physician immediately. Avoid alcoholic beverages. Never give anything by mouth to an unconscious person.
Section 5	**Fire Fighting Measures**
Precautions	Do not use or store near heat, sparks, or flames. Do not smoke when using. Vapors may accumulate in low places and may cause flash fires.
Special Fire Fighting Procedures	**For Small Fires:** Use dry chemical, CO_2, water, or foam extinguisher. **For Large Fires:** Evacuate area and call Fire Department immediately.
Section 6	**Accidental Release Measures**
Spill or Leak Procedures	Remove all sources of ignition and ventilate area. Stop leak if it can be done without risk. Personnel cleaning up the spill should wear appropriate personal protective equipment, including respirators if vapor concentrations are high. Soak up spill with absorbent material such as sand, earth, or other noncombustible material. Put absorbent material in covered, labeled metal containers. Contaminated absorbent material may pose the same hazards as the spilled product.
Section 7	**Handling and Storage**
Precautions	**Handling & Storage:** Keep away from heat, sparks and flames; store in cool, dry place. **Other:** Containers, even empties, will retain residue and flammable vapors.
Section 8	**Exposure Controls/Personal Protection**
Protective Equipment Types	**Eyes:** Safety glasses with side shields. **Respiratory:** NIOSH-approved canister respirator in absence of adequate ventilation. **Gloves:** Rubber gloves are suitable for normal use of the product. For long exposures to pure solvents, chemical-resistant gloves may be required. **Other:** Eye wash and safety shower should be available.
Ventilation	**Local Exhaust:** Open doors & windows. Exhaust ventilation capable of maintaining emissions at the point of use below PEL. If used in enclosed area, use exhaust fans. Exhaust fans should be explosion-proof or set up in a way that flammable concentrations of solvent vapors are not exposed to electrical fixtures or hot surfaces.
Section 9	**Physical and Chemical Properties**

NFPA Hazard Signal	Health 2	Stability 1	Flammability 3	Special None
HMIS Hazard Signal	Health 3	Stability 1	Flammability 4	Special None
Boiling Point	151 °F/66 °C			
Melting Point	N/A			
Vapor Pressure	145 mmHg @ 20°C			
Vapor Density (Air = 1)	2.5			
Volatile Components	70–80%			
Solubility In Water	Negligible			
PH	N/A			
Specific Gravity	0.95 ± 0.015			

Table S–1 Concluded

Evaporation Rate	(BUAC = 1) = 5.5 – 8.0
Appearance	Clear Liquid
Odor	Ether-Like
Will Dissolve In	Tetrahydrofuran
Material Is	Liquid

Table S–1 Concluded

Trenches

It is often necessary to dig trenches to bury conduits. Under some conditions, these trenches can be deep enough to bury a person if a cave-in should occur. Safety regulations for the shoring of trenches are found in OSHA Standard 1926, Subpart P, App C, titled "Timber Shoring for Trenches." These procedures and regulations are federally mandated and must be followed. Some general safety rules also should be followed:

1. Do not walk close to trenches unless it is necessary. This can cause the dirt to loosen and increase the possibility of a cave-in.
2. Do not jump over trenches.
3. Place barricades around trenches (*Figure S–2*).
4. Use ladders to enter and exit trenches.

Figure S–2 Place a barricade around a trench and use a ladder to enter and exit the trench.

Confined Spaces

Confined spaces have a limited means of entrance or exit (*Figure S–3*). They can be very hazardous workplaces, often containing atmospheres that are extremely harmful or deadly. Confined spaces are very difficult to ventilate because of their limited openings. It is often necessary for a worker to wear special clothing and use a separate air supply to work in one. OSHA 1910.146 lists rules and regulations for working in a confined space that requires a permit for entry. In addition, many industries have written procedures that must be followed when working in confined spaces. Some general guidelines for working in confined spaces include the following:

1. Determine whether the confined space has the potential to develop into a hazardous atmosphere. If so, the atmosphere must be tested to determine if it is suitable for entry. If oxygen levels are not between 19.5% and 23.5%, it is unsafe to enter. This should be continuously monitored throughout the work task.

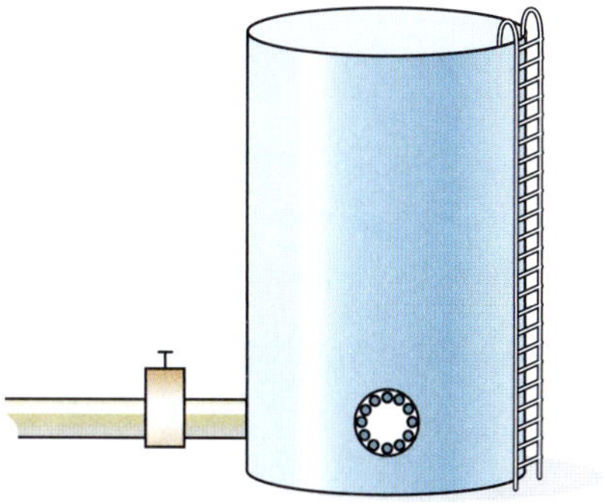

Figure S–3 A confined space is any space having a limited means of entrance or exit, is not intended for continuous occupancy, and is large enough for a person to enter and perform work.

2. Have a person stationed outside the confined space to watch the person or persons working inside. The outside person should stay in voice or visual contact with the inside workers at all times.
3. The outside person should never enter the space, even in an emergency, but should contact the proper emergency personnel. If they should enter the space and become incapacitated, there would be no one available to call for help.
4. Use only electric equipment and tools that are approved for the atmosphere found inside the confined area. It may be necessary to obtain a hot work permit to operate tools that have open brushes and that spark when they are operated.
5. As a general rule, a retrieval system should be in place to perform a confined space rescue. This often entails a person working in a confined space wearing a harness with a lanyard that extends to the outside person, so the outside person could pull them to safety, if necessary.

Lockout and Tagout Procedures

Lockout and tagout procedures are required to prevent someone from energizing a piece of equipment without the knowledge of the people who shut it down to work on it. This could apply to switches, circuit breakers, valves, and any other source of energy that has the potential to cause bodily harm to an employee. OSHA has three different sections that reference lockout/tagout and control of hazardous energy. OSHA 1910.333(b)(2)(iii) requires that a tag similar to the one shown in *Figure S–4* be placed on the isolation point for the piece of equipment being serviced; this section also requires a lockout device be applied to the isolation device to physically prevent the re-energizing of the equipment. The person performing the work places the lock on the equipment and keeps the key in their possession. Some tasks might require that multiple people have a vested interest in making sure that the equipment is not turned back on until the job is finished. A device that permits the use of multiple padlocks and a safety tag is shown in *Figure S–5*. Another strategy for ensuring de-energization can include the use of a group lockout box and a well-orchestrated procedure that establishes an employee or person in charge of overseeing the safety of all the employees working under the lockout procedure. Violating lockout and tagout procedures is considered an extremely serious offense in most industries, often resulting in immediate termination of employment, and can even result in prosecution. As a general rule, there are no first-time warnings.

After locking out and tagging a piece of equipment, it should be tested to make certain it is truly de-energized before working on it. For a mechanical lockout, this might mean trying to restart the machine to verify non-operation. However, more verification is needed for an electrical procedure. The

Figure S–4 Safety tag used to ensure no unauthorized operation of the equipment.

Figure S–5 The equipment can be locked out by several different people.

three-step procedure referenced at the beginning of this Unit is generally recommended for making certain that a piece of electric equipment is de-energized. An adequately rated voltage tester or voltmeter must be used to test the circuit elements for absence of voltage. Again, the procedure is as follows:

1. Test the voltage tester or voltmeter on a known energized circuit to make certain the tester is working properly.
2. Test the circuit you intend to work on with the voltage tester or voltmeter to make sure it is truly de-energized.
3. Test the voltage tester or voltmeter on a known energized circuit to make sure the tester is still working properly.

This procedure helps reduce the possibility of a faulty piece of equipment indicating that a circuit is de-energized when it is not.

S–4 Personal Protective Equipment

Maintenance and construction workers alike are usually required to wear certain articles of protective clothing, dictated by the environment of the work area and the job being performed.

Head Protection

Some type of head protection is required on almost any work site. An example of a Type E hard hat that might be worn by an electrician, made of nonconductive plastic, is shown in *Figure S–6*. It has a pair of safety goggles attached that can be used when desired or necessary.

Figure S–6 An example of a nonconductive, Type E hard hat with attached safety goggles.

Eye Protection

Eye protection is another piece of safety gear required on almost all worksites. Manufacturers require that eye protection, and sometimes hard hats, be worn by anyone, not just employees, entering the production area. Eye protection can come in different forms, ranging from the goggles shown in *Figure S–6* to the safety glasses with side shields shown in *Figure S–7*. Common safety glasses may or may not be prescription glasses, but almost all provide side protection (*Figure S–7*). Sometimes a full-face shield may be required.

Figure S–7 Safety glasses provide side protection.

Hearing Protection

OSHA 1910.95 includes requirements concerning hearing protection and the need to develop a hearing conservation program. The need for hearing protection is based on the ambient sound level of the worksite or the industrial location. Workers are usually required to wear some type of hearing protection when working in certain areas, usually in the form of earplugs or earmuffs.

Arc-Rated Clothing

Special clothing made of arc-rated material is required to be worn by electricians, engineers, and any personnel who will be exposed to an arc-flash hazard and are within the arc-flash boundary. **Arc-rated clothing** is often also required for maintenance personnel who work with high-power sources such as transformer installations and motor-control centers. An arc flash in a motor-control center can easily set a person's clothes on fire. The typical motor-control center can sometimes release enough incident energy during an arc flash to injure or kill a person 20 to 30 feet away.

Gloves

Another common article of personal protective equipment is rubber insulating gloves. Electricians often wear rubber insulating gloves with protectors when it is necessary to work on or near energized circuits (*Figure S–8*). These gloves are rated for a certain amount of voltage and must be selected based on the voltage of the system to be worked on. They must be inspected for holes or tears before they are used and air tested to help find any potential damage to the gloves; after all, if it can leak an air molecule, you can count on an electron getting through. These gloves are also periodically dielectrically tested to ensure they are still able to provide the required insulation capabilities. The protectors (*Figure S–9*) are used over the top of the rubber gloves to help protect against cuts and punctures.

Fall Protection Harnesses

Fall protection harnesses provide protection from falling. They buckle around the upper body with leg, shoulder, and chest straps, and the back has a heavy metal D-ring (*Figure S–10*). A lanyard is attached to the D-ring and secured to a stable structure above the worker. Lanyards can be made of different types of materials and can be fixed length or retractable. If the worker falls, the lanyard limits the distance they can drop. A fall protection harness should be worn:

1. when working more than 6 feet above the ground or floor in construction activities or 4 feet above the ground or floor in general industry work
2. when working near a hole or drop-off with a leading edge that has not been addressed
3. when working on high scaffolding that does not have guardrails in place

A fall protection harness is shown in *Figure S–11*.

National Safety Apparel, Inc.

Figure S–8 Rubber insulating gloves are rated to protect the wearer from voltages up to the voltage rating of their class of gloves.

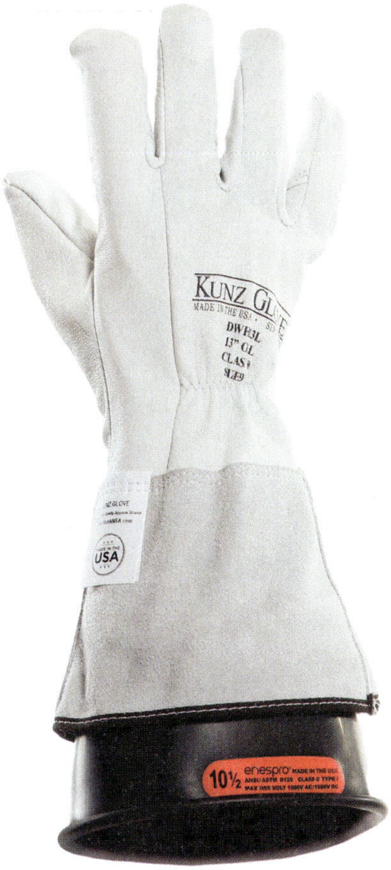

National Safety Apparel, Inc.

Figure S–9 Protector gloves help protect rubber insulating gloves from damage like cuts or punctures during work.

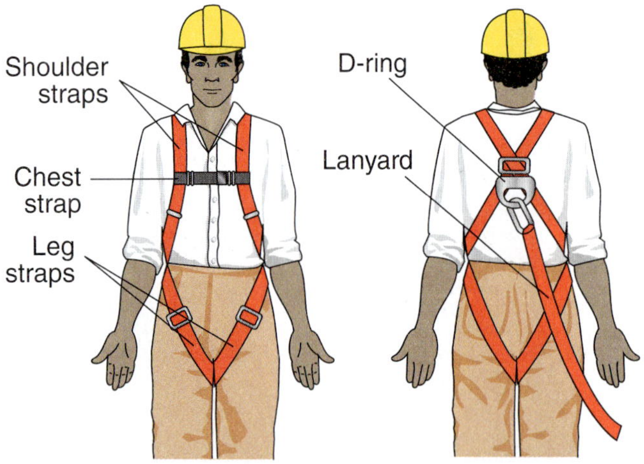

Figure S–10 Fall protection harnesses buckle around the upper body and legs.

Figure S–11 Example of a fall protection harness.

S–5 Ladders and Scaffolds

It is often necessary to work in an elevated location. When this is the case, ladders or scaffolds are employed. **Scaffolds** generally provide the safest elevated working platforms. They are commonly assembled on the worksite from standard sections (*Figure S–12*). The bottom sections usually contain adjustable feet that can be used to level the sections. Two end sections are connected by X braces that form a rigid work platform (*Figure S–13*). Sections of scaffolding are stacked on top of each other to reach the desired height.

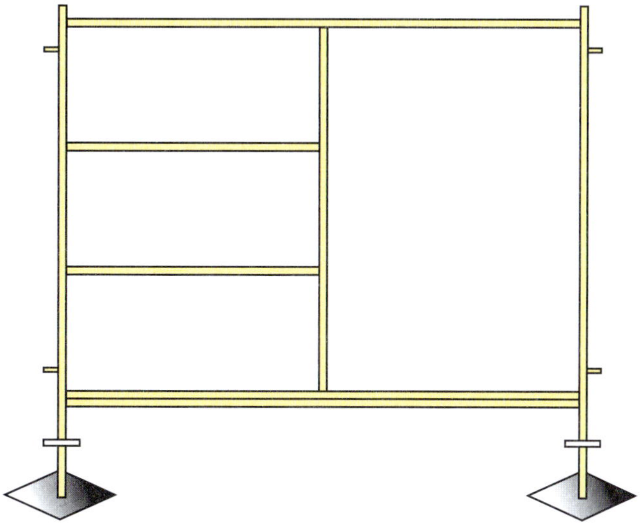

Figure S–12 Example of a section of scaffolding.

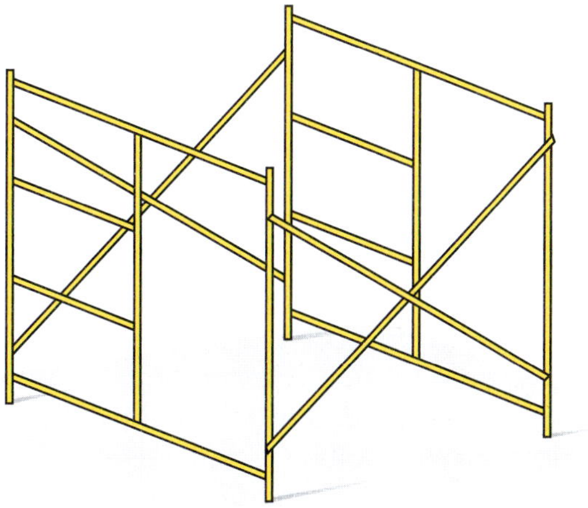

Figure S–13 X braces connect scaffolding sections together and add rigidity to the scaffold.

Rolling Scaffolds

Rolling scaffolds are used in areas that contain level floors, such as inside a building. The major difference between a rolling scaffold and those discussed previously is that a rolling scaffold is equipped with wheels on the bottom section that permit it to be moved from one position to another. However, keep in mind that it is usually prohibited to move a scaffold while a person is still on the scaffolding. The wheels usually contain a mechanism that permits them to be locked after the scaffold is rolled to the desired location.

Hanging or Suspended Scaffolds

Hanging or suspended scaffolds are suspended by cables from a support structure. They are generally used on the sides of buildings to raise and lower workers by using hand cranks or electric motors.

Straight Ladders

Ladders can be divided into two main types: straight and step. Straight ladders are constructed by placing rungs between two parallel rails and can be either a fixed length or an adjustable length (*Figure S–14*). They generally contain safety feet on one end that help prevent the ladder from slipping. Ladders used for electrical work are usually wood or fiberglass; aluminum ladders are avoided because they conduct electricity. Regardless of the type of ladder used, you should check its load capacity before using it. This information is found on the side of the ladder. Load capacities of 200, 250, and 300 pounds are common. Do not use a ladder that does not have enough load capacity to support your weight plus the weight of your tools and the weight of any object you are taking up the ladder with you.

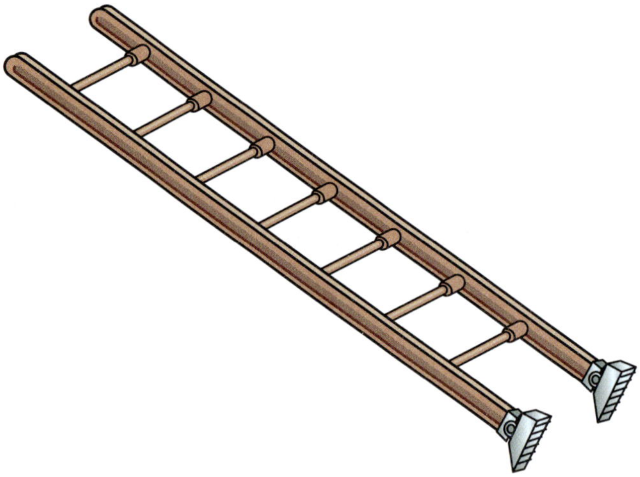

Figure S–14 Example of a straight ladder.

Straight ladders should be placed against the side of a building or other structure at an angle of approximately 76° (*Figure S–15*). This can be accomplished by moving the base of the ladder away from the structure a distance equal to one-fourth the height of the ladder. If the ladder is 20 feet high, it should be placed 5 feet from the base of the structure. If the ladder is to provide access to the top of the structure, it should extend 3 feet above the structure. A general check to see if the angle is correct is to stand with your toes just touching the bottom of the ladder and reach your hands straight out. If you can just touch the ladder without leaning forward, the angle is likely going to be close to the proper angle. You always want to check by measuring the rise and run of the ladder before ascending.

Step Ladders

Step ladders are self-supporting, constructed of two sections hinged at the top (*Figure S–16*). The front section has two rails and steps, the rear portion has two rails and braces. Like straight ladders, step ladders are designed to withstand a certain load capacity. Always check the load capacity before using a ladder. As a general rule, ladder manufacturers recommend that the top step not be used because of the danger of becoming unbalanced and falling. Many people mistakenly think the top step is the top or

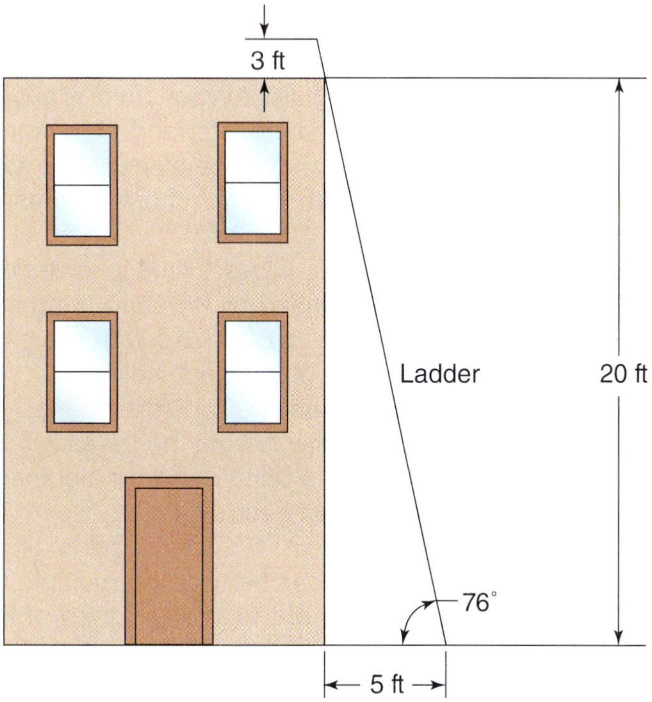

Figure S–15 A straight ladder should be placed at an angle of approximately 76°.

cap of the ladder, but it is actually the last step before the ladder top. The top of the ladder is not a step at all but is rather a platform for staging tools and materials for those working on the ladder. However, be careful not to leave tools on the top of the ladder when not on the ladder as they can fall during a move and injure the person moving the ladder.

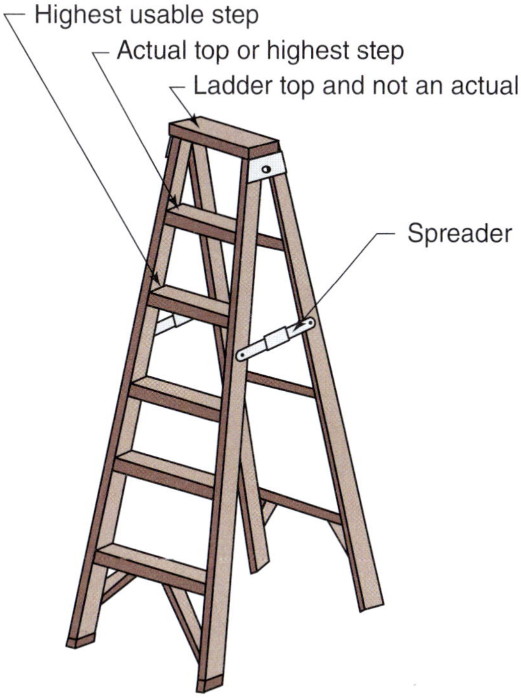

Figure S–16 Example of a step ladder.

S–6 Fires

For a fire to burn, it must have three things: fuel, heat, and oxygen. Fuel is anything that can burn, including materials such as wood, paper, cloth, combustible dusts, and even some metals. Different materials require different amounts of heat for combustion to take place. If the temperature of any material is below its combustion temperature, it will not burn. Oxygen must be present for combustion to take place. If a fire is denied oxygen, it will extinguish.

Fires are divided into four classes: A, B, C, and D. Class A fires involve common combustible materials such as wood or paper. They are often extinguished by lowering the temperature of the fuel below the combustion temperature. Class A fire extinguishers often use water to extinguish a fire. A fire extinguisher listed as Class A only should never be used on an electrical fire.

Class B fires involve fuels such as grease, combustible liquids, or gases. A Class B fire extinguisher generally employs carbon dioxide (CO_2), which greatly lowers the temperature of the fuel and deprives the fire of oxygen. Carbon dioxide extinguishers are often used on electrical fires because they coat surrounding equipment with a dry powder and do not destroy it.

Class C fires involve energized electric equipment. A Class C fire extinguisher usually uses a dry powder to smother the fire. Many fire extinguishers can be used on multiple types of fires; for example, an extinguisher labeled ABC could be used on any of the three classes of fire. The important thing to remember is never to use an extinguisher on a fire for which it is not rated. Using a Class A extinguisher filled with water on an electrical fire could be fatal.

Class D fires consist of burning metal. Spraying water on some burning metals can actually cause the fire to increase. Class D extinguishers place a powder on top of the burning metal that forms a crust to cut off the oxygen supply to the metal. Some metals cannot be extinguished by placing powder on them, in which case the powder should be used to help prevent the fire from spreading to other combustible materials.

Probably the single greatest cause of electrical fires is poor or loose connections. Poor connections create resistance, and resistance causes heat. Electricity is a form of pure energy. The electrical unit of watts measures the amount of electrical energy converted into some other form. When resistance is present, electrical energy is converted into thermal energy or heat. The typical soldering irons used for connecting electronic components generally range from 15 to 30 watts (*Figure S–17*). Even a small amount of resistance can cause an extreme amount of heat in a high current load. Assume that a connection has developed 0.25 ohm of resistance in a circuit with a current draw of 25 amperes. The heat dissipated at the connection can be determined using Ohm's law:

$$P = I^2R$$

$$P = 25 \times 25 \times 0.25$$

$$P = 156.25 \text{ watts}$$

This amount of heat can start a fire in many different types of combustible materials.

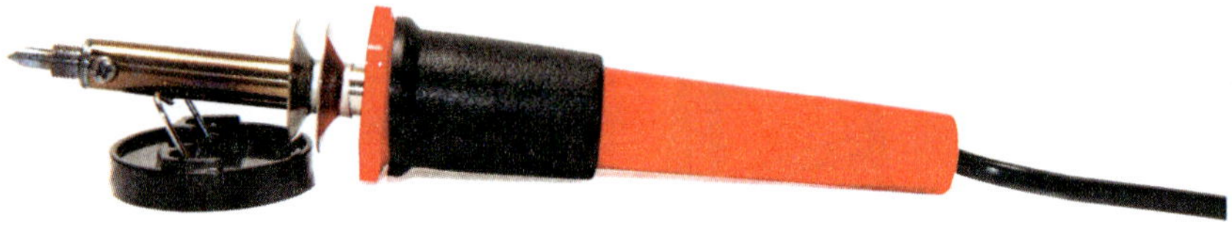

Figure S–17 Example of a soldering iron used to connect electronic components.

Once a poor connection begins overheating, it causes the metal of the connector and terminal to oxidize and develop even more resistance, causing more damage. The result of a poor connection at a terminal is shown in *Figure S–18*. The conductor connected to the terminal eventually melted and was burned off (*Figure S–19*).

Figure S–18 Burned terminal caused by a poor connection.

Figure S–19 Burned wire caused by a poor connection.

S–7 Ground-Fault Circuit Interrupters

Ground-fault circuit interrupters (GFCIs) are used to prevent people from being electrocuted. However, they will not prevent a person from receiving a shock; they only seek to limit the amount of current that is allowed to flow through the body. They work by sensing the amount of current flow on both the ungrounded (hot) and grounded (neutral) conductors supplying power to a device. In theory, the amount of current in the two conductors should be equal but opposite in polarity (*Figure S–20*). In this example, a current of 10 amperes flows in both the hot and neutral conductors.

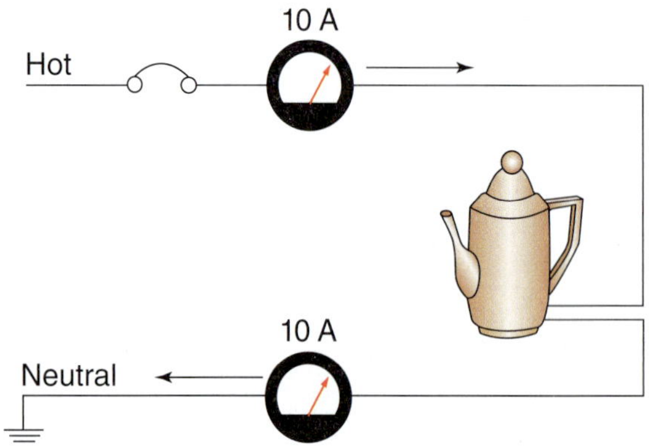

Figure S–20 The current in both the "hot" and neutral conductors should be the same but flowing in opposite dircctions.

A ground fault occurs when a path to ground other than the intended path is established (*Figure S–21*). Assume that a person comes in contact with a defective electric appliance. If the person is grounded, a current path can be established through the person's body. In the example shown in *Figure S–21*, it is assumed that a current of 0.1 ampere is flowing through the person. This means the hot conductor now has a current of 10.1 amperes, but the neutral conductor has a current of only 10 amperes. The GFCI is designed to detect this current difference to protect personnel by opening the circuit when it detects a current difference of approximately 5 milliamperes (0.005 ampere). The *National Electrical Code® (NEC®)* lists locations where ground-fault circuit interrupter protection is required as well as certain equipment types that require this protection. The *National Electrical Code* and *NEC* are registered trademarks of the National Fire Protection Association, Quincy, MA.

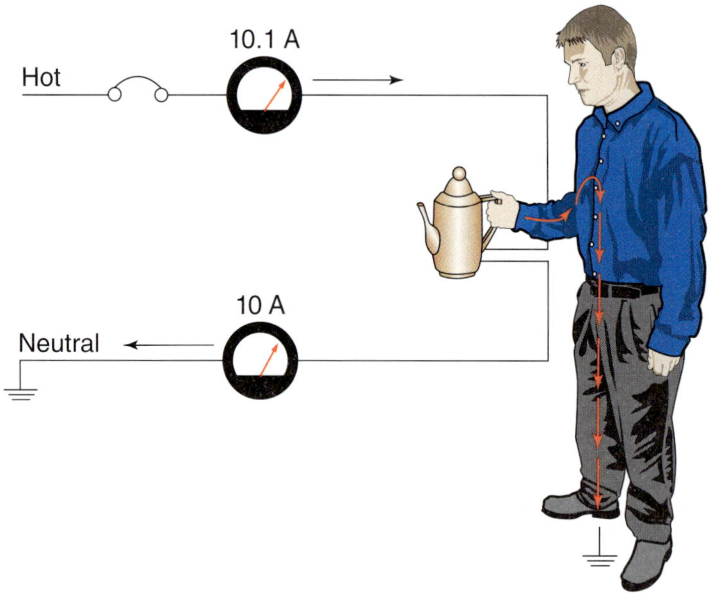

Figure S–21 A ground fault occurs when a path to ground other than the intended path is established.

GFCI Devices

Several devices can be used to provide GFCI protection, including the GFCI-type circuit breaker (*Figure S–22*). The circuit breaker provides GFCI protection for an entire circuit, so any device connected to the circuit is protected. A second method of protection, GFCI-type receptacles (*Figure S–23*), provides protection at the point of attachment. They have some advantages over the GFCI-type circuit breaker. They can be connected

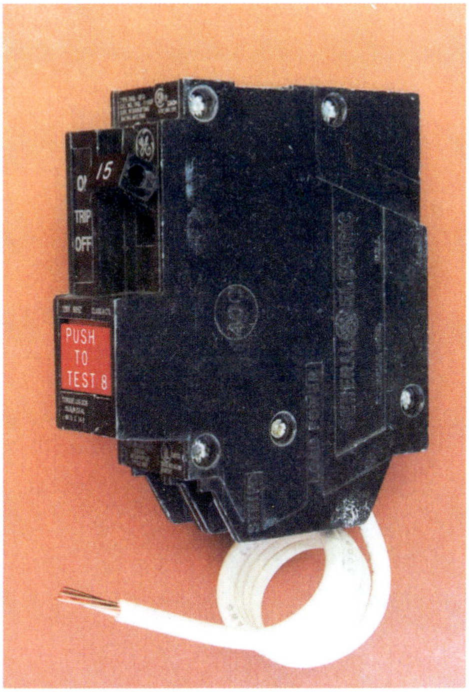

Figure S–22 GFCI-type circuit breaker.

Figure S–23 GFCI-type receptacle.

so they protect only the devices connected to them and do not protect any other outlets on the same circuit, or they can be connected so they provide protection to other outlets. Another advantage is that, because they are located at the point of attachment for the device, there is no stray capacitance loss between the panel box and the equipment being protected. A third type of GFCI device is the GFCI extension cord (*Figure S–24*). It can be connected into any standard receptacle outlet, and any devices connected to it are then protected by a GFCI device.

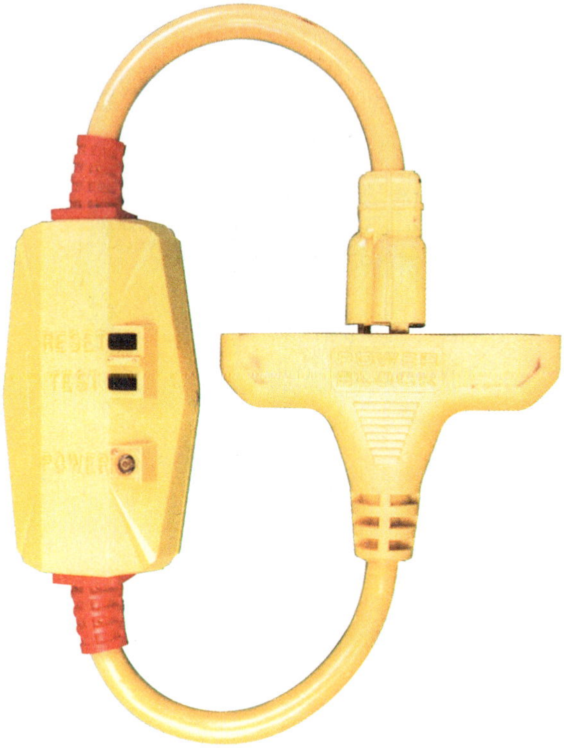

Figure S–24 Pigtail-type GFCI device.

S–8 Arc-Fault Circuit Interrupters

Arc-fault circuit interrupters (AFCIs) are similar to ground-fault circuit interrupters in that they are designed to protect people from a particular hazard. Whereas the GFCI is designed to protect against electrocution, the AFCI is intended to protect against fire. Studies have shown that one-third of electrical-related fires are caused by an arc-fault condition. At present, the *National Electrical Code* requires that AFCIs be used on all 120-volt, single-phase, 10-, 15-, and 20-ampere circuits installed in dwelling units supplying power to kitchens, family rooms, dining rooms, living rooms, parlors, libraries, dens, bedrooms, sunrooms, recreation rooms, closets, hallways, laundry areas, or similar areas.

An arc fault is a plasma flame that can develop temperatures in excess of 6000°C (10,832°F). Arc faults occur when an intermittent gap between two conductors or a conductor and ground permits current to "jump" between the two conductive surfaces. There are two basic types of arc faults: the parallel and the series.

Parallel Arc Faults

Parallel arc faults are caused by two conductors becoming shorted together (*Figure S–25*). A prime example of this is when the insulation of a lamp cord or extension cord has become damaged and permits the two conductors to short together. The current in this type of fault is limited by the resistance of the conductors in the circuit ***and not by the load***. The current in this type of fault is generally much higher than the rated current of a typical thermomagnetic circuit breaker. A continuous short will usually cause the circuit breaker to trip almost immediately because it will activate the magnetic part of the circuit breaker, but an intermittent short may take some time to heat the thermal part of the circuit breaker enough to cause it to trip open. Thermal/magnetic-type circuit breakers are generally effective in protecting against this type of arc fault, but cords with small conductors, such as lamp and small extension cords, can add enough resistance to the circuit to permit the condition to exist long enough to produce sufficient heat to start a fire.

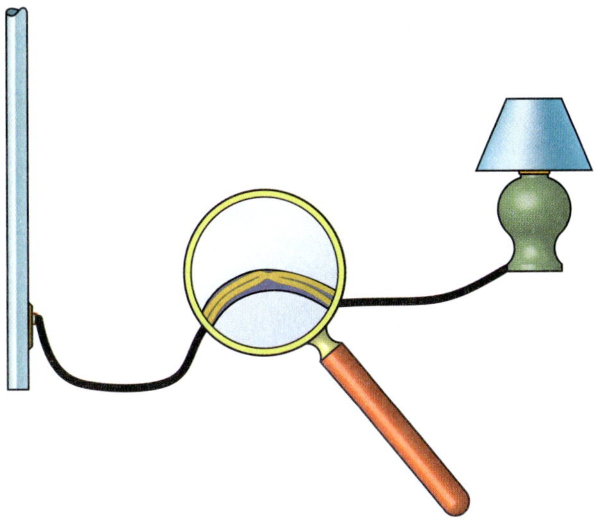

Figure S–25 Parallel arc faults are caused by two conductors touching.

Parallel arc faults can be more hazardous than series arc faults because they generate a greater amount of heat. Arc faults of this type often cause hot metal to be ejected into combustible material. Parallel arc faults, however, generally produce peak currents that are well above the normal current rating of a circuit breaker. This permits the electronic circuits in the AFCI to detect them very quickly and trip the breaker in a fraction of a second.

Series Arc Faults

Series arc faults are generally caused by loose connections. A loose screw on an outlet terminal or an improperly made wire nut connection are prime examples of this type of problem. They are called series arc faults because the circuit contains some type of current-limiting resistance connected in series with the arc (*Figure S–26*). Although the amount of electrical energy converted into heat is less than that of a parallel arc fault, series arc faults are dangerous because the current is limited by the load. This keeps the current below the thermal and magnetic trip rating of a common thermomagnetic circuit breaker. Because the peak arc current is never greater than the normal steady current flow due to the arcing current being the circuit current itself, series arcing is more difficult to detect than parallel arcing.

When the current of an arc remains below the normal range of a common thermomagnetic circuit breaker, it cannot provide protection. For example, if a hair dryer normally has a current draw of 12 amperes, but the receptacle outlet has a loose connection at the terminal screw, the current supplying the hair dryer will bridge the gap between the conductor and the termination point. This will lead to the termination point overheating and potentially igniting building materials, even though the current never exceeds the circuit rating. A special type of protective device is needed to interrupt the circuit when loose connections or broken conductors are present.

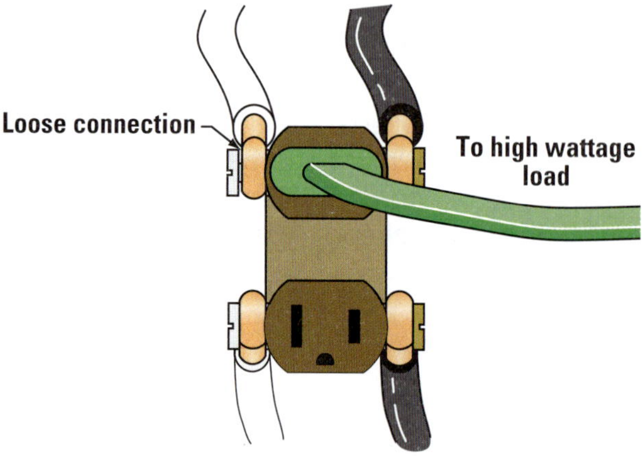

Figure S–26 Series arc faults are generally caused by bad connections.

Arc-Fault Detection

There are conditions where arcing in an electric circuit is normal, such as these:

- Turning a light switch on or off
- Switching on or off of a motor relay
- Using motors that contain a commutator and brushes

The AFCI is designed to be able to distinguish between normally occurring arcs and an arc fault. An arc caused by a toggle switch being used to turn a light on or off will produce a current spike of short duration, as shown in *Figure S–27*. An arc fault, however, is an intermittent connection and will generally produce current spikes of various magnitudes and lengths of time (*Figure S–28*).

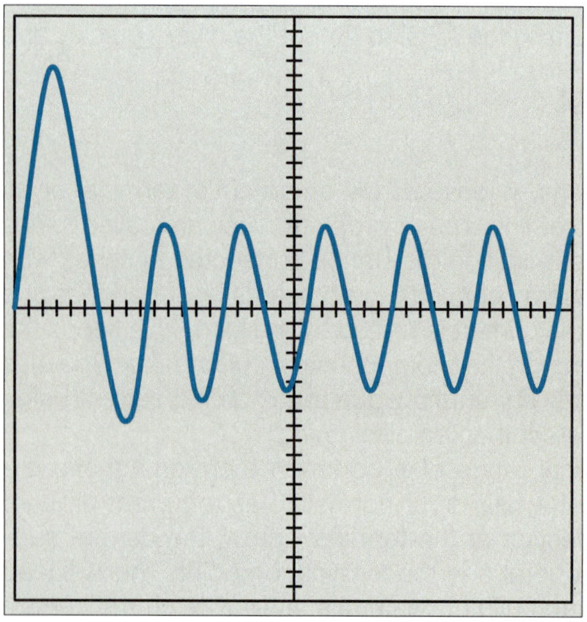

Figure S–27 Current spike produced by turning a light on or off.

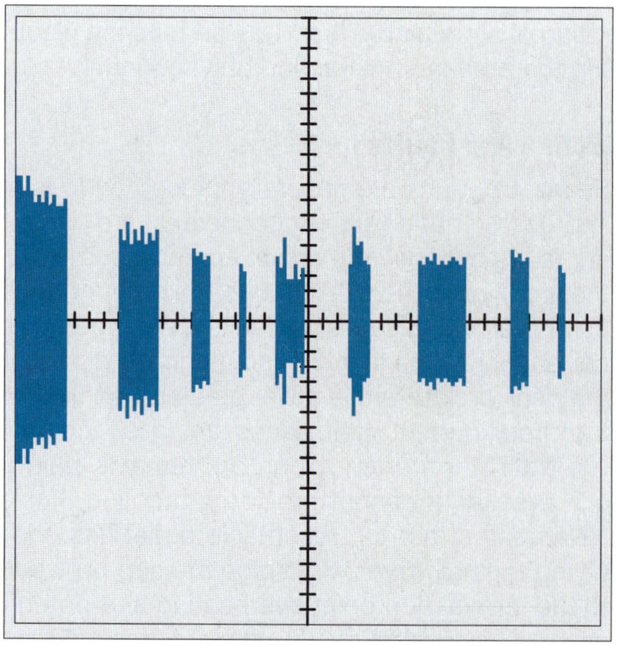

Figure S–28 Waveform produced by typical arc fault.

For an AFCI to determine the difference between a normally occurring arc and an arc fault, a microprocessor and other related electronic components are employed to detect these differences. The AFCI contains current and temperature sensors as well as a microprocessor and nonvolatile (retains its information when power is switched off) memory. The current and temperature sensors permit the AFCI to operate as a normal circuit breaker in the event of a circuit overload or short circuit. The microprocessor continuously monitors the current and compares the waveform to information stored in the memory. The microprocessor is monitoring the current for the magnitude, duration, and length of time between pulses, not for a particular waveform.

Connecting an Arc-Fault Circuit Interrupter

The AFCI is connected in the same manner as a ground-fault circuit breaker. The AFCI contains a white pigtail (*Figure S–29*) that is connected to the neutral bus bar in the panel box. Both the neutral and hot or ungrounded conductors of the branch circuit are connected to the AFCI-type circuit breaker. The circuit breaker contains a silver-colored and a brass-colored screw. The neutral or white wire of the branch circuit is inserted under the silver screw, and the black wire is inserted under the brass screw (*Figure S–30*). A rocker switch located on the front of the AFCI permits the breaker to be tested for both short and arc condition. In addition to the manual test switch, the microprocessor performs a self-test about once every 10 minutes.

Figure S–29 Arc-fault circuit interrupter-type circuit breaker.

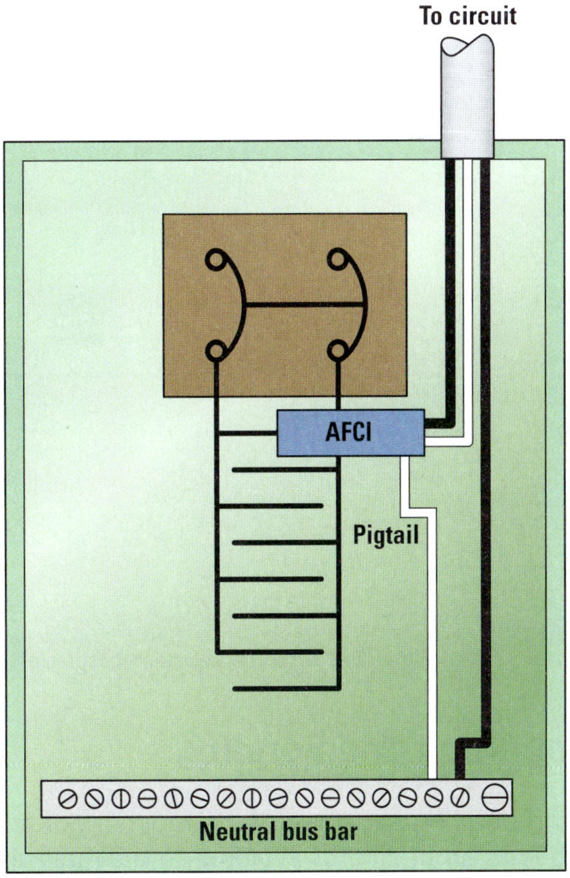

Figure S–30 The arc-fault circuit interrupter connects in the same manner as a ground-fault interrupter.

Tamper-Resistant Receptacles

Tamper-resistant receptacles contain a plastic mechanism that covers the outlet openings (*Figure S–31*). The covering will retract and permit entry only when a device with two blades, such as a male plug, is inserted. The covering will not retract if an object is pressed into only one opening. Tamper-resistant receptacles are used in areas where small children are present, such as day-care facilities, a child's playroom, and nurseries.

Figure S–31 Tamper-resistant receptacle.

S–9 Grounding and Bonding

Grounding and bonding are some of the most important safety considerations in the electrical field. The *NEC* requires grounding of systems to maintain a consistent and stable voltage-to-ground by holding the system at an earth potential. Bonding often gets lumped in with grounding as all normally non-current-carrying conductive surfaces are bonded together to carry this earth potential throughout the system. However, bonding also serves a critical safety function by creating a low-impedance conductive path back to the power source that facilitates the automatic operation of overcurrent protective devices to de-energize a circuit during ground-fault conditions. The equipment grounding conductor is an example of a conductor that both provides a low-resistance path for fault current to return to the service and connects the equipment to ground. Many electric appliances are provided with a three-wire cord. The third prong is connected to the case of the appliance and forces the case to exist at ground potential. If an ungrounded conductor comes in contact with the case, the equipment grounding conductor conducts the current directly back to the service and causes a high level of current to flow in the circuit which causes the overcurrent protection to operate and interrupt the circuit automatically. The third prong on a plug should never be cut off or defeated. Grounding requirements are far too numerous to list in this chapter, but *NEC 250* covers the requirements for the grounding of electrical systems.

Summary

- Never work on an **energized circuit**. Always establish an **electrically safe work condition**.
- The most important rule of safety is to think before you act.
- Avoid horseplay.
- Do not work alone, if possible.
- Learn first aid and CPR.
- A current of 100 to 200 milliamperes can cause death.
- The mission of OSHA is to ensure safe and healthy workplaces in the United States.
- Avoid using alcohol and drugs in the workplace.
- Do not walk close to trenches unless it is necessary.
- Do not jump over trenches.
- Place barricades around trenches.
- Use ladders to enter and exit trenches.
- When working in confined spaces, an outside person should keep in constant contact with people inside the space.
- Lockout and tagout procedures are used to prevent someone from energizing until the work is finished.
- Scaffolds generally provide the safest elevated working platforms.
- The bottom of a straight ladder should be placed at a distance from the wall that is equal to one-fourth the height of the ladder.
- Fires can be divided into four classes: Class A is common items such as wood and paper; Class B is grease, liquids, and gases; Class C is energized electric equipment; and Class D is metals.
- GFCIs are used to protect people from electrocution.
- GFCI protectors open the circuit when approximately 5 milliamperes of ground-fault current are sensed.
- AFCIs protect against electrical fires by sensing an arc-fault condition. They employ a microprocessor to sense an arc-fault condition.
- *NEC Article 250* lists requirements for grounding electrical systems.
- Poor connections are one of the leading causes of electrical fires.

Review Questions

1. What one strategy does OSHA mandate above all others with regard to minimizing the risk of injury from an electrical hazard?

   A) Personal protective equipment

   B) De-energization

   C) Work practices

   D) Job planning

2. True or False: The meter must be tested on a known voltage source both before and after it is used
   to verify de-energization.

3. At what voltage does OSHA establish that the voltage is high enough to pose a threat to a person if they make contact with energized parts?

   A) 100 volts

   B) 30 volts

   C) 120 volts

   D) 50 volts

4. At what current level does the heart run the risk of going into ventricular fibrillation?

   A) 75 milliamperes

   B) 20 milliamperes

   C) 5 milliamperes

   D) 30 milliamperes

5. The Occupational Safety and Health Administration was created by Congress in 1971 and serves a mission to:

   A) provide personal protective equipment.

   B) develop industry consensus standards.

   C) ensure safe and healthful workplaces in the United States.

   D) guarantee a workplace free from all hazards.

6. True or False: OSHA is only concerned with electrical hazards within the workplace.

7. What special type of clothing are workers required to wear when performing energized tasks on high-power equipment such as motor-control centers and switchgear?

   A) Arc-rated clothing

   B) Flame-resistant clothing

   C) Limited flammability clothing

   D) Long sleeve cotton work shirt

8. What type of personal protective equipment is commonly worn by electricians to prevent shock when they must perform work on or near electrical equipment that has not been de-energized?

   A) Safety glasses

   B) Hearing protection

   C) A Class E hard hat

   D) Rubber insulating gloves with protectors

9. True or False: A properly installed and maintained GFCI will prevent an individual from experiencing any shock at all if they contact a protected circuit.

10. AFCIs operate by monitoring the connected circuit for signs that electrical current is arcing between conductors in either a parallel or a series arcing fault or, as in the case with combination type, both series and parallel. In a series arc, how much of the circuit current is involved in the arc fault?

    A) Roughly 80% of the circuit current

    B) 100% of the normal circuit current

    C) None of the normal circuit current

    D) Roughly 300–400% of the circuit current

11. True or False: The equipment grounding conductor provides a low-impedance path for fault current to return to the service.

12. It is important for the equipment grounding conductor to allow a high amount of fault current to flow in a ground fault to facilitate the operation of:

    A) electrical equipment.

    B) sensitive electronic equipment.

    C) overcurrent protection.

    D) portable tools.

Unit

1

Atomic Structure

Key Terms

Alternating current (AC)
Atom
Atomic number
Attraction
Bidirectional
Conductor
Direct current (DC)
Electron
Electron orbit
Element
Insulators
Matter
Molecules
Negative
Neutron
Nucleus
Positive
Proton
Repulsion
Semiconductors
Unidirectional
Valence electrons

Why You Need to Know

Atoms are the building blocks of the universe, and all matter is composed of them. One of the smallest components of the atom is the electron. This tiny little particle in our universe can be credited with many of the largest innovations and accomplishments of all humankind! All electrical quantities, such as voltage, current, and watts, are based on other electrical units that are derived from the measurement of electrons. A basic understanding of electron flow will remove the "mystery" of electricity and will start you on a path to a further understanding of electrical theory. This unit explains

- how electricity is produced and how those sources are divided into alternating current (AC) and direct current (DC) for utilization.
- why some materials are conductors and others are insulators.
- why a conductor becomes warm as current flows through it.

Outline

Objectives

After studying this unit, you should be able to

- describe the basic building blocks of an atom.
- demonstrate how key principles of Coulomb's law affect the action of particles with respect to other particles.
- describe the importance of valence shell electrons to electricity.
- describe how electrons orbit the nucleus of an atom.
- discuss the differences between conductors, semiconductors, and insulators.
- explain methods for producing an electrical charge and the various effects of electricity.

Preview

Electricity is the driving force that provides most of the power for the industrialized world. It is used to light homes, cook meals, heat and cool buildings, and drive motors, and it even powers many of the automobiles on the road today. Today's workforce, regardless of their chosen field, can benefit from an understanding of how electricity works as our world is ever evolving and becoming more and more based on the power of the electron!

Electric sources are divided into two basic types: **direct current (DC)** and **alternating current (AC)**. Direct current is **unidirectional**, which means it flows in only one direction. The first part of this text is mainly devoted to the study of direct current. Alternating current is **bidirectional**, which means it reverses its direction of flow at regular intervals. The latter part of this text is devoted mainly to the study of alternating current.

1–1 Early History of Electricity

Although the practical use of electricity has become common only within the past hundred years, it has been known as a force for much longer. The Greeks were the first to discover electricity about 2500 years ago. They noticed that when amber was rubbed with other materials, it became charged with an unknown force that had the power to attract objects such as dried leaves, feathers, bits of cloth, or other lightweight materials. The Greeks called amber *elektron*. The word *electric* was derived from it and meant "to be like amber" or to have the ability to attract other objects.

This mysterious force remained little more than a curious phenomenon until about 2000 years later, when other people began to conduct experiments. In the early 1600s, William Gilbert discovered that amber was not the only material that could be charged to attract other objects. He called materials that could be charged *electriks* and materials that could not be charged *nonelektriks*.

About 300 years ago, a few men began to study the behavior of various charged objects. In 1733, a Frenchman named Charles DuFay found that a piece of charged glass would repel some charged objects and attract others. These men soon learned that the force of **repulsion** was just as important as the force of **attraction**. From these experiments, two lists were developed (*Figure 1–1*). It was determined that any material in list A would attract any material in list B, that all materials in list A would repel each other, and that all materials in list B would repel each other (*Figure 1–2*). Various names were suggested for the materials in lists A and B. Any opposite-sounding names could have been chosen, such as east and west, north and south, left and right. Benjamin Franklin named the materials in list A **positive** and the materials in list B **negative**. These names are still used today. The first item in each list was used as a standard for determining whether a charged object was positive or negative. Any object repelled by a piece of glass rubbed on silk would have a positive charge, and any item repelled by a hard rubber rod rubbed on wool would have a negative charge.

List A	List B
Glass (rubbed on silk)	Hard rubber (rubbed on wool)
Glass (rubbed on wool or cotton)	Block of sulfur (rubbed on wool or fur)
Mica (rubbed on cloth)	Most kinds of rubber (rubbed on cloth)
Asbestos (rubbed on cloth or paper)	Sealing wax (rubbed on silk, wool, or fur)
Stick of sealing wax (rubbed on wool)	Mica (rubbed on dry wool)
	Amber (rubbed on cloth)

Figure 1–1 List of charged materials.

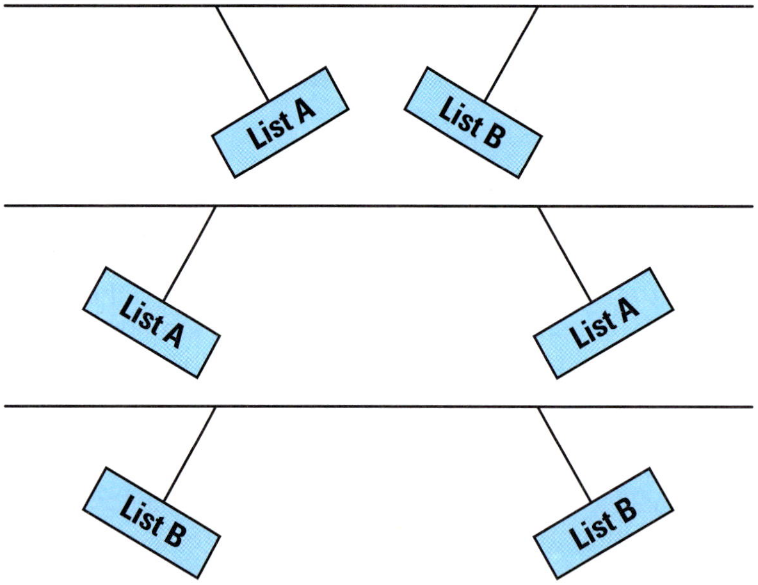

Figure 1–2 Unlike charges attract and like charges repel.

1–2 Atoms

Understanding electricity necessitates starting with the study of atoms. The **atom** is the basic building block of the universe. All **matter** is made from a combination of atoms. Matter is any substance that has mass and occupies space. Matter can exist in any of three states: solid, liquid, or gas. Water, for example, can exist as a solid in the form of ice, as a liquid, or as a gas in the form of steam (*Figure 1–3*). An **element** is a substance that cannot be chemically divided into two or more simpler substances. A table listing both natural and artificial elements is shown in *Figure 1–4*. An atom is the smallest part of an element. The three principal parts of an atom are the **electron**, the **neutron**, and the **proton**.

Although most atoms contain these three principal parts, the smallest atom, hydrogen, does not contain a neutron (*Figure 1–5*). Hydrogen contains one proton and one electron. The smallest atom that contains neutrons is helium (*Figure 1–6*). Helium contains two protons, two neutrons, and two electrons. Protons and neutrons are made up of smaller particles called *quarks*.

Figure 1–3 Water can exist in three states: liquid, solid, and gas. The state it exists in depends on both temperature and pressure.

ATOMIC NUMBER	NAME	VALENCE ELECTRONS	SYMBOL	ATOMIC NUMBER	NAME	VALENCE ELECTRONS	SYMBOL	ATOMIC NUMBER	NAME	VALENCE ELECTRONS	SYMBOL
1	Hydrogen	1	H	37	Rubidium	1	Rb	73	Tantalum	2	Ta
2	Helium	2	He	38	Strontium	2	Sr	74	Tungsten	2	W
3	Lithium	1	Li	39	Yttrium	2	Y	75	Rhenium	2	Re
4	Beryllium	2	Be	40	Zirconium	2	Zr	76	Osmium	2	Os
5	Boron	3	B	41	Niobium	1	Nb	77	Iridium	2	Ir
6	Carbon	4	C	42	Molybdenum	1	Mo	78	Platinum	1	Pt
7	Nitrogen	5	N	43	Technetium	2	Tc	79	Gold	1	Au
8	Oxygen	6	O	44	Ruthenium	1	Ru	80	Mercury	2	Hg
9	Fluorine	7	F	45	Rhodium	1	Rh	81	Thallium	3	Tl
10	Neon	8	Ne	46	Palladium	–	Pd	82	Lead	4	Pb
11	Sodium	1	Na	47	Silver	1	Ag	83	Bismuth	5	Bl
12	Magnesium	2	Ma	48	Cadmium	2	Cd	84	Polonium	6	Po
13	Aluminum	3	Al	49	Indium	3	In	85	Astatine	7	At
14	Silicon	4	Si	50	Tin	4	Sn	86	Radon	8	Rd
15	Phosphorus	5	P	51	Antimony	5	Sb	87	Francium	1	Fr
16	Sulfur	6	S	52	Tellurium	6	Te	88	Radium	2	Ra
17	Chlorine	7	Cl	53	Iodine	7	I	89	Actinium	2	Ac
18	Argon	8	A	54	Xenon	8	Xe	90	Thorium	2	Th
19	Potassium	1	K	55	Cesium	1	Cs	91	Protactinium	2	Pa
20	Calcium	2	Ca	56	Barium	2	Ba	92	Uranium	2	U
21	Scandium	2	Sc	57	Lanthanum	2	La				
22	Titanium	2	Ti	58	Cerium	2	Ce		Artificial Elements		
23	Vanadium	2	V	59	Praseodymium	2	Pr				
24	Chromium	1	Cr	60	Neodymium	2	Nd	93	Neptunium	2	Np
25	Manganese	2	Mn	61	Promethium	2	Pm	94	Plutonium	2	Pu
26	Iron	2	Fe	62	Samarium	2	Sm	95	Americium	2	Am
27	Cobalt	2	Co	63	Europium	2	Eu	96	Curium	2	Cm
28	Nickel	2	Ni	64	Gadolinium	2	Gd	97	Berkelium	2	Bk
29	Copper	1	Cu	65	Terbium	2	Tb	98	Californium	2	Cf
30	Zinc	2	Zn	66	Dysprosium	2	Dy	99	Einsteinium	2	E
31	Gallium	3	Ga	67	Holmium	2	Ho	100	Fermium	2	Fm
32	Germanium	4	Ge	68	Erbium	2	Er	101	Mendelevium	2	Mv
33	Arsenic	5	As	69	Thulium	2	Tm	102	Nobelium	2	No
34	Selenium	6	Se	70	Ytterbium	2	Yb	103	Lawrencium	2	Lw
35	Bromine	7	Br	71	Lutetium	2	Lu				
36	Krypton	8	Kr	72	Hafnium	2	Hf				

Figure 1–4 Table of both natural and artificial elements and associated information such as atomic number and symbol.

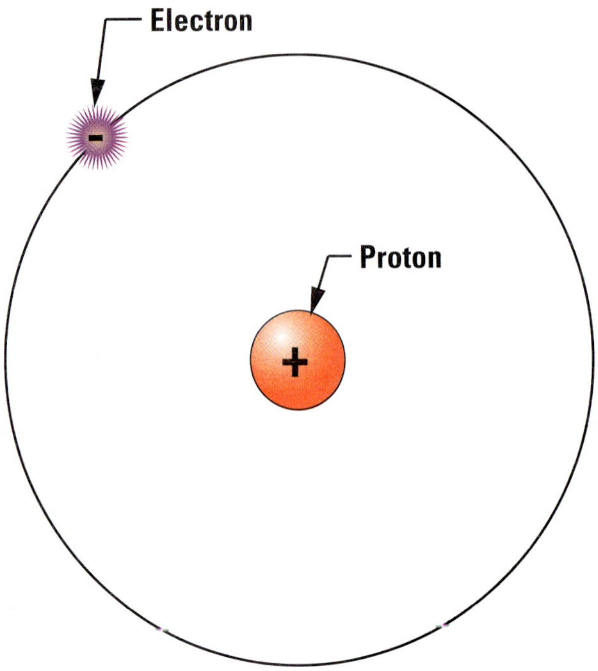

Figure 1–5 Hydrogen contains one proton and one electron.

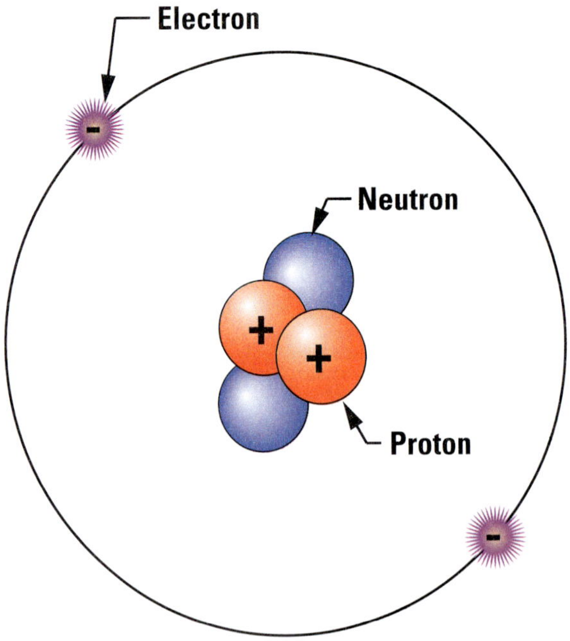

Figure 1–6 Helium contains two protons, two neutrons, and two electrons.

Notice that the proton has a positive charge, the electron a negative charge, and the neutron no charge. The neutrons and the protons combine to form the **nucleus** of the atom. Because the neutron has no charge, the nucleus has a net positive charge. The number of protons in the nucleus determines what kind of element an atom is. Oxygen, for example, contains 8 protons in its nucleus, and gold contains 79. The **atomic number** of an element is the same as the number of protons in the nucleus. These protons produce lines of force that extend outward in all directions (*Figure 1–7*). The nucleus may or may not contain as many neutrons as protons. For example, an atom of helium contains 2 protons and 2 neutrons in its nucleus, whereas an atom of copper contains 29 protons and 35 neutrons (*Figure 1–8*).

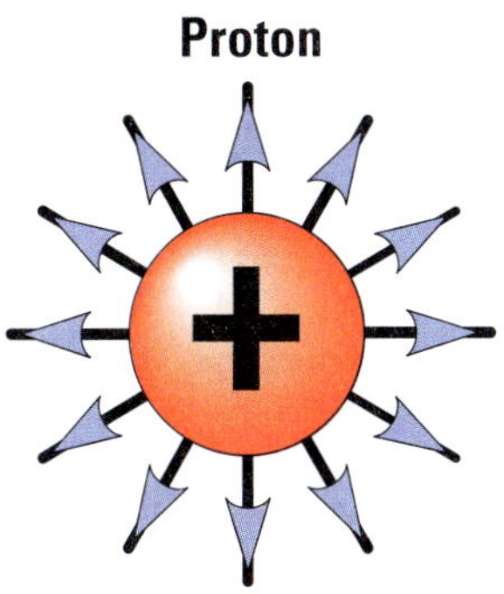

Figure 1–7 The proton produces lines of force that extend outward.

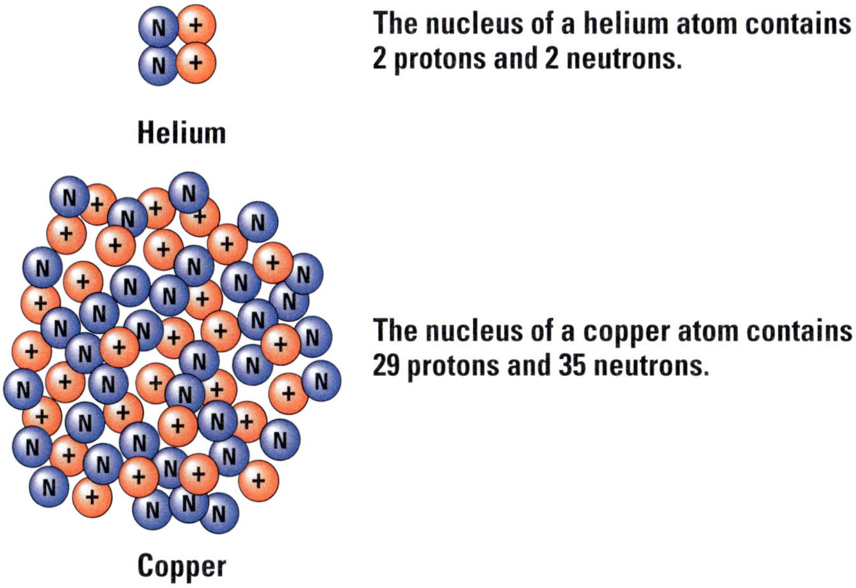

Figure 1–8 The nucleus may or may not contain the same number of protons and neutrons.

The electron orbits the outside of the nucleus. The latest scientific measurements suggest that protons and neutrons weigh about 1838 times more than the electron and that the electron is approximately 1000 times smaller than a proton. *Figure 1–5* shows a representative image of the size relationship between these two particles; however, this is not drawn to scale. Previously, it was believed that the electron was about three times larger than the proton. Many scientists believe that it is almost impossible to accurately measure the size of these particles and that the actual size is of no real importance. It is the characteristics of the particles that are of importance. Because the electron exhibits a negative charge, the lines of force come in from all directions (*Figure 1–9*).

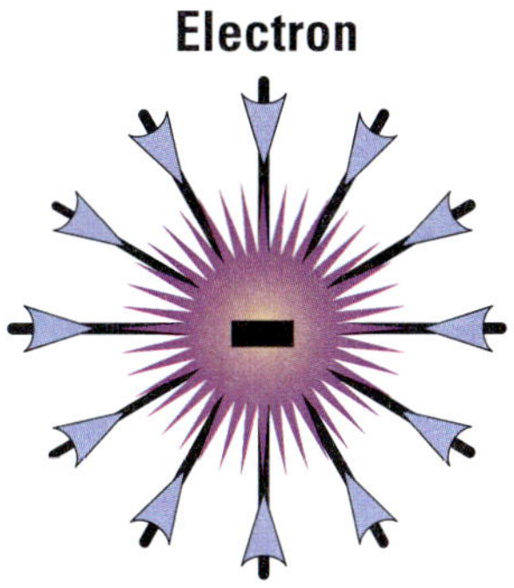

Figure 1–9 The electron produces lines of force that come inward.

1–3 Coulomb's Law

Understanding atoms necessitates first understanding a basic law of physics that states that **opposite charges attract and like charges repel**. In *Figure 1–10*, which illustrates this principle, charged particles are suspended from strings. Notice that the two particles that contain opposite charges are attracted to each other. The two positively charged particles repel each other, as do the two negatively charged particles. The reason for this is that lines of force can never cross each other. The outward-going lines of force of a positively charged object combine with the inward-going lines of force of a negatively charged object (*Figure 1–11*). This combination produces an attraction between the two objects. If two objects with like charges come close to each other, the lines of force repel them (*Figure 1–12*). Because the nucleus has a net positive charge and the electron has a negative charge, the electron is attracted to the nucleus.

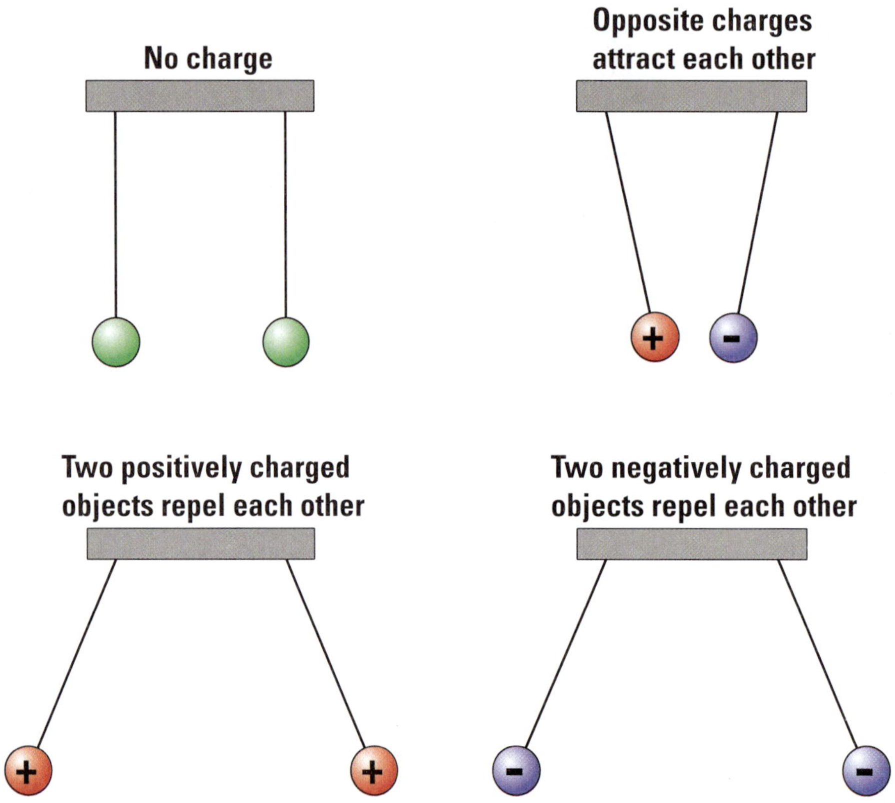

Figure 1–10 Unlike charges attract and like charges repel.

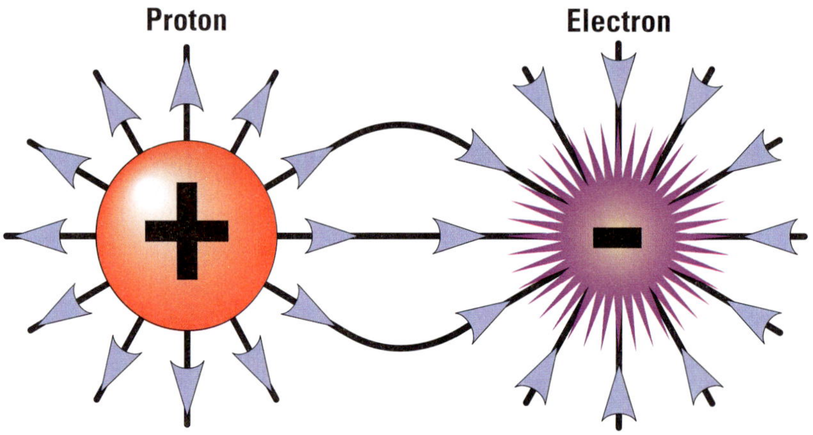

Figure 1–11 Unlike charges attract each other.

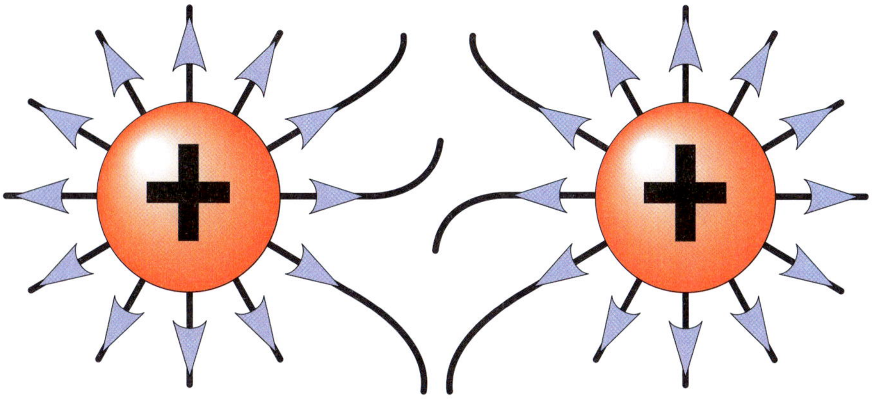

Figure 1–12 Like charges repel each other.

Because the nucleus of an atom is formed from the combination of protons and neutrons, one may ask why the protons of the nucleus do not repel each other because they all have the same charge. Two theories attempt to explain this. The first theory asserted that the force of gravity held the nucleus together. Neutrons, like protons, are extremely massive particles. It was first theorized that the gravitational attraction caused by their mass overcame the repelling force of the positive charges. By the mid-1930s, however, it was known that the force of gravity could not hold the nucleus together. According to Coulomb's law, the electromagnetic force in helium is about 1.1×10^{36} times greater than the gravitational force as determined by Newton's law. In 1947, the Japanese physicist Hideki Yukawa identified a subatomic particle that acts as a mediator to hold the nucleus together. The particle is a quark known as a *gluon*. The force of the gluon is about 10^{2} times stronger than the electromagnetic force.

1–4 Structure of the Atom

In 1808, an English scientist named John Dalton proposed that all matter was composed of atoms. Although the assumptions that Dalton used to prove his theory were later found to be factually incorrect, the idea that all matter is composed of atoms was adopted by most of the scientific world. Then, in 1897, J.J. Thomson discovered the electron. Thomson determined that electrons have a negative charge and that they have very little mass compared to the atom. He proposed that atoms have a large, positively charged massive body with negatively charged electrons scattered throughout it. Thomson also proposed that the negative charge of the electrons exactly balanced the positive charge of the large mass, causing the atom to have a net charge of zero. Thomson's model of the atom proposed that electrons existed in a random manner within the atom, much like firing BBs from a BB gun into a slab of cheese. This was referred to as the plum pudding model of the atom.

In 1913, a Danish scientist named Neils Bohr presented the most widely accepted theory concerning the structure of an atom. In the Bohr model, electrons exist in specific or "allowed" orbits around the nucleus in much the same way that planets orbit the Sun (*Figure 1–13*). The orbit in which the electron exists is determined by the electron's mass times its speed times the radius of the orbit. These factors must equal the positive force of the nucleus. In theory, there can be an infinite number of allowed orbits. Because the number of protons in the atom is equal to the number of electrons in the atom, the overall net charge of the atom is zero. This is the stable condition of the atom.

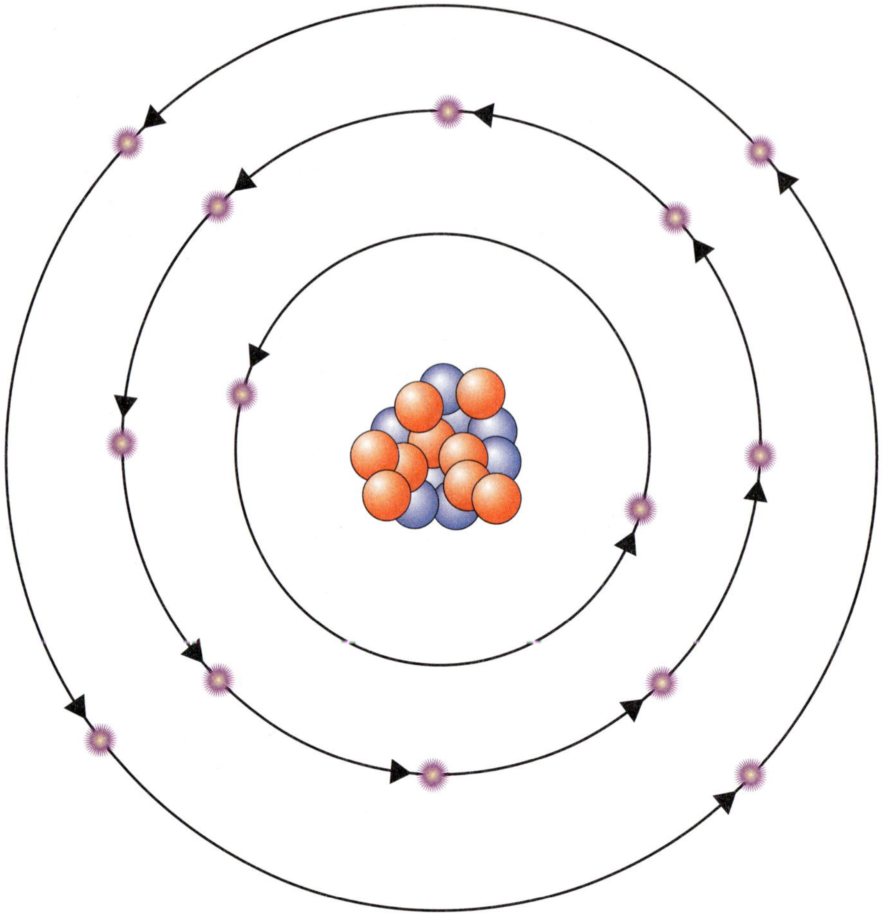

Figure 1–13 Bohr's model of the atom proposed that electrons orbit the nucleus in much the same way that planets orbit the Sun.

When an electron receives enough energy from some other source, it "quantum jumps" into a higher allowed orbit. Electrons, however, tend to return to a lower allowed orbit. When this occurs, the electron emits the excess energy as a single photon of electromagnetic energy. This movement of electrons caused by the external energy source is the basis of electric current flow. This will be described further in Section 1–6 as it relates to valence electrons.

1–5 Electron Orbits

Each **electron orbit** of an atom contains a set number of electrons (*Figure 1–14*). The number of electrons that can be contained in any one orbit, or shell, is found by the formula ($2N^2$). The letter *N* represents the number of the orbit or shell. For example, the first orbit can hold no more than 2 electrons:

$$2 \times (1)^2 \text{ or}$$

$$2 \times 1 = 2$$

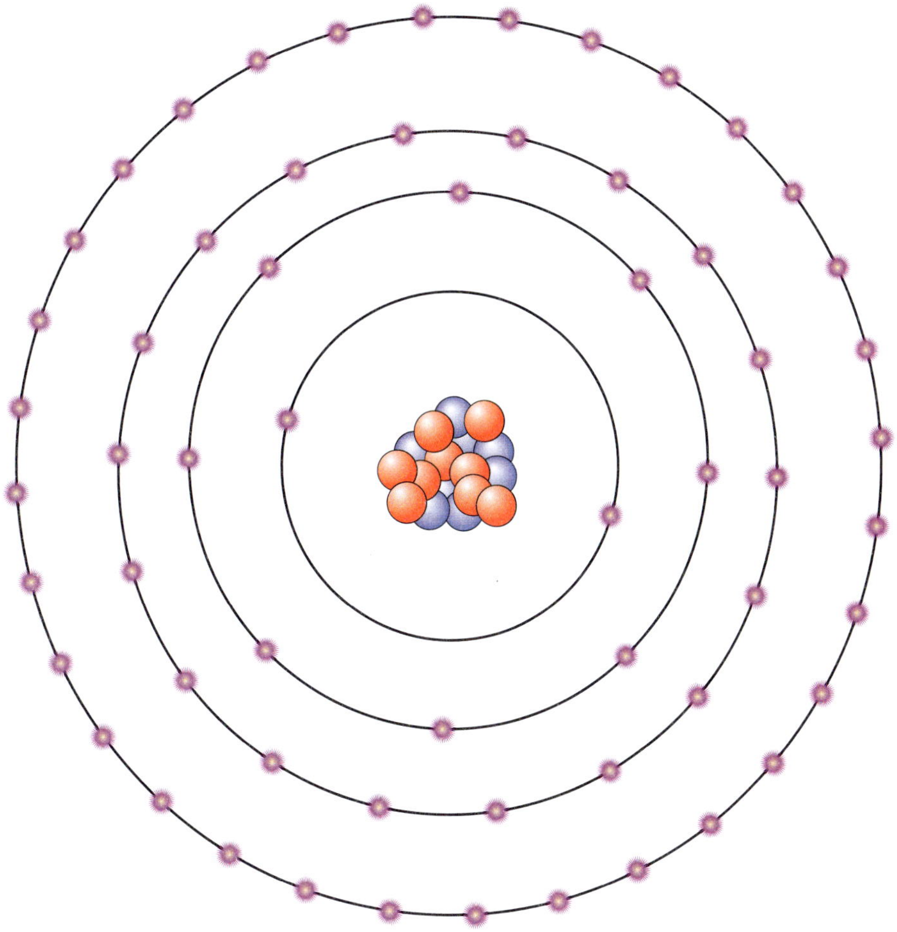

Figure 1–14 Electron orbits.

The second orbit can hold no more than 8 electrons:

$$2 \times (2)^2 \text{ or}$$

$$2 \times 4 = 8$$

The third orbit can contain no more than 18 electrons:

$$2 \times (3)^2 \text{ or}$$

$$2 \times 9 = 18$$

The fourth and fifth orbits cannot hold more than 32 electrons. Thirty-two is the maximum number of electrons that can be contained in any orbit:

$$2 \times (4)^2 \text{ or}$$

$$2 \times 16 = 32$$

Although atoms are often drawn flat, as illustrated in *Figure 1–14*, electrons orbit the nucleus in a spherical fashion, as shown in *Figure 1–15*. Electrons travel at such a high rate of speed that they form a shell around the nucleus. For this reason, electron orbits are often referred to as *shells*.

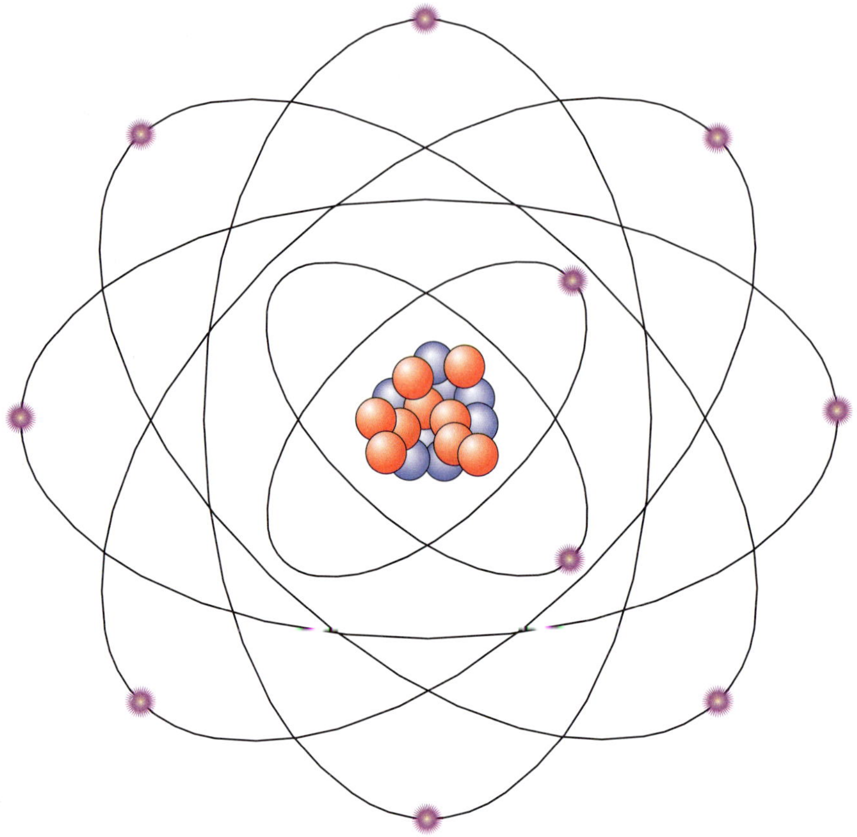

Figure 1–15 Electrons orbit the nucleus in a spherical fashion.

1–6 Valence Electrons

The outer shell of an atom is known as the *valence shell*. Any electrons located in the outer shell of an atom are known as **valence electrons** (*Figure 1–16*). The valence shell of an atom cannot hold more than 8 electrons. Atoms with 8 electrons in the valence shell are very stable. Elements such as neon, argon, krypton, xenon, and radon all contain 8 valence electrons. These elements are called noble gases and are inert. Atoms with less than 8 electrons in the valence shell can join with other atoms to share valence electrons to form stable molecules or compounds when the shared valence electrons effectively add up to the desired 8 outer electrons. The joining of different atoms to form molecules and compounds is described in Section 1–10. Atoms with less than 8 electrons are less stable, and an external energy force can cause these valence electrons to escape the attraction of the nucleus. When an electron escapes the hold of the nucleus, it becomes a free electron, and the atom ends up with a net positive charge. This is called ionization, and the positive atom is called a positive ion. In a similar manner, the free electron may lose energy and fall into orbit around a neutral atom, causing that atom to have a net negative charge. That atom is called a negative ion. This explains why some materials can be made to have a positive charge, while others have a negative charge. Atoms with one, two, or three valence electrons are the least stable, and they can be made to give up these valence electrons with less energy applied. The valence electrons are of primary concern in the study of electricity because these electrons explain much of electrical theory. A **conductor**, for instance, is made from a material that contains between one and three valence electrons. Conductors are materials that permit electrons to flow through them easily. When an atom has only one or two valence electrons, these electrons are loosely held by the atom and are easily given up for current flow. Silver, copper, and gold all contain one valence electron and are excellent conductors of electricity. Silver is the best natural conductor of electricity, followed by copper, gold, and aluminum. An atom of copper is shown in *Figure 1–17*. Although it is known that atoms containing few valence electrons are the best conductors, it is not known why some of these materials are better conductors than others. Copper, gold, platinum, and silver all contain only one valence electron. Silver, however, conducts electricity more readily than any of the others. Aluminum, which contains three valence electrons, is a better conductor than platinum, which contains only one valence electron.

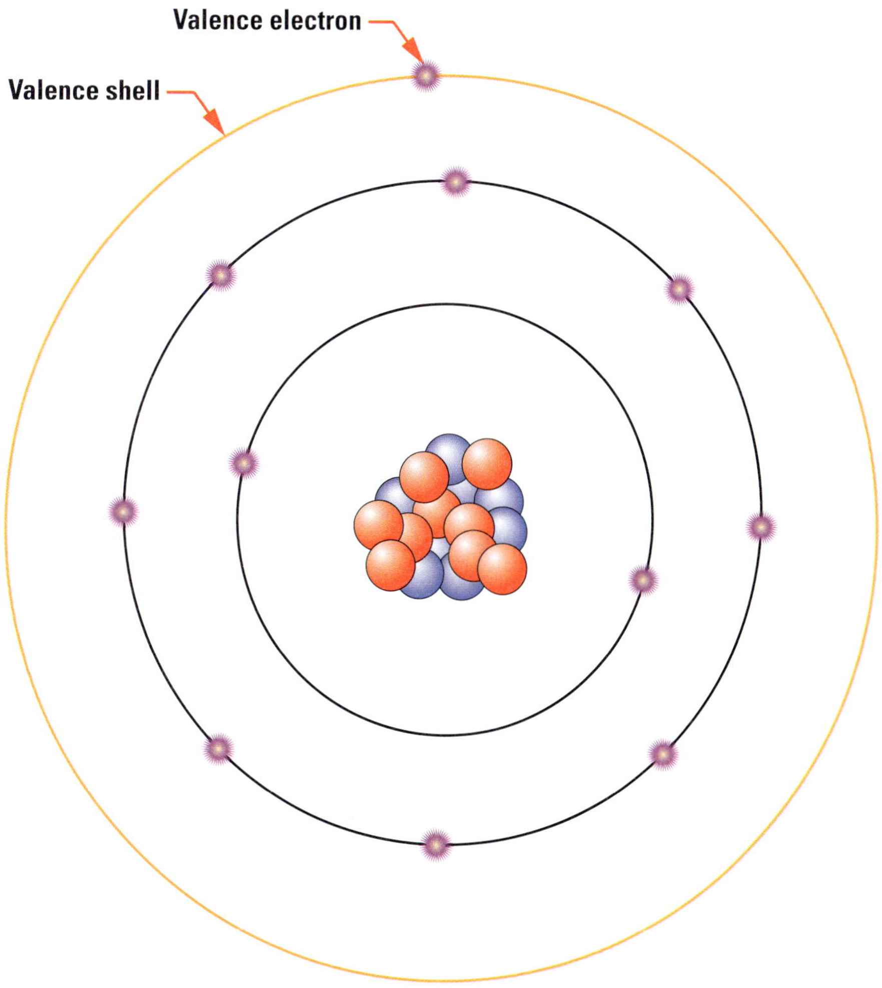

Figure 1–16 The electrons located in the outer orbit of an atom are valence electrons.

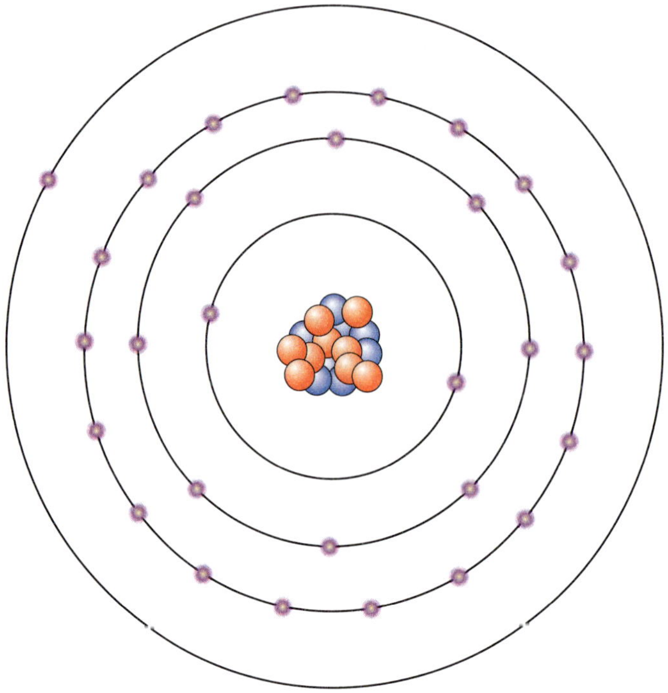

Figure 1–17 A copper atom contains 29 electrons and has 1 valence electron.

1–7 The Bump Theory

Electrical current is the flow of electrons between atoms of a conductive material. There are several theories concerning how electrons are made to flow through a conductor. One theory is generally referred to as the *bump theory*. It states that current flow is produced when an electron from one atom knocks electrons of another atom out of orbit. *Figure 1–18* illustrates this action. When an atom contains only one valence electron, that electron is easily given up when struck by another electron. The striking electron gives its energy to the electron being struck. The striking electron may settle into orbit around the atom, and the electron that was struck moves off to strike another electron. This same effect can be seen in the game of pool. If the moving cue ball strikes a stationary ball exactly right, the energy of the cue ball is given to the stationary ball. The stationary ball then moves off with the cue ball's energy, and the cue ball stops moving (*Figure 1–19*). The additional energy causes the electron to move out of orbit and become a free electron. After traveling a short distance, the electron enters the valence orbit of a different atom. When it returns to orbit, some or all of the gained energy is released in the form of heat, which is why conductors become warm when current flows through them. If too much current flows in a conductor, it may become hot enough to cause a fire.

If an atom containing two valence electrons is struck by a moving electron, the energy of the striking electron is divided between the two valence electrons (*Figure 1–20*). If the valence electrons are knocked out of orbit, they contain only half the energy of the striking electron. This effect can also be seen in the game of pool (*Figure 1–21*). If a moving cue ball strikes two stationary balls at the same time, the energy of the cue ball is divided between the two stationary balls. Both stationary balls will move but with only half the energy of the cue ball.

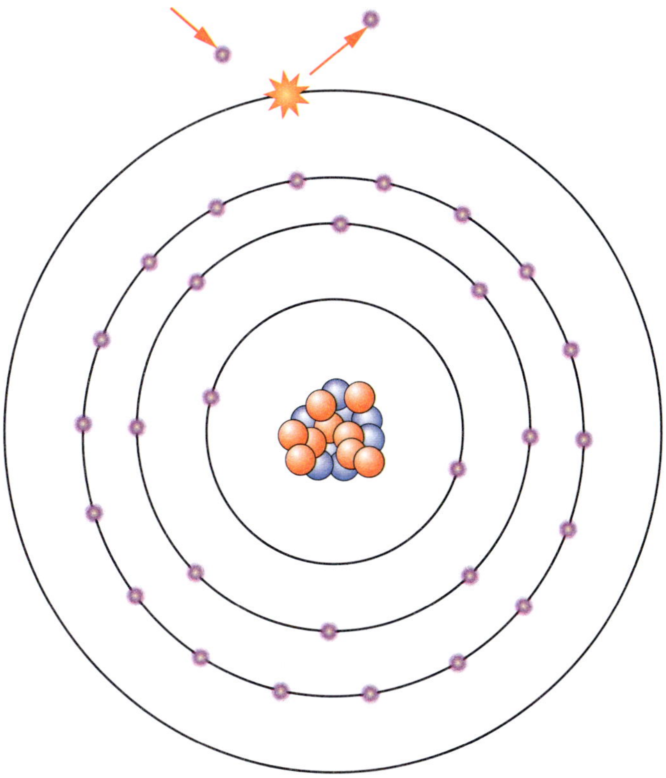

Figure 1–18 An electron of one atom knocks an electron of another atom out of orbit.

Figure 1–19 The energy of the cue ball is given to the ball being struck.

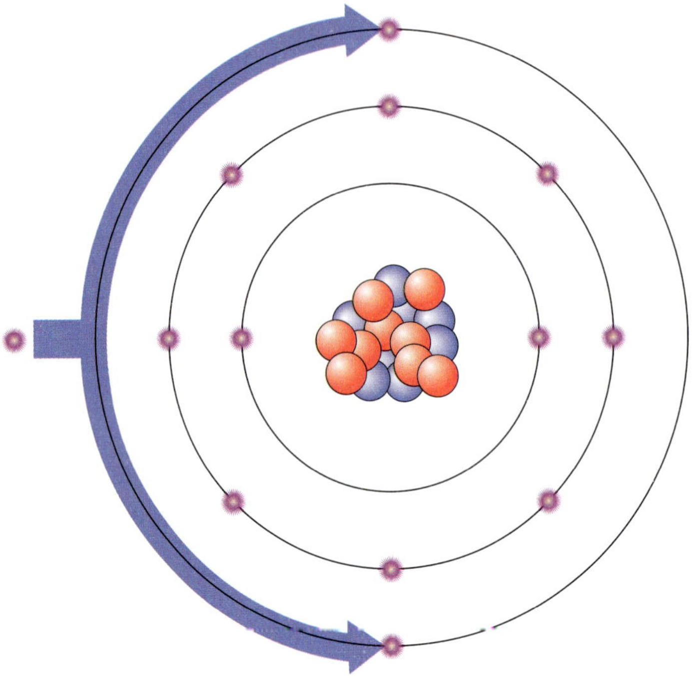

Figure 1–20 The energy of the striking electron is divided.

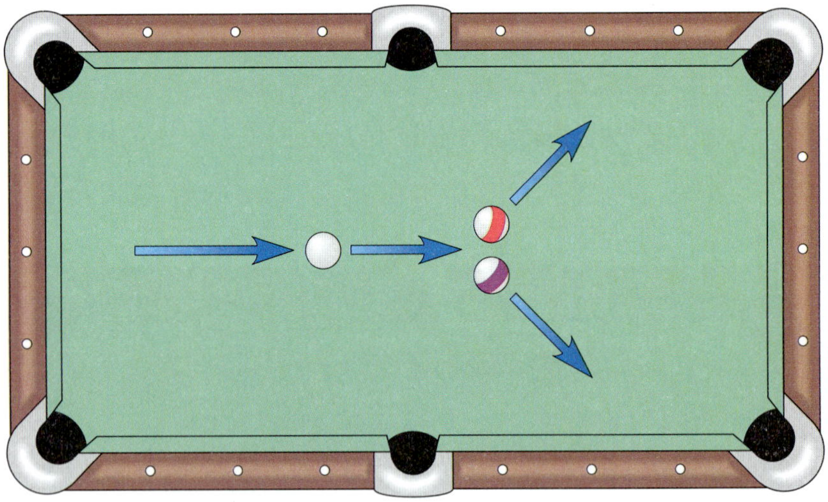

Figure 1–21 The energy of the cue ball is divided between the two other balls.

1–8 Power Sources

Many theories on how electricity is produced state that electric power sources produce a positive terminal and a negative terminal. The negative terminal is created by causing an excess of electrons to form at that terminal, and the positive terminal is created by removing a large number of electrons from that terminal, leaving an imbalance of positively charged protons (*Figure 1–22*). Different methods can be employed to produce the excess of electrons at one terminal and deficiency of electrons at the other, but when a circuit is completed between the two terminals, negative electrons are pushed from the negative terminal into the conductive material and are attracted to the positive terminal (*Figure 1–23*). The greater the difference in the number of electrons between the negative and positive terminals, the greater the force of repulsion and attraction. Different methods may be employed to cause a difference of potential. The battery, illustrated in *Figures 1–22* and *1–23*, develops a difference in potential by chemical reaction. Generators develop a difference in potential by cutting lines of magnetic flux with conductors.

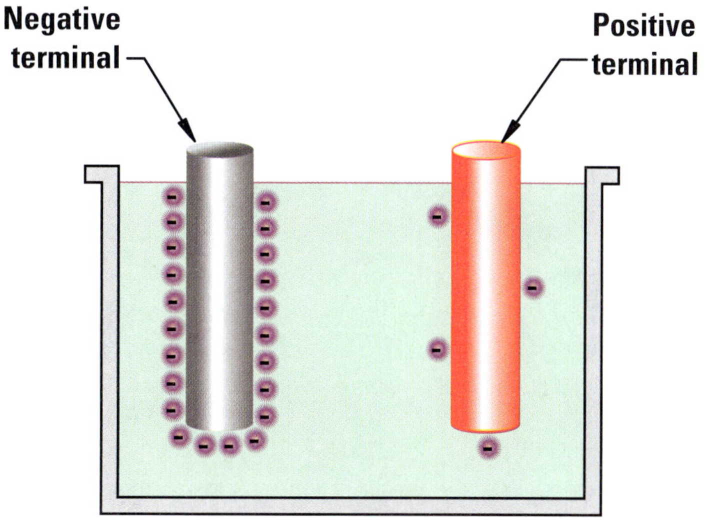

Figure 1–22 All electrical power sources produce a positive and a negative terminal.

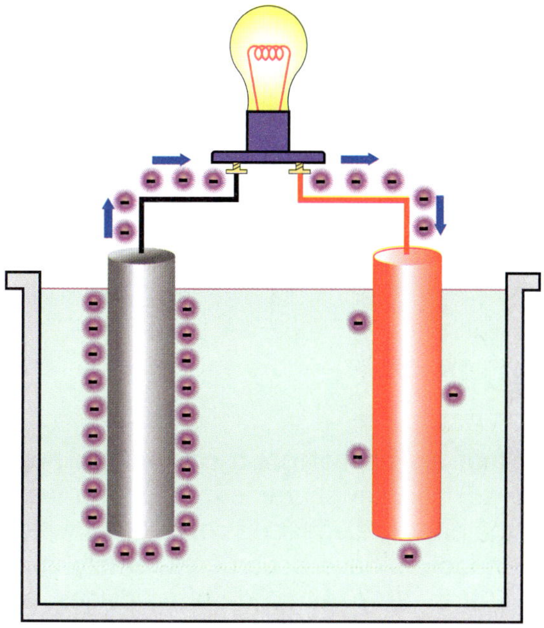

Figure 1–23 Completing a circuit between the positive and negative terminals causes electrons to be repelled from the negative terminal and attracted to the positive terminal.

1–9 Insulators

Materials containing seven or eight valence electrons are known as **insulators**. Insulators are materials that resist the flow of electricity. When the valence shell of an atom is full or almost full, the electrons are held tightly and are not given up easily. Some good examples of insulator materials are rubber, plastic, glass, and wood. *Figure 1–24* illustrates what happens when a moving electron strikes an atom containing eight valence electrons. The energy of the moving electron is divided so many times that it has little effect on the atom. Any atom that has seven or eight valence electrons is extremely stable and does not easily give up an electron.

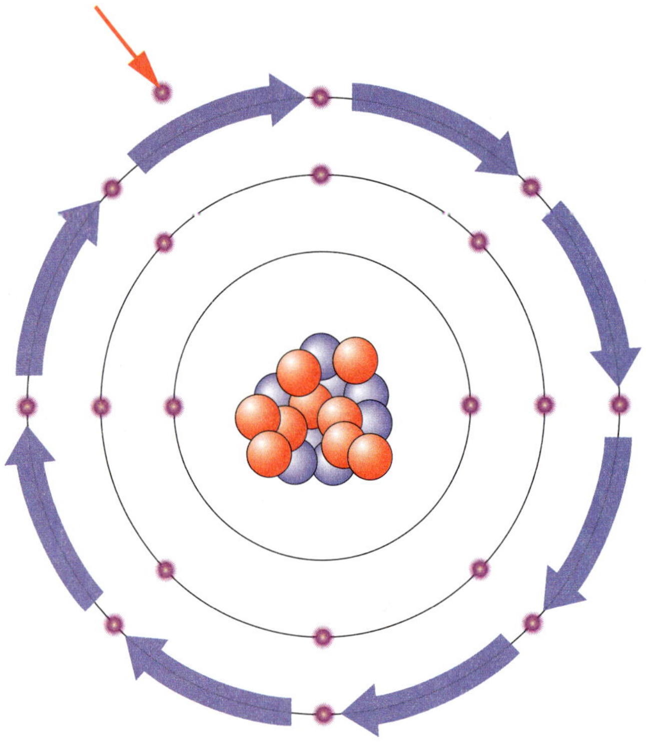

Figure 1–24 The energy of the striking electron is divided among the eight valence electrons of an atom with insulator properties.

1–10 Semiconductors

Semiconductors are materials that are neither good conductors nor good insulators. They contain four valence electrons (*Figure 1–25*) and are characterized by the fact that, as they are heated, their resistance decreases. Heat has the opposite effect on conductors, whose resistance *increases* with an increase in temperature. Semiconductors have become extremely important in the electrical industry since the invention of the transistor in 1947. All solid-state devices, such as diodes, transistors, and integrated circuits, are made from combinations of semiconductor materials. The two most common materials used in the production of electronic components are silicon and germanium. Of the two, silicon is used more often because of its ability to withstand heat. Before any pure semiconductor can be used to construct an electronic device, it must be mixed or "doped" with an impurity to create either a positive or negative state in the silicon.

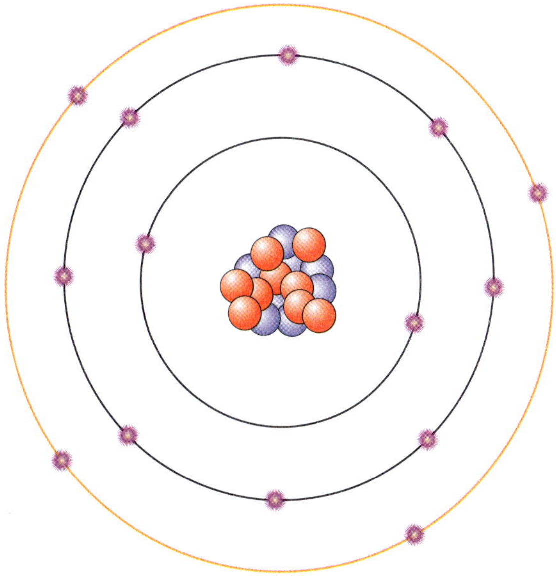

Figure 1–25 Semiconductors contain four valence electrons.

1–11 Molecules

Although all matter is made from atoms, atoms should not be confused with **molecules**, which are the smallest part of a compound. Water, for example, is a compound, not an element. The smallest particle of water is a molecule made up of two atoms of hydrogen and one atom of oxygen, H_2O (*Figure 1–26*). Notice that the two hydrogen atoms share their single valence electron with the six valence electrons in the oxygen atom to effectively make a stable valence shell of eight electrons. Thus, water is a stable molecule. If the molecule of water is broken apart, it becomes two hydrogen atoms and one oxygen atom and is no longer water.

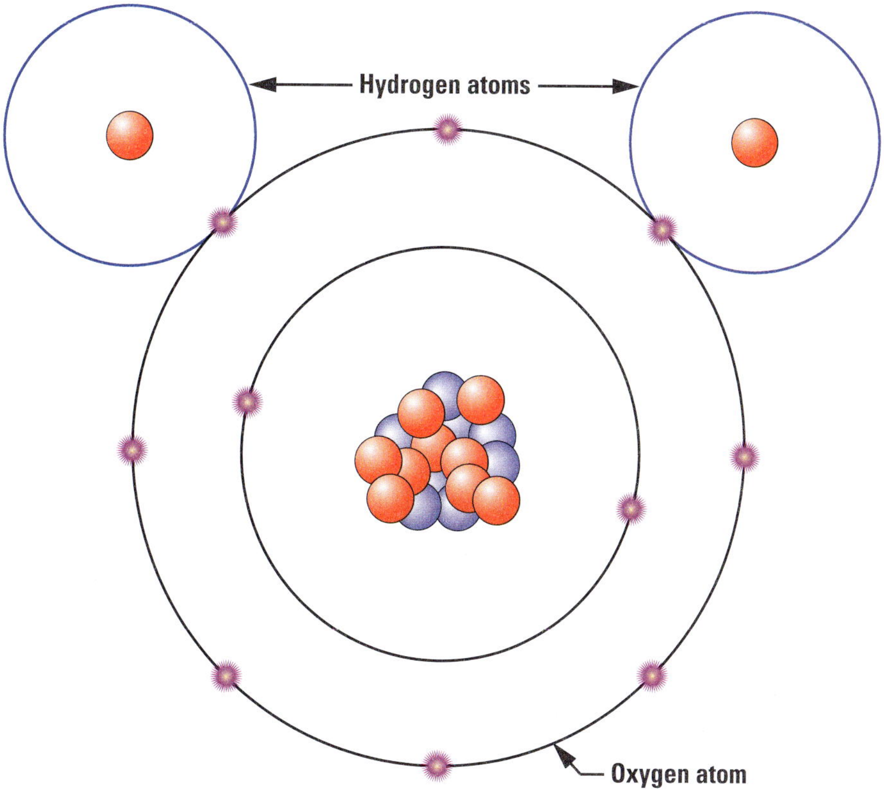

Figure 1–26 A water molecule.

1–12 Methods of Producing Electricity

So far in this unit, it has been discussed that electricity is a flow of electrons. There are six basic methods for producing electricity:

1. Magnetism
2. Chemical action
3. Pressure
4. Heat
5. Friction
6. Light

Of the six methods listed, magnetism is the most common method used to produce electricity. Electromagnetic induction is the operating principle of all generators and alternators. These principles are covered in more depth later in this text.

The second most common method of producing electricity is chemical action. The chemical production of electricity involves the movement of entire ions instead of just electrons. The principles of conduction in liquids are discussed in Units 11 and 12.

The production of electricity by pressure involves the striking, bending, or twisting of certain crystals. This effect is referred to as the *piezo* electric effect. The word *piezo* is derived from a Greek word meaning "pressure."

Producing electricity with heat is referred to as the *Seebeck* effect. The Seebeck effect is the operating principle of thermocouples. Thermocouples are discussed in Unit 12.

Static charges are probably the best example of producing electricity by friction. A static charge occurs when certain materials are rubbed together, and electrons are transferred from one object to the other. Static electricity is discussed in Unit 3.

Producing electricity from light involves the use of particles called *photons*. In theory, photons are massless particles of pure energy. Photons can be produced when electrons are forced to change to a lower energy level. This is the operating principle of gas-filled lights such as sodium vapor, mercury vapor, and so on. Electricity can be produced by photons when they strike a semiconductor material. The energy of the photon is given to an electron, forcing it to move out of orbit. This is the operating principle of *photovoltaic* devices called *solar cells*. Solar cells are discussed in Unit 12. Other photo-operated devices are *photoemissive* and *photoconductive*. Photoemissive devices include photodiodes, phototransistors, photo–silicon-controlled rectifiers, and so on. These devices are generally used to sense light when the speed of operation is imperative. Photoconductive devices change resistance with a change of light. The most common photoconductive device is the cad cell (*Figure 1–27*). Cad cells exhibit a resistance of about 50 ohms (Ω) in direct sunlight and several hundred thousand ohms in darkness. Cad cells are often used to create devices that can be used to turn on outside lighting when it becomes dark outside.

Figure 1–27 Cad cell.

1–13 Electrical Effects

With the exception of friction, electricity can be used to cause the same effects that produce it:

1. Magnetism
2. Chemical reactions
3. Pressure
4. Heat
5. Light

Anytime electric current flows through a conductor, a magnetic field is created around the conductor. This principle is discussed in Unit 4.

Electricity can be used to produce certain chemical reactions, such as electroplating. Electroplating is accomplished by placing a base metal and a pure metal in a chemical solution. An object can be copper plated, for example, by placing a base metal object and a piece of pure copper in a solution of cuprous cyanide. The object to be plated is connected to the negative electrode, and the pure copper is connected to the positive electrode. Atoms of copper are transferred through the solution and deposited on the object to be plated.

Another example of electricity producing chemical reactions can be seen in the process of *electrolysis*. Electrolysis is the process of separating elements electrically. These principles are discussed in Unit 11.

Just as the twisting or bending of certain crystals can produce electricity, electricity can cause certain crystals to bend or twist. When electricity is applied to a certain size and shape of quartz crystal, the crystal vibrates at a certain rate. This principle has been used in crystal radios for many years. If an electric current is applied to a piece of Rochelle salt crystal, the crystal vibrates. This is the operating principle of a crystal earphone (*Figure 1–28*).

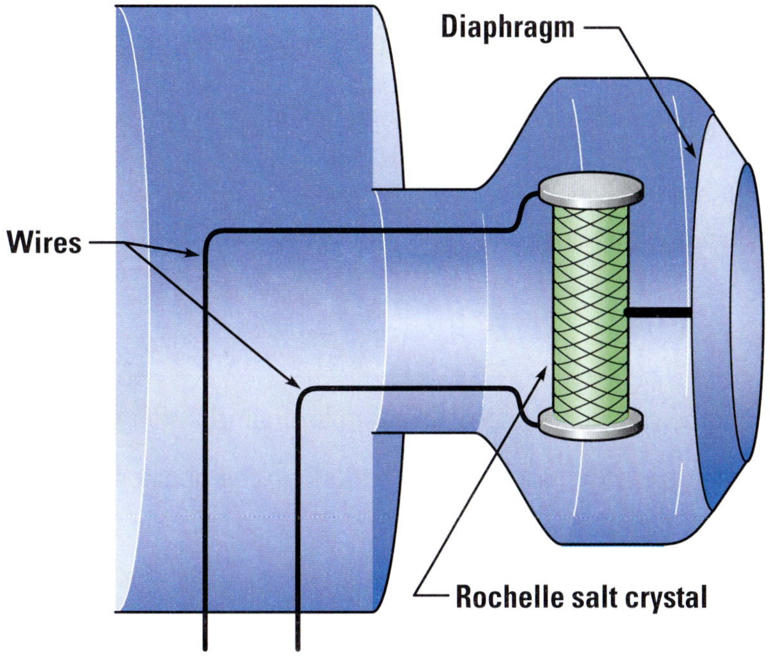

Figure 1–28 Producing sound with electricity.

As discussed previously in this unit, when electrons enter a valence orbit, heat is often produced. This is the reason that conductors become warm as current flows through them. This is also the operating principle of many heat-producing devices such as electric ranges, electric irons, electric heaters, and so on.

Light is produced when electrons move to a lower orbit and produce a photon. When electric current is applied to certain conductors, they not only become hot, but they also emit photons of light. Incandescent lamps are an example of this principle. When electric current is applied to the filament of an incandescent lamp, most of the electrical energy is converted into heat, but part of it produces photons of light. Incandescent lamps, however, are very inefficient. The typical 100-watt lamp produces about 95 watts of heat and 5 watts of light. Other lighting sources such as sodium vapor, mercury vapor, and fluorescent are much more efficient. Some semiconductor devices can be used to produce light without heat. Light-emitting diodes are a good example of these devices and are far more efficient than other types of light-producing technology.

Summary

- The atom is the smallest part of an element.
- The three principal parts of an atom are the proton, the electron, and the neutron.
- Protons have a positive charge, electrons a negative charge, and neutrons no charge.
- Valence electrons are located in the outer orbit, or valence shell, of an atom.
- Conductors are materials that provide an easy path for electrons to flow between atoms.
- Conductors are made from materials that contain from one to three valence electrons.
- Insulators are materials that do not provide an easy path for the flow of electrons.
- Insulators are generally made from materials containing seven or eight valence electrons.
- Semiconductors contain four valence electrons.
- Semiconductors are used in the construction of all solid-state devices, such as diodes, transistors, and integrated circuits.
- A molecule is the smallest part of a compound.
- Six basic methods for producing electricity are magnetism, chemical action, light, heat, pressure, and friction.
- Five basic effects that can be caused by electricity are magnetism, chemical reactions, light, heat, and pressure.
- A photon is a massless particle of pure energy.
- Photons can be produced when electrons move from one energy level to another.

Review Questions

1. True or False: The three principal parts of an atom are named the proton, neutron, and electron.
2. Which of the following parts of an atom can be found in the nucleus? (Select all that apply.)
   A) Electron
   B) Valence shell
   C) Proton
   D) Neutron

3. True or False: Coulomb's law provides for an understanding that like-charged particles will be attracted to each other.
4. When outward-going lines of force that surround a positively charged particle draw near another particle with inward-going lines of force, the resulting action can best be described as:
   A) repelling particles.
   B) attracting particles.
   C) nuclear fusion.
   D) electricity.
5. The movement of electrons between the valence shells of atoms due to an external force being applied is the basis for which concept?
   A) Coulomb's law
   B) Nuclear fusion
   C) Nuclear fission
   D) Electrical current flow
6. True or False: Atoms with 8 electrons in their valence shell allow for the ready transmission of electrons between atoms and thus encourage the flow of electricity.
7. True or False: Electrons orbit the nucleus of an atom at various intervals called electron orbits or shells.
8. What is the maximum number of electrons that can orbit the nucleus of an atom in any given electron shell?
   A) 32
   B) 8
   C) 18
   D) 2
9. How many valence electrons are generally contained in materials used for insulators?
   A) 2
   B) 18
   C) 8
   D) 3
10. True or False: Conductor material is generally made up of atoms that have valence electron counts of 1 or 2 electrons.
11. Which of the following methods is commonly used to produce an electrical current in a generator or alternator?
    A) Chemical action
    B) Magnetism
    C) Pressure
    D) Friction

12. Which of the following are often the result of electrical current flowing through a material? (Select all that apply.)

    A) Magnetism

    B) Heat

    C) Moisture

    D) Light

Unit 2 Electrical Quantities and Ohm's Law

Key Terms

Ampere
British thermal unit
Complete path
Conventional current flow theory
Coulomb
Electromotive force
Electron flow theory
Equipment grounding conductor
Horsepower
Impedance
Joule
Neutral conductor
Ohm
Ohm's law
Potential difference
Power
Resistance
Volt
Watt

Why You Need to Know

Anyone working in the electrical field must know and understand the basic units used to measure electric power. To accomplish this, it is important to understand how electricity works and is evaluated. In order to work with electric components and control devices in the field, you must know the electrical measuring terms and how to apply these terms. This unit presents

- the difference between voltage, current, resistance, and watts. These are the basic units of measurement in the electrical field. You will need to learn how each applies to the field of electricity.
- how electrical current flows within a conductive material. Understanding the theory of how current moves within a circuit is a cornerstone of grasping how a circuit will work. This comes in very useful in diagnostic work later!
- Ohm's law. This is the basis for all electrical calculations. Ohm's law is a method of mathematically determining electrical quantities when other quantities are known. It is possible, for example, to determine the amount of current that will flow in a circuit if the resistance in the circuit and the voltage applied to it are known.
- a discussion of the similarity of electricity and water systems.

Outline

Objectives

After studying this unit, you should be able to

- discuss key terms related to measuring electrical circuit parameters.
- analyze common theories on how electricity flows in conductive materials.
- employ Ohm's law to calculate expected values of electrical measurements within a circuit.
- discuss different types of electrical circuits.

Preview

Electricity has a standard set of values. Before one can work with electricity, one must know these values, understand what they mean, and know how they apply. Because the values of electrical measurement have been standardized, they are the same for everyone who uses them. For example, carpenters use a standard system for measuring length, such as the inch, foot, meter, or centimeter. Imagine what a house would look like if it was constructed by two carpenters who used different lengths of measure for an inch or a foot. The same holds true for people who work with electricity. The standards of measurement must be the same for everyone. Meters are calibrated to indicate the same quantity of current flow or voltage or resistance. A volt, an ampere, a watt, or an ohm is the same anywhere electricity is used in the world.

2–1 The Coulomb

The **coulomb (C)** is a measure of electric charge. The letter *C* is used to represent the coulomb. An electron, designated as q_e, has a charge of 1.6×10^{-19} C. To produce a charge of 1 C would require 6.25×10^{18} or 6,250,000,000,000,000,000 electrons. To better understand the number of electrons contained in one coulomb, think of comparing 1 second to 200 billion years. The coulomb can be thought of as a quantity measure of electrons similar to a quart, gallon, or liter. It requires a certain amount of liquid to equal a liter, just as it requires a certain number of electrons to equal a coulomb.

The coulomb is named after a French scientist who lived in the 1700s named Charles Augustin de Coulomb. Coulomb experimented with electrostatic charges and developed a law dealing with the attraction and repulsion of these forces. The law, known as *Coulomb's law or Coulomb's Inverse-Square Law, proposes that the force of electrostatic attraction or repulsion is directly proportional to the product of the two charges and inversely proportional to the square of the distance between them*. It is the Système Internationale (SI) unit of electric charge. A coulomb is defined as the charge transferred by a current of 1 ampere in 1 second.

2–2 The Ampere

The **ampere (A)** is named after André Ampère, a French scientist who lived from the late 1700s to the early 1800s. Ampère is most famous for his work dealing with electromagnetism, which is discussed in a later unit. The ampere is equal to 1 coulomb per second. Notice that the definition of an ampere involves a quantity measurement, the coulomb, and a time measurement, the second ($I = q/t$). An ampere is the measurement of how much electricity is flowing in a given circuit. One ampere of current flows through a wire when 1 coulomb of electrons flows past a given point in a conductor in 1 second (*Figure 2–1*). The ampere is a rate measurement of how much electricity is flowing through a circuit. It is comparable to gallons per minute or gallons per second in systems that carry fluids (*Figure 2–2*). The letter *I*, which

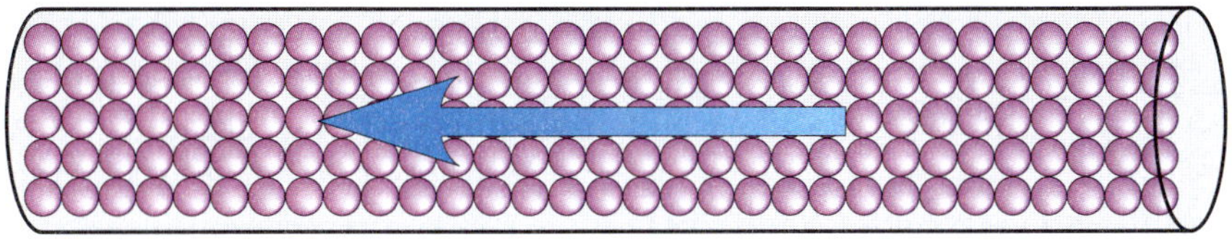

Figure 2–1 One ampere equals one coulomb of electrons flowing past a given point in a conductor in 1 second.

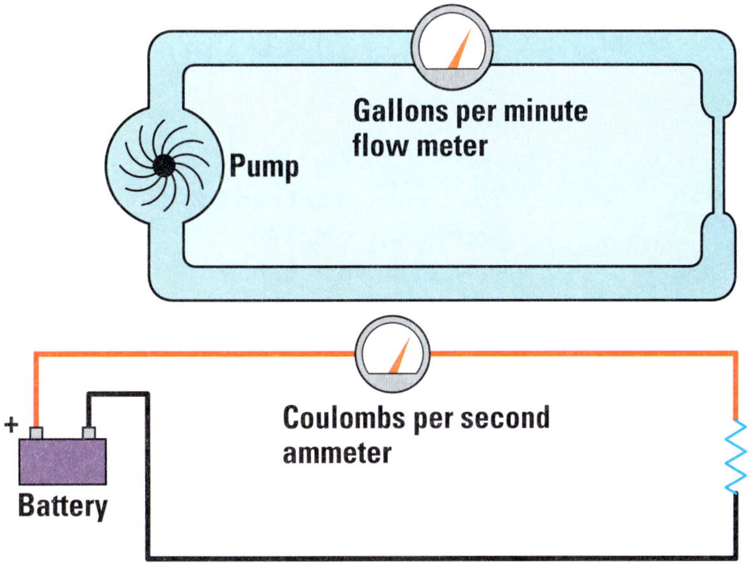

Figure 2–2 Amperes (current) in an electric circuit can be compared to gallons per minute flow rate in a water system.

stands for intensity of current, and the letter *A*, which stands for ampere, are both used to represent current flow in various algebraic formulas. This text uses the letter *I* in formulas to represent current.

2–3 The Electron Flow Theory

There are two theories concerning electrical current flow. One theory is known as the **electron flow theory** and states that because electrons are negative particles, current flows from the most negative point in the circuit to the most positive. The electron flow theory is more widely accepted as being correct and is used throughout this text.

2–4 The Conventional Current Flow Theory

The second theory, known as the **conventional current flow theory**, was defined by Benjamin Franklin. It is older than the electron flow theory and states that current flows from the most positive point to the most negative. Although it has been established almost to a certainty that the electron flow theory is correct, the conventional current flow theory is still widely used for several reasons. Most electronic circuits use the negative terminal as ground or common reference point. When the negative terminal is used as ground, the positive terminal is considered to be above ground, or hot. It is easier for most people to think of something flowing down rather than up, or from a point above ground to ground. An automobile electric system is a good example of this type of circuit. Most people consider the positive battery terminal to be the hot terminal.

Many people who work in the electronics field prefer the conventional current flow theory because all the arrows on the semiconductor symbols point in the direction of conventional current flow. If the electron flow theory is used, it must be assumed that current flows against the arrow (*Figure 2–3*). Another reason that many people prefer using the conventional current flow theory is that most electronic schematics are drawn in a manner that assumes that current flows from the more positive to the more negative source. In *Figure 2–4,* the positive voltage point is shown at the top of the schematic and the negative (ground) is shown at the bottom. When tracing the flow of current through a circuit, most people find it easier to go from top to bottom than from bottom to top.

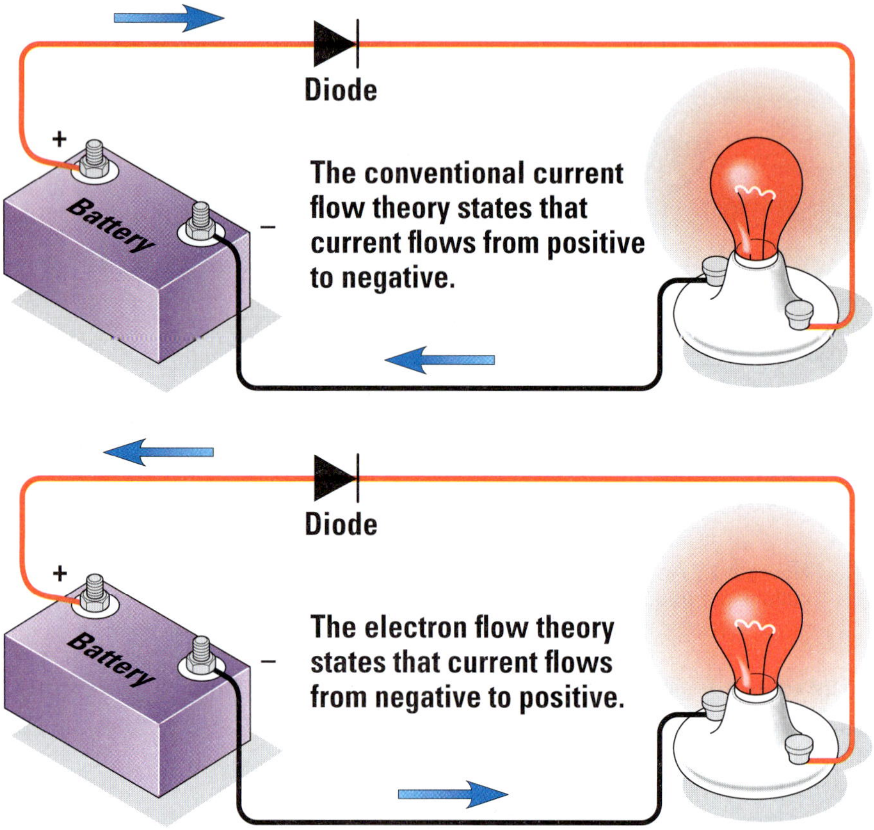

Figure 2–3 Depiction of both the conventional current flow theory and the electron flow theory.

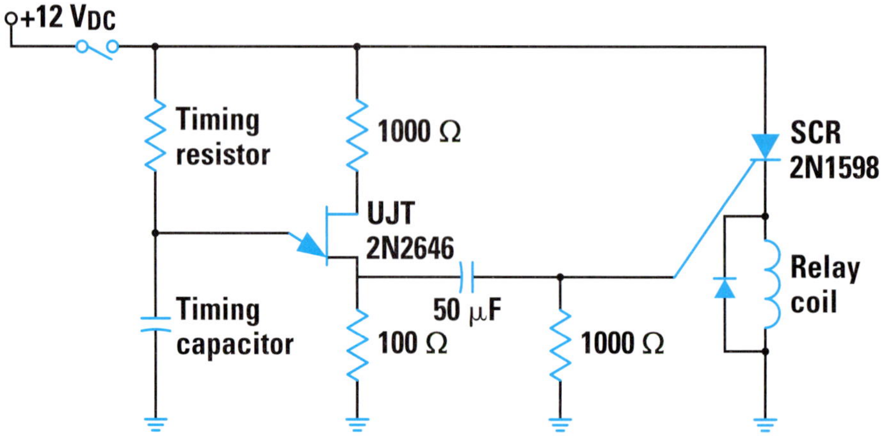

Figure 2–4 Example schematic drawing of an on-delay timer lending more usefulness to conventional current flow theory.

2–5 Speed of Current

no

To determine the speed of current flow through a wire, one must first establish exactly what is being measured. As stated previously, current is a flow of electrons through a conductive substance. Assume for a moment that it is possible to remove a single electron from a wire and identify it by painting it red. If it were possible to observe the progress of the identified electron as it moved from atom to atom, it would be seen that a single electron moves rather slowly (*Figure 2–5*). It is estimated that a single electron moves at a rate of about 3 inches per hour at 1 ampere of current flow. However, this must not be mistaken to mean that current moves slowly through a conductor. Think of how quickly a light turns on when you operate the switch.

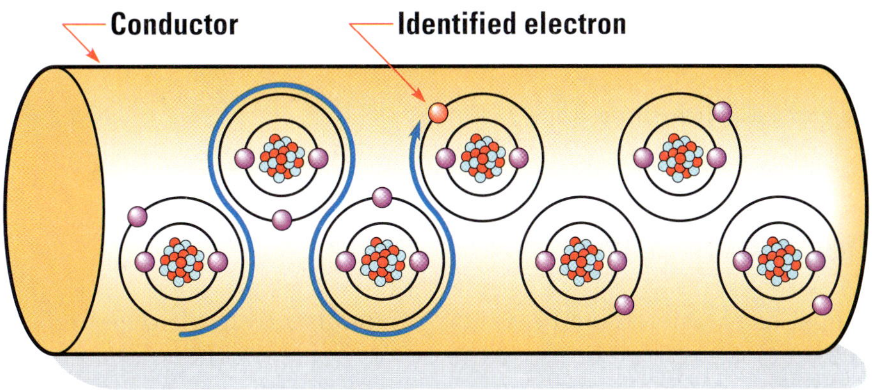

Figure 2–5 Electrons moving from atom to atom.

Another factor that must be considered is whether the circuit is DC, AC, or radio waves. Radio waves move at the speed of light, which is approximately 186,000 miles per second or 300,000,000 meters per second. The velocity of AC through a conductor is less than the speed of light because magnetic fields travel more slowly in material dielectrics than they do through free air. The formula shown can be used to calculate the wavelength of a signal traveling through a conductor. Wavelength is the distance that current travels during one AC cycle. Wavelength is discussed in more detail in Unit 13.

$$L = \frac{984\ V}{f}$$

where

$$L = \text{length in feet}$$

$$V = \text{velocity factor}$$

$$f = \text{frequency in megahertz (MHz)}$$

The velocity factor is determined by the type of conductor. *Table 2–1* gives the velocity factor for several different types of coaxial cables and parallel conductors.

How many feet would a 5-MHz signal travel through a conductor with a velocity factor of 0.66 during one AC cycle?

$$L = \frac{984 \times 0.66}{5}$$

$$L = 129.888\ \text{feet}$$

Description or Type Number	Velocity Factor	Characteristic Impedance	Capacitance per Foot
		Coaxial Cable	
RG-8A/U	0.66	53	29.5 pF
RG-58A/U	0.66	53	28.5 pF
RG-17A/U	0.66	50	30 pF
RG-11A/U	0.66	75	20.5 pF
RG59A/U	0.66	73	21 pF
		Parallel Conductors	
Air Insulated	0.975	200–600	–
214–023	0.71	75	20 pF
214–056	0.82	300	5.8 pF
214–076	0.84	300	3.9 pF
214–022	0.85	300	3 pF

Table 2–1 Data for Different Types of Transmission Lines

2–6 Basic Electric Circuits

A **complete path** must exist before current can flow through a circuit (*Figure 2–6*). A complete circuit is often referred to as a closed circuit because the power source, conductors, and load form a closed loop. In *Figure 2–6,* an incandescent lamp is used as the load. The lamp provides resistance to the flow

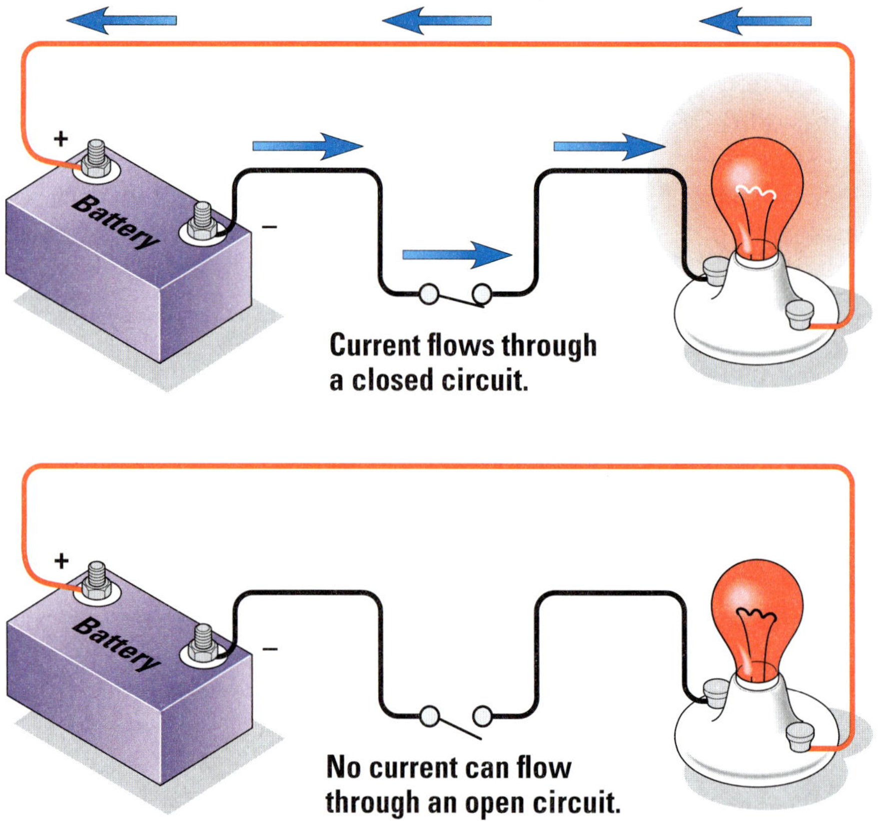

Figure 2–6 Current flows only through a closed circuit.

of current in the circuit and limits how much will flow. If the switch is opened, this introduces nearly infinite resistance to the circuit and there is no longer a closed loop, and no current can flow. This is often referred to as an incomplete, or open, circuit.

Another example of a closed circuit is the short circuit, which has very little or no resistance. It generally occurs when the supply and return conductors from the power source become connected; this is usually unintentional (*Figure 2–7*). In this example, an unintentional current path has been established that bypasses the load. Because the load is what limits the flow of current, when it is bypassed, an excessive amount of current can flow as most of the circuit resistance has now been removed from the equation. Short circuits generally result in an opened fuse or tripped circuit breaker. If the circuit has not been protected by a fuse or circuit breaker, a short circuit can seriously damage equipment, melt wires, and start fires.

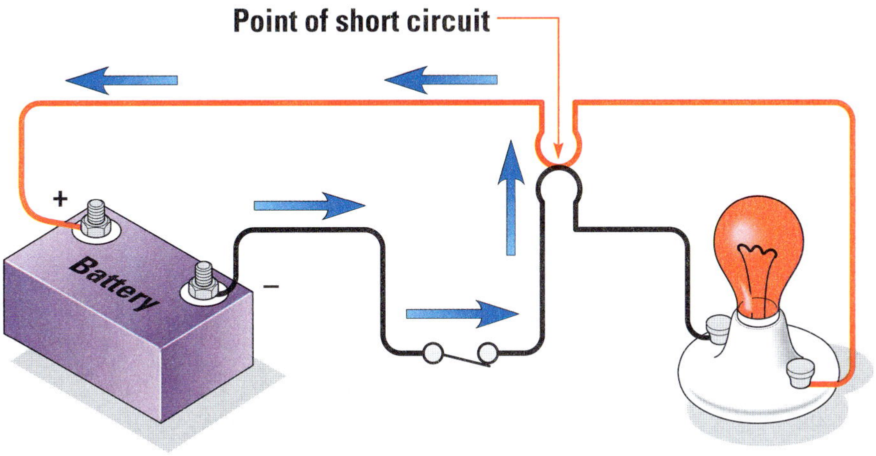

Figure 2–7 The positive and negative legs in this circuit have made contact, leaving very low resistance in this circuit and resulting in excessive amounts of current.

Another type of circuit, one that can be similar to a short circuit, is a ground-fault circuit. Ground-fault circuits can also cause an excessive amount of current to flow. These types of faults in the circuit occur when the ungrounded circuit conductor makes contact with an alternate path back to the power source that is not a normal current-carrying circuit conductor. Most circuits contain an extra conductor called the **equipment grounding conductor.** A typical 120-volt appliance circuit is shown in *Figure 2–8*. In this circuit, the ungrounded, or hot, conductor is connected to the fuse or circuit breaker. The hot conductor supplies power to the load. The grounded conductor,

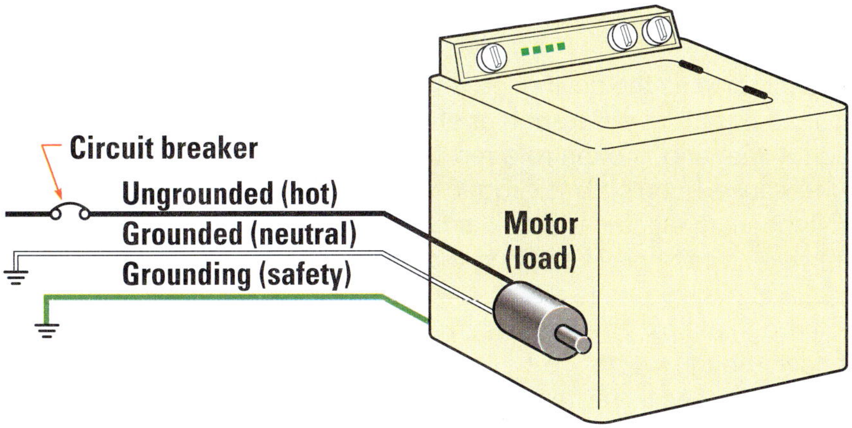

Figure 2–8 Example of a 120-V appliance circuit.

or **neutral conductor** in this case, provides the return path and completes the circuit back to the power source. The equipment grounding conductor is generally connected to the frame of the appliance and other normally noncurrent carrying metal parts to provide a low-resistance fault-current path back to the source and to establish a consistent reference to ground. Although both the neutral and equipment grounding conductors are connected at the power source, the equipment grounding conductor is not considered to be a circuit conductor because current will flow through the equipment grounding conductor only when a circuit fault develops. In normal operation, current flows through the hot and neutral conductors only.

The equipment grounding conductor is used to help prevent a shock hazard in the event that the ungrounded, or hot, conductor comes in contact with the case or frame of the appliance (*Figure 2–9*). This condition can occur in several ways. In this example, assume that the motor winding becomes damaged and makes a connection to the frame of the motor. Because the frame of the motor is connected to the frame of the appliance, the equipment grounding conductor provides a return path for current to flow on. If enough current flows, the hope is that the circuit breaker will open. Without an equipment grounding conductor connected to the frame of the appliance, the frame would become energized, and anyone touching the case and a grounded point, such as a water line, would complete the circuit. The resulting shock could be fatal. For this reason, the equipment grounding conductor prong of a plug should never be cut off or bypassed.

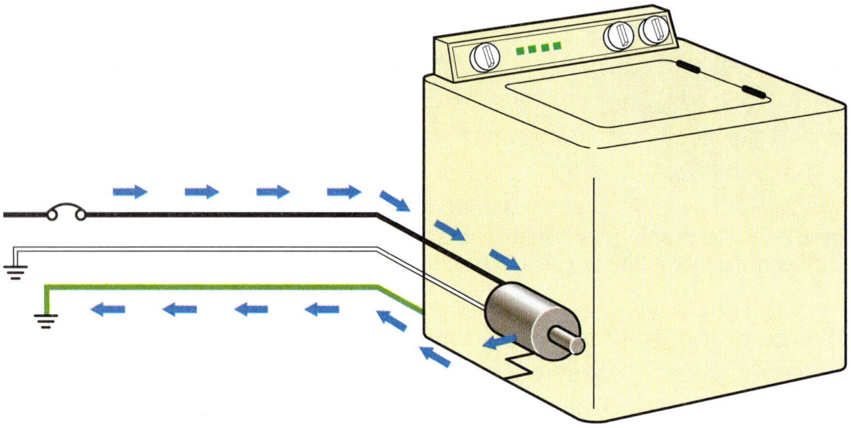

Figure 2–9 The equipment grounding conductor provides a low-resistance path for fault current to return to the source.

2–7 The Volt

Voltage can best be described as the driving force that causes current to flow between two points. Voltage is also referred to as **potential difference** or **electromotive force (EMF).** It is the force that pushes the electrons through a wire and is often referred to as electrical pressure. A **volt (V)** is the amount of potential necessary to cause 1 coulomb to produce 1 joule of work. One thing to remember is that voltage cannot flow. Voltage in an electrical circuit is like pressure in a water system (*Figure 2–10*). To say that voltage flows through a circuit is like saying that pressure flows through a pipe. Pressure can push water through a pipe, and it is correct to say that water flows through a pipe, but it is not correct to say that pressure flows through a pipe. The same is true for voltage. Voltage pushes current through a wire, but voltage cannot flow through a wire.

Voltage is often thought of as the potential to do something. For this reason, it is frequently referred to as potential or potential difference, especially in older publications and service manuals. Voltage must be present before current can flow, just as pressure must be present before water can flow.

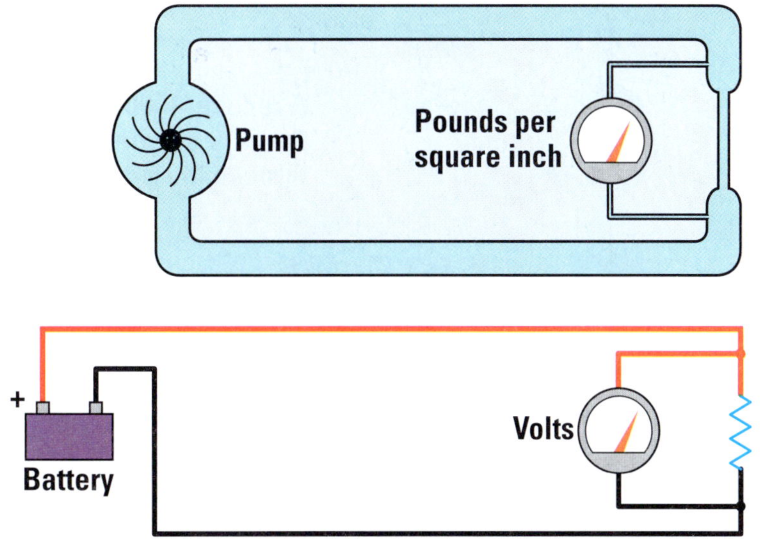

Figure 2–10 Voltage in an electric circuit can be compared to pressure in a water system.

A voltage, or potential, of 120 volts is present at a common wall outlet, but there is no current flow until some device is connected and a complete circuit exists. The same is true in a water system. Pressure is present, but water cannot flow until the valve is opened and a path is provided to a region of lower pressure. The letter E, which stands for EMF, or the letter V, which stands for volt, can be used to represent voltage in an algebraic formula. This text uses the letter E to represent voltage in an algebraic formula.

2–8 The Ohm

An **ohm** is the unit of **resistance** to current flow. It was named after the German scientist Georg S. Ohm. The symbol used to represent an ohm, or resistance, is the Greek letter omega (Ω). The letter *R*, which stands for resistance, is used to represent ohms in algebraic formulas. An ohm is the amount of resistance that allows 1 ampere of current to flow when the applied voltage is 1 volt. Without resistance, every electric circuit would be a short circuit. All electric loads, such as heating elements, lamps, motors, transformers, and so on, are measured in ohms. In a water system, a reducer can be used to control the flow of water; in an electric circuit, a resistor can be used to control the flow of electrons (*Figure 2–11*).

To understand the effect of resistance on an electric circuit, imagine a person running along a beach. As long as the runner stays on the hard, compact sand, they can run easily along the beach. Likewise, current can flow easily through a good conductive material, such as a copper wire. Now imagine that the runner wades out into the water until it is knee deep. They will no longer be able to run along the beach as easily because of the resistance of the water. Now imagine that the runner wades out into the water until it is waist deep. The ability to run along the beach will be hindered to an even greater extent because of the increased resistance of the water against the runner's body. The same is true for resistance in an electric circuit. The higher the resistance, the greater the hindrance to current flow.

Green Tip

Larger wire sizes will result in less resistance. Less resistance reduces the voltage drop and consequently the amount of power loss due to heating in the conductor.

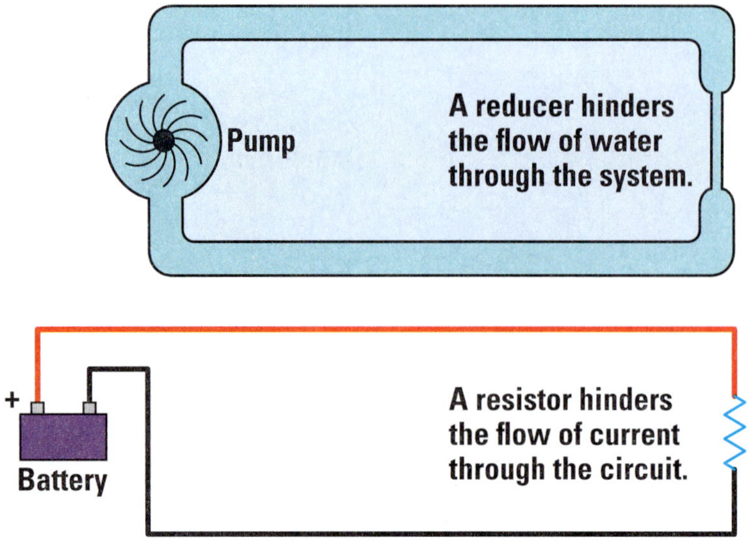

Figure 2–11 A resistor in an electric circuit can be compared to a reducer in a water system.

Another fact an electrician should be aware of is that any time current flows through resistance, heat is produced (*Figure 2–12*). That is why a wire becomes warm when current flows through it. The elements of an electric range become hot, and the filament of an incandescent lamp becomes extremely hot because of resistance.

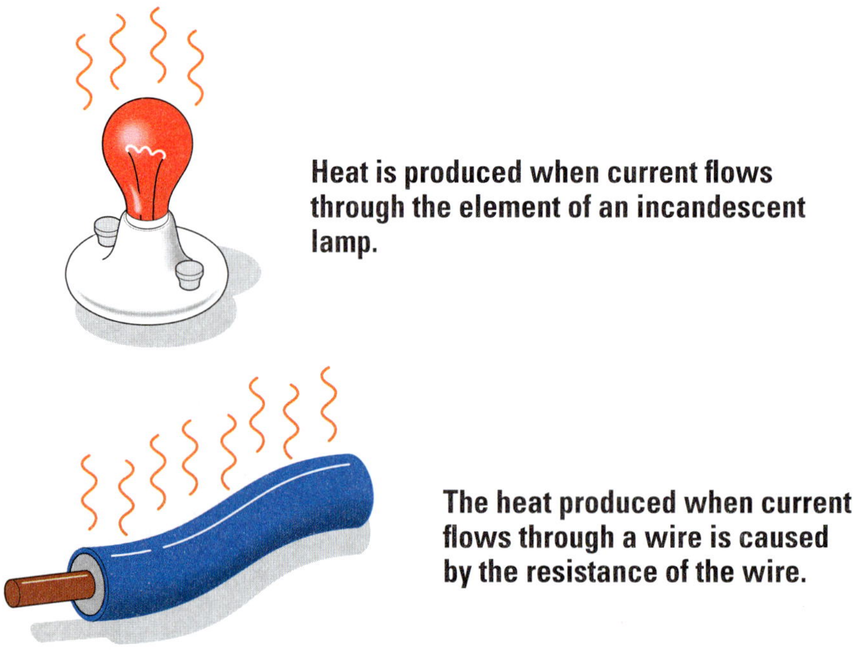

Figure 2–12 Heat is produced when current flows through resistance.

Another term similar in meaning to resistance is **impedance.** Impedance is most often used in calculations of AC rather than DC. Impedance is discussed to a greater extent later in this text.

2–9 The Watt

Wattage is a measure of the amount of power in a circuit. This is often described as the amount of work the circuit will perform. The **watt (W)** was named in honor of the English scientist James Watt. In an algebraic formula, wattage is generally represented either by the letter *P*, for power, or *W*, for watts. It is proportional to the amount of voltage and the amount of current flow (P = E × I). To understand watts, return to the example of the water system. Assume that a water pump has a pressure of 120 pounds per square inch (PSI) and causes a flow rate of 1 gallon per second. Now assume that this water is used to drive a turbine, as shown in *Figure 2–13.* The turning force, or torque, developed by the turbine is proportional to the amount of water flow and the pressure forcing it against the turbine blades. If the pressure is increased and the flow rate remains constant, the water will strike the turbine blades with greater force and the torque will increase. If the pressure remains constant and a greater volume of water is permitted to flow, the turbine blades will be struck by more pounds of water in the same amount of time and torque will again increase. As you can see, the torque developed by the turbine is proportional to both the pressure and the flow rate of the water.

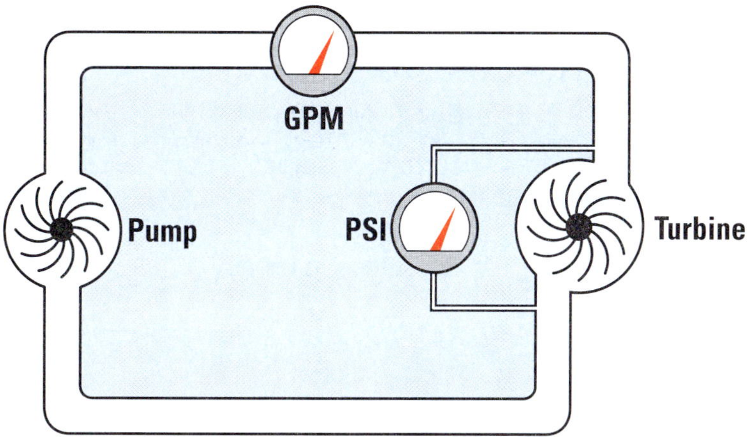

Figure 2–13 Power produced in a circuit is equal to the flow rate times the pressure in the circuit.

The power of an electric circuit is very similar. *Figure 2–14* shows a resistor connected to a circuit with a voltage of 120 volts and a current flow of 1 ampere. The resistor shown represents an electric heating element. When 120 volts force a current of 1 ampere through it, the heating element will produce 120 watts of heat (120 V × 1 A = 120 W). If the voltage is increased to 240 volts but the current remains constant, the element will produce 240 watts of heat (240 V × 1 A = 240 W). If the voltage remains at 120 volts but the current is increased to 2 amperes, the heating element will again produce 240 watts (120 V × 2 A = 240 W). Notice that the amount of power used by the heating element is determined by the amount of current flow and the voltage driving it.

An important concept concerning **power** in an electric circuit is that before true power, or watts, can exist, there must be some type of energy change or conversion. Electricity is a form of pure energy. In accordance with physical laws, energy cannot be created or destroyed, but its form can be changed. The watt is a measure of the amount of electrical energy converted into some other form. It makes no difference if electrical energy is converted into thermal energy in the form of heat or if it is converted into mechanical energy supplied by an electric motor. Some form of energy conversion must take place before watts, or true power, can exist.

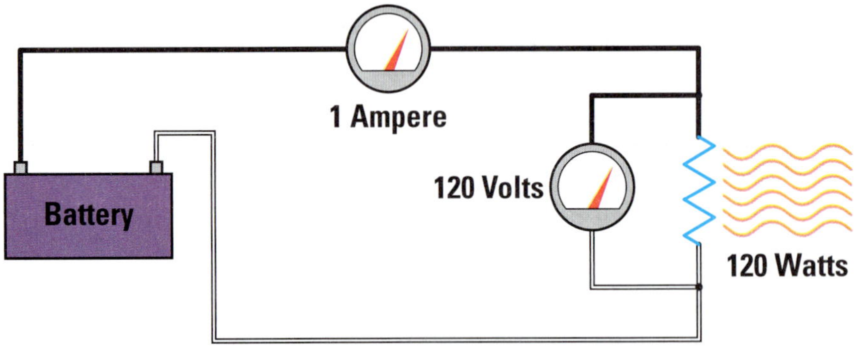

Figure 2–14 Amperes times volts equals watts.

Formulas that may be used to determine power in a circuit are

$$P = EI$$

$$P = I^2R$$

$$P = \frac{E^2}{R}$$

where

P = power or watts
I = current or amps
E = volts
R = resistance in ohms

2–10 Other Measures of Power

The watt is not the only unit of power measure. Many years ago, James Watt decided that to sell his steam engines, he would have to rate their power in terms the average person could understand. He decided to compare his steam engines to the horses he hoped his engines would replace. After experimenting, Watt found that the average horse working at a steady rate could do 550 foot-pounds of work per second. A foot-pound (ft-lb) is the amount of force required to raise a 1-pound weight 1 foot. This rate of doing work is the definition of a **horsepower (hp):**

$$1 \text{ hp} = 550 \text{ ft-lb/s}$$

Horsepower can also be expressed as 33,000 foot-pounds per minute (550 ft-lb/s × 60 s/min = 33,000 ft-lb/min):

$$1 \text{ hp} = 33{,}000 \text{ ft-lb/min}$$

It was later calculated that the amount of electric energy needed to produce one horsepower was 746 watts:

$$1 \text{ hp} = 746 \text{ W}$$

Another measure of energy frequently used in the English system of measure is the **British thermal unit (BTU).** A BTU is defined as the amount of heat required to raise the temperature of

1 pound of water 1 degree Fahrenheit. In the metric system, the calorie is used instead of the BTU to measure heat. A calorie is the amount of heat needed to raise the temperature of 1 gram of water 1 degree Celsius. The **joule** is the SI equivalent of the watt. A joule is defined as 1 newton-meter. A newton is a force of 100,000 dynes, or about $3\frac{1}{2}$ ounces, and a meter is about 39 inches. The joule can also be expressed as the amount of work done by 1 coulomb flowing through a potential of 1 volt or as the amount of work done by 1 watt for 1 second:

$$1 \text{ joule} = 1 \text{ wattsecond}$$

$$1 \text{ watt} = 1 \text{ joule/s}$$

The chart in *Figure 2–15* gives some common conversions for different quantities of energy. These quantities can be used to calculate different values. Examples 2-1 and 2-2 show how conversion values such as those found in *Figure 2-15* can be used to determine useful information.

1 horsepower =	746 watts
1 horsepower =	550 ft-lb/s
1 watt =	0.00134 horsepower
1 watt =	3.412 BTU/hr
1 wattsecond =	1 joule
1 BTU-hr =	0.293 Whr
1 cal/s =	4.19 watts
1 ft-lb/s =	1.36 watts
1 BTU =	1050 joules
1 joule =	0.2388 cal
1 cal =	4.187 joules

Figure 2–15 Common power units and conversion rates.

Example 2–1

An elevator must lift a load of 4000 lb to a height of 50 ft in 20 s. How much horsepower is required to operate the elevator?

Solution

Find the amount of work that must be performed, and then convert that to horsepower:

$$4000 \text{ lb} \times 50 \text{ ft} = 200{,}000 \text{ ft-lb}$$

$$\frac{200{,}000 \text{ ft-lb}}{20 \text{ s}} = 10{,}000 \text{ ft-lb/s}$$

$$\frac{10{,}000 \text{ ft-lb/s}}{550 \text{ ft-lb/s}} = 18.18 \text{ hp}$$

Example 2–2

A water heater contains 40 gallons of water. Water weighs 8.34 lb per gallon. The present temperature of the water is 68°F. The water must be raised to a temperature of 160°F in 1 hour. How much power will be required to raise the water to the desired temperature?

Solution

First, determine the weight of the water in the tank because a BTU is the amount of heat required to raise the temperature of 1 pound of water 1 degree Fahrenheit:

$$40 \text{ gal} \times 8.34 \text{ lb per gal} = 333.6 \text{ lb}$$

The second step is to determine how many degrees of temperature the water must be raised. This amount will be the difference between the present temperature and the desired temperature:

$$160°\text{F} - 68°\text{F} = 92°\text{F}$$

The amount of heat required in BTU will be the product of the pounds of water and the desired increase in temperature:

$$333.6 \text{ lb} \times 92°\text{F} = 30{,}691.2 \text{ lb-degrees or BTU}$$

$$1 \text{ Whr} = 3.412 \text{ BTU/hr}$$

Therefore,

$$\frac{30{,}691.2 \text{ BTU}}{3.412 \text{ BTU/hr}} = 8995.1 \text{ Whr}$$

2–11 Ohm's Law

In its simplest form, **Ohm's law** states that *it takes 1 volt to push 1 ampere through 1 ohm.* Ohm discovered that all electric quantities are proportional to each other and can therefore be expressed as mathematical formulas. He found that if the resistance of a circuit remained constant and the voltage increased, there was a corresponding proportional increase of current. Similarly, if the resistance remained constant and the voltage decreased, there would be a proportional decrease of current. He also found that if the voltage remained constant and the resistance increased, there would be a decrease of current, and if the voltage remained constant and the resistance decreased, there would be an increase of current. This finding led Ohm to the conclusion that *in a DC circuit, the current is directly proportional to the voltage and inversely proportional to the resistance.*

Because Ohm's law is a statement of proportion, it can be expressed as an algebraic formula when standard values such as the volt, the ampere, and the ohm are used. The three basic Ohm's law formulas are

$$E = I \times R$$

$$I = \frac{E}{R}$$

$$R = \frac{E}{I}$$

where

E = EMF, or voltage (in volts)

I = intensity of current, or amperage (in amperes)

R = resistance (in ohms)

The first formula states that the voltage can be found if the current and resistance are known. Voltage is equal to amperes multiplied by ohms. For example, assume that a circuit has a resistance of 50 ohms and a current flow through it of 2 amperes. The voltage connected to this circuit is 100 volts.

$$E = I \times R$$

$$E = 2\text{ A} \times 50\ \Omega$$

$$E = 100\text{ V}$$

The second formula states that the current can be found if the voltage and resistance are known. In the example shown, 120 volts are connected to a resistance of 30 ohms. The amount of current flow will be 4 amperes.

$$I = \frac{E}{R}$$

$$I = \frac{120\text{ V}}{30\ \Omega}$$

$$I = 4\text{ A}$$

The third formula states that if the voltage and current are known, the resistance can be found. Assume that a circuit has a voltage of 240 volts and a current flow of 10 amperes. The resistance in the circuit is 24 ohms.

$$R = \frac{E}{I}$$

$$R = \frac{240\text{ V}}{10\text{ A}}$$

$$R = 24\ \Omega$$

Figure 2–16 shows a simple chart that can be a great help when trying to remember an Ohm's law formula. To use the chart, cover the unit of measurement that is being determined. For example, if the voltage, E, is to be found, cover the E on the chart. The chart now shows the remaining letters IR (*Figure 2–17*); thus, E = I × R. The same method reveals the formulas for current (I) and resistance (R).

A larger chart, which shows the formulas needed to find watts as well as voltage, amperage, and resistance, is shown in *Figure 2–18*. The letter P (power) is used to represent the value of watts. Notice that this chart is divided into four sections and that each section contains three different formulas. To use this chart, select the section containing the quantity to be found, and then choose the proper formula from the given quantities. See examples 2-3, 2-4, and 2-5 for examples of how the chart found in *Figure 2-18* can be used to determine missing values when certain values are known.

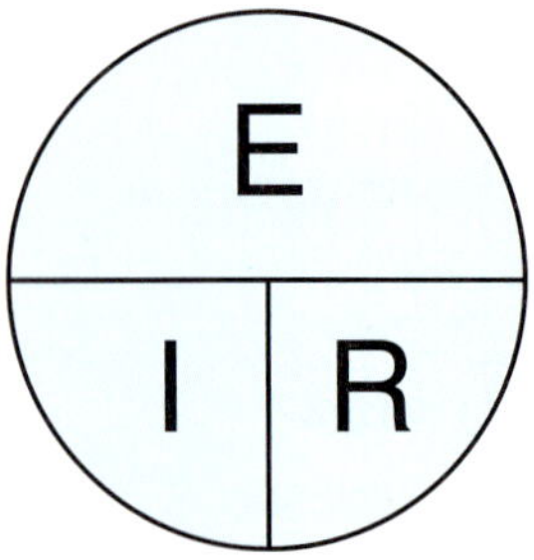

Figure 2–16 Chart for finding the values of voltage, current, and resistance.

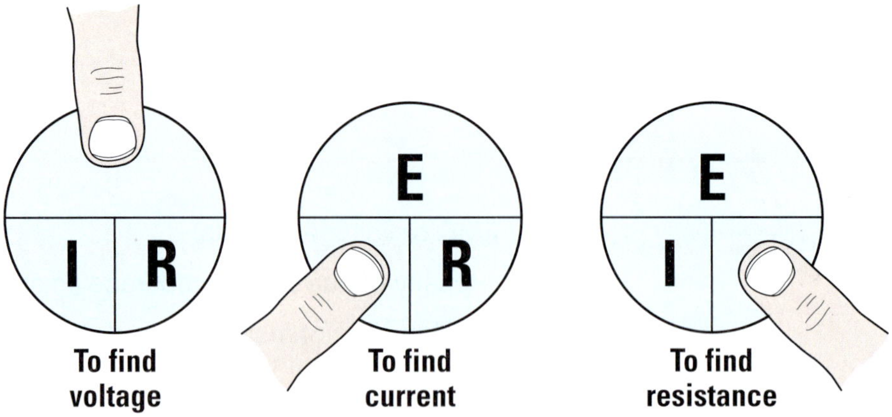

Figure 2–17 Using the Ohm's law chart.

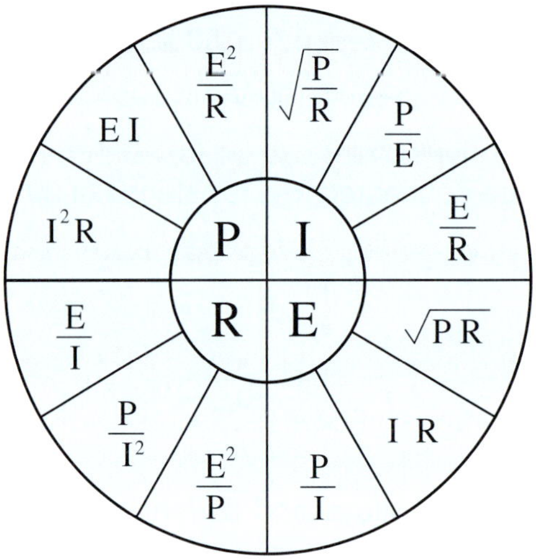

Figure 2–18 Formula chart for finding the values of voltage, current, resistance, and power.

Example 2–3

An electric iron is connected to 120 V and has a current draw of 8 A. How much power is used by the iron?

Solution

The quantity to be found is watts, or power. The known quantities are voltage and amperage. The proper formula to use is shown in *Figure 2–19.*

$$P = EI$$

$$P = 120\text{ V} \times 8\text{ A}$$

$$P = 960\text{ W}$$

Example 2–3 (continued)

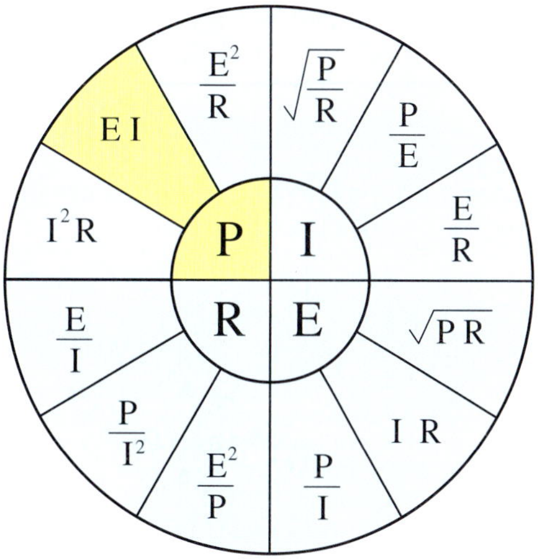

Figure 2–19 Finding power when voltage and current are known.

Example 2–4

An electric hair dryer has a power rating of 1000 W. How much current will it draw when connected to 120 V?

Solution

The quantity to be found is amperage, or current. The known quantities are power and voltage. To solve this problem, choose the formula shown in *Figure 2–20*.

$$I = \frac{P}{E}$$

$$I = \frac{1000\ W}{120\ V}$$

$$I = 8.333\ A$$

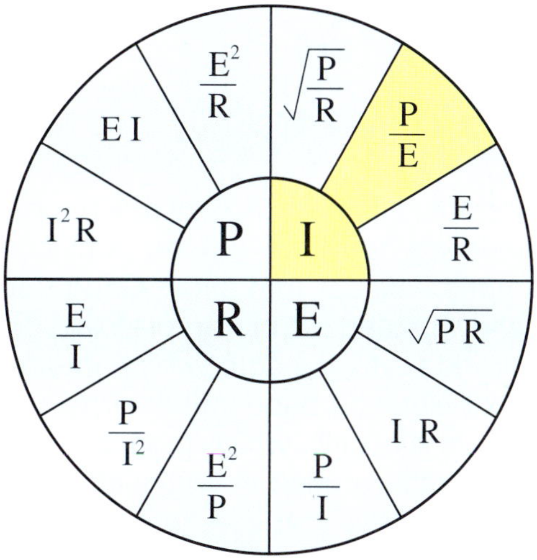

Figure 2–20 Finding current when power and voltage are known.

Example 2–5

An electric hotplate has a power rating of 1440 W and a current draw of 12 A. What is the resistance of the hotplate?

Solution

The quantity to be found is resistance, and the known quantities are power and current. Use the formula shown in *Figure 2–21*.

$$R = \frac{P}{I^2}$$

$$R = \frac{1440\text{ W}}{12\text{ A} \times 12\text{ A}}$$

$$R = \frac{1440\text{ W}}{144\text{ A}}$$

$$R = 10\ \Omega$$

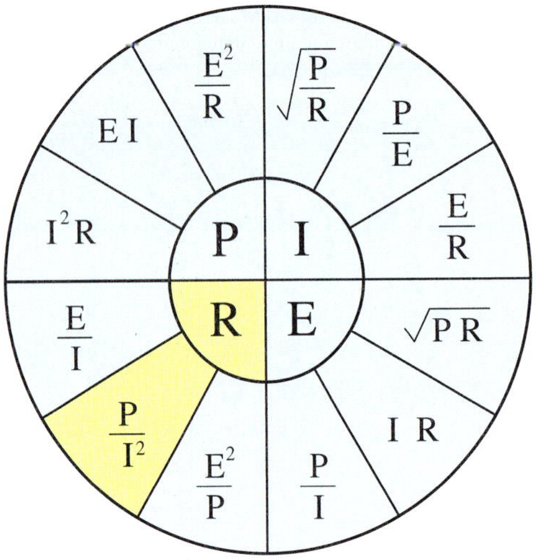

Figure 2–21 Finding resistance when power and current are known.

2–12 Metric Prefixes

Metric prefixes are used in the electrical field just as they are in most other scientific fields. A special type of notation, known as *engineering notation,* is used in electrical measurements. Engineering notation is similar to scientific notation except that engineering notation is in steps of 1000 instead of 10. The chart in *Figure 2–22* shows standard metric prefixes. The first step above the base unit is deka, which means 10. The second prefix is hecto, which means 100, and the third prefix is kilo, which means 1000. The first prefix below the base unit is deci, which means $\frac{1}{10}$; the second prefix is centi, which means $\frac{1}{100}$; and the third is milli, which means $\frac{1}{1000}$.

The chart in *Figure 2–23* shows prefixes used in engineering notation. The first prefix above the base unit is kilo, or 1000; the second prefix is mega, or 1,000,000; and the third prefix is giga, or 1,000,000,000. Notice that each prefix is 1000 times greater than the previous prefix. The chart also shows that the first prefix below the base unit is milli, or $\frac{1}{1000}$; the second is micro, represented by the Greek mu (μ), or $\frac{1}{1{,}000{,}000}$; and the third is nano, or $\frac{1}{1{,}000{,}000{,}000}$.

Kilo	1000
Hecto	100
Deka	10
Base Unit	1
Deci	1/10, or 0.1
Centi	1/100, or 0.01
Milli	1/1000, or 0.001

Figure 2–22 Standard metric prefixes.

Engineering unit	Symbol	Multiply by	
Tera	T	1,000,000,000,000	X 10^{12}
Giga	G	1,000,000,000	X 10^{9}
Mega	M	1,000,000	X 10^{6}
Kilo	k	1,000	X 10^{3}
Base unit		1	
Milli	m	0.001	X 10^{-3}
Micro	μ	0.000,001	X 10^{-6}
Nano	n	0.000,000,001	X 10^{-9}
Pico	p	0.000,000,000,001	X 10^{-12}

Figure 2–23 Standard prefixes of engineering notation.

Metric prefixes are used in almost all scientific measurements for ease of notation. It is much simpler to write a value such as 10 MΩ than it is to write 10,000,000 ohms, or to write 0.5 ns than to write 0.000,000,000,5 second. Once the metric prefixes have been learned, measurements such as 47 kilohms (kΩ) or 50 milliamperes (mA) become commonplace to the technician.

The SI System

Note that the term *metric* is commonly used to indicate a system that employs measurements that increase or decrease in steps of 10. The prefixes just discussed are commonly referred to as *metric* units of measure. These prefixes are actually part of the SI (Système Internationale) system that was adopted for use in the United States in 1960 by the 11th General Conference on Weights and Measures (abbreviated CGPM from the official French name *Conference Generale des Poids et Mesures*). The intention of the SI system is to provide a worldwide standard of weights and measures. The SI system uses seven base and two supplementary units that are regarded as dimensionally independent (*Figure 2–24*). From these base and supplementary units, other units have been derived. Some of the units commonly used in the electrical field are shown in *Figure 2–25*.

Quantity	Unit	Symbol
Length	Meter	m
Mass	Kilogram	kg
Time	Second	s
Electric current	Ampere	A
Thermodynamic temperature	Kelvin	K
Amount of substance	Mole	mol
Luminous intensity	Candela	cd
Phase angle	Radian	rad
Solid angle	Sterdian	sr

Figure 2–24 SI base and supplementary units.

Quantity	Unit	Symbol	Formula
Frequency	Hertz	Hz	I/s
Force	Newton	N	$(kg \cdot m)/s^2$
Pressure, stress	Pascal	Pa	N/m^2
Energy, work, quantity of heat	Joule	J	$N \cdot m$
Power, radiant flux	Watt	W	J/s
Quantity of electricity, charge	Coulomb	C	$A \cdot s$
Electric potential, electromotive force	Volt	V	W/A
Capacitance	Farad	F	C/V
Electric resistance	Ohm	Ω	V/A
Conductance	Siemens	S	A/V
Magnetic flux	Weber	Wb	$V \cdot s$
Magnetic flux density	Tesla	T	Wb/m^2
Inductance	Henry	H	Wb/A
Luminous flux	Lumen	lm	$cd \cdot sr$
Illuminance	Lux	lx	lm/m^2

Figure 2–25 Derived SI units.

Summary

- A coulomb is a measure of electrical charge.
- An ampere (A) is 1 coulomb per second.
- The letter *I*, which stands for intensity of current flow, is normally used in Ohm's law formulas.
- Voltage is referred to as electric pressure, potential difference, or electromotive force. An E or a V can be used to represent voltage in Ohm's law formulas.
- An ohm (Ω) is a measurement of resistance (R) in an electric circuit.
- The watt (W) is a measurement of power in an electric circuit. It is represented by either a W or a P (power) in Ohm's law formulas.

- Electric measurements are generally expressed in engineering notation.
- Engineering notation differs from scientific notation in that it uses steps of 1000 instead of steps of 10.
- Before current can flow, there must be a complete circuit.
- A short circuit has little or no resistance.
- An open circuit has infinite resistance.

Review Questions

1. Which of the following best describes the term "voltage" as it applies in an electrical circuit?
   A) The amount of electricity flowing in the conductor
   B) The pressure or driving force that pushes and pulls electrons through a circuit
   C) The amount of work performed by the electrons flowing through the circuit
   D) The opposition to electrons flowing through the circuit
2. Which of the following best describes the term "resistance" as it applies in an electrical circuit?
   A) The amount of electricity flowing in the conductor
   B) The pressure or driving force that pushes and pulls electrons through a circuit
   C) The amount of work performed by the electrons flowing through the circuit
   D) The opposition to electrons flowing through the circuit
3. Which of the following best describes the term "current" as it applies in an electrical circuit?
   A) The amount of electricity flowing in the conductor
   B) The pressure or driving force that pushes and pulls electrons through a circuit
   C) The amount of work performed by the electrons flowing through the circuit
   D) The opposition to electrons flowing through the circuit
4. Which of the following best describes the term "power" as it applies in an electrical circuit?
   A) The amount of electricity flowing in the conductor
   B) The pressure or driving force that pushes and pulls electrons through a circuit
   C) The amount of work performed by the electrons flowing through the circuit
   D) The opposition to electrons flowing through the circuit
5. True or False: The conventional current flow theory states that current flows in a circuit from the negative side of the voltage source to the positive side of the supply.
6. A circuit has a total resistance of 25 ohms. How much voltage must be applied to the circuit to see a current flow value of 4 amperes?
   A) 6.25 V
   B) 10 V
   C) 100 V
   D) 400 V

7. How much resistance is present in a circuit when connected to a 12-volt battery and 2.5 amperes of current is measured flowing from the terminals of the battery?

   A) 5.2 ohms

   B) 2.8 ohms

   C) 10.6 ohms

   D) 4.8 ohms

8. True or False: Current can flow in a circuit that has been opened by the operation of a switch.

9. In the electron flow theory, which side of the voltage source does current initiate from?

   A) Positive

   B) Negative

10. How much work can be expected to be performed by a circuit with 5 ohms of resistance that is connected to a battery with a voltage of 12 volts?

    A) 28.8 W

    B) 2.4 W

    C) 60 W

    D) 12 W

11. A short circuit is a type of closed circuit that allows current to bypass the load. This leads to a circuit with very low ____________ and results in an extremely high amount of ____________. (Select all that apply.)

    A) voltage, current

    B) ohms, amperes

    C) resistance, ohms

    D) resistance, current

12. A 1500W heater is connected to a circuit operating at 240 volts. If the heater is operating at maximum capacity, will the current in the circuit be enough to trip a 15-ampere circuit breaker or fuse?

    A) Yes, the heater draws more than 15A and will trip the circuit breaker.

    B) Unsure, there is not enough information to determine how much current the heater will draw.

    C) No, the heater will not draw enough current to cause the circuit breaker to trip.

Practical Applications

1. You are an electrician on the job. The electrical blueprint shows that eight 500-W lamps are to be installed on the same circuit. The circuit voltage is 277 V and is protected by a 20-A circuit breaker. A circuit can be loaded to only 80% of its rating when the load is considered continuous. Is a 20-A circuit large enough to carry this load?

2. You have been sent to a new home. The homeowner reports that sometimes the electric furnace trips the 240-V, 60-A circuit breaker connected to it. Upon examination, you find that the furnace contains three 5000-W heating elements designed to turn on in stages. For example, when the thermostat calls for heat, the first 5000-W unit turns on. After some period of time, the second unit will turn on, and then, after another time delay, the third unit will turn on. What do you think the problem is, and what would be your recommendation for correcting it? Explain your answer.

3. You are an electrician installing the wiring in a new home. The homeowner wants a ceiling fan with a light kit to be installed in five different rooms. Each fan contains a light kit that can accommodate four 60-watt lamps. Each fan motor draws a current of 1.8 amperes when operated on high speed. It is assumed that each fan will likely operate at maximum current for more than three hours at a time and, therefore, must be considered a continuous load. The fans are to be connected to a 15-ampere circuit. Because the fans are considered a continuous load, the circuit must be rated at least to 125% of the continuous connected load. How many fans can be connected to a single 15-ampere circuit? How many circuits will be required to supply power to all five fans?

4. A homeowner is installing a swimming pool. You have been asked to install a circuit to operate a 600-watt underwater light and a circulating pump. The motor nameplate reveals that the pump has a current draw of 8.5 amperes. The devices are considered a continuous load. Can the power to operate both of these devices be supplied by a single 20-ampere circuit?

Practice Problems

Ohm's Law

Fill in the missing values.

Volts (E)	Amperes (I)	Ohms (R)	Watts (P)
153 V	0.056 A		
	0.65 A	470 Ω	
24 V			124 W
	0.00975 A		0.035 W
		6.8 kΩ	0.86 W
460 V		72 Ω	
48 V	1.2 A		
	154 A	0.8 Ω	
277 V			760 W
	0.0043 A		0.0625 W
		130 kΩ	0.0225 W
96 V		2.2 kΩ	

Unit 3

Static Electricity

Key Terms

Electroscope

Electrostatic charges

Lightning

Lightning arrestor

Lightning bolts

Lightning rods

Nuisance static charges

Precipitators

Selenium

Static

Thundercloud

Useful static charges

Why You Need to Know

Many processes and devices use static electricity in a productive way. Electronic air cleaners and paint spray operations employ static electricity. The concepts of static electricity and how it can be used or prevented are important to anyone in the electrical field. To fully understand static electricity, this unit presents

- a demonstration of electron flow and how, by adding electrons, an object is negatively charged, and how, by removing electrons, an object becomes positively charged, and
- an explanation of electron flow in thunderclouds and lightning. This natural element is probably the greatest example of static electricity, and in this unit, you can see how it applies to the theory of positive and negative charges in nature.

Outline

Objectives

After studying this unit, you should be able to

- explain the basic principles of static electricity.
- describe how an electroscope can be used to determine charges.
- apply static electricity principles to lightning.
- explain how static electricity can be a nuisance.
- explain how static electricity can be useful in daily use.

Preview

Static electric charges occur often in everyday life. Almost everyone has received a shock after walking across a carpet and then touching a metal object or after sliding across a car seat and touching the door handle. Almost everyone has combed their hair with a hard rubber or plastic comb and then used the comb to attract small pieces of paper or other lightweight objects. Static electric charges cause clothes to stick together when they are taken out of a clothes dryer. Lightning is without doubt the greatest display of a static electric discharge.

Static electricity can harm sensitive electronic components. Components subject to damage by static electricity are often packaged in a special wrapping. Locations that contain a significant number of sensitive electronics, such as data centers, will often employ electrostatic flooring or localized mats, and workers will also need to wear special shoes that bleed off this build up of static electrical charge.

3–1 Static Electricity

Static electricity is caused by an imbalance of electrical charges (positive and negative). The imbalance occurs, then an object either gains or loses electrons. Objects that gain electrons exhibit a negative charge, and objects that lose electrons exhibit a positive charge. This electrical differential will exist until electrons on the negatively charged object move to the positively charged object. When the electrons are equal on each object, there will no longer be a static charge.

Although static charges can be a nuisance (*Figure 3–1*) or even dangerous, they can also be beneficial. The manufacture of sandpaper relies on the application of static electricity. Grains of sand receive a

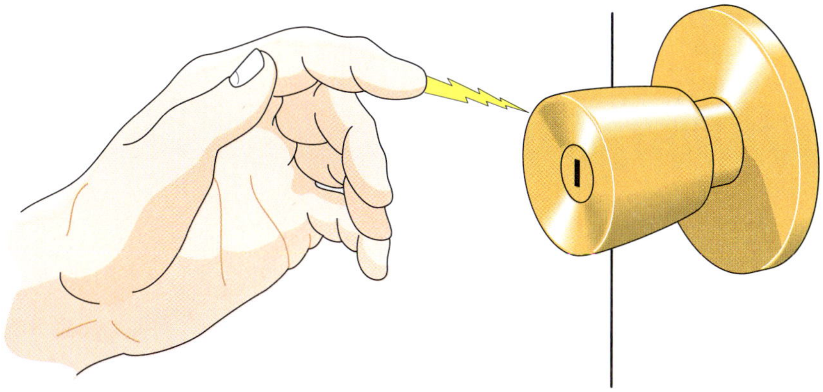

Figure 3–1 Static electric charges can cause a painful shock.

static charge to make them stand apart and expose a sharper edge (*Figure 3–2*). Electronic air filters—**precipitators**—use static charges to attract small particles of smoke, dust, and pollen (*Figure 3–3*). The precipitator uses a high-voltage DC power supply to provide a set of wires with a positive charge and a set of plates with a negative charge. As a blower circulates air through the unit, small particles receive a positive charge as they move across the charged wires. The charged particles are then attracted to the negative plates. The negative plates hold the particles until the unit is turned off and the plates are cleaned.

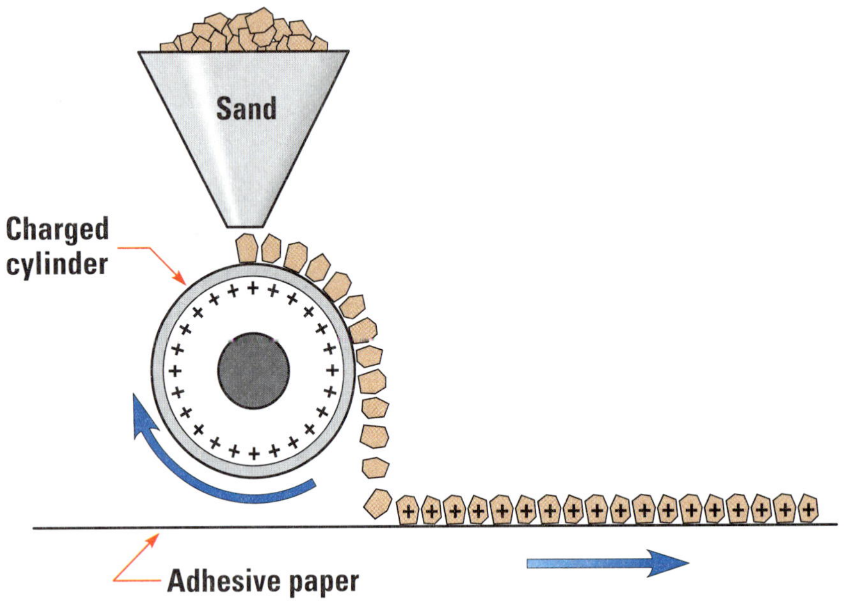

Figure 3–2 Grains of sand receive a charge to help them stand apart.

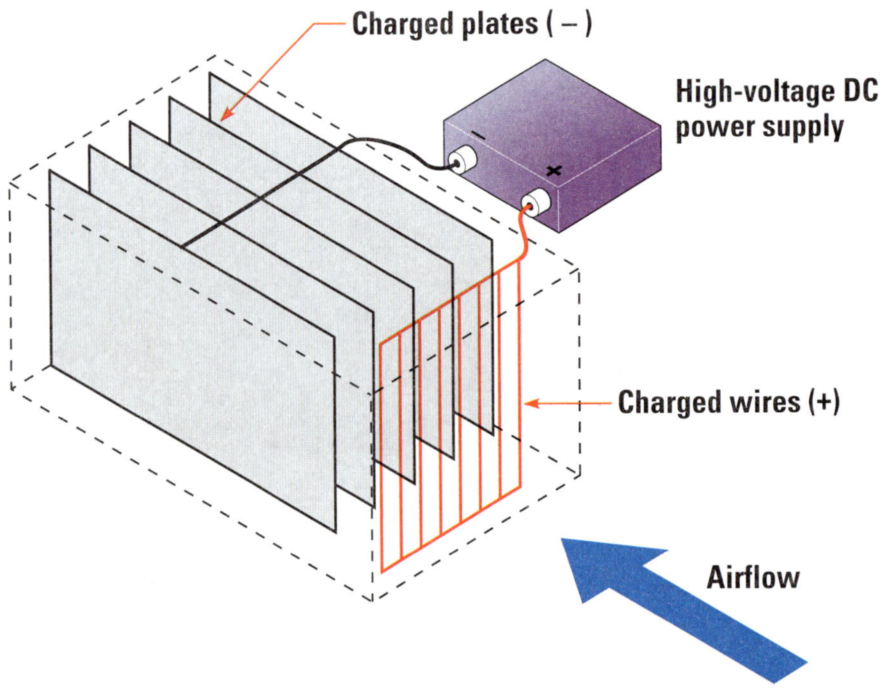

Figure 3–3 Electronic air cleaner.

The word **static** means not moving or sitting still. Static electricity refers to electrons that are sitting still and not moving. Static electricity is, therefore, a charge and not a current. **Electrostatic charges** are built up on insulator materials because insulators are the only materials that can hold the electrons stationary and keep them from flowing to a different location. A static charge can be built up on a conductor only if the conductor is electrically insulated from surrounding objects. A static charge can be either positive or negative. If an object has a lack of electrons, it has a positive charge; if it has an excess of electrons, it has a negative charge.

3–2 Charging an Object

The charge that accumulates on an object is determined by the materials used to produce the charge. If rubber is rubbed on a piece of wool, the wool deposits excess electrons on the rubber and gives it a negative charge. If glass is rubbed on a piece of wool, electrons are removed from the glass, producing a positive charge on the rod (*Figure 3–4*).

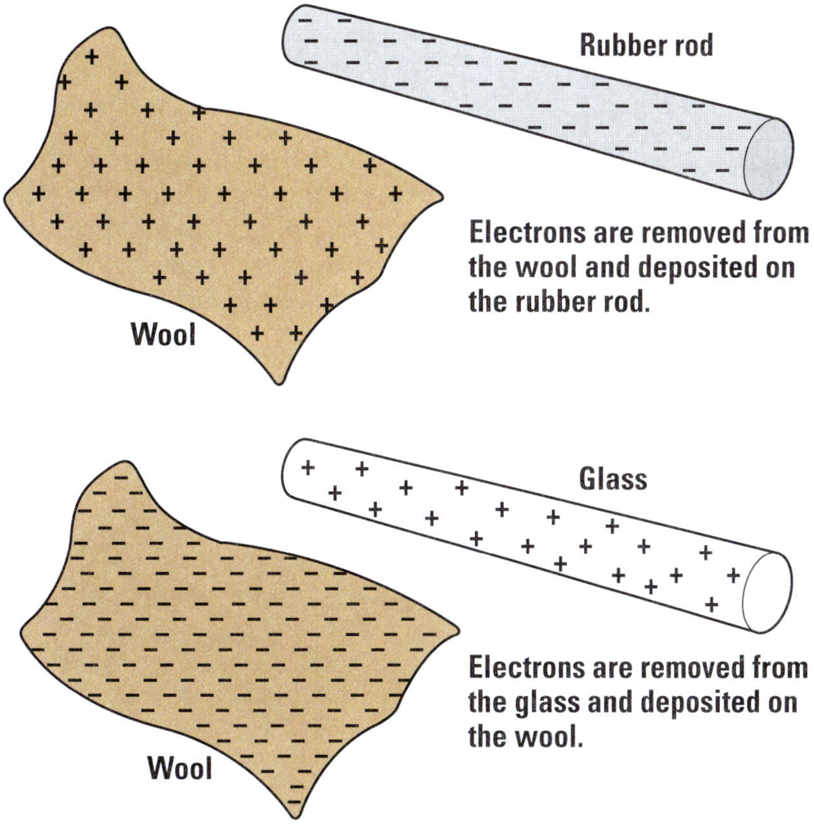

Figure 3–4 Producing a static charge.

3–3 The Electroscope

An early electric instrument that can be used to determine the polarity of the electrostatic charge of an object is the **electroscope** (*Figure 3–5*). An electroscope is a metal ball attached to the end of a metal rod. The other end of the rod is attached to two thin metal leaves. The metal leaves are inside a transparent container to permit the action of the leaves to be seen and to prevent air currents from affecting their movement. The metal rod is insulated from the box.

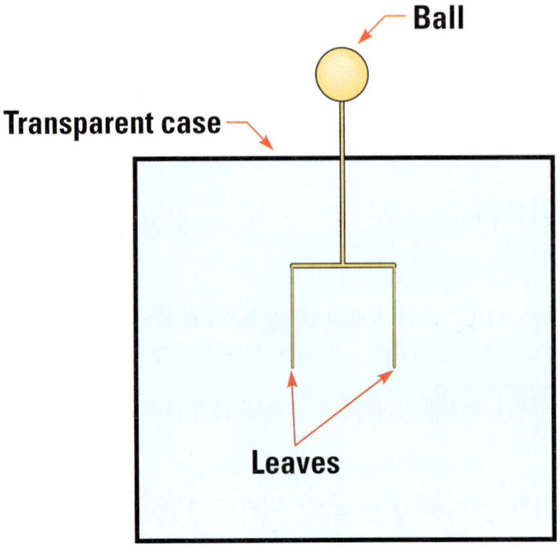

Figure 3–5 An example of an electroscope.

Before the electroscope can be used, it must first be charged. This is done by touching the ball with an object that has a known charge. For this example, assume that a rubber rod has been given a negative charge by friction with a wool cloth. When the rubber rod is wiped against the metal ball, excess electrons are deposited on the metal surface of the electroscope. Because both of the metal leaves now have an excess of electrons, they repel each other, as shown in *Figure 3–6*.

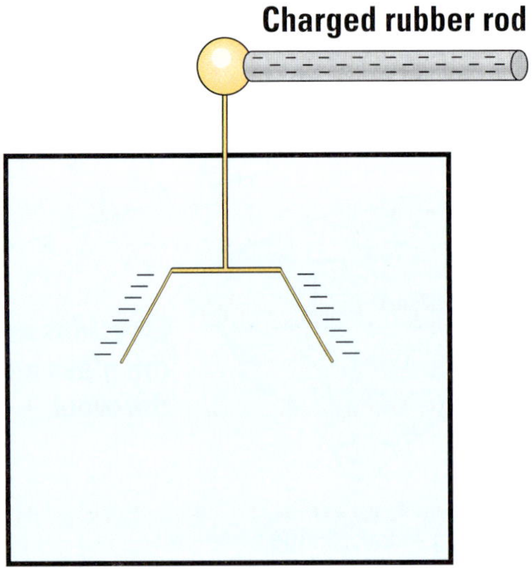

Figure 3–6 The electroscope is charged with a known static charge.

Testing an Object

A charged object can now be tested to determine whether it has a positive or negative polarity. Assume that a ballpoint pen is charged by rubbing the plastic body through a person's hair. Now, bring the pen close to but not touching the ball, and observe the action of the leaves. If the pen has taken on a negative charge, the leaves will move farther apart, as shown in *Figure 3–7*. The field caused by the negative electrons on the pen repels electrons from the ball. These electrons move down the rod to the leaves, causing the leaves to become more negative and to repel each other more, forcing the leaves to move farther apart.

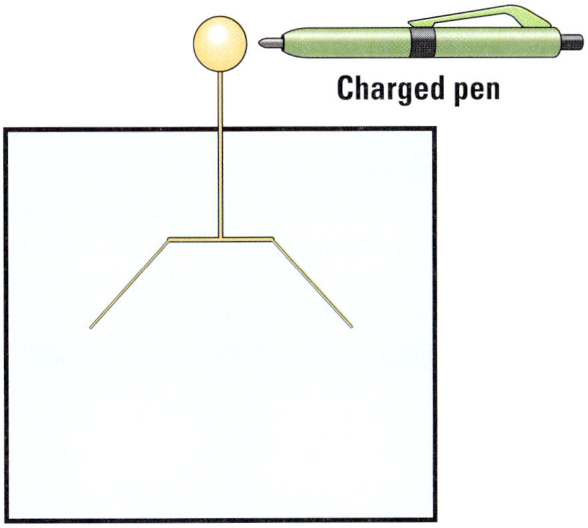

Figure 3–7 The leaves are deflected farther apart, indicating that the object has a negative charge.

If the pen has a positive charge, the leaves will move closer together when the pen is moved near the ball (*Figure 3–8*). This action is caused by the positive field of the pen attracting electrons. When electrons are attracted away from the leaves, they become less negative and move closer together. If the electroscope is charged with a positive charge in the beginning, a negatively charged object will cause the leaves to move closer together, and a positively charged object will cause the leaves to move farther apart.

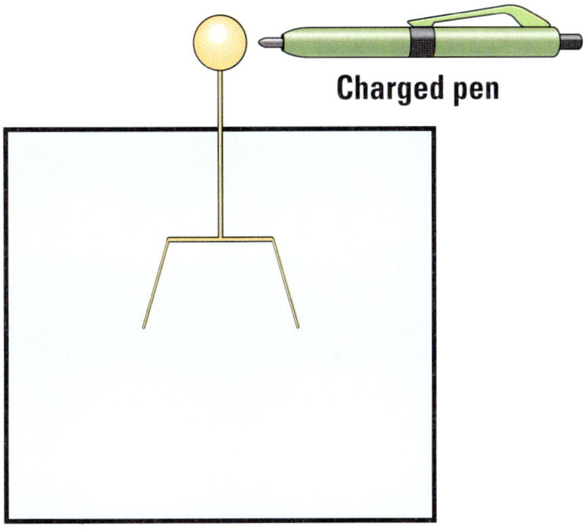

Figure 3–8 The leaves move closer together, indicating that the object has a positive charge.

3–4 Static Electricity in Nature

When static electricity occurs in nature, it can be harmful. The best example of natural static electricity is **lightning**. A static charge builds up in clouds that contain a large amount of moisture as they move through the air. It is theorized that the movement causes a static charge to build up on the surface of drops of water. Large drops become positively charged, and small drops become negatively charged. *Figure 3–9* illustrates a typical **thundercloud**. Notice that both positive and negative charges can be contained in the same cloud. Most lightning discharges, or **lightning bolts**, occur within the cloud. Lightning discharges can also take place between different clouds, between a cloud and the ground, and between the ground and the cloud (*Figure 3–10*). Whether a lightning bolt travels from the cloud to the ground or from the ground to the cloud is determined by which contains the negative and which contains the positive charge. Current in static discharge events likely flows from negative to positive. If a cloud is negative and an object on the ground is positive, the lightning discharge travels from the cloud to the ground. If the cloud has a positive charge and the object on the ground has a negative charge, the discharge travels from the ground to the cloud. According to the National Weather Service (NWS), a typical lightning bolt has an average voltage of about 300,000,000 volts!

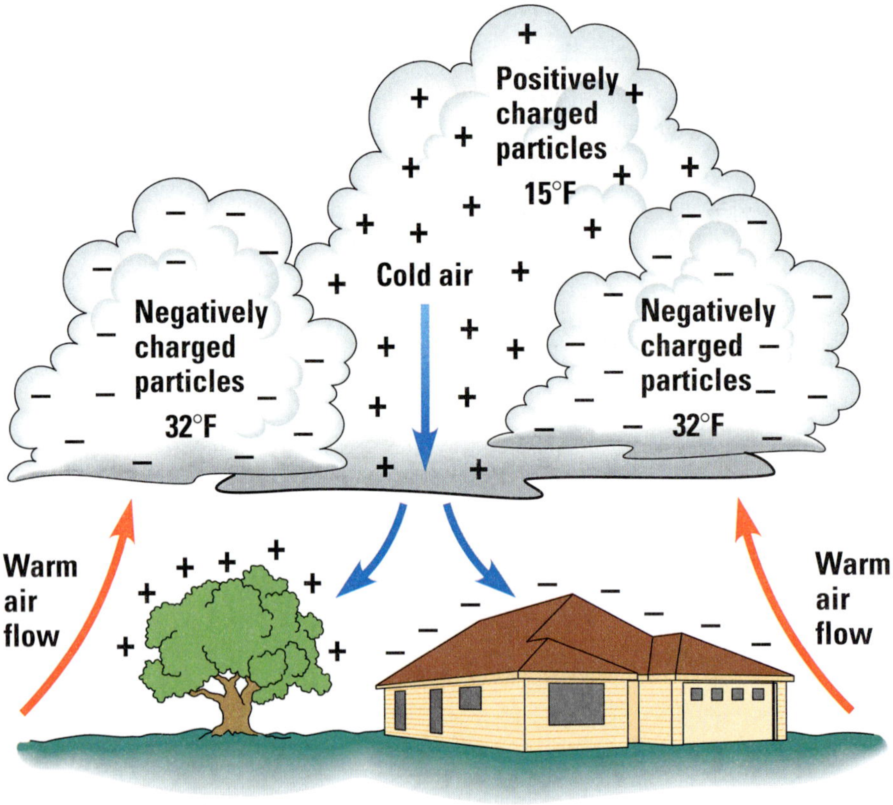

Figure 3–9 The typical thundercloud contains both negatively and positively charged particles.

Lightning Protection

Lightning rods are sometimes used to help protect objects from lightning. Lightning rods work by providing an easy path to ground for current flow. If the protected object is struck by a lightning bolt,

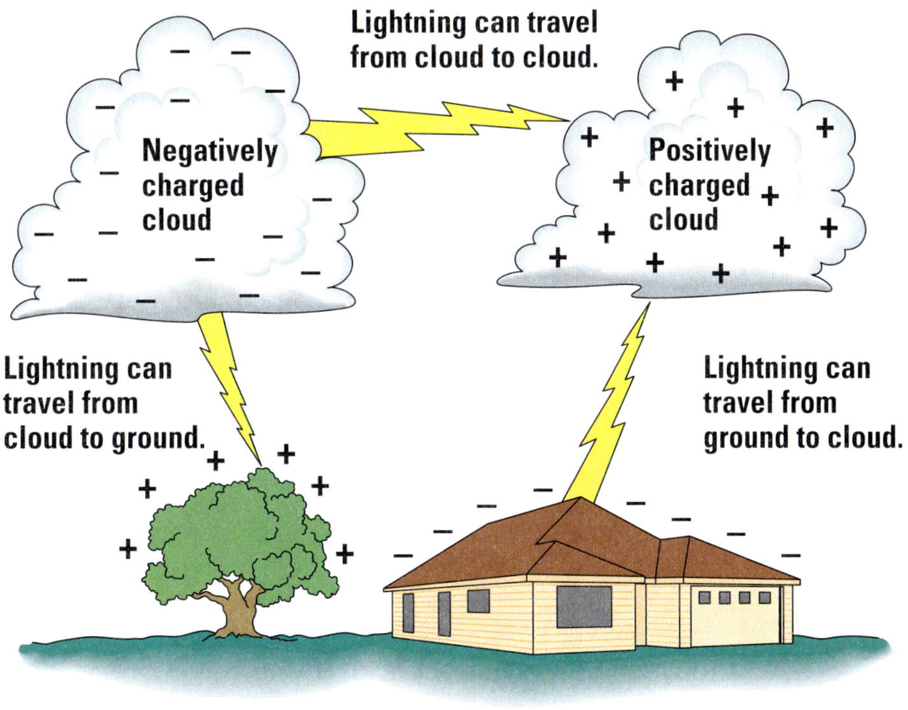

Figure 3–10 Lightning travels from negative to positive.

the lightning rod bleeds the lightning discharge to ground before the protected object can be harmed (*Figure 3–11*). The concept of the first lightning rods was initially detailed by Benjamin Franklin, and a common style of lightning rod still bears his name today, the Franklin Type Lightning Rod.

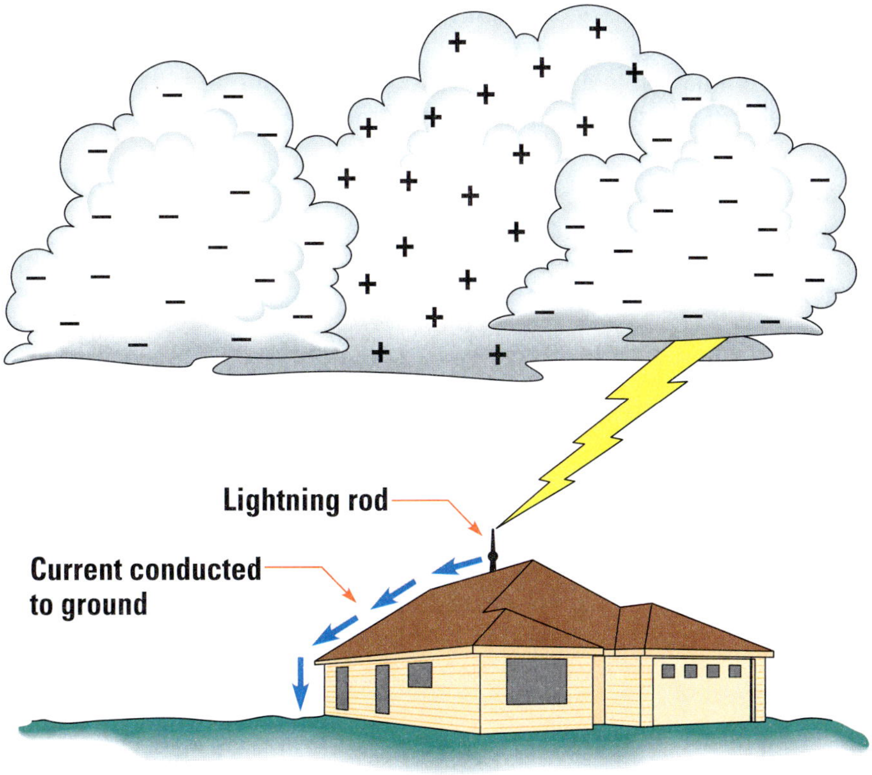

Figure 3–11 A lightning rod and down conductor provide an easy, low resistance path to ground.

Another device used for lightning protection is the **lightning arrestor.** The lightning arrestor works in a manner very similar to that of the lightning rod except that it is not designed to be struck by lightning itself, and it does not provide a direct path to ground. The lightning arrestor is grounded at one end, and the other end is brought close to but not touching the object to be protected. If the protected object is struck, the high voltage of the lightning arcs across to the lightning arrestor and bleeds to ground.

Power lines are often protected by lightning arrestors that exhibit a very high resistance at the normal voltage of the line. If the power line is struck by lightning, the increase of voltage causes the resistance of the arrestor to decrease and conduct the lightning discharge to ground.

3–5 Nuisance Static Charges

Static charges are sometimes a nuisance. Some examples of **nuisance static charges** are listed here:

1. **The static charge that accumulates on automobiles as they move through dry air.** These static charges can cause dangerous conditions under certain circumstances. For that reason, trucks carrying flammable materials such as gasoline or propane use a drag chain. One end of the drag chain is attached to the frame of the vehicle, and the other end drags the ground. The chain is used to provide a path to ground while the vehicle is moving and to prevent a static charge from accumulating on the body of the vehicle.
2. **The static charge that accumulates on a person's body as he or she walks across a carpet.** This charge can cause a painful shock when a metal object is touched, and it discharges in the form of an electric spark. Most carpets are made from man-made materials that are excellent insulators, such as nylon. In the winter, the heating systems of most dwellings remove moisture from the air and cause the air to have low humidity. The dry air combined with an insulating material provides an excellent setting for the accumulation of a static charge. This condition can generally be eliminated by the installation of a humidifier. A simple way to prevent the painful shock of a static discharge is to hold a metal object, such as a key or coin, in one hand. Touch the metal object to a grounded surface, and the static charge will arc from the metal object to ground instead of from your finger to ground.
3. **The static charge that accumulates on clothes in a dryer.** This static charge is caused by the clothes moving through the dry air. The greatest static charges generally are built up on man-made fabrics because they are the best insulators and retain electrons more readily than natural fabrics such as cotton or wool.

3–6 Useful Static Charges

Not all static charges are a nuisance. Some examples of **useful static charges** follow:

1. Static electricity is often used in spray painting. A high-voltage grid is placed in front of the spray gun. This grid has a positive charge. The object to be painted has a negative charge (*Figure 3–12*). As the droplets of paint pass through the grid, the positive charge causes electrons to be removed from the paint droplets. The positively charged droplets are attracted to the negatively charged object. This static charge helps prevent waste of the paint and, at the same time, produces a uniform finish.
2. Another device that depends on static electricity is the dry copy machine. The copy machine uses an aluminum drum coated with **selenium** (*Figure 3–13*). Selenium is a semiconductor material that changes its conductivity with a change of light intensity. When selenium is in the

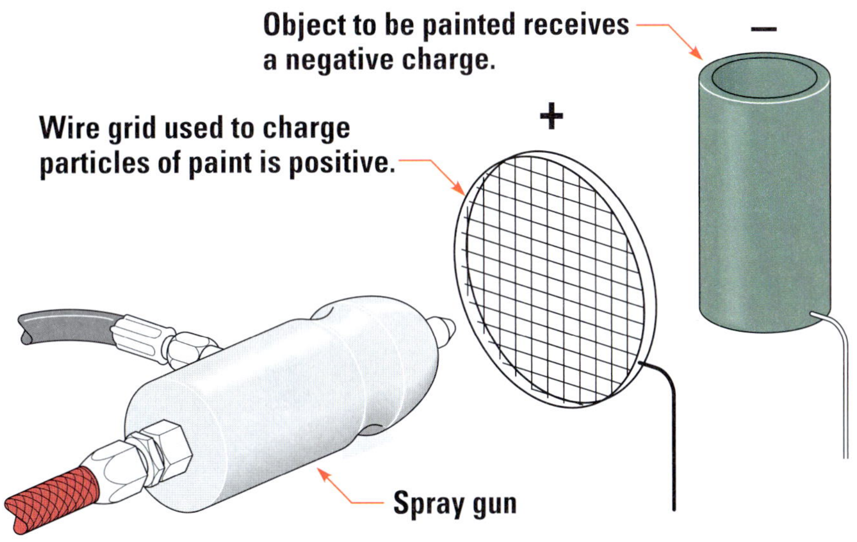

Figure 3–12 Static electric charges are often used in spray painting.

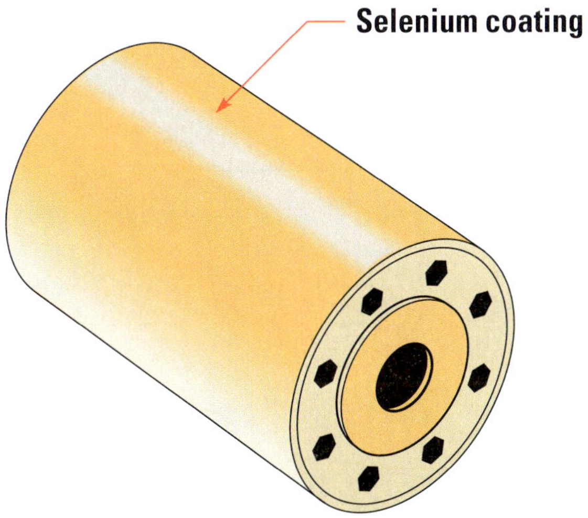

Figure 3–13 The drum of a copy machine is coated with selenium.

presence of light, it has a very high conductivity. When it is in darkness, it has a very low conductivity. However, most modern office copy machines are now digital and no longer utilize this outdated technology.

A high-voltage wire located near the drum causes the selenium to have a positive charge as it rotates (*Figure 3–14*). The drum is in darkness when it is charged. An image of the material to be copied is reflected on the drum by a system of lenses and mirrors (*Figure 3–15*). The light portions of the paper reflect more light than the dark portions. When the reflected light strikes the drum, the conductivity of the selenium increases greatly, and negative electrons from the aluminum drum neutralize the selenium charge at that point. The dark area of the paper causes the drum to retain a positive charge.

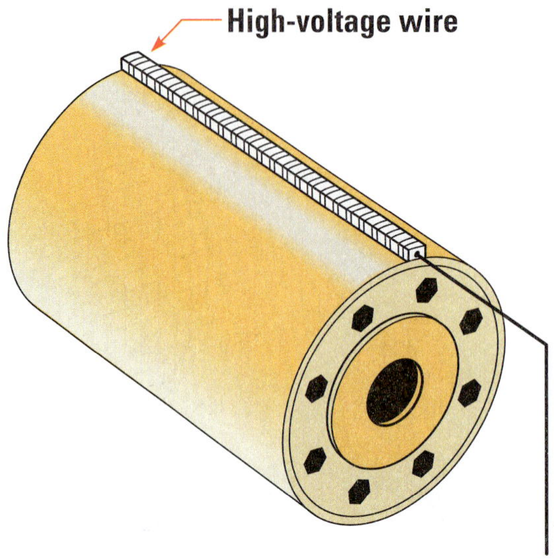

Figure 3–14 The drum receives a positive charge.

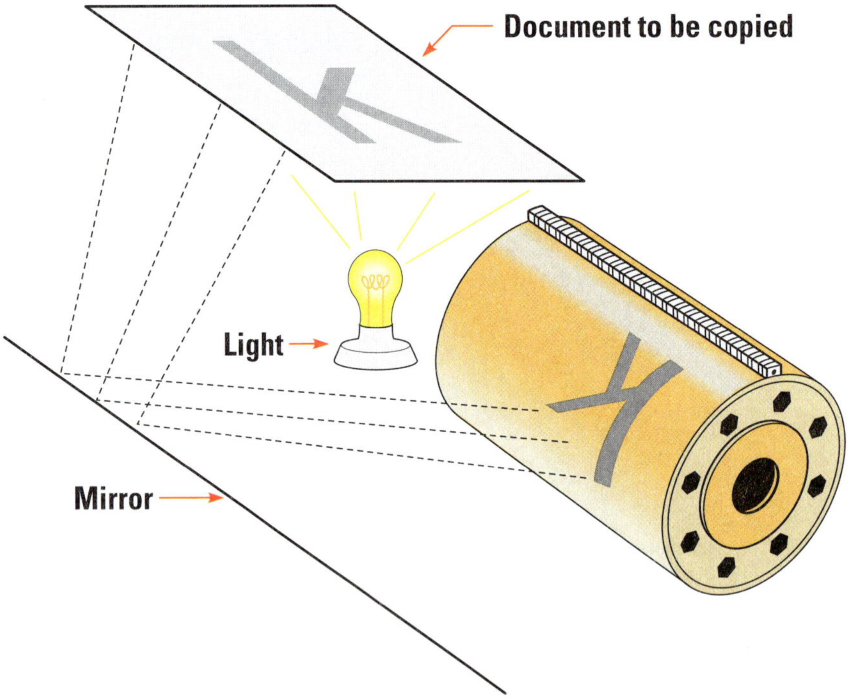

Figure 3–15 The image is transferred to the selenium drum.

A dark powder that has a negative charge is applied to the drum (*Figure 3–16*). The powder is attracted to the positively charged areas on the drum. The powder on the neutral areas of the drum falls away.

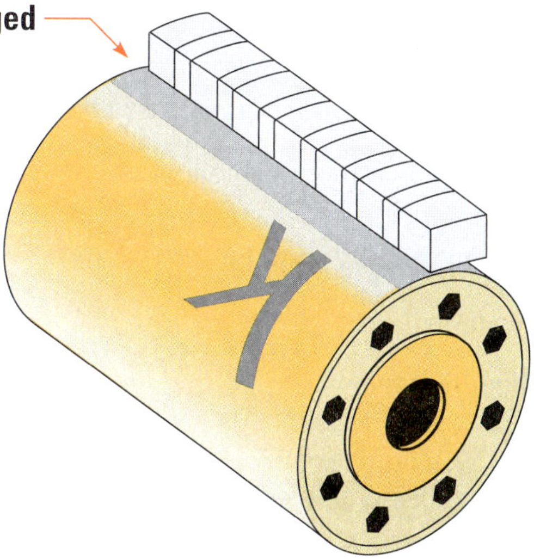

Figure 3–16 Negatively charged powder is applied to the positively charged drum.

A piece of positively charged paper passes under the drum (*Figure 3–17*) and attracts the powder from the drum. The paper then passes under a heating element, which melts the powder into the paper and causes the paper to become a permanent copy of the original.

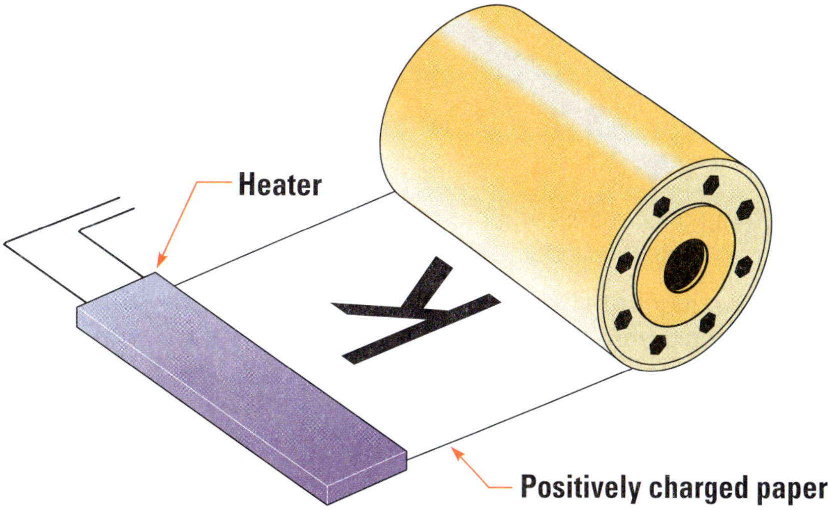

Figure 3–17 The negatively charged powder is attracted to the positively charged paper.

Summary

- The word *static* means not moving.
- An object can be positively charged by removing electrons from it.
- An object can be negatively charged by adding electrons to it.
- An electroscope is a device used to determine the polarity of an object.
- Static charges accumulate on insulator materials.
- Lightning is an example of a natural static charge.

Review Questions

1. True or False: The term static refers to something that is not moving or that is sitting still.
2. For an object to be charged, it must exhibit which of the following electrical properties to hold the electrons stationary?
   A) Be of a grounded conductive material
   B) Be a source of voltage
   C) Be of an insulating material
   D) Be magnetic
3. What is the initial step in using an electroscope to determine the charge of an object?
   A) The electroscope must be charged itself.
   B) The object must be charged.
   C) The metal leaves of the electroscope must be removed.
   D) The electroscope must be rubbed with a wool cloth.
4. On a negatively pre-charged electroscope, the metal leaves will behave in which of the following ways when a positively charged object nears the instrument?
   A) The blades will move farther apart.
   B) The blades will attract and get closer.
   C) There will be no movement of the blades.
   D) The blades will spin like a fan.
5. True or False: Lightning only occurs when clouds moving through the air develop enough of a negative charge that the electrons can travel to Earth and go to ground.
6. Lightning "bolts" can occur both between clouds and between the cloud and ground. Which of the following values would best describe the electrical qualities of a lightning bolt?
   A) Millions of amperes flowing
   B) Millions of volts
   C) Thousands of volts
   D) Thousands of amperes
7. The chain being dragged by a truck carrying LP or propane gas serves which of the following purposes?
   A) Let the driver know that they are still in contact with the road
   B) Make noise to warn other drivers of the truck's location
   C) Prevent the tank from discharging static electricity
   D) Prevent the tank from accumulating a static charge on the frame of the vehicle

8. True or False: Carpets are fantastic insulators and, therefore, will not allow a static charge to build up when a person walks across the floor.
9. Static electricity is often used in which of the following tasks to improve the process?
   A) Spray painting of objects that require uniform coverage
   B) The drum of one of the original photocopy machines
   C) The manufacture of sandpaper
   D) All of these
10. True or False: Electronic air filters utilize static electricity to attract small particles like smoke, dust, and pollen and trap them in the filter or precipitator.

Unit 4 Magnetism

Key Terms

Ampere-turns
Demagnetized
Electromagnets
Electron spin patterns
Flux
Flux density
Left-hand rule
Lines of flux
Lodestones
Magnetic domains
Magnetic molecules
Magnetomotive force (mmf)
Permanent magnets
Permeability
Reluctance
Residual magnetism
Saturation

Why You Need to Know

Magnetism and electricity are inseparable. If you have one, the other can be produced. The number of devices that operate on magnetism is almost endless. Everything from a simple compass to the largest electric motor in industry operates on magnetism. This unit presents those basic principles and discusses the most common terms used to measure magnetism:

- flux density,
- reluctance,
- ampere-turns, and
- the left-hand rule regarding electromagnets.

These concepts appear many times in the study of electricity.

Outline

Objectives

After studying this unit, you should be able to

- identify the properties of permanent magnets.
- differentiate between the axis poles of the earth and the magnetic poles of the earth.
- describe the operation of electromagnets.
- determine the polarity of an electromagnet.
- describe properties outlined in the electron theory of magnetism.
- identify the different systems used to measure magnetism.
- identify how working with magnetism can produce sound in a speaker.

Preview

Magnetism is one of the most important phenomena in the study of electricity. It is the force used to produce most of the electrical power in the world. The force of magnetism has been known for over 2000 years. It was first discovered by the Greeks when they noticed that a certain type of stone was attracted to iron. This stone was first found in Magnesia in Asia Minor and was named magnetite.

4–1 The Earth Is a Magnet

The first compass was invented when it was noticed that a piece of magnetite, a type of stone that is attracted to iron, placed on a piece of wood floating in water, always aligned itself north and south (*Figure 4–1*). Because they are always able to align themselves north and south, natural magnets became known as "leading stones" or **lodestones**. The reason that the lodestone aligned itself north and south is because the earth itself contains magnetic poles. *Figure 4–2* illustrates the positions of the true North and South Poles, or the axis, of the earth and the positions of the magnetic poles. Notice that *magnetic* north is not located at the true North Pole of the earth. This is the reason that navigators must distinguish between true north and magnetic north. The angular difference between the two is known as the *angle of declination*. This magnetic field surrounds the entire planet and converges at the poles.

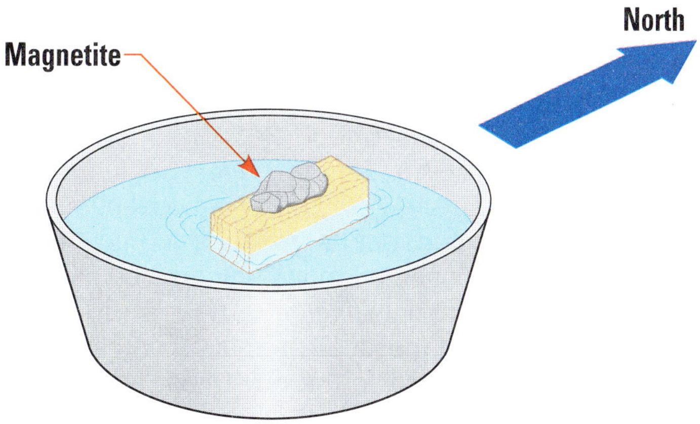

Figure 4–1 The first compass.

Also, notice that the magnetic north pole is located near the southern polar axis, and the magnetic south pole is located near the northern polar axis. The reason that the *geographic* poles (axes) are called north and south is because the north pole of a compass needle points in the direction of the north geographic pole. Because unlike magnetic poles attract, the north magnetic pole of the compass needle is attracted to the south magnetic pole of the earth.

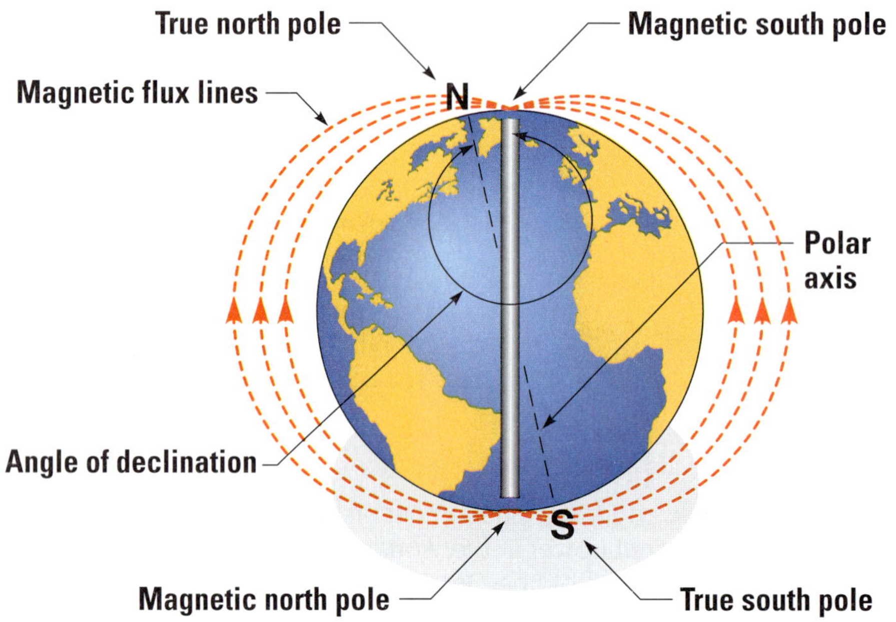

Figure 4–2 The earth is a magnet.

4–2 Permanent Magnets

Permanent magnets are magnets that do not require any power or force to maintain their field. They are an excellent example of one of the basic laws of magnetism that states that *energy is required to create a magnetic field, but no energy is required to maintain a magnetic field*. Man-made permanent magnets are much stronger and can retain their magnetism longer than natural magnets.

The Electron Theory of Magnetism

Only three substances actually form natural magnets: iron, nickel, and cobalt. Why these materials form magnets has been the subject of complex scientific investigations, resulting in an explanation of magnetism based on **electron spin patterns**. It is believed that electrons spin on their axes as they orbit the nucleus of the atom. This spinning motion causes each electron to become a tiny permanent magnet. Although all electrons spin, they do not all spin in the same direction. In most atoms, electrons that spin in opposite directions tend to form pairs (*Figure 4–3*). Because the electron pairs spin in opposite directions, their magnetic effects cancel each other out as far as having any effect on distant objects. In a similar manner, two horseshoe magnets connected together would be strongly attracted to each other but would have little effect on surrounding objects (*Figure 4–4*).

An atom of iron contains 26 electrons; of these, 22 are paired and spin in opposite directions, canceling each other's magnetic effect. In the next to the outermost shell, however, 4 electrons are not paired and spin in the same direction. These 4 electrons account for the magnetic properties of iron. At a temperature of 1420°F, or 771.1°C, the electron spin patterns rearrange themselves, and iron loses its magnetic properties.

When the atoms of most materials combine to form molecules, they arrange themselves in a manner that produces a total of eight valence electrons. The electrons form a spin pattern that cancels the magnetic field of the material. When the atoms of iron, nickel, and cobalt combine, however, the magnetic field is not canceled. Their electrons combine so they share valence electrons in such a way that their spin patterns are in the same direction, causing their magnetic fields to add instead of cancel.

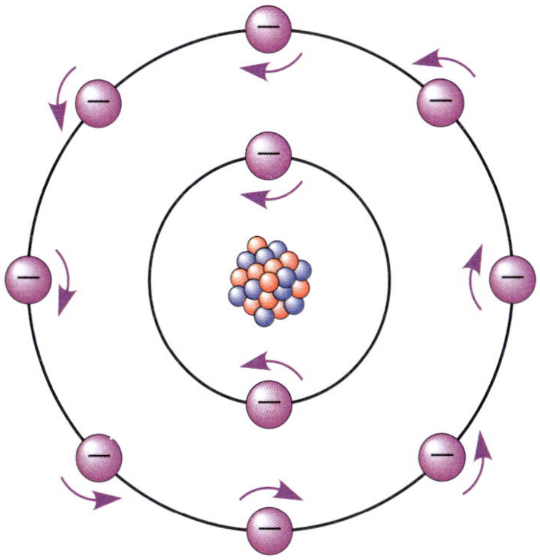

Figure 4–3 Electron pairs generally spin in opposite directions.

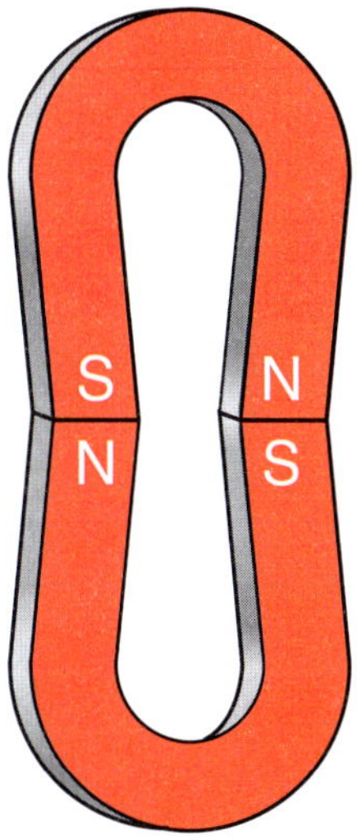

Figure 4–4 Two horseshoe magnets attract each other.

The additive effect forms regions in the molecular structure of the metal called **magnetic domains** or **magnetic molecules.** These magnetic domains act like small permanent magnets.

A piece of nonmagnetized metal has its molecules in a state of disarray (*Figure 4–5*). When the metal is magnetized, its molecules align themselves in an orderly pattern (*Figure 4–6*). In theory, each molecule of a magnetic material is itself a small magnet. If a permanent magnet is cut into pieces, each piece is a separate magnet (*Figure 4–7*).

Figure 4–5 The atoms are disarrayed in a piece of nonmagnetized metal.

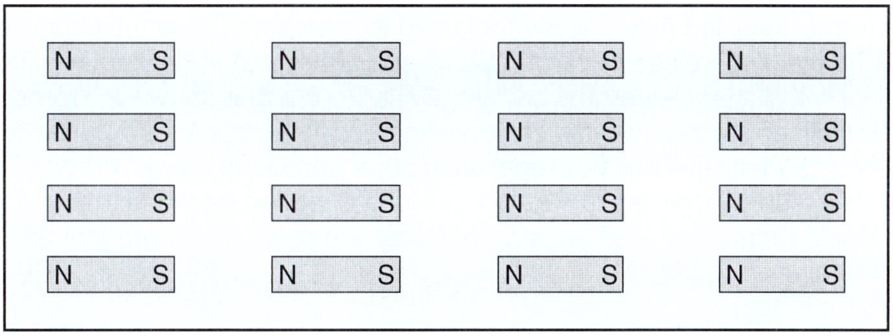

Figure 4–6 The atoms are aligned in an orderly fashion in a piece of magnetized metal.

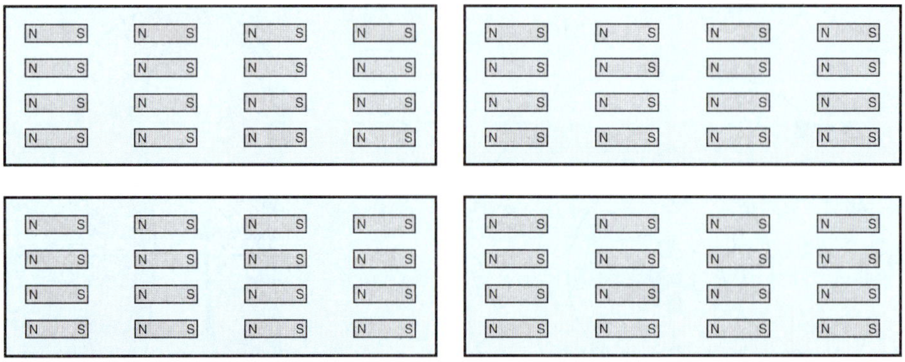

Figure 4–7 When a magnet is cut apart, each piece becomes a separate magnet.

Magnetic Materials

Magnetic materials can be divided into three basic classifications:

- *Ferromagnetic materials* are metals that are easily magnetized. Examples of these materials are iron, nickel, cobalt, and manganese.
- *Paramagnetic materials* are metals that can be magnetized but not as easily as ferromagnetic materials. Some examples of paramagnetic materials are platinum, titanium, and chromium.
- *Diamagnetic materials* are either metal or nonmetal materials that cannot be magnetized. The magnetic lines of force tend to go around them instead of through them. Some examples of these materials are copper, brass, and antimony.

Some of the best materials for the production of permanent magnets are alloys. One of the best permanent magnet materials is Alnico 5, which is made from a combination of aluminum, nickel, cobalt, copper, and iron. Another type of permanent magnet material is made from a combination of barium ferrite and strontium ferrite. Ferrites can have an advantage in some situations because they are insulators and not conductors. They have a resistivity of approximately 1,000,000 ohm-centimeters. Barium ferrite and strontium ferrite can be powdered. The powder is heated to the melting point and then rolled and heat treated. This treatment changes the grain structure and magnetic properties of the material. The new type of material has a property more like stone than metal and is known as a *ceramic magnet*. Ceramic magnets can be powdered and mixed with rubber, plastic, or liquids. Ceramic magnetic materials mixed with liquids can be used to make magnetic ink, which is used on checks. Another frequently used magnetic material is iron oxide, which is used to make magnetic recording tape and computer disks.

Magnetic Lines of Force

Magnetic lines of force are called **flux**. The symbol used to represent flux is the Greek letter phi (Φ). Flux lines can be seen by placing a piece of cardboard on a magnet and sprinkling iron filings on the cardboard. The filings will align themselves in a pattern similar to the one shown in *Figure 4–8*. The pattern produced by the iron filings forms a two-dimensional figure, but the flux lines actually surround the entire magnet (*Figure 4–9*). Magnetic **lines of flux** repel each other and never cross. Although magnetic lines of flux do not flow, it is assumed they are in a north-to-south direction because that is the direction that a compass needle would point. Flux lines are stronger near the poles of the magnet and become weaker as they move farther away.

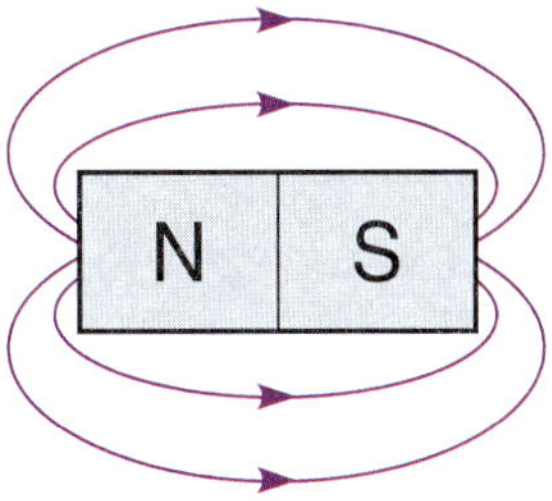

Figure 4–8 Magnetic lines of force are called lines of flux.

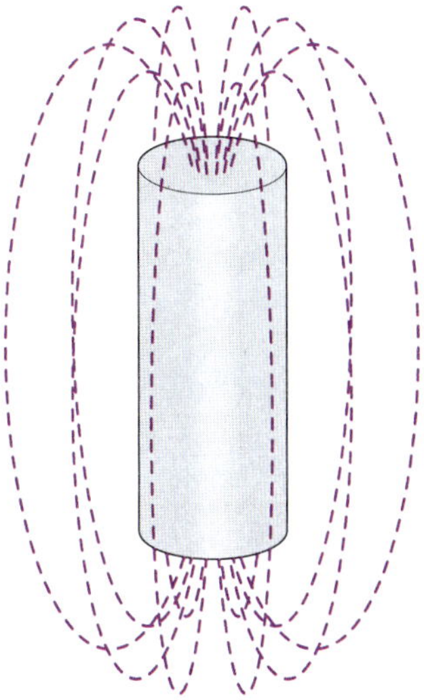

Figure 4–9 Magnetic lines of flux surround the entire magnet.

A basic law of magnetism states that *unlike poles attract and like poles repel. Figure 4–10* illustrates how the magnetic lines of flux interact when two magnets with their north and south poles facing each other are placed in close proximity. *Figure 4–11* illustrates what happens when two magnets with like poles facing each other are pushed closer together.

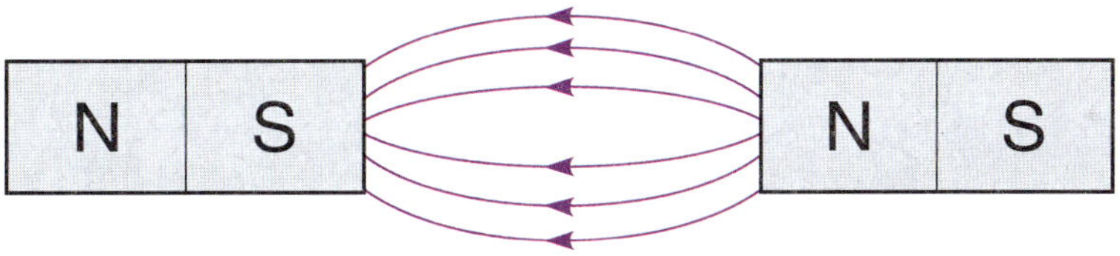

Figure 4–10 Magnetic lines of flux between two magnets when opposite poles are facing each other.

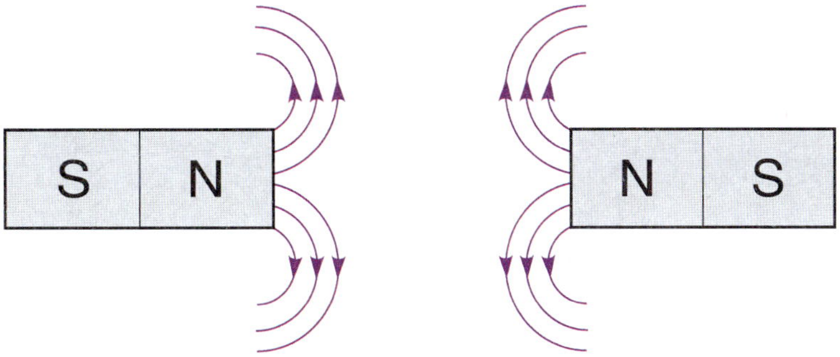

Figure 4–11 Magnetic lines of flux repel when like poles are brought near each other.

If the opposite poles of two magnets are brought close to each other, they are attracted to each other and will want to attach themselves together (*Figure 4–12*). If like poles of two magnets are brought together, they repel each other, and it will take a force stronger than the strength of the magnetism to force the magnets to make contact.

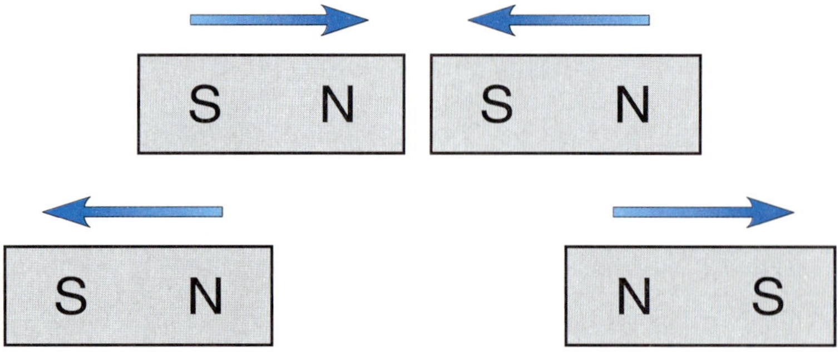

Figure 4–12 Opposite poles of a magnet attract, and like poles repel.

4–3 Electromagnetics

Whenever an electric current flows through a conductor, a magnetic field is formed around the conductor. This is one of the basic principles of electricity when we study the physics behind it (*Figure 4–13*). **Electromagnets** depend on electric current flow to produce a magnetic field. They are generally designed to produce a magnetic field only as long as the current is flowing; they do not retain their magnetism when the current flow stops If the conductor is wound into a coil (*Figure 4–14*), the magnetic lines of flux add to produce a stronger magnetic field. A coil with 10 turns of wire produces a magnetic field that is ten times as strong as the magnetic field around a single conductor.

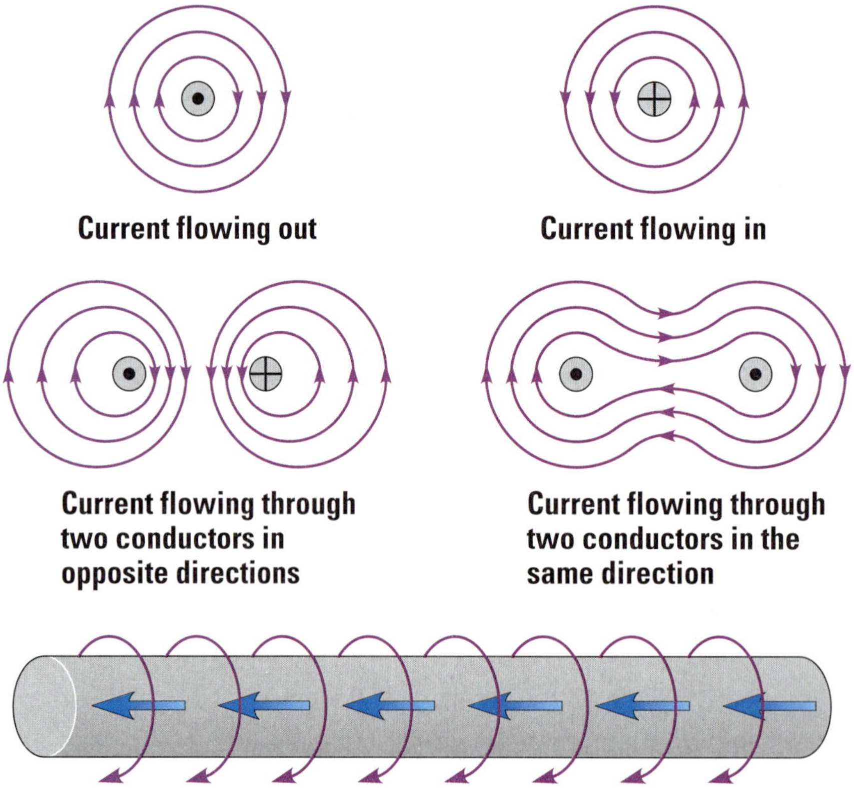

Figure 4–13 Current flowing through a conductor produces a magnetic field around the conductor.

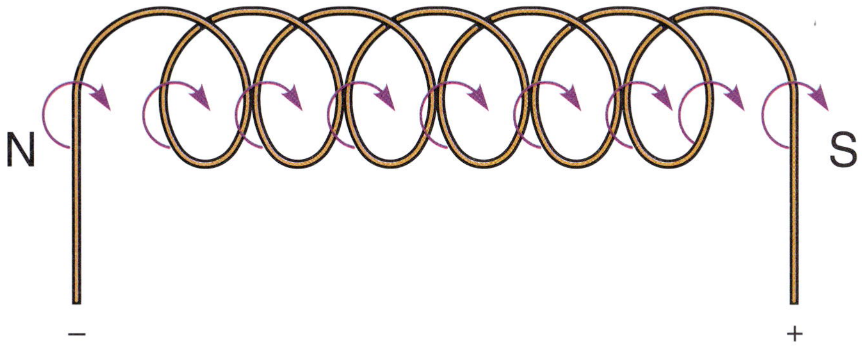

Figure 4–14 Winding the wire into a coil increases the strength of the magnetic field.

Another factor that affects the strength of an electromagnetic field is the amount of current flowing through the wire. An increase in current flow causes an increase in magnetic field strength. The two factors that determine the number of flux lines produced by an electromagnet are the number of turns of wire and the amount of current flowing through the wire. The strength of an electromagnet is proportional to its **ampere-turns.** Ampere-turns are determined by multiplying the number of turns of wire by the current flow.

Core Material

Coils can be wound around any type of material to form an electromagnet. The base material is called the *core material*. When a coil is wound around a nonmagnetic material such as wood or plastic, it is known as an *air-core magnet*. When a coil is wound around a magnetic material such as iron or soft steel, it is known as an *iron-core magnet*. The addition of magnetic material to the center of the coil can greatly increase the strength of the magnet. If the core material causes the magnetic field to become ten times stronger, the core material has a **permeability** of 10 (*Figure 4–15*). Permeability is a measure of a material's ability to become magnetized. The number of flux lines produced is proportional to the ampere-turns. The magnetic core material provides an easy path for the flow of magnetic lines in much the same way a conductor provides an easy path for the flow of electrons. This increased permeability permits the flux lines to be concentrated in a smaller area, which increases the number of flux lines per square inch or per square centimeter.

Another common magnetic measurement is **reluctance.** Reluctance is resistance to magnetism. A material such as soft iron or steel has a high permeability and low reluctance because it is easily magnetized. A material such as copper has a low permeability and high reluctance.

If the current flow in an electromagnet is continually increased, the magnet eventually reaches a point where its strength increases only slightly with an increase in current. When this condition occurs, the magnetic material is at a point of **saturation**. Saturation occurs when all the molecules of the magnetic material are lined up. Think of magnetic saturation like pouring water into a bucket; once the bucket is full, it simply cannot hold any more water. If it became necessary to construct a stronger magnet, a larger piece of core material would be required.

When the current flow through the coil of a magnet is stopped, some magnetism may be left in the core material. The amount of magnetism left in a material after the magnetizing force has stopped is called **residual magnetism**. If the residual magnetism of a piece of core material is hard to remove, the material has a high coercive force. *Coercive force* is a measure of a material's ability to retain magnetism. A high coercive force is desirable in materials that are intended to be used as permanent magnets. A low coercive force is generally desirable for materials intended to be used as electromagnets. Another term that is used to describe a material's ability to retain magnetism is *retentivity*.

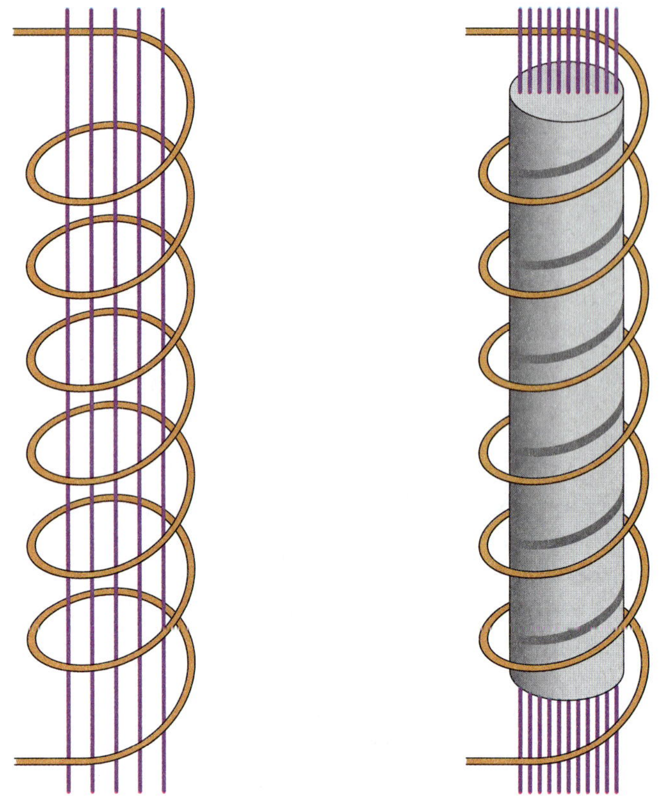

Figure 4–15 An iron core increases the number of flux lines per square inch.

4–4 Magnetic Measurements

The terms used to measure the strength of a magnetic field are determined by the system that is being used. Three different systems are used to measure magnetism: the English system, the CGS system, and the MKS or SI system.

The English System

In the English system of measure, magnetic strength is measured in a term called **flux density**. Flux density is measured in lines per square inch. The Greek letter phi (**Φ**) is used to denote flux, and the letter *B* is used to represent flux density. The following formula is used to determine flux density:

$$\text{B (flux density)} = \frac{\Phi \text{ (flux lines)}}{\text{A (area)}}$$

In the English system, the term used to describe the total force producing a magnetic field, or flux, is **magnetomotive force (mmf)**. Magnetomotive force can be computed using the following formula:

$$\text{mmf} = \Phi \times \text{rel (reluctance)}$$

The following formula can be used to determine the strength of the magnet:

$$\text{Pull (in pounds)} = \frac{B \times A}{72{,}000{,}000}$$

where

B = flux density in lines per square inch

A = area of the magnet.

The CGS System

In the CGS (centimeter-gram-second) system of measurement, one magnetic line of force is known as a *maxwell*. A gauss represents a magnetic force of 1 maxwell per square centimeter. A gauss is also equal to 6.4516 lines of flux per square inch. In the English or SI system, magnetomotive force is measured in ampere-turns. In the CGS system, gilberts are used to represent the same measurement. Because the main difference between these two systems of measurement is that they use different units of measure, a conversion factor can be employed to help convert one set of units to another:

1 gilbert = 1.256 ampere-turns

1 ampere-turn = 0.7905 gilbert

In the CGS system, a standard called the *unit magnetic pole* is used. In *Figure 4–16,* two magnets are separated by a distance of 1 centimeter. These magnets repel each other with a force of 1 dyne. The dyne is a very weak unit of force. One dyne is equal to $\frac{1}{27{,}800}$ of an ounce, or it requires 27,800 dynes to equal a force of 1 ounce. When the two magnets separated by a distance of 1 centimeter exert a force of 1 dyne, they are considered to be a unit magnetic pole. Magnetic force can then be determined using the formula:

$$\text{Force (in dynes)} = \frac{M_1 \times M_2}{D^2}$$

where

M_1 = strength of first magnet in unit magnetic poles

M_2 = strength of second magnet in unit magnetic poles

D = distance between the poles in centimeters

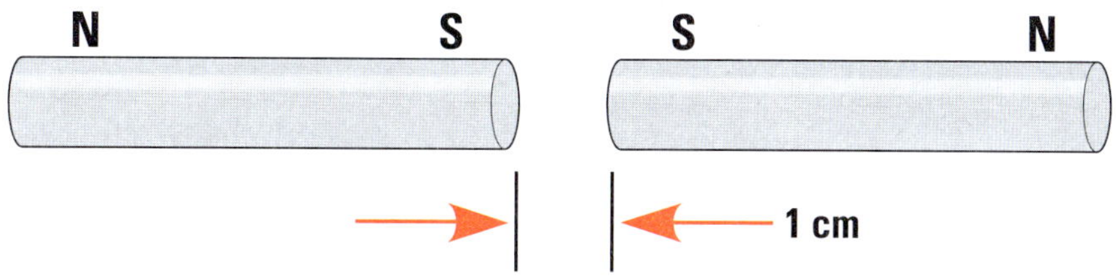

Figure 4–16 A unit magnetic pole produces a force of one dyne.

The MKS or SI System

The MKS or SI system employs the units of measure of the MKS (meter-kilogram-second) system. In the SI system, the unit of force is the newton. One newton is equal to 0.2248 pounds, or it requires 4.448 newtons to equal a force of 1 pound. The weber is used to measure magnetic flux. One weber equals 100,000,000 lines of flux or 10^8 maxwells. The main unit of measure in the SI system is the tesla. One tesla is equal to one weber per square meter. *Table 4–1* lists different magnetic units of measure and the conversion factor to other magnetic units of measure.

1 gauss	=	6.4516 lines/sq. in.
1 gauss	=	0.0001 tesla
1 tesla	=	10,000 gauss
1 tesla	=	64,516 lines/sq. in.
1 tesla	=	1 Weber/sq. meter
1 weber	=	100,000,000 lines of flux
1 line/sq. in.	=	0.155 gauss
1 line/sq. in.	=	0.0000155 tesla
1 gilbert	=	0.7959949 ampere–turns
1 ampere–turn	=	1.256637 gilberts

Table 4–1 Magnetic Units of Measure

Magnetic Polarity

The polarity of an electromagnet can be determined using the **left-hand rule.** When the fingers of the left hand are placed around the windings in the direction of electron current flow, the thumb points to the north magnetic pole (*Figure 4–17*). If the direction of current flow is reversed, the polarity of the magnetic field also reverses.

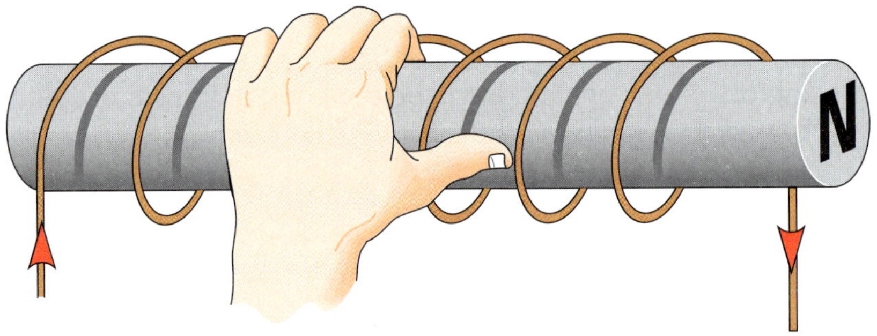

Figure 4–17 The left-hand rule can be used to determine the polarity of an electromagnet.

4–5 Working with Magnetism

When an object is to be **demagnetized**, its molecules must be rearranged from their magnetized state into a state of disarray as if they are in a nonmagnetized material. This can be done by placing the object in the field of a strong electromagnet connected to an AC line. Because the magnet is connected to AC, the polarity of the magnetic field reverses each time the current changes direction. The molecules of the object to be demagnetized are, therefore, aligned first in one direction and then in the other. If the object is pulled away from the AC magnetic field, the effect of the field becomes weaker as the object is moved farther away (*Figure 4–18*). The weakening of the magnetic field causes the molecules of the object to be left in a state of disarray. The ease or difficulty with which an object can be demagnetized depends on the strength of the AC magnetic field and the coercive force of the object.

An object can be demagnetized in two other ways (*Figure 4–19*). If a magnetized object is struck, the vibration often causes the molecules to rearrange themselves in a disordered fashion. It may be necessary to strike the object several times. Heating also demagnetizes an object. When the temperature becomes high enough, the molecules rearrange themselves in a disordered fashion.

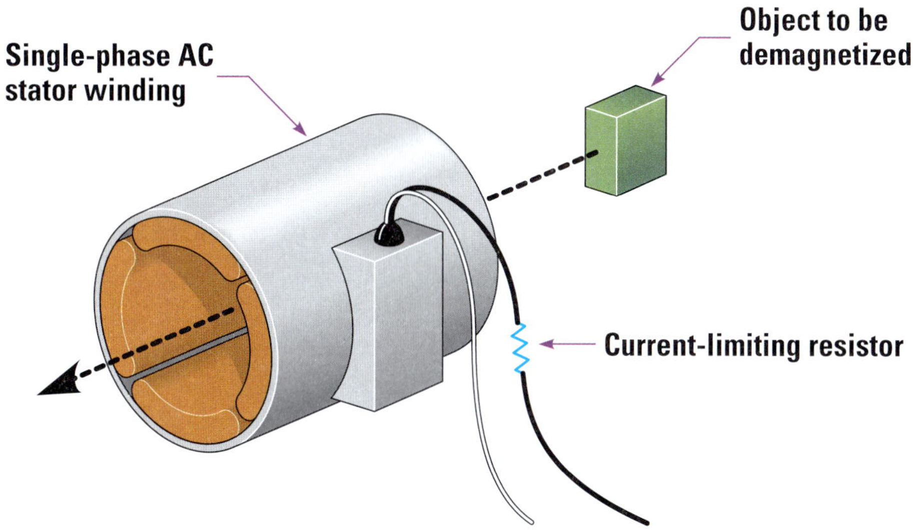

Figure 4–18 Demagnetizing an object.

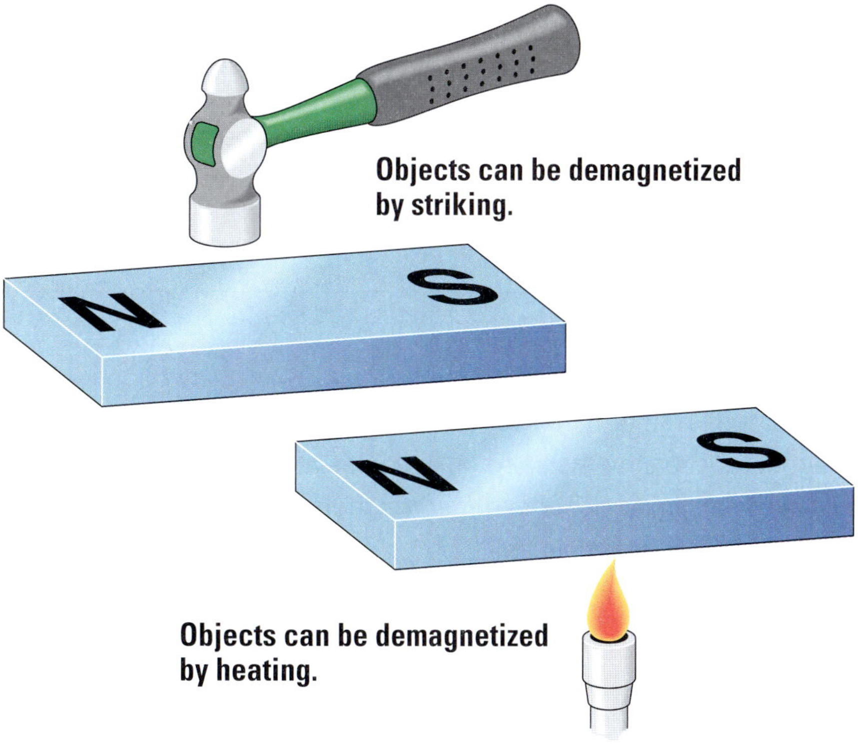

Figure 4–19 Other methods for demagnetizing objects.

Magnetic Devices

A list of devices that operate on magnetism would be very long indeed. Some of the more common devices are electromagnets, measuring instruments, inductors, transformers, and motors.

The Speaker

The speaker is a common device that operates on the principle of magnetism (*Figure 4–20*). The speaker produces sound by moving a cone; the movement causes a displacement of air. The tone is determined by how fast the cone vibrates. Low or bass sounds are produced by vibrations in the range of 20 cycles per second. High sounds are produced when the speaker vibrates in the range of 20,000 cycles per second.

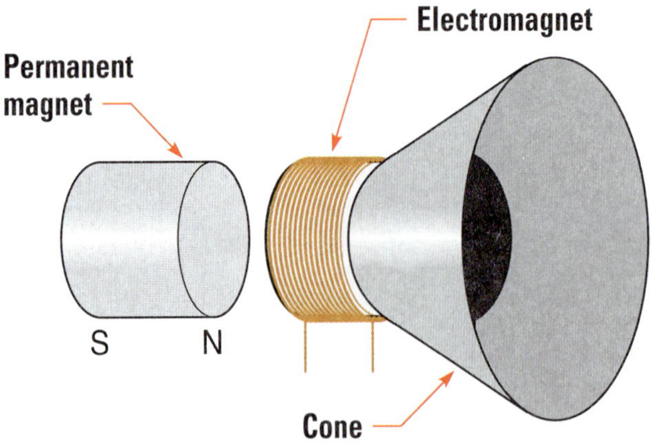

Figure 4–20 A speaker uses both an electromagnet and a permanent magnet.

The speaker uses two separate magnets. One is a permanent magnet, and the other is an electromagnet. The permanent magnet is held stationary, and the electromagnet is attached to the speaker cone. When current flows through the coil of the electromagnet, a magnetic field is produced. The polarity of the field is determined by the direction of the current flow. When the electromagnet has a north polarity, it is repelled from the permanent magnet, causing the speaker cone to move outward and displace air. When the current flow reverses through the coil, the electromagnet has a south polarity and is attracted to the permanent magnet. The speaker cone then moves inward and again displaces air. The number of times per second that the current through the coil reverses determines the tone of the speaker.

Summary

- Early natural magnets were known as lodestones.
- The Earth has a north and a south magnetic pole.
- The magnetic poles of the Earth and the axis poles are not the same.
- Like poles of a magnet repel each other, and unlike poles attract each other.
- Some materials have the ability to become better magnets than others.
- Three basic types of magnetic materials are
  a. Ferromagnetic
  b. Paramagnetic
  c. Diamagnetic
- When current flows through a wire, a magnetic field is created around the wire.
- The direction of current flow through the wire determines the polarity of the magnetic field.
- The strength of an electromagnet is determined by the ampere-turns.
- The type of core material used in an electromagnet can increase its strength.
- Three different systems are used to measure magnetic values:
  a. The English system
  b. The CGS system
  c. The SI system
- An object can be demagnetized by placing it in an AC magnetic field and pulling it away, by striking, and by heating.

Review Questions

1. Energy is required to create a magnetic field, but no energy is required to perform which of these after the magnetic field is created?
   A) Generate electricity
   B) Maintain the magnetic field
   C) Demagnetize the magnet
   D) Reverse polarity of the magnetic field
2. Diamagnetic is a term used to describe which of the following?
   A) Easily magnetized
   B) An alloy that can be magnetized electrically
   C) Ceramic magnets
   D) Material that cannot be magnetized
3. True or False: True north and magnetic north are the same direction, and no additional consideration was necessary for early global travelers.
4. The angular difference between true north and magnetic north is known as what?
   A) Angle of differentiation
   B) Angle of magnetic lines
   C) Angle of declination
   D) Angle of true squares
5. Electromagnets only produce a magnetic field when which of the following events happen?
   A) Current flows.
   B) Heat is applied.
   C) The object is struck.
   D) Another magnet is nearby.
6. True or False: The strength of an electromagnet is directly proportional to the number of turns of wire around the core material and the amperes flowing within the conductor. This is referred to as ampere-turns.
7. What is the name of the rule that is often used to determine the polarity of an electromagnet?
   A) The thumb rule.
   B) The one-hand rule
   C) The rule of polarity
   D) The left-hand rule
8. If the direction of current flow in an electromagnet is suddenly reversed, what will happen to the magnet?
   A) The magnetic field will double in intensity.
   B) The magnet will reverse polarity.

C) The magnetic field will decrease in intensity.

D) Nothing will happen.

9. Which system of magnet strength measurement utilizes the unit of measurement called flux density and the Greek letter phi as the symbol?

   A) The CGS system

   B) The SI system

   C) The English system

   D) The magnetic measurement system

10. True or False: The newton is used to measure the strength of a magnet in the SI system of measurement.

11. Which of the following materials will not form a natural magnet?

    A) Iron

    B) Nickel

    C) Cobalt

    D) Platinum

12. True or False: The electron theory of magnetism theorizes that the magnetic properties of a material can be explained through the electron spin patterns. Electrons spinning in the opposite direction add to, and spinning in the same direction will cancel out, the magnetic fields.

13. Which of the following statements best summarizes how a speaker uses magnetism to produce an audible sound?

    A) An electromagnet repels or attracts a permanent magnet to deflect a cone and displace air to produce sound waves.

    B) When current is run through a permanent magnet, it produces a sound applicable to frequency.

    C) The cone of the speaker funnels the electrical signal to the permanent magnet and produces a sound.

    D) Electromagnets make noise when current is applied.

14. True or False: The tone of the sound produced by a speaker is determined by the number of times the speaker moves in and out in a second.

Practical Applications

Industry often employs the use of an electromagnet to pick up metal objects. These large magnets are called round horseshoe magnets. They contain a center core surrounded by an outer core. A direct current winding of magnetic wire is wound around the center core. An illustration of a round horseshoe magnet that has been cut in half is shown in *Figure 4–21*. When direct current flows through the winding, the magnetic field produces the polarities as shown. The dimensions of the magnet are shown in *Figure 4–22*. It is assumed that the magnetic field developed has a flux density of 2 webers per square inch. Determine the maximum pulling force of the magnet.

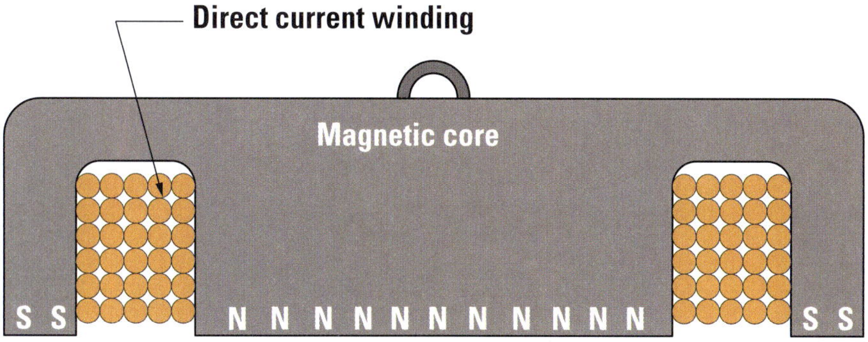

Figure 4–21 A round horseshoe magnet contains wire wound around a magnetic core that produces an electromagnet when current is flowing through the wire.

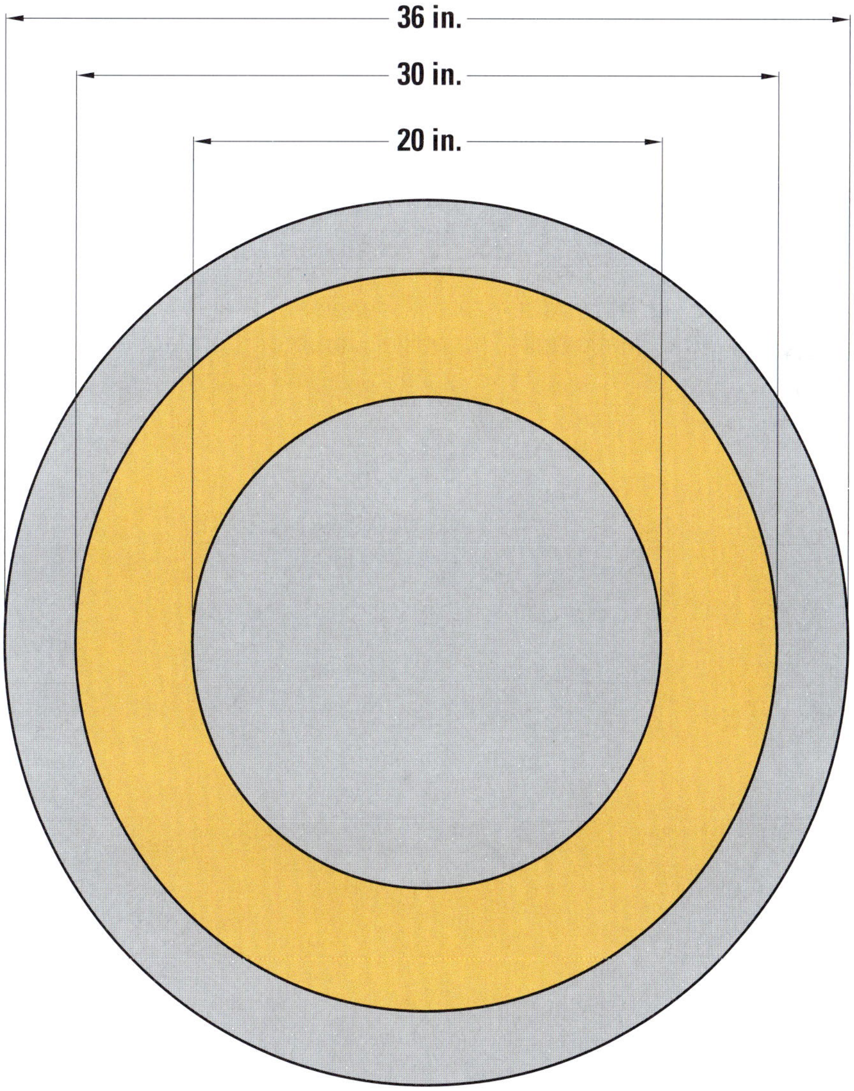

Figure 4–22 Dimensions of a round horseshoe magnet.

Solution

The first step is to determine the surface area of the magnet. The maximum area of the magnet can be determined by multiplying the square of the radius by the number pi. The magnet has a diameter of 36 inches. The radius is one-half the diameter.

$$A = \pi r^2$$

$$A = 3.1416 \times 18 \times 18$$

$$A = 1017.89 \text{ sq. in.}$$

The next dimension is a circle 30 inches in diameter.

$$A = 3.1416 \times 15 \times 15$$

$$A = 706.86 \text{ sq. in.}$$

The surface area of the outer ring can be determined by subtracting these two dimensions.

$$1017.89 - 706.86 = 311.03 \text{ sq. in.}$$

The area of the center part of the magnet can be determined using the formula for the area of a circle.

$$A = 3.1416 \times 10 \times 10$$

$$A = 314.16 \text{ sq. in.}$$

The total surface area of the magnet is the sum of the two areas.

$$311.03 + 314.16 = 626.19 \text{ sq. in.}$$

The problem stated that the magnet produces a flux density of 2 webers per square inch. A weber is equal to 100,000,000 lines of magnetic flux. The pulling force of the magnet can be determined using the formula:

$$\text{Pounds pull} = \frac{B \times A}{72{,}000{,}000}$$

$$\text{Pounds pull} = \frac{200{,}000{,}000 \times 626.19}{72{,}000{,}000}$$

$$\text{Pounds pull} = 1739.42$$

Unit 5 Resistors

Key Terms

Carbon film resistor

Color code

Composition carbon resistor

Fixed resistors

Metal film resistor

Metal glaze resistor

Multiturn variable resistors

Pot

Potentiometer

Rheostat

Short circuit

Tolerance

Variable resistor

Voltage divider

Wire-wound resistors

Why You Need to Know

Resistors are one of the basic components of an electrical circuit. Resistance is also one of the basic units of measurement in electrical theory, and we will be using the resistance provided by resistors extensively in the application of concepts such as Ohm's law. It is important to learn how resistors are employed in circuits and the effect resistance has on how the circuit operates. To understand resistors, this unit covers

- how the values of different types of resistors are determined.
- how some resistors use bands of color while others employ markings.
- how some resistors are intended to dissipate large amounts of heat and should be mounted a particular way.

Outline

5–1 Uses of Resistors

5–2 Fixed Resistors

5–3 Color Code

5–4 Standard Resistance Values of Fixed Resistors

5–5 Power Ratings

5–6 Variable Resistors

5–7 Schematic Symbols

Objectives

After studying this unit, you should be able to

- describe the role resistors play in a circuit.
- discuss the major types of fixed resistors.
- apply the resistor color code to determine the resistance value of a fixed resistor.
- analyze whether a resistor is operating within its power rating.
- describe how a variable resistor is used as a potentiometer.

Preview

Resistors are one of the most common components found in electric circuits. The unit of measure for resistance (R) is the *ohm,* which was named after a German scientist named Georg S. Ohm. The symbol used to represent resistance is the Greek letter omega (Ω). Resistors come in various sizes, types, and ratings to accommodate the needs of almost any circuit applications.

5–1 Uses of Resistors

Resistors, or resistive loads, limit the flow of current through the circuit. In *Figure 5–1,* a 30-ohm resistor is connected to a 15-volt battery. The current in this circuit is limited to a value of 0.5 ampere.

$$I = \frac{E}{R}$$

$$I = \frac{15\ V}{30\ \Omega}$$

$$I = 0.5\ A$$

Figure 5–1 The resistance in this circuit limits the current that can flow.

If this resistor were not present, the circuit current would be limited only by the resistance of the conductor, which would be very low, and a large amount of current would flow. For example, a 16 AWG solid copper conductor has a resistance of around 5 ohms per 1000 feet of material. If the total length of conductor in the circuit is 50 feet, the resistance of the wire is around 0.25 ohms. When the wire is connected across the 15-volt power source, a current of 75 amperes would try to flow through the circuit (15 V/0.0.25 Ω = 60 A). This is commonly known as a **short circuit**.

A secondary function of resistors is to produce a **voltage divider**. The three resistors shown in *Figure 5–2* are connected in series with a 17.5-volt battery. If the leads of a voltmeter were connected between different points in the circuit, it would indicate the following voltages:

A to B, 1.5 V

A to C, 7.5 V

A to D, 17.5 V

B to C, 6 V

B to D, 16 V

C to D, 10 V

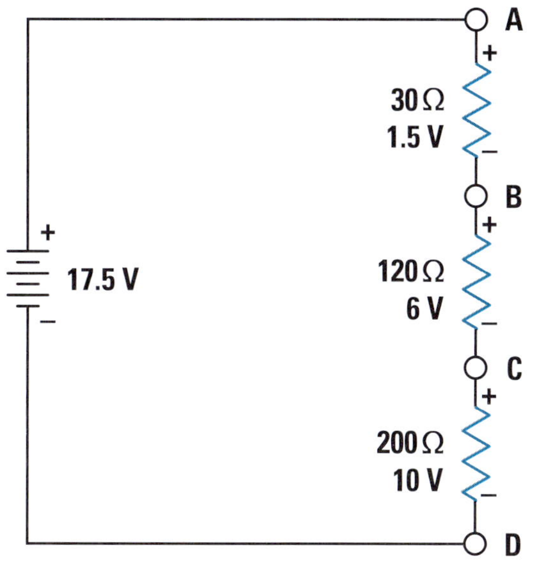

Figure 5–2 Resistors used as a voltage divider to obtain a specific voltage drop across a component.

By connecting resistors of the proper value, almost any voltage desired can be obtained. Voltage dividers were used to a large extent in vacuum-tube circuits many years ago. Voltage divider circuits are still used in applications involving field-effect transistors (FETs) and in multirange voltmeter circuits.

5–2 Fixed Resistors

Fixed resistors have only one ohmic value, which cannot be changed or adjusted. There are several different types of fixed resistors. One of the most common types of fixed resistors is the **composition carbon resistor**. Carbon resistors are made from a compound of carbon graphite and a resin-bonding material. The proportions of carbon and resin material determine the value of resistance. This compound is enclosed in a case of nonconductive material with connecting leads (*Figure 5–3*).

Carbon resistors are very popular for most applications because they are inexpensive and readily available. They are made in standard values that range from about 1 ohm to about 22 megaohms (MΩ), and they can be obtained in power ratings of ⅛, ¼, ½, 1, and 2 watts. The power rating of the resistor is indicated by its size. A ½-watt resistor is approximately ⅜ inch in length and ⅛ inch in diameter.

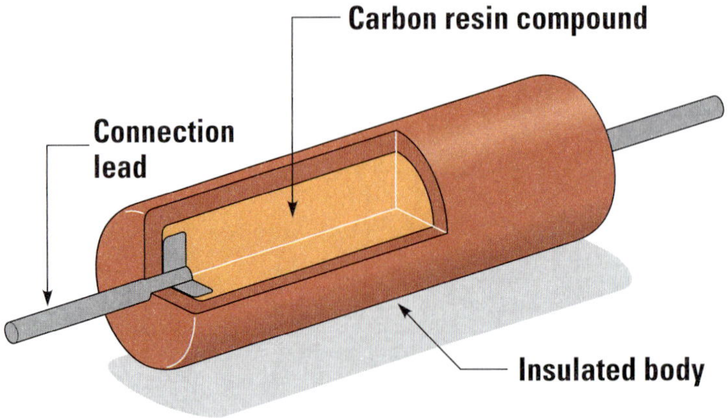

Figure 5–3 Cut away example of a composition carbon resistor.

A 2-watt resistor has a length of approximately 11⁄16 inch and a diameter of approximately 5⁄16 inch (*Figure 5–4*). The 2-watt resistor is larger than the ½-watt or 1-watt because it must have a larger surface area to be able to dissipate more heat. Although carbon resistors have a lot of desirable characteristics, they have one characteristic that is not desirable. Carbon resistors will change their value with age or if they are overheated. Carbon resistors generally increase instead of decrease in value.

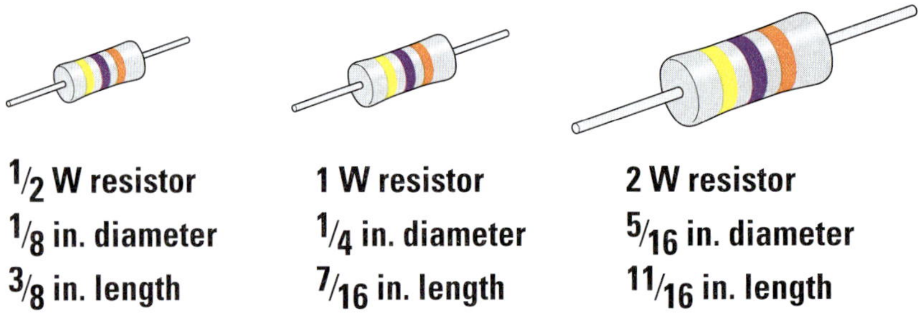

Figure 5–4 Power rating is indicated by size.

Metal Film Resistors

Another type of fixed resistor is the **metal film resistor**. Metal film resistors are constructed by applying a film of metal to a ceramic rod in a vacuum (*Figure 5–5*). The resistance is determined by the type of metal used to form the film and the thickness of the film. Typical thicknesses for the film are from 0.00001 to 0.00000001 inch. Leads are then attached to the film coating and the entire assembly is covered with a coating. These resistors are superior to carbon resistors in several respects. Metal film resistors do not change their value with age, and their **tolerance** is generally better than that of carbon resistors. Tolerance indicates the plus and minus limits of a resistor's ohmic value. Carbon resistors commonly have a tolerance range of 20%, 10%, or 5%. Metal film resistors generally range in tolerance from 2% to 0.1%. The disadvantage of the metal film resistor is that it costs more.

Carbon Film Resistors

Another type of fixed resistor that is constructed in a similar manner is the **carbon film resistor**. This resistor is made by coating a ceramic rod with a film of carbon instead of metal. Carbon film resistors are less expensive to manufacture than metal film resistors and can have a higher tolerance rating than composition carbon resistors.

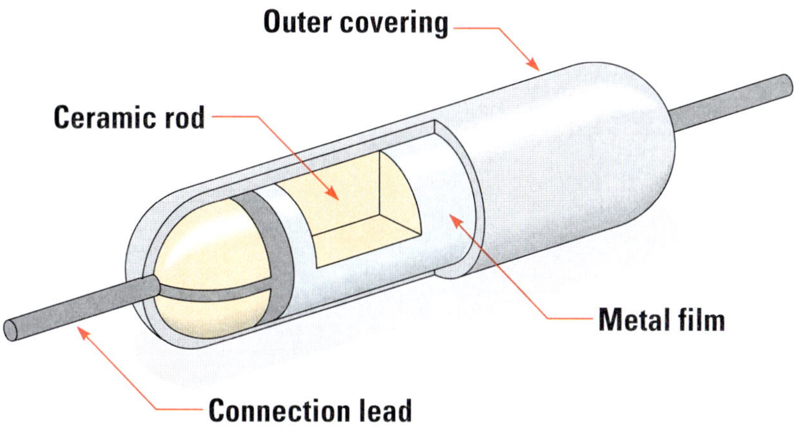

Figure 5–5 Metal film resistor.

Metal Glaze Resistors

The **metal glaze resistor** is also a fixed resistor, similar to the metal film resistor. This resistor is made by combining metal with glass. The compound is then applied to a ceramic base as a thick film. The resistance is determined by the amount of metal used in the compound. Tolerance ratings of 2% and 1% are common.

Wire-Wound Resistors

Wire-wound resistors are fixed resistors that are made by winding a piece of resistive wire around a ceramic core (*Figure 5–6*). The resistance of a wire-wound resistor is determined by three factors:

1. The type of material used to make the resistive wire
2. The diameter of the wire
3. The length of the wire

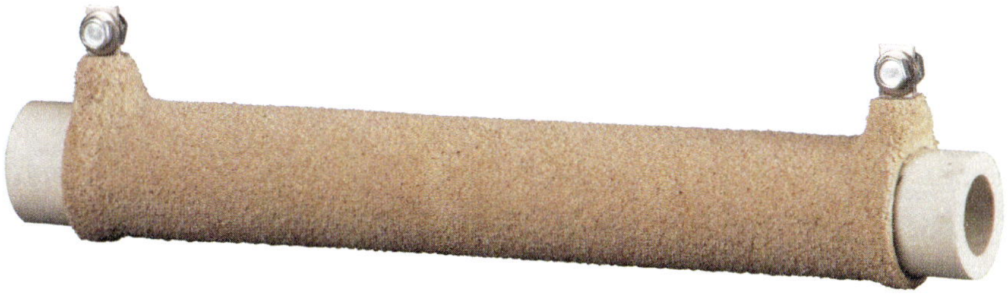

Figure 5–6 Wire-wound resistor. Courtesy of POWEROHM RESISTORS, INC.

Wire-wound resistors can be found in various case styles and sizes. These resistors are generally used when a high-power rating is needed. Wire-wound resistors can operate at higher temperatures than any other type of resistor. An adjustable wire-wound resistor that has a hollow center is shown in *Figure 5–7*. This type of resistor should be mounted vertically and not horizontally. The center of the resistor is hollow for a very good reason. When the resistor is mounted vertically, the heat from the resistor produces a chimney effect and causes air to circulate through the center (*Figure 5–8*). This increase in airflow dissipates heat at a faster rate to help keep the resistor from overheating. The disadvantage of wire-wound resistors is that they are expensive and generally require a large amount of space for mounting. They can also exhibit an amount of inductance in circuits that operate at high frequencies. This added inductance can cause problems for the rest of the circuit. Inductance is covered in later units.

Figure 5–7 Wire-wound resistor with hollow core. Courtesy of POWEROHM RESISTORS, INC.

Figure 5–8 Airflow helps cool the resistor.

5–3 Color Code

The value of a resistor can often be determined by the **color code**. Many resistors have bands of color that are used to determine the resistance value, tolerance, and in some cases, reliability. The color bands represent numbers. Each color represents a different numerical value (*Table 5–1*). The chart shown in *Figure 5–9* lists the colors and the number values assigned to each. The resistor shown beside the color chart illustrates how to determine the value of a resistor. Resistors can have from three to five bands of color. Resistors that have a tolerance of ±20% have only three color bands. Most resistors

Color	Value	Tolerance	
Black	0	No color	± 20%
Brown	1	Silver	± 10%
Red	2	Gold	± 5%
Orange	3	Red	± 2%
Yellow	4	Brown	± 1%
Green	5		
Blue	6		
Violet	7		
Gray	8		
White	9		

Table 5–1 Colors and Numeric Values

contain four bands of color. For resistors with tolerances that range from ±10% to ±2%, the first two color bands represent number values. The third color band is the multiplier. This means to combine the first two numbers and multiply the resulting two-digit number by the power of 10 indicated by the value of the third band. The fourth band indicates the tolerance. For example, assume a resistor has color bands of brown, green, red, and silver (*Figure 5–10*). The first two bands represent the

First Band	Second Band	Multiplier Band	Resistance Ω
Brown	Black	Black	10
		Brown	100
		Red	1000
		Orange	10000
		Yellow	0.1 Meg.
		Green	1.0 Meg.
		Blue	10.0 Meg.
	Brown	Black	11
		Brown	110
		Red	1100
		Orange	11000
		Yellow	0.11 Meg.
		Green	1.1 Meg.
		Blue	11.0 Meg.
	Red	Black	12
		Brown	120
		Red	1200
		Orange	12000
		Yellow	0.12 Meg.
		Green	1.2 Meg.
		Blue	12.0 Meg.
	Orange	Black	13
		Brown	130
		Red	1300
		Orange	13000
		Yellow	0.13 Meg.
		Green	1.3 Meg.
		Blue	13.0 Meg.
	Green	Black	15
		Brown	150
		Red	1500
		Orange	15000
		Yellow	0.15 Meg.
		Green	1.5 Meg.
		Blue	15.0 Meg.
	Blue	Black	16
		Brown	160
		Red	1600
		Orange	16000
		Yellow	0.16 Meg.
		Green	1.6 Meg.
		Blue	16.0 Meg.
	Gray	Black	18
		Brown	180
		Red	1800
		Orange	18000
		Yellow	0.18 Meg.
		Green	1.8 Meg.
		Blue	18.0 Meg.
Yellow	Orange	Gold	4.3
		Black	43
		Brown	430
		Red	4300
		Orange	43000
		Yellow	0.43 Meg.
		Green	4.3 Meg.
	Violet	Gold	4.7
		Black	47
		Brown	470
		Red	4700
		Orange	47000
		Yellow	0.47 Meg.
		Green	4.7 Meg.
Green	Brown	Gold	5.1
		Black	51
		Brown	510
		Red	5100
		Orange	51000
		Yellow	0.51 Meg.
		Green	5.1 Meg.
	Blue	Gold	5.6
		Black	56
		Brown	560
		Red	5600
		Orange	56000
		Yellow	0.56 Meg.
		Green	5.6 Meg.
Blue	Red	Gold	6.2
		Black	62
		Brown	620
		Red	6200
		Orange	62000
		Yellow	0.62 Meg.
		Green	6.2 Meg.
	Gray	Gold	6.8
		Black	68
		Brown	680
		Red	6800
		Orange	68000
		Yellow	0.68 Meg.
		Green	6.8 Meg.
Violet	Green	Gold	7.5
		Black	75
		Brown	750
		Red	7500
		Orange	75000
		Yellow	0.75 Meg.
		Green	7.5 Meg.
Gray	Red	Gold	8.2
		Black	82
		Brown	820
		Red	8200
		Orange	82000
		Yellow	0.82 Meg.
		Green	8.2 Meg.
White	Brown	Gold	9.1
		Black	91
		Brown	910
		Red	9100
		Orange	91000
		Yellow	0.91 Meg.
		Green	9.1 Meg.

Figure 5–9 Resistor color code chart.

First Band	Second Band	Multiplier Band	Resistance Ω
Red	Black	Black	20
		Brown	200
		Red	2000
		Orange	20000
		Yellow	0.2 Meg.
		Green	2.0 Meg.
		Blue	20.0 Meg.
	Red	Black	22
		Brown	220
		Red	2200
		Orange	22000
		Yellow	0.22 Meg.
		Green	2.2 Meg.
		Blue	22.0 Meg.
	Yellow	Black	24
		Brown	240
		Red	2400
		Orange	24000
		Yellow	0.24 Meg.
		Green	2.4 Meg.
	Violet	Gold	2.7
		Black	27
		Brown	270
		Red	2700
		Orange	27000
		Yellow	0.27 Meg.
		Green	2.7 Meg.

First Band	Second Band	Multiplier Band	Resistance Ω
Orange	Black	Gold	3.0
		Black	30
		Brown	300
		Red	3000
		Orange	30000
		Yellow	0.30 Meg.
		Green	3.0 Meg.
	Orange	Gold	3.3
		Black	33
		Brown	330
		Red	3300
		Orange	33000
		Yellow	0.33 Meg.
		Green	3.3 Meg.
	Blue	Gold	3.6
		Black	36
		Brown	360
		Red	3600
		Orange	36000
		Yellow	0.36 Meg.
		Green	3.6 Meg.
	White	Gold	3.9
		Black	39
		Brown	390
		Red	3900
		Orange	39000
		Yellow	0.39 Meg.
		Green	3.9 Meg.

Standard Color Code

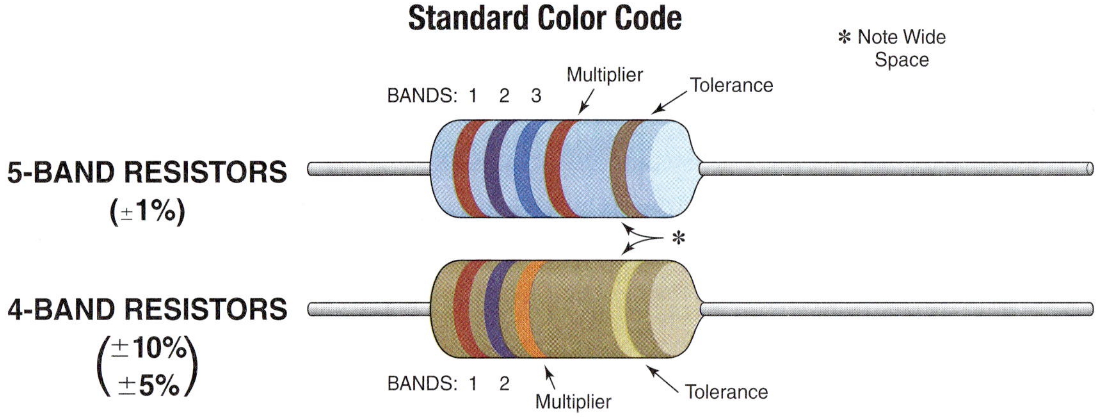

Band 1 1st Digit	
Color	**Digit**
Black	0
Brown	1
Red	2
Orange	3
Yellow	4
Green	5
Blue	6
Violet	7
Gray	8
White	9

Band 2 2nd Digit	
Color	**Digit**
Black	0
Brown	1
Red	2
Orange	3
Yellow	4
Green	5
Blue	6
Violet	7
Gray	8
White	9

Band 3 (if used) 3rd Digit	
Color	**Digit**
Black	0
Brown	1
Red	2
Orange	3
Yellow	4
Green	5
Blue	6
Violet	7
Gray	8
White	9

Multiplier	
Color	**Multiplier**
Black	1
Brown	10
Red	100
Orange	1,000
Yellow	10,000
Green	100,000
Blue	1,000,000
Silver	0.01
Gold	0.1

Resistance Tolerance	
Color	**Tolerance**
Silver	±10%
Gold	± 5%
Brown	± 1%

Figure 5–9 Concluded. Courtesy of Heathkit Educational Systems

numbers 1 and 5 (brown is 1 and green is 5). The third band is red, which has a number value of 2. The number 15 should be multiplied by 10^2, or 100. The value of the resistor is 1500 ohm. Another method, which is simpler to understand, is to add the number of zeros indicated by the multiplier band to the combined first two numbers. The multiplier band in this example is red, which has a numeric value of 2. Add two zeros to the first two numbers. The number 15 becomes 1500.

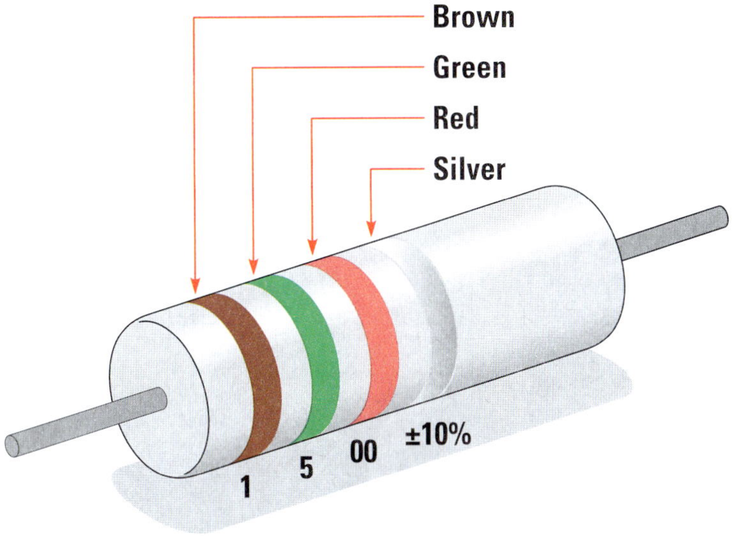

Figure 5–10 Determining resistor values using the color code.

The fourth band is the tolerance band. The tolerance band in this example is silver, which means $\pm 10\%$. This resistor should be 1500 ohm plus or minus 10%. To determine the value limits of this resistor, find 10% of 1500 ohm:

$$1500\ \Omega \times 0.10 = 150\ \Omega$$

The value can range from $1500\ \Omega + 10\%$, or $1500\ \Omega + 150\ \Omega = 1650\ \Omega$, to $1500\ \Omega = 10\%$ or $1500\ \Omega - 150\ \Omega = 1350\ \Omega$.

Resistors that have a tolerance of $\pm 1\%$, as well as some military resistors, contain five bands of color.

Gold and Silver as Multipliers

The colors gold and silver are generally found in the fourth band of a resistor, but they can be used in the multiplier band also. When the color gold is used as the multiplier band, it means to divide the combined first two numbers by 10. If silver is used as the multiplier band, it means to divide the combined first two numbers by 100. For example, assume a resistor has color bands of orange, white, gold, and gold. The value of this resistor is 3.9 ohm with a tolerance of $\pm 5\%$ (orange = 3; white = 9; gold means to divide 39 by 10 = 3.9; and gold in the fourth band means $\pm 5\%$ tolerance). Examples 5-1 and 5-2 demonstrate how to use the bands on a resistor to determine its resistance value.

Example 5–1

The resistor shown in *Figure 5–11* contains the following bands of color:

First band = brown

Second band = black

Third band = black

Fourth band = brown

Fifth band = brown

Example 5–1 (continued)

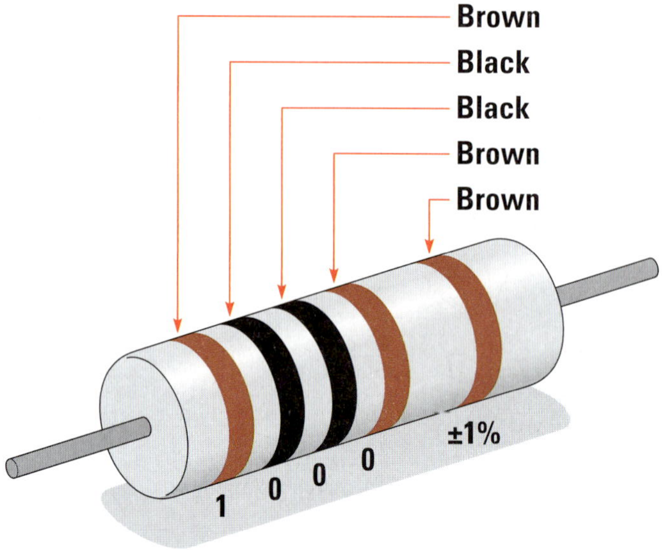

Figure 5–11 Determining the value of a ±1% resistor.

Solution

The brown fifth band indicates that this resistor has a tolerance of ±1%. To determine the value of a 1% resistor, the first three bands are numbers, and the fourth band is the multiplier. In this example, the first band is brown, which has a number value of 1. The next two bands are black, which represents a number value of 0. The fourth band is brown, which means add one 0 to the first three numbers. The value of this resistor is 1000 Ω ±1%.

Example 5–2

A five-band resistor has the following color bands:

First band = red

Second band = orange

Third band = violet

Fourth band = red

Fifth band = brown

Solution

The first three bands represent number values. Red is 2, orange is 3, and violet is 7. The fourth band is the multiplier; in this case, red represents 2. Add two zeros to the number 237. The value of the resistor is 23,700 Ω. The fifth band is brown, which indicates a tolerance of ±1%.

Military resistors often have five bands of color also. These resistors are read in the same manner as a resistor with four bands of color. The fifth band can represent different things. A fifth band of orange or yellow is used to indicate reliability. Resistors with a fifth band of orange have a reliability good enough to be used in missile systems, and a resistor with a fifth band of yellow can be used in space-flight equipment. A military resistor with a fifth band of white indicates the resistor has solderable leads.

Resistors with tolerance ratings ranging from 0.5% to 0.1% generally have their values printed directly on the resistor.

5–4 Standard Resistance Values of Fixed Resistors

Fixed resistors are generally produced in standard values. The higher the tolerance value, the fewer resistance values available. Standard resistor values are listed in the chart shown in *Figure 5–12.* In the column under 10%, only 12 values of resistors are listed. These standard values, however, can be multiplied by factors of 10. Notice that one of the standard values listed is 33 ohm. There are also standard values in 10% resistors of 0.33, 3.3, 330, 3300, 33,000, 330,000, and 3,300,000 ohm. The 2% and 5% column shows 24 resistor values, and the 1% column lists 96 values. All the values listed in the chart can be multiplied by factors of 10 to obtain other resistance values.

STANDARD RESISTANCE VALUES (Ω)									
0.1% 0.25% 0.5%	1%	0.1% 0.25% 0.5%	1%	0.1% 0.25% 0.5%	1%	0.1% 0.25% 0.5%	1%	0.1% 0.25% 0.5%	1%
10.0	10.0	17.2	–	29.4	29.4	50.5	–	86.6	86.6
10.1	–	17.4	17.4	29.8	–	51.1	51.1	87.6	–
10.2	10.2	17.6	–	30.1	30.1	51.7	–	88.7	88.7
10.4	–	17.8	17.8	30.5	–	52.3	52.3	89.8	–
10.5	10.5	18.0	–	30.9	30.9	53.0	–	90.9	90.9
10.6	–	18.2	18.2	31.2	–	53.6	53.6	92.0	–
10.7	10.7	18.4	–	31.6	31.6	54.2	–	93.1	93.1
10.9	–	18.7	18.7	32.0	–	54.9	54.9	94.2	–
11.0	11.0	18.9	–	32.4	32.4	55.6	–	95.3	95.3
11.1	–	19.1	19.1	32.8	–	56.2	56.2	96.5	–
11.3	11.3	19.3	–	33.2	33.2	56.9	–	97.6	97.6
11.4	–	19.6	19.6	33.6	–	57.6	57.6	98.8	–
11.5	11.5	19.8	–	34.0	34.0	58.3	–		
11.7	–	20.0	20.0	34.4	–	59.0	59.0		
11.8	11.8	20.3	–	34.8	34.8	59.7	–		
12.0	–	20.5	20.5	35.2	–	60.4	60.4		
12.1	12.1	20.8	–	35.7	35.7	61.2	–		
12.3	–	21.0	21.0	36.1	–	61.9	61.9		
12.4	12.4	21.3	–	36.5	36.5	62.6	–		
12.6	–	21.5	21.5	37.0	–	63.4	63.4		
12.7	12.7	21.8	–	37.4	37.4	64.2	–	2%,5%	10%
12.9	–	22.1	22.1	37.9	–	64.9	64.9	10	10
13.0	13.0	22.3	–	38.3	38.3	65.7	–	11	–
13.2	–	22.6	22.6	38.8	–	66.5	66.5	12	12
13.3	13.3	22.9	–	39.2	39.2	67.3	–	13	–
13.5	–	23.2	23.2	39.7	–	68.1	68.1	15	15
13.7	13.7	23.4	–	40.2	40.2	69.0	–	16	–
13.8	–	23.7	23.7	40.7	–	69.8	69.8	18	18
14.0	14.0	24.0	–	41.2	41.2	70.6	–	20	–
14.2	–	24.3	24.3	41.7	–	71.5	71.5	22	22
14.3	14.3	24.6	–	42.2	42.2	72.3	–	24	–
14.5	–	24.9	24.9	42.7	–	73.2	73.2	27	27
14.7	14.7	25.2	–	43.2	43.2	74.1	–	30	–
14.9	–	25.5	25.5	43.7	–	75.0	75.0	33	33
15.0	15.0	25.8	–	44.2	44.2	75.9	–	36	–
15.2	–	26.1	26.1	44.8	–	76.8	76.8	39	39
15.4	15.4	26.4	–	45.3	45.3	77.7	–	43	–
15.6	–	26.7	26.7	45.9	–	78.7	78.7	47	47
15.8	15.8	27.1	–	46.4	46.4	79.6	–	51	–
16.0	–	27.4	27.4	47.0	–	80.6	80.6	56	56
16.2	16.2	27.7	–	47.5	47.5	81.6	–	62	–
16.4	–	28.0	28.0	48.1	–	82.5	82.5	68	68
16.5	16.5	28.4	–	48.7	48.7	83.5	–	75	–
16.7	–	28.7	28.7	49.3	–	84.5	84.5	82	82
16.9	16.9	29.1	–	49.9	49.9	85.6	–	91	–

Figure 5–12 Standard resistance values.

5–5 Power Ratings

Resistors also have a power rating in watts that should not be exceeded, or the resistor will be damaged. See Example 5-3 for an example of how to determine if a resistor is properly applied within its power rating. In a purely resistive circuit, the amount of heat that will be given off by the resistor can be determined by the use of one of the following Ohm's law formulas for power:

$$P = I^2R$$

$$P = \frac{E^2}{R}$$

$$P = EI$$

Example 5–3

The resistor shown in Figure 5–13 has a value of 100 Ω and a power rating of ½ W. If the resistor is connected to a 10-V power supply, will it be damaged?

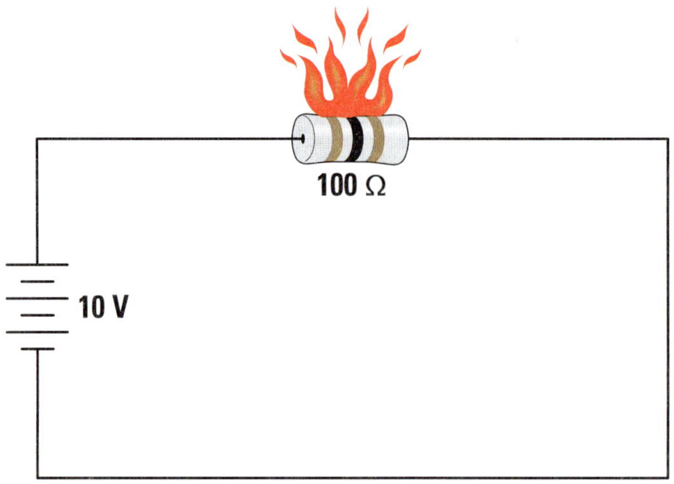

Figure 5–13 Exceeding the power rating causes damage to the resistor.

Solution

Using the formula $P = \frac{E^2}{R}$, determine the amount of heat that will be dissipated by the resistor.

$$P = \frac{E^2}{R}$$

$$P = \frac{10\text{ V} \times 10\text{ V}}{100\ \Omega}$$

$$P = \frac{100\text{ V}}{100\ \Omega}$$

$$P = 1\text{ W}$$

Because the resistor has a power rating of ½ W and the amount of heat that will be dissipated is 1 W, the resistor will be damaged.

5–6 Variable Resistors

A **variable resistor** is a resistor whose values can be changed or varied over a range. Variable resistors can be obtained in different case styles and power ratings. *Figure 5–14* illustrates how a variable resistor is constructed. In this example, a resistive wire is wound in a circular pattern, and a sliding tap makes contact with the wire. The value of resistance can be adjusted between one end of the resistive wire and the sliding tap. If the resistive wire has a total value of 100 ohms, the resistor can be set between the values of 0 and 100 ohms.

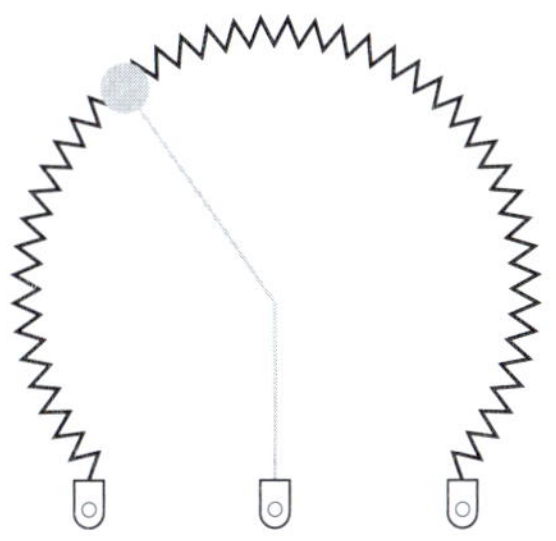

Figure 5–14 Variable resistor.

A variable resistor with three terminals is shown in *Figure 5–15*. This type of resistor has a wiper arm inside the case that makes contact with the resistive element. The full resistance value is between the two outside terminals, and the wiper arm is connected to the center terminal. The resistance between the center terminal and either of the two outside terminals can be adjusted by turning the shaft and changing the position of the wiper arm. Wire-wound variable resistors of this type can be obtained also (*Figure 5–16*). The advantage of the wire-wound type is a higher power rating.

Figure 5–15 Three-terminal variable resistors. © DeSerg/Shutterstock

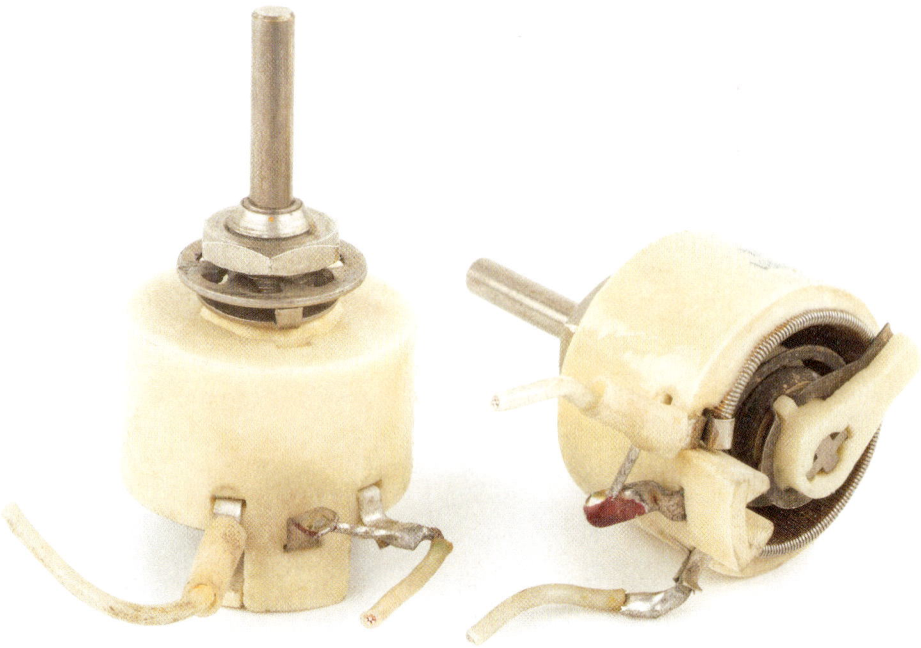

Figure 5–16 Wire-wound variable resistor. © SeDmi/Shutterstock.com

The resistors shown in *Figure 5–15* can be adjusted from their minimum to maximum value by turning the control approximately three-quarters of a turn. In some types of electric equipment, this range of adjustment may be too coarse to allow for sensitive adjustments. When this becomes a problem, a multiturn resistor (*Figure 5–17*) can be used. **Multiturn variable resistors** operate by moving the wiper arm with a screw of some number of turns. They generally range from 3 turns to 10 turns. If a 10-turn variable resistor is used, it will require 10 turns of the control knob to move the wiper from one end of the resistor to the other end instead of three-quarters of a turn.

Figure 5–17 Multiturn variable resistor.

Variable Resistor Terminology

Variable resistors are known by several common names. The most popular name is **pot**, which is shortened from the word **potentiometer**. Another common name is **rheostat**. A rheostat is actually a variable resistor that has two terminals. They are used to adjust the current in a circuit to a certain value.

A potentiometer is a variable resistor that has three terminals. Potentiometers can be used as rheostats by using only two of their three terminals. A potentiometer is a variable resistor used to provide a variable voltage, as shown in *Figure 5–18*. In this example, one end of a variable resistor is connected to +12 volts, and the other end is connected to ground. The middle terminal, or wiper, is connected to the positive terminal of a voltmeter, and the negative lead is connected to ground. If the wiper is moved to the upper end of the resistor, the voltmeter will indicate a potential of 12 volts. If the wiper is moved to the bottom, the voltmeter will indicate a value of 0 volts. The wiper can be adjusted to provide any value of voltage between 12 and 0 volts.

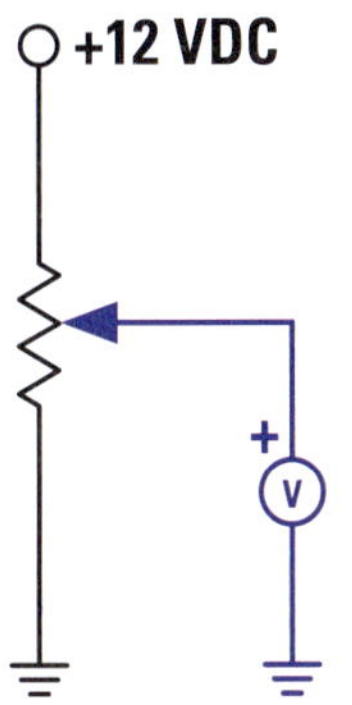

Figure 5–18 Variable resistor used as a potentiometer.

5–7 Schematic Symbols

Electrical schematics use symbols to represent the use of a resistor. *Figure 5–19* illustrates several schematic symbols used to represent both fixed and variable resistors.

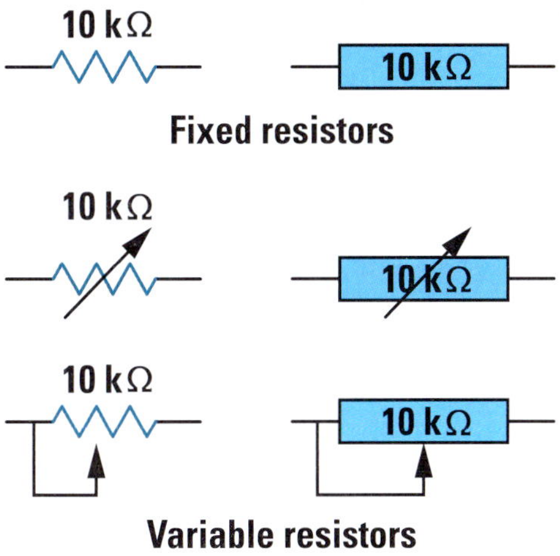

Figure 5–19 Schematic symbols used to represent resistors.

Summary

- Resistors are used in two main applications: as voltage dividers and to limit the flow of current in a circuit.
- The value of fixed resistors cannot be changed.
- There are several types of fixed resistors, such as composition carbon, metal film, and wire-wound.
- Carbon resistors change their resistance with age or if overheated.
- Metal film resistors never change their value but are more expensive than carbon resistors.
- The advantage of wire-wound resistors is their high power ratings.
- Resistors typically have bands of color to indicate their resistance value and tolerance.
- Resistors are produced in standard values. The number of values between 0 and 100 ohms is determined by the tolerance.
- Variable resistors can change their value within the limit of their full value.
- A potentiometer is a variable resistor used as a voltage divider.

Review Questions

1. True or False: A resistive load in a circuit will allow unlimited current to flow in the circuit.
2. Which of the following best describes a secondary use of resistors in an electrical circuit?
   A) A current divider
   B) A current multiplier
   C) A voltage divider
   D) A voltage multiplier
3. Which of the following fixed resistor types is constructed by using a carbon graphite compound and a resin-bonding material?
   A) Metal-film resistor
   B) Composition carbon resistor
   C) Carbon-film resistor
   D) Wire-wound resistor
4. True or False: If it were desirable to use a resistor that did not change its resistance value over time, a metal-film resistor would be a good choice.
5. A resistor has color bands of orange, blue, yellow, and gold. What is the tolerance of this resistor?
   A) 10%
   B) 1%
   C) 15%
   D) 5%

6. If gold or silver are used in the multiplier band on the resistor, does this indicate multiplying or dividing the first two numbers by either 10 or 100?
   A) Divide
   B) Multiply
   C) Neither, the values stay the same
   D) Gold and silver are never in the multiplier band
7. A resistor has a power rating of 500 W and has a resistance value of 25 Ω. What is the maximum amount of current that can flow through this resistor without exceeding the power rating?
   A) 20 amperes
   B) 5 amperes
   C) 2 amperes
   D) 50 amperes
8. A resistor has a power rating of 1000 W and is connected to an applied voltage of 120 V. What is the minimum resistance value of this resistor to stay operating within the power rating?
   A) 1000 Ω
   B) 144 Ω
   C) 14.4 Ω
   D) 28.8 Ω
9. A potentiometer is a variable resistor that utilizes how many terminals?
   A) Two
   B) Three
   C) Four
   D) Five
10. True or False: For sensitive electronic equipment where small incremental adjustments are required, it might be desirable to utilize a multi-turn variable resistor if a standard is too coarse.

Practical Applications

1. You are an electrician on the job. It is decided that the speed of a large DC motor is to be reduced by connecting a resistor in series with its armature. The DC voltage applied to the motor is 250 V, and the motor has a full-load armature current of 50 A. Your job is to reduce the armature current to 40 A at full load by connecting the resistor in series with the armature. What value of resistance should be used, and what is the power rating of the resistor?

---

2. You are working on an electronic circuit. The circuit current is 5 mA. A resistor is marked with the following bands: brown, black, red, gold. A voltmeter measures a voltage drop of 6.5 V across the resistor. Is this resistor within its tolerance rating?

3. A homeowner uses a 100-watt incandescent lamp as a heater in an outside well pump house to protect the pump from freezing in cold weather. Unfortunately, however, the lamp can burn out and leave the pump unprotected. You have been asked to install a heater that will not burn out and leave the pump unprotected. You have available a 100-watt, 150-ohm wire-wound resistor. Can this resistor be connected to the 120-volt source without damage to the resistor? If so, what would be the power output of the resistor?

4. You have determined that a 4700-ohm, ½-watt resistor on an electronic circuit board is defective. Assuming room permits, can the resistor be replaced with a 4700-ohm, 1-watt resistor without damage to the rest of the board, or will the higher-wattage resistor generate excessive heat that could damage other components?

Practice Problems

Resistors

Fill in the missing values.

1st Band	2nd Band	3rd Band	4th Band	Value	% Tolerance
Red	Yellow	Brown	Silver		
				6800 Ω	5
Orange	Orange	Orange	Gold		
				12 Ω	2
Brown	Green	Silver	Silver		
				1.8 MΩ	10
Brown	Black	Yellow	None		
				10 kΩ	5
Violet	Green	Black	Red		
				4.7 kΩ	20
Gray	Red	Green	Red		
				5.6 Ω	2

Section 2 Basic Electric Circuits

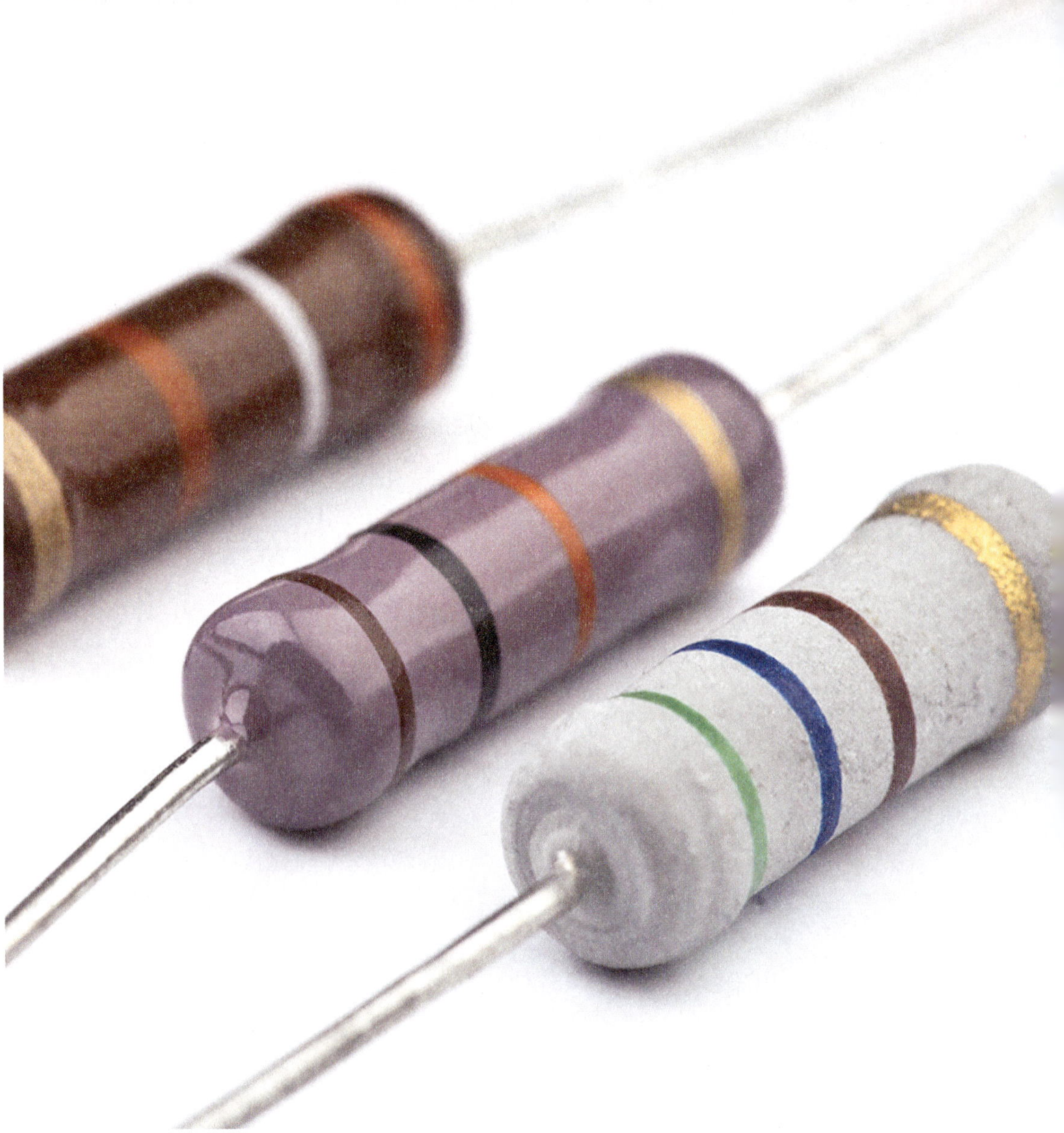

© K. Miri Photography/Shutterstock.com

Unit 6 Series Circuits

Key Terms

Chassis ground

Circuit breakers

Earth ground

Fuses

Ground point

Series circuit

Voltage drop

Voltage polarity

Why You Need to Know

It is important to learn the rules governing the values of resistance, voltage, current, and power in a series circuit. Knowledge of series circuits is essential in understanding more complex circuits that are encountered throughout an electrician's career. There are some series circuits in everyday use, such as airport runway lighting circuits, but knowledge of series circuits is most useful in understanding how components in combination circuits relate to each other. Combination circuits contain components or branches that are connected in both series and parallel.

Outline

Objectives

After studying this unit, you should be able to

- describe how electrical current behaves in a series circuit.
- describe how voltage is applied throughout a series circuit.
- describe how resistance behaves in a series circuit.
- apply three basic rules for solving electrical values of series circuits.
- calculate values of voltage, current, resistance, and power for series circuits using Ohm's law.
- explain how resistors in series can function as a voltage divider.
- identify voltage polarity and how using a ground reference impacts the circuit.

Preview

Electric circuits can be divided into three major types: series, parallel, and combination. Combination circuits are circuits that contain both series and parallel paths. The first type discussed in this unit is the series circuit.

6–1 Series Circuits

A **series circuit** is a circuit that has only one path for current to flow (*Figure 6–1*). Because there is only one path for current to flow, the current is the same at any point in the circuit. In a series circuit, an electron leaves the negative terminal of the battery and must flow through each resistor before it can complete the circuit to the positive battery terminal.

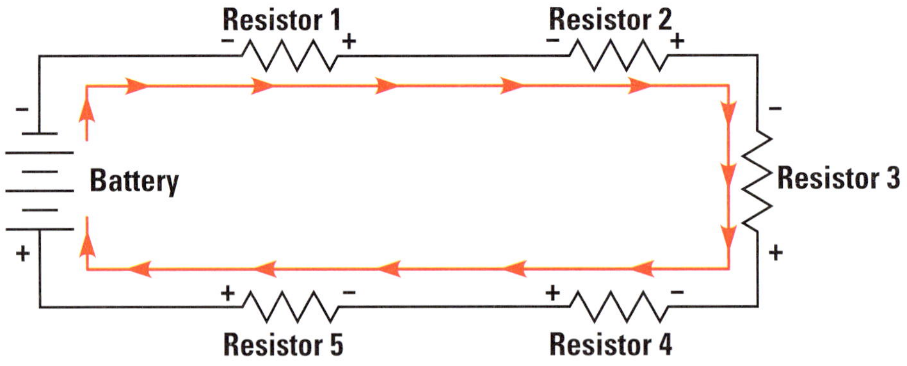

Figure 6–1 A series circuit has only one path for current flow.

One of the best examples of a series-connected device is a fuse or circuit breaker (*Figure 6–2*). Because **fuses** and **circuit breakers** are connected in series with the rest of the circuit, all the circuit current must flow through them. If the current becomes excessive, the fuse or circuit breaker will open and disconnect the rest of the circuit from the power source.

Voltage Drops in a Series Circuit

Voltage is the electromotive force that pushes the electrons through a resistance. The amount of current that will flow in a series circuit is determined by the applied voltage and total resistance. If a voltmeter is connected across a resistor (*Figure 6–3*), the amount of voltage necessary to push the current through

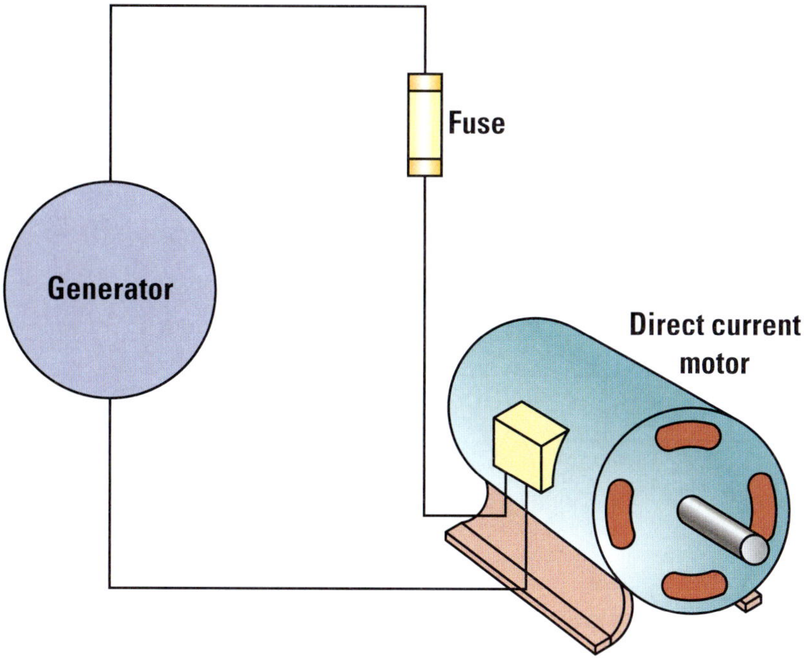

Figure 6–2 All the current must flow through the fuse.

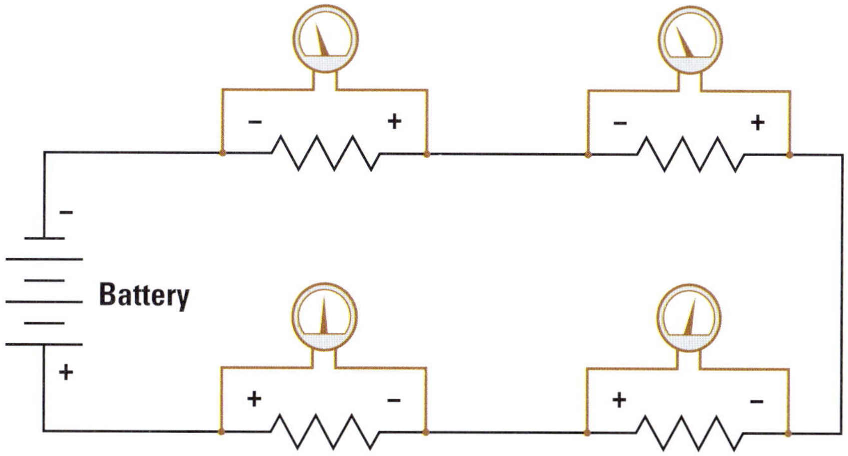

Figure 6–3 The sum of the individual voltage drops in a series circuit will equal the applied circuit voltage.

that single resistor is indicated by the meter. This amount is known as **voltage drop**. It is similar to pressure drop in a water system. *In a series circuit, the sum of all the voltage drops across all the resistors will equal the voltage applied to the circuit.* The amount of voltage drop across each resistor is proportional to its resistance value and the amount of current flowing in the circuit.

In the circuit shown in *Figure 6–4*, four resistors are connected in series. It is assumed that all four resistors have the same value. The circuit is connected to a 24-volt battery. Because all the resistors have the same value, the voltage drop across each will be 6 volts (24 V/4 resistors = 6 V). Note: Since voltage drop is proportional to current and resistance, all four resistors will have the same voltage drop across them if they all have the same resistance value. The circuit shown in *Figure 6–5* illustrates a series circuit comprised of resistors of different values. Notice that the voltage drop across each resistor is proportional to its resistance. Also notice that the sum of the voltage drops is equal to the applied voltage of 24 volts.

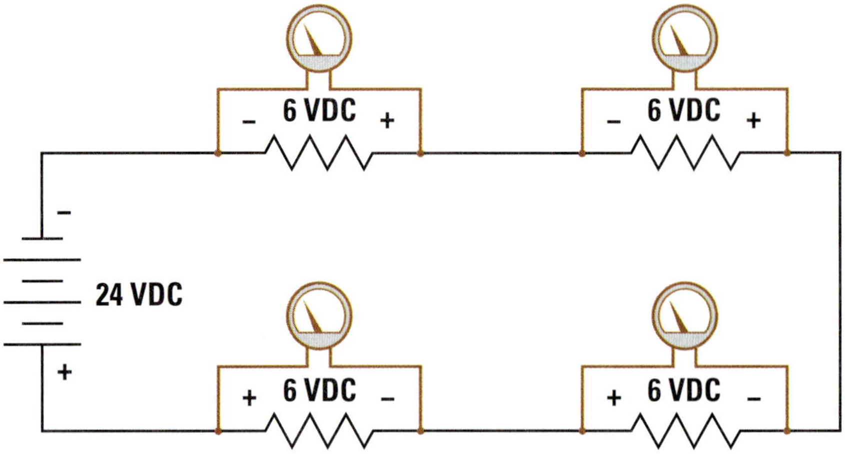

Figure 6–4 The voltage drop across each resistor is proportional to its resistance.

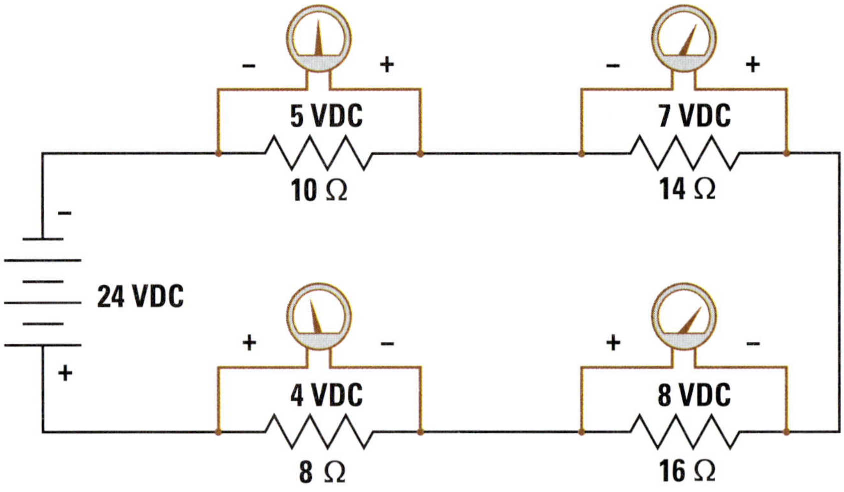

Figure 6–5 Series circuit with four resistors having different voltage drops.

Resistance in a Series Circuit

Because only one path exists for the current to flow through a series circuit, it must flow through each resistor in the circuit (*Figure 6–1*). Each resistor limits or impedes the flow of current in the circuit. Therefore, the total amount of resistance to current flow in a series circuit is equal to the sum of the resistances in that circuit, and the amount of current flowing in the circuit is inversely proportional to the resistance in the circuit.

6–2 Calculating Series Circuit Values

Three rules can be used with Ohm's law for finding values of voltage, current, resistance, and power in any series circuit:

1. The current is the same at any point in the circuit.
2. The total resistance is the sum of the individual resistors.
3. The applied voltage is equal to the sum of the voltage drops across all the resistors.

The circuit shown in *Figure 6–6* shows the values of current flow, voltage drop, and resistance for each of the resistors. The total resistance (R_T) of the circuit can be found by adding the values of the three resistors—resistance adds:

$$R_T = R_1 + R_2 + R_3$$

$$R_T = 20\ \Omega + 10\ \Omega + 30\ \Omega$$

$$R_T = 60\ \Omega$$

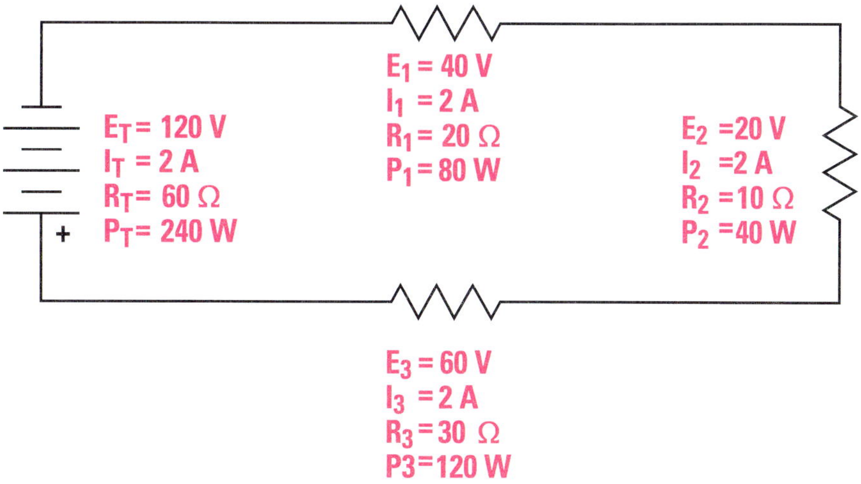

Figure 6–6 Series circuit values.

The amount of current flow in the circuit can be found by using Ohm's law:

$$I = \frac{E}{R}$$

$$I = \frac{120\ V}{60\ \Omega}$$

$$I = 2\ A$$

A current of 2 amperes flows through each resistor in the circuit:

$$I_T = I_1 = I_2 = I_3$$

Because the amount of current flowing through resistor R_1 is known, the voltage drop across each resistor can be found using Ohm's law:

$$E_1 = I_1 \times R_1$$

$$E_1 = 2\ A \times 20\ \Omega$$

$$E_1 = 40\ V$$

In other words, it takes 40 volts to push 2 amperes of current through 20 ohms of resistance. If a voltmeter were connected across resistor R_1, it would indicate a value of 40 volts (*Figure 6–7*). The voltage drop across resistors R_2 and R_3 can be found in the same way:

$$E_2 = I_2 \times R_2$$

$$E_2 = 2\ A \times 10\ \Omega$$

$$E_2 = 20\ V$$

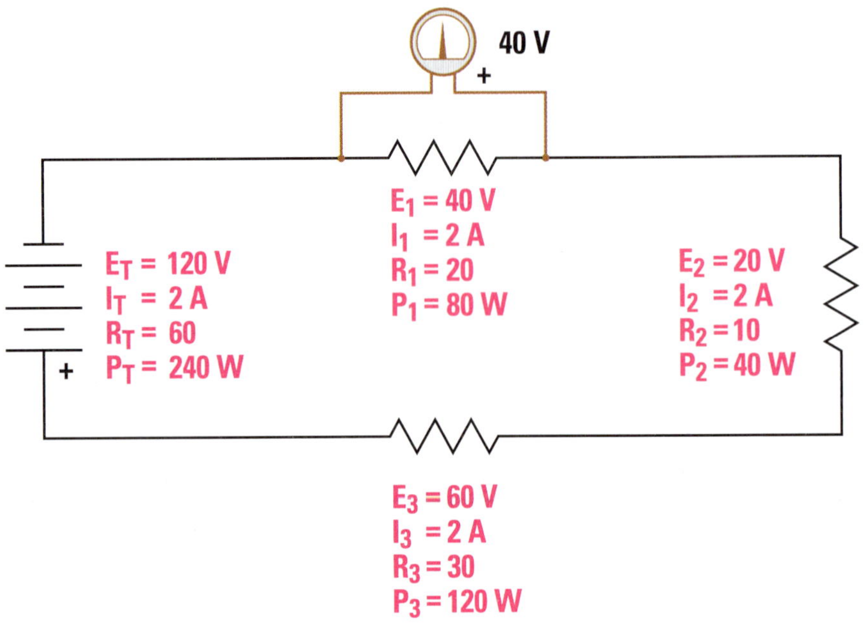

Figure 6–7 The voltmeter indicates a voltage drop of 40 volts.

$$E_3 = I_3 \times R_3$$
$$E_3 = 2\text{ A} \times 30\ \Omega$$
$$E_3 = 60\text{ V}$$

If the voltage drop across all the resistors is added, it equals the total applied voltage (E_T):

$$E_T = E_1 + E_2 + E_3$$
$$E_T = 40\text{ V} + 20\text{ V} + 60\text{ V}$$
$$E_T = 120\text{ V}$$

The final value to determine in the circuit is power, or in other words, the usefulness of the circuit. Power is easy to calculate as it is the product of voltage and current and, therefore, follows the properties of both of these values in a circuit. Since current remains the same throughout a series circuit and voltage drops add up to equal the total applied voltage, power values are also additive in a series circuit. The power seen at each individual component will add up to equal the total power performed by the circuit. Power can be found using the following formulas:

$$P_T = I_T \times E_T \text{ or } P_T = P_1 + P_2 + P_3$$

And the power at each individual component can be found using the circuit current and the voltage drop across that given component. Since these values are all known for the circuit shown in *Figure 6–6*, it becomes easy to calculate the power values for each resistor:

$$P_1 = I_1 \times E_1$$
$$P_1 = 2\text{ A} \times 40\text{ V}$$
$$P_1 = 80\text{ W}$$
$$P_2 = I_2 \times E_2$$
$$P_2 = 2\text{ A} \times 20\text{ V}$$

$$P_2 = 40\ \text{W}$$
$$P_3 = I_3 \times E_3$$
$$P_3 = 2\ \text{A} \times 60\ \text{V}$$
$$P_3 = 120\ \text{W}$$
$$P_T = I_T \times E_T \text{ or } P_T = P_1 + P_2 + P_3$$
$$P_T = 2\ \text{A} \times 120\ \text{V}$$
$$P_T = 240\ \text{W}$$
$$P_T = 80\ \text{W} + 40\ \text{W} + 120\ \text{W}$$
$$P_T = 240\ \text{W}$$

The same can be done for the circuit shown in *Figure 6–7* because the voltage and current values are known throughout the circuit.

Solving Circuits

In the following examples, circuits that have missing values are shown. The missing values can be found by using the rules for series circuits and Ohm's law.

Example 6–1

The first step in finding the missing values in the circuit shown in *Figure 6–8* is to find the total resistance (R_T). This can be done using the second rule of series circuits, which states that resistances add to equal the total resistance of the circuit:

$$R_T = R_1 + R_2 + R_3 + R_4$$
$$R_T = 100\ \Omega + 250\ \Omega + 150\ \Omega + 300\ \Omega$$
$$R_T = 800\ \Omega$$

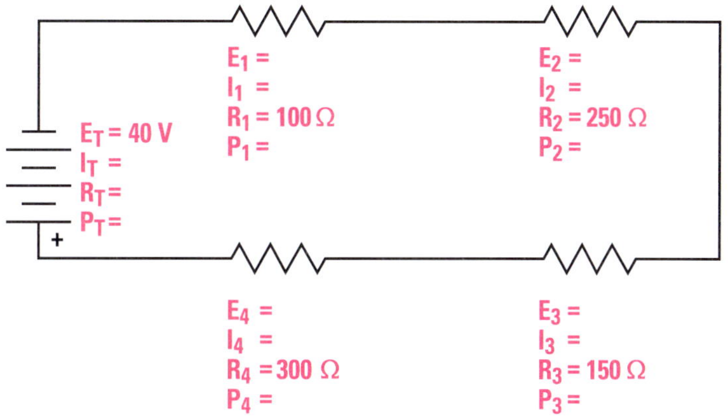

Figure 6–8 Series circuit, Example 1.

Now that the total voltage and the total resistance are known, the current flow through the circuit can be found using Ohm's law:

$$I = \frac{E}{R}$$

Example 6–1 (continued)

$$I = \frac{40\text{ V}}{800\ \Omega}$$

$$I = 0.050\text{ A}$$

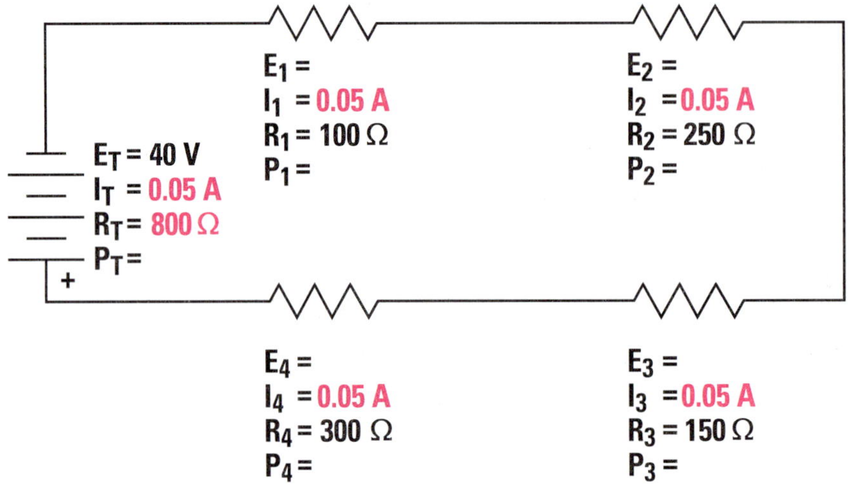

Figure 6–9 The current is the same at any point in a series circuit.

The first rule of series circuits states that current remains the same at any point in the circuit. Therefore, 0.050 A flows through each resistor in the circuit (*Figure 6–9*). The voltage drop across each resistor can now be found using Ohm's law (*Figure 6–10*):

$$E_1 = I_1 \times R_1$$

$$E_1 = 0.050\text{ A} \times 100\ \Omega$$

$$E_1 = 5\text{ V}$$

$$E_2 = I_2 \times R_2$$

$$E_2 = 0.050\text{ A} \times 250\ \Omega$$

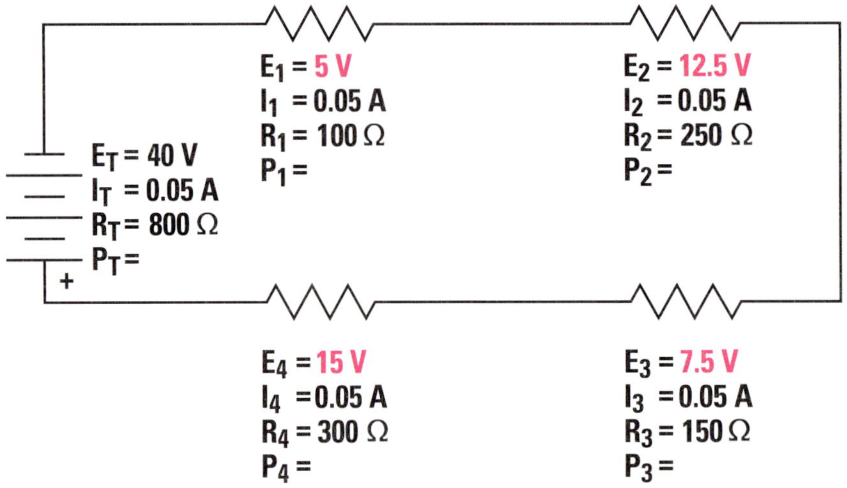

Figure 6–10 The voltage drop across each resistor can be found using Ohm's law.

Example 6–1 (continued)

$$E_2 = 12.5\ V$$

$$E_3 = I_3 \times R_3$$

$$E_3 = 0.050\ A \times 150\ \Omega$$

$$E_3 = 7.5\ V$$

$$E_4 = I_4 \times R_4$$

$$E_4 = 0.050\ A \times 300\ \Omega$$

$$E_4 = 15\ V$$

Several formulas can be used to determine the amount of power dissipated (converted into heat) by each resistor. It makes no difference which of the power formulas are used because all will give the same answer when the same values of voltage, current, and resistance are substituted in the formula. The power dissipation of resistor R_1 will be found using the formula:

$$P_1 = E_1 \times I_1$$

$$P_1 = 5\ V \times 0.05\ A$$

$$P_1 = 0.25\ W$$

The amount of power dissipation for resistor R_2 will be calculated using the formula:

$$P_2 = \frac{E_2^2}{R_2}$$

$$P_2 = \frac{156.25\ V^2}{250\ \Omega}$$

$$P_2 = 0.625\ W$$

The amount of power dissipation for resistor R_3 will be calculated using the formula:

$$P_3 = I_3^2 \times R_3$$

$$P_3 = 0.0025\ A^2 \times 150\ \Omega$$

$$P_3 = 0.375\ W$$

The amount of power dissipation for resistor R_4 will be found using the formula:

$$P_4 = E_4 \times I_4$$

$$P_4 = 15\ V \times 0.05\ A$$

$$P_4 = 0.75\ W$$

A good rule to remember when calculating values of electric circuits is that *the total power used in a circuit is equal to the sum of the power used by all parts.* That is, the total power can be found in any kind of circuit—series, parallel, or combination—by adding the power dissipation of all the parts. The total power for this circuit can be found using the formula:

$$P_T = P_1 + P_2 + P_3 + P_4$$

$$P_T = 0.25\ W + 0.625\ W + 0.375\ W + 0.75\ W$$

$$P_T = 2\ W$$

Example 6–1 (continued)

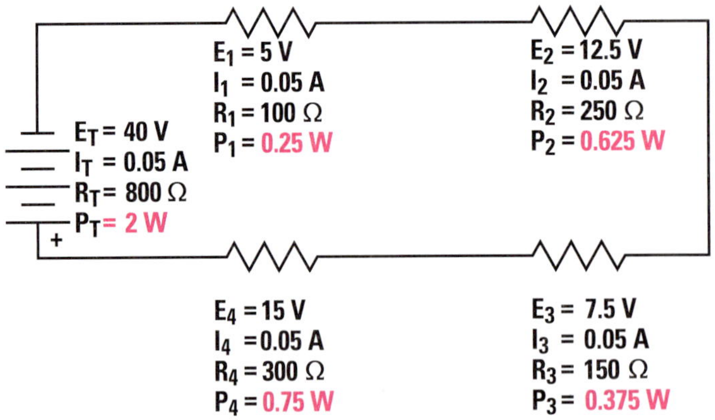

Figure 6–11 The final values for the circuit in Example 1.

Now that all the missing values have been found (*Figure 6–11*), the circuit can be checked by using the third rule of series circuits, which states that voltage drops add to equal the applied voltage:

$$E_T = E_1 + E_2 + E_3 + E_4$$

$$E_T = 5\text{ V} + 12.5\text{ V} + 7.5\text{ V} + 15\text{ V}$$

$$E_T = 40\text{ V}$$

Example 6–2

The second circuit to be solved is shown in *Figure 6–12*. In this circuit, the total resistance is known, but the value of resistor R_2 is not. The second rule of series circuits states that resistances add to equal the total resistance of the circuit. Because the total resistance is known, the missing resistance of R_2 can be

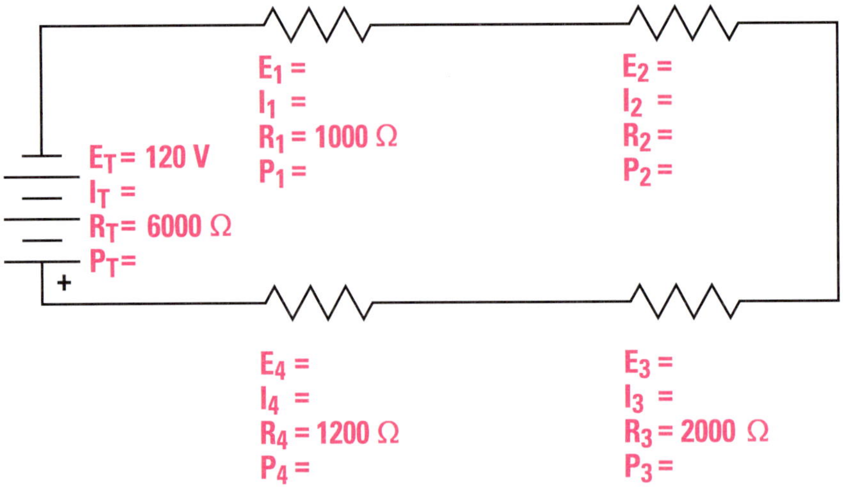

Figure 6–12 Series circuit, Example 2.

Example 6–2 (continued)

found by adding the values of the other resistors and subtracting their sum from the total resistance of the circuit (*Figure 6–13*):

$$R_2 = R_T - (R_1 + R_3 + R_4)$$

$$R_2 = 6000\ \Omega - (1000\ \Omega + 2000\ \Omega + 1200\ \Omega)$$

$$R_2 = 6000\ \Omega - 4200\ \Omega$$

$$R_2 = 1800\ \Omega$$

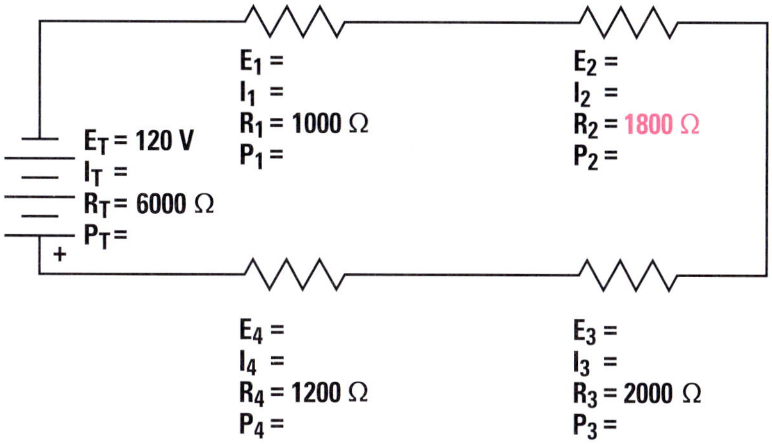

Figure 6–13 The missing resistor value.

The amount of current flow in the circuit can be found using Ohm's law:

$$I = \frac{E}{R}$$

$$I = \frac{120\ V}{6000\ \Omega}$$

$$I = 0.020\ A$$

Because the amount of current flow is the same through all elements of a series circuit (*Figure 6–14*), the voltage drop across each resistor can be found using Ohm's law (*Figure 6–15*):

$$E_1 = I_1 \times R_1$$

$$E_1 = 0.020\ A \times 1000\ \Omega$$

$$E_1 = 20\ V$$

$$E_2 = I_2 \times R_2$$

$$E_2 = 0.020\ A \times 1800\ \Omega$$

$$E_2 = 36\ V$$

$$E_3 = I_3 \times R_3$$

$$E_3 = 0.020\ A \times 2000\ W$$

$$E_3 = 40\ V$$

$$E_4 = I_4 \times R_4$$

$$E_4 = 0.020\ A \times 1200\ \Omega$$

$$E_4 = 24\ V$$

Example 6–2 (continued)

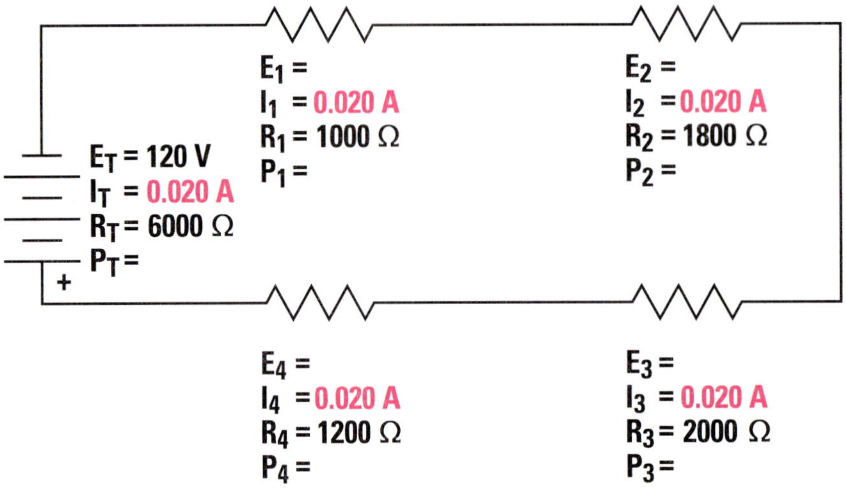

Figure 6–14 The current is the same through each circuit element.

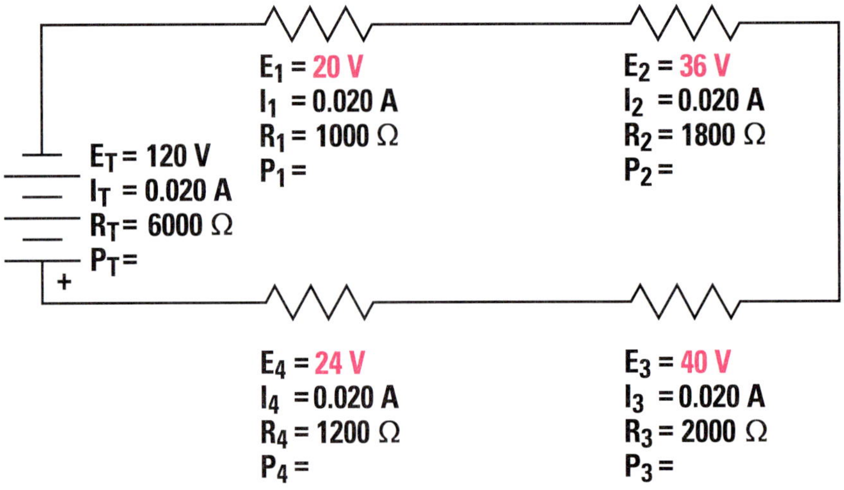

Figure 6–15 Voltage drops across each resistor.

The third rule of series circuits can be used to check the answers:

$$E_T = E_1 + E_2 + E_3 + E_4$$

$$E_T = 20\text{ V} + 36\text{ V} + 40\text{ V} + 24\text{ V}$$

$$E_T = 120\text{ V}$$

The amount of power dissipation for each resistor in the circuit can be calculated using the same method used to solve the circuit in Example 1. The power dissipated by resistor R_1 is calculated using the formula:

$$P_1 = E_1 \times I_1$$

$$P_1 = 20\text{ V} \times 0.02\text{ A}$$

$$P_1 = 0.4\text{ W}$$

Example 6–2 (continued)

The amount of power dissipation for resistor R_2 is found by using the formula:

$$P_2 = \frac{E_2^{\,2}}{R_2}$$

$$P_2 = \frac{1296\ V^2}{1800\ \Omega}$$

$$P_2 = 0.72\ W$$

The power dissipation of resistor R_3 is found using the formula:

$$P_3 = I_3^{\,2} \times R_3$$

$$P_3 = 0.0004\ A^2 \times 2000\ \Omega$$

$$P_3 = 0.8\ W$$

The power dissipation of resistor R_4 is calculated using the formula:

$$P_4 = E_4 \times I_4$$

$$P_4 = 24\ V \times 0.02\ A$$

$$P_4 = 0.48\ W$$

The total power is calculated using the formula:

$$P_T = E_T \times I_T$$

$$P_T = 120\ V \times 0.02\ A$$

$$P_T = 2.4\ W$$

The circuit with all calculated values is shown in *Figure 6–16.*

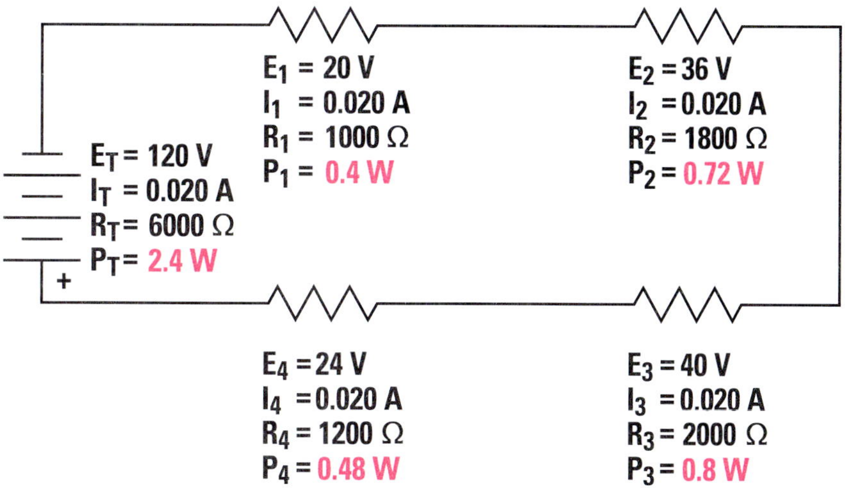

Figure 6–16 The remaining unknown values for the circuit in Example 2.

Example 6–3

In the circuit shown in *Figure 6–17*, resistor R_1 has a voltage drop of 6.4 V, resistor R_2 has a power dissipation of 0.102 W, resistor R_3 has a power dissipation of 0.154 W, resistor R_4 has a power dissipation of 0.307 W, and the total power consumed by the circuit is 0.768 W.

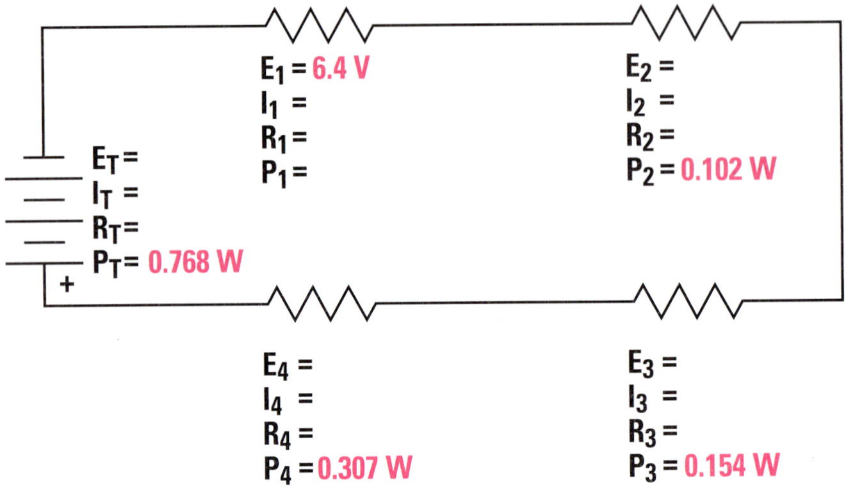

Figure 6–17 Series circuit, Example 3.

The only value that can be found with the given quantities is the amount of power dissipated by resistor R_1. Because the total power is known and the power dissipated by the three other resistors is known, the power dissipated by resistor R_1 can be found by subtracting the power dissipated by resistors R_2, R_3, and R_4 from the total power used in the circuit:

$$P_1 = P_T - (P_2 + P_3 + P_4)$$

or

$$P_1 = P_T - P_2 - P_3 - P_4$$

$$P_1 = 0.768\text{ W} - 0.102\text{ W} - 0.154\text{ W} - 0.307\text{ W}$$

$$P_1 = 0.205\text{ W}$$

Now that the amount of power dissipated by resistor R_1 and the voltage drop across R_1 are known, the current flow through resistor R_1 can be found using the formula:

$$I = \frac{P}{E}$$

$$I = \frac{0.205\text{ W}}{6.4\text{ V}}$$

$$I = 0.032\text{ A}$$

Because the current in a series circuit must be the same at any point in the circuit, it must be the same through all circuit components (*Figure 6–18*).

Now that the power dissipation of each resistor and the amount of current flowing through each resistor are known, the voltage drop of each resistor can be calculated (*Figure 6–19*):

$$E_2 = \frac{P_2}{I_2}$$

Example 6–3 (continued)

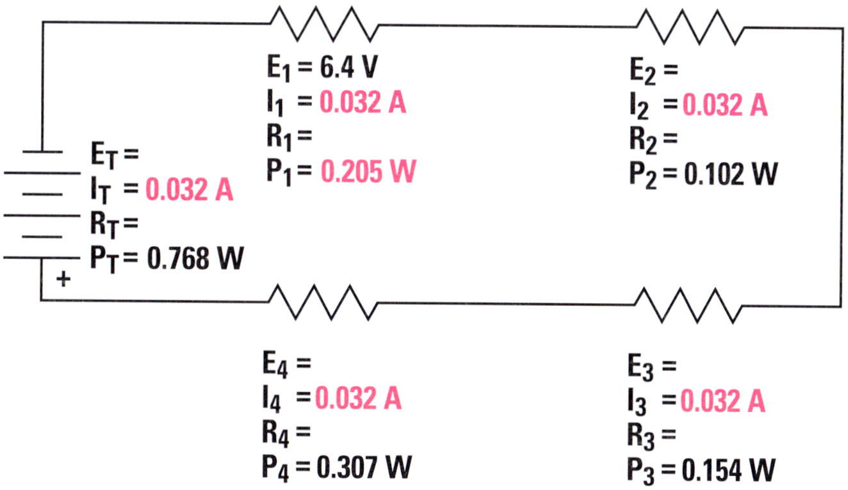

Figure 6–18 The current flow in the circuit in Example 3.

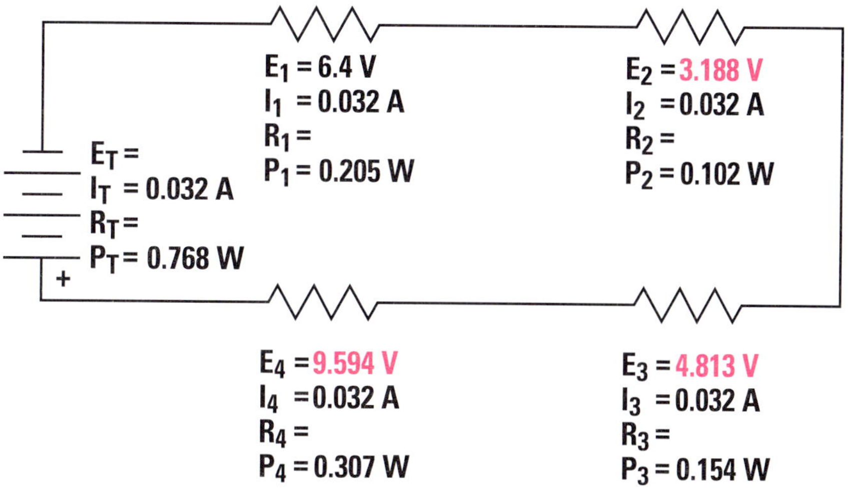

Figure 6–19 Voltage drops across each resistor.

$$E_2 = \frac{0.102 \text{ W}}{0.032 \text{ A}}$$

$$E_2 = 3.188 \text{ V}$$

$$E_3 = \frac{P_3}{I_3}$$

$$E_3 = \frac{0.154 \text{ W}}{0.032 \text{ A}}$$

$$E_3 = 4.813 \text{ V}$$

$$E_4 = \frac{P_4}{I_4}$$

Example 6–3 (continued)

$$E_4 = \frac{0.307\ \text{W}}{0.032\ \text{A}}$$

$$E_4 = 9.594\ \text{V}$$

Ohm's law can now be used to find the resistance value of each resistor in the circuit (*Figure 6–20*):

$$R_1 = \frac{E_1}{I_1}$$

$$R_1 = \frac{6.4\ \text{V}}{0.032\ \text{A}}$$

$$R_1 = 200\ \Omega$$

$$R_2 = \frac{E_2}{I_2}$$

$$R_2 = \frac{3.188\ \text{V}}{0.032\ \text{A}}$$

$$R_2 = 99.625\ \Omega$$

$$R_3 = \frac{E_3}{I_3}$$

$$R_3 = \frac{4.813\ \text{V}}{0.032\ \text{A}}$$

$$R_3 = 150.406\ \Omega$$

$$R_4 = \frac{E_4}{I_4}$$

$$R_4 = \frac{9.594\ \text{V}}{0.032\ \text{A}}$$

$$R_4 = 299.813\ \Omega$$

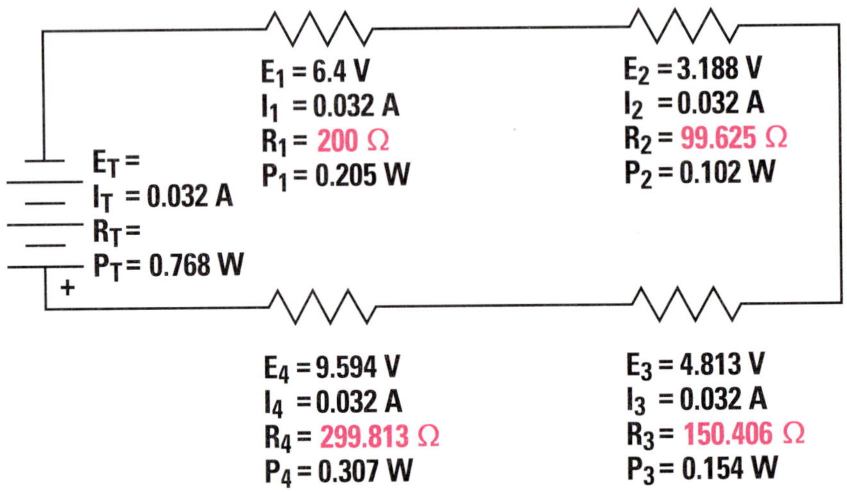

Figure 6–20 The resistance value of each resistor.

Example 6–3 (continued)

The voltage applied to the circuit can be found by adding the voltage drops across the resistors (*Figure 6–21*):

$$E_T = E_1 + E_2 + E_3 + E_4$$
$$E_T = 6.4\ V + 3.188\ V + 4.813\ V + 9.594\ V$$
$$E_T = 23.995\ V$$

The total resistance of the circuit can be found in a similar manner (*Figure 6–21*). The total resistance is equal to the sum of all the resistive elements in the circuit:

$$R_T = R_1 + R_2 + R_3 + R_4$$
$$R_T = 200\ \Omega + 99.625\ \Omega + 150.406\ \Omega + 299.813\ \Omega$$
$$R_T = 749.844\ \Omega$$

If Ohm's law is used to determine total voltage and total resistance, slightly different answers are produced:

$$E_T = \frac{P_T}{I_T}$$
$$E_T = \frac{0.768\ W}{0.032\ A}$$
$$E_T = 24\ V$$
$$R_T = \frac{E_T}{I_T}$$
$$R_T = \frac{24\ V}{0.032\ A}$$
$$R_T = 750\ \Omega$$

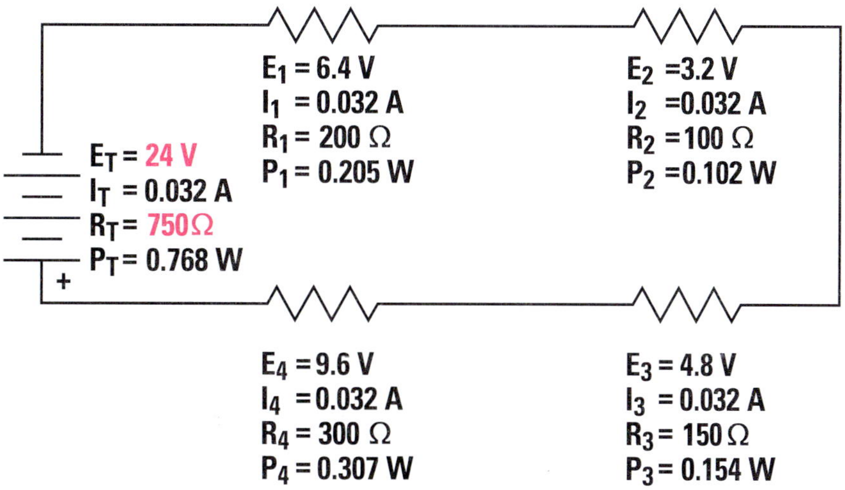

Figure 6–21 The applied voltage and the total resistance.

Example 6–3 (continued)

These slight variances are easily dismissed as being a result of rounding the values of each of the components. Although there is a small difference between the answers, they are within 1% of each other. This small difference has very little effect on the operation of the circuit, and most electric measuring instruments cannot measure values this accurately anyway.

6–3 Voltage Dividers

One common use for series circuits is the construction of voltage dividers. A voltage divider works on the principle that the sum of the voltage drops across a series circuit must equal the applied voltage. Voltage dividers are used to provide different voltages between certain points (*Figure 6–22*). If a voltmeter is connected between Point A and Point B, a voltage of 20 volts will be seen. If the voltmeter is connected between Point B and Point D, a voltage of 80 volts will be seen.

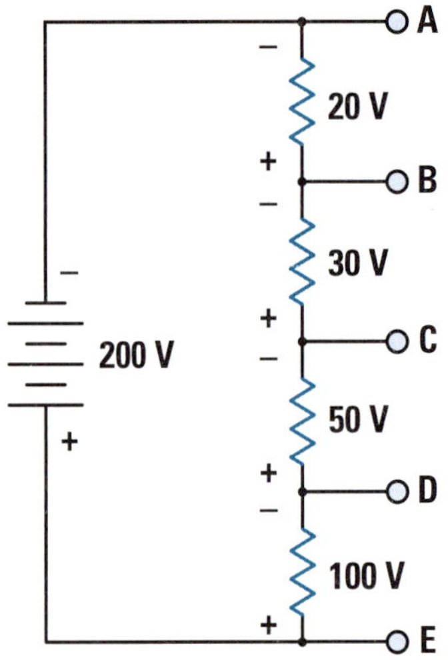

Figure 6–22 Series circuit used as a voltage divider.

Voltage dividers can be constructed to produce any voltage desired. For example, assume that a voltage divider is connected to a source of 120 volts and produces voltage drops of 36 volts, 18 volts, and 66 volts. Notice that the sum of the voltage drops equals the applied voltage. The next step is to decide how much current is to flow through the circuit. Because there is only one path for current flow, the current will be the same through all the resistors. In this circuit, a current flow of 15 milliamperes (0.015 A) is measured. The resistance value of each resistor can now be determined:

$$R = \frac{E}{I}$$

$$R_1 = \frac{36\ V}{0.015\ A}$$

$$R_1 = 2.4\ k\Omega\ (2400\ \Omega)$$

$$R_2 = \frac{18\ \text{V}}{0.015\ \text{A}}$$

$$R_2 = 1.2\ \text{k}\Omega\ (1200\ \Omega)$$

$$R_3 = \frac{66\ \text{V}}{0.015\ \text{A}}$$

$$R_3 = 4.4\ \text{k}\Omega\ (4400\ \Omega)$$

Transposing Ohm's Law to Find Voltage Drop

Another method of determining the voltage drop across series elements is to get creative with Ohm's law. Because the current flow through a series circuit is the same at all points in the circuit, the voltage drop across any particular resistance is equal to the total circuit current times the value of that resistor:

$$E_X = I_T \times R_X$$

The total circuit current is proportional to the source voltage (E_T) and the total resistance of the circuit:

$$I_T = \frac{E_T}{R_T}$$

If the value of I_T is substituted for E_T/R_T in the previous formula, the expression now becomes:

$$E_X = \left(\frac{E_T}{R_T}\right) R_X$$

If the formula is rearranged, it becomes what is known as the general voltage divider formula, a derivation of Ohm's law:

$$E_X = \left(\frac{R_X}{R_T}\right) E_T$$

The voltage drop across any series component (E_X) can be calculated by substituting the value of R_X for the resistance value of that component when the source voltage and total resistance are known.

Example 6–4

Three resistors are connected in series to a 24-V source. Resistor R_1 has a resistance of 200 Ω, resistor R_2 has a value of 300 Ω, and resistor R_3 has a value of 160 Ω. What is the voltage drop across each resistor?

Solution

Find the total resistance of the circuit:

$$R_T = R_1 + R_2 + R_3$$

$$R_T = 200\ \Omega + 300\ \Omega + 160\ \Omega$$

$$R_T = 660\ \Omega$$

Now use the voltage divider formula to calculate the voltage drop across each resistor:

$$E_1 = \left(\frac{R_1}{R_T}\right) E_T$$

Example 6–4 (continued)

$$E_1 = \left(\frac{200\ \Omega}{660\ \Omega}\right) 24\ V$$

$$E_1 = 7.273\ V$$

$$E_2 = \left(\frac{R_2}{R_T}\right) E_T$$

$$E_2 = \left(\frac{300\ \Omega}{660\ \Omega}\right) 24\ V$$

$$E_2 = 10.909\ V$$

$$E_3 = \left(\frac{R_3}{R_T}\right) E_T$$

$$E_3 = \left(\frac{160\ \Omega}{660\ \Omega}\right) 24\ V$$

$$E_3 = 5.818\ V$$

6–4 Voltage Polarity and Ground as a Reference

It is often necessary to know the polarity of the voltage dropped across a resistor. **Voltage polarity** can be determined by observing the direction of current flow through the circuit. In the circuit shown in *Figure 6–22*, it will be assumed that the current flows from the negative terminal of the battery to the positive terminal. Point A is connected to the negative battery terminal, and Point E is connected to the positive terminal. If a voltmeter is connected across Terminals A and B, Terminal B will be positive with respect to A. If a voltmeter is connected across Terminals B and C, however, Terminal B will be negative with respect to Terminal C. Notice that Terminal B is closer to the negative terminal of the battery than Terminal C is. Consequently, electrons flow through the resistor in a direction that makes Terminal B more negative than C. Terminal C would be negative with respect to Terminal D for the same reason.

Using Ground as a Reference

Two symbols are used to represent ground (*Figure 6–23*). The symbol shown in *Figure 6–23(A)* is an **earth ground** symbol. It symbolizes a **ground point** that is made by physically driving an object such as a rod or a pipe into the ground or connecting to a conductive object in contact with Earth such as the rebar in the footing of a house. The symbol shown in *Figure 6–23(B)* symbolizes a **chassis ground**. This is a point that is used as a common connection for other parts of a circuit, but it is not actually driven into the ground. Although the symbol shown in *Figure 6–23(B)* is the accepted symbol for a chassis ground, the symbol shown in *Figure 6–23(A)* is often used to represent a chassis ground also.

Figure 6–23 Common symbols used to show a ground reference point in a circuit.

An excellent example of using a chassis ground as a common connection can be found in the electric system of an automobile. The negative terminal of the battery is connected to the frame or chassis of the vehicle. The frame of the automobile is not connected directly to earth ground, however. In the case of an automobile electric system, the chassis of the vehicle is used as the negative side of the circuit. An electric circuit using ground as a common connection point is shown in *Figure 6–24*. This circuit is an electronic burglar alarm. Notice the numerous ground points in the schematic. In practice, when the circuit is connected, all the ground points will be connected together.

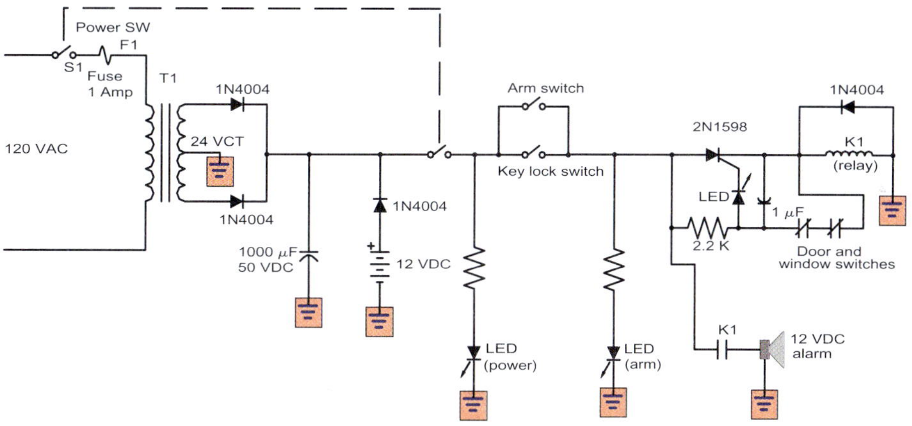

Figure 6–24 Burglar alarm with battery backup.

In voltage divider circuits, ground is often used to provide a common reference point to produce voltages that are above and below ground (*Figure 6–25*). An above-ground voltage is a voltage that is positive with respect to ground. A below-ground voltage is negative with respect to ground. In *Figure 6–25*, one terminal of a zero-center voltmeter is connected to ground. If the probe is connected to Point A, the pointer of the voltmeter gives a negative indication for voltage. If the probe is connected to Point B, the pointer indicates a positive voltage.

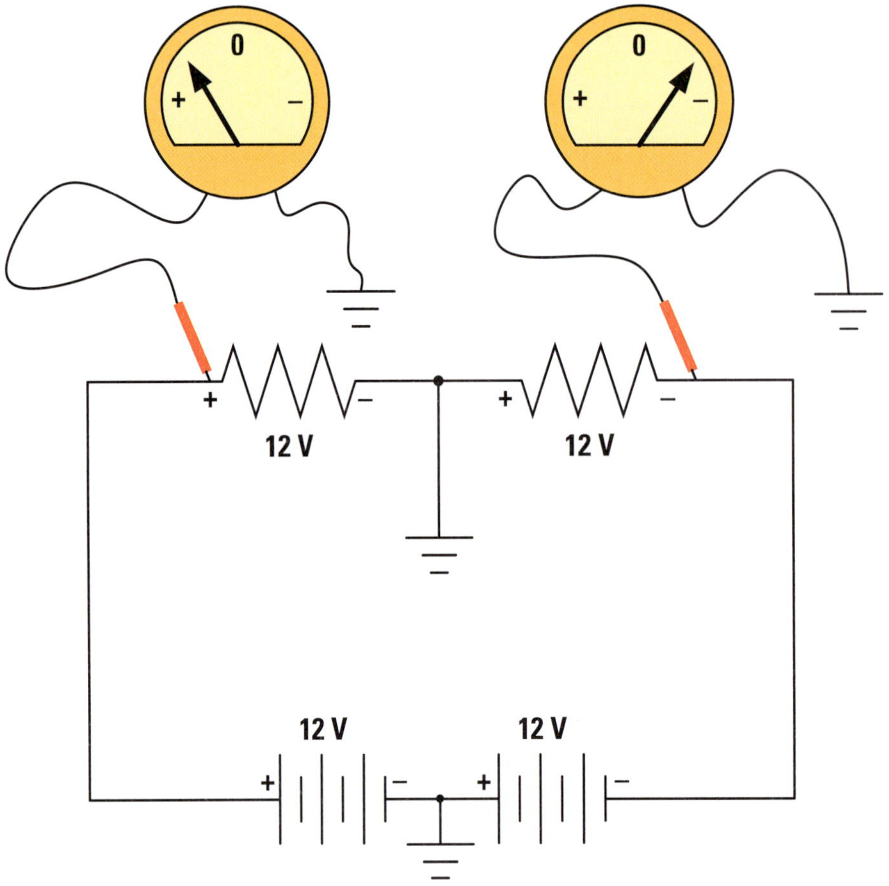

Figure 6–25 A common ground used to produce above- and below-ground voltage.

Summary

- Series circuits have only one path for current flow.
- The individual voltage drops in a series circuit can be added to equal the applied voltage.
- The current is the same at any point in a series circuit.
- The individual resistors can be added to equal the total resistance of the circuit.
- Fuses and circuit breakers are connected in series with the devices they are intended to protect.
- The total power in any circuit is equal to the sum of the power dissipated by all parts of the circuit.
- When the source voltage and the total resistance are known, the voltage drop across each element can be calculated using the general voltage divider formula.

Review Questions

1. True or False: In a series circuit, there is only a single path for electrical current to flow, and therefore, the entire current for the circuit will flow through each component.
2. Circuit breakers and fuses are connected in which of the following circuit configurations?

   A) Series
   B) Parallel
   C) Either series or parallel
   D) Neither series nor parallel

3. A voltmeter connected across, or in parallel with, a component in an electrical circuit will read the value of voltage applied to that specific component. Which of the following best describes what this value is called?

   A) Total voltage

   B) Voltage drop

   C) Voltage divider

   D) Voltage adder

4. True or False: The sum of all voltages measured across individual components of a series circuit will add together to equal the total circuit current.

5. A series circuit has individual resistor values of 200 Ω, 86 Ω, 91 Ω, 180 Ω, and 150 Ω. Which of the following is the total resistance of the circuit?

   A) 710 Ω

   B) 507 Ω

   C) 557 Ω

   D) 707 Ω

6. Resistance in a series circuit is additive and affects the total current flowing in the circuit. Which of the following best describes this relationship?

   A) The current is directly proportional to the amount of resistance in a series circuit.

   B) The current is not affected by the overall resistance in a series circuit.

   C) The current is inversely proportional to the resistance in a circuit.

   D) The current flowing in a series circuit is split among the individual resistors in the circuit.

7. A series circuit contains five resistors. The total voltage applied to the circuit is 120 V. Four resistors have voltage drops of 35 V, 28 V, 22 V, and 15 V. What is the voltage drop of the fifth resistor?

   A) 35 volts

   B) 20 volts

   C) 15 volts

   D) 10 volts

8. A series circuit contains eight resistors. The total resistance in the circuit is 240 volts, and the total resistance is 240 Ω. This results in a total circuit current of 1 ampere. What is the current value through resistor 7?

   A) 0.125 A

   B) 8 A

   C) 240 A

   D) 1 A

9. A series circuit has an applied total voltage of 240 volts. At one of the resistors, the voltage drop is measured at 48 volts, and the power at that resistor is measured at 96 watts. What is the total resistance in the circuit?

   A) 120 Ω

   B) 480 Ω

   C) 240 Ω

   D) Not enough information is provided.

10. A series circuit contains four resistors of the same value and has an applied total voltage of 60 volts. The total resistance in the circuit is eight times greater than the number of volts dropped across each resistor. How much power is dissipated at one of the resistors?

    A) 15 W

    B) 30 W

    C) 7.5 W

    D) Not enough information is provided.

11. Voltage dividers can be used to provide which of the following functions?

    A) Voltage dividers are commonly used to achieve a desired circuit current.

    B) Voltage dividers can be used to achieve any desired voltage in a series circuit.

    C) Voltage dividers are used to change the speed of large industrial motors.

    D) Voltage dividers are used as a risk mitigation strategy by safety professionals.

12. True or False: Understanding the polarity of a power source such as a battery is critical when using a ground reference as only the negative terminal should be connected to the ground reference point.

Practical Applications

1. A 12-V DC automobile head lamp is to be used on a fishing boat with a 24-V power system. The head lamp is rated at 50 W. A resistor is to be connected in series with the lamp to permit it to operate on 24 V. What should be the resistance and power rating of the resistor?

2. Three wire-wound resistors have the following values: 30 Ω, 80 Ω, and 100 Ω. Each resistor has a voltage rating of 100 V. If these three resistors are connected in series, can they be connected to a 240-V circuit without damage to the resistors? Explain your answer.

3. You are an electrician working in an industrial plant. A circuit contains eight incandescent lamps connected in series across 480 volts. One lamp has burned out, and you must determine which one is defective. You have available a voltmeter, ammeter, and ohmmeter. Which meter would you use to determine which lamp is defective in the shortest possible time? Explain how you would use this meter and why.

4. A direct current motor is connected to a 250-volt DC supply. The armature has a current draw of 165 amperes when operating at full load. You have been assigned the task of connecting two resistors in the armature circuit to provide speed control for the motor. When both resistors are connected in the circuit, the armature current is to be limited to 50% of the full-load current draw. When only one resistor is connected in the circuit, the armature current is to be limited to 85% of full-load current. Determine the ohmic value and minimum power rating of each resistor. Refer to *Figure 6–26.* When both switches S_1 and S_2 are open (off), both resistors are connected in the armature circuit, limiting current to 50% of its normal value. When switch S_1 is closed, it causes the current to bypass resistor 1. Resistor 2 now limits the current to 85% of the full-load current. When both switches S_1 and S_2 are closed, all resistance is bypassed, and the armature is connected to full power.

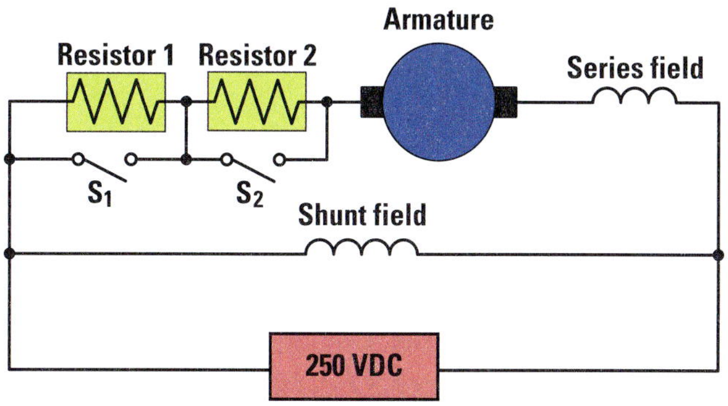

Figure 6–26 Determine the resistance and power rating of the two series resistors.

Practice Problems

Series Circuits

1. Using the three rules for series circuits and Ohm's law, solve for the missing values.

E_T 120 V	E_1	E_2	E_3	E_4	E_5
I_T	I_1	I_2	I_3	I_4	I_5
R_T	R_1 430 Ω	R_2 360 Ω	R_3 750 Ω	R_4 1000 Ω	R_5 620 Ω
P_T	P_1	P_2	P_3	P_4	P_5

E_T	E_1	E_2	E_3 11 V	E_4	E_5
I_T	I_1	I_2	I_3	I_4	I_5
R_T	R_1	R_2	R_3	R_4	R_5
P_T 0.25 W	P_1 0.03 W	P_2 0.0825 W	P_3	P_4 0.045 W	P_5 0.0375 W

E_T 340 V	E_1 44 V	E_2 94 V	E_3 60 V	E_4 40 V	E_5
I_T	I_1	I_2	I_3	I_4	I_5
R_T	R_1	R_2	R_3	R_4	R_5
P_T	P_1	P_2	P_3	P_4	P_5 0.204 W

2. Use the general voltage divider formula to calculate the values of voltage drop for the following series-connected resistors. Assume a source voltage of 120 V.

$R_1 = 1K\ \Omega$	$R_2 = 2.2K\ \Omega$	$R_3 = 1.8K\ \Omega$	$R_4 = 1.5K\ \Omega$
$E_1 =$ _____ V	$E_2 =$ _____ V	$E_3 =$ _____ V	$E_4 =$ _____ V

3. Find all missing values for the circuit shown in *Figure 6–27*.
4. Find all missing values for the circuit shown in *Figure 6–28*.

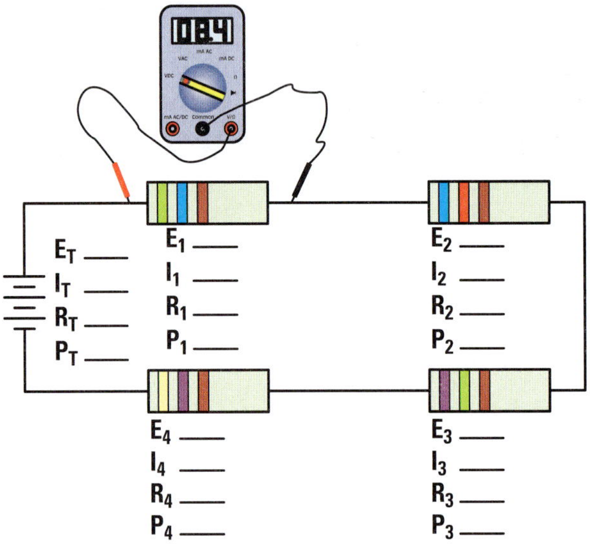

Figure 6–27 Determine resistor values using the color code, and determine all missing electrical values.

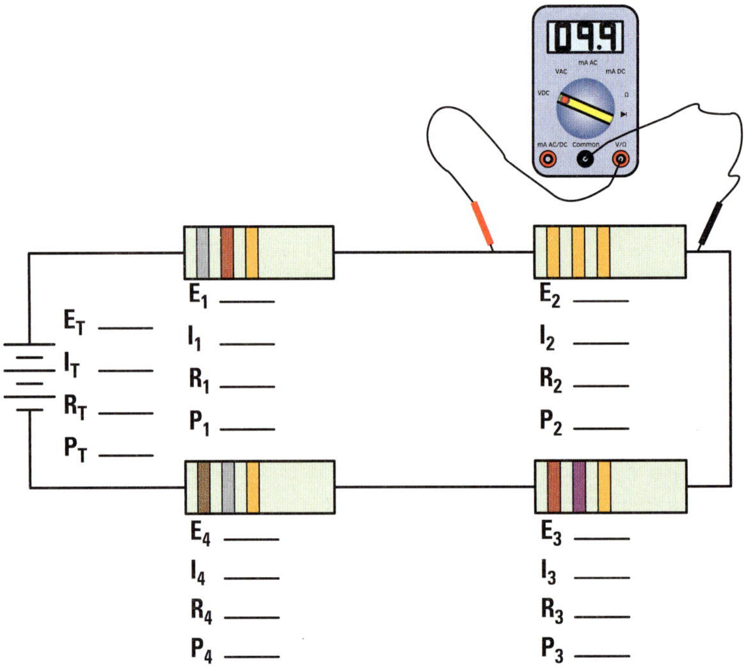

Figure 6–28 Determine resistor values using the color code, and determine all missing electrical values.

Unit 7 Parallel Circuits

Key Terms

Circuit branch

Current dividers

Load

Parallel circuits

Why You Need to Know

Most electrical equipment installed in our daily lives is connected using a parallel circuit configuration. Learning the rules concerning voltage, current, resistance, and power in parallel circuits is important for understanding how electric circuits work. In this unit, you will use Ohm's law to

- understand the rules governing parallel circuits to determine different electrical values.
- understand how voltage remains constant while current changes in a parallel circuit.
- learn how and why total resistance decreases when resistors are connected in parallel.

Although working Ohm's law problems often becomes tedious, it is one of the best ways to demonstrate the relationship between voltage, current, resistance, and power in a parallel circuit.

Outline

Objectives

After studying this unit, you should be able to

- describe the characteristics of parallel circuits.
- describe how voltage is applied throughout a parallel circuit.
- explain how resistance is affected by parallel circuits.
- describe how electrical current flows in a parallel circuit.
- employ Ohm's law for solving electrical values of parallel circuits.
- explain how a parallel circuit acts as a current divider.

Preview

Parallel circuits are the type of circuit that is most commonly used to supply the electrical equipment we use every day. Most devices such as lights and receptacles in homes and office buildings are connected in parallel. Imagine if the lights in your home were wired in series. All the lights in the home would have to be turned on for any light to operate, and if one were to burn out, all the lights would go out. Receptacle circuits would be far more of an inconvenience than what we experience today. If receptacles were connected in series, some device would need to be connected to each receptacle for the circuit to work.

7–1 Parallel Circuit Values

Total Current

Parallel circuits are circuits that have more than one path for current flow (*Figure 7–1*). If it is assumed that current leaves Terminal A and returns to Terminal B, it can be seen that the electrons can take three separate paths. In *Figure 7–1*, 6 amperes of current leave Terminal A. Three amperes flow through Resistor R_1, and 3 amperes flow to Resistors R_2 and R_3. At the junction of Resistors R_2 and R_3, 2 ampere flows through Resistor R_2, and 1 ampere flows to Resistor R_3. Notice that the power supply, Terminals A and B, must furnish all the current that flows through each individual resistor, or **circuit branch**.

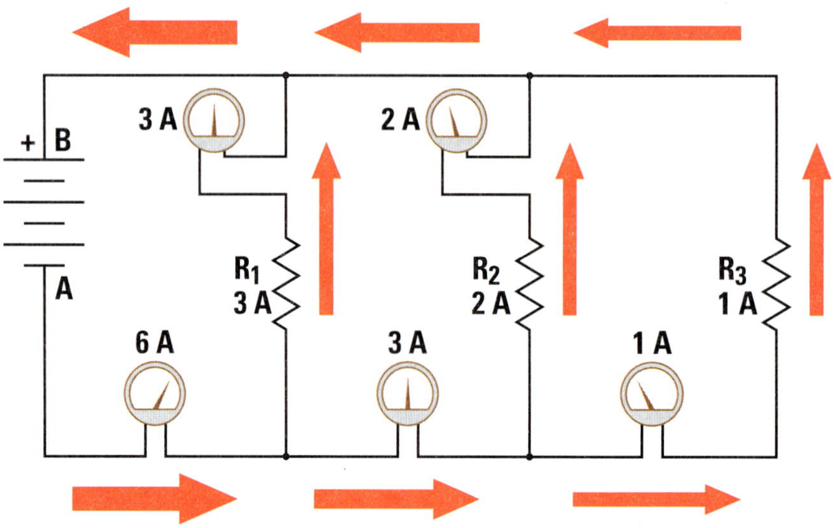

Figure 7–1 Parallel circuits provide more than one path for current flow.

One of the rules for parallel circuits states that *the total current flow in the circuit is equal to the sum of the currents through all the branches.* This is similar to the way voltage drops across each component are added together in a series circuit to equal the applied voltage. Notice that the amount of current leaving the source *must* return to the source. See Examples 7-1 through 7-3 for different ways to find all unknown values in parallel circuits.

Voltage Drop

Figure 7–2 shows another parallel circuit and gives the values of voltage, current, and resistance for each individual resistor or branch. Notice that the voltage drop across all three resistors is the same. If the circuit is traced, it can be seen that each resistor is connected directly to the power source. A second rule for parallel circuits states that *the voltage drop across any branch of a parallel circuit is the same as the applied voltage.* For this reason, most electric circuits in homes are connected in parallel. Each lamp and receptacle is supplied with 120 volts (*Figure 7–3*).

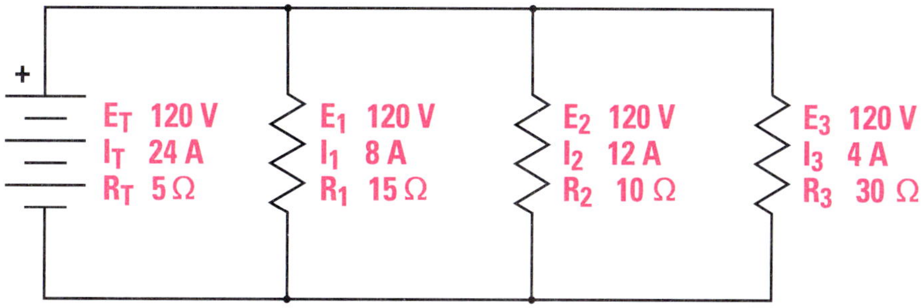

Figure 7–2 Parallel circuit values.

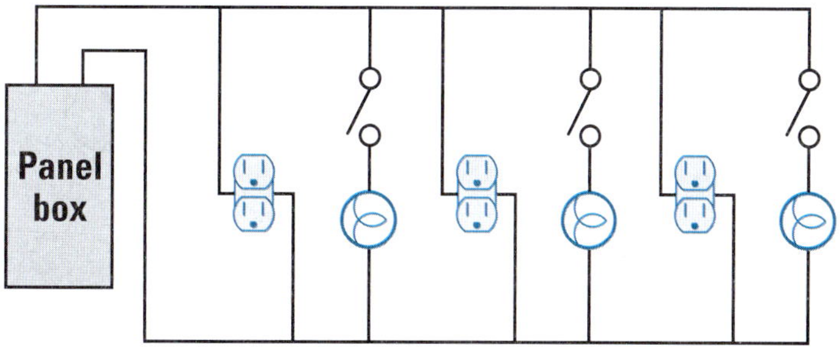

Figure 7–3 Lights and receptacles are connected in parallel.

Total Resistance

Resistance in a parallel circuit behaves nothing like it does in a series circuit where the total resistance was equal to the sum of all individual resistance values. *In a parallel circuit, the total resistance is always less than the resistance of the lowest value resistor, or branch, in the circuit.* In the circuit shown in *Figure 7–4,* three separate resistors have values of 15 ohms, 10 ohms, and 30 ohms. The total resistance of the circuit, however, is 5 ohms. Each time another element is connected in parallel, it creates another path for current to take, and there is less overall opposition to the flow of current through the entire circuit. Imagine a water system consisting of a holding tank, a pump, and return lines to the tank (*Figure 7–5*). Although large return pipes have less resistance to the flow of water than small pipes, the small pipes do provide a return path to the holding tank. Each time another return path is added, regardless of size, there is less overall resistance to flow, and the pump can push more water back into the tank.

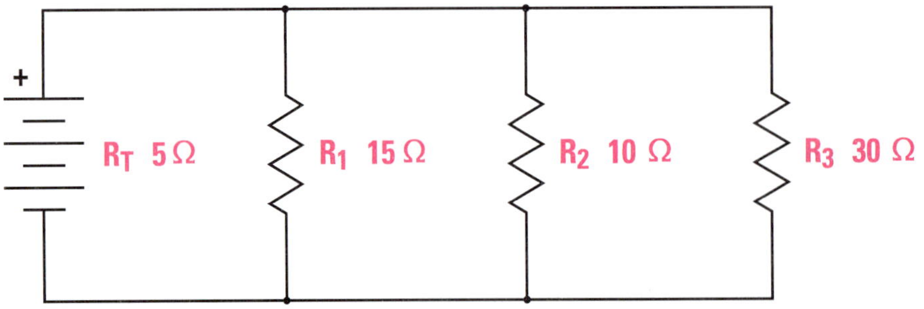

Figure 7–4 Total resistance is always less than the resistance of any single branch.

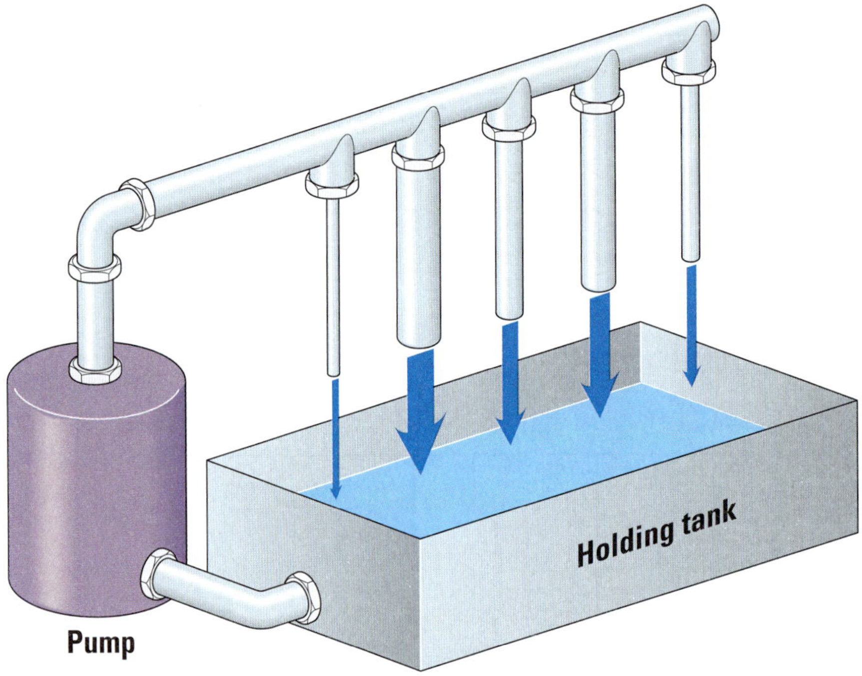

Figure 7–5 Each new path reduces the total resistance to the flow of water.

That concept often causes confusion concerning how much **load** is connected to a circuit. Students often associate an increase of resistance with an increase of load. And while adding equipment to an existing circuit will increase the load on the circuit, it is an increase of current, not resistance. The total resistance will actually decrease. In laboratory exercises, students can observe the circuit current increase each time a new parallel element is connected to the circuit. If they equate the increase of resistance elements in the circuit to an increase of current, they might conclude that as they add resistance to the circuit, it increases the current it takes to power the circuit. That conclusion is completely contrary to Ohm's law, which states that an increase of resistance must cause a proportional decrease of current. Even the name "resistance" implies that it is making it more difficult for current to flow.

7–2 Parallel Resistance Formulas

Resistors of Equal Value

Three formulas can be used to determine the total resistance of a parallel circuit. The first formula can be used only when all the resistors in the circuit are of equal value. This formula states that *when all resistors are of equal value, the total resistance is equal to the value of one individual resistor, or branch, divided by the number (N) of resistors or branches.*

$$R_T = \frac{R}{N}$$

For example, assume that three resistors, each having a value of 24 ohms, are connected in parallel (*Figure 7–6*). The total resistance of this circuit can be found by dividing the resistance of one single resistor by the total number of resistors:

$$R_T = \frac{R}{N}$$

$$R_T = \frac{24\ \Omega}{3}$$

$$R_T = 8\ \Omega$$

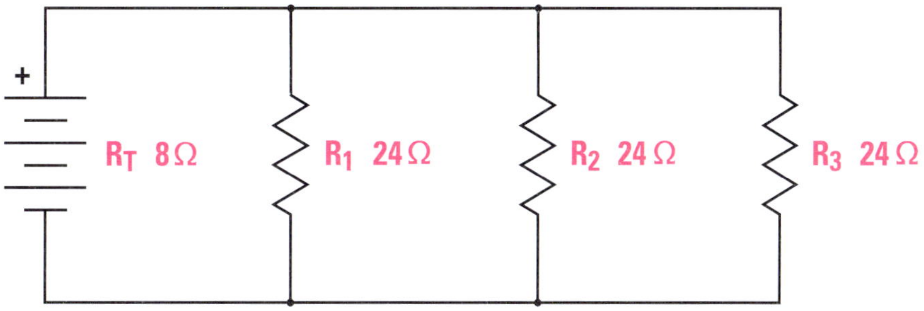

Figure 7–6 Finding the total resistance when all resistors have the same value.

Product over Sum

The second formula used to determine the total resistance in a parallel circuit divides the product of pairs of resistors by their sum sequentially until only one pair is left. This is commonly referred to as the product-over-sum method for finding total resistance.

$$R_T = \frac{R_1 \times R_2}{R_1 + R_2}$$

In the circuit shown in *Figure 7–7,* three branches having resistors with values of 20 ohms, 30 ohms, and 60 ohms are connected in parallel. To find the total resistance of the circuit using the product-over-sum method, find the total resistance of any two branches in the circuit (*Figure 7–8*):

$$R_T = \frac{R_2 \times R_3}{R_2 + R_3}$$

$$R_T = \frac{30\ \Omega \times 60\ \Omega}{30\ \Omega + 60\ \Omega}$$

$$R_T = \frac{1800\ \Omega}{90\ \Omega}$$

$$R_T = 20\ \Omega$$

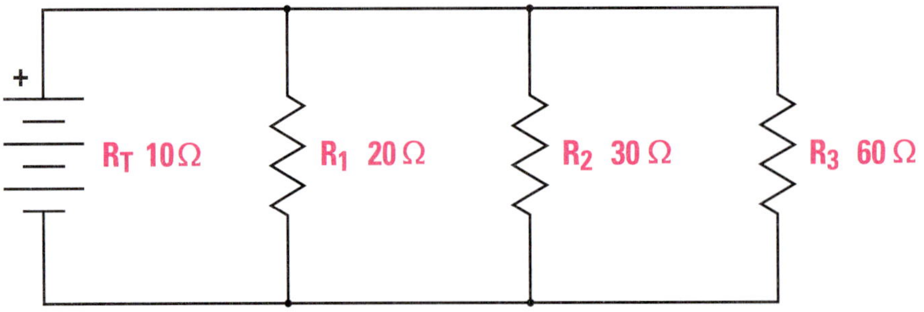

Figure 7–7 Finding the total resistance of a parallel circuit by dividing the product of two resistors by their sum.

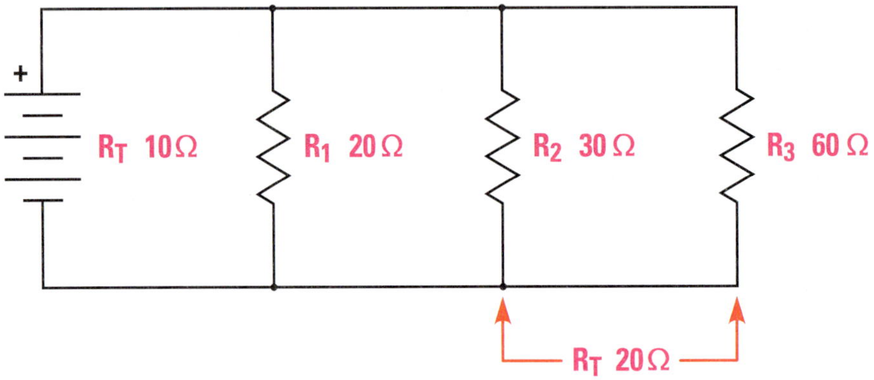

Figure 7–8 The total resistance of resistors R_2 and R_3.

The total resistance of the last two resistors in the circuit is 20 ohms. This 20 ohms, however, is connected in parallel with the original R_1, a 20-ohms resistor. The total resistance of the last two resistors is now substituted for the value of R_2 in the formula, and the value of the first resistor is substituted for the value of R_1 (*Figure 7–9*):

$$R_T = \frac{R_1 \times R_2}{R_1 + R_2}$$

$$R_T = \frac{20\ \Omega \times 20\ \Omega}{20\ \Omega + 20\ \Omega}$$

$$R_T = \frac{400\ \Omega}{40\ \Omega}$$

$$R_T = 10\ \Omega$$

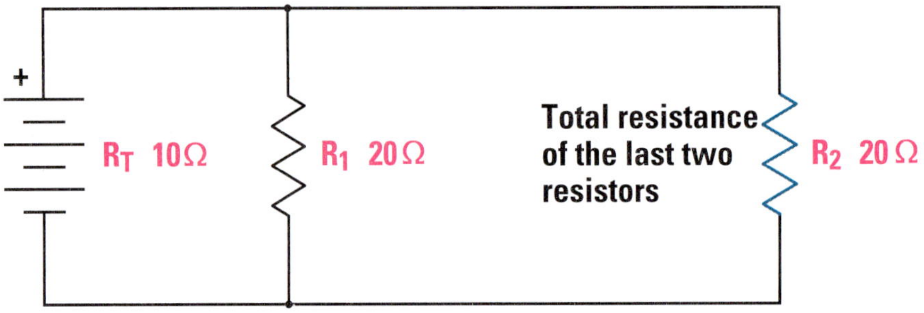

Figure 7–9 The total value of the first two resistors is used as Resistor 2.

Reciprocal Formula

The third formula used to find the total resistance of a parallel circuit is:

$$\frac{1}{R_T} = \frac{1}{R_1} + \frac{1}{R_2} + \frac{1}{R_3} + \frac{1}{R_N}$$

Notice that this formula actually finds the reciprocal of the total resistance instead of the total resistance. To make the formula equal to the total resistance, it can be rewritten as follows:

$$R_T = \frac{1}{\frac{1}{R_1} + \frac{1}{R_2} + \frac{1}{R_3} + \frac{1}{R_N}}$$

The value R_N stands for the number of resistors in the circuit. If the circuit has 25 resistors connected in parallel, for example, the last resistor in the formula would be R_{25}.

This formula is known as the reciprocal formula. The reciprocal of any number is that number divided into 1. The reciprocal of 4, for example, is 0.25 because $1/4 = 0.25$. Another rule of parallel circuits is that *the total resistance of a parallel circuit is the reciprocal of the sum of the reciprocals of the individual branches.* A modified version of this formula is used in several different applications to find values other than resistance. Some of those other formulas are covered later.

In *Figure 7–10,* three resistors having values of 150 ohms, 300 ohms, and 100 ohms are connected in parallel. The total resistance can be found using the reciprocal formula:

$$R_T = \frac{1}{\frac{1}{R_1} + \frac{1}{R_2} + \frac{1}{R_3}}$$

$$R_T = \frac{1}{\frac{1}{150\ \Omega} + \frac{1}{300\ \Omega} + \frac{1}{100\ \Omega}}$$

$$R_T = \frac{1}{(0.006667 + 0.003333 + 0.01)\frac{1}{\Omega}}$$

$$R_T = \frac{1}{(0.02)\frac{1}{\Omega}}$$

$$R_T = 50\ \Omega$$

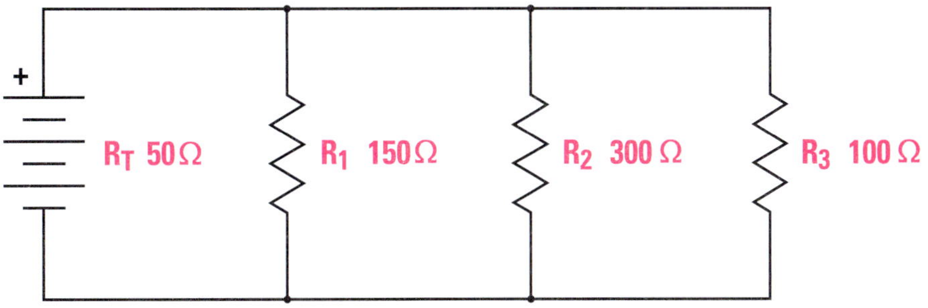

Figure 7–10 Finding the total resistance using the reciprocal method.

To find the total resistance of the previous example using a scientific calculator, press the following keys. Note that the calculator automatically carries each answer to the maximum number of decimal places. This increases the accuracy of the answer.

[1] [5] [0] [1/x] [+] [3] [0] [0] [1/x] [+] [1] [0] [0] [1/x] [=] [1/x]

Note that this is intended to illustrate how total parallel resistance can be determined using many scientific calculators. Some calculators may require a different key entry or pressing the equal key at the end.

Example 7–1

In the circuit shown in *Figure 7–11,* three resistors having values of 300 Ω, 200 Ω, and 600 Ω are connected in parallel. The total current flow through the circuit is 0.6 A. Find all the missing values in the circuit.

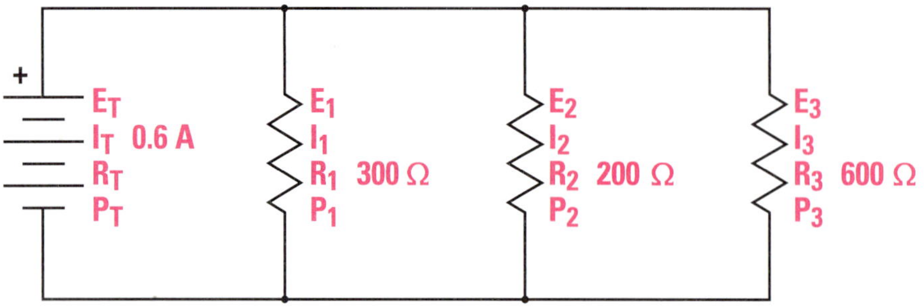

Figure 7–11 Parallel circuit, Example 1.

Solution

The first step is to find the total resistance of the circuit. The reciprocal formula is used:

$$R_T = \frac{1}{\frac{1}{R_1} + \frac{1}{R_2} + \frac{1}{R_3}}$$

$$R_T = \frac{1}{\frac{1}{300\ \Omega} + \frac{1}{200\ \Omega} + \frac{1}{600\ \Omega}}$$

$$R_T = \frac{1}{(0.00333 + 0.0050 + 0.00167)\frac{1}{\Omega}}$$

$$R_T = \frac{1}{(0.01)\frac{1}{\Omega}}$$

$$R_T = 100\ \Omega$$

Now that the total resistance of the circuit is known, the voltage applied to the circuit can be found by using the total current value and Ohm's law:

$$E_T = I_T \times R_T$$

$$E_T = 0.6\ \text{A} \times 100\ \Omega$$

$$E_T = 60\ \text{V}$$

One of the rules for parallel circuits states that the voltage drops across all the parts of a parallel circuit are the same as the total voltage. Therefore, the voltage drop across each resistor is 60 V (*Figure 7–12*):

$$E_T = E_1 = E_2 = E_3$$

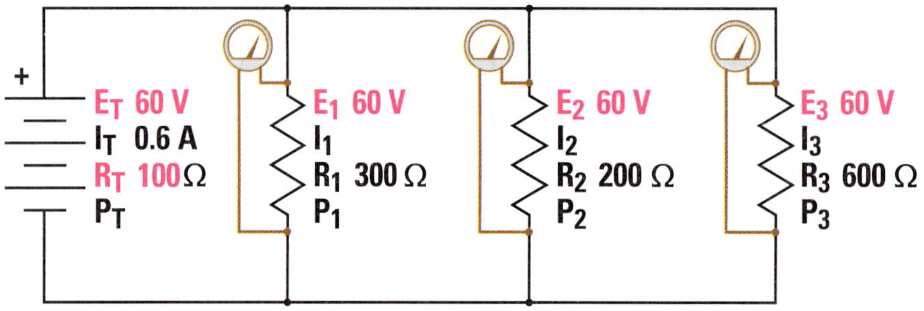

Figure 7–12 The voltage drop is the same across all branches of a parallel circuit.

Example 7–1 (continued)

Because the voltage drop and the resistance of each resistor are known, Ohm's law can be used to determine the amount of current flow through each resistor (*Figure 7–13*):

$$I_1 = \frac{E_1}{R_1}$$

$$I_1 = \frac{60\ V}{300\ \Omega}$$

$$I_1 = 0.2\ A$$

$$I_2 = \frac{E_2}{R_2}$$

$$I_2 = \frac{60\ V}{200\ \Omega}$$

$$I_2 = 0.3\ A$$

$$I_3 = \frac{E_3}{R_3}$$

$$I_3 = \frac{60\ V}{600\ \Omega}$$

$$I_3 = 0.1\ A$$

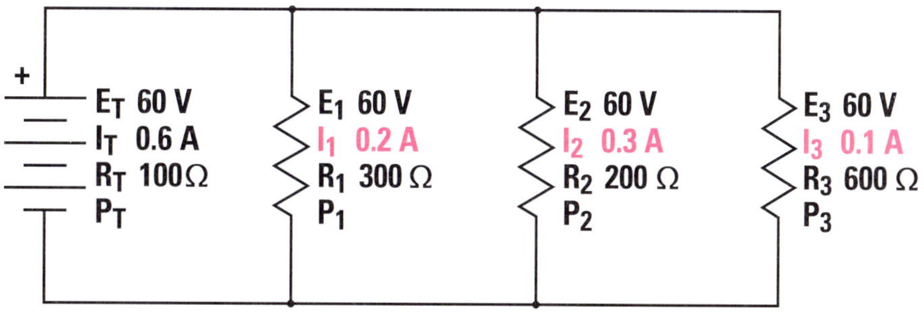

Figure 7–13 Ohm's law is used to calculate the amount of current through each branch.

The amount of power (W) used by each resistor can be found by using Ohm's law. A different formula is used to find the amount of electrical energy converted into heat by each of the resistors:

$$P_1 = \frac{E_1^2}{R_1}$$

$$P_1 = \frac{60\ V \times 60\ V}{300\ \Omega}$$

$$P_1 = \frac{3600\ V^2}{300\ \Omega}$$

$$P_1 = 12\ W$$

$$P_2 = I_2^2 \times R_2$$

$$P_2 = 0.3\ A \times 0.3\ A \times 200\ \Omega$$

$$P_2 = 0.09\ A^2 \times 200\ \Omega$$

$$P_2 = 18\ W$$

Example 7–1 (continued)

$$P_3 = E_3 \times I_3$$

$$P_3 = 60\ V \times 0.1\ A$$

$$P_3 = 6\ W$$

In Unit 6, it was stated that the total amount of power in a circuit is equal to the sum of the power used by all the parts. This is true for any type of circuit. Therefore, the total amount of power used by this circuit can be found by taking the sum of the power used by all the resistors (*Figure 7–14*):

$$P_T = P_1 + P_2 + P_3$$

$$P_T = 12\ W + 18\ W + 6\ W$$

$$P_T = 36\ W$$

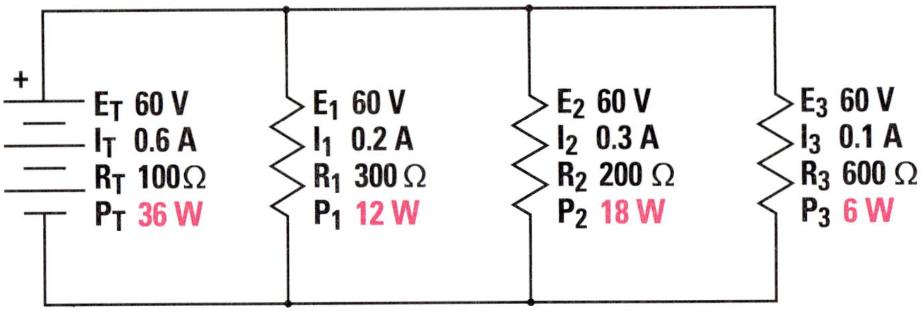

Figure 7–14 The total amount of power used and the amount used by each component in the circuit.

Example 7–2

In the circuit shown in *Figure 7–15,* three resistors are connected in parallel. Two of the resistors have values of 900 Ω and 1800 Ω. The value of Resistor R_2 is unknown. The total resistance of the circuit is 300 Ω. Resistor R_2 has a current flow of 0.2 A. Find the missing circuit values.

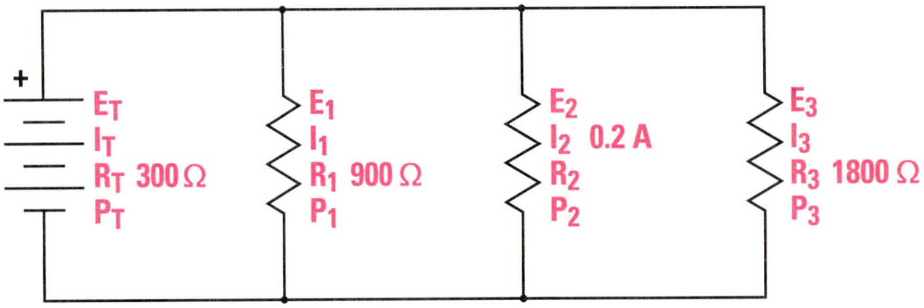

Figure 7–15 Parallel circuit, Example 2.

Solution

The first step in solving this problem is to find the missing resistor value. This can be done by transposing the reciprocal formula as shown:

$$\frac{1}{R_2} = \frac{1}{R_T} - \frac{1}{R_1} - \frac{1}{R_3}$$

Example 7–2 (continued)

or

$$R_2 = \frac{1}{\frac{1}{R_T} - \frac{1}{R_1} - \frac{1}{R_3}}$$

Using the basic principles of algebra, we can transpose the original resistance reciprocal formula to solve for the missing resistance. The reciprocal of any individual resistor is equal to the reciprocal of the difference between the reciprocal of the total resistance and the sum of the reciprocals of the other resistors in the circuit:

$$R_2 = \frac{1}{\frac{1}{R_T} - \frac{1}{R_1} - \frac{1}{R_3}}$$

$$R_2 = \frac{1}{\frac{1}{300\ \Omega} - \frac{1}{900\ \Omega} - \frac{1}{1800\ \Omega}}$$

$$R_2 = \frac{1}{(0.003333 - 0.001111 - 0.0005556)\frac{1}{\Omega}}$$

$$R_2 = \frac{1}{(0.001666)\frac{1}{\Omega}}$$

$$R_2 = 600\ \Omega$$

Now that the resistance of Resistor R_2 has been found, the voltage drop across Resistor R_2 can be determined using the current flow through the resistor and Ohm's law (*Figure 7–16*):

$$E_2 = I_2 \times R_2$$

$$E_2 = 0.2\ A \times 600\ \Omega$$

$$E_2 = 120\ V$$

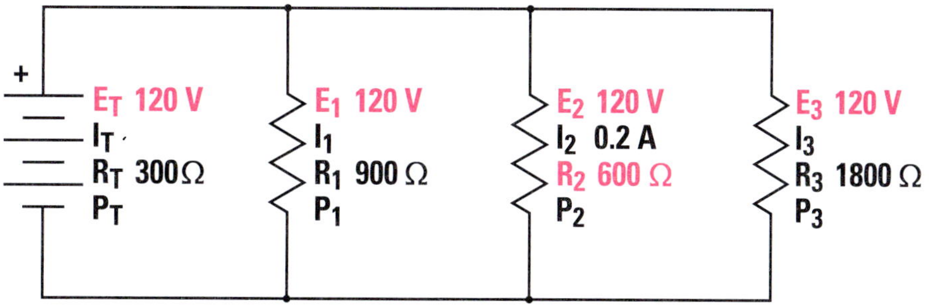

Figure 7–16 The missing resistor and voltage values allow the remaining values to be calculated.

If 120 V is dropped across Resistor R_2, the same voltage is dropped across each component of the circuit:

$$E_2 = E_T = E_1 = E_3$$

Example 7–2 (continued)

Once two values of any component are known, the rest of the values can be determined. Because the voltage drop across each part of the circuit is known and the resistance is known, the current flow through each branch can be determined using Ohm's law (*Figure 7–17*):

$$I_T = \frac{E_T}{R_T}$$

$$I_T = \frac{120\ V}{300\ \Omega}$$

$$I_T = 0.4\ A$$

$$I_1 = \frac{E_1}{R_1}$$

$$I_1 = \frac{120\ V}{900\ \Omega}$$

$$I_1 = 0.1333\ A$$

$$I_3 = \frac{E_3}{R_3}$$

$$I_3 = \frac{120\ V}{1800\ \Omega}$$

$$I_3 = 0.0667\ A$$

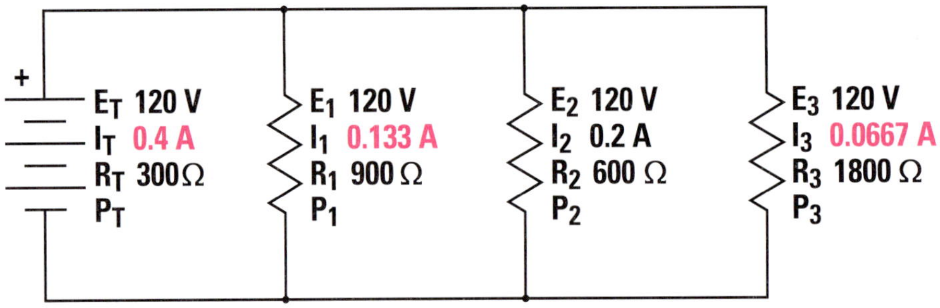

Figure 7–17 Determining the current using Ohm's law.

The amount of power used by each resistor can be found using Ohm's law (*Figure 7–18*):

$$P_1 = \frac{E_1^2}{R_1}$$

$$P_1 = \frac{120\ V \times 120\ V}{900\ \Omega}$$

$$P_1 = \frac{14{,}400\ V^2}{900\ \Omega}$$

$$P_1 = 16\ W$$

$$P_2 = I_2^2 \times R_2$$

$$P_2 = 0.2\ A \times 0.2\ A \times 600\ \Omega$$

$$P_2 = 0.04\ A^2 \times 600\ \Omega$$

$$P_2 = 24\ W$$

Example 7–2 (continued)

$$P_3 = E_3 \times I_3$$

$$P_3 = 120\ V \times 0.0667\ A$$

$$P_3 = 8.004\ W$$

$$P_T = E_T \times I_T$$

$$P_T = 120\ V \times 0.4\ A$$

$$P_T = 48\ W$$

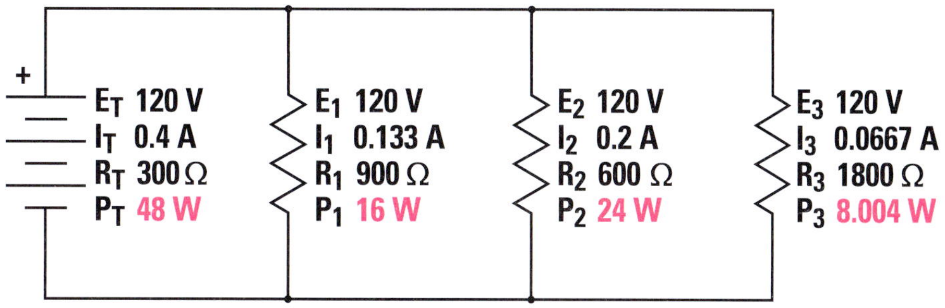

Figure 7–18 The values of power for the circuit in Example 2.

Example 7–3

In the circuit shown in *Figure 7–19,* three resistors are connected in parallel. Resistor R_1 is producing 0.075 W of heat, R_2 is producing 0.45 W of heat, and R_3 is producing 0.225 W of heat. The circuit has a total current of 0.05 A.

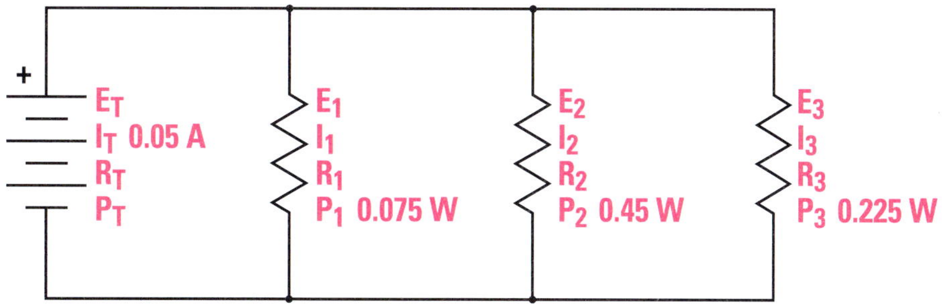

Figure 7–19 Parallel circuit, Example 3.

Solution

Because the amount of power dissipated by each resistor is known, the total power for the circuit can be found by finding the sum of the power used by the components:

$$P_T = P_1 + P_2 + P_3$$

$$P_T = 0.075\ W + 0.45\ W + 0.225\ W$$

$$P_T = 0.75\ W$$

Example 7–3 (continued)

Now that the amount of total current and total power for the circuit are known, the applied voltage can be found using Ohm's law (*Figure 7–20*):

$$E_T = \frac{P_T}{I_T}$$

$$E_T = \frac{0.75\ \text{W}}{0.05\ \text{A}}$$

$$E_T = 15\ \text{V}$$

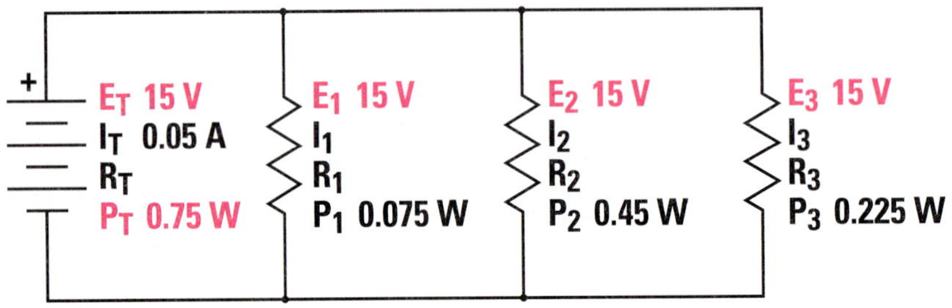

Figure 7–20 The applied voltage for the circuit can be determined from total power and circuit current.

Because voltage must be equal on all branches of a parallel circuit, the amount of current flow through each resistor can now be found using Ohm's law (*Figure 7–21*):

$$I_1 = \frac{P_1}{E_1}$$

$$I_1 = \frac{0.075\ \text{W}}{15\ \text{V}}$$

$$I_1 = 0.005\ \text{A}$$

$$I_2 = \frac{P_2}{E_2}$$

$$I_2 = \frac{0.45\ \text{W}}{15\ \text{V}}$$

$$I_2 = 0.03\ \text{A}$$

$$I_3 = \frac{P_3}{E_3}$$

$$I_3 = \frac{0.225\ \text{W}}{15\ \text{V}}$$

$$I_3 = 0.015\ \text{A}$$

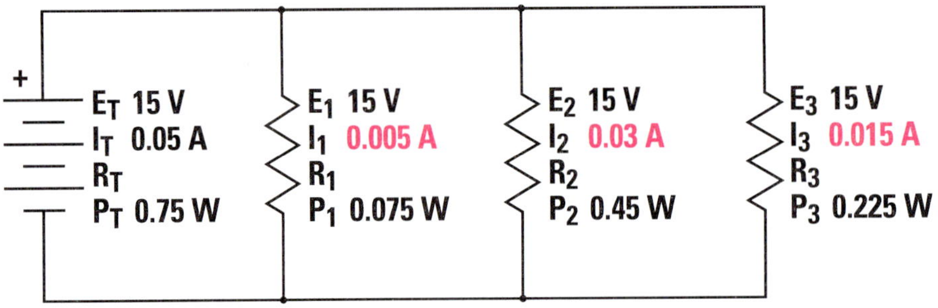

Figure 7–21 The current through each branch.

Example 7–3 (continued)

All resistance values for the circuit can now be found using Ohm's law (*Figure 7–22*):

$$R_1 = \frac{E_1}{I_1}$$

$$R_1 = \frac{15\ V}{0.005\ A}$$

$$R_1 = 3000\ \Omega$$

$$R_2 = \frac{E_2}{I_2}$$

$$R_2 = \frac{15\ V}{0.03\ A}$$

$$R_2 = 500\ \Omega$$

$$R_3 = \frac{E_3}{I_3}$$

$$R_3 = \frac{15\ V}{0.015\ A}$$

$$R_3 = 1000\ \Omega$$

$$R_T = \frac{E_T}{I_T}$$

$$R_T = \frac{15\ V}{0.05\ A}$$

$$R_T = 300\ \Omega$$

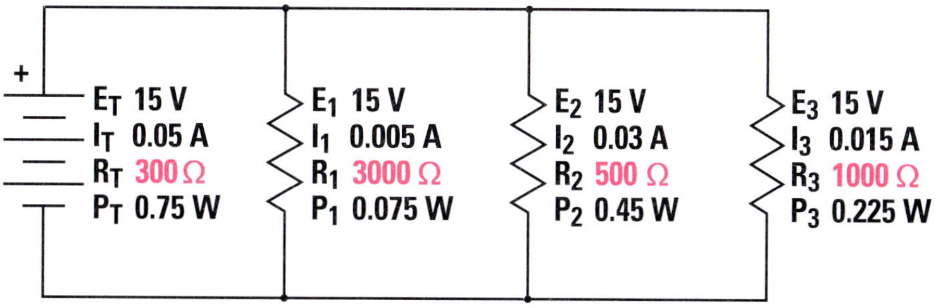

Figure 7–22 The remaining values for the circuit.

7–3 Current Dividers

All parallel circuits are **current dividers** (*Figure 7–23*). This is a similar function to how resistors in series act as voltage dividers. As previously discussed in this unit, the sum of the currents in a parallel circuit must equal the total current. Assume that a current of 1 ampere enters the circuit at Point A. This 1 ampere of current will divide between Resistors R_1 and R_2 and then recombine at Point B. The amount of current that flows through each resistor is inversely proportional to the resistance value. A greater amount of current will flow through a low-value resistor, and less current will flow through a high-value resistor.

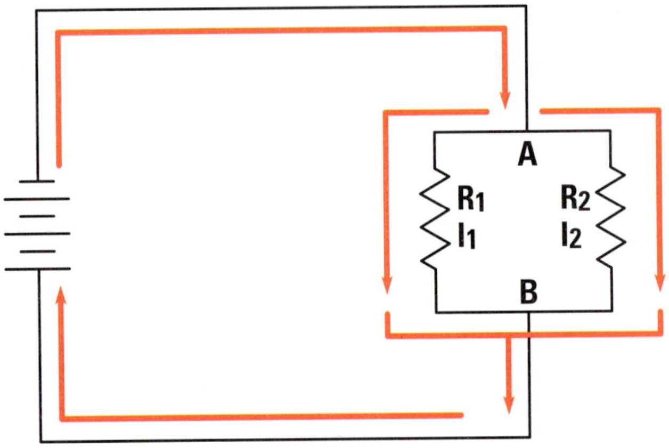

Figure 7–23 Current dividing over two branches of a parallel circuit.

In a purely parallel circuit, the voltage applied to each branch will be equal (*Figure 7–24*). Therefore, the current flow through any branch can be calculated by dividing the source voltage (E_T) by the resistance of that branch. The current flow through Branch 1 can be calculated using the formula: $\mathbf{I_1 = E_T \div R_1}$**:**

$$I_1 = \frac{E_T}{R_1}$$

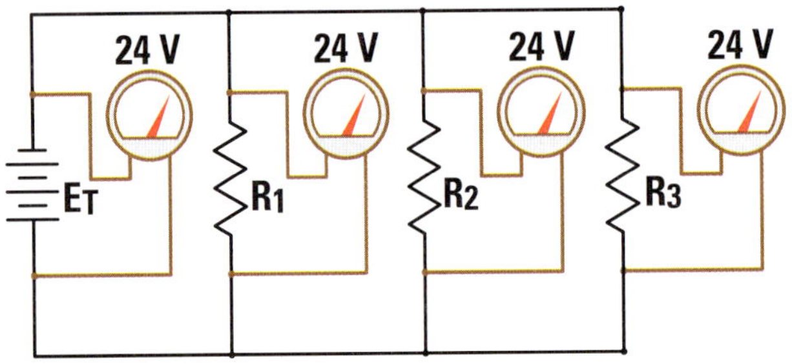

Figure 7–24 The voltage is the same across all branches of a parallel circuit.

It is also true that the total circuit voltage is equal to the product of the total circuit current and the total circuit resistance:

$$E_T = I_T \times R_T$$

If the value of E_T is substituted for ($I_T \times R_T$) in the previous formula, it becomes:

$$I_1 = \frac{I_T \times R_T}{R_1}$$

If the formula is rearranged and the values of I_1 and R_1 are substituted for I_X and R_X, it becomes what is generally known as *the current divider formula:*

$$I_X = \left(\frac{R_T}{R_X}\right) I_T$$

This formula can be used to calculate the current flow through any branch by substituting the values of I_X and R_X for the branch values when the total circuit current and the resistance are known. In the circuit shown in *Figure 7–25,* Resistor R_1 has a value of 1200 ohms, Resistor R_2 has a value of 300 ohms, and Resistor R_3 has a value of 120 ohms, producing a total resistance of 80 ohms for the circuit. It is assumed that a total current of 2 amperes flows in the circuit. The amount of current flow through Resistor R_1 can be found using the formula:

$$I_1 = \left(\frac{R_T}{R_1}\right)I_T$$

$$I_1 = \left(\frac{80\ \Omega}{1200\ \Omega}\right)2\ \text{A}$$

$$I_1 = 0.133\ \text{A}$$

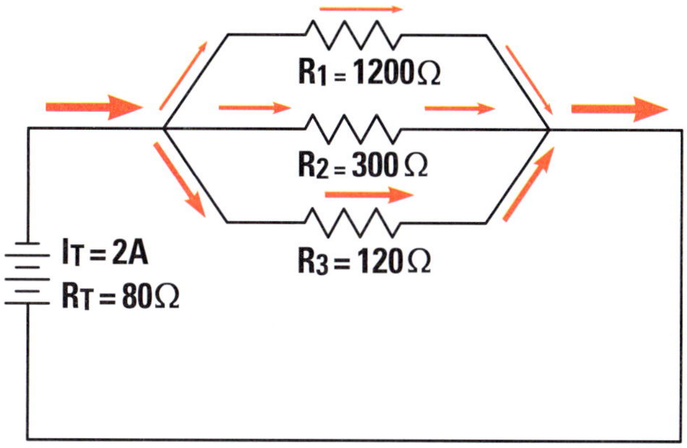

Figure 7–25 The current divides through each branch of a parallel circuit depending on the resistance in each branch.

The current flow through each of the other resistors can be found by substituting in the same formula:

$$I_2 = \left(\frac{R_T}{R_2}\right)I_T$$

$$I_2 = \left(\frac{80\ \Omega}{300\ \Omega}\right)2\ \text{A}$$

$$I_2 = 0.533\ \text{A}$$

$$I_3 = \left(\frac{R_T}{R_3}\right)I_T$$

$$I_3 = \left(\frac{80\ \Omega}{120\ \Omega}\right)2\ \text{A}$$

$$I_3 = 1.333\ \text{A}$$

Summary

- A parallel circuit is characterized by the fact that it has more than one path for current flow.
- The three rules for solving parallel circuits are as follows:
  a. The total current is the sum of the currents through all of the branches of the circuit.
  b. The voltage across any part of the circuit is the same as the total voltage.
  c. The total resistance is the reciprocal of the sum of the reciprocals of each individual branch.

- Circuits in homes are connected in parallel.
- The total power in a parallel circuit is equal to the sum of the power dissipation of all the components.
- The amount of current flow through each branch of a parallel circuit is inversely proportional to its resistance.

Review Questions

1. Which of the following best describes the major difference between a series and a parallel circuit?

   A) A series circuit has multiple resistors, and a parallel circuit only has one resistor.

   B) A parallel circuit has multiple paths for current to flow, and a series circuit only has a single path.

   C) A parallel circuit has a current that remains the same throughout the circuit as the total.

   D) A series circuit sees the total resistance decrease as resistors are added to the circuit.

2. True or False: Parallel circuits are the type of circuit most commonly used to supply power to the electrical equipment we use every day.

3. In a parallel circuit, what is the relationship between the voltage on each individual branch and the voltage applied to the circuit?

   A) The total applied voltage is equal to the sum of the voltage drops on each branch.

   B) The total applied voltage is always greater than the voltage drop across any single branch of a parallel circuit.

   C) The total applied voltage in a parallel circuit is equal to the voltage dropped across individual branches.

   D) Both the total applied voltage is equal to the sum of the voltage drops on each branch and the total applied voltage is always greater than the voltage drop across any single branch of a parallel circuit

4. True or False: Voltage in a parallel circuit will be different on each branch because the resistance on each branch varies.

5. In a parallel circuit, what happens to overall resistance when another branch is added to the circuit? Select the answer that best describes the result.

   A) Adding another branch to the circuit will result in an increase in overall resistance.

   B) Adding another branch to the circuit adds a resistor and therefore increases the resistance.

   C) Adding another resistor limits the amount of current that can flow, which means more resistance.

   D) Adding another branch allows for an additional path for current to take and therefore lowers the overall resistance in the circuit.

6. Total resistance in a parallel circuit can be determined by using one of three methods. All methods will result in a total resistance value that is less than the smallest resistor connected in parallel. Which of the following is not one of the three methods?

   A) The reciprocal method

   B) The additive method

   C) Product over sum method

   D) Resistors of equal value method

7. A parallel circuit has three branches. An ammeter is connected in series with the output of the power supply and indicates a total current flow of 3.0 A. If Branch 1 has a current flow of 0.9 A and Branch 2 also has a current flow of 0.9 A, what is the current flow through Branch 3?

   A) Branch three will also have 0.9 A.

   B) Branch three will measure 1.2 A.

   C) Branch three will measure 3.0 A.

   D) The current on the third branch depends on what the resistance is, and, therefore, there is not enough information to answer.

8. In a parallel circuit with four branches of equal resistors, how much of the total current will flow across each branch?

   A) Each branch will see 25% of the total current.

   B) Each branch will see 50% of the total current.

   C) Each branch will see 100% of the total current.

   D) Not enough information is provided.

9. Four resistors having values of 270 Ω, 300 Ω, 510 Ω, and 430 Ω are connected in parallel. If the current flowing through the 300 Ω resistor is measured at 0.40 A with an ammeter, what is the total amount of current flowing through the entire circuit? Round to two decimal places.

   A) 0.40 A

   B) 0.08 A

   C) 1.36 A

   D) 1.56 A

10. A circuit contains a 1200-Ω, a 2200-Ω, and a 3300-Ω resistor connected in parallel. The circuit has a total current flow of 0.19 A. Which of the following nominal voltage systems is most likely applied to the circuit? Select the best answer.

    A) 125 V

    B) 120 V

    C) 240 V

    D) 277 V

11. In a parallel circuit with only two branches, if a current value is known for one branch and the total circuit current is also known, which of the following methods would give you the missing current amount on the remaining branch? Select the best answer.

    A) $I_T = I_1 \times R_2$

    B) $I_2 = E_T \times I_1$

    C) $I_2 = E_1 - I_1$

    D) $I_2 = I_T - I_1$

12. True or False: All parallel circuits will act as current dividers.

Practical Applications

1. You have been hired by a homeowner to install a ceiling fan and light kit in a living room. The living room luminaire (light fixture) being used at present contains two 60-W lamps. After locating the circuit in the panelboard, you find that the circuit is protected by a 15-A circuit breaker and is run with 14 AWG copper wire. After turning on all lights connected to this circuit, you have a current draw of 8.6 A and a voltage of 120 V. The ceiling fan light kit contains four 13-W LED lamps, and the fan has a maximum current draw of 1.6 A. Can this fan be connected to the existing living room circuit?

---

2. You are employed in a large industrial plant. A 480-V, 5000-W heater is used to melt lead in a large tank. It has been decided that the heater is not sufficient to raise the temperature of the lead to the desired level. A second 5000-W heater is to be installed on the same circuit. What will be the circuit current after installation of the second heater? If the maximum a continuous load can draw for current on a branch circuit is 80% of the overcurrent protective device rating, what is the minimum standard-size circuit breaker that must be used for this circuit?

---

3. You are an electrician. You have been asked by a homeowner to install a lighted mirror in a bathroom. The mirror contains eight 40-watt lamps. Upon checking the service panel, you discover that the bathroom circuit is connected to a single 120-volt, 20-ampere circuit breaker. At present, the circuit supplies power to an electric wall heater rated at 1000 watts, a ceiling exhaust fan with a heat lamp feature, and a light fixture over the mirror. The fan motor has a full-load current draw of 3.2 amperes, and the light kit contains a 250-watt heat lamp. The light fixture presently installed over the mirror contains four 60-watt lamps. The homeowner asked whether the present light fixture over the mirror can be replaced by the lighted mirror. Assuming all loads are purely resistive, can the present circuit supply the power needed to operate all the loads without overloading the circuit? The loads on this circuit are not considered continuous.

---

4. A car lot uses incandescent lamps to supply outside lighting during the night. There are three strings of lamps connected to a single-pole 120-volt, 20-ampere branch circuit. Each string contains eight lamps. What is the largest standard lamp that can be used without overloading the circuit? Standard size lamps are 25 watts, 40 watts, 60 watts, 75 watts, and 100 watts.

Practice Problems

Parallel Circuits

Using the rules for parallel circuits and Ohm's law, solve for the missing values.

1.

E_T	E_1	E_2	E_3	E_4
I_T 0.942 A	I_1	I_2	I_3	I_4
R_T	R_1 680 Ω	R_2 820 Ω	R_3 470 Ω	R_4 330 Ω
P_T	P_1	P_2	P_3	P_4

2.

E_T	E_1	E_2	E_3	E_4
I_T 0.00639 A	I_1	I_2 0.00139 A	I_3 0.00154 A	I_4 0.00115 A
R_T	R_1	R_2	R_3	R_4
P_T	P_1 0.640 W	P_2	P_3	P_4

3.

E_T	E_1	E_2	E_3	E_4
I_T	I_1	I_2	I_3 3.2 A	I_4
R_T 3.582 Ω	R_1 16 Ω	R_2 10 Ω	R_3	R_4 20 Ω
P_T	P_1	P_2	P_3	P_4

4.

E_T	E_1	E_2	E_3	E_4
I_T	I_1	I_2	I_3	I_4
R_T	R_1 82 kΩ	R_2 75 kΩ	R_3 56 kΩ	R_4 62 kΩ
P_T 3.436 W	P_1	P_2	P_3	P_4

5. A parallel circuit contains the following resistor values:

   $R_1 = 360\ \Omega$ $R_2 = 470\ \Omega$ $R_3 = 300\ \Omega$

   $R_4 = 270\ \Omega$ $I_T = 0.05$ A

   Find the following missing values:

   R_T = ______________ Ω I_1 = ______________ A I_2 = ______________ A

   I_3 = ______________ A I_4 = ______________ A

6. A parallel circuit contains the following resistor values:

   $R_1 = 270K\ \Omega$ $R_2 = 360K\ \Omega$ $R_3 = 430K\ \Omega$

   $R_4 = 100K\ \Omega$ $I_T = 0.006$ A

   Find the following missing values:

   R_T = ______________ Ω I_1 = ______________ A I_2 = ______________ A

   I_3 = ______________ A I_4 = ______________ A

7. Find all missing values for the circuit shown in *Figure 7–26.*

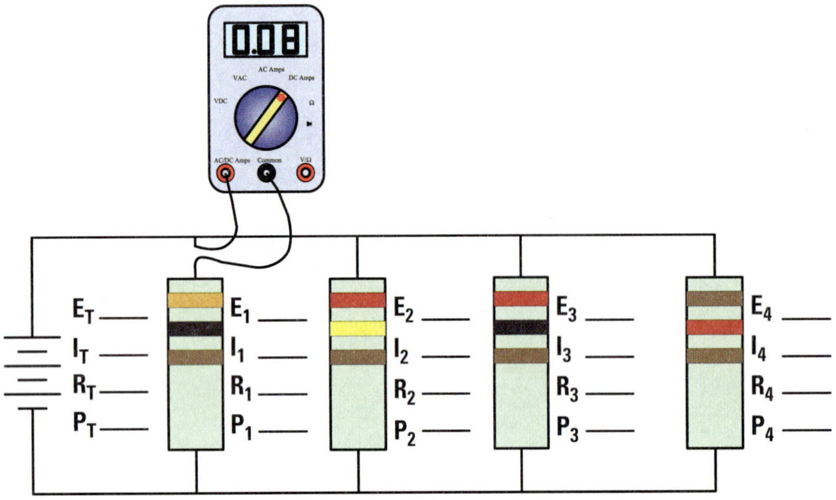

Figure 7–26 Determine all unknown values in this parallel circuit.

8. Find all missing values for the circuit shown in *Figure 7–27*.

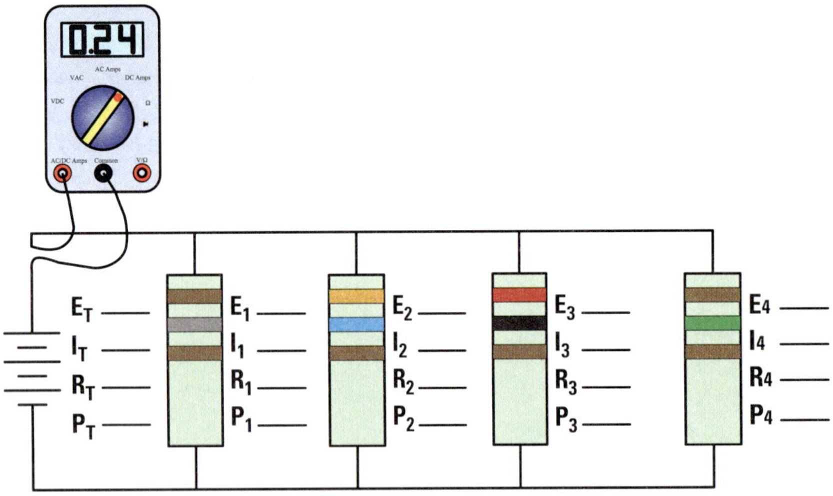

Figure 7–27 Determine all unknown values in this parallel circuit.

Unit 8 Combination Circuits

Key Terms

Combination circuit

Node

Parallel block

Redraw

Reduce

Simple parallel circuit

Trace the current path

Why You Need to Know

Understanding how values of voltage, current, and resistance relate to each other in combination circuits is essential to understanding how all electric circuits work. Although it is not a common practice for electricians on the job to use a calculator to determine values of voltage and current, they must have an understanding of how these electrical values relate to each other in different kinds of circuits. An understanding of Ohm's law and basic circuits is also essential for being able to troubleshoot problems that occur in an electric circuit. In this unit

- the rules for series circuits are employed for those components connected in series, and parallel rules are used for components connected in parallel.
- you are shown how to reduce a complex combination circuit to a simple series or parallel circuit and then apply circuit rules and Ohm's law to determine the values of voltage, current, and power for the different components.

Outline

8–1 Combination Circuits

8–2 Solving Combination Circuits

8–3 Simplifying the Circuit

Objectives

After studying this unit, you should be able to

- explain how current flows in a combination circuit.
- simplify combination circuits using the rules for series and parallel circuits.
- determine the missing electrical values for each component of a combination circuit using Ohm's law.

Preview

Combination circuits contain a combination of both series and parallel elements. To determine which components are in parallel and which are in series, trace the flow of current through the circuit. Remember that a series circuit has only one path for current flow, and a parallel circuit has more than one path for current flow.

8–1 Combination Circuits

A simple combination circuit is shown in *Figure 8–1.* To identify the series and parallel elements, **trace the current path.** All the current in the circuit must flow through the first resistor, R_1, and R_1 is therefore in series with the rest of the circuit. When the current reaches the junction point of R_2 and R_3, however, it splits. A junction point such as this can be referred to as a **node.** Part of the current flows through R_2, and part flows through R_3. These two resistors are in parallel. Because this circuit contains both series and parallel elements, it is a **combination circuit.**

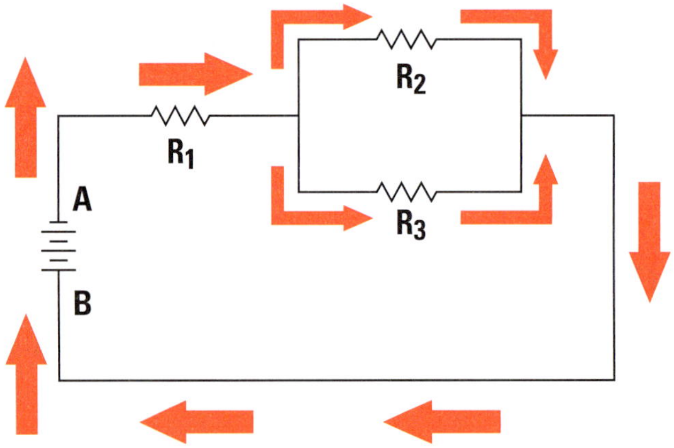

Figure 8–1 A simple combination circuit.

8–2 Solving Combination Circuits

The circuit shown in *Figure 8–2* contains four resistors with values of 325, 275, 150, and 250 ohms. The circuit has a total current flow of 1 ampere. To determine which resistors are in series and which are in parallel, trace the path for current flow through the circuit. When the path of current flow is traced, it can be seen that current can flow by two separate paths from the negative terminal to the positive terminal. One path is through R_1 and R_2, and the other path is through R_3 and R_4. These two paths are parallel. However, the same current will flow through both R_1 and R_2. So, these two resistors are in series with each other. The same is true for R_3 and R_4.

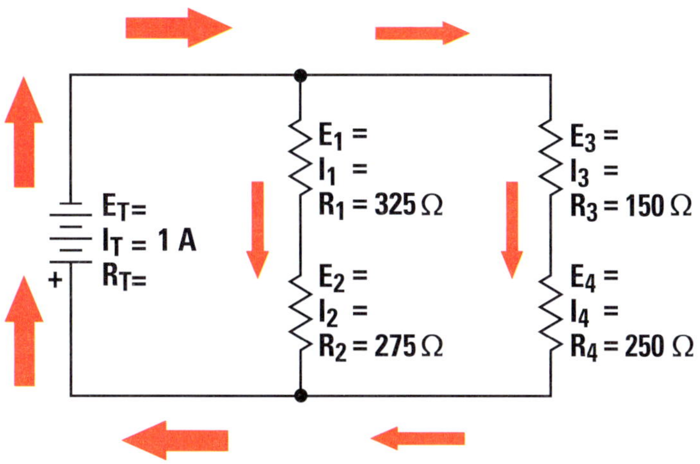

Figure 8–2 Tracing the current paths through a combination circuit.

To solve the unknown values in a combination circuit, use series circuit rules for those sections of the circuit that are connected in series and parallel circuit rules for those sections connected in parallel. The circuit rules are as follows:

Series Circuits

1. The current is the same at any point in the circuit.
2. The total resistance is the sum of the individual resistances.
3. The applied voltage is equal to the sum of the voltage drops across all the resistors.

Parallel Circuits

1. The voltage drop across any branch of a parallel circuit is the same as the applied voltage.
2. The total current flow is equal to the sum of the currents through all of the circuit branches.
3. The total resistance is equal to the reciprocal of the sum of the reciprocals of the branch resistances.

8–3 Simplifying the Circuit

The circuit shown in *Figure 8–2* can be reduced or simplified to a **simple parallel circuit** (*Figure 8–3*). Because R_1 and R_2 are connected in series, their values can be added to form one equivalent resistor,

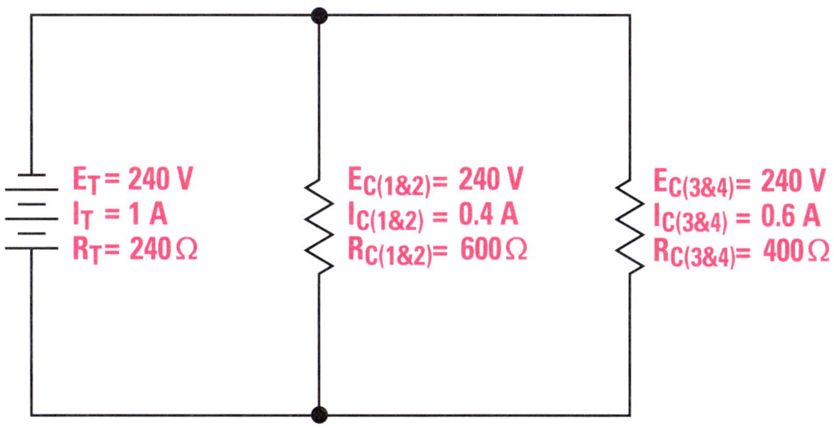

Figure 8–3 Simplifying the combination circuit.

called the combined resistor $R_{C(1\&2)}$. The same is true for R_3 and R_4. Their values are added to form $R_{C(3\&4)}$. Now that the circuit has been reduced to a simple parallel circuit, the total resistance can be found:

$$R_T = \frac{1}{\frac{1}{R_{C(1\&2)}} + \frac{1}{R_{C(3\&4)}}}$$

$$R_T = \frac{1}{\frac{1}{600\ \Omega} + \frac{1}{400\ \Omega}}$$

$$R_T = \frac{1}{(0.0016667 + 0.0025)\frac{1}{\Omega}}$$

$$R_T = \frac{1}{(0.0041667)\frac{1}{\Omega}}$$

$$R_T = 240\ \Omega$$

Now that the total resistance has been found, the other circuit values can be calculated. The applied voltage can be found using Ohm's law:

$$E_T = I_T \times R_T$$

$$E_T = 1\ A \times 240\ \Omega$$

$$E_T = 240\ V$$

One of the rules for parallel circuits states that the voltage is the same across each branch of the circuit. For this reason, the voltage drops across $R_{c(1\&2)}$ and $R_{c(3\&4)}$ are the same. Because the voltage drop and the resistance are known, Ohm's law can be used to find the current flow through each branch:

$$I_{C(1\&2)} = \frac{E_{C(1\&2)}}{R_{C(1\&2)}}$$

$$I_{C(1\&2)} = \frac{240\ V}{600\ \Omega}$$

$$I_{C(1\&2)} = 0.4\ A$$

$$I_{C(3\&4)} = \frac{E_{C(3\&4)}}{R_{C(3\&4)}}$$

$$I_{C(3\&4)} = \frac{240\ V}{400\ \Omega}$$

$$I_{C(3\&4)} = 0.6\ A$$

These values can now be used to solve the missing values in the original circuit. The values of voltage and current that apply to $R_{c(1\&2)}$ will apply to R_1 and R_2. Since R_1 and R_2 are connected in series, the current is the same through both resistors. Since 0.4 ampere of current flows through both R_1 and R_2, the voltage drop across each can be calculated using Ohm's law:

$$E_1 = I_1 \times R_1$$

$$E_1 = 0.4\ A \times 325\ \Omega$$

$$E_1 = 130\ A$$

$$E_2 = I_2 \times R_2$$

$$E_2 = 0.4\ \text{A} \times 275\ \Omega$$

$$E_2 = 110\ \text{V}$$

These values of voltage and current can now be added to the circuit in *Figure 8–2* to produce the circuit shown in *Figure 8–4.*

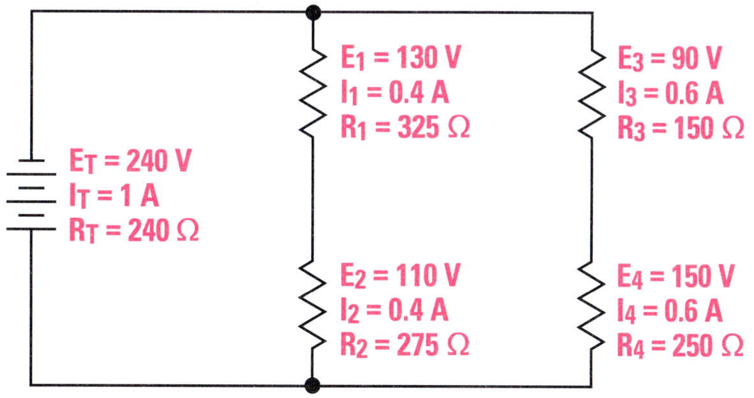

Figure 8–4 All the missing values for the combination circuit.

Because R_3 and R_4 are placed in the same configuration as R_1 and R_2, the process for determining these values is the same. The voltage across these two resistors can now be calculated using Ohm's law:

$$E_3 = I_3 \times R_3$$

$$E_3 = 0.6\ \text{A} \times 150\ \Omega$$

$$E_3 = 90\ \text{V}$$

$$E_4 = I_4 \times R_4$$

$$E_4 = 0.6\ \text{A} \times 250\ \Omega$$

$$E_4 = 150\ \text{V}$$

Examples 8-1 and 8-2 demonstrate the process for simplifying combination circuits to solve for unknown values.

Example 8–1

Solve the combination circuit shown in *Figure 8–5*.

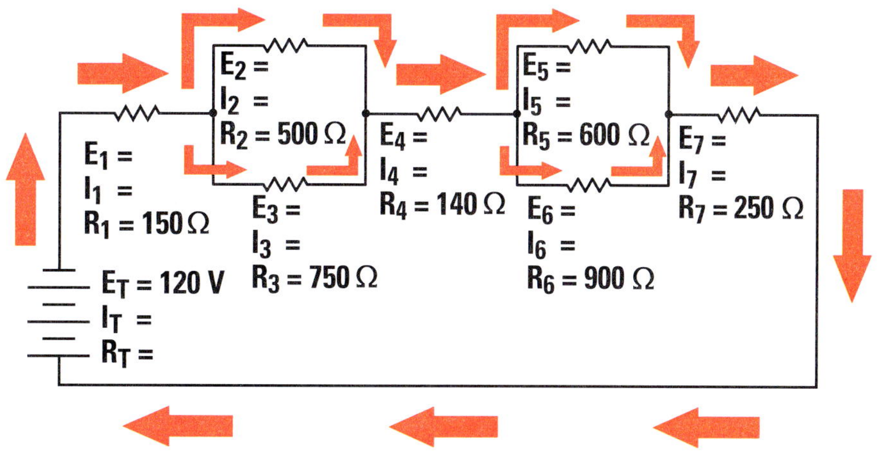

Figure 8–5 Tracing the flow of current through the combination circuit.

Example 8–1 (continued)

Solution

The first step in finding the missing values is to trace the current path through the circuit to determine which resistors are in series and which are in parallel. All the current must flow through R_1. R_1 is, therefore, in series with the rest of the circuit. When the current reaches the junction of R_2 and R_3, it divides, and part flows through each resistor. R_2 and R_3 are in parallel. All the current must then flow through R_4, which is connected in series, to the junction of R_5 and R_6. The current path is divided between these two resistors, and R_5 and R_6 are connected in parallel. All the circuit current must then flow through R_7.

The next step in solving this circuit is to **reduce** it to a simpler circuit. If the total resistance of the first **parallel block** formed by R_2 and R_3 is found, this block can be replaced by a single resistor:

$$R_{(2\&3)} = \frac{1}{\frac{1}{R_2} + \frac{1}{R_3}}$$

$$R_{(2\&3)} = \frac{1}{\frac{1}{500\ \Omega} + \frac{1}{750\ \Omega}}$$

$$R_{(2\&3)} = \frac{1}{(0.002 + 0.0013333)\frac{1}{\Omega}}$$

$$R_{(2\&3)} = \frac{1}{(0.0033333)\frac{1}{\Omega}}$$

$$R_{(2\&3)} = 300\ \Omega$$

The equivalent resistance of the second parallel block can be calculated in the same way:

$$R_{(5\&6)} = \frac{1}{\frac{1}{R_5} + \frac{1}{R_6}}$$

$$R_{(5\&6)} = \frac{1}{\frac{1}{600\ \Omega} + \frac{1}{900\ \Omega}}$$

$$R_{(5\&6)} = \frac{1}{(0.0016667 + 0.0011111)\frac{1}{\Omega}}$$

$$R_{(5\&6)} = \frac{1}{(0.00277778)\frac{1}{\Omega}}$$

$$R_{(5\&6)} = 360\ \Omega$$

Now that the total resistance of the second parallel block is known, you can **redraw** the circuit as a simple series circuit as shown in *Figure 8–6*. The first parallel block has been replaced with a single resistor of 300 Ω labeled $R_{c(2\&3)}$, and the second parallel block has been replaced with a single 360-Ω resistor labeled $R_{c(5\&6)}$. Ohm's law can be used to find the missing values in this series circuit.

One of the rules for series circuits states that the total resistance of a series circuit is equal to the sum of the individual resistances. R_T can be calculated by adding the resistances of all the resistors:

$$R_T = R_1 + R_{c(2\&3)} + R_4 + R_{c(5\&6)} + R_7$$

$$R_T = 150\ \Omega + 300\ \Omega + 140\ \Omega + 360\ \Omega + 250\ \Omega$$

$$R_T = 1200\ \Omega$$

Example 8–1 (continued)

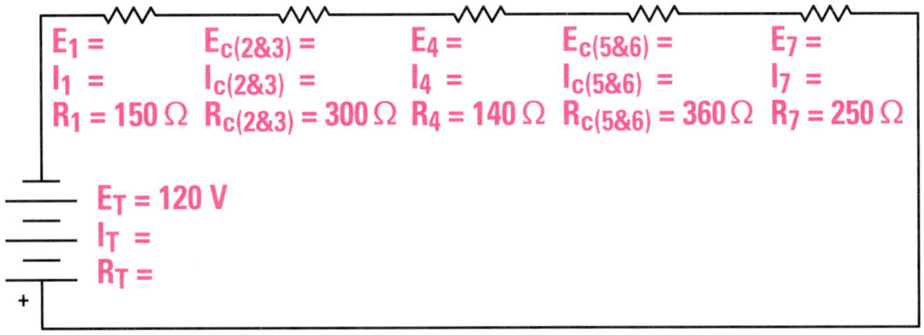

Figure 8–6 Simplifying the combination circuit makes the missing values easier to determine.

Because the total voltage and total resistance are known, the total current flow through the circuit can be calculated:

$$I_T = \frac{E_T}{R_T}$$

$$I_T = \frac{120\text{ V}}{1200\ \Omega}$$

$$I_T = 0.1\text{ A}$$

The first rule of series circuits states that the current is the same at any point in the circuit. The current flow through each resistor is therefore 0.1 A. The voltage drop across each resistor can now be calculated using Ohm's law:

$$E_1 = I_1 \times R_1$$

$$E_1 = 0.1\text{ A} \times 150\ \Omega$$

$$E_1 = 15\text{ V}$$

$$E_{c(2\&3)} = I_{c(2\&3)} \times R_{c(2\&3)}$$

$$E_{c(2\&3)} = 0.1\text{ A} \times 300\ \Omega$$

$$E_{c(2\&3)} = 30\text{ V}$$

$$E_4 = I_4 \times R_4$$

$$E_4 = 0.1\text{ A} \times 140\ \Omega$$

$$E_4 = 14\text{ V}$$

$$E_{c(5\&6)} = I_{c(5\&6)} \times R_{c(5\&6)}$$

$$E_{c(5\&6)} = 0.1\text{ A} \times 360\ \Omega$$

$$E_{c(5\&6)} = 36\text{ V}$$

$$E_7 = I_7 \times R_7$$

$$E_7 = 0.1\text{ A} \times 250\ \Omega$$

$$E_7 = 25\text{ V}$$

The series circuit with all solved values is shown in *Figure 8–7*. These values can now be used to solve missing parts in the original circuit.

Example 8–1 (continued)

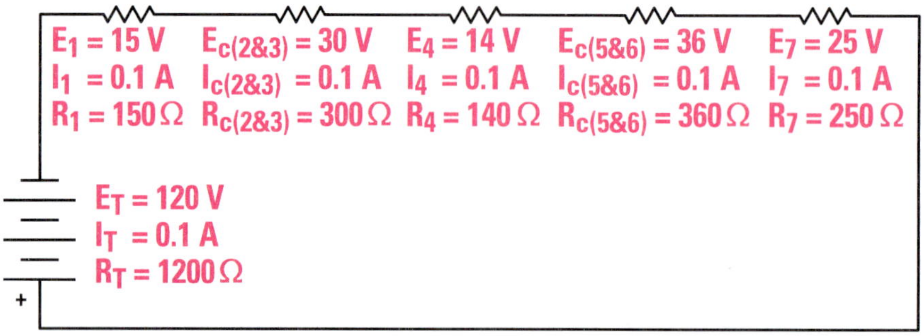

Figure 8–7 The simplified circuit with all values solved.

$R_{c(2\&3)}$ is actually the parallel block containing R_2 and R_3. The values for $R_{c(2\&3)}$, therefore, apply to this parallel block. One of the rules for a parallel circuit states that the voltage drop of a parallel circuit is the same at any point in the circuit. Because 30 V are dropped across $R_{c(2\&3)}$, the same 30 V are dropped across R_2 and R_3 (*Figure 8–8*). The current flow through these resistors can now be calculated using Ohm's law:

$$I_2 = \frac{E_2}{R_2}$$

$$I_2 = \frac{30\text{ V}}{500\ \Omega}$$

$$I_2 = 0.06\text{ A}$$

$$I_3 = \frac{E_3}{R_3}$$

$$I_3 = \frac{30\text{ V}}{750\ \Omega}$$

$$I_3 = 0.04\text{ A}$$

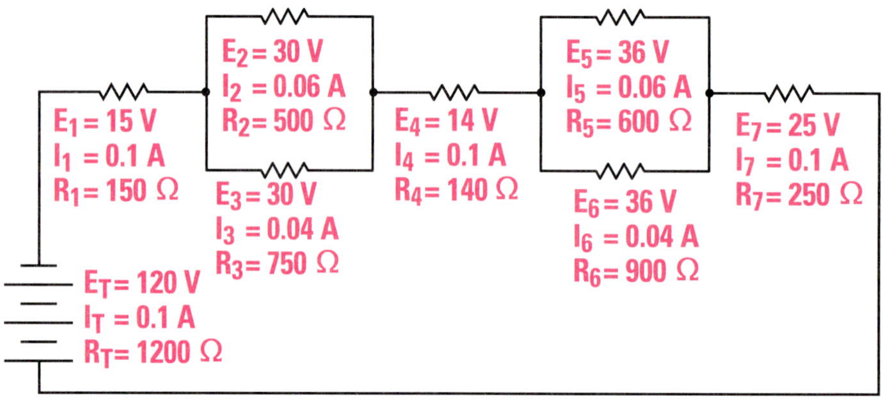

Figure 8–8 All values solved for the combination circuit.

Example 8–1 (continued)

The values of $R_{c(5\&6)}$ can be applied to the parallel block composed of R_5 and R_6. $E_{c(5\&6)}$ is 36 volts. This is the voltage drop across R_5 and R_6. The current flow through these two resistors can be calculated using Ohm's law:

$$I_5 = \frac{E_5}{R_5}$$

$$I_5 = \frac{36\ V}{600\ \Omega}$$

$$I_5 = 0.06\ A$$

$$I_6 = \frac{E_6}{R_6}$$

$$I_6 = \frac{36\ V}{900\ \Omega}$$

$$I_6 = 0.04\ A$$

Example 8–2

Both of the preceding circuits were solved by first determining which parts of the circuit were in series and which were in parallel. The circuits were then reduced to a simple series or parallel circuit. This same procedure can be used for any combination circuit. The circuit shown in *Figure 8–9* must be reduced to a simpler circuit first. Once the values of the simple circuit are found, they can be placed back in the original circuit to find other values.

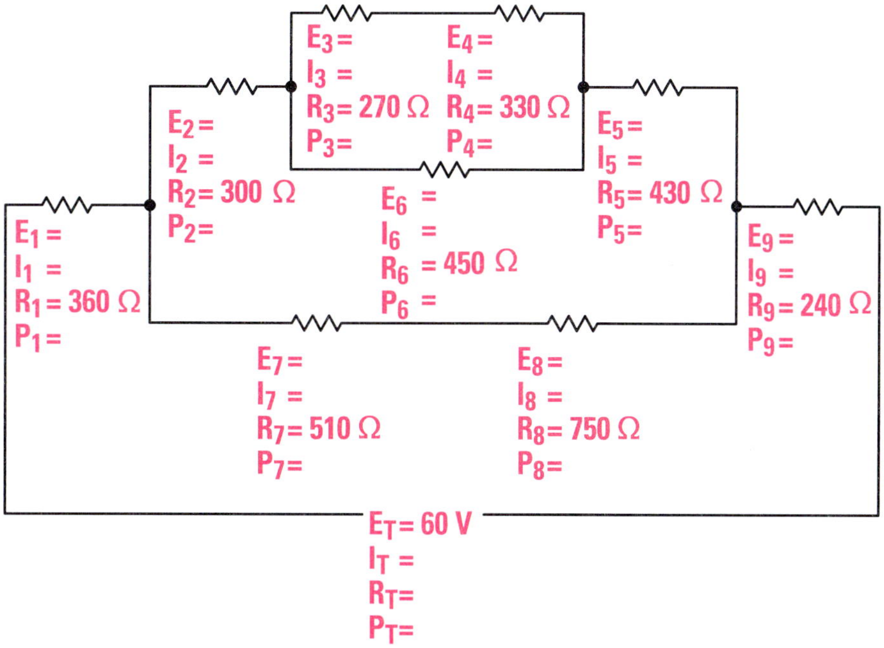

Figure 8–9 A complex combination circuit.

Example 8–2 (continued)

Solution

The first step is to reduce the top part of the circuit to a single resistor. This part consists of R_3 and R_4. Because these two resistors are connected in series with each other, their values can be added to form one single resistor. This combination will form R_{c1} (*Figure 8–10*):

$$R_{c1} = R_3 + R_4$$

$$R_{c1} = 270\ \Omega + 330\ \Omega$$

$$R_{c1} = 600\ \Omega$$

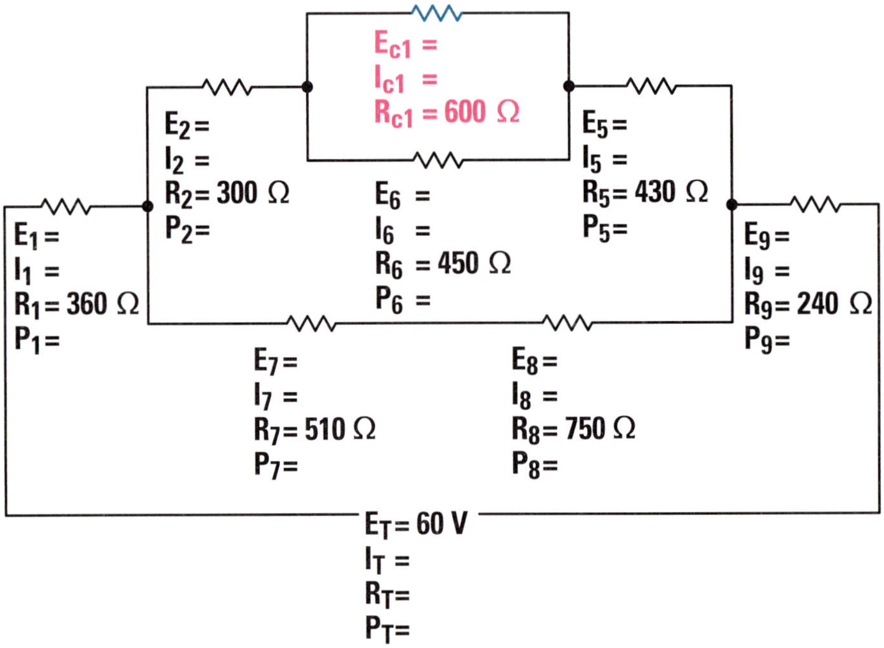

Figure 8–10 R_1 and R_2 are combined to form R_{c1}.

The top part of the circuit is now formed by R_{c1} and R_6. These two resistors are in parallel with each other. If their total resistance is calculated, they can be changed into one single resistor with a value of 257.143 Ω. This combination becomes R_{c2} (*Figure 8–11*):

$$R_T = \frac{1}{\frac{1}{600\ \Omega} + \frac{1}{450\ \Omega}}$$

$$R_T = 257.143\ \Omega$$

The top of the circuit now consists of R_2, R_{c2}, and R_5. These three resistors are connected in series with each other. They can be combined to form R_{c3} by adding their resistances together (*Figure 8–12*):

$$R_{c3} = R_2 + R_{c2} + R_5$$

$$R_{c3} = 300\ \Omega + 257.143\ \Omega + 430\ \Omega$$

$$R_{c3} = 987.143\ \Omega$$

Example 8–2 (continued)

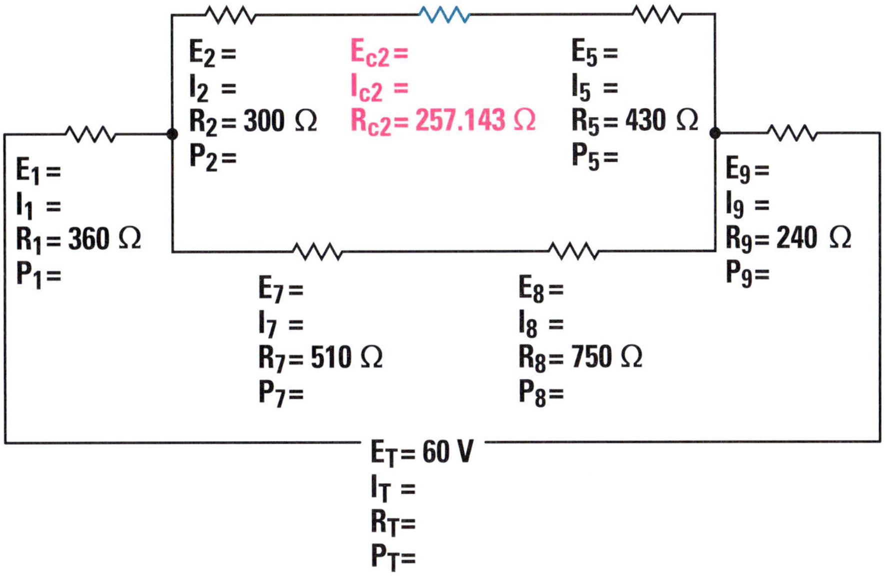

Figure 8–11 R_{c1} and R_6 are combined to form R_{c2}.

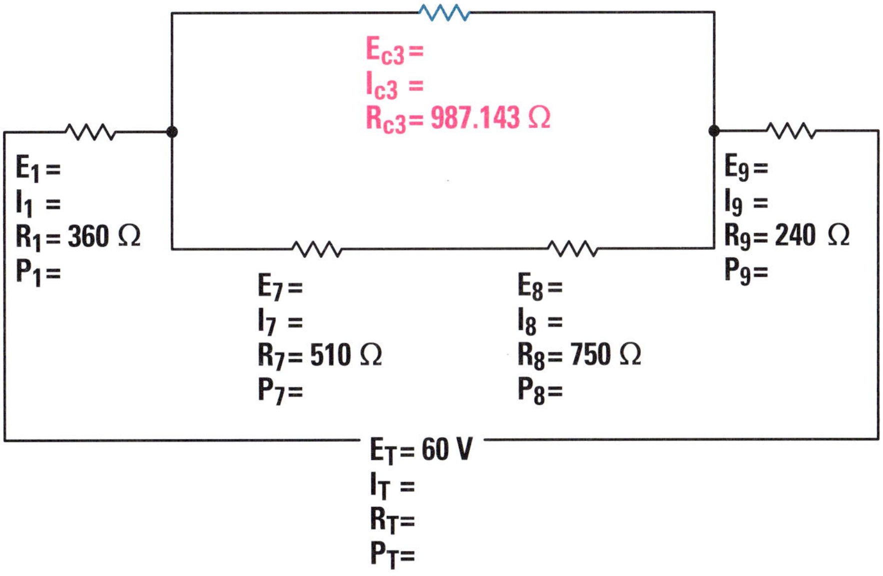

Figure 8–12 R_2, R_{c2}, and R_5 are combined to form R_{c3}.

R_7 and R_8 are also connected in series with each other. These two resistors are added to form R_{c4} (*Figure 8–13*):

$$R_{c4} = R_7 + R_8$$

$$R_{c4} = 510\ \Omega + 750\ \Omega$$

$$R_{c4} = 1260\ \Omega$$

Example 8–2 (continued)

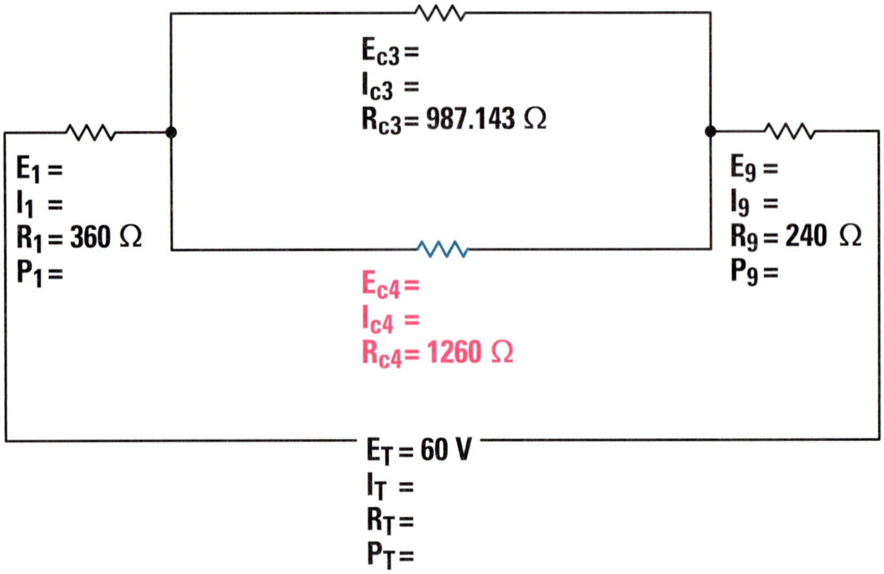

Figure 8–13 R_7 and R_8 are combined to form R_{c4}.

R_{c3} and R_{c4} are connected in parallel with each other. The total resistance can be calculated to form R_{c5} (*Figure 8–14*):

$$R_{c5} = \frac{1}{\frac{1}{987.143\ \Omega} + \frac{1}{1260\ \Omega}}$$

$$R_{c5} = 553.503\ \Omega$$

The circuit has now been reduced to a simple series circuit containing three resistors. The total resistance of the circuit can be calculated by adding R_1, R_{c5}, and R_9:

$$R_T = R_1 + R_{c5} + R_9$$

$$R_T = 360\ \Omega + 553.503\ \Omega + 240\ \Omega$$

$$R_T = 1153.503\ \Omega$$

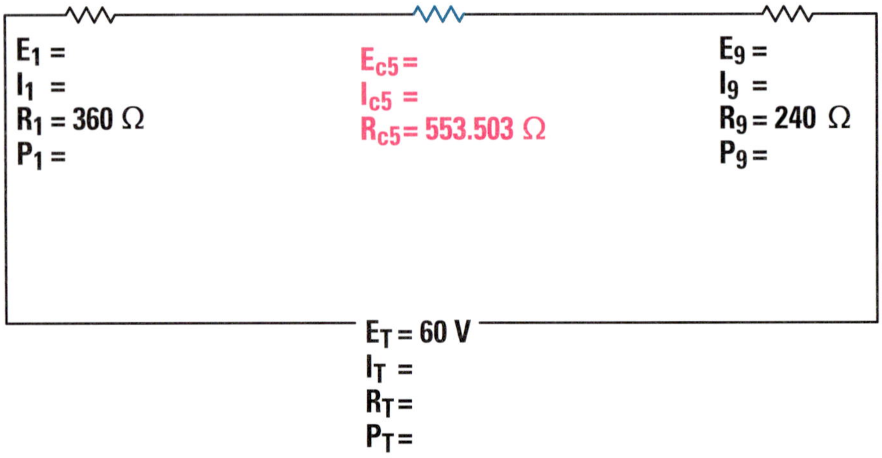

Figure 8–14 R_{c3} and R_{c4} are combined to form R_{c5}.

Example 8–2 (continued)

Now that the total resistance and total voltage are known, the total circuit current and total circuit power can be calculated using Ohm's law:

$$I_T = \frac{E_T}{R_T}$$

$$I_T = \frac{60\text{ V}}{1153.503\ \Omega}$$

$$I_T = 0.052\text{ A}$$

$$P_T = E_T \times I_T$$

$$P_T = 60\text{ V} + 0.052\text{ A}$$

$$P_T = 3.12\text{ W}$$

Ohm's law can now be used to find the missing values for R_1, R_{c5}, and R_9 (*Figure 8–15*):

$$E_1 = I_1 \times R_1$$

$$E_1 = 0.052\text{ A} \times 360\ \Omega$$

$$E_1 = 18.72\text{ V}$$

$$P_1 = E_1 \times I_1$$

$$P_1 = 18.72\text{ V} \times 0.052\text{ A}$$

$$P_1 = 0.973\text{ W}$$

$$E_{c5} = I_{c5} \times R_{c5}$$

$$E_{c5} = 0.052\text{ A} \times 553.503\ \Omega$$

$$E_{c5} = 28.782\text{ V}$$

$$E_9 = I_9 \times R_9$$

$$E_9 = 0.052\text{ A} \times 240\ \Omega$$

$$E_9 = 12.48\text{ V}$$

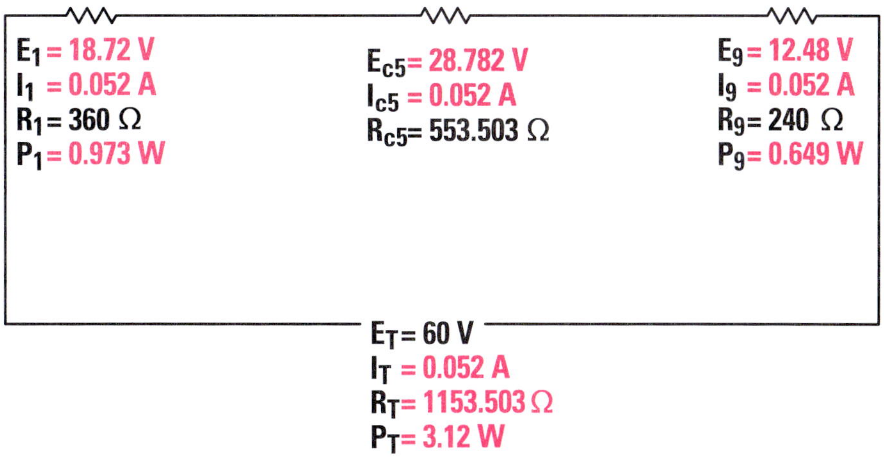

Figure 8–15 Missing values are found for the first part of the circuit.

Example 8–2 (continued)

$$P_9 = E_9 \times I_9$$

$$P_9 = 12.48 \text{ V} \times 0.052 \text{ A}$$

$$P_9 = 0.649 \text{ W}$$

Remember that R_{c5} is actually the combination of R_{c3} and R_{c4} (Figure 8–13). The values of R_{c5} therefore apply to R_{c3} and R_{c4}. Since these two resistors are connected in parallel with each other, the voltage drop across them will be the same. Each will have the same voltage drop as R_{c5} (*Figure 8–16*). Ohm's law can now be used to find the remaining values of R_{c3} and R_{c4}:

$$I_{c4} = \frac{E_{c4}}{R_{c4}}$$

$$I_{c4} = \frac{28.782 \text{ V}}{1260 \ \Omega}$$

$$I_{c4} = 0.0228 \text{ A}$$

$$I_{c3} = \frac{E_{c3}}{R_{c3}}$$

$$I_{c3} = \frac{28.782 \text{ V}}{987.143 \ \Omega}$$

$$I_{c3} = 0.0292 \text{ A}$$

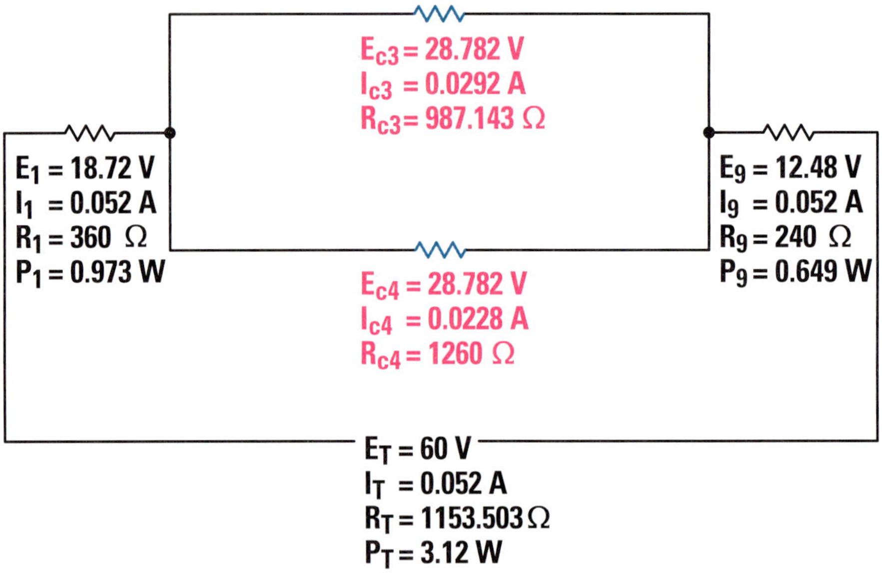

Figure 8–16 The values for R_{c3} and R_{c4}.

Also, R_{c4} is the combination of R_7 and R_8, and the values of R_{c4} apply to R_7 and R_8. Because R_7 and R_8 are connected in series with each other, the current flow from R_{c4} will be the same through both (*Figure 8–17*). Ohm's law can now be used to calculate the remaining values for these two resistors:

$$E_7 = I_7 \times R_7$$

$$E_7 = 0.0228 \text{ A} \times 510 \ \Omega$$

$$E_7 = 11.628 \text{ V}$$

$$P_7 = E_7 \times I_7$$

Example 8–2 (continued)

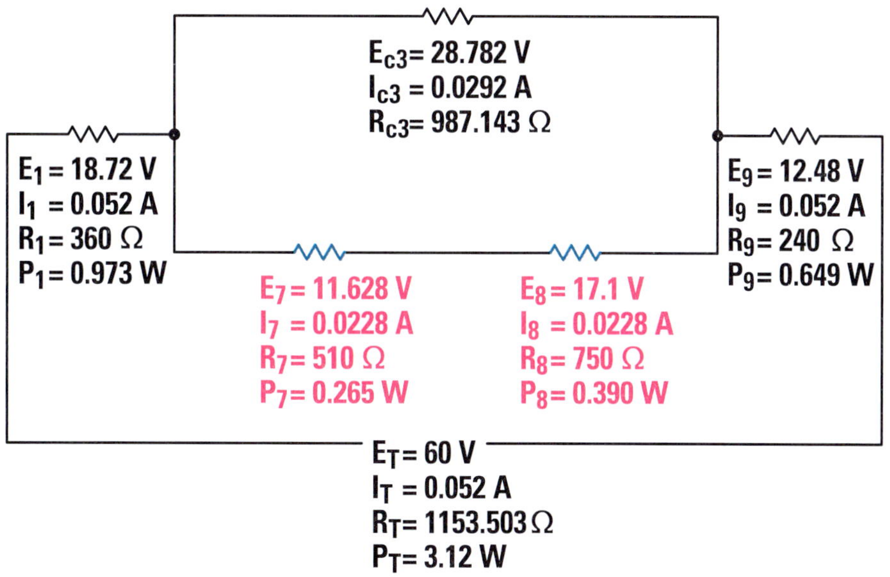

Figure 8–17 The values for R_7 and R_8.

$$P_7 = 11.628\text{ V} \times 0.0228\text{ A}$$

$$P_7 = 0.265\text{ W}$$

$$E_8 = I_8 \times R_8$$

$$E_8 = 0.0228\text{ A} \times 750\ \Omega$$

$$E_8 = 17.1\text{ V}$$

$$P_8 = E_8 \times I_8$$

$$P_8 = 17.1\text{ V} \times 0.0228\text{ A}$$

$$P_8 = 0.390\text{ W}$$

R_{c3} is the combination of R_2, R_{c2}, and R_5. Because these resistors are connected in series with each other, the current flow through each will be the same as the current flow through R_{c3}. The remaining values can now be calculated using Ohm's law (*Figure 8–18*):

$$E_{c2} = I_{c2} \times R_{c2}$$

$$E_{c2} = 0.0292\text{ A} \times 257.143\ \Omega$$

$$E_{c2} = 7.509\text{ V}$$

$$E_2 = I_2 \times R_2$$

$$E_2 = 0.0292\text{ A} \times 300\ \Omega$$

$$E_2 = 8.76\text{ V}$$

$$P_2 = E_2 \times I_2$$

$$P_2 = 8.76\text{ A} \times 0.292\text{ A}$$

$$P_2 = 0.256\text{ W}$$

Example 8–2 (continued)

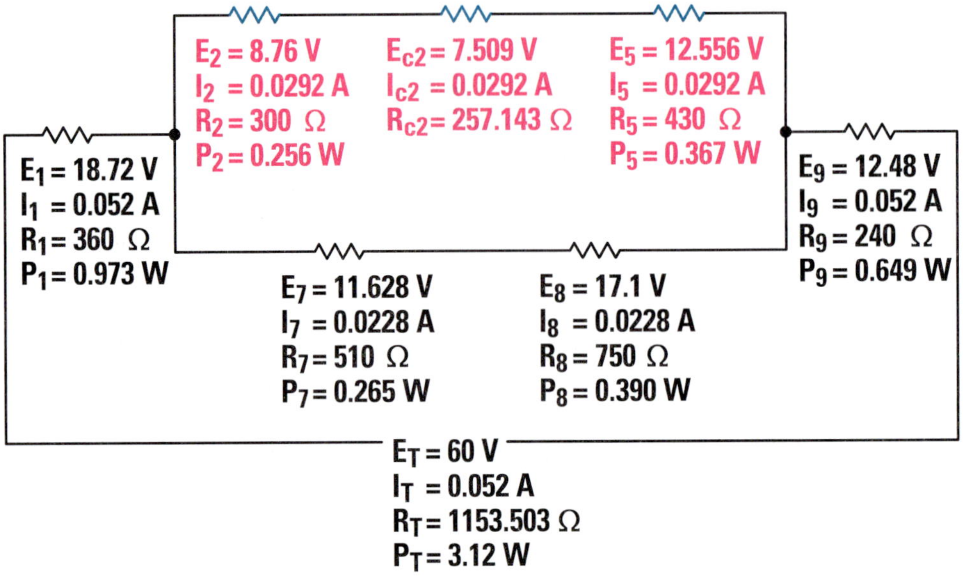

Figure 8–18 Determining values for R_1, R_{c2}, and R_5.

$$E_5 = I_5 \times R_5$$

$$E_5 = 0.0292\ A \times 430\ \Omega$$

$$E_5 = 12.556\ V$$

$$P_5 = E_5 \times I_5$$

$$P_5 = 12.556\ V \times 0.0292\ A$$

$$P_5 = 0.367\ W$$

R_{c2} is the combination of R_{c1} and R_6. R_{c1} and R_6 are connected in parallel and will therefore have the same voltage drop as R_{c2}. Ohm's law can be used to calculate the remaining values for R_{c2} and R_6 (*Figure 8–19*):

$$I_{c1} = \frac{E_{c1}}{R_{c1}}$$

$$I_{c1} = \frac{7.509\ V}{600\ \Omega}$$

$$I_{c1} = 0.0125\ A$$

$$I_6 = \frac{E_6}{R_6}$$

$$I_6 = \frac{7.509\ V}{450\ \Omega}$$

$$I_6 = 0.0167\ A$$

$$P_6 = E_6 \times I_6$$

$$P_6 = 7.509\ V \times 0.0167\ A$$

$$P_6 = 0.125\ W$$

Example 8–2 (continued)

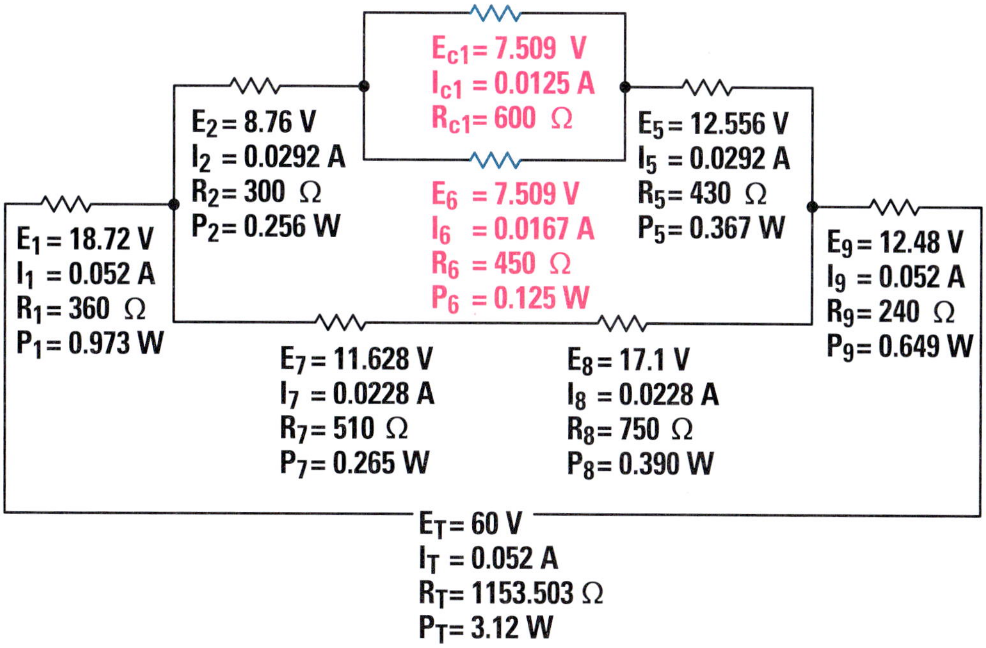

Figure 8–19 The values of R_{c1} and R_6.

R_{c1} is the combination of R_3 and R_4. Because these two resistors are connected in series, the amount of current flow through R_{c1} will be the same as the flow through R_3 and R_4. The remaining values of the circuit can now be found using Ohm's law (*Figure 8–20*):

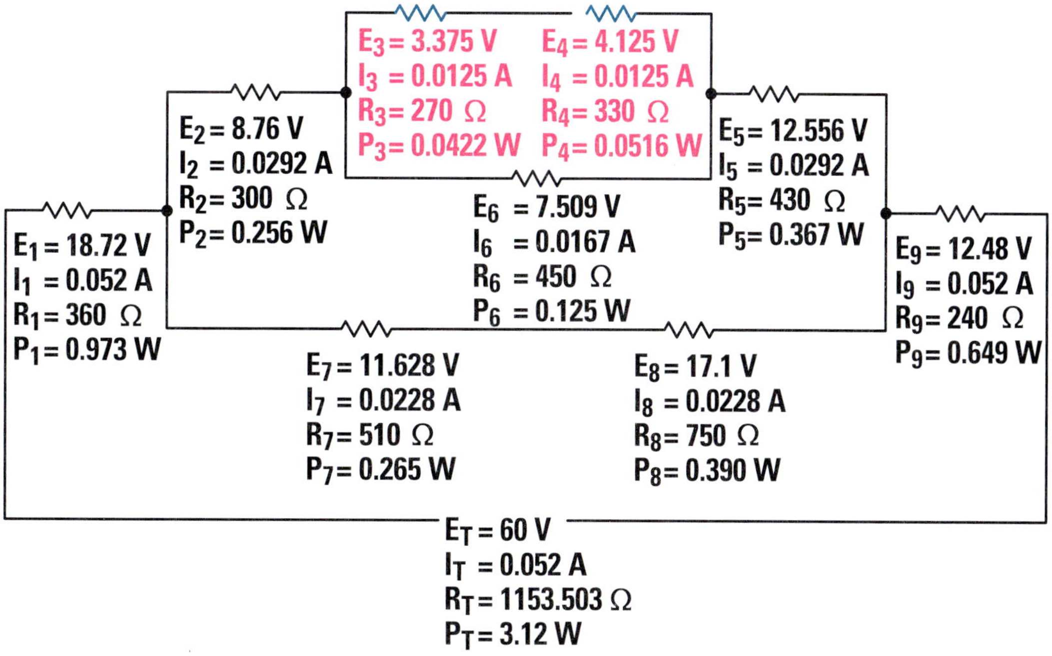

Figure 8–20 The values for R_3 and R_4.

Example 8–2 (continued)

$$E_3 = I_3 \times R_3$$
$$E_3 = 0.0125\ A \times 270\ \Omega$$
$$E_3 = 3.375\ V$$
$$P_3 = E_3 \times I_3$$
$$P_3 = 3.375\ V \times 0.0125\ A$$
$$P_3 = 0.0422\ W$$
$$E_4 = I_4 \times R_4$$
$$E_4 = 0.0125\ A \times 330\ \Omega$$
$$E_4 = 4.125\ V$$
$$P_4 = E_4 \times I_4$$
$$P_4 = 4.125\ V \times 0.0125\ A$$
$$P_4 = 0.0516\ W$$

Summary

- Combination circuits contain both series and parallel branches.
- The three rules for series circuits are as follows.
  a. The current is the same at any point in the circuit.
  b. The total resistance is the sum of the individual resistances.
  c. The applied voltage is equal to the sum of the voltage drops across all the resistors.
- The three rules for parallel circuits are as follows.
  a. The voltage drop across any branch of a parallel circuit is the same as the applied voltage.
  b. The total current is the sum of the individual currents through each path in the circuit (current adds).
  c. The total resistance is the reciprocal of the sum of the reciprocals of the branch resistances.
- When solving combination circuits, it often requires the circuit to be reduced to simpler circuits.

Review Questions

1. Refer to *Figure 8–21* to select the correct answer. If all four resistors are of equal resistance, what percentage of the total circuit current will flow through R_3?

   A) 100%

   B) 75%

   C) 50%

   D) 33%

2. True or False: In a combination circuit, resistors and conductors that are in series with the power supply will see the full amount of circuit current flow through them.
3. When tracing out a circuit to follow the potential paths for current to take, if the current can split between two resistors, those resistors are in __________, and if all current must pass through both resistors, those resistors are in __________.

   A) series, parallel

   B) parallel, series

   C) parallel, parallel

   D) series, series
4. Referring to *Figure 8–9*, how much of the total circuit current will flow through resistor R_9 as a percentage of the total current flow in the circuit?

   A) 100%

   B) 75%

   C) 50%

   D) 33%
5. True or False: When solving for the missing values in a series-parallel combination circuit, the first step is to simplify the circuit into either a purely series circuit or a purely parallel circuit by combining individual elements throughout the circuit.
6. Referring to *Figure 8–5*, when simplifying this circuit, which of the following methods will best assist in redrawing this circuit in the simplest form?

   A) Additive voltage drops across series resistors

   B) Additive currents in parallel branches

   C) Product over sum method for parallel resistance

   D) Additive resistance in a series circuit
7. In a combination circuit, if the first resistor in the circuit encountered when tracing current from the negative terminal of the battery is in series, the simplified circuit will be which of the following?

   A) A parallel circuit

   B) A series circuit

   C) A combination circuit

   D) Need more information
8. When simplifying parallel sections of a combination circuit, what do we know about the voltage in the section?

   A) Voltage drops in a parallel section are equal.

   B) Voltage drops in a parallel section are additive.

   C) Voltage drops in a parallel section are inversely added.

   D) Voltage drops in a parallel section are not needed.

9. A segment of a combination circuit consists of three parallel branches in series with a single 100-ohm resistor. Each parallel branch has two resistors in series with each other. If the total voltage applied to this segment is 240 volts and the power dissipated by the resistor in series is 16 watts, what is the voltage applied to a single branch of the parallel section?

   A) 120 volts

   B) 60 volts

   C) 200 volts

   D) Not enough information is provided.

10. A combination circuit has two parallel branches with one resistor each in series with two resistors, one upstream and one downstream of the parallel branches. If all resistors are the same resistance value, what is the resistance value if the total voltage applied to the circuit is 125 volts and it results in a total current flow of 5 amperes?

   A) 20 ohms

   B) 15 ohms

   C) 10 ohms

   D) 5 ohms

11. A combination circuit consists of a resistor in series with three parallel resistors which are then in series with another single resistor which is in series with three additional parallel resistors. If the current through one of the parallel resistors is measured to be exactly one-third of the current flowing through one of the series resistors, which of the following statements is true?

   A) All six parallel resistors have the same resistance.

   B) The current is also one-third the total in the two other resistors in that parallel segment.

   C) The voltage across the parallel branch is one-third of the total voltage.

   D) Not enough information is provided.

12. A combination circuit has two parallel branches in series with two resistors, one upstream and one downstream of the parallel branches. If the voltage applied to the circuit is 120 volts and the voltage measured across the parallel branches is 60 volts, what is the total current flowing in the circuit?

   A) 0.5 amperes

   B) 0.25 amperes

   C) 1.0 amperes

   D) Not enough information is provided.

Practical Applications

1. Two resistors are connected in series. One resistor has a resistance of 1000 Ω, and the other resistor value is not known. The circuit is connected to a 50-V power supply, and there is a current of 25 mA flowing in the circuit. If a 1500-Ω resistor is connected in parallel with the unknown resistor, how much current will flow in this circuit?

2. The hot water for the heating system for a small factory is supplied by a single boiler. The boiler is located in a small building separate from the factory. The power supplied to the boiler is 240-volt single-phase without a neutral conductor. At the present time, the small building housing the boiler has no inside lighting. The business owner desires that four 100-watt lamps be installed inside the building. All lamps are to be connected to a single switch. The lamps are to be connected to the present 240-volt service inside the building. How would you accomplish this task? Explain your answer.

Practice Problems

Series-Parallel Circuits

Refer to the circuit shown in *Figure 8–21* to solve the following problems.

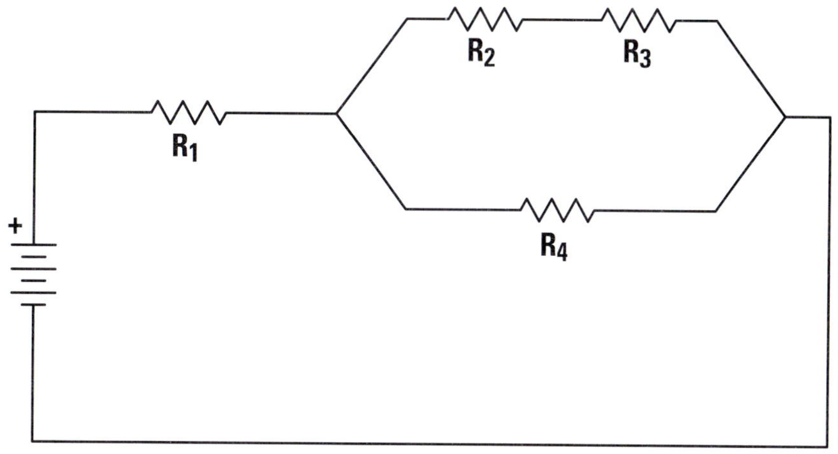

Figure 8–21 A combination series-parallel circuit with four resistors.

1. Find the unknown values in the circuit if the applied voltage is 75 V and the resistors have the following values:

$R_1 = 1.5\ k\Omega$	$R_2 = 910\ \Omega$	$R_3 = 2\ k\Omega$	$R_4 = 3.6\ k\Omega$	
I_T ______	E_1 ______	E_2 ______	E_3 ______	E_4 ______
R_T ______	I_1 ______	I_2 ______	I_3 ______	I_4 ______

2. Find the unknown values in the circuit if the applied voltage is 350 V and the resistors have the following values:

$R_1 = 22\ k\Omega$	$R_2 = 18\ k\Omega$	$R_3 = 12\ k\Omega$	$R_4 = 30\ k\Omega$	
I_T ______	E_1 ______	E_2 ______	E_3 ______	E_4 ______
R_T ______	I_1 ______	I_2 ______	I_3 ______	I_4 ______

3. Find the unknown values in the circuit if the applied voltage is 18 V and the resistors have the following values:

$R_1 = 82\ \Omega$	$R_2 = 160\ \Omega$	$R_3 = 220\ \Omega$	$R_4 = 470\ \Omega$	
I_T ______	E_1 ______	E_2 ______	E_3 ______	E_4 ______
R_T ______	I_1 ______	I_2 ______	I_3 ______	I_4 ______

Parallel-Series Circuits

Refer to the circuit shown in *Figure 8–22* to solve the following problems.

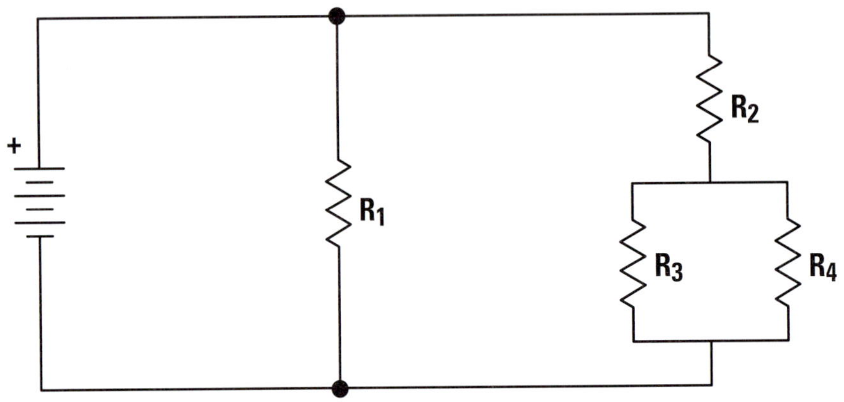

Figure 8–22 A parallel-series circuit.

4. Find the unknown values in the circuit if the total current is 0.8 A and the resistors have the following values:

$R_1 = 1.5\ k\Omega$	$R_2 = 910\ \Omega$	$R_3 = 2\ k\Omega$	$R_4 = 3.6\ k\Omega$	
E_T ______	E_1 ______	E_2 ______	E_3 ______	E_4 ______
R_T ______	I_1 ______	I_2 ______	I_3 ______	I_4 ______

5. Find the unknown values in the circuit if the total current is 0.65 A and the resistors have the following values:

$R_1 = 22\ k\Omega$	$R_2 = 15\ k\Omega$	$R_3 = 22\ k\Omega$	$R_4 = 33\ k\Omega$	
E_T ______	E_1 ______	E_2 ______	E_3 ______	E_4 ______
R_T ______	I_1 ______	I_2 ______	I_3 ______	I_4 ______

6. Find the unknown values in the circuit if the total current is 1.2 A and the resistors have the following values:

$R_1 = 75\ \Omega$	$R_2 = 47\ \Omega$	$R_3 = 220\ \Omega$	$R_4 = 160\ \Omega$	
E_T ______	E_1 ______	E_2 ______	E_3 ______	E_4 ______
R_T ______	I_1 ______	I_2 ______	I_3 ______	I_4 ______

Combination Circuits

Refer to the circuit shown in *Figure 8–23* to solve the following problems.

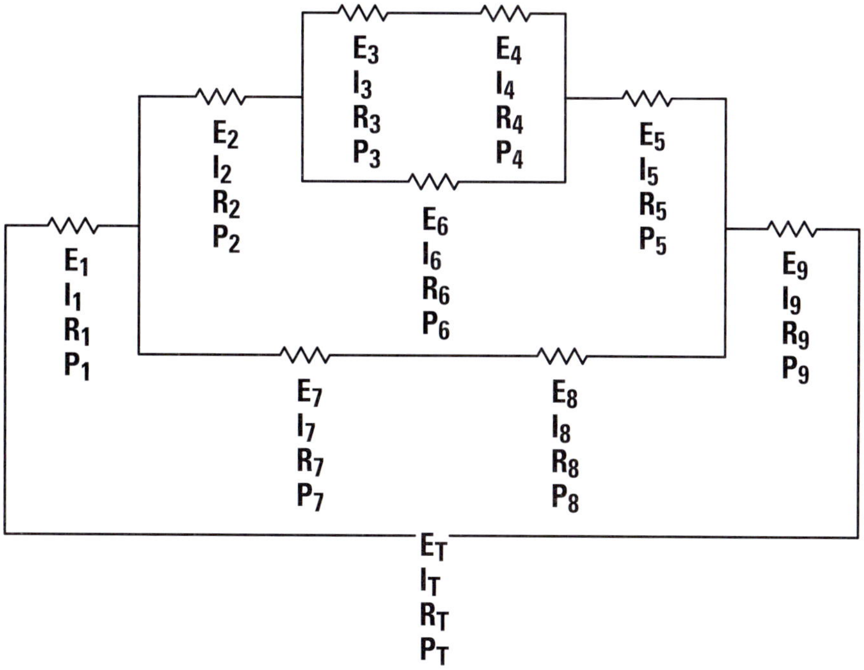

Figure 8–23 A combination circuit.

7.

E_T 250 V	E_1	E_2	E_3	E_4
I_T	I_1	I_2	I_3	I_4
R_T	R_1 220 Ω	R_2 500 Ω	R_3 470 Ω	R_4 280 Ω
P_T	P_1	P_2	P_3	P_4

E_5	E_6	E_7	E_8	E_9
I_5	I_6	I_7	I_8	I_9
R_5 400 Ω	R_6 500 Ω	R_7 350 Ω	R_8 450 Ω	R_9 300 Ω
P_5	P_6	P_7	P_8	P_9

8.

E_T	E_1	E_2	E_3	E_4 1.248 V
I_T	I_1	I_2	I_3	I_4
R_T	R_1	R_2	R_3	R_4
P_T 0.576 W	P_1 0.0806 W	P_2 0.0461 W	P_3 0.00184 W	P_4

E_5	E_6	E_7	E_8	E_9
I_5	I_6	I_7	I_8	I_9
R_5	R_6	R_7	R_8	R_9
P_5 0.0203 W	P_6 0.00995 W	P_7 0.0518 W	P_8 0.0726 W	P_9 0.288 W

Section 3 Test Instruments and Wire Sizes

Unit 9 Test Instruments

Key Terms

Ammeter

Ammeter shunt

Analog meters

Ayrton shunt

Clamp-on ammeter

Current transformer

d'Arsonval movement

digital multimeters

Galvanometers

Moving-coil meter

Multirange ammeters

Multirange voltmeters

Ohmmeter

Oscilloscope

Voltmeter

Wattmeter

Why You Need to Know

Test instruments are the eyes, and often the lifeline, of the electrician. An understanding of how these devices operate is critical to anyone working in the electrical field and a requirement for anyone deemed an electrically qualified person by their employer. They provide the electrician with the ability to evaluate problems and identify solutions to keep electrical equipment up and running. They also enable technicians to correctly determine electrical values of voltage, current, resistance, power, and many others. In this unit you will learn

- how different types of test instruments are constructed. This knowledge will aid you in knowing how they are to be connected in a circuit and how to interpret their readings.
- the differences between analog and digital meters, as well as how to interpret the waveforms shown on an oscilloscope. Oscilloscopes must be used to observe specific voltage patterns that would render a conventional voltmeter useless.

Outline

9–1 Analog Meters
9–2 The Voltmeter
9–3 The Ammeter
9–4 The Ohmmeter
9–5 Digital Meters
9–6 The Oscilloscope
9–7 The Wattmeter

Objectives

After studying this unit, you should be able to

- explain how an analog meter functions using a d'Arsonval meter movement.
- explain how to connect a voltmeter to a circuit to measure voltage.
- discuss how to connect an ammeter to a circuit to measure current flow.
- describe how resistance in a circuit is measured using an ohmmeter.
- explain the operating principles of digital meters.
- identify different waveforms shown on the display of an oscilloscope.
- explain how to connect a wattmeter into a circuit to measure power.

Preview

Anyone desiring to work in the electrical and electronics field must become proficient with the common test instruments used to measure electrical quantities. These instruments are the voltmeter, the ammeter, and the ohmmeter. Without meters, it would be nearly impossible to make meaningful interpretations of what is happening in a circuit. Meters can be divided into two general types: analog and digital.

9–1 Analog Meters

Analog meters are characterized by the use of a pointer and scale to indicate their value (*Figure 9–1*). It should be noted that analog meters are not commonly used any longer outside of very specific tasks that require one. The most common type of meter movement is the **d'Arsonval movement,** shown in *Figure 9–2*. This type of movement is often referred to as a **moving-coil meter.** A coil of wire is suspended between the poles of a permanent magnet. The coil is suspended either by jeweled movements similar to those used in watches or by taut bands. The taut-band type offers less turning friction than the jeweled movement. These meters can be made to operate on very small amounts of current and often are referred to as **galvanometers.**

Principle of Operation

Analog meters operate on the principle that like magnetic poles repel each other. As current passes through the coil, a magnetic field is created around the coil. The direction of current flow through the meter is such that the polarity of the magnetic pole created around the coil is the same as that of the permanent magnet. This similar polarity causes the coil to be deflected away from the pole of the magnet. A spring is used to retard the turning of the coil. The distance the coil turns against the spring is proportional to the strength of the magnetic field developed in the coil. If a pointer is added to the coil and a scale is placed behind the pointer, a meter movement is created.

Because the turning force of this meter depends on the repulsion of magnetic fields, it will operate on DC only. If AC is connected to the moving coil, the magnetic polarity will change 60 times per second, and the net turning force will be zero. For this reason, a DC voltmeter will indicate zero if connected to an AC line. When this type of movement is to be used to measure AC values, the current must be rectified, or changed into DC, before it is applied to the meter (*Figure 9–3*).

Courtesy Triplett Test Equipment and Tools

Figure 9–1 A classic example of an analog meter.

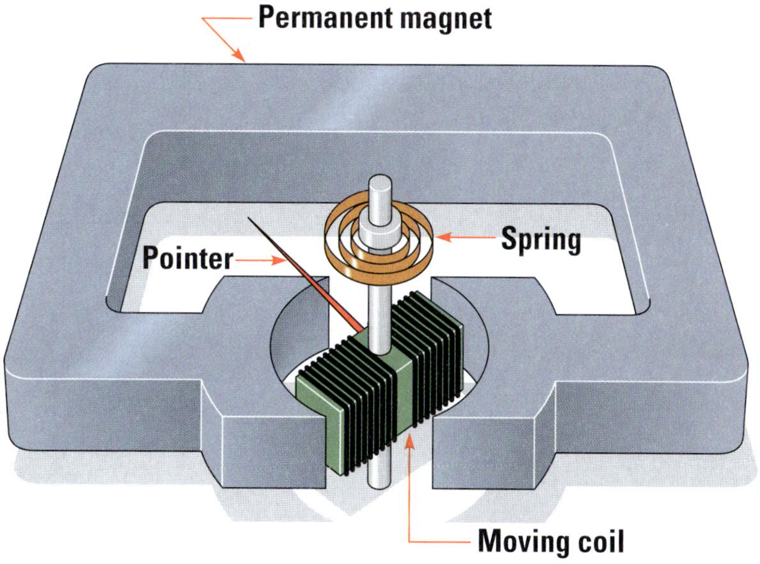

Figure 9–2 Basic d'Arsonval meter movement.

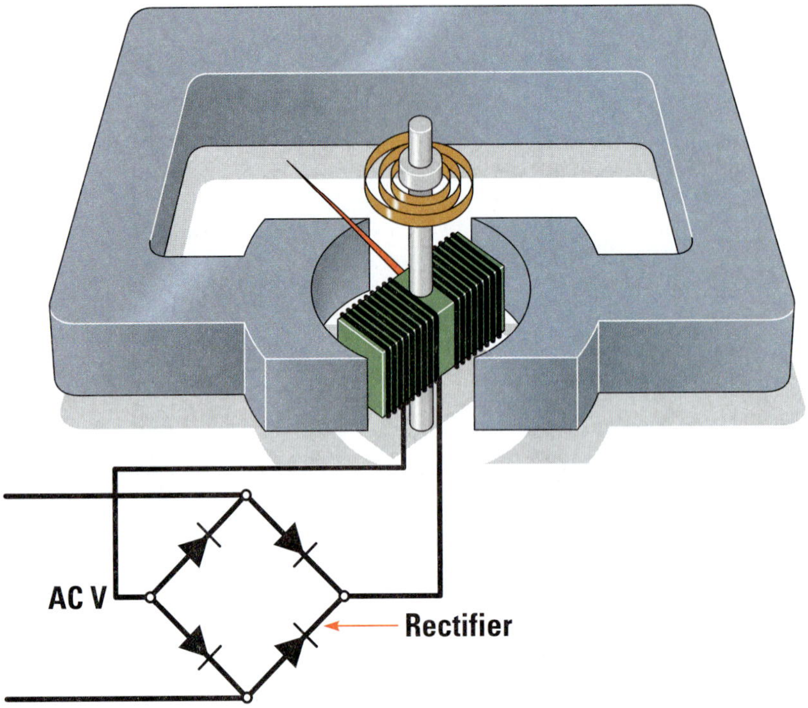

Figure 9–3 Rectifier changes AC voltage into DC voltage.

9–2 The Voltmeter

The **voltmeter** is designed to be connected directly across the source of power. Another way of explaining this connection is that you are connecting the voltmeter in parallel with the power source. *Figure 9–4* shows a voltmeter being used to test the voltage of a battery. Notice that the leads of the meter are connected directly across the source of voltage. A voltmeter can be connected directly across the power source because it has a very high internal resistance connected in series with the voltmeter movement (*Figure 9–5*). However, be aware that most modern meters will have multimeter functionality, and other settings do not have this same internal resistance. The industrial standard for a voltmeter is 20,000 ohms per volt for DC and 5000 ohms per volt for AC. Assume the voltmeter shown in *Figure 9–5* is an AC meter and has a full-scale range of 300 volts. The meter circuit (meter plus resistor) would therefore have a resistance of 1,500,000 ohms (300 V $\times$ 5000 Ω per volt $=$ 1,500,000 Ω).

Figure 9–4 A voltmeter connects directly across the power source.

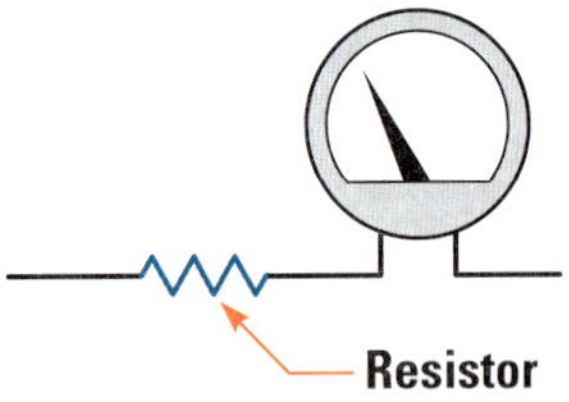

Figure 9–5 A resistor connects in series with the meter.

Calculating the Resistor Value

If a person wanted to determine whether their meter meets this industrial standard, a simple check can be performed to calculate the internal resistance of the meter. But before the resistor value can be calculated, the operating characteristics of the meter must be known. The meter manufacturer lists that the meter requires a current of 50 microamperes and a voltage of 1 volt to deflect the pointer full scale. These are known as the *full-scale values* of the meter.

When the meter and resistor are connected to a source of voltage, their combined voltage drop must be 300 volts. Because the meter has a voltage drop of 1 volt, the resistor must have a drop of 299 volts. The resistor and meter are connected in series with each other. In a series circuit, the current flow will be the same in all parts of the circuit. If 50 microamperes are required to deflect the meter full scale, then the resistor must have a current of 50 microamperes flowing through it when it has a voltage drop of 299 volts. The value of resistance can now be calculated using Ohm's law:

$$R = \frac{E}{I}$$

$$R = \frac{299\,V}{0.000050\,A}$$

$$R = 5.98\ M\Omega\ (5{,}980{,}000\ \Omega)$$

Multirange Voltmeters

Most voltmeters are **multirange voltmeters,** which means that they are designed to use one meter movement to measure several ranges of voltage. For example, one meter may have a selector switch that permits full-scale ranges to be selected. These ranges may be 3 volts full scale, 12 volts full scale, 30 volts full scale, 60 volts full scale, 120 volts full scale, 300 volts full scale, and 600 volts full scale. Meters are made with that many scales so they will be as versatile as possible. Many of the digital options on the market today are also going to be auto-ranging and have no need for the selector switch to contain anything other than voltage, current, or resistance. Some even automatically choose between AC and DC systems. Digital meters are covered later in this unit. The meter shown in *Figure 9–6* has multirange selection for voltage.

Courtesy of Triplett Corporation

Figure 9–6 Example of an analog multimeter with multirange selection.

The selector switch changes the resistance value that is inserted in the circuit to either increase or decrease the voltage range. *Figure 9–7* is an example of a selector switch that has four range settings for full-scale voltage: 30, 60, 300, and 600 volts. Notice that when the higher voltage settings are selected, more resistance is inserted in the circuit.

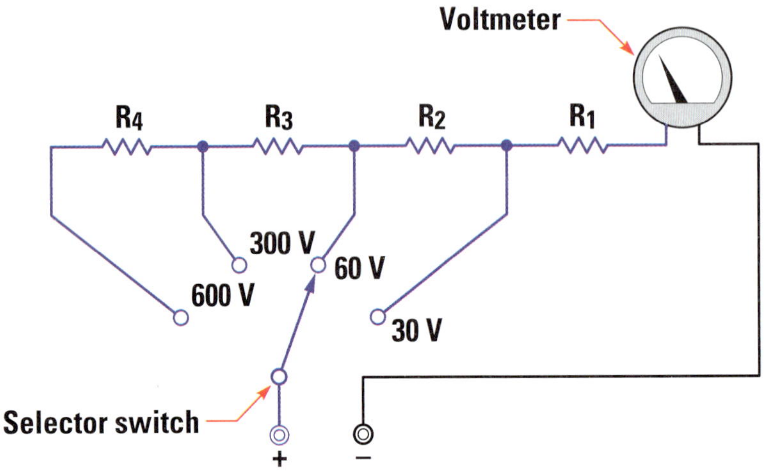

Figure 9–7 Example of how a rotary selector switch is used to change the full-range voltage setting.

Calculating the Resistor Values

The values of the four resistors shown in *Figure 9–7* can be determined using Ohm's law. Assume that the full-scale values of the meter are 50 microamperes and 1 volt. The first step is to determine the value for R_1, which is used to provide a full-scale value of 30 volts. R_1, therefore, must have a voltage drop of 29 volts when a current of 50 microamperes is flowing through it.

$$R_1 = \frac{E}{I}$$

$$R_1 = \frac{29\,V}{0.000050\,A}$$

$$R_1 = 598\ K\Omega\ (580{,}000\ \Omega)$$

The remaining resistance values can be determined using the total circuit current of 50 microamperes and a voltage drop equal to the difference between range settings. For example, the voltage drop used to determine the resistance needed for R_3 would be 240 volts since the range is between 300 V and 60 V. The rest is simple Ohm's law.

Reading a Meter

Learning to read the scale of an analog multimeter takes time and practice. Most people use meters every day without thinking about it. A common example of how people use meters every day is the instrument cluster in our vehicles. Of course, as our cars and trucks become more and more advanced, these instrument clusters change configuration from time to time, but the basics remain: speedometer, fuel gauge, tachometer, and temperature. The speedometer often also utilizes multiple scales. One scale is in miles per hour, and the other is in kilometers per hour. We know how to select the correct scale based on what the speed limit signs list along our route. This is similar to how an analog multimeter scale is constructed.

The meter scale shown in *Figure 9–8* has several scales used to measure different quantities and values. The top of the scale is used to measure resistance, or ohms. Notice that the scale begins on the left at infinity and ends at zero on the right. Ohmmeters are covered later in this unit. The second scale is labeled AC–DC and is used to measure voltage. Notice that this scale has three different full-scale values. The top scale is 0–300, the second scale is 0–60, and the third scale is 0–12. The scale used is determined by the setting of the range control switch. The third set of scales is labeled "AC." This scale is used to measure AC voltage when the range control switch is set at 3 volts.

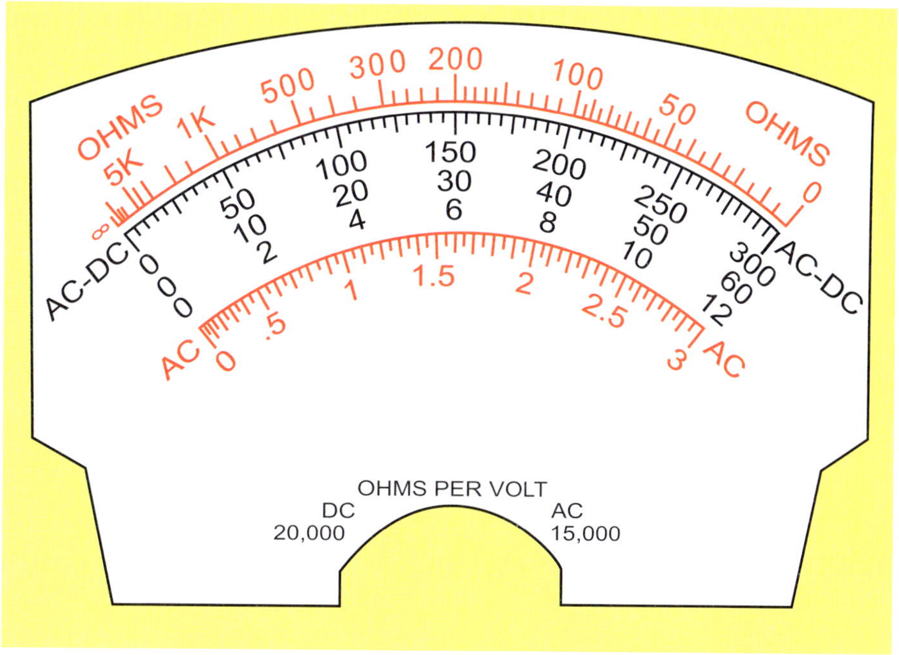

Figure 9–8 An example of an analog meter scale on a meter that measures resistance and voltage.

Reading a Voltmeter

Analog voltmeter scales use the primary numbers, such as 3, 6, and 12 and are read in multiples of 10 of these numbers. Because the numbers are in multiples of 10, it is easy to multiply or divide the readings in your head by moving a decimal point. Remember that any number can be multiplied by 10 by moving the decimal point one place to the right, and any number can be divided by 10 by moving the decimal point one place to the left. For example, if the selector switch was set to permit the meter to indicate a voltage of 30 volts full scale, the 300-volt scale would be used, and the reading would be divided by 10. The reading can be divided by 10 by moving the decimal point one place to the left. In *Figure 9–9*, the pointer is indicating a value of 250. If the selector switch is set for 30 volts full scale, moving the decimal point two places to the left will give a reading of 25 volts. If the selector switch were set for a full-scale value of 300 volts, the meter shown in *Figure 9–9* would be indicating a value of 250 volts.

Now assume that the meter has been set to have a full-scale value of 600 volts. Referring to *Figure 9–10,* the pointer is indicating a value of 44 out of a scale of 60 V total. Because the full-scale value of the meter is set for 600 volts, multiply the reading on the meter by 10. The correct reading becomes 440 volts.

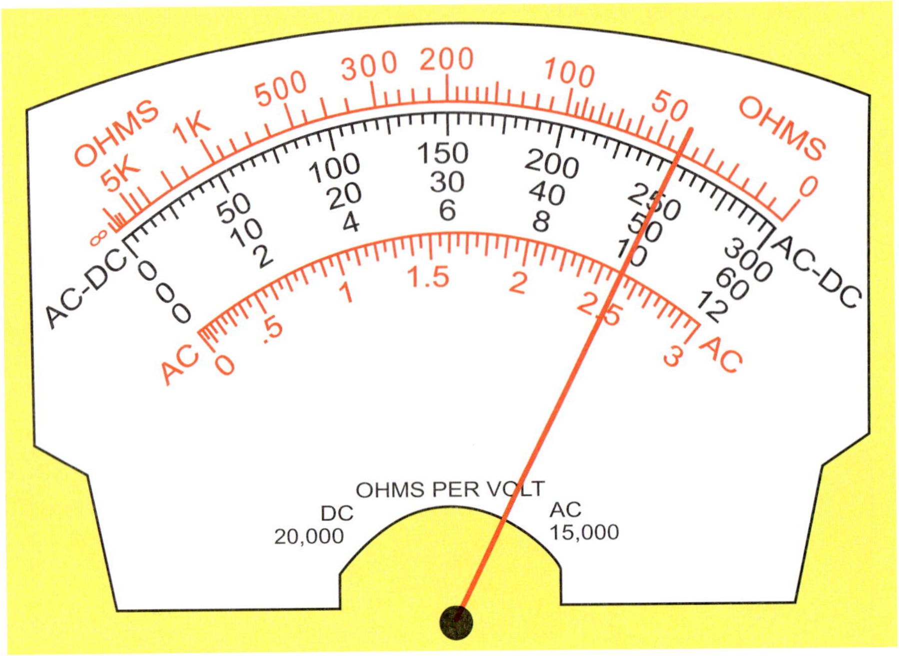

Figure 9–9 The meter indicates a value of 250 V based on a full-scale of 300 V.

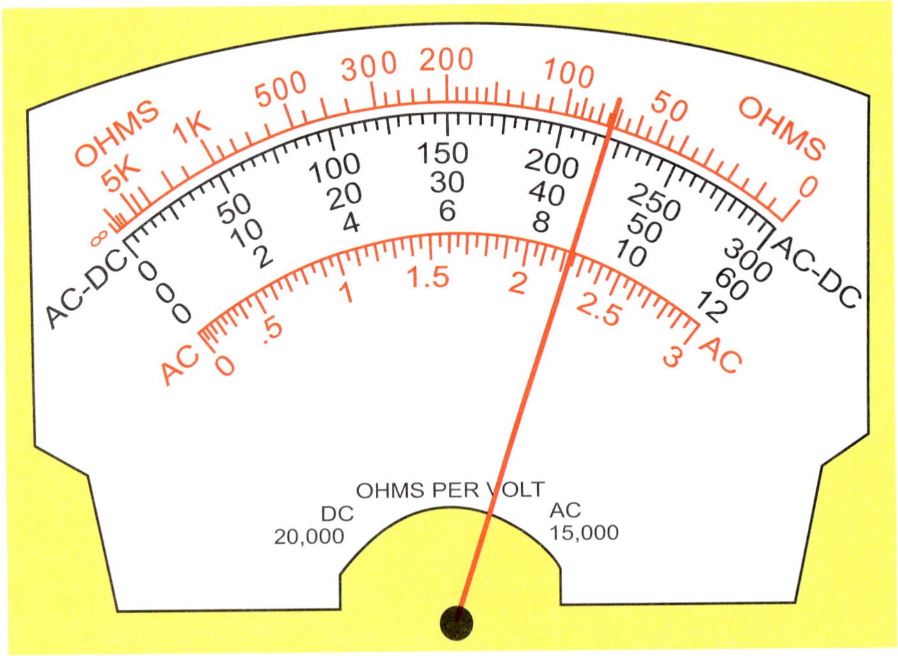

Figure 9–10 The meter indicates a value of 440 based on a full-scale of 600 V.

Three distinct steps should be followed when reading an analog meter. These steps can be helpful for someone who has not had a great deal of experience reading one of these devices. The steps are as follows:

1. *Determine what the meter is set to indicate.* Is the meter set to read a value of DC voltage, DC current, AC voltage, AC current, or ohms? It is impossible to read a meter if you do not know what the meter is used to measure.

2. *Determine which scale of the meter will apply to the measurement.* The advantage of a multimeter is that it can measure a wide range of values and quantities. After it has been determined what quantity the meter is set to measure, it must then be determined what range of the meter is set to and which corresponding scale will be used.

   There is a great deal of difference in reading when the meter is set to indicate a value of 600 volts full scale and when it is set for 30 volts full scale.
3. *Read the meter.* The last step is to determine what the meter is indicating. It may be necessary to determine the value of the hash marks on the meter face for the range for which the selector switch is set. If the meter in *Figure 9–8* is set for 300 volts full scale, each hash mark has a value of 5 volts. If the full-scale value of the meter is 60 volts, however, each hash mark has a value of 1 volt.

9–3 The Ammeter

Caution

The **ammeter**, unlike the voltmeter, is a very low-impedance device. The ammeter is used to measure current and must be connected in series with the load to permit the load to limit the current flow (*Figure 9–11*).

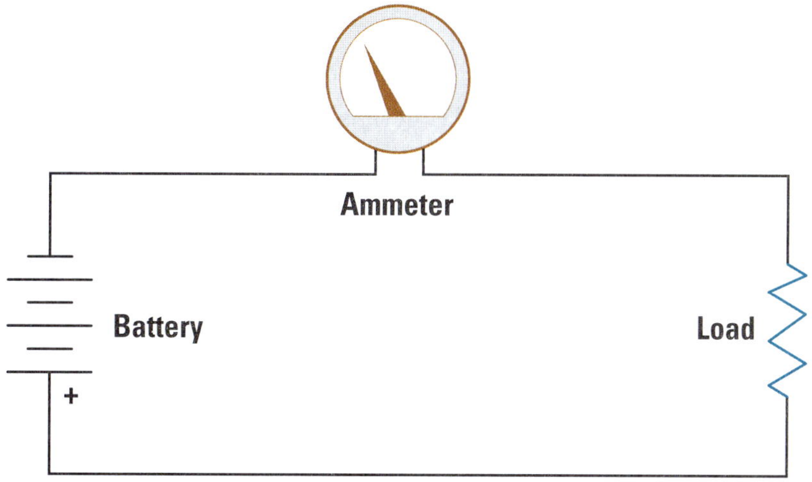

Figure 9–11 A directly connected ammeter must be connected in series with the load.

An analog ammeter has a typical impedance of less than 0.1 ohm. If this meter is connected in parallel with the power supply, the full circuit voltage will be applied, and the impedance of the ammeter is the only thing to limit the amount of current flow in the circuit. Assume that an ammeter with a resistance of 0.1 ohm is connected across a 240-volt AC line. The current flow in this circuit would be 2400 amperes (240 V/0.1 Ω = 2400 A). Improperly connected ammeters have been the source of an arc flash in several instances. It is imperative that ammeters connected directly to circuit conductors be connected in series, as shown in *Figure 9–11*. *Figure 9–12* shows an ammeter of this type that might be installed integral to equipment.

Figure 9–12 Example of one type of in-line ammeter.

Ammeter Shunts

DC ammeters are constructed by connecting a common moving-coil type of meter across a shunt. An **ammeter shunt** is a low-resistance device used to conduct most of the circuit current away from the meter movement. Because the meter movement is connected in parallel with the shunt, the voltage drop across the shunt is the voltage applied to the meter. Most ammeter shunts are manufactured to have a voltage drop of 50 millivolts (mV). If a 50-millivolt meter movement is connected across the shunt as shown in *Figure 9–13*, the pointer will move to the full-scale value when the rated current of the shunt is flowing. In the example shown, the ammeter shunt is rated to have a 50-millivolt drop when a 10-ampere current is flowing in the circuit. Because the meter movement has a full-scale voltage of 50 millivolts, it will indicate the full-scale value when 10 amperes of current are flowing through the shunt. An ammeter shunt is shown in *Figure 9–14*.

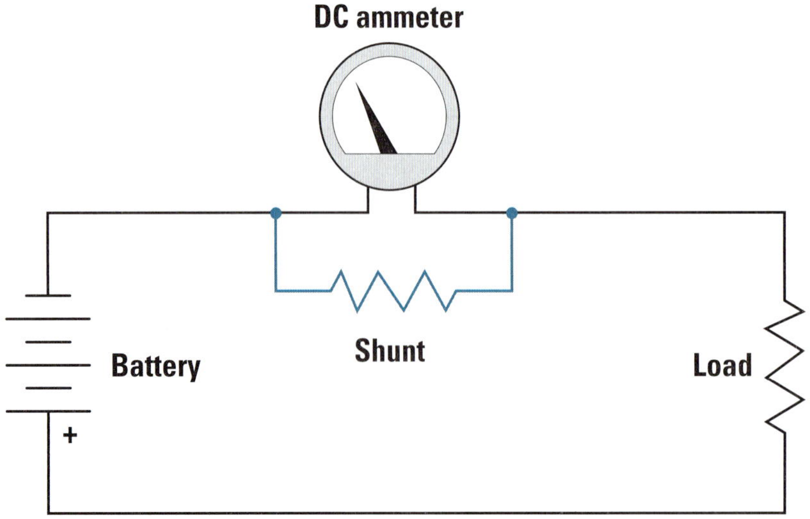

Figure 9–13 A shunt is used to set the value of the ammeter.

Ra3m/Shutterstock.com

Figure 9–14 An example of an ammeter shunt.

Ammeter shunts can be purchased to indicate different values. If the same 50-millivolt movement is connected across a shunt designed to drop 50 millivolts when 100 amperes of current flow through it, the meter will have a full-scale value of 100 amperes.

The resistance of an ammeter shunt can be calculated using Ohm's law. The resistance of a shunt designed to have a voltage drop of 50 millivolts when 100 amperes of current flow through it is:

$$R = \frac{E}{I}$$

$$R = \frac{0.050\,V}{100\,A}$$

$$R = 0.0005\ \Omega\text{, or } 0.5\ m\Omega$$

In the preceding problem, no consideration was given to the electrical values of the meter movement. The reason is that the amount of current needed to operate the meter movement is so small compared with the 100-ampere circuit current that it could have no meaningful effect on the resistance value of the shunt. When calculating the value for a low-current shunt, however, the meter values must be taken into consideration. For example, assume the meter has a voltage drop of 50 millivolts (0.050 V) and requires a current of 1 milliampere (0.001 A) to deflect the meter full scale. Using Ohm's law, it can be found that the meter has an internal resistance of 50 ohms (0.050 V/0.001 A = 50 Ω). Now assume that a shunt is to be constructed that will permit the meter to have a full-scale value of 10 milliamperes. If a total of 10 milliamperes is to flow through the circuit and 1 milliampere must flow through the meter, then 9 milliamperes must flow through the shunt (*Figure 9–15*). Because the shunt must have a voltage drop of 50 millivolts when 9 milliamperes of current are flowing through it, its resistance must be 5.555 ohms (0.050 V/0.009 A = 5.555 Ω).

Multirange Ammeters

Multirange ammeters are designed to operate on more than one range by connecting the meter movement to different shunts. ***When a multirange ammeter is used, care must be taken to ensure a shunt is always in place around the meter when the meter is connected to the circuit.*** Disconnection

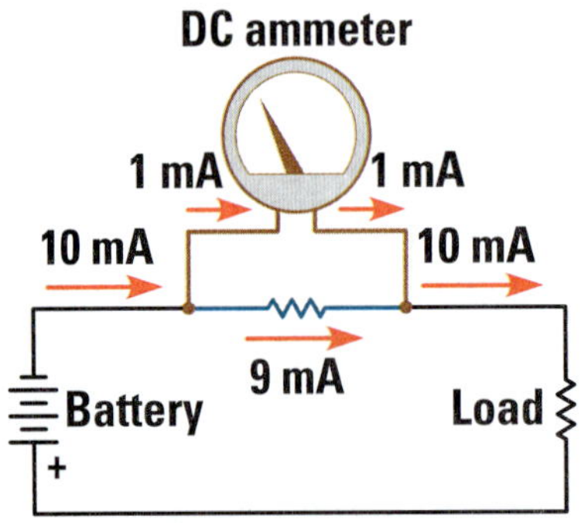

Figure 9–15 The total current is divided between the meter and the shunt.

would cause the meter movement to be inserted in series with the circuit, and full-circuit current would flow through the meter movement. Two basic methods are used for connecting shunts to a meter movement. One method is to use a make-before-break switch. This type of switch is designed so that it will make contact with the next shunt before it breaks connection with the shunt to which it is connected (*Figure 9–16*). This method does, however, present a problem—contact resistance. Notice in *Figure 9–16* that the rotary switch is in series with the shunt resistors. This arrangement causes the contact resistance to be added to the shunt resistance and can cause inaccuracy in the meter reading.

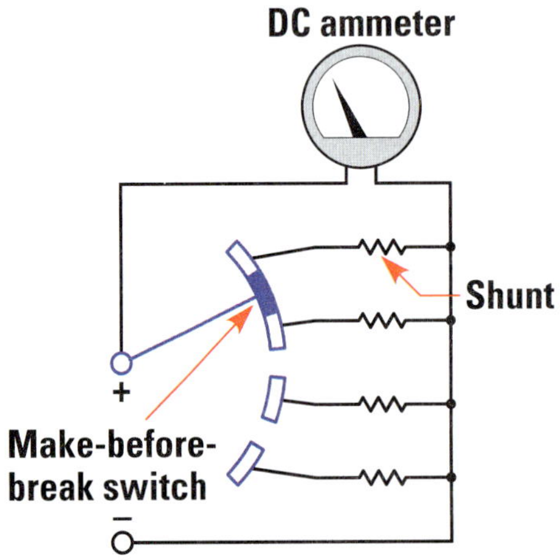

Figure 9–16 A make-before-break switch is used to change meter shunts without ever fully disconnecting a shunt from the circuit.

The Ayrton Shunt

The second method of connecting a shunt to a meter movement is to use an **Ayrton shunt** (*Figure 9–17*). In this type of shunt, connection is made to different parts of the shunt, and the meter movement is never disconnected from the shunt. Also notice that the switch connections are made external to the shunt and meter. This arrangement prevents contact resistance from affecting the accuracy of the meter.

When an Ayrton shunt is used, the resistors are connected in parallel with the meter on some ranges and in series with the meter for other ranges. In this example, the meter movement has full-scale values of 50 millivolts, 1 milliampere, and 50 ohms of resistance. The shunt will permit the meter to have full-scale current values of 100 milliamperes, 500 milliamperes, and 1 ampere.

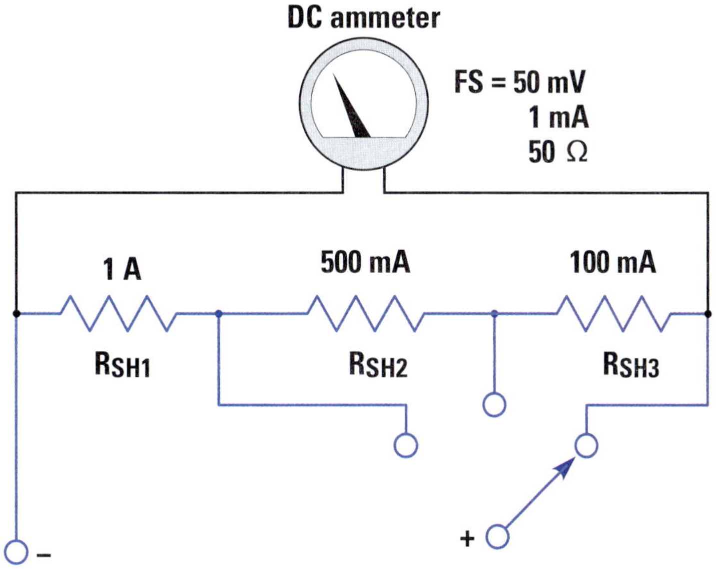

Figure 9–17 An Ayrton shunt.

When the range switch in *Figure 9–18* is set in the 100-milliampere position, all three shunt resistors are connected in series with each other and are in parallel with the meter movement.

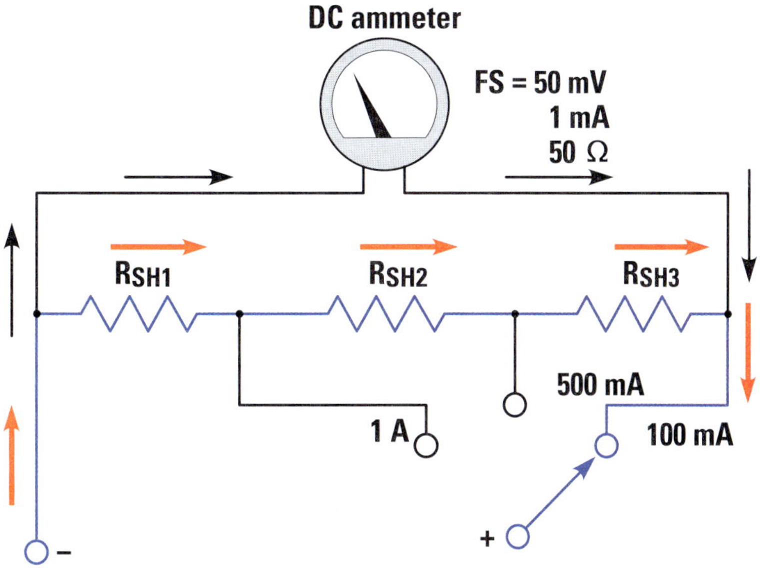

Figure 9–18 The meter movement is in parallel with all three shunt resistors connected in series.

When the selector switch is changed to the 500-milliampere position, R_{SH1} and R_{SH2} are connected in series with each other and are in parallel with the meter movement. R_{SH3} is connected in series with the meter movement, and together, they are in parallel with the other two resistors (*Figure 9–19*). This configuration changes again when the selector switch is moved to the 1 ampere full scale position. In this position, both resistors R_{SH2} and R_{SH3} are connected in series with the meter movement. This leaves only resistor R_{SH1} to be connected in parallel to the meter movement (*Figure 9–20*).

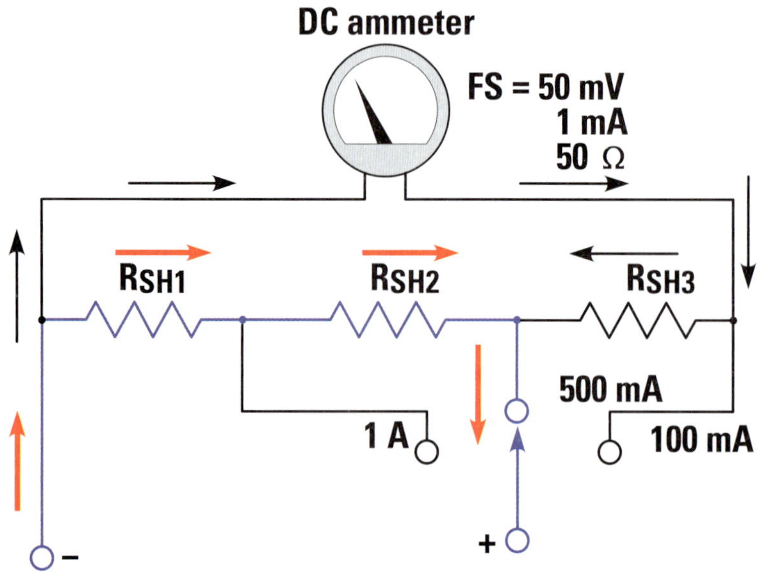

Figure 9–19 Current path through the meter and shunt for a full-scale value of 500 milliamperes.

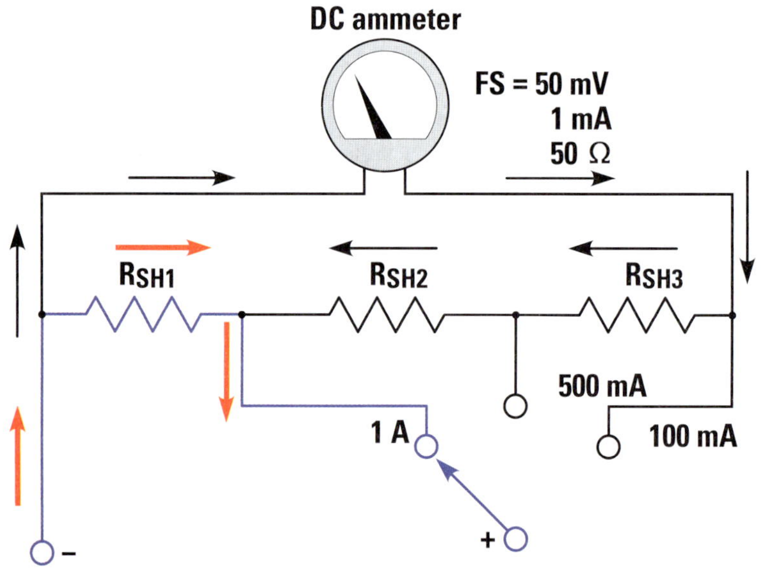

Figure 9–20 Current path through shunt and meter movement for a full-scale value of 1 ampere.

However, it should be noted that the Ayrton shunt is not used for large current values and is seldom used for currents above 10 amperes. An ammeter with an Ayrton shunt is shown in *Figure 9–21*.

AC Ammeters

Most AC ammeters use a **current transformer** instead of shunts to change the scale values since shunts can only be used to increase their range but cannot be used to decrease the range. This type of ammeter is shown in *Figure 9–22*. The primary of the transformer is connected in series with the load, and the ammeter is connected to the secondary of the transformer. Notice that the range of the meter is changed by selecting different taps on the secondary of the current transformer. The different taps on the transformer

Figure 9–21 An example of a DC ammeter circuit utilizing an Ayrton shunt.

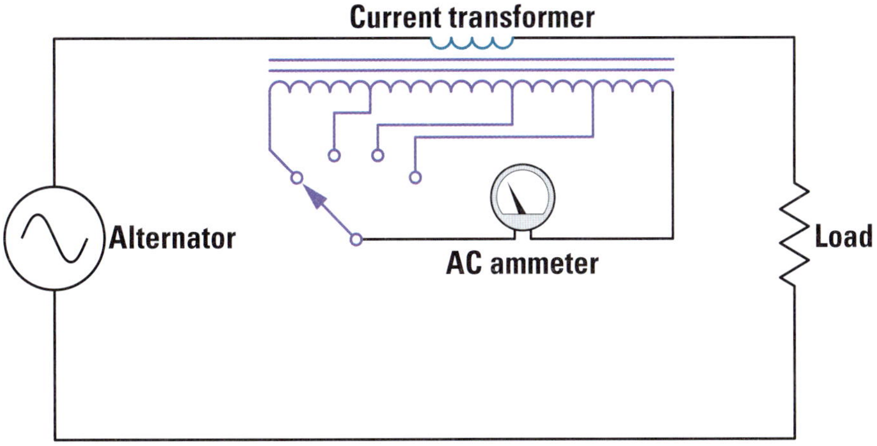

Figure 9–22 A current transformer is often used to change the range of an AC ammeter.

provide different ratios of the number of turns of wire in the primary as compared to the number of turns of wire in the secondary. This adjusts the output on the secondary of the current transformer.

Current Transformers (CTs)

When a large amount of AC current must be measured, such as when measuring how much current a large building uses, a different type of current transformer is installed around the circuit conductors. These transformers have ratios that start at 200:5 and can have ratios of several thousand to five. These current transformers, generally referred to in industry as CTs, have a standard secondary current rating of 5 amperes AC. They are designed to be operated with a 5-ampere AC ammeter connected directly to their secondary winding, which produces a short circuit. CTs are designed to operate with the secondary winding shorted.

> **Caution**
>
> The secondary winding of a CT should never be opened when there is power applied to the primary. This will cause the transformer to produce a step-up in voltage that could be high enough to kill someone who comes in contact with it.

A type of current transformer that is often used in this capacity is a toroid transformer. A toroid transformer is constructed with a hollow core similar to that of a doughnut, hence the nickname these have been given in several areas of the country (*Figure 9–23*). When current transformers are used, the circuit conductor is inserted through the opening in the transformer (*Figure 9–24*). If the circuit has multiple conductors, as in a three-phase circuit, there will be multiple transformers installed. The circuit conductor acts as the primary winding of the current transformer and consists of a winding with only a single turn.

Figure 9–23 An example of a toroid-style current transformer.

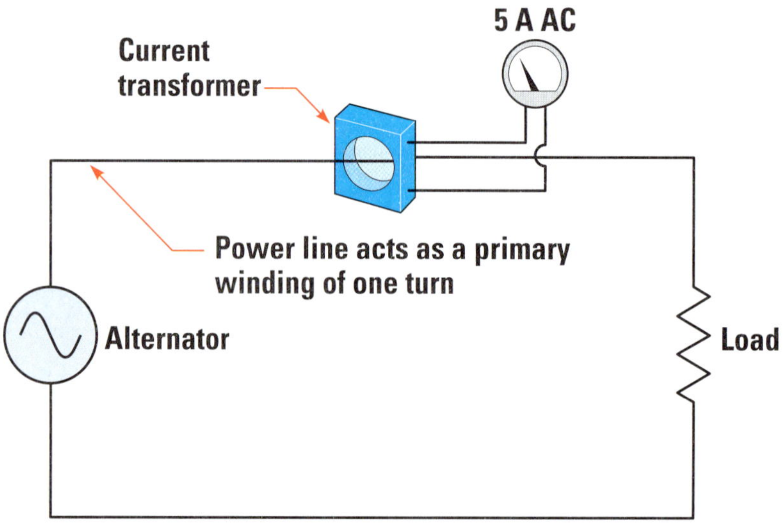

Figure 9–24 Toroid current transformer used with an ammeter in AC circuits when higher amounts of current are flowing in the circuit.

The ratio of the current transformer input to output can be changed by looping the power wire through the opening in the transformer to produce a primary winding of more than one turn. For example, assume a current transformer has a ratio of 600:5. If the primary power wire is inserted through the opening, it will require a current of 600 amperes to induce the 5 amperes needed to deflect the meter full scale. If the primary power conductor is looped around and inserted through the window a second time, the primary now contains two turns of wire instead of one (*Figure 9–25*). It now requires only 300 amperes of current flow in the primary to deflect the meter full scale. If the primary conductor is looped through the opening a third time, it will require only 200 amperes of current flow to deflect the meter full scale. This principle allows us to essentially set our own current range or scale based on how much current is expected to flow in the circuit conductors.

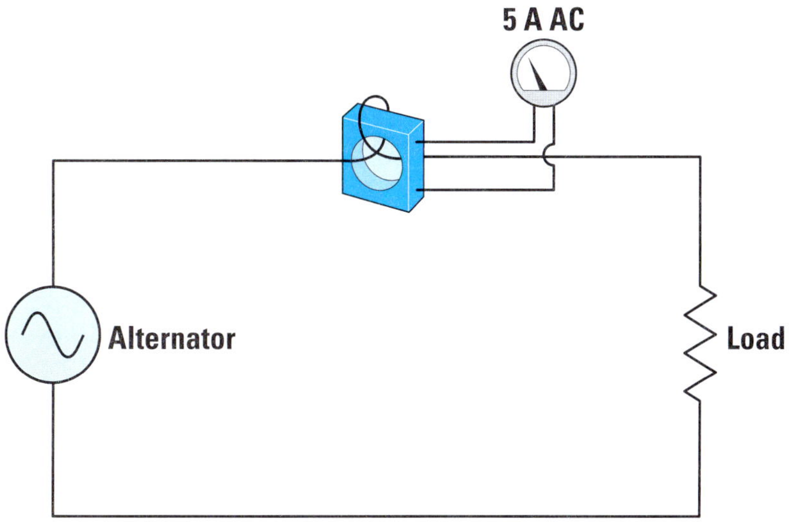

Figure 9–25 The primary conductor loops through the CT to produce a second turn, which changes the ratio between primary current and secondary current.

Clamp-On Ammeters

Many technicians and electricians will use a **clamp-on ammeter** (*Figure 9–26*) when measuring current in the field. This meter derives its name because the jaw is clamped around one of the circuit conductors

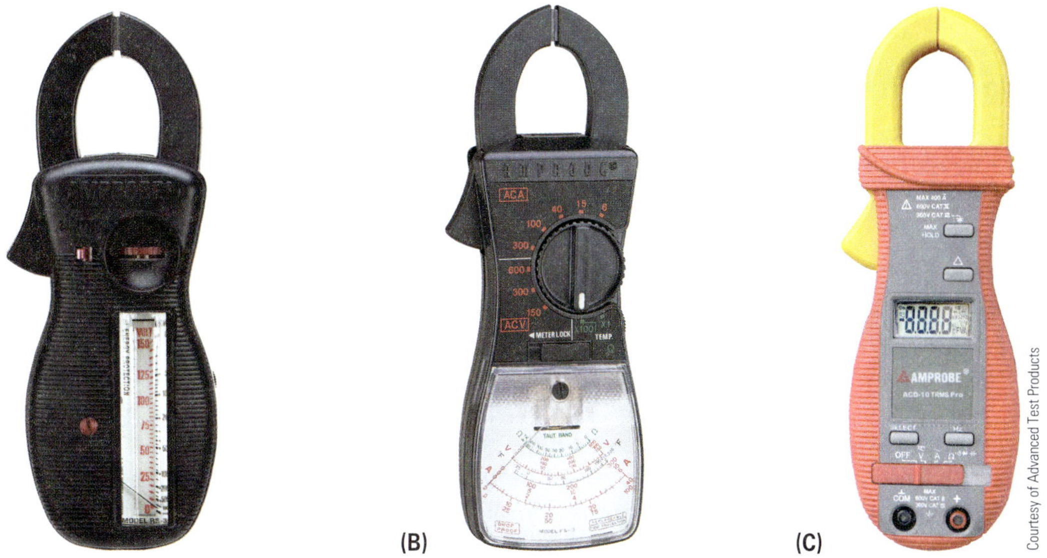

Figure 9–26 **(A)** Analog-type clamp-on ammeter with vertical scale. **(B)** Analog-type clamp-on ammeter with flat scale. **(C)** Clamp-on ammeter with digital scale.

supplying power to the load (*Figure 9–27*). If the meter is clamped around more than one line, the electro-magnetic fields of the separate conductors will cancel out and the meter will indicate zero.

Figure 9–27 The clamp-on ammeter connects around only one conductor in the circuit.

Many AC clamp-on ammeters use a current transformer to operate. The jaw of the meter is part of the core material of the transformer. When the meter is connected around the current-carrying wire, the changing magnetic field produced by the AC induces a voltage into the current transformer. The strength and frequency of the magnetic field determine the amount of voltage induced in the current transformer. Because 60 hertz is a standard frequency throughout the country, the amount of induced voltage is proportional to the strength of the magnetic field. The induced voltage then causes a certain amount of current to flow based on how much resistance is in the secondary circuit of the meter. It is important to remember the trick often used with power CTs to adjust their range by wrapping the conductor through the CT a second time (*Figure 9–25*). Doing this with a clamp-on ammeter will cause the reading of the meter to be double the actual current in the circuit. This procedure should not be used with a clamp-on ammeter unless it is specifically called out by the meter manufacturer and performed by a qualified technician.

DC–AC Clamp-On Ammeters

Most clamp-on ammeters that can measure both DC and AC system currents do not operate on the same principle as the current transformer. Current transformers depend on induction, which means that the current in the circuit conductor must change direction periodically to provide a change of magnetic field polarity. It is the continuous change of field strength and direction that permits the current transformer to operate. The current in a DC circuit is unidirectional and does not change polarity and, therefore, would not permit the current transformer to operate.

DC–AC clamp-on ammeters (*Figure 9–28*) use the *Hall effect* as the basic principle of operation. The Hall effect was discovered by Edward H. Hall at Johns Hopkins University in 1879. Hall originally used a piece of pure gold to produce the Hall effect, but today, a semiconductor material is used because it has better operating characteristics and is less expensive. The device is often referred to as a *Hall generator. Figure 9–29* illustrates the operating principle of the Hall generator. As current from the constant-current generator flows through the semiconductor material, a balance is observed. Current through the middle of the semiconductor produces no potential difference. However, if an external magnetic force pulls or pushes the current flowing through the semiconductor material, a potential difference can be measured across the semiconductor material.

Courtesy of Advanced Test Products

Figure 9–28 DC–AC clamp-on ammeter.

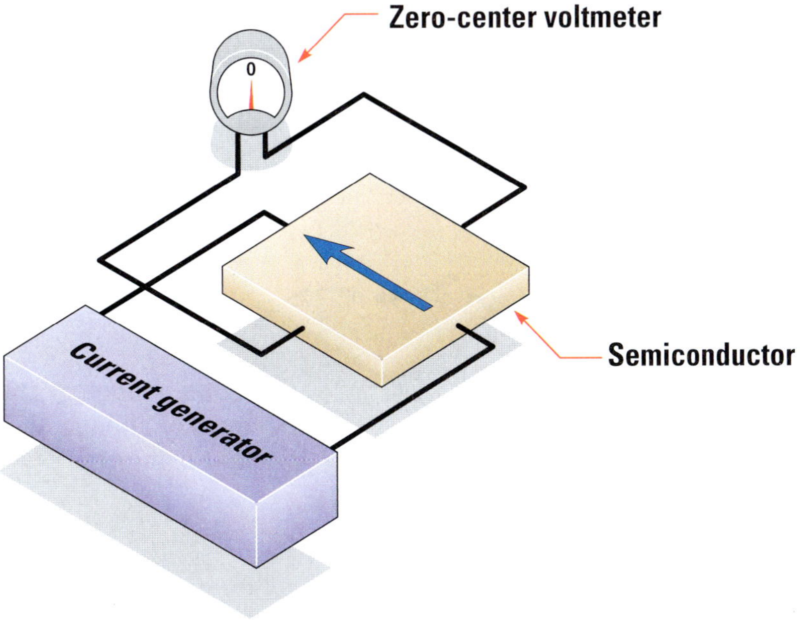

Figure 9–29 Basic Hall generator.

What might generate a force that can cause the electrons to flow through the semiconductor in other than that equilibrium state? If a conductor with current flowing through it is brought into close proximity, the magnetic field generated by the current will cause the electrons to move to one side or the other of the semiconductor material, and, thus, a potential difference is created. It does not matter what the polarity is since each will cause a deflection and, therefore, this Hall generator works with both a DC or an AC conductor as the source of electron deflection. *Figures 9–30 and 9–31* show how either polarity of a magnetic field in the vicinity of the semiconductor material will cause the desired deflection of the electrons.

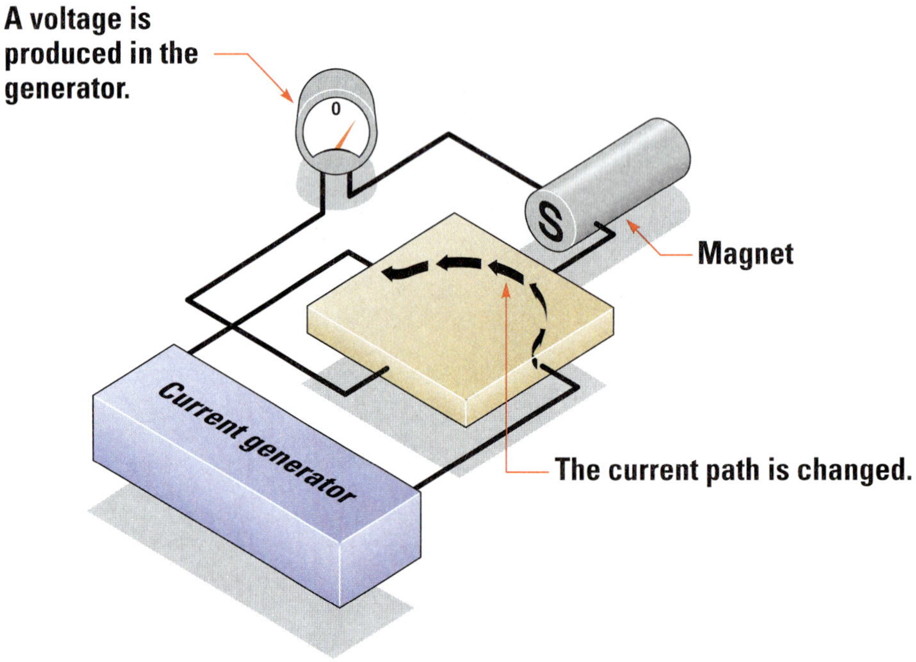

Figure 9–30 The presence of a magnetic field causes the Hall generator to produce a voltage.

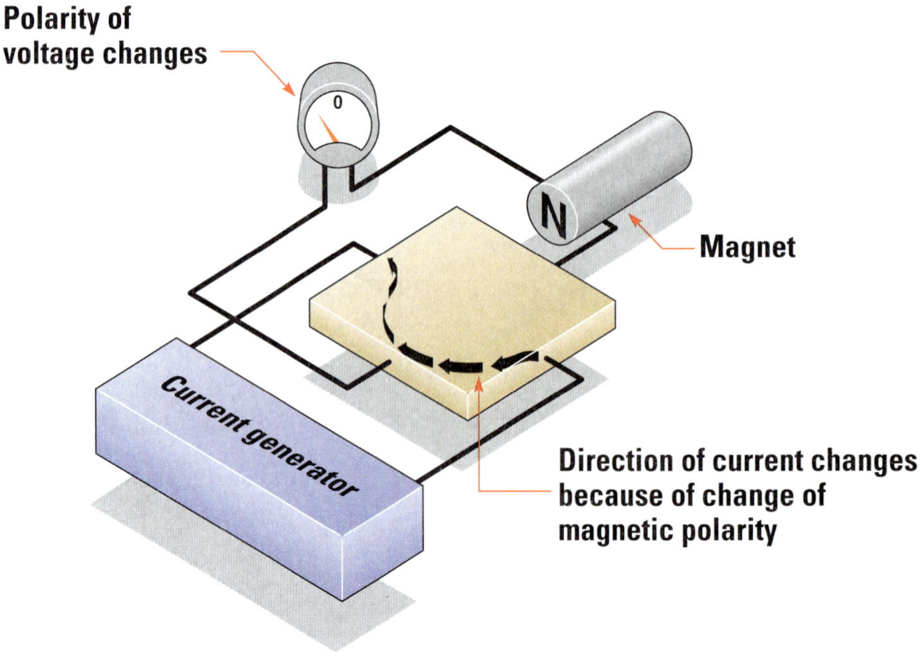

Figure 9–31 If the magnetic field polarity changes, the polarity of the voltage changes.

9–4 The Ohmmeter

Resistance is measured by using an **ohmmeter**. It is common to see an ohmmeter as one of the options in a multifunction testing device, such as a multimeter. When using an analog meter, such as a volt-ohmmilliammeter (VOM), the ohmmeter has the only scale that is nonlinear. The scale numbers increase in value as they progress from right to left. There are two basic types of analog ohmmeters—the series and the shunt. The series ohmmeter is used to measure high values of resistance, and the shunt type is used to measure low values of resistance. Regardless of the type used, the meter must provide its own power source to measure resistance. The power is provided by batteries located inside the instrument.

The Series Ohmmeter

A schematic for a basic series ohmmeter is shown in *Figure 9–32*. The meter movement is placed in series with two resistors, R_1 and R_2. Resistor R_1 is a fixed resistance type, and R_2 is an adjustable resistor, so the total circuit resistance can be adjusted as the strength of the battery degrades. The resistance values are specifically chosen so the current supplied by the battery is equal to the current required for a full-scale deflection on the meter movement.

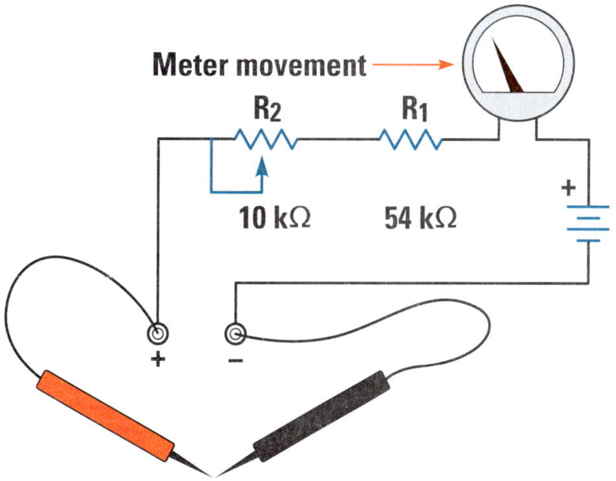

Figure 9–32 An example of a basic series ohmmeter circuit.

When resistance is to be measured, the meter must first be zeroed. This is done with the ohms-adjust control, the variable resistor located on the front of the meter. To zero the meter, connect the leads (*Figure 9–32*) and turn the ohms-adjust knob until the meter indicates zero at the far right end of the scale (*Figure 9–33*), or in other words, a full-scale meter movement. When the leads are separated, the meter will again indicate infinity resistance at the left side of the scale since there will be no current flowing in the circuit and, therefore, no meter movement deflection. When the leads are connected across a resistance, the meter will again go up the scale. Because resistance has been added to the circuit, less than full deflection current will flow, and the meter will indicate some value other than zero. *Figure 9–34* shows a meter indicating a resistance of 150 ohms when the range setting is on R×1.

Ohmmeters can have different range settings, such as R×1, R×100, R×1000, or R×10,000. These different scales can be obtained by adding different values of resistance in the meter circuit and resetting the meter to zero. *An ohmmeter should always be readjusted to zero when the scale is changed.* On the R×1 setting, the resistance is measured straight off the resistance scale located at the top of the meter. If the range is set for R×1000, however, the reading must be multiplied by 1000. The ohmmeter

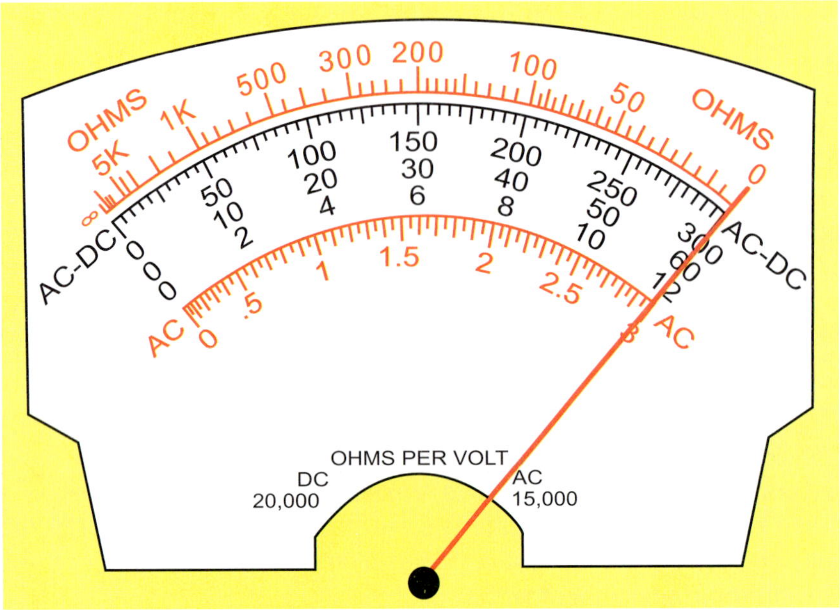

Figure 9–33 Ensure the ohmmeter reads zero by adjusting the variable resistor.

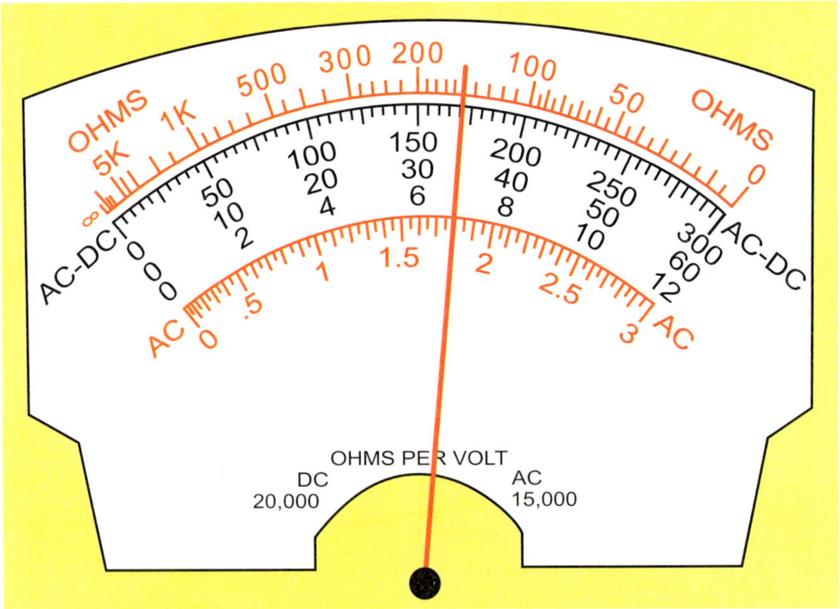

Figure 9–34 The ohmmeter only has enough current to move to 150 on the scale.

reading shown in *Figure 9–34* would be indicating a resistance of 150,000 ohms if the range had been set for R×1000. Notice that the ohmmeter scale is read backward from the other scales. Zero ohms is located on the far right side of the scale, and maximum ohms is located on the far left side.

Shunt-Type Ohmmeters

The shunt-type ohmmeter is used for measuring low values of resistance. It operates on the same basic principle as an ammeter shunt. When using a shunt-type ohmmeter, place the unknown value of resistance in parallel with the meter movement. This placement causes part of the circuit current to bypass the meter (*Figure 9–35*).

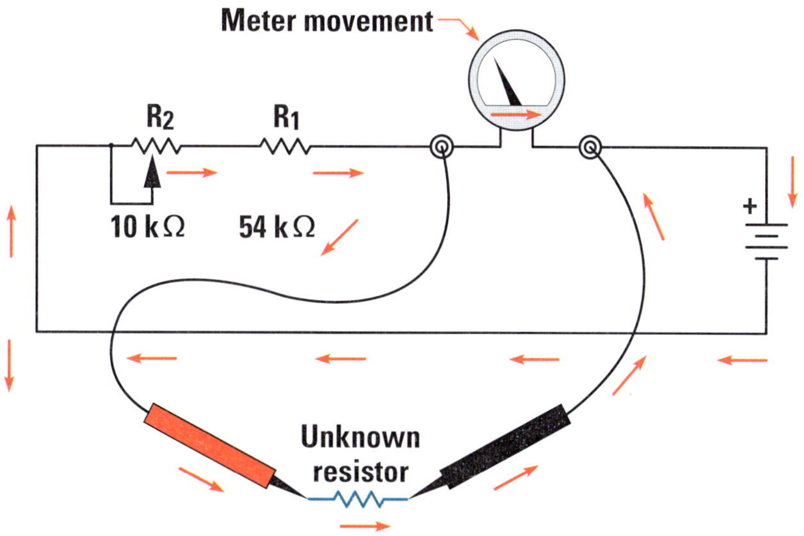

Figure 9–35 An example of a shunt-type ohmmeter.

9–5 Digital Meters

Digital Ohmmeters

Digital ohmmeters display the resistance in figures instead of using a meter movement and scale. When using a digital ohmmeter, care must be taken to notice the scale indication on the meter. For example, most digital meters will display a *K* on the scale to indicate kilohms or an M to indicate megohms (kilo means 1000 and mega means 1,000,000). If the meter is showing a resistance of 0.200 kilohms, it means 0.200 × 1000, or 200 ohms. If the meter indicates 1.65 megohms, it means 1.65 × 1,000,000, or 1,650,000 ohms.

Appearance is not the only difference between analog and digital ohmmeters. Their operating principle is different also. Analog meters operate by measuring the amount of current change in the circuit when an unknown value of resistance is added. Digital ohmmeters measure resistance by measuring the amount of voltage drop across an unknown resistance. In the circuit shown in *Figure 9–36*, a constant-current generator is used to supply a known amount of current to a resistor, R_X. It will be assumed that the amount of current supplied is 1 milliampere. The voltage dropped across the resistor is proportional to the resistance of the resistor and the amount of current flow. For example, assume the value of

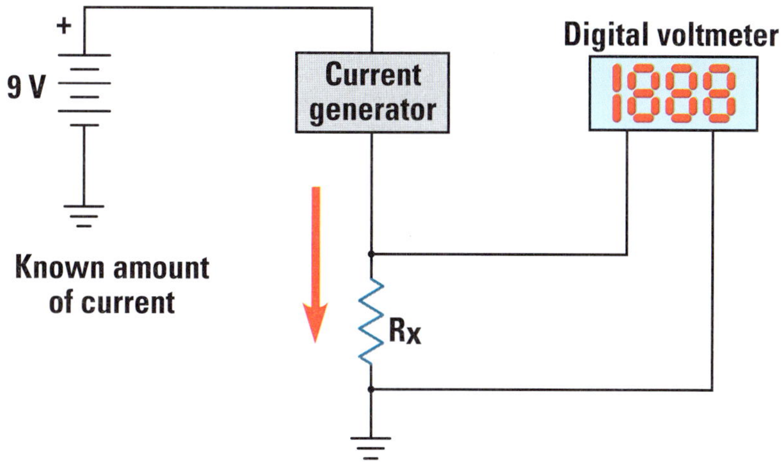

Figure 9–36 Digital ohmmeters operate by measuring the voltage drop across a resistor when a known amount of current flows through it.

the unknown resistor is 4700 ohms. The voltmeter would indicate a drop of 4.7 volts when 1 milliampere of current flowed through the resistor. The scale factor of the ohmmeter can be changed by changing the amount of current flow through the resistor.

The ohmmeter, whether digital or analog, must never be connected to a circuit when the power is turned on. Because the ohmmeter uses its own internal power supply, it is designed for a very low operating voltage. Connecting a meter to power when it is set in the ohms position can possibly damage or destroy the meter. Most modern digital meters have built-in safeguards, but it is still poor practice to use them on energized circuits and in every case, it is against the manufacturer's recommendations.

Digital Multimeters

Most test instruments used in the field today are what are called **digital multimeters**. The most apparent difference between digital meters and analog meters is that digital meters display their reading in discrete digits instead of with a pointer and scale. A digital multimeter is shown in *Figure 9–37*. Some digital meters have a range switch similar to the range switch used with analog meters. This switch sets the full-range value of the meter. However, many options available today are what are called "auto-ranging" and only require the user to know if the system is AC or DC voltage. Some don't even require the user to choose the system type and can auto-select either AC or DC voltage.

Courtesy of Advanced Test Products

Figure 9–37 An example of a digital multimeter.

Digital meters do not have the same loading effect on a circuit that analog meters do. This loading effect is a direct result of how the analog meter introduces internal resistance aspects that change the circuit characteristics. Most digital meters have an input impedance of about 10 megohms on all ranges. The input circuit impedance limits the current through the meter. This impedance is accomplished by using field-effect transistors (FETs) and a voltage divider circuit. A simple schematic for such a circuit is shown in *Figure 9–38*. Notice that the meter input is connected across 10 megohms of resistance regardless of the range setting of the meter. If this meter is used to measure the voltage of the 24-volt

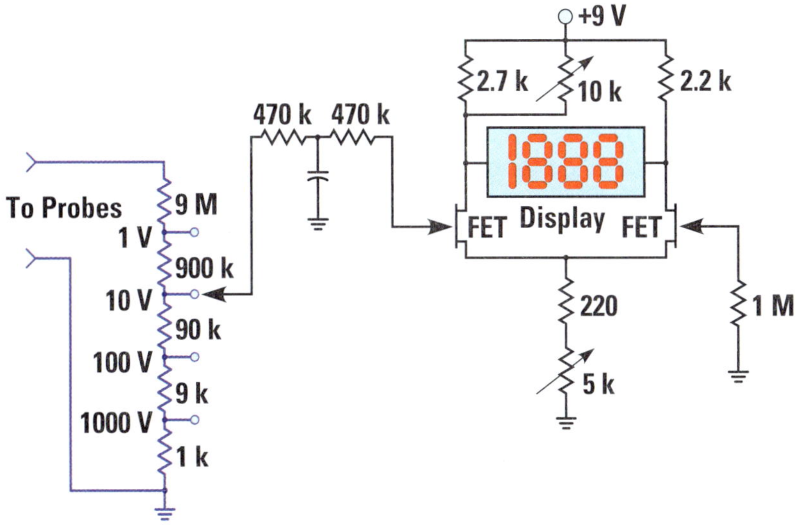

Figure 9–38 A simplified version of a digital voltmeter diagram.

circuit, a current of 2.4 microamperes will flow through the meter. This is not enough current to upset the rest of the circuit, and voltage measurements can be made accurately.

Another device used to test voltage is often referred to as a voltage tester. Historically, these devices have been a low-impedance type tester. Early versions often used a solenoid and spring as the indicator of voltage. The higher the voltage, the more the action. You might hear the phrase "Wiggy" used from time to time to refer to these types of testers. These were often desired as they did not require a battery to work and functioned off of the power in the circuit. However, the low impedance of these devices often caused problems if they were used incorrectly. These have been replaced in today's workplace with digital voltage testers that often look similar to appeal to the nostalgia of the "Wiggy." *Figure 9–39* shows an example of one of the modern devices that utilizes a system of LED lights to indicate the voltage system range of the circuit. Many of these digital devices can also test for continuity.

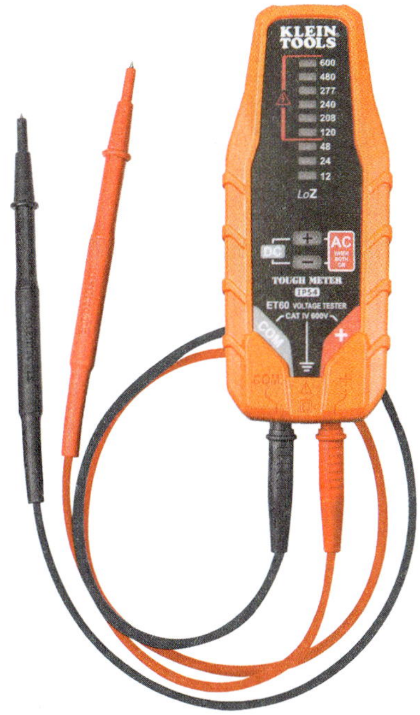

Courtesy Klein Tools, Inc

Figure 9–39 An example of a low-impedance AC/DC voltage tester.

9–6 The Oscilloscope

Electronic control systems often produce voltage pulses that are not detectable to a digital multimeter or even the analog VOM. In many instances, it is necessary to know not only the amount of voltage present at a particular point but also the length or duration of the pulse and its frequency. Some pulses may be less than 1 volt and last for only a millisecond. A VOM or digital multimeter would be useless for measuring such a pulse. It is therefore necessary to use an **oscilloscope** to learn what is actually happening in the circuit.

The oscilloscope is a powerful tool in the hands of a trained technician. The first thing to understand is that an *oscilloscope* is a voltmeter. It does not measure current, resistance, or watts. The oscilloscope measures voltage during a period of time and produces a two-dimensional image.

Voltage Range Selection

The oscilloscope is divided into two main sections. One section is the voltage section, and the other is the time base. The display of the oscilloscope is divided by vertical and horizontal lines (*Figure 9–40*). The space between these lines is called a division. The display in *Figure 9–40* has 8 divisions on the *Y*, or vertical, axis and 10 divisions on the *X*, or horizontal, axis. Voltage is measured on the vertical, or *Y*, axis of the display, and time is measured on the horizontal, or *X*, axis.

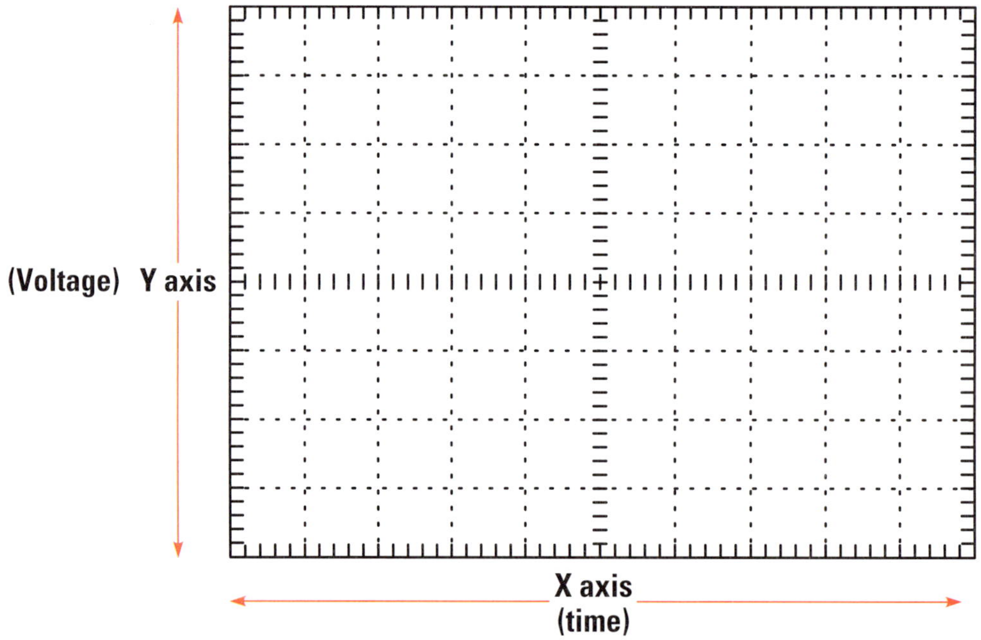

Figure 9–40 Voltage is measured on the vertical, or *Y*, axis and time is measured on the horizontal, or *X*, axis.

Oscilloscopes can be divided into two main types: analog and digital. Analog oscilloscopes have been used for years and many are still in use; however, digital oscilloscopes are rapidly taking their place. Analog scopes generally employ some type of control knob to change their range of operation, *Figure 9–41*. The setting indicates the volts per division instead of volts full scale. The settings in *Figure 9–41* indicate that Channel 1 is set for 0.2 volt per division, and Channel 2 is set for 0.5 volt per division.

Digital oscilloscopes often indicate their setting on the display instead of marking them on the face of the oscilloscope. Voltage control knobs for a four-channel digital oscilloscope are shown

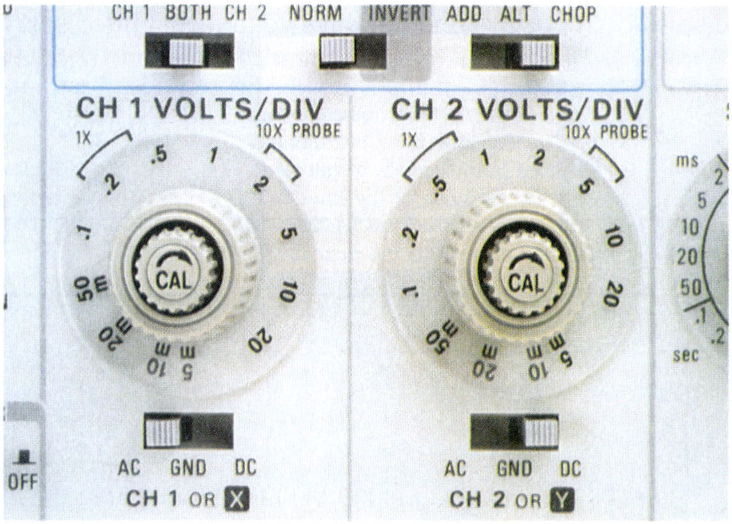

Courtesy of Tektronix, Inc.

Figure 9–41 Voltage control of an analog oscilloscope.

in *Figure 9–42*. The display of a typical digital oscilloscope is shown in *Figure 9–43*. In the lower left-hand corner of the display, the notation CH1 200mV can be seen. This indicates that the voltage range has been set for 200 millivolts per division.

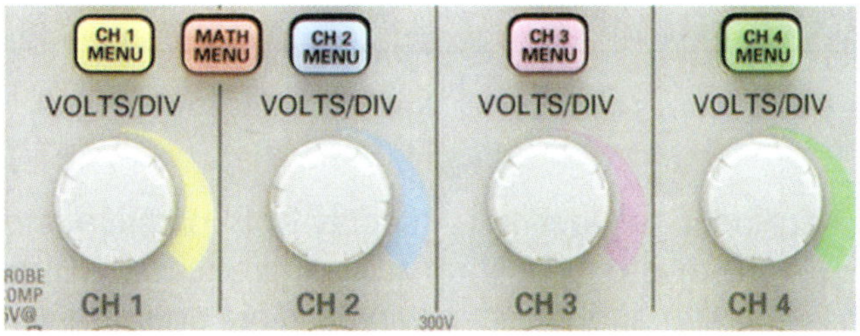

Courtesy of Tektronix, Inc.

Figure 9–42 Voltage control knobs of a four-channel digital oscilloscope.

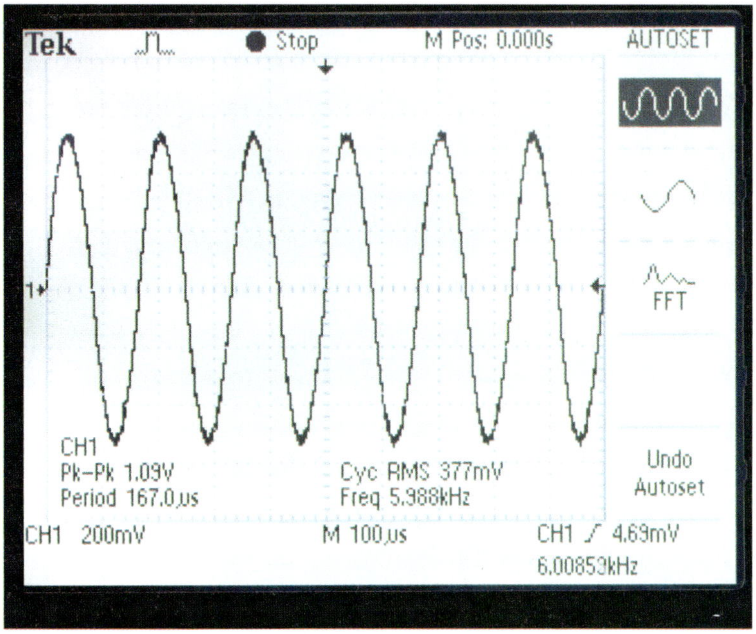

Courtesy of Tektronix, Inc

Figure 9–43 Display of a digital oscilloscope.

Oscilloscopes can display both positive and negative voltages. In the display shown in *Figure 9–44,* assume that a value of 0 volts has been set at the center line. The voltage shown at Position A is positive with respect to 0, and the voltage at Position B is negative with respect to 0. If the oscilloscope were set for a value of 2 volts per division, the value at Point A would be 6 volts positive with respect to 0, and the value at Point B would be 6 volts negative with respect to 0.

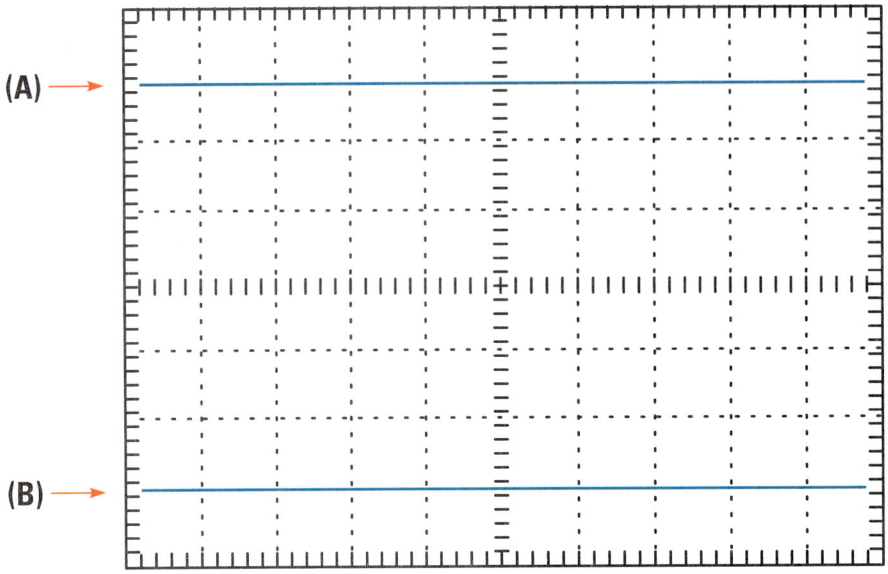

Courtesy of Tektronix, Inc.

Figure 9–44 The oscilloscope displays both positive and negative voltages.

Another example of the oscilloscope's ability to display both positive and negative voltage values is shown in *Figure 9–45*. The waveform is basically an AC square wave. Notice that the voltage peaks at the leading edge of both the negative and positive waves. This could never be detected with a digital multimeter or VOM.

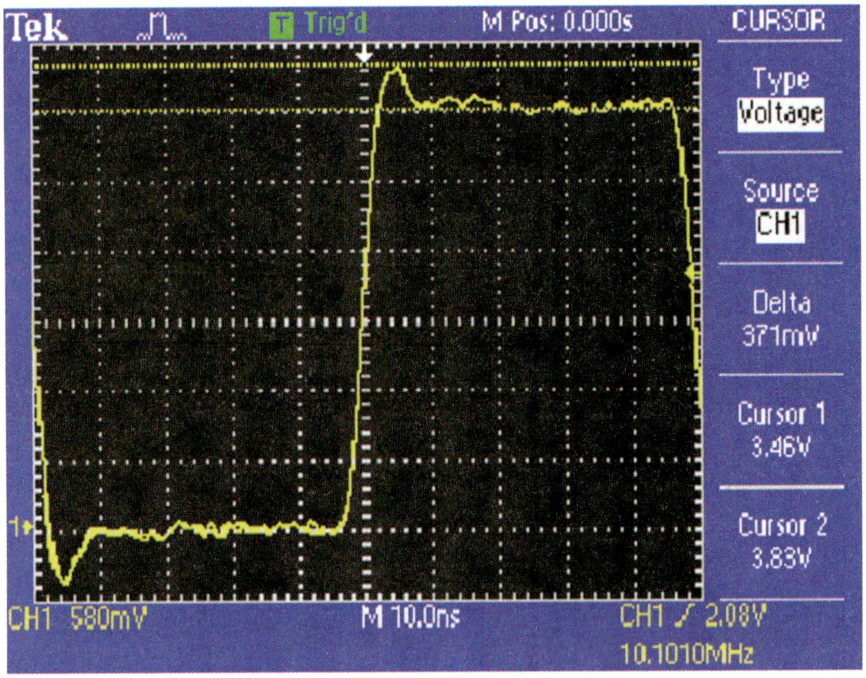

Courtesy of Tektronix, Inc.

Figure 9–45 Square AC waveform observed with an oscilloscope.

Many oscilloscopes also have the ability to display more than one voltage at a time. Each voltage is generally referred to as a trace or channel. The oscilloscope shown in *Figure 9–46* has the ability to display four voltages or four traces. Many digital-type oscilloscopes have the ability to display a different color for each trace.

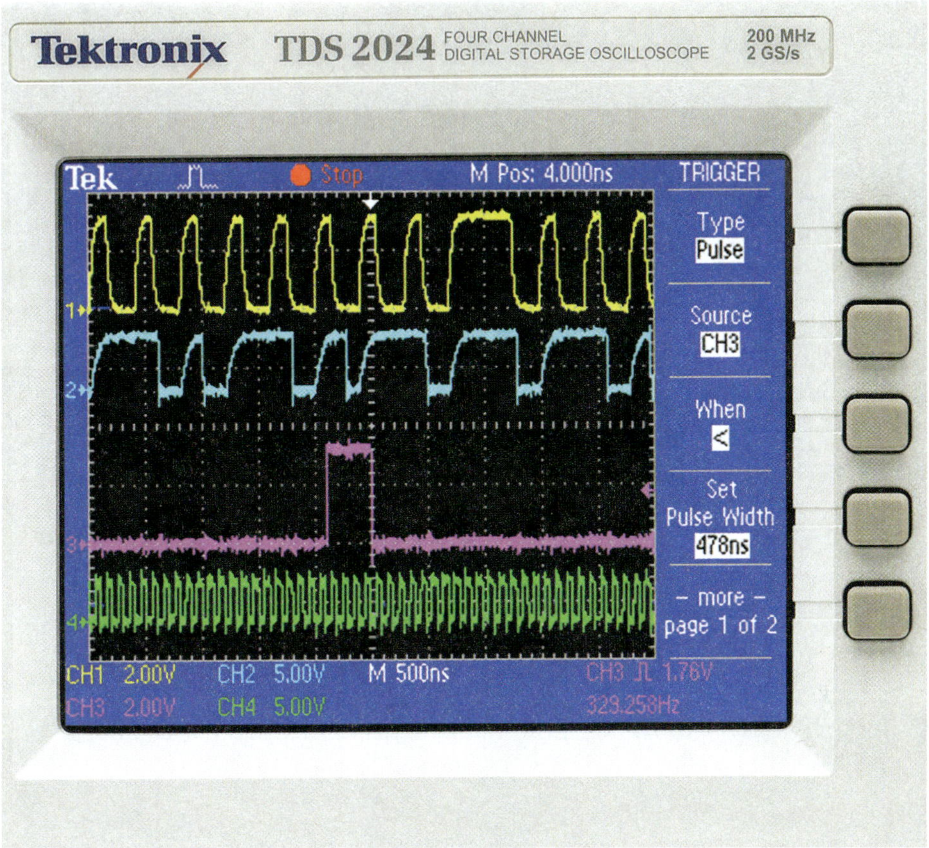

Courtesy of Tektronix, Inc.

Figure 9–46 An example of a four-trace or four-channel oscilloscope.

The Time Base

The next section of the oscilloscope to discuss is the time base. The time base is calibrated in seconds per division and has ranges from seconds to microseconds. The controls for these are very similar in use to the voltage selection controls. Analog-type oscilloscopes use a range selection switch similar to the one shown in *Figure 9–47*. The time base selection control for a typical digital-type oscilloscope is shown in *Figure 9–48*. Digital oscilloscopes generally indicate the time base setting on the display as shown in *Figure 9–43*. The lower middle section of the display shows M 100 microseconds, which indicates that the scope is set for 100 microseconds per division. With the time base set at this value, it would take the trace 1000 microseconds, or one millisecond, to sweep across the face of the display.

Measuring Frequency

Because the oscilloscope has the ability to display the voltage with respect to time, it is possible to calculate the frequency of the waveform. The frequency (f) of an AC waveform can be found by determining the inverse of the time (t) it takes to complete one cycle ($f = 1/t$). For example, the time base

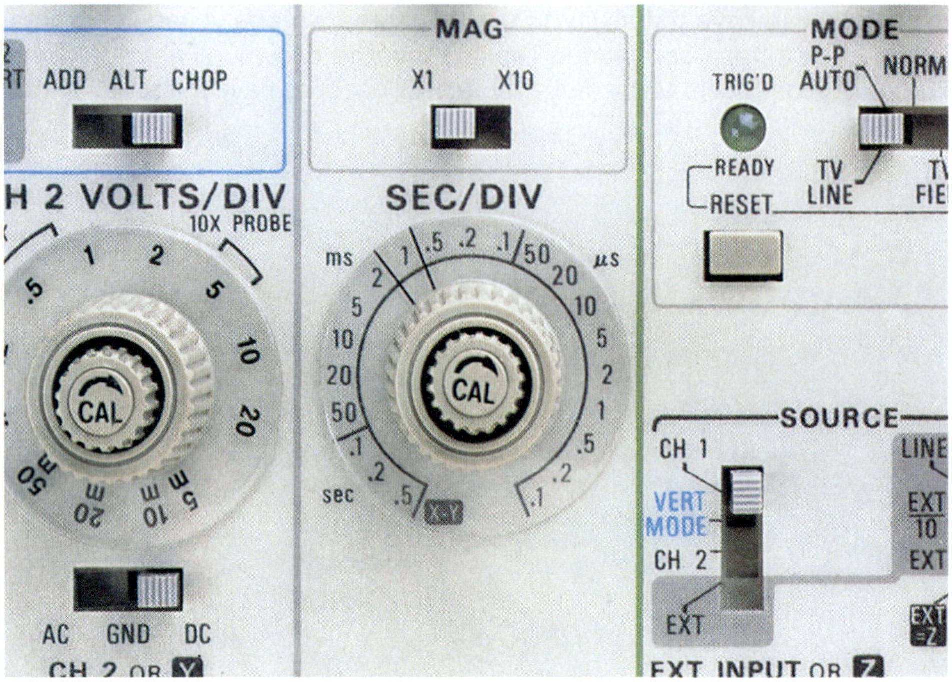

Courtesy of Tektronix, Inc.

Figure 9–47 Example of the time base of an analog oscilloscope.

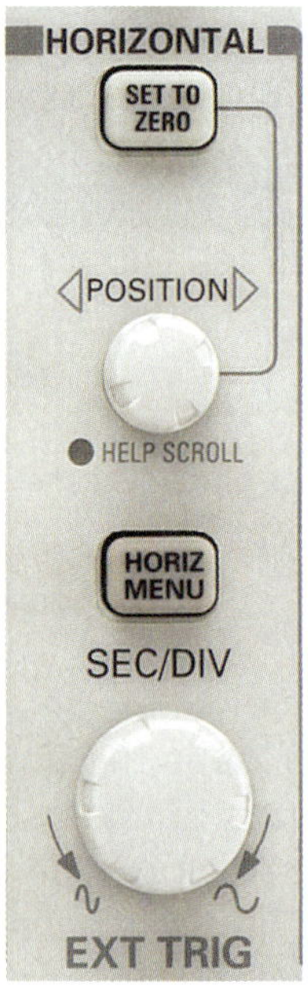

Courtesy of Tektronix, Inc.

Figure 9–48 Example of the time base control for a digital oscilloscope.

in *Figure 9–43* is set at 100 microseconds per division. The AC sine wave being displayed completes one complete cycle in 167 microseconds. The frequency is 5988 hertz, or 5.988 kilohertz (1/0.000167). The oscilloscope display in *Figure 9–43* displays these values. Many oscilloscopes have the ability to measure frequency automatically and display the value for you.

Attenuated Probes

Most oscilloscopes use a probe that acts as an attenuator. An attenuator is a device that divides or makes smaller the input signal (*Figure 9–49*). An attenuated probe is used to permit higher voltage readings than are normally possible. For example, most attenuated probes are 10 to 1. This means that if the voltage range switch is set for 5 volts per division, the display would indicate 50 volts per division. If the voltage range switch is set for 2 volts per division, each division on the display has a value of 20 volts per division.

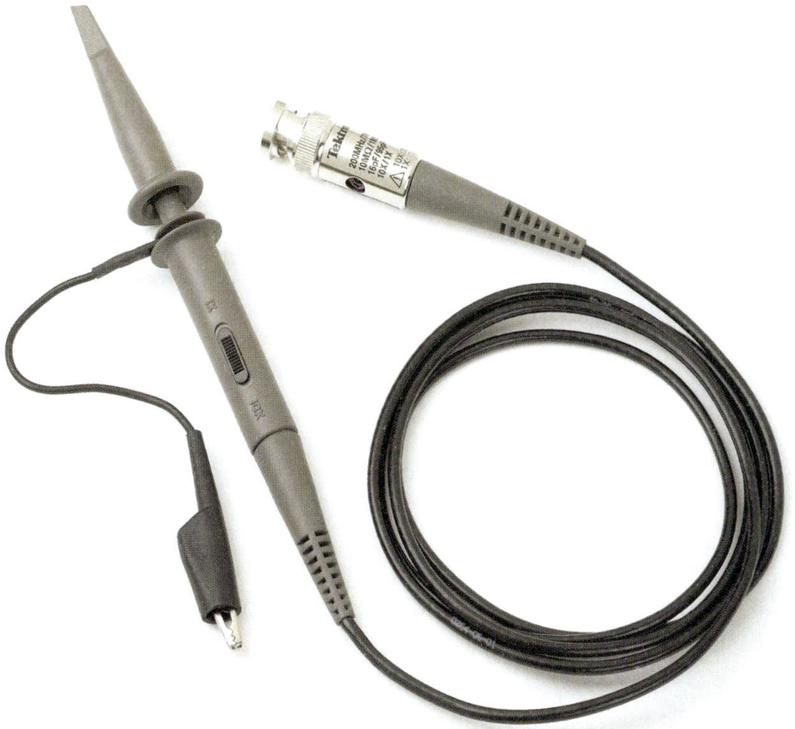

Courtesy of Tektronix, Inc.

Figure 9–49 Oscilloscope attenuated probe.

Probe attenuators are made in different styles by different manufacturers. On some probes, the attenuator is in the probe head itself, whereas on others, the attenuator is located at the scope input. Regardless of the type of attenuated probe used, it may have to be compensated or adjusted. In fact, probe compensation should be checked frequently. Different manufacturers use different methods for compensating their probes, so it is generally necessary to follow the procedures given in the operator's manual for the probe being used.

Oscilloscope Controls

The following is a list of common controls found on the oscilloscope. Refer to the oscilloscope shown in *Figure 9–50*.

1. **POWER.** The power switch is used to turn the oscilloscope ON or OFF.
2. **BEAM FINDER.** This control is used to locate the position of the trace if it is off the display. The BEAM FINDER button will indicate the approximate location of the trace. The position controls are then used to move the trace back on the display.
3. **PROBE ADJUST** (sometimes called calibrate). This is a reference voltage point used when compensating the probe. Most probe adjust points produce a square wave signal of about 0.5 volt.

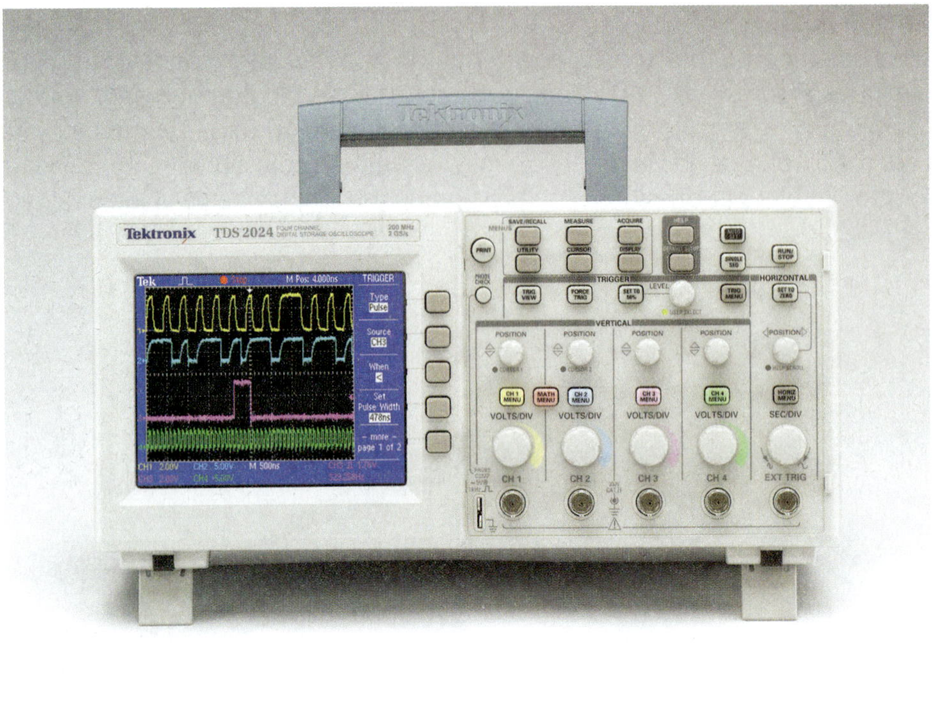

Figure 9–50 An example of a digital oscilloscope.

4. **INTENSITY** and **FOCUS.** The INTENSITY control adjusts the brightness of the trace. A bright spot should never be left on the display because it will burn a spot on the face of the cathode ray tube (CRT). This burned spot results in permanent damage to the CRT. The FOCUS control sharpens the image of the trace.
5. **VERTICAL POSITION.** This is used to adjust the trace up or down on the display. If a dual-trace oscilloscope is being used, there will be two vertical POSITION controls. (A dual-trace oscilloscope contains two separate traces that can be used separately or together.)
6. **CH 1–BOTH–CH 2.** This control determines which channel of a dual-trace oscilloscope is to be used or whether they are both to be used at the same time.
7. **ADD–ALT.–CHOP.** This control is active only when both traces are being displayed at the same time. The ADD adds the two waves together. *ALT*. stands for alternate. This alternates the sweep between Channel 1 and Channel 2. The CHOP mode alternates several times during one sweep. This generally makes the display appear more stable. The CHOP mode is generally used when displaying two traces at the same time.
8. **AC–GND–DC.** The AC is used to block any DC voltage when only the AC portion of the voltage is to be seen. For instance, assume an AC voltage of a few millivolts is riding on a DC voltage of several hundred volts. If the voltage range is set high enough that 100 V DC can be seen on the display, the AC voltage cannot be seen. The AC section of this switch inserts a capacitor in series with the probe. The capacitor blocks the DC voltage and permits the AC voltage to pass. Because the 100 V DC has been blocked, the voltage range can be adjusted for millivolts per division, which will permit the AC signal to be seen.

   The GND section of the switch stands for ground. This section grounds the input so the sweep can be adjusted for 0 volt at any position on the display. The ground switch grounds at the scope and does not ground the probe. This permits the ground switch to be used when the probe is connected to a live circuit. The DC section permits the oscilloscope to display all of the voltage, both AC and DC, connected to the probe.

9. **HORIZONTAL POSITION.** This control adjusts the position of the trace from left to right.
10. **AUTO–NORMAL.** This determines whether the time base will be triggered automatically or operated in a free-running mode. If this control is operated in the NORM setting, the trigger signal is taken from the line to which the probe is connected. The scope is generally operated with the trigger set in the AUTO position.
11. **LEVEL.** The LEVEL control determines the amplitude the signal must be before the scope triggers.
12. **SLOPE.** The SLOPE permits selection as to whether the trace is triggered by a negative or positive waveform.
13. **INT.–LINE–EXT.** *INT*. stands for internal. The scope is generally operated in this mode. In this setting, the trigger signal is provided by the scope. In the LINE mode, the trigger signal is provided from a sample of the line. *EXT*, or external, mode permits the trigger pulse to be applied from an external source.

These are not all the controls shown on the oscilloscope in *Figure 9–50,* but they are the major controls. Most oscilloscopes contain these controls.

Interpreting Waveforms

The ability to interpret the waveforms on the display of the oscilloscope takes time and practice. When using the oscilloscope, one must keep in mind that the display shows the voltage with respect to time.

In *Figure 9–51*, it is assumed that the voltage range has been set for 0.5 volt per division, and the time base is set for 2 milliseconds per division. It is also assumed that 0 volt has been set on the center line of the display. The waveform shown is a square wave. The display shows that the voltage rises in the positive direction to a value of 1.48 volts and remains there for 2 milliseconds. The voltage then drops to 1.48 volts negative and remains there for 2 milliseconds before going back to positive. Because the voltage changes between positive and negative, it is an AC voltage. The length of one cycle is 4 milliseconds. The frequency is therefore 250 hertz (1/0.004 s = 250 Hz).

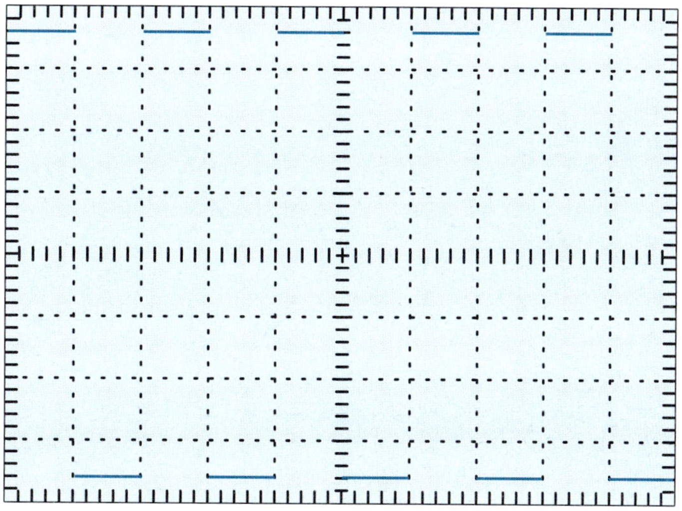

Figure 9–51 A 250 hertz AC square waveform.

In *Figure 9–52*, the oscilloscope has been set for 50 millivolts per division and 20 microseconds per division. The display shows a voltage that is negative to the probe's ground lead and has a peak value of 150 millivolts. The waveform lasts for 20 microseconds and produces a frequency of 50 kilohertz

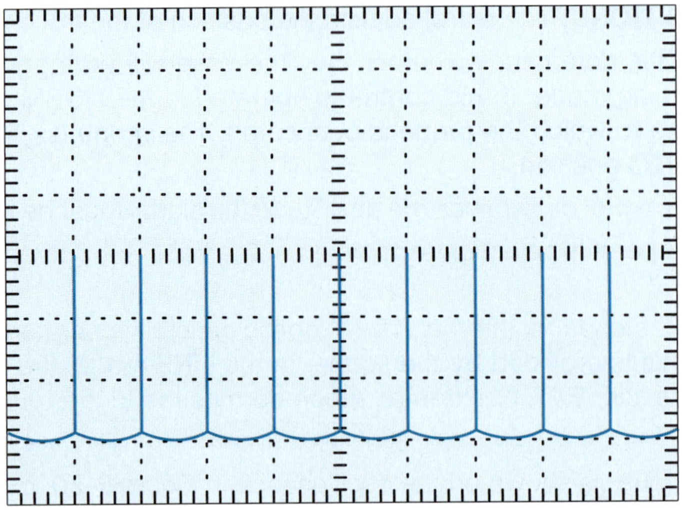

Figure 9–52 A 50 kHz pulsating DC waveform.

(1/0.000020 s = 50,000 Hz). The voltage is DC because it never crosses the zero reference and goes in the positive direction. This type of voltage is called *pulsating DC*.

In *Figure 9–53*, assume the oscilloscope has been set for a value of 50 volts per division and 4 milliseconds per division. The waveform shown rises from 0 volts to about 45 volts in a period of about 1.5 milliseconds. The voltage gradually increases to about 50 volts in the space of 1 millisecond and then rises to a value of about 100 volts in the next 2 milliseconds. The voltage then decreases to 0 in the next 4 milliseconds. It then increases to a value of about 10 volts in 0.5 millisecond and remains at that level for about 8 milliseconds. This is one complete cycle for the waveform. The length of the one cycle is about 16.6 milliseconds, which is a frequency of 60.2 hertz. (1/0.0166). The voltage is DC because it remains positive and never drops below the 0 line.

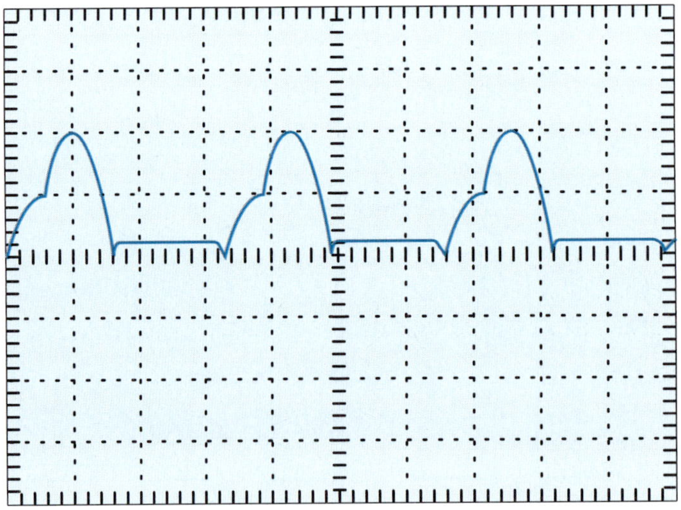

Figure 9–53 A 60 Hz chopped DC waveform.

The oscilloscope is the only means by which many of the waveforms and voltages found in electronic circuits can be understood. Consequently, the oscilloscope is the single most valuable piece of equipment a technician can use in determining the characteristics of sensitive electronics.

9–7 The Wattmeter

The **wattmeter** is used to measure true power in a circuit. There are two basic types of wattmeters, dynamic and electronic. Dynamic wattmeters differ from d'Arsonval-type analog meters in that they do not contain a permanent magnet. They contain an electromagnet and a moving coil (*Figure 9–54*). The electromagnets are connected in series with the load in the same manner that an ammeter is connected. The moving coil has resistance connected in series with it and is connected directly across the power source in the same manner as a voltmeter (*Figure 9–55*).

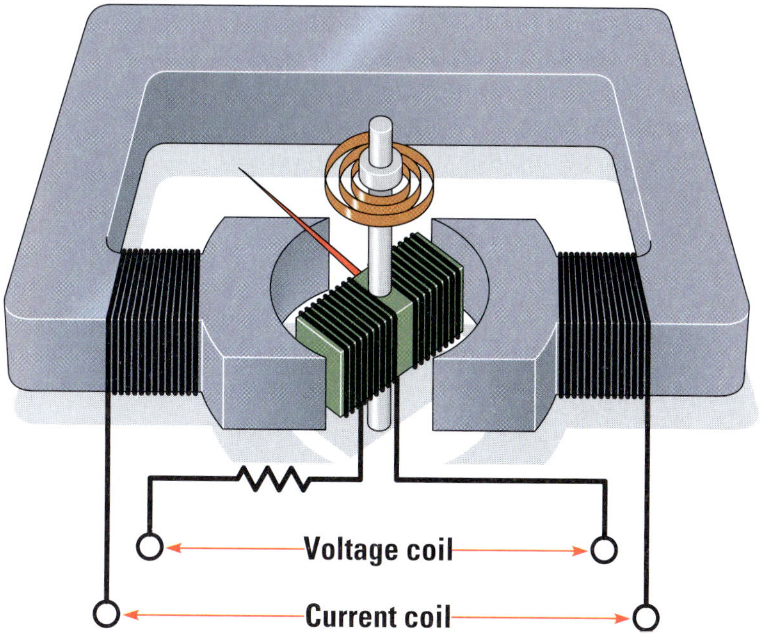

Figure 9–54 The dynamic wattmeter contains two coils—one for voltage and the other for current.

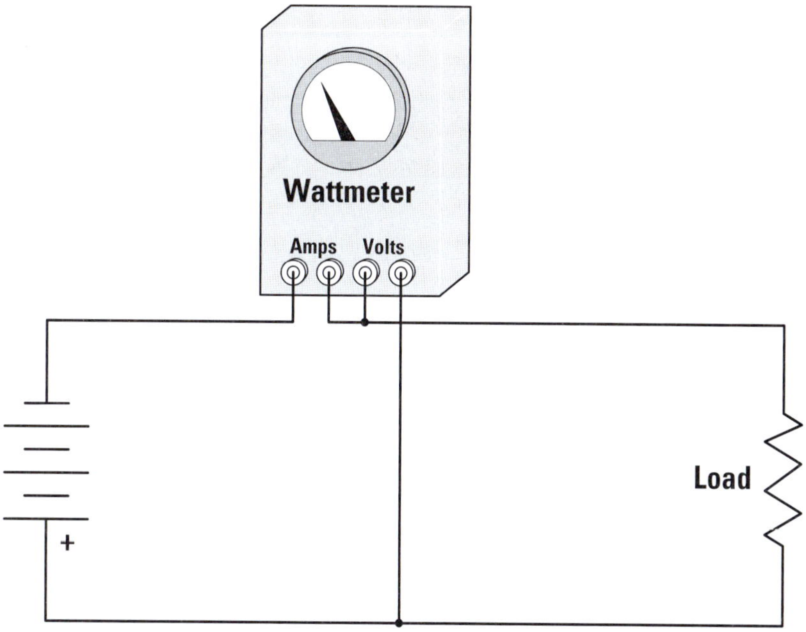

Figure 9–55 The current section of the wattmeter is connected in series with the load, and the voltage section is connected in parallel with the load.

Because the electromagnet is connected in series with the load, the current flow through the load determines the magnetic field strength of the stationary magnet. The magnetic field strength of the moving coil is determined by the amount of line voltage. The turning force of the coil is proportional to the strength of these two magnetic fields. The deflection of the meter against the spring is proportional to the amount of current flow and voltage.

Because the wattmeter contains an electromagnet instead of a permanent magnet, the polarity of the magnetic field is determined by the direction of current flow. The same is true of the polarity of the moving coil connected across the source of voltage. If the wattmeter is connected into an AC circuit, the polarity of the two coils will reverse at the same time, producing a continuous torque. For this reason, the wattmeter can be used to measure power in either a DC or an AC circuit. However, if the connection of the stationary coil or the moving coil is reversed, the meter will attempt to read backward.

Dynamic-type wattmeters are being replaced by wattmeters that contain electronic circuitry to determine true power. They are less expensive and generally more accurate than the dynamic type. Like dynamic wattmeters, electronic-type meters contain amperage terminals that connect in series with the load and voltage terminals that connect in parallel with the load. Portable-type wattmeters often have terminals labeled "power in" and "power out." Connection to the current and voltage section of the meter is made inside the meter (*Figure 9–56*). Analog-type electronic wattmeters use a standard d'Arsonval-type movement to indicate watts. The electronic circuit determines the true power of the circuit and then supplies the appropriate power to the meter movement. Wattmeters with digital displays are also available and are often the standard method for measuring power.

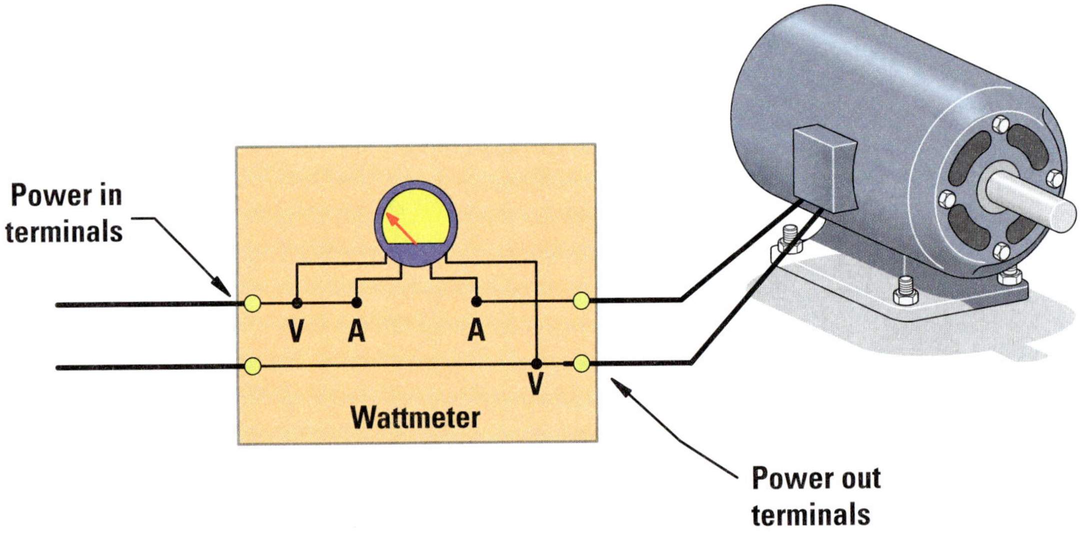

Figure 9–56 Portable wattmeters often make connection to the voltage and current terminals inside the meter.

Recording Meters

On occasion, it becomes necessary to make a recording of an electrical value over a long period of time. Such an ability is often desired when studying the power quality of a system. Recording meters produce a graph of metered values during a certain length of time. They are used to detect spike voltages, currents of short duration, or sudden drops in voltage, current, or power. Recording meters can show the amount of voltage or current, its duration, and the time of occurrence. Some meters have the ability to store information in memory over a period of several days. This information can be recalled later by the service technician. Several types of recording meters are shown in *Figure 9–57*.

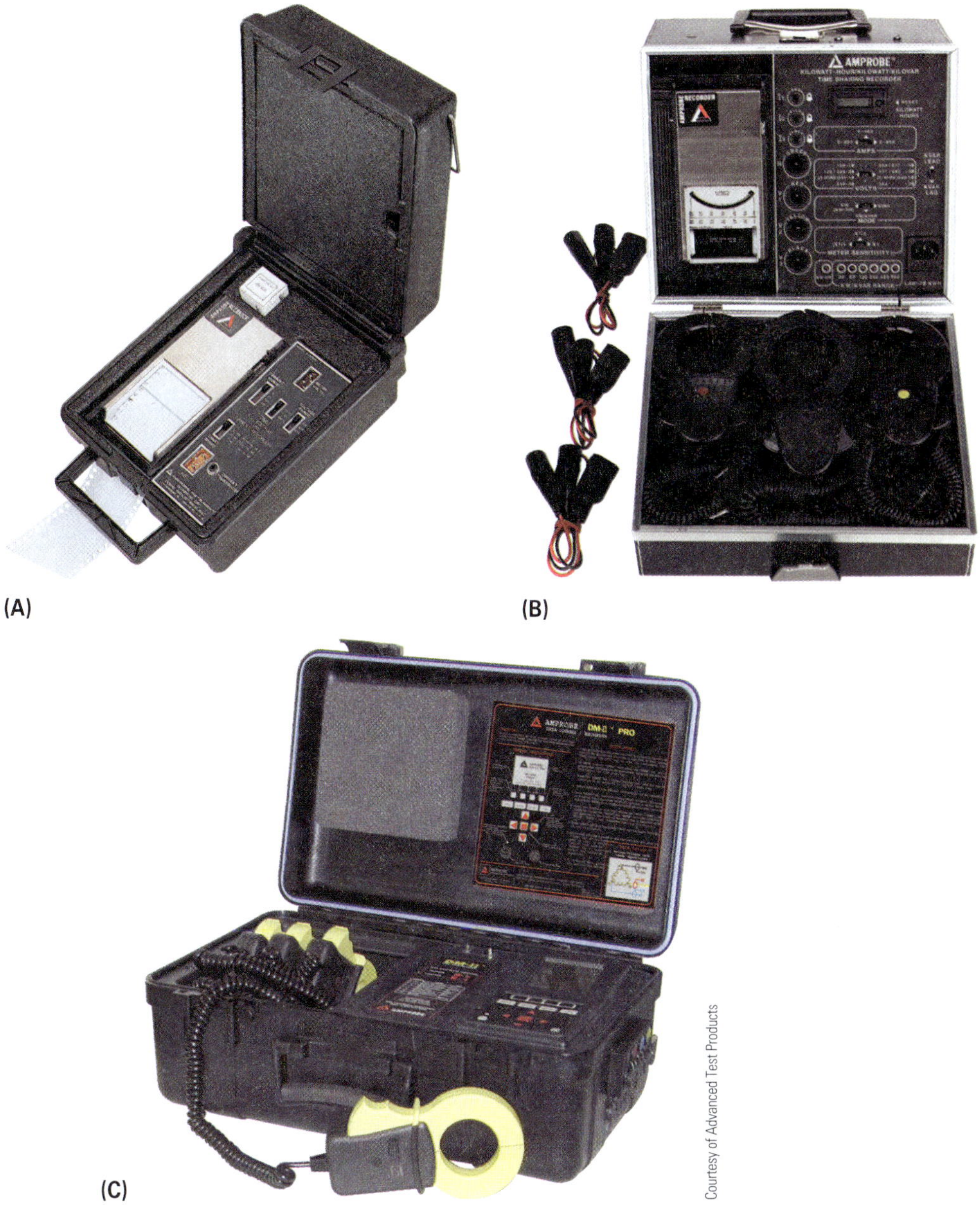

(A) (B) (C)

Courtesy of Advanced Test Products

Figure 9–57 **(A)** Single-line recording volt-ammeter. **(B)** A kilowatt-kiloVAR recording meter. **(C)** A single-phase or three-phase voltage and current recording meter.

Summary

- The d'Arsonval type of meter movement is based on the principle that like magnetic fields repel.
- The d'Arsonval movement operates only on DC.
- Voltmeters have a high resistance and are designed to be connected directly across the power line.
- The steps to reading a meter are as follows

    a. Determine what the meter is set to indicate.

    b. Determine which full-scale value of the meter will apply to the measurement.

    c. Read the meter.

- Ammeters have a low resistance and must be connected in series with a load to limit the flow of current.
- Shunts are used to change the value of DC ammeters.
- AC ammeters use a current transformer to change the range setting.
- Clamp-on ammeters measure the flow of current by measuring the strength of the magnetic field around a conductor.
- Ohmmeters are used to measure the resistance in a circuit.
- Ohmmeters contain an internal power source, generally batteries.
- Ohmmeters must never be connected to a circuit that has power applied to it.
- Digital multimeters display their value in digits instead of using a meter movement.
- Digital multimeters generally have an input impedance of 10 megohms on all ranges.
- The oscilloscope measures the amplitude of voltage with respect to time.
- The frequency of a waveform can be determined by finding the inverse of the time of one cycle ($f = 1/t$).
- Wattmeters contain a stationary coil and a movable coil.
- The stationary coil of a wattmeter is connected in series with the load, and the moving coil is connected to the line voltage.
- The turning force of the dynamic wattmeter is proportional to the strength of the magnetic field of the stationary coil and the strength of the magnetic field of the moving coil.
- Digital ohmmeters measure resistance by measuring the voltage drop across an unknown resistor when a known amount of current flows through it.

Review Questions

1. Which of the following determines how far a d'Arsonval meter movement will turn?

   A) The selector switch

   B) The strength of the magnetic field in the coil

   C) Gravitational forces

   D) The amount of voltage that flows in the circuit

2. Since the principle of operation for a d'Arsonval meter movement is based on repelling magnetic fields, which of the following statements are true? (Select all that apply.)

   A) The meter movement functions with DC voltage.

   B) The meter movement functions with AC voltage.

   C) The meter movement functions with inverted AC voltage.

   D) The meter movement functions with rectified AC voltage that has been changed into DC.

3. Which of the following is true when connecting a voltmeter to a circuit to measure the voltage drop across a circuit component?

   A) The voltmeter is connected in series with the circuit component.

   B) The circuit must be de-energized before connection.

   C) The voltmeter is connected in parallel to the circuit component.

   D) The voltmeter is connected both in series and parallel to the circuit component.

4. True or False: A voltmeter has an internal resistance that is significantly greater than that of the circuit being measured.

5. Which of the following are true when connecting an ammeter to a circuit to measure the current flowing in the circuit?

   A) The ammeter is connected in both series and parallel to the circuit component being measured.

   B) The circuit must be de-energized to measure current.

   C) The ammeter is connected in parallel to the circuit.

   D) The ammeter is connected in series to the circuit.

6. Which two methods are used to allow a DC ammeter to have multirange functionality?

   A) Make-before-break switch; Ayrton shunt

   B) Make-before-break switch; voltage divider circuit

   C) Ayrton shunt; voltage divider circuit

   D) Voltage divider circuit; low-resistance shunt

7. Which of the following is true when connecting an ohmmeter to a circuit to measure the resistance of a circuit component?

   A) The ohmmeter is connected in series with the circuit component to be measured.

   B) The circuit must be de-energized to measure resistance.

   C) The ohmmeter is connected in both series and parallel to the circuit component.

   D) The ohmmeter is connected in parallel to energized electrical components to read their resistance value.

8. The selector switch on an ohmmeter is set to the setting labeled MΩ. What must the value indicated be multiplied by to find the actual number of ohms being detected by the ohmmeter?

   A) 100

   B) 1000

   C) 100,000

   D) 1,000,000

9. True or False: A square waveform shown on an oscilloscope display is one where the voltage is displayed in only the positive polarity, and the waveform never crosses the zero point.

10. A waveform observed on the display of an oscilloscope stays in a positive polarity with a frequency of 50 kHz. Which of the following best describes what type of waveform is being observed?

    A) A pulsating AC waveform

    B) An inverted AC waveform

    C) A pulsating DC waveform

    D) An AC square waveform

11. Which of the following is true when connecting a wattmeter to measure power within a circuit?
    A) The wattmeter is connected both in series and in parallel with the circuit.
    B) The wattmeter is connected only in series to the circuit.
    C) The wattmeter is connected only in parallel to the circuit.
    D) The wattmeter can only measure power on a de-energized circuit.
12. True or False: Electronic-type wattmeters are similar to dynamic wattmeters in that they have current terminals that connect in series with the load and have voltage terminals that can connect in parallel with the load.
13. Which of the following best describes how a digital ohmmeter determines the resistance when measuring across a component in a circuit?
    A) Digital meters measure the change in current
    B) Digital meters measure the voltage flowing in the circuit
    C) Digital meters measure the amount of voltage drop
    D) Digital meters measure the power dissipated
14. Digital multimeters use which of the following to display their reading?
    A) Discrete digits
    B) A digital pointer and scale
    C) A solenoid and indicator arrow
    D) None of the above

Practical Applications

You are an electrician on the job. You have been given a multimeter that has the following AC voltage ranges: 30, 60, and 150. The meter states that it has a resistance of 5000 Ω/V. You need to be able to measure a voltage of 277 volts. How much resistance should be inserted in series with the meter to make the 30-volt range indicate a full-scale value of 300 volts?

Unit 10 Determining Conductor Sizes

Key Terms

Ambient air temperature

American Wire Gauge (AWG)

Ampacity

Circular mil

Correction factor

Damp locations

Dry locations

Insulation

Maximum operating temperature

MEGGER

Mil-foot

National Electrical Code

Parallel conductors

Wet locations

Why You Need to Know

The ability to select the proper conductor for an installation based on how much current a conductor can carry is a basic necessity for anyone working in the electrical field. This unit

- explains how the amount of current a conductor is permitted to carry is not the same as selecting the proper wire for an installation and describes the differences.
- differentiates the various types of wire insulation and the appropriate use of each based on ambient temperatures.
- explains the method for using tools such as a MEGGER when determining the resistance of wire insulation.
- discusses how conductor length and size impact resistance and how to select the required conductor size.
- provides the tools for determining the ampacity of conductors when applying correction factors for wiring in a raceway.
- explains that, as a general rule, electricians select conductors based on their ampacity as specified in the *NFPA 70®: National Electrical Code® (NEC®)*. However, there are instances where length or some unique condition or situation is being employed. In those instances, wire size and type are chosen by determining the maximum voltage drop and calculating the resistance of the wire. This unit explains how to determine wire size using the *NEC* and by calculating wire resistance.

Outline

Objectives

After studying this unit, you should be able to

- apply NEC ampacity tables in conductor selection.
- compare conductor insulation types and characteristics.
- apply correction factors to conductor ampacity values based on conditions of use.
- describe how conductor length affects resistance.
- explain how parallel conductors must be installed.

Preview

The size of the conductor needed for a particular application can be determined by several methods. The ***NFPA 70: National Electrical Code* (NEC)** is used throughout industry to determine the conductor size for most applications. Typically, the circuit conductors are sized based on their ability to safely carry the circuit current without overheating. This property is defined in *NEC Article 100* as the conductor's ***ampacity***. Ampacity is defined as "*the maximum current, in amperes, that a conductor can carry continuously under the conditions of use without exceeding its temperature rating.*" It is imperative that an electrician be familiar with the application of *Code* tables. However, in some circumstances, the conductor selected will not be able to adequately supply the load, as in the case of extremely long conductor runs where the resistance of the conductor itself affects the circuit. In these instances, the electrician should know how to determine the conductor size needed by calculating maximum voltage drop and resistance of the conductor.

10–1 Conductor Sizes and Ampacity

The **American Wire Gauge (AWG)** was standardized in 1857 and is used mainly in the United States for the diameters of round, solid, nonferrous electrical wire. The gauge size is important for determining the current-carrying capacity of a conductor. Gauge sizes are determined by the number of draws necessary to produce a given diameter of wire, the higher the number of draws, the smaller the diameter of the wire. Electrical wire is made by drawing it through a succession of dies (*Figure 10–1*).

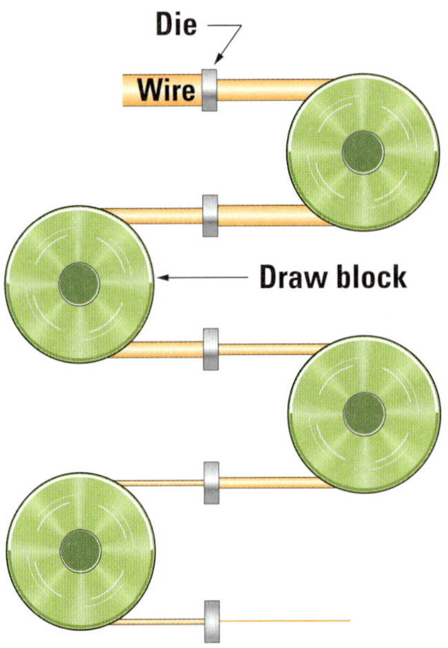

Figure 10–1 Copper is drawn through a succession of dies to produce the desired conductor diameter.

Each time a wire passes through a die, it is wrapped around a draw block several times. The draw block provides the pulling force necessary to draw the material through the die. A 24 AWG wire would be drawn through 24 dies, each having a smaller diameter. In the field, conductor size can be determined with a wire gauge (*Figure 10–2*). One side of the wire gauge lists the AWG size of the wire (*Figure 10–3*). The opposite side of the wire gauge indicates the diameter of the wire in thousandths of an inch (*Figure 10–4*). When determining conductor size, first remove the insulation from around the conductor. The slots in the wire gauge, not the holes behind the slots, are used to determine the size (*Figure 10–5*).

Figure 10–2 Wire gauge.

Figure 10–3 One side of the wire gauge is marked with the AWG size.

Figure 10–4 The other side of the wire gauge lists the diameter of the wire in thousandths of an inch.

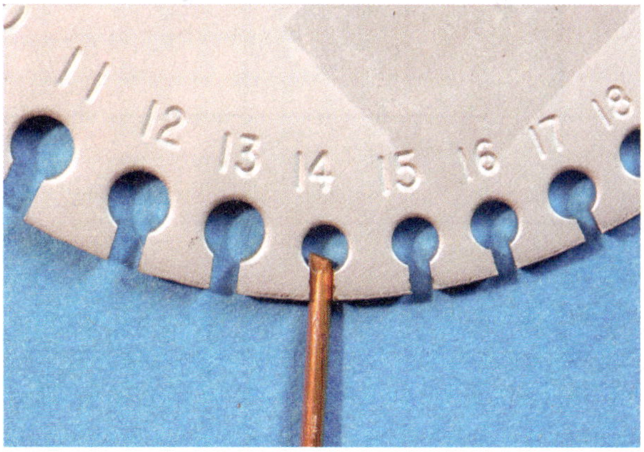

Figure 10–5 The slot in the wire gauge is used to measure conductor size.

The largest AWG size is 4/0, which has an area of 211,600 circular mils. Conductors with a larger area are measured in kilos, or thousands, of circular mils. The next largest conductor past 4/0 is 250 thousand circular mils (250 kcmil). Conductors can be obtained up to 2000 kcmil. In practice, conductors this large are often difficult to install, and, therefore, it is usually desirable to install parallel runs of smaller conductors instead.

Using the *NEC* Ampacity Tables

NEC Article 310 deals with conductors for general wiring and *Tables 310.16* through *310.21* list conductor ampacities according to certain parameters of the circuit. The tables are divided by the type of insulation used on the conductor and the conductor material. Each of these tables lists the parameters upon which it is based. *Table 310.16* (*Figure 10–6*) lists the ampacity of conductors based on not more than three current-carrying conductors in raceway or cable or directly buried in the earth. These values are also based on an **ambient** (surrounding) **air temperature** of 30°C (86°F). *Table 310.17* (*Figure 10–7*) lists ampacities of conductors in free air based on an ambient temperature of 30°C. *Table 310.18* (*Figure 10–8*) and *Table 310.19* apply to conductors with specific insulation types and are generally used for high-temperature locations as indicated by an ambient temperature of 40°C (104°F). The heading at the top of each table lists a different set of conditions to which they apply.

Table 310.16 Ampacities of insulated Conductors with Not More Than Three Current-Carrying Conductors in Raceway, Cable, or Earth (Directly Buried)

Size AWG or kcmil	Temperature Rating of Conductor [SeeTable 310.4(I)]						Size AWG or kcmil
	60°C (140°F)	75°C (167°F)	90°C (194°F)	60°C (140°F)	75°C (167°F)	90°C (194°F)	
	Types TW, UF	Types RHW, THHW, THW, THWN, XHHW, XHWN, USE, ZW	Types TBS, SA, SIS, FEP, FEPB, MI, PFA, RHH, RHW-2, THHN, THHW, THW-2, THWN-2, USE-2, XHH, XHHW, XHHW-2, XHWN, XHWN-2 XHHN, Z, ZW-2	Types TW, UF	Types RHW, THHW, THW, THWN, XHHW, XHWN, USE	Types TBS, SA, SIS, THHN, THHW, THW-2, THWN-2 RHH, RHW-2, USE-2, XHH, XHHW, XHHW-2, XHWN, XHWN-2, XHHN	
	COPPER			ALUMINUM OR COPPER-CLAD ALUMINUM			
18*	-	-	14	-	-	-	-
16*	-	-	18	-	-	-	-
14*	15	20	25	-	-	-	-
12*	20	25	30	15	20	25	12*
10*	30	35	40	25	30	35	10*
8	40	50	55	35	40	45	8
6	55	65	75	40	50	55	6
4	70	85	95	55	65	75	4
3	85	100	115	65	75	85	3
2	95	115	130	75	90	100	2
1	110	130	145	85	100	115	1
1/0	125	150	170	100	120	135	1/0
2/0	145	175	195	115	135	150	2/0
3/0	165	200	225	130	155	175	3/0
4/0	195	230	260	150	180	205	4/0
250	215	255	290	170	205	230	250
300	240	285	320	195	230	260	300
350	260	310	350	210	250	280	350
400	280	335	380	225	270	305	400
500	320	380	430	260	310	350	500
600	350	420	475	285	340	385	600
700	385	460	520	315	375	425	700
750	400	475	535	320	385	435	750
800	410	490	555	330	395	445	800
900	435	520	585	355	425	480	900
1000	455	545	615	375	445	500	1000
1250	495	590	665	405	485	545	1250
1500	525	625	705	435	520	585	1500
1750	545	650	735	455	545	615	1750
2000	555	665	750	470	560	630	2000

Notes:
1. Section 310.15 (B) shall be referenced for ampacity correction factors where the ambient temperature is other than 30ºC (86º F).
2. Section 310.15 (C) (I) shall be referenced for more than three current-carrying conductors.
3. Section 310.16 shall be referenced for conditions of use.
* Section 240.4 (D) shall be referenced for conductor overcurrent protection limitations, except as modified elsewhere in the *Code*.

Figure 10–6 *NEC Table 310.16.*

Table 310.17 Ampacities of Single-Insulated Conductors in Free Air

Size AWG or kcmil	Temperature Rating of Conductor [See Table 310.4(I)]						Size AWG or kcmil
	60°C (140°F)	75°C (167°F)	90°C (194°F)	60°C (140°F)	75°C (167°F)	90°C (194°F)	
	Types TW, UF	Types RHW, THHW, THW, THWN, XHHW, XHWN, ZW	Types TBS, SA, SIS, FEP, FEPB, MI, PFA, RHH, RHW-2, THHN, THHW, THW-2 THWN-2, USE-2, XHH, XHHW, XHHW-2, XHWN, XHWN-2, XHHN, Z, ZW-2	Types TW, UF	Types RHW, THHW, THW, THWN, XHHW, XHWN	Types TBS, SA, SIS, THHN, THHW, THW-2, THWN-2, RHH, RHW-2, USE-2, XHH, XHHW, XHHW-2, XHWN, XHWN-2 XHHN	
	COPPER			ALUMINUM OR COPPER-CLAD ALUMINUM			
18	-	-	18	-	-	-	-
16	-	-	24	-	-	-	-
14*	25	30	35	-	-	-	-
12*	30	35	40	25	30	35	12*
10*	40	50	55	35	40	45	10*
8	60	70	80	45	55	60	8
6	80	95	105	60	75	85	6
4	105	125	140	80	100	115	4
3	120	145	165	95	115	130	3
2	140	170	190	110	135	150	2
1	165	195	220	130	155	175	1
1/0	195	230	260	150	180	205	1/0
2/0	225	265	300	175	210	235	2/0
3/0	260	310	350	200	240	270	3/0
4/0	300	360	405	235	280	315	4/0
250	340	405	455	265	315	355	250
300	375	445	500	290	350	395	300
350	420	505	570	330	395	445	350
400	455	545	615	355	425	480	400
500	515	620	700	405	485	545	500
600	575	690	780	455	545	615	600
700	630	755	850	500	595	670	700
750	655	785	885	515	620	700	750
800	680	815	920	535	645	725	800
900	730	870	980	580	700	790	900
1000	780	935	1055	625	750	845	1000
1250	890	1065	1200	710	855	965	1250
1500	980	1175	1325	795	950	1070	1500
1750	1070	1280	1445	875	1050	1185	1750
2000	1155	1385	1560	960	1150	1295	2000

Notes:
1. Section 310.15(B) shall be referenced for ampacity correction factors where the ambient temperature is other than 30ºC (86ºF).
2. Section 310.17 shall be referenced for conditions of use.
3. Section 240.4(D) shall be referenced for conductor overcurrent protection limitations, except as modified elsewhere in the *Code*.

Figure 10–7 *NEC Table 310.17.*

10–2 Factors That Determine Ampacity

Conductor Material

One of the factors that determines the amount of current that a conductor can safely carry is the material from which it is made. The ampacity tables list values for copper, aluminum, and copper-clad aluminum conductors. For example, refer to *Table 310.16.* The left half of the table is ampacity values for copper conductors and the right half of the table is for aluminum and copper-clad aluminum conductors. The table indicates that a copper conductor can carry a bit more current than an aluminum or copper-clad aluminum conductor of the same size and insulation type without exceeding the temperature rating of the insulation. An 8 AWG copper conductor with Type TW insulation is rated to carry a maximum of 40 amperes. An 8 AWG aluminum conductor with Type TW insulation is rated to carry only 35 amperes.

Insulation Type

Another factor that determines the amount of current a conductor can safely carry is the type of insulation used. This is the basis for the term ampacity. The current-carrying capacity of a conductor in

Table 310.18 Ampacities of Insulated Conductors with Not More Than Three Current-Carrying Conductors in Raceway or Cable

	Temperature Rating of Conductors [See Table 310.4(I)]				
	150°C (302°F)	200°C (392°F)	250°C (482°F)	150°C (302°F)	
	Type Z	Types FEP, FEPB, PFA, SA	Types PFAH, TFE	Type Z	
Size AWG or kcmil	COPPER		NICKEL OR NICKEL-COATED COPPER	ALUMINUM OR COPPER-CLAD ALUMINUM	Size AWG or kcmil
14	34	36	39	-	14
12	43	45	54	30	12
10	55	60	73	44	10
8	76	83	93	57	8
6	96	110	117	75	6
4	120	125	148	94	4
3	143	152	166	109	3
2	160	171	191	124	2
1	186	197	215	145	1
1/0	215	229	244	169	1/0
2/0	251	260	273	198	2/0
3/0	288	297	308	227	3/0
4/0	332	346	361	260	4/0

Notes:
1. Section 310.15(B) shall be referenced for ampacity correction factors where the ambient temperature is other than 40ºC (104ºF).
2. Section 310.15(C)(I) shall be referenced for more than three current-carrying conductors.
3. Section 310.18 shall be referenced for conditions of use.

Figure 10–8 *NEC Table 310.18.*

the NEC is solely based on not exceeding the temperature rating of the insulation. Different insulation types can permit higher current values because some types of insulation can withstand more heat than others. The **insulation** is the nonconductive covering around the conductor material (*Figure 10–9*). The voltage rating of the conductor is also determined by the type of insulation. The amount of voltage a particular type of insulation can withstand without breaking down is determined by the type of material it is made of and its thickness. *NEC Table 310.4(1)* lists information concerning the different types of insulation. The table is divided into columns that list the trade name; identification letters; **maximum operating temperature;** whether the insulation can be used in a wet, damp, or dry location; material; thickness; and outer covering. A good thing to remember is that insulation materials that contain the letter *W*, such as RHW, THW, and THWN must be used if the installation is in a wet location.

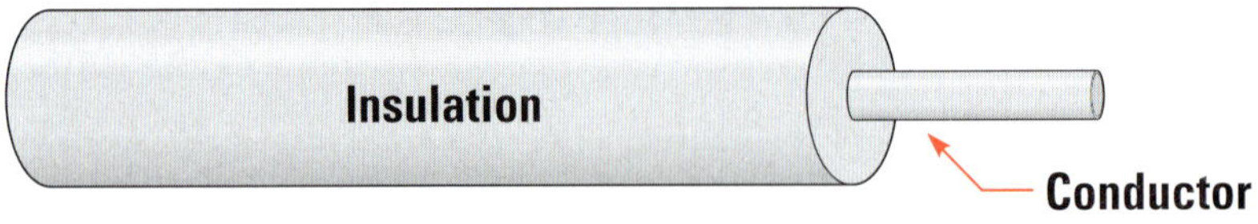

Figure 10–9 Insulation around conductor.

Example 10–1

Find the maximum operating temperature of Type RHW insulation. (Note: Refer to the *NEC.*)

Solution

Find Type RHW in the second column of *Table 310.4(1).* The third column lists a maximum operating temperature of 75°C, or 167°F.

Example 10–2

Can Type THHN insulation be used in wet locations?

Solution

Locate Type THHN insulation in the second column of *Table 310.4(1)*. The fourth column indicates that this insulation can be used in **dry** and **damp locations**. This type of insulation cannot be used in **wet locations**. For an explanation of the difference between damp and wet locations, consult "locations" in *Article 100* of the *NEC*.

Condition of Use Factors

Not every installation is going to directly follow the parameters of these critical ampacity tables in the *NEC*. When the installation conditions or conditions of use differ from what the tables are based on, it becomes necessary to correct or adjust the ampacity value of conductors listed in the tables to fit the actual installation conditions, One of the main factors that determines the amount of current a conductor can carry safely is the ambient, or surrounding, air temperature. *Table 310.16,* for example, lists the ampacity of conductors in a raceway or cable when the ambient air temperature is 30°C(86°F). If these conductors are used in a location that has a higher or lower ambient temperature, the ampacity of the conductor must be corrected accordingly due to a change in the ability of the conductor to handle heat generated when current flows through the material. Temperature **correction factors** can be found in *Table 310.15(B)(1)(1)* and *310.15(B)(1)(2)* depending on which ambient temperature the original value was based on. *Table 310.15(B)(1)(1)* is for correcting conductor ampacity values if the original value was based on an ambient temperature of 30°C (86°F). The ampacity of conductors in *Tables 310.16* and *310.17* are based on an ambient air temperature of 30°C (86°F). The correction factors for conductors rated at 40°C are found in *Table 310.15(B)(1)(1)*. The ampacity of conductors in *Tables 310.18* and *310.19* are based on an ambient temperature of 40°C (104°F). The correction factors found in Table *310.15(B)(1)(2)* are shown in *Figure 10–10.*

Table 310.15(B)(1)(1) Ambient Temperature Correction Factors Based on 30°C (86°F)

For ambient temperatures other than 30°C (86°F), multiply the ampacities specified in the ampacity tables by the appropriate correction factor shown below.

Ambient Temperature (°C)	Temperature Rating of Conductor			Ambient Temperature (°F)
	60°C	75°C	90°C	
10 or less	1.29	1.20	1.15	50 or less
11–15	1.22	1.15	1.12	51–59
16–20	1.15	1.11	1.08	60–68
21–25	1.08	1.05	1.04	69–77
26–30	1.00	1.00	1.00	78–86
31–35	0.91	0.94	0.96	87–95
36–40	0.82	0.88	0.91	96–104
41–45	0.71	0.82	0.87	105–113
46–50	0.58	0.75	0.82	114–122
51–55	0.41	0.67	0.76	123–131
56–60	-	0.58	0.71	132–140
61–65	-	0.47	0.65	141–149
66–70	-	0.33	0.58	150–158
71–75	-	-	0.50	159–167
76–80	-	-	0.41	168–176
81–85	-	-	0.29	177–185

Note: Table 310.15(B)(1)(1) shall be used with Table 310.16 and Table 310.17 as required.

Figure 10–10 *NEC Table 310.15(B)(1)(1)* lists correction factors for ampacities based on an original 30°C (86°F) ambient temperature.

More Than Three Current-Carrying Conductors in a Raceway or Cable

Tables 310.16 and *310.18* are also ampacity values based on a maximum of three current-carrying conductors in a raceway or cable. If a raceway or cable is to contain more than three current-carrying conductors, the ampacity of the conductors must be adjusted because each additional conductor creates additional heat in the raceway or cable, impedes a conductor's ability to properly dissipate the heat, and results in a higher temperature inside the raceway or cable. *NEC Table 310.15(C)(1)* (*Figure 10–11*) lists these adjustment factors. If the raceway is used in an area with a greater ambient temperature than that listed in the appropriate wire table, the temperature correction factor must also be applied.

Table 310.15(C)(I) Adjustment Factors for More Than Three Current-Carrying Conductors

Number of Conductors*	Percent of Values in Table 310.16 Through Table 310.19 as Adjusted for Ambient Temperature if Necessary
4–6	80
7–9	70
10–20	50
21–30	45
31–40	40
41 and above	35

*Number of conductors is the total number of conductors in the raceway or cable, including spare conductors. The count shall be adjusted in accordance with 310.15(E) and (F). The count shall not include conductors that are connected to electrical components that cannot be simultaneously energized.

Figure 10–11 *NEC Table 310.15(C)(1).*

Temperature Limitations of Terminations

Using the *NEC* to determine the amount of current a conductor can carry safely is not always the same thing as determining the proper size conductor to use for a particular application. Several factors must be considered when selecting a conductor for a specific installation, and ampacity is only one of them. One factor is whether the load is continuous or not. The *NEC* defines a continuous load as one where the maximum current is expected to continue for three hours or more. Most industrial motor and lighting loads, for example, would be considered continuous loads. *NEC 210.19(A)(1)* states that conductors must have an ampacity not less than the noncontinuous load plus 125% of the continuous load. Basically, the ampacity of a conductor must be 125% greater than the current required to supply a certain continuous load. This helps to ensure the conductor's ability to dissipate the heat created by the continuous current flow in the circuit.

Another factor that affects the selection of a conductor is the temperature rating of the equipment to which the conductor is connected. *NEC 110.14(C)* lists the requirements on how the termination temperature limitations can impact the temperature rating of the conductor insulation. *The section* states that the temperature rating of a conductor must be selected so as not to exceed the lowest temperature rating of any connected conductor, termination, or device. Because the termination temperature rating is often unknown, *NEC 110.14(C)(1)(a)* states that terminations for circuits rated at 100 amperes or less, or that are marked for 14 AWG through 1 AWG conductors, are to be used with conductors with 60°C (140°F) rated insulation. This does not mean that conductors with a higher temperature rating cannot be used, but their ampacity must not exceed the ampacity of the same-sized conductor in the 60°C column. There are two exceptions to this rule. Conductors with a higher temperature ratings can be used at the higher temperature rating if the termination at the equipment is listed and identified for such a rating. The other exception is for motors that are marked with *NEMA* (National Electrical Manufacturers Association) code letters B, C, or D. *NEC 110.14(C)(1)(a)(4)* permits conductors for motors with these code letters to be selected from the 75°C column.

The requirements of *NEC 110.14(C)* also require that the ampacities used in determining termination ratings be selected from *Table 310.16. NEC Table 310.17,* for example, lists the current-carrying capacity of conductors located in free air. Because the conductors are generally terminated inside an enclosure, however, they must be chosen from a table that lists the ampacity of conductors inside an enclosure.

For circuits with a current of 100 A or greater or that are marked for conductors larger than 1 AWG, *NEC 110.14(C)(1)(b)* requires conductors with an insulation rating of 75°C (167°F). This section also allows higher temperature conductors to be used if the equipment is listed and identified for use with such conductors or as long as the ampacity of the conductor does not exceed the ampacity of the same size and material conductor in the 75°C (167°F) column.

Example 10–3

What is the maximum ampacity of a 4 AWG copper conductor with Type THWN insulation in a raceway or cable installed in an area with an ambient temperature of 43°C?

Solution

Determine the ampacity of a 4 AWG copper conductor with Type THWN insulation from the wire table. Type THWN insulation is located in the second column of *Table 310.16.* The table lists an ampacity of 85 A for this conductor. Locate 43°C in the far left-hand column of the correction factor chart shown in *Figure 10–10*; 43°C falls between 41°C and 45°C. Follow across to the 75°C column. The chart lists a correction factor of 0.82. The ampacity of the conductor in the above wire table is to be multiplied by the correction factor:

$$85\text{ A} \times 0.82 = 69.7\text{ A}$$

Example 10–4

What is the maximum ampacity of a 1/0 AWG copper-clad aluminum conductor with Type RHH insulation if the conductor is installed in rigid metallic conduit and installed in an area with an ambient air temperature of 100°F?

Solution

Locate the column that contains Type RHH insulation in the copper-clad aluminum section of *Table 310.16.* The table indicates a maximum ampacity of 135 A for this conductor. The chart shown in *Figure 10–10* is used to determine the correction factor for this temperature. Fahrenheit degrees are located in the far right-hand column of the chart; 100°F falls between 97°F and 104°F. The correction factor for this temperature is 0.91. Multiply the ampacity of the conductor by this factor:

$$135\text{ A} \times 0.91 = 122.85\text{ A}$$

Example 10–5

Twelve 14 AWG copper conductors with Type RHW insulation are to be run in intermediate metal conduit. The conduit is installed in an area that has an ambient temperature of 110°F. What is the maximum ampacity of these conductors per the *NEC* based on their conditions of use?

Solution

Find the ampacity of a 14 AWG copper conductor with Type RHW insulation. Type RHW insulation is located in the second column of *Table 310.16*. A 14 AWG copper conductor has an ampacity of 20 A. Next, use the correction factor for an ambient temperature of 110°F in *Table 310.15(B)(1)(1)* shown in *Figure 10–10*. A correction factor of 0.82 will be used:

$$20\text{ A} \times 0.82 = 16.4\text{ A}$$

The correction factor located in *Table 310.15(C)(1)* must now be used. The table indicates a correction factor of 50% when 10 through 20 conductors are run in a raceway:

$$16.4\text{ A} \times 0.50 = 8.2\text{ A}$$

$$16.4\text{ A} \times 0.50 = 8.2\text{ A}$$

Each 14 AWG conductor has a maximum current rating of 8.2 A.

Example 10–6

Assume that a motor with a full-load current rating of 28 A is to be connected with copper conductors that have Type THW insulation. The motor is located in an area with an ambient temperature of 30°C. The motor is not marked with NEMA code letters, and the termination temperature is not known. What size conductors should be used? (Hint: The *NEC* Article 430 requires motor conductors to be sized at 125% of the full-load current of the motor.)

Solution

First, multiply the full-load current rating by 125%:

$$28\text{ A} \times 1.25 = 35\text{ A}$$

Refer to *NEC Table 310.16*. Type THW insulation is located in the 75°C column. Although this conductor is located in the 75°C column, the ampacity of the conductor must not exceed that listed in the 60°C column because the termination temperature rating is not known, and the motor does not contain *NEMA* code letters. However, if adjustment or correction factors were required, the derating can start from the higher temperature rating but must default to the lower value if the adjustment and correction factors do not reduce the value to lower than the 60°C column. The nearest wire size without going under 35 A is an 8 AWG.

Example 10–7

Assume that a bank of heating resistors is rated at 28 kW and is connected to 240 V. The resistors operate at maximum capacity for more than three hours at a time and are located in an area with an ambient temperature of 86°F. Aluminum conductors with Type THWN-2 insulation are to be used to connect the heaters. What size conductors should be used?

Solution

Determine the amperage of the load using Ohm's law:

$$I = \frac{P}{E}$$

$$I = \frac{28{,}000\ \text{W}}{240\ \text{V}}$$

$$I = 116.667\ \text{A}$$

Because the load is continuous, the conductors must have an ampacity 125% greater than the load current.

$$I = 116.667\ \text{A} \times 1.25$$

$$I = 145.834\ \text{A or } 146\ \text{A}$$

Refer to *NEC Table 310.16*. Type THWN-2 insulation is located in the 90°C column. The conductor ampacity, however, must not exceed the value listed in the 75°C column. The nearest size aluminum conductor without going less than 146 A is 3/0 AWG.

Example 10–8

An electric annealing oven is located in an area with an ambient temperature of 125°F. The oven contains a 50-kW electric heating element and is connected to 480 volts, single-phase. The conductors are to be copper with type THHN insulation. The termination temperature is not known. The furnace is not expected to operate at maximum output for more than three hours continuously. What size conductor should be employed to make this connection?

Solution

Determine the amount of current needed to operate the furnace.

$$I = \frac{P}{E}$$

$$I = \frac{50{,}000}{480}$$

$$I = 104.167\ \text{amperes}$$

Section 210.19(A) requires branch circuit conductors to have an ampacity not less than the larger of two concepts. The first concept accounts for the continuous nature of a load and requires the conductor to be able to handle 125% of the continuous load plus any noncontinuous load.

Because the load is considered noncontinuous, the conductor must be large enough to supply the load:

$$104.167 \times 1.00 = 104.167\ \text{amperes}$$

Example 10–8 (continued)

The second concept that must be considered is will the ampacity of the conductor be sufficient for the load after we apply correction and adjustment factors. Since we only have a high ambient temperature to consider in this example, the next step is to apply the correction factor for temperature. The correction factor for a temperature of 125°F is 0.76, as shown in *Figure 10–10*. To determine the minimum size conductor required to supply 104.167 amperes at 125°F, divide the current by the correction factor:

$$104.167 \div 0.76 = 137.06 \text{ amperes}$$

Therefore, the temperature correction factor is going to require the larger conductor.

Because the termination temperature is not known and the current is over 100 amperes, the conductor ampacity must not exceed that listed in the 75°C column of *NEC Table 310.16*. This means we can apply the temperature correction factor to the 90°C column ampacity as long as the final ampacity is less than the ampacity listed in the 75°C column for that size of conductor. Since our conductor must have an initial ampacity that is greater than 137.06 amperes to still be sufficient for the load, we can select a 1 AWG copper THHN conductor with an ampacity of 145 amperes. However, we need to verify the corrected ampacity is not greater than the ampacity listed in the 75°C column.

$$145 \times 0.76 = 110.2 \text{ amperes}$$

This verifies that the corrected ampacity of the conductor is less than the 130 amperes listed in the 75°C column and is sufficient to supply the 104.167 ampere load. The correct size conductor for this application is a 1 AWG copper conductor.

10–3 Determining Conductor Sizes and Resistance

Although the ampacity tables in the *NEC* are used to determine the proper size conductor for most installations, there are instances in which these tables are not used.

Longer Circuit Lengths

Another situation in which it becomes necessary to calculate the appropriate size conductor is when the conductor becomes excessively long. Keep in mind that the ampacity values in the *NEC* tables are based on not overheating the conductor insulation; they do not take into account how the resistance of the conductor material might affect the operation of the circuit. When the circuit length becomes long enough, the resistance of the conductor can affect the load because it is essentially a resistor in series with the load. As we discussed in previous units, series connected resistors cause a voltage drop and reduce the voltage available at the load. In these instances, the correct conductor size must be determined by calculating a conductor size that will have little to no effect on the operation of the load.

All wire contains resistance. As wire is added to a circuit, it has the effect of adding resistance in series with the load (*Figure 10–12*). Four factors determine the resistance of a length of wire:

1. **The type of material from which the wire is made.** Different types of material have different wire resistances. A copper conductor will have less resistance than an aluminum conductor of the same size and length. An aluminum conductor will have less resistance than a piece of iron wire the same size and length.
2. **The diameter of the conductor.** The larger the diameter, the less resistance it will have. A large-diameter pipe, for example, will have less resistance to the flow of water than a small-diameter

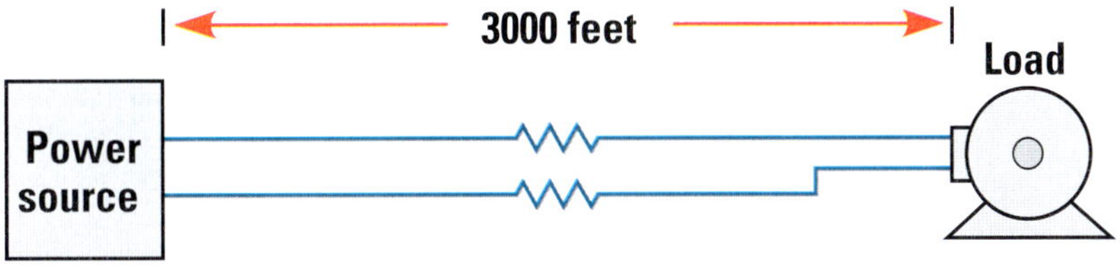

Figure 10–12 Long conductors act as a resistor in series with the load.

pipe (*Figure 10–13*). The cross-sectional area of round wire is measured in circular mils. One mil equals 0.001 inch. A **circular mil** is the cross-sectional area of the wire in mils squared. For example, assume a wire has a diameter of 0.064 inch. Sixty-four thousandths should be written as a whole number, not as a decimal or a fraction (64^2 [64 mil × 64 mil] = 4096 CM).

Figure 10–13 A larger water pipe will have less resistance to the flow of water than a smaller diameter pipe.

3. **The length of the conductor**. The longer the conductor, the more resistance it will have. Adding length to a conductor has the same effect as connecting resistors in series.
4. **The temperature of the conductor**. Most conductive materials will increase in resistance with an increase of temperature. Some exceptions to this rule are carbon, silicon, and germanium. If the coefficient of temperature for a particular material is known, its resistance at different temperatures can be calculated. Materials that increase their resistance with an increase in temperature have a positive coefficient of temperature. Materials that decrease their resistance with an increase in temperature have a negative coefficient of temperature.

In the English system of measure, the resistivity of a material is measured in ohms per **mil-foot**. A mil-foot is a piece of wire 1 foot long and 1 mil in diameter (*Figure 10–14*). A chart showing the resistivity of different materials at 20°C is shown in *Figure 10–15*. Notice the wide range of resistances for different materials. The temperature coefficient of the different types of conductors is listed also.

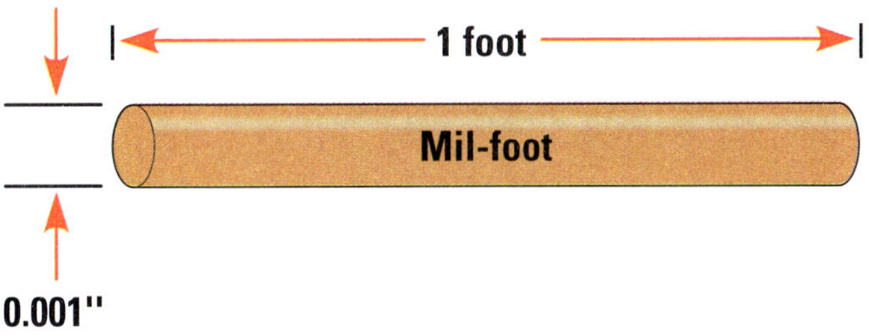

Figure 10–14 A mil-foot describes a conductor that is 1 foot long and one thousandth of an inch in diameter.

Resistivity (K) of Materials

Material	Ω per CM ft at 20°C	Temp. coeff. (Ω per °C)
Aluminum	17	0.004
Carbon	22,000 apprx.	−0.0004
Constantan	295	0.000002
Copper	10.4	0.0039
Gold	14	0.004
Iron	60	0.0055
Lead	126	0.0043
Manganin	265	0.000000
Mercury	590	0.00088
Nichrome	675	0.0002
Nickel	52	0.005
Platinum	66	0.0036
Silver	9.6	0.0038
Tungsten	33.8	0.005

Figure 10–15 Resistivity of materials.

Calculating Resistance

Now that a standard measure of resistance for different types of materials is known, the resistance of different lengths and sizes of these materials can be calculated. The formula for calculating resistance of a certain length, size, and type of wire is:

$$R = \frac{K \times L}{CM}$$

where

R = resistance of the wire

K = ohms-CM per foot

L = length of wire in feet

CM = circular mil (area of the wire)

This formula can be converted to calculate other values in the formula such as to find the SIZE of conductor required:

$$CM = \frac{K \times L}{R}$$

to find the LENGTH of conductor:

$$L = \frac{R \times CM}{K}$$

Example 10–9

Find the resistance of a piece of 6 AWG copper wire 550 ft long. Assume a temperature of 20°C. The formula to be used is:

$$R = \frac{K \times L}{CM}$$

Solution

The value for K can be found in the table in *Figure 10–15*. The table indicates a value of 10.4 Ω-CM per foot for a copper conductor. The length, L, was given as 550 ft, and the circular mil area of 6 AWG wire is listed as 26,250 in the table shown in *Figure 10–16*.

$$R = \frac{10.4\ \Omega\text{-CM/ft} \times 550\ \text{ft}}{26{,}250\ \text{CM}}$$

$$R = \frac{5720\ \Omega\text{-CM}}{26{,}250\ \text{CM}}$$

$$R = 0.218\ \Omega$$

Example 10–10

An aluminum wire 2250 ft long cannot have a resistance greater than 0.2 Ω. What size aluminum wire must be used?

Solution

To find the size of wire, use

$$CM = \frac{K \times L}{R}$$

$$CM = \frac{17\ \Omega\text{-CM/ft} \times 2250\ \text{ft}}{0.2\ \Omega}$$

$$CM = \frac{38{,}250\ \Omega\text{-CM}}{0.2\ \Omega}$$

$$CM = 191{,}250$$

The nearest standard size conductor for this installation can be found in the American Wire Gauge table. Because the resistance cannot be greater than 0.2 Ω, the conductor cannot be smaller than 191,250 circular mils. The nearest standard conductor size is 0000 AWG, with a circular mil area of 211,600 circular mils.

The following examples illustrate when it might be necessary to determine the size of a conductor using methods other than selecting the size from the *NEC* ampacity tables.

B & S Gauge No.	Diam. in Mils	Area in Circular Mils	Ohms per 1000 Ft (ohms per 100 meters)						Pounds per 1000 Ft (kg per 100 meters)			
			Copper*		Copper*		Aluminum		Copper		Aluminum	
			68°F	(20°C)	167°F	(75°C)	68°F	(20°C)				
0000	460	211,600	0.049	(0.016)	0.0596	(0.0195)	0.0804	(0.0263)	640	(95.2)	195	(29.0)
000	410	167,800	0.0618	(0.020)	0.0752	(0.0246)	0.101	(0.033)	508	(75.5)	154	(22.9)
00	365	133,100	0.078	(0.026)	0.0948	(0.031)	0.128	(0.042)	403	(59.9)	122	(18.1)
0	325	105,500	0.0983	(0.032)	0.1195	(0.0392)	0.161	(0.053)	320	(47.6)	97	(14.4)
1	289	83,690	0.1239	(0.0406)	0.151	(0.049)	0.203	(0.066)	253	(37.6)	76.9	(11.4)
2	258	66,370	0.1563	(0.0512)	0.191	(0.062)	0.256	(0.084)	201	(29.9)	61.0	(9.07)
3	229	52,640	0.1970	(0.0646)	0.240	(0.079)	0.323	(0.106)	159	(23.6)	48.4	(7.20)
4	204	41,740	0.2485	(0.0815)	0.302	(0.099)	0.408	(0.134)	126	(18.7)	38.4	(5.71)
5	182	33,100	0.3133	(0.1027)	0.381	(0.125)	0.514	(0.168)	100	(14.9)	30.4	(4.52)
6	162	26,250	0.395	(0.129)	0.481	(0.158)	0.648	(0.212)	79.5	(11.8)	24.1	(3.58)
7	144	20,820	0.498	(0.163)	0.606	(0.199)	0.817	(0.268)	63.0	(9.37)	19.1	(2.84)
8	128	16,510	0.628	(0.206)	0.764	(0.250)	1.03	(0.338)	50.0	(7.43)	15.2	(2.26)
9	114	13,090	0.792	(0.260)	0.963	(0.316)	1.30	(0.426)	39.6	(5.89)	12.0	(1.78)
10	102	10,380	0.999	(0.327)	1.215	(0.398)	1.64	(0.538)	31.4	(4.67)	9.55	(1.42)
11	91	8,234	1.260	(0.413)	1.532	(0.502)	2.07	(0.678)	24.9	(3.70)	7.57	(1.13)
12	81	6,530	1.588	(0.520)	1.931	(0.633)	2.61	(0.856)	19.8	(2.94)	6.00	(0.89)
13	72	5,178	2.003	(0.657)	2.44	(0.80)	3.29	(1.08)	15.7	(2.33)	4.80	(0.71)
14	64	4,107	2.525	(0.828)	3.07	(1.01)	4.14	(1.36)	12.4	(1.84)	3.80	(0.56)
15	57	3,257	3.184	(1.044)	3.98	(1.27)	5.22	(1.71)	9.86	(1.47)	3.00	(0.45)
16	51	2,583	4.016	(1.317)	4.88	(1.60)	6.59	(2.16)	7.82	(1.16)	2.40	(0.36)
17	45.3	2,048	5.06	(1.66)	6.16	(2.02)	8.31	(2.72)	6.20	(0.922)	1.90	(0.28)
18	40.3	1,624	6.39	(2.09)	7.77	(2.55)	10.5	(3.44)	4.92	(0.713)	1.50	(0.22)
19	35.9	1,288	8.05	(2.64)	9.79	(3.21)	13.2	(4.33)	3.90	(0.580)	1.20	(0.18)
20	32	1,022	10.15	(3.33)	12.35	(4.05)	16.7	(5.47)	3.09	(0.459)	0.94	(0.14)
21	28.5	810	12.8	(4.2)	15.6	(5.11)	21.0	(6.88)	2.45	(0.364)	0.745	(0.11)
22	25.4	642	16.1	(5.3)	19.6	(6.42)	26.5	(8.69)	1.95	(0.290)	0.591	(0.09)
23	22.6	510	20.4	(6.7)	24.8	(8.13)	33.4	(10.9)	1.54	(0.229)	0.468	(0.07)
24	20.1	404	25.7	(8.4)	31.2	(10.2)	42.1	(13.8)	1.22	(0.181)	0.371	(0.05)
25	17.9	320	32.4	(10.6)	39.4	(12.9)	53.1	(17.4)	0.97	(0.14)	0.295	(0.04)
26	15.9	254	40.8	(13.4)	49.6	(16.3)	67.0	(22.0)	0.77	(0.11)	0.234	(0.03)
27	14.2	202	51.5	(16.9)	62.6	(20.5)	84.4	(27.7)	0.61	(0.09)	0.185	(0.03)
28	12.6	160	64.9	(21.3)	78.9	(25.9)	106	(34.7)	0.48	(0.07)	0.147	(0.02)
29	11.3	126.7	81.8	(26.8)	99.5	(32.6)	134	(43.9)	0.384	(0.06)	0.117	(0.02)
30	10	100.5	103.2	(33.8)	125.5	(41.1)	169	(55.4)	0.304	(0.04)	0.092	(0.01)
31	8.93	79.7	130.1	(42.6)	158.2	(51.9)	213	(69.8)	0.241	(0.04)	0.073	(0.01)
32	7.95	63.2	164.1	(53.8)	199.5	(65.4)	269	(88.2)	0.191	(0.03)	0.058	(0.01)
33	7.08	50.1	207	(68)	252	(82.6)	339	(111)	0.152	(0.02)	0.046	(0.01)
34	6.31	39.8	261	(86)	317	(104)	428	(140)	0.120	(0.02)	0.037	(0.01)
35	5.62	31.5	329	(108)	400	(131)	540	(177)	0.095	(0.01)	0.029	
36	5	25	415	(136)	505	(165)	681	(223)	0.076	(0.01)	0.023	
37	4.45	19.8	523	(171)	636	(208)	858	(281)	0.0600	(0.01)	0.0182	
38	3.96	15.7	660	(216)	802	(263)	1080	(354)	0.0476	(0.01)	0.0145	
39	3.53	12.5	832	(273)	1012	(332)	1360	(446)	0.0377	(0.01)	0.0115	
40	3.15	9.9	1049	(344)	1276	(418)	1720	(564)	0.0299	(0.01)	0.0091	
41	-											
42	2.5											
43	-											
44	1.97											

* Resistance figures are given for standard annealed copper. For hard-drawn copper add 2%

Figure 10–16 American Wire Gauge table.

Example 10–11

A manufacturing plant has a cooling pond located 400 ft from the plant. Six pumps are used to circulate water between the pond and the plant. In cold weather, however, the pumps can freeze and fail to supply water. The plant owner decides to connect electric-resistance heaters to each pump to prevent this problem. The six heaters are connected to a two-conductor cable from the plant at a junction box *(Figure 10–17)*. The heaters operate on 480 V and have a total current draw of 50 A. What size copper conductors should be used to supply power to the heaters if the voltage drop at the junction box is to be kept to 3% of the applied voltage? Assume an average ambient temperature of 20°C.

Example 10–11 (continued)

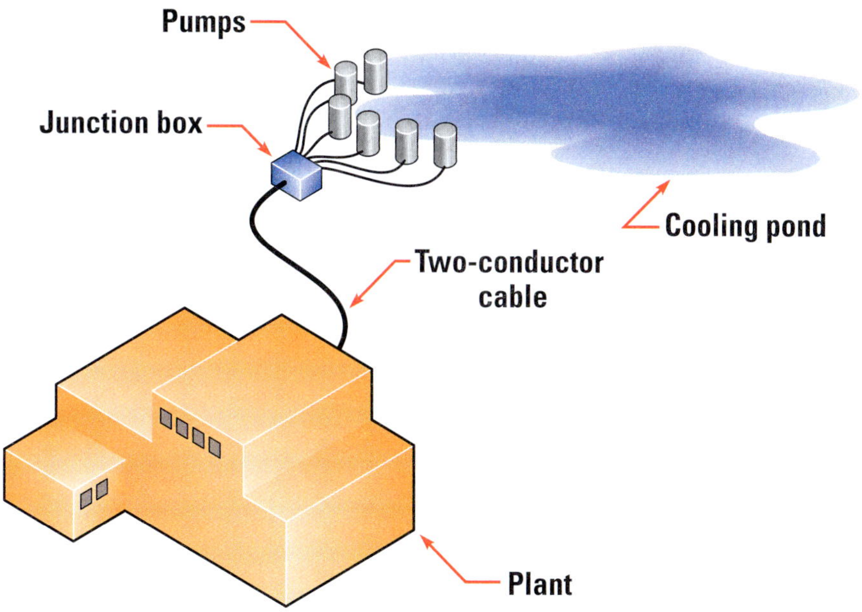

Figure 10–17 Lengthy circuits can pose an issue for voltage drop in circuits.

Solution

The first step in the solution to this problem is to determine the maximum amount of resistance the conductors can have without producing a voltage drop greater than 3% of the applied voltage. The maximum amount of voltage drop can be calculated by multiplying the applied voltage by 3%:

$$480\ \text{V} \times 0.03 = 14.4\ \text{V}$$

The maximum amount of resistance can now be calculated using Ohm's law:

$$R = \frac{E}{I}$$

$$R = \frac{14.4\ \text{V}}{50\ \text{A}}$$

$$R = 0.288\ \Omega$$

The distance to the pond is 400 ft. Because two conductors are used, the resistance of both conductors must be considered. Two conductors 400 ft long will have the same resistance as one conductor 800 ft long. For this reason, a length of 800 ft is used in the formula:

$$CM = \frac{K \times L}{R}$$

$$CM = \frac{10.4\ \Omega\text{-CM/ft} \times 800\ \text{ft}}{0.288\ \Omega}$$

$$CM = \frac{8{,}320\ \Omega\text{-CM}}{0.288\ \text{V}}$$

$$CM = 28{,}888.9$$

For this installation, 4 AWG conductors will be used to keep the voltage drop under 3%.

Example 10–12

The next problem concerns conductors used in a three-phase system. Assume that a motor is located 500 ft from its power source and operates on 560 V. When the motor starts, it has a current draw of 168 A. The voltage drop at the motor terminals cannot be permitted to be greater than 5% of the source voltage during starting. What size aluminum conductors should be used for this installation?

Solution

The solution to this problem is very similar to the solution in the previous example. First, find the maximum voltage drop that can be permitted at the load by multiplying the source voltage by 5%:

$$E = 560\ V \times 0.05$$

$$E = 28\ V$$

The second step is to determine the maximum amount of resistance of the conductors. To calculate this value, the maximum voltage drop is divided by the starting current of the motor:

$$R = \frac{E}{I}$$

$$R = \frac{28\ V}{168\ A}$$

$$R = 0.167\ \Omega$$

The next step is to calculate the length of the conductors. In the previous example, the lengths of the two conductors were added to find the total amount of wire resistance. In a single-phase system, each conductor must carry the same amount of current. During any period of time, one conductor is supplying current from the source to the load, and the other conductor completes the circuit by permitting the same amount of current to flow from the load to the source.

In a balanced three-phase circuit, three currents are 120° out of phase with each other *(Figure 10–18)*. These three conductors share the flow of current between source and load. In *Figure 10–18*, two lines labeled A and B have been drawn through the three current waveforms. Notice that at position A, the current flow in phase 1 is maximum and in a positive direction. The current flow in phases 2 and 3 is less than maximum and in a negative direction. This condition corresponds to the example shown in *Figure 10–19*. Notice that maximum current is flowing in only one conductor. Less than maximum current is flowing in the other two conductors.

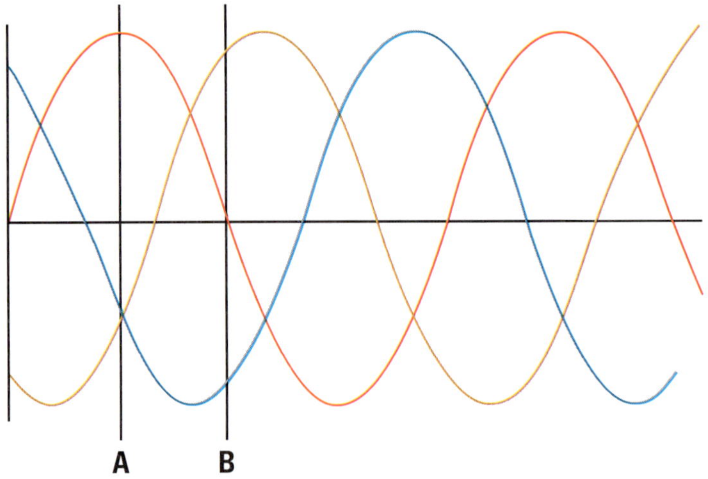

Figure 10–18 The line currents in a three-phase system are 120° out of phase with each other.

Example 10–12 (continued)

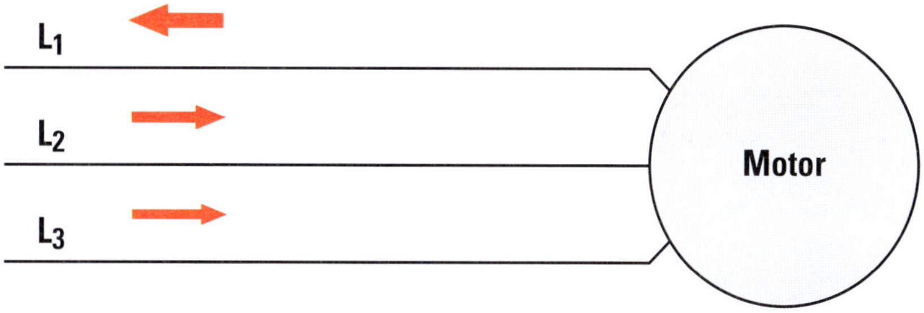

Figure 10–19 Current flows from lines 2 and 3 to line 1.

Observe the line marking position B in *Figure 10–18.* The current flow in phase 1 is zero, and the currents flowing in phases 2 and 3 are in opposite directions and less than maximum. This condition of current flow is illustrated in *Figure 10–20.* Notice that only two of the three phase lines are conducting current and that the current in each line is less than maximum.

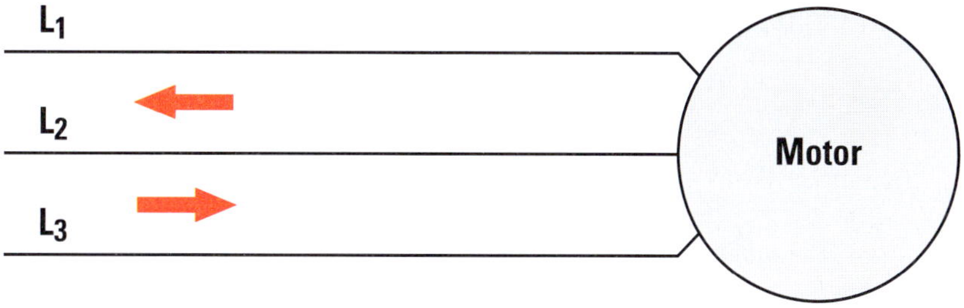

Figure 10–20 Current flows from line 3 to line 2.

Because the currents flowing in a three-phase system are never maximum at the same time, and at other times, the current is divided between two phases, the total conductor resistance will not be the sum of two conductors. To calculate the resistance of conductors in a three-phase system, a demand factor of 0.866 is used.

In this problem, the motor is located 500 ft from the source. The conductor length is calculated by doubling the length of one conductor and then multiplying by 0.866:

$$L = 500 \text{ ft} \times 2 \times 0.866$$

$$L = 866 \text{ ft}$$

Now that all the factors are known, the size of the conductor can be calculated using the formula:

$$CM = \frac{K \times L}{R}$$

where

$$K = \frac{17\ \Omega\text{-CM}}{\text{ft}}$$

$$L = 866 \text{ ft}$$

Example 10–12 (continued)

$$R = 0.167\ \Omega$$

$$CM = \frac{17\ \Omega\text{-CM/ft} \times 866\ \text{ft}}{0.167\ \Omega}$$

$$CM = \frac{14{,}722\ \Omega\text{-CM}}{0.167\ \Omega}$$

$$CM = 88{,}155.7$$

Three 1/0 AWG conductors will be used.

Calculating Voltage Drop

Sometimes it is necessary to calculate the voltage drop of an installation when the circuit length, size of the conductor, and current are known. The following formula, which combines the conductor resistance formula above with Ohm's law, can be used to find the voltage drop of conductors used on a single-phase system:

$$E_D = \frac{2\ KIL}{CM}$$

where

E_D = voltage drop

K = ohms per mil ft

I = circuit current

L = length of conductor in ft

CM = circular mil area of the conductor

Example 10–13

A single-phase motor is located 250 ft from its power source. The conductors supplying power to the motor are 10 AWG copper. The motor has a full-load current draw of 24 A. What is the voltage drop across the conductors when the motor is in operation?

Solution

$$E_D = \frac{2 \times 10.4\ \Omega\text{-CM/ft} \times 24\ \text{A} \times 250\ \text{ft}}{10{,}380\ \text{CM}}$$

$$E_D = 12.023\ \text{V}$$

A slightly different formula can be used to calculate the voltage drop on a three-phase system. Instead of multiplying KIL by 2, multiply KIL by the square root of 3:

$$E_D = \frac{\sqrt{3}\ KIL}{CM}$$

Example 10–14

A three-phase motor is located 175 ft from its source of power. The conductors supplying power to the motor are 1/0 AWG aluminum. The motor has a full-load current draw of 88 A. What is the voltage drop across the conductors when the motor is operating at full load?

Solution

$$E_D = \frac{1.732 \times 17\ \Omega\text{-CM/ft} \times 88\ \text{A} \times 175}{105{,}500\ \text{CM}}$$

$$E_D = 4.298\ \text{V}$$

Coefficient of Temperature

The temperature of a conductor can greatly affect its resistance. *Figure 10–15* lists the ohms per circular mil-foot (K) at 20°C for various materials. The resistance of a material is generally given at 20°C because it is the standard used in the *American Engineers Handbook* and is considered a standard throughout the United States. The temperature coefficient can be used to determine the resistance of a material at different temperatures. Most conductors will increase in resistance with an increase in temperature. Semiconductor materials such as silicon, germanium, and carbon will exhibit a decrease in resistance with an increase in temperature. These materials have a negative coefficient of temperature.

Example 10–15

Determine the ohms-per-mil foot at 75°C for a copper conductor.

Solution

Use the formula:

$$R = R_{ref}\,[1 + \alpha(T - T_{ref})]$$

where

R = Conductor resistance at temperature "T"
R_{ref} = Conductor resistance at reference temperature (20°C in this example)
α = Coefficient of resistance for the conductor material
T = Conductor temperature in °C
T_{ref} = Reference temperature that α is specified at for the conductor material.

$$R = 10.4[1 + 0.0039(75 - 20)]$$

$$R = 10.4[1 + 0.0039(55)]$$

$$R = 10.4[1 + 0.2145]$$

$$R = 10.4[1.2145]$$

$$R = 12.63$$

At a temperature of 75°C, copper would have a resistance of 12.63 ohms-per-mil foot.

10–4 Parallel Conductors

Under certain conditions, it may become necessary or advantageous to connect conductors in parallel. One such condition for **parallel conductors** is when the conductor is very large, as in the earlier example, where it was calculated that the conductors supplying a motor 2500 feet from its source would have to be 500 kcmil. A 500-kcmil conductor is very large and difficult to handle. Therefore, it may be preferable to use smaller conductors in parallel for this installation. The *NEC* lists five conditions that must be met when conductors are connected in parallel *(310.10 (G))*. These conditions are listed here:

1. **The conductors must be the same length.**
2. **The conductors must be made of the same material. For example, all parallel conductors must be either copper or aluminum. It is not permissible to use copper for one of the conductors and aluminum for the other.**
3. **The conductors must have the same circular mil area.**
4. **The conductors must use the same type of insulation.**
5. **The conductors must be terminated or connected in the same manner.**

In the example, the actual conductor size needed was calculated to be 440,778.443 circular mils. This circular mil area could be obtained by connecting two 250-kcmil conductors in parallel for each phase or three 000 (3/0) conductors in parallel for each phase. [Note: Each 000 (3/0) conductor has an area of 167,800 circular mils. This is a total of 503,400 CM.]

Another example of when it may be necessary to connect wires in parallel is when conductors of a large size must be run in a raceway. However, extra care is required when planning parallel conductor installations. One of the factors that must be considered is making sure that all circuit conductors are installed in the same raceway or cable as required by *NEC 300.3(B)*. This is a difficult requirement to meet when installing several conductors per phase. Another factor is to ensure that the conductors do not induce currents or otherwise heat any ferrous metal enclosures or raceways as is required in *NEC 300.20*. When conductors of a single phase are installed in ferrous metal raceways, as shown in *Figure 10–21,* a magnetic field is produced around the conductor. In an AC circuit, the current continuously changes direction and magnitude, which causes the magnetic field to continuously expand and contract through the wall of the metal raceway or cable (*Figure 10–22*). Remember from previous units that what is required to induce a current to flow in a conductor is a magnetic field and motion of the field with respect to the conductor. This expanding and contracting action of the magnetic field induces a current called an *eddy current* into the metal of the raceway or cable.

Eddy currents are currents that are induced into metals. They tend to move in a circular fashion similar to the eddies of a river, hence the name eddy currents (*Figure 10–23*). Eddy currents can produce enough heat to melt the insulation surrounding the conductors. All metal raceways and cables can have eddy current induction, but those made of magnetic or ferrous materials, such as steel, have an added problem with hysteresis loss. Hysteresis loss is caused by molecular friction (*Figure 10–24*). As the direction of the magnetic field reverses, the molecules of the metal are magnetized with the opposite polarity and swing

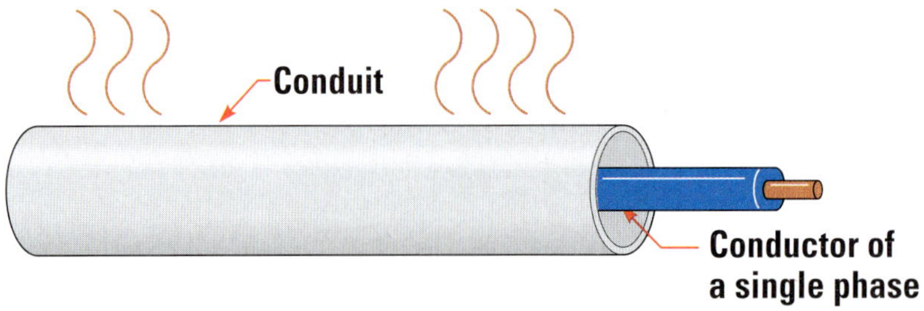

Figure 10–21 Conductors of a single phase cause heat to be produced in the raceway or cable.

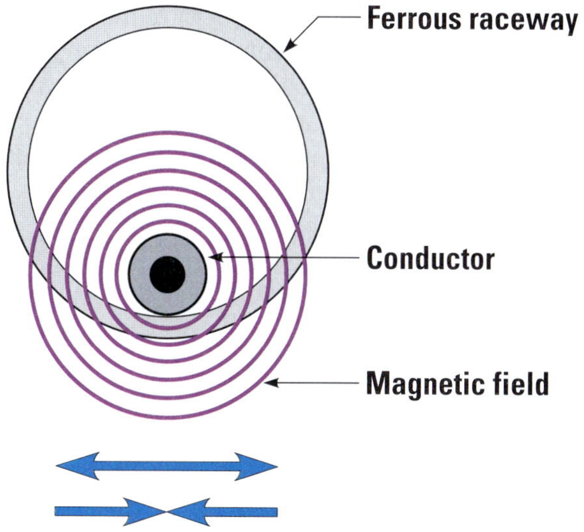

Figure 10–22 The magnetic field expands and contracts through the wall of the raceway.

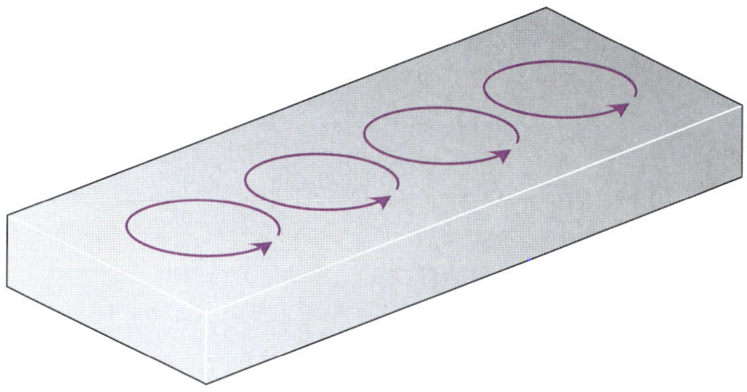

Figure 10–23 Eddy currents are circular currents induced in ferrous raceways and enclosures.

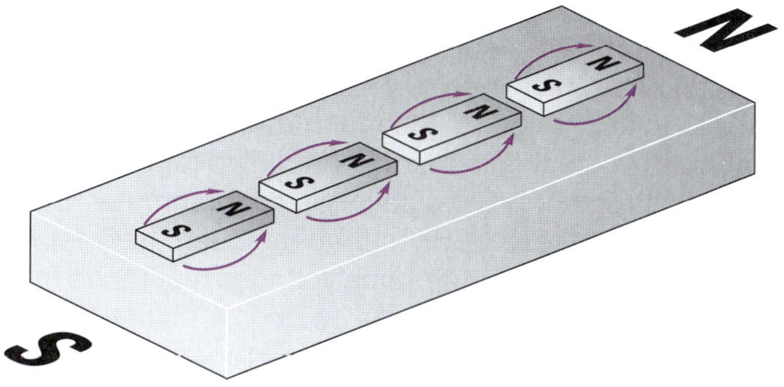

Figure 10–24 Constant changing of the polarity of molecules in the ferrous material causes hysteresis losses from friction.

to realign themselves. This continuous aligning and realigning of the molecules produces heat caused by friction. Hysteresis losses become greater with an increase in frequency.

To correct this problem, a conductor of each phase must be run in each raceway or cable (*Figure 10–25*). When all three phases are contained in the same raceway, cable, or enclosure, the magnetic fields of the separate conductors cancel each other, resulting in no current being induced.

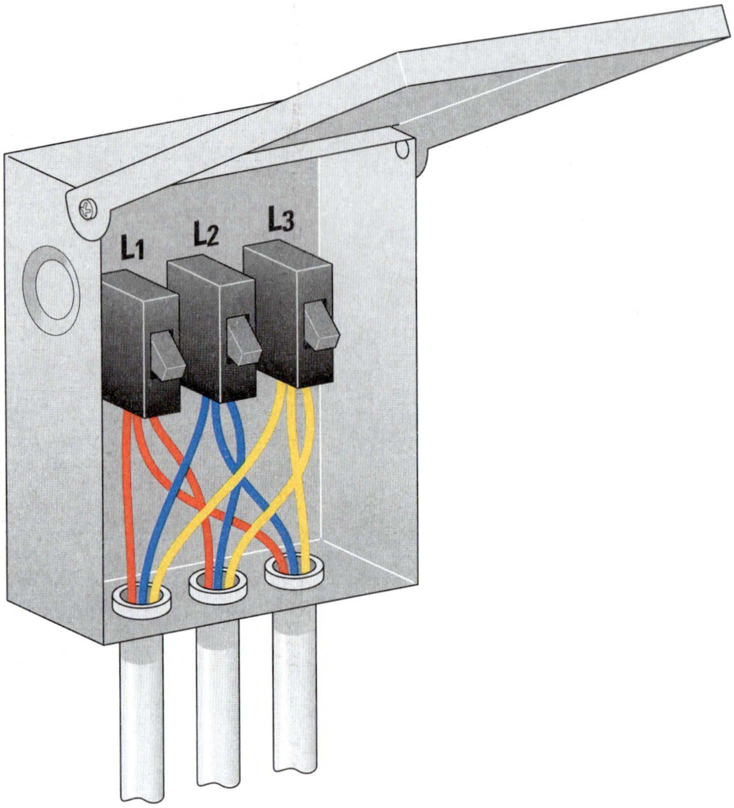

Figure 10–25 Each raceway contains a conductor from each phase. The magnetic fields cancel each other out.

Testing Wire Installations

After the conductors have been installed, it is common practice to test the installation for ground-faults and short circuits that might have resulted from the process of installing them in a raceway. This test requires a specialized piece of test equipment that not only measures resistance in millions of ohms but also provides a high enough voltage to ensure that the insulation will not break down when rated line voltage is applied to the conductors. Most ohmmeters operate with a maximum voltage that ranges from 1.5 volts to about 9 volts; this is hardly enough voltage to test the insulation integrity. To test wire insulation, a special instrument is used called a megohmmeter. Sometimes this is also called a **MEGGER** after a popular brand of these devices that has been used for decades.

The MEGGER is a megohmmeter that can produce voltages that range from about 250 to 5000 volts depending on the model of the meter and the range setting. An older model of a MEGGER is shown in *Figure 10–26*. This example contains a hand crank that is connected to the rotor of a brushless AC generator. An advantage of this particular instrument is that it does not require the use of batteries. A range-selector switch permits the meter to be used as a standard ohmmeter or as a megohmmeter. When it is used as a megohmmeter, the selector switch permits the test voltage to be selected.

Courtesy of Megger®

Figure 10–26 An example of a hand-crank MEGGER.

Megohmmeters can also be obtained in battery-operated models (*Figure 10–27*). The major advantage to digital and battery powered models is that they can be compact, lightweight, and particularly useful when it becomes necessary to test the dielectric of a capacitor. Another advantage to an electronic megohmmeter is that they can also be incorporated into test instruments that have much more functionality than just testing the resistance of conductor insulation.

Figure 10–27 An example of a digital multimeter that includes a megohmmeter.

Conductors are generally tested for two conditions, short circuits and ground-faults. Short circuits are current paths that exist between two circuit conductors. To test conductors for short circuits, the megohmmeter is connected across two conductors at a time (*Figure 10–28*). The circuit is tested at rated voltage or slightly higher. The test instrument indicates the resistance between the two conductors. Because both conductors are insulated, the resistance between them should be extremely high. Each conductor should be tested against every other conductor in the installation.

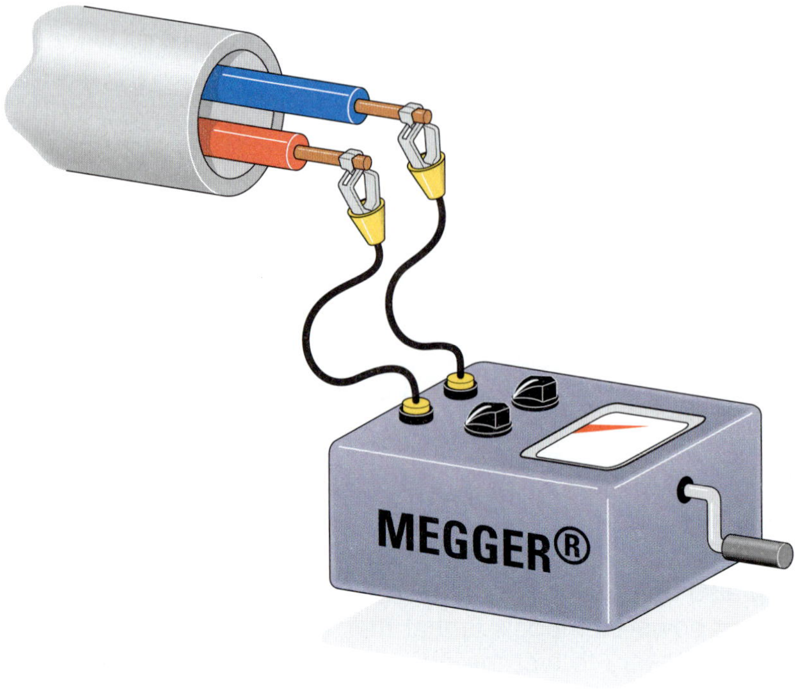

Figure 10–28 Testing for short circuits with a megohmmeter.

To test the installation for ground-faults, one lead of the test instrument is connected to the enclosure or raceway (*Figure 10–29*). The other lead is connected to the conductor being tested. The conductor should be tested at rated voltage or slightly higher. Each conductor should be tested. It is also important to take these measurements when the conductors are not connected to the load, otherwise, the reading you get will be that of the load, and if the test requires a higher voltage, such as a 5kV test, there could be significant damage to the equipment.

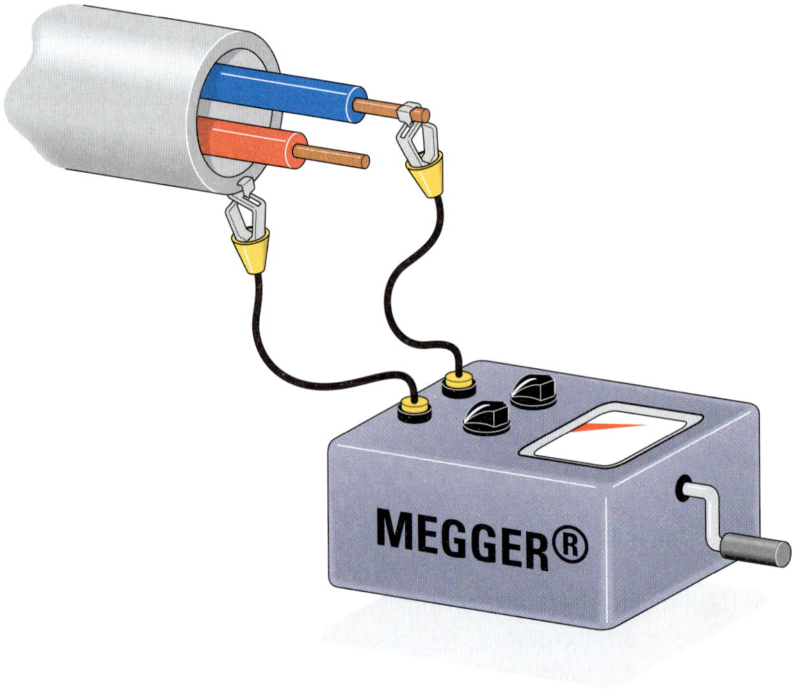

Figure 10–29 Testing for ground-faults with a megohmmeter.

Summary

- The *NEC* is used to determine the conductor size for most installations.
- The resistance of a wire is determined by four factors:
  a. the type of material from which the conductor is made,
  b. the length of the conductor,
  c. the area of the conductor, and
  d. the temperature of the conductor.
- When conductors are used in high-temperature locations, their ampacity must be corrected to account for such temperatures.
- When there are more than three conductors in a raceway, their ampacity must be adjusted to account for the added heat produced by the additional conductors in the raceway.
- A conductor's ampacity is dependent on the type of insulation around the conductor because this value is about not exceeding the insulation's temperature rating.

Review Questions

Use of the *NEC* will be required to answer some of the following questions.

1. Which of the following tables in Article 310 of the NEC contains ampacity values for conductors in a raceway or cable based on an ambient air temperature of 30°C?
   A) Table 310.17
   B) Table 310.18
   C) Table 310.16
   D) Table 310.15(B)(1)(1)
2. What is the ampacity value of a 4 AWG copper conductor with THWN insulation installed in free air with an ambient temperature of 30°C before the application of any adjustment or correction factors or any considerations for termination limitations?
   A) 105 A
   B) 125 A
   C) 70 A
   D) 85 A
3. True or False: The letter W in the conductor insulation type, such as in THWN, stands for weatherproof insulation.
4. Which of the following tables in Article 310 of the NEC contains conductor characteristics and applications such as whether a conductor can be installed in a wet location?
   A) Table 310.4(1)
   B) Table 310.4(2)
   C) Table 310.12
   D) Table 310.15(B)(1)(2)

5. What size THHN conductor is needed to supply a current of 72 amperes if the conductor is installed in a raceway with a total of six conductors and terminates in equipment rated for 75°C?
   A) 1 AWG
   B) 1/0 AWG
   C) 3 AWG
   D) 4 AWG
6. True or False: Conductors installed in ambient temperatures different than the ambient temperature of the ampacity table must have their ampacity corrected.
7. What percentage of the original ampacity must a conductor's ampacity be reduced to when installed in a raceway with eight total current carrying conductors?
   A) 80%
   B) 70%
   C) 60%
   D) 50%
8. The ampacity of a conductor with Type XHHW-2 insulation will be multiplied by which of the following ambient temperature correction factors when exposed to an ambient temperature of 125°F?
   A) 0.76
   B) 0.67
   C) 0.71
   D) 0.82
9. True or False: Voltage drop within a conductor has little to no effect on a circuit, especially when the conductors are over 250 feet in length.
10. Which of the following materials would result in the lowest voltage drop in a circuit that stretched over 500 feet in length?
    A) Aluminum
    B) Copper
    C) Gold
    D) Nickel
11. Which of the following is not a condition that the NEC requires for parallel conductor installations?
    A) They must be the same length.
    B) They must have the same circular mil area.
    C) They must have the same insulation type.
    D) They must be the same color.
12. True or False: When installing conductors in parallel, it is important to ensure that the installation does not induce currents or otherwise heat any ferrous metal raceway system or parts.

Practical Applications

1. You have been hired by a company to connect an outside lighting system to a panel box located in the main office building. The conductors are to have Type THWN-2 insulation, and the wire is copper. The termination temperature of the connections is not known. The lighting system is rated at 20 kW. The voltage is 240 V. Because the lights are expected to operate at maximum current for more than three hours at a time, they are considered a continuous load. Using the ***NEC***, determine the conductor size needed to make this connection.

---

2. You are a journeyman electrician in an industrial plant. You have been assigned the task of connecting a 25-kW electric heater to a 240-V, single-phase panel. The heater is located in an area with an ambient temperature of 90°F. The conductors are to be copper with Type THW insulation. The heater is expected to operate for more than three hours at a time at maximum output. What size conductors would you use to connect this heater per the ***NEC***?

---

3. You are an electrician working in an industrial plant. The company is constructing a maintenance shop located 225 feet from the main plant electrical system. You are to install a 400-ampere, three-phase panelboard in the new maintenance shop. The service is supplied by the existing plant service. The voltage is 480 volts, and the voltage drop is to be kept at 3% or below at full load. The calculated full load for this maintenance shop is 365 amperes, including the 125% calculation for continuous loads. The conductors are to be aluminum. What size conductors should be installed for the new service? Express your answer to the nearest standard wire size. (Note: Determine the wire size by calculating length, resistance, and voltage drop as opposed to using the ***National Electrical Code***.)

---

4. Using the above example, determine the conductor size needed for the 400-ampere panelboard in accordance with the ampacity tables in the ***National Electrical Code***. Assume that the conductors have type THWH insulation. Does this meet the required size for voltage drop?

Practice Problems

Refer to the AWG table in Figure 10–16 and the applicable conductor ampacity tables in the NEC.

1. A 2 AWG copper conductor is 450 ft long. What is the resistance of this wire? Assume the ambient temperature to be 20°C.
2. An 8 AWG conductor is 500 ft long and has a resistance of 1.817 Ω. The ambient temperature is 20°C. Of what material is the wire made?
3. Three 500-kcmil copper conductors with Type RHH insulation are to be used in an area that has an ambient temperature of 58°C. What is the maximum current-carrying capacity of these conductors without exceeding the temperature rating of the RHH insulation?
4. Eight 10 AWG aluminum conductors with Type THWN insulation are installed in a single conduit. What is the maximum current-carrying capacity of these conductors based on the requirements of the *NEC*? Assume an ambient temperature of 30°C.
5. A 50-hp DC motor is connected to 250 volts. The motor does not contain a NEMA code. The motor is expected to operate at maximum current for more than three hours at a time. The conductors are to be copper with type RHW-2 insulation. The motor is located in an area with an

ambient temperature of 112°F. The two conductors are to be run in conduit. What size conductors should be used to connect this motor? Assume the motor operates at 100% efficiency and at 100% of its 50 hp rating. (Note: 1 hp = 746 watts.)

6. A bank of electric heaters has a power rating of 25 kW. The heaters are connected to 480 volts. The heaters will operate continuously and at maximum current draw for more than three hours at a time. The conductors supplying the heaters are to be aluminum with type THWN-2 insulation. The heaters are located in an area with an ambient temperature of 48°C. There will be six conductors in the conduit. What size conductors should be used for this installation?
7. Determine the resistance of a 16 AWG copper conductor located in an area with an ambient temperature of 188.6°F (87°C) with a length of 250 feet (76.2 m).
8. Determine the ohms-per-mil-foot of an aluminum conductor located in an area with a temperature of 104°F (40°C).
9. Use the *NEC* to determine the ampacity of the following conductors.

Size	Material	Insulation	Ambient Temperature	Conductors in Raceway	Amps
#10 AWG	Copper	RHW	44°C	3	
350 kcmil	Copper	XHH	128°F	6	
#2 AWG	Aluminum	TW	86°F	2	
3/0 AWG	Aluminum	XHHW-2	38°C	9	
500 kcmil	Copper	THWN	48°C	6	
#6 AWG	Copper	THW-2	150°F	3	
2/0 AWG	Aluminum	UF	86°F	12	
750 kcmil	Aluminum	RHW-2	34°F	6	

Section 4

Small Sources of Electricity

Taras Verkhovynets/Shutterstock

Unit 11 Conduction in Liquids and Gases

Key Terms

Acids

Alkalis

Anode

Arc

Cathode

Copper sulfate

Cuprous cyanide

Electrolysis

Electrolytes

Electron impact

Electroplating

Ion

Ionization potential

Metallic salt

Sulfuric acid

X-rays

Why You Need to Know

To better understand how batteries produce electricity, how fluorescent lights work, or how it is possible to electroplate objects, it is necessary to understand how electricity is conducted in gases and liquids. In this unit we will

- examine how electricity is conducted in gases and liquids and how it is different from conduction through metals such as silver, copper, or aluminum. While metals rely on the conduction of electrons, gases and liquids rely on the movement of ions.
- explain the process of separating elements electrically.
- describe what X-rays are and how they are used.
- define the ionization potential.

Outline

Objectives

After studying this unit, you should be able to

- describe the process by which ions are made.
- explain how ions make electrical current flow easier.
- summarize how electrical current can travel through a gas.
- explain how electrical current travels through a liquid.
- apply the principles of ionization to several industrial processes.

Preview

The conduction of electric current is generally thought of as electrons moving through a conductive material like copper or aluminum. However, many processes depend on electric current flowing through a gas or liquid. Batteries, for example, would not work if conduction could not take place through a liquid or electrolyte, and fluorescent lighting operates on the principle of conduction through a gas. Conduction through a gas or liquid does not depend on the flow of individual electrons as is the case with electrical conductors. Conduction in gases and liquids depends on the movement of ions.

11–1 The Ionization Process: Magnesium and Chlorine

Simply put, an **ion** is a charged atom. Atoms that have a deficiency of electrons are known as positive ions. Atoms that have gained extra electrons are known as negative ions.

An example of how ionization occurs can be seen by the combination of two atoms, magnesium and chlorine. The magnesium atom contains two valence electrons and is considered to be a metal. Chlorine contains seven valence electrons and is considered a nonmetal (*Figure 11–1*).

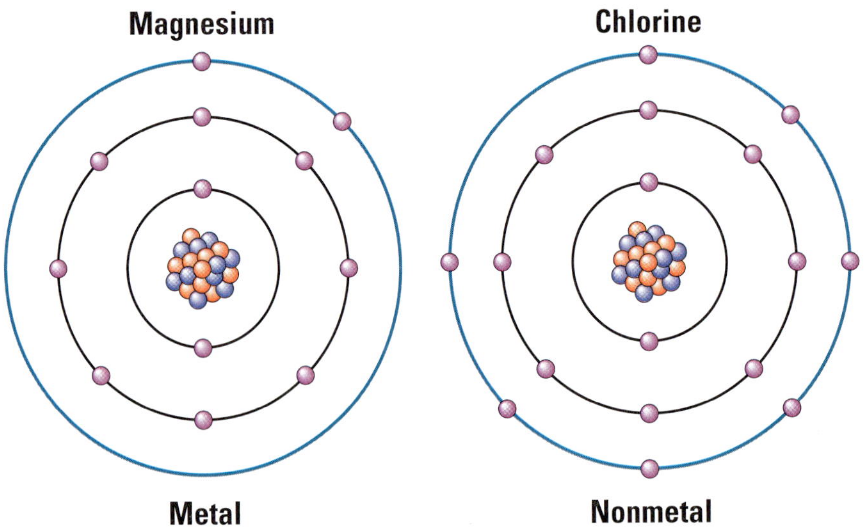

Figure 11–1 Simplified diagrams of magnesium and chlorine atoms.

When magnesium is heated in the presence of chlorine gas, a magnesium atom combines with two chlorine atoms to form a **metallic salt** called magnesium chloride (*Figure 11–2*). When this process occurs, the magnesium atom gives up its two valence electrons to the two chlorine atoms. These atoms are no longer called atoms; they are now called ions. Because the magnesium atom gave up two electrons, it has a positive charge and has become a positive ion. The two chlorine atoms have each gained an electron and are now negative ions.

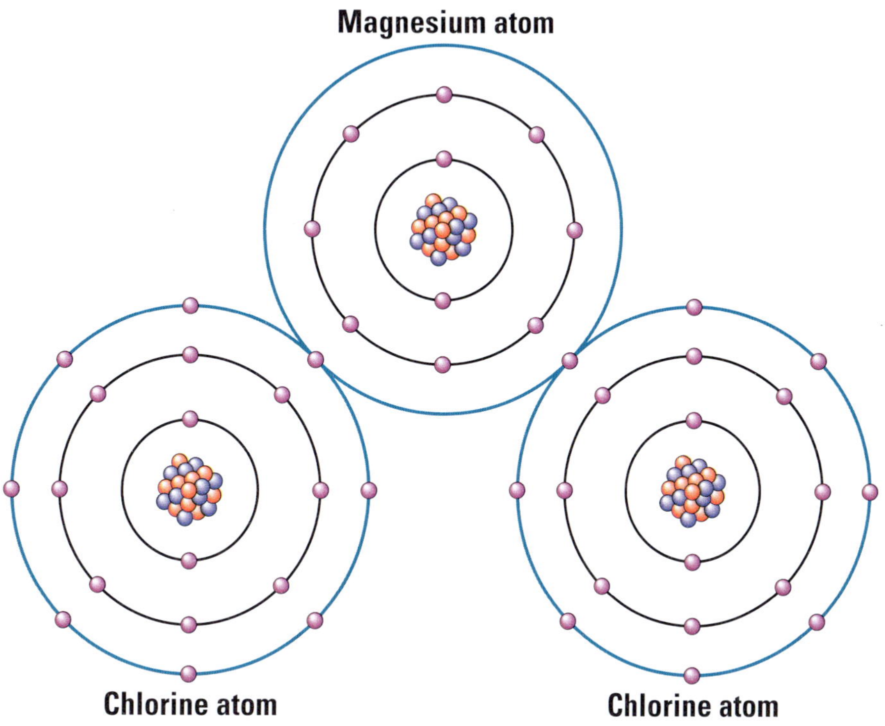

Figure 11–2 Magnesium and chlorine atoms combine to form magnesium chloride.

Magnesium chloride is similar to table salt, which is sodium chloride. Both are formed by combining a metal, sodium or magnesium, with chlorine. When salts like these are mixed with a nonconducting liquid, such as distilled water, the water becomes conductive. A simple illustration can demonstrate conduction through a liquid. In *Figure 11–3*, copper electrodes have been placed in a glass. One electrode is

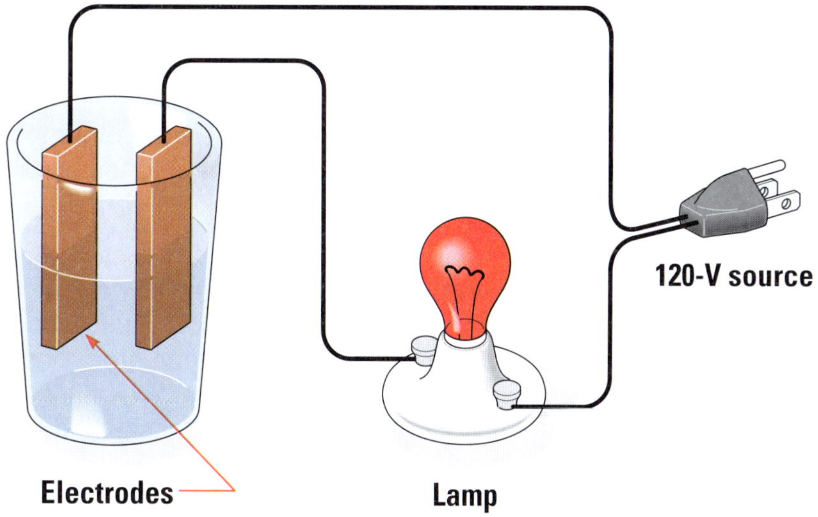

Figure 11–3 Distilled water does not conduct electricity, and the lamp will not light up.

connected to a lamp. The other is connected to one side of a 120-volt circuit. The other side of the lamp is connected to the other side of the circuit. The lamp acts as a load to prevent the circuit from drawing excessive current and to indicate that current is flowing in the circuit. If distilled water is poured into the glass, nothing will happen because there are no ions in the water to carry the electrical current from one electrode to the other. However, once salt is added to the water, the lamp begins to glow (*Figure 11–4*). The lamp glows dimly at first and increases in brightness as more salt is added to the water and the number of ions in the water increases. The glow continues to brighten until the solution becomes saturated with salt, and the limiting factor becomes the resistance of the filament in the lamp.

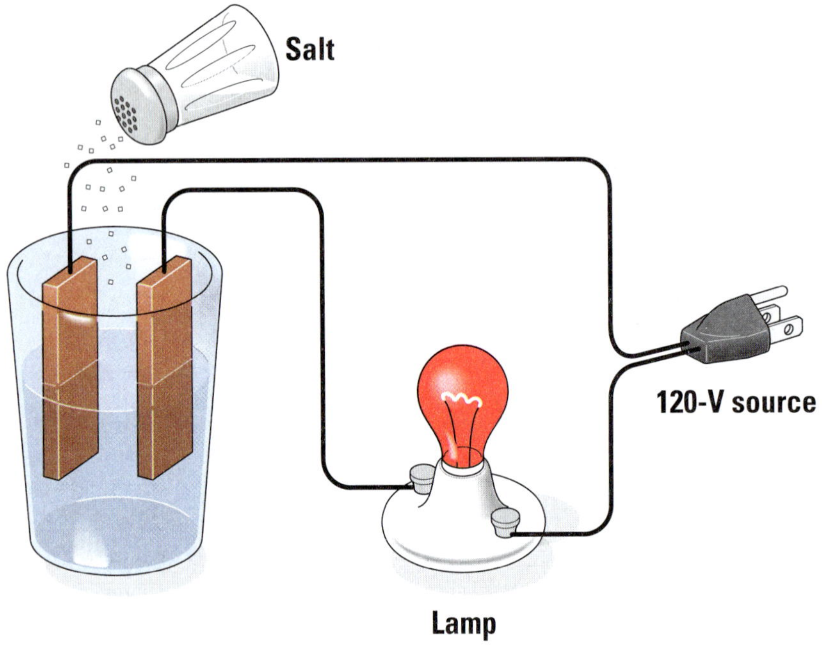

Figure 11–4 Adding salt to the water causes the lamp to glow.

Table salt is not the only compound that can be used to promote conduction in a liquid. **Acids, alkalis,** and other types of metallic salts can be used. The solutions created by adding these compounds to a liquid are referred to as **electrolytes.**

11–2 Other Types of Ions

The ions discussed so far are formed by charging a single atom. There are other types of ions, however, that are formed from groups of atoms. These ions are extremely useful to the electric industry.

Sulfuric Acid

One of the most useful ion compounds is **sulfuric acid.** It is the electrolyte that is commonly used in the lead-acid batteries used in automobiles. The chemical formula for sulfuric acid, H_2SO_4, indicates that one molecule of sulfuric acid contains two atoms of hydrogen, one atom of sulfur, and four atoms of oxygen. Sulfuric acid in its pure form is not useful as an electrolyte. When it is combined with water, H_2O, the molecules separate into ions of H^+, H^+, and SO_4^{--}. The plus sign indicates that each of two hydrogen atoms has lost its electrons and is now a positive ion. The SO_4^{--} is called a sulfate ion and indicates that the two electrons lost by the hydrogen atoms are connected to the sulfur atom and four oxygen atoms. The two extra electrons actually hold the sulfate ion together because they are shared in the valence electron shells of the sulfur and oxygen atoms.

Copper Sulfate

Another useful compound is **copper sulfate,** $CuSO_4$. In its natural form, copper sulfate is in the form of blue crystals. When it is mixed with water, it dissolves and forms two separate ions, Cu^{++} and SO_4^{--}. The Cu^{++} is called a cupric ion and indicates that a copper atom has lost two of its electrons. The SO_4^{--} is the same sulfate ion formed when sulfuric acid is mixed with water. This copper sulfate solution is used in copper electroplating processes.

11–3 Electroplating

Electroplating is the process of depositing atoms of one type of metal on another. Several factors are true in an electroplating process as follows:

1. ***The electrolyte solution must contain ions of the metal to be plated.***
2. ***Metal ions are positively charged ions.***
3. ***The object to be plated must be connected to the negative power terminal.*** The negative terminal is called the **cathode** and refers to the terminal where electrons enter the circuit.
4. ***Direct current is used as the power source.***
5. ***The positive terminal is made of the same metal that is to form the coating.*** The positive terminal is referred to as the **anode** and refers to the terminal where electrons leave the circuit.

An example of the electroplating process is shown in *Figure 11–5*. Note that the object to be plated has been connected to the negative battery terminal, or cathode, and the copper bar has been connected to the positive terminal, or anode. Both objects are submerged in a **cuprous cyanide** solution, which ionizes into positive copper ions (Cu^+) and negative cyanide ions (CN^-). When power is applied to the circuit, the copper ions in the solution move toward the object to be plated, and the cyanide ions move toward the copper bar. When the copper ion contacts the object, it receives an electron and becomes a neutral copper atom. These neutral copper atoms form a copper covering over the object to be plated. The thickness of the coating is determined by the length of time the process is permitted to continue.

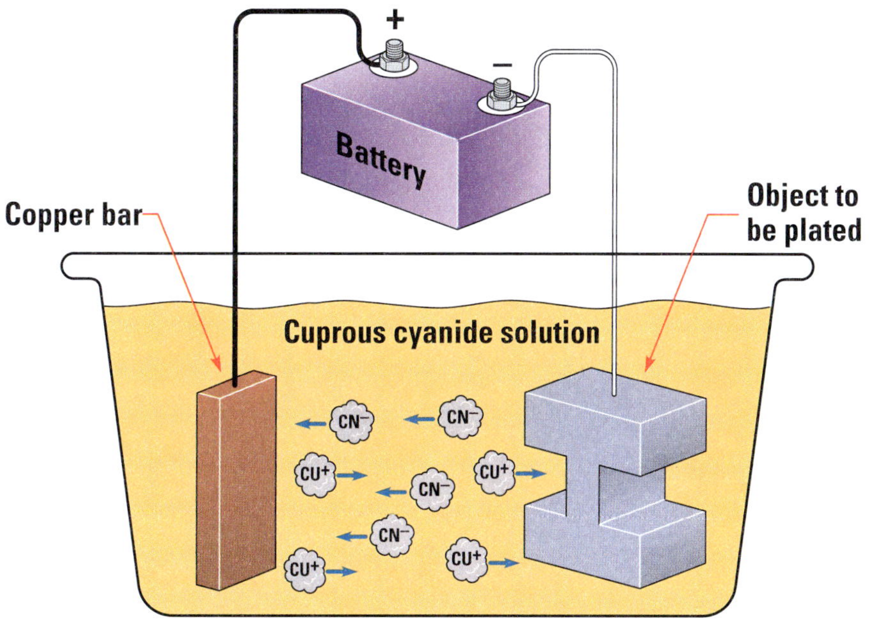

Figure 11–5 The electroplating process transfers positively charged metal ions onto the surface of the negatively charged cathode material.

When the negatively charged cyanide ions contact the positive copper bar, copper atoms on the surface of the bar lose an electron and become positive ions. The copper ion is attracted into the solution and flows toward the object to be plated. Just as many copper ions are being formed as are being attracted to the object to be plated. The solution, therefore, remains at a constant strength.

One of the most common uses for electroplating is the production of pure metals. Impure, or raw, copper is connected to the positive terminal, and pure copper is connected to the negative terminal. Because only pure atoms of copper become ions, impurities remain in solution or never leave the positive plate. Impure copper has a high resistance and is not suitable for use as an electrical conductor. Wire manufacturers require electrolytically purified copper because of its low-resistance characteristics.

11–4 Electrolysis

The term **electrolysis** refers to the process of separating elements electrically. Elements are often chemically combined with other elements in naturally occurring compounds and must be separated to obtain the pure element. Aluminum is a prime example of this process. Aluminum oxide is an abundant compound found in the Earth's crust, but pure aluminum is very rare, and extracting aluminum from aluminum oxide was once difficult and expensive. Obtaining pure aluminum became much less expensive when a process was discovered for separating the aluminum and oxygen atoms electrically. In this process, electrons are removed from oxygen ions and returned to aluminum ions. The aluminum ions then become aluminum atoms.

11–5 Conduction in Gases

At atmospheric pressure, air is an excellent insulator. In *Figure 11–6,* two electrodes are separated by an air gap. The electrodes are connected to a variable-voltage power supply. Assume that the output voltage of the power supply is zero when the switch is turned on. As the output voltage is increased, the ammeter will remain at zero until the voltage reaches a high enough potential difference to allow the electrons to travel through the air. Once this voltage level has been reached, an **arc** will be established across the electrodes and current will begin to flow (*Figure 11–7*). The amount of current is still going to be limited by the impedance of the circuit. However, once an arc has been established, the resistance of the air gap changes as the air becomes ionized and conduction is maintained by ionized gases. One

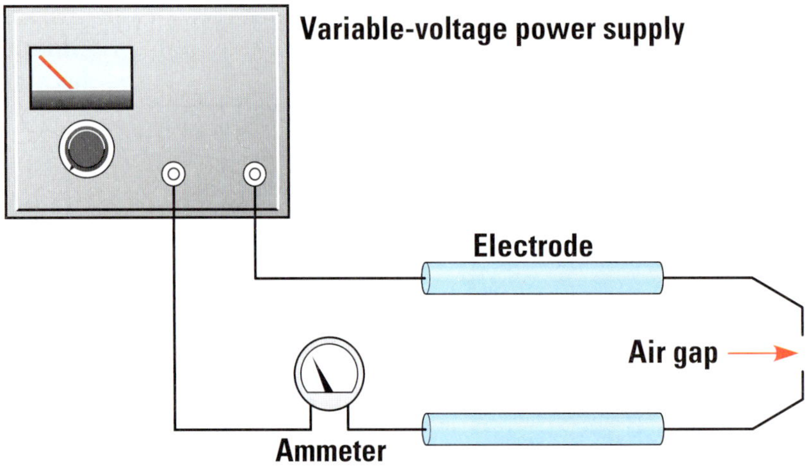

Figure 11–6 Air gap between electrodes acts as an insulator under normal conditions.

factor that determines the voltage required is the distance between the two conductors. The farther apart they are, the higher the voltage must be. When using an electric arc welder, the arc is first established by touching the electrode to the object to be welded and then withdrawing the electrode once conduction begins. Recall that once an arc has been established, the amount of voltage required to maintain it is reduced.

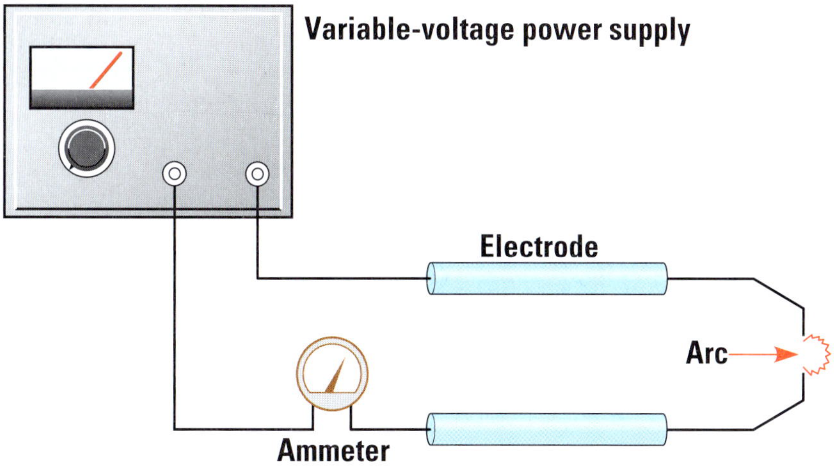

Figure 11–7 When the voltage becomes high enough, an arc jumps across the air gap and creates conductive air.

Ionization Potential in a Gas

The amount of voltage, or potential, an electron must possess to cause ionization is called the **ionization potential.** The following factors determine the amount of voltage required to cause conduction in a gas-filled envelope.

1. *Atmospheric pressure*. The amount of air pressure greatly influences the voltage required to reach ionization potential. Atmospheric pressure is 14.7 pounds per square inch (PSI) at sea level. If the pressure is increased, such as in an automobile engine, the amount of voltage required to reach ionization potential increases greatly. The ignition system of most automobiles produces voltages that range from 20,000 volts to 70,000 volts.

   If the atmospheric pressure is reduced, the amount of voltage required to reach ionization potential is reduced. This is the principle of operation of the gas-filled tubes that have been used in industry for many years. Gas-filled tubes operate with lower-than-normal atmospheric pressure inside the tube.
2. *The type of gas in the surrounding atmosphere*. The type of gas in the atmosphere can greatly influence the amount of voltage required to cause ionization. Sodium vapor, for example, requires a potential of approximately 5 volts. Mercury vapor requires 10.4 volts; neon, 21.5 volts; and helium, 24.5 volts. Before an electron can ionize an atom of mercury, it must possess a potential difference of 10.4 volts.

Electron Impact

Conduction in a gas is different from conduction in a liquid or metal. The most important factor in the ionization of a gas is **electron impact.** The process begins when an electron is freed from the negative terminal inside a gas-filled tube or envelope and begins to travel toward the positive terminal (*Figure 11–8*). The electron is repelled from the negative terminal and attracted to the positive terminal. The speed the electron attains is proportional to the applied voltage and the distance it travels before

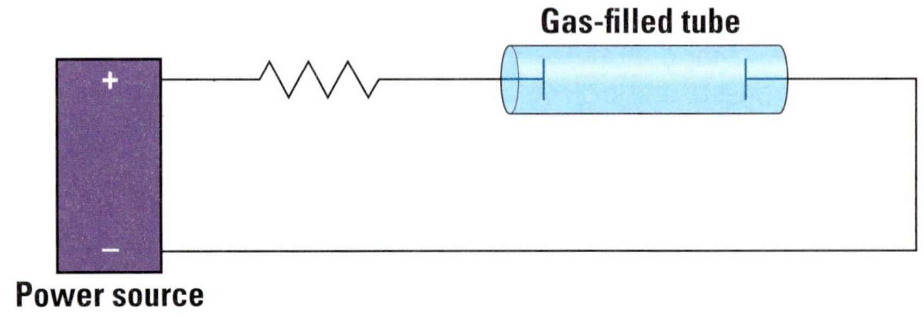

Figure 11–8 Conduction in a gas tube is made possible by electrons traveling through the tube.

striking a gas molecule. If the electron's speed (potential) is great enough, it will liberate other electrons from the gas molecule, which, in turn, flow toward the positive terminal and strike other molecules (*Figure 11–9*). These electrons are very energetic and are the main source of current flow in a gas. When electrons are removed from the molecules, the molecules become positive ions similar to those in a liquid. These positive ions are attracted to the negative terminal, where they receive electrons and become neutral molecules again.

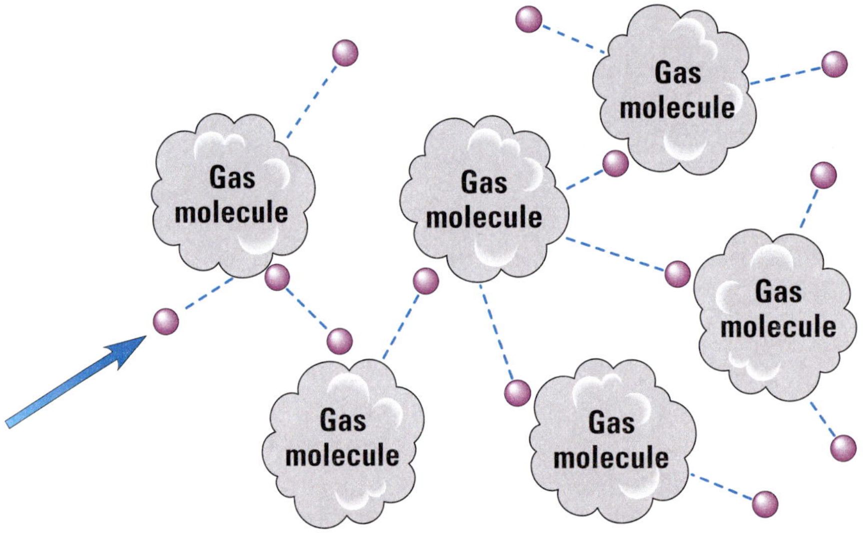

Figure 11–9 Electrons colliding with gas molecules free other electrons from the gas molecules that can carry the action through the tube.

If the electron does not possess enough potential to cause ionization, it bounces off and begins traveling toward the positive terminal again. The pressure inside the tube determines the density of the gas molecules (*Figure 11–10*). If the pressure is too high, the gas molecules will be so dense that the electron cannot gain enough kinetic energy to cause another electron to leave its gas molecule and create

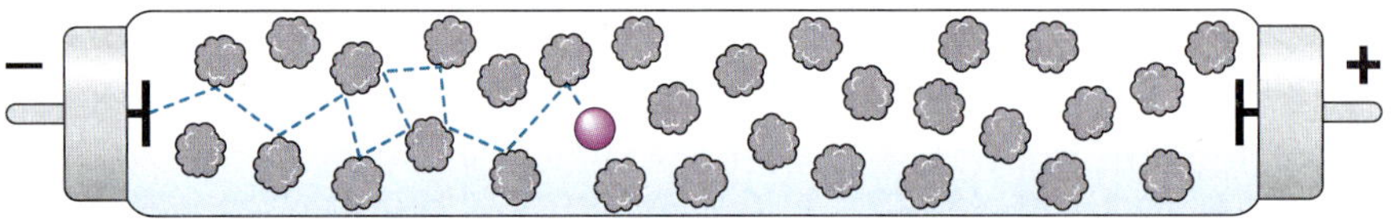

Figure 11–10 Under high pressure, the electron can travel only a short distance between impacts with gas molecules and does not cause ionization.

ionization. In this case, the electron will bounce from one molecule to another until it finally reaches the positive terminal. If the pressure is low, however, the electron can travel a great enough distance to attain ionization potential (*Figure 11–11*).

Figure 11–11 Under low pressure, the electron can travel a greater distance before striking a gas atom.

Useful Applications

Ionization in a gas can be very useful, such as in the production of light. When an electron strikes a gas molecule that does not have enough energy to release other electrons, the molecule will emit light to rid itself of the excess energy. The color of the light is a characteristic of the type of gas. Some examples of this type of lighting are sodium vapor, mercury vapor, and neon.

Another useful device that operates on the principle of conduction in a gas was invented in 1895 by German mechanical engineer and physicist Wilhelm Roentgen. During experimentation in his laboratory, he discovered that when cathode rays struck metal or glass, a new kind of radiation was produced. These new rays passed through wood, paper, and air without effect. They passed through thin metal better than through thick metal and through flesh easier than through bone. Roentgen named these new rays **X-rays.** Today, X-rays are used by doctors throughout the world. Another practical use of this concept was in the early versions of television.

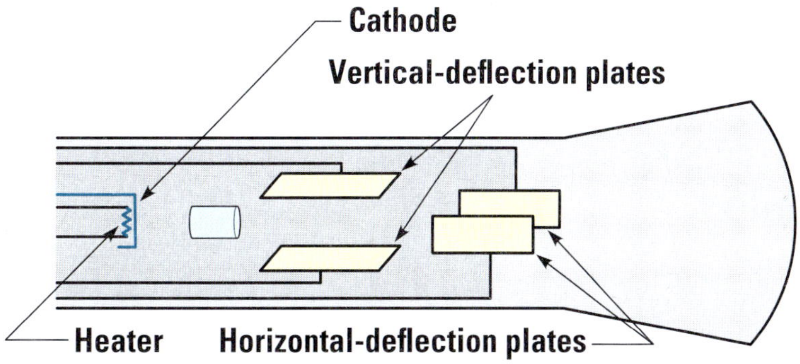

Figure 11–12 Example of how a cathode ray tube deflects the electrons flowing within the tube.

11–6 Ionization in Nature

Ionization of gases often occurs naturally as well. One of the most familiar phenomena caused by ionization of gas is the aurora borealis, or northern lights. During an expedition to the arctic, Kristian Birkeland hypothesized that disturbances in the sun cause large amounts of electrons and high-speed protons to be given off and blown into space. When these highly energetic particles reach the earth, they strike the very thin layers of the upper atmosphere and cause ionization to occur. The different colors are caused by different layers of the atmosphere and varying concentrations of gases being ionized.

Another example of natural ionization is the buildup in the atmosphere of charged particles that are not strong enough to cause lightning. They will, however, produce a continuous discharge of light around objects such as the masts of ships or church steeples. This glowing discharge is often referred to as Saint Elmo's fire by sailors.

Summary

- Electrical conduction in a liquid is caused by the movement of charged atoms called ions.
- Pure or distilled water is an insulator and becomes a conductor if an acid, an alkali, or a metallic salt is added.
- Solutions of acids, alkalis, and metallic salts are referred to as electrolytes.
- Electroplating is the process of depositing the atoms of one type of metal onto another.
- Several factors that affect the electroplating of metals are
  a. the electrolyte solution must contain ions of the metal to be plated.
  b. metal ions are always positively charged.
  c. the object to be plated must be connected to the negative power terminal.
  d. direct current is used as the power source.
  e. the positive terminal is made of the same metal that forms the coating.
- The negative terminal of a device is called the cathode.
- The positive terminal of a device is called the anode.
- One of the most common uses for electroplating is the production of pure metals.
- The term *electrolysis* refers to the process of separating elements electrically.
- Conduction through a gas depends on the ionization of gas molecules.
- The amount of voltage required to cause ionization of a gas is called the ionization potential.
- Gas can be ionized more easily when it is at a low pressure than when it is at a high pressure.
- The ionization potential is different for different gases.
- The most important factor in the ionization of a gas is electron impact.

Review Questions

1. When an element such as magnesium gives up its valence electrons, the Mg atoms become ions. What type of ion does Mg become after such an event?
   A) A negatively charged ion
   B) A positively charged ion
   C) A neutral charged ion
   D) A pure ion
2. True or False: An atom with an excess of electrons in its valence shell is considered a negative ion.
3. Electrical conduction in a liquid or gas depends on which of the following?
   A) Movement of ions within the electrolyte or gas
   B) The material makeup of the ion
   C) The salt content of the liquid
   D) Transfer of electrons between the valence shells

4. For electrical conduction to occur through a gas, what must be high enough in order to reach ionization potential?
   A) Atmospheric pressure
   B) Resistance of the gas
   C) Applied voltage
   D) Density of the gas
5. Which of the following is a factor in how electricity can travel within a gas environment?
   A) Distance between conductors
   B) Atmospheric pressure
   C) Type of gas involved
   D) All of these
6. At atmospheric pressure, which of the following terms best describes the conductive properties of air?
   A) An excellent insulator
   B) An excellent conductor
   C) A semiconductor
   D) A superconductor
7. True or False: Electricity will not flow in pure or distilled water.
8. Which of the following best describes the conductivity of a liquid when salt is added to the solution?
   A) The solution becomes an insulator.
   B) The solution becomes a partial conductor.
   C) The solution becomes more conductive.
   D) The solution becomes a gas.
9. True or False: Electroplating uses the concept of ionization to chemically change one type of metal into a different type of material using AC power systems.
10. Ionization can be used to separate elements that are chemically combined or naturally occurring compounds. The process for separating these elements is referred to by which of the following terms?
    A) Electroplating
    B) Electrolysis
    C) Ionization potential
    D) Electrolyting

Unit 12 Batteries and Other Sources of Electricity

Key Terms

Battery
Cell
Current capacity
Electromotive series of metals
Hydrometer
Internal resistance
Load test
Nickel-cadmium (nicad) cell
Piezoelectricity
Primary cell
Secondary cell
Specific gravity
Thermal runaway
Thermocouple
Voltaic cell
Voltaic pile

Why You Need to Know

Methods other than conductors cutting through magnetic fields can be employed to produce electricity. Although magnetic devices such as generators and alternators are certainly the greatest producers of electricity, they are not the only ones. Batteries are common as a way to store energy in our electrical systems and power items we use every day. Understanding how they operate is essential to understanding how electricity works. This unit

- discusses different types of batteries and how to identify their current capacity for correct use.
- discusses how some batteries can be recharged and others cannot, and why you must understand those differences.
- explains why different types of batteries produce different voltages and what determines the amount of current a battery can produce.
- illustrates how to calculate the voltage and current capacity of batteries by connecting them in series and/or parallel.
- discusses solar cells, thermocouples, and piezoelectricity.

Outline

Objectives

After studying this unit, you should be able to

- explain how batteries were discovered.
- describe the two main types of voltaic cells.
- determine voltages produced by different types of primary voltaic cells.
- determine proper series or parallel connections for batteries to achieve desired voltage and current characteristics.
- describe the discharging and charging cycles of secondary type voltaic cells.
- explain how other methods to produce electricity are used in practice.

Preview

Most of the electric power in the world is produced by large rotating machines called alternators or generators. However, these machines are not the only way we are able to power our world. Several options exist that produce usable electricity without motion and magnetism. The most common of these alternative power sources is the battery. Batteries are used for everything from starting or powering automobiles and operating power tools to powering personal computers and cellphones. Batteries have even found a place in providing an energy buffer to our electrical grid to offset large demands on the power grid. They range in size from small enough to fit into a hearing aid to large enough battery energy storage systems that occupy entire floors of high-rise buildings.

12–1 History of the Battery

In 1791, Italian physician, physicist, biologist and philosopher Luigi Galvani was conducting experiments in anatomy using dissected frog legs preserved in a salt solution. Galvani suspended the frog legs by means of a copper wire. He noticed that when he touched the leg with an iron scalpel, the leg would twitch. Galvani realized that the twitch was caused by electricity, but he thought that the electricity was produced by the muscular contraction of the frog's leg. This was the first recorded incident of electricity being produced by chemical action. However, archeologists have found evidence that chemical-based energy storage devices might have been in use in ancient civilizations, like Egypt, long before Galvani's experiment.

In 1800, Italian physicist and chemist Alessandro Volta repeated Galvani's experiment. Volta, however, concluded that the electricity was produced by the chemical action of the copper wire, the iron scalpel, and the salt solution. Further experiments led Volta to produce the first practical battery (*Figure 12–1*). The battery was constructed using zinc and silver discs separated by a piece of cardboard soaked in brine, or saltwater. Volta called his battery a **voltaic pile** because it was a series of individual cells connected together. Each cell produced a certain amount of voltage depending on the materials used to make the cell. A **battery** is actually several cells connected together, although the word *battery* is often used in reference to a single cell. The schematic symbols for an individual **cell** and for a battery are shown in *Figure 12–2*.

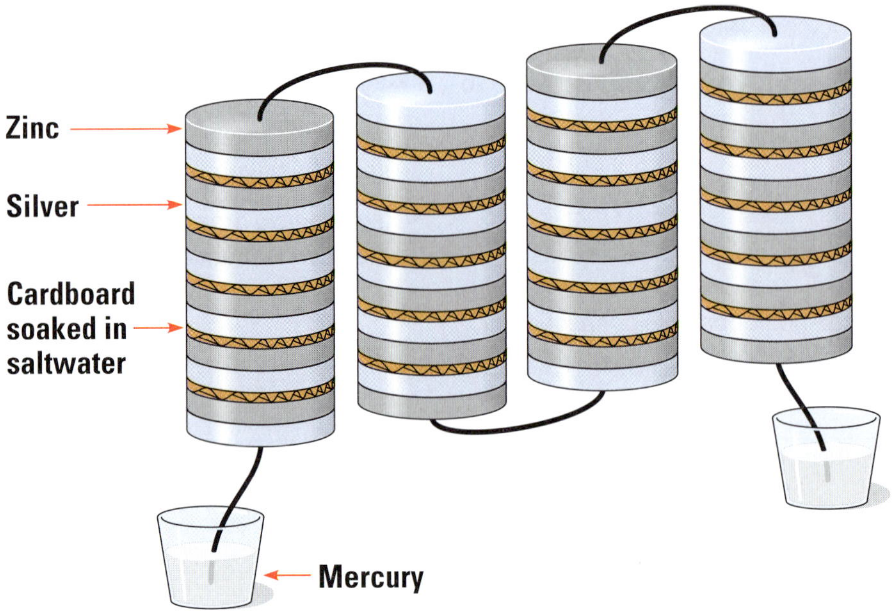

Figure 12–1 The first practical example of a battery, known as a voltaic pile.

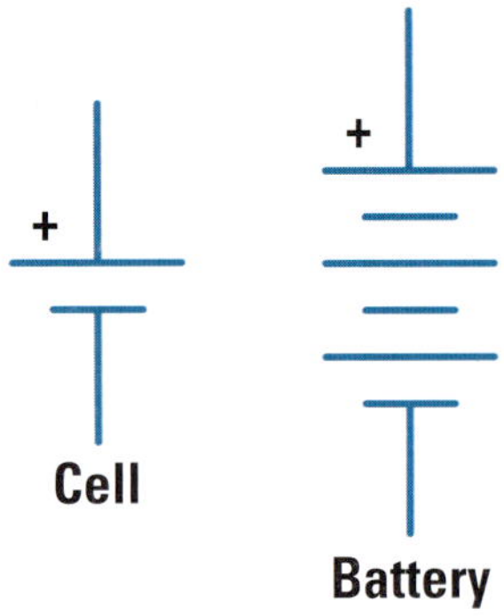

Figure 12–2 Schematic symbols used to represent an individual cell and a battery.

12–2 Voltaic Cells

A **voltaic cell** can be constructed using virtually any two unlike metals and an acid, alkaline, or salt solution. A very simple cell can be constructed as shown in *Figure 12–3*. In this example, a copper wire is inserted into one end of a potato, and an aluminum wire is inserted into the other. The acid in the potato acts as the electrolyte. If a high-impedance voltmeter is connected to the wires, a small voltage can be measured.

Another example of a simple voltaic cell is shown in *Figure 12–4*. In this example, two coins of different metal, a nickel and a penny, are separated by a piece of paper. The paper has been wetted with saliva from a person's mouth. The saliva contains an acid or alkali that acts as the electrolyte. When a high-impedance voltmeter is connected across the two coins, a small voltage is produced.

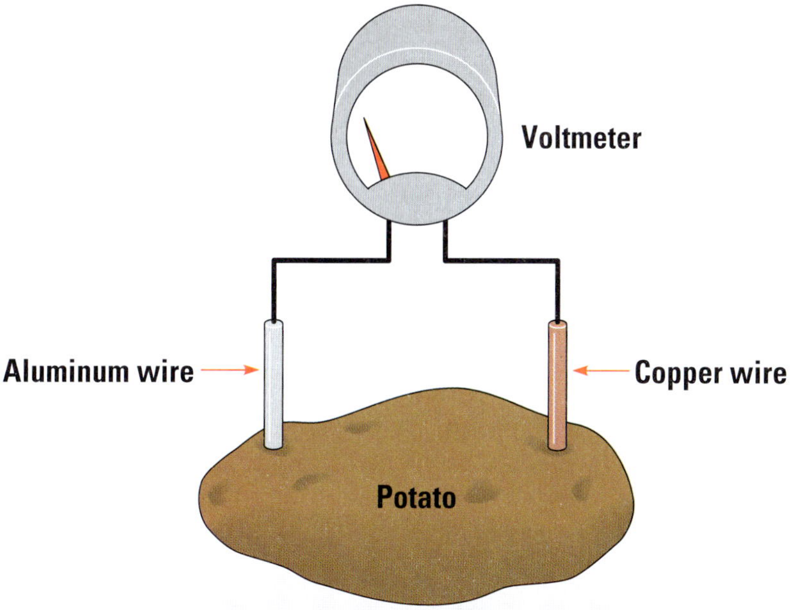

Figure 12–3 Simple voltaic cell constructed from a potato and two different metals.

Figure 12–4 A simple voltaic cell constructed from two coins and a piece of paper.

12–3 Cell Voltage

The amount of voltage produced by an individual cell is determined by the materials from which it is made. When a voltaic cell is constructed, the plate metals are chosen on the basis of how easily one metal will give up electrons compared to the other. A special list of metals called the **electromotive series of metals** is shown in *Figure 12–5*. The table in this figure lists metals in the order of their ability to accept or receive electrons. The metals at the top accept electrons more easily than those at the bottom. The farther apart the metals are on the list, the higher the voltage developed by the cell. One of the first practical cells to be constructed was the zinc-copper cell. This cell uses zinc and copper as the active metals and a solution of water and hydrochloric acid as the electrolyte. Notice that zinc is located closer to the top of the list than copper. Because zinc accepts electrons more readily than copper, zinc is the negative electrode, and copper is the positive.

Although it is possible to construct a cell from virtually any two unlike metals and an electrolyte solution, not all combinations are practical. Some metals corrode rapidly when placed in an electrolyte solution, and some produce chemical reactions that cause a buildup of resistance. In practice, relatively few metals can be used to produce a practical cell.

Electromotive Series of Metals (Partial List)
Lithium
Potassium
Calcium
Sodium
Magnesium
Aluminum
Manganese
Zinc
Chromium
Iron
Cadmium
Cobalt
Nickel
Tin
Lead
Antimony
Copper
Silver
Mercury
Platinum
Gold

Figure 12–5 A partial list of the electromotive series of metals.

The table in *Figure 12–6* lists common cells. The table is divided into two sections. One section lists primary cells, and the other lists secondary cells. A **primary cell** is a cell that *cannot be recharged*. The chemical reaction of a primary cell causes one of the electrodes to be eaten away as power is produced.

Cell	Negative Plate	Positive Plate	Electrolyte	Volts per Cell
		Primary Cells		
Carbon-zinc (Leclanché)	Zinc	Carbon, manganese dioxide	Ammonium chloride	1.5
Alkaline	Zinc	Manganese dioxide	Potassium hydroxide	1.5
Mercury	Zinc	Mercuric oxide	Potassium hydroxide	1.35
Silver-zinc	Zinc	Silver oxide	Potassium hydroxide	1.6
Zinc-air	Zinc	Oxygen	Potassium hydroxide	1.4
Edison-lalande	Zinc	Copper oxide	Sodium hydroxide	0.8
		Secondary Cells		
Lead-acid	Lead	Lead dioxide	Dilute sulfuric acid	2.2
Nickel-iron (Edison)	Iron	Nickel oxide	Potassium hydroxide	1.4
Nickel-cadmium	Cadmium	Nickel hydroxide	Potassium hydroxide	1.2
Silver-zinc	Zinc	Silver oxide	Potassium hydroxide	1.5
Silver-cadmium	Cadmium	Silver oxide	Potassium hydroxide	1.1

Figure 12–6 Examples of both primary and secondary voltaic cell compositions.

When a primary cell becomes discharged, it should be replaced with a new cell. A **secondary cell** *can be recharged*. The recharging process is covered later in this unit.

12–4 Primary Cells

A simplified example of a primary cell is the zinc-copper cell shown in *Figure 12–7*. This cell consists of two electrodes (terminals that conduct electricity into or away from a conducting substance), one zinc and one copper, suspended in a solution of water and hydrochloric acid. This cell will produce approximately 1.08 volts. The electrolyte contains positive hydrogen ions (H^+), which attract electrons from the zinc atoms. Zinc atoms have two valence electrons. When the electrons bond with the H^+ ions, they become neutral hydrogen (H) atoms. The neutral H atoms combine to form H_2 molecules of hydrogen gas. This hydrogen can be seen bubbling away in the electrolyte solution. Removing the valence shell of the zinc atoms creates positively charged zinc ions (Zn^{++}). The zinc ions are attracted to the negative chlorine (Cl^-) ions in the electrolyte solution.

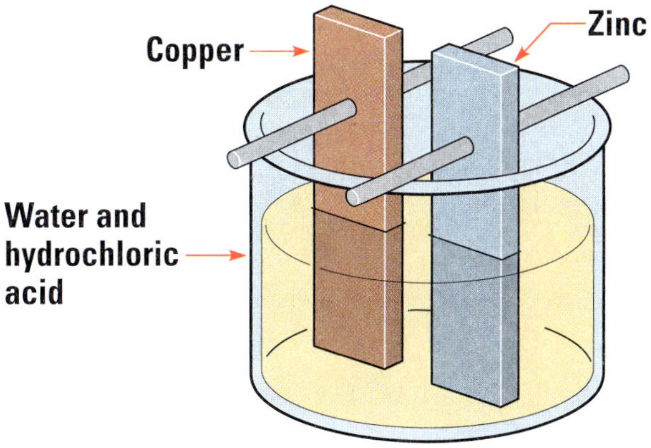

Figure 12–7 Zinc-copper cell.

The copper electrode provides another source of electrons that can be attracted by the H^+ ions. When a circuit is completed between the zinc and copper electrodes (*Figure 12–8*), electrons are attracted from the zinc electrode to replace the electrons in the copper electrode that are attracted into the solution. After some period of time, the zinc electrode dissolves as a result of the zinc ions being distributed in the solution.

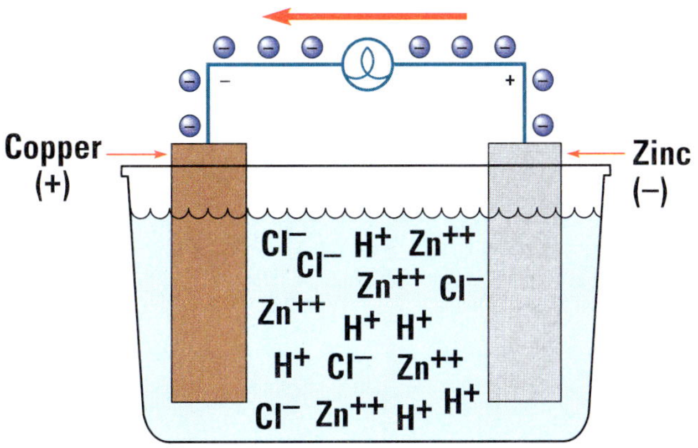

Figure 12–8 Electrons flow from the zinc to the copper electrode.

The Carbon-Zinc Cell

One of the first practical primary cells was invented by Leclanché. It is known as the carbon-zinc cell, or the Leclanché cell. Leclanché used a carbon rod as an electrode instead of copper and a mixture of ammonium chloride, manganese dioxide, and granulated carbon as the electrolyte. The mixture was packed inside a zinc container that acted as the negative electrode (*Figure 12–9*). This cell is often referred to as a dry cell because the electrolyte is actually a paste instead of a liquid. The use of a paste permits the cell to be used in any position without spilling the electrolyte. As the cell is discharged, the zinc container is eventually dissolved. When the container has dissolved, the cell should be discarded immediately and replaced. Failure to discard the cell can result in damage to equipment.

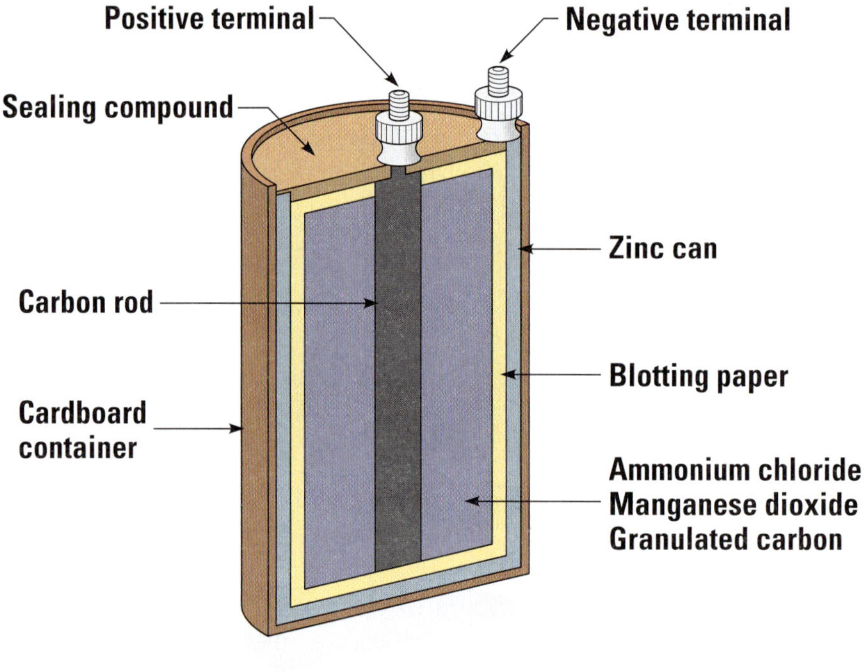

Figure 12–9 Carbon-zinc cell.

Alkaline Cells

Another primary cell very similar to the carbon-zinc cell is the alkaline cell. This cell uses a zinc can as the negative electrode and manganese dioxide, MnO_2, as the positive electrode. The major electrolytic ingredient is potassium hydroxide. Like the carbon-zinc cell, the alkaline cell contains a paste electrolyte and is considered a dry cell also. The voltage developed is the same as in the carbon-zinc cell, 1.5 volts. The major advantage of the alkaline cell is longer life. The average alkaline cell can supply from three to five times the energy of a carbon-zinc cell, depending on the discharge rate when the cell is being used. Although some special alkaline cells can be recharged, the number of charge and discharge cycles is limited. In general, the alkaline cell is considered to be a nonrechargeable cell. A D-size alkaline-manganese cell is shown in *Figure 12–10*. Several different sizes and case styles of alkaline and mercury cells are shown in *Figure 12–11*.

Figure 12–10 Alkaline-manganese cell.

Figure 12–11 Different sizes and case styles of primary cells.

One of the devices shown in *Figure 12–11* is a true battery and not a cell. The battery is rectangular and has two snap-on type terminal connections at the top. This battery actually contains six individual cells rated at 1.5 volts each. The cells are connected in series and provide a terminal voltage of 9 volts. The internal makeup of the battery is shown in *Figure 12–12*.

Figure 12–12 Cross-sectional view of a 9-V battery.

Button Cells

Another common type of primary cell is the button cell. The button cell is so named because it resembles a button. Button cells are commonly used in cameras, watches, hearing aids, and handheld calculators. Most button cells are constructed using mercuric oxide as the anode, zinc as the cathode, and potassium hydroxide as the electrolyte (*Figure 12–13*). Although these cells are more expensive than other cells, they have a high energy density and a long life. The mercury-zinc cell has a voltage of 1.35 volts.

Another type of button cell that is less common because of its higher cost is the silver-zinc cell. This cell is the same as the mercury-zinc cell except that it uses silver oxide as the cathode material instead of mercuric oxide. The silver-zinc cell does have one distinct advantage over the mercury-zinc cell. The silver-zinc cell develops a voltage of 1.6 volts compared with the 1.35 volts developed by the mercury-zinc cell. This higher voltage can be of major importance in some electronic circuits.

A variation of the silver-zinc cell uses divalent silver oxide as the cathode material instead of silver oxide. The chemical formula for silver oxide is Ag_2O. The chemical formula for divalent silver oxide is Ag_2O_2. One of the factors that determines the energy density of a cell is the amount of oxygen in the cathode material. Although divalent silver oxide contains twice as much oxygen as silver oxide, it does not produce twice the energy density. The increased energy density is only about 10% to 15%. In addition to divalent silver oxide only having a marginal increase in energy density, the compound is less stable, too, which can cause the cell to have a shorter shelf life.

Lithium Cells

Lithium cells should probably be referred to as the lithium system because there are several different types of lithium cells. Lithium cells can have voltages that range from 1.9 volts to 3.6 volts depending on the material used to construct the cell. Lithium is used as the anode material because of its high affinity for oxygen. In fact, for many years, lithium had to be handled in an airless and moisture-free environment because of its extreme reactivity with oxygen. Several different cathode materials can be used depending on the desired application. One type of lithium cell uses a solid electrolyte of lithium hydroxide. The use of a solid electrolyte produces a highly stable compound, resulting in a shelf life that is measured in decades. This cell produces a voltage of 1.9 volts but has an extremely low current capacity. The output

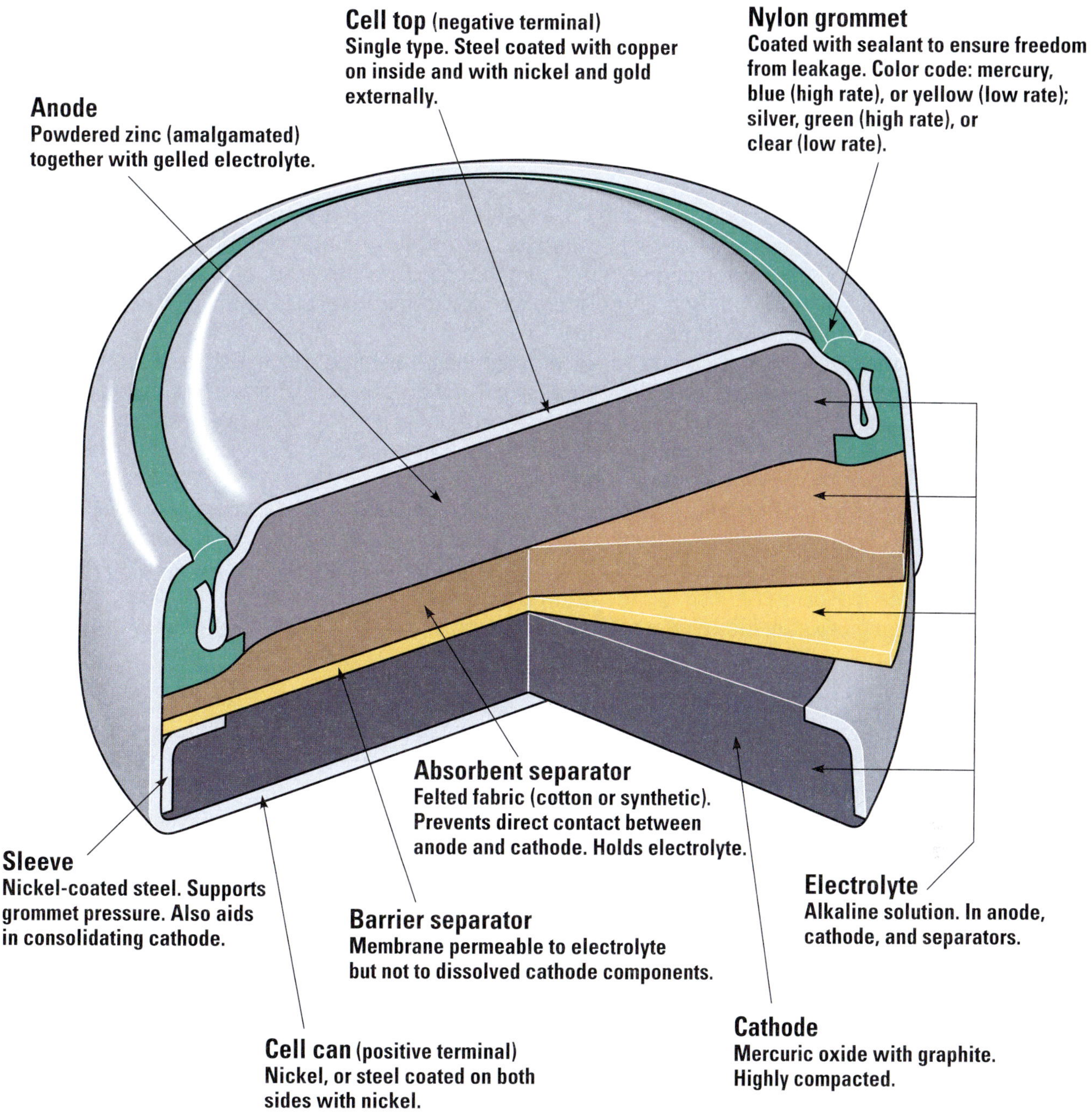

Figure 12–13 Cut-away view of different parts of a mercury-zinc button cell.

current of the lithium cell is measured in microamperes (millionths of an ampere), and they are used to power watches with liquid crystal displays and to maintain memory circuits in computers.

Other lithium cells use liquid electrolytes and can provide current outputs comparable to alkaline-manganese cells. Another lithium system uses a combination electrolyte-cathode material. One of these is sulfur dioxide, SO_2. This combination produces a voltage of 2.9 volts. In another electrolyte-cathode system, sodium chloride, NaCl, is used. This combination provides a terminal voltage of 3.6 volts.

Although lithium cells are generally considered to be primary cells, many types are rechargeable, such as lithium-ion cells. The amount of charging current, however, can be critical for these cells. An incorrect amount of charging current can cause the cell to go into what is called thermal runaway. **Thermal runaway** is an event wherein the cell is damaged in such a way that it begins an uncontrollable self-heating event that leads to the destruction of the battery and uncontrollable fire. These types of cells are also very susceptible to catastrophic failure if damaged. This is a major area of concern since lithium battery systems are the battery of choice used in powering electric vehicles. For more information on the dangers of lithium battery fires, the National Fire Protection Association (NFPA) has several resources at www.nfpa.org.

Current Capacity and Cell Ratings

The amount of current a particular type of cell can deliver is determined by the surface area of its plates. A D cell can deliver more current than a C cell, and a C cell can deliver more current than an AA cell. The amount of energy a cell can deliver is called its **current capacity**. To determine a cell's current capacity, several factors must be considered, such as the type of cell, the rate of current flow, the voltage, and the length of time involved. Primary cells are generally limited by size and weight and therefore do not contain a large amount of energy.

One of the common ratings for primary cells is the milliampere-hour (mAh). A milliampere (mA) is 1/1000 of an ampere. Therefore, if a cell can provide a current of 1 milliampere for 1 hour, it will have a rating of 1 milliampere-hour. An average D-size alkaline cell has a capacity of approximately 10,000 milliampere-hours. Some simple calculations would reveal that this cell should be able to supply 100 milliamperes of current for a period of 100 hours, or 200 milliamperes of current for 50 hours.

Another common measure of a primary cell's current capacity is watt-hours or, as many of us are familiar with in our monthly electric bill, kilowatt-hours (kWh). If the alkaline-manganese cell just discussed has a voltage of 1.5 volts, its watt-hour capacity would be 15 watt-hours (10,000 mAh $\times$ 1.5 V $=$ 15,000 mWh, or 15 Watt-hr). The chart in *Figure 12–14* shows the watt-hour capacity for several sizes of alkaline-manganese cells. The chart lists the cell size, the volume of the cell, and the energy density of the cells in watt-hours per cubic inch.

Alkaline-Manganese Cell Size	Volume in Cubic Inches	Watt-Hours per Cubic Inch
D cell	3.17	4.0
C cell	1.52	4.2
AA cell	0.44	4.9
AAA cell	0.20	5.1

Figure 12–14 Energy density for different size alkaline-manganese cells.

The amount of energy a cell contains depends not only on the volume of the cell but also on the type materials that make up the cell. The chart in *Figure 12–15* compares the energy densities for different types of materials used in different cells.

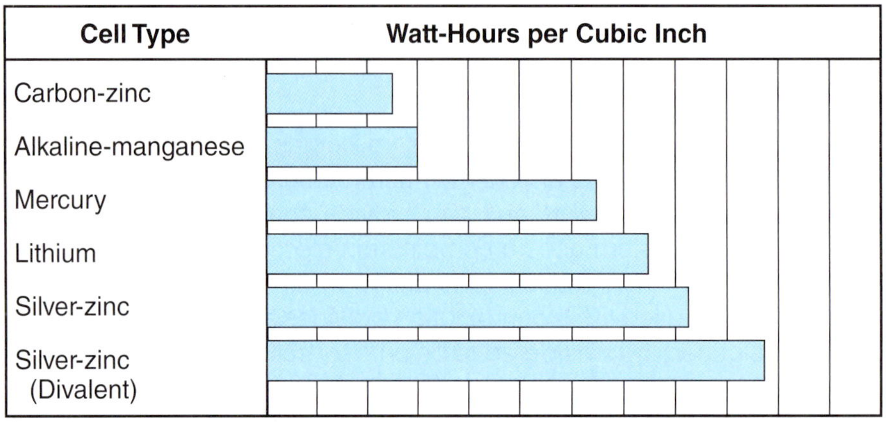

Figure 12–15 Energy density of cells utilizing different anode and cathode materials.

Internal Resistance

Batteries actually have two voltage ratings, one at no load and the other at normal load. The cell's rated voltage at normal load is the one used. The no-load voltage of a cell is greater because of the **internal resistance** of the cell. All cells have some amount of internal resistance. For example, the alkaline cell discussed previously had a rating of 10,000 milliampere-hours, or 10 ampere-hours. Theoretically, the cell should be able to deliver 10 amperes of current for one hour or 20 amperes of current for a half-hour. In practice, the cell could not deliver 10 amperes of current even under a short-circuit condition due to the limiting factor of the internal resistance of the cell. *Figure 12–16* illustrates what happens when a DC ammeter is connected directly across the terminals of a D-size alkaline cell. Assume that the ammeter indicates a current flow of 4.5 amperes and that the terminal voltage of the cell has dropped to 0.5 volt. By applying Ohm's law, it can be determined that the cell has an internal resistance of 0.111 ohm ($0.5\ \text{V}/4.5\ \text{A} = 0.111\ \Omega$).

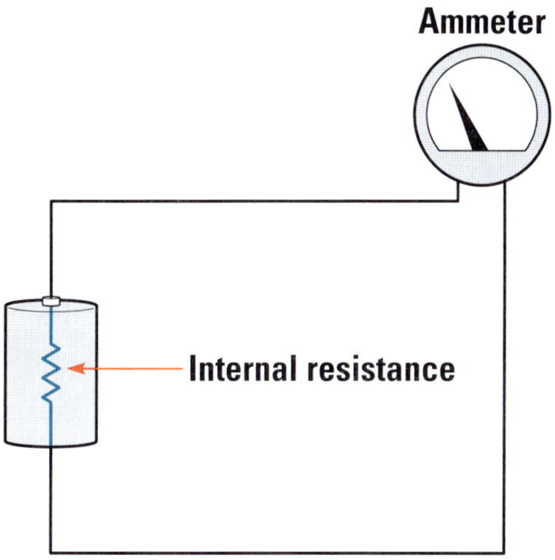

Figure 12–16 Available fault current in a battery is limited by the cell's internal resistance.

As the cell ages and energy is used, the electrodes and electrolyte begin to deteriorate. This causes them to become less conductive, which results in an increase of internal resistance. As the internal resistance increases, the terminal voltage decreases.

12–5 Secondary Cells

The secondary cell is characterized by the fact that once its stored energy has been depleted, it can be recharged. One of the most common types of secondary cells is the lead-acid cell, and it has been used in battery energy storage systems for years. A string of individual cells connected to form one battery is shown in *Figure 12–17*. A single lead-acid cell consists of one plate made of pure lead, Pb; a second plate of lead dioxide, PbO_2; and an electrolyte of dilute sulfuric acid, H_2SO_4 with a specific gravity that can range from 1.215 to 1.28, depending on the application and the manufacturer (*Figure 12–18*). **Specific gravity** is a measure of the amount of acid contained in the water. Water has a specific gravity of 1.000. The specific gravity of a cell is sometimes measured during routine maintenance using a device called a **hydrometer** (*Figure 12–19*).

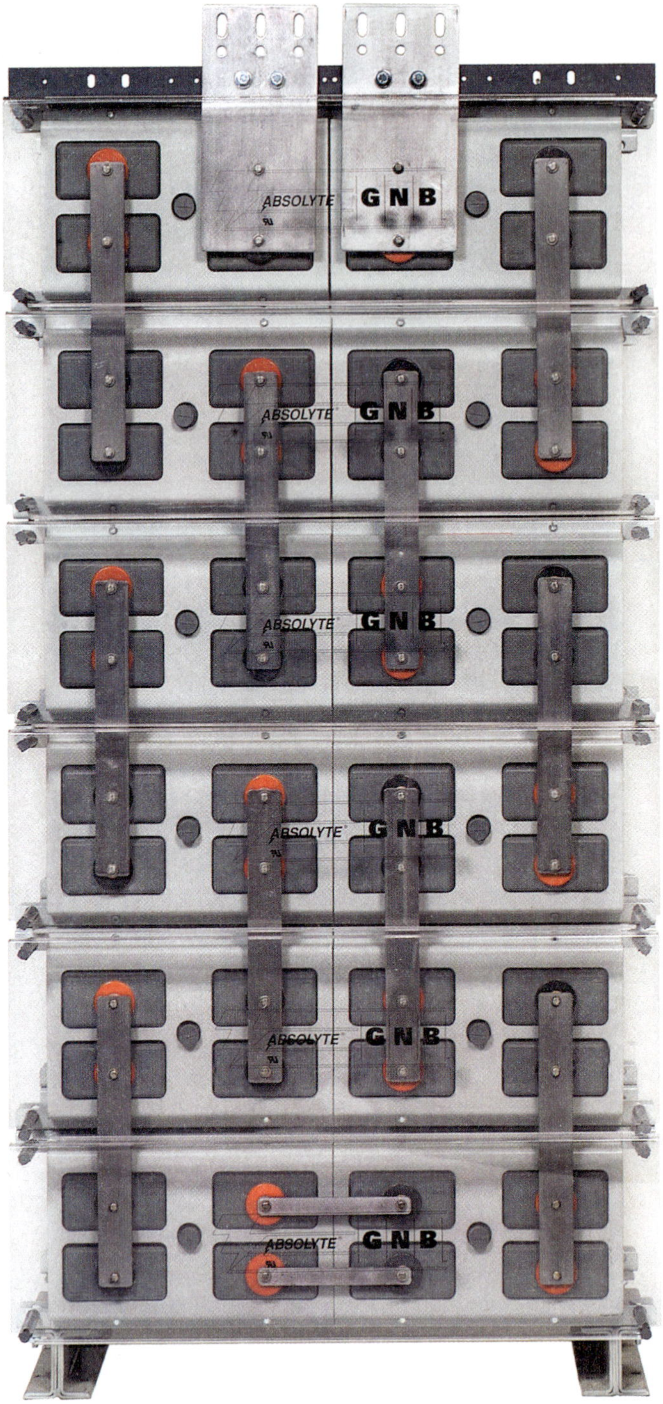

Figure 12–17 Lead-acid storage batteries. (Courtesy of GNB Industrial Power)

Discharge Cycle

When a complete circuit is connected between the positive and negative terminals, the battery begins to discharge its stored energy through the circuit and powers any connected loads such as a lamp or motor (*Figure 12–20*). This process is called the discharge cycle. Each of the lead atoms on the surface of the negative plate loses two electrons to become a Pb^{++} ion. These positive ions attract SO_4^{--} ions from the electrolyte. As a result, a layer of lead sulfate, $PbSO_4$, forms on the negative lead plate.

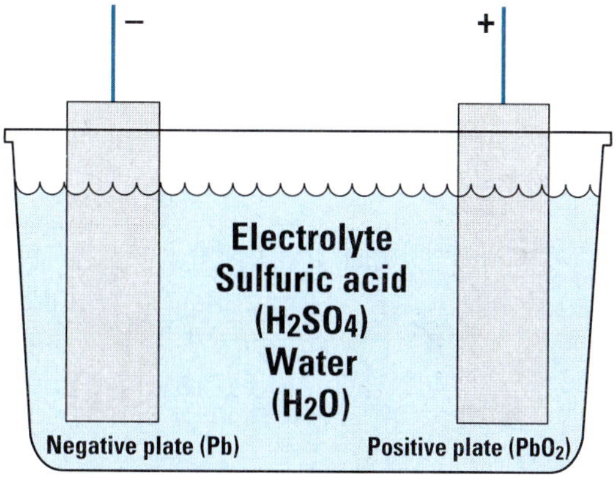

Figure 12–18 Basic lead-acid cell.

Figure 12–19 Hydrometer. (© feelphoto2521/Shutterstock.com.)

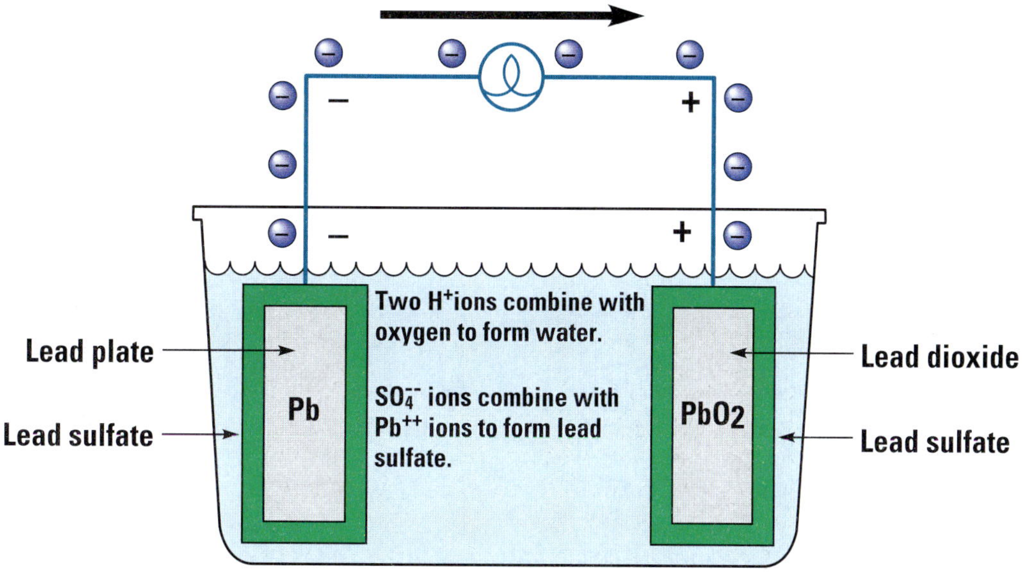

Figure 12–20 The discharge cycle of a lead-acid cell when connected to a complete circuit.

The positive plate is composed of lead dioxide, PbO_2. Each of the Pb atoms lacks four electrons that were given to the oxygen atom when the compound was formed and the Pb became a Pb^{++++} ion. Each of the Pb^{++++} ions takes two electrons from the load circuit to become a Pb^{++} ion. The Pb^{++} ions cannot hold the oxygen atoms that are released into the electrolyte solution and combine with hydrogen atoms to form molecules of water, H_2O. The remaining Pb^{++} ions combine with SO_4^{--} ions and form a layer of lead sulfate around the positive plate.

As the cell is discharged, two H^+ ions contained in the electrolyte combine with an oxygen atom liberated from the lead dioxide to form water. For this reason, the hydrometer can be used to test the specific gravity of the cell to determine the state of charge. The more discharged the cell becomes, the more water is formed in the electrolyte and the lower the specific gravity reading becomes.

Charging Cycle

The secondary cell can be recharged by reversing the chemical action that occurred during the discharge cycle. This process, called the charging cycle, is accomplished by connecting an external power source, such as a DC power supply, to the cell. The positive output of the power supply connects to the positive terminal of the cell, and the negative output of the power supply connects to the negative terminal of the cell (*Figure 12–21*).

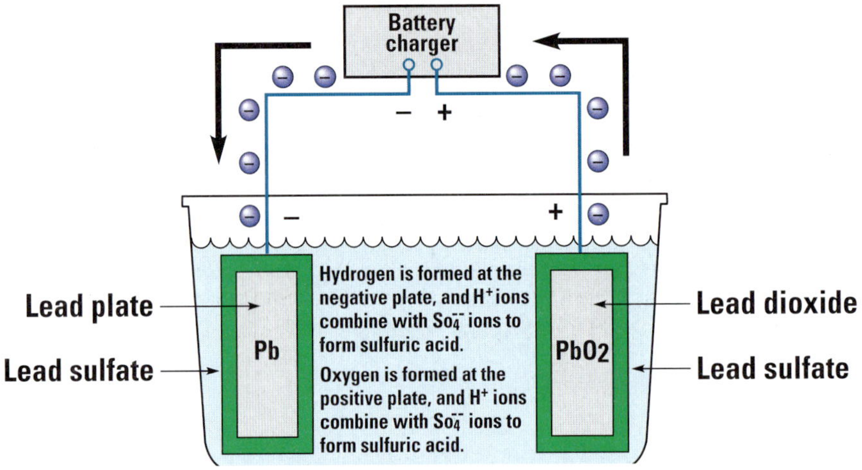

Figure 12–21 A lead-acid cell being charged by an external power source.

The terminal voltage of the power supply must be greater than that of the cell or battery. As current flows through the cell, hydrogen is produced at the negative plate, and oxygen is produced at the positive plate. If the cell is in a state of discharge, both plates are covered with a layer of lead sulfate. The H^+ ions move toward the negative plate and combine with SO_4^{--} ions to form new molecules of sulfuric acid. As a result, Pb^{++} ions are left at the plate. These ions combine with electrons being supplied by the power supply and again become neutral lead atoms.

At the same time, water molecules break down at the positive plate. The hydrogen atoms combine with SO_4^{--} ions in the electrolytic solution to form sulfuric acid. The oxygen atom recombines with the lead dioxide to form a Pb^{++} ion. As electrons are removed from the lead dioxide by the power supply, the Pb^{++} ions become Pb^{++++} ions.

As the cell is charged, sulfuric acid is again formed in the electrolyte. The hydrometer can again be used to determine the state of charge. When the cell is fully charged, the electrolyte should be back to its original strength and specific gravity.

Most lead-acid cells contain multiple plates (*Figure 12–22*). One section of plates is connected together to form a single positive plate, while the other section forms a single negative plate. This arrangement increases the surface area and thus the current capacity of the cell.

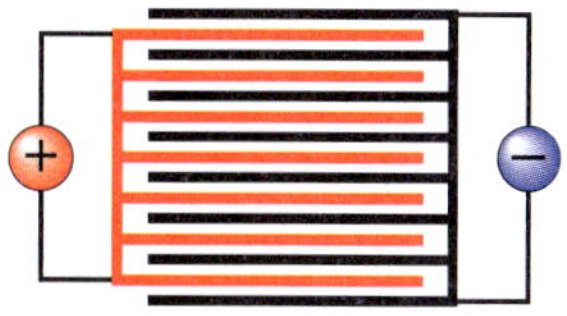

Figure 12–22 Multiple plates increase the surface area and the amount of current the cell can produce.

Cautions Concerning Charging

Theoretically, a secondary cell should be able to be discharged and recharged indefinitely. In practice, however, it cannot. If the cell is overcharged, that is, the amount of charge current is too great, the lead sulfate does not have a chance to dissolve back into the electrolyte and become acid. High charging current or mechanical shock can cause large flakes of lead sulfate to break away from the plates and fall to the bottom of the cell (*Figure 12–23*). These flakes can no longer recombine with the H^+ ions to become sulfuric acid. Therefore, the electrolyte is permanently weakened. If the flakes build up to a point where they touch the plates, they can cause an internal short circuit and the cell can no longer operate.

Caution

Overcharging can also cause hydrogen gas to form. Because hydrogen is the most explosive element known, sparks or open flames should be kept away from batteries or cells, especially during the charging process. Overcharging also causes excess heat that can permanently damage the cell. The accepted temperature limit for most lead-acid cells is 110°F (43°C).

The proper amount of charging current can vary from one type of battery to another, and manufacturers' specifications should be followed when possible. A general rule concerning charging current is that the current should not be greater than 1/10 the ampere-hour capacity. An 80-ampere-hour battery, for instance, should not be charged with a current greater than 8 amperes.

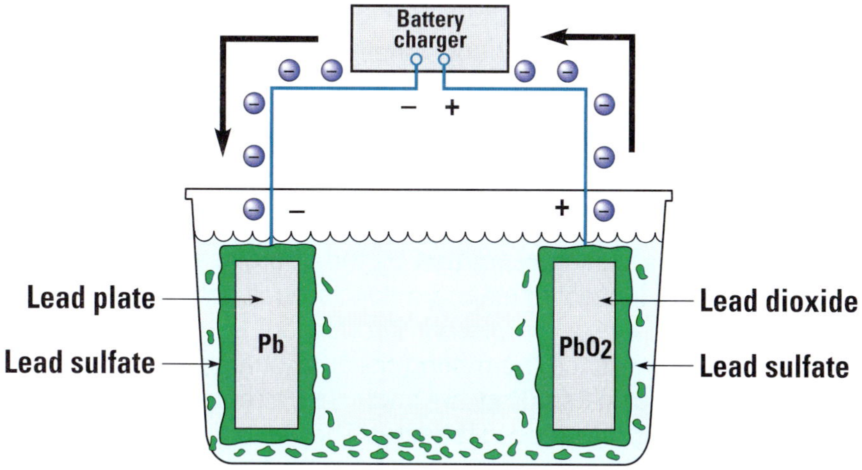

Figure 12–23 Improper charging can cause lead sulfate to flake off and fall to the bottom of the cell.

Sealed Lead-Acid Batteries

Sealed lead-acid batteries are commonly used in energy storage systems. They come in different sizes, voltage ratings, ampere-hour ratings, and case styles (*Figure 12–24*). One type of sealed lead-acid battery is referred to as a gel cell because the sulfuric acid electrolyte is suspended in an immobilized gelatin state. This treatment prevents spillage and permits the battery to be used in any position. Gel cells use a cast grid constructed of lead-calcium, which is free of antimony. The calcium is used to add strength to the grid. The negative plate is actually a lead paste material, and the positive plate is made of lead-dioxide paste. A one-way pressure-relief valve set to open at 2–6 PSI is used to vent any gas buildup during charging.

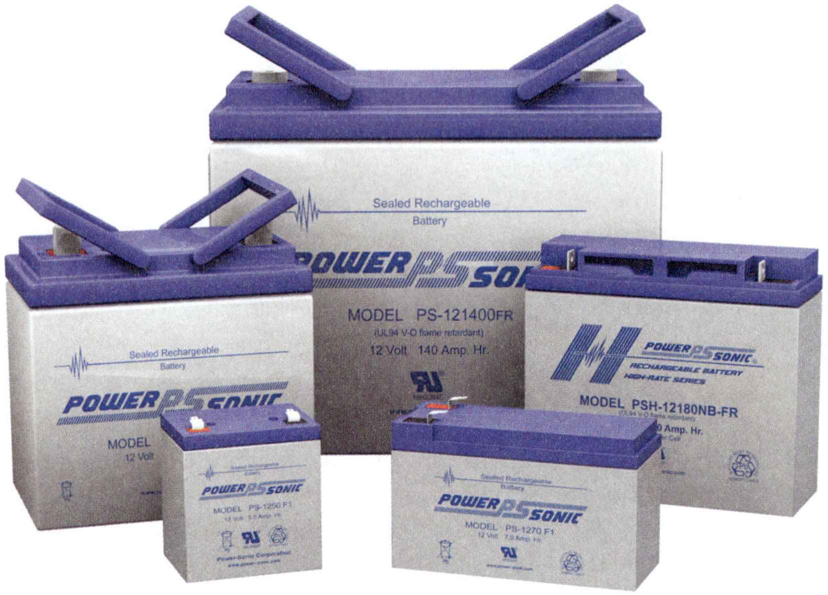

Figure 12–24 Sealed lead-acid and nicad cells and batteries. (Courtesy of Power-Sonic Corporation)

AGM Batteries

An AGM (Absorbed Glass Mat) battery is a lead-acid battery that does not spill sulfate or degrade like the common flooded cell battery. Several methods can be used to manufacture AGM batteries. Some contain spiral wound cells composed of continuous sheets of pure lead and lead dioxide separated by a fiberglass mesh that holds the electrolyte in suspension and provides separation between the plates (*Figure 12–25*). The mesh is a micro-fiberglass blend that feels like fine cotton and is capable of holding the electrolyte in a uniform suspension. The spiral design keeps the electrolyte in constant contact with the lead plates, resulting in higher efficiency during both charging and discharging than liquid-filled batteries. Spiral-wound batteries can be identified by the shape of the battery case (*Figure 12–26*). Other types of AGM batteries contain flat plates separated by the fiberglass mesh containing the electrolyte. The shape of these batteries are identical to standard flooded-cell batteries.

Some advantages of AGM batteries over common flooded-cell batteries are that they exhibit a longer life span, have a lower internal resistance (permitting them to charge faster and deliver more current), and have a longer shelf life when not in use. They are much less affected by shock and vibration because the plates and fiberglass mat are tightly pressed together. They also produce much less hydrogen during the charge cycle than flooded-cell batteries and are more resilient in extreme temperature environments. However, those performance benefits do come at a higher cost than a typical flooded-cell battery. AGM batteries can be obtained that are specifically designed for vehicle start-stop duty in vehicles equipped with such features and deep cycle duty for boats and RVs. AGM batteries are also favored by vehicle owners who have a need for large auxiliary power supplies for tools such as snow removal equipment, winch motors, and high-performance sound systems. Although AGM batteries typically cost more than flooded-cell batteries, their advantages over flooded-cell batteries can often offset the cost difference.

Figure 12–25 AGM battery with spiral-wound cells. (Courtesy of OPTIMA Batteries)

Figure 12–26 Spiral-cell batteries can be identified by the shape of the battery case. (Courtesy of OPTIMA Batteries)

Ratings for Lead-Acid Batteries

One of the most common ratings for lead-acid batteries is the ampere-hour rating. The ampere-hour rating for lead-acid batteries is determined by measuring the battery's ability to produce current for a 20-hour period at 80°F (26.6°C). A battery with the ability to produce a current of 4 amperes for 20 hours would have a rating of 80 ampere-hours.

Another common battery rating, especially for automotive batteries, is cold-cranking amperes. This rating has nothing to do with the ampere-hour rating of the battery. Cold-cranking amperes are the maximum amount of initial current the battery can supply at 68°F (20°C).

Testing Lead-Acid Batteries

It is sometimes necessary to test the state of charge or condition of a lead-acid battery. This is often done as a maintenance procedure for battery energy storage systems. The state of charge can often be tested with a hydrometer, as previously described. As batteries continuously charge and discharge, however, the specific gravity will begin to remain lower than the original specifications after the battery has been charged. When this happens, it is an indication that the battery has lost part of its materials because of lead sulfate flaking off the plates and falling to the bottom of the battery.

Another standard test for lead-acid batteries is the **load test**. This test will probably reveal more information concerning the condition of the battery than any other test. To perform a load test, the amount of test current should be three times the ampere-hour capacity. The voltage should not drop below 80% of the terminal voltage for a period of three minutes. For example, an 80-ampere-hour 12-volt battery is to be load tested. The test current will be 240 amperes (80 A × 3), and the voltage should not drop below 9.6 volts (12 V × 0.80) for a period of 3 minutes.

12–6 Other Secondary Cells

Nickel-Iron Batteries (Edison Battery)

The nickel-iron cell is often referred to as the Edison cell or Edison battery. The nickel-iron battery was developed in 1899 for use in electric cars being built by the Edison Company. The negative plate is a nickel-steel grid that contains powdered iron. The positive plates are nickel tubes that contain nickel oxides and nickel hydroxides. The electrolyte is a solution of 21% potassium hydroxide.

The nickel-iron cell weighs less than lead-acid cells but has a lower energy density. The greatest advantage of the nickel-iron cell is its ability to withstand deep discharges and to recover without harm to the cell. The nickel-iron cell can also be left in a state of discharge for long periods of time without harm. Because these batteries need little maintenance, they are sometimes found in portable and emergency lighting equipment. They are also used to power electric mine locomotives and electric forklifts.

The nickel-iron battery does have two major disadvantages. One is high cost. These batteries cost several times more than comparable lead-acid batteries. The second disadvantage is high internal resistance. Nickel-iron batteries do not have the ability to supply the large initial currents needed to start gasoline or diesel engines.

Nickel-Cadmium Batteries

The **nickel-cadmium (nicad) cell** was first developed in Sweden by Junger and Berg in 1898. The positive plate is constructed of nickel hydroxide mixed with graphite. The graphite is used to increase conductivity. The negative plate is constructed of cadmium oxide, and the electrolyte is potassium hydroxide with a specific gravity of approximately 1.2. Nicad batteries have extremely long life spans. On average, they can be charged and discharged about 2000 times. Nicad batteries can be purchased in a variety of case styles (*Figure 12–27*).

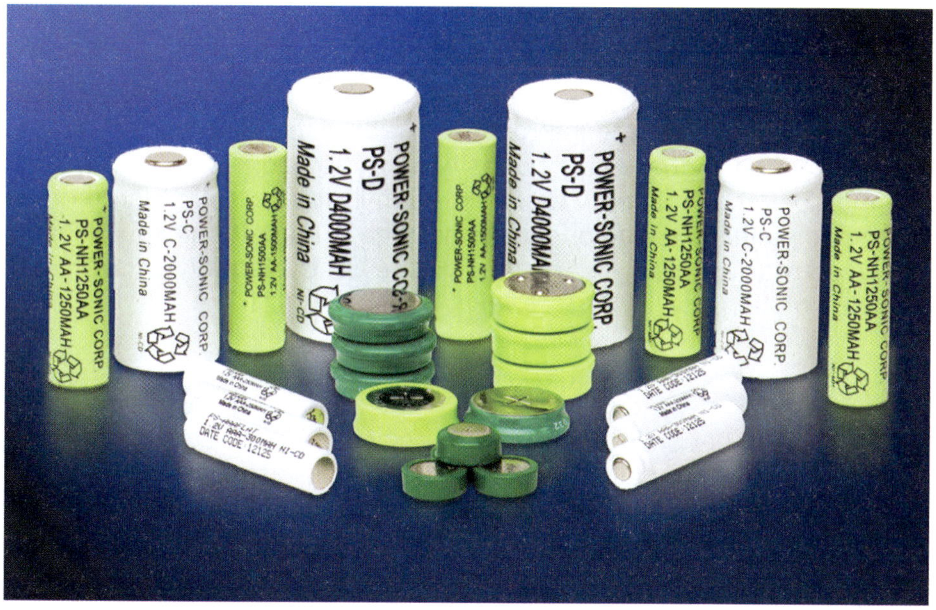

Figure 12–27 Nickel-cadmium batteries. (Courtesy of Power-Sonic Corporation)

Nickel-cadmium batteries have the ability to produce large amounts of current, similar to the lead-acid battery, but do not experience the voltage drop associated with the lead-acid battery (*Figure 12–28*).

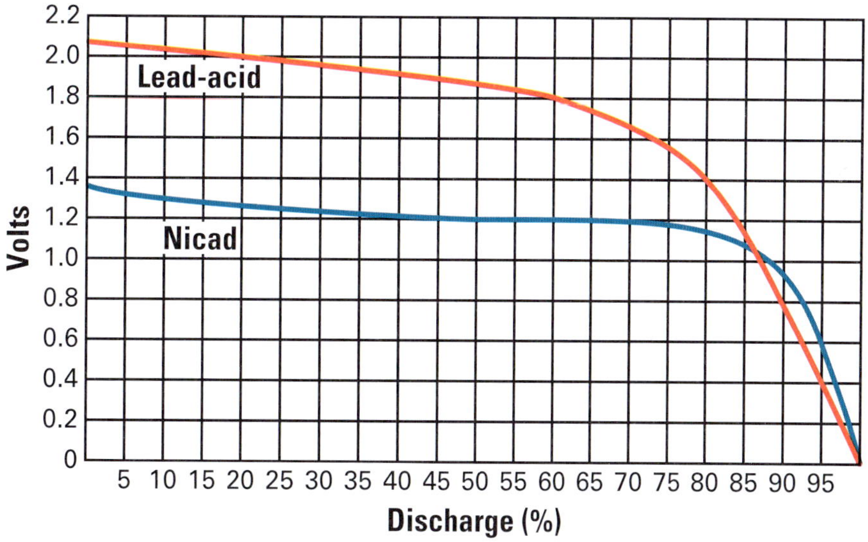

Figure 12–28 Examples of discharge curves for nicad and lead-acid cells.

Nickel-cadmium batteries do have some disadvantages:

1. The nicad battery develops only 1.2 volts per cell as compared with 1.5 volts for carbon-zinc and alkaline primary cells or 2 volts for lead-acid cells.
2. Nicad batteries cost more initially than lead-acid batteries.
3. Nicad batteries remember their charge-discharge cycles. If they are used at only low currents and are permitted to discharge through only part of a cycle and are then recharged, over a long period of time, they will develop a characteristic curve to match this cycle.

Nickel-Metal Hydride Cells

Nickel-metal hydride cells (Ni-MH) are similar to nickel-cadmium cells in many respects. Both exhibit a voltage of 1.2 volts per cell, and both have very similar charge and discharge curves. Nickel-metal hydride cells exhibit some improved characteristics over nickel-cadmium cells, however. They have about a 40% higher energy density, and they do not exhibit as great a problem of charge/discharge memory accumulation. Nickel-metal hydride cells are also more environmentally friendly than nickel-cadmium cells. The positive electrode is made of nickel oxyhydroxide (NiOOH), and the negative electrode is made of metal hydride. The electrolyte is an aqueous (watery) potassium-hydroxide solution.

Lithium-Ion Cells

Lithium-ion cells are quickly becoming the most popular type of cell in use today. These cells are used in nearly every aspect of our daily lives. They can be used for small portable equipment such as laptop computers, cell phones, and power tools but are versatile enough to be found powering large equipment such as electric vehicles and electric excavating equipment. Lithium-ion cells can be quickly recharged and offer a very high energy density for their size and weight. They exhibit a voltage of 3.6 volts per cell, which is the same voltage that can be obtained by connecting three nickel-cadmium (Ni-Cd) or three nickel-metal hydride (Ni-MH) cells in series. Lithium-ion cells have a weight-energy density ratio that is about three times greater than nickel-cadmium cells. They also exhibit a rather flat discharge curve (*Figure 12–29*), which makes them an ideal choice for electronic devices that require a constant voltage. Unlike nickel-cadmium cells and to some extent nickel-metal hydride cells, lithium-ion cells do not have the problem of memory accumulation. They can be recharged to their full capacity each time they are charged. In fact, lithium-ion cells are healthiest when they are in a state of constant charge and discharge so the electrons continue to flow through the cell.

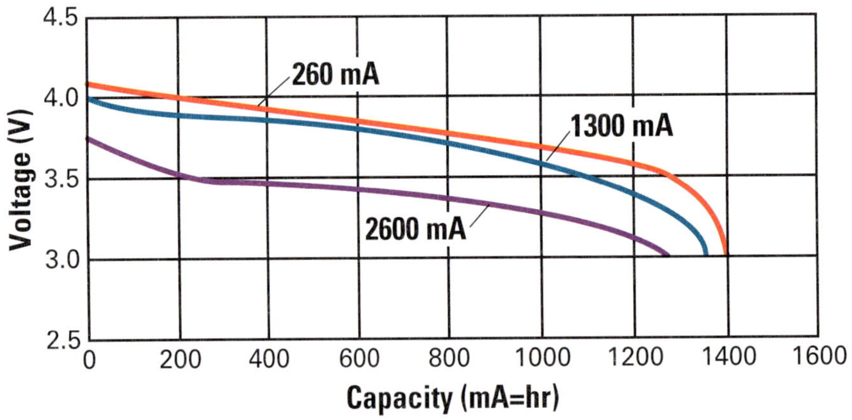

Figure 12–29 Examples of discharge curves for various lithium-ion cells.

Lithium-ion cells can safely be recharged because they do not contain metallic lithium. The positive electrode or anode is made of lithium metallic oxide, and the negative electrode or cathode is made of carbon. These cells work by transferring lithium ions between the cathode and anode during discharging and charging. Although lithium-ion cells are safe to recharge, they do require a special charger. Chargers for these cells generally produce a constant voltage and constant current. Overcurrent or overvoltage during charging can cause early deterioration of the cell.

12–7 Series and Parallel Battery Connections

When batteries or cells are connected in series, their voltages add, and their current capacities remain the same. In *Figure 12–30*, four batteries, each having a voltage of 12 volts and 60 ampere-hours, are connected in series. Connecting them in series has the effect of maintaining the surface area of the plates and increasing the number of cells. The connection shown in *Figure 12–30* will have an output voltage of 48 volts and a current capacity of 60 ampere-hours.

Caution

Batteries of different voltages should never be connected in parallel. The batteries could be damaged or one of the batteries could explode.

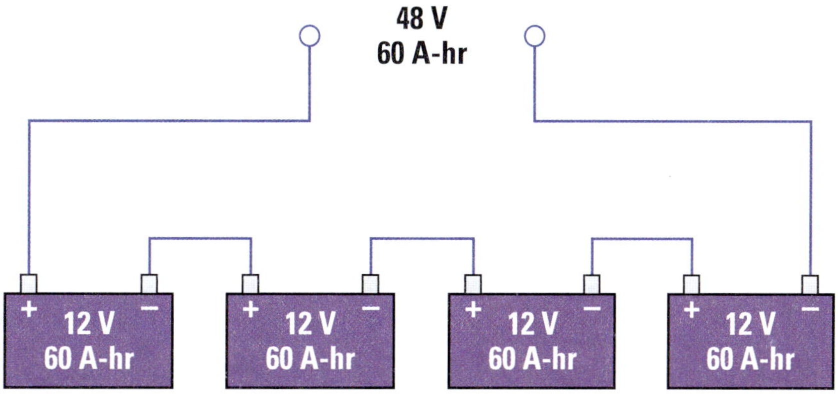

Figure 12–30 When batteries are connected in series, their voltages add, and their ampere-hour capacities remain the same.

Connecting batteries or cells in parallel (*Figure 12–31*), however, has the opposite effect and increases the area of the plates. In this example, the same four batteries are connected in parallel. The output voltage remains 12 volts, but the ampere-hour capacity has increased to 240 ampere-hours.

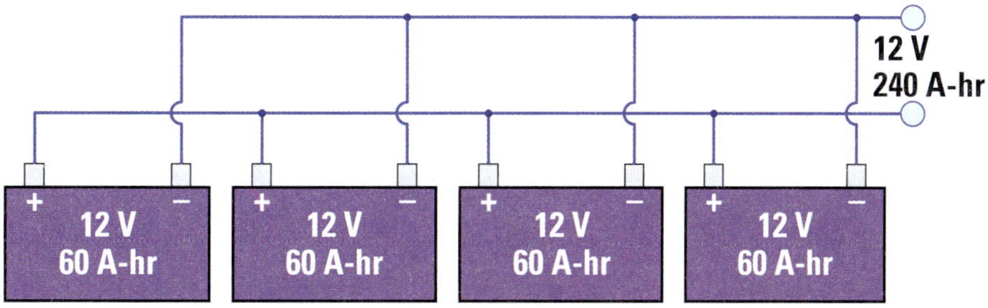

Figure 12–31 When batteries are connected in parallel, their voltages remain the same and their ampere-hour capacities add.

Batteries can also be connected in a series-parallel combination. In *Figure 12–32*, the four batteries have been connected in such a manner that the output has a value of 24 volts and 120 ampere-hours. To make this connection, the four batteries were divided into two groups of two batteries each. The batteries of each group were connected in series to produce an output of 24 volts at 60 ampere-hours. These two groups were then connected in parallel to provide an output of 24 volts at 120 ampere-hours.

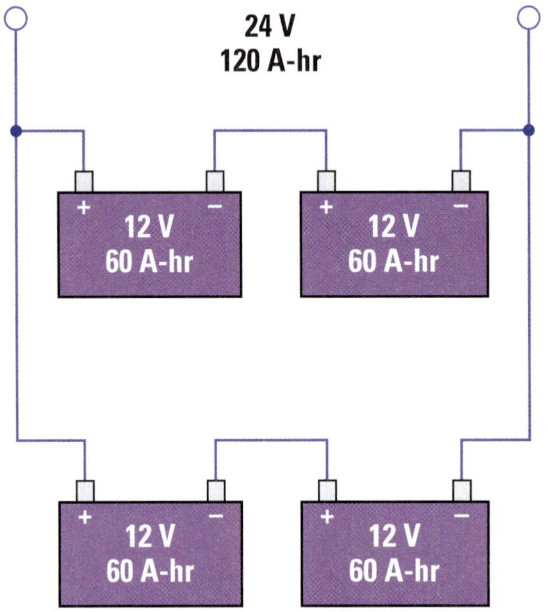

Figure 12–32 Series-parallel battery system connection.

12–8 Other Small Sources of Electricity

Solar Photovoltaic Cells

Although batteries are the largest source of electricity after alternators and generators, they are not the only source. One source of electricity is the photovoltaic cell or, as it is more commonly known, the solar cell. Solar cells are constructed from a combination of P- and N-type semiconductor material. Semiconductors are made from materials that contain four valence electrons. The most common semiconductor materials are silicon and germanium. Impurities must be added to the pure semiconductor material to form P- and N-type materials. If a material containing three valence electrons is added to a pure semiconductor material, P-type material is formed. P-type material has a lack of electrons. N-type material is formed when a material containing five valence electrons is added to a pure semiconductor material. N-type material has an excess of electrons.

Light is composed of particles called photons. A photon is a small package of pure energy that contains no mass. When photons strike the surface of the photocell, the energy contained in the photon is given to a free electron. This additional energy causes the electron to cross the junction between the two types of semiconductor material and produce a voltage (*Figure 12–33*).

The amount of voltage produced by a solar cell is determined by the material from which it is made. Silicon solar cells produce an open-circuit voltage of 0.5 volt per cell in direct sunlight. The amount of current a cell can deliver is determined by the surface area of the cell. Because the solar cell produces a voltage in the presence of light, the schematic symbol for a solar cell is the same as

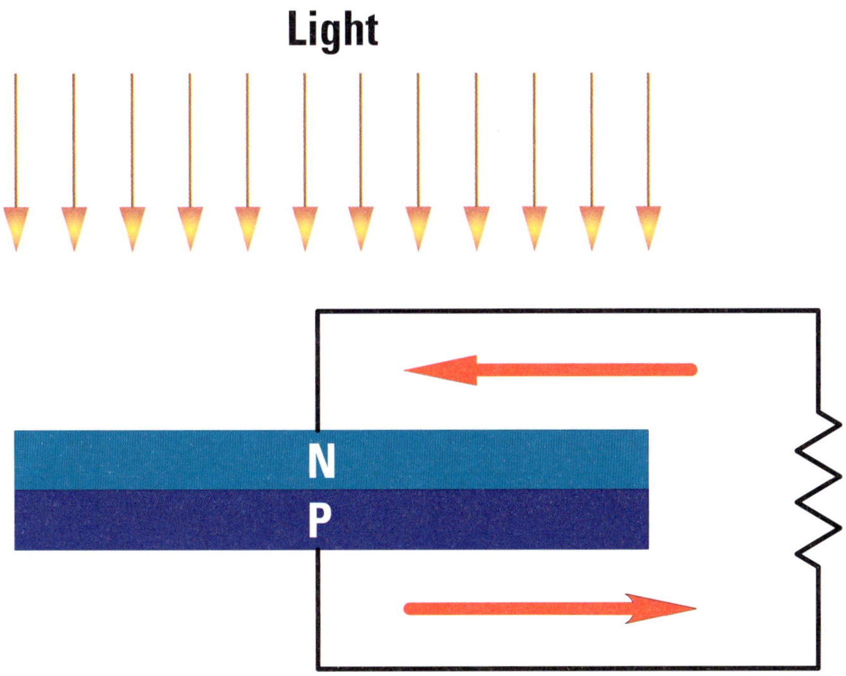

Figure 12–33 Solar cells are formed by bonding P- and N-type semiconductor materials.

that used to represent a single voltaic cell with the addition of an arrow to indicate it is receiving light (*Figure 12–34*).

It is often necessary to connect solar cells in series or parallel or both to obtain desired amounts of voltage and current, just as we saw previously with battery systems. For example, assume that an array of photovoltaic cells is to be used to charge a 12-volt lead-acid battery. The charging voltage is to be 14 volts, and the charging current should be 0.5 ampere. Now assume that each cell produces 0.5 volt with a current capacity of 0.25 ampere. In order to produce 14 volts, it will be necessary to connect 28 solar cells in series. An output of 14 volts with a current capacity of 0.25 ampere will be produced. To produce an output of 14 volts with a current capacity of 0.5 ampere, it

Green Tip

Photovoltaic cells can produce electricity without the use of any fossil fuels. Arrays are often used to charge batteries that in turn supply power to rural locations where access to power lines is not available.

Green Tip

Solar collectors are a very efficient method of heating water. Many countries throughout the world depend almost entirely on solar energy for heating residential water.

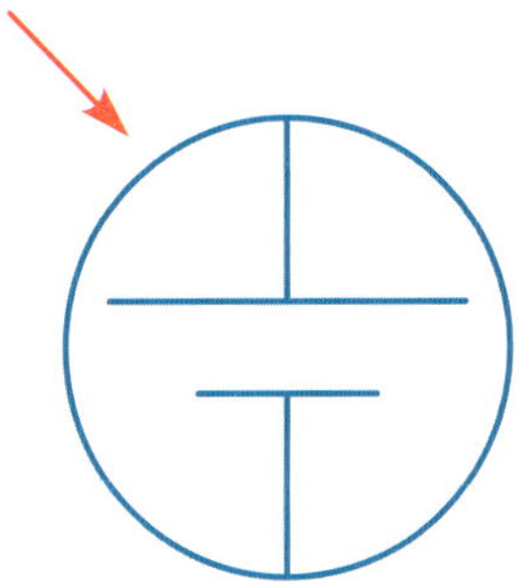

Figure 12–34 Schematic symbol for a solar cell with an arrow to indicate it receives light input to produce electricity.

will be necessary to connect a second set of 28 cells in series and parallel this set with the first set (*Figure 12–35*).

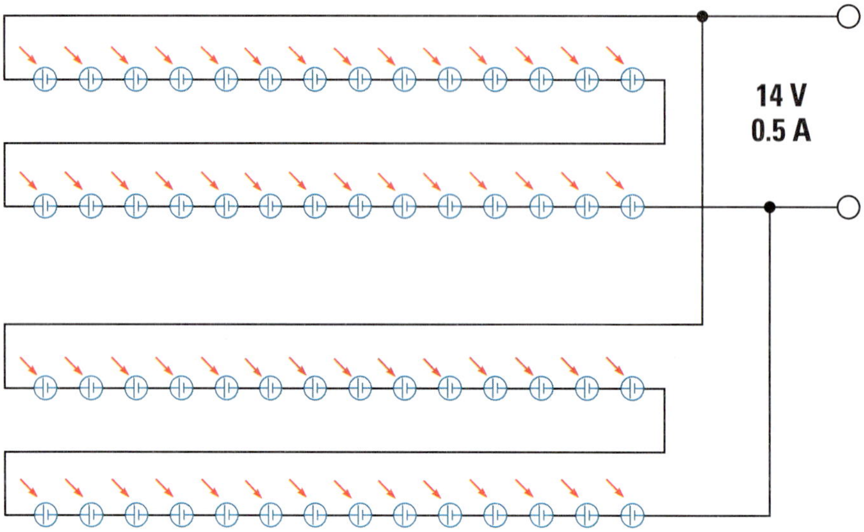

Figure 12–35 Series-parallel connection of solar photovoltaic cells.

Thermocouples

In 1822, a German scientist named Seebeck discovered that when two dissimilar metals are joined at one end and that junction is heated, a voltage is produced (*Figure 12–36*). This is known as the Seebeck effect. The device produced by the joining of two dissimilar metals for the purpose of producing electricity with heat is called a **thermocouple**. The amount of voltage produced by a thermocouple is determined by

1. the type of materials used to produce the thermocouple.
2. the temperature difference between the junction and the open ends.

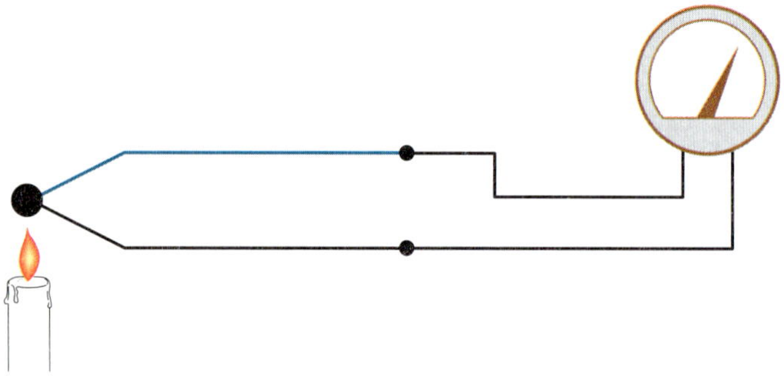

Figure 12–36 When a thermocouple is heated, a voltage will be produced.

The chart in *Figure 12–37* shows common types of thermocouples. The different metals used in the construction of thermocouples are shown as well as the normal temperature ranges of the thermocouples.

Type	Material		Degrees F	Degrees C
J	Iron	Constantan	–328 to +32 +32 to +1432	–200 to 0 0 to 778
K	Chromel	Alumel	–328 to +32 +32 to +2472	–200 to 0 0 to 1356
T	Copper	Constantan	–328 to +32 +32 to 752	–200 to 0 0 to 400
E	Chromel	Constantan	–328 to +32 +32 to 1832	–200 to 0 0 to 1000
R	Platinum 13% rhodium	Platinum	+32 to +3232	0 to 1778
S	Platinum 10% rhodium	Platinum	+32 to +3232	0 to 1778
B	Platinum 6% rhodium	Platinum	+32 to +3092	0 to 1700

Figure 12–37 Thermocouple chart.

The amount of voltage produced by a thermocouple is small, generally on the order of millivolts (1 mV = 0.001 V). The polarity of the voltage of some thermocouples is determined by the temperature. For example, a type J thermocouple produces 0 volt at about 32°F (0°C). At temperatures above 32°F, the iron wire is positive, and the constantan wire is negative. At temperatures below 32°F, the iron wire becomes negative, and the constantan wire becomes positive. At a temperature of 300°F (148.9°C), a type J thermocouple will produce a voltage of about 7.9 millivolts. At a temperature of −300°F, it will produce a voltage of about −7.9 millivolts.

Because thermocouples produce such low voltages, they are often connected in series, as shown in *Figure 12–38*. This connection is referred to as a thermopile. Thermocouples and thermopiles are generally used for measuring temperature and are sometimes used to detect the presence of a pilot light

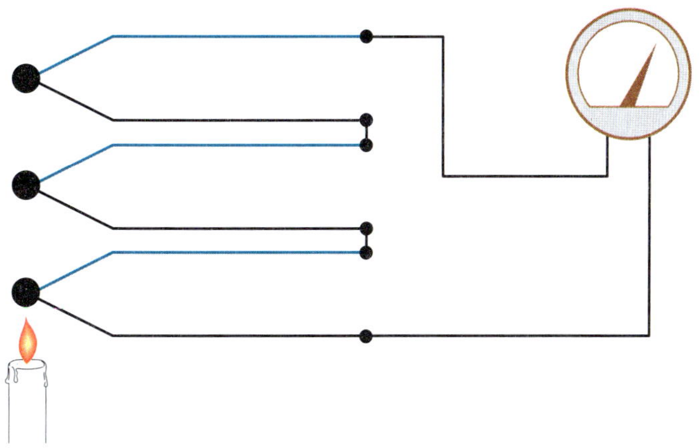

Figure 12–38 A thermopile is a series connection of thermocouples.

in appliances that operate with natural gas. The thermocouple is heated by the pilot light. The current produced by the thermocouple is used to produce a magnetic field, which holds a gas valve open to permit gas to flow to the main burner. If the pilot light goes out, the thermocouple ceases to produce current, and the valve closes (*Figure 12–39*).

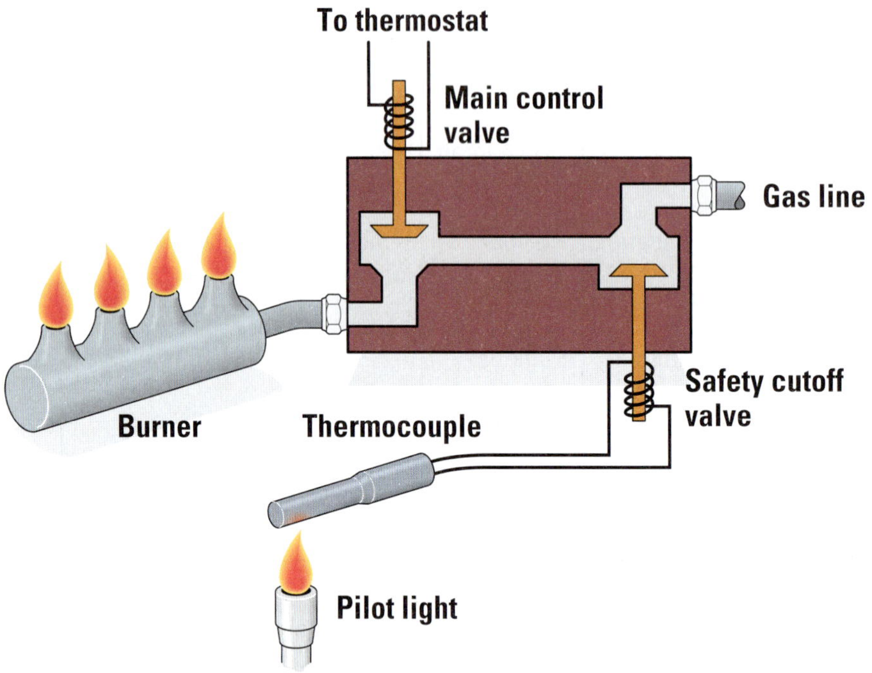

Figure 12–39 A thermocouple provides power to the safety cutoff valve.

Piezoelectricity

Piezoelectricity is produced by some materials when they are placed under pressure. The pressure can be caused by compression, twisting, bending, or stretching. Rochelle salt (sodium potassium tartrate) is often used as the needle, or stylus, of a phonograph or record player. Grooves in the vinyl record cause the crystal to be twisted back and forth, producing an alternating voltage. This voltage is amplified and is heard as music or speech. Rochelle salt crystals are also used as the pickup for microphones. The vibrations caused by sound waves produce stress on the crystals. This stress causes the crystals to produce an alternating voltage that can be amplified.

Another crystal used to produce the piezoelectric effect is barium titanate. Barium titanate can actually produce enough voltage and current to flash a small neon lamp when struck by a heavy object (*Figure 12–40*). Industry uses the crystal's ability to produce voltage in transducers for sensing pressure and mechanical vibration of machine parts.

Quartz crystal has been used for many years as the basis for crystal oscillators. In this application, an AC voltage close to the natural mechanical vibration frequency of a slice of quartz is applied to opposite surfaces of the crystal. This causes the quartz to vibrate at its natural frequency. The frequency is extremely constant for a particular slice of quartz. Quartz crystals have been used to change the frequency range of two-way radios for many years.

Fuel Cells

Another device that employs chemical processes to produce electricity is the fuel cell. Similar to the voltaic cells in battery systems, the fuel cell transfers electrons from one material to another. An important difference between the voltaic cell and the fuel cell, however, is that the material giving the electrons

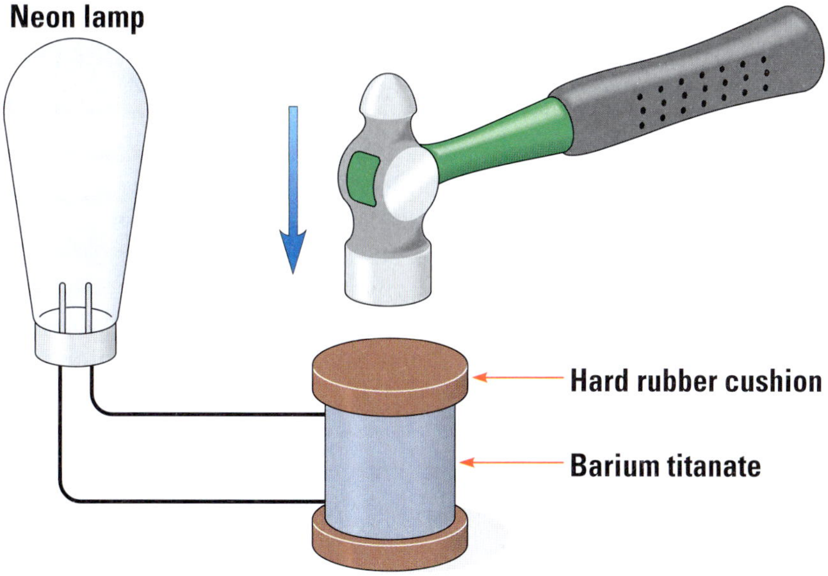

Figure 12–40 Voltage produced by the piezoelectric effect.

and the material accepting the electrons are not consumed in the process. Fuel cells require high-purity fuel and a catalytic surface for the reaction to take place. They also require auxiliary equipment such as gas containers and pressure controls. While some vehicle manufacturers have produced fuel-cell electric vehicles (FCEVs), research is still under way to better develop the fuel cells as an alternative power source for automobiles. A common type of fuel cell is the hydrogen-oxygen cell (*Figure 12–41*).

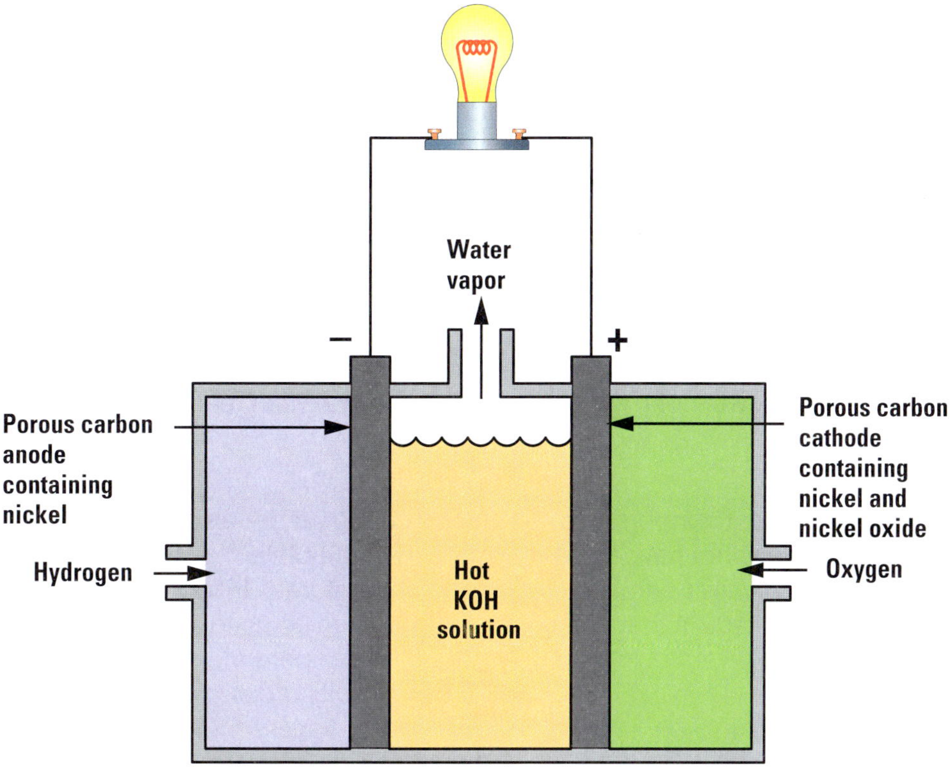

Figure 12–41 Basic hydrogen-oxygen fuel cell.

The cell contains hollow, porous carbon electrodes immersed in a potassium hydroxide solution. Hydrogen gas is supplied to one electrode, and oxygen gas is supplied to the other. The carbon electrodes contain metal or metal oxide that acts as a catalyst. A catalyst is a substance that aids the chemical reaction. In this case, the catalyst helps hydrogen molecules, which are pairs of hydrogen atoms, to separate into single atoms that combine with the negatively charged hydroxide ions of the electrolyte. The combination of the hydrogen and hydroxide ion forms a molecule of water, H_2O, with one electron left over. The excess electrons are attracted to the carbon electrode that is supplied with oxygen. The hydroxide ions are deposited at the cathode at the same rate as they are used up at the anode. The result is that hydrogen and oxygen are converted into water.

The alkaline-hydrogen-oxygen fuel cell just discussed was one of the very early types introduced in the 1960s. There are other types of fuel cells today that offer different characteristics and advantages.

The PEM Fuel Cell

Polymer Electrolyte Membrane (PEM) fuel cells are often called proton exchange membrane fuel cells. These cells offer high power density combined with low weight and volume compared with other types of fuel cells. PEM fuel cells employ a solid polymer as the electrolyte. The electrodes are porous carbon combined with a platinum catalyst. They require only hydrogen, oxygen from the air, and water to operate. They do not need corrosive fluids like some other fuel cells. PEM fuel cells operate at a relatively low temperature, around 176°F (80°C).

Direct Methanol Fuel Cells

Direct methanol fuel cells are powered by pure methanol mixed with steam. The methanol-steam mixture is fed directly to the fuel cell anode. These fuel cells have an advantage in that they do not have the fuel storage problems of cells that rely on pure hydrogen. Because methanol is a liquid, it is much easier to store. Methanol also has a higher energy density than hydrogen, although it does not contain as much energy as gasoline or diesel fuel.

Phosphoric Acid Fuel Cells

Phosphoric acid fuel cells use liquid phosphoric acid as an electrolyte. The acid is contained in a Teflon-bonded silicon carbide matrix. The electrodes are composed of porous carbon containing a platinum catalyst. This cell is considered the first generation of modern fuel cells and has been used commercially for stationary power generation and to power large vehicles such as city buses.

Molten Carbonate Fuel Cells

Molten carbonate fuel cells are being developed for natural gas and coal-fired power plants. These fuel cells operate at very high temperatures, typically 1200°F (650°C) and above. The high operating temperature has an advantage in that nonprecious metals can be used as a catalyst in the anode and cathode electrodes, resulting in much lower cost. Molten carbonate fuel cells can reach efficiencies as high as 60%.

Solid Oxide Fuel Cells

Solid oxide fuel cells use a hard nonporous ceramic compound as the electrolyte. Because the electrolyte is a solid, the fuel cell does not have to be constructed in the plate-like configuration typical of other types of fuel cells. These fuel cells operate at very high temperatures, 1830°F (1000°C). The high operating temperature has an advantage in that nonprecious metals can be used as a catalyst in the electrodes. Also, the high operating temperature of the solid oxide fuel cell permits reclaiming of fuels internally, enabling the use of a variety of fuels, which helps reduce the cost of adding a reformer to the system. Solid oxide fuel cells have an efficiency of 50% to 60%.

Green Tip

Fuel cells can be used to produce electricity with very little harmful effect to the environment.

Summary

- The first modern practical battery was invented by Alessandro Volta in 1800.
- A voltaic cell converts chemical energy into electrical energy.
- A battery is a group of cells connected together.
- Primary cells cannot be recharged.
- Secondary cells can be recharged.
- The amount of voltage produced by a cell is determined by the materials from which it is made.
- A voltaic cell can be constructed from almost any two unlike metals and an electrolyte of acid, alkali, or salt.
- Voltaic cells depend on the movement of ions in solution to produce electricity.
- The amount of current a cell can provide is determined by the area of the cell plates.
- Primary cells are often rated in milliampere-hours or watt-hours.
- The amount of energy a battery can contain is called its capacity or energy density.
- Secondary cells can be recharged by reversing the current flow through them.
- A hydrometer is a device for measuring the specific gravity of the electrolyte.
- When lead-acid batteries are charged, hydrogen gas is produced.
- Nickel-cadmium batteries have a life span of about 2000 charge-discharge cycles.
- When cells are connected in series, their voltages add, and their ampere-hour capacities remain the same.
- When cells are connected in parallel, their voltages remain the same, and their ampere-hour capacities add.
- Cells or batteries of different voltages should never be connected in parallel.
- Photovoltaic cells produce electricity in the presence of light.
- The amount of voltage produced by a photovoltaic cell is determined by the materials from which it is made.
- The amount of current produced by a photovoltaic cell is determined by the surface area.
- Thermocouples produce a voltage when the junction of two unlike metals is heated.
- The amount of voltage produced by a thermocouple is determined by the type of materials used and the temperature difference between the ends of the junction.
- The voltage polarity of some thermocouples is determined by the temperature.
- Thermocouples connected in series to produce a higher voltage are called a thermopile.
- Some crystals can produce a voltage when placed under pressure.
- The production of voltage by the application of pressure is called the piezoelectric effect.

Review Questions

1. A battery cell that can be discharged and recharged is known as which of the following types of battery cells?

   A) A primary cell

   B) A re-useable cell

   C) A secondary cell

   D) A replacement cell

2. In primary cells, which of the following is a true statement?
   A) The electrode material in a primary cell is consumed during the discharging process.
   B) The electrolyte is always a liquid in primary cells.
   C) A primary cell is always rechargeable with higher voltages.
   D) Primary cells always have a zinc anode.
3. True or False: Early experiments produced one of the first recorded batteries by using zinc and silver plates separated by cardboard soaked in brine.
4. Lithium cells produce a cell voltage that varies based on the materials used in the cell. Which of the following best describes the range that one might expect to see from a lithium cell?
   A) Lithium cells range between 0.9 V and 1.2 V per cell.
   B) Lithium cells range between 1.35 V and 2 V per cell.
   C) Lithium cells range between 5 V and 7.2 V per cell.
   D) Lithium cells range between 1.9 V and 3.6 V per cell.
5. A cell has a voltage of 1.5 V, and you need to construct a battery to supply a 9 V motor. How many cells must be connected together to supply the motor, and how must they be connected?
   A) 9 cells connected in series
   B) 6 cells connected in series
   C) 6 cells connected in parallel
   D) 9 cells connected in parallel
6. Cell current capacity is measured in milliampere-hours. How can cells be connected so that only this current capacity is increased?
   A) Connect the cells in series
   B) Connect the cells in both series and parallel
   C) Connect the cells in parallel
   D) Any of these
7. Alkaline cells are one of the most common types of primary cells available. How many volts is a single alkaline cell able to produce?
   A) 1.5 volts
   B) 2.2 volts
   C) 6 volts
   D) 9 volts
8. What are the names of the two scientists that are often given credit for discovering that electricity can be produced by a chemical action?
   A) Galileo and Sir Isaac Newton
   B) Albert Einstein and Robert Oppenheimer
   C) Thomas Edison and Nikolai Tesla
   D) Luigi Galvani and Alessandro Volta

9. Assume that individual photovoltaic cells are able to produce 0.5 V per cell and 0.25 A per cell. If you are to build an array capable of supplying 120 volts and 10 amperes, how many cells must be connected in series?

   A) 480 cells

   B) 240 cells

   C) 40 cells

   D) 20 cells

10. In the array from the previous question, refer to the series connected cells as a string. If the minimum number of cells needed for a string were used, how many strings would need to be connected in parallel to achieve the target of 10 A at 120 V?

    A) 480 strings

    B) 240 strings

    C) 40 strings

    D) 20 strings

11. During the discharge cycle of a secondary cell, such as a lead-acid cell, uses a chemical reaction to create an electrical charge. Which of the following best describes what happens to the specific gravity of the electrolyte as the cell discharges?

    A) The specific gravity decreases.

    B) The specific gravity increases.

    C) The specific gravity remains the same.

    D) Not enough information is provided.

12. True or False: During the charging cycle of a secondary cell, an external power source must be applied with a voltage that is greater than that of the battery.

13. A fuel cell is similar to a battery cell in that they both use a chemical process to transfer electrons from one material to another. What is one of the main differences between fuel cells and battery cells?

    A) Fuel cells still use gasoline.

    B) Fuel cells have higher emissions.

    C) Fuel cells do not consume the anode and cathode materials.

    D) Fuel cells use only dry-type electrolytes.

14. True or False: In a hydrogen-oxygen fuel cell, the byproduct of the chemical reaction is water.

Practical Applications

1. Your job is to order and connect lead-acid cells used to supply an uninterruptible power supply (UPS). The battery output voltage must be 126 V and have a current capacity of not less than 250 A-hr. Each cell has a rating of 2 V and 100 A-hr. How many cells must be ordered, and how would you connect them?

2. A bank of nickel-cadmium cells is used as the emergency lighting supply for a hospital. There are 100 cells connected in series, and each has an A-hr rating of 120 A-hr. The bank has to be replaced, and the manufacturer is no longer supplying nickel-cadmium cells. It was decided to replace the cells with lead-acid cells. How many lead-acid cells will be required if each has an A-hr rating of 60 A-hr, and how should they be connected?

3. An office building uses a bank of 63 lead-acid cells connected in series with a capacity of 80 amp-hours each to provide battery backup for their computers. The lead-acid cells are to be replaced with nickel-metal hydride cells with a capacity of 40 amp-hours each. How many nickel-metal hydride cells will be required to replace the lead-acid cells, and how should they be connected?

Unit 13 Magnetic Induction

Key Terms

Eddy current
Electromagnetic induction
Exponential curve
Henry (H)
Hysteresis loss
Left-hand rule
Lenz's law
Metal oxide varistor (MOV)
R-L time constant
Speed
Strength of magnetic field
Turns of wire
Voltage spike
Weber (Wb)

Why You Need to Know

Magnetic induction is one of the most important concepts in how electricity shapes our daily lives. Devices such as generators, alternators, motors, and transformers operate on this principle. This unit

- explains why current cannot rise in an inductor instantly.
- describes what determines the amount of induced voltage and its polarity.
- describes what a voltage spike is, when it occurs, and how to control it.
- illustrates how an induced current produces a magnetic field around a coil.
- explains how the amount of voltage induced in a conductor relates to magnetic induction and lines of flux.
- discusses time constants and how the rise time of current in an inductor can be determined.
- introduces the exponential curve. Understanding the exponential curve is important because so many things in both science and nature operate on this principle. Not only does the current in an inductor rise at an exponential rate, but also clothes hung on a line dry at an exponential rate.

Outline

Objectives

After studying this unit, you should be able to

- explain how electromagnetic induction produces electric current in a conductor.
- determine the amount and polarity of an induced voltage using principles of magnetic induction.
- determine direction of induced current using Fleming's left-hand rule.
- explain how Lenz's law applies to circuits.
- describe how current rises in a DC coil.
- identify how to protect equipment from voltage spikes.

Preview

Electromagnetic induction is one of the most important concepts in the electrical field. It is one of the basic operating principles underlying electrical generation, transformation, and utilization. It is imperative that anyone desiring to work in the electrical field understand how this principle works.

13–1 Electromagnetic Induction

In Unit 4, it was stated that one of the basic laws of electricity is that whenever current flows through a conductor, a magnetic field is created around the conductor (*Figure 13–1*). The direction of the current flow determines the polarity of the magnetic field, and the amount of current determines the strength of the magnetic field.

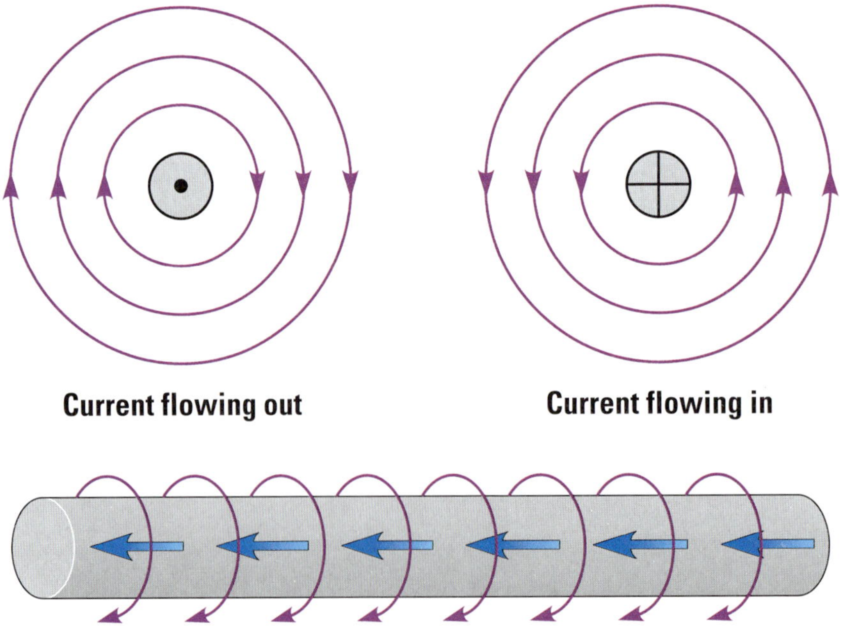

Figure 13–1 Current flowing through a conductor produces a magnetic field around the conductor.

This basic law in reverse is the principle of **electromagnetic induction,** which states that ***whenever a conductor cuts through magnetic lines of flux, a voltage is induced into the conductor.*** The conductor in *Figure 13–2* is connected to a zero-center microammeter, creating a complete circuit. When the conductor is moved downward through the magnetic lines of flux, the induced voltage causes electrons to flow in the direction indicated by the arrows. This flow of electrons causes the pointer of the meter to be deflected from the zero-center position.

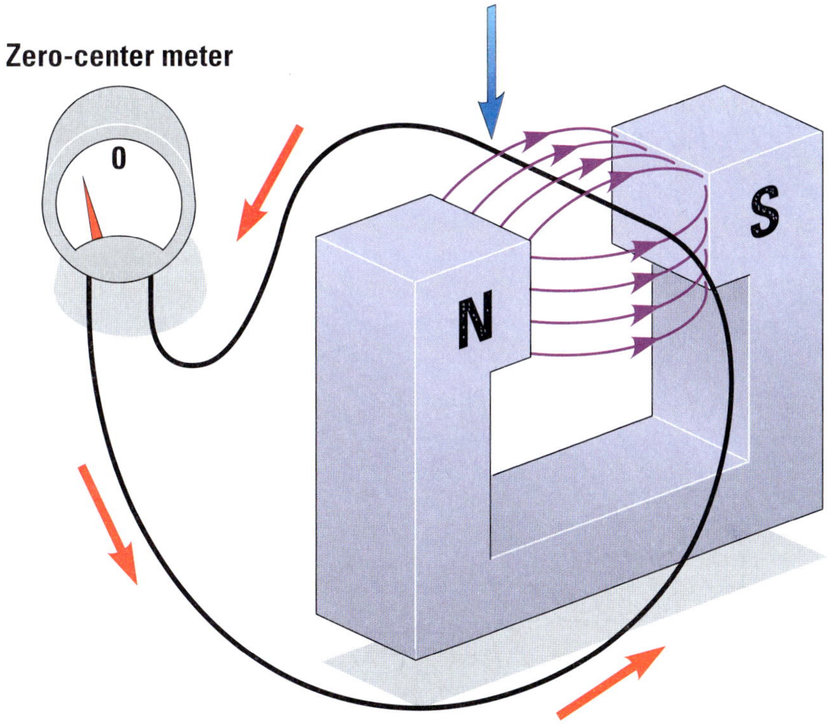

Figure 13–2 Voltage is induced when a conductor cuts magnetic lines of flux.

If the conductor is moved upward, the polarity of induced voltage is reversed, and the current flows in the opposite direction (*Figure 13–3*). Consequently, the pointer is deflected in the opposite direction.

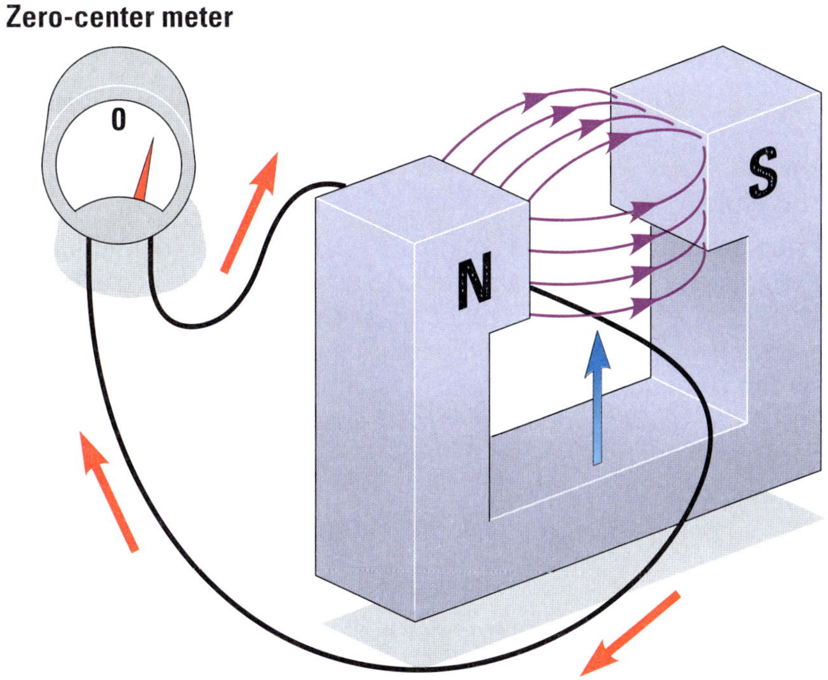

Figure 13–3 Reversing the direction of movement reverses the polarity of the voltage.

The polarity of the induced voltage can also be changed by reversing the polarity of the magnetic field (*Figure 13–4*). In this example, the conductor is again moved downward through the lines of flux, but the polarity of the magnetic field has been reversed. Therefore, the polarity of the induced voltage is the opposite of that in *Figure 13–2,* and the pointer of the meter is deflected in the opposite direction. It can be concluded that ***the polarity of the induced voltage is determined by the polarity of the magnetic field in relation to the direction of movement.***

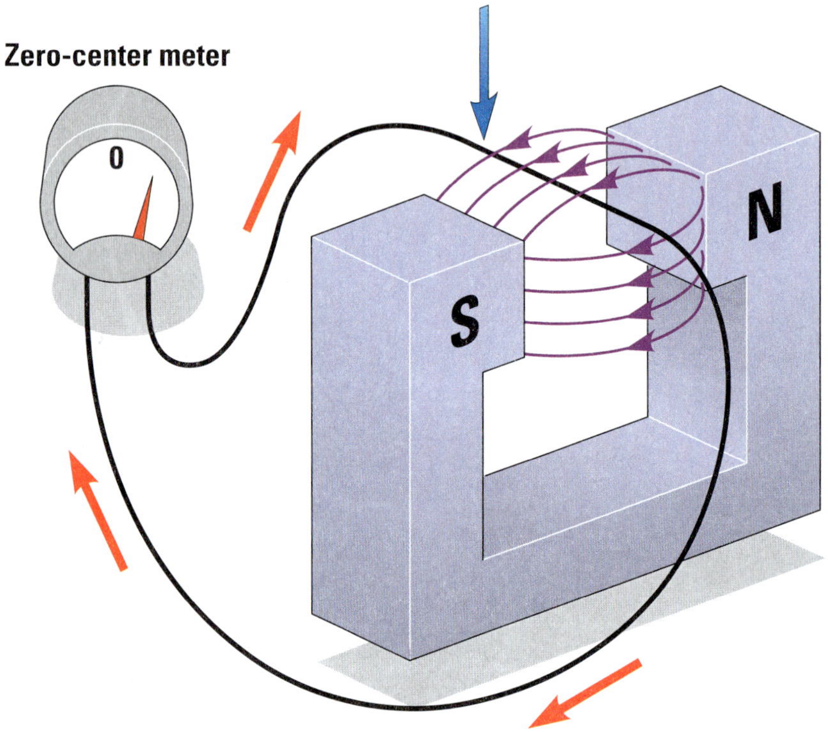

Figure 13–4 Reversing the polarity of the magnetic field reverses the polarity of the voltage.

Fleming's Left-Hand Rule

Fleming's **left-hand rule** can be used to determine the relationship of the motion of the conductor in a magnetic field to the direction of the induced current. To use the left-hand rule, place the thumb, forefinger, and center finger at right angles to each other, as shown in *Figure 13–5*. ***The forefinger points in the direction of the field flux,*** assuming that magnetic lines of force are in a direction of north to south. ***The thumb points in the direction of thrust,*** or movement of the conductor, ***and the center finger shows the direction of the current induced into the armature.*** An easy method of remembering which finger represents which quantity follows:

THumb = **TH**rust

Forefinger = **F**lux

Center finger = **C**urrent

The left-hand rule can be used to clearly illustrate that if the polarity of the magnetic field is changed or if the direction of armature rotation is changed, the direction of induced current also changes.

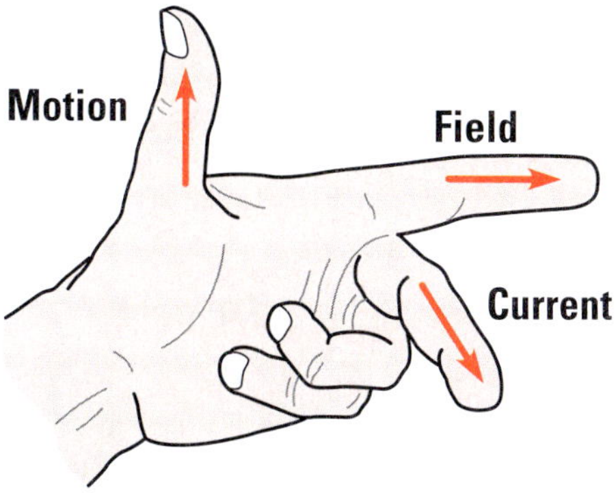

Figure 13–5 Fleming's left-hand rule.

Moving Magnetic Fields

The important factors concerning electromagnetic induction are a conductor, a magnetic field, and relative motion. In practice, it is often desirable to move the magnet instead of the conductor. Most AC generators or alternators operate on this principle. In *Figure 13–6,* a coil of wire is held stationary while a magnet is moved through the coil. As the magnet is moved, the lines of flux cut through the windings of the coil and induce a voltage into them.

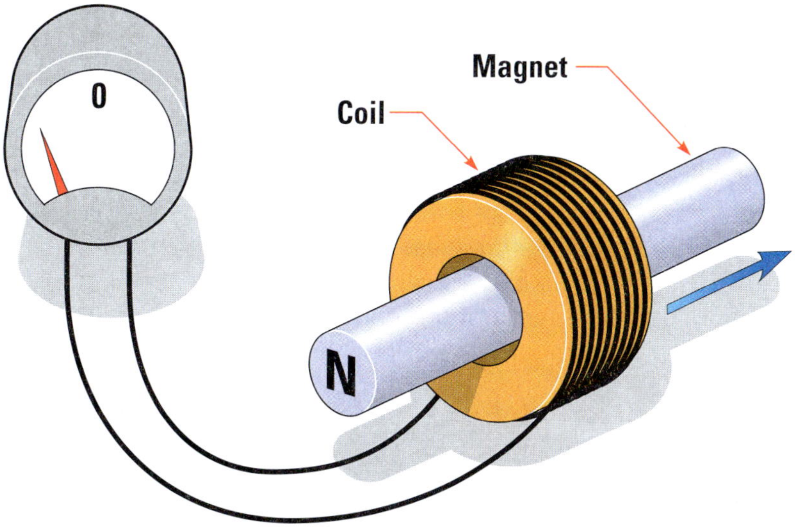

Figure 13–6 Voltage can also be induced by moving a magnetic field near a stationary conductor.

Determining the Amount of Induced Voltage

Three factors determine the amount of voltage that will be induced in a conductor as follows:

1. The number of **turns of wire**
2. The **strength of the magnetic field** (flux density)
3. The **speed** of the cutting action

In order to induce 1 volt in a conductor, the conductor must cut 100,000,000 lines of magnetic flux in 1 second. In magnetic measurement, 100,000,000 lines of flux are equal to 1 **weber (Wb).** Therefore, if a conductor cuts magnetic lines of flux at a rate of 1 weber per second, a voltage of 1 volt is induced.

A simple one-loop generator is shown in *Figure 13–7*. The loop is attached to a rod that is free to rotate. This assembly is suspended between the poles of two stationary magnets. When an external force is applied to the rod and the loop is turned, the conductor cuts through magnetic lines of flux and a voltage is induced into the conductor. The assembly that consists of the rod and the conductor windings is often referred to as the rotor.

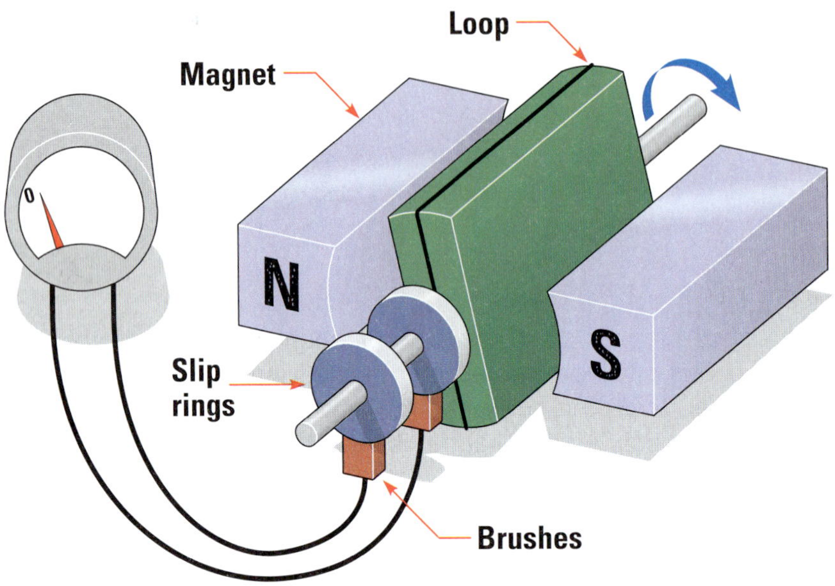

Figure 13–7 A simplified example of a single-loop generator.

If the speed of rotation is increased, the conductor cuts more lines of flux per second, and the amount of induced voltage increases. If the speed of rotation remains constant and the strength of the magnetic field is increased, there will be more lines of flux per square inch. When there are more lines of flux, the number of lines cut per second increases, and the induced voltage increases. If more turns of wire are added to the loop (*Figure 13–8*), more flux lines are cut per second, and the amount of induced voltage increases again. Adding more turns of the conductor on the rotor has the effect of connecting single conductors in series, and the amount of induced voltage in each conductor adds.

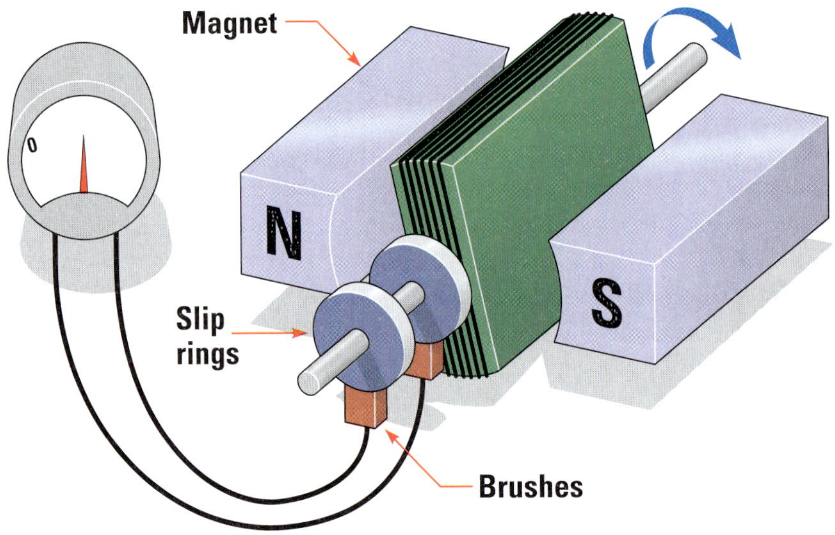

Figure 13–8 Increasing the number of conductor turns or windings increases the induced voltage.

13–2 Lenz's Law

When a voltage is induced in a coil and there is a complete circuit, current flows through the coil (*Figure 13–9*). When current flows through the coil, a magnetic field is created around the coil. This magnetic field develops a polarity opposite that of the moving magnet. The magnetic field developed by the induced current acts to attract the moving magnet and pull it back inside the coil.

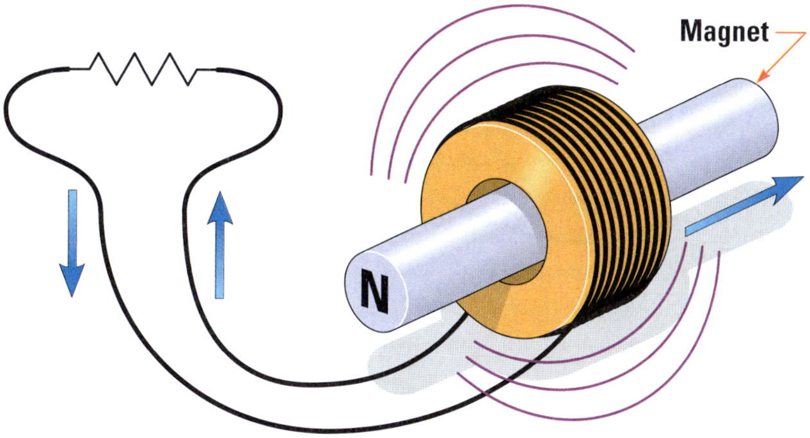

Figure 13–9 An induced current produces a magnetic field around the coil.

If the direction of motion is reversed, the polarity of the induced current is reversed, and the magnetic field created by the induced current again opposes the motion of the magnet. This principle was first noticed by Russian physicist Heinrich Lenz many years ago and is summarized in **Lenz's law,** which states that ***an induced voltage or current opposes the motion that causes it.*** From this basic principle, other laws concerning inductors have been developed. One is that ***inductors always oppose a change of current.*** The coil in *Figure 13–10,* for example, has no induced voltage and therefore no induced current. If the magnet is moved toward the coil, however, magnetic lines of flux begin to cut the conductors of the coil, and a current is induced in the coil. The induced current causes magnetic lines of flux to expand outward around the coil (*Figure 13–11*). As this expanding magnetic field cuts through the conductors of the coil, a voltage is induced in the coil. The polarity of the voltage is such that it opposes the induced current caused by the moving magnet. When the movement of the magnet stops, magnetic lines of force no longer cut the conductors in the coil and there is no induced voltage and, therefore, no induced current. As long as the permanent magnet remains inside the coil, a magnetic field surrounds the coil, but because there is no movement, there is no cutting action.

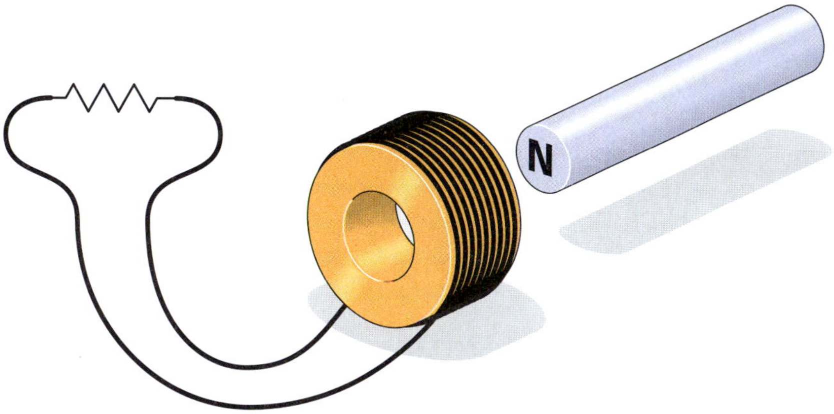

Figure 13–10 No current flows through the coil.

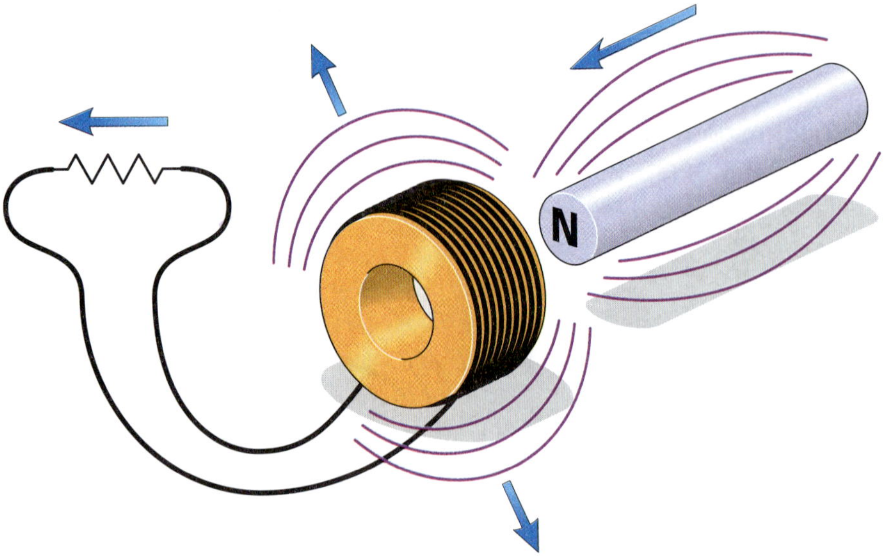

Figure 13–11 Induced current produces a magnetic field around the coil.

If the magnet is moved away, the magnetic field around the coil collapses and induces a voltage in the coil (*Figure 13–12*). Because the direction of movement of the collapsing field has been reversed, the induced voltage is opposite in polarity, forcing the current to flow in the opposite direction. The induced voltage again opposes a change of current.

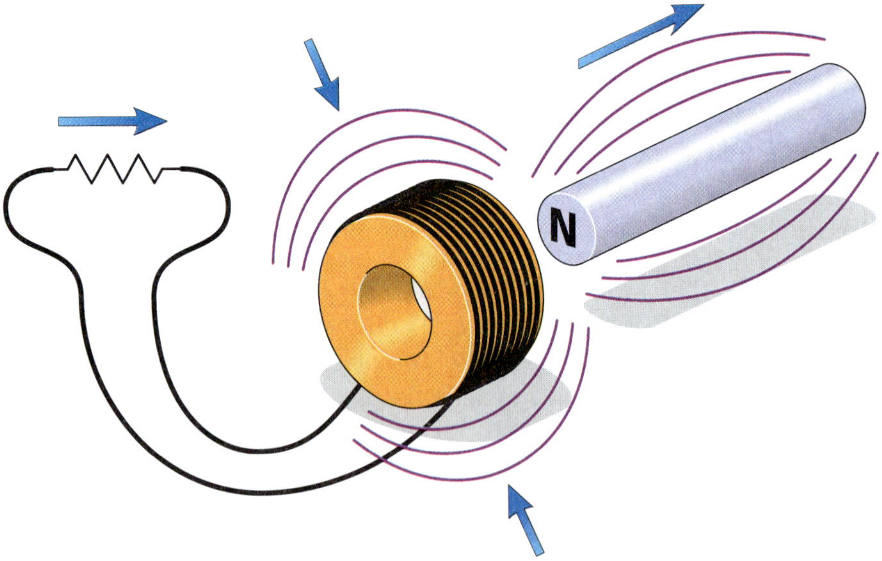

Figure 13–12 The induced voltage forces current to flow in the opposite direction.

13–3 Inductance in a Coil

When a resistive load is suddenly connected to a DC power source (*Figure 13–13*), the current instantly rises to its maximum value. The resistor shown in *Figure 13–13* has a value of 10 ohms and is connected to a 20-volt source. When the switch is closed, the current instantly rises to a value of 2 amperes (20 V/10 Ω = 2 A).

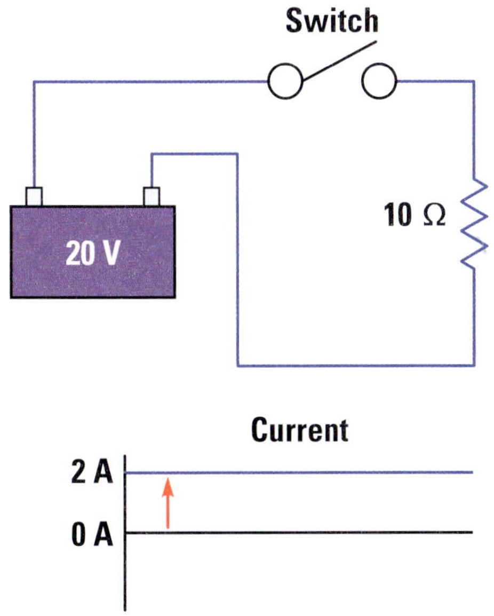

Figure 13–13 Total circuit current flows instantly in a resistive circuit.

If the resistor is replaced with a coil that has a wire resistance of 10 ohms and the switch is closed, the current cannot instantly rise to its maximum value of 2 amperes (*Figure 13–14*). As current begins to flow through the coil, the expanding magnetic field cuts through the conductors, inducing a voltage into them. In accordance with Lenz's law, the induced voltage is opposite in polarity to the applied voltage. The induced voltage therefore acts like a resistance to hinder the flow of current through the coil (*Figure 13–15*).

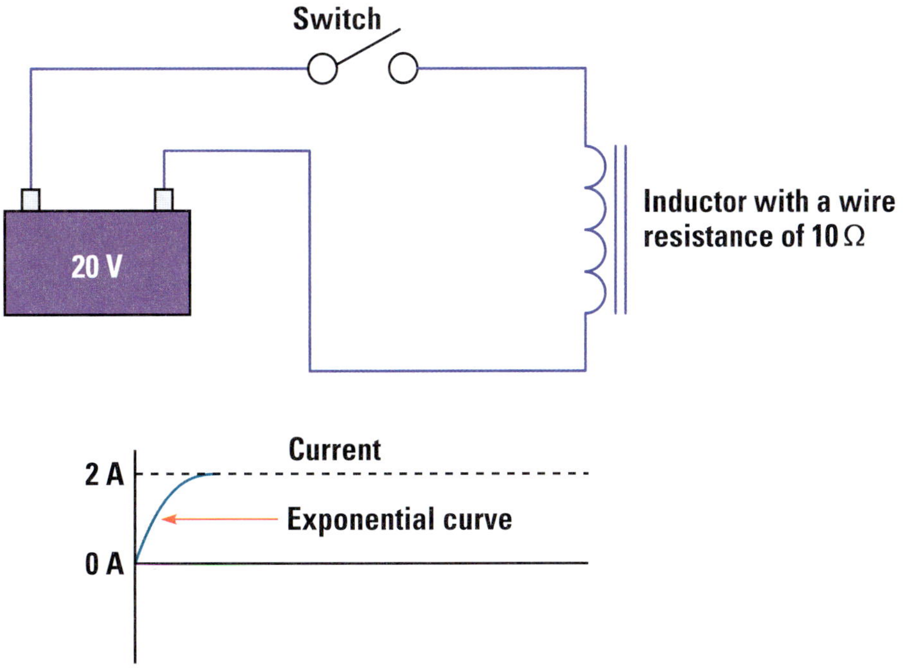

Figure 13–14 Current rises through an indicator quickly initially then gradually equalizes as it approaches full current.

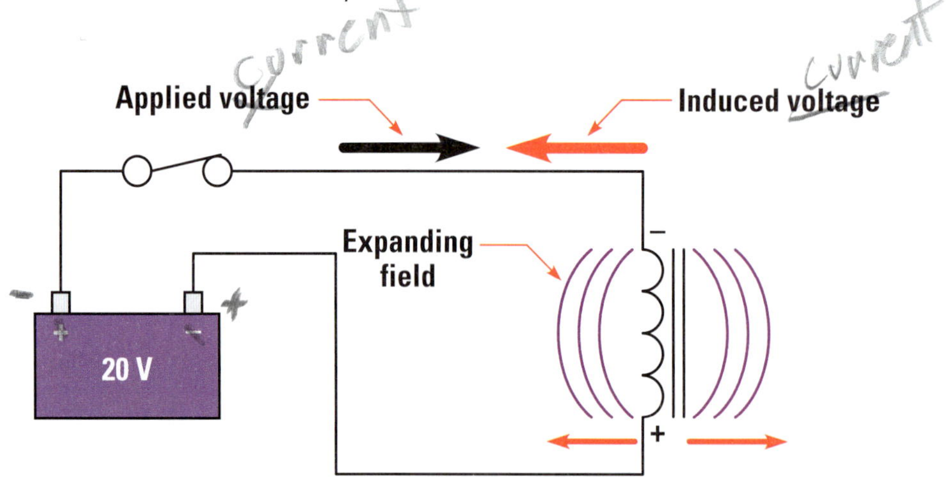

Figure 13–15 The applied voltage is opposite in polarity to the induced voltage.

The induced voltage is proportional to the rate of change of current (speed of the cutting action). When the switch is first closed, current flow through the coil tries to rise instantly. This extremely fast rate of current change induces maximum voltage in the coil. As the current flow approaches its maximum Ohm's law value—in this example, 2 amperes—the rate of change becomes less and the amount of induced voltage decreases. When the current reaches its Ohm's law value of 2 amperes, there is no longer a change of current, and the magnetic field becomes stationary. Because the magnetic field is no longer moving, there is no induced voltage to oppose the applied voltage. Current flow is now limited by the resistance of the wire used in the construction of the coil.

The Exponential Curve

The **exponential curve** describes the rate of certain occurrences. The curve is divided into five time constants. During each time constant, the current rises an amount equal to 63.2% of some stated value. An exponential curve is shown in *Figure 13–16*. In this example, current must rise from zero to a value of 1.5 amperes at an exponential rate. In this example, 100 milliseconds are required for the current to rise to its full value. Because the current requires a total of 100 milliseconds to rise to its full value, each time constant is 20 milliseconds (100 ms/5 time constants = 20 ms per time constant). During the first time constant, the current rises from 0 to 63.2% of its total value, or 0.984 ampere (1.5 A × 0.632 = 0.948 A). The remaining current is now 0.552 ampere (1.5 A − 0.948 A = 0.552 A).

During the second time constant, the current rises 63.2% of the remaining value, or 0.349 ampere (0.552 A × 0.632 = 0.349 A). At the end of the second time constant, the current has reached a total value of 1.297 amperes (0.948 A + 0.349 A = 1.297 A). The remaining current is now 0.203 ampere (1.5 A − 1.297 A = 0.203 A). During the third time constant, the current rises 63.2% of the remaining 0.203 ampere, or 0.128 ampere (0.203 A × 0.632 = 0.128 A). The total current at the end of the third time constant is 1.425 amperes (1.297 A + 0.128 A = 1.425 A).

Because the current increases at a rate of 63.2% during each time constant, it is theoretically impossible to reach the total value of 1.5 amperes. After five time constants, however, the current has reached approximately 99.3% of the maximum value and for all practical purposes is considered to be complete.

The exponential curve can often be found in nature. If clothes are hung on a line to dry, they will dry at an exponential rate. Another example of the exponential curve can be illustrated by draining water from a bucket. In *Figure 13–17,* if the stopper is removed and water is permitted to drain from the bucket for a period of time equal to one time constant for draining the bucket, during that first time constant, 63.2% of the water in the bucket will drain out. If the stopper is again removed for one time constant, 63.2% of the water remaining in the bucket will drain out. Each time the stopper is removed for a period of one time constant, the bucket will lose 63.2% of its remaining water until eventually reaching a point where all the water can be considered drained from the bucket.

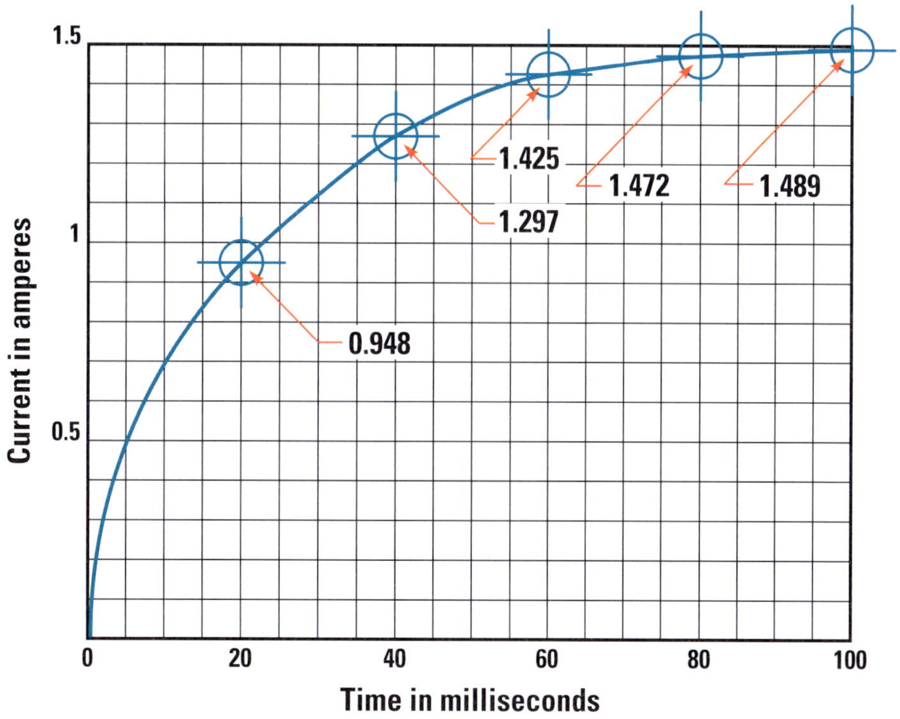

Figure 13–16 Example of the current rise curve.

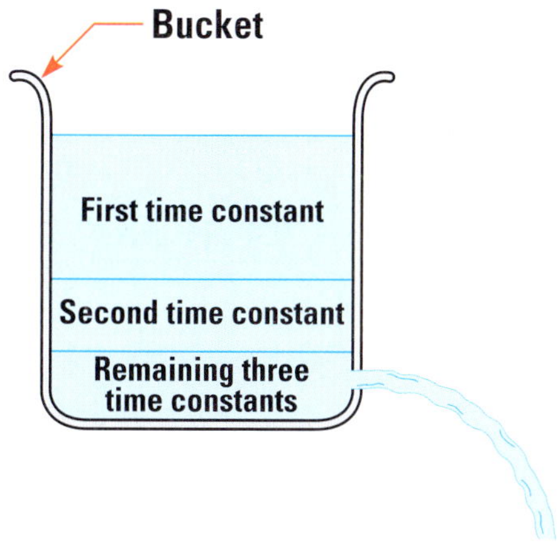

Figure 13–17 Water flows from a bucket at an exponential rate similar to current change in a coil.

Measuring Inductance

The unit of measurement for inductance is the **henry (H),** named to honor an American scientist Joseph Henry, and inductance is represented by the letter *L,* used to honor the Russian scientist Heinrich Lenz. ***A coil has an inductance of 1 henry when a current change of 1 ampere per second results in an induced voltage of 1 volt.***

The amount of inductance exhibited by a coil is determined by its physical properties and construction. A coil wound on a nonmagnetic core material such as wood or plastic is referred to as an *air-core* inductor. If the coil is wound on a core made of magnetic material such as silicon steel or soft iron, it is referred to as an *iron-core* inductor. Iron-core inductors produce more inductance with fewer turns than air-core inductors because of the good magnetic path provided by the core material. Iron-core inductors cannot be used for high-frequency applications, however, because of **eddy current** loss and **hysteresis loss** in the core material.

Another factor that determines inductance is how far the windings are separated from each other. If the turns of wire are far apart, they will have less inductance than turns wound closer together (*Figure 13–18*).

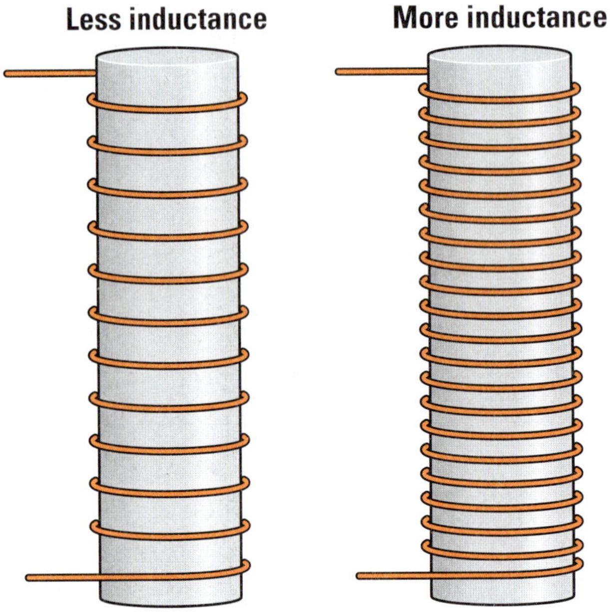

Figure 13–18 Inductance is determined by the physical construction of the coil, such as the coil winding spacing.

The inductance of a coil can be determined using the formula

$$L = \frac{0.4\ \pi N^2\ \mu A}{l}$$

where

L = inductance in henrys

π = 3.1416

N = number of turns of wire

μ = permeability of the core material

A = cross-sectional area of the core

l = length of the core

The formula indicates that the inductance is proportional to the number of turns of wire, the type of core material used, and the cross-sectional area of the core but inversely proportional to core length. An inductor is basically an electromagnet that changes its polarity at regular intervals. Because the inductor is an electromagnet, the same factors that affect magnets affect inductors. The permeability of the core material is just as important to an inductor as it is to any other electromagnet. Flux lines pass through a core material with a high permeability (such as silicon, steel, or soft iron) better than through a material with a low permeability (such as brass, copper, or aluminum). Once the core material has become saturated, however, the permeability value becomes approximately 1, and an increase in turns of wire has only a small effect on the value of inductance.

R-L Time Constants

The time necessary for a current in an inductor to reach its full Ohm's law value, called the **R-L time constant,** can be computed using the formula

$$T = \frac{L}{R}$$

where

$$T = \text{time in seconds}$$
$$L = \text{inductance in henrys}$$
$$R = \text{resistance in ohms}$$

This formula computes the time of one time constant.

The current build-up rate for an inductor at different time constants is shown in *Table 13–1.* The table gives the decimal fraction of the maximum current for each time constant.

R-L Current Buildup Table

Time Constant	*Multiplication Factor*
1	$0.632 \times I_{Max}$
2	$0.865 \times I_{Max}$
3	$0.950 \times I_{Max}$
4	$0.981 \times I_{Max}$
5	$0.992 \times I_{Max}$

Table 13–1 Rate of current decay when power is interrupted to an inductor.

Example 13–1

A coil has an inductance of 1.5 H and a wire resistance of 6 Ω. If the coil is connected to a battery of 3 V, how long will it take the current to reach its full Ohm's law value of 0.5 A (3 V/6 Ω = 0.5 A)?

Solution

To find the time of one time constant, use the formula:

$$T = \frac{L}{R}$$
$$T = \frac{1.5\ \text{H}}{6\ \Omega}$$
$$T = 0.25\ \text{s}$$

The time for one time constant is 0.25 s. Because five time constants are required for the current to reach its full value of 0.5 A, 0.25 s will be multiplied by 5:

$$0.25\ \text{s} \times 5 = 1.25\ \text{s}$$

Example 13–2

An inductor has a maximum current flow of 1.25 amperes. When power is first applied to the inductor, how much current will flow at the end of the third time constant?

$$1.25 \times 0.950 = 1.1875 \text{ A}$$

When power is interrupted to an inductor, the current will decay at an exponential rate also. The R-L current decay rate is shown in *Table 13–2.*

R-L Current Decay Table	
Time Constant	*Multiplication Factor*
1	$0.368 \times I_{Max}$
2	$0.135 \times I_{Max}$
3	$0.050 \times I_{Max}$
4	$0.019 \times I_{Max}$
5	$0.008 \times I_{Max}$

Table 13–2 Rate of current change in an inductor when power is applied.

Example 13–3

An inductor has a maximum current of 4 amperes. When power is interrupted, how much current will flow at the end of the second time constant?

$$4 \times 0.135 = 0.54 \text{ A}$$

13–4 Induced Voltage Spikes

A **voltage spike** may occur when the current flow through a coil suddenly stops. Current in the coil will also decrease at an exponential rate (*Figure 13–19*). If a complete circuit exists when the power is interrupted, there is little or no problem. In the circuit shown in *Figure 13–20*, a resistor and inductor are connected in parallel. When the switch is closed, the battery supplies current to both. When the switch is opened, the magnetic field surrounding the inductor collapses and induces a voltage into the inductor. The induced voltage attempts to keep current flowing in the same direction.

Recall that inductors oppose a change of current. The amount of current flow and the time necessary for the flow to stop is determined by the resistor and the properties of the inductor. The amount of voltage produced by the collapsing magnetic field is determined by the maximum current in the circuit and the total resistance in the circuit. In the circuit shown in *Figure 13–20,* assume that the inductor has an internal wire resistance of 6 ohms and the resistor has a resistance of 100 ohms. Also assume that when the switch is closed, a current of 2 amperes will flow through the inductor. These spikes can be on the order of hundreds or even thousands of volts and can damage circuit components, especially in circuits containing solid-state devices such as diodes, transistors, and integrated circuits. The amount of induced voltage can be determined if the inductance of the coil, the amount of current change, and the amount of time change are known by using the formula.

$$EMF = -L\left(\frac{\Delta I}{\Delta t}\right)$$

where

L = inductance in henrys (A negative sign is placed in front of the L because the induced voltage is always opposite in polarity to the voltage that produces it.)

ΔI = change of current

Δt = change of time

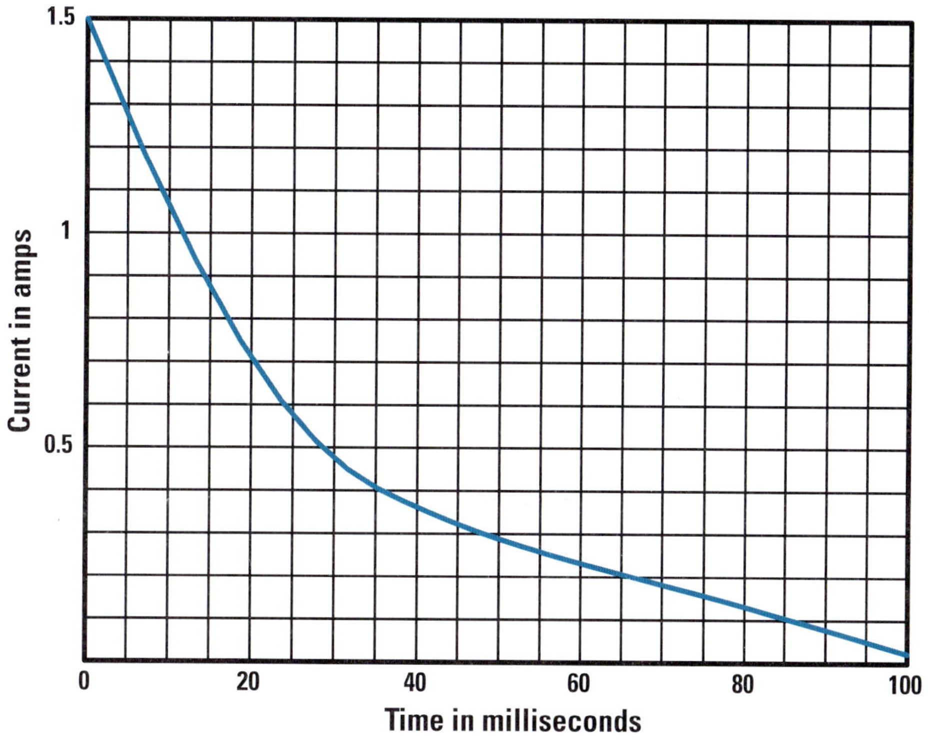

Figure 13–19 Current flow through an inductor decreases at an exponential rate.

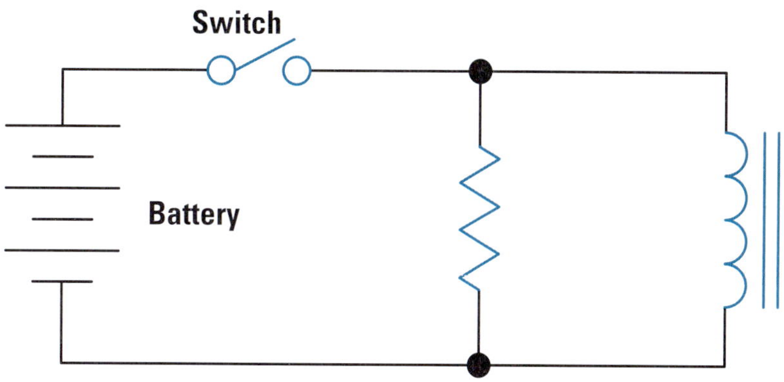

Figure 13–20 The resistor helps prevent voltage spikes caused by the inductor.

Assume that a 1.5-henry inductor has a current flow of 2.5 amperes. When a switch is opened, the current changes from 2.5 amperes to 0.5 amperes in 0.005 second. How much voltage is induced across the inductor?

$$\text{EMF} = -L\left(\frac{\Delta I}{\Delta t}\right)$$

$$\text{EMF} = -1.5\ \text{H}\left(\frac{2\ \text{A}}{0.005\ \text{s}}\right)$$

$$\text{EMF} = -1.5\ \text{H} \times 400$$

$$\text{EMF} = -600\ \text{V}$$

When the switch is opened, the resistor and inductor create their own series circuit (*Figure 13–21*).

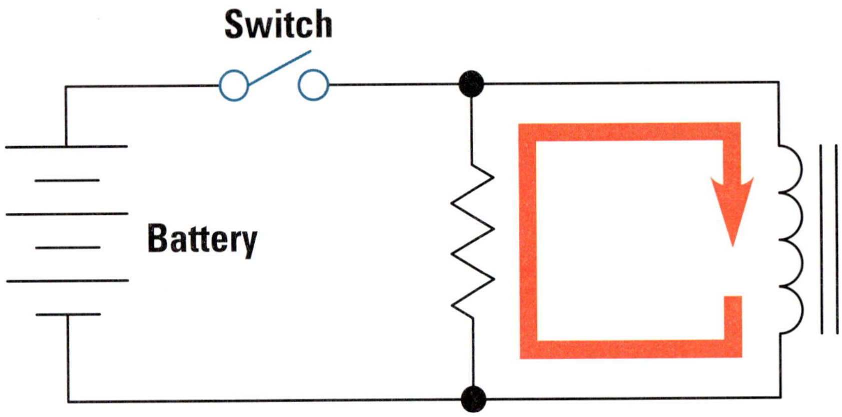

Figure 13–21 When the switch is opened, a series path is formed by the resistor and inductor.

Another device often used to prevent induced voltage spikes when the current flow through an inductor is stopped is the diode (*Figure 13–22*). The diode is an electronic component that operates like an electric one-way valve. The diode permits current to flow through it in only one direction. The diode is connected in parallel with the inductor in such a manner that when voltage is applied to the circuit, the diode is reverse-biased and acts like an open switch. When the diode is reverse-biased, no current flows through it.

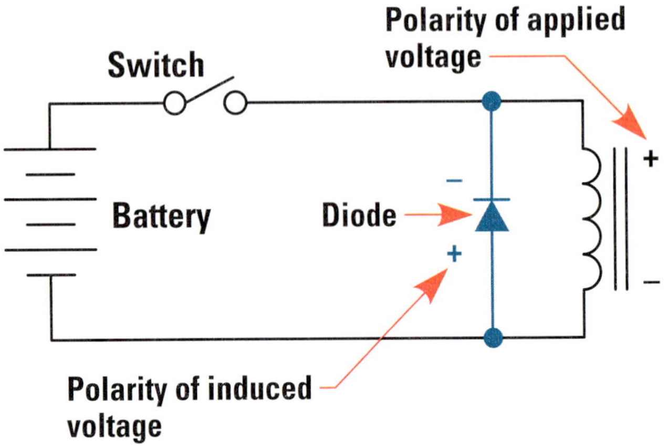

Figure 13–22 A diode can be used to prevent damage from induced voltage spikes.

When the switch is opened, the induced voltage produced by the collapsing magnetic field is opposite in polarity to the applied voltage. The diode then becomes forward-biased and acts like a closed switch. Current can now flow through the diode and complete a circuit back to the inductor. A silicon diode has a forward voltage drop of approximately 0.7 volts regardless of the current flowing through it. The diode and inductor are now a series-connected circuit. Remember that in a series circuit, the applied or source voltage must equal the sum of the voltage drops in the circuit. In this case, the diode is the total connected load. Because the diode will not permit a voltage drop greater than about 0.7 volt, the source voltage is limited to 0.7 volt also. The circuit energy is dissipated as heat by the diode. The diode can be used for this purpose in DC circuits only; it cannot be used for this purpose in AC circuits.

A device that can be used for spike suppression in either DC or AC circuits is the **metal oxide varistor (MOV).** The MOV is a bidirectional device, which means that it allows current to flow in either direction and can therefore be used in AC circuits. The MOV is an extremely fast-acting solid-state component that exhibits a change of resistance when the voltage reaches a certain point. Assume that the MOV shown in *Figure 13–23* has a voltage rating of 140 volts and that the voltage applied to the circuit is 120 volts. When the switch is closed and current flows through the circuit, a magnetic field is established around the inductor (*Figure 13–24*). As long as the voltage applied to the MOV is less than 140 volts, it will exhibit an extremely high resistance, in the range of several hundred thousand ohms.

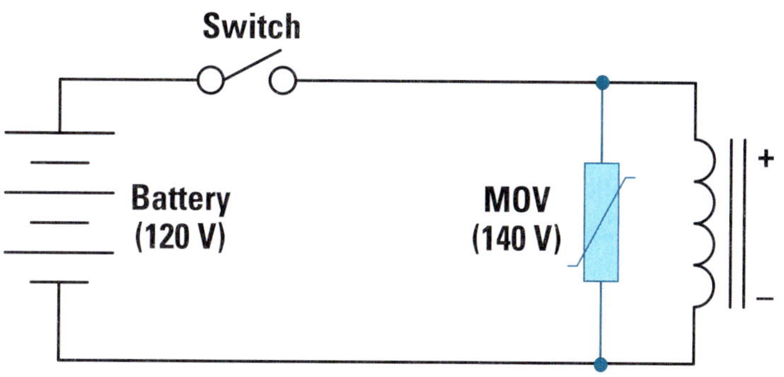

Figure 13–23 Metal oxide varistor used in parallel to a coil to suppress a voltage spike.

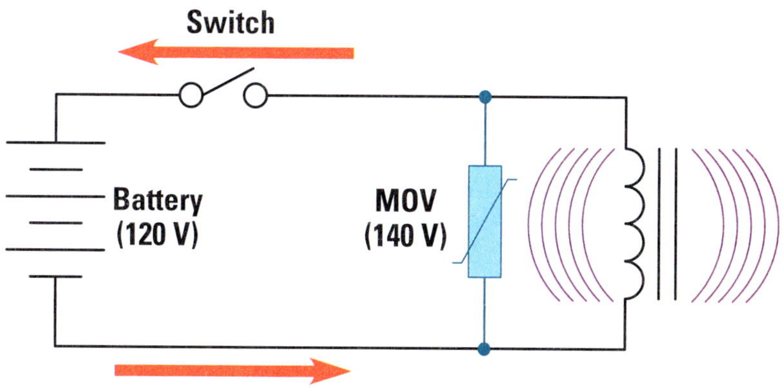

Figure 13–24 When the switch is closed, a magnetic field is established around the inductor.

When the switch is opened, current flow through the coil suddenly stops and the magnetic field collapses. This sudden collapse of the magnetic field causes an extremely high voltage to be induced in the coil. When this induced voltage reaches 140 volts, however, the MOV suddenly changes from a high resistance to a low resistance, preventing the voltage from becoming greater than 140 volts (*Figure 13–25*).

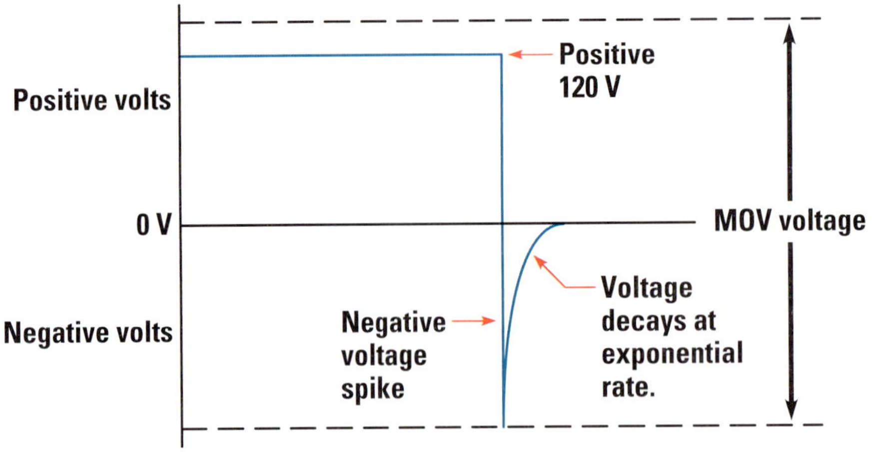

Figure 13–25 The MOV prevents the spike from becoming too high.

Metal oxide varistors are extremely fast acting. They can typically change resistance values in less than 20 nanoseconds (ns). They are often found connected across the coils of relays and motor starters in control systems to prevent voltage spikes from being induced back into the line. They are also found in the surge protectors used to protect many home appliances, such as televisions, stereos, and computers.

If nothing is connected in the circuit with the inductor when the switch opens, the induced voltage can become extremely high. In this instance, the resistance of the circuit is the air gap of the switch contacts, which is practically infinite. The inductor will attempt to produce any voltage necessary to prevent a change of current. Inductive voltage spikes can reach thousands of volts. This is the principle of operation of many high-voltage devices, such as the ignition systems of many automobiles.

Another device that uses the collapsing magnetic field of an inductor to produce a high voltage is the electric fence charger (*Figure 13–26*). The switch is constructed in such a manner that it pulses on and off. When the switch closes, current flows through the inductor and a magnetic field is produced around the inductor. When the switch opens, the magnetic field collapses and induces a high voltage across the inductor. If anything or anyone standing on the ground touches the fence, a circuit is completed through the object or person and the ground. The coil is generally constructed of many turns of very small wire. This construction provides the coil with a high resistance and limits current flow when the field collapses.

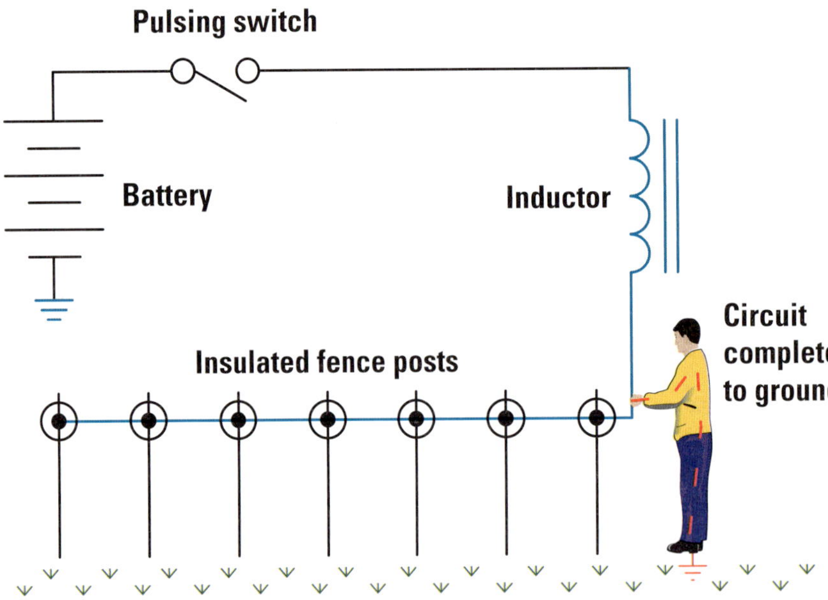

Figure 13–26 An inductor is used to produce a high voltage for an electric fence.

Summary

- When current flows through a conductor, a magnetic field is created around the conductor.
- When a conductor cuts through a magnetic field, a voltage is induced in the conductor.
- The polarity of the induced voltage is determined by the polarity of the magnetic field in relation to the direction of motion.
- Three factors that determine the amount of induced voltage are

  a. the number of turns of wire.

  b. the strength of the magnetic field.

  c. the speed of the cutting action.
- One volt is induced in a conductor when magnetic lines of flux are cut at a rate of 1 weber per second.
- Induced voltage is always opposite in polarity to the applied voltage.
- Inductors oppose a change of current.
- Current rises in an inductor at an exponential rate.
- An exponential curve is divided into five time constants.
- Each time constant is equal to 63.2% of some value.
- Inductance is measured in units called henrys (H).
- A coil has an inductance of 1 henry when a current change of 1 ampere per second results in an induced voltage of 1 volt.
- Air-core inductors are inductors wound on cores of nonmagnetic material.
- Iron-core inductors are wound on cores of magnetic material.
- The amount of inductance an inductor will have is determined by the number of turns of wire and the physical construction of the coil.
- Inductors can produce extremely high voltages when the current flowing through them is stopped.
- Two devices used to help prevent large spike voltages are the resistor and the diode.

Review Questions

1. True or False: When a conductive material cuts through magnetic lines of flux, a voltage is induced into the conductor.

2. The polarity of the induced voltage produced by a conductor cutting through magnetic lines of flux is determined by which of the following?

   A) The conductor material

   B) The speed of the conductor

   C) The direction of the conductor relative to the magnetic field

   D) The strength of the magnetic field

3. The amount of voltage induced into a conductor is dependent upon which of the following? (Select all that apply.)

   A) The number of turns of wire

   B) The conductor material

   C) The speed of the conductor cutting through lines of flux

   D) The strength of the magnetic field

4. True or False: One volt will be induced into a conductor when 100,000,000 lines of magnetic flux are cut per rotation.

5. An induced voltage will always oppose which of the following?

   A) The motion that produced it

   B) The current in the conductor

   C) The circuit voltage

   D) The changing temperature

6. When current flows in a coil, the increasing strength of the resulting magnetic field around the coil induces a current back into the coil. Which of the following statements best describes this process?

   A) The induced current is additive to the current flowing in the coil.

   B) The induced current has little to no effect on the coil.

   C) The induced current opposes the voltage applied to the coil.

   D) The induced current opposes the current flowing in the coil.

7. True or False: When DC voltage is applied to a coil and current begins to flow, the rising current induces an opposing current in the coil that acts similar to resistance and slows the initial rise in current through the coil.

8. The current rise in a circuit containing a coil will follow an exponential curve. This curve can be divided into five time constants. Each time constant is what percentage of the remaining current until the circuit reaches maximum current?

   A) 50%

   B) 63.2%

   C) 75%

   D) Not enough information

9. Which of the following can help to prevent a voltage spike from the coil when the current flow through a coil suddenly stops?

   A) A magnet

   B) A second coil

   C) A capacitor

   D) A diode

10. True or False: A metal oxide varistor helps mitigate the potential high voltage spikes from a suddenly collapsing circuit current by increasing its internal resistance and opposing current flow.
11. When using Fleming's left-hand rule to determine which direction current will flow in the conductor, what does the thumb represent?
    A) The movement of the conductor
    B) The polarity of the magnetic lines of flux
    C) The direction of current flow
    D) The speed of conductor motion
12. Using Fleming's left-hand rule, determine which direction current will flow if the magnetic lines of flux are in a direction towards your body and the conductor is moving downward.
    A) Current will flow away from your body.
    B) Current will flow to your right.
    C) Current will flow to your left.
    D) Current will travel towards your body.

Practical Applications

You are an electrician working in a plant that uses programmable controllers to perform much of the logic for the motor controls. The plant produces highly flammable chemicals, so the programmable controllers use a 24-volt DC output that is intrinsically safe in a hazardous area. You are having trouble with the output modules of the programmable controllers that are connected to DC control relays. These outputs are going bad at an unusually high rate, and you suspect that spike voltages produced by the relay coils are responsible. To test your theory, you connect an oscilloscope to an output that is operating a DC relay coil and watch the display when the programmable controller turns the relay off. You discover that there is a high voltage spike produced by the relay. What device would you use to correct this problem, and how would you install it?

Section 5 Basics of Alternating Current

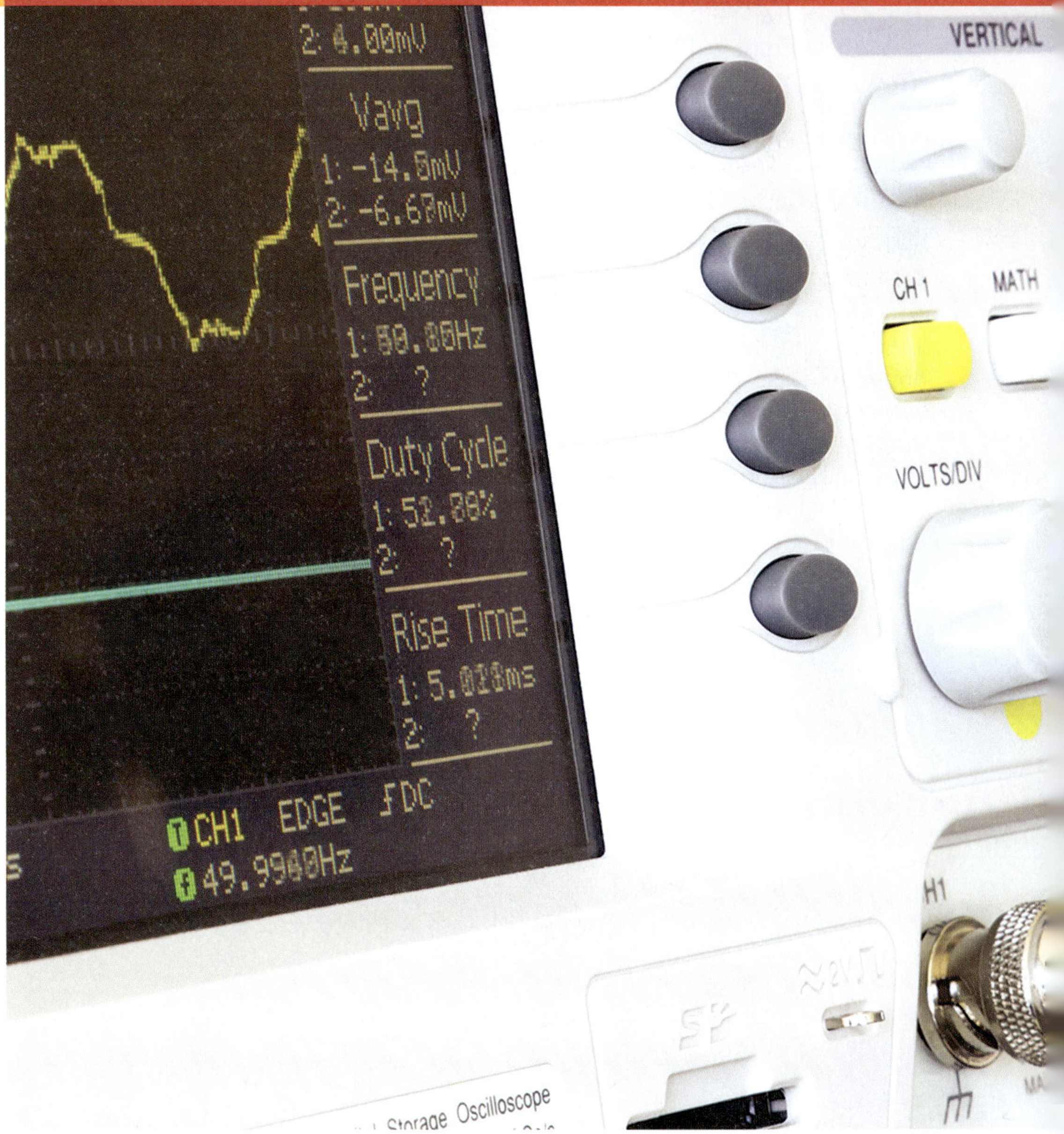

© Iudinko/Shutterstock.com

Unit

14 Basic Trigonometry and Vectors

Key Terms

Adjacent

Cosine

Hypotenuse

Opposite

Parallelogram method

Pythagoras

Pythagorean theorem

Right angle

Right triangle

Sine

Tangent

Vector

Vector addition

Why You Need to Know

In DC circuits, the values within the circuit are easily calculated because the voltage in the circuit stays at a constant value, and the polarity remains unchanged. This always keeps the voltage and current in sync with each other. Alternating-current or AC circuits, however, do not enjoy this same benefit as the polarity of the voltage source is constantly changing and can lead to the current being in either a leading or lagging position at given points in the system. This relationship is determined based on what makes up the circuit and can be predicted using some basic right triangle trigonometry. This unit

- explains how the phase difference between voltage and current is determined by the type of load or loads connected to the circuit.
- explains how voltage and current relationships in AC circuits are based on right triangles.
- describes several different methods, all of them involving vectors, that can be employed to determine these phase relationships.
- illustrates how the Pythagorean theorem provides the formula needed for electricians to understand 90° angles and how to calculate voltages and current in AC circuits.

Outline

14–1 Right Triangles

14–2 The Pythagorean Theorem

14–3 Sines, Cosines, and Tangents

14–4 Formulas

14–5 Practical Application

Objectives

After studying this unit, you should be able to

- describe characteristics that are unique to a right triangle.
- describe the relationship between sides of a right triangle using the Pythagorean theorem.
- calculate unknown values of a right triangle using the Pythagorean theorem.
- explain the relationship between the angles of a right triangle and the length of its sides.
- calculate the unknown values of a right triangle using sines, cosines, and tangents.
- employ vector addition methods as an alternative to right triangle calculations.

Preview

Before beginning the study of AC electricity, a brief discussion of right triangles and the mathematical functions involving them is appropriate. Many AC formulas are based on right triangles because, depending on the type of load, the voltage and current in an AC circuit can be out of phase with each other by approximately 90°. The exact amount of the out-of-phase condition is determined by different factors, which are covered in later units.

14–1 Right Triangles

A **right triangle** is a triangle that contains one angle that is a 90° angle (*Figure 14–1*). The **hypotenuse** is the longest side of a right triangle and is always opposite the **right angle**. The sum of the three angles of any triangle always equals 180°. If one angle is 90°, the sum of the other two angles must equal 90°. Several thousand years ago, a Greek mathematician named **Pythagoras** made some interesting discoveries concerning triangles that contain right angles. One of these discoveries was that the sum of the squares of the two non-hypotenuse sides equals the square of the hypotenuse. Because we know this relationship between the three sides of a right triangle, we can determine the length of unknown sides if we know the other two sides. For example, assume a right triangle has one side 3 feet long and the other side is 4 feet long (*Figure 14–2*). If the side that is 3 feet long is squared, it produces a value of 9 square feet (3 ft × 3 ft = 9 sq. ft). If the side that is 4 feet long is squared, it produces

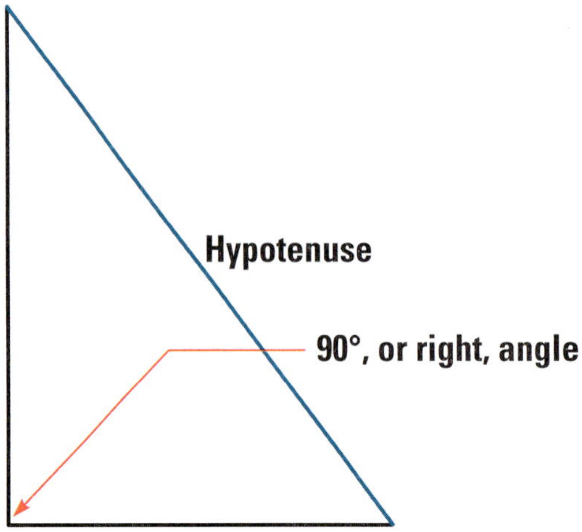

Figure 14–1 A right triangle.

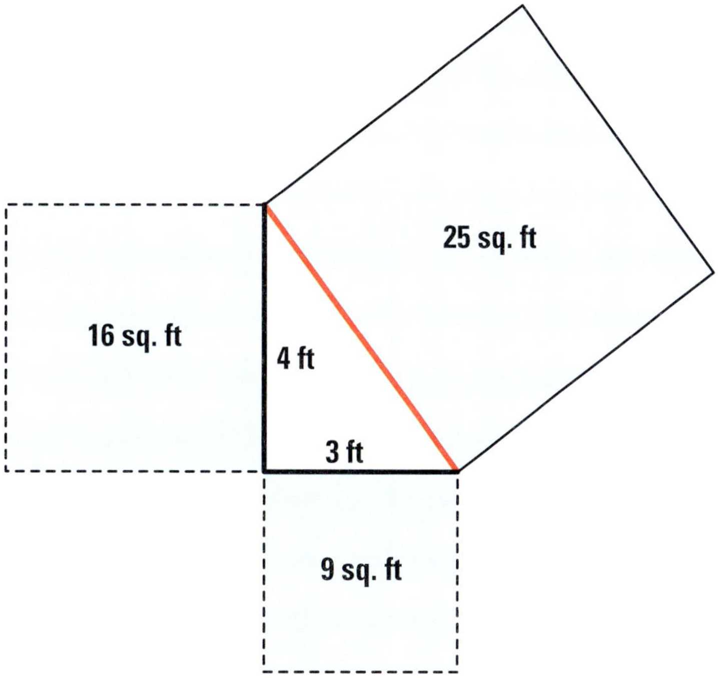

Figure 14–2 The Pythagorean theorem states that the sum of the squares of the sides of a right triangle is equal to the square of the hypotenuse.

a value of 16 square feet (4 ft × 4 ft = 16 sq. ft). The sum of these two values will equal the value of the hypotenuse length squared. In this instance, the square of the hypotenuse has a value of 25 square feet (9 sq. ft + 16 sq. ft = 25 sq. ft). If we find the square root of this value, we can determine that the length of the hypotenuse is 5 feet.

14–2 The Pythagorean Theorem

From this knowledge concerning the relationship of the length of the sides to the length of the hypotenuse, Pythagoras derived a formula known as the **Pythagorean theorem:**

$$c^2 = a^2 + b^2$$

where

c = the length of the hypotenuse

a = the length of one side

b = the length of the other side

If the lengths of the two sides are known, the length of the hypotenuse can be found using the formula:

$$c = \sqrt{a^2 + b^2}$$

It is also possible to determine the length of one of the sides if the length of the other side and the length of the hypotenuse are known. Because the sum of the squares of the two sides equals the square of the hypotenuse, the square of one side equals the difference between the square of the hypotenuse and the square of the other side.

Example 14–1

The triangle shown in *Figure 14–3* has a hypotenuse (C) 18 in. long. One side (B) is 7 in. long. What is the length of the second side (A)?

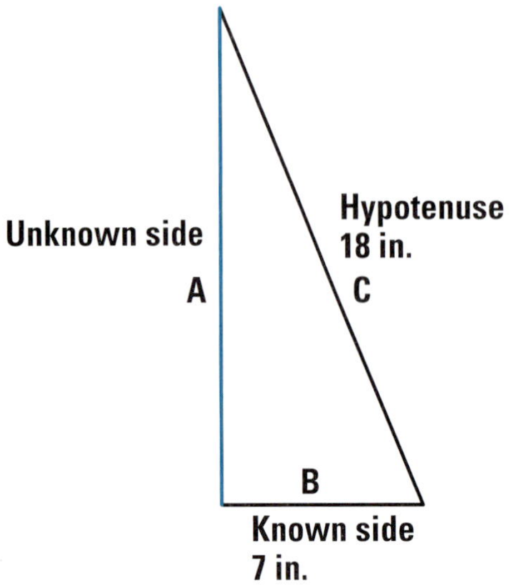

Figure 14–3 Using the Pythagorean theorem to find the length of one side.

Solution

Transpose the formula $c^2 = a^2 + b^2$ to find the value of a. This can be done by subtracting b^2 from both sides of the equation. The result is the formula:

$$a^2 = c^2 - b^2$$

Now take the square root of each side of the equation so the answer is equal to a, not a^2:

$$a = \sqrt{c^2 - b^2}$$

The formula can now be used to find the length of the unknown side:

$$a = \sqrt{(18 \text{ in.})^2 - (7 \text{ in.})^2}$$

$$a = \sqrt{324 \text{ in.}^2 - 49 \text{ in.}^2}$$

$$a = \sqrt{275 \text{ in.}^2}$$

$$a = 16.583 \text{ in.}$$

14–3 Sines, Cosines, and Tangents

A second important concept concerning right triangles is the relationship of the length of the sides to the angles. Because one angle is always 90°, only the two angles that are not right angles are of concern. Another important fact concerning triangles is that the sum of the angles of any triangle equals 180°; because one angle of a right triangle is always 90°, the sum of the two other angles must equal 90°.

It was discovered long ago that the number of degrees in each of these two angles is proportional to the lengths of the three sides. This relationship is expressed as the sine, cosine, or tangent of a particular angle. The function used is determined by which sides are known and which of the two angles is to be found.

When using sines, cosines, or tangents, the sides are designated the *hypotenuse*, the **opposite,** and the **adjacent.** The hypotenuse is always the longest side of a right triangle, but the opposite and adjacent sides are determined by which of the two angles is to be found. In *Figure 14–4*, a right triangle has its sides labeled A, B, and hypotenuse. The two unknown angles are labeled X and Y. To determine which side is opposite an angle, draw a line bisecting the angle. This bisect line would intersect the opposite side. In *Figure 14–5*, side A is opposite angle X, and side B is opposite angle Y. If side A is opposite angle X, then side B is adjacent to angle X, and if side B is opposite angle Y, then side A is adjacent to angle Y (*Figure 14–6*).

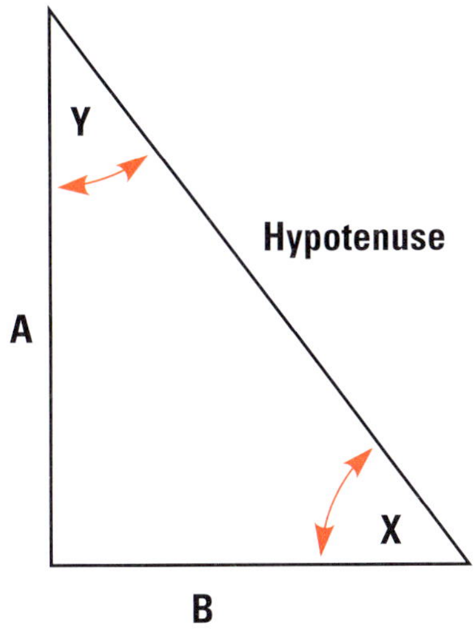

Figure 14–4 Determining which side is opposite and which is adjacent.

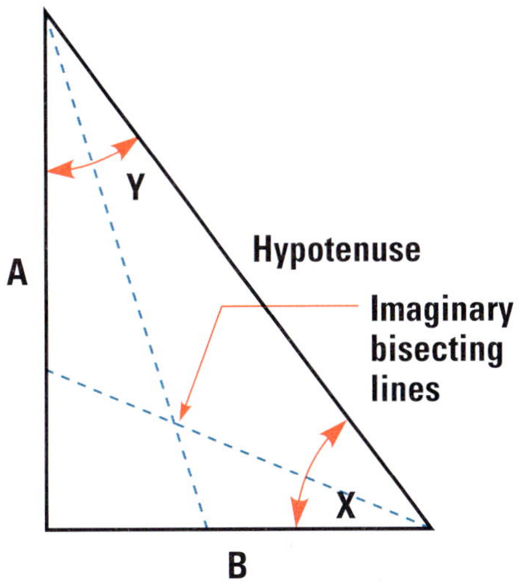

Figure 14–5 Side A is opposite angle X, and side B is opposite angle Y.

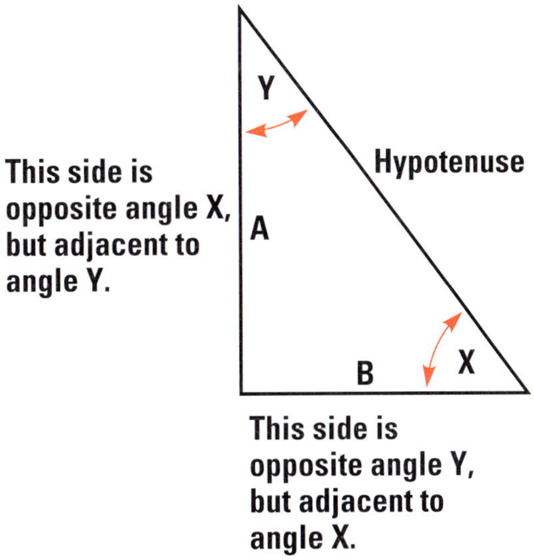

Figure 14–6 Determining opposite and adjacent sides.

The **sine** of an angle is equal to the ratio of the opposite side to the hypotenuse:

$$\text{sine} \angle = \frac{\text{opposite}}{\text{hypotenuse}}$$

The **cosine** of an angle is equal to the ratio of the adjacent side to the hypotenuse:

$$\text{cosine} \angle = \frac{\text{adjacent}}{\text{hypotenuse}}$$

The **tangent** of an angle is equal to the ratio of the opposite side to the adjacent side:

$$\text{tangent} \angle = \frac{\text{opposite}}{\text{adjacent}}$$

A simple saying is often used to help remember the relationship of the trigonometric function and the side of the triangle. This saying is **SOH—CAH—TOA.** In this saying, SOH stands for ***S**ine-**O**pposite-**H**ypotenuse,* CAH stands for ***C**osine-**A**djacent-**H**ypotenuse*, and TOA stands for ***T**angent-**O**pposite-**A**djacent*.

Once the sine, cosine, or tangent of the angle has been determined, the angle can be found by using the trigonometric functions on a scientific calculator.

Example 14–2

The triangle in *Figure 14–7* has a hypotenuse 14 in. long, and side A is 9 in. long. How many degrees are in angle X?

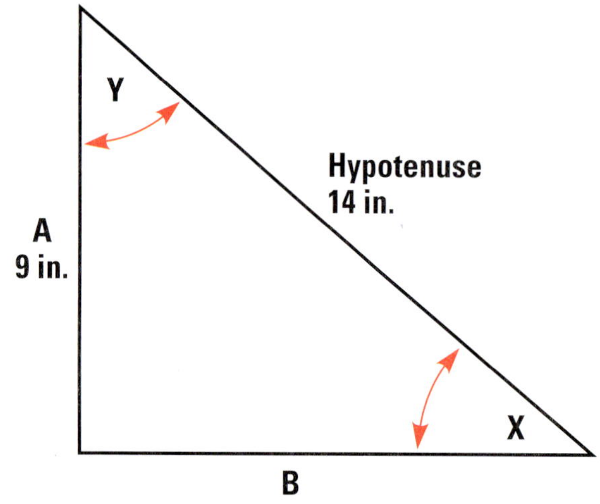

Figure 14–7 Finding the angles using trigonometric functions.

Solution

Because the lengths of the hypotenuse and the opposite side are known, the sine function is used to find the angle:

$$\text{sine} \angle X = \frac{\text{opposite}}{\text{hypotenuse}}$$

$$\text{sine} \angle X = \frac{9}{14}$$

$$\text{sine} \angle X = 0.643$$

Example 14–2 (continued)

Note that 0.643 is the sine of the angle, not the angle. To find the angle, use the SIN function on a scientific calculator. Also note that the calculator must be set to operate in degrees, not radians or grads. If you are using a scientific calculator, it will be necessary to use the ARC SIN function or INV SIN (SIN^{-1}) function to find the answer. If the number 0.643 is entered and the SIN key is pressed, an answer of 0.0112 results. This is the sine of a 0.643° angle. If the sine, cosine, or tangent of an angle is known, the ARC key or INV key (depending on the manufacturer of the calculator) must be pressed before the SIN, COS, or TAN key is pressed:

$$\text{SIN}^{-1} \text{ or ARCSIN } 0.643 = 40°$$

Example 14–3

Using the same triangle (*Figure 14–7),* determine the number of degrees in angle Y.

Solution

In this example, the lengths of the hypotenuse and the adjacent side are known. The cosine function can be used to find the angle:

$$\text{cosine } \angle Y = \frac{\text{adjacent}}{\text{hypotenuse}}$$

$$\text{cosine } \angle Y = \frac{9}{14}$$

$$\text{cosine } \angle Y = 0.643$$

To find what angle corresponds to the cosine of 0.643, use the trigonometric tables in Appendices A and B or the COS function of a scientific calculator:

$$\text{COS}^{-1} \text{ or ARC COS } 0.643 = 50°$$

14–4 Formulas

Some formulas that can be used to find the angles and lengths of different sides follow:

$$\sin \angle\theta = \frac{O}{H} \qquad \cos \angle\theta = \frac{A}{H} \qquad \tan \angle\theta = \frac{O}{A}$$

$$\text{Adj.} = \cos \angle\theta \times \text{Hyp.}$$

$$\text{Opp} = \sin \angle\theta \times \text{Hyp.}$$

$$\text{Hyp.} = \frac{O}{\sin \angle\theta}$$

$$\text{Adj.} = \frac{O}{\tan \angle\theta}$$

$$\text{Opp.} = \text{Adj.} \times \tan \angle\theta$$

$$\text{Hyp.} = \frac{A}{\cos \angle\theta}$$

14–5 Practical Application

Although the purpose of this unit is to provide preparation for the study of AC circuits, basic trigonometry can provide answers to other problems that may be encountered on the job. Assume that it is necessary to know the height of a tall building (*Figure 14–8*). Now assume that the only tools available to make this measurement are a 1-foot ruler, a tape measure, and a scientific calculator. To make the measurement, find a relatively flat area in the open sunlight. Hold the ruler upright and measure the shadow cast by the sun (*Figure 14–9*). Assume the length of the shadow to be 7.5 inches. Using the length of the shadow as one side of a right triangle and the ruler as the other side, the angle of the sun can now be determined.

Figure 14–8 Trigonometry can be used to estimate the height of a building.

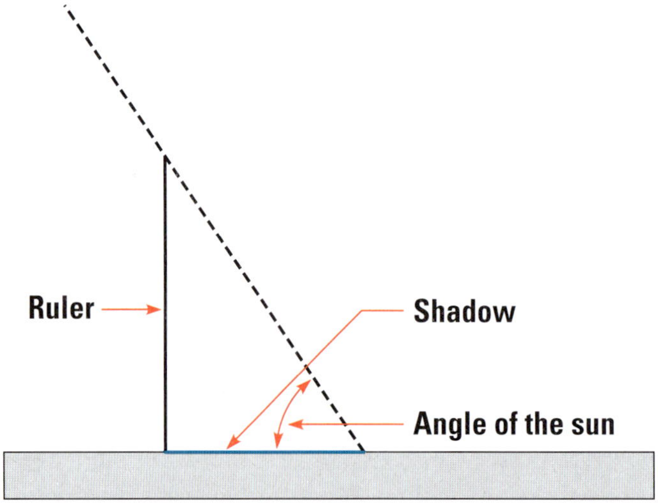

Figure 14–9 Determining the angle of the sun.

The two known sides are the opposite and the adjacent. The tangent function corresponds to these two sides. The angle of the sun is

$$\tan \angle\theta = \frac{O}{A}$$

$$\tan \angle\theta = \frac{12}{7.5}$$

$$\tan \angle\theta = 1.6$$

$$= 57.99^\circ$$

The tape measure can now be used to measure the shadow cast by the building on the ground. Assume the length of the shadow to be 35 feet. If the height of the building is used as one side of a right triangle and the shadow is used as the other side, the height can be found because the angle of the sun is known (*Figure 14–10*).

$$\text{Opp.} = \text{Adj.} \times \tan \angle\theta$$

$$\text{Opp.} = 35 \text{ ft} \times \tan 57.99^\circ$$

$$\text{Opp.} = 35 \text{ ft} \times 1.6$$

$$\text{Opp.} = 56 \text{ ft}$$

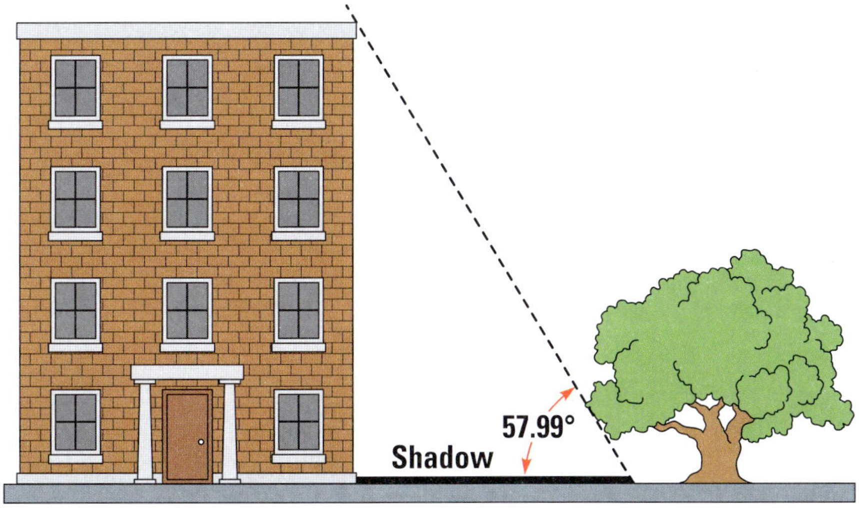

Figure 14–10 Measuring the building's shadow helps determine the adjacent side to the sun's angle.

Vectors

Using the right triangle is just one method of graphically showing how angular quantities can be added. Another method of illustrating this concept is with the use of vectors. A **vector** is a line that indicates both magnitude and direction. The magnitude is indicated by its length, and the direction is indicated by its angle of rotation from 0°. Vectors should not be confused with *scalars*, which are used to represent magnitude only and do not take direction into consideration. Imagine, for example, that you are in a strange city, and you ask someone for directions to a certain building. If the person said, "Walk three blocks," that would be a scalar because it contains only the magnitude, three blocks. If the person said, "Walk three blocks south," it would be a vector because it contains both the magnitude, three blocks, and the direction, south.

Zero degrees is indicated by a horizontal line. An arrow is placed at one end of the line to indicate direction. The magnitude can represent any quantity such as inches, meters, miles, volts, amperes, ohms, and power. A vector with a magnitude of 5 at an angle of 0° is shown in *Figure 14–11.* Vectors rotate in a counterclockwise direction. Assume that a vector with a magnitude of 3 is to be drawn at a 45° angle from the first vector (*Figure 14–12*). Now assume that a third vector with a magnitude of 4 is to be drawn in the direction of 120° (*Figure 14–13*). Notice that the direction of the third vector is referenced from the horizontal 0° line and not from the second vector line, which was drawn at an angle of 45°.

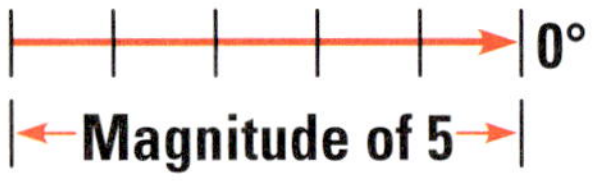

Figure 14–11 A vector with a magnitude of 5 and an angle of 0°.

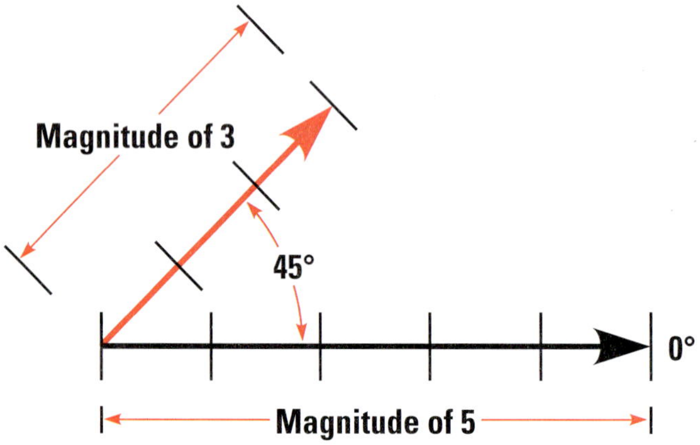

Figure 14–12 A second vector with a magnitude of 3 and a direction of 45°.

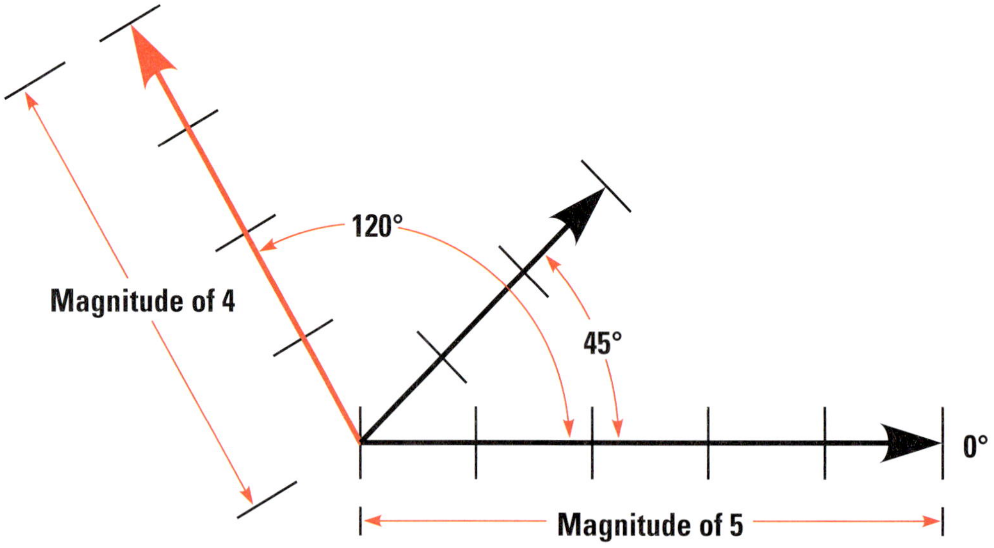

Figure 14–13 A third vector with a magnitude of 4 and a direction of 120° is added.

Adding Vectors Because vectors are used to represent quantities such as volts, amperes, ohms, power, and so on, they can be added, subtracted, multiplied, and divided. In electrical work, however, addition is the only function needed, so it is the only one discussed. Several methods can be used to add vectors. Regardless of the method used, because vectors contain both magnitude and direction, they must be added with a combination of geometric and algebraic addition. This method is often referred to as **vector addition.**

One method is to connect the starting point of one vector to the end point of another. This method works especially well when all vectors are in the same direction. Consider the circuit shown in *Figure 14–14*. In this circuit, two batteries, one rated at 6 volts and the other rated at 4 volts, are connected in such a manner that their voltages add. It should be noted that if all the quantities are in the same direction, they become scalars instead of vectors because only magnitude is considered. If vector addition is used, the starting point of one scalar is placed at the end point of the other. Notice that the sum of the two scalar quantities is equal to the sum of the two voltages, 10 volts. Another example of this type of vector addition is shown in *Figure 14–15*. In this example, three resistors are connected in series. The first resistor has a resistance of 80 ohms; the second, a resistance of 50 ohms; and the third, a resistance of 30 ohms. Because there is no phase angle shift of voltage or current, the impedance is 160 ohms, which is the sum of the resistances of the three resistors.

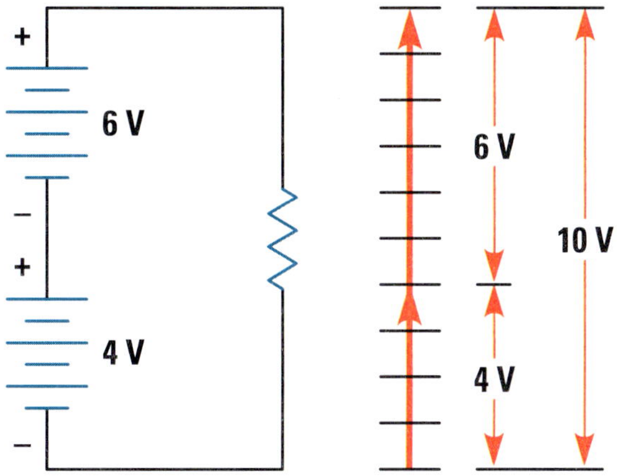

Figure 14–14 Adding vectors in the same direction considers only their magnitude.

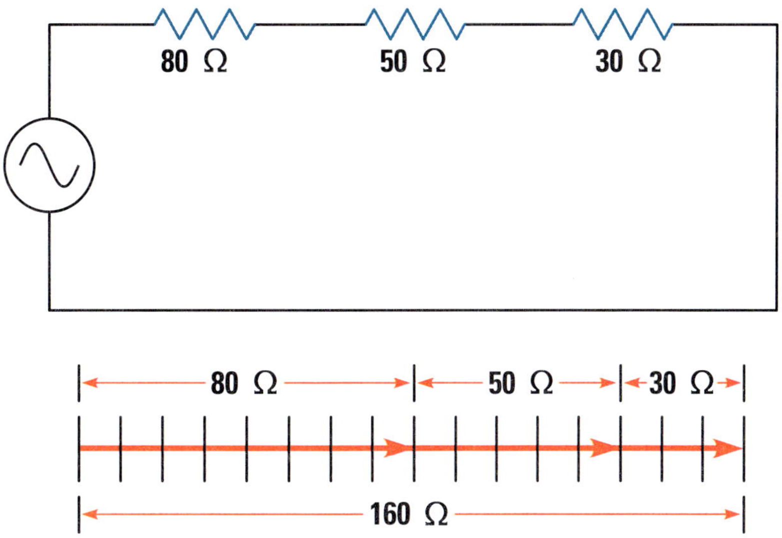

Figure 14–15 Addition of series resistors is an example of adding vectors with the same directional value.

Adding Vectors with Opposite Directions To add vectors that are exactly opposite in direction (180° apart), subtract the magnitude of the smaller vector from the magnitude of the larger vector. The remainder is the magnitude of a vector with the same direction as the larger of the original vectors. If one of the batteries in *Figure 14–14* were reversed, the two voltages would oppose each other (*Figure 14–16*). This means that 4 volts of the 6-volt battery A would have to be used to overcome the voltage of battery B. The resultant would be a vector with a magnitude of 2 volts in the same direction as the 6-volt battery. In algebra, this is the same operation as adding a positive number and a negative number ($+6 + [-4] = +2$). When the -4 is brought out of brackets, the equation becomes $6 - 4 = 2$.

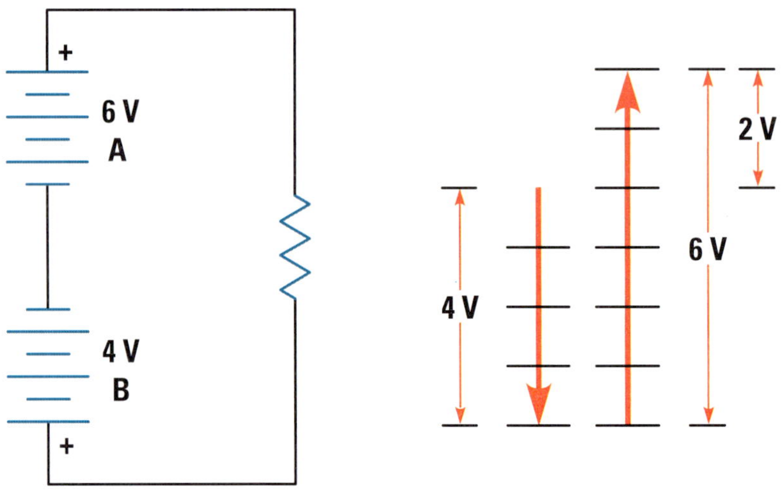

Figure 14–16 Adding vectors with opposite directions.

Adding Vectors of Different Directions Vectors that have directions other than 180° from each other can also be added. *Figure 14–17* illustrates the addition of a vector with a magnitude of 4 and a direction of 15° to a vector with a magnitude of 3 and a direction of 60°. The addition is made by connecting the starting point (tail) of the second vector to the ending point (tip) of the first vector. This is often referred to as the "tip-to-tail" method of vector addition. The resultant is drawn from

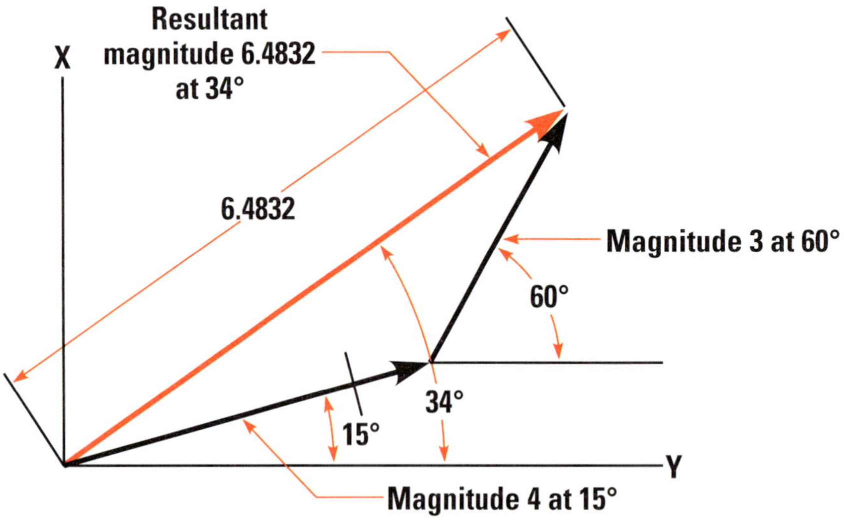

Figure 14–17 Adding two vectors with different directions using the tip-to-tail method.

the starting point of the first vector to the ending point of the second. It is possible to add several different vectors using this method. *Figure 14–18* illustrates the addition of several different vectors and the resultant.

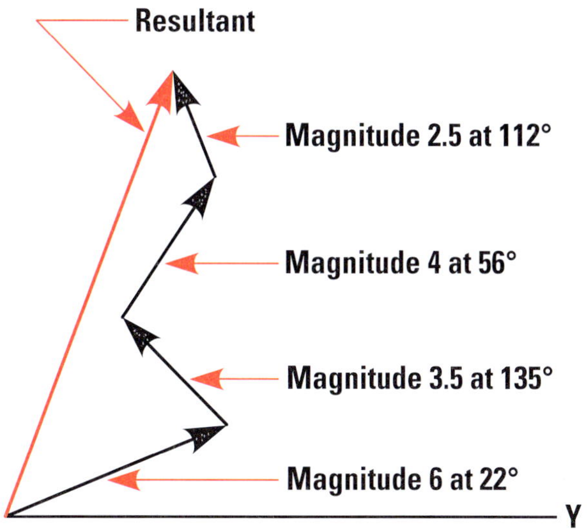

Figure 14–18 Adding several vectors with different magnitudes and directions.

The Parallelogram Method of Vector Addition The **parallelogram method** can also be used to find the resultant of two vectors that originate at the same point. A parallelogram is a four-sided figure whose opposite sides form parallel lines. A rectangle, for example, is a parallelogram with 90° angles. Assume that a vector with a magnitude of 24 and a direction of 26° is to be added to a vector with a magnitude of 18 and a direction of 58°. Also assume that the two vectors originate from the same point (*Figure 14–19*). To find the resultant of these two vectors, form a parallelogram using the vectors as two of the sides. The parallelogram is formed by connecting the tale of one vector to the head of the other. The resultant is drawn from the corner of the parallelogram where the two vectors intersect to the opposite corner (*Figure 14–20*).

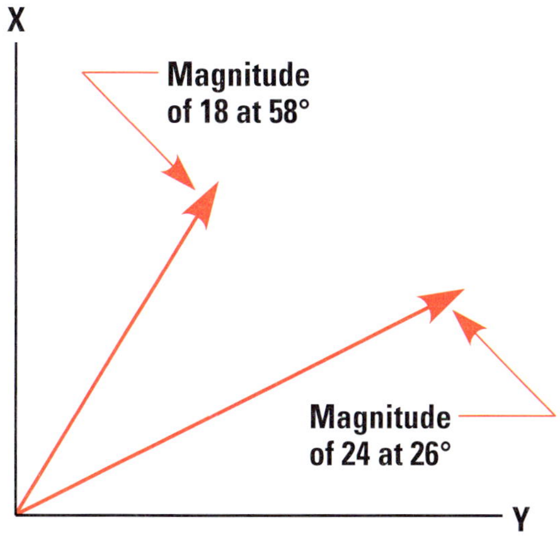

Figure 14–19 Vectors that originate from the same point.

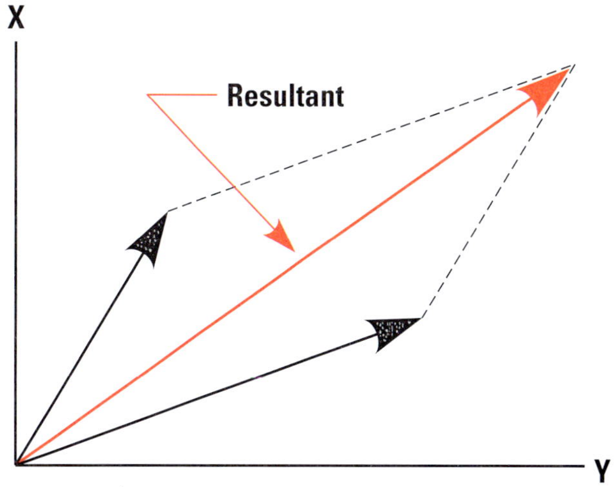

Figure 14–20 The parallelogram method of vector addition.

Another example of the parallelogram method of vector addition is shown in *Figure 14–21*. In this example, one vector has a magnitude of 50 and a direction of 32°. The second vector has a magnitude of 60 and a direction of 120°. The resultant is found by using the two vectors as two of the sides of a parallelogram.

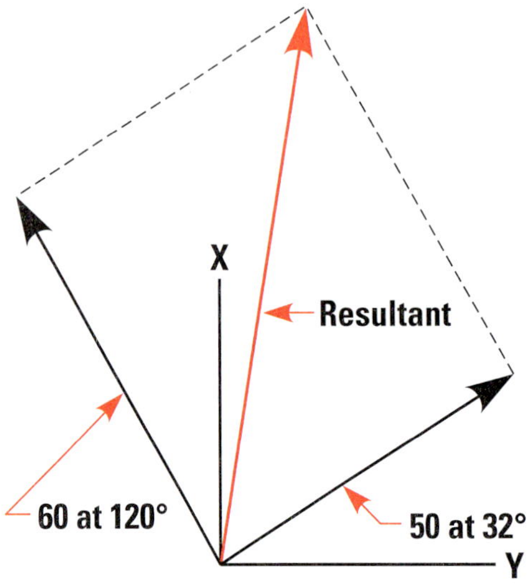

Figure 14–21 The parallelogram method of vector addition, second example.

Summary

- Several AC formulas are based on right triangle trigonometry.
- A right triangle is a triangle that contains a 90°, or right, angle.
- The Pythagorean theorem states that the sum of the squares of the sides of a right triangle equals the square of the hypotenuse.
- The sum of the three angles of any triangle is always 180°.
- The relationship of the length of the sides of a right triangle to the number of degrees in its angles can be expressed as the sine, cosine, or tangent of a particular angle.
- The sine function of an angle is the ratio of the opposite side to the hypotenuse.
- The cosine function of an angle is the ratio of the adjacent side to the hypotenuse.
- The tangent function of an angle is the ratio of the opposite side to the adjacent side.
- The hypotenuse is always the longest side of a right triangle.
- A simple saying that can be used to help remember the relationship of the trigonometric functions to the sides of a right triangle is *SOH–CAH–TOA.*
- Vectors are lines that indicate both magnitude and direction.
- Scalars indicate magnitude only.

Review Questions

Refer to *Figure 14–22* to answer the following questions.

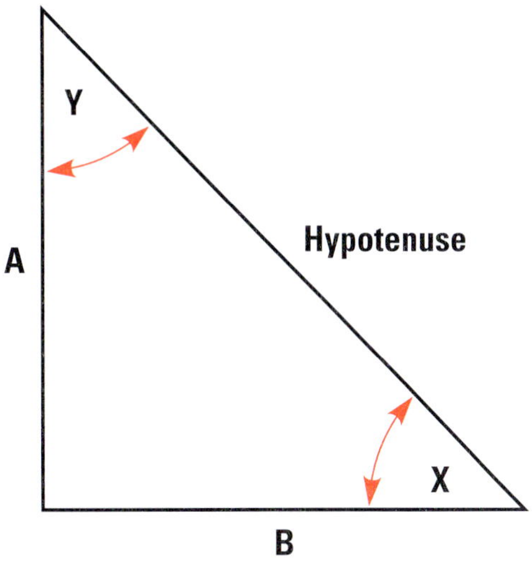

Figure 14–22 Finding the values of a right triangle.

1. A right triangle will always have which of the following characteristics?
   A) An obtuse angle
   B) A 90° angle
   C) Three equal sides
   D) A total amount of degrees that equals 360°
2. True or False: The total amount of degrees in a right triangle is 180°, including a right angle of 90° and two acute angles that add together to equal the remaining 90°.
3. In a right triangle, the sum of the squares of the two sides that form the 90° angle will be equal to which of the following values?
   A) The square of the hypotenuse
   B) 90°
   C) The hypotenuse
   D) The sine of angle Φ
4. True or False: The hypotenuse is always the shortest side of a right triangle.
5. If side A of the triangle in Figure 14–22 has a length of 18.5 ft and side B has a length of 28 ft, what is the length of the hypotenuse?
   A) 38 feet
   B) 28 feet
   C) 33.6 feet
   D) 21 feet

6. The hypotenuse of a right triangle is 25 inches and one of the sides is 15 inches, what is the length of the unknown side?
   A) 29.2 inches
   B) 10 inches
   C) 400 inches
   D) 20 inches
7. Which of the following trigonometric functions is used to find the angle if the length of the hypotenuse and of the adjacent side are known?
   A) Sine
   B) Cosine
   C) Tangent
   D) Pythagorean theorem
8. The sine of an angle is equal to what relationship between the sides of a right triangle?
   A) The ratio of the adjacent side to the hypotenuse
   B) The ratio of the opposite to the adjacent sides
   C) The ratio of the hypotenuse to the angle
   D) The ratio of the opposite side to the hypotenuse
9. Side A in Figure 14–22 has a length of 12 m and is adjacent to angle Y, which is 12°. What is the length of side B?
   A) 2.6 m
   B) 2.4 m
   C) 11.7 m
   D) 56.4 m
10. The hypotenuse in Figure 14–22 has a length of 65 in., and side A has a length of 31 in. What is angle X?
    A) 61.5°
    B) 24.5°
    C) 41.3°
    D) 28.5°
11. Which of the following describes a method to add vectors that involves drawing the vectors in a manner that starts one vector at the end of the previous vector and draws the resulting vector by connecting the start of the first vector to the end of the last vector?
    A) The parallelogram method
    B) The connect the dots method

C) The tip-to-tail method

D) The vector drawing method

12. True or False: The parallelogram method can be used to find the resulting vector when adding two vectors by creating a four-sided figure where the two vectors being added form two connecting sides and equal imaginary sides are drawn opposite. The resulting vector is drawn diagonally across the four-sided figure to reach the opposite corner from the starting point.

Practical Applications

1. A parking lot is 275 ft by 200 ft (*Figure 14–23*). A trade size 4 conduit has been buried 2 ft beneath the concrete. The conduit runs at an angle from one corner of the parking lot to the other. Two pull boxes have been placed in the conduit run. The pull boxes are equally spaced along the length of conduit. Your job is to pull conductors through the conduit from one end of the run to the other.

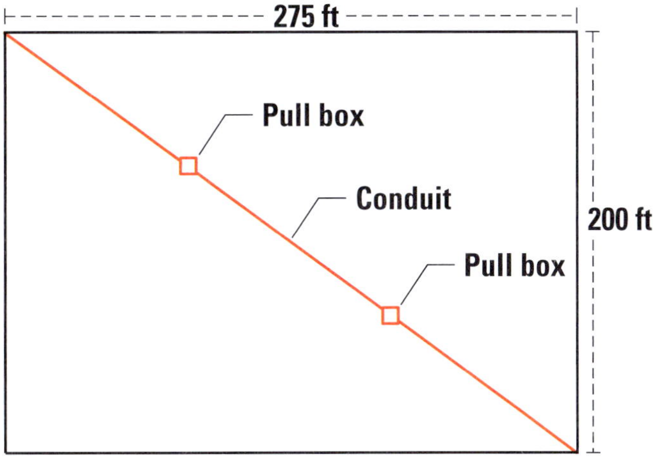

Figure 14–23 Determining length.

A) Allowing 12 ft of extra conductor, what is the length of each conductor? The conductors are not to be spliced.

B) How much rope will be required to pull these conductors through the conduit, allowing an extra 5 ft at one end that is to be connected to the pulling device and 3 ft extra that is to be used to connect the conductors to the rope?

2. You desire to add a solar heating system to your swimming pool to extend the length of time you can use the pool. For maximum efficiency, the collector should be placed on an angle that will permit the sun's rays to strike the collector at a 90° angle (*Figure 14–24*). When the sun is at its highest point of the day, you place a yardstick perpendicular to a flat concrete surface. The stick casts a shadow 3.5 ft long. At what angle with respect to the ground should the solar collector be placed?

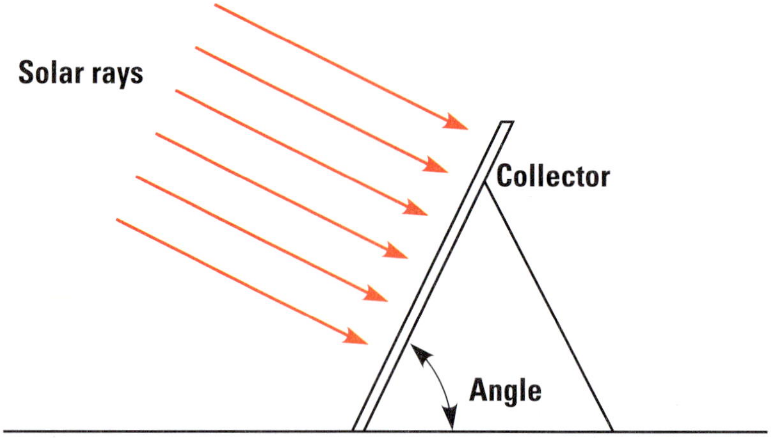

Figure 14–24 The sun's rays should strike the face of the solar collector at a 90° angle.

Practice Problems

Basic Trigonometry

Refer to *Figure 14–22* to find the missing values in the following chart.

∠X	∠Y	Side A	Side B	Hyp.
40°				8
	33°	72		
		38		63
52°			14	
	42°		156	

Unit

15 Alternating Current

Key Terms

Amplitude

Average value

Cycle

Effective value

Frequency

Hertz (Hz)

In phase

Linear wave

Oscillators

Peak value

Resistive loads

Ripple

RMS (root-mean-square) value

Sine wave

Skin effect

Triangle wave

True power

Why You Need to Know

Most of the electric power used in daily life around the world is alternating-current or AC power. In this unit, you will learn about different types of waveforms, different methods employed to measure and calculate AC values, and the relationship of voltage and current in a circuit that contains pure resistance. This unit will also

- introduce the sine wave. The most common of all AC waveforms is the sine wave because it is produced by a rotating machine. All rotating machines operate on the principle of the sine wave. The pistons in an internal combustion engine, for example, travel up and down inside the cylinder in a sine wave pattern because they are connected to a rotating crankshaft.
- explain that the main advantage for the use of AC is that AC voltage can be transformed easily, and DC voltage cannot.
- present the concepts of eddy currents and skin effect that act as resistance to the flow of current in an AC circuit.
- explain the differences between peak, RMS, and average voltage values. These are very important concepts and should be understood by anyone working in the electrical field. Some electric components such as capacitors and solid-state devices are very sensitive to voltage. Many of these devices list a peak rating of voltage instead of the RMS value. It then becomes imperative that you know how to determine the peak value from the RMS value.

Outline

Objectives

After studying this unit, you should be able to

- explain the differences between DC and AC.
- determine instantaneous values of voltage and current for a sine wave using trigonometric functions.
- explain the relationship between peak, RMS, and average values of voltage and current.
- describe the phase relationship of voltage and current in a pure resistive circuit.
- explain how true power is produced in an AC circuit.
- describe how the skin effect influences current flow in an AC circuit.

Preview

Most of the electric power utilized around the world is AC. It is used to operate everything from home appliances, such as television sets, computers, microwave ovens, and electric toasters, to the largest motors found in industry. AC has several advantages over DC that make it the more popular choice for the large-scale production of electric power. However, there are several advancements in technology that are quickly making DC systems more viable.

15–1 Advantages of AC

Probably the single greatest advantage of AC is the fact that AC can be transmitted over great distances without significant voltage drop, and DC cannot. A transformer permits AC voltage to be stepped up or down. Voltage can be stepped up for the purpose of transmission and then stepped back down when it is to be used by some device. Transmission voltages of 69, 138, and 345 kilovolts are common. The advantage of high-voltage transmission is that less current is required to produce the same amount of power. The reduction of current permits smaller wires to be used, which results in a savings of material.

In the very early days of electric power generation, Thomas Edison, an American inventor, proposed powering the country with low-voltage DC. He reasoned that low-voltage DC was safer for people to use than higher-voltage AC. A Serbian immigrant named Nikola Tesla, however, argued that DC was impractical for large-scale applications. The disagreement was finally settled at the 1904 World's Fair held in St. Louis, Missouri. The 1904 World's Fair not only introduced the first ice cream cone and the first iced tea but was also the first World's Fair to be lighted with "electric candles." At that time, the only two companies capable of providing electric lighting for the World's Fair were the Edison Company, headed by Thomas Edison, and the Westinghouse Company, headed by George Westinghouse, a close friend of Nikola Tesla. The Edison Company submitted a bid of over $1 per lamp to light the fair with low-voltage DC. The Westinghouse Company submitted a bid of less than 25 cents per lamp to light the fair using higher-voltage AC. This set the precedent for how electric power would be supplied throughout the world.

15–2 AC Waveforms

Square Waves

AC differs from DC in that AC reverses its direction of flow at periodic intervals (*Figure 15–1*). AC waveforms can vary depending on how the current is produced. One waveform frequently encountered is the square wave (*Figure 15–2*). It is assumed that the oscilloscope in *Figure 15–2* has been adjusted

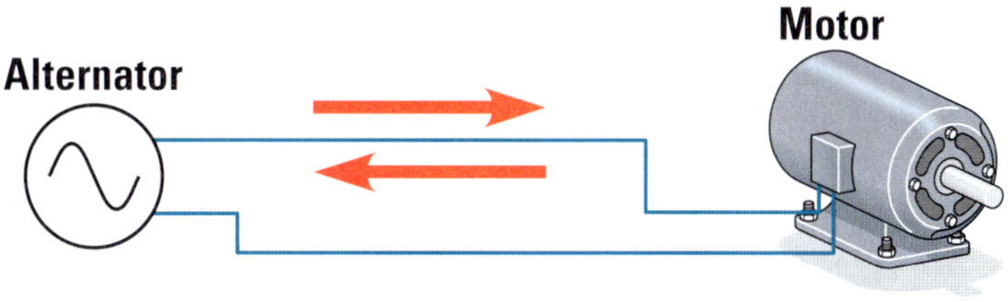

Figure 15–1 AC current alternates the direction of flow in the circuit.

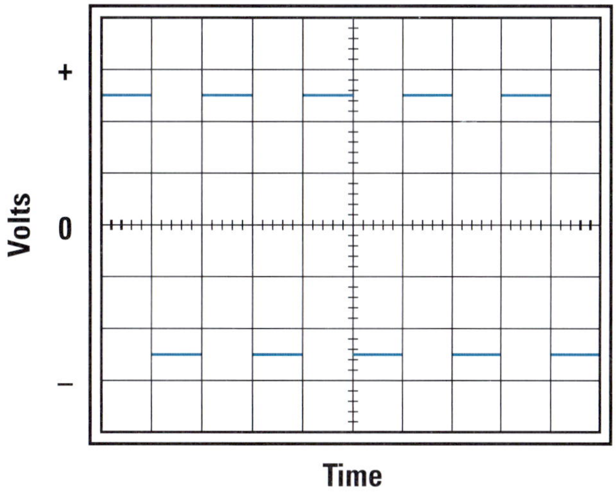

Figure 15–2 Example of a square wave AC.

so that 0 volts is represented by the center horizontal line. The waveform shows that the voltage is in the positive direction for some length of time and then changes polarity. The voltage remains negative for some length of time and then changes back to positive again. Each time the voltage reverses polarity, the current flow through the circuit changes direction. A square wave could be produced by a simple single-pole double-throw switch connected to two batteries as shown in *Figure 15–3*. Each time the switch position is changed, current flows through the resistor in a different direction. Although this circuit will produce a square wave AC, it is not practical. Square waves are generally produced

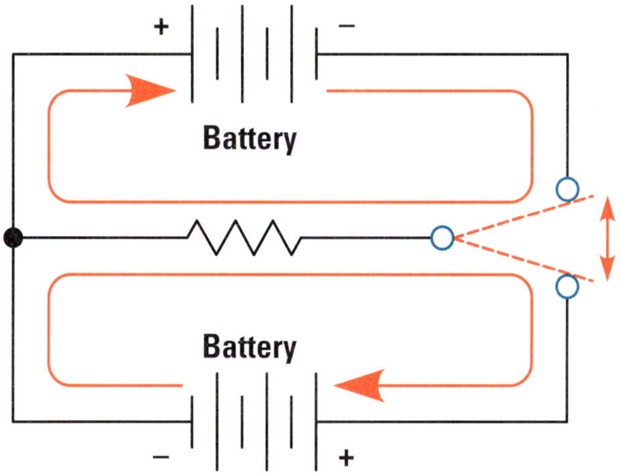

Figure 15–3 Square wave AC produced with a switch and two batteries.

by electronic devices called **oscillators.** The schematic diagram of a simple square wave oscillator is shown in *Figure 15–4*. In this circuit, two bipolar transistors are used as switches to reverse the direction of current flow through the windings of the transformer.

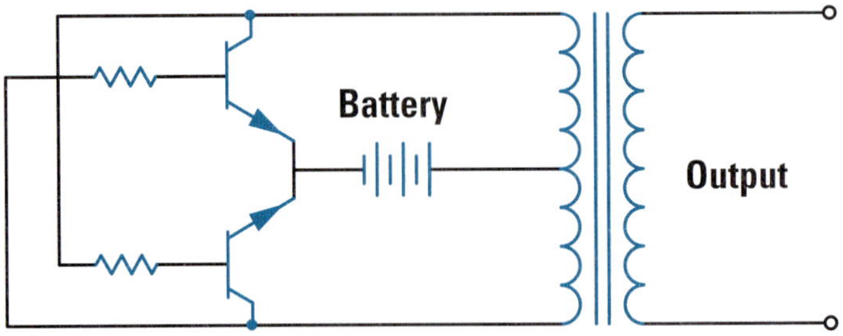

Figure 15–4 Square wave oscillator.

Triangle Waves

Another common AC waveform is the **triangle wave** shown in *Figure 15–5*. The triangle wave is a **linear wave,** one in which the voltage rises at a constant rate with respect to time. Linear waves form straight lines when plotted on a graph. For example, assume that the waveform shown in *Figure 15–5* reaches a maximum positive value of 100 volts after 2 milliseconds. The voltage will be 25 volts after 0.5 milliseconds, 50 volts after 1 milliseconds, and 75 volts after 1.5 milliseconds.

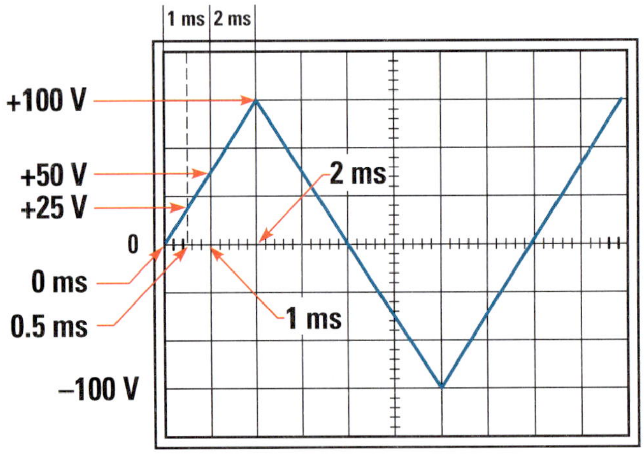

Figure 15–5 Example of a triangular AC waveform.

Sine Waves

The most common of all AC waveforms is the **sine wave** (*Figure 15–6*). Sine waves are produced by using rotating machines to generate electricity. The sine wave contains a total of 360 electric degrees. It reaches its peak positive voltage at 90°, returns to a value of 0 volts at 180°, increases to its maximum negative voltage at 270°, and returns to 0 volts at 360°. Each complete waveform of 360° is called a **cycle.** The number of complete cycles that occur in one second is called the **frequency.** Frequency is measured in **hertz (Hz).** The most common frequency in the United States and Canada is 60 hertz. This means that the voltage increases from zero to its maximum value in the positive direction, returns to zero, increases to its maximum value in the negative direction, and returns to zero 60 times each second.

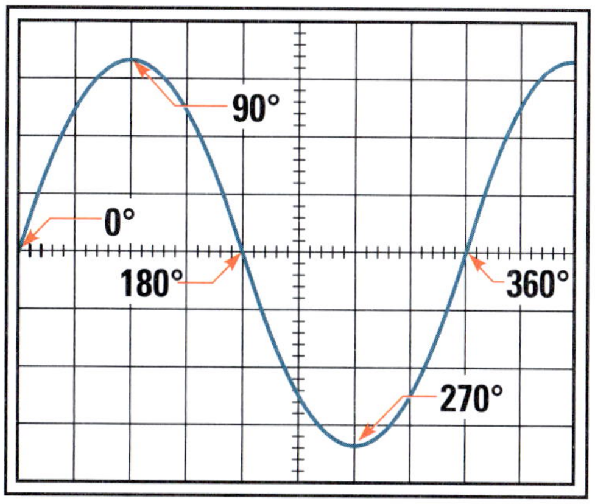

Figure 15–6 Example of a sine wave.

The amount of time necessary for a waveform to complete one cycle can be determined by using the formula:

$$T = \frac{1}{f}$$

where

T = time in seconds

f = frequency in Hz

Example 15–1

An AC waveform has a frequency of 1200 Hz. What is the amount of time necessary to complete one cycle?

Solution

$$T = \frac{1}{1200}$$

$$T = 0.000833 \text{ S or } 0.833 \text{ mS}$$

If the time of one cycle is known, the frequency can be determined using the formula:

$$f = \frac{1}{T}$$

Example 15–2

An AC waveform completes a cycle every 0.04 millisecond. What is the frequency of the waveform?

Solution

$$f = \frac{1}{0.00004}$$

$$f = 25{,}000 \text{ Hz or } 25 \text{ kHz}$$

Sine waves are so named because the voltage at any point along the waveform is equal to the maximum, or peak, value times the sine of the angle of rotation. *Figure 15–7* illustrates one-half of a loop of wire cutting through lines of magnetic flux. The flux lines are shown with equal spacing between each line, and the arrow denotes the arc of the loop as it cuts through the lines of flux. Notice the number of flux lines that are cut by the loop during the first 30° of rotation. Now notice the number of flux lines that are cut during the second and third 30° of rotation. Because the loop is cutting the flux lines at an angle, it must travel a greater distance between flux lines during the first degrees of rotation. Consequently, fewer flux lines are cut per second, which results in a lower induced voltage. Recall that 1 volt is induced in a conductor when it cuts lines of magnetic flux at a rate of 1 weber per second (Wb/s). One weber is equal to 100,000,000 lines of flux.

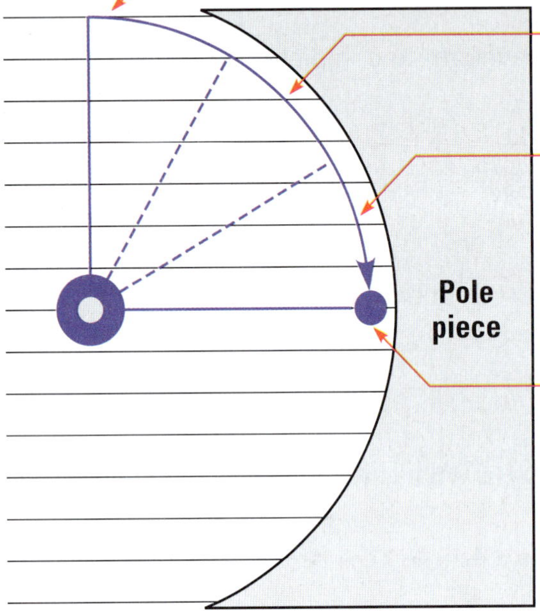

Figure 15–7 As the loop approaches 90° of rotation, the flux lines are cut at a faster rate.

When the loop has rotated 90°, its direction of motion is perpendicular to the flux lines and is cutting them at the maximum rate, which results in the highest, or peak, voltage being induced in the loop. The voltage at any point during the rotation is equal to the maximum induced voltage times the sine of the angle of rotation. For example, if the induced voltage after 90° of rotation is 100 volts, the voltage after 30° of rotation will be 50 volts because the sine of a 30° angle is 0.5 ($100 \times 0.5 = 50$ V). The induced voltage after 45° of rotation is 70.7 volts because the sine of a 45° angle is 0.707 ($100 \times 0.707 = 70.7$ V). A sine wave showing the instantaneous voltage values after different degrees of rotation is shown in *Figure 15–8*. The instantaneous voltage value is the value of voltage at any instant on the waveform.

The following formula can be used to determine the instantaneous value at any point along the sine wave:

$$E_{(INST)} = E_{(MAX)} \times \sin \angle\theta$$

where

$E_{(INST)}$ = the voltage at any point on the waveform

$E_{(MAX)}$ = the maximum, or peak, voltage

$\sin \angle\theta$ = the sine of the angle of rotation

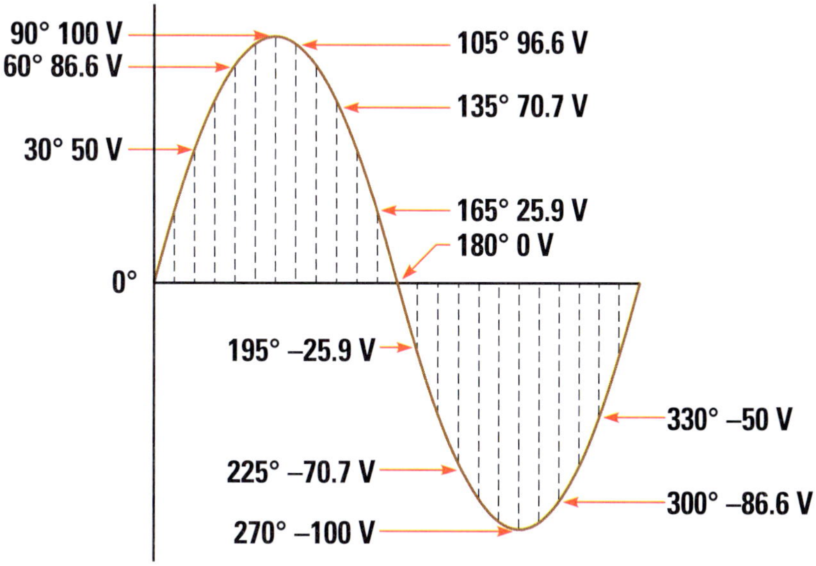

Figure 15–8 Instantaneous values of voltage plotted at various points along a sine wave.

Example 15–3

A sine wave has a maximum voltage of 138 V. What is the voltage after 78° of rotation?

Solution

$$E_{(INST)} = E_{(MAX)} \sin \angle\theta$$

$$E_{(INST)} = 138 \times 0.978 \text{ (sin of } 78^\circ\text{)}$$

$$E_{(INST)} = 134.964 \text{ V}$$

The formula can be changed to find the maximum value if the instantaneous value and the angle of rotation are known or to find the angle if the maximum and instantaneous values are known:

$$E_{(MAX)} = \frac{E_{(INST)}}{\sin \angle\theta}$$

$$\sin \angle\theta = \frac{E_{(INST)}}{E_{(MAX)}}$$

Example 15–4

A sine wave has an instantaneous voltage of 246 V after 53° of rotation. What is the maximum value the waveform will reach?

Solution

$$E_{(MAX)} = \frac{E_{(INST)}}{\sin \angle\theta}$$

$$E_{(MAX)} = \frac{246 \text{ V}}{0.799}$$

$$E_{(MAX)} = 307.885 \text{ V}$$

Example 15–5

A sine wave has a maximum voltage of 350 V. At what angle of rotation will the voltage reach 53 V?

Solution

$$\sin \angle\theta = \frac{E_{(INST)}}{E_{(MAX)}}$$

$$\sin \angle\theta = \frac{53\ V}{350\ V}$$

$$\sin \angle\theta = 0.151$$

Note: 0.151 is the *sine* of the angle, not the angle.

$$\angle\theta = 8.685$$

15–3 Sine Wave Values

Several measurements of voltage and current are associated with sine waves. These measurements are peak to peak, peak, RMS, and average. A sine wave showing peak-to-peak, peak, and RMS measurements is shown in *Figure 15–9*.

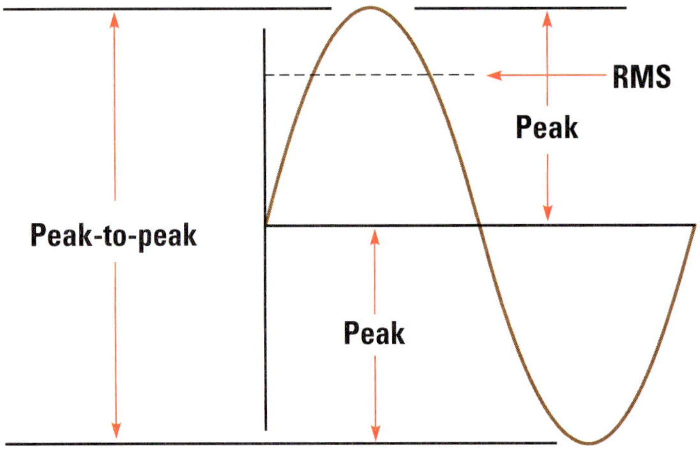

Figure 15–9 Peak-to-peak, RMS, and peak values of a sine wave.

Peak-to-Peak and Peak Values

The peak-to-peak value is measured from the maximum value in the positive direction to the maximum value in the negative direction. The peak-to-peak value is often the simplest measurement to make when using an oscilloscope.

The **peak value,** or **amplitude,** is measured from zero to the highest value obtained in either the positive or negative direction. The peak value is one-half of the peak-to-peak value.

RMS Values

In *Figure 15–10,* a 100-volt battery is connected to a 100-ohm resistor. This connection will produce 1 ampere of current flow, and the resistor will dissipate 100 watts of power in the form of heat. An AC

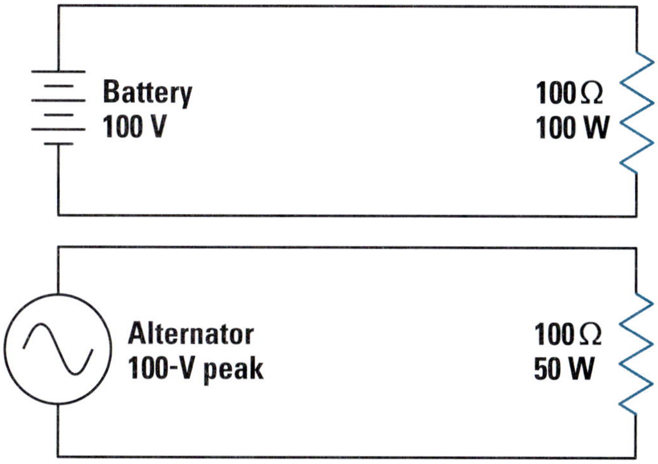

Figure 15–10 A DC circuit produces higher power than a sine wave AC circuit.

alternator that produces a peak voltage of 100 volts is also shown connected to a 100-ohm resistor. A peak current of 1 ampere will flow in the circuit, but the resistor will dissipate only 50 watts in the form of heat. The reason is that the voltage produced by a pure source of DC, such as a battery, is one continuous value, and the AC source varies over the same time frame (*Figure 15–11*). The AC sine wave begins at zero, increases to the maximum value, and decreases back to zero during an equal period of time. Because the sine wave has a value of 100 volts for only a short period of time and is less than 100 volts during the rest of the half-cycle, it cannot produce as much power as 100 volts of DC.

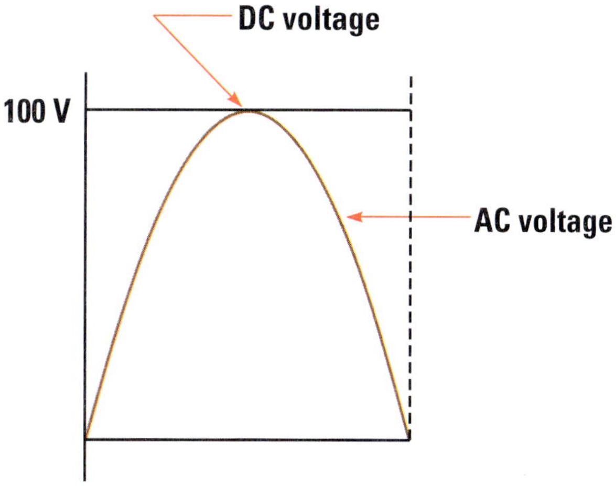

Figure 15–11 The DC voltage remains at a constant value during a half cycle of AC voltage.

The solution to this problem is to use a value of AC voltage that will produce the same amount of power as a like value of DC voltage. This AC value is called the **RMS,** or **effective, value.** It is the value indicated by almost all AC voltmeters and ammeters. ***RMS stands for root-mean-square, which is an abbreviation for the square root of the mean of the square of the instantaneous currents.*** The RMS value can be found by dividing the peak value by the square root of 2 (1.414) or by multiplying the peak value by 0.707 (the reciprocal of 1.414). The formulas for determining the RMS and peak values are:

$$\text{RMS} = \text{peak} \times 0.707$$

$$\text{peak} = \text{RMS} \times 1.414$$

Example 15–6

A sine wave has a peak value of 354 V. What is the RMS value?

Solution

$$RMS = peak \times 0.707$$

$$RMS = 354\ V \times 0.707$$

$$RMS = 250.278\ V$$

Example 15–7

An AC voltage has a value of 120 V RMS. What is the peak value of voltage?

Solution

$$peak = RMS \times 1.414$$

$$peak = 120\ V \times 1.414$$

$$peak = 169.68\ V$$

When the RMS values of voltage and current are used, the result is the same amount of power as the values of DC voltage and current. If 100 V RMS is applied to a 100-Ω resistor, the resistor will produce 100 W of heat. AC voltmeters and ammeters indicate the RMS value, not the peak value. Oscilloscopes, however, display the peak-to-peak value of voltage. All values of AC voltage and current used from now on in this text are RMS values unless otherwise stated.

Average Values

Average values of voltage and current are actually DC values. The average value must be found when a sine wave AC voltage is changed into DC with a rectifier (*Figure 15–12*). The rectifier shown

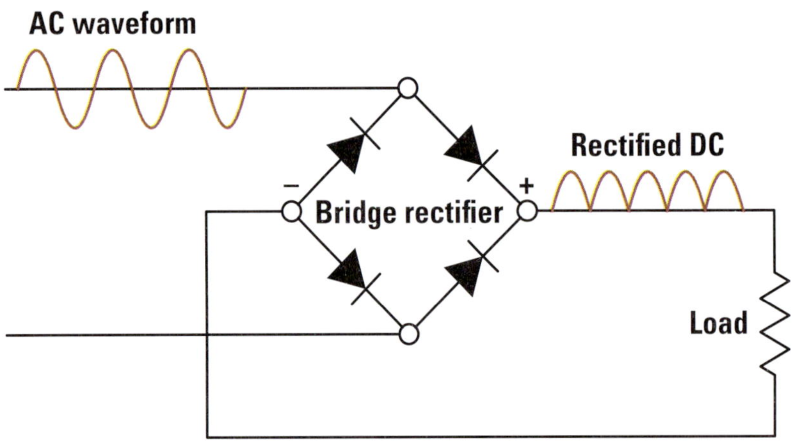

Figure 15–12 The bridge rectifier changes AC voltage into DC voltage.

is a bridge-type rectifier that produces full-wave rectification. This means that both the positive and negative half of the AC waveform are changed into the same polarity, which simulates DC power. The average value is the amount of voltage that would be indicated by a DC voltmeter if it were connected across the load resistor. The average voltage is proportional to the peak, or maximum, value of the waveform and to the length of time it is on compared with the length of time it is off (*Figure 15–13*). Notice in *Figure 15–13* that the voltage waveform turns on and off, but it never changes polarity. The current, therefore, never reverses direction. This is called pulsating DC. The pulses are often referred to as **ripples.** The average value is the voltage that will produce the same amount of power as a nonpulsating source of voltage such as a battery (*Figure 15–14*). For a full-wave rectified sine wave, the average value of voltage is found by multiplying the peak value by 0.637 or by multiplying the RMS value by 0.9.

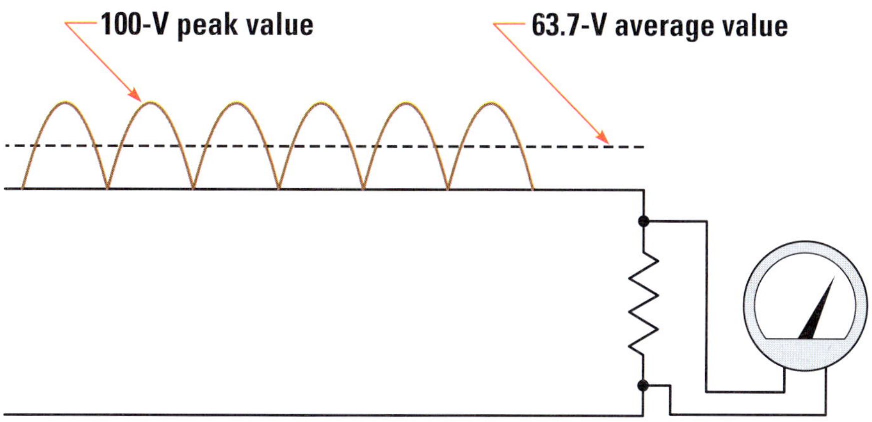

Figure 15–13 A DC voltmeter indicates the average value of a pulsating waveform.

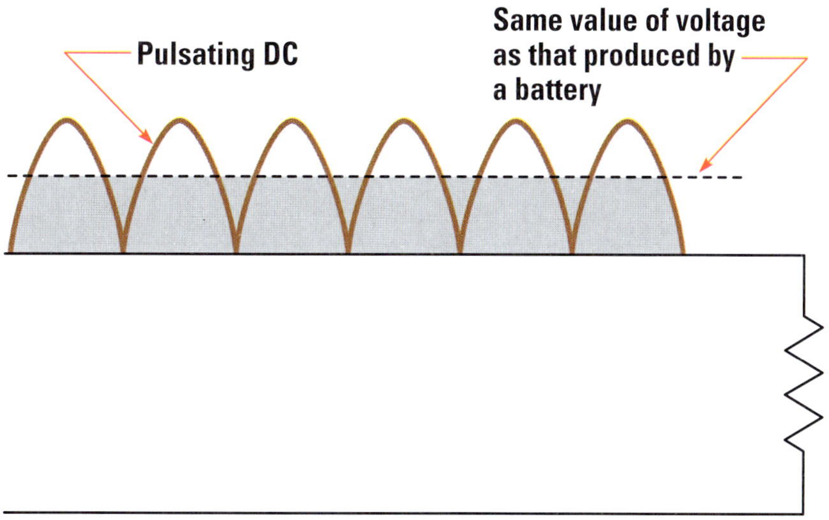

Figure 15–14 The average value of pulsating DC produces the same amount of power as a nonpulsating source of voltage.

Example 15–8

An AC sine wave with an RMS value of 120 V is connected to a full-wave rectifier. What is the average DC voltage?

Solution

The problem can be solved in one of two ways. The RMS value can be changed into peak, and then the peak value can be changed to the average value:

$$\text{peak} = \text{RMS} \times 1.414$$

$$\text{peak} = 120\text{ V} \times 1.414$$

$$\text{peak} = 169.68\text{ V}$$

$$\text{average} = \text{peak} \times 0.637$$

$$\text{average} = 169.68\text{ V} \times 0.637$$

$$\text{average} = 108.086\text{ V}$$

The second method of determining the average value is to multiply the RMS value by 0.9:

$$\text{average} = \text{RMS} \times 0.9$$

$$\text{average} = 120\text{ V} \times 0.9$$

$$\text{average} = 108\text{ V}$$

The conversion factors given are for full-wave rectification. If a half-wave rectifier is used (*Figure 15–15*), only one-half of the AC waveform is converted into DC. To determine the average voltage for a half-wave rectifier, multiply the peak value by 0.637 or the RMS value by 0.9 and then divide the product by 2. Because only half of the AC waveform has been converted into DC, the average voltage will be only half that of a full-wave rectifier (*Figure 15–16*).

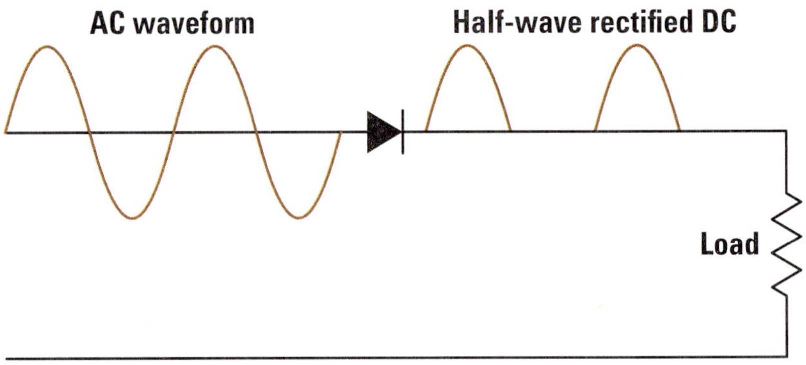

Figure 15–15 A half-wave rectifier converts only one-half of the AC waveform into DC.

Example 15–8 (continued)

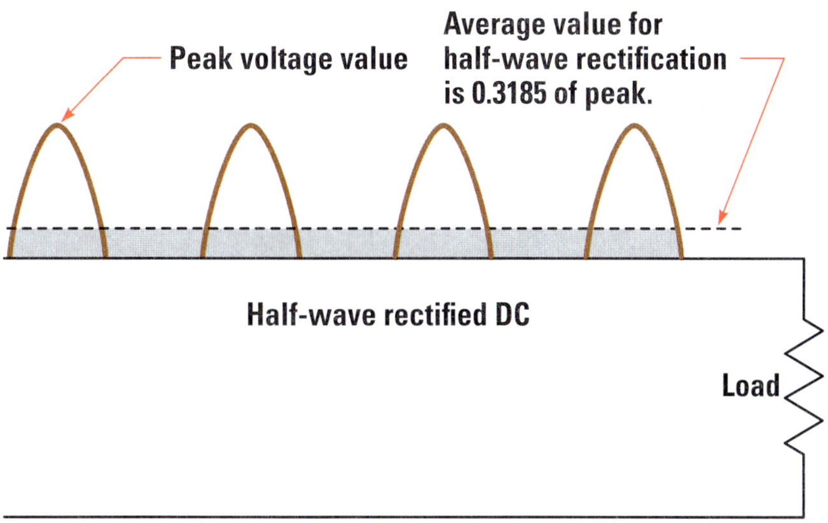

Figure 15–16 The average value for a half-wave rectifier is only half that of a full wave.

Example 15–9

A half-wave rectifier is connected to 277 VAC. What is the average DC voltage?

Solution

$$\text{average} = \text{RMS} \times \frac{0.9}{2}$$

$$\text{average} = 277\text{ V} \times \frac{0.9}{2}$$

$$\text{average} = 124.65\text{ V}$$

Wavelength

When current flows through a conductor, a magnetic field is created around the conductor. When high-frequency AC is used to produce a magnetic field, the field will radiate away from the source at the speed of light, which is approximately 186,000 miles per second or 300,000,000 meters per second (*Figure 15–17*). Wavelength is the distance the radiating field or signal can travel during one cycle of alternating voltage. Wavelength is generally represented by the Greek letter lambda (λ) and can be computed using the formula:

$$\lambda = \frac{\text{speed of light in meters per second}}{\text{frequency}}$$

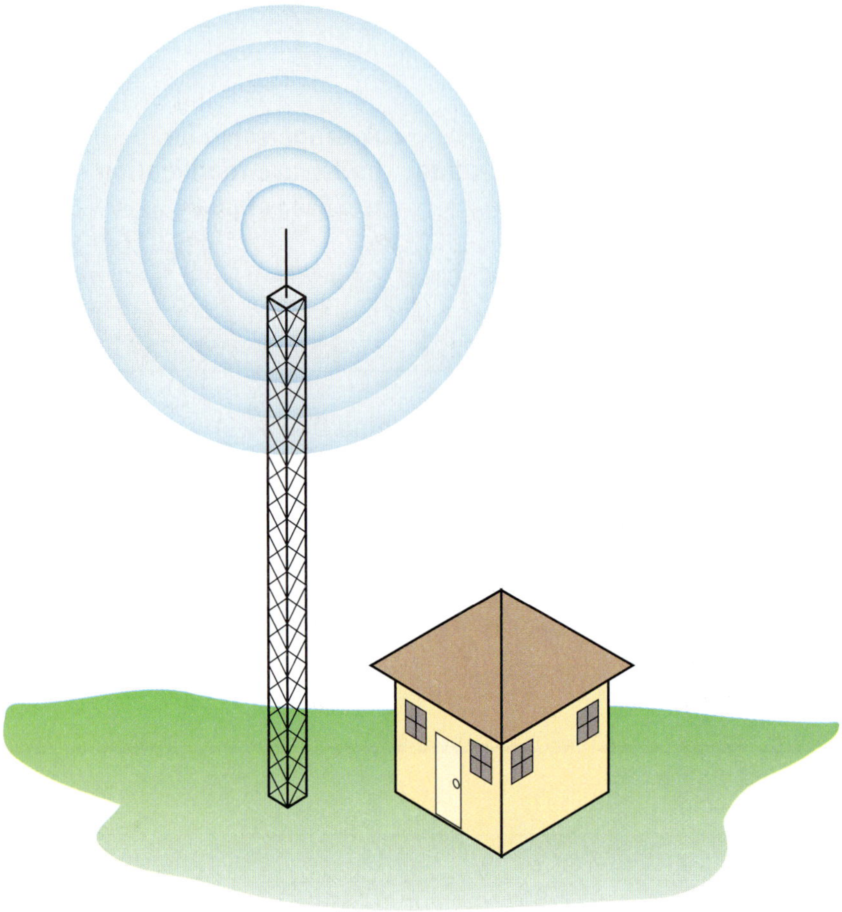

Figure 15–17 High-frequency AC causes a magnetic field to radiate at the speed of light.

Assume that a radio transmitter is operating at a frequency of 100 MHz. What is the wavelength of the signal?

$$\lambda = \frac{300{,}000{,}000 \text{ m/s}}{100{,}000{,}000 \text{ Hz}}$$

$$\lambda = 3 \text{ m}$$

At a frequency of 100 MHz, the signal will travel 9.843 ft (3 m) during one cycle (*Figure 15–18*).

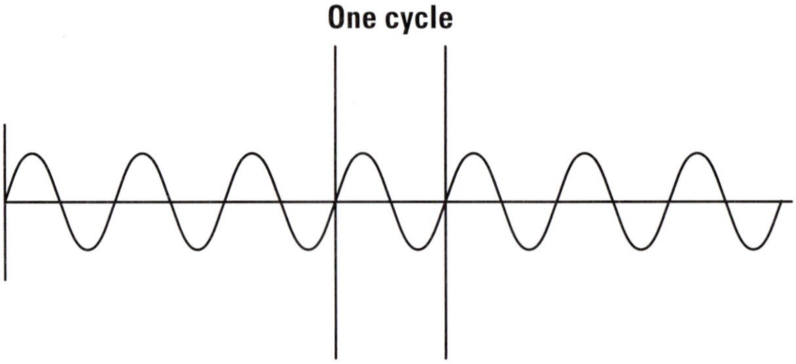

Figure 15–18 Wavelength is the distance traveled by the signal during one cycle.

15–4 Resistive Loads

In DC circuits, there is only one basic type of load, which is resistive. Even motor loads appear to be resistive because there is a conversion of electrical energy into mechanical energy. In this type of load, the **true power,** or watts, is the product of the volts times the amperes. In AC circuits, the type of load can vary depending on several factors. AC loads are generally described as being resistive, inductive, or capacitive, depending on the phase-angle relationship of voltage and current and the amount of true power produced by the circuit. Inductive and capacitive loads are discussed in later units. **Resistive loads** are loads that contain pure resistance, such as electric heating equipment and incandescent lighting. The following two characteristics hold true for resistive loads:

1. They produce heat.
2. The current and voltage are in phase with each other.

Any time a circuit contains resistance, electrical energy is changed into heat.

When an AC voltage is applied to a resistor, the current flow through the resistor is a copy of the voltage with regard to the time or frequency but not necessarily in magnitude (*Figure 15–19*). The current rises and falls at the same rate as the voltage and reverses the direction of flow when the voltage reverses polarity. In this condition, the current is said to be **in phase** with the voltage. When an AC circuit contains pure resistance, all the Ohm's law formulas discussed in Unit 2 can be employed to find values of voltage, current, resistance, and power. These formulas are also shown in Appendix B.

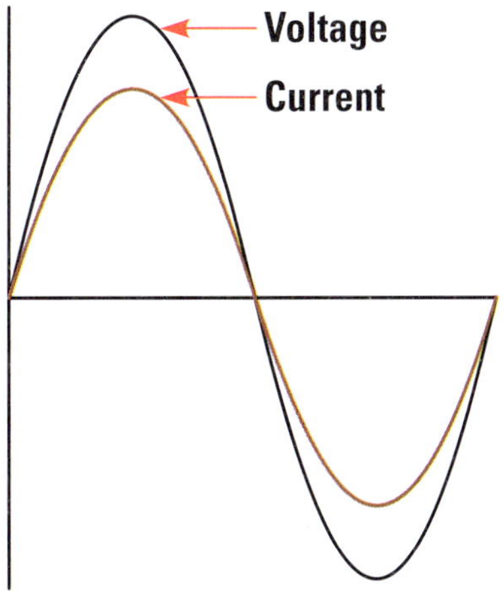

Figure 15–19 In a pure resistive circuit, the voltage and current are in phase with each other.

15–5 Power in an AC Circuit

True power, or watts, can be produced only when both current and voltage are either positive or negative. When like signs are multiplied, the product is positive (+ × + = +, or − × − = +), and when unlike signs are multiplied, the product is negative (+ × − = −). Because the current and voltage are either both positive or both negative at the same time, the product, watts, is always positive (*Figure 15–20*).

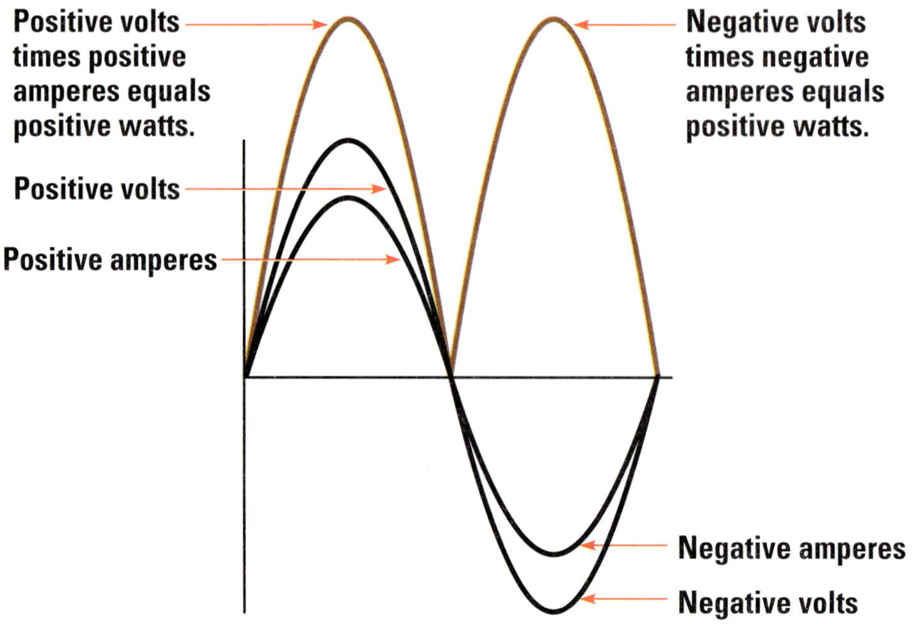

Figure 15–20 Power stays positive in a purely resistive AC circuit.

15–6 Skin Effect in AC Circuits

When current flows through a conductor connected to a source of DC power, the electrons flow through the entire conductor (*Figure 15–21*). The conductor offers some amount of ohmic resistance to the flow of electrons, depending on the type of material from which the conductor is made, its length, and its diameter. If that same conductor were connected to a source of AC, the resistance of the conductor to the flow of current would be slightly higher because of a phenomenon called the **skin effect.** The AC induces eddy currents into the conductor. These eddy currents cause the electrons to be repelled toward the outer surface of the conductor (*Figure 15–22*). Forcing the electrons toward the outer surface of the conductor has the same effect as decreasing the diameter of the conductor, which increases conductor resistance. The skin effect is proportional to the frequency. As the frequency increases, the

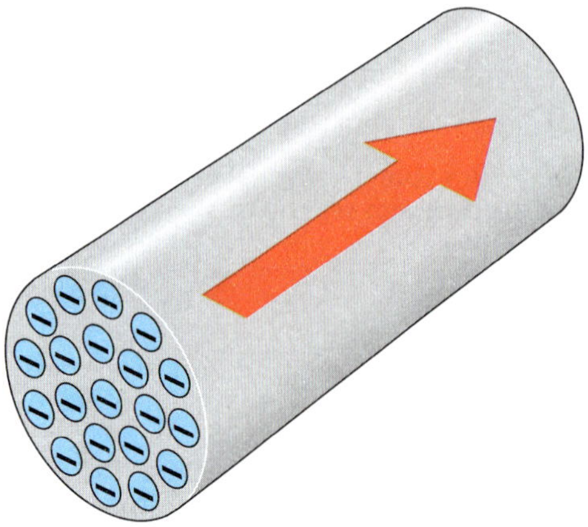

Figure 15–21 In a DC circuit, no eddy currents are produced, and the electrons travel through the entire cross-section of the conductor.

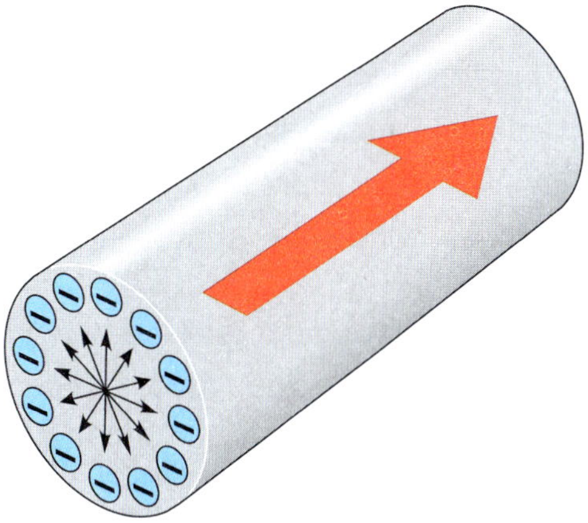

Figure 15–22 In an AC circuit, the electrons are forced to the outside of the conductor by eddy currents. This is called the skin effect.

skin effect increases. At low frequencies, such as 60 hertz, the skin effect has very little effect on the resistance of a conductor and generally is not taken into consideration when computing the size of wire needed for a particular circuit. At high frequencies, however, the skin effect can have a great effect on the operation of the circuit.

To help overcome the problem of skin effect in high-frequency circuits, conductors with a large amount of surface area must be used. Grounding a piece of equipment operating at 60 hertz, for example, may be as simple as using a grounding rod and a piece of 6 AWG copper conductor. Grounding a piece of equipment operating at 20 megahertz, however, may require the use of wide copper tape or a wide, flat, braided cable. Braided cable is less affected by skin effect because it contains many small conductors, which provide a large amount of surface area.

Green Tip

High-frequency circuits should employ conductors that have a very large surface area, such as braided cable, to help reduce the increased conductor resistance due to skin effect.

Summary

- Most of the electric power generated in the world is AC.
- AC can be stepped up and down easily, and DC cannot.
- AC power reverses its direction of current flow at periodic intervals.
- The most common AC waveform is the sine wave.
- There are 360° in one complete sine wave.
- One complete waveform is called a cycle.
- The number of complete cycles that occur in 1 second is called the frequency.
- Sine waves are produced by rotating machines.
- Frequency is measured in hertz (Hz).
- The instantaneous voltage at any point on a sine wave is equal to the peak, or maximum, voltage times the sine of the angle of rotation.
- The peak-to-peak voltage is the amount of voltage measured from the positive-most peak to the negative-most peak.

- The peak value is the maximum amount of voltage attained by the waveform.
- The RMS value of voltage will produce as much power as a like amount of DC voltage.
- The average value of voltage is used when an AC sine wave is changed into DC.
- The current and voltage in a pure resistive circuit are in phase with each other.
- True power, or watts, can be produced only when current and voltage are both positive or both negative.
- Resistance in AC circuits is characterized by the fact that the resistive part will produce heat.
- There are three basic types of AC loads: resistive, inductive, and capacitive.
- The electrons in an AC circuit are forced toward the outside of the conductor by eddy current induction in the conductor itself.
- Skin effect is proportional to frequency.
- Skin effect can be reduced by using conductors with a large surface area.

Review Questions

1. Which of the following best describes the fundamental difference between AC and DC electricity?

   A) DC current flow produces a magnetic field around the conductor.

   B) AC current is always greater than DC.

   C) DC current has greater voltage drop in conductors.

   D) AC current reverses its direction of flow at periodic intervals.

2. True or False: Direct current systems facilitate the transmission of electricity over great distances because they are capable of being transformed into higher voltages and then stepped back down to more useable voltages.

3. An AC system with a sine waveform is the most common type of AC waveform. Which of the following best describes how to determine any value of instantaneous voltage depending on the angle of rotation of the generating source?

   A) $E_{inst} = E_{max} \times \sin \Phi$

   B) $E_{inst} = E_{max} \times \cos \Phi$

   C) $E_{inst} = E_{max} \times \sin^{-1} \Phi$

   D) $E_{inst} = E_{max} \times \tan \Phi$

4. If an oscilloscope shows the voltage at 70° of rotation to be 447 volts, what is the peak-to-peak voltage of the system?

   A) 475.7 volts

   B) 1,306.9 volts

   C) 951.4 volts

   D) 2613.8 volts

5. A digital multimeter indicates that it reads RMS values of AC systems. If the multimeter indicates a voltage of 465 V, what voltage is produced as the peak value?
   A) 657.5 V
   B) 418.5 V
   C) 591.9 V
   D) 480 V
6. True or False: In an AC circuit, the current value that has the same impact on the power in a circuit as if the source was DC is known as the root-mean-square or RMS value.
7. In a purely resistive AC circuit, which of the following best describes the phase relationship between voltage and current?
   A) The voltage and current are 180° out of phase with each other.
   B) The voltage and current are 0° out of phase with each other.
   C) Current leads applied voltage in a resistive circuit.
   D) Voltage leads current in a resistive circuit.
8. Which of the following applies to resistive loads connected to AC circuits?
   A) The current is in phase with the voltage.
   B) When voltage is at peak, current is at peak.
   C) When voltage is in negative polarity, current is in negative polarity.
   D) All of these are correct.
9. True power, watts, is created in a circuit when which of the following statements are true?
   A) When current leads the voltage by 90°
   B) When voltage and current are opposite polarity
   C) When voltage and current are either both positive or both negative
   D) When only voltage is positive
10. True or False: The power waveform does not follow the same path as voltage and current because power is always in a positive direction since it is only produced when current and voltage are of the same polarity.
11. Skin effect in the conductors of an AC circuit is caused by which of the following?
   A) The resistance of the conductor material
   B) Eddy currents caused by constantly changing polarity
   C) The amount of applied voltage
   D) The diameter of the conductor

12. Skin effect forces the current in a conductor to travel on the outside of the wire, thus effectively reducing the current carrying capabilities of the conductor. Which of the following can be used to minimize the impact of skin effect?

    A) Using a conductor with a large amount of surface area such as a conductive tape or braided conductor

    B) A copper #6 grounding conductor

    C) Using aluminum versus copper conductors

    D) Using insulated conductors

Practical Applications

1. You are an electrician working on an overhead crane. The crane uses a large electromagnet to pick up large metal pipes. The magnet must have a minimum of 200 VDC to operate properly. The crane has an AC source of 240 V. You are given four diodes that have a peak voltage rating of 400 V each. These diodes are to be used to form a bridge rectifier to convert the AC voltage into DC voltage. Is the voltage rating of the diodes sufficient? To the nearest volt, what will be the DC output voltage of the bridge rectifier?

2. You are a journeyman electrician working in a large office building. The fluorescent lighting system is operated at 277 V. You have been instructed to replace the existing light ballasts with a new electronic type that is more efficient. The manufacturer of the ballast states that the maximum peak operating voltage for the ballast is 350 V. Will the new electronic ballast operate on the building's lighting system without harm?

Practice Problems

Refer to the AC formulas in Appendix B to answer the following questions.

Sine Wave Values

Fill in all the missing values.

Peak Volts	Inst. Volts	Degrees
347	208	
780		43.5
	24.3	17.6
224	5.65	
48.7		64.6
	240	45
87.2	23.7	
156.9		82.3
	62.7	34.6
1256	400	
15,720		12
	72.4	34.8

$$E_{(INST)} = E_{(MAX)} \times \sin \angle\theta$$

$$E_{(MAX)} = \frac{E_{(INST)}}{\sin \angle\theta}$$

$$\sin \angle\theta = \frac{E_{(INST)}}{E_{(MAX)}}$$

Peak, RMS, and Average Values

Fill in all the missing values.

Peak	RMS	Full-Wave Rectified Average
12.7		
	53.8	
		164.2
1235		
	240	
		16.6
339.7		
	12.6	
		9
123.7		
	74.8	
		108

Section

6 Alternating Current (AC) Circuits Containing Inductance

Unit 16 Inductance in AC Circuits

Key Terms

Current lags voltage

Induced voltage

Inductance (L)

Inductive reactance (X_L)

Quality (Q)

Reactance

Reactive power

Volt-amperes-reactive (VARs)

Why You Need to Know

Inductive loads are one of the three major types of loads found in alternating current circuits. It is important to understand the impact of an inductive load on a circuit and how current lags voltage when induction is applied in an AC circuit. This unit

- explains how properties other than resistance can limit the flow of current.
- introduces another electrical measurement called impedance. Impedance is the total current-limiting effect in an AC circuit and can be comprised of more than one element, such as resistance and inductance. Without an understanding of inductance, you will not be able to understand many of the concepts to follow in later units.

Outline

16–1 Inductance

16–2 Inductive Reactance

16–3 Inductors Connected in Series and Parallel Circuits

16–4 Voltage and Current Relationships in an Inductive Circuit

16–5 Power in an Inductive Circuit

16–6 Q of an Inductor

Objectives

After studying this unit, you should be able to

- explain how inductance influences an AC circuit.
- describe how inductance opposes current flow with inductive reactance.
- calculate inductive reactance and inductance using principles of induction.
- explain the relationship of voltage and current in a purely inductive circuit.
- apply principles for series or parallel circuits when inductors are involved.
- differentiate between true power (Watts) in a resistive circuit and reactive power (VARs) in an inductive circuit.
- describe how the quality (Q) of an inductor is determined.

Preview

This unit discusses the effects of inductance on AC circuits. The unit explains how current is limited in an inductive circuit as well as the effect inductance has on the relationship between voltage and current.

16–1 Inductance

Inductance (L) is one of the primary types of load in AC circuits. Some amount of inductance is present in all AC circuits because of the continually expanding and collapsing magnetic field created by current flowing through the conductor (*Figure 16–1*). The amount of inductance of a single conductor is extremely small, and, in most instances, it is not considered in circuit calculations. Circuits are generally considered to contain inductance when they contain a coil; inductance *is* considered in these types of circuit calculations. Loads such as dishwashers, exhaust fans, and air conditioner units are inductive because they all contain coils of wire in the motor.

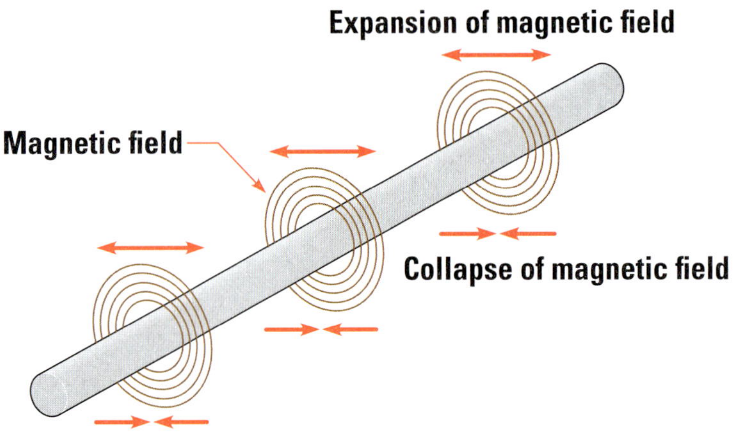

Figure 16–1 A continually changing magnetic field induces a voltage into any conductor.

In Unit 13, it was discussed that whenever current flows through a coil of wire, a magnetic field is created around the wire (*Figure 16–2*). If the amount of current decreases, the magnetic field collapses (*Figure 16–3*). Recall from Unit 13 several facts concerning inductance:

1. When magnetic lines of flux cut through a coil, a voltage is induced in the coil.
2. An induced voltage is always opposite in polarity to the applied voltage. This is often referred to as counter-electromotive force (CEMF).
3. The amount of induced voltage is proportional to the rate of change of current.
4. An inductor opposes a change of current.

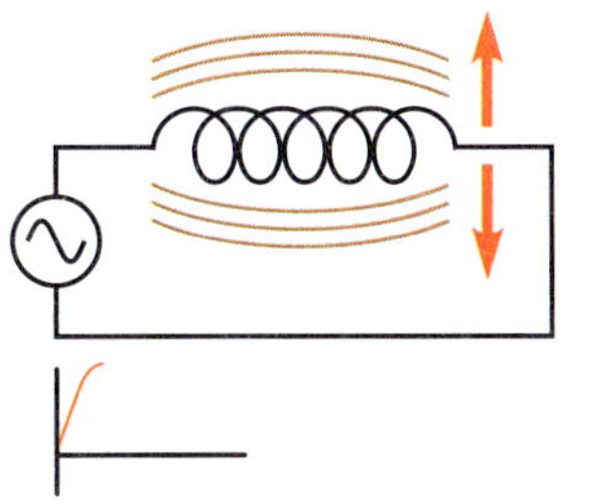

Figure 16–2 As current flows through a coil, a magnetic field is created around the coil.

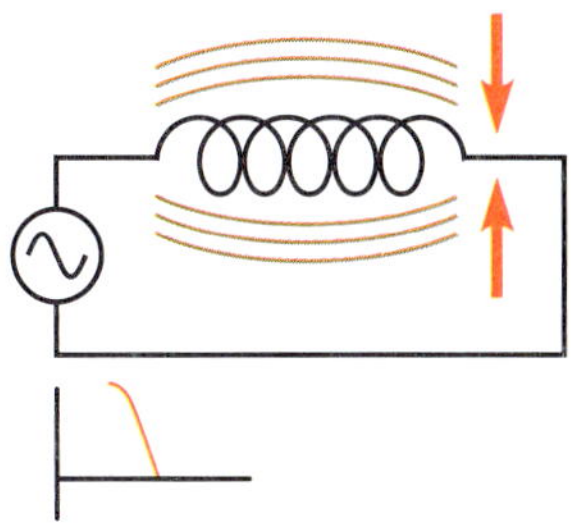

Figure 16–3 As current flow decreases, the magnetic field collapses.

When the inductors in *Figures 16–2 and 16–3* are connected to an AC power source, the magnetic field continually increases, decreases, and reverses polarity. Because the magnetic field continually changes magnitude and direction, a voltage is continually being induced in the coil. This **induced voltage** is 180° out of phase with the applied voltage and is always in opposition to the applied voltage (*Figure 16–4*). Because the induced voltage is always in opposition to the applied voltage, the effective applied voltage is reduced by the induced voltage. For example, assume an inductor is connected to a 120-volt AC line. Now assume that the inductor has an induced voltage of 116 volts. Because the induced voltage subtracts from the applied voltage, there are only 4 volts to push current through the wire resistance of the coil.

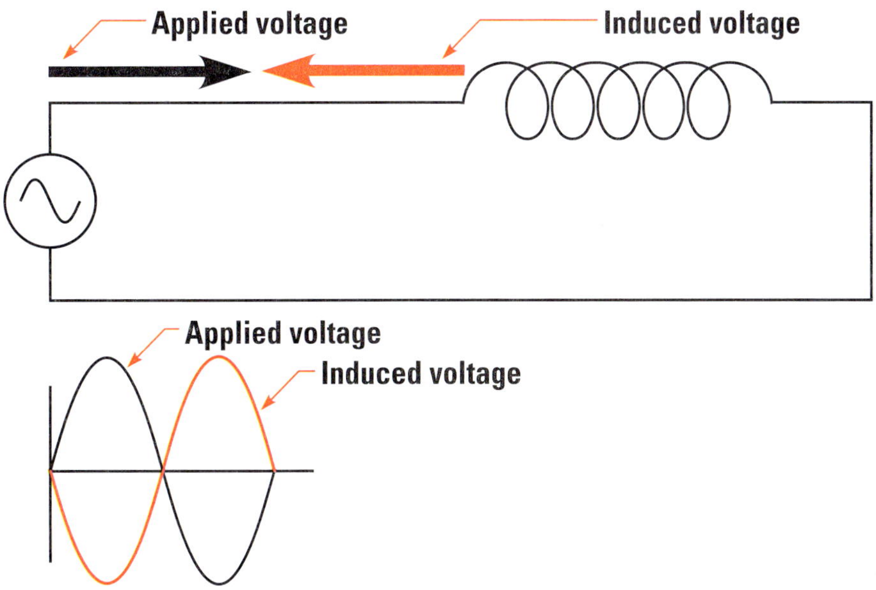

Figure 16–4 The applied voltage and induced voltage are 180 degrees out of phase with each other.

Calculating the Induced Voltage

The amount of induced voltage in an inductor can be calculated if the resistance of the wire in the coil and the amount of circuit current are known. For example, assume that an ohmmeter is used to measure the actual amount of resistance in a coil, and the coil is found to contain 6 ohms of wire resistance (*Figure 16–5*). Now assume that the coil is connected to a 120-volt AC circuit, and an ammeter measures a current flow of 0.8 ampere (*Figure 16–6*). Ohm's law can now be used to determine the amount of voltage necessary to push 0.8 ampere of current through 6 ohms of resistance:

$$E = I \times R$$

$$E = 0.8\ A \times 6$$

$$E = 4.8\ V$$

Because only 4.8 volts are needed to push the current through the wire resistance of the inductor, the remainder of the 120 volts is used to overcome the coil's induced voltage of 119.904 volts ($\sqrt{(120\ V)^2 - (4.8\ V)^2} = 119.904$ volts).

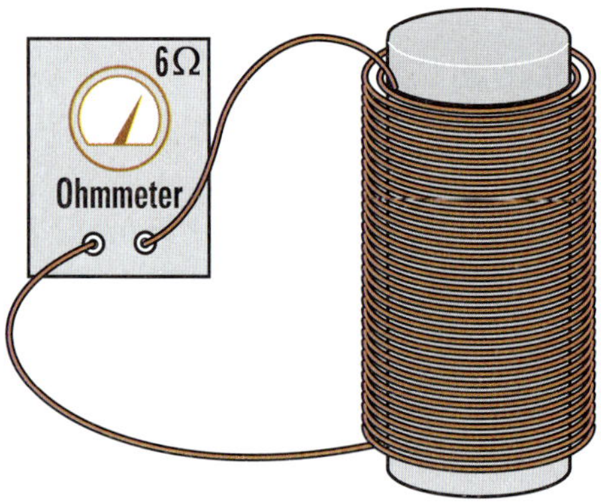

Figure 16–5 Measuring the resistance of a coil.

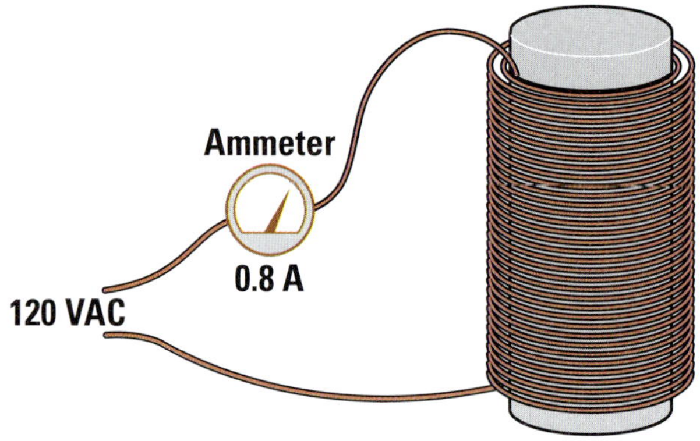

Figure 16–6 Measuring circuit current with an ammeter.

16–2 Inductive Reactance

Notice that the induced voltage is able to limit the flow of current through the circuit in a manner similar to that of resistance. This induced voltage is *not* resistance, but it can limit the flow of current just as resistance does. This current-limiting property of the inductor is called **reactance** and is symbolized by the letter *X*. Because this reactance is caused by inductance, it is called **inductive reactance** and is symbolized by $\mathbf{X_L}$. Inductive reactance is measured in ohms just as resistance is and can be calculated when the values of inductance and frequency are known. The following formula can be used to find inductive reactance:

$$X_L = 2\ \pi\ fL$$

where

X_L = inductive reactance

2 = a constant

π = 3.1416

f = frequency in hertz (Hz)

L = inductance in henrys (H)

The inductive reactance, in ohms, is proportional to the three factors that determine induced voltage:

1. The *number* of turns of wire
2. The *strength* of the magnetic field
3. The *speed* of the cutting action (relative motion between the inductor and the magnetic lines of flux)

Factors such as the size of wire used, the number of turns, how close the turns are to each other, and the type of core material determine the amount of inductance (in henrys, H) of the coil (*Figure 16–7*). The speed of the cutting action is proportional to the frequency (hertz). An increase of frequency causes the magnetic lines of flux to cut the conductors at a faster rate and thus produces a higher induced voltage or more inductive reactance.

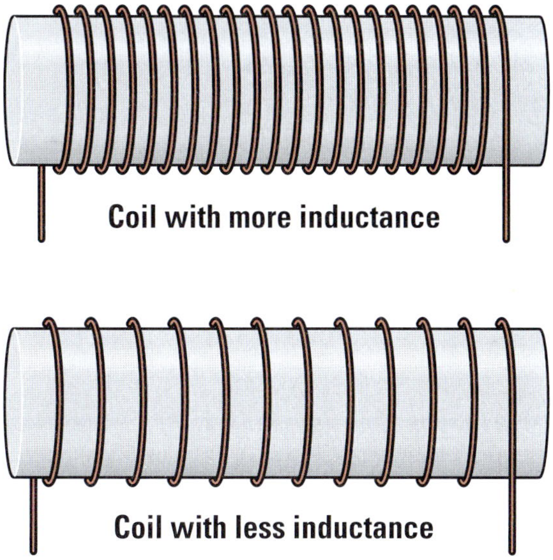

Figure 16–7 Coils with turns closer together produce more inductance than coils with turns farther apart.

Example 16–1

The inductor shown in *Figure 16–8* has an inductance of 0.8 H and is connected to a 120-V, 60-Hz line. How much current will flow in this circuit if the wire resistance of the inductor is negligible?

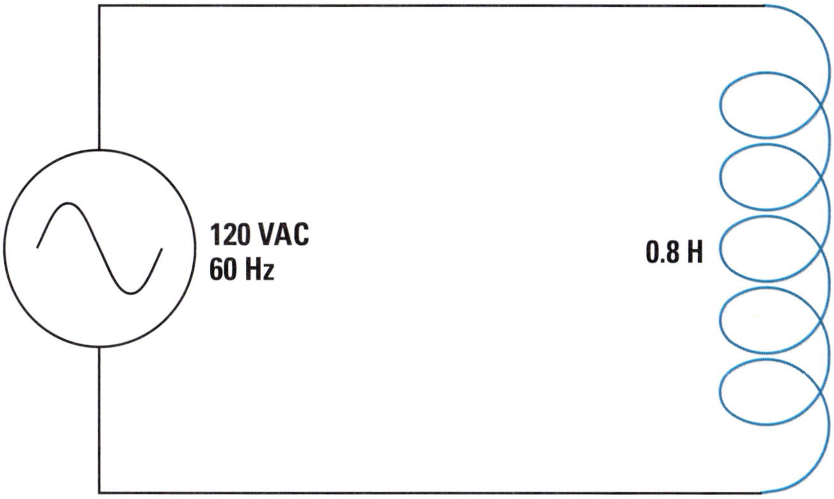

Figure 16–8 Circuit current is limited by inductive reactance.

Example 16–1 (continued)

Solution

The first step is to determine the amount of inductive reactance of the inductor:

$$X_L = 2\pi fL$$

$$X_L = 2 \times 3.1416 \times 60 \text{ Hz} \times 0.8$$

$$X_L = 301.594\ \Omega$$

Because inductive reactance is the current-limiting property of this circuit, it can be substituted for the value of R in an Ohm's law formula:

$$I = \frac{E}{X_L}$$

$$I = \frac{120 \text{ V}}{301.594\ \Omega}$$

$$I = 0.398 \text{ A}$$

If the amount of inductive reactance is known, the inductance of the coil can be determined using the formula

$$L = \frac{X_L}{2\pi f}$$

Example 16–2

Assume an inductor with a negligible internal wire resistance is connected to a 36-V, 400-Hz line. If the circuit has a current flow of 0.2 A, what is the inductance of the inductor?

Solution

The first step is to determine the inductive reactance of the circuit:

$$X_L = \frac{E}{I}$$

$$X_L = \frac{36 \text{ V}}{0.2 \text{ A}}$$

$$X_L = 180\ \Omega$$

Now that the inductive reactance is known, the inductance can be determined:

$$L = \frac{X_L}{2\pi f}$$

$$L = \frac{180\ \Omega}{2 \times 3.1416 \times 400 \text{ Hz}}$$

$$L = 0.0716 \text{ H}$$

Example 16–3

An inductor with negligible internal wire resistance is connected to a 480-V, 60-Hz line. An ammeter indicates a current flow of 24 A. How much current will flow in this circuit if the frequency is increased to 400 Hz?

Solution

The first step in solving this problem is to determine the amount of inductive reactance in the circuit. Because the resistance of the wire used to make the inductor is negligible, the current is limited by inductive reactance. The inductive reactance can be found by substituting X_L for R in an Ohm's law formula:

$$X_L = \frac{E}{I}$$

$$X_L = \frac{480\ V}{24\ A}$$

$$X_L = 20\ \Omega$$

Now that the inductive reactance is known, the inductance of the coil can be found using the formula:

$$L = \frac{X_L}{2\pi f}$$

Note: When using a frequency of 60 hertz, $2 \times \pi \times 60 = 376.992$. To simplify calculations, this value is generally rounded to 377. Because 60 hertz is the most common frequency used throughout the United States, 377 can be memorized and substituted for $2 \times \pi \times 60$.

$$L = \frac{20\ \Omega}{377\ Hz}$$

$$L = 0.053\ H$$

Because the inductance of the coil is determined by its physical construction, it does not change when connected to a different frequency. Now if the inductance of the coil is known, the inductive reactance at 400 hertz can be calculated:

$$X_L = 2\pi fL$$

$$X_L = 2 \times 3.1416 \times 400\ Hz \times 0.053$$

$$X_L = 133.204\ \Omega$$

The amount of current flow can now be found by substituting the value of inductive reactance for resistance in an Ohm's law formula:

$$I = \frac{E}{X_L}$$

$$I = \frac{480\ V}{133.204\ \Omega}$$

$$I = 3.603\ A$$

Schematic Symbols

The schematic symbol used to represent an inductor depicts a coil of wire. Different types of inductor symbols are shown in *Figure 16–9*. The symbols shown with the two parallel lines represent iron-core inductors, and the symbols without the parallel lines represent air-core inductors.

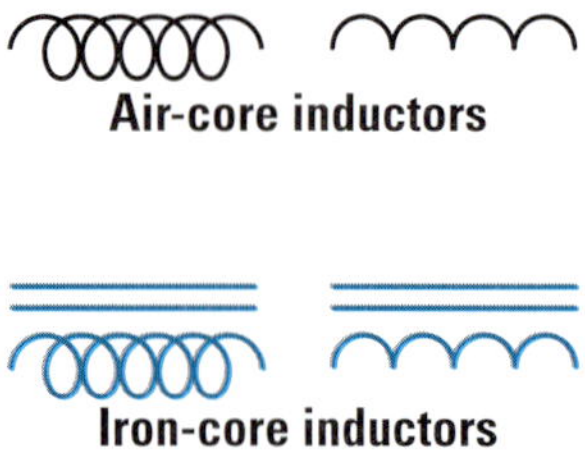

Figure 16–9 Schematic symbols for inductors.

16–3 Inductors Connected in Series and Parallel Circuits

Inductors or coils in series behave very similar to resistors in series. (*Figure 16–10*). The total inductance of the circuit (L_T) equals the sum of the inductances of all the inductors:

$$L_T = L_1 + L_2 + L_3$$

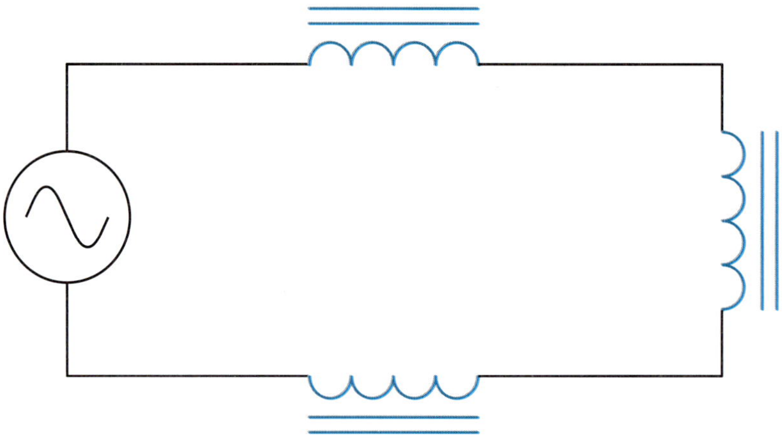

Figure 16–10 Inductors connected in series.

The total inductive reactance (X_{LT}) of inductors connected in series equals the sum of the inductive reactance of the individual inductors:

$$X_{LT} = X_{L1} + X_{L2} + X_{L3}$$

Example 16–4

Three inductors are connected in series. Inductor 1 has an inductance of 0.6 H, Inductor 2 has an inductance of 0.4 H, and Inductor 3 has an inductance of 0.5 H. What is the total inductance of the circuit?

Solution

$$L_T = 0.6\text{ H} + 0.4\text{ H} + 0.5\text{ H}$$

$$L_T = 1.5\text{ H}$$

Example 16–5

Three inductors are connected in series. Inductor 1 has an inductive reactance of 180 Ω, Inductor 2 has an inductive reactance of 240 Ω, and Inductor 3 has an inductive reactance of 320 Ω. What is the total inductive reactance of the circuit?

Solution

$$X_{LT} = 180\ \Omega + 240\ \Omega + 320\ \Omega$$

$$X_{LT} = 740\ \Omega$$

Inductors Connected in Parallel

The same similarity to resistors holds true for parallel circuits (*Figure 16–11*). The total inductance can be found in a manner similar to finding the total resistance of a parallel circuit. The reciprocal of the total inductance is equal to the sum of the reciprocals of all the inductors:

$$\frac{1}{L_T} = \frac{1}{L_1} + \frac{1}{L_2} + \frac{1}{L_3}$$

or

$$L_T = \frac{1}{\frac{1}{L_1} + \frac{1}{L_2} + \frac{1}{L_3}}$$

Another formula that can be used to find the total inductance of parallel inductors is the product-over-sum formula. This formula is used to find the total inductance of two inductors connected in parallel:

$$L_T = \frac{L_1 \times L_2}{L_1 + L_2}$$

If the values of all the inductors are the same, total inductance can be found by dividing the inductance of one inductor by the total number of inductors:

$$L_T = \frac{L}{N}$$

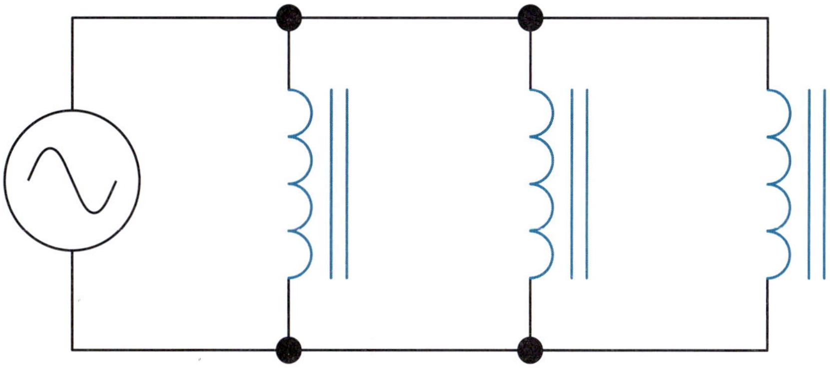

Figure 16–11 Inductors connected in parallel.

Similar formulas can be used to find the total inductive reactance of inductors connected in parallel:

$$\frac{1}{X_{LT}} = \frac{1}{X_{L1}} + \frac{1}{X_{L2}} + \frac{1}{X_{L3}}$$

or

$$X_{LT} = \frac{1}{\frac{1}{X_{L1}} + \frac{1}{X_{L2}} + \frac{1}{X_{L3}}}$$

or

$$X_{LT} = \frac{X_{L1} \times X_{L2}}{X_{L1} + X_{L2}}$$

or

$$X_{LT} = \frac{X_L}{N}$$

Example 16–6

Three inductors are connected in parallel. Inductor 1 has an inductance of 2.5 H, Inductor 2 has an inductance of 1.8 H, and Inductor 3 has an inductance of 1.2 H. What is the total inductance of this circuit?

Solution

$$L_T = \frac{1}{\frac{1}{2.5\ \text{H}} + \frac{1}{1.8\ \text{H}} + \frac{1}{1.2\ \text{H}}}$$

$$L_T = \frac{1}{(1.788)}$$

$$L_T = 0.559\ \text{H}$$

16–4 Voltage and Current Relationships in an Inductive Circuit

In Unit 15, it was discussed that when current flows through a purely resistive circuit, the current and voltage are in phase with each other, meaning that when voltage hits its peak in either polarity, current hits its corresponding peak at the same time. ***In a purely inductive circuit, however, the current lags voltage by 90°***. How the current and applied voltage can become 90° out of phase with each other can best be explained by comparing the relationship of the current and induced voltage (*Figure 16–12*). Recall that the induced voltage is proportional to the rate of change of the current (speed of cutting action). At the beginning of the waveform, the current is shown at its maximum value in the negative direction. At this time, the current is not changing, so induced voltage is zero. As the magnitude of the current begins to decrease in value, the magnetic field produced by the flow of current decreases or collapses and begins to induce a voltage into the coil as it cuts through the conductors (*Figure 16–13*).

The greatest rate of change in current occurs when the current passes from negative through zero and begins to increase in the positive direction (*Figure 16–13*). Because the current is changing at the greatest

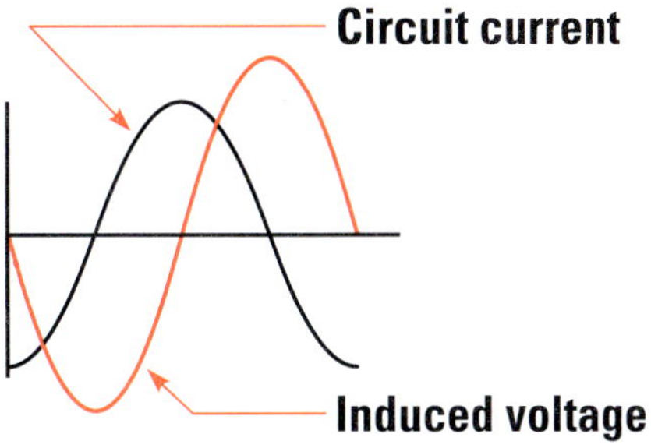

Figure 16–12 Induced voltage is proportional to the rate of change of current.

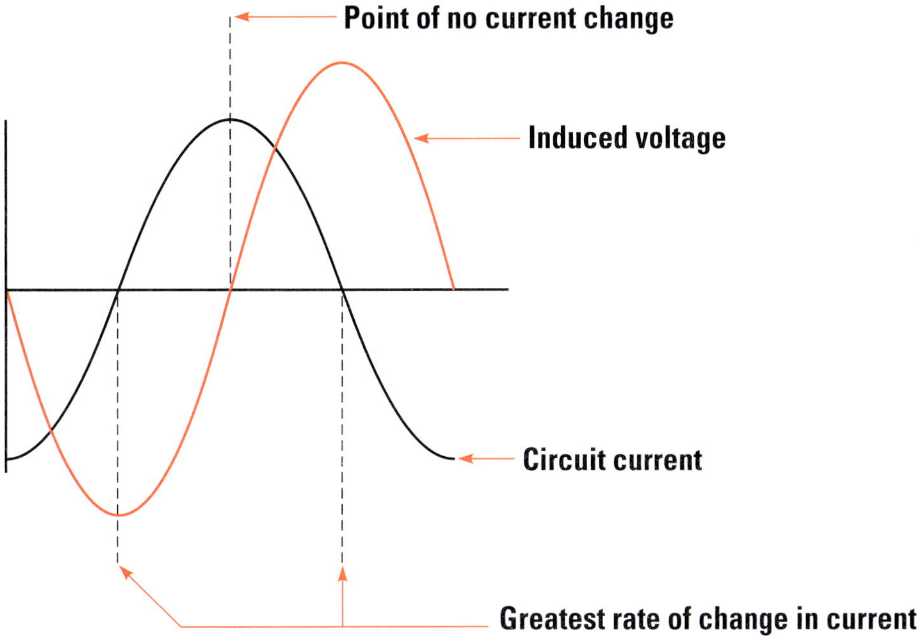

Figure 16–13 No voltage is induced when the current is not changing, such as at peak current.

rate, the induced voltage waveform is at its peak. As current approaches its peak value in the positive direction, the rate of change decreases, causing a decrease in the induced voltage. The induced voltage will again be zero when the current reaches its peak value, and the magnetic field stops expanding.

The current flowing through the inductor is leading the induced voltage by 90°. Because the induced voltage is 180° out of phase with the applied voltage, the current lags the applied voltage by 90° (*Figure 16–14*).

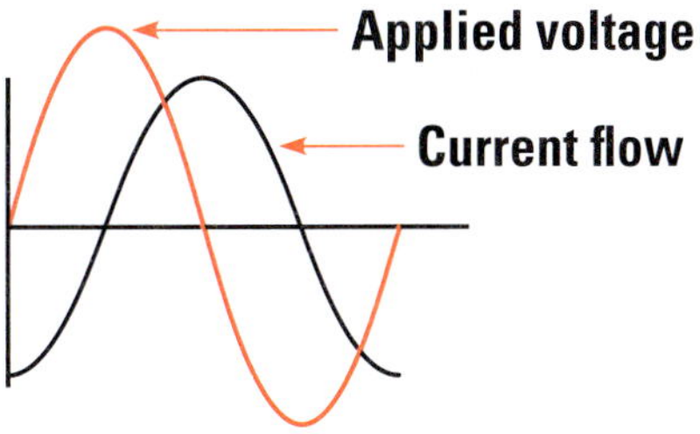

Figure 16–14 The current lags the applied voltage by 90°.

16–5 Power in an Inductive Circuit

In a purely resistive circuit, the true power, or watts, is equal to the product of the voltage and current. In a purely inductive circuit, however, no true power, or watts, is produced. Recall that voltage and current must both be either positive or negative before true power can be produced. Because the voltage and current are 90° out of phase with each other in a purely inductive circuit, the current and voltage will be at different polarities 50% of the time and at the same polarity 50% of the time. During the period of time that the current and voltage have the same polarity, power is being given to the circuit in the form of creating a magnetic field. When the current and voltage are opposite in polarity, power is being given back to the circuit as the magnetic field collapses and induces a voltage back into the circuit. Because power is stored in the form of a magnetic field and then given back, no power is used by the inductor. Any power used in an inductor is caused by losses such as the resistance of the wire used to construct the inductor, generally referred to as I^2R losses, eddy current losses, and hysteresis losses.

The current and voltage waveform in *Figure 16–15* has been divided into four sections: A, B, C, and D. During the first time period, indicated by A, the current is negative, and the voltage is positive. During this period, energy is being given to the circuit as the magnetic field collapses. During the second time period, B, both the voltage and current are positive. Power is being used to produce the magnetic field. In the third time period, C, the current is positive, and the voltage is negative. Power is again being given back to the circuit as the field collapses. During the fourth time period, D, both the voltage and current are negative. Power is again being used to produce the magnetic field. If the amount of power used to produce the magnetic field is subtracted from the power given back, the net result will be zero power produced in the purely inductive circuit.

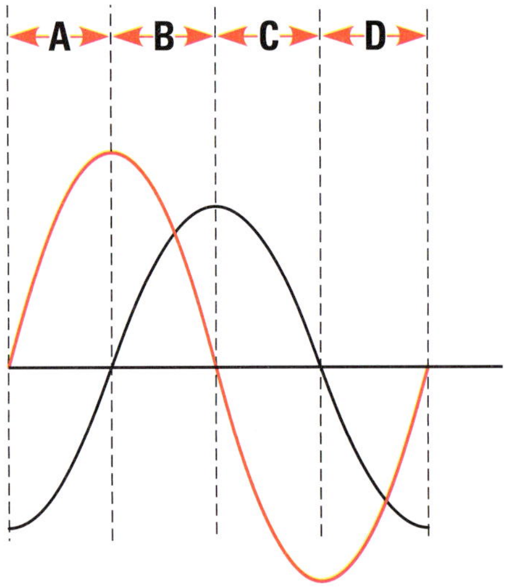

Figure 16–15 Voltage and current relationships during different parts of a cycle.

The formula for determining power in any type of electrical circuit is:

$$P = E \times I \times \text{Cos} \angle \theta$$

where

P = power in watts

E = volts

I = current

$\angle\theta$ = angle theta (Angle theta is the phase angle difference between the applied voltage and the current.)

In a purely resistive AC circuit, the voltage and current are in phase with each other. Therefore, angle theta is zero degrees (0°). The cosine of 0° is 1. In a purely inductive circuit, angle theta is 90°. The cosine of 90° is 0. It also stands to reason that with every variation of current and voltage being somewhere between 0° and 90° out of phase with each other, the cosine of angle theta is somewhere between 0 and 1.

Reactive Power

Although essentially no true power is being used, except by previously mentioned losses, a new electrical measurement called **volt-amperes-reactive (VARs)** is used to measure the **reactive power** in a purely inductive circuit. VARs can be calculated in the same way as watts except that inductive values are substituted for resistive values in the formulas. VARs is equal to the amount of current flowing through an inductive circuit multiplied by the voltage applied to the inductive part of the circuit. Several formulas for calculating VARs are:

$$\text{VARs} = E_L \times I_L$$

$$\text{VARs} = \frac{E_L^2}{X_L}$$

$$\text{VARs} = I_L^2 \times X_L$$

where

E_L = voltage applied to an inductor

I_L = current flow through an inductor

X_L = inductive reactance

Although VARs are not true power, they do represent real values of voltage and current. Inductors are given a VAR or kVAR rating. This rating is similar to the wattage rating of resistors and should not be ignored.

Example 16–7

An inductor is connected to a 480-volt circuit with a current flow of 125 amperes. What is the minimum kVAR rating for the inductor?

$$\text{kVAR} = \frac{E_{(\text{inductive})} \times I_{(\text{inductive})}}{1000}$$

$$\text{kVAR} = \frac{480 \times 125}{1000}$$

$$\text{kVAR} = 60$$

Solution

The inductor in this example should have a minimum rating of 60 kVARs. An inductor with a rating less than 60 kVARs would probably be destroyed in a short period of time.

16–6 Q of an Inductor

So far in this unit, it has been generally assumed that an inductor has no resistance and that inductive reactance is the only current-limiting factor. Since inductors are actually coils of wire, they all contain some amount of internal resistance. Inductors actually appear to be a coil connected in series with some amount of resistance (*Figure 16–16*). The amount of resistance compared with the inductive reactance determines the **quality (Q)** of the coil. Inductors that have a higher ratio of inductive reactance to resistance are considered to be inductors of higher quality. An inductor constructed with a large wire will have a low wire resistance and therefore a higher Q (*Figure 16–17*). Inductors constructed with many turns of small wire have a much higher resistance and therefore a lower Q. To determine the Q of an inductor, divide the inductive reactance by the resistance:

$$Q = \frac{X_L}{R}$$

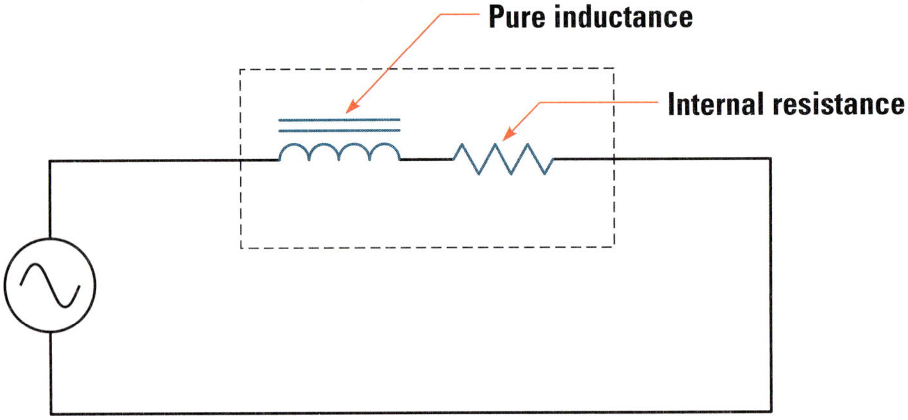

Figure 16–16 Inductors contain internal resistance that acts as a resistor in series with the coil.

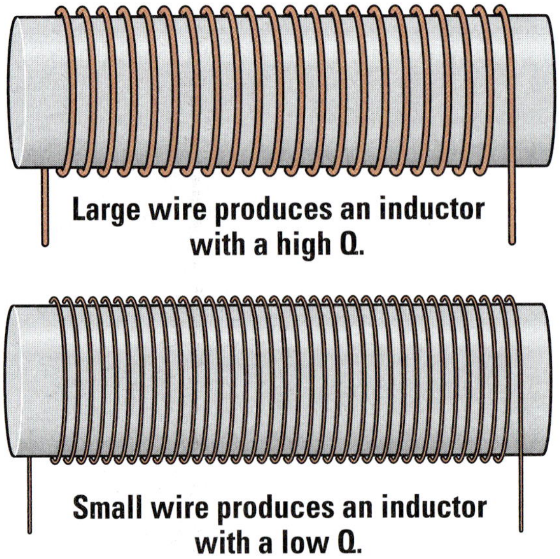

Figure 16–17 The Q of an inductor is a ratio of inductive reactance compared to resistance. The letter Q stands for quality.

Although every inductor will have some amount of internal resistance, inductors that have a Q of 10 or greater are generally considered to be pure inductors. Once the ratio of inductive reactance becomes 10 times as great as resistance, the amount of resistance is considered negligible. For example, assume an inductor has an inductive reactance of 100 ohms and a wire resistance of 10 ohms. The inductive reactive component in the circuit is 90° out of phase with the resistive component. This relationship can be better illustrated by using a right triangle (*Figure 16–18*). The total current-limiting effect of the inductor is a combination of the inductive reactance and resistance. This total current-limiting effect is called impedance and is symbolized by the letter *Z*. The total impedance of the circuit is represented by the hypotenuse of the right triangle formed by the inductive reactance and the resistance. To calculate the value of impedance for the coil, the inductive reactance and the resistance must be added. However, remember they are 90° out of phase with each other. Therefore, resistance and inductive reactance form the legs of a right triangle and the impedance forms the hypotenuse. Using the Pythagorean theorem, discussed in Unit 14, we can calculate the value of impedance:

$$Z = \sqrt{R^2 + X_L^2}$$

$$Z = \sqrt{10^2 + 100^2}$$

$$Z = \sqrt{10,100}$$

$$Z = 100.499\ \Omega$$

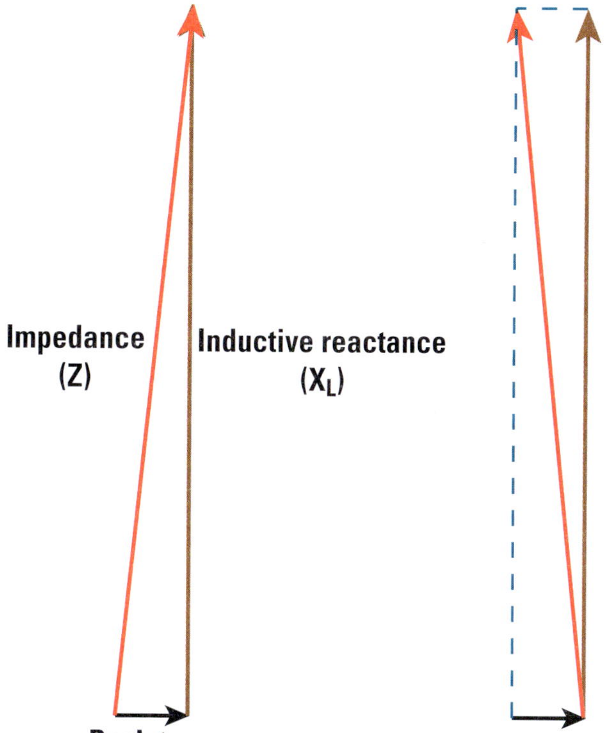

Figure 16–18 Coil impedance is a combination of wire resistance and inductive reactance.

Notice that the value of total impedance for the inductor is only 0.5 ohm greater than the value of inductive reactance.

If it should become necessary to determine the true inductance of an inductor, the resistance of the wire must be taken into consideration. Assume that an inductor is connected to a 480-volt, 60-Hz power source and that an ammeter indicates a current flow of 0.6 ampere. Now assume that an ohmmeter measures 150 Ω of wire resistance in the inductor. What is the inductance of the inductor?

To determine the inductance, it will be necessary to first determine the amount of inductive reactance as compared to the wire resistance. The total current-limiting value (impedance) can be found with Ohm's law:

$$Z = \frac{E}{I}$$

$$Z = \frac{480}{0.6}$$

$$Z = 800\ \Omega$$

The total current-limiting effect of the inductor is 800 Ω. This value is a combination of both the inductive reactance of the inductor and the wire resistance. Because the resistive part and the inductive reactance part of the inductor are 90° out of phase with each other, they form the legs of a right triangle with the impedance forming the hypotenuse of the triangle (*Figure 16–19*). To determine the amount of inductive reactance, use the following formula:

$$X_L = \sqrt{Z^2 - R^2}$$

$$X_L = \sqrt{800^2 - 150^2}$$

$$X_L = 785.812\ \Omega$$

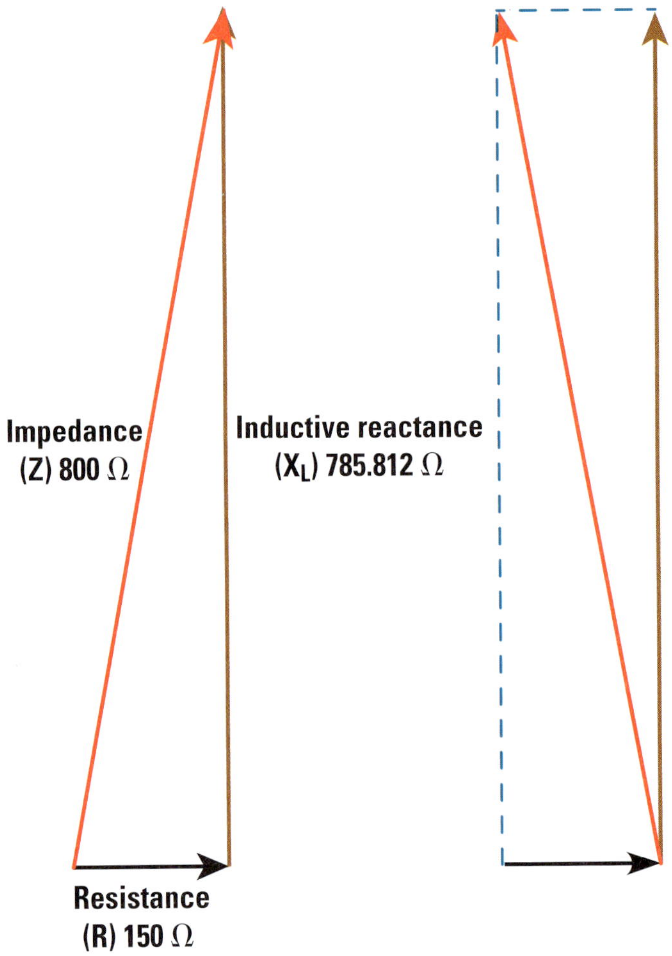

Figure 16–19 The inductive reactance forms one leg of a right triangle.

Now that the amount of inductive reactance has been determined, the inductance, in henrys, can be calculated using the formula:

$$L = \frac{X_L}{2\pi f}$$

$$L = \frac{785.812}{2 \times 3.1416 \times 60}$$

$$L = 2.084 \text{ H}$$

Summary

- The magnitude of induced voltage is proportional to the rate of change of current.
- Induced voltage is always opposite in polarity to the applied voltage.
- Inductive reactance is a counter-voltage or counter-EMF that limits the flow of current, as does resistance.
- Inductive reactance is measured in ohms.
- Inductive reactance is proportional to the inductance of the coil and the frequency of the line.
- Inductive reactance is symbolized by X_L.
- Inductance is measured in henrys (H) and is symbolized by the letter *L*.
- When inductors are connected in series, the total inductance is equal to the sum of all the inductors.
- When inductors are connected in parallel, the reciprocal of the total inductance is equal to the sum of the reciprocals of all the inductors.
- The current lags the applied voltage by 90° in a pure inductive circuit.
- All inductors contain some amount of resistance internal to the conductor material.
- The Q of an inductor is the ratio of the inductive reactance to the resistance.
- Inductors with a Q of 10 are generally considered to be "pure" inductors.
- Purely inductive circuits contain no true power or watts.
- Reactive power is measured in VARs.
- VARs is an abbreviation for volt-amperes-reactive.

Review Questions

1. Which of the following best describes how inductance impacts current flow in an AC circuit?

   A) Inductance boosts the current flow in a conductor by inducing a complimentary voltage in the circuit.

   B) Inductance has no effect on current flow in an AC circuit.

   C) Inductance opposes current flow by inducing an opposing voltage at 180° to the applied voltage.

   D) Inductors only affect a circuit due to the resistance of the coil material.

2. Which of the following is not a true statement about inductance in an AC circuit?
   A) When magnetic lines of flux cut across the conductors of a coil, a voltage is induced in the coil.
   B) The amount of voltage induced into the coil is proportional to the magnitude of applied voltage.
   C) An inductor opposes a change in current.
   D) The induced voltage is always opposite in polarity to the applied voltage.
3. The opposition to current flow caused by an inductor is referred to as inductive reactance and is measured in ohms. Which of the following is not a determining factor in the strength of a circuit's inductive reactance?
   A) The resistance of the wire
   B) The number of turns of wire in the coil
   C) The strength of the magnetic field
   D) The speed at which the magnetic field cuts across the conductor
4. True or False: Inductance is measured by the amount of henrys a coil contains, but the overall effect on a circuit is the reactance.
5. A 120-V, 60-Hz circuit containing only an inductor with negligible internal resistance has 2 amperes of current flowing. What is the inductance value of this inductor?
   A) 240 mH
   B) 22.6 kH
   C) 0.64 H
   D) 160 mH
6. An inductor provides 188.5 Ω of inductive reactance when connected to a 480-V, 60-Hz circuit. If this same coil is used on a 400-V, 50-Hz circuit, what will be the total opposition to current flow the inductor provides?
   A) 188.5 Ω
   B) 300 Ω
   C) 157.1 Ω
   D) 243.7 Ω
7. Which of the following statements is true in a purely inductive AC circuit?
   A) The current is in phase with the voltage.
   B) The current lags the voltage by 90°.
   C) The current is 180° out of phase with the voltage.
   D) The current leads the voltage by 90°.
8. True or False: The induced voltage is 180° out of phase with the applied voltage because the current lags the applied voltage by 90°.

9. A 240-V, 60-Hz circuit contains three inductors with inductances of 30, 40, and 50 mH. What is the total circuit current that will flow in this circuit if the inductors are connected in series?
   A) 45.24 amperes
   B) 2.0 amperes
   C) 5.31 milliamperes
   D) 5.31 amperes
10. A 240-V, 60-Hz circuit contains three inductors with inductances of 30, 40, and 50 mH. What is the total circuit current that will flow in this circuit if the inductors are connected in parallel?
   A) 18.8 amperes
   B) 50 milliamperes
   C) 49.9 amperes
   D) 5.31 amperes
11. Power in a purely inductive circuit is best described by which of the following statements?
   A) Reactive power in a purely inductive circuit is called VARs.
   B) Reactive power in a purely inductive circuit is called watts.
   C) Watts and VARs are interchangeable and can be added to get the total power.
   D) VARs are the amount of true power consumed by the circuit.
12. True or False: In a purely inductive circuit, Watts are created by the inductor and used as the measurement for power in these circuits.
13. The quality (Q) of an inductor is a ratio of which of the following properties of an inductor?
   A) Induction to resistance
   B) Inductive reactance to resistance
   C) Resistance to reactive power
   D) Inductance to reactive power
14. Inductor Q values are typically going to be lower when the conductor that forms the coil of the inductor meets which of the following statements?
   A) The coil uses a large wire with a low resistance.
   B) The inductor has no magnetic core material.
   C) The inductor is connected to a low frequency.
   D) The coil uses a small wire with a high resistance.

Practical Applications

1. You are working as an electrician installing fluorescent lights. You notice that the lights were made in Europe and that the ballasts are rated for operation on a 50-Hz system. Will these ballasts be harmed by overcurrent if they are connected to 60 Hz? If there is a problem with these lights, what will be the most likely cause of the trouble?

2. You have the task of ordering a replacement coil for one that has become defective. The information on the nameplate has been painted over and cannot be read. The machine that contains the inductor operates on 480 V at a frequency of 60 Hz. An identical machine is still in operation. A clamp-on ammeter indicates a current of 18 A, and a voltmeter indicates a voltage drop across the inductor of 324 V in the machine that is still in operation. After turning off the power and locking out the panel, you disconnect the inductor in the operating machine and measure a wire resistance of 1.2 Ω with an ohmmeter. Using the identical inductor in the operating machine as an example, what inductance value should you order, and what would be the minimum VAR rating of the inductor? Should you be concerned with the amount of wire resistance in the inductor when ordering? Explain your answers.

Practice Problems

Inductive Circuits

1. Fill in all the missing values. Refer to the following formulas:

$$X_L = 2\pi fL$$

$$L = \frac{X_L}{2\pi f}$$

$$f = \frac{X_L}{2\pi L}$$

Inductance (H)	Frequency (Hz)	Inductive Reactance (Ω)
1.2	60	
0.085		213.628
	1000	4712.389
0.65	600	
3.6		678.584
	25	411.459
0.5	60	
0.85		6408.849
	20	201.062
0.45	400	
4.8		2412.743
	1000	40.841

2. What frequency must be applied to a 33-mH inductor to produce an inductive reactance of 99.526 Ω?
3. An inductor is connected to a 120-volt, 60-Hz line and has a current flow of 4 amperes. An ohmmeter indicates that the inductor has a wire resistance of 12 Ω. What is the inductance of the inductor?
4. A 0.75-henry inductor has a wire resistance of 90 Ω. When connected to a 60-Hz power line, what is the total current-limiting effect of the inductor?
5. An inductor has a current flow of 3 amperes when connected to a 240-volt, 60-Hz power line. The inductor has a wire resistance of 15 Ω. What is the Q of the inductor?

Unit 17 Resistive-Inductive Series Circuits

Key Terms

Angle theta ($\angle\theta$)

Apparent power (VA)

Power factor (PF)

Total current (I_T)

Wattless power

Why You Need to Know

In previous units, you learned that current and voltage are in phase with each other in a purely resistive circuit and that current and voltage are 90° out of phase with each other in a purely inductive circuit. This unit

- describes what happens when resistive and inductive elements are combined in the same circuit. Although there are some applications for connecting resistors and inductors in series with each other, more often, you will encounter devices that only appear to have both elements connected in series with each other. An example of this is AC motors and transformers. Motors and transformers are inductive because they contain windings or coils. At no load, these devices appear to be only inductive, and current is limited by inductive reactance. As load is added, electrical energy is converted to some other form, and they appear to become more resistive.
- explains what power factor is and how to correct for its impact. Without a knowledge of what happens when inductance and resistance are connected in the same circuit, you will not be able to understand such concepts as power factor and power factor correction.

Outline

Objectives

After studying this unit, you should be able to

- explain the relationship between resistance and inductance in an AC series circuit.
- determine the power factor in an R-L series circuit.
- solve for missing total circuit values in an R-L series circuit.
- find the phase angle for current and voltage in an R-L series circuit.
- determine unknown values at the resistance in R-L series circuits.
- determine unknown values at the inductor in R-L series circuits.

Preview

This unit covers the relationship of resistance and inductance used in the same circuit when the resistors and inductors are connected in series. Concepts such as total circuit impedance, power factor, and vector addition will be important factors. Although it is true that some circuits are basically purely resistive or purely inductive, many circuits contain a combination of both resistive and inductive elements.

17–1 Solving R-L Series Circuits

Remember, when a purely resistive load is connected to an AC circuit, the voltage and current are in phase with each other. When a purely inductive load is connected to an AC circuit, the voltage and current are 90° out of phase with each other (*Figure 17–1*). When a circuit containing both resistance, R, and inductance, L, is connected to an AC circuit, the voltage and current will be out of phase with each other by some amount between 0° and 90°. ***The exact amount of phase angle difference is determined by the ratio of resistance to inductive reactance in the circuit.*** In the following example, a series circuit containing 30 ohms of resistance (R) and 40 ohms of inductive reactance (X_L) is connected to a 240-volt, 60-hertz line (*Figure 17–2*). It is assumed that the inductor has negligible internal resistance; Q is above 10. The following unknown values will be calculated:

Z—total circuit impedance

I—current flow

E_R—voltage drop across the resistor

P—watts (true power)

L—inductance of the inductor

E_L—voltage drop across the inductor

VARs—reactive power

VA—apparent power

PF—power factor

$\angle\theta$—the angle at which the voltage and current are out of phase with each other

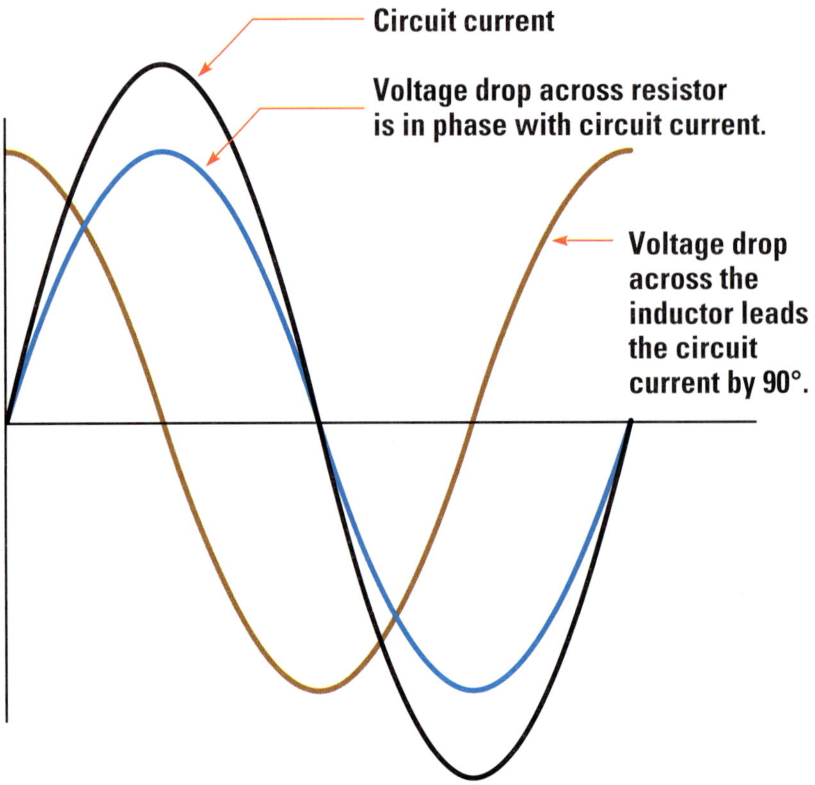

Figure 17–1 Relationship of resistive and inductive current with voltage.

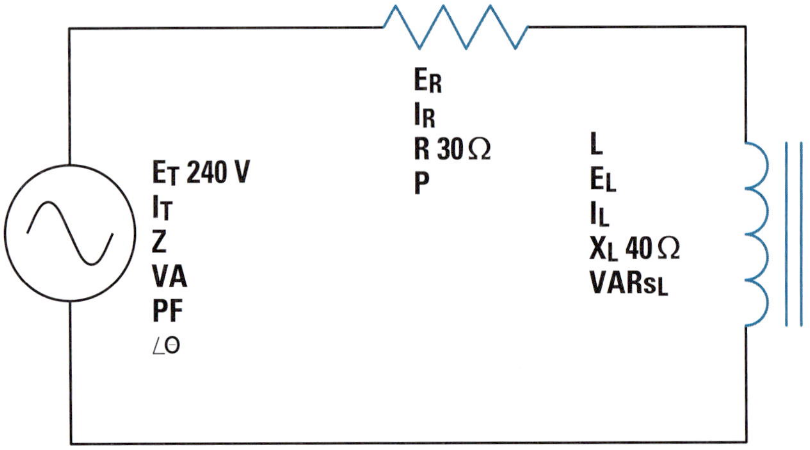

Figure 17–2 R-L series circuit supplied by an AC source.

Impedance

In Unit 16, impedance was defined as a measure of the part of the circuit that impedes, or hinders, the flow of current. It is measured in ohms and symbolized by the letter *Z*. In this circuit, total circuit impedance is a combination of resistance and inductive reactance.

In a series circuit, the total resistance is equal to the sum of the individual resistors. In this instance, however, the total impedance (Z) is the sum of the resistance and the inductive reactance. It would first appear that the sum of these two quantities should be 70 ohms ($30\ \Omega + 40\ \Omega = 70\ \Omega$). In practice, however, the resistive part of the circuit and the reactive part of the circuit are out of phase with each other by 90°. To find the sum of these two quantities, vector addition must be used. Because these two quantities are 90° out of phase with each other, the resistive and inductive reactance form the two legs of a right triangle, and the impedance is the hypotenuse (*Figure 17–3*). They can be added vectorially or by way of Pythagorean theorem.

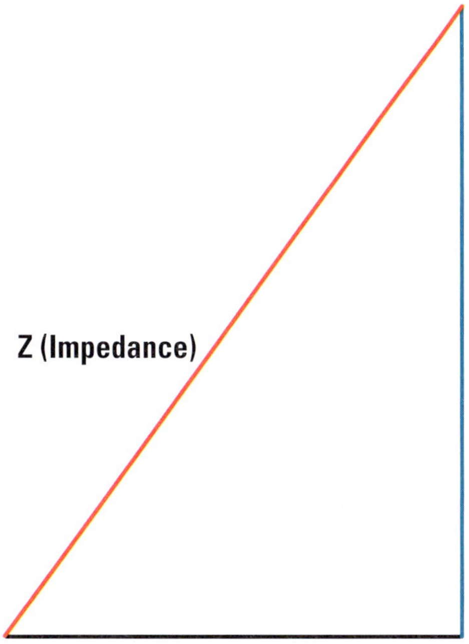

Figure 17–3 The combination of resistance, inductive reactance, and total circuit impedance for a right triangle.

The total circuit impedance (Z) can be calculated using the Pythagorean theorem formula:

$$Z = \sqrt{R^2 + X_L^2}$$

$$Z = \sqrt{(30\ \Omega)^2 + (40\ \Omega)^2}$$

$$Z = \sqrt{900\ \Omega^2 + 1600\ \Omega^2}$$

$$Z = \sqrt{2500\ \Omega^2}$$

$$Z = 50\ \Omega$$

To find the impedance of the circuit in *Figure 17–3* using vector addition, connect the starting point of one vector to the ending point of the other. Because resistance and inductive reactance are 90° out of phase with each other, the two vectors must be placed at a 90° angle. If the resistive vector has a magnitude of 30 ohms and the inductive vector has a magnitude of 40 ohms, the resultant vector (impedance) will have a magnitude of 50 ohms (*Figure 17–4*). Notice that the result is the same as that found using the right triangle.

The parallelogram method of vector addition can also be used to find the total impedance. The resistance forms a vector with a magnitude of 30 ohms and a direction of 0°. The inductive reactance forms a vector with a magnitude of 40 ohms and a direction of 90°. When lines are extended to form a parallelogram and a resultant is drawn, the resultant will have a magnitude of 50 ohms, which is the total impedance of the circuit (*Figure 17–5*). Some students of electricity find the right triangle concept easier to understand, and others find vectors more helpful. For this reason, this text uses both methods to help explain the relationship of voltage, current, and power in AC circuits.

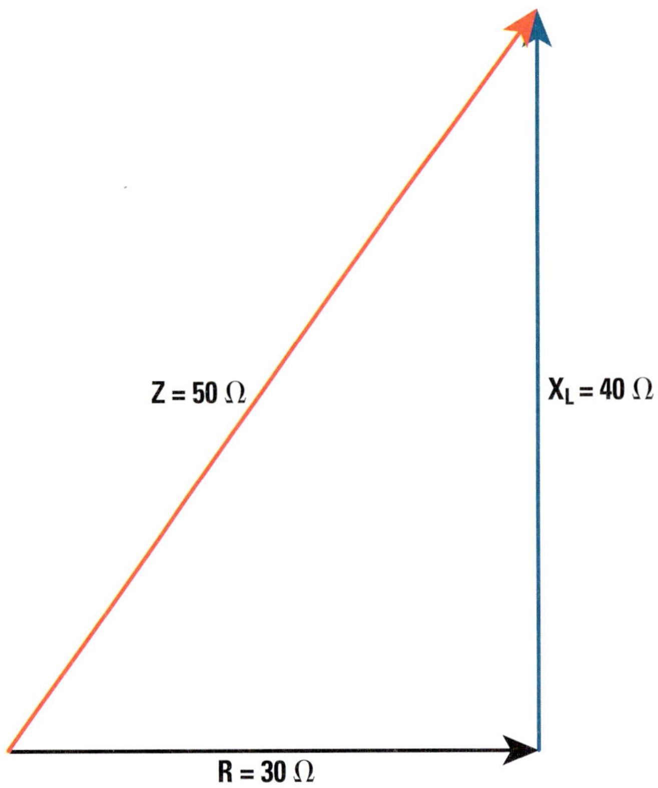

Figure 17–4 Determining impedance using tip-to-tail vector addition.

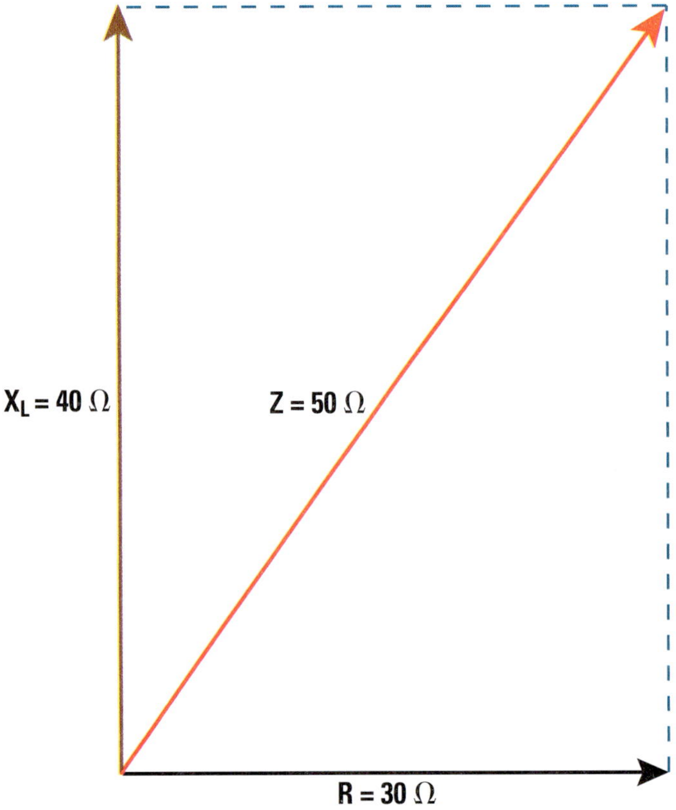

Figure 17–5 Finding the total circuit impedance using the parallelogram method of vector addition.

Total Current

One of the primary laws for all series circuits is that the current will be the same in any part of the circuit. Because the impedance is the total current-limiting component of the circuit, it can be used to replace R in an Ohm's law formula. The **total current (I_T)** flow through the circuit can be calculated by dividing the total applied voltage by the total current-limiting factor. Total current can be found by using the formula

$$I_T = \frac{E_T}{Z}$$

The total current for the circuit in *Figure 17–2,* then, is:

$$I_T = \frac{240\ V}{50\ \Omega}$$

$$I_T = 4.8\ A$$

In a series circuit, the current is the same at any point in the circuit. Therefore, 4.8 amperes of current flow through both the resistor and the inductor. These values can be added to the circuit, as shown in *Figure 17–6*.

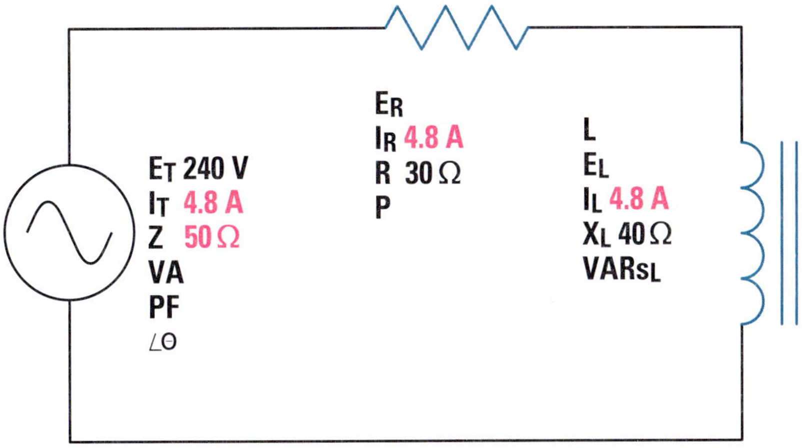

Figure 17–6 Total voltage divided by total circuit impedance equals total circuit current.

17–2 Values at the Resistor

Now that the amount of current flowing through the resistor is known, the voltage drop across the resistor (E_R) can be calculated using the formula:

$$E_R = I_R \times R$$

The voltage drop across the resistor in our circuit is:

$$E_R = 4.8\ A \times 30\ \Omega$$

$$E_R = 144\ V$$

Notice that the amount of voltage dropped across the resistor was found using quantities that pertained only to the resistive part of the circuit. Just like in a series circuit with several resistors, you wouldn't try to determine the voltage drop across resistor R_2 by using the resistance of R_4. The amount of voltage dropped across the resistor could not be found using a formula such as:

$$E_R = I_R \times X_L$$

or

$$E_R = I_R \times Z$$

However, total impedance and total inductive reactance can be used with vector addition or the Pythagorean theorem to find total resistance in the circuit. For example, the resistive quantity that is measured in ohms is resistance (R). If the impedance and inductive reactance of a circuit were known, they could be used with the following formula (Pythagorean theorem) to find the circuit resistance:

$$R = \sqrt{Z^2 - X_L^2}$$

Note: Refer to the Resistive-Inductive Series Circuits section of the AC formulas listed in Appendix B.

Watts

True power (P) for the circuit can be calculated by using any of the watts formulas with pure resistive parts of the circuit. Watts (W) can be calculated, for example, by multiplying the voltage dropped across the resistor (E_R) by the current flow through the resistor (I_R), by squaring the voltage dropped across the resistor and dividing by the resistance of the resistor, or by squaring the current flow through the resistor and multiplying by the resistance of the resistor. Watts cannot be calculated by using any values in the circuit that are not attributed to the resistive load, such as by multiplying the total voltage (E_T) by the total current flow. Recall that true power, or watts, can be produced only during periods of time when the voltage and current are both positive or both negative.

In an R-L series circuit, the current is the same through both the resistor and the inductor. The voltage dropped across the resistor, however, is in phase with the current, and the voltage dropped across the inductor is 90° out of phase with the current (*Figure 17–7*). Because true power, or watts, can be produced only when the current and voltage are both positive or both negative, only resistive parts of the circuit can produce watts.

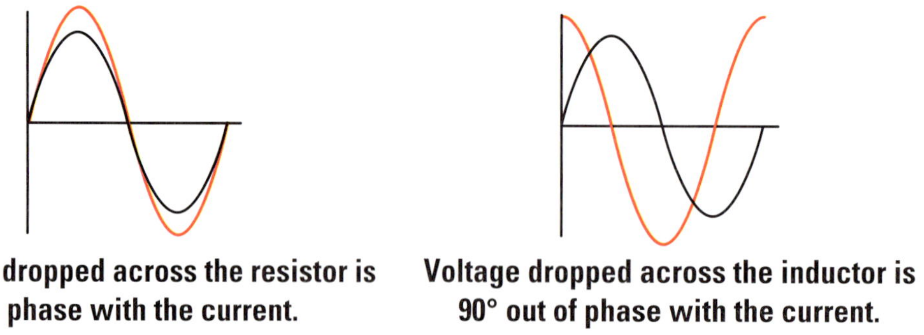

Figure 17–7 Relationship of current and voltage in an R-L series circuit.

The formula used in this example is:

$$P = E_R \times I_R$$

$$P = 144\ \text{V} \times 4.8\ \text{A}$$

$$P = 691.2\ \text{W}$$

17–3 Values at the Inductor

The amount of inductance can be calculated using the formula:

$$L = \frac{X_L}{2\pi f}$$

$$L = \frac{40\ \Omega}{377}$$

$$L = 0.106\ \text{H}$$

Voltage Drop Across the Inductor

The voltage drop across the inductor (E_L) can be calculated using the formula:

$$E_L = I_L \times X_L$$

$$E_L = 4.8\ \text{A} \times 40\ \text{W}$$

$$E_L = 192\ \text{V}$$

Notice that only inductive quantities were used to find the voltage drop across the inductor.

Calculating the Reactive Power

Volt-amperes-reactive (VARs) is the amount of reactive power in the circuit. VARs should not be confused with watts, which is true power. VARs represents the product of the volts and amperes that are 90° out of phase with each other, such as the voltage dropped across the inductor and the current flowing through the inductor. Recall that true power can be produced only during periods of time when the voltage and current are both positive or both negative (*Figure 17–8*). During these periods, the power is being stored in the form of a magnetic field. During the periods that voltage and current have opposite signs, the power is returned to the circuit. For this reason, VARs is often referred to as **wattless power.** It can be calculated in a manner similar to that for watts except that reactive values of voltage and current are used instead of resistive values. In this example, the following formula used is:

$$\text{VARs} = I_L^2 \times X_L$$

$$\text{VARs} = (4.8\ \text{A})^2 \times 40\ \Omega$$

$$\text{VARs} = 921.6$$

17–4 Total Circuit Values

Although the total applied voltage in this circuit is known (240 volts), the total voltage is also equal to the sum of the voltage drops, just as it is in any other series circuit. Because the voltage dropped across the resistor is in phase with the current and the voltage dropped across the inductor is 90° out of phase with the current, tip-to-tail vector addition can be used to verify the total is indeed 240 volts. The total voltage will be the hypotenuse of a right triangle, and the resistive and inductive voltage drops will form the legs of the right triangle (*Figure 17–9*). This relationship of voltage drops can also be represented using the parallelogram method of vector addition, as shown in *Figure 17–10*. The

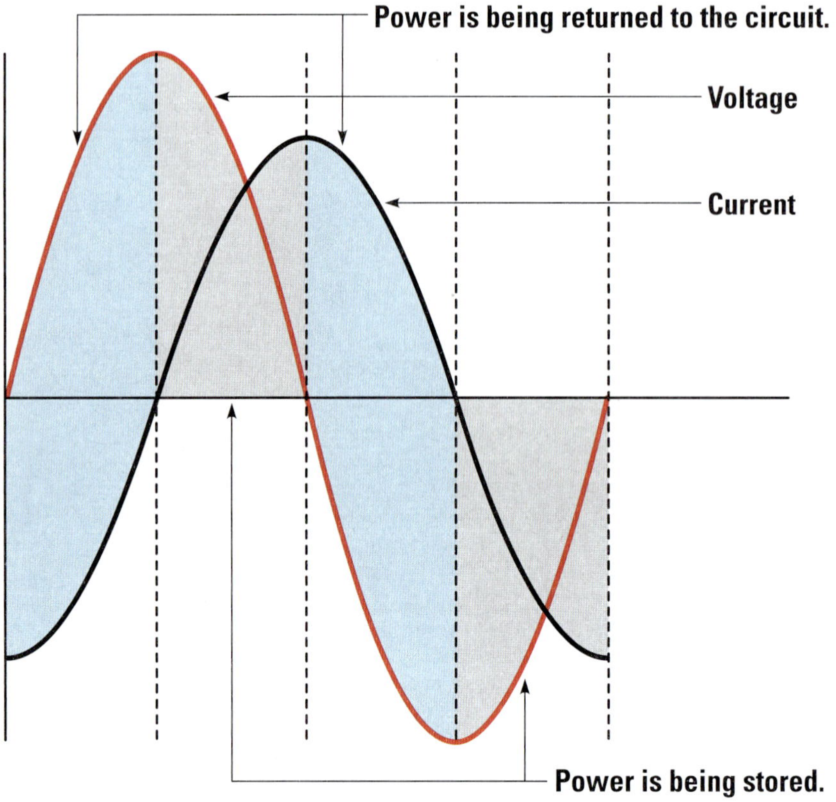

Figure 17–8 Power is stored and then returned to the circuit in equal and offsetting quadrants.

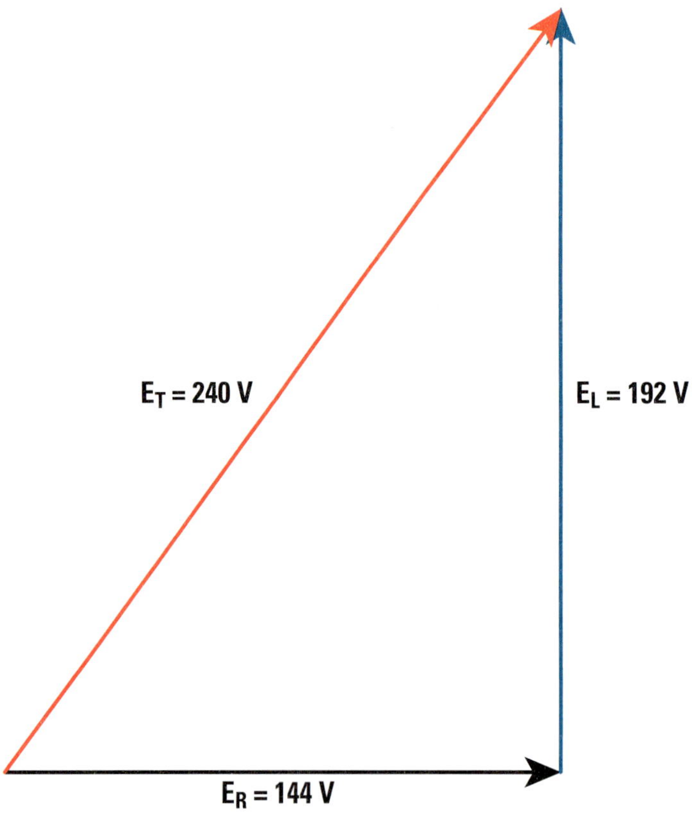

Figure 17–9 Tip-to-tail vector addition of E_L and E_R voltage drops in an R-L series circuit to find E_T.

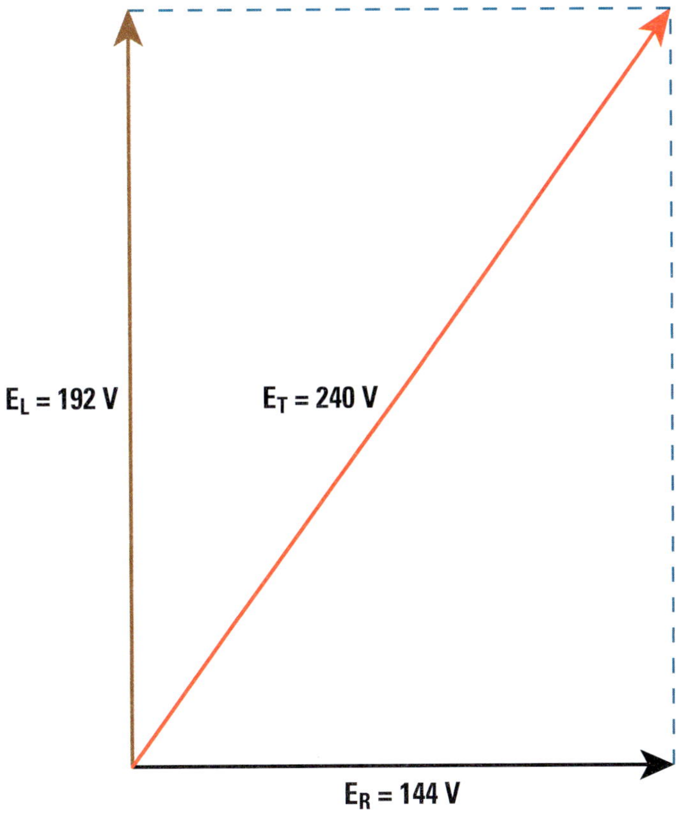

Figure 17–10 Using the parallelogram method of vector addition to solve for total voltage in an R-L series circuit.

following Pythagorean formulas can also be used to solve for a missing voltage value if the other two voltage values are known:

$$E_T = \sqrt{E_R^2 + E_L^2}$$

$$E_R = \sqrt{E_T^2 - E_L^2}$$

$$E_L = \sqrt{E_T^2 - E_R^2}$$

Note: Refer to the Resistive-Inductive Series Circuits section of the AC formulas listed in Appendix B.

Calculating the Apparent Power

Volt-amperes (VA) is the apparent power of the circuit. It can be calculated in a manner similar to watts or VARs, except that only total values of the circuit are used. It is called **apparent power (VA)** because it is the value that would be found if a voltmeter and ammeter were used to measure the circuit voltage and current and then these measured values were multiplied together (*Figure 17–11*). In this example, the following formula is used:

$$VA = E_T \times I_T$$

$$VA = 240\text{ V} \times 4.8\text{ A}$$

$$VA = 1152$$

The apparent power can also be found using vector addition in a manner similar to that for impedance or total voltage. Because true power, or watts, is a pure resistive component and VARs is a pure reactive component, they form the legs of a right triangle. The apparent power is the hypotenuse of this triangle (*Figure 17–12*). This relationship of the three power components can also be plotted using the

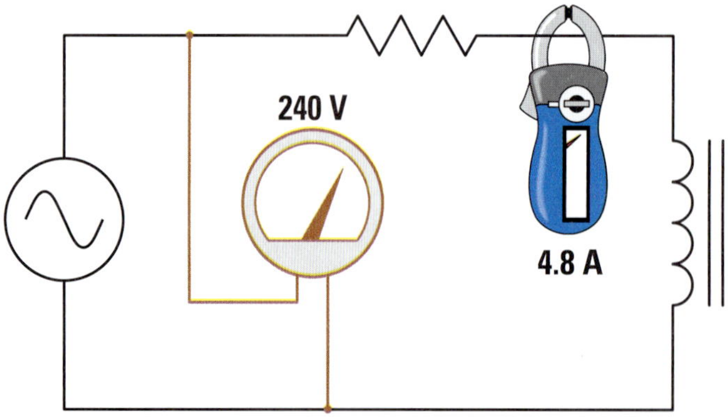

Figure 17–11 Apparent power is the product of measured voltage and current values (240 V × 4.8 A = 1152 VA).

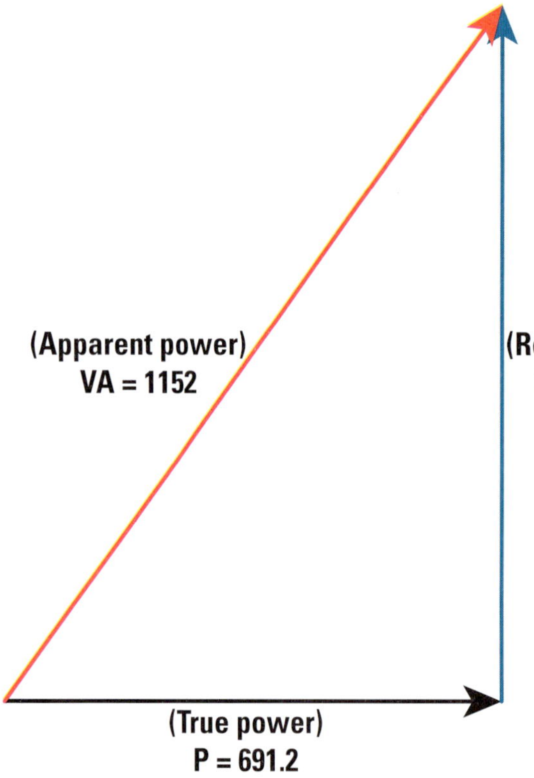

Figure 17–12 Tip-to-tail vector addition of true power (watts) and reactive power (VARs) to find apparent power (volt-amperes) in an R-L series circuit.

parallelogram method (*Figure 17–13*). The following Pythagorean formulas can be used to calculate the values of apparent power, true power, and reactive power when the other two values are known:

$$VA = \sqrt{P^2 + VARs^2}$$

$$P = \sqrt{VA^2 - VARs^2}$$

$$VARs = \sqrt{VA^2 - P^2}$$

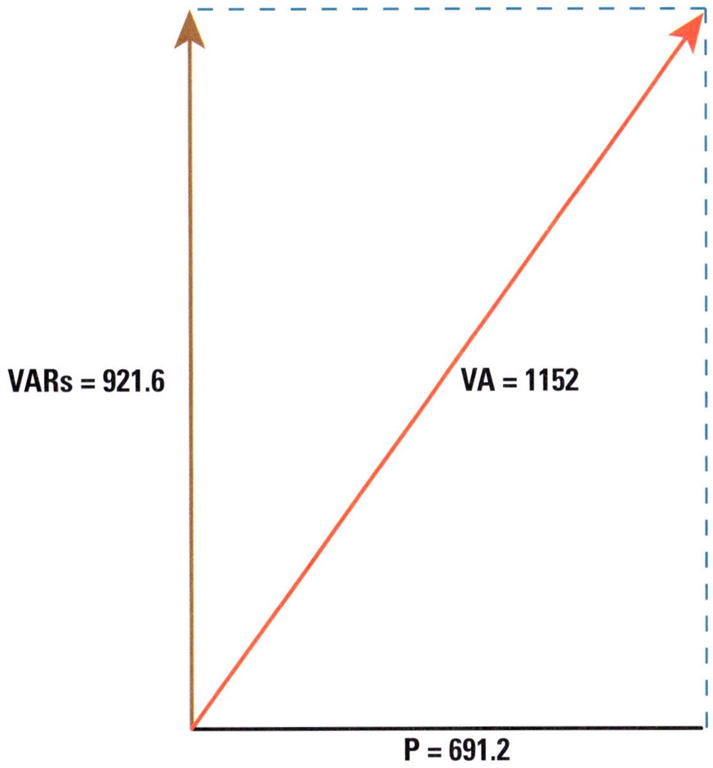

Figure 17–13 Using the parallelogram method to plot the relationship of volt-amperes, watts, and VARs.

17–5 Power Factor

Power factor (PF) is a ratio of the true power to the apparent power. It can be calculated by dividing any resistive value by its like total value. For example, power factor can be calculated by dividing the voltage drop across the resistor by the total circuit voltage, by dividing resistance by impedance, or by dividing watts by volt-amperes:

$$PF = \frac{E_R}{E_T}$$

$$PF = \frac{R}{Z}$$

$$PF = \frac{P}{VA}$$

Power factor is generally expressed as a percentage. The decimal fraction calculated from the division will, therefore, be changed to a percent by multiplying it by 100%. In this circuit, the formula used is

$$PF = \frac{P}{VA}$$

$$PF = \frac{691.2\ W}{1152\ VA}$$

$$PF = 0.6 \times 100\%, \text{ or } 60\%$$

Note that in a series circuit, the power factor cannot be calculated using current because current is the same in all parts of the circuit.

Power factor is very important in an industrial application. Most power companies charge a substantial surcharge when the power factor drops below a certain percentage. The reason for this is that electric power is sold on the basis of true power, or watt-hours, consumed. The power company, however, must supply the apparent power. Assume that an industrial plant has a power factor of 60% and is consuming 5 megawatts of power. At a power factor of 60%, the power company must supply 8.333 megavolt-amperes (5 MW/0.6 = 8.333 MVA). If the power factor were to be corrected to 95%, the power company would have to supply only 5.263 megavolt-amperes to furnish the same amount of power to the plant.

Angle Theta

The angular displacement by which the voltage and current are out of phase with each other is referred to as **angle theta ($\angle\theta$).** Because the power factor is the ratio of true power to apparent power, the phase angle of voltage and current is formed between the resistive leg of the right triangle and the hypotenuse (*Figure 17–14*). The resistive leg of the triangle is adjacent to the angle, and the total leg is the hypotenuse. The trigonometric function in SOH—CAH—TOA that corresponds to the adjacent side and the hypotenuse is the cosine. The cosine of angle theta is watts divided by volt-amperes. Watts divided by volt-amperes is also the power factor. Therefore, the cosine of angle theta ($\angle\theta$) is the power factor (PF):

$$\cos \angle\theta = \text{PF}$$

$$\cos \angle\theta = 0.6$$

$$\angle\theta = 53.13^\circ$$

Angle theta can also be determined using the impedance triangle or voltage triangle for a particular circuit. Each of these triangles is proportional to each other and will provide the same angular result for $\angle\theta$.

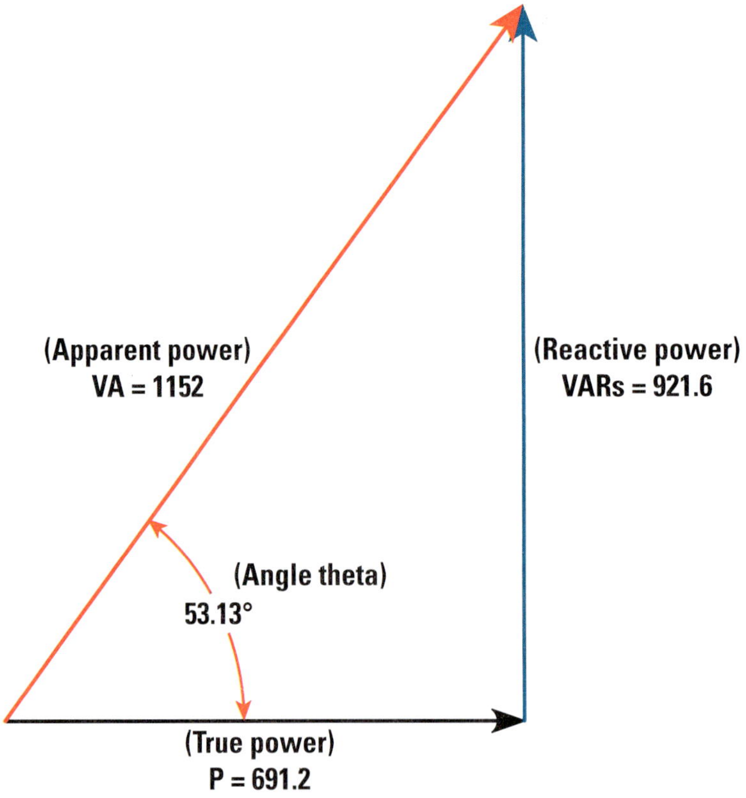

Figure 17–14 The angle theta is the "out of phase" relationship of true power to apparent power.

The vectors formed using the parallelogram method of vector addition can also be used to find angle theta as shown in *Figure 17–15*. Notice that the hypotenuse (volt-amperes) and the adjacent resistive leg (watts) are again used to determine angle theta.

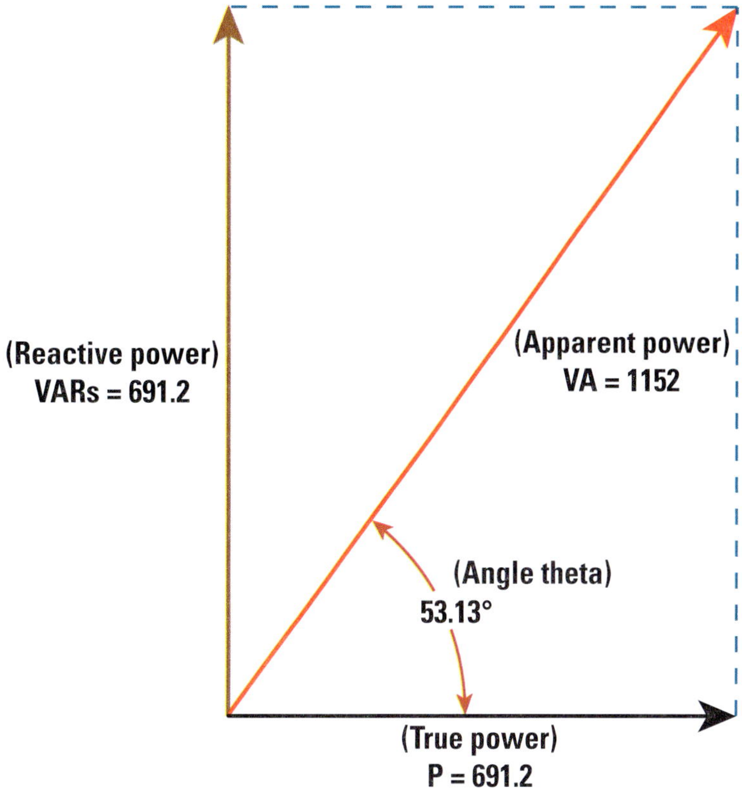

Figure 17–15 The angle theta can be found using vectors provided by the parallelogram method.

Because this circuit contains both resistance and inductance, the current is lagging the voltage by 53.13° (*Figure 17–16*). Angle theta can also be determined by using any of the other trigonometric functions:

$$\sin \angle\theta = \frac{\text{VARs}}{\text{VA}}$$

$$\tan \angle\theta = \frac{\text{VARs}}{\text{P}}$$

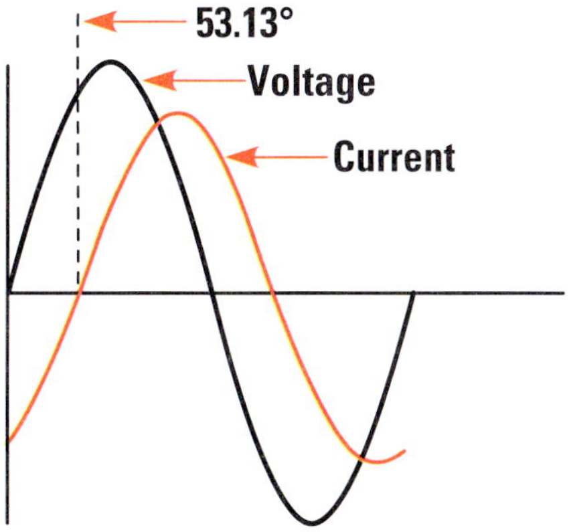

Figure 17–16 Current and voltage are 53.13° out of phase with each other.

Now that all the unknown values have been calculated, they can be filled in as shown in *Figure 17–17*.

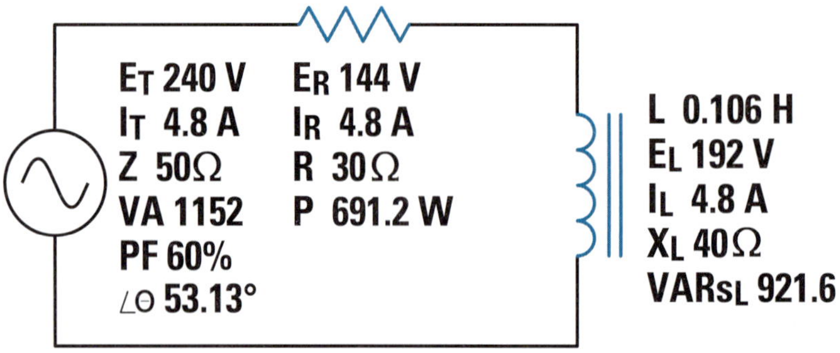

Figure 17–17 Solving for all unknown circuit values.

Example 17–1

Two resistors and two inductors are connected in series (*Figure 17–18*). The circuit is connected to a 130-V, 60-Hz line. The first resistor has a power dissipation of 56 W, and the second resistor has a power dissipation of 44 W. One inductor has a reactive power of 152 VARs, and the second has a reactive power of 88 VARs. Find the unknown values in this circuit.

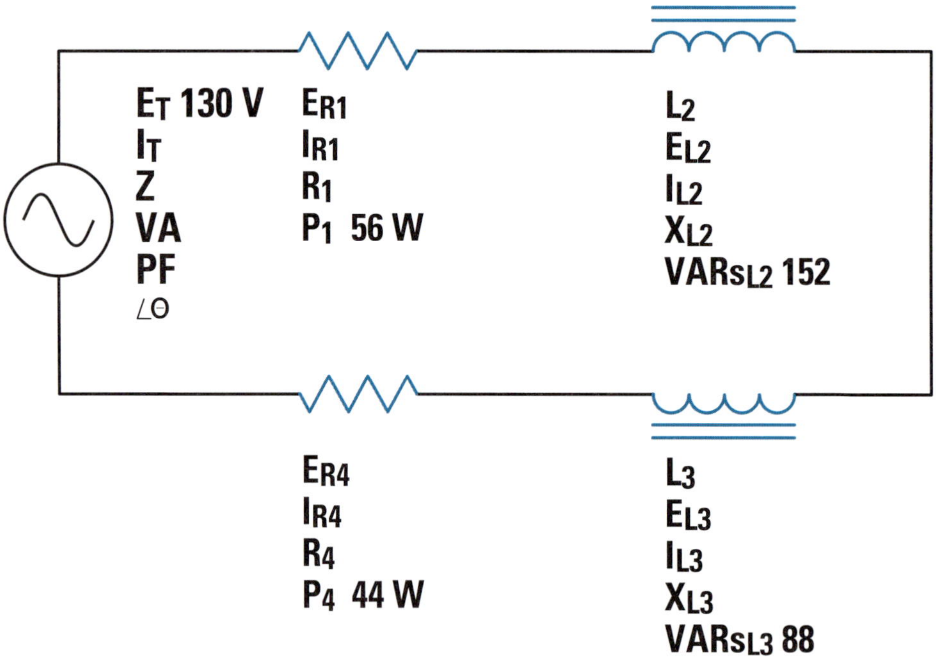

Figure 17–18 An R-L series circuit containing two resistors and two inductors. Find all of the unknown values.

Example 17–1 (continued)

Solution

The first step is to find the total amount of true power and the total amount of reactive power. The total amount of true power can be calculated by adding the values of the resistors together. A vector diagram of these two values would reveal that they both have a direction of 0°.

$$P_T = P_1 + P_4$$

$$P_T = 56\text{ W} + 44\text{ W}$$

$$P_T = 100\text{ W}$$

The total reactive power in the circuit can be found in the same manner. Like watts, the VARs are both inductive and are therefore in the same direction. The total reactive power will be the sum of the two reactive power ratings:

$$VARs_T = VARs_{L2} + VARs_{L3}$$

$$VARs_T = 152 + 88$$

$$VARs_T = 240$$

Apparent Power

Now that the total amount of true power and the total amount of reactive power are known, the apparent power (VA) can be calculated using Pythagorean addition:

$$VA = \sqrt{P_T^2 + VARs_T^2}$$

$$VA = \sqrt{(100\text{ W})^2 + (240\text{ VARs})^2}$$

$$VA = 260$$

The parallelogram method of vector addition is shown in *Figure 17–19* for this calculation.

Total Circuit Current

Now that the apparent power is known, the total circuit current can be found using the applied voltage and Ohm's law:

$$I_T = \frac{VA}{E_T}$$

$$I_T = \frac{260\text{ VA}}{130\text{ V}}$$

$$I_T = 2\text{ A}$$

Because this is a series circuit, the current will be the same at all points in the circuit. The known values added to the circuit are shown in *Figure 17–20*.

Other Circuit Values

Now that the total circuit current has been found, other values can be calculated using Ohm's law.

Example 17–1 (continued)

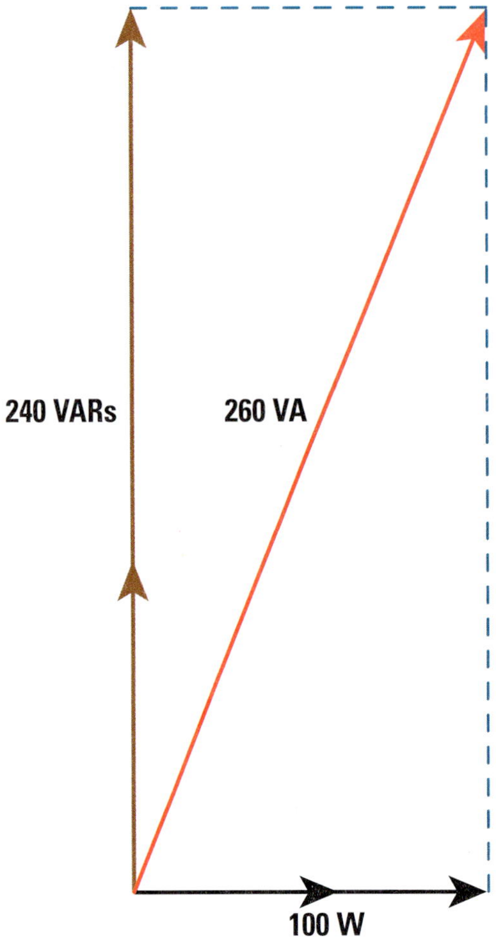

Figure 17–19 Adding the true and reactive power vectors in the circuit to solve for the unknown apparent power vector.

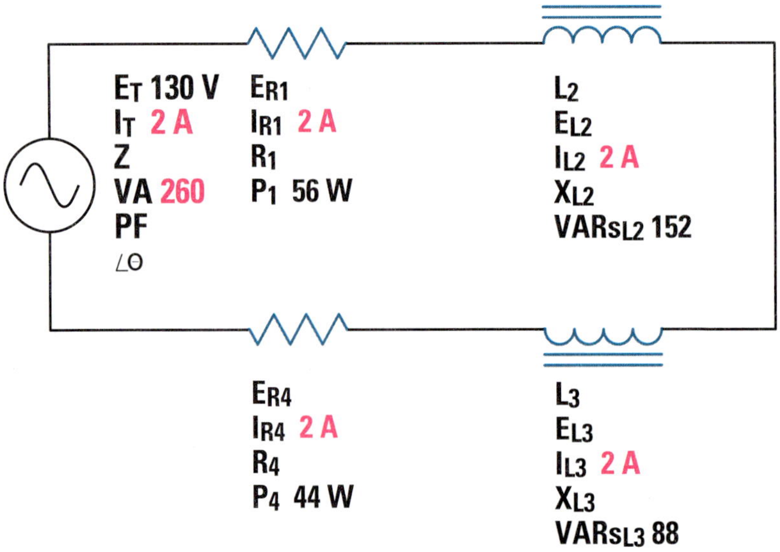

Figure 17–20 Adding circuit current values.

Example 17–1 (continued)

Impedance

$$Z = \frac{E_T}{I_T}$$

$$Z = \frac{130\ V}{2\ A}$$

$$Z = 65\ \Omega$$

Power Factor

$$PF = \frac{W_T}{VA}$$

$$PF = \frac{100\ W}{260\ VA}$$

$$PF = 38.462\%$$

Angle Theta

The cosine of angle theta is the power factor. A vector diagram showing this relationship is shown in *Figure 17–21*.

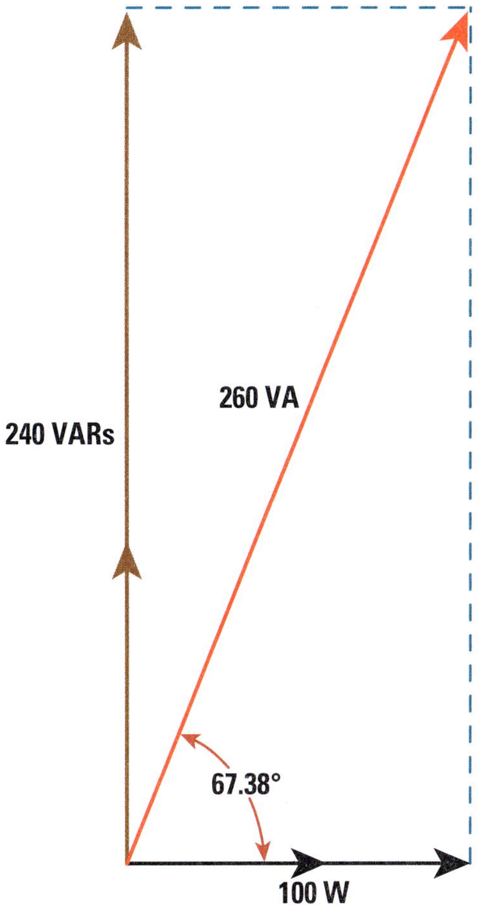

Figure 17–21 Angle theta for the circuit.

Example 17–1 (continued)

$$\cos \angle\theta = PF$$

$$\cos \angle\theta = 0.38462$$

$$\angle\theta = 67.38°$$

The relationship of voltage and current for this circuit is shown in *Figure 17–22*.

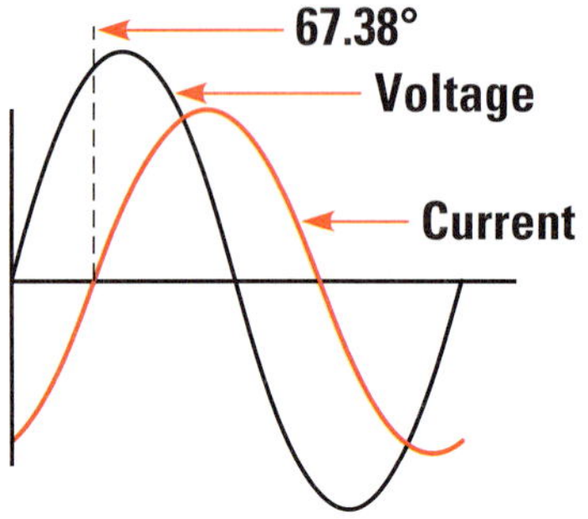

Figure 17–22 Voltage and current are 67.38° out of phase with each other.

E_{R1}

$$E_{R1} = \frac{P_1}{I_{R1}}$$

$$E_{R1} = \frac{56\ W}{2\ A}$$

$$E_{R1} = 28\ V$$

R_1

$$R_1 = \frac{E_{R1}}{I_{R1}}$$

$$R_1 = \frac{28\ V}{2\ A}$$

$$R_1 = 14\ \Omega$$

E_{L2}

$$E_{L2} = \frac{VARs_{L2}}{I_{L2}}$$

$$E_{L2} = \frac{152\ VARs}{2\ A}$$

$$E_{L2} = 76\ V$$

Example 17–1 (continued)

X_{L2}

$$X_{L2} = \frac{E_{L2}}{I_{L2}}$$

$$X_{L2} = \frac{76\ V}{2\ A}$$

$$X_{L2} = 38\ \Omega$$

L_2

$$L_2 = \frac{X_{L2}}{2\pi f}$$

$$L_2 = \frac{38\ \Omega}{377}$$

$$L_2 = 0.101\ H$$

E_{L3}

$$E_{L3} = \frac{VARs_{L3}}{I_{L3}}$$

$$E_{L3} = \frac{88\ VARs}{2\ A}$$

$$E_{L3} = 44\ V$$

X_{L3}

$$X_{L3} = \frac{E_{L3}}{I_{L3}}$$

$$X_{L3} = \frac{44\ V}{2\ A}$$

$$X_{L3} = 22\ \Omega$$

L_3

$$L_3 = \frac{X_{L3}}{2\pi f}$$

$$L_3 = \frac{22\ \Omega}{377}$$

$$L_3 = 0.0584$$

E_{R4}

$$E_{R4} = \frac{P_4}{I_{R4}}$$

$$E_{R4} = \frac{44\ W}{2\ A}$$

$$E_{R4} = 22\ V$$

Example 17–1 (continued)

R_4

$$R_4 = \frac{E_{R4}}{I_{R4}}$$

$$R_4 = \frac{22\text{ V}}{2\text{ A}}$$

$$R_4 = 11\ \Omega$$

The complete circuit with all values is shown in *Figure 17–23*.

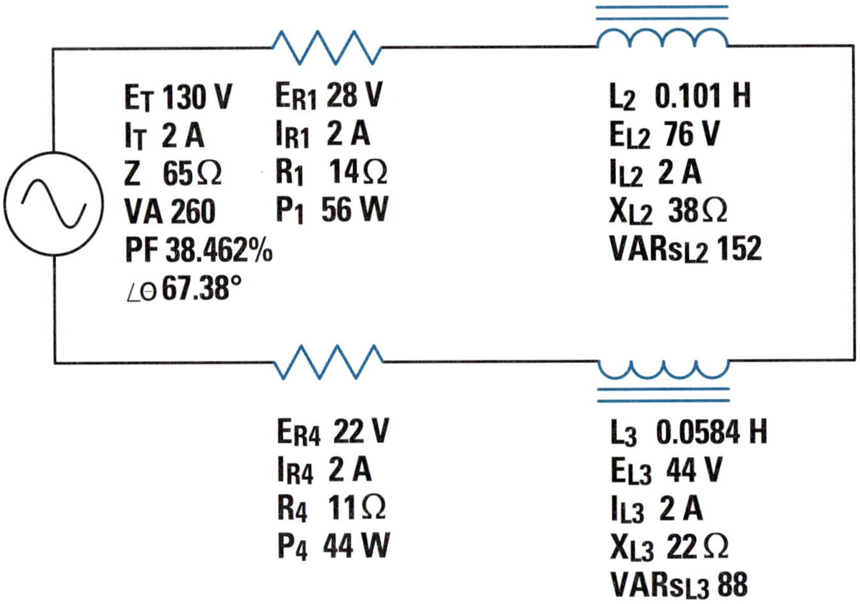

Figure 17–23 All values for the circuit have been determined.

Many people working in the electrical field encounter R-L series circuits almost every day in ways they do not realize. A good example of this is an electric motor. As far as the circuit is concerned, a motor is an R-L series circuit. The motor winding is basically an inductor. It is a coil of wire surrounded by an iron core. The wire does have resistance, however, and because there is only one path for current flow through the circuit, it is a series circuit (*Figure 17–24*). When a motor is operated at no load, it will have a large inductive component compared to resistance. As load is added to the motor, electrical energy is converted into kinetic energy, causing an increase in true power or watts. This has the effect of increasing the amount of circuit resistance. When a motor is operated at full load, the circuit has become an R-L series circuit with more resistance than inductance.

Another example of an R-L series circuit is a luminaire (lighting fixture) that depends on the ionization of gas such as fluorescent, mercury vapor, and sodium vapor. When a gas is ionized, it basically becomes a short circuit. A ballast is used to prevent excessive current flow when the gas ionizes. The ballast is connected in series with the lamp (*Figure 17–25*). When the gas in the lamp ionizes, the inductive reactance of the ballast limits current. The lamp appears as a resistive load after ionization takes place. Without an understanding of the relationship of inductance and resistance in a series circuit, it would be impossible to understand how this everyday electric device actually works.

Example 17–1 (continued)

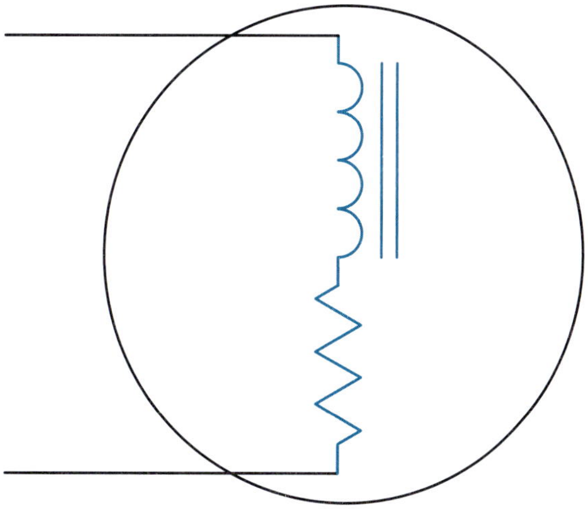

Figure 17–24 A motor winding is an R-L series circuit because wire resistance is in series with coil inductive reactance.

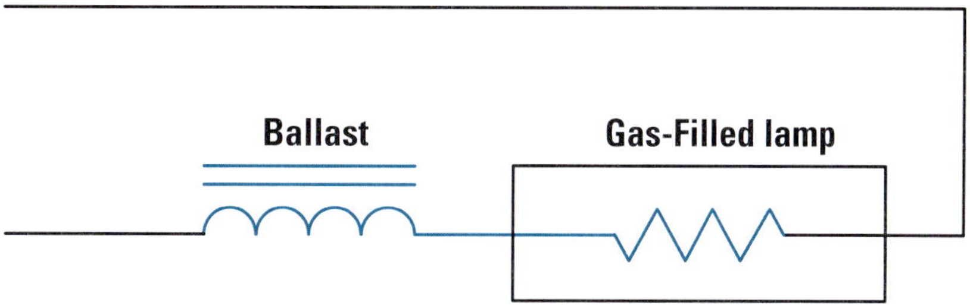

Figure 17–25 A ballast uses inductive reactance to prevent excessive current flow when the gas ionizes and current flows.

Summary

- In a purely resistive circuit, the voltage and current are in phase with each other.
- In a purely inductive circuit, the voltage and current are 90° out of phase with each other.
- In an R-L series circuit, the voltage and current will be out of phase with each other by some value between 0° and 90°.
- The amount the voltage and current are out of phase with each other is determined by the ratio of resistance to inductive reactance in the circuit.
- Total circuit values include total voltage, E_T; total current, I_T; volt-amperes, VA; and impedance, Z.
- Resistive values include voltage drop across the resistor, E_R; current flow through the resistor, I_R; resistance, R; and watts, P.
- Inductive values include inductance of the inductor, L; voltage drop across the inductor, E_L; current through the inductor, I_L; inductive reactance, X_L; and inductive VARs, $VARs_L$.
- Angle theta measures the phase angle difference between the applied voltage and total circuit current.
- The cosine of angle theta is equal to the power factor.
- Power factor is a ratio of true power to apparent power.

Review Questions

Use the formulas provided in Appendix B under Resistive-Inductive Series Circuits to solve the following problems.

1. True or False: In an R-L series circuit, the phase angle difference is determined by the ratio of resistance to current in the circuit.
2. Total circuit impedance in an R-L series circuit is the combination of resistance and inductive reactance. Which of the following best describes this relationship?
   A) Resistance and inductive reactance can simply be added together to get impedance.
   B) Resistance and inductive reactance must be added vectorially to determine impedance.
   C) Resistance and inductance can be added vectorially to determine total impedance.
   D) Resistance can simply be added to inductance to determine impedance.
3. Power factor is the ratio between which two factors in an R-L series circuit?
   A) Resistance and current
   B) Current and voltage
   C) Inductive reactance and voltage
   D) Watts and apparent power
4. An R-L series circuit has a total circuit current of 2 amperes when 120 volts are applied. If 90 volts are measured across the resistor in the circuit, what is the power factor?
   A) 75%
   B) 80%
   C) 85%
   D) 90%
5. In an R-L series circuit supplied by a 60 V source at 50 Hz, the total circuit current is measured at 4 amperes. If the value of the resistor is fixed at 10 Ω, what is the inductance value of the inductor?
   A) 5 Ω
   B) 31.8 mH
   C) 15.9 mH
   D) 10 Ω
6. If a series R-L circuit is connected to a voltage source at 120V and the apparent power in the circuit is 2400 VA, how much resistance is in the circuit if there are 1600 VARs?
   A) 7.21 Ω
   B) 4.00 Ω
   C) 4.47 Ω
   D) 6.00 Ω
7. True or False: The cosine of angle theta is the power factor of the circuit.

8. What is the angular difference or "angle theta" in an R-L series circuit that has an apparent power of 300 VA and true power of 150 W?

   A) 75°

   B) 60°

   C) 45°

   D) 30°

9. If an R-L series circuit contains a resistor of 60 Ω and an inductive reactance of 80 Ω when connected to an applied voltage of 100 V at 50 Hz, how much true power is used in the circuit?

   A) 600 watts

   B) 6000 watts

   C) 6 watts

   D) 60 watts

10. A series R-L circuit has a power factor of 80%. If the total circuit current is 2 amperes and the reactive power in the circuit is 300 VARs, how much voltage is dropped across the inductor?

    A) 200 V

    B) 250 V

    C) 150 V

    D) 400 V

11. True or False: The parallelogram method of vector addition can be used to find apparent power when Watts and VARs are known.

12. A series R-L circuit contains a resistor of 25 Ω and has a measured circuit current of 1.5 amperes. What is the apparent power and applied voltage if the reactive power in the circuit 139.05 VARs?

    A) 200 VA and 100 volts

    B) 150 VA and 100 volts

    C) 37.5 VA and 25 volts

    D) 75 VA and 50 volts

Practical Applications

1. An AC electric motor is connected to a 240-V, 60-Hz source. A clamp-on ammeter with a peak hold function reveals that the motor has an inrush current of 34 A when the motor is first started. Your job is to reduce the inrush current to a value of 20 A by connecting a resistor in series with the motor. The resistor will be shunted out of the circuit after the motor is started. Using an ohmmeter, you find that the motor has a wire resistance of 3 Ω. How much resistance should be connected in series with the motor to reduce the starting current to 20 A?

---

2. You are an electrician working in an industrial plant. Your task is to connect an inductor to a 480-V, 60-Hz line. To determine the proper conductor and fuse size for this installation, you need to know the amount of current the inductor will draw from the line. The nameplate on the inductor indicates that it has an inductance of 0.1 H. An ohmmeter reveals that it has a wire resistance of 10 Ω. How much current should this inductor draw when connected to the line?

Practice Problems

Refer to the circuit shown in *Figure 17–2* and the Resistive-Inductive Series Circuits section of the AC formulas listed in Appendix B.

1. Assume that the circuit shown in *Figure 17–2* is connected to a 480-V, 60-Hz line. The inductor has an inductance of 0.053 H, and the resistor has a resistance of 12 Ω.

E_T 480 V	E_R ______	E_L ______
I_T ______	I_R ______	I_L ______
Z ______	R 12 Ω	X_L ______
VA ______	P ______	$VARs_L$ ______
PF ______	$\angle\theta$ ______	L 0.053 H

2. Assume that the voltage drop across the resistor, E_R, is 78 V, that the voltage drop across the inductor, E_L, is 104 V, and the circuit has a total impedance, Z, of 20 Ω. The frequency of the AC voltage is 60 Hz.

E_T ______	E_R 78 V	E_L 104 V
I_T ______	I_R ______	I_L ______
Z 20 Ω	R ______	X_L ______
VA ______	P ______	$VARs_L$ ______
PF ______	$\angle\theta$ ______	L ______

3. Assume the circuit shown in *Figure 17–2* has an apparent power of 144 VA and a true power of 115.2 W. The inductor has an inductance of 0.15915 H, and the frequency is 60 Hz.

E_T ______	E_R ______	E_L ______
I_T ______	I_R ______	I_L ______
Z ______	R ______	X_L ______
VA 144	P 115.2 W	$VARs_L$ ______
PF ______	$\angle\theta$ ______	L 0.15915 H

4. Assume the circuit shown in *Figure 17–2* has a power factor of 78%, an apparent power of 374.817 VA, and a frequency of 400 Hz. The inductor has an inductance of 0.0382 H.

E_T ______	E_R ______	E_L ______
I_T ______	I_R ______	I_L ______
Z ______	R ______	X_L ______
VA 374.817	P ______	$VARs_L$ ______
PF 78%	$\angle\theta$ ______	L 0.0382 H

5. In an R-L series circuit, E_T = 240 volts, R = 60 Ω, and X_L = 75 W. Find E_L.
6. In an R-L series circuit, E_T = 208 volts, R = 2.4 kΩ, and X_L = 1.5 kW. Find PF.
7. In an R-L series circuit, E_T = 120 volts, R = 35 Ω, and X_L = 48 W. Find reactive power.

8. In an R-L series circuit, the apparent power is 560 VA, and PF = 62%. Find reactive power.
9. An R-L series circuit is connected to a 60-Hz, 208-volt power line. A clamp-on ammeter indicates a current of 18 amperes. The inductor has a reactive power of 1860 VARs. What is the value of R?
10. In an R-L series circuit, Z = 88 Ω, and R = 32 Ω. Find X_L.
11. In an R-L series circuit, apparent power = 450 VA, and reactive power = 224 VARs. Find true power.
12. In an R-L series circuit, $\angle\theta = 22°$, and true power = 94 watts. Find reactive power.
13. An R-L series circuit contains two resistors and two inductors. The resistors have values of 120 Ω and 300 Ω. The inductors have reactive values of 220 Ω and 470 Ω. Find impedance.
14. An R-L series circuit contains two resistors and two inductors. The resistors dissipate powers of 96 watts and 125 watts. The inductors have reactive powers of 100 VARs and 78 VARs. What is the power factor?
15. An R-L series circuit contains two resistors and two inductors. The resistors are 68 Ω and 124 Ω. The inductors have inductive reactances of 44 Ω and 225 Ω. The total voltage is 240 volts. Find the voltage drop across the 124-Ω resistor.
16. An R-L series circuit contains two resistors and two inductors. The resistors are 86 kΩ and 68 kΩ. The inductors have inductive reactances of 24 kΩ and 56 kΩ. The total voltage is 480 volts. Find the voltage drop across the 56-kΩ inductor.

Unit 18 Resistive-Inductive Parallel Circuits

Key Terms

Current flow through the inductor (I_L)

Current flow through the resistor (I_R)

Why You Need to Know

Many times, resistive and inductive loads are connected in parallel with each other. This can be seen in every kind of electrical environment. Incandescent lights in a home are resistive loads, but they are often connected in parallel with a motor or transformer. A common heat-pump system in a home is a perfect example of this type of circuit. The compressor and blower motors are inductive loads, but the electric heat strips are resistive loads. To understand how these different types of loads affect the circuit, this unit will

- illustrate how to calculate the effect on voltage, current, and impedance when a parallel circuit contains both resistance and inductance.
- explain how to apply the relationship of power factor to true power in a parallel circuit containing resistive-inductive factors.

Outline

Objectives

After studying this unit, you should be able to

- describe the flow of current through a parallel circuit containing resistance and inductance.
- explain how the relationship between inductive reactance and resistance affects the phase angle in a parallel circuit.
- calculate power factor and phase angle using true power and apparent power.
- calculate total impedance in an R-L parallel circuit using vector addition.
- determine missing values in an R-L parallel circuit using Ohm's law.

Preview

This unit discusses circuits that contain resistance and inductance connected in parallel with each other. Mathematical calculations are used to show the relationship of current and voltage on the entire circuit and the relationship of current through different branches of the circuit.

18–1 Resistive-Inductive Parallel Circuits

A circuit containing a resistor and an inductor connected in parallel is shown in *Figure 18–1*. Because the voltage applied to any device in parallel must be the same, the voltage applied to the resistor and inductor must be in phase and have the same value. The current flow through the inductor will be 90° out of phase with the voltage, and the current flow through the resistor will be in phase with the voltage (*Figure 18–2*). This configuration produces a phase angle difference of 90° between the current flow through a pure inductive load and a pure resistive load (*Figure 18–3*). The vectors in *Figure 18–3* illustrate that the resistive current and voltage are in phase with each other, and the inductive current is 90° out of phase with both the voltage and resistive current.

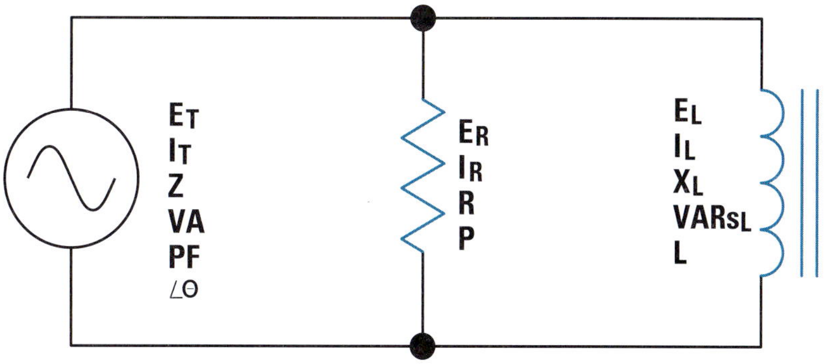

Figure 18–1 A circuit with a resistor in parallel with a coil.

The amount of phase angle shift between the total circuit current and voltage is determined by the ratio of the amount of resistance to the amount of inductive reactance. The circuit power factor is still determined by the ratio of true power to apparent power.

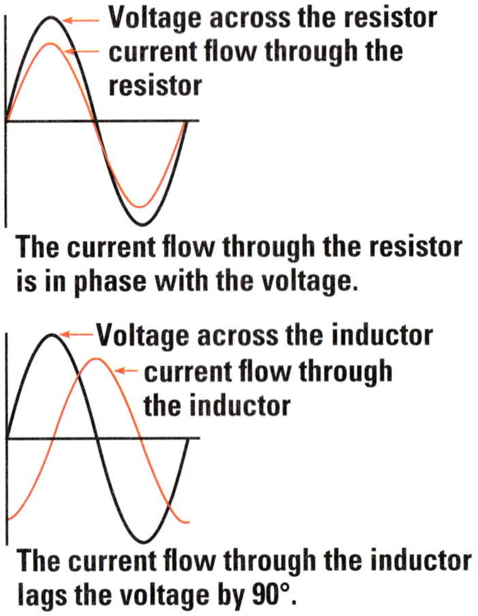

Figure 18–2 Relationship of voltage and current in an R-L parallel circuit.

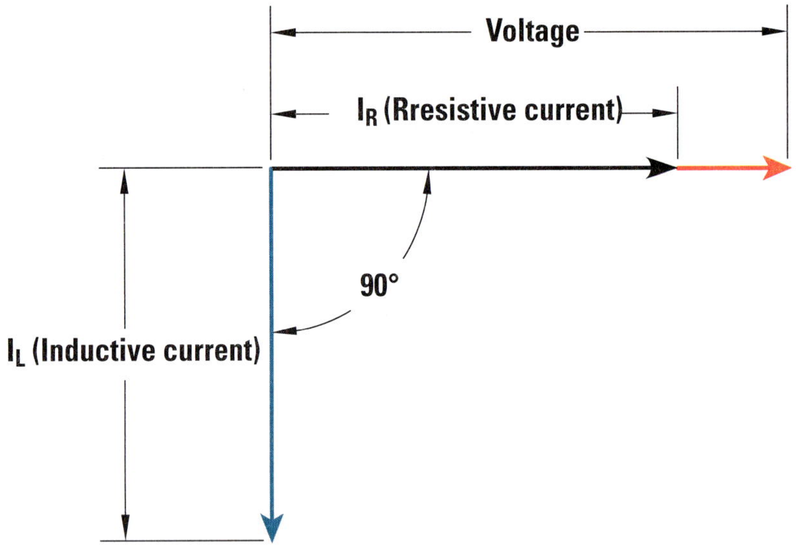

Figure 18–3 Currents through the resistive and inductive loads are 90° out of phase with each other in an R-L parallel circuit.

18–2 Calculating Circuit Values

In the circuit shown in *Figure 18–4*, a resistance of 15 ohms is connected in parallel with an inductive reactance of 20 ohms. The circuit is connected to a voltage of 240 volts AC and a frequency of 60 hertz. In this example problem, the following circuit values will be calculated:

I_R—current flow through the resistor

P—watts (true power)

I_L—current flow through the inductor

VARs—reactive power

I_T—total circuit current

Z—total circuit impedance

VA—apparent power

PF—power factor

$\angle\theta$—the angle the voltage and current are out of phase with each other

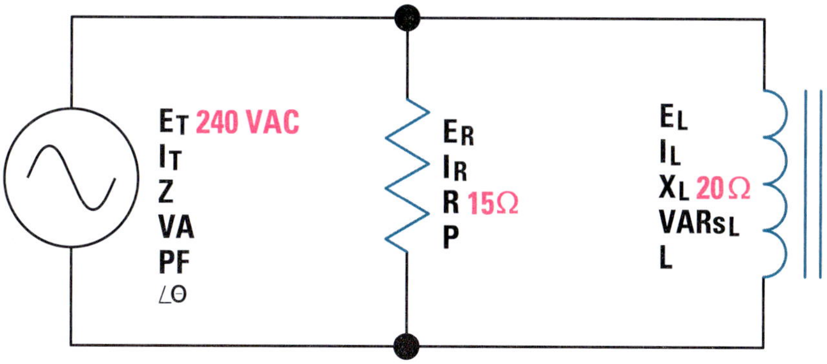

Figure 18–4 Typical R-L parallel circuit.

Resistive Current

In a parallel circuit, the voltage is the same across each branch of the circuit. Therefore, 240 volts are applied across both the resistor and the inductor. Because the amount of voltage applied to the resistor is known, the amount of **current flow through the resistor (I_R)** can be calculated:

$$I_R = \frac{E}{R}$$

$$I_R = \frac{240\ V}{15\ \Omega}$$

$$I_R = 16\ A$$

Watts

True power (P), or watts (W), can be calculated using any of the watts formulas and resistive values. The amount of true power in this circuit can be calculated:

$$P = E_R \times I_R$$

$$P = 240\ V \times 16\ A$$

$$P = 3840\ W$$

Inductive Current

Because the voltage applied to the inductor is known, the current flow can be found by dividing the voltage by the inductive reactance. The amount of **current flow through the inductor (I_L)** can be calculated:

$$I_L = \frac{E}{X_L}$$

$$I_L = \frac{240\ V}{20\ \Omega}$$

$$I_L = 12\ A$$

VARs

The amount of reactive power (VARs) can also be calculated:

$$VARs = E_L \times I_L$$

$$VARs = 240\ V \times 12\ A$$

$$VARs = 2880$$

Inductance

Because the frequency and the inductive reactance are known, the inductance of the coil can be found using the formula for inductance:

$$L = \frac{X_L}{2\pi f}$$

$$L = \frac{20\ \Omega}{377}$$

$$L = 0.0531\ H$$

Total Current

The total current (I_T) flow through the circuit can be calculated by adding the current flow through the resistor and the inductor. However, because these two currents are 90° out of phase with each other, they must be added vectorially. If these current values were plotted on a graph, you could see that they form a right triangle similar to how we solved for missing values in an R-L series circuit (*Figure 18–5*). Notice that the current through the resistor and inductor forms the two right angle legs of the triangle, and the total circuit current is represented by the hypotenuse. The Pythagorean theorem can be used to find the total circuit current:

$$I_T = \sqrt{I_R^2 + I_L^2}$$

$$I_T = \sqrt{(16\ A)^2 + (12\ A)^2}$$

$$I_T = \sqrt{256\ A^2 + 144\ A^2}$$

$$I_T = \sqrt{400\ A^2}$$

$$I_T = 20\ A$$

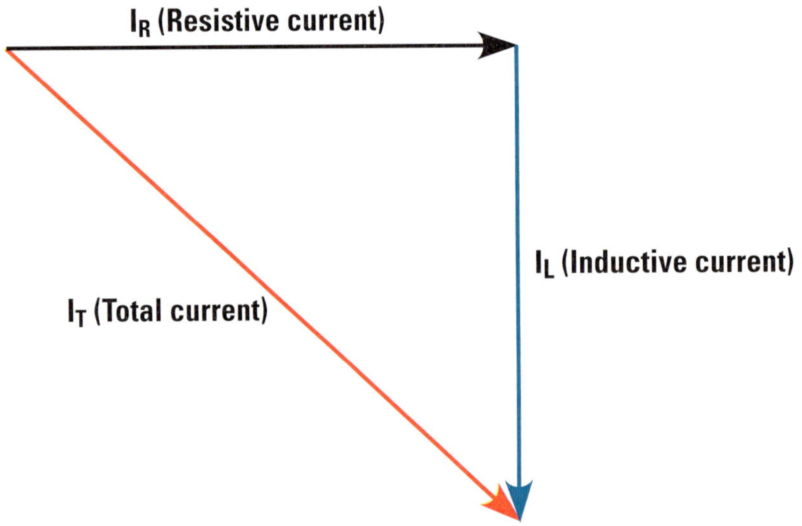

Figure 18–5 Relationship of resistive, inductive, and total current in an R-L parallel circuit.

The parallelogram method for plotting the total current can also be used and is shown in *Figure 18–6*.

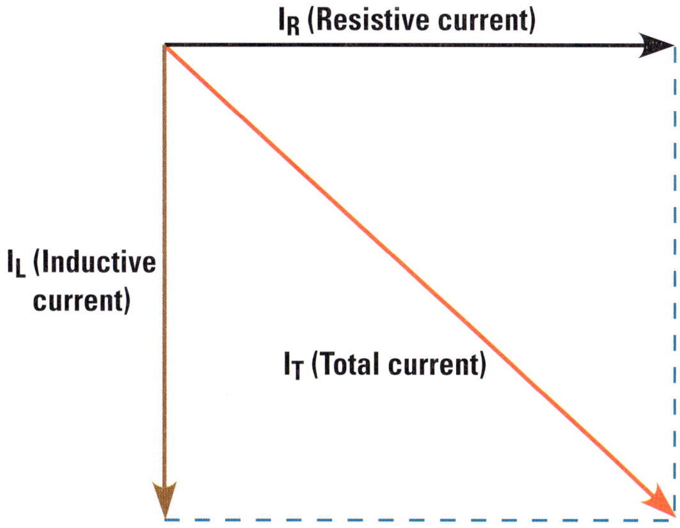

Figure 18–6 Using the parallelogram method to solve for total circuit current.

Impedance

Now that the total current and total voltage are known, the total impedance (Z) can be calculated by substituting Z for R in an Ohm's law formula. The total impedance of the circuit can be calculated using the formula:

$$Z = \frac{E}{I_T}$$

$$Z = \frac{240\ \text{V}}{20\ \text{A}}$$

$$Z = 12\,\Omega$$

The value of impedance can also be found if total current and voltage are not known. Remember that in parallel circuits, the reciprocal of the total resistance is equal to the sum of the reciprocals of each resistor. This same rule can be amended to determine total impedance in an R-L parallel circuit. Resistance and inductive reactance are 90° out of phase with each other; therefore, they must be added vectorially as well. The initial formula is formed by combining the reciprocal resistance formula with the Pythagorean theorem:

Helpful Hint

It should be noted that in an R-L parallel circuit, vectors cannot be used to plot the impedance because the values of resistance and inductive reactance do not add as they do in a series circuit.

$$\left(\frac{1}{Z}\right)^2 = \left(\frac{1}{R}\right)^2 + \left(\frac{1}{X_L}\right)^2$$

This formula states that the square of the reciprocal of the impedance is equal to the sum of the squares of the reciprocals of resistance and inductive reactance. To remove the square from the reciprocal of the impedance, take the square root of both sides of the equation:

$$\frac{1}{Z} = \sqrt{\left(\frac{1}{R}\right)^2 + \left(\frac{1}{X_L}\right)^2}$$

Notice that the formula can now be used to find the reciprocal of the impedance, not the impedance. To change the formula so it is equal to the impedance, take the reciprocal of both sides of the equation:

$$Z = \frac{1}{\sqrt{\left(\frac{1}{R}\right)^2 + \left(\frac{1}{X_L}\right)^2}}$$

Numeric values can now be substituted in the formula to find the impedance of the circuit:

$$Z = \frac{1}{\sqrt{\left(\frac{1}{15\ \Omega}\right)^2 + \left(\frac{1}{20\ \Omega}\right)^2}}$$

$$Z = \frac{1}{\sqrt{(0.004444 + 0.0025)\frac{1}{\Omega^2}}}$$

$$Z = \frac{1}{\sqrt{0.006944\frac{1}{\Omega^2}}}$$

$$Z = \frac{1}{0.08333\frac{1}{\Omega}}$$

$$Z = 12\ \Omega$$

Another formula that can be used to determine the impedance of resistance and inductive reactance connected in parallel is:

$$Z = \frac{R \times X_L}{\sqrt{R^2 + X_L^2}}$$

Substituting the same values for resistance and inductive reactance in this formula will result in the same answer:

$$Z = \frac{15\ \Omega \times 20\ \Omega}{\sqrt{(15\ \Omega)^2 + (20\ \Omega)^2}}$$

$$Z = \frac{300\ \Omega}{\sqrt{625\ \Omega^2}}$$

$$Z = \frac{300\ \Omega}{25\ \Omega}$$

$$Z = 12\ \Omega$$

Apparent Power

The apparent power (VA) can be calculated by multiplying the applied circuit voltage by the total current flow, or, if these values are unknown, we can use true power and VARs. The relationship of volt-amperes, watts, and VARs is similar in an R-L parallel circuit to that in an R-L series circuit. The power values will always add in any type of circuit because power is comprised of voltage and current. Just like in an R-L series circuit, the true power and reactive power are 90° out of phase with each other. However, in an R-L parallel circuit, it is the current that is out of phase, not the voltage as in an R-L series circuit. We will still draw a right triangle between watts and VARs with apparent power (VA) as the hypotenuse (*Figure 18–7*).

$$VA = E_T \times I_T$$

$$VA = 240\text{ V} \times 20\text{ A}$$

$$VA = 4800$$

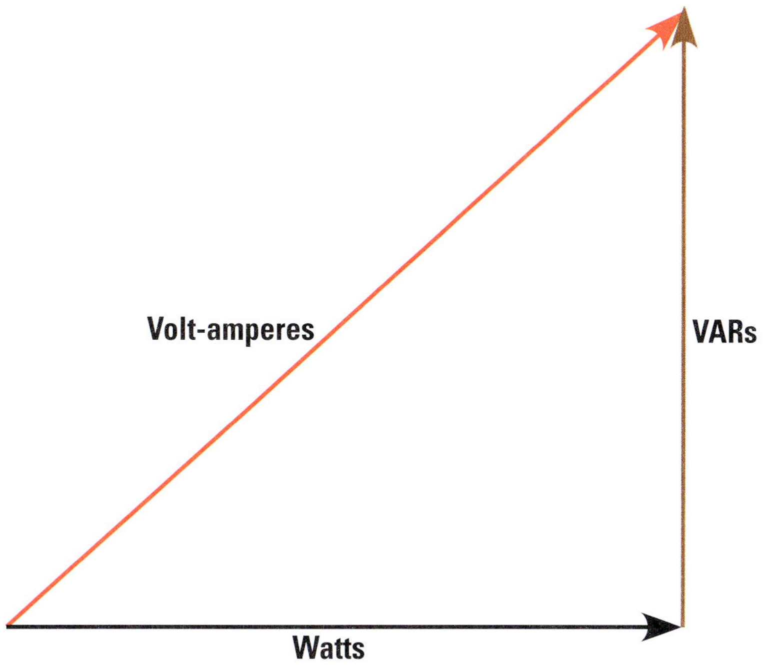

Figure 18–7 Relationship of apparent power (volt-amperes), true power (watts), and reactive power (VARs) in an R-L parallel circuit.

Power Factor

Power factor (PF) in an R-L parallel circuit is the relationship of apparent power to the true power, just as it was in the R-L series circuit. However, in an R-L series circuit, power factor could also be calculated by dividing the voltage dropped across the resistive component of the circuit by the total, or applied, voltage. In a parallel circuit, the voltage is the same, but the currents are different. Therefore, power factor can be calculated by dividing the current flow through the resistive parts of the circuit by the total circuit current:

$$PF = \frac{I_R}{I_T}$$

Power factor can also be determined using the impedance and resistance. In a circuit that contains resistance and reactance connected in parallel, there are two formulas that can be used to determine power factor. Because the impedance of a parallel circuit will be less than the resistance, the formula must be modified from the one used in a circuit containing resistance and reactance connected in series. One formula divides the impedance by the resistance:

$$PF = \frac{Z}{R}$$

The second formula determines power factor by finding the reciprocal of the resistance divided by the impedance. Either formula will produce the same answer:

$$PF = \frac{1}{\left(\frac{R}{Z}\right)}$$

Example 18–1

Find the power factor of a circuit with a resistance of 8 Ω and an impedance of 5 Ω.

Solution

$$PF = \frac{5}{8} \qquad PF = 0.625 \quad \text{or} \quad 62.5\%$$

$$PF = \frac{1}{\left(\frac{8}{5}\right)} \qquad PF = \frac{1}{1.6} \qquad PF = 0.625 \quad \text{or} \quad 62.5\%$$

The circuit power factor in this example can now be calculated because the apparent power and true power are both known:

$$PF = \frac{P}{VA} \times 100$$

$$PF = \frac{3840\ \text{W}}{4800\ \text{VA}} \times 100$$

$$PF = 0.80\text{, or } 80\%$$

Angle Theta

The cosine of angle theta is also still equal to the power factor, and, therefore, angle theta can be found by determining the inverse cosine of the power factor, if power factor alone is known. The same formula can be used to determine the power factor of any type of AC circuit.

$$\cos \angle\theta = 0.80$$

$$\angle\theta = 36.87°$$

A right triangle diagram using apparent power, true power, and reactive power is shown in *Figure 18–8*. Notice that angle theta is the angle produced by the apparent power and the true power. The relationship of current and voltage for this circuit is shown in *Figure 18–9*. Because voltage is the same across both the resistor and inductive load, voltage does not need a vector diagram or right triangle to illustrate the relationship; it is the same throughout the circuit. The circuit with all values is shown in *Figure 18–10*.

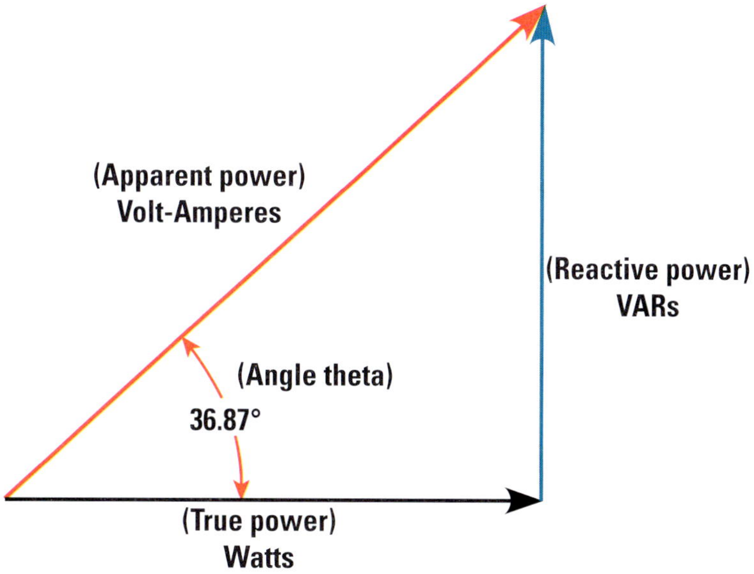

Figure 18–8 Angle theta.

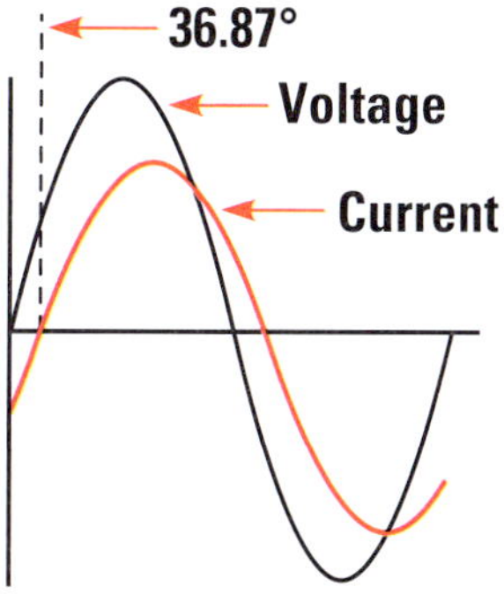

Figure 18–9 The current is 36.87° out of phase with the voltage.

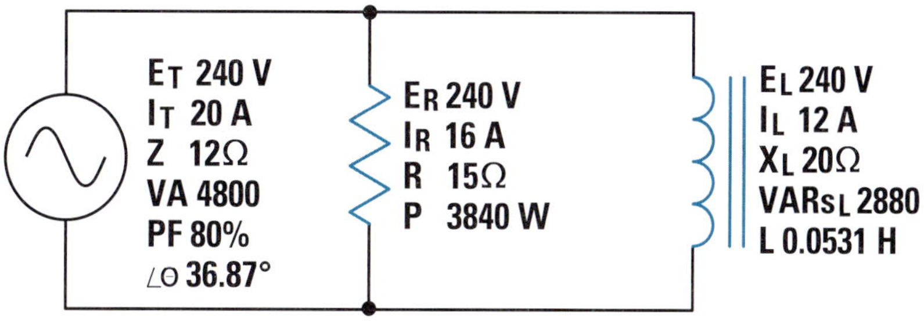

Figure 18–10 All values have been found.

Example 18–2

In this circuit, one resistor is connected in parallel with two inductors (*Figure 18–11*). The frequency is 60 Hz. The circuit has an apparent power of 6120 VA, the resistor has a resistance of 45 Ω, the first inductor has an inductive reactance of 40 Ω, and the second inductor has an inductive reactance of 60 Ω. It is assumed that both inductors have a Q greater than 10, and their resistance is negligible. Find the following missing values:

Z—total circuit impedance

I_T—total circuit current

E_T—applied voltage

E_R—voltage drop across the resistor

I_R—current flow through the resistor

P—watts (true power)

E_{L1}—voltage drop across the first inductor

I_{L1}—current flow through the first inductor

$VARs_{L1}$—reactive power of the first inductor

L_1—inductance of the first inductor

E_{L2}—voltage drop across the second inductor

I_{L2}—current flow through the second inductor

$VARs_{L2}$—reactive power of the second inductor

L_2—inductance of the second inductor

$\angle\theta$— the angle that voltage and current are out of phase with each other

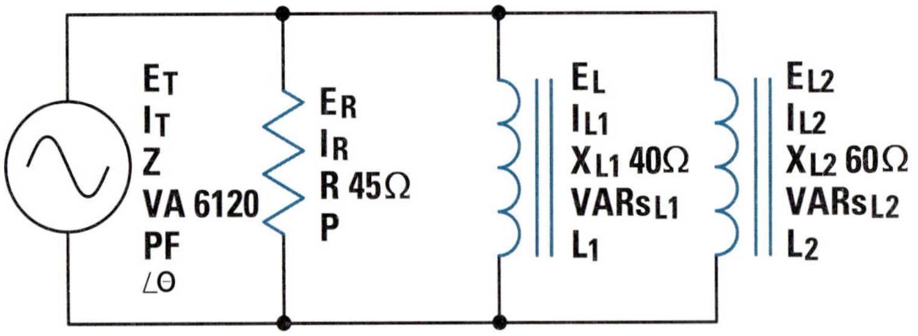

Figure 18–11 Example circuit 2.

Solution

Impedance

Before it is possible to calculate the impedance of the circuit, the total amount of inductive reactance for the circuit must be found. Because these two inductors are connected in parallel, the reciprocal of their inductive reactance values must be added. This will give the reciprocal of the total inductive reactance:

$$\frac{1}{X_{LT}} = \frac{1}{X_{L1}} + \frac{1}{X_{L2}}$$

To find the total inductive reactance, take the reciprocal of both sides of the equation:

$$X_{LT} = \frac{1}{\frac{1}{X_{L1}} + \frac{1}{X_{L2}}}$$

Example 18–2 (continued)

Refer to the formulas for pure inductive circuits shown in the AC formulas section of Appendix B. Numeric values can now be substituted into the formula to find the total inductive reactance:

$$X_{LT} = \frac{1}{\frac{1}{40\ \Omega} + \frac{1}{60\ \Omega}}$$

$$X_{LT} = \frac{1}{(0.025 + 0.01667)\frac{1}{\Omega}}$$

$$X_{LT} = 24\ \Omega$$

Now that the total amount of inductive reactance for the circuit is known, the impedance can be calculated using the formula:

$$Z = \frac{1}{\sqrt{\left(\frac{1}{R}\right)^2 + \left(\frac{1}{X_{LT}}\right)^2}}$$

$$Z = \frac{1}{\sqrt{\left(\frac{1}{45\ \Omega}\right)^2 + \left(\frac{1}{24\ \Omega}\right)^2}}$$

$$Z = 21.176\ \Omega$$

E_T

Now that the circuit impedance and the apparent power are known, the applied voltage can be calculated using the formula:

$$E_T = \sqrt{VA \times Z}$$

$$E_T = \sqrt{6120\ VA \times 21.176\ \Omega}$$

$$E_T = 360\ V$$

E_R, E_{L1}, E_{L2}

In a parallel circuit, the voltage is the same across any leg or branch. Therefore, 360 V is dropped across the resistor, the first inductor, and the second inductor:

$$E_R = 360\ V$$

$$E_{L1} = 360\ V$$

$$E_{L2} = 360\ V$$

I_T

The total current of the circuit can now be calculated using the formula:

$$I_T = \frac{E_T}{Z}$$

$$I_T = \frac{360\ V}{21.176\ \Omega}$$

$$I_T = 17\ A$$

Example 18–2 (continued)

The remaining values of the circuit can be found using Ohm's law. Refer to the Resistive-Inductive Parallel Circuits listed in the AC formula section in Appendix B.

I_R

$$I_R = \frac{E_R}{R}$$

$$I_R = \frac{360\ V}{45\ \Omega}$$

$$I_R = 8\ A$$

P

$$P = E_R \times I_R$$

$$P = 360\ V \times 8\ A$$

$$P = 2880\ W$$

I_{L1}

$$I_{L1} = \frac{E_{L1}}{X_{L1}}$$

$$I_{L1} = \frac{360\ V}{40\ \Omega}$$

$$I_{L1} = 9\ A$$

$VARs_{L1}$

$$VARs_{L1} = E_{L1} \times I_{L1}$$

$$VARs_{L1} = 360\ V \times 9\ A$$

$$VARs_{L1} = 3240$$

L_1

$$L_1 = \frac{X_{L1}}{2\pi f}$$

$$L_1 = \frac{40\ \Omega}{377}$$

$$L_1 = 0.106\ H$$

Example 18–2 (continued)

I_{L2}

$$I_{L2} = \frac{E_{L2}}{X_{L2}}$$

$$I_{L2} = \frac{360\ V}{60\ \Omega}$$

$$I_{L2} = 6\ A$$

$VARs_{L2}$

$$VARs_{L2} = E_{L2} \times I_{L2}$$

$$VARs_{L2} = 360\ V \times 6\ A$$

$$VARs_{L2} = 2160\ VARs$$

L_2

$$L_2 = \frac{X_{L2}}{2\pi f}$$

$$L_2 = \frac{60\ \Omega}{377}$$

$$L_2 = 0.159\ H$$

PF

$$PF = \frac{P}{VA}$$

$$PF = \frac{2880\ W}{6120\ VA}$$

$$PF = 47.06\%$$

$\angle\theta$

$$\cos \angle\theta = PF$$

$$\cos \angle\theta = 0.4706$$

$$\angle\theta = 61.93^\circ$$

A vector diagram illustrating angle theta is shown in *Figure 18–12.* The vectors used are those for apparent power, true power, and reactive power. The phase relationship of voltage and current for this circuit is shown in *Figure 18–13*, and the circuit with all completed values is shown in *Figure 18–14.*

Example 18–2 (continued)

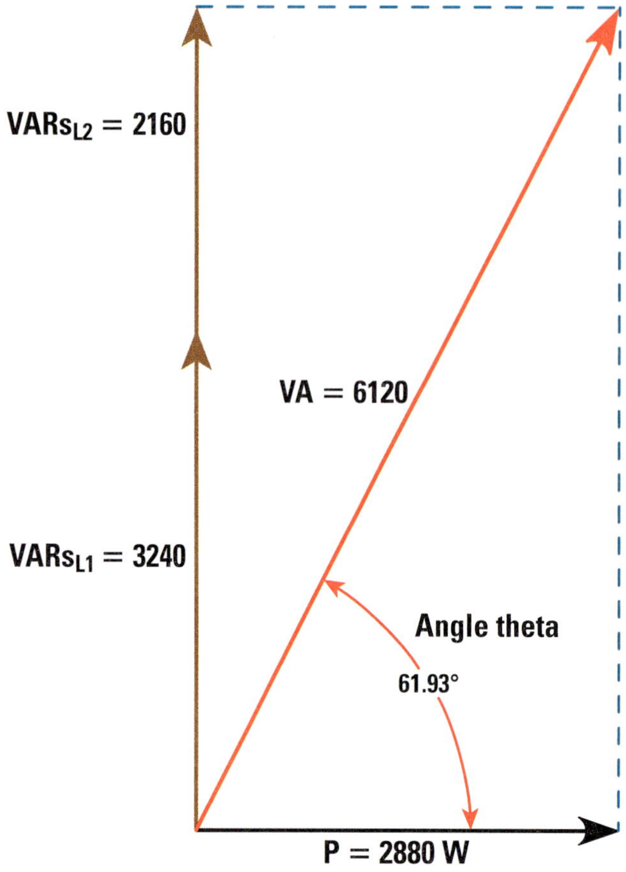

Figure 18–12 Angle theta determined by power vectors.

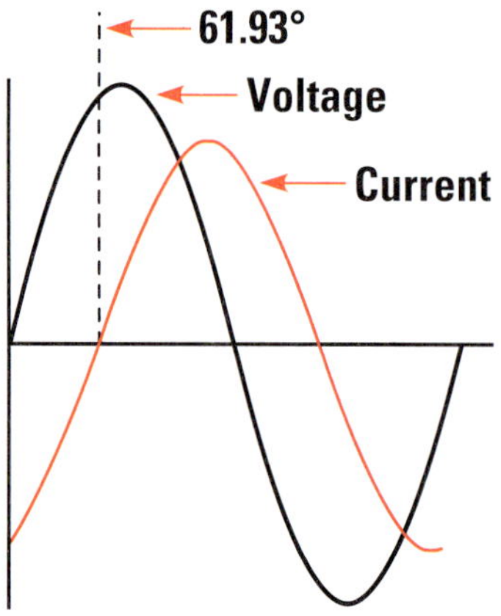

Figure 18–13 Voltage and current are 61.93° out of phase with each other.

Example 18–2 (continued)

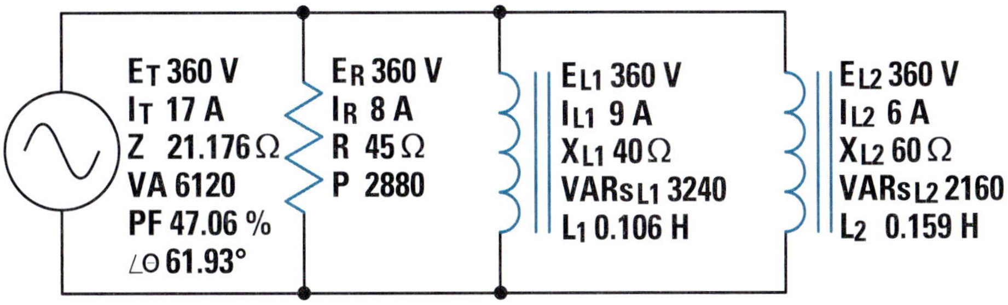

Figure 18–14 Completed values for example circuit 2.

R-L parallel circuits exist throughout the electrical field. Many circuits contain both inductive and resistive loads connected to the same branch circuit. A ceiling fan, for example, often has a light kit attached to the fan. The fan is an inductive load, and if the light kit uses incandescent lamps, these lamps are a resistive load. Branch circuits can also contain incandescent lamps, fluorescent lights, and motors connected on the same circuit. All these loads are connected in parallel.

Summary

- The voltage applied across each branch in a parallel circuit must be the same.
- The current flowing through resistive parts of the circuit will be in phase with the voltage.
- The current flowing through inductive parts of the circuit will lag the voltage by 90°.
- The total current in a parallel circuit is equal to the sum of the individual branch currents but must be added vectorially because the current through the inductive branch of the circuit is 90° out of phase with the current flowing through the resistive branch.
- Total impedance in an R-L parallel circuit can be determined using a method that combines the Pythagorean theorem with the reciprocal method for adding resistance in parallel circuits.
- Apparent power, true power, and reactive power will always be additive in any kind of circuit. However, because true power and reactive power are 90° out of phase with each other, power adds vectorially.

Review Questions

1. When an inductor and resistor are connected in parallel, how many degrees out of phase are the current flow through the resistor and the inductor? Select the most correct answer.

   A) 65°

   B) 78°

   C) 90°

   D) 180°

2. In an R-L parallel circuit with one resistive branch and one inductive branch, which of the following statements describes the relationship between current and voltage in the circuit?
   A) The applied voltage lags the circuit current by 90°.
   B) The phase angle between voltage and current is determined by the ratio of R and X_L.
   C) The applied voltage leads the circuit current by 90°.
   D) The voltage and current are in phase with each other.
3. Inductive reactance is directly proportional to the frequency of an AC circuit. As the frequency increases, so does the inductive reactance. Which of the following best describes the impact on the phase angle between voltage and current?
   A) The phase angle is not impacted.
   B) The phase angle decreases as X_L increases.
   C) The phase angle increases as R increases.
   D) The phase angle increases as X_L increases.
4. True or False: As resistance increases in an R-L parallel circuit, the angle of phase shift between voltage and current decreases.
5. Which of the following best describes power factor?
   A) Power factor is the ratio of watts to volt-amperes.
   B) Power factor is the ratio of ohms to amperes.
   C) Power factor is the ratio of VARs to volt-amperes.
   D) Power factor is the ratio of volts to volt-amperes.
6. Power factor is equal to cosine of angle theta, and angle theta can be determined by finding which function of the power factor?
   A) Angle theta is the sine of power factor.
   B) Angle theta is equal to the tangent of power factor.
   C) Angle theta is equal to the cos-1 of power factor.
   D) Angle theta is equal to the sin-1 of power factor.
7. An R-L parallel circuit has 30 ohms of inductive reactance and 40 ohms of resistance. What is the total circuit impedance?
   A) 32 ohms
   B) 78 ohms
   C) 50 ohms
   D) 24 ohms
8. An R-L parallel circuit has a voltage supply of 240 volts, and it is measured that 5 amperes of current is flowing. If the resistor has a resistance of 171 ohms, which of the following is most accurate for inductive reactance?
   A) 24 ohms
   B) 50 ohms

C) 75 ohms

D) 48 ohms

9. If an R-L parallel circuit has an applied voltage of 240 volts and a total current of 50 amperes, what is the resistance value if the power factor is 80%?

   A) 6 ohms

   B) 8 ohms

   C) 10 ohms

   D) 12 ohms

10. True or False: Voltage drop through the inductive branch of an R-L parallel circuit can be determined if true power and resistance in the circuit are known.

Practical Applications

1. A small space heater of 500 W is connected in parallel with an inductive load, such as a blower fan. A clamp-on ammeter reveals a total circuit current of 7 A. What is the inductance of the load connected in parallel with the space heater? Assume an applied voltage of 120 V and a frequency of 60 Hz.

2. You are working on a residential heat pump. The heat pump is connected to a 240-V, 60-Hz power line. The compressor has a current draw of 34 amperes when operating. The compressor has a power factor of 70%. The backup strip heat is rated at 10 kW. You need to know the amount of total current draw that will occur if the strip heat comes on while the compressor is operating.

Practice Problems

Refer to the circuit shown in *Figure 18–1*. Use the AC formulas in the Resistive-Inductive Parallel Circuits section of Appendix B.

1. Assume that the circuit shown in *Figure 18–1* is connected to a 60-Hz line and has a total current flow of 34.553 A. The inductor has an inductance of 0.02122 H, and the resistor has a resistance of 14 W.

E_T ______	E_R ______	E_L ______
I_T 34.553 A ______	I_R ______	I_L ______
Z ______	R 14 Ω ______	X_L ______
VA ______	P ______	$VARs_L$ ______
PF ______	$\angle\theta$ ______	L 0.02122 H ______

2. Assume that the current flow through the resistor, I_R, is 15 A; the current flow through the inductor, I_L, is 36 A; and the circuit has an apparent power of 10,803 VA. The frequency of the AC voltage is 60 Hz.

E_T ______	E_R ______	E_L ______
I_T ______	I_R 15 A ______	I_L 36 A ______
Z ______	R ______	X_L ______
VA 10,803 ______	P ______	$VARs_L$ ______
PF ______	$\angle\theta$ ______	L ______

3. Assume that the circuit in *Figure 18–1* has an apparent power of 144 VA and a true power of 115.2 W. The inductor has an inductance of 0.15915 H, and the frequency is 60 Hz.

E_T ______	E_R ______	E_L ______
I_T ______	I_R ______	I_L ______
Z ______	R ______	X_L ______
VA 144 ______	P 115.2 ______	$VARs_L$ ______
PF ______	$\angle\theta$ ______	L 0.15915 H ______

4. Assume that the circuit in *Figure 18–1* has a power factor of 78%, an apparent power of 374.817 VA, and a frequency of 400 Hz. The inductor has an inductance of 0.0382 H.

E_T ______	E_R ______	E_L ______
I_T ______	I_R ______	I_L ______
Z ______	R ______	X_L ______
VA 374.817 ______	P ______	$VARs_L$ ______
PF 78% ______	$\angle\theta$ ______	L 0.0382 H ______

5. In an R-L parallel circuit, R = 240 Ω, and X_L = 360 Ω. Find impedance.
6. In an R-L parallel circuit, I_T = 0.25 amps, and I_R = 0.125 amps. The inductor has a reactive power of 75 VARs. What is E_T?
7. In an R-L parallel circuit, E_T = 120 volts, R = 120 Ω, and X_L = 150 Ω. Find I_L.
8. In an R-L parallel circuit, E_T = 48 volts, I_T = 0.25 amps, and R = 320 Ω. Find X_L.
9. In an R-L parallel circuit, E_T = 240 volts, R = 560 Ω, and X_L = 330 Ω. Find reactive power.
10. In an R-L parallel circuit, E_T = 240 volts, R = 560 Ω, and X_L = 330 Ω. Find apparent power.
11. In an R-L parallel circuit, E_T = 208 volts, R = 2.4 k Ω, and X_L = 1.8 k Ω. Find I_T.
12. In an R-L parallel circuit, E_T = 480 volts, R = 16 Ω, and X_L = 24 Ω. Find PF.
13. In an R-L parallel circuit, I_T = 1.25 amps, R = 1.2 k Ω, and X_L = 1 k Ω. Find I_R.
14. In an R-L parallel circuit, true power = 4.6 watts, and reactive power = 5.4 VARs. What is the apparent power?
15. An R-L parallel circuit is connected to 240 volts at 60 Hz. The resistor has a resistance of 68 Ω, and the inductor has an inductive reactance of 48 Ω. What is the reactance of the inductor?
16. An R-L parallel circuit has an applied voltage of 208 volts and a total current of 2 amperes. The resistor has a value 180 Ω. Find I_L.

Section 7 AC Circuits Containing Capacitors

Sochillplanets/Shutterstock.com

Unit 19 Capacitors

Key Terms

Capacitor
Dielectric
Dielectric constant
Dielectric stress
Electrolytic
Exponential
Farad
HIPOT
JAN (Joint Army-Navy) standard
Leakage current
Nonpolarized capacitors
Polarized capacitors
RC time constant
Surface area
Variable capacitors

Why You Need to Know

Capacitors are one type of electrical device that is used in nearly every aspect of our lives today. These devices can be found in everything from computers to LED drivers and can range in size from tiny devices that help us communicate with our friends and family to large scale capacitors that help correct the power factor in some of the largest manufacturing facilities across the globe. This unit will

- explain how capacitors are constructed, how they are charged and discharged, and the differences between different types of capacitors and their markings.
- describe how capacitance is measured and the importance of the voltage rating.
- explain how capacitors store energy in an electrostatic field and how current can flow only when the capacitors are charging or discharging.

Outline

19–1 Introduction to Capacitors
19–2 Charging Capacitors
19–3 Capacitors in Circuits
19–4 Capacitor Applications
19–5 Capacitor Types and Markings
19–6 Testing Capacitors

Objectives

After studying this unit, you should be able to

- identify unique characteristics of capacitors that determine capacitance.
- explain the process of how a capacitor builds and holds a charge.
- differentiate between capacitors connected in parallel circuits and series circuits.
- differentiate between polarized and nonpolarized capacitors.
- describe the meaning and purpose of common capacitor markings
- identify adverse situations related to capacitors using test methods.

Preview

Capacitors perform a variety of jobs such as power factor correction, storing an electric charge to produce a large current pulse, timing circuits, and electronic filters. Capacitors can be nonpolarized or polarized depending on the application. Nonpolarized capacitors can be used in both AC and DC circuits, whereas polarized capacitors can be used in DC circuits only. Both types will be introduced and discussed in this unit.

19–1 Introduction to Capacitors

Caution

Capacitors store electrical charge and have the capabilities to supply significant amounts of current. Under certain conditions, a capacitor can have enough stored energy to initiate ventricular fibrillation. Extreme caution must be taken when handling these devices so as not to be seriously injured. For more information on how to ensure a capacitor is safe to handle and how to place these devices in an electrically safe condition, see Article 360 in NFPA 70E®: Standard for Electrical Safety in the Workplace®.

Capacitors ***are devices that oppose a change of voltage.*** The simplest type of capacitor is constructed by separating two metal plates by some type of insulating material called the **dielectric** (*Figure 19–1*). *Figure 19–2* shows the capacitor symbol. Three factors determine the capacitance of a capacitor:

1. The surface area of the plates
2. The distance between the plates
3. The type of dielectric used

The greater the **surface area** of the plates, the more capacitance a capacitor will have. If a capacitor is connected to a DC source (*Figure 19–3*), electrons are removed from the plate connected to the positive battery terminal and are deposited on the plate connected to the negative terminal. This flow of current continues until a voltage equal to the battery voltage is established across the plates of the capacitor (*Figure 19–4*). When these two voltages become equal, the flow of electrons stops. The capacitor is now charged. If the battery is disconnected from the capacitor, the capacitor will remain charged as long as there is no path by which the electrons can move from one plate to the other (*Figure 19–5*). A good rule to remember concerning a capacitor and current flow is that ***current can flow only during the period of time that a capacitor is either charging or discharging.***

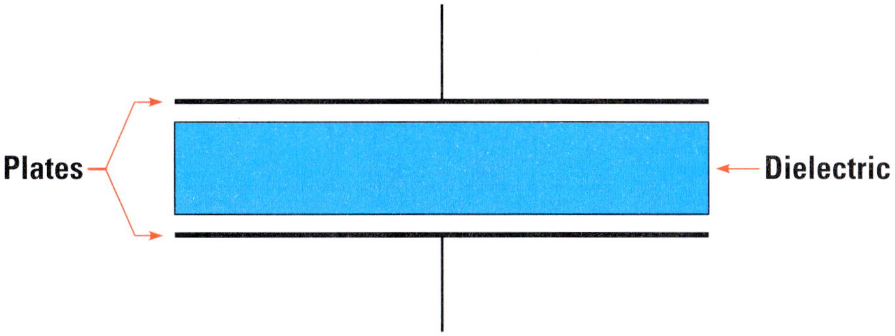

Figure 19–1 A capacitor is made by separating two metal plates with a dielectric.

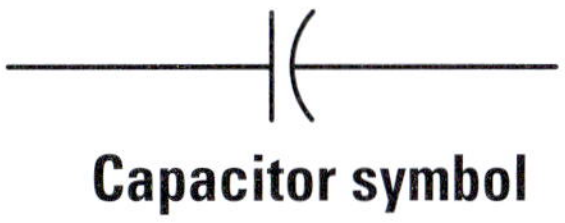

Figure 19–2 Schematic symbol generally used to indicate a capacitor.

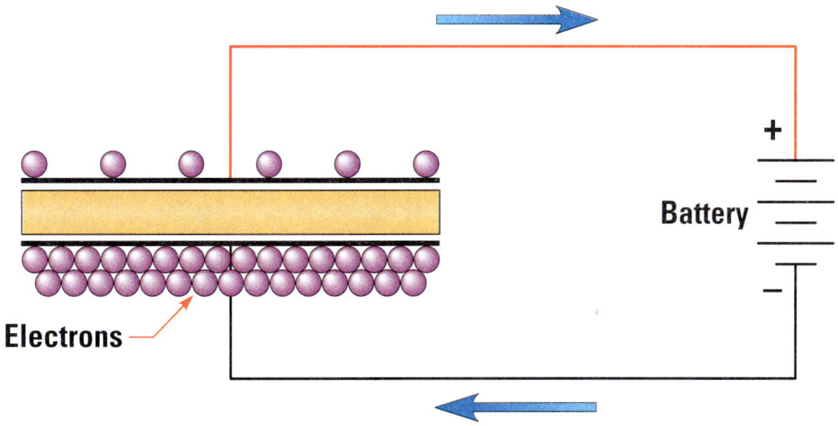

Figure 19–3 A capacitor can be charged by removing electrons from one plate and depositing them on the other plate.

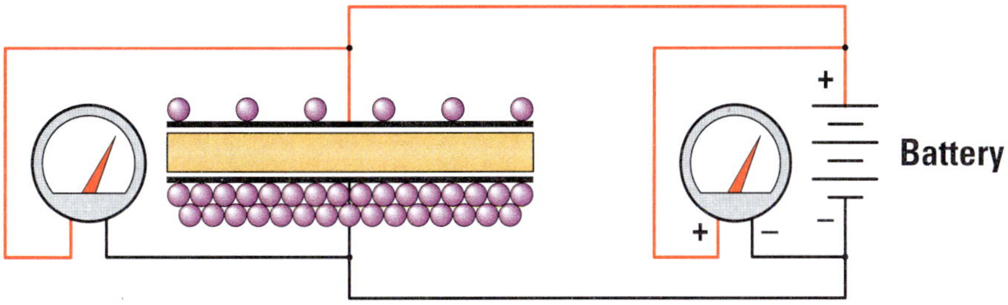

Figure 19–4 Current flows until the voltage across the capacitor is equal to the voltage of the battery.

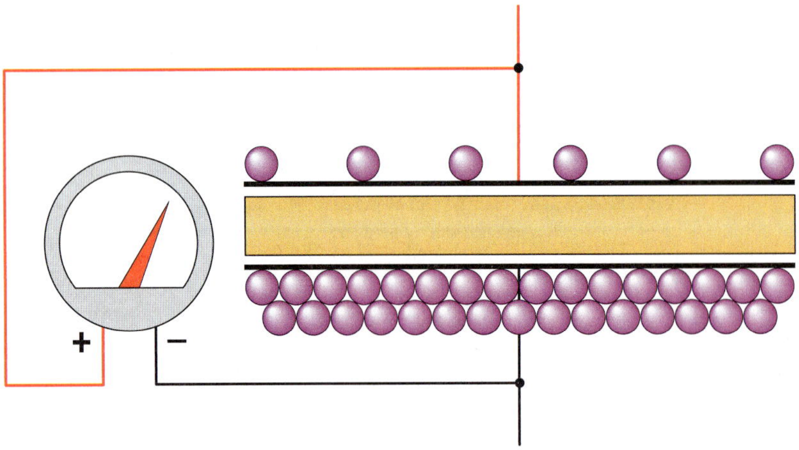

Figure 19–5 The capacitor remains charged after the battery is removed from the circuit.

In theory, it should be possible for a capacitor to remain in a charged condition forever. In actual practice, however, it cannot. No dielectric is a perfect insulator, and electrons eventually move through the dielectric from the negative plate to the positive, causing the capacitor to discharge (*Figure 19–6*). This current flow through the dielectric is called **leakage current** and is proportional to the resistance of the dielectric and the charge across the plates. If the dielectric of a capacitor becomes weak, it will permit an excessive amount of leakage current to flow. A capacitor in this condition is often referred to as a *leaky capacitor.*

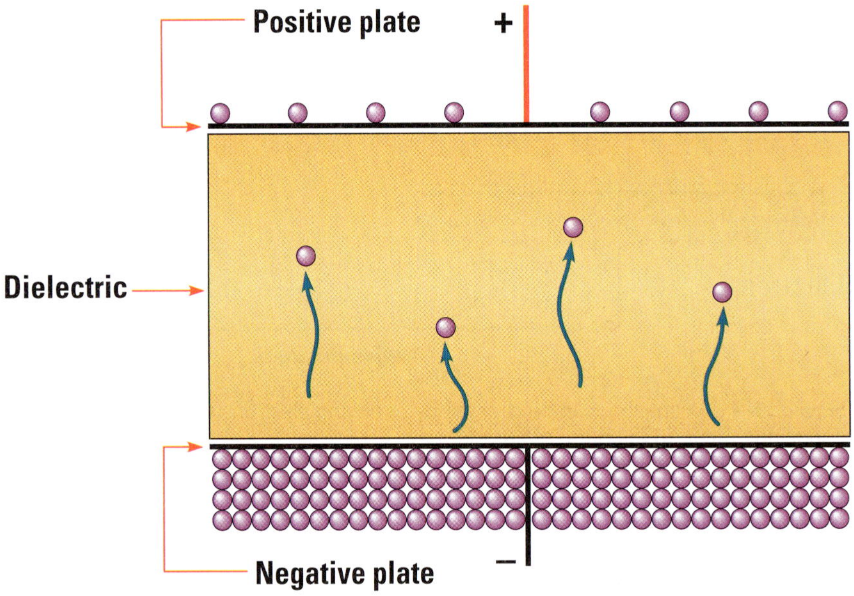

Figure 19–6 Electrons eventually leak through the dielectric. This flow of electrons is known as leakage current.

19–2 Charging Capacitors

Two other factors that determine capacitance are the type of dielectric used and the distance between the plates. To understand these concepts, it is necessary to understand how a capacitor stores energy. In previous units, it was discussed that an inductor stores energy in the form of a magnetic field. A capacitor stores energy in an electrostatic field.

The term *electrostatic* refers to electric charges that are static or stationary. They are very similar to the static electric charges that form on objects that are good insulators, as discussed in Unit 3. The electrostatic field is formed when electrons are removed from one plate and deposited on the other.

Dielectric Stress

When a capacitor is not charged, the atoms of the dielectric are uniform as shown in *Figure 19–7*. The valence electrons orbit the nucleus in a circular pattern. When the capacitor becomes charged, however, a potential exists between the plates of the capacitor. The plate with the lack of electrons has a positive charge, and the plate with the excess of electrons has a negative charge. Because electrons are negative particles, they are repelled from the negative plate and attracted to the positive plate. This attraction causes the electron orbit to become stretched, as shown in *Figure 19–8.*

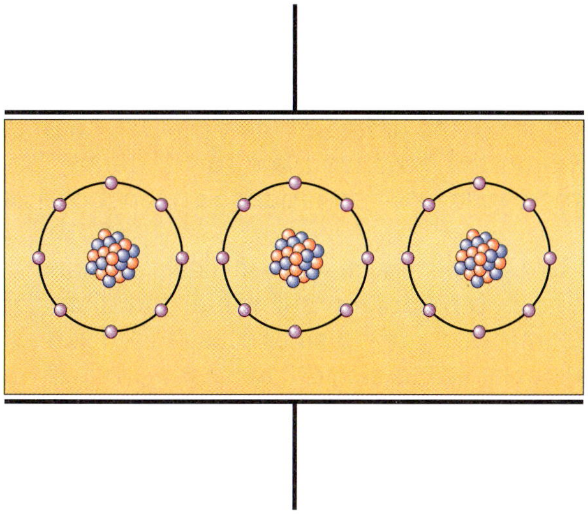

Figure 19–7 Atoms in the dielectric material of an uncharged capacitor.

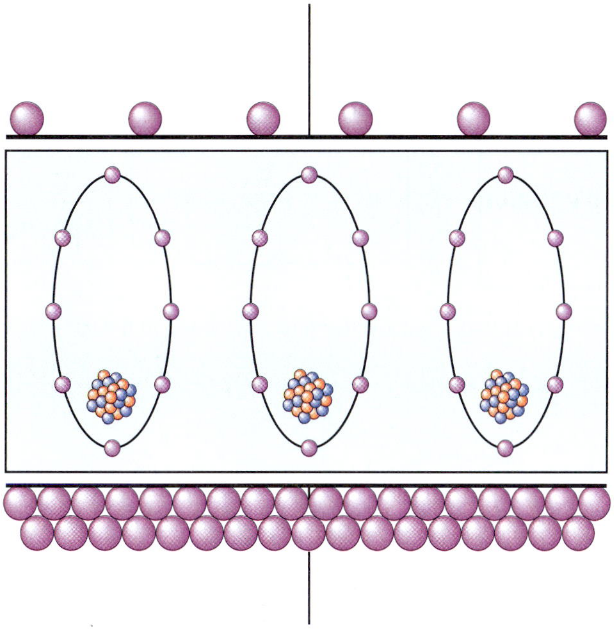

Figure 19–8 Atoms in the dielectric material of a charged capacitor become stretched or "stressed."

This stretching of the atoms of the dielectric is called **dielectric stress.** Placing the atoms of the dielectric under stress has the same effect as drawing back a bowstring with an arrow and holding it; that is, it stores energy.

The amount of dielectric stress is proportional to the voltage or potential difference between the plates. Remember that voltage is the pressure in an electrical circuit. The greater the voltage, the greater the pressure and, therefore, the dielectric stress. If the voltage becomes too great, the dielectric will break down and permit current to flow between the plates. At this point, the capacitor develops a short circuit path for electrons to flow through the dielectric material and has become "shorted." Capacitors have a voltage rating that should not be exceeded. The voltage rating indicates the maximum amount of voltage the dielectric is intended to withstand without breaking down. The amount of voltage applied to a capacitor is critical to its life span. Capacitors operated above their voltage rating will fail relatively quickly. Inversely, capacitors operated well below their rated voltage will not experience the same level of stress and will have a much longer lifespan.

The energy of the capacitor is stored in the dielectric in the form of an electrostatic charge. It is this electrostatic charge that permits the capacitor to produce extremely high currents under certain conditions. If the leads of a capacitor are shorted together, it releases all the stored energy in an instant. When the leads are shorted, the atoms of the dielectric snap back to their normal position. Shorting causes the electrons on the negative plate to be literally blown off and attracted to the positive plate. Some very large capacitors can produce currents of thousands of amperes for short periods of time.

This principle was used to operate the electronic flash of many cameras before LED camera flashes were introduced. Some photographers still prefer the older technology to today's modern options. Electronic flash attachments contain a small glass tube filled with a gas called xenon. Xenon produces a very bright white light similar to sunlight when the gas is ionized. However, a large amount of power is required. A battery capable of directly ionizing the xenon would be very large and expensive and would need a voltage around 500 volts. The simple circuit shown in *Figure 19–9* can be used to overcome the problem. In this circuit, two small 1.5-volt batteries are all it takes to charge the capacitor to the peak value of the voltage waveform. When the switch is closed, the capacitor suddenly discharges through the xenon tube and supplies the power needed to ionize the gas. The downside to this system is that it takes several seconds to store enough energy in the capacitor to ionize the gas in the tube.

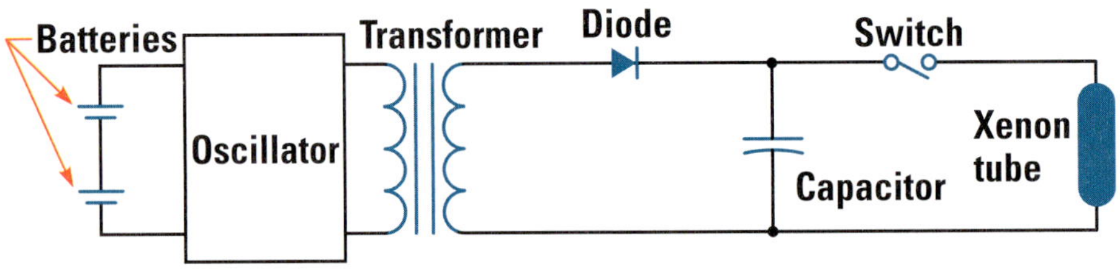

Figure 19–9 Energy is stored in a capacitor and discharged to the camera flash.

Dielectric Constant

Because much of the capacitor's energy is stored in the dielectric, the type of dielectric material used is important in determining the amount of capacitance a capacitor will have. Materials are assigned a number called the **dielectric constant.** Vacuum is assigned the number 1 and is used as a reference. Air has a dielectric constant of approximately 1.006. For example, assume that a capacitor uses vacuum as the dielectric and is found to have a capacitance of 1 microfarad (1 μF). Now assume that some material is placed between the plates without changing the spacing and the capacitance value becomes

5 microfarad. This material has a dielectric constant of 5. A chart showing the dielectric constant of different materials is shown in *Figure 19–10*.

Material	Dielectric constant
Air	1.006
Bakelite	4.0–10.0
Castor oil	4.3–4.7
Cellulose acetate	7.0
Ceramic	1200
Dry paper	3.5
Hard rubber	2.8
Insulating oils	2.2–4.6
Lucite	2.4–3.0
Mica	6.4–7.0
Mycalex	8.0
Paraffin	1.9–2.2
Porcelain	5.5
Pure water	81
Pyrex glass	4.1–4.9
Rubber compounds	3.0–7.0
Teflon	2
Titanium dioxide compounds	90–170

Figure 19–10 Dielectric constants for several different dielectric materials.

Capacitor Ratings

The basic unit of capacitance is the **farad,** and it is symbolized by the letter F. It is named after a famous scientist named Michael Faraday. ***A capacitor has a capacitance of one farad when a change of 1 volt across its plates results in a movement of 1 coulomb:***

$$Q = C \times V$$

where

Q = charge in coulombs

C = capacitance in farads

V = charging voltage

Although the farad is the basic unit of capacitance, it is seldom used because it is an extremely large amount of capacitance. The following formula can be used to determine the capacitance of a capacitor when the area of the plates, the dielectric constant, and the distance between the plates are known:

$$C = \frac{K \times A}{4.45\ D}$$

where

C = capacitance in pF (picofarads)

K = dielectric constant (F/in.)

A = area of one plate in sq. in.

D = distance between the plates in in.

Example 19–1

What would be the plate area of a 1-F (farad) capacitor if air is used as the dielectric and the plates are separated by a distance of 1 in.?

Solution

The first step is to convert the preceding formula to solve for area:

$$A = \frac{C \times 4.45 \times D}{K}$$

$$A = \frac{1{,}000{,}000{,}000{,}000 \times 4.45 \times 1}{1.006}$$

$$A = 4{,}447{,}000{,}000{,}000 \text{ sq. in.}$$

$$A = 1107.7 \text{ sq. miles}$$

Because the basic unit of capacitance is so large, other units such as the microfarad (μF), nanofarad (nF), and picofarad (pF) are generally used:

$$\mu F = \frac{1}{1{,}000{,}000} (1 \times 10^{-6}) \text{ of a farad}$$

$$nF = \frac{1}{1{,}000{,}000{,}000} (1 \times 10^{-9}) \text{ of a farad}$$

$$pF = \frac{1}{1{,}000{,}000{,}000{,}000} (1 \times 10^{-12}) \text{ of a farad}$$

19–3 Capacitors in Circuits

Connecting capacitors in parallel (*Figure 19–11*) has the same effect as increasing the plate area of one capacitor. Unlike inductance and resistance, capacitance adds linearly in a parallel circuit because capacitors provide current to the circuit and current in parallel is additive. In the example shown, three capacitors having a capacitance of 20, 30, and 60 microfarads are connected in parallel. The total capacitance of this connection is:

$$C_T = C_1 + C_2 + C_3$$

$$C_T = 20\ \mu F + 30\ \mu F + 60\ \mu F$$

$$C_T = 110\ \mu F$$

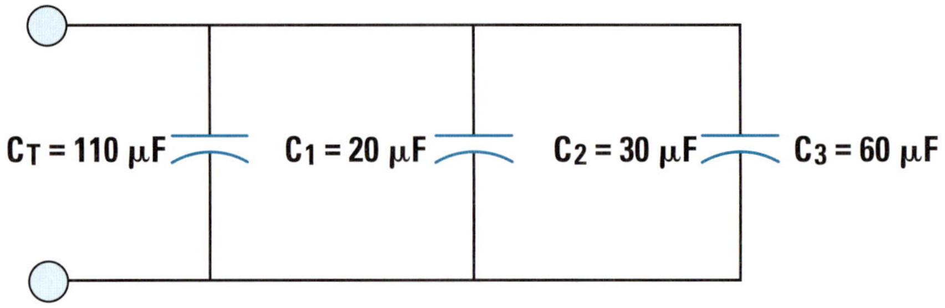

Figure 19–11 Capacitors connected in parallel.

Capacitors Connected in Series

Connecting capacitors in series (*Figure 19–12*) has the effect of increasing the distance between the plates, thus reducing the total capacitance of the circuit. The total capacitance can be calculated in a manner similar to that for calculating parallel resistance. The following formulas can be used to find the total capacitance when capacitors are connected in series:

$$C_T = \frac{1}{\frac{1}{C_1} + \frac{1}{C_2} + \frac{1}{C_3}}$$

or

$$C_T = \frac{C_1 \times C_2}{C_1 + C_2}$$

or

$$C_T = \frac{C}{N}$$

where

C = capacitance of one capacitor

N = number of capacitors connected in series

Note: The last formula can be used only when all the capacitors connected in series have the same value.

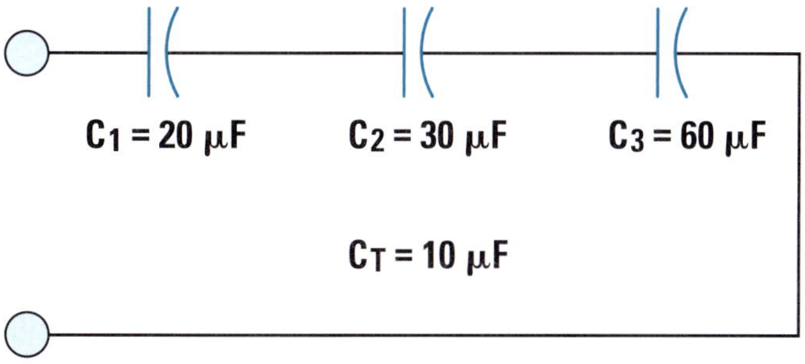

Figure 19–12 Capacitors connected in series.

Example 19–2

What is the total capacitance of three capacitors connected in series if C_1 has a capacitance of 20 μF, C_2 has a capacitance of 30 μF, and C_3 has a capacitance of 60 μF?

Solution

$$C_T = \frac{1}{\frac{1}{C_1} + \frac{1}{C_2} + \frac{1}{C_3}}$$

$$C_T = \frac{1}{\frac{1}{20\ \mu F} + \frac{1}{30\ \mu F} + \frac{1}{60\ \mu F}}$$

$$C_T = 10\ \mu F$$

Capacitive Charge and Discharge Rates

Capacitors charge and discharge at an **exponential** rate. A charge curve for a capacitor is shown in *Figure 19–13*. The curve is divided into five time constants, and during each time constant, the voltage changes by an amount equal to 63.2% of the maximum amount that it can change. In *Figure 19–13,* it is assumed that a capacitor is to be charged to a total of 100 volts. At the end of the first time constant, the voltage has reached 63.2% of 100, or 63.2 volts. At the end of the second time constant, the voltage reaches 63.2% of the remaining voltage, or 86.4 volts. This pattern continues until the capacitor has been charged to 100 volts.

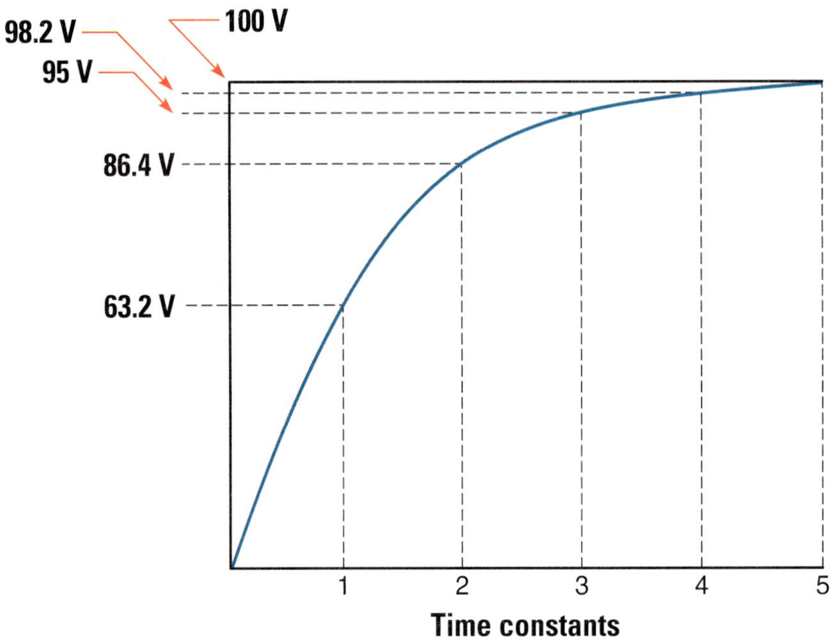

Figure 19–13 Example of a curve showing the rate at which a capacitor is charged.

The capacitor discharges in the same manner (*Figure 19–14*). At the end of the first time constant, the voltage will decrease by 63.2% of its charged value. In this example, the voltage decreases from

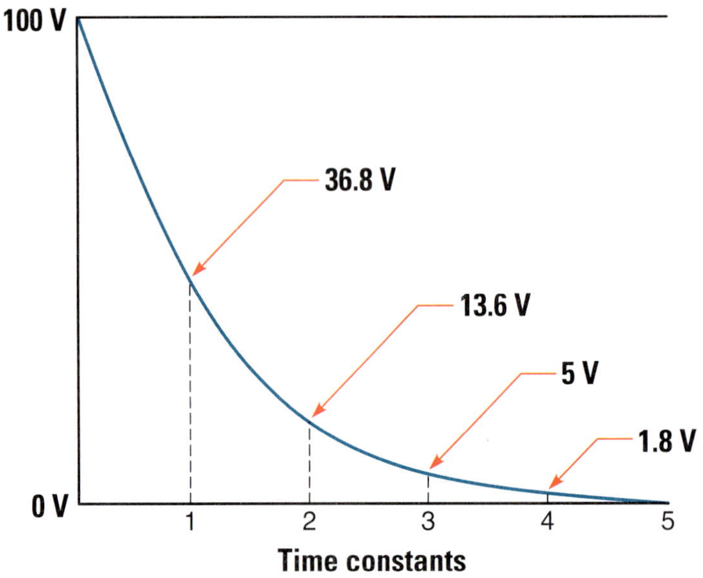

Figure 19–14 Capacitor discharge curve shows the rate at which capacitors discharge.

100 volts to 36.8 volts in the first time constant. At the end of the second time constant, the voltage will drop to 13.6 volts, and by the end of the third time constant, the voltage will drop to 5 volts. The voltage will continue to drop at this rate until it reaches approximately 0 after five time constants.

RC Time Constants

When a capacitor is connected in a circuit with a resistor (*Figure 19-15*), the amount of time needed to charge the capacitor—known as the **RC time constant**—can be determined by the following formula:

$$\tau = R \times C$$

where

τ = the time for one time constant in seconds

R = resistance in ohms

C = capacitance in farads

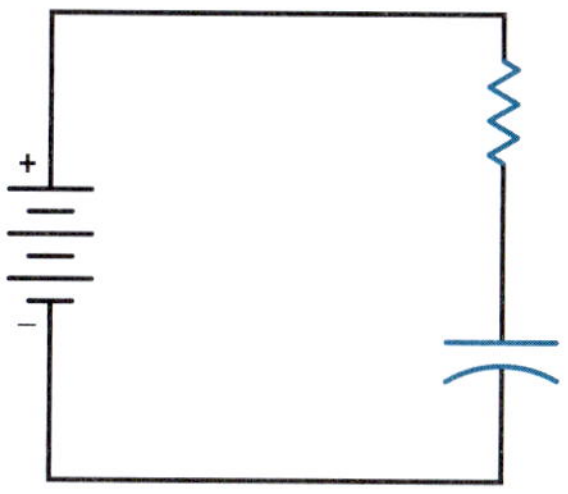

Figure 19–15 Capacitor charging times can be determined by calculating the RC time constant.

The Greek letter τ (tau) is used to represent the time for one time constant. It is not unusual, however, for the letter *t* to be used to represent time.

Example 19–3

How long will it take the capacitor shown in *Figure 19–14* to charge if it has a value of 50 μF and the resistor has a value of 100 kW?

Solution

$$\tau = R \times C$$

$$\tau = 0.000050\ \text{F} \times 100{,}000\ \Omega$$

$$\tau = 5\ \text{s}$$

The formula is used to find the time for one time constant. Five time constants are required to charge the capacitor:

$$\text{Total time} = 5\ \text{s} \times 5\ \text{time constants}$$

$$\text{Total time} = 25\ \text{s}$$

Example 19–4

How much resistance should be connected in series with a 100-pF capacitor to give it a total charge time of 0.2 s?

Solution

Change the preceding formula to solve for resistance:

$$R = \frac{\tau}{C}$$

The total charge time is to be 0.2 s. The value of τ is therefore 0.2/5 = 0.04 s. Substitute these values in the formula:

$$R = \frac{0.04\ \text{s}}{100^{\times 10^{-12}}\ \text{F}}$$

$$R = 400\ \text{M}\Omega$$

Example 19–5

A 500-kW resistor is connected in series with a capacitor. The total charge time of the capacitor is 15 s. What is the capacitance of the capacitor?

Solution

Change the base formula to solve for the value of capacitance:

$$C = \frac{\tau}{R}$$

Because the total charge time is 15 s, the time of one time constant will be 3 s (15 s/5 = 3 s):

$$C = \frac{3\ \text{s}}{500{,}000\ \Omega}$$

$$C = 0.000006\ \text{F}$$

or

$$C = 6\ \mu\text{F}$$

19–4 Capacitor Applications

Capacitors are found in nearly every aspect of our electrified world today. One application where we commonly find capacitors is in applications where they are used as a counter measure to induction in a circuit and to correct power factor. This is possible as capacitance has the opposite effect in a circuit from inductance by causing current to lead voltage instead of lag as in inductive circuits. They are also used extensively alongside solid-state electronics to produce effects such as phase shifts for silicon-controlled rectifier (SCR) and Triac circuits and filtering pulsating DC, and they are often used in RC timing circuits. (SCRs and Triacs are solid-state electronic devices used throughout industry to control high-current circuits.) Capacitors are also used extensively in electronic circuits for control of frequency and pulse generation. The type of capacitor used is dictated by the circuit application.

Nonpolarized Capacitors

Capacitors can be divided into two basic groups: nonpolarized and polarized. **Nonpolarized capacitors** are often referred to as *AC capacitors* because they are not sensitive to polarity connection. Nonpolarized capacitors can be connected to either DC or AC circuits without harm to the capacitor. Nonpolarized capacitors are constructed by separating metal plates by some type of dielectric (*Figure 19–1*). These capacitors can be obtained in many different styles and case types (*Figure 19–16*).

Figure 19–16 Examples of some commonly used nonpolarized capacitors. Vishay Intertechnology, Inc.

A common type of AC capacitor called the paper capacitor or oil-filled paper capacitor is often used in motor circuits and for power factor correction (*Figure 19–17*). It derives its name from the type of

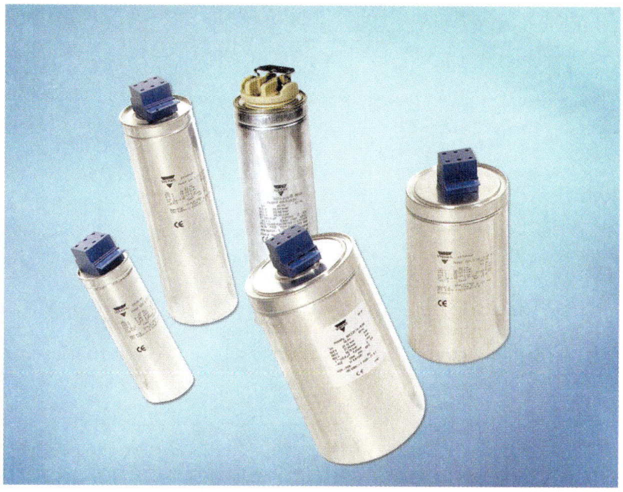

Figure 19–17 Examples of oil-filled paper capacitors.
Vishay Intertechnology, Inc.

dielectric used. This capacitor is constructed by separating plates made of metal foil with thin sheets of paper soaked in a dielectric oil (*Figure 19–18*). These capacitors are sometimes used as the run or starting capacitors for single-phase motors, if one is needed. Many manufacturers of oil-filled capacitors identify one terminal with an arrow, a painted dot, or a stamped dash in the capacitor can (*Figure 19–19*). This identified terminal marks the connection to the plate that is located nearer to the metal container or can. It has long been known that when a capacitor's dielectric breaks down and permits a short circuit to ground, the plate nearer to the outside case most often becomes grounded. For this reason, it is generally desirable to connect the identified capacitor terminal to the line side instead of to the motor start winding.

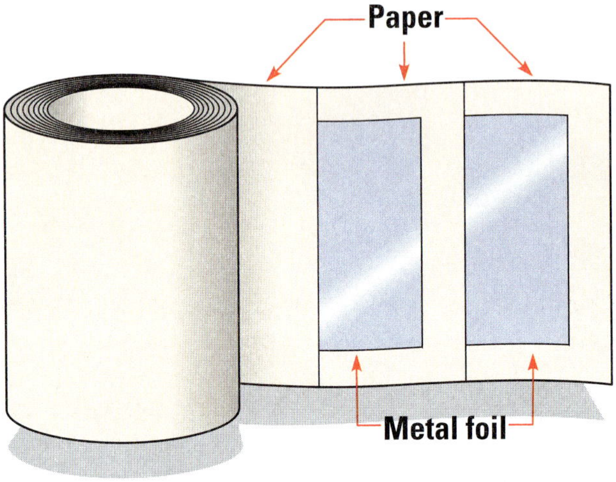

Figure 19–18 Example of how an oil-filled paper capacitor is constructed.

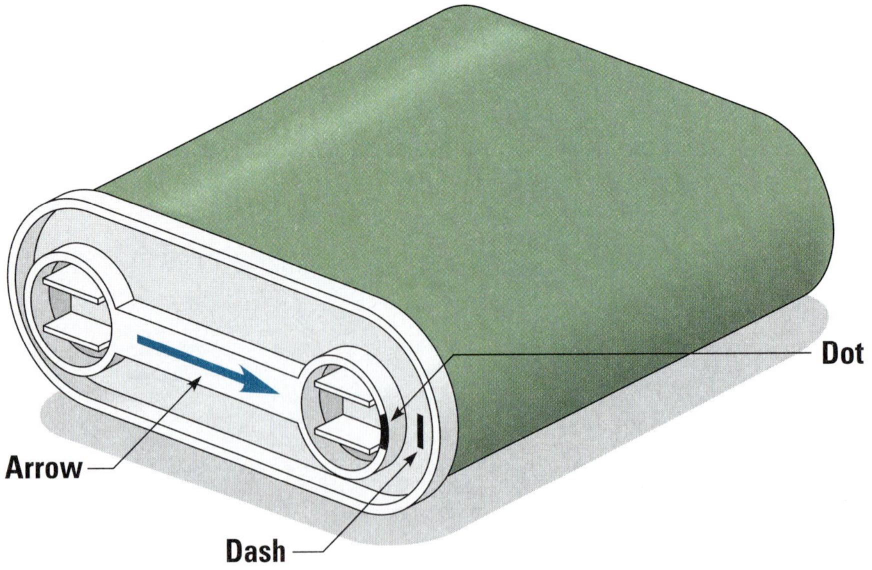

Figure 19–19 The terminal closest to the capacitor case is often marked by the manufacturer.

In *Figure 19–20,* a capacitor has been connected in such a manner that the terminal marked as being closest to the case is connected to the start winding of a single-phase motor. If the capacitor developed an internal fault to the case, a current path returning to the power source still remains through the motor

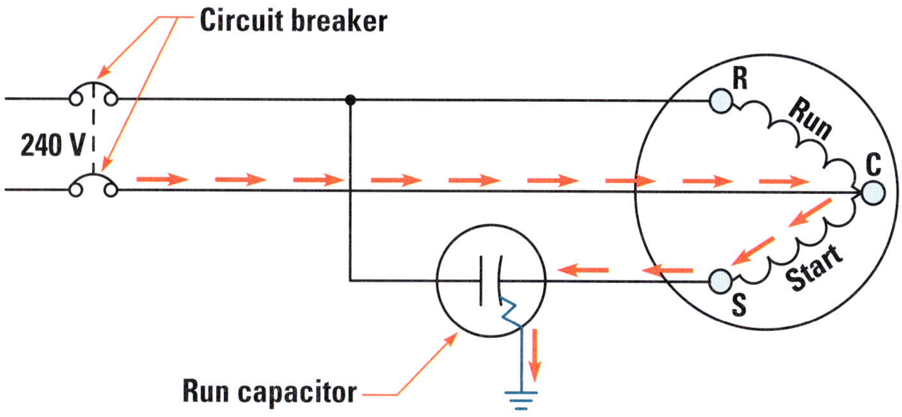

Figure 19–20 Identified capacitor terminal connected to motor start winding (incorrect connection).

start winding and the circuit equipment grounding conductor pathway. The start winding is an inductive-type load, and inductive reactance will limit the value of current flowing in the circuit. Because the flow of current is limited, it will delay the circuit breaker or fuse from opening the circuit and could result in injury or damage to property.

In *Figure 19–21,* the run capacitor has been connected in the circuit in such a manner that the identified terminal is connected to the line side. If the capacitor should develop a fault to the case, the current path will now be from the circuit conductor to the capacitor case and onto the equipment grounding conductor pathway, if one is installed. The current then has a pathway that bypasses the motor windings altogether. When the capacitor is connected in this manner, there is nothing limiting the current flow, and it facilitates the automatic operation of the fuse or circuit breaker.

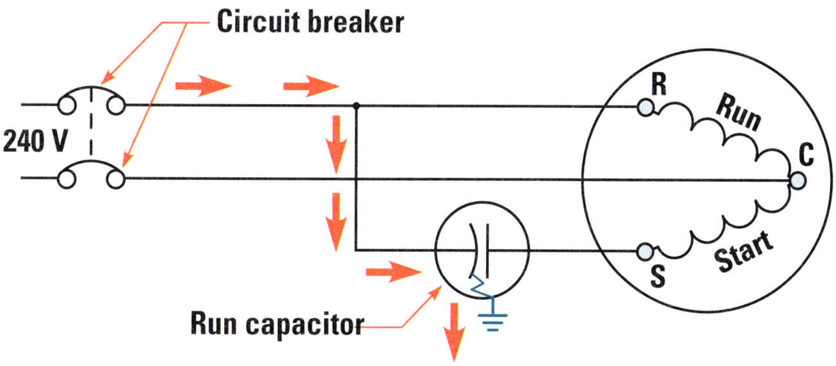

Figure 19–21 Identified terminal connected to the line bypasses the motor during a ground fault (correct connections).

Polarized Capacitors

Polarized capacitors are generally referred to as **electrolytic** capacitors. These capacitors are sensitive to the polarity they are connected to and have one terminal identified as positive or negative (*Figure 19–22*). Polarized capacitors can be used in DC circuits only. If their polarity connection is reversed, the capacitor can be damaged and will sometimes explode. The advantage of electrolytic capacitors is that they can have very high capacitance in a small case.

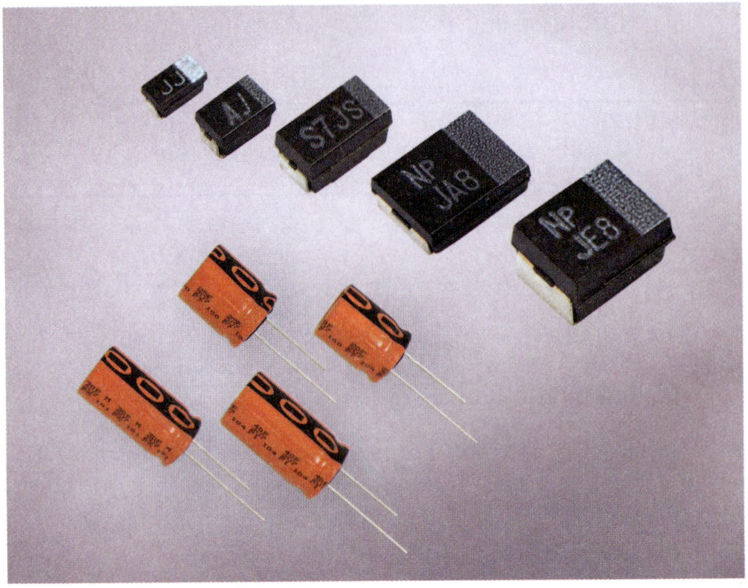

Figure 19–22 Examples of different types of polarized capacitors. Vishay Intertechnology, Inc.

The two basic types of electrolytic capacitors are the wet type and the dry type. The wet-type electrolytic capacitor (*Figure 19–23*) has a positive plate made of aluminum foil. The negative plate is actually an electrolyte made from a borax solution. A second piece of aluminum foil is placed in contact with the electrolyte and becomes the negative terminal. When a source of DC is connected to the capacitor, the borax solution forms an insulating oxide film on the positive plate. This film is only a few molecules thick and acts as the insulator to separate the plates. The capacitance is very high because the distance between the plates is so small.

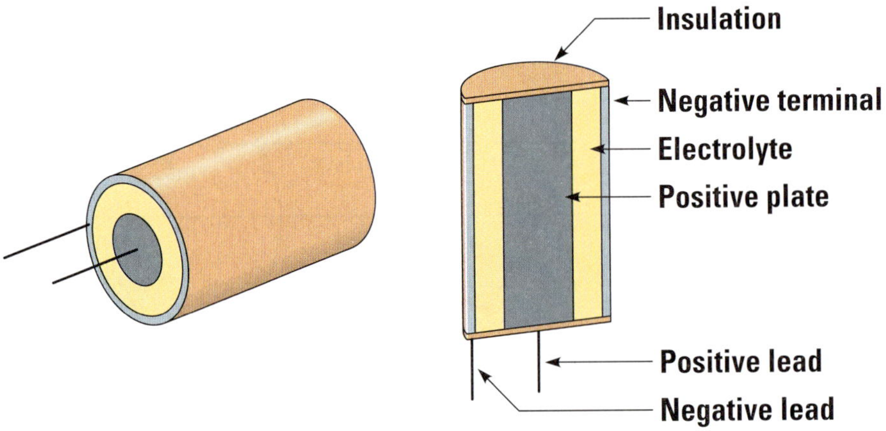

Figure 19–23 Example of how a wet-type electrolytic capacitor is constructed.

If the polarity of the wet-type electrolytic capacitor becomes reversed, the oxide insulating film dissolves and the capacitor becomes shorted. If the polarity connection is corrected, the film reforms and restores the capacitor.

AC Electrolytic Capacitors

This ability of the wet-type electrolytic capacitor to be shorted and then reformed is the basis for a special type of nonpolarized electrolytic capacitor called the AC electrolytic capacitor. This capacitor is used as the starting capacitor for many small single-phase motors, as the run capacitor in many ceiling fan motors, and for low-power electronic circuits when a nonpolarized capacitor with a high capacitance is required. The AC electrolytic capacitor is made by connecting two wet-type electrolytic capacitors together inside the same case (*Figure 19–24*). In the example shown, the two wet-type electrolytic capacitors have their negative terminals connected. When AC is applied to the leads, one capacitor will be connected to reverse polarity and become shorted. The other capacitor will be connected to the correct polarity and will form. During the next half cycle, the polarity changes and forms the capacitor that was shorted and shorts the other capacitor. An AC electrolytic capacitor is shown in *Figure 19–25*.

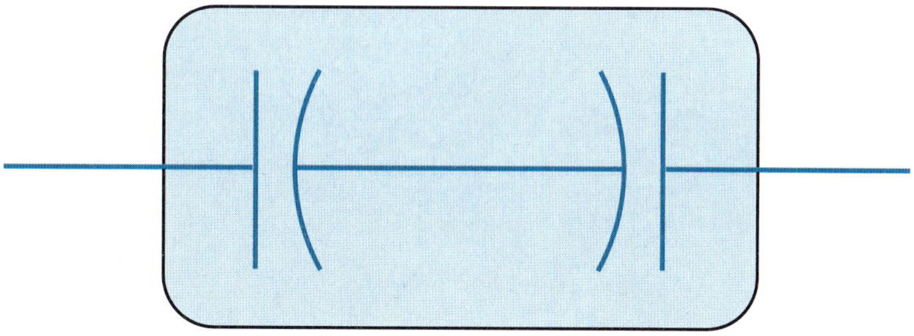

Figure 19–24 Two wet-type electrolytic capacitors connect to form an AC electrolytic capacitor.

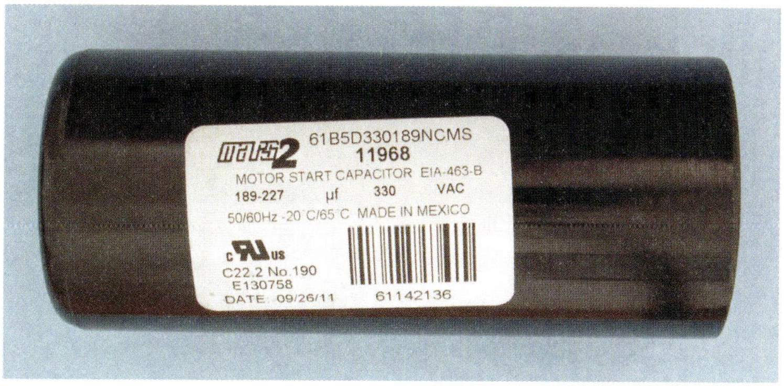

Figure 19–25 An AC electrolytic capacitor.

Dry-Type Electrolytic Capacitors

The dry-type electrolytic capacitor is very similar to the wet type except that gauze is used to hold the borax solution. This prevents the capacitor from leaking. Although the dry-type electrolytic capacitor has the advantage of being relatively leak proof, it does have one disadvantage. If the polarity connection should be reversed and the oxide film is broken down, it will not reform when connected to the proper polarity. Reversing the polarity of a dry-type electrolytic capacitor permanently damages the capacitor. Dry-type capacitors should never be connected directly to an AC circuit without the use of a device such as a diode to prevent current flowing in the incorrect polarity.

Variable Capacitors

Variable capacitors are constructed in such a manner that their capacitance value can be changed over a certain range. They generally contain a set of movable plates, which are connected to a shaft, and a set of stationary plates (*Figure 19–26*). The movable plates interlock with the stationary plates to increase or decrease the capacitance value. Because air is used as the dielectric and the plate area is relatively small, variable capacitors are generally rated in picofarads. Another type of small variable capacitor is called the trimmer capacitor (*Figure 19–27*). This capacitor has one movable plate and one stationary plate. The capacitance value is changed by turning an adjustment screw that moves the movable plate closer to or farther away from the stationary plate. *Figure 19–28* shows schematic symbols used to represent variable capacitors.

Figure 19–26 Example of a type of variable capacitor. © SeDmi/Shutterstock.com

Figure 19–27 Example of different trimmer capacitors. Vishay Intertechnology, Inc.

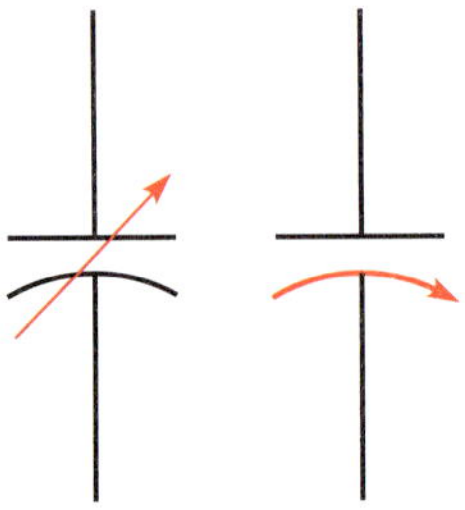

Figure 19–28
Examples of variable capacitor symbols.

19–5 Capacitor Types and Markings

Different types of capacitors are marked in different ways. Large AC oil-filled paper capacitors generally have their capacitance and voltage values written on the capacitor. The same is true for most electrolytic and small nonpolarized capacitors. Some smaller types of capacitors even use a color code system similar to that of resistors to indicate capacitance and voltage rating. However, this is not common, and there is no standard color code for this.

Two methods for color coding mica capacitors that do follow a standard are the EIA (Electronic Industries Association) standard and the **JAN (Joint Army-Navy) standard.** The JAN standard is used for electronic components intended for military use. When the EIA standard is employed, the first dot is white. In some instances, the first dot may be silver instead of white. This indicates that the capacitor's dielectric is paper instead of mica. When the JAN standard is used, the first dot is black. The second and third dots represent digits, the fourth dot is the multiplier, the fifth dot is the tolerance, and the sixth dot indicates classes A to E of temperature and leakage coefficients. However, neither method is common as most manufacturers opt to mark the capacitor with specific values.

Temperature Coefficients

The temperature coefficient indicates the amount of capacitance change with temperature. Temperature coefficients are listed in parts per million (ppm) per degree Celsius. A positive temperature coefficient indicates that the capacitor will increase its capacitance with an increase in temperature. A negative temperature coefficient indicates that the capacitance will decrease with an increase in temperature.

Ceramic Capacitors

Another capacitor that often uses color codes is the ceramic capacitor (*Figure 19–29*). This capacitor generally has one band that is wider than the others. The wide band indicates the temperature coefficient, and the other bands are the first and second digits, multiplier, and tolerance.

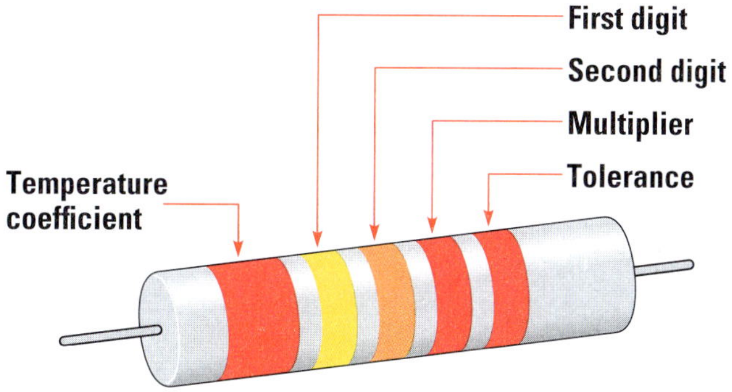

Color	Number	Multiplier	Tolerance: Over 10 pF	Tolerance: 10 pF or Less	Temp. Coeff.
Black	0	1	20%	2.0 pF	0
Brown	1	10	1%		N30
Red	2	100	2%		N80
Orange	3	1000			N150
Yellow	4				N220
Green	5				N330
Blue	6		5%	0.5 pF	N470
Violet	7				N750
Gray	8	0.01		0.25 pF	P30
White	9	0.1	10%	1.0 pF	P500

Figure 19–29 Typical color coding for ceramic capacitors.

Dipped Tantalum Capacitors

A dipped tantalum capacitor is shown in *Figure 19–30*. This capacitor has the general shape of a match head but is somewhat larger. Color bands and dots determine the value, tolerance, and voltage. The capacitance value is given in picofarads.

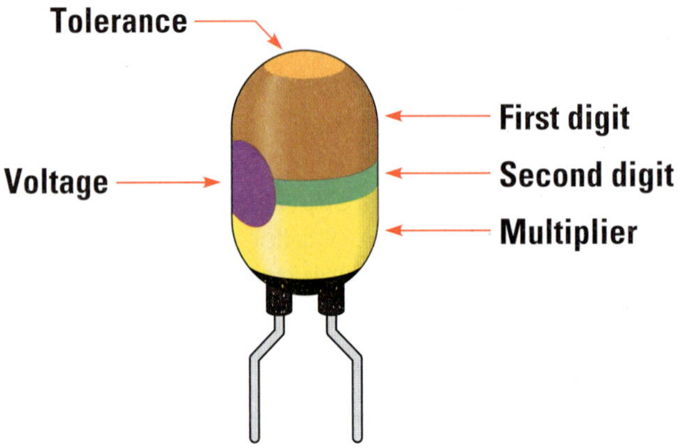

Color	Number	Multiplier	Tolerance (%)	Voltage
			No dot 20	
Black	0			4
Brown	1			6
Red	2			10
Orange	3			15
Yellow	4	10,000		20
Green	5	100,000		25
Blue	6	1,000,000		35
Violet	7	10,000,000		50
Gray	8			
White	9			3
Gold			5	
Silver			10	

Figure 19–30 Example of a dipped tantalum capacitor.

Film Capacitors

Not all capacitors use color codes to indicate values. Some capacitors use numbers and letters. A film-type capacitor is shown in *Figure 19–31*. This capacitor is marked 105 K. The value can be read as follows:

1. The first two numbers indicate the first two digits of the value.
2. The third number is the multiplier. Add the number of zeros to the first two numbers indicated by the multiplier. In this example, add five zeros to 10. The value is given in picofarads (pF). This capacitor has a value of 1,000,000 pF or 1 μF.
3. The K is the tolerance. In this example, K indicates a tolerance of $\pm$10%.

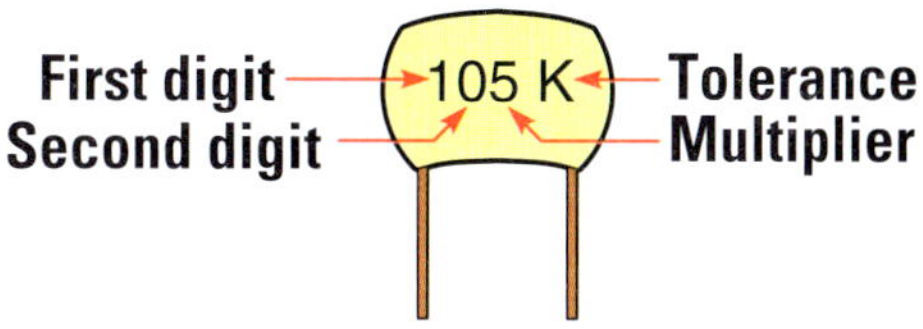

Number	Multiplier	Tolerance		
			10 pF or Less	Over 10 pF
0	1	B	0.1 pF	
1	10	C	0.25 pF	
2	100	D	0.5 pF	
3	1000	F	1.0 pF	1%
4	10,000	G	2.0 pF	2%
5	100,000	H		3%
6		J		5%
7		K		10%
8	0.01	M		20%
9	0.1			

Figure 19–31 Film-type capacitors.

19–6 Testing Capacitors

Testing capacitors is difficult at best. Small electrolytic capacitors are generally tested for shorts with an ohmmeter. If the capacitor is not shorted, it should be tested for leakage using a variable DC power supply and a microammeter (*Figure 19–32*). When rated voltage is applied to the capacitor, the microammeter should indicate zero current flow.

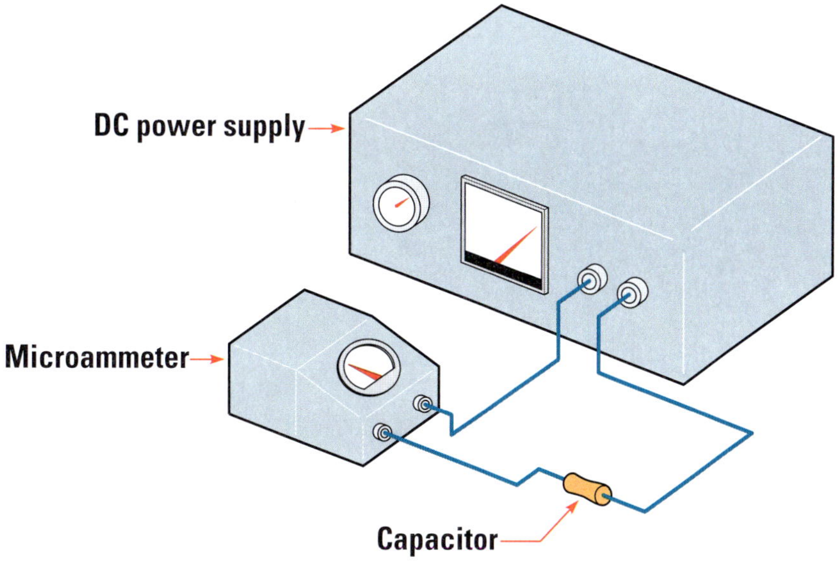

Figure 19–32 Testing a capacitor for leakage.

Large AC oil-filled capacitors can be tested in a similar manner. To test the capacitor accurately, two measurements must be made. One is to measure the capacitance value of the capacitor to determine if it is the same or approximately the same as the rated value. The other is to test the strength of the dielectric.

The first test should be made with an ohmmeter. With the power disconnected, connect the terminals of an ohmmeter directly across the capacitor terminals (*Figure 19–33*). This test determines if the dielectric is shorted. If the capacitor is shorted, the ohmmeter will read a low resistance value. If the capacitor is not shorted, the ohmmeter will display a high value of resistance to indicate an open loop.

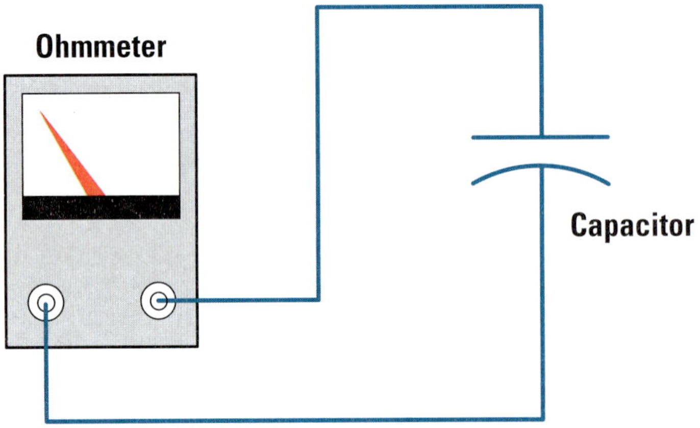

Figure 19–33 Testing the capacitor with an ohmmeter.

If the ohmmeter test is successful, the dielectric must be tested at its rated voltage. This is called a dielectric strength test. To make this test, a dielectric test set must be used (*Figure 19–34*). This device is often referred to as a **HIPOT** because of its ability to produce a high voltage or high potential. The dielectric test set contains a variable voltage control, a voltmeter, and a microammeter. To use the HIPOT, connect its terminal leads to the capacitor terminals. Increase the output voltage until the rated voltage is applied to the capacitor. The microammeter indicates any current flow between the plates of the dielectric. If the capacitor is good, the microammeter should indicate zero current flow.

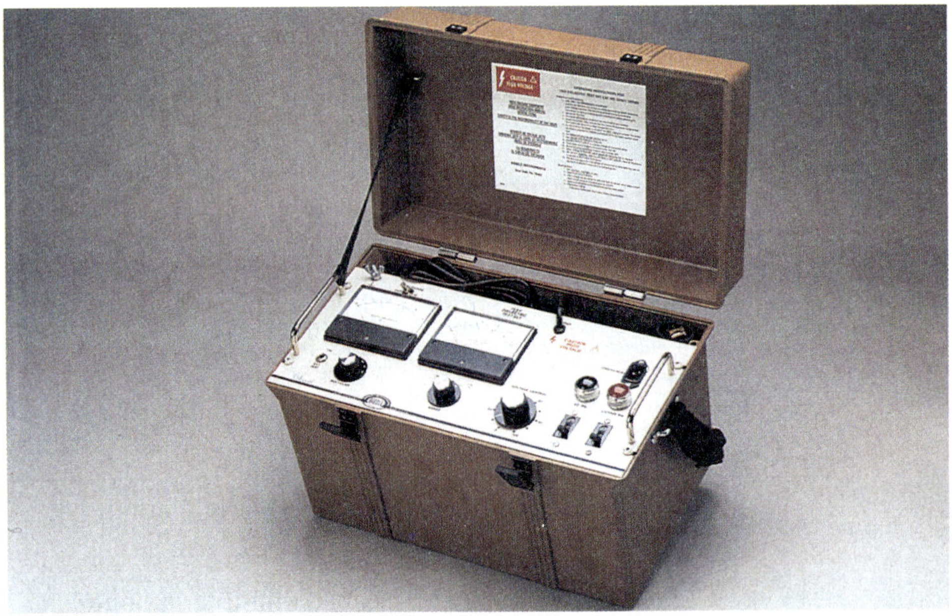

Figure 19–34 A dielectric test set. Courtesy of Megger®

The capacitance value must be measured to determine if there are any open plates in the capacitor. To measure the capacitance value of the capacitor, connect some value of AC voltage across the plates of the capacitor (*Figure 19–35*). This voltage must not be greater than the rated capacitor voltage. Then, measure the amount of current flow in the circuit. Now that the voltage and current flow are known, the capacitive reactance of the capacitor can be calculated using the formula:

$$X_C = \frac{E}{I}$$

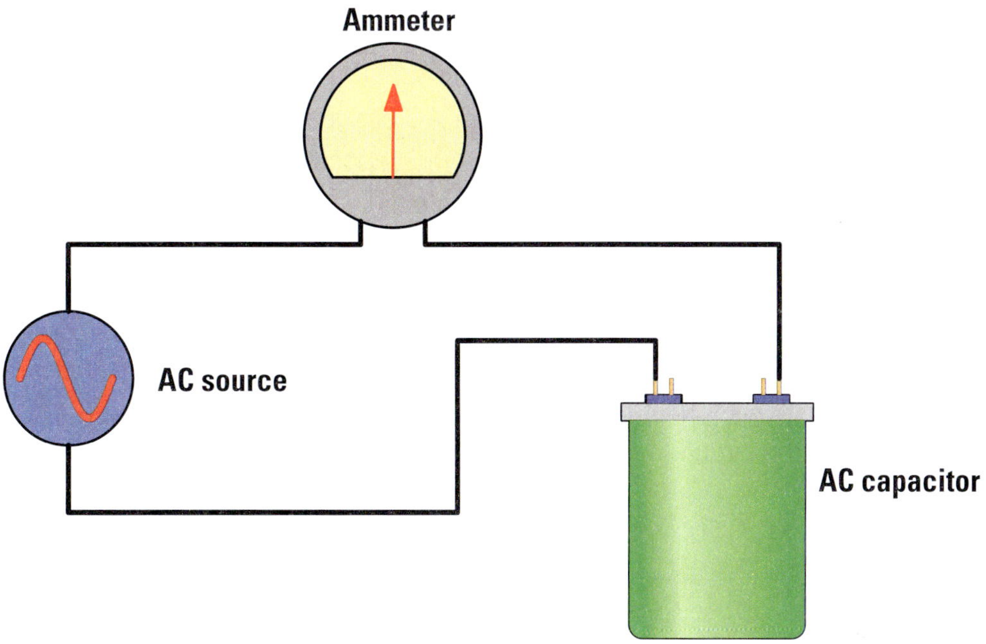

Figure 19–35 Determining the capacitance value.

After the capacitive reactance has been determined, the capacitance can be calculated using the formula:

$$C = \frac{1}{2\pi f X_C}$$

Note: Capacitive reactance is measured in ohms and limits current flow in a manner similar to inductive reactance. Capacitive reactance is covered in detail in Unit 20.

Summary

- Capacitors are devices that oppose a change of voltage.
- Three factors that determine the capacitance of a capacitor are
  a. the surface area of the plates.
  b. the distance between the plates.
  c. the type of dielectric used.
- A capacitor stores energy in an electrostatic field.
- Current can flow only during the time a capacitor is charging or discharging.

- Capacitors charge and discharge at an exponential rate.
- The basic unit of capacitance is the farad (F).
- Capacitors are generally rated in microfarads (μF), nanofarads (nF), or picofarads (pF).
- When capacitors are connected in parallel, their capacitance values add.
- When capacitors are connected in series, the reciprocal of the total capacitance is equal to the sum of the reciprocals of all the capacitors.
- The charge and discharge times of a capacitor are proportional to the amount of capacitance and resistance in the circuit.
- Five time constants are required to charge or discharge a capacitor.
- Nonpolarized capacitors are often called AC capacitors.
- Nonpolarized capacitors can be connected to DC or AC circuits.
- Polarized capacitors are often referred to as electrolytic capacitors.
- Polarized capacitors can be connected to DC circuits only.
- There are two basic types of electrolytic capacitors, the wet type and the dry type.
- Wet-type electrolytic capacitors can be reformed if reconnected to the correct polarity.
- Dry-type electrolytic capacitors will be permanently damaged if connected to the incorrect polarity.
- Capacitors are often marked with color codes or with numbers and letters.
- To test a capacitor for leakage, a microammeter should be connected in series with the capacitor and rated voltage applied to the circuit.

Review Questions

1. Which of the following factors determine the capacitance of a capacitor? (Select all that apply.)

   A) The surface area of the plates

   B) The frequency of the AC source

   C) The type of dielectric used

   D) The system voltage

2. The dielectric material of a capacitor can best be described by which of the following statements?

   A) A chemical material that produces electricity

   B) A conductive metal used for the plates

   C) The outer coating of the capacitor

   D) An insulating material between the plates

3. True or False: A capacitor stores energy in the form of a magnetic field.

4. As a capacitor becomes charged, the plates develop a potential difference between them. Which of the following terms is a result of the different charges and is like drawing back the bowstring of a bow?

   A) Dielectric constant

   B) Dielectric stress

   C) Leakage current

   D) Polarized capacitor

5. Connecting capacitors in series has an overall effect on the circuit. Which of the following statements best describes the impact on a circuit if capacitors are connected in series?
   A) Capacitance in series adds together.
   B) Capacitance in series increases the resistance.
   C) Capacitance in series adds similarly to resistance in parallel.
   D) Capacitance in series decreases the total current.
6. Four capacitors having values of 20, 50, 40, and 60 μF are connected in parallel. What is the total capacitance of this circuit?
   A) 8.95 μF
   B) 42.5 μF
   C) 100 μF
   D) 170 μF
7. True or False: Nonpolarized capacitors are often referred to as AC capacitors because they are not sensitive to which polarity the connection is.
8. Polarized capacitors are sensitive to the polarity of the connection and will typically have one terminal identified. Which of the following is another name for these types of capacitors?
   A) Electrolytic capacitors
   B) Dielectric capacitors
   C) Variable capacitors
   D) Dry-type capacitors
9. The Electronic Industries Association employs a color-coding system for mica capacitors. Which of the following is an indication that the color-coding is following the EIA standard?
   A) The first dot is black.
   B) The first dot is silver.
   C) The first dot is white.
   D) The first dot is gold.
10. A positive temperature coefficient indicates that an increase in temperature will have what type of impact on the capacitance?
    A) The capacitance will decrease.
    B) The capacitance will increase.
    C) The capacitance will remain the same.
    D) There is not enough information provided.
11. Which of the following tests can help indicate if a capacitor is experiencing leakage current through the dielectric material?
    A) Test for leakage with an ohmmeter
    B) Test for leakage by connecting a wattmeter

C) Test for leakage using an AC power supply and microammeter

D) Test for leakage using a DC power supply and microammeter

12. What is being tested if the device used to test the capacitor is also known as a HIPOT?

A) Testing the dielectric material

B) Testing the integrity of the plates

C) Testing the frequency of the system

D) Testing for capacitive reactance

Practical Applications

1. You are changing the starting relay on a central air-conditioning unit when you notice that the identifying mark on the compressor-run capacitor is connected to the run winding of the compressor. Should you change the capacitor connection so the identifying mark is facing the line side, or is it correct as connected? Explain your answer.

---

2. You are an electrician working in an industrial plant. You discover that the problem with a certain machine is a defective capacitor. The capacitor is connected to a 240-volt AC circuit. The information on the capacitor reveals that it has a capacitance value of 10 mF and a voltage rating of 240 VAC. The only 10-mF AC capacitor in the storeroom is marked with a voltage rating of 350 VDC. Can this capacitor be used to replace the defective capacitor? Explain your answer.

---

3. You find that a 25-μF capacitor connected to 480 VAC is defective. The storeroom has no capacitors with a 480-VAC rating. However, you find two capacitors rated at 50 μF and 370 VAC. Can these two capacitors be connected in such a manner that they can replace the defective capacitor? If yes, explain how they are connected and why the capacitors will not be damaged by the lower voltage rating. If no, explain why they cannot be used without damaging the capacitor.

Practice Problems

RC Time Constants

1. Fill in all the missing values. Refer to the formulas that follow.

Resistance	Capacitance	Time constant	Total time
150 kW	100 μF		
350 kW			35 s
	350 pF	0.05 s	
	0.05 μF		10 s
1.2 MW	0.47 μF		
	12 μF	0.05 s	
86 kW			1.5 s
120 kW	470 pF		
	250 nF		100 ms
	8 μF		150 μs
100 kW		150 ms	
33 kW	4 μF		

$$\tau = RC$$

$$R = \frac{\tau}{C}$$

$$C = \frac{\tau}{R}$$

$$\text{Total time} = \tau \times 5$$

2. Two capacitors having values of 80 μF and 60 μF are connected in series. What is the total capacitance?
3. Three capacitors having values of 120, 20, and 60 μF are connected in parallel. What is the total capacitance?
4. Three capacitors having values of 2.2 μF, 280 nF, and 470 pF are connected in parallel. What is the total capacitance?
5. A 470-μF capacitor is connected in series with a 120-kW resistor. How long will it take the capacitor to charge completely?

Unit 20 Capacitance in AC Circuits

Key Terms

Capacitive reactance (X_C)

Out of phase

Voltage rating

Why You Need to Know

Capacitors in AC circuits act in the exact opposite manner of inductors in almost every respect. Although a capacitor is technically an open circuit, you will see how current appears to flow through it. This unit will

- illustrate how to calculate the current-limiting effect of a capacitor.
- discuss the different voltage ratings that are listed on nonpolarized capacitors.
- illustrate how to connect the capacitor into an AC circuit and calculate capacitance charge and discharge rates.
- discuss the effects of frequency in a capacitance circuit.

Outline

20–1 Capacitors in an AC Circuit

20–2 Voltage, Current, and Power in a Purely Capacitive Circuit

20–3 Capacitor Quality and Voltage Ratings

20–4 Effects of Frequency in a Capacitive Circuit

20–5 Capacitors Connected in Series and Parallel

Objectives

After studying this unit, you should be able to

- illustrate how a capacitor behaves when connected to an AC power source.
- demonstrate the relationship between voltage, current, and power in a purely capacitive AC circuit.
- describe how the quality and voltage rating of a capacitor impact a circuit.
- analyze how the frequency impacts a capacitor's effect in an AC circuit.
- calculate missing circuit values for purely capacitive circuits containing both series and parallel capacitors.

Preview

In Unit 19, it was discussed that a basic capacitor is composed of two metal plates separated by an insulating material called the dielectric. Because there is no complete circuit between the plates, current cannot flow through the capacitor. The only time that current can flow is during the period of time that the capacitor is being charged or discharged.

20–1 Capacitors in an AC Circuit

When a capacitor (*Figure 20–1*) is connected to an AC circuit, current appears to flow through the capacitor. The reason is that in an AC circuit, the current continually changes direction and polarity, and the capacitor is in a constant state of charging and discharging. To understand this concept, consider the pump and tank system shown in *Figure 20–2*. Two tanks are connected to a common pump. Assume Tank A is full, and Tank B is empty. Now assume that the pump pumps water from Tank A to Tank B. When Tank B becomes full, the pump reverses and pumps the water from Tank B back into Tank A. Each time a tank becomes filled, the pump reverses and pumps water back into the other tank. Notice that water is continually flowing in this circuit, but there is not a complete circuit because there is no direct connection between the two tanks.

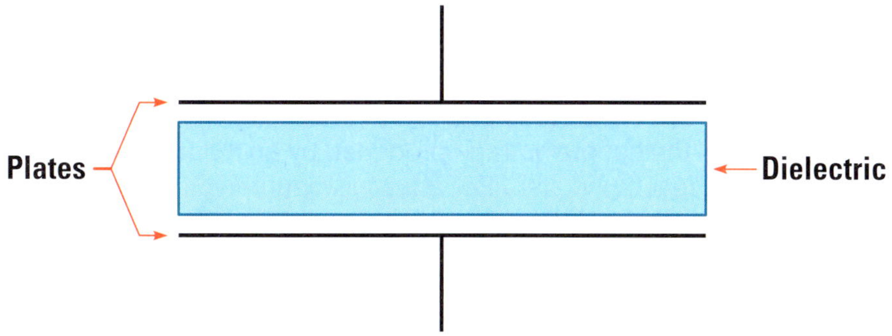

Figure 20–1 A basic capacitor consists of two plates separated by a dielectric material.

A similar action takes place when a capacitor is connected to an AC circuit (*Figure 20–3*). In this circuit, the AC power supply charges one plate of the capacitor positive and the other plate negative. During the next half cycle, the voltage changes polarity and the capacitor discharges and recharges to the opposite polarity. As long as the voltage continues to increase, decrease, and change polarity,

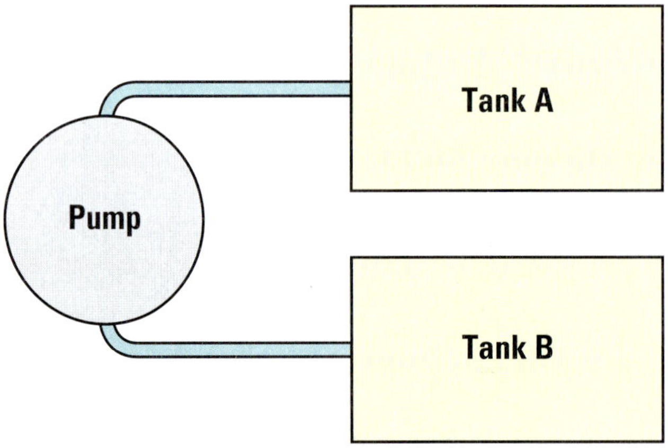

Figure 20–2 Water appears to flow continuously, even without a complete circuit.

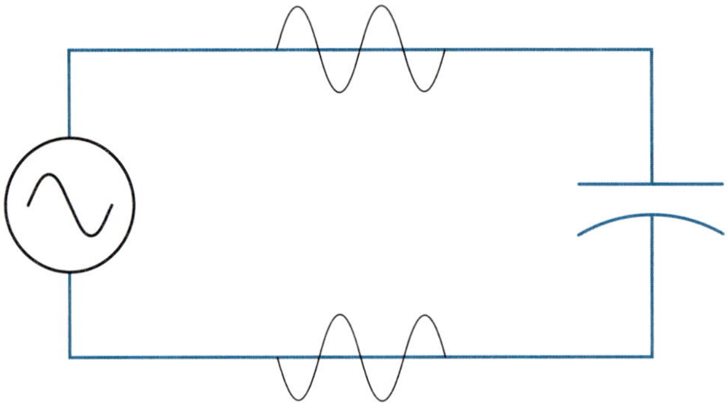

Figure 20–3 A capacitor in an AC circuit is continuously charging and discharging.

current flows from one plate of the capacitor to the other. If an ammeter were placed in the circuit, it would indicate current flowing in the circuit, appearing as though current is flowing through the capacitor.

Capacitive Reactance

As the capacitor is charged, an impressed voltage is developed across its plates as an electrostatic charge is built up (*Figure 20–4*). The impressed voltage is the voltage provided by the electrostatic charge. This impressed voltage opposes the applied voltage and limits the flow of current in the circuit. This countervoltage is similar to the countervoltage produced by an inductor. The countervoltage developed by the capacitor is called reactance also. Because this countervoltage is caused by capacitance, it is called **capacitive reactance (X_C)** and is also measured in ohms. The formula for finding capacitive reactance is:

$$X_C = \frac{1}{2\pi f C}$$

where

X_C = capacitive reactance

π = 3.1416

f = frequency in hertz

C = capacitance in farads

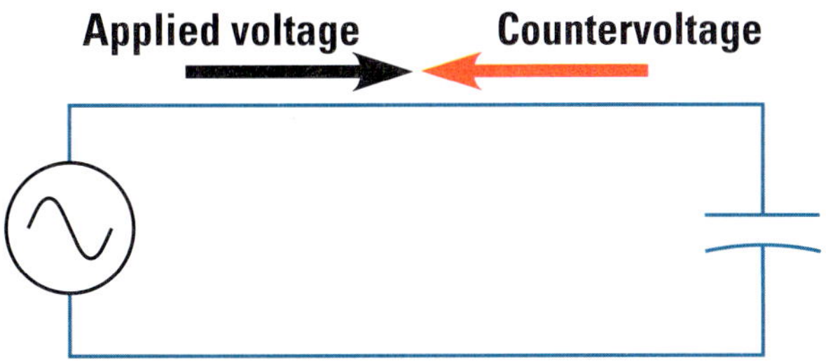

Figure 20–4 Countervoltage limits the flow of current.

Example 20–1

A 35-μF capacitor is connected to a 120-V, 60-Hz line. How much current will flow in this circuit?

Solution

The first step is to calculate the capacitive reactance. Recall that the value of C in the formula is in Farads. This means we must convert the capacitive units being used, in this case μF, to Farads:

$$X_C = \frac{1}{2 \times 3.1416 \times 60 \times (35 \times 10^{-6})}$$

$$X_C = 75.788\ \Omega$$

Once the value of capacitive reactance is known, it can be used like resistance in an Ohm's law formula. Because capacitive reactance is the current-limiting factor, it will replace the value of R:

$$I = \frac{E}{X_C}$$

$$I = \frac{120\ \text{V}}{75.788\ \Omega}$$

$$I = 1.583\ \text{A}$$

Calculating Capacitance

If the value of capacitive reactance is known, the capacitance of the capacitor can be found by transposing the capacitive reactance formula:

$$C = \frac{1}{2\pi f X_C}$$

Example 20–2

A capacitor is connected into a 480-V, 60-Hz circuit. An ammeter indicates a current flow of 2.6 A. What is the capacitance value of the capacitor?

Example 20–2 (continued)

Solution

The first step is to calculate the value of capacitive reactance. Because capacitive reactance, like resistance, limits current flow, it can be substituted for R in an Ohm's law formula:

$$X_C = \frac{E}{I}$$

$$X_C = \frac{480\ \text{V}}{2.6\ \text{A}}$$

$$X_C = 184.615\ \Omega$$

Now that the capacitive reactance of the circuit is known, the value of capacitance can be found:

$$C = \frac{1}{2\pi f X_C}$$

$$C = \frac{1}{2 \times 3.1416 \times 60\ \text{Hz} \times 184.615\ \Omega}$$

$$C = \frac{1}{69{,}598.378}$$

$$C = 0.000014368\ \text{F} = 14.368\ \mu\text{F}$$

20–2 Voltage, Current, and Power in a Purely Capacitive Circuit

Earlier in this text, it was shown that the current in a purely resistive circuit is in phase with the applied voltage and that current in a purely inductive circuit lags the applied voltage by 90°. In this unit, it will be shown that in a purely capacitive circuit, the current will *lead* the applied voltage by 90°.

When a capacitor is connected to an AC, the capacitor charges and discharges at the same rate and time as the applied voltage. The charge in coulombs is equal to the capacitance of the capacitor times the applied voltage ($Q = C \times V$). When the applied voltage is zero, the charge in coulombs and impressed voltage is also zero. When the applied voltage reaches its maximum value, positive or negative, the charge in coulombs and impressed voltage also reaches its maximum (*Figure 20–5*). The impressed voltage follows the same curve as the applied voltage.

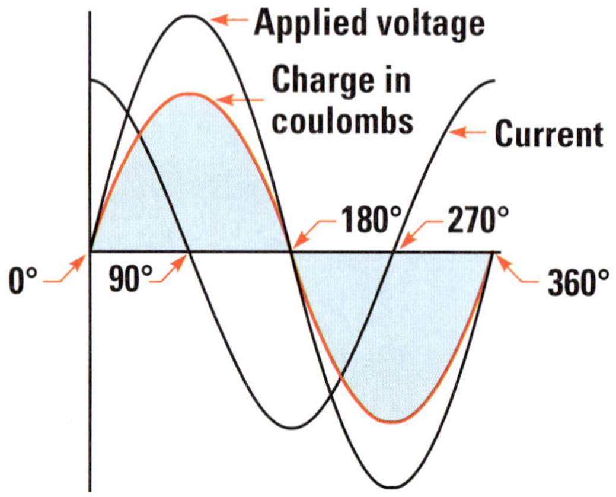

Figure 20–5 Capacitive current leads the applied voltage by 90°.

In the waveform shown, voltage and charge are both zero at 0°. Because there is no charge on the capacitor, there is no opposition to current flow, which is shown to be maximum. As the applied voltage increases from zero toward its positive peak at 90°, the capacitor begins to charge at the same time. The charge produces an impressed voltage across the plates of the capacitor that opposes the flow of current. The impressed voltage is 180° **out of phase** with the applied voltage (*Figure 20–6*). When the applied voltage reaches 90° in the positive direction, the charge reaches maximum, the impressed voltage reaches peak in the negative direction, and the current flow is zero.

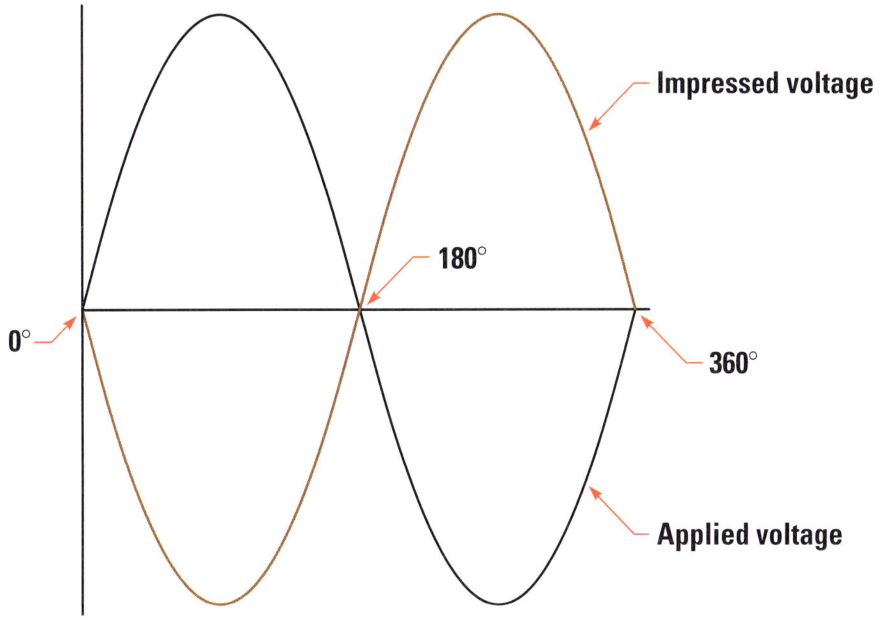

Figure 20–6 The impressed voltage is 180° out of phase with applied voltage.

As the applied voltage begins to decrease, the capacitor begins to discharge, causing the current to flow in the opposite or negative direction. When the applied voltage and charge reach zero at 180°, the impressed voltage is zero also, and the current flow is maximum in the negative direction. As the applied voltage and charge increase in the negative direction, the increase of the impressed voltage across the capacitor again causes the current to decrease. The applied voltage and charge reach maximum negative after 270° of rotation. The impressed voltage reaches maximum positive, and the current has decreased to zero (*Figure 20–7*). As the applied voltage decreases from its maximum negative value, the capacitor again begins to discharge. This causes the current to flow in the positive direction. The current again reaches its maximum positive value when the applied voltage and charge reach zero after 360° of rotation.

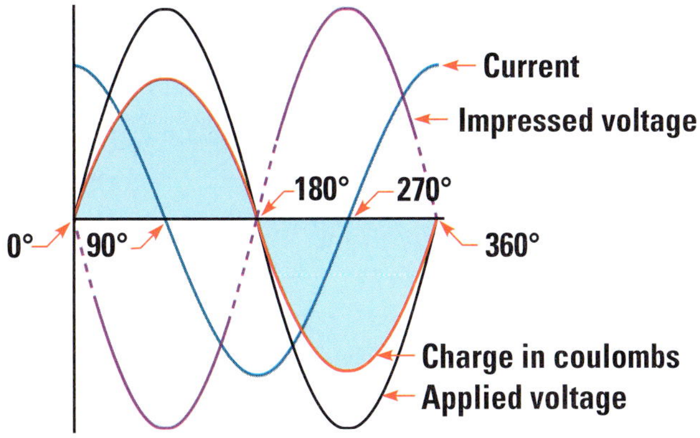

Figure 20–7 Applied and impressed voltage, current, and charge relationships for a purely capacitive circuit.

Power in a Purely Capacitive Circuit

Because the current flow in a purely capacitive circuit leads the applied voltage by 90°, the voltage and current have the same polarity for half the time during one cycle and have opposite polarities the other half of the time (*Figure 20–8*). During the period of time that the voltage and current have the same polarity, energy is being stored in the capacitor in the form of an electrostatic field. When the voltage and current have opposite polarities, the capacitor is discharging, and the energy is returned to the circuit. When the values of current and voltage for one full cycle are added, the sum equals zero just as it does with purely inductive circuits. Therefore, there is no true power, or watts, produced in a purely capacitive circuit.

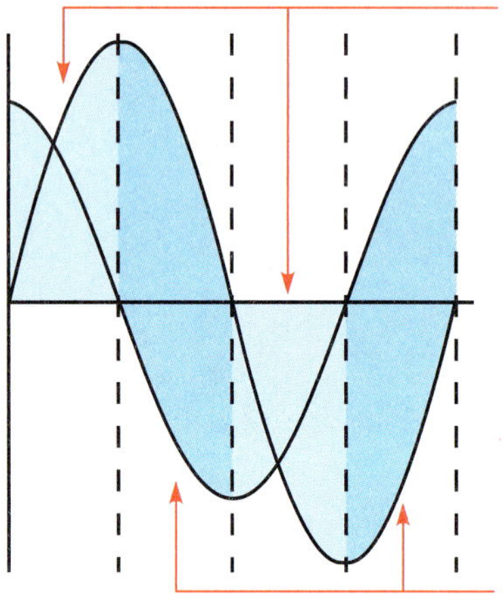

Figure 20–8 In a purely capacitive circuit, the power required to charge the capacitor is returned to the circuit when the capacitor discharges.

The power value for a capacitor is reactive power and is measured in VARs, just as it is for an inductor. Inductive VARs and capacitive VARs are 180° out of phase with each other, however (*Figure 20–9*). To distinguish between inductive and capacitive VARs, inductive VARs are labeled as $VARs_L$, and capacitive VARs are shown as $VARs_C$.

As with other types of AC circuits, power in a capacitive circuit is equal to the voltage times the current, times the cosine of angle theta:

$$P = E \times I \times \text{Cos } \angle\theta$$

In a purely capacitive circuit, the voltage and current are 90° out of phase with each other. The cosine of 90° is 0.

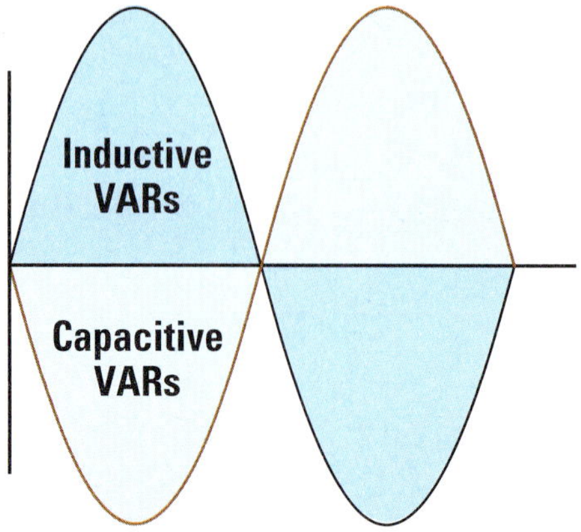

Figure 20–9 Inductive VARs and capacitive VARs are 180° out of phase with each other.

20–3 Capacitor Quality and Voltage Ratings

The quality (Q) of a capacitor is generally very high. As with inductors, it is a ratio of the resistance to capacitive reactance:

$$Q = \frac{R}{X_C}$$

The R value for a capacitor is generally very high because it is the equivalent resistance of the dielectric between the plates of the capacitor. If a capacitor is leaky, however, the dielectric will appear to be a much lower resistance and the Q rating will decrease.

The **voltage rating** of a capacitor is actually the voltage rating of the dielectric. ***Voltage rating is extremely important with regard to the life of the capacitor and should never be exceeded.*** Unfortunately, there are no set standards concerning how voltage ratings are marked. It is not unusual to see capacitors marked VOLTS AC, VOLTS DC, PEAK VOLTS, and WVDC (WORKING VOLTS DC). The voltage rating of electrolytic or polarized capacitors is always given in DC volts. The voltage rating of nonpolarized capacitors, however, can be given as AC or DC volts.

If a nonpolarized capacitor has a voltage rating given in AC volts, the voltage indicated is the RMS value. If the voltage rating is given as PEAK or as DC volts, it indicates the peak value of AC volts. If a capacitor is to be connected to an AC circuit, it is necessary to calculate the peak value if the voltage rating is given as DC volts.

Example 20–3

An AC oil-filled capacitor has a voltage rating of 300 VDC. Will the voltage rating of the capacitor be exceeded if the capacitor is connected to a 240-V, 60-Hz line?

Solution

The DC voltage rating of the capacitor indicates the peak value of voltage when connected to an AC circuit. To determine whether the voltage rating will be exceeded, find the peak value of 240 V by multiplying by 1.414:

$$\text{Peak} = 240 \text{ V} \times 1.414$$

$$\text{Peak} = 339.36 \text{ V}$$

The answer is that the capacitor voltage rating will be exceeded.

20–4 Effects of Frequency in a Capacitive Circuit

One of the factors that determines the capacitive reactance of a capacitor is the frequency. Capacitive reactance is inversely proportional to frequency. As the frequency increases, the capacitive reactance decreases. *Table 20–1* shows the capacitive reactance for different values of capacitance at different frequencies. Frequency has an effect on capacitive reactance because the capacitor charges and discharges faster at higher frequencies. Recall that current is a rate of electron flow. A current of 1 ampere is 1 coulomb per second:

$$I = \frac{C}{t}$$

where

I = current

C = charge in coulombs

t = time in seconds

	Capacitive Reactance			
Capacitance	30 Hz	60 Hz	400 Hz	1000 Hz
10 pF	530.515 M Ω	265.258 M Ω	39.789 M Ω	15.915 M Ω
350 pF	15.156 M Ω	7.579 M Ω	1.137 M Ω	454.727 k Ω
470 nF	11.286 k Ω	5.644 k Ω	846.567 k Ω	338.627 k Ω
750 nF	7.074 k Ω	3.537 k Ω	530.515 k Ω	212.206 Ω
1 μF	5.305 k Ω	2.653 k Ω	397.886 Ω	159.155 Ω
25 μF	212.206 Ω	106.103 Ω	15.915 Ω	6.366 Ω

Table 20–1 Capacitive Reactance Is Inversely Proportional to Frequency

Assume that a capacitor is connected to a 30-hertz line, and 1 coulomb of charge flows each second. If the frequency is doubled to 60 hertz, 1 coulomb of charge will flow in 0.5 second because the capacitor is being charged and discharged twice as fast (*Figure 20–10*). This means that in a period of 1 second, 2 coulombs of charge will flow. Because the capacitor is being charged and discharged at a faster rate, the opposition to current flow is decreased.

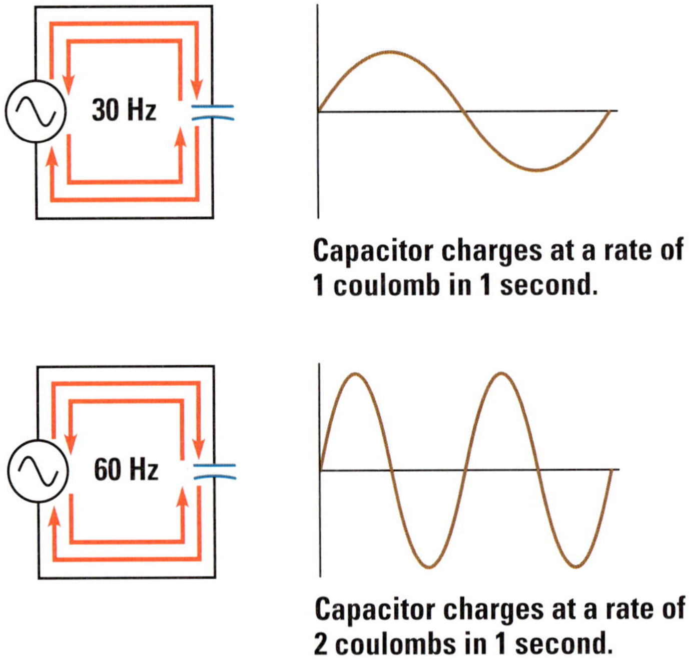

Figure 20–10 The rate of charging and discharging increases with frequency and allows more current to flow.

20–5 Capacitors Connected in Series and Parallel

Example 20–4

Three capacitors with values of 10, 30, and 15 μF are connected in series to a 480-V, 60-Hz line (*Figure 20–11*). Find the following circuit values:

X_{C1}—capacitive reactance of the first capacitor
X_{C2}—capacitive reactance of the second capacitor
X_{C3}—capacitive reactance of the third capacitor
X_{CT}—total capacitive reactance for the circuit
C_T—total capacitance for the circuit
I_T—total circuit current
E_{C1}—voltage drop across the first capacitor
$VARs_{C1}$—reactive power of the first capacitor
E_{C2}—voltage drop across the second capacitor
$VARs_{C2}$—reactive power of the second capacitor
E_{C3}—voltage drop across the third capacitor
$VARs_{C3}$—reactive power of the third capacitor
$VARs_{CT}$—total reactive power for the circuit

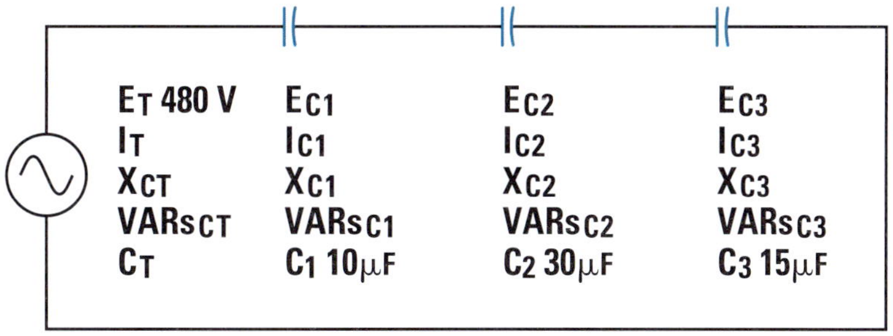

Figure 20–11 Capacitors connected in series.

Solution

Because the frequency and the capacitance of each capacitor are known, the capacitive reactance for each capacitor can be found using the formula:

$$X_C = \frac{1}{2\pi fC}$$

Recall that the value for C in the formula is in farads and the capacitors in this problem are rated in microfarads.

$$X_{C1} = \frac{1}{2\pi fC}$$

$$X_{C1} = \frac{1}{377 \times 0.000010}$$

$$X_{C1} = 265.252\ \Omega$$

Example 20–4 (continued)

$$X_{C2} = \frac{1}{2\pi fC}$$

$$X_{C2} = \frac{1}{377 \times 0.000030}$$

$$X_{C2} = 88.417\ \Omega$$

$$X_{C3} = \frac{1}{2\pi fC}$$

$$X_{C3} = \frac{1}{377 \times 0.000015}$$

$$X_{C3} = 176.835\ \Omega$$

Because there is no phase angle shift among the three capacitive loads, the total capacitive reactance is the sum of the capacitive reactance of each capacitor (*Figure 20–12*):

$$X_{CT} = X_{C1} + X_{C2} + X_{C3}$$

$$X_{CT} = 265.252\ \Omega + 88.417\ \Omega + 176.835\ \Omega$$

$$X_{CT} = 530.504\ \Omega$$

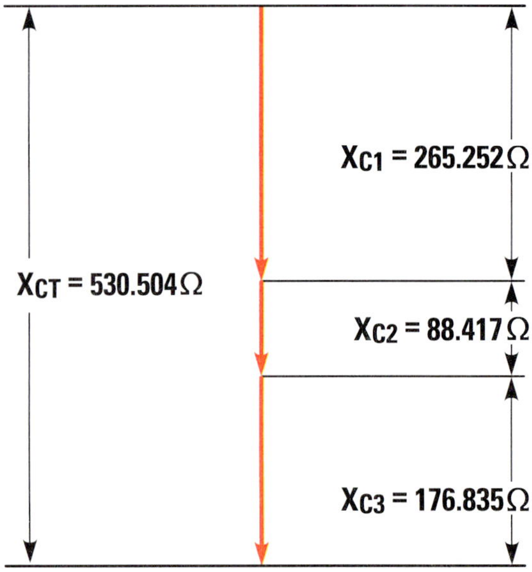

Figure 20–12 Vector sum for total capacitive reactance.

The total capacitance of a series circuit can be calculated in a manner similar to that used for calculating parallel resistance. Refer to the Purely Capacitive Circuits section of Appendix B. Total capacitance in this circuit is calculated using the formula:

$$C_T = \frac{1}{\frac{1}{C_1} + \frac{1}{C_2} + \frac{1}{C_3}}$$

$$C_T = \frac{1}{\frac{1}{10\ \mu F} + \frac{1}{30\ \mu f} + \frac{1}{15\ \mu F}}$$

Example 20–4 (continued)

$$C_T = \frac{1}{0.2\frac{1}{\mu F}}$$

$$CT = 5\ \mu F$$

The total capacitance can also be determined using the total capacitive reactance:

$$C = \frac{1}{2\pi f X_c}$$

$$C = \frac{1}{377 \times 530.504}$$

$$C = 0.000005\ F \text{ or } 5\ \mu F$$

The total current can be found by using the total capacitive reactance to substitute for R in an Ohm's law formula:

$$I_T = \frac{E_{CT}}{X_{CT}}$$

$$I_T = \frac{480\ V}{530.504\ \Omega}$$

$$I_T = 0.905\ A$$

Because the current is the same at any point in a series circuit, the voltage drop across each capacitor can now be calculated using the capacitive reactance of each capacitor and the current flowing through it.

$$E_{C1} = I_{C1} \times X_{C1}$$

$$E_{C1} = 0.905 \times 265.25$$

$$E_{C1} = 240.051\ V$$

$$E_{C2} = I_{C2} \times X_{C2}$$

$$E_{C2} = 0.905 \times 88.417$$

$$E_{C2} = 80.017\ V$$

$$E_{C3} = I_{C3} \times X_{C3}$$

$$E_{C3} = 0.905 \times 176.83$$

$$E_{C3} = 160.031\ V$$

Now that the voltage drops of the capacitors are known, the reactive power of each capacitor can be found.

$$VARs_{C1} = E_{C1} \times I_{C1}$$

$$VARs_{C1} = 240.051 \times 0.905$$

$$VARs_{C1} = 217.246$$

$$VARs_{C2} = E_{C2} \times I_{C2}$$

$$VARs_{C2} = 80.017 \times 0.905$$

$$VARs_{C2} = 72.415$$

Example 20–4 (continued)

$$VARs_{C3} = E_{C3} \times I_{C3}$$

$$VARs_{C3} = 160.031 \times 0.905$$

$$VARs_{C3} = 144.828$$

Power, whether true, apparent, or reactive power, will always be additive in any circuit. The total reactive power in this circuit can be found by taking the sum of all the VARs for the capacitors or by using total values of voltage and current and Ohm's law:

$$VARs_{CT} = VARs_{C1} + VARs_{C2} + VARs_{C3}$$

$$VARs_{CT} = 217.248 + 72.415 + 144.828$$

$$VARs_{CT} = 434.491$$

The circuit with all calculated values is shown in *Figure 20–13.*

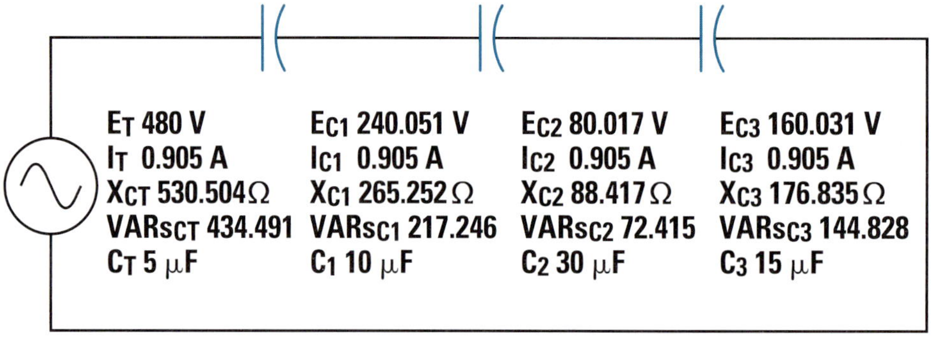

Figure 20–13 Series circuit 1 with all values.

Parallel Capacitors

Example 20–5

Three capacitors having values of 50, 75, and 20 μF are connected in parallel to a 60-Hz line. The circuit has a total reactive power of 787.08 VARs (*Figure 20–14*). Find the following unknown values:

X_{C1}—capacitive reactance of the first capacitor

X_{C2}—capacitive reactance of the second capacitor

X_{C3}—capacitive reactance of the third capacitor

X_{CT}—total capacitive reactance for the circuit

E_T—total applied voltage

I_{C1}—current flow through the first capacitor

$VARs_{C1}$—reactive power of the first capacitor

I_{C2}—current flow through the second capacitor

$VARs_{C2}$—reactive power of the second capacitor

Example 20–5 (continued)

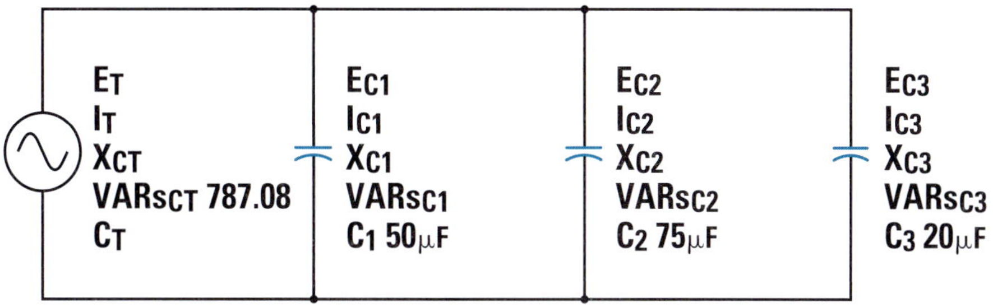

Figure 20–14 Capacitors connected in parallel.

I_{C3}—current flow through the third capacitor

$VARs_{C3}$—reactive power of the third capacitor

I_T—total circuit current

Because the frequency of the circuit and the capacitance of each capacitor are known, the capacitive reactance of each capacitor can be calculated using the formula:

$$X_C = \frac{1}{2\pi f C}$$

Note: Refer to the Purely Capacitive Circuits section of Appendix B.

$$X_{C1} = \frac{1}{377 \times 0.000050\ \text{F}}$$

$$X_{C1} = 53.05\ \Omega$$

$$X_{C2} = \frac{1}{377 \times 0.000075\ \text{F}}$$

$$X_{C2} = 35.367\ \Omega$$

$$X_{C3} = \frac{1}{377 \times 0.000020\ \text{F}}$$

$$X_{C3} = 132.626\ \Omega$$

The total capacitive reactance can be found in a manner similar to finding the resistance of parallel resistors:

$$X_{CT} = \frac{1}{\frac{1}{X_{C1}} + \frac{1}{X_{C2}} + \frac{1}{X_{C3}}}$$

$$X_{CT} = \frac{1}{\frac{1}{53.05\ \Omega} + \frac{1}{35.367\ \Omega} + \frac{1}{132.626\ \Omega}}$$

$$X_{CT} = \frac{1}{0.05467\ \frac{1}{\Omega}}$$

$$X_{CT} = 18.292\ \Omega$$

Example 20–5 (continued)

Now that the total capacitive reactance of the circuit is known and the total reactive power is known, the voltage applied to the circuit can be found using the formula:

$$E_T = \sqrt{VARs_{CT} \times X_{CT}}$$

$$E_T = \sqrt{787.08\ VARs \times 18.292\ \Omega}$$

$$E_T = 119.989\ V \text{ or approx. } 120\ V$$

Refer to the formulas provided in Appendix B for Purely Capacitive Circuits for other formulas used to determine electrical values.

In a parallel circuit, the voltage must be the same across each branch of the circuit. Therefore, 120 V is applied across each capacitor.

Now that the circuit voltage is known, the amount of total current for the circuit and the amount of current in each branch can be found using Ohm's law:

$$I_{CT} = \frac{E_{CT}}{X_{CT}}$$

$$I_{CT} = \frac{119.989\ V}{18.292\ \Omega}$$

$$I_{CT} = 6.56\ A$$

$$I_{CT} = \frac{E_{C1}}{X_{C1}}$$

$$I_{C1} = \frac{119.989\ V}{53.05\ \Omega}$$

$$I_{C1} = 2.262\ A$$

$$I_{C2} = \frac{E_{C2}}{X_{C2}}$$

$$I_{C2} = \frac{119.989\ V}{35.367\ \Omega}$$

$$I_{C2} = 3.393\ A$$

$$I_{C3} = \frac{E_{C3}}{X_{C3}}$$

$$I_{C3} = \frac{119.989\ V}{132.626\ \Omega}$$

$$I_{C3} = 0.905\ A$$

The amount of reactive power for each capacitor can now be calculated using Ohm's law:

$$VARs_{C1} = E_{C1} \times I_{C1}$$

$$VARs_{C1} = 119.989\ V \times 2.262\ A$$

$$VARs_{C1} = 271.415$$

$$VARs_{C2} = E_{C2} \times I_{C2}$$

$$VARs_{C2} = 119.989\ V \times 3.393\ A$$

$$VARs_{C2} = 407.123$$

Example 20–5 (continued)

$$VARs_{C3} = E_{C3} \times I_{C3}$$

$$VARs_{C3} = 119.987\ V \times 0.905\ A$$

$$VARs_{C3} = 108.590$$

To make a quick check of the circuit values, add the VARs for all the capacitors and see if they equal the total circuit VARs:

$$VARs_{CT} = VARs_{C1} + VARs_{C2} + VARs_{C3}$$

$$VARs_{CT} = 271.415 + 407.123 + 108.590$$

$$VARs_{CT} = 787.128$$

The slight difference in answers is caused by the rounding off of values. The circuit with all values is shown in *Figure 20–15.*

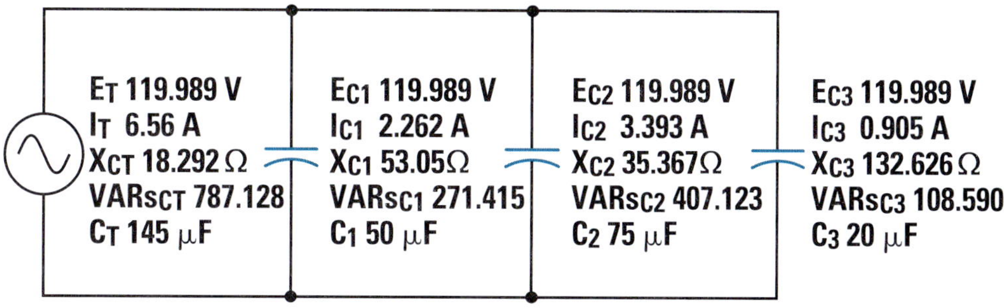

Figure 20–15 Parallel capacitive circuit with completed values.

Summary

- When a capacitor is connected to an AC circuit, current appears to flow through a capacitor because of the continuous increase and decrease of voltage and because of the continuous change of polarity in an AC circuit.
- The current flow in a purely capacitive circuit is limited by capacitive reactance.
- Capacitive reactance is inversely proportional to the capacitance of the capacitor and the frequency of the AC line.
- Capacitive reactance is measured in ohms.
- In a purely capacitive circuit, the current leads the applied voltage by 90°.
- There is no true power, or watts, in a purely capacitive circuit.
- Capacitive power is reactive and is measured in VARs, similar to inductive power.
- Capacitive and inductive VARs are 180° out of phase with each other.
- The Q of a capacitor is the ratio of the resistance to the capacitive reactance.
- Capacitor voltage ratings are given as volts AC, peak volts, and volts DC.
- A DC voltage rating for an AC capacitor indicates the peak value of voltage.

Review Questions

1. When a capacitor is connected to an AC circuit, which of the following most accurately describes how current flows in the circuit?
   A) Current flows through a capacitor from positive to negative.
   B) Current flows through a capacitor from negative to positive.
   C) Current only appears to flow in a capacitor.
   D) Current flows through the dielectric material.
2. Which of the following factors determines the capacitive reactance of a capacitor?
   A) The voltage applied to the circuit
   B) The frequency of the applied voltage
   C) The polarity of the capacitor
   D) The current flowing in the capacitor
3. True or False: In a purely capacitive circuit, the voltage will lag the current by 90 degrees.
4. The power produced in a purely capacitive circuit is measured in which of the following measurements?
   A) $VARs_C$
   B) Volt-amperes
   C) $VARs_L$
   D) Watts
5. Which of the following statements about the quality of a capacitor is most accurate?
   A) The quality of a capacitor is generally very high.
   B) The quality of a capacitor is based on the resistance of the dielectric material.
   C) A leaky capacitor will experience a decrease in quality.
   D) All of these are correct.
6. If a nonpolarized capacitor's voltage rating is given in DC volts but the circuit is an AC circuit, what must we do to determine if the circuit voltage exceeds the voltage rating of the capacitor?
   A) We must add a diode in the system.
   B) We must treat the DC voltage as peak AC and find RMS voltage for AC.
   C) We must determine the positive and negative terminal.
   D) We must treat DC voltage as RMS and find peak voltage for AC.

7. True or False: As frequency increases, the capacitive reactance in the circuit will decrease because the charge and discharge cycle happens faster at higher frequencies.
8. If a capacitor is connected to a 50-hertz circuit at 240 volts and shows a current of 5 amperes, which of the following statements accurately represents the results of connecting the same capacitor to a circuit with a 240-volt, 60-hertz supply?

   A) The current will be less than 5 amperes.

   B) The current will remain unchanged.

   C) The current will be greater than 5 amperes.

   D) Not enough information is provided.
9. A circuit contains 20, 40, and 60 μF capacitors. If it is connected to a 480-volt, 60-hertz source, what is the total current in the circuit?

   A) 5 amperes

   B) 16.4 amperes

   C) 18.3 amperes

   D) 21.7 amperes
10. A circuit contains two identical capacitors in series with each other. The applied voltage is 120 V at 60 hertz, and the total current is 3 amperes. What is the capacitance value of one capacitor?

    A) 132.6 μF

    B) 66.3 μF

    C) 97.4 μF

    D) 153.4 μF

Practical Applications

1. You are working as an electrician in an industrial plant. You are given an AC oil-filled capacitor to install on a 480-V, 60-Hz AC line. The capacitor has the following marking: (15 μF 600 VDC). Will this capacitor be damaged if it is installed? Explain your answer.

---

2. You are working in an industrial plant. You have been instructed to double the capacitance connected to a machine. The markings on the capacitor, however, are not visible. The capacitor is connected to 560 volts, and an ammeter indicates a current of 6 amperes flowing to the capacitor. What size capacitor should be connected in parallel with the existing capacitor? What is the minimum AC voltage rating of the new capacitor? What is the minimum DC voltage rating of the new capacitor? What is the minimum kVAR size that can be used in this installation?

Practice Problems

Capacitive Circuits

1. Fill in all the missing values. Refer to the formulas that follow.

$$X_C = \frac{1}{2\pi f C}$$

$$C = \frac{1}{2\pi f X_C}$$

$$f = \frac{1}{2\pi C X_C}$$

Capacitance	X_C	Frequency
38 μF		60 Hz
	78.8 Ω	400 Hz
250 pF	4.5 kΩ	
234 μF		10 kHz
	240 Ω	50 Hz
10 μF	36.8 Ω	
560 nF		2 MHz
	15 kΩ	60 Hz
75 μF	560 Ω	
470 pF		200 kHz
	6.8 kΩ	400 Hz
34 μF	450 Ω	

2. A 4.7-μF capacitor is connected to a 60-Hz power source. What is the capacitive reactance of the capacitor?
3. A capacitor is connected to a 208-volt, 60-Hz power source. An ammeter indicates a current flow of 0.28 amperes. What is the capacitance of the capacitor?
4. A 0.47-μF capacitor is connected to a 240-volt power source. An ammeter indicates a current of 0.2 ampere. What is the frequency of the power source?
5. Three capacitors having capacitance values of 20, 40, and 50 μF are connected in parallel to a 60-Hz power line. An ammeter indicates a circuit current of 8.6 amperes. How much current is flowing through the 40-μF capacitor?
6. A capacitor has a capacitive reactance of 300 Ω when connected to a 60-Hz power line. What is the capacitive reactance if the frequency is increased to 100 Hz?
7. A pure capacitive circuit is connected to a 480-volt, 60-Hz power source. An ammeter indicates a current flow of 24 amperes. The circuit current must be reduced to 16 amperes by connecting a second capacitor in series with the first. What is the value of the existing capacitor? What value capacitor should be connected in series with the original capacitor to limit the circuit current to 16 amperes?

Unit 21 Resistive-Capacitive Series Circuits

Key Terms

Capacitance (C)

Reactive power ($VARs_c$)

Total voltage (E_T)

Voltage drop across the capacitor (E_C)

Voltage drop across the resistor (E_R)

Why You Need to Know

The relationship of voltage, current, impedance, and power when resistance and capacitance are connected in series with each other follows a similar but different relationship from that of resistive-inductive series circuits. This unit

- describes the effect of voltage, current, power, and impedance in series circuits that contain both resistance and capacitance.
- illustrates how to calculate voltage drop across a resistor and how voltage and current are out of phase in an amount less than 90°.

Outline

Objectives

After studying this unit, you should be able to

- break down the relationship between resistance and capacitance in an RC series circuit connected to an AC power source.
- calculate unknown circuit values to determine how much true power is produced in an RC series circuit connected to an AC power source.
- calculate unknown circuit values to determine how much reactive power is produced in an RC series circuit connected to an AC power source.
- calculate remaining unknown values to determine the apparent power, power factor, and phase angle between current and voltage in an RC series circuit.

Preview

In this unit, the relationship of voltage, current, impedance, and power in a resistive-capacitive series circuit is discussed. As with any series circuit, the current flow will be the same through all parts of the circuit. This unit explores the effect of voltage drop across each component; the relationship of resistance, reactance, and impedance; and the differences between true power, reactive power, and apparent power.

21–1 Resistive-Capacitive Series Circuits

When a purely capacitive load is connected to an AC circuit, the voltage and current are 90° out of phase with each other, with the current leading the voltage. When both resistance and capacitance are connected to an AC circuit, the voltage and current will be out of phase with each other by some amount between 0° and 90°. The exact amount of phase angle difference is determined by the ratio of resistance to capacitance. Resistive-capacitive series circuits are similar to resistive-inductive series circuits, covered in Unit 17. Other than changing a few formulas, the procedure for solving circuit values is essentially the same.

In the following example, a series circuit containing 12 ohms of resistance and 16 ohms of capacitive reactance is connected to a 240-volt, 60-hertz line (*Figure 21–1*). The following unknown values are calculated:

Z—total circuit impedance

I—total current

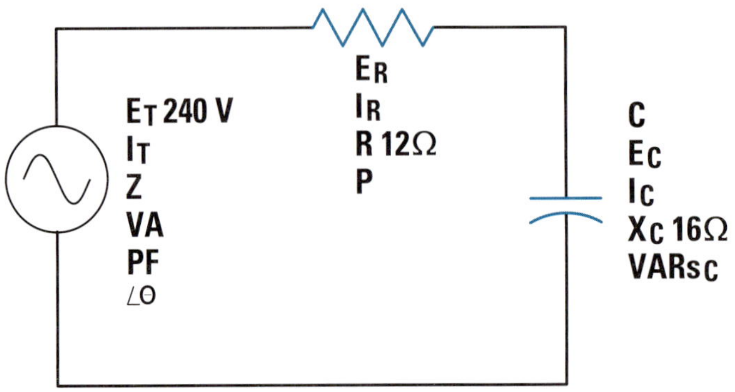

Figure 21–1 An incomplete resistive-capacitive series circuit.

E_R—voltage drop across the resistor

P—watts (true power)

C—capacitance

E_C—voltage drop across the capacitor

$VARs_C$—volt-amperes-reactive (reactive power)

VA—volt-amperes (apparent power)

PF—power factor

$\angle\theta$—angle theta (the angle the voltage and current are out of phase with each other)

21–2 Calculating True Power (Watts)

If we are to determine the amount of true power in the circuit, the first step is to determine how much current will flow based on circuit voltage and the total impedance of the circuit. The total impedance (Z) is the total current-limiting element in the circuit. It is a combination of both resistance and capacitive reactance. Because this is a series circuit, the current-limiting elements must be added. However, because resistance and capacitive reactance are 90° out of phase with each other, they will form a right triangle with impedance being the hypotenuse (*Figure 21–2*). A parallelogram vector diagram illustrating this relationship is shown in *Figure 21–3*. Impedance can be calculated using the following formula:

$$Z = \sqrt{R^2 = X_C^{\ 2}}$$

$$Z = \sqrt{(12\ \Omega)^2 + (16\ \Omega)^2}$$

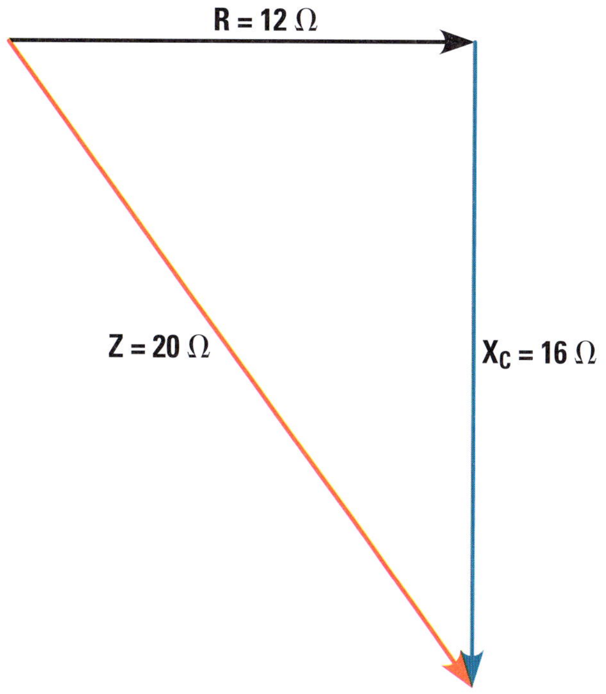

Figure 21–2 Resistance and capacitive reactance are 90° out of phase and can be illustrated using a right triangle.

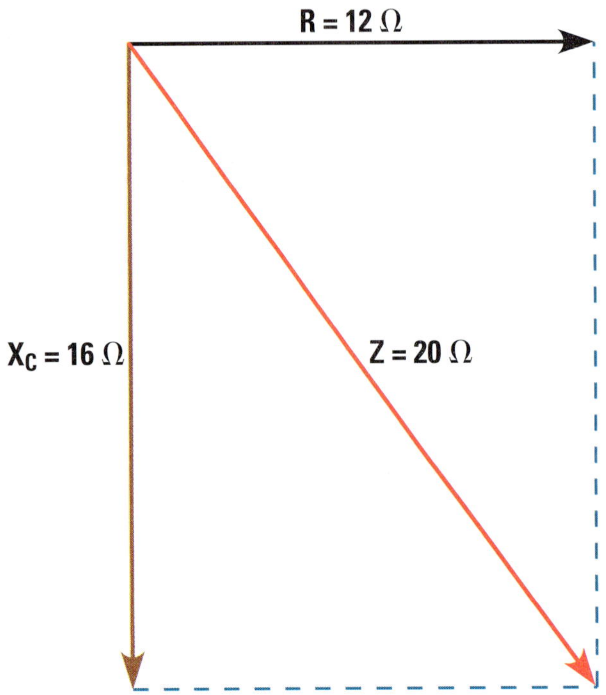

Figure 21–3 Impedance vector for example circuit 1.

$$Z = \sqrt{144\ \Omega^2 + 256\ \Omega^2}$$

$$Z = \sqrt{400\ \Omega^2}$$

$$Z = 20\ \Omega$$

Total Current

Now that the total impedance of the circuit is known, the total current flow (I_T) can be calculated using the formula:

$$I_T = \frac{E}{Z}$$

$$I_T = \frac{240\ V}{20\ \Omega}$$

$$I_T = 12\ A$$

Voltage Drop Across the Resistor

In a series circuit, the current is the same at any point in the circuit. Therefore, 12 amperes of current flow through both the resistor and the capacitor. The **voltage drop across the resistor (E_R)** can be calculated by using Ohm's law:

$$E_R = I \times R$$

$$E_R = 12\ A \times 12\ \Omega$$

$$E_R = 144\ V$$

True Power

True power (P) for the circuit can be calculated by using any of the power formulas as long as values that apply only to the resistive part of the circuit are used. Recall that current and voltage must be in phase with each other for true power to be produced. The formula used in this example is:

$$P = E_R \times I$$

$$P = 144\ V \times 12\ A$$

$$P = 1728\ W$$

21–3 Calculating Reactive Power (VARs)

The amount of **capacitance (C)** can be calculated using the following formula:

$$C = \frac{1}{2\pi f X_c}$$

$$C = \frac{1}{377 \times 16\ \Omega}$$

$$C = \frac{1}{6032}$$

$$C = 0.0001658\ F = 165.8\ \mu F$$

Voltage Drop Across the Capacitor

The **voltage drop across the capacitor (E_C)** can be calculated using the following formula:

$$E_C = I \times X_C$$

$$E_C = 12\ A \times 16\ \Omega$$

$$E_C = 192\ V$$

Total Voltage

Although the amount of **total voltage (E_T)** applied to the circuit is given as 240 volts in this circuit, it is possible to calculate the total voltage if it is not known by adding together the voltage drop across the resistor and the voltage drop across the capacitor. In a series circuit, the voltage drops across the resistor and capacitor are 90° out of phase with each other, and vector addition must be used. These two values form the legs of a right triangle, and the total applied voltage forms the hypotenuse (*Figure 21–4*). The total voltage can be calculated using the following formula:

$$E_T = \sqrt{E_R^2 + E_C^2}$$

$$E_T = \sqrt{(144\ V)^2 + (192\ V)^2}$$

$$E_T = 240\ V$$

A parallelogram vector diagram illustrating the voltage relationships for this circuit is shown in *Figure 21–5* as an alternative to the right triangle.

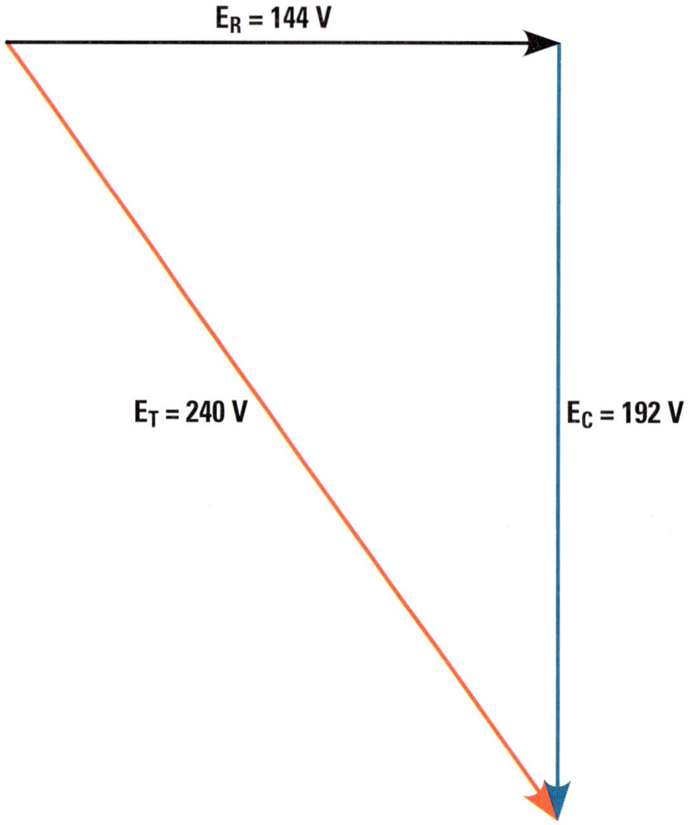

Figure 21–4 The voltage drops across the resistor (E_R) and capacitor (E_C) are 90° out of phase with each other.

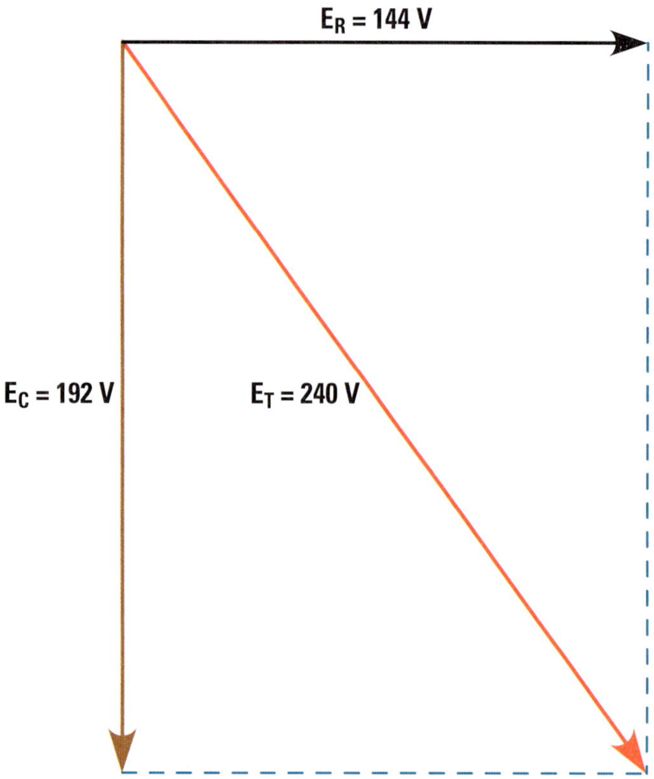

Figure 21–5 Parallelogram formed by voltage vectors for the example circuit.

Reactive Power

The **reactive power ($VARs_C$)** in the circuit can be calculated in a manner similar to that used for true power except that reactive values of voltage and current are used instead of resistive values.
In this example, the formula used is:

$$VARs_C = E_C \times I$$

$$VARs_C = 192\ V \times 12\ A$$

$$VARs_C = 2304$$

21–4 Calculating Apparent Power and Power Factor

The apparent power (VA) of the circuit can be calculated in a manner similar to that used for watts or $VARs_C$, except that total values of voltage and current are used. In this example, the formula used is:

$$VA = E_T \times I$$

$$VA = 240\ V \times 12\ A$$

$$VA = 2880$$

The apparent power can also be determined by the right triangle or vector addition of the true power and reactive power (*Figure 21–6*).

$$VA = \sqrt{P^2 + VARs_C{}^2}$$

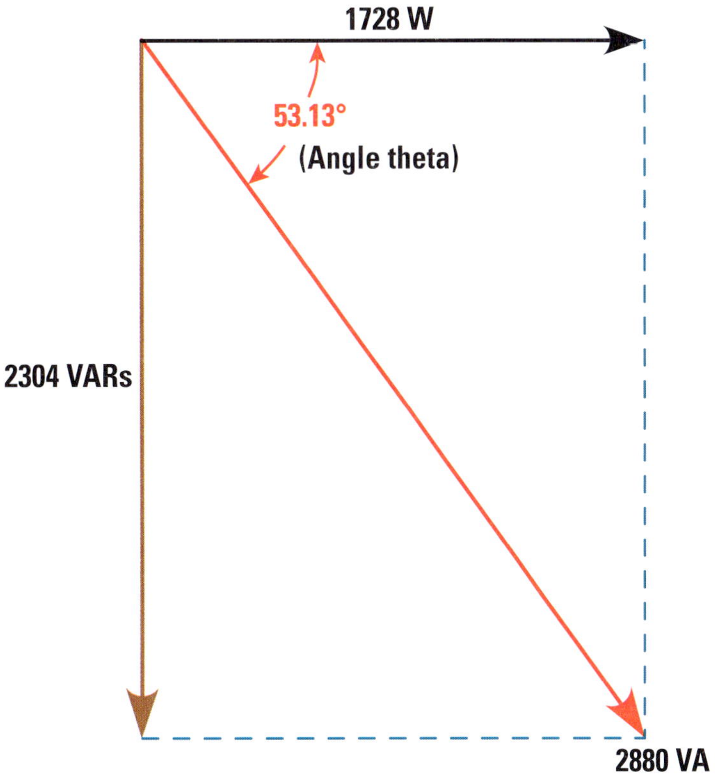

Figure 21–6 Parallelogram method to illustrate power vectors for the example circuit.

Power Factor

Power factor (PF) is a ratio of the true power to the apparent power. It can be calculated by dividing any resistive value by its like total value. In this circuit, the power values are used:

$$PF = \frac{P}{VA}$$

$$PF = \frac{1728\ W}{2880\ VA}$$

$$PF = 0.6 \times 100, \text{ or } 60\%$$

Angle Theta

The power factor of a circuit is the cosine of the phase angle. Because the power factor of this circuit is 0.6, angle theta ($\angle\theta$) is

$$COS\ \angle\theta = PF$$

$$COS\ \angle\theta = 0.6$$

$$\angle\theta = 53.13°$$

In this circuit, the current leads the applied voltage by 53.13° (*Figure 21–7*).

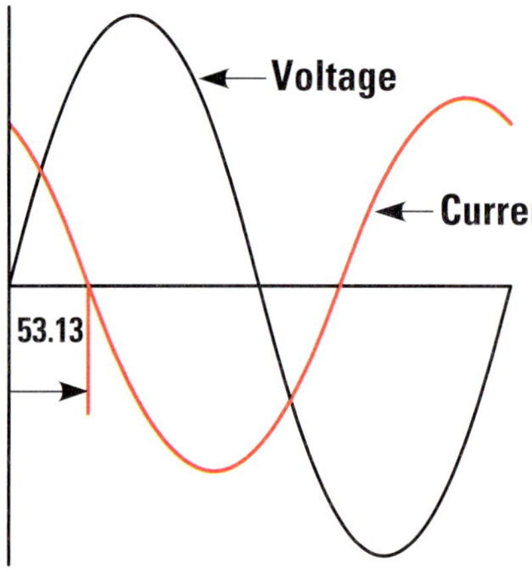

Figure 21–7 The current leads the voltage by 53.13°.

Example 21–1

One resistor and two capacitors are connected in series to a 42.5-V, 60-Hz line (*Figure 21–8*). Capacitor 1 has a reactive power of 3.75 VARs, the resistor has a true power of 5 W, and the second capacitor has a reactive power of 5.625 VARs. Find the following values:

VA—apparent power

I_T—total circuit current

Z—impedance of the circuit

PF—power factor

$\angle\theta$—angle theta

Example 21–1 (continued)

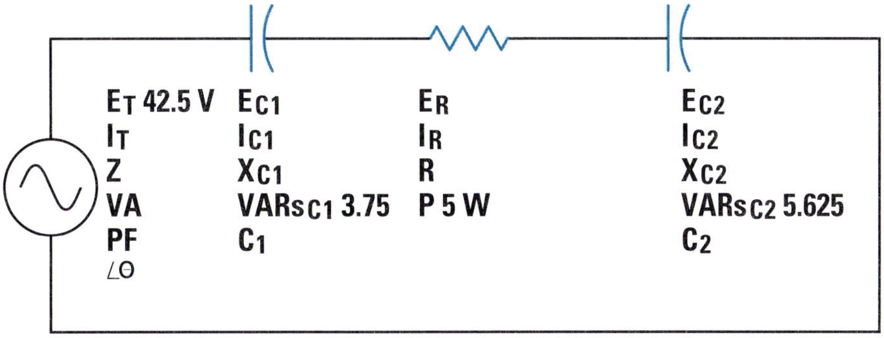

Figure 21–8 RC series circuit for example 21–1.

E_{C1}—voltage drop across the first capacitor

X_{C1}—capacitive reactance of the first capacitor

C_1—capacitance of the first capacitor

E_R—voltage drop across the resistor

R—resistance of the resistor

E_{C2}—voltage drop across the second capacitor

X_{C2}—capacitive reactance of the second capacitor

C_2—capacitance of the second capacitor

Solution

Because the reactive power of the two capacitors is known and the true power of the resistor is known, the apparent power can be found using the formula:

$$VA = \sqrt{P^2 + VARs_C^{\ 2}}$$

In this circuit, $VARs_C$ is the sum of the VARs of the two capacitors. A power triangle for this circuit is shown in *Figure 21–9*.

$$VA = \sqrt{(5\ W)^2 + (3.75\ VARs + 5.625\ VARs)^2}$$

$$VA = \sqrt{25\ W^2 + 87.891\ VARs^2}$$

$$VA = 10.625$$

Now that the apparent power and applied voltage are known, the total circuit current can be calculated using the following formula:

$$I_T = \frac{VA}{E_T}$$

$$I_T = \frac{10.625\ VA}{42.5\ V}$$

$$I_T = 0.25\ A$$

Example 21–1 (continued)

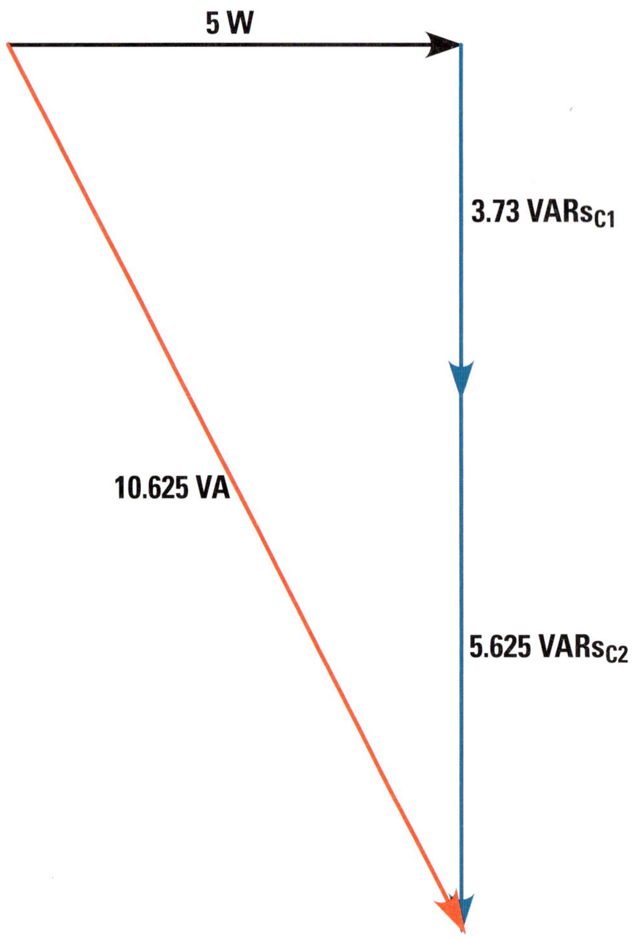

Figure 21–9 Power triangle for example circuit.

In a series circuit, the current will be the same at any point in the circuit. Therefore, I_{C1}, I_R, and I_{C2} will all have a value of 0.25 A.

The impedance of the circuit can now be calculated using the formula:

$$Z = \frac{E_T}{I_T}$$

$$Z = \frac{42.5\ V}{0.25\ A}$$

$$Z = 170\ \Omega$$

The power factor can be calculated using the following formula:

$$PF = \frac{P}{VA}$$

$$PF = \frac{5\ W}{10.625\ VA}$$

$$PF = 0.4706, \text{ or } 47.06\%$$

Example 21–1 (continued)

The cosine of angle theta is the power factor:

$$\cos \angle\theta = 0.4706$$
$$\angle\theta = 61.93^\circ$$

The vector diagram shown in *Figure 21–10* illustrates the relationship of angle theta to the reactive power, true power, and apparent power.

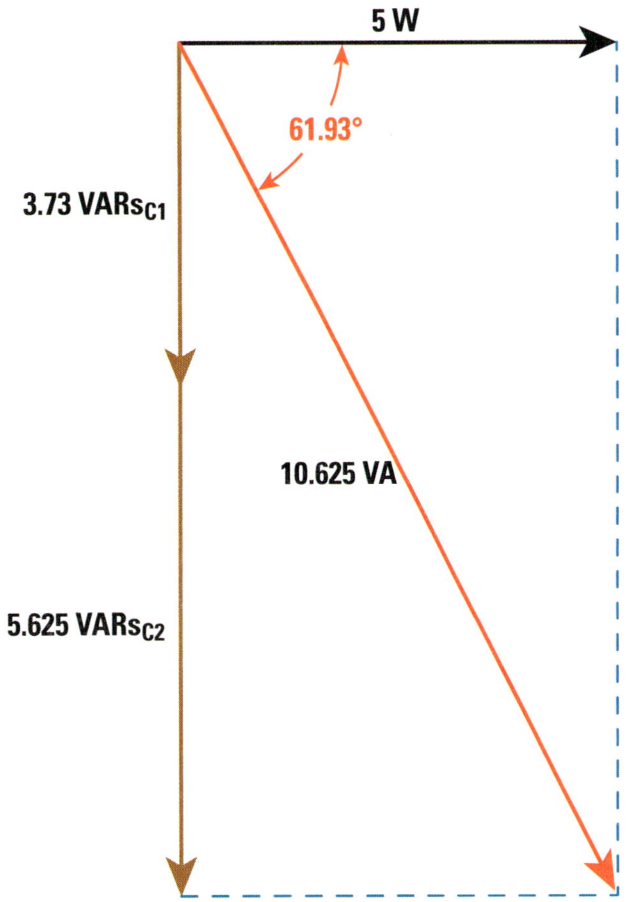

Figure 21–10 Vector relationship of reactive, true, and apparent power.

Now that the current through each circuit element is known and the power of each element is known, the voltage drop across each element can be calculated:

$$E_{C1} = \frac{VARs_{C1}}{I_{C1}}$$
$$E_{C1} = \frac{3.75 \text{ VARs}}{0.25 \text{ A}}$$
$$E_{C1} = 15 \text{ V}$$

Example 21–1 (continued)

$$E_R = \frac{P}{I_R}$$

$$E_R = \frac{5\ W}{0.25\ A}$$

$$E_R = 20\ V$$

$$E_{C2} = \frac{VARs_{C2}}{I_{C2}}$$

$$E_{C2} = \frac{5.625\ VARs}{0.25\ A}$$

$$E_{C2} = 22.5\ V$$

The capacitive reactance of the first capacitor is:

$$X_{C1} = \frac{E_{C1}}{I_{C1}}$$

$$X_{C1} = \frac{15\ V}{0.25\ A}$$

$$X_{C1} = 60\ \Omega$$

The capacitance of the first capacitor is:

$$C_1 = \frac{1}{2\pi f X_{C1}}$$

$$C_1 = \frac{1}{377 \times 30\ \Omega}$$

$$C_1 = 0.0000442\ F,\ \text{or}\ 44.2\ \mu F$$

The resistance of the resistor is:

$$R = \frac{E_R}{I_R}$$

$$R = \frac{20\ V}{0.25\ A}$$

$$R = 80\ \Omega$$

The capacitive reactance of the second capacitor is:

$$X_{C2} = \frac{E_{C2}}{I_{C2}}$$

$$X_{C2} = \frac{22.5\ V}{0.25\ A}$$

$$X_{C2} = 90\ \Omega$$

The capacitance of the second capacitor is:

$$C_2 = \frac{1}{2\pi f X_{C2}}$$

$$C_2 = \frac{1}{377 \times 90\ \Omega}$$

$$C_2 = 0.0000295\ F,\ \text{or}\ 29.5\ \mu F$$

Example 21–1 (continued)

The completed circuit with all values is shown in *Figure 21–11*.

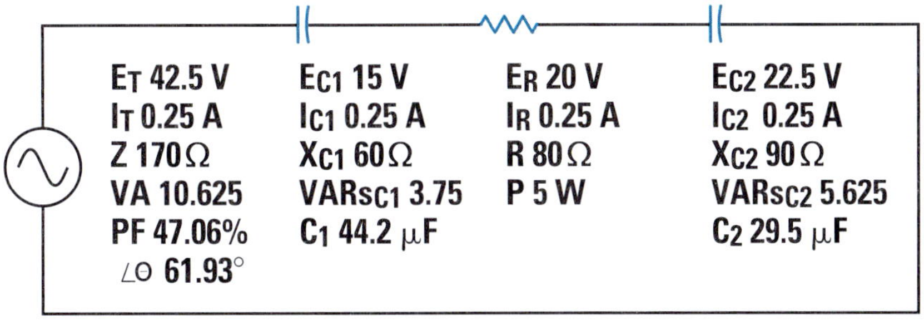

Figure 21–11 Example 21–1 circuit with all completed values.

Example 21–2

A small indicating lamp has a rating of 2 W when connected to 120 V. The lamp must be connected to a voltage of 480 V at 60 Hz. A capacitor will be connected in series with the lamp to reduce the circuit current to the proper value. What value of capacitor will be needed to perform this job?

Solution

The first step is to determine the amount of current the lamp will normally draw when connected to a 120-V line:

$$I_{LAMP} = \frac{P}{E}$$

$$I_{LAMP} = \frac{2\text{ W}}{120\text{ V}}$$

$$I_{LAMP} = 0.0167\text{ A}$$

The next step is to determine the amount of voltage that must be dropped across the capacitor when a current of 0.01667 A flows through it. Because the voltage dropped across the resistor and the voltage dropped across the capacitor are 90° out of phase with each other, vectors must be used to determine the voltage drop across the capacitor (*Figure 21–12*). The voltage drop across the capacitor can be calculated using the formula:

$$E_C = \sqrt{E_T^2 - E_R^2}$$

$$E_C = \sqrt{(480\text{ V})^2 - (120\text{ V})^2}$$

$$E_C = \sqrt{216{,}000\text{ V}^2}$$

$$E_C = 464.758\text{ V}$$

Example 22–2 (continued)

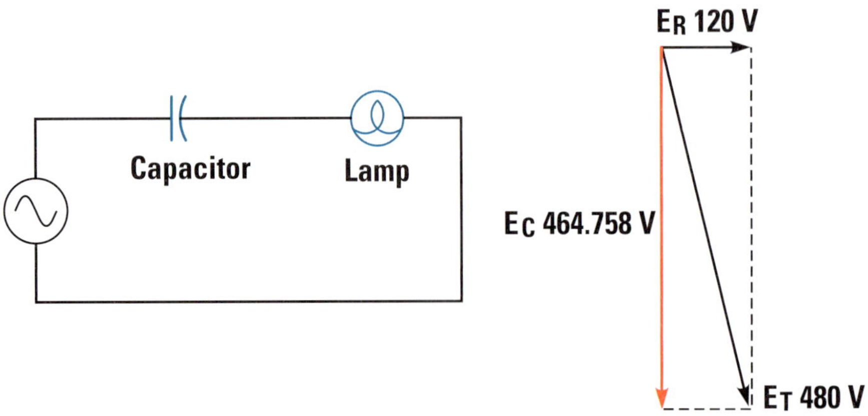

Figure 21–12 Determining voltage drop across the capacitor.

Now that the voltage drop across the capacitor and the amount of current flow are known, the capacitive reactance can be calculated:

$$X_C = \frac{E_C}{I}$$

$$X_C = \frac{464.758\text{ V}}{0.0167\text{ A}}$$

$$X_C = 27{,}829.82\ \Omega$$

The amount of capacitance needed to produce this capacitive reactance can now be calculated using the formula:

$$C = \frac{1}{2\pi f X_C}$$

$$C = \frac{1}{377 \times 27{,}829.82\ \Omega}$$

$$C = 0.0000000953\text{ F, or } 95.3\text{ nF}$$

Summary

- In a purely capacitive circuit, the voltage and current are 90° out of phase with each other.
- In a purely resistive circuit, the voltage and current are in phase with each other.
- In a circuit containing resistance and capacitance, the voltage and current will be out of phase with each other by some amount between 0°and 90°.
- The amount of phase angle difference between voltage and current in an RC series circuit is the ratio of resistance to capacitive reactance.
- In a series circuit, the current flow through all components is the same. Therefore, the voltage drops across the resistive and capacitive parts become out of phase with each other.
- True power can be produced only by resistive components of the circuit.
- Power factor is the ratio of true power to apparent power.

Review Questions

Refer to the formulas in the Resistive-Capacitive Series Circuits section of Appendix B.

1. In a purely capacitive circuit, the current leads the voltage by 90°. Which of the following statements is true for a circuit that has a mixture of capacitance and resistance?

   A) The current will lag the voltage by some amount between 0° and 90°.

   B) The current will be in phase with the voltage.

   C) The current will remain leading the voltage by 90°.

   D) The current will lead the voltage by some amount between 0° and 90°.

2. In an RC series circuit, which of the following best describes the impact on the phase angle that is made by an increase in resistance?

   A) As resistance increases, the phase angle decreases.

   B) As resistance increases, the phase angle increases.

   C) As resistance increases, the phase angle remains unchanged.

   D) As resistance increases, the phase angle is 90°.

3. An RC series circuit is supplied by a 120-volt supply at 60 hertz. The resistance in the circuit is 30 Ω and is in series with a 53.05-μf capacitor. How many watts are produced in this circuit?

   A) 62 watts

   B) 212 watts

   C) 127 watts

   D) 103 watts

4. An RC series circuit produces 1500 watts of heat. If the capacitive reactance is 18.73 ohms and the resistance is 15 ohms, what is the applied voltage to the circuit?

   A) 120 volts

   B) 240 volts

   C) 480 volts

   D) 600 volts

5. An RC series circuit is supplied by a 50-hertz source and has a current of 5 amperes through the resistor. If the capacitor is marked as 25 μF, how many VARs will it produce?

   A) 3183 VARs

   B) 2872 VARs

   C) 637 VARs

   D) Not enough information

6. An RC series circuit is supplied by a 120-volt, 30-hertz supply, and a volt meter reads 67 volts across the resistor. If an ammeter reads 2 amps downstream from the resistor, what is the value of capacitance in the circuit?

   A) 49.8 μF

   B) 78.4 μF

   C) 106.6 μF

   D) 129.3 μF

7. A 120-volt, 60-hertz circuit contains a 100-μF capacitor and a 20-ohm resistor connected in series. What is the power factor of this circuit?

   A) 50%

   B) 60%

   C) 70%

   D) 80%

8. A 120-volt, 60-hertz circuit contains a 100-μF capacitor and a 20-ohm resistor connected in series. What is the phase angle between voltage and current for this circuit?

   A) 45°

   B) 60°

   C) 37°

   D) 53°

9. True or False: The voltage drop through the capacitor lags the circuit current by 90°.
10. True or False: In an RC series circuit, the total current is equal to the vector sum of the current through the capacitor and the current through the resistor.
11. True or False: VARs in an RC series circuit can be determined if the circuit frequency, current, and the rating of the capacitor are known.
12. True or False: The power factor in an RC series circuit is the ratio of watts to apparent power.

Practice Problems

Refer to the formulas in the Resistive-Capacitive Series Circuits section of Appendix B and *Figure 21–1*.

1. Assume that the circuit shown in *Figure 21–1* is connected to a 480-V, 60-Hz line. The capacitor has a capacitance of 165.782 μF, and the resistor has a resistance of 12 Ω. Find the missing values.

E_T 480 V ______	E_R ______	E_C ______
IT ______	I_R ______	I_C ______
Z ______	R 12 Ω ______	X_C ______
VA ______	P ______	$VARs_C$ ______
PF ______	$\angle\theta$ ______	C 165.782 μF ______

2. Assume that the voltage drop across the resistor, ER, is 78 V; the voltage drop across the capacitor, EC, is 104 V; and the circuit has a total impedance, Z, of 20 Ω. The frequency of the AC voltage is 60 Hz. Find the missing values.

E_T ______	E_R 78 V ______	E_C 104 V ______
I_T ______	I_R ______	I_C ______
Z 20 Ω ______	R ______	X_C ______
VA ______	P ______	$VARs_C$ ______
PF ______	$\angle\theta$ ______	C ______

3. Assume the circuit shown in *Figure 21–1* has an apparent power of 432 VA and a true power of 345.6 Ω. The capacitor has a capacitance of 15.8919 μF, and the frequency is 60 Hz. Find the missing values.

E_T ______	E_R ______	E_C ______
I_T ______	I_R ______	I_C ______
Z ______	R ______	X_C ______
VA 432 ______	P 345.6 Ω ______	$VARs_C$ ______
PF ______	$\angle\theta$ ______	C 15.8919 μF ______

4. Assume the circuit in *Figure 21–1* has a power factor of 68%, an apparent power of 300 VA, and a frequency of 400 Hz. The capacitor has a capacitance of 4.7125 μF. Find the missing values.

E_T ______	E_R ______	E_C ______
I_T ______	I_R ______	I_C ______
Z ______	R ______	X_C ______
VA 300 ______	P ______	$VARs_C$ ______
PF 68% ______	$\angle\theta$ ______	C 4.7125 μF ______

5. In a series RC circuit, $E_T = 240$ volts, $R = 60\ \Omega$, and $X_C = 85\ \Omega$. Find E_C.
6. In a series RC circuit, $E_T = 120$ volts, $R = 124\ \Omega$, and $X_C = 64\ \Omega$. Find reactive power.
7. In a series RC circuit, $E_T = 208$ volts, $I_T = 2.4$ amperes, and $R = 45\ \Omega$. Find the power factor.
8. In a series RC circuit, $E_T = 460$ volts, and $\angle\theta = 44°$. Find E_C.
9. In a series RC circuit, $E_T = 240$ volts at 60 Hz. An ammeter indicates a total current of 0.75 amperes. The resistor has a value of 140 Ω. What is the capacitance of the capacitor?
10. In a series RC circuit, the apparent power is 4250 VA, and the reactive power is 2125 VARs. What is the true power?

Unit 22 Resistive-Capacitive Parallel Circuits

Key Terms

Circuit impedance (Z)

Current flow through the capacitor (I_C)

Phase angle shift

Total circuit current (I_T)

Why You Need to Know

The relationship of voltage, current, impedance, and power when resistance and capacitance are connected in parallel with each other is very similar to resistive-inductive parallel circuits. Although it may seem that the units on resistive-capacitive series circuits and resistive-capacitive parallel circuits are a repeat of the information covered previously, they are an important step toward understanding what happens when elements of resistance, inductance, and capacitance are combined into the same circuit. This unit

- discusses parallel circuits that contain both resistance and capacitance and the effect on voltage, current, power, and impedance.
- illustrates how to calculate circuit current and voltage and their relationship in a resistive-capacitive parallel circuit.

Outline

Objectives

After studying this unit, you should be able to

- analyze the operation of RC parallel circuits.
- interpret how capacitance impacts phase angle shift in an RC parallel circuit.
- calculate voltage, current, impedance, and power values in an RC parallel circuit.
- calculate power factor and phase angle for an RC parallel circuit.

Preview

This unit discusses the relationship of different electrical quantities such as voltage, current, impedance, and power in a circuit that contains both resistance and capacitance elements connected in parallel. Because all components connected in parallel will have the same voltage applied to them, the current flow through different components will be additive but out of phase with each other. The effect this condition has on other circuit quantities is explored.

22–1 Operation of RC Parallel Circuits

When resistance and capacitance are connected in parallel, the voltage across all the devices will be in phase and will have the same value. The current flow through the capacitor, however, will lead the applied voltage by 90° (*Figure 22–1*). The amount of **phase angle shift** between the total circuit current and voltage is determined by the ratio of the amount of resistance to the amount of capacitive reactance. The circuit power factor is still determined by the ratio of true power to apparent power.

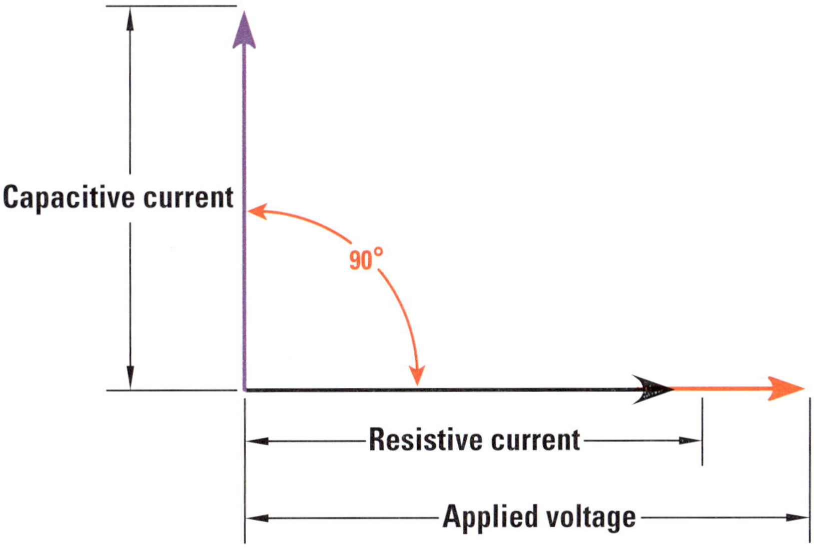

Figure 22–1 Current flow through the capacitor is 90° out of phase with both the current flow through the resistor and the applied voltage.

22–2 Calculating Circuit Values

Example 22–1

In the RC parallel circuit shown in *Figure 22–2*, assume that a resistance of 30 Ω is connected in parallel with a capacitive reactance of 20 Ω. The circuit is connected to a voltage of 240 VAC and a frequency of 60 Hz. Calculate the following circuit values:

I_R—current flow through the resistor

P—watts (true power)

I_C—current flow through the capacitor

$VARs_C$—volt-amperes reactive (reactive power)

C—capacitance of the capacitor

I_T—total circuit current

Z—total impedance of the circuit

VA—volt-amperes (apparent power)

PF—power factor

$\angle\theta$—angle theta (phase angle of voltage and current)

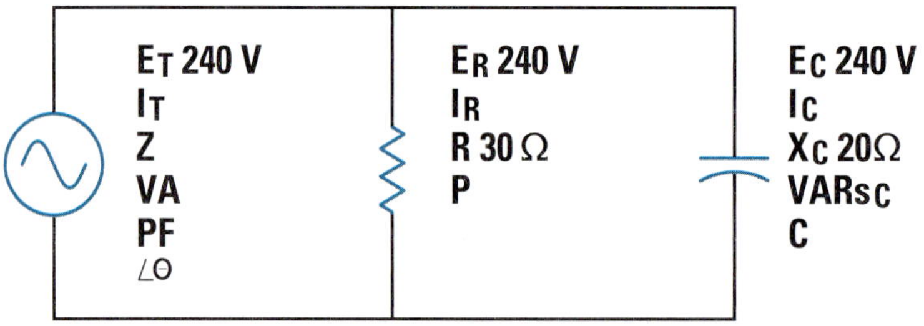

Figure 22–2 Resistive-capacitive parallel example circuit.

Solution

Resistive Current

The amount of current flow through the resistor (I_R) can be calculated by using the formula:

$$I_R = \frac{E}{R}$$

$$I_R = \frac{240\ V}{30\ \Omega}$$

$$I_R = 8\ A$$

True Power

The amount of total true power (P) in the circuit can be determined by using any of the values associated with the pure resistive part of the circuit. In this example, true power is found using the Ohm's law power formula:

$$P = E \times I_R$$

$$P = 240\ V \times 8\ A$$

$$P = 1920\ W$$

Example 22–1 (continued)

Capacitive Current

The amount of **current flow through the capacitor (I_C)** can be calculated using the following formula:

$$I_C = \frac{E}{X_C}$$

$$I_C = \frac{240\ V}{20\ \Omega}$$

$$I_C = 12\ A$$

Reactive Power

The amount of reactive power ($VARs_C$) can be found using any of the values applicable to the capacitive branch. In this example, $VARs_C$ is calculated using the following formula:

$$VARs_C = E \times I_C$$

$$VARs_C = 240\ V \times 12\ A$$

$$VARs_C = 2880$$

Capacitance

The capacitance of the capacitor can be calculated using the formula for capacitance:

$$C = \frac{1}{2\pi f X_C}$$

$$C = \frac{1}{377 \times 20\ \Omega}$$

$$C = \frac{1}{7540}$$

$$C = 0.0001326\ F = 132.6\ \mu F$$

Total Current

The voltage is the same across all legs of a parallel circuit, but the current is additive. However, the current flow through the resistor is in phase with the voltage, and the current flow through the capacitor is leading the voltage by 90° (*Figure 22–3*). The 90° difference in capacitive and resistive current forms a right triangle as shown in *Figure 22–4*. Because these two currents are connected in parallel, vector addition must be used to find the total current flow in the circuit (*Figure 22–5*). The **total circuit current (I_T)** flow can be calculated by using the following formula:

$$I_T = \sqrt{I_R^2 + I_C^2}$$

$$I_T = \sqrt{(8A)^2 + (12\ A)^2}$$

$$I_T = \sqrt{64\ A^2 + 144\ A^2}$$

$$I_T = \sqrt{208\ A^2}$$

$$I_T = 14.422\ A$$

Example 22–1 (continued)

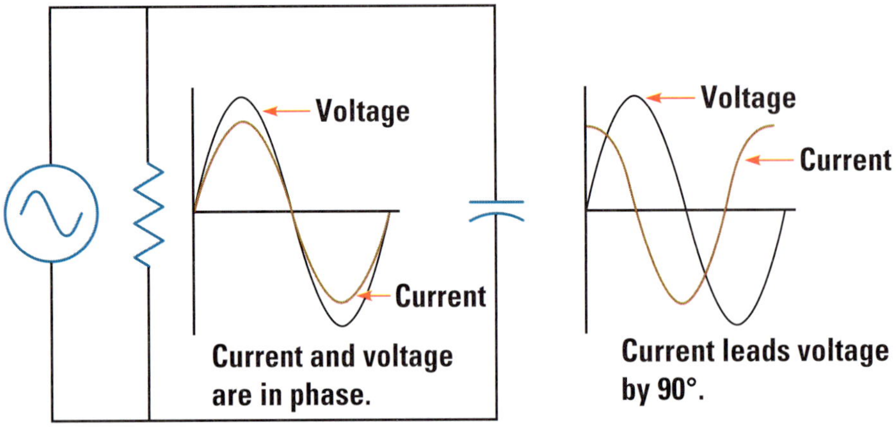

Figure 22–3 Phase relationship of current and voltage in an RC parallel circuit.

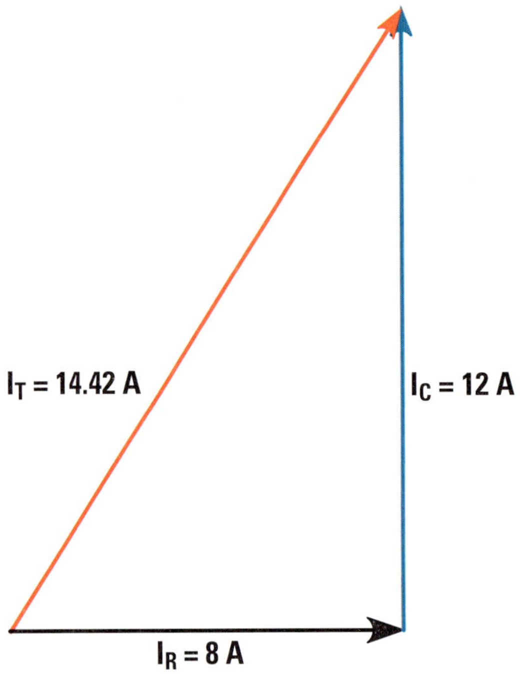

Figure 22–4 Resistance current and capacitive current are 90° out of phase with each other and form a right triangle.

Impedance

The total **circuit impedance (Z)** can be found by using any of the total values and substituting Z for R in an Ohm's law formula. The total impedance of this circuit is calculated using the formula:

$$Z = \frac{E}{I_T}$$

$$Z = \frac{240 \text{ V}}{14.422 \text{ A}}$$

$$Z = 16.641 \ \Omega$$

Example 22–1 (continued)

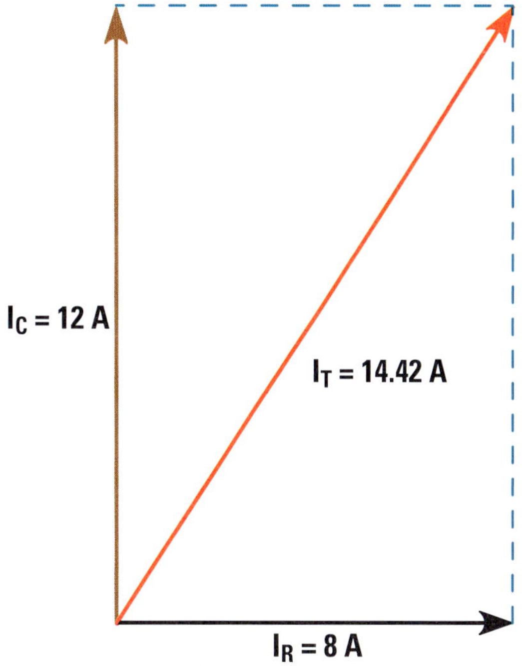

Figure 22–5 Vector addition (parallelogram method) can be used to find total current.

The impedance can also be found by adding the reciprocals of the resistance and capacitive reactance. This formula is used when resistance and capacitance are connected in parallel and cannot be used when the two elements are connected in series. Because the resistance and capacitive reactance are 90° out of phase with each other, vector addition must be used:

$$Z = \frac{1}{\sqrt{\left(\frac{1}{R}\right)^2 + \left(\frac{1}{X_C}\right)^2}}$$

Another formula that can be used to determine the impedance in a circuit that contains both resistance and capacitive reactance connected in parallel is:

$$Z = \frac{R \times X_C}{\sqrt{R^2 + X_C^{\,2}}}$$

Apparent Power

The apparent power (VA) can be calculated by multiplying the circuit voltage by the total current flow:

$$VA = E \times I_T$$

$$VA = 240\ V \times 14.422\ A$$

$$VA = 3461.28$$

Example 22–1 (continued)

Power Factor

The power factor (PF) is the ratio of true power to apparent power. The circuit power factor can be calculated using the formula:

$$PF = \frac{W}{VA} \times 100$$

$$PF = \frac{1920}{3461.28}$$

$$PF = 0.5547\text{, or } 55.47\%$$

Angle Theta

The cosine of angle theta ($\angle\theta$) is equal to the power factor:

$$\cos \angle\theta = 0.5547$$

$$\angle\theta = 56.31°$$

A vector diagram of apparent, true, and reactive power is shown in *Figure 22–6*. Angle theta is the angle developed between the apparent power and the true power. The complete circuit with all values is shown in *Figure 22–7*.

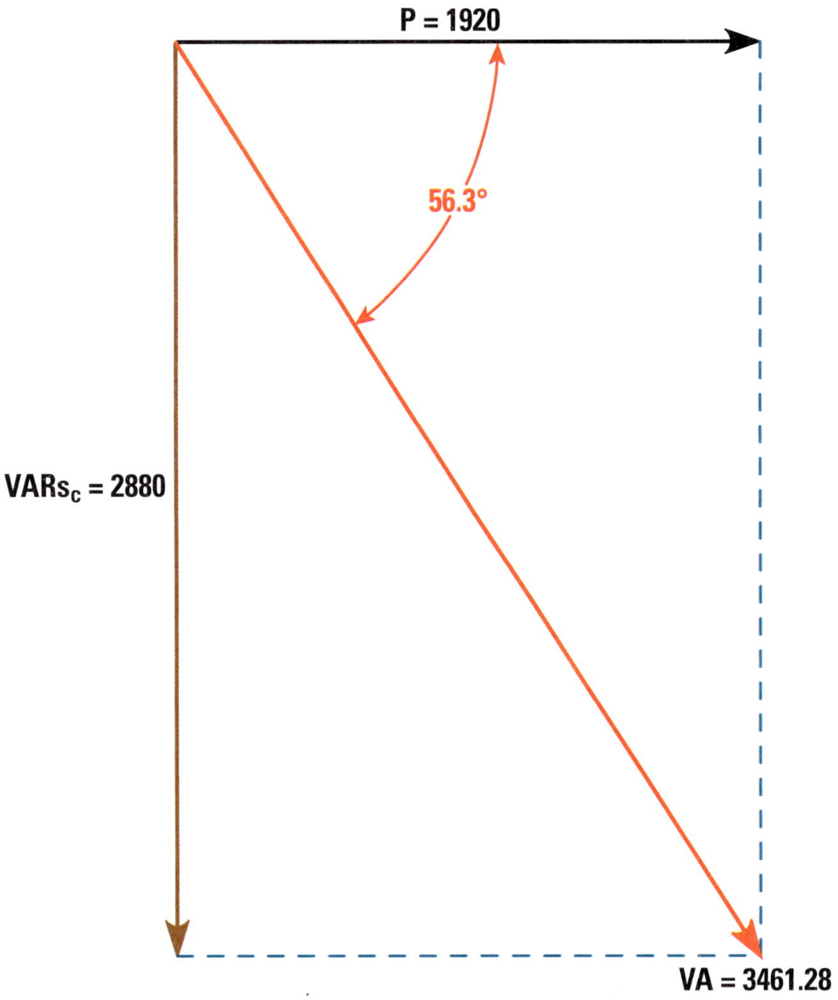

Figure 22–6 Vector relationship of apparent, true, and reactive power.

Example 22–1 (continued)

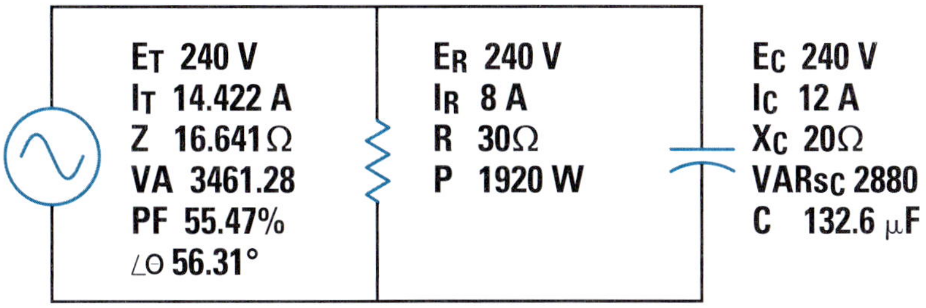

Figure 22–7 Example circuit completed with all calculated values.

Example 22–2

In this circuit, a resistor and a capacitor are connected in parallel to a 400-Hz line. The power factor is 47.1%, the apparent power is 4086.13 VA, and the capacitance of the capacitor is 33.15 μF (*Figure 22–8*). Find the following unknown values:

$\angle\theta$—angle theta

P—true power

$VARs_C$—capacitive VARs

X_C—capacitive reactance

E_C—voltage drop across the capacitor

I_C—capacitive current

E_R—voltage drop across the resistor

I_R—resistive current

R—resistance of the resistor

E_T—applied voltage

I_T—total circuit current

Z—impedance of the circuit

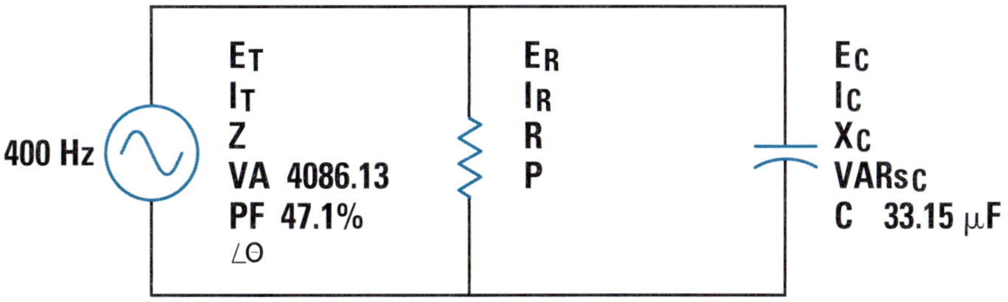

Figure 22–8 Example 22–2 circuit.

Example 22–2 (continued)

Solution

Angle Theta

The power factor is the cosine of angle theta. To find angle theta, change the power factor from a percentage into a decimal fraction by dividing by 100:

$$PF = \frac{47.1}{100}$$
$$PF = 0.471$$
$$\cos\angle\theta = 0.471$$
$$\angle\theta = 61.9°$$

True Power

The power factor is determined by the ratio of true power to apparent power:

$$PF = \frac{P}{VA}$$

This formula can be changed to calculate the true power when the power factor and apparent power are known (refer to the Resistive-Capacitive Parallel Circuits section of Appendix B).

$$P = VA \times PF$$
$$P = 4086.13 \times 0.471$$
$$P = 1924.47\ W$$

Reactive Power

The apparent power, true power, and reactive power form a right triangle as shown in *Figure 22–9*. Because these powers form a right triangle, the Pythagorean theorem can be used to find the leg of the triangle represented by the reactive power:

$$VARs_C = \sqrt{VA^2 - P^2}$$
$$VARs_C = \sqrt{4086.13^2 - 1924.47^2}$$
$$VARs_C = \sqrt{12{,}992{,}873.6}$$
$$VARs_C = 3604.56$$

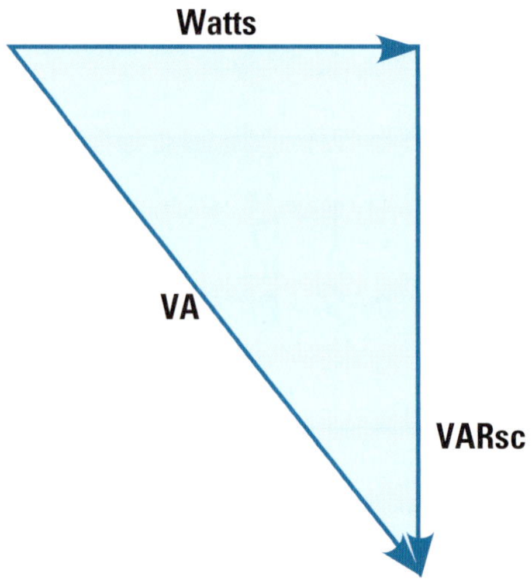

Figure 22–9 Right triangle formed by the apparent, true, and reactive powers.

Example 22–2 (continued)

A vector diagram showing the relationship of apparent power, true power, reactive power, and angle theta is shown in *Figure 22–10*.

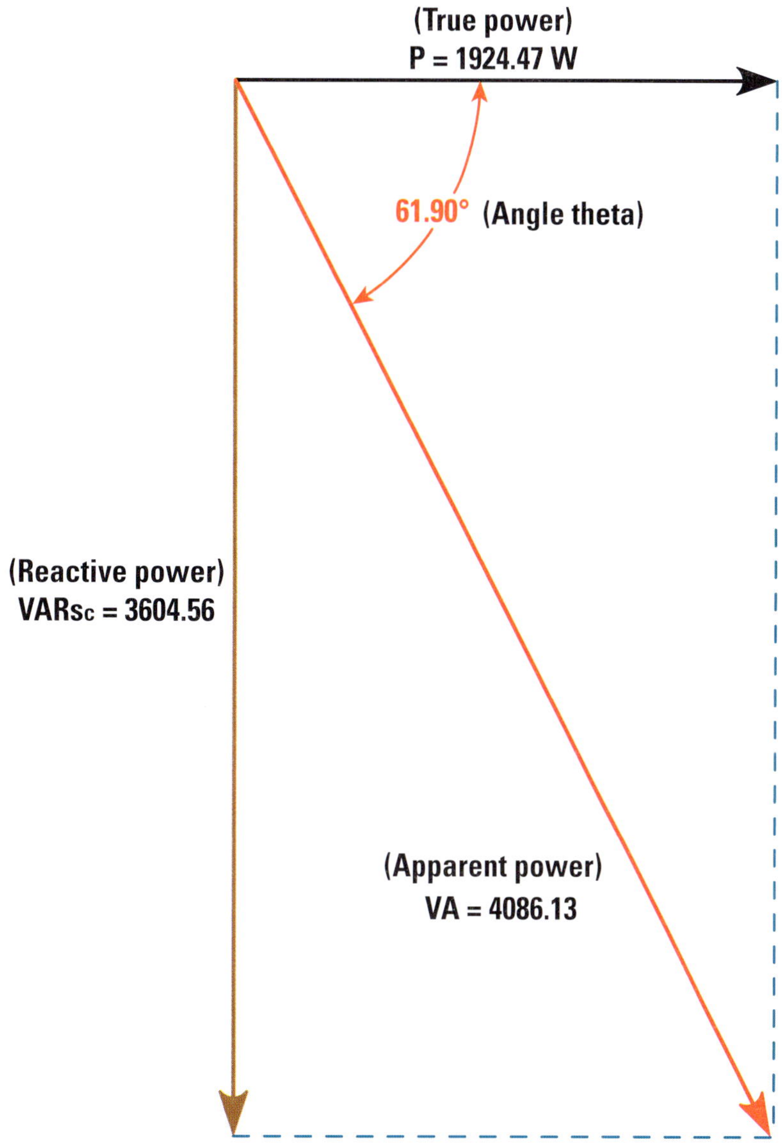

Figure 22–10 Vector diagram of apparent power, true power, reactive power, and angle theta.

Capacitive Reactance

Because the capacitance of the capacitor and the frequency are known, the capacitive reactance can be found using the formula:

$$X_C = \frac{1}{2\pi fC}$$

$$X_C = \frac{1\text{Hz}}{2 \times 3.1416 \times 400\text{ Hz} \times 0.00003315\text{ F}}$$

$$X_C = 12.003\ \Omega$$

Example 22–2 (continued)

Voltage Drop Across the Capacitor

$$E_C = \sqrt{VARs_C \times X_C}$$

$$E_C = \sqrt{3604.56\ VARs \times 12.003\ \Omega}$$

$$E_C = \sqrt{43{,}265.53}$$

$$E_C = 208\ V$$

Note: This equation is found in Appendix B in the section Resistive-Capacitive Parallel Circuits.

E_R and E_T

The voltage is the same across all branches of a parallel circuit. Therefore, if 208 V is applied across the capacitive branch, 208 V must be the total voltage of the circuit as well as the voltage applied across the resistive branch:

$$E_T = 208\ V$$

$$E_R = 208\ V$$

I_C

The amount of current flowing in the capacitive branch can be calculated using the formula:

$$I_C = \frac{E_C}{X_C}$$

$$I_C = \frac{208\ V}{12\ \Omega}$$

$$I_C = 17.331\ A$$

I_R

The amount of current flowing through the resistor can be calculated using the formula:

$$I_R = \frac{P}{E_R}$$

$$I_R = \frac{1924.47\ W}{208}$$

$$I_R = 9.25\ A$$

Resistance

The amount of resistance can be calculated using the formula:

$$R = \frac{E_R}{I_R}$$

$$R = \frac{208\ V}{9.25\ A}$$

$$R = 22.49\ \Omega$$

Example 22–2 (continued)

Total Current

The total current can be calculated using Ohm's law or by vector addition because both the resistive and capacitive currents are known. Vector addition is used in this example:

$$I_T = \sqrt{I_R^2 + I_C^2}$$

$$I_T = \sqrt{(9.25\ A)^2 + (17.33\ A)^2}$$

$$I_T = \sqrt{385.89\ A^2}$$

$$I_T = 19.64\ A$$

Impedance

The impedance of the circuit is calculated using the product over sum formula adjusted for resistive–capacitive parallel circuits:

$$Z = \frac{R \times X_C}{\sqrt{R^2 + X_C^2}}$$

$$Z = \frac{22.49\ \Omega \times 12.003\ \Omega}{\sqrt{(22.49\ \Omega)^2 + (12.003\ \Omega)^2}}$$

$$Z = \frac{269.947\ \Omega}{\sqrt{649.872\ \Omega^2}}$$

$$Z = \frac{269.947\ \Omega}{25.493\ \Omega}$$

$$Z = 10.589\ \Omega$$

The complete circuit with all values is shown in *Figure 22–11*.

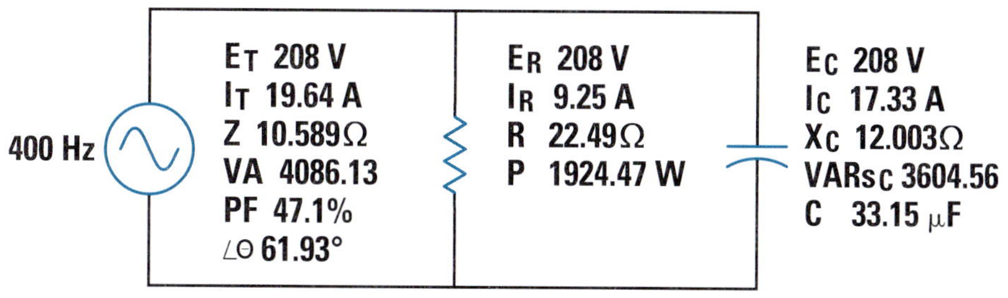

Figure 22–11 Example Circuit 2 with completed values.

A vector diagram showing the relationship of resistive current, capacitive current, and total current is shown in *Figure 22–12*.

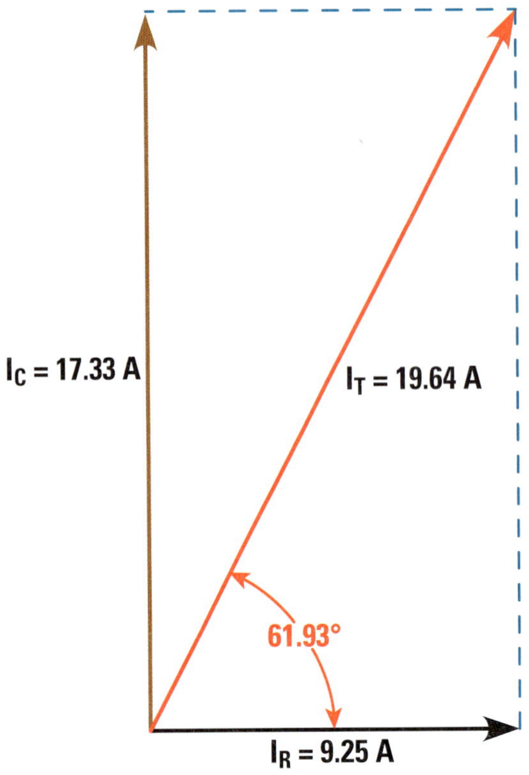

Figure 22–12 The total circuit current leads the resistive current and voltage by 61.93°.

Summary

- The current flow in the resistive part of the circuit is in phase with the voltage.
- The current flow in the capacitive part of the circuit leads the voltage by 90°.
- The amount the current and voltage are out of phase with each other is determined by the ratio of resistance to capacitive reactance.
- The voltage is the same across any leg of a parallel circuit.
- The circuit power factor is the ratio of true power to apparent power.

Review Questions

1. True or False: In an RC parallel circuit, the current through both the resistor and capacitor will be equal to the total current in the circuit.
2. Power factor in an RC parallel circuit is the ratio of which of the following statements?

   A) Capacitive reactance to apparent power

   B) True power to apparent power

   C) Resistance to capacitive reactance

   D) Reactive power to apparent power

3. Which of the following statements best describes the impact that the addition of capacitance has on the shift in phase angle between the applied voltage and the overall circuit current?
   A) Phase angle increases as capacitance is lowered.
   B) Phase angle increases as resistance is added.
   C) Phase angle increases as capacitance is added.
   D) Phase angle increases as resistance is lowered.
4. In an RC parallel circuit, how many degrees out of phase is the current through the capacitor from the current through the resistor?
   A) 0°
   B) 30°
   C) 60°
   D) 90°
5. A capacitor and a resistor are connected in parallel to a 120-V, 60-Hz line. The resistor has a resistance of 40 Ω, and the capacitor has a capacitance of 132.6 μF. What is the total current flow through the circuit?
   A) 6.7 amperes
   B) 9 amperes
   C) 12.3 amperes
   D) Not enough information to solve
6. A resistor and capacitor are connected in parallel to a 208-V, 60-Hz power source. An ammeter indicates a total circuit current of 46 amps, and a wattmeter indicates a true power of 4975 W. What is the resistance of the circuit?
   A) 108 ohms
   B) 4.5 ohms
   C) 8.7 ohms
   D) 23.9 ohms
7. In the previous question, what is the capacitance of the capacitor connected in parallel with the resistor?
   A) 39.3 μF
   B) 501 μF
   C) 5.29 μF
   D) 4.52 μF
8. A resistor and capacitor are connected in parallel to a 277-V, 60-Hz power source. The resistor has a value of 50 Ω, and the capacitor has a value of 40 μF. Which of the following is the approximate circuit power factor?
   A) Not enough information
   B) 75%
   C) 90%
   D) 80%

9. In the previous question, which of the following best represents the phase angle difference?
   A) 37°
   B) 45°
   C) 32°
   D) 35°
10. True or False: The power factor is the ratio of true power to apparent power in an RC parallel circuit.

Practice Problems

Refer to the formulas in the Resistive-Capacitive Parallel Circuits section of Appendix B and to *Figure 22–2*.

1. Assume that the circuit shown in *Figure 22–2* is connected to a 60-Hz line and has a total current flow of 10.463 A. The capacitor has a capacitance of 132.626 μF, and the resistor has a resistance of 14 Ω. Find the missing values.

E_T ______	E_R ______	E_C ______
I_T 10.463 A	I_R ______	I_C ______
Z ______	R 14 Ω	X_C ______
VA ______	P ______	$VARs_C$ ______
PF ______	∠θ ______	C 132.626 μF

2. Assume that the circuit is connected to a 400-Hz line and has a total impedance of 21.6 Ω. The resistor has a resistance of 36 Ω, and the capacitor has a current flow of 2 A through it. Find the missing values.

E_T ______	E_R ______	E_C ______
I_T ______	I_R ______	I_C 2 A
Z 21.6 Ω	R 36 Ω	X_C ______
VA ______	P ______	$VARs_C$ ______
PF ______	∠θ ______	C ______

3. Assume that the circuit shown in *Figure 22–2* is connected to a 600-Hz line and has a current flow through the resistor of 65.6 A and a current flow through the capacitor of 124.8 A. The total impedance of the circuit is 2.17888 Ω. Find the missing values.

E_T ______	E_R ______	E_C ______
I_T ______	I_R 65.6 A	I_C 124.8 A
Z 2.17888 Ω	R ______	X_C ______
VA ______	P ______	$VARs_C$ ______
PF ______	∠θ ______	C ______

4. Assume that the circuit shown in *Figure 22–2* is connected to a 1000-Hz line and has a true power of 486.75 W and a reactive power of 187.5 VARs. The total current flow in the circuit is 7.5 A. Find the missing values.

E_T ______	E_R ______	E_C ______
I_T 7.5 A	I_R ______	I_C ______
Z ______	R ______	X_C ______
VA ______	P 486.75 W	$VARs_C$ 187.5
PF ______	$\angle\theta$ ______	C ______

5. In an RC parallel circuit, R = 3.6 kΩ, and X_C = 4.7 kΩ. Find Z.
6. In an RC parallel circuit, I_R = 0.6 amperes, R = 24 Ω, and X_C = 33 Ω. Find I_C.
7. In an RC parallel circuit, E_T = 120 volts, I_T = 1.2 amperes, and R = 240 Ω. Find X_C.
8. In an RC parallel circuit, the apparent power is 3400 VA, and $\angle\theta = 58°$. Find reactive power.
9. In an RC parallel circuit, the true power is 780 watts, and the reactive power is 560 VARs. Find the power factor.
10. In an RC parallel circuit, E_T = 7.5 volts at 1 kHz. The circuit current is 2.214 amperes. R = 4 Ω. What is the value of the capacitor connected in the circuit?

Section 8 AC Circuits Containing Resistance-Inductance-Capacitance

© I. Pilon/Shutterstock.com

Unit 23 Resistive-Inductive-Capacitive Series Circuits

Key Terms

Bandwidth

Lagging power factor

Leading power factor

Resonance

Why You Need to Know

This unit combines the elements of resistance, inductance, and capacitance in the same circuit. RLC series circuits are generally employed in a particular application. An RLC series resonant circuit, for example, can produce a very large increase in the voltage across a particular component. In this unit, we will

- discuss the relationship between voltage, current, power, and impedance in series circuits containing resistance, inductance, and capacitance.
- determine the voltage and current phase relationship in the RLC series circuit.
- calculate the values needed to produce a resonant circuit at a particular frequency.
- introduce the concept of bandwidth when dealing with resonant circuits.

Outline

Objectives

After studying this unit, you should be able to

- explain the operational characteristics of an RLC series circuit.
- describe how to achieve RLC series circuit resonance.
- calculate unknown values of voltage, current, impedance, and power for an RLC series circuit.
- calculate power factor and phase angle in an RLC series circuit.

Preview

Circuits containing resistance, inductance, and capacitance connected in series are presented in this unit. Electrical quantities for voltage drop, impedance, and power are calculated for the total circuit values and for the individual components. Circuits that become resonant at a certain frequency are presented as well as the effect a resonant circuit has on electrical quantities such as voltage, current, and impedance.

23–1 RLC Series Circuits

When an AC circuit contains elements of resistance, inductance, and capacitance connected in series, the ***current is the same*** through all components, but the ***voltages dropped across the elements are out of phase*** with each other. The voltage dropped across the resistance is in phase with the current, the voltage dropped across the inductor leads the current by 90°, and the voltage dropped across the capacitor lags the current by 90° (*Figure 23–1*). Notice how in this illustration, the voltage drop across the inductor and the voltage drop across the capacitor are 180° out of phase with each other. This will be important as we explore this relationship further. An RLC series circuit is shown in *Figure 23–2*. The ratio between resistance, inductive reactance, and capacitive reactance determines how much the applied

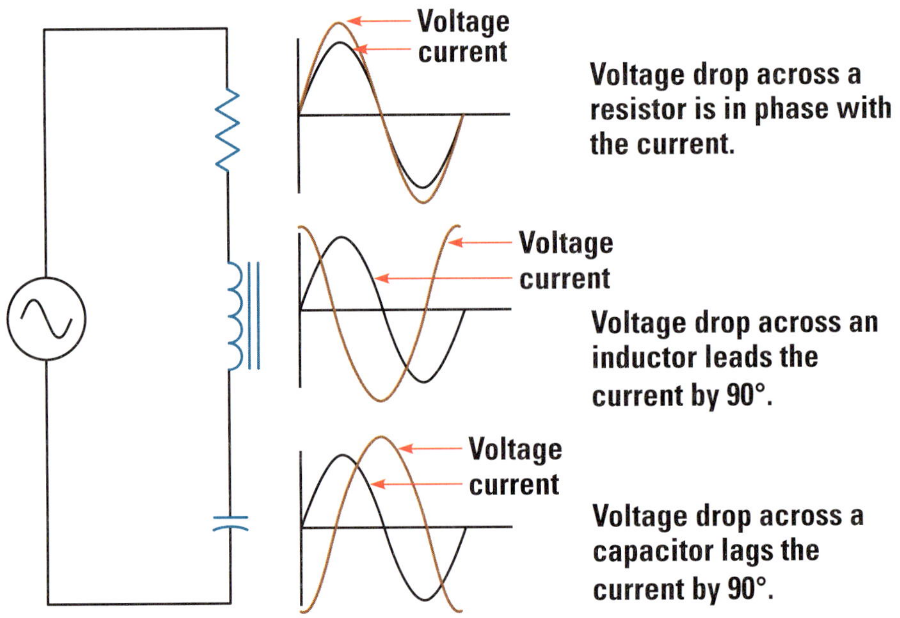

Figure 23–1 Voltage and current relationship in the individual components of an RLC series circuit.

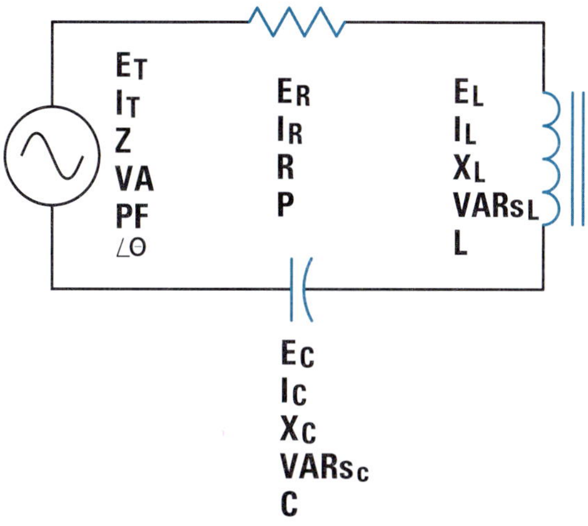

Figure 23–2 Resistive-inductive-capacitive series circuit.

voltage will either lead or lag the circuit current. If the circuit contains more inductive VARs than capacitive VARs, the current lags the applied voltage and the power factor is a **lagging power factor.** If there are more capacitive VARs than inductive VARs, the current leads the voltage, and the power factor is a **leading power factor.**

Because inductive reactance and capacitive reactance are 180° out of phase with each other, they subtract from each other in an AC circuit. This subtraction causes the total circuit impedance to be the vector addition of resistance and the difference of X_L and X_C. If the inductive and capacitive reactance are equal ohmic values but in opposite directions, the total circuit impedance will be equal to the resistance because the reactive vector equals 0 ohms. When Ohm's law is applied to the circuit values, note that the voltage drops developed across the inductor or capacitor can be higher than the applied voltage.

Example 23–1

Assume that the circuit shown in *Figure 23–2* has an applied voltage of 240 Ω at 60 Hz, the resistor has a value of 12 Ω, the inductor has an inductive reactance of 24 Ω, and the capacitor has a capacitive reactance of 8 Ω. Find the following unknown values:

Z—impedance of the circuit

I_T—circuit current

E_R—voltage drop across the resistor

P—true power (watts)

L—inductance of the inductor

E_L—voltage drop across the inductor

$VARs_L$—reactive power of the inductor

C—capacitance

E_C—voltage drop across the capacitor

$VARs_C$—reactive power of the capacitor

VA—volt-amperes (apparent power)

PF—power factor

$\angle\theta$—angle theta

Example 23–1 (continued)

Solution

Total Impedance

The impedance of the circuit is the sum of resistance, inductive reactance, and capacitive reactance. Because inductive reactance and capacitive reactance are 180° out of phase with each other, vector addition must be used to find their sum. This method results in the smaller of the two reactive values being subtracted from the larger (*Figure 23–3*). The smaller value is eliminated, and the larger value is reduced by the amount of the smaller value. After subtracting 8 Ω of capacitive reactance from 24 Ω of inductive reactance, the total reactive value is 16 Ω of remaining inductive reactance. The total impedance is the hypotenuse formed by the resulting right triangle (*Figure 23–4*). The impedance is calculated by using the formula:

$$Z = \sqrt{R^2 + (X_L - X_C)^2}$$

$$Z = \sqrt{(12\ \Omega)^2 + (24\ \Omega - 8\ \Omega)^2}$$

$$Z = \sqrt{(12\ \Omega)^2 + (16\ \Omega)^2}$$

$$Z = \sqrt{144\ \Omega + 256\ \Omega}$$

$$Z = \sqrt{400\ \Omega}$$

$$Z = 20\ \Omega$$

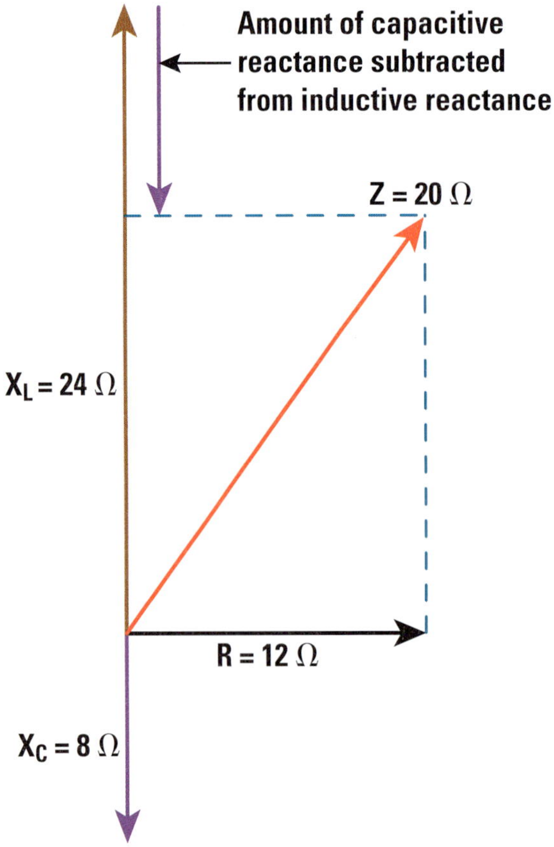

Figure 23–3 Vector addition must be used to determine total circuit impedance.

Example 23–1 (continued)

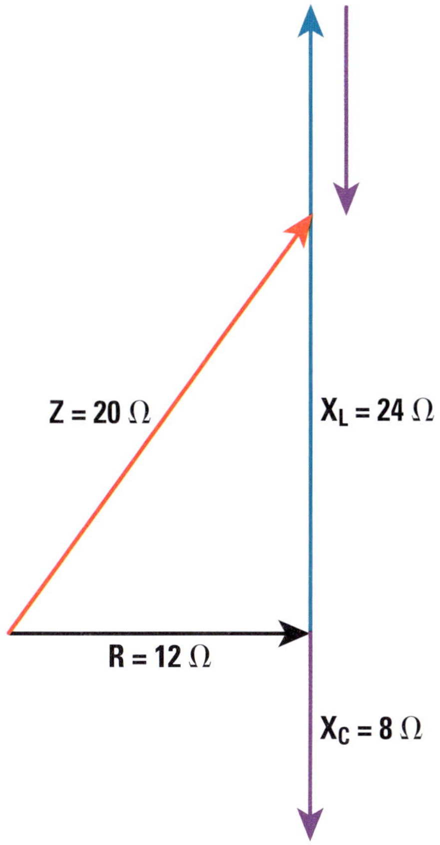

Figure 23–4 Right triangle formed by combining resistance, capacitive reactance, and inductive reactance to find total circuit impedance.

In the preceding formula, there is no need to make sure the smaller reactance value is subtracted from the larger. Even if the larger value is subtracted from the smaller value, it will yield the correct result because the difference is squared, and the square of a negative or positive number is always positive. For example, assume that an RLC series circuit contains a resistor with a value of 10 Ω, an inductor with an inductive reactance of 30 Ω, and a capacitor with a capacitive reactance of 54 Ω. When these values are substituted in the previous formula, the difference between the inductive reactance and capacitive reactance is a negative number:

$$Z = \sqrt{R^2 + (X_L - X_C)^2}$$

$$Z = \sqrt{(10\ \Omega)^2 + (30\ \Omega - 54\ \Omega)^2}$$

$$Z = \sqrt{(10\ \Omega)^2 + (-24\ \Omega)^2}$$

$$Z = \sqrt{(100 + 576\ \Omega)^2}$$

$$Z = \sqrt{676\ \Omega}$$

$$Z = 26\ \Omega$$

Example 23–1 (continued)

Current

The total current flow through the circuit can now be calculated using the formula:

$$I_T = \frac{E_T}{Z}$$

$$I_T = \frac{240\ V}{20\ \Omega}$$

$$I_T = 12\ A$$

In a series circuit, the current flow is the same at any point in the circuit. Therefore, 12 A flow through each of the circuit components.

Resistive Voltage Drop

The voltage drop across the resistor can be calculated using the formula:

$$E_R = I_R \times R$$

$$E_R = 12\ A \times 12\ \Omega$$

$$E_R = 144\ V$$

Watts

The true power of the circuit can be calculated using any of the pure resistive values. In this example, true power is found using the formula:

$$P = E_R \times I$$

$$P = 144\ V \times 12\ A$$

$$P = 1728\ W$$

Inductance

The amount of inductance in the circuit can be calculated using the formula:

$$L = \frac{X_L}{2\pi f}$$

$$L = \frac{24\ \Omega}{377}$$

$$L = 0.0637\ H$$

Voltage Drop Across the Inductor

The amount of voltage drop across the inductor can be calculated using the formula:

$$E_L = I \times X_L$$

$$E_L = 12\ A \times 24\ \Omega$$

$$E_L = 288\ V$$

Notice that the voltage drop across the inductor is greater than the applied voltage.

Example 23–1 (continued)

Inductive VARs

The amount of reactive power of the inductor can be calculated by using inductive values:

$$VARs_L = E_L \times I$$

$$VARs_L = 288\ V \times 12\ A$$

$$VARs_L = 3456$$

Capacitance

The amount of capacitance in the circuit can be calculated by using the formula:

$$C = \frac{1}{2\pi f X_C}$$

$$C = \frac{1}{377 \times 8\ \Omega}$$

$$C = \frac{1}{3016}$$

$$C = 0.000331565\ F, \text{ or } 331.565\ \mu F$$

Voltage Drop Across the Capacitor

The voltage dropped across the capacitor can be calculated using the formula:

$$E_C = I \times X_C$$

$$E_C = 12\ A \times 8\ \Omega$$

$$E_C = 96\ V$$

Capacitive VARs

The amount of capacitive VARs can be calculated using the formula:

$$VARs_C = E_C \times I$$

$$VARs_C = 96\ V \times 12\ A$$

$$VARs_C = 1152$$

Apparent Power

The VAs (apparent power) can be calculated by multiplying the applied voltage and the circuit current:

$$VA = E_T \times I$$

$$VA = 240\ V \times 12\ A$$

$$VA = 2880$$

The apparent power can also be found by vector addition of true power, inductive VARs, and capacitive VARs (*Figure 23–5*). As with the addition of resistance, inductive reactance, and capacitive reactance, inductive VARs ($VARs_L$), and capacitive VARs ($VARs_C$) are 180° out of phase with each other. The result is

Example 23–1 (continued)

the elimination of the smaller and a reduction of the larger. The following formula can be used to determine apparent power:

$$VA = \sqrt{P^2 + (VARs_L - VARs_C)^2}$$

$$VA = \sqrt{(1728\ W)^2 + (3456\ VARs_L - 1152\ VARs_C)^2}$$

$$VA = \sqrt{(1728\ W)^2 + (2304\ VARs)^2}$$

$$VA = \sqrt{8{,}294{,}400}$$

$$VA = 2880$$

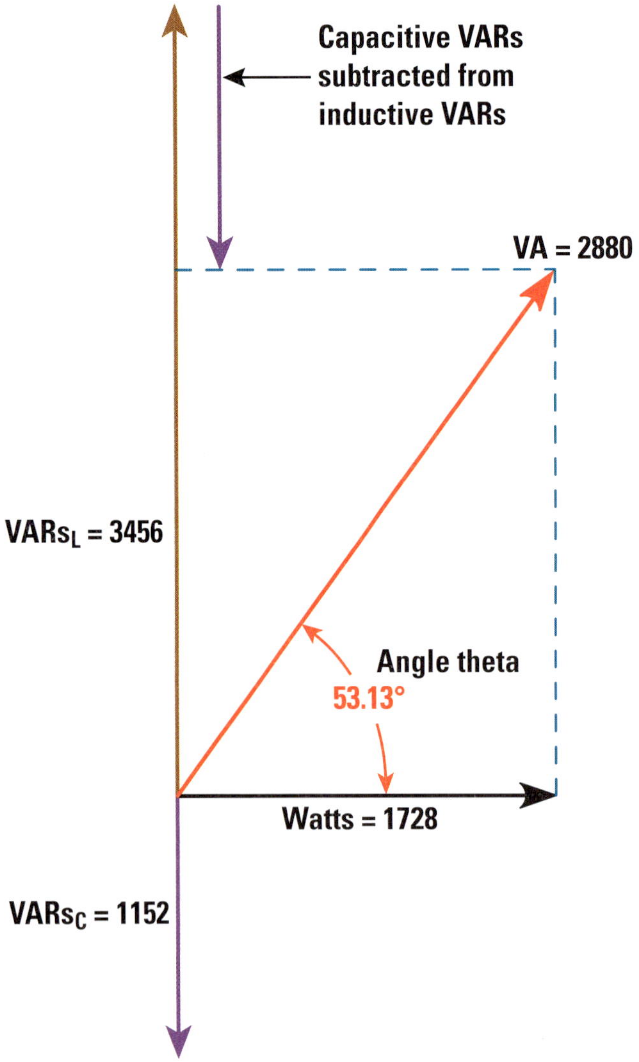

Figure 23–5 Vector addition of true power and reactive power to find the total apparent power.

Example 23–1 (continued)

Power Factor

The power factor can be calculated by dividing the true power of the circuit by the apparent power. The answer is multiplied by 100 to change the decimal into a percent:

$$PF = \frac{W}{VA} \times 100$$

$$PF = \frac{1728\ W}{2880\ VA} \times 100$$

$$PF = 0.06 \times 100$$

$$PF = 60\%$$

Angle Theta

The power factor is the cosine of angle theta:

$$COS\ \angle\theta = 0.06$$

$$\angle\theta = 53.13°$$

The circuit with all calculated values is shown in *Figure 23–6*.

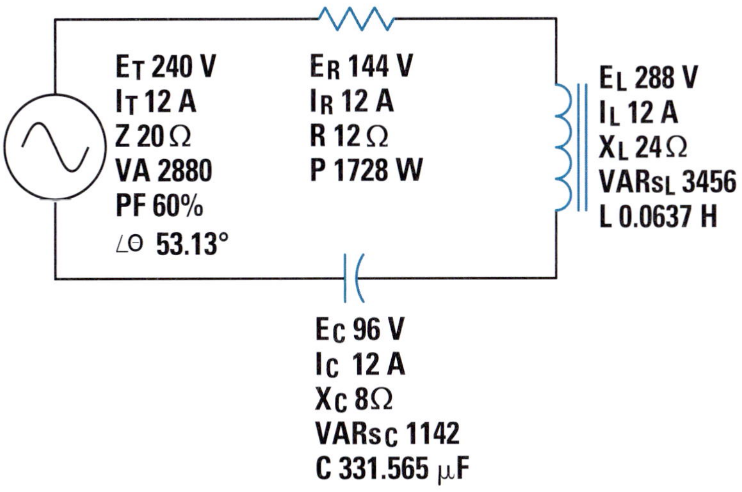

Figure 23–6 Example circuit with calculated values.

Example 23–2

An RLC series circuit contains a capacitor with a capacitance of 66.3 μF, an inductor with an inductance of 0.0663 H, and a resistor with a value of 8 Ω connected to a 120-V, 60-Hz line (*Figure 23–7*). How much current will flow in this circuit?

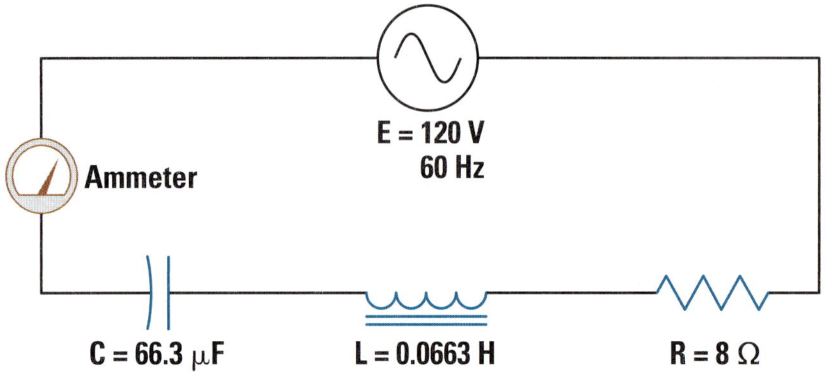

Figure 23–7 Example RLC series circuit.

Solution

The first step in solving this problem is to find the values of capacitive and inductive reactance:

$$X_C = \frac{1}{2\pi fC}$$

$$X_C = \frac{1}{377 \times 0.0000663\ \text{F}}$$

$$X_C = 40\ \Omega$$

$$X_L = 2\pi fL$$

$$X_L = 377 \times 0.0663\ \text{H}$$

$$X_L = 25\ \Omega$$

Now that the capacitive and inductive reactance values are known, the circuit impedance can be found using the formula:

$$Z = \sqrt{R^2 + (X_L - X_C)^2}$$

$$Z = \sqrt{(8\ \Omega)^2 + (25\ \Omega - 40\ \Omega)^2}$$

$$Z = \sqrt{64\ \Omega + 225\ \Omega}$$

$$Z = 17\ \Omega$$

Now that the circuit impedance is known, the current flow can be found using Ohm's law:

$$I_T = \frac{E_T}{Z}$$

$$I_T = \frac{120\ \text{V}}{17\ \Omega}$$

$$I_T = 7.05\ \text{A}$$

Example 23–3

The RLC series circuit shown in *Figure 23–8* contains an inductor with an inductive reactance of 62 Ω and a capacitor with a capacitive reactance of 38 Ω. The circuit is connected to a 208-V, 60-Hz line. How much resistance should be connected in the circuit to limit the circuit current to a value of 8 A?

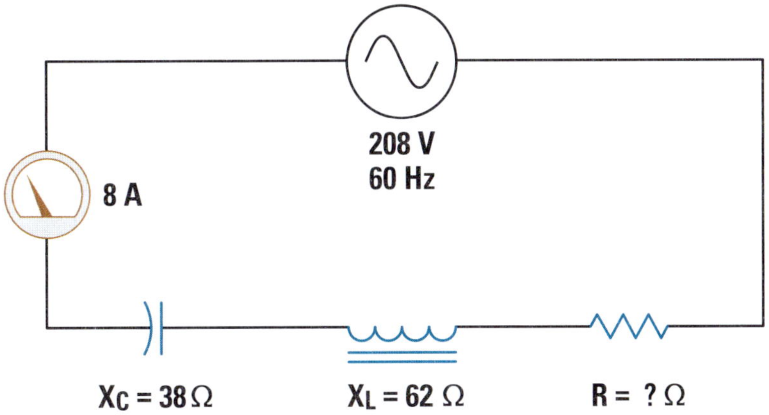

Figure 23–8 Example RLC series circuit.

Solution

The first step is to determine the total impedance necessary to limit the circuit current to a value of 8 A:

$$Z = \frac{E_T}{I_T}$$

$$Z = \frac{208\ V}{8\ A}$$

$$Z = 26\ \Omega$$

The formula for finding impedance in an RLC series circuit can now be adjusted to find the missing resistance value (refer to the Resistive-Inductive-Capacitive Series Circuits section of Appendix B):

$$R = \sqrt{Z^2 - (X_L - X_C)^2}$$

$$R = \sqrt{(26\ \Omega)^2 - (62\ \Omega - 38\ \Omega)^2}$$

$$R = \sqrt{(26\ \Omega)^2 - (24\ \Omega)^2}$$

$$R = \sqrt{100\ \Omega^2}$$

$$R = 10\ \Omega$$

23–2 Series Resonant Circuits

When an inductor and capacitor are connected in series (*Figure 23–9*), there is one frequency at which the inductive reactance and capacitive reactance become equal. The reason for this is that, as frequency increases, inductive reactance increases and capacitive reactance decreases. The point at which the two reactive values become equal and cancel each other is called **resonance.** Resonant circuits are used to provide great increases of current and voltage at the resonant frequency.

The following formula can be used to determine the resonant frequency when the values of inductance (L) and capacitance (C) are known:

$$f_R = \frac{1}{2\pi\sqrt{LC}}$$

where

f_R = frequency at resonance

L = inductance in henrys

C = capacitance in farads

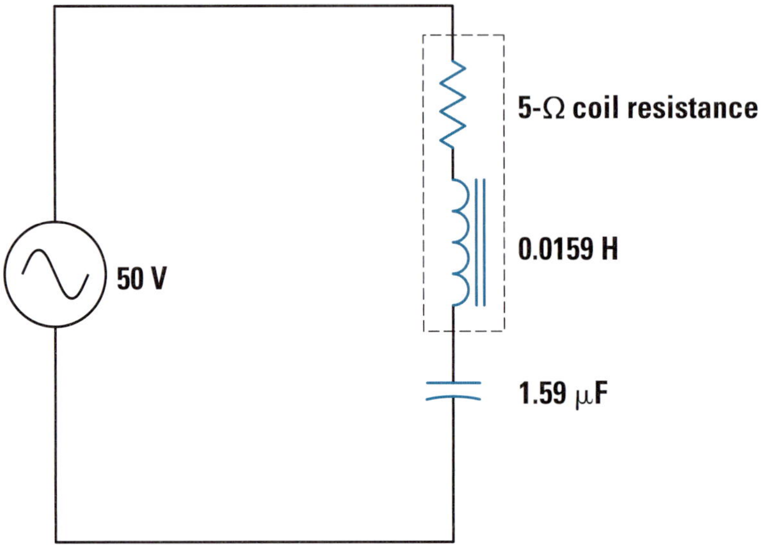

Figure 23–9 Example of an LC series circuit.

In the circuit shown in *Figure 23–9,* an inductor has an inductance of 0.0159 henry and a wire resistance in the coil of 5 ohms. The capacitor connected in series with the inductor has a capacitance of 1.59 microfarads. This circuit reaches resonance at 1000 hertz, when both the inductor and capacitor produce reactive values of 100 ohms. At this point, the two values are equal and opposite in direction, and the only current-limiting factor in the circuit is the 5 ohms of wire resistance in the coil (*Figure 23–10*).

During the period of time the circuit is not at resonance, current flow is limited by the combination of inductive reactance and capacitive reactance. At 600 hertz, the inductive reactance is 59.942 ohms, and the capacitive reactance is 166.829 ohms. The total circuit impedance is:

$$Z = \sqrt{R^2 + (X_L - X_C)^2}$$

$$Z = \sqrt{(5\ \Omega)^2 + (59.942\ \Omega - 166.829\ \Omega)^2}$$

$$Z = 107\ \Omega$$

If 50 volts are applied to the circuit, the current flow will be 0.467 ampere (50 V/107 Ω).

If the frequency is greater than 1000 hertz, the inductive reactance increases, and the capacitive reactance decreases. At a frequency of 1400 hertz, for example, the inductive reactance has become 139.864 ohms, and the capacitive reactance has become 71.498 Ω. The total impedance of the circuit at this point is 68.549 ohms. The circuit current is 0.729 ampere (50 V/68.549 Ω).

When the circuit reaches resonance, the current suddenly increases to 10 amperes because the only current-limiting factor is the 5 ohms of wire resistance (50 V/5 Ω =10 A). A graph illustrating the effect of current in a resonant circuit is shown in *Figure 23–11*.

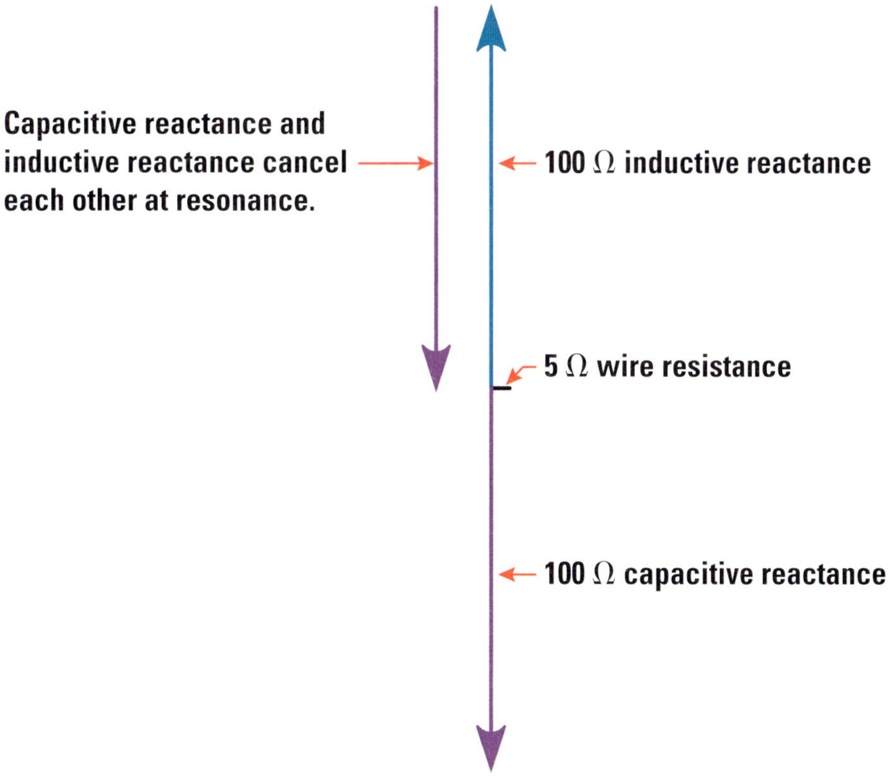

Figure 23–10 Inductive reactance and capacitive reactance become equal at the resonant frequency.

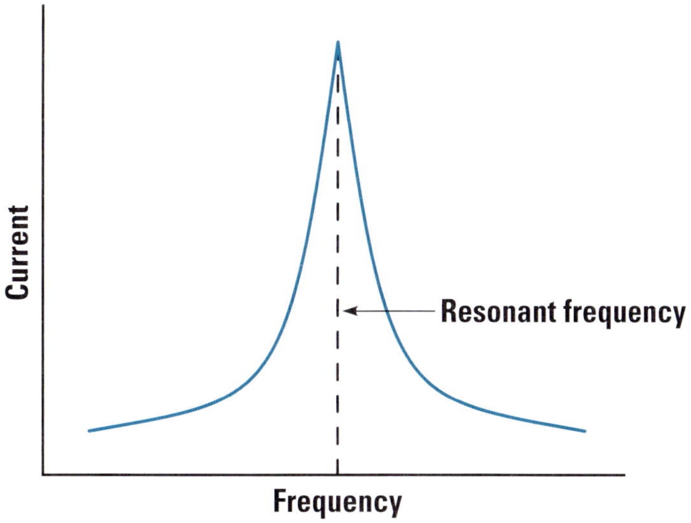

Figure 23–11 The current increases sharply at the resonant frequency.

Although inductive and capacitive reactance cancel each other at resonance, each is still a real value. In this example, both the inductive reactance and capacitive reactance have an ohmic value of 100 ohms at the resonant frequency. The voltage drop across each component is proportional to the reactance and the amount of current flow. Theoretically, if voltmeters were connected across each component, a voltage of 1000 volts would be seen ($10 \text{ A} \times 100 \ \Omega = 1000 \text{ V}$) (*Figure 23–12*). In reality, the voltmeter connected across the inductor would differ very slightly from the capacitor because of the resistance in the coil wire.

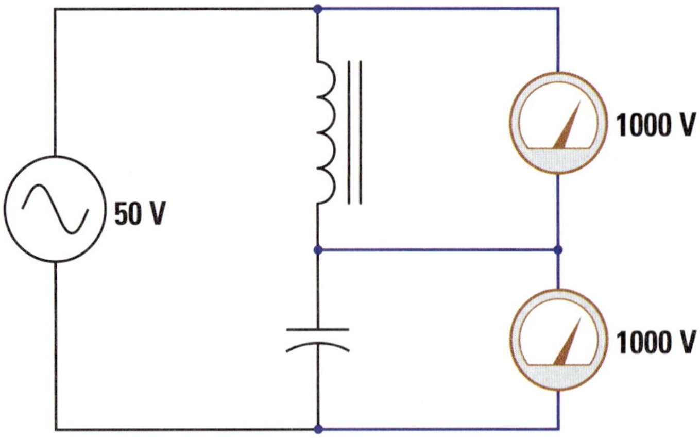

Figure 23–12 The voltage drops across the inductor and capacitor increase at the resonant frequency.

Bandwidth

Bandwidth is the range within the limits of a band. The rate that current increases and decreases is proportional to the quality (Q) of the components in the circuit:

$$B = \frac{f_R}{Q}$$

where

B = bandwidth

Q = quality of circuit components

f_R = frequency at resonance

High-Q components result in a sharp increase and decrease of current as illustrated by the curve of *Figure 23–11*. Not all series resonant circuits produce as sharp an increase or decrease of current as illustrated in *Figure 23–11*. The term used to describe this rate of increase or decrease is bandwidth. Bandwidth is a measurement of the frequency range at which the current is at a value of 0.707 of the maximum current value (*Figure 23–13*):

$$B = f_2 - f_1$$

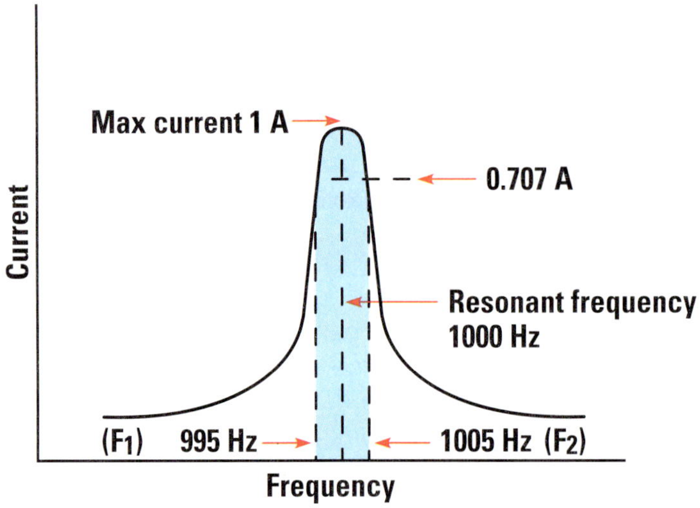

Figure 23–13 A narrow-band resonant circuit is produced by high-Q inductors and capacitors.

Assume the circuit producing the curve in *Figure 23–13* reaches resonance at a frequency of 1000 hertz. Also assume that the circuit reaches a maximum value of 1 ampere at resonance. The bandwidth of this circuit can be determined by finding the lower and upper frequencies on either side of 1000 hertz at which the current reaches a value of 0.707 of the maximum value. In this illustration, that is 0.707 ampere (1 A × 0.707 = 0.707 A). The lower frequency value is determined to be 995 hertz, and the upper value is determined to be 1005 hertz. This circuit has a bandwidth of 10 hertz (1005 Hz − 995 Hz = 10 Hz). When resonant circuits are constructed with components that have a relatively high Q, the difference between the two frequencies is small because the resistance value is relatively small compared to the reactive value. These circuits are said to have a *narrow* bandwidth.

In a resonant circuit using components with a lower Q rating, the current does not increase as sharply, as shown in *Figure 23–14*. In this circuit, the resonant frequency is 1200 hertz, and the maximum current flow at resonance is 0.8 ampere. The bandwidth is determined by the difference between the two frequencies at which the current is at a value of 0.566 ampere (0.8 A × 0.707 = 0.566 A). It is determined that the lower frequency is 1150 hertz, and the upper frequency is 1250 hertz. This circuit has a bandwidth of 100 hertz (1250 Hz − 1150 Hz = 100 Hz). This circuit is said to have a *wide* bandwidth.

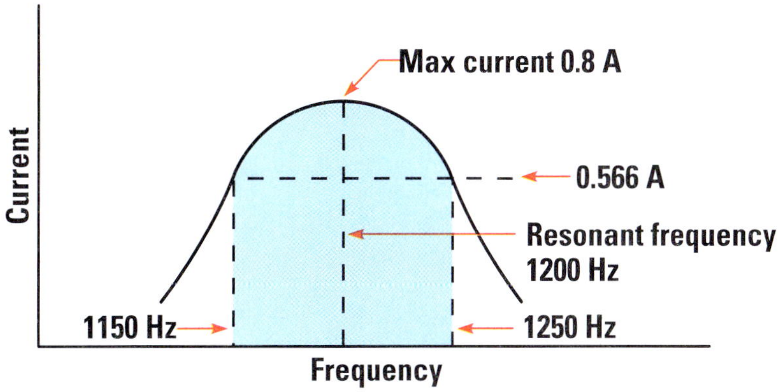

Figure 23–14 A wide-band resonant circuit is produced by low-Q inductors and capacitors.

Determining Resonant Values of L and C

There are times when it is necessary to determine what value of inductance or capacitance is needed to resonate with an existing value for inductance or capacitance. The two formulas shown can be used to determine these values for a series or parallel resonant circuit:

$$L = \frac{1}{4\pi^2 f_R^2 C} \quad \text{or} \quad C = \frac{1}{4\pi^2 f_R^2 L}$$

where

$\pi = 3.1416$

f_R = resonant frequency

C = capacitance in farads

L = inductance in henrys

What value of capacitance would be needed to resonate at a frequency of 1200 Hz with a 0.2-henry inductor?

$$C = \frac{1}{4\pi^2 f_R^2 L}$$

$$C = \frac{1}{4 \times 3.1416^2 \times (1200\ \text{Hz})^2 \times 0.2\ \text{H}}$$

$$C = \frac{1}{4 \times 9.869 \times 1{,}440{,}000\ \text{Hz}^2 \times 0.2\ \text{H}}$$

$$C = \frac{1}{11{,}369{,}088}$$

$$C = 0.000{,}000{,}088 \text{ farads or } 0.088\ \mu\text{F or } 88\ \text{nF}$$

What value of inductance is needed to resonate with a 50-microfarad capacitor at 400 hertz?

$$L = \frac{1}{4\pi^2 f_R^2 C}$$

$$L = \frac{1}{4 \times 3.1416^2 \times (400\ \text{Hz})^2 \times 0.000{,}050\ \text{F}}$$

$$L = \frac{1}{4 \times 9.869 \times 160{,}000\ \text{Hz}^2 \times 0.000{,}050\ \text{F}}$$

$$L = \frac{1}{315.808}$$

$$L = 0.003166 \text{ henry, or } 3.166\ \text{mH}$$

Summary

- The voltage dropped across the resistor in an RLC series circuit will be in phase with the current.
- The voltage dropped across the inductor in an RLC series circuit will lead the current by 90°.
- The voltage dropped across the capacitor in an RLC series circuit will lag the current by 90°.
- Vector addition can be used in an RLC series circuit to find values of total voltage, impedance, and apparent power.
- In an RLC circuit, inductive and capacitive values are 180° out of phase with each other. Adding them results in the elimination of the smaller value and a reduction of the larger value.
- LC resonant circuits increase the current and voltage drop at the resonant frequency.
- Resonance occurs when inductive reactance and capacitive reactance become equal.
- The rate the current increases is proportional to the Q of the circuit components.
- Bandwidth is determined by calculating the upper and lower frequencies at which the current reaches a value of 0.707 of the maximum value.
- Bandwidth is inversely proportional to the Q of the components in the circuit.

Review Questions

1. Which of the following statements best describes the phase angle relationship of current and the voltage dropped across a purely resistive load in an RLC series circuit?
   A) Current and voltage dropped across the resistor are 180° out of phase.
   B) Current and voltage dropped across the resistor are 90° out of phase.
   C) Current and voltage dropped across the resistor are between 0° and 90° out of phase.
   D) Current and voltage dropped across the resistor are in phase with each other.
2. Which of the following statements best describes the phase angle relationship of current and the voltage dropped across the inductive load in an RLC series circuit?
   A) The voltage dropped across the inductive load lags the current by 90°.
   B) The voltage dropped across the inductive load leads the current by 90°.
   C) The current through the inductive load leads the voltage dropped by 90°.
   D) The current and voltage dropped across the inductive load are in phase.
3. Which of the following statements best describes the phase angle relationship of current and the voltage dropped across the capacitive load in an RLC series circuit?
   A) The voltage dropped across the capacitive load lags the current by 90°.
   B) The voltage dropped across the capacitive load leads the current by 90°.
   C) The current through the capacitive load lags the voltage dropped by between 0° and 90°.
   D) The voltage dropped across the capacitive load is 180° out of phase with the current.
4. RLC series circuit resonance is achieved when which of the following conditions are true in the circuit?
   A) Resonance in an RLC series circuit is when resistance equals inductive or capacitive reactance.
   B) Resonance is when capacitance and inductance are equal.
   C) Resonance is when inductive reactance and capacitive reactance cancel each other out.
   D) Resonance is when the voltage dropped across resistive, inductive, and capacitive elements is equal.
5. True or False: All RLC series circuits will exhibit resonance at a specific frequency as the capacitive reactance and inductive reactance will be equal but opposite.
6. Which of the following best describes the impact of achieving resonance in an RLC series circuit?
   A) The circuit will see a voltage spike because induced voltage adds to applied voltage.
   B) The circuit will see an increase in total circuit current as only resistance opposes current flow.
   C) The circuit will see a decrease in current flow as impedance increases at resonance.
   D) The circuit experiences no impact from being at resonance.

7. An AC circuit has a frequency of 400 Hz. A 16 Ω resistor, a 0.0119 H inductor, and a 16.6 μF capacitor are connected in series. What is the total impedance of the circuit?

   A) 17.1 ohms

   B) 24 ohms

   C) 29.9 ohms

   D) 69.9 ohms

8. If the applied voltage to the circuit in question #7 is measured with a voltmeter at 222.3 volts, what is the voltage that will be dropped across the resistor component in the circuit?

   A) 388.7 volts

   B) 312 volts

   C) 189.3 volts

   D) 208 volts

9. If the frequency of the circuit in question #7 were to be modified to achieve resonance, what frequency would the circuit need to be changed to in order to reach this state?

   A) 60 hertz

   B) 125 hertz

   C) 358 hertz

   D) 50 hertz

10. Calculate the power factor of the circuit in question #7 using the original frequency and values.

    A) 84.5%

    B) 94.5%

    C) 89.7%

    D) 92.3%

11. Using the power factor determined in question #10, what is the phase angle between voltage and current?

    A) 19.1°

    B) 26.2°

    C) 22.6°

    D) 32.3°

Practical Applications

You are an electrician working in a plant. A series resonant circuit is to be used to produce a high voltage at a frequency of 400 Hz. The inductor has an inductance of 15 mH and a wire resistance of 2 Ω. How much capacitance should be connected in series with the inductor to produce a resonant circuit? The voltage supplied to the circuit is 240 V at 400 Hz. What is the minimum voltage rating of the capacitor?

Practice Problems

Refer to the Resistive-Inductive-Capacitive Series Circuits section of Appendix B and to *Figure 23–2*. Find all the missing values in the following problems.

1. The circuit shown in *Figure 23–2* is connected to a 120-V, 60-Hz line. The resistor has a resistance of 36 Ω, the inductor has an inductive reactance of 100 Ω, and the capacitor has a capacitive reactance of 52 Ω.

E_T 120 V ________	E_R ________	E_L ________	E_C ________
I_T ________	I_R ________	I_L ________	I_C ________
Z ________	R 36 Ω	X_L 100 Ω	X_C 52 Ω
VA ________	P ________	$VARs_L$ ________	$VARs_C$ ________
PF ________	$\angle\theta$ ________	L ________	C ________

2. The circuit is connected to a 400-Hz line with an applied voltage of 35.678 V. The resistor has a true power of 14.4 W, and there are 12.96 inductive VARs and 28.8 capacitive VARs.

E_T 35.678 V ________	E_R ________	E_L ________	E_C ________
I_T ________	I_R ________	I_L ________	I_C ________
Z ________	R ________	X_L ________	X_C ________
VA ________	P 14.4 W	$VARs_L$ 12.96	$VARs_C$ 28.8
PF ________	$\angle\theta$ ________	L ________	C ________

3. The circuit is connected to a 60-Hz line. The apparent power in the circuit is 29.985 VA, and the power factor is 62.5%. The resistor has a voltage drop of 14.993 V, the inductor has an inductive reactance of 60 Ω, and the capacitor has a capacitive reactance of 45 Ω.

E_T ________	E_R 14.993 V	E_L ________	E_C ________
I_T ________	I_R ________	I_L ________	I_C ________
Z ________	R ________	X_L 60 Ω	X_C 45 Ω
VA 29.985	P ________	$VARs_L$ ________	$VARs_C$
PF 62.5%	$\angle\theta$ ________	L ________	C ________

4. This circuit is connected to a 1000-Hz line. The resistor has a voltage drop of 185 V, the inductor has a voltage drop of 740 V, and the capacitor has a voltage drop of 444 V. The circuit has an apparent power of 51.8 VA.

E_T ________	E_R 185 V	E_L 740 V	E_C 444 V
I_T ________	I_R ________	I_L ________	I_C ________
Z ________	R ________	X_L ________	X_C ________
VA 51.8 ________	P ________	$VARs_L$ ________	$VARs_C$ ________
PF ________	$\angle\theta$ ________	L ________	C ________

5. A series RLC circuit contains a 4-kΩ resistor, an inductor with an inductive reactance of 3.5 kΩ, and a capacitor with a capacitive reactance of 2.4 kΩ. A 120-VAC, 60-Hz power source is connected to the circuit. How much voltage is dropped across the inductor?

6. A series RLC circuit contains a resistor with a true power of 18 watts, an inductor with a reactive power of 24 VARs, and a capacitor with a reactive power of 34 VARs. What is the circuit power factor?
7. Is the power factor in question #6 a leading or lagging power factor? Explain your answer.
8. A series RLC circuit contains a resistor with a resistance of 8 Ω, an inductor with a reactance of 12 Ω, and a capacitor with a reactance of 16 Ω. $E_R = 17.8$ volts. Find E_C.
9. A series RLC circuit has an applied voltage of 240 volts and an apparent power of 600 VA. The circuit power factor is 62%. Find the value of the resistor.
10. A series RLC circuit is connected to a 60-Hz power line. The resistor has a value of 240 Ω. The inductor has an inductance of 0.796 henrys, and the capacitor has a capacitance of 5.89 μF. Find Z.

Unit 24 Resistive-Inductive-Capacitive Parallel Circuits

Key Terms

Power factor correction

Tank circuits

Unity

Why You Need to Know

RLC parallel circuits are used throughout the electrical field. They are employed in electronic equipment as filters to separate one frequency from another. They are used in industry to produce large increases in current flow for induction heating applications and to correct the power factor of inductive loads in an effort to reduce the amperes needed to supply a particular load. RLC parallel resonant circuits are often referred to as "tank" circuits. Tank circuits are the principle used in many motor-control applications as proximity sensors that detect the presence or absence of metal. In this unit, we will

- discuss the relationship between voltage, current, power, and impedance in parallel circuits that contain resistance, capacitance, and inductance.
- give a step-by-step procedure for determining and correcting power factor to the desired percentage.

Outline

Objectives

After studying this unit, you should be able to

- explain the operational characteristics of an RLC parallel circuit.
- describe how to achieve RLC parallel circuit resonance.
- calculate unknown values of voltage, current, impedance, and power for an RLC parallel circuit.
- calculate power factor and phase angle in an RLC parallel circuit.
- calculate power factor correction in AC circuits.

Preview

Circuits containing elements of resistance, inductance, and capacitance connected in parallel are found in nearly every aspect of our daily lives. Electrical quantities of current, impedance, and power must be calculated for the entire circuit as well as for individual components during critical activities such as troubleshooting and diagnostic work. Parallel resonant circuits, or tank circuits, and their effects on voltage, current, and impedance also play an important role and will be introduced in this unit.

24–1 RLC Parallel Circuits

When an AC circuit contains elements of resistance, inductance, and capacitance connected in parallel, the ***voltage dropped across each element is the same. The currents flowing through each branch, however, will have different values and be out of phase with each other*** (*Figure 24–1*). The current flowing through a purely resistive parallel branch will be in phase with the applied voltage. The current flowing through a purely inductive parallel branch will lag the applied voltage by 90 electrical degrees, and the current flowing through a purely capacitive parallel branch will lead the voltage by 90 electrical degrees. The phase angle difference between the applied voltage and the total current is determined by the ratio of resistance, inductive reactance, and capacitive reactance connected in parallel. As with an RLC series circuit, if the inductive VARs are greater than the capacitive VARs, the current will lag the voltage and the power factor will be lagging. If the capacitive VARs are greater, the current will lead the voltage and the power factor will be leading.

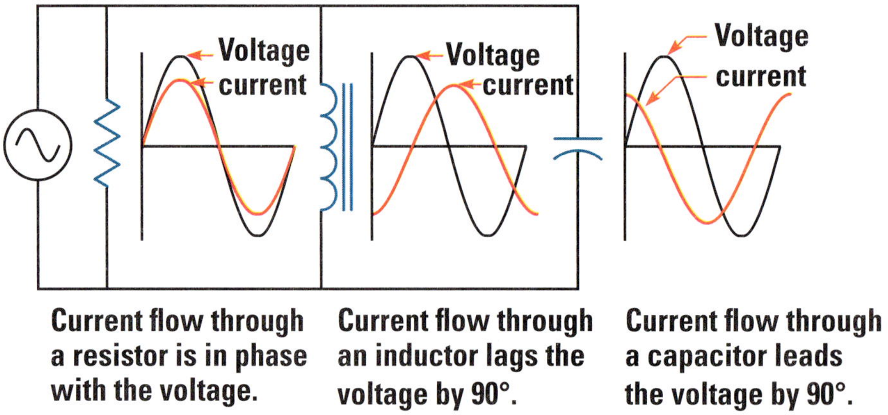

Figure 24–1 Voltage and current relationship in an RLC parallel circuit. The voltage is the same across each branch, but the currents will be out of phase.

Example 24–1

Assume that the RLC parallel circuit shown in *Figure 24–2* is connected to a 240-V, 60-Hz line. The resistor has a resistance of 12 Ω, the inductor has an inductive reactance of 8 Ω, and the capacitor has a capacitive reactance of 16 Ω. Complete the following unknown values:

Z—impedance of the circuit

I_T—total circuit current

I_R—current flow through the resistor

P—true power (watts)

L—inductance of the inductor

I_L—current flow through the inductor

$VARs_L$—reactive power of the inductor

C—capacitance

I_C—current flow through the capacitor

$VARs_C$—reactive power of the capacitor

VA—volt-amperes (apparent power)

PF—power factor

$\angle\theta$—angle theta

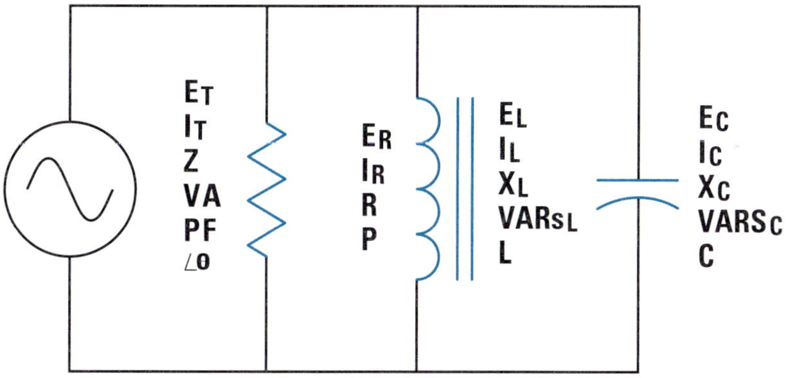

Figure 24–2 Example RLC parallel circuit.

Solution

Impedance

The impedance of the circuit is the reciprocal of the sum of the reciprocals of the current opposition values of each of the branches. Because these values are out of phase with each other, vector addition must be used:

$$Z = \frac{1}{\sqrt{\left(\frac{1}{R}\right)^2 + \left(\frac{1}{X_L} - \frac{1}{X_C}\right)^2}}$$

$$Z = \frac{1}{\sqrt{\left(\frac{1}{12\ \Omega}\right)^2 + \left(\frac{1}{8\ \Omega} - \frac{1}{16\ \Omega}\right)^2}}$$

Example 24–1 (continued)

$$Z = \frac{1}{\sqrt{(0.006944 + 0.003906)\frac{1}{\Omega}}}$$

$$Z = \frac{1}{\sqrt{(0.01085)\frac{1}{\Omega}}}$$

$$Z = \frac{1}{\sqrt{(0.10416)\frac{1}{\Omega}}}$$

$$Z = 9.601\ \Omega$$

To find the total impedance of the previous example using a scientific calculator, press the following keys. Note that the calculator automatically carries each answer to the maximum number of decimal places. This increases the accuracy of the answer.

[1] [2] [1/x] [x²] [+] [(] [8] [1/x] [−] [1] [6] [1/x] [)] [x²] [=] [√] [1/x]

Some calculators may require a different key entry or pressing the equal key at the end.

Another formula that can be used to determine the total impedance of a circuit containing resistance, inductive reactance, and capacitive reactance is:

$$Z = \frac{R \times X}{\sqrt{R^2 + X^2}}$$

where

$$X = \frac{X_L \times X_C}{X_L + X_C}$$

In this formula, X_L is a positive number, and X_C is a negative number. Therefore, Z will be either positive or negative depending on whether the circuit is more inductive (positive) or capacitive (negative). To find the total impedance of this circuit using this formula, first determine the value of X:

$$X = \frac{X_L \times X_C}{X_L + X_C}$$

$$X = \frac{8\ \Omega \times (-16\ \Omega)}{8\ \Omega + (-16\ \Omega)}$$

$$X = \frac{-128\ \Omega^2}{-8\ \Omega}$$

$$X = 16\ \Omega$$

Now that the value of X has been determined, the impedance can be calculated using the formula:

$$Z = \frac{R \times X}{\sqrt{R^2 + X^2}}$$

$$Z = \frac{12\ \Omega \times 16\ \Omega}{\sqrt{(12\ \Omega)^2 + (16\ \Omega)^2}}$$

$$Z = \frac{192\ \Omega}{\sqrt{400\ \Omega}}$$

$$Z = \frac{192\ \Omega}{20\ \Omega}$$

$$Z = 9.6\ \Omega$$

Example 24–1 (continued)

Resistive Current

The next unknown value to be found is the current flow through the resistor. This can be calculated by using the formula:

$$I_R = \frac{E}{R}$$

$$I_R = \frac{240\ V}{12\ \Omega}$$

$$I_R = 20\ A$$

True Power

The true power, or watts (W), can be calculated using the formula:

$$P = E \times I_R$$

$$P = 240\ V \times 20\ A$$

$$P = 4800\ W$$

Inductive Current

The amount of current flow through the inductor can be calculated using the formula:

$$I_L = \frac{E}{X_L}$$

$$I_L = \frac{240\ V}{8\ \Omega}$$

$$I_L = 30\ A$$

Inductive VARs

The amount of reactive power, or VARs, produced by the inductor can be calculated using the formula:

$$VARs_L = E \times I_L$$

$$VARs_L = 240\ V \times 30\ A$$

$$VARs_L = 7200$$

Inductance

The amount of inductance in the circuit can be calculated using the formula:

$$L = \frac{X_L}{2\pi f}$$

$$L = \frac{8\ \Omega}{377}$$

$$L = 0.0212\ H$$

Capacitive Current

The current flow through the capacitor can be calculated using the formula:

$$I_C = \frac{E}{X_C}$$

$$I_C = \frac{240\ V}{16\ \Omega}$$

$$I_C = 15\ A$$

Example 24–1 (continued)

Capacitance

The amount of circuit capacitance can be calculated using the formula:

$$C = \frac{1}{2\pi f X_C}$$

$$C = \frac{1}{377 \times 16\ \Omega}$$

$$C = 0.000165782\ \text{F} = 165.782\ \mu\text{F}$$

Capacitive VARs

The capacitive VARs can be calculated using the formula:

$$VARs_C = E \times I_C$$

$$VARs_C = 240 \times 15$$

$$VARs_C = 3600$$

Total Circuit Current

The amount of total current flow in the circuit can be calculated by vector addition of the current flowing through each branch of the circuit (*Figure 24–3*). The inductive current is 180° out of phase with the capacitive current. These two currents subtract from each other, resulting in the elimination of the

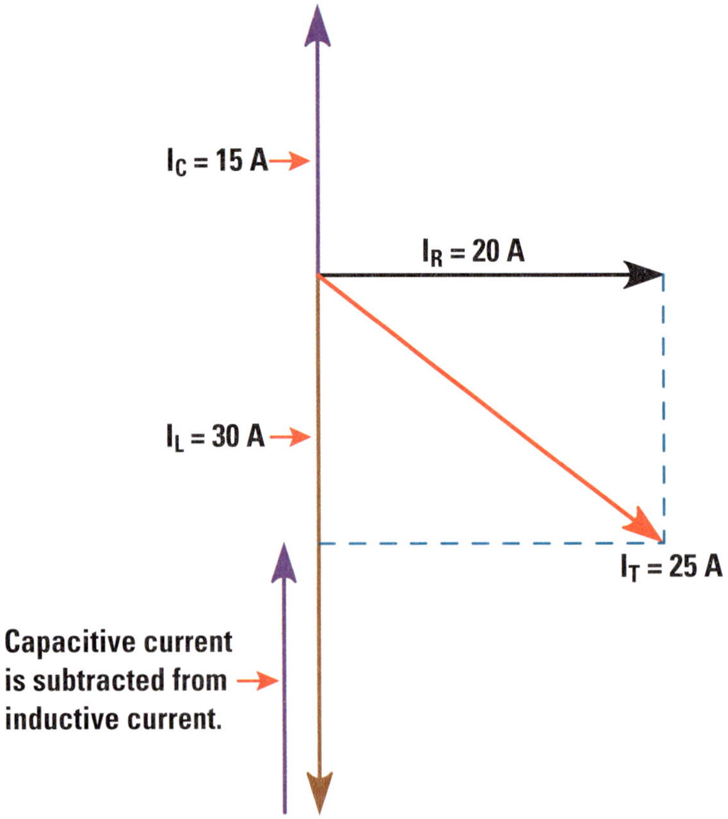

Figure 24–3 Vector diagram of resistive, inductive, and capacitive currents in example circuit.

Example 24–1 (continued)

smaller and reduction of the larger. The total circuit current is the hypotenuse of the resulting right triangle (*Figure 24–4*). The following formula can be used to find total circuit current:

$$I_T = \sqrt{I_R^2 + (I_L - I_C)^2}$$

$$I_T = \sqrt{(20\ A)^2 + (30\ A - 15\ A)^2}$$

$$I_T = \sqrt{(20\ A)^2 + (15\ A)^2}$$

$$I_T = \sqrt{400\ A + 225\ A}$$

$$I_T = \sqrt{625\ A}$$

$$I_T = 25\ A$$

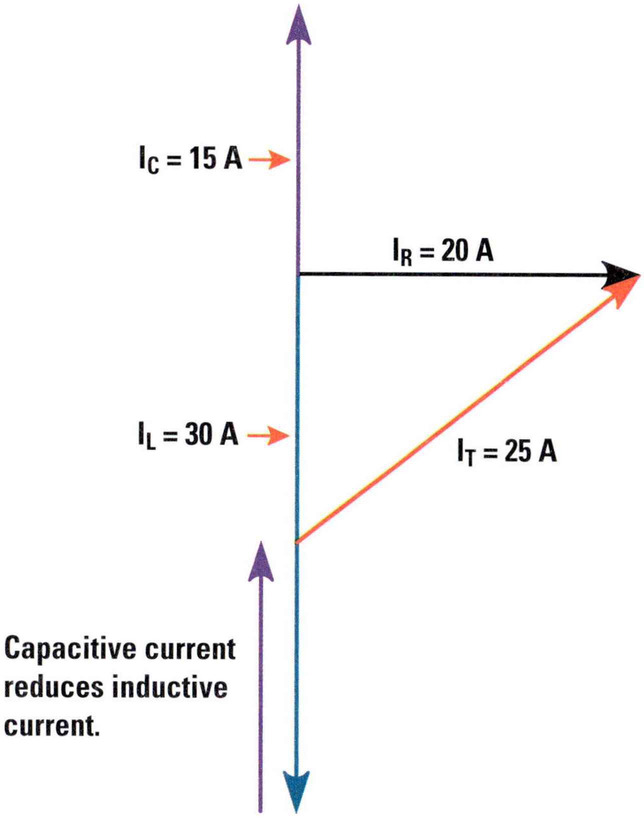

Figure 24–4 Inductive and capacitive currents cancel each other.

The total current could also be calculated by using the value of impedance found earlier in the problem:

$$I_T = \frac{E}{Z}$$

$$I_T = \frac{240\ V}{9.6\ \Omega}$$

$$I_T = 25\ A$$

Example 24–1 (continued)

Apparent Power

Now that the total circuit current has been calculated, the apparent power, or VAs, can be found using the formula:

$$VA = E \times I_T$$

$$VA = 240\ V \times 25\ A$$

$$VA = 6000$$

The apparent power can also be found by vector addition of the true power and reactive power:

$$VA = \sqrt{P^2 + (VARs_L - VARs_C)^2}$$

Power Factor

The power factor can now be calculated using the formula:

$$PF = \frac{P}{VA} \times 100$$

$$PF = \frac{4800\ W}{6000\ VA} \times 100$$

$$PF = 0.80 \times 100$$

$$PF = 80\%$$

Angle Theta

The power factor is the cosine of angle theta. Angle theta is therefore:

$$\cos \angle\theta = 0.80$$

$$\angle\theta = 36.87°$$

The circuit with all calculated values is shown in *Figure 24–5*.

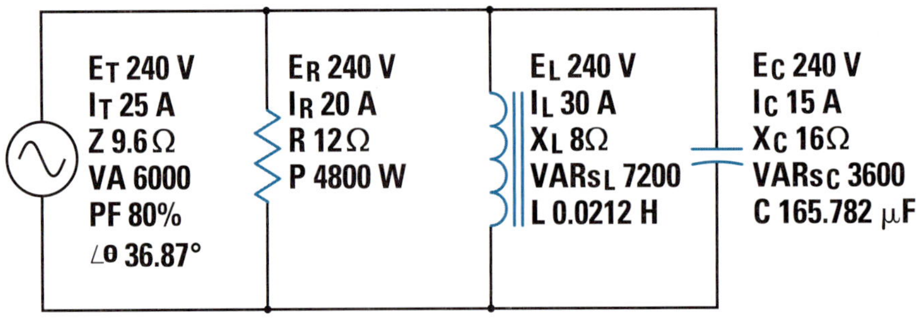

Figure 24–5 Example circuit with all calculated values.

The phase angle relationship between voltage and current from the previous example is shown in *Figure 24–6*. Because the circuit has more inductive VARs than capacitive VARs, the current lags the voltage by 36.87°.

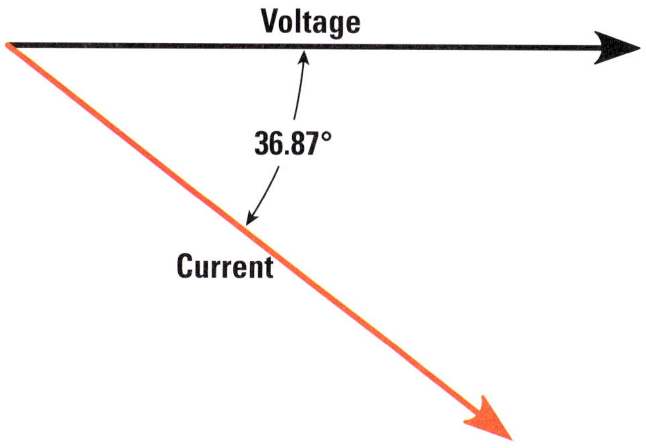

Figure 24–6 The current lags the voltage by 36.87°.

Example 24–2

In the circuit shown in *Figure 24–7*, a resistor, an inductor, and a capacitor are connected to a 1200-Hz power source. The resistor has a resistance of 18 Ω, the inductor has an inductance of 9.76 mH (0.00976 H), and the capacitor has a capacitance of 5.5 μF. What is the impedance of this circuit?

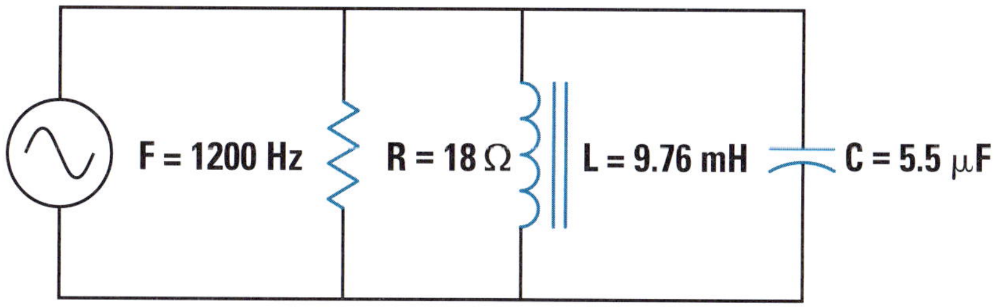

Figure 24–7 Example RLC parallel circuit.

Solution

The first step in finding the impedance of this circuit is to find the values of inductive and capacitive reactance:

$$X_L = 2\pi fL$$

$$X_L = 2 \times 3.1416 \times 1200\text{ Hz} \times 0.00976\text{ H}$$

$$X_L = 73.589\ \Omega$$

$$X_C = \frac{1}{2\pi fC}$$

$$X_C = \frac{1}{2 \times 3.1416 \times 1200\text{ Hz} \times 0.0000055\text{ F}}$$

$$X_C = 24.114\ \Omega$$

Example 24–2 (continued)

The total impedance of the circuit can now be calculated using the formula:

$$Z = \frac{1}{\sqrt{\left(\frac{1}{R}\right)^2 + \left(\frac{1}{X_L} - \frac{1}{X_C}\right)^2}}$$

$$Z = \frac{1}{\sqrt{\left(\frac{1}{18\ \Omega}\right)^2 + \left(\frac{1}{74.589\ \Omega} - \frac{1}{24.114\ \Omega}\right)^2}}$$

$$Z = \frac{1}{\sqrt{(0.0555^2 + (0.0134 - 0.0415)^2)\frac{1}{\Omega^2}}}$$

$$Z = \frac{1}{\sqrt{(0.00387)}\frac{1}{\Omega}}$$

$$Z = 16.075\ \Omega$$

24–2 Parallel Resonant Circuits

When the values of inductive reactance and capacitive reactance become equal, they are said to be resonant. In a parallel circuit, inductive current and capacitive current cancel each other at the resonant frequency because they are 180° out of phase with each other. This produces a minimum line current at the point of resonance. Although the line current can be close to 0, high current can flow in the loop formed by the parallel connection of the inductor and capacitor at the resonant frequency. An LC parallel circuit is shown in *Figure 24–8*. LC parallel circuits are often referred to as **tank circuits**. In the example circuit, the inductor has an inductance of 0.0398 henry and a wire resistance of 10 ohms. The capacitor has a capacitance of 0.442 microfarad. This circuit will reach resonance at 1200 hertz, when both the capacitor and inductor exhibit reactance values of 300 ohms each.

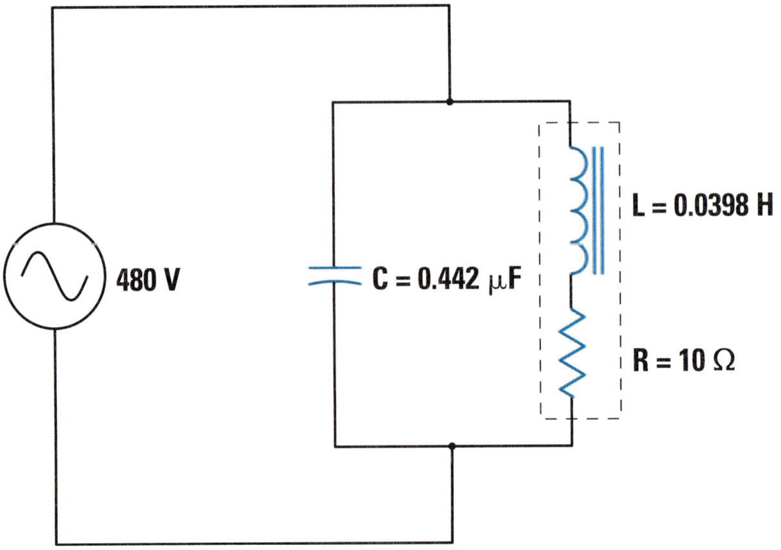

Figure 24–8 Parallel resonant circuit.

Calculating the values for a parallel resonant circuit is a bit more involved than calculating the values for a series resonant circuit. In theory, when a parallel circuit reaches resonance, the total circuit current should reach zero, and total circuit impedance should become infinite because the capacitive current and inductive current cancel each other. In practice, the quality (Q) of the circuit components determines total circuit current and, therefore, total circuit impedance. Because capacitors generally have an extremely high Q by their very nature, the Q of the inductor is the determining factor (*Figure 24–9*):

$$Q = \frac{I_{TANK}}{I_{LINE}}$$

or

$$Q = \frac{X_L}{R}$$

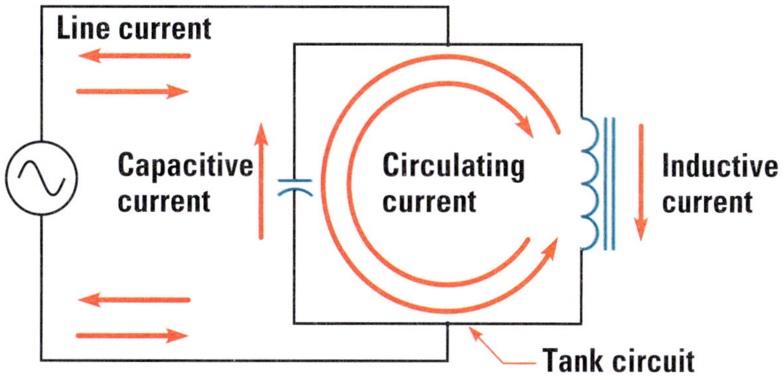

Figure 24–9 The amount of current circulating inside the tank is equal to the product of the line current and the Q of the circuit.

In the example shown in *Figure 24–8,* the inductor has an inductive reactance of 300 ohms at the resonant frequency and a wire resistance of 10 ohms. The Q of this inductor is 30 at the resonant frequency ($Q = X_L/R$). To determine the total circuit current at resonance, it is first necessary to determine the amount of current flow through each of the components at the resonant frequency. Because this is a parallel circuit, the inductor and capacitor will have the alternator voltage of 480 volts applied to them. At the resonant frequency of 1200 hertz, both the inductor and capacitor will have a current flow of 1.6 amperes (480 V/300 Ω= 1.6 A). The total current flow in the circuit will be the in-phase current caused by the wire resistance of the coil (*Figure 24–10*). This value can be calculated by dividing the circulating current inside the LC parallel loop by the Q of the circuit. The total current in this circuit will be 0.0533 ampere (1.6 A/30 = 0.0533 A). Now that the total circuit current is known, the total impedance at resonance can be found using Ohm's law:

$$Z = \frac{E}{I_T}$$

$$Z = \frac{480\ \text{V}}{0.0533\ \text{A}}$$

$$Z = 9006.63\ \Omega$$

Another method of calculating total current for a tank circuit is to determine the true power in the circuit caused by the resistance of the coil. The resistive part of a coil is considered to be in series with the reactive part (*Figure 24–8*). Because this is a series connection, the current flow through the inductor

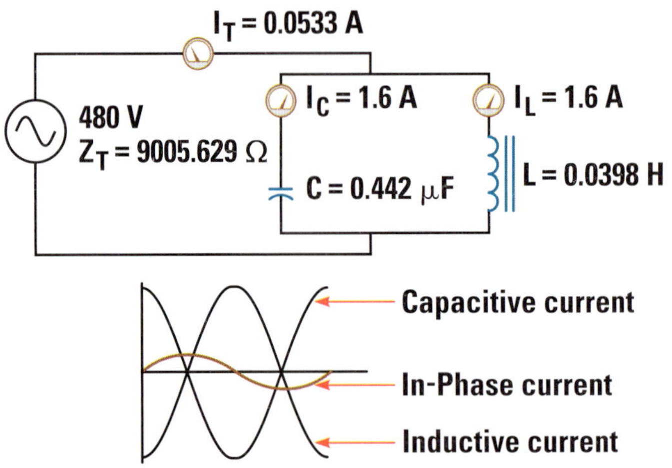

Figure 24–10 Total current is equal to the in-phase current.

is the same for both the reactive and resistive elements. The coil has a resistance of 10 ohms. The true power produced by the coil can be calculated by using the formula:

$$P = I^2R$$

$$P = (1.6\ A)^2 \times (10\ \Omega)$$

$$P = 25.6\ W$$

Now that the true power is known, the total circuit current can be found using Ohm's law:

$$I_T = \frac{P}{E}$$

$$I_T = \frac{25.6\ W}{480\ V}$$

$$I_T = 0.0533\ A$$

Graphs illustrating the decrease of current and increase of impedance in a parallel resonant circuit are shown in *Figure 24–11*.

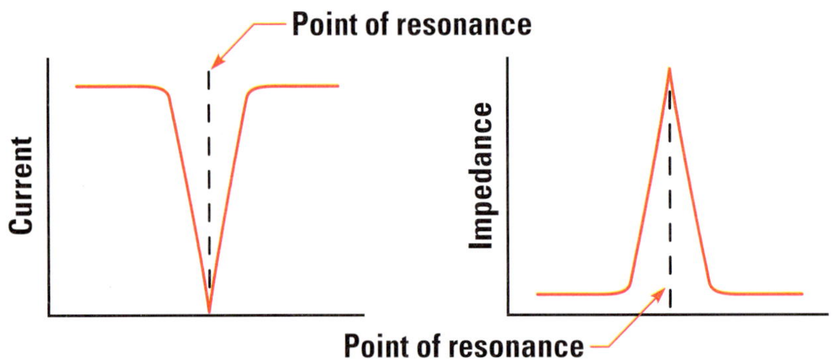

Figure 24–11 Total current and impedance curves of an LC parallel circuit at resonance.

Bandwidth

The bandwidth for a parallel resonant circuit is determined in a manner similar to that used for a series resonant circuit. The bandwidth of a parallel circuit is determined by calculating the frequency on either side of resonance at which the impedance is 0.707 of maximum. As in series resonant circuits, the Q of the parallel circuit determines the bandwidth. Circuits that have a high Q will have a narrow bandwidth, and circuits with a low Q will have a wide bandwidth (*Figure 24–12*).

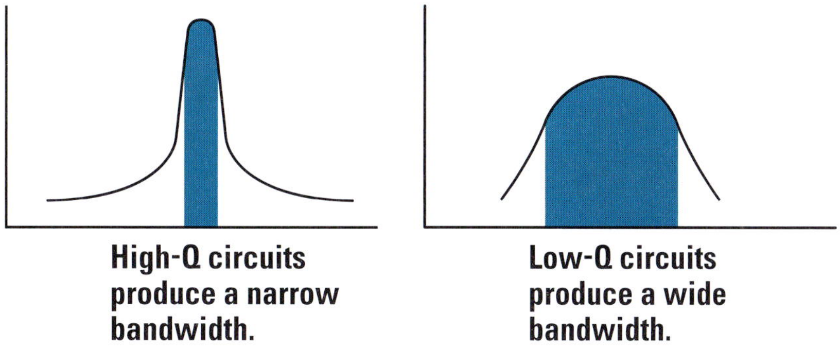

Figure 24–12 The Q of the circuit determines the bandwidth.

24–3 RLC Parallel Circuit Applications

Induction Heating

The tank circuit is often used when a large amount of current flow is needed. Recall that the formula for the Q of a parallel resonant circuit is:

$$Q = \frac{I_{TANK}}{I_{LINE}}$$

If this formula is rearranged, it can be seen that the current circulating inside the tank is equal to the line current times the Q of the circuit:

$$I_{TANK} = I_{LINE} \times Q$$

A high-Q circuit can produce an extremely high current inside the tank with very little line current. An example of this is an induction heater used to heat pipe for tempering (*Figure 24–13*). In this example, the coil around the pipe forms an inductor. A capacitor is selected that will create resonance at the circuit's frequency and is connected in parallel with the coil around the pipe. The pipe is heated by eddy current induction. Assume that the coil has a Q of 10. If this circuit has a total current of 100 amperes, then 1000 amperes of current flow in the tank. This 1000 amperes is used to heat the pipe. Because the pipe acts as a core for the inductor, the inductance of the coil changes when the pipe is not in the coil. Therefore, the circuit is resonant only during the times that the pipe is in the coil.

Induction heaters of this type have another advantage over other methods that heat pipe with flames produced by oil- or gas-fired furnaces. When induction heating is used, the resonant frequency can be changed by adding or subtracting capacitance in the tank circuit. This ability to control the frequency greatly affects the tempering of metal. If the frequency is relatively low, 400 hertz or less, the metal is heated evenly. If the frequency is increased to 1000 hertz or greater, skin effect causes most of the heating effect to localize at the surface of the metal. This localization at the surface permits a hard coating to develop at the outer surface of the metal without greatly changing the temper of the inside of the metal (*Figure 24–14*).

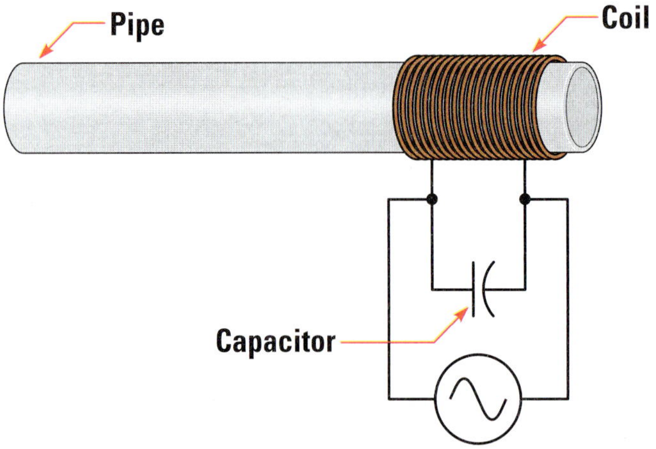

Figure 24–13 Induction heating system for tempering metal pipe.

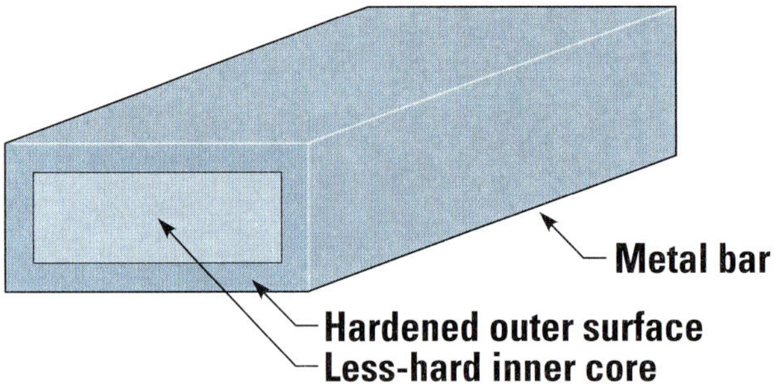

Figure 24–14 Frequency controls depth of heat penetration.

Power Factor Correction

Another very common application for LC parallel circuits is the **power factor correction**. Assume that a motor is connected to a 240-volt single-phase line with a frequency of 60 Hz (*Figure 24–15*). An ammeter indicates a current flow of 10 amperes, and a wattmeter indicates a true power of 1630 watts when the motor is at full load. In this problem, the existing power factor will be determined, and then the amount of capacitance needed to correct the power factor will be calculated.

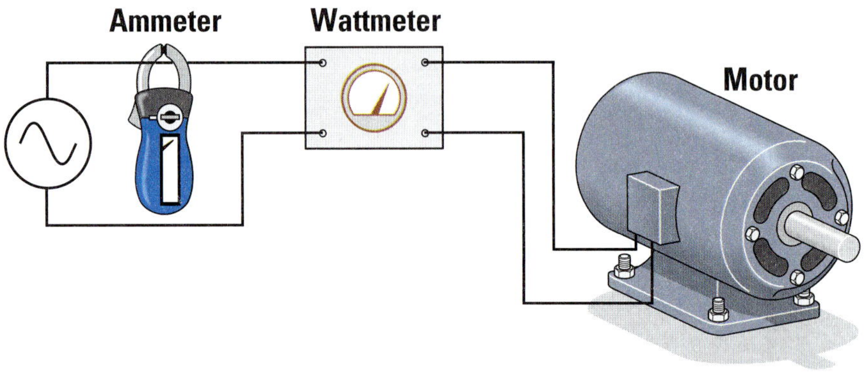

Figure 24–15 Determining motor power factor.

Although an AC motor is an inductive device, when it is loaded, it must produce true power to overcome the load connected to it. For this reason, the motor appears to be a resistance connected in series with an inductance (*Figure 24–16*). Recall that true power, or watts, is produced only when electrical energy is converted into some other form. A resistor produces true power because it converts electrical energy into thermal (heat) energy. In the case of a motor, electrical energy is being converted into both thermal energy and kinetic (mechanical) energy. When a motor is operated at a no-load condition, the current is relatively small in comparison with the full-load current. At no load, most of the current is used to magnetize the iron core of the stator and rotor. This current is inductive and is 90° out of phase with the voltage. The only true power produced at no load is caused by motor losses, such as eddy currents being induced into the iron core, the heating effect caused by the resistance of the wire in the windings, hysteresis losses, and the small amount of mechanical energy required to overcome the losses of bearing friction and windage. At no load, the motor would appear to be a circuit containing a large amount of inductance and a small amount of resistance (*Figure 24–17*).

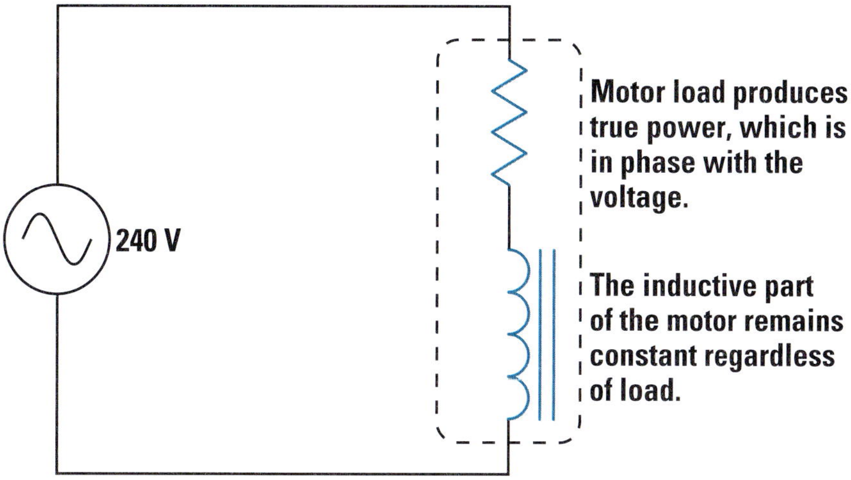

Figure 24–16 Equivalent motor circuit.

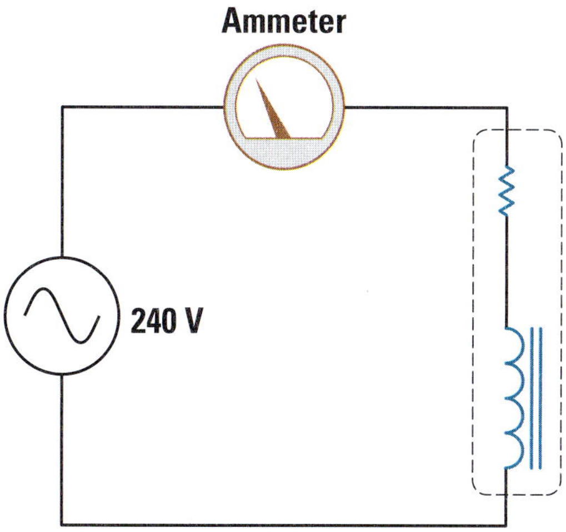

Figure 24–17 At no load, the motor appears to be an R-L series circuit with a large amount of inductance and a small amount of resistance.

As load is added to the motor, more electrical energy is converted into kinetic energy to drive the load. The increased current used to produce the mechanical energy is in phase with the voltage. This causes the circuit to appear to be more resistive. By the time the motor reaches full load, the circuit appears to be more resistive than inductive (*Figure 24–18*). Notice that as load is added or removed, only the resistive value of the motor changes, which means that once the power factor has been corrected, it will remain constant regardless of the motor load.

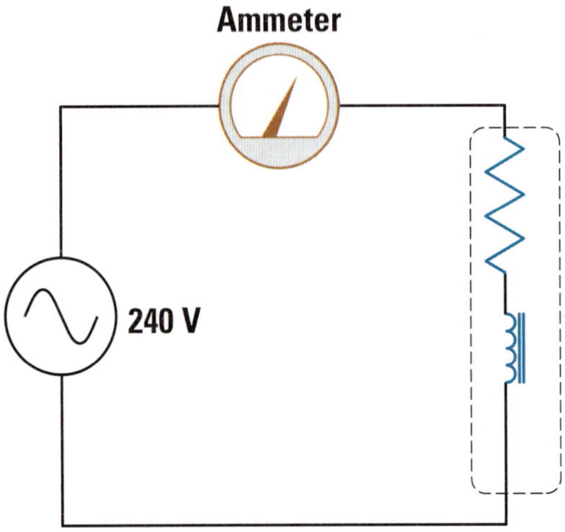

Figure 24–18 At full load, the motor appears to be an R-L series circuit with a large amount of resistance and a smaller amount of inductance.

When determining the amount of capacitance needed to correct power factor, it is helpful to use a step-by-step procedure. The first step in this procedure is to determine the apparent power of the circuit:

$$VA = E \times I$$

$$VA = 240 \times 10$$

$$VA = 2400$$

Once true power and apparent power are known, the power factor can be calculated using the formula:

$$PF = \frac{P}{VA}$$

$$PF = \frac{1630}{2400}$$

$$PF = 0679 \textit{ or } 67.9\%$$

The next step is to determine the amount of reactive power in the circuit. The reactive part of the circuit can be determined by finding the reactive power produced by the inductance. Inductive VARs can be calculated using the formula:

$$VARs_L = \sqrt{VA^2 - P^2}$$

$$VARs_L = \sqrt{2400^2 - 1630^2}$$

$$VARs_L = 1761.562$$

To correct the power factor to 100% or **unity**, an equal amount of capacitive VARs would be connected in parallel with the motor. In actual practice, however, it is generally not considered practical to correct the power factor to unity or 100%. It is common practice to correct motor power factor to a value somewhere between 95% and 100%. The next step in this example is to determine the amount of apparent power necessary to produce a power factor of 95%:

$$VA = \frac{P}{PF}$$

$$VA = \frac{1630}{0.95}$$

$$VA = 1715.789$$

Next, determine the amount of inductive VARs that would result in an apparent power of 1715.789 VA. The inductive VARs needed to produce this amount of apparent power can now be determined using the formula:

$$VARs_L = \sqrt{VA^2 - P^2}$$

$$VARs_L = \sqrt{1715.789^2 - 1630^2}$$

$$VARs_L = 535.754$$

Then, we must determine the amount of capacitive VARs needed to reduce the inductive VARs from 1761.562 to 535.754. To find the capacitive VARs needed to produce a total reactive power of 535.754 in the circuit, subtract the amount of reactive power needed from the present amount:

$$VARs_C = VARs_{L(present)} - VARs_{L(needed)}$$

$$VARs_C = 1761.562 - 535.754$$

$$VARs_C = 1225.808$$

Then, it is possible to determine the amount of capacitive reactance that would produce 1225.808 capacitive VARs. To determine the capacitive reactance needed to produce the required reactive power at 240 volts, the following formula can be used:

$$X_C = \frac{E^2}{VARs_C}$$

$$X_C = \frac{240^2}{1225.808}$$

$$X_C = 46.989\ \Omega$$

The last step in this example is to determine the amount of capacitance needed to produce a capacitive reactance of 46.989 Ω. The amount of capacitance needed to produce the required capacitive reactance at 60 Hz can be calculated using the formula:

$$C = \frac{1}{2\pi f X_C}$$

$$C = \frac{1}{377 \times 46.989}$$

$$C = 56.45\ \mu F$$

The power factor will be corrected to 95% when a capacitor with a capacitance of 56.45 μF is connected in parallel with the motor (*Figure 24–19*).

Green Tip

Power factor correction results in less power loss on conductors by lowering the amount of current flow through the conductor. Reduced current flow results in less power loss due to heating the conductor.

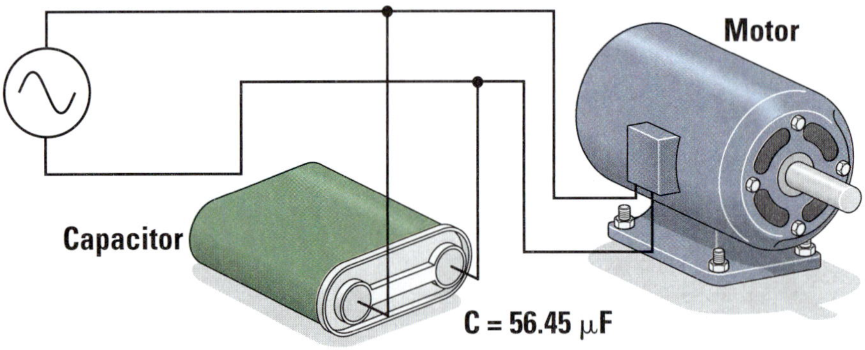

Figure 24–19 Adding a power factor correction capacitor corrects the power factor to 95%.

Summary

- The voltage applied to all branches of an RLC parallel circuit will be the same.
- The current flow in the resistive branch will be in phase with the voltage.
- The current flow in the inductive branch will lag the voltage by 90°.
- The current flow in the capacitive branch will lead the voltage by 90°.
- Angle theta for the circuit is determined by the amounts of inductance and capacitance.
- An LC resonant parallel circuit is often referred to as a tank circuit.
- When an LC parallel circuit reaches resonance, the line current drops, and the total impedance increases.
- When an LC parallel circuit becomes resonant, the total circuit current is determined by the amount of pure resistance in the circuit.
- Total circuit current and total impedance in a resonant tank circuit are proportional to the Q of the circuit.
- Motor power factor can be corrected by connecting capacitance in parallel with the motor. The amount of capacitive VARs added must be the same as the amount of inductive VARs that must be removed to reach the desired power factor.

Review Questions

1. Which of the following statements is true with respect to the relationship between voltage and current in an RLC parallel circuit?

   A) The current through each branch is in phase with voltage applied.

   B) The current through each branch is equal to the total circuit current.

   C) The voltage on each branch will be the same, and currents will be different and out of phase with each other.

   D) The voltages on each branch will be out of phase with each other.

2. In an RLC parallel circuit, which of the following statements is true with respect to power factor in a circuit where the capacitive VARs are greater than the inductive VARs?

   A) Power factor is leading.

   B) Power factor is lagging.

   C) Power factor is at unity.

   D) There is not enough information.

3. In an RLC parallel circuit that achieves resonance, which of the following statements is true?

   A) Inductive and capacitive reactance cancel each other.

   B) Inductive and capacitive current values cancel each other.

   C) Inductive and capacitive voltage drops cancel each other.

   D) Total circuit impedance equals the resistance value only.

4. True or False: In a parallel circuit that contains only an inductor and a capacitor, the total circuit current will be nearly infinite at resonance frequency because the reactive impedances cancel out.

5. An RLC parallel circuit contains a resistor with a resistance of 16 Ω, an inductor with an inductive reactance of 8 Ω, and a capacitor with a capacitive reactance of 20 Ω. What is the total impedance of this circuit?

   A) 10.24 ohms

   B) 4.21 ohms

   C) 20 ohms

   D) 44.7 ohms

6. An RLC parallel circuit has a current of 38 A flowing through the resistor, 22 A flowing through the inductor, and 7 A flowing through the capacitor. What is the total circuit current?

   A) 67 amperes

   B) 98.23 amperes

   C) 52.31 amperes

   D) 40.85 amperes

7. An RLC parallel circuit contains a resistor with a resistance of 40 Ω, an inductor with an inductive reactance of 28 Ω, and a capacitor with a capacitive reactance of 12 Ω. What is the power factor for this circuit?

   A) 17.4%

   B) 46.5%

   C) 85.2%

   D) 215%

8. In the previous question, what is the phase angle?
   A) 80°
   B) 31.5°
   C) 62.3°
   D) 180°
9. A 132.6-mH inductor produces an inductive reactance of 50Ω at 60 Hz. What value of capacitance will be needed to produce a resonant circuit at this frequency?
   A) 53.05 μF
   B) 24.72 μF
   C) 194.3 μF
   D) 5.305 μF
10. An AC motor is connected to a 560-V, 60-Hz line. The motor has a current draw at full load of 53 A. A wattmeter indicates a true power of 18,700 W. Find the power factor of the motor and the amount of capacitance that should be connected in parallel with the motor to correct the power factor to 100%, or unity.
    A) 132.46 μF
    B) 156.24 μF
    C) 175.31 μF
    D) 194.95 μF

Practical Applications

A single-phase AC motor is connected to a 240-V, 60-Hz supply. A clamp-on ammeter indicates the motor has a current draw of 15 A at full load. A wattmeter connected to the motor indicates a true power of 2.2 kW. What is the power factor of the motor, and how much capacitance is needed to correct the power factor to 95%?

Practice Problems

Refer to the Resistive-Inductive-Capacitive Parallel Circuits section of Appendix B and *Figure 24–2*. Find all the missing values in the following problems.

1. The circuit in *Figure 24–2* is connected to a 120-V, 60-Hz line. The resistor has a resistance of 36 Ω, the inductor has an inductive reactance of 40 Ω, and the capacitor has a capacitive reactance of 50 Ω.

E_T 120 V	E_R ______	E_L ______	E_C ______
I_T ______	I_R ______	I_L ______	I_C ______
Z ______	R 36 Ω	X_L 40 Ω	X_C 50 Ω
VA ______	P ______	$VARs_L$ ______	$VARs_C$ ______
PF ______	$\angle\theta$ ______	L ______	C ______

2. The circuit in *Figure 24–2* is connected to a 400-Hz line with a total current flow of 22.267 A. There is a true power of 3840 W, and the inductor has a reactive power of 1920 VARs. The capacitor has a reactive power of 5760 VARs.

E_T ______	E_R ______	E_L ______	E_C ______
I_T 22.267 A	I_R ______	I_L ______	I_C ______
Z ______	R ______	X_L ______	X_C ______
VA ______	P 3840 W	$VARs_L$ 1920	$VARs_C$ 5760
PF ______	$\angle\theta$ ______	L ______	C ______

3. The circuit in *Figure 24–2* is connected to a 60-Hz line. The apparent power in the circuit is 48.106 VA. The resistor has a resistance of 12 W. The inductor has an inductive reactance of 60 Ω, and the capacitor has a capacitive reactance of 45 W.

E_T ______	E_R ______	E_L ______	E_C ______
I_T ______	I_R ______	I_L ______	I_C ______
Z ______	R 12 Ω	X_L 60 Ω	X_C 45 Ω
VA 48.106	P ______	$VARs_L$ ______	$VARs_C$ ______
PF ______	$\angle\theta$ ______	L ______	C ______

4. The circuit in *Figure 24–2* is connected to a 1000-Hz line. The resistor has a current flow of 60 A, the inductor has a current flow of 150 A, and the capacitor has a current flow of 70 A. The circuit has a total impedance of 4.8 W.

E_T ______	E_R ______	E_L ______	E_C ______
I_T ______	I_R 60 A	I_L 150 A	I_C 70 A
Z 4.8 Ω	R ______	X_L ______	X_C ______
VA ______	P ______	$VARs_L$ ______	$VARs_C$ ______
PF ______	$\angle\theta$ ______	L ______	C ______

5. In an RLC parallel circuit, the resistor has a resistance of 24 kΩ, the inductor has a reactance of 36 kΩ, and the capacitor has a reactance of 14 kΩ. Find Z.

6. In an RLC parallel circuit, the resistor has a resistance of 60 Ω, the inductor has a reactance of 180 Ω, and the capacitor has a reactance of 80 Ω. Find the circuit power factor.

7. In an RLC parallel circuit, the true power is 260 watts. The inductor has a reactive power of 360 VARs, and the capacitor has a reactive power of 760 VARs. By how many degrees are the voltage and current out of phase with each other?

8. An RLC parallel circuit has an apparent power of 400 VA. The inductor has a reactive power of 450 VARs, and the capacitor has a reactive power of 200 VARs. Find the true power or watts.

9. In an RLC parallel circuit, the resistor has a value of 12 Ω and a current flow of 40 amperes. What is the voltage drop across the capacitor (E_C)?

10. An RLC parallel circuit is connected to 240 volts. The resistor has a power dissipation of 600 watts. If the circuit power factor is 75%, what is the total circuit current (I_T)?

Unit 25 Surge Protection and Lightning Protection

Key Terms

Capacitive reactance (X_C)

Capacitor charge rate

Electrical noise

High-frequency occurrences

Inductive reactance (X_L)

Induced voltage

Isolation transformer

Joule (J)

Lightning arrester

Metal-oxide varistor (MOV)

Noise-to-signal ratio

Thermal fuse

Voltage spike

Voltage surge

Why You Need to Know

Surge protectors are basically filter circuits used to protect electrical equipment from sudden power surges and voltage spikes that occur on power lines. These surges and spikes, called transients, can do great damage to electronic devices such as computers, televisions, electronic circuit boards contained in heating and cooling systems, programmable logic controllers, and variable frequency motor speed controllers used throughout industry.

Outline

25–1 Introduction to Voltage Surges and Spikes

25–2 Surge and Spike Protection Methods

25–3 Surge Protector Ratings

25–4 Lightning Protection Methods

Objectives

After studying this unit, you should be able to

- differentiate between voltage surges and voltage spikes.
- determine the frequency of a voltage surge.
- identify several methods to mitigate the impact of voltage surges and spikes.
- describe how surge protection devices are rated.
- differentiate between lightning arrester devices and lightning protection systems.

25–1 Introduction to Voltage Surges and Spikes

Voltage surges and spikes are one of the leading causes of electronic device failure and can be responsible for damage to transformers, motors, and other electrical equipment. A **voltage surge** is defined as a dramatic increase in voltage lasting up to 50 microseconds. A **voltage spike** is defined as a dramatic increase in voltage lasting 2 nanoseconds or less. If a voltage surge is high enough and lasts long enough, it can cause excess heat great enough to destroy semiconductor devices, overheat transformer and motor windings, and damage the dielectric of capacitors. Although voltage spikes do not last as long, the extremely high voltage can also damage or destroy semiconductor devices. Because voltage surges and spikes last for such a short period of time, they are **high-frequency occurrences**. The frequency in Hertz is the reciprocal of the time in seconds. The frequency of a voltage surge lasting for 50 microseconds is 20 kHz:

$$f = \frac{1}{T}$$

where

f = frequency in Hertz

T = time in seconds

$$f = \frac{1}{0.000050}$$

$$f = 20{,}000 \quad \text{or} \quad 20\text{ k}$$

A voltage spike lasting 2 nanoseconds has a frequency of 500 MHz:

$$f = \frac{1}{0.000{,}000{,}002}$$

$$f = 500{,}000{,}000 \quad \text{or} \quad 500\text{ MHz}$$

The fact that surges and spikes operate at a high frequency affects the methods that can be used to help eliminate and protect against them.

25–2 Surge and Spike Protection Methods

Capacitors are one of the typical components employed to help minimize the damage caused by voltage spikes and surges. Capacitors exhibit two characteristics that help reduce or eliminate voltage spikes and surges. One is that the current limiting effect of a capacitor is affected by the frequency. The current limiting effect of a capacitor is called **capacitive reactance (X_C)**. Capacitive reactance is inversely proportional to the amount of capacitance and the frequency of the applied voltage:

$$X_C = \frac{1}{2\pi fc}$$

where

X_C = capacitive reactance in ohms

π = 3.1416

f = frequency in hertz

C = capacitance in farads

A 1000-picofarad capacitor connected to a 60-Hz source would exhibit a capacitive reactance of 2.65 megohms:

$$X_C = \frac{1}{2 \times 3.1416 \times 60 \times 1000 \times 10^{-12}}$$

$$X_C = 2{,}652{,}520\ \Omega$$

If the frequency of a voltage spike is 500 MHz, the capacitive reactance becomes 0.318 ohms:

$$X_C = \frac{1}{2 \times 3.1416 \times 500 \times 10^{6} \times 1000 \times 10^{-12}}$$

$$X_C = 0.3183\ \Omega$$

If a 1000-picofarad capacitor is connected in parallel with a load that is connected to a 60-Hz power source and has a voltage of 120 volts, the current flow through the capacitor is 45.24 microamperes:

$$I = \frac{E}{X_C}$$

$$I = \frac{120}{2{,}652{,}250}$$

$$I = 0.00004542\ \text{A} \quad \text{or} \quad 45.24\ \mu\text{A}$$

A current flow of 45.24 microamperes is probably not enough to affect the operation of the circuit. If a voltage spike having a duration of 2 nanoseconds should occur, however, the capacitive reactance changes to 0.318 ohm and, therefore, almost all of the energy from the spike is conducted through the capacitive parallel branch. This branch is often connected to ground so the voltage spike energy is dissipated before it reaches the load (*Figure 25–1*).

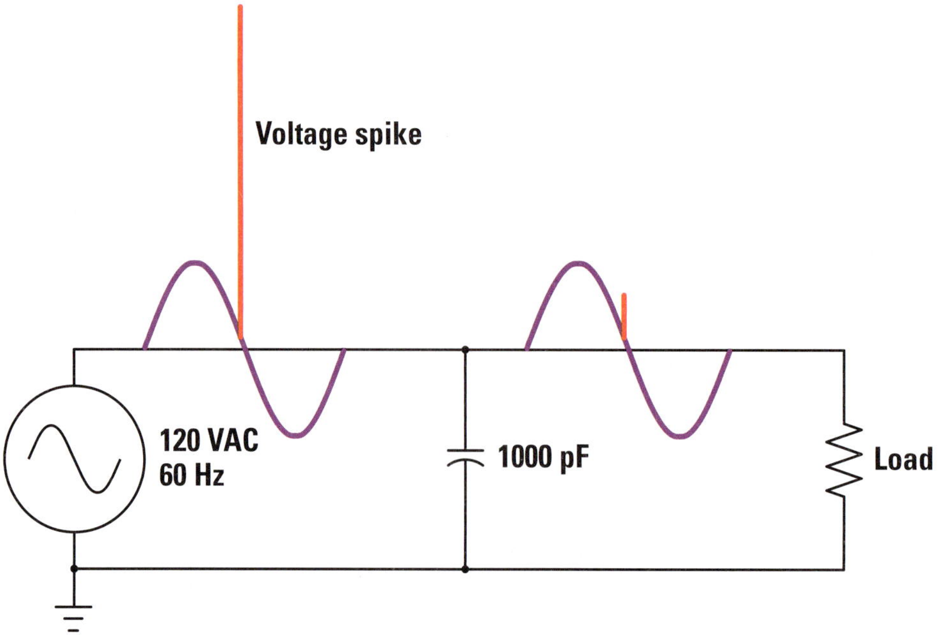

Figure 25–1 A capacitor is connected in parallel with the load to help shunt spike voltages to ground.

Capacitor Charge Rate

The second characteristic of capacitors that enables them to reduce voltage surges and spikes is the fact that capacitors oppose a change of voltage. When power is applied to a capacitor, it begins to charge to the peak value of the applied voltage. It cannot, however, charge instantly. As the capacitor charges, the voltage across the plates opposes the applied voltage. The greater the voltage becomes across the plates, the more it opposes the applied voltage. The capacitor charges at an exponential rate, or on an exponential curve **capacitor charge rate**. An exponential curve is shown in *Figure 25–2.* If it is assumed that the curve will reach a maximum of 100 volts at the end of the fifth time constant, the voltage will have only reached a value of 63.2 volts after the first time constant. Each subsequent time constant sees the capacitor charge to another 63.2% of the remaining voltage to reach 100 volts.

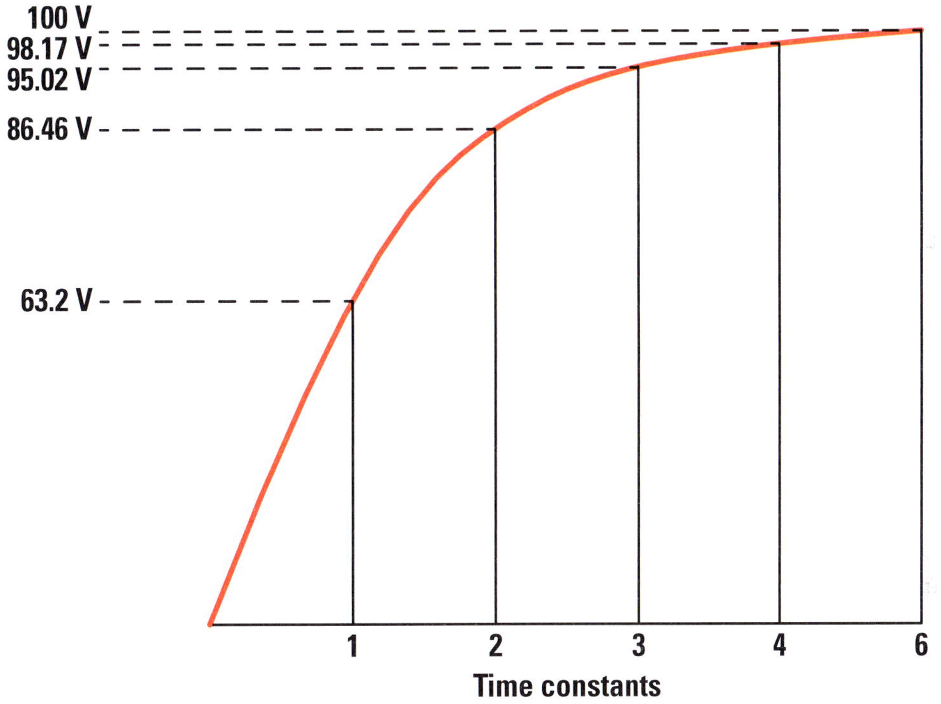

Figure 25–2 An exponential curve showing the charging rate to 100 volts.

During the first time constant, the voltage increased 63.2% of the full 100 volts. During the second time constant, the voltage increased 63.2% of the remaining voltage from 63.2 volts to 100 volts, or 63.2% of 36.8 volts ($100 - 63.2 = 36.8$)($36.8 \times 0.632 = 23.26$). During the third time constant, the voltage increases 63.2% of the voltage that remained from 86.46 volts to 100 volts ($100 - 86.46 = 13.54$) ($13.45 \times 0.632 = 8.56$). In theory, the voltage can never reach the maximum value by increasing 63.2% of the remaining value for each time constant. After five time constants, however, the voltage has reached about 99.9% of the maximum value and the capacitor is considered charged. Because voltage spikes and surges have such short duration, the time needed to charge the capacitor at an exponential rate greatly reduces the magnitude of a voltage spike.

Inductors

Like capacitors, inductors are frequently used to reduce damage caused by spikes and surges in the system. Capacitors oppose a change of voltage and are, therefore, used to minimize the effect of the high levels of voltage. Inductors, on the other hand, oppose a change of current. Current cannot rise through an inductor instantly. As the current increases, a magnetic field expands around the inductor, inducing a voltage into the coil (*Figure 25–3*). The **induced voltage** is always 180° opposite the applied voltage and therefore always opposes a change of current flow through the coil. Because the induced voltage always opposes a change of current flow, it acts like a resistance to the increase of current. The amount of voltage induced in the coil is proportional to the rate of change of current, forcing the current to rise at an exponential rate through an inductor. However, capacitors are connected in parallel with the load as shown in *Figure 25–1*. Inductors are connected in series with the load, as shown in *Figure 25–4*.

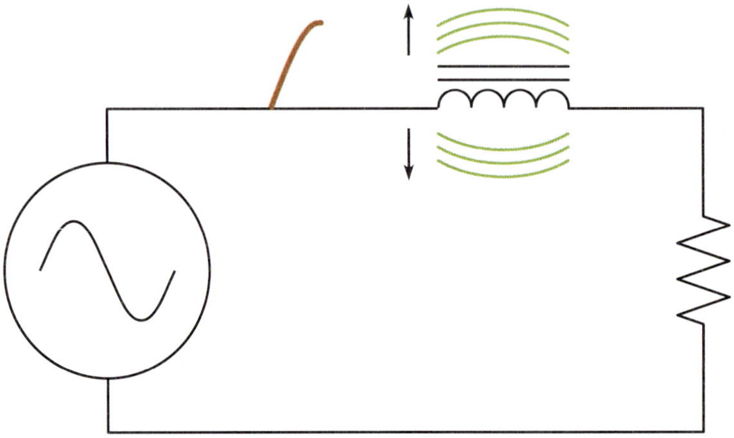

Figure 25–3 As current increases through an inductor, a magnetic field expands around it.

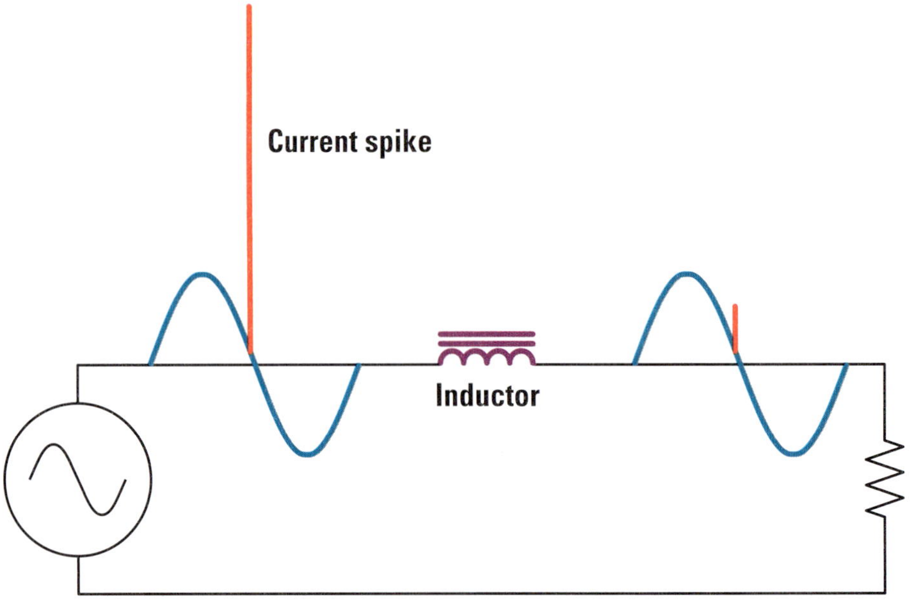

Figure 25–4 Inductors are connected in series with the load.

The current-limiting property of an inductor is called **inductive reactance (X_L)**. Inductive reactance is directly proportional to the inductance of the inductor and the frequency of the applied voltage. Frequency determines the rate of change of the current—the higher the frequency, the greater the opposition to current flow.

$$X_L = 2\pi fL$$

where

$$X_L = \text{inductive reactance in ohms}$$
$$\pi = 3.1416$$
$$f = \text{frequency in hertz}$$
$$L = \text{inductance in henrys}$$

Although capacitive reactance *decreases* with an increase in frequency, inductive reactance *increases* with an increase in frequency. A 5-mH inductor connected to a 60-Hz power source would have an inductive reactance of 1.88 ohms.

$$X_L = 2 \times 3.1416 \times 60 \times 5 \times 10^{-3}$$
$$X_L = 1.88496\ \Omega$$

If a current surge with a duration of 50 microseconds should occur, the frequency becomes 20k Hz. The inductive reactance of the 5-mH inductor becomes 628.32 ohms:

$$X_L = 2 \times 3.1416 \times 20{,}000 \times 5 \times 10^{-3}$$
$$X_L = 628.32\ \Omega$$

Isolation Transformers

Isolation transformers are another device that can greatly reduce the effects of spikes and surges. An **isolation transformer** is a transformer that has the primary and secondary winding physically and electrically separated from each other. They are magnetically coupled, not electrically coupled. The primary winding of a transformer is the winding connected to the power source. The secondary winding is the winding connected to the load. Although transformers are not true inductors, they are inductive devices. As such, current cannot rise instantly; it rises at an exponential rate. The isolation transformer also has an advantage in that it can separate the load connected to the secondary winding from ground. Ground is generally considered to be the zero reference point. Ground connections are not always zero, however, and voltage spikes and surges can travel to electrical equipment via the ground connection. Removing a piece of equipment from ground can increase its protection against voltage spikes and surges (*Figure 25–5*).

Electrical Noise

Many surge and spike suppressors include components to help reduce or eliminate the effects of electrical noise in the circuit. **Electrical noise** is a random fluctuation in an electrical signal and is commonly associated with electronic devices. It can appear as a rider signal on a main power signal (*Figure 25–6*). Electrical noise is so named because it often causes a hum or hissing sound in audio equipment, generally heard in silent spaces between songs. The term **noise-to-signal ratio** is often used to describe the amount of interference caused by the unwanted signal. A noise-to-signal ratio of 100 to 1 would indicate that the desired signal is 100 times greater than the electrical noise. A

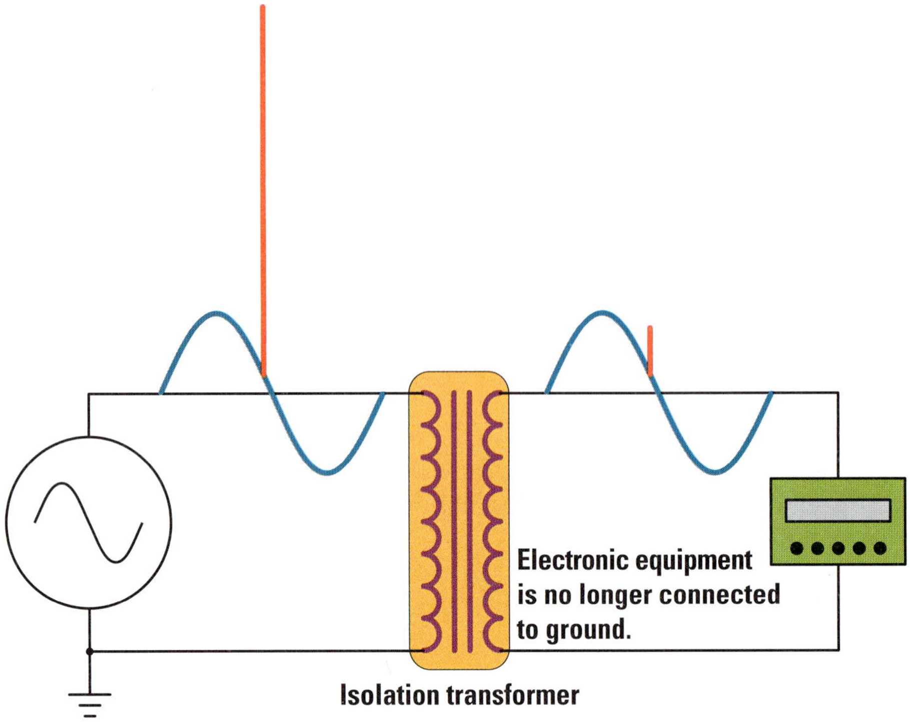

Figure 25–5 Isolation transformers reduce voltage and current spikes. They also isolate electronic equipment from the ground reference.

noise-to-signal ratio of 1 to 2 would indicate that the unwanted noise is twice as strong as the desired signal. In the worst cases, the electrical noise can become so powerful that it completely overrides the desired signal (*Figure 25–7*). The components most often used to reduce electrical noise are inductors and capacitors.

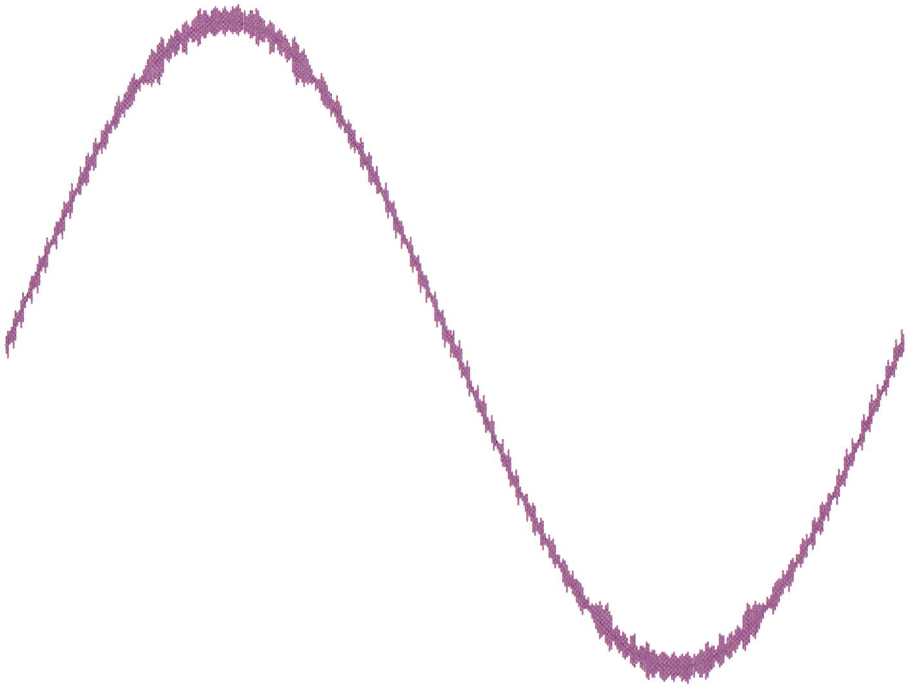

Figure 25–6 Electrical noise can appear as a rider signal on a waveform.

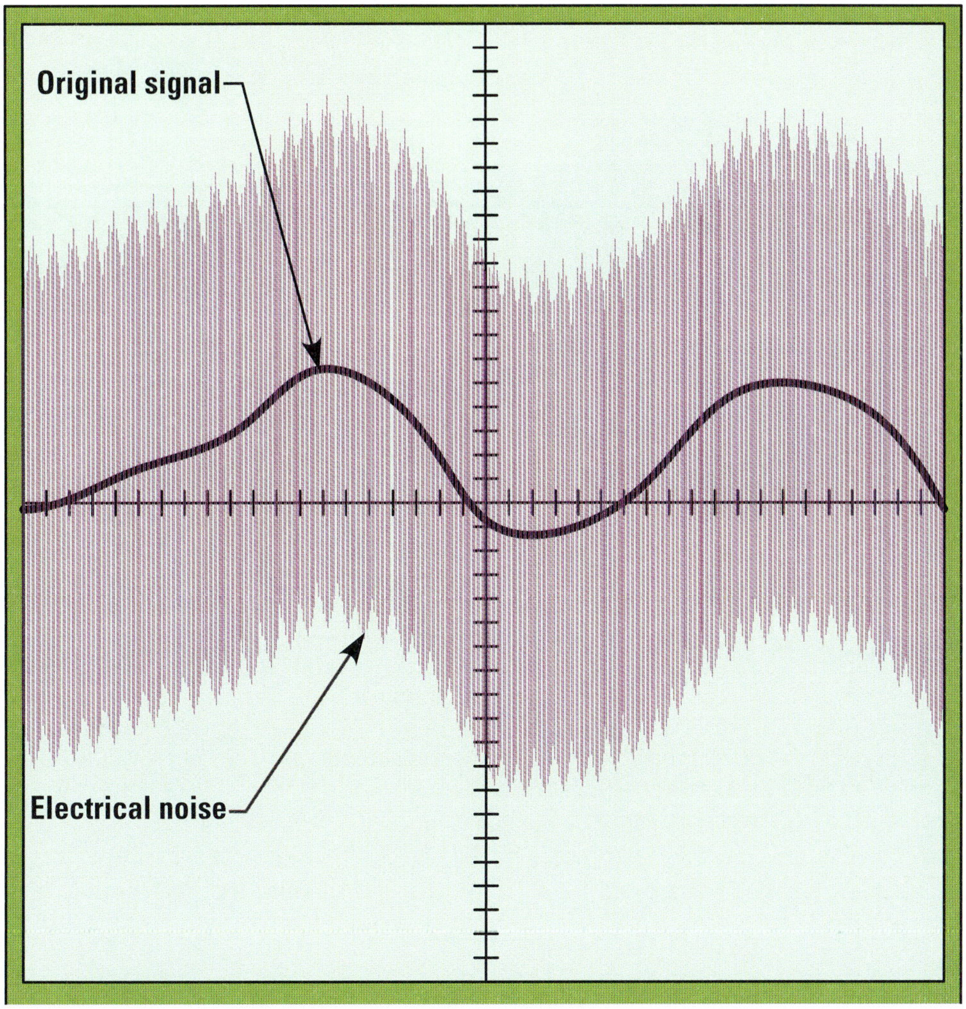

Figure 25–7 Electrical noise can become so great that it completely overrides the original signal.

Metal-Oxide Varistor (MOV)

The **metal-oxide varistor (MOV)** is one of the major electronic components employed in surge and spike suppressors (*Figure 25–8*). The schematic symbol representing an MOV is shown in *Figure 25–9*. The word *varistor* is derived from the term *variable resistor.* MOVs are voltage-activated variable resistors. They are bidirectional, which means that they will conduct regardless of polarity. If the applied voltage is below their voltage rating, they exhibit a very high resistance. When their voltage rating is exceeded, their resistance decreases dramatically to a few ohms. An MOV intended to protect a 120-VAC circuit will generally have RMS (root-mean-square) voltage ratings of 130 to 140 volts. As long as the circuit voltage is below the voltage rating of the MOV, the resistance remains very high. If a spike voltage rises above the MOV voltage rating, the resistance decreases rapidly and prevents the spike voltage from becoming greater than the voltage rating (*Figure 26–10*). The energy of the spike is dissipated as heat by the MOV. The typical rise time of a voltage spike is 8 microseconds (8 μs). The typical response times for MOVs are 40 to 60 nanoseconds (40 to 60 ns), which is much faster than the rise time of the voltage spike.

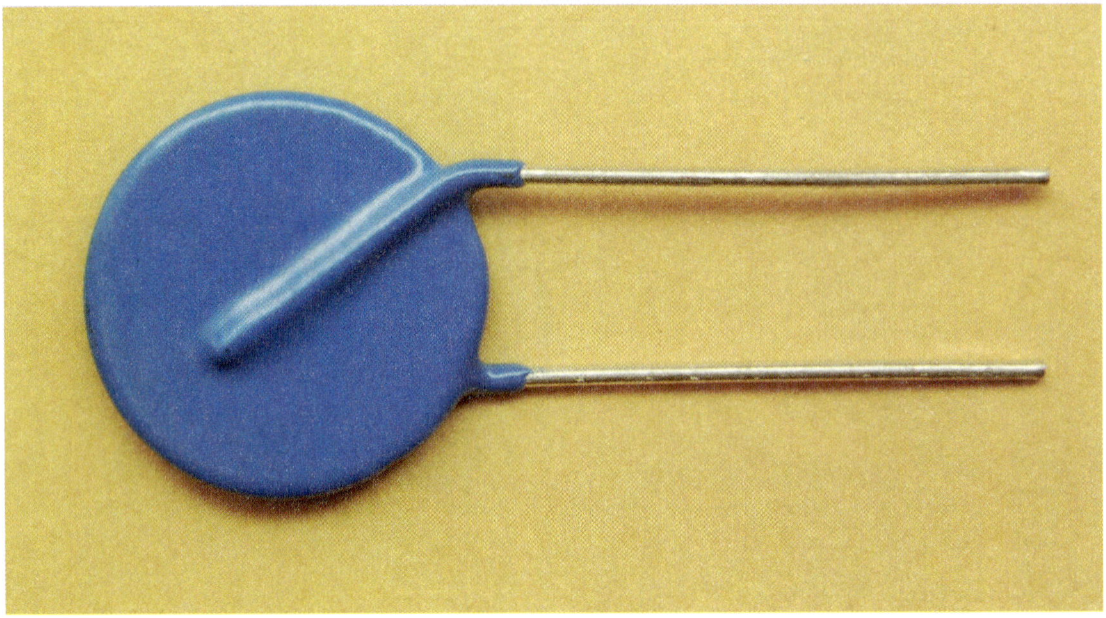

Figure 25–8 Metal-oxide varistor.

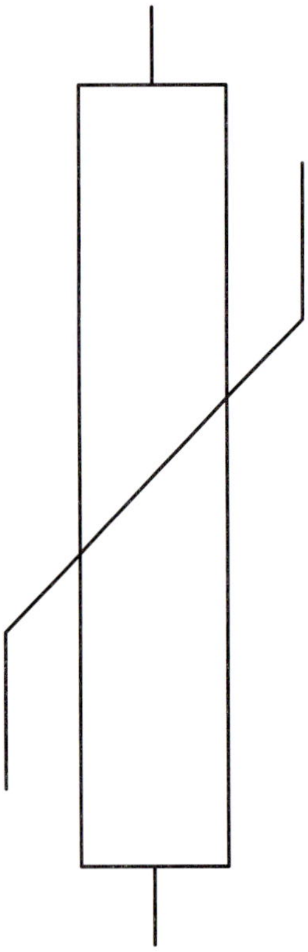

Figure 25–9 Symbol used to represent an MOV.

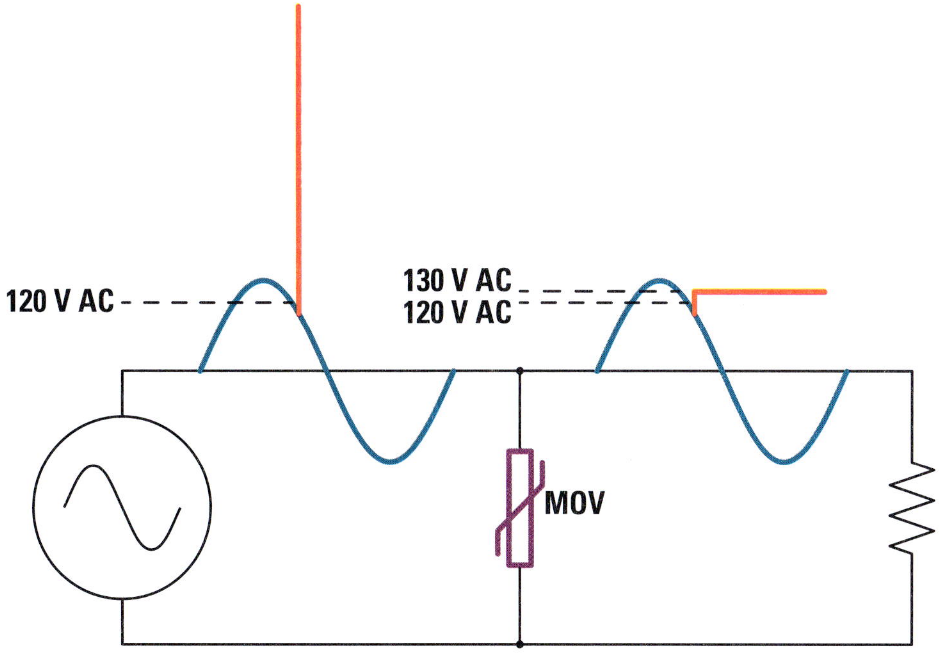

Figure 25–10 The MOV minimizes the effect of a spike or surge on the circuit.

Metal-oxide varistors are designed to conduct the current caused by excessive voltages for very short periods of time, typically from 8 to 20 microseconds. This is generally not a concern with the voltage spikes and surges created on power utility lines by the starting and stopping of motors and other equipment. Large voltage surges caused by lightning strikes, however, may damage or destroy the MOV. The voltage clamping ability of the MOV may be degraded by continual and/or excessive voltage surges and spikes. Typical electrical characteristics for an MOV are shown in *Figure 25–11.*

Absolute maximum ratings	
RMS voltage	130 V AC
DC voltage	175 V DC
Peak current	6500 A (8 x 20 μs)
Energy	70 joules
Electrical characteristics	
Varistor voltage (minimum)	184 V
Varistor voltage (normal)	200 V
Varistor voltage (maximum)	246 V
Typical capacitance	1900 pf
Rated power	1 W

Figure 25–11 Electrical ratings of a typical MOV.

Increasing the amount of energy that can be conducted to ground can be accomplished by either using an MOV with a higher rating or by connecting several MOVs in parallel (*Figure 25–12*).

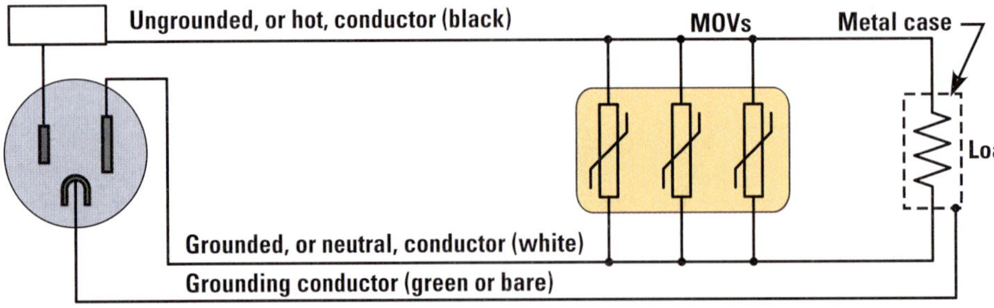

Figure 25–12 MOVs are often connected in parallel to increase the power rating.

Thermal Fuses

Another component often used in voltage surge and spike suppression circuits is the thermal fuse (*Figure 25–13*). **Thermal fuses** are used in many types of electrical circuits, from appliances to electronic equipment. Heat-producing appliances such as coffee makers and hair dryers are generally protected with a thermal fuse. Thermal fuses are sensitive to heat instead of current. They open the circuit in the event of an unsafe thermal condition instead of an unsafe amount of current flow. Once a thermal fuse has opened, like a standard fuse, it must be replaced.

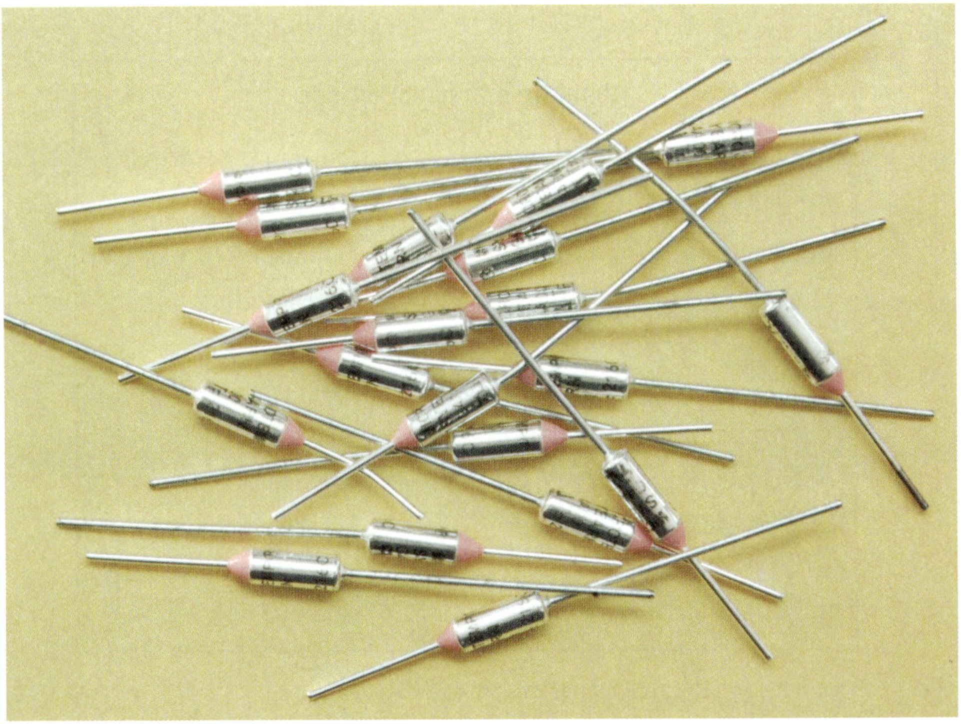

Figure 25–13 Thermal fuses.

25–3 Surge Protector Ratings

Surge and spike protectors are generally rated in joules. The joule is a metric unit of measure. One **joule (J)** is the amount of energy expended, or work done, in applying a force of 1 Newton through a distance of 1 meter or passing a current of 1 ampere through a resistance of 1 ohm for 1 second. The joule can also be defined as the amount of energy necessary to produce 1 watt of power for 1 second. Therefore, 1 joule is equal to 1 watt-second. Basically, the higher the rating in joules, the more protection offered by the surge protection device (SPD). Some surge protection devices may be as simple as connecting 1 or 2 MOVs in parallel with an outlet (*Figure 25–14*). An SPD circuit board that uses multiple MOVs connected in parallel is shown in *Figure 25–15.* SPDs with a higher joule rating contain multiple elements similar to the circuit shown in *Figure 25–16.* A circuit board of this type is shown in *Figure 25–17.* Although these SPDs are more costly because of the components involved, they offer a much greater degree of protection against voltage spikes and surges. An SPD of this type is shown in *Figure 25–18.*

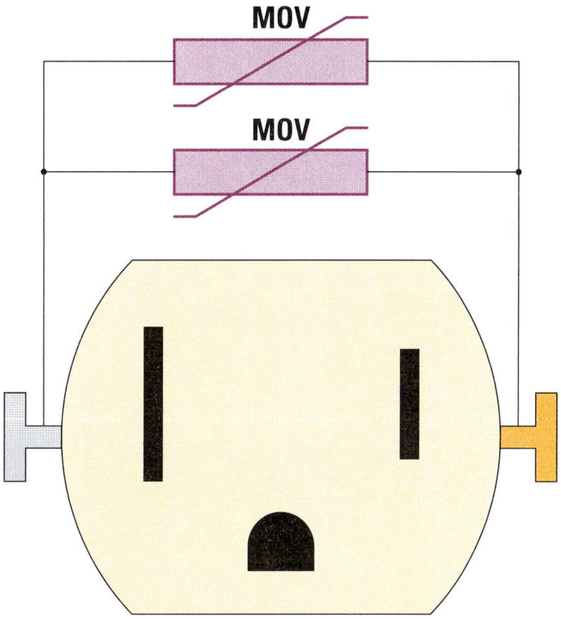

Figure 25–14 A simple SPD may connect a couple of MOVs across an outlet.

Figure 25–15 SPD circuit board containing multiple MOVs connected in parallel.

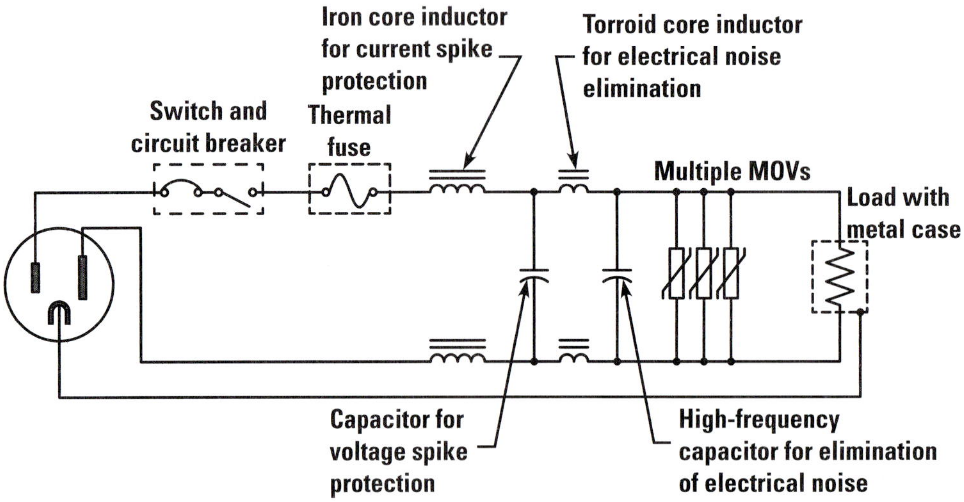

Figure 25–16 Schematic of an SPD circuit containing multiple components used to eliminate spikes, surges, and electrical noise.

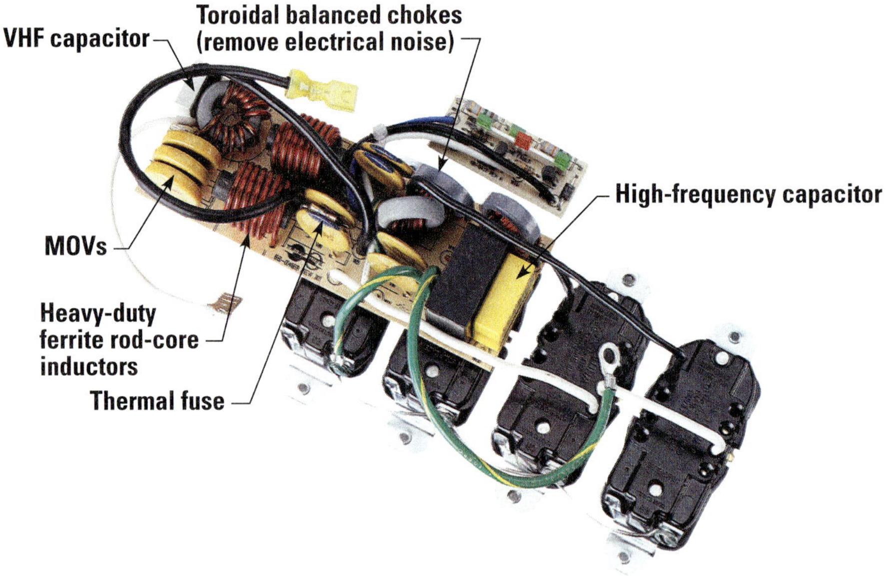

Figure 25–17 Circuit board of an SPD offering a high joule rating. (Courtesy of Tripp Lite)

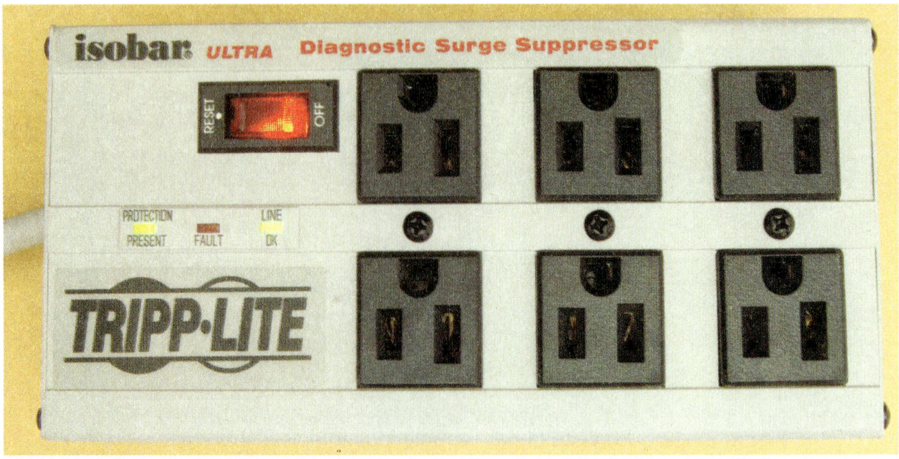

Figure 25–18 Type III SPD offering a high degree of protection.

25–4 Lightning Protection Methods

Another type of surge protector is the lightning arrester. **Lightning arresters** are designed to protect against much higher voltages and currents than standard surge protectors. Lightning strikes are typically hundreds of thousands of volts and can produce momentary currents in the range of 50,000 to 60,000 amperes. Some lightning arresters are simple spark gap devices that operate when an extremely high voltage spike occurs (*Figure 25–19*). At normal power-line voltage, the resistance of the spark gap is too high to conduct current. In the event of a lightning strike, the voltage becomes great enough to cause an arc across the spark gap and conduct the energy of the lightning strike to ground via a grounding electrode, such as a grounding rod.

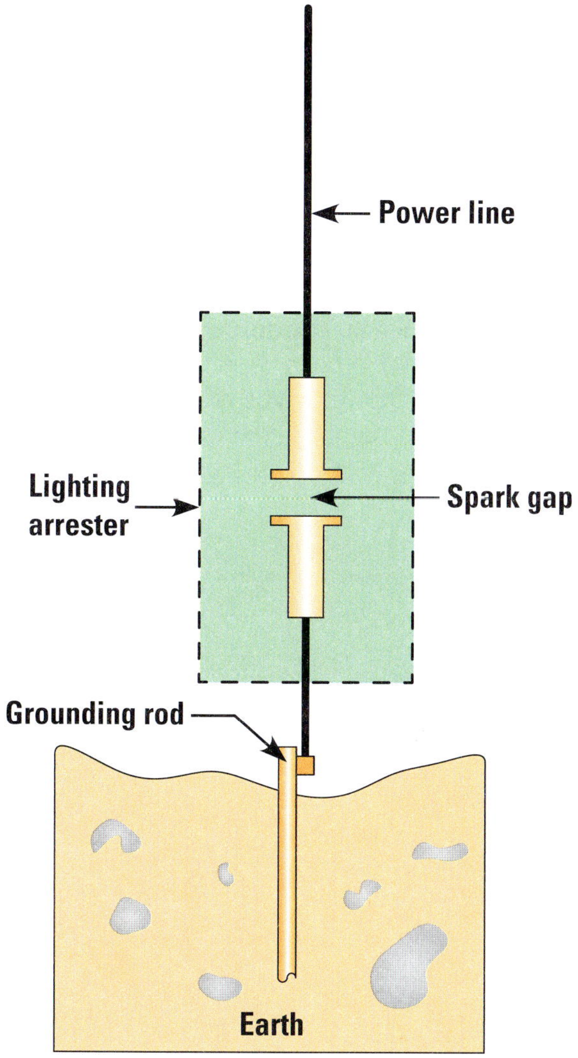

Figure 25–19 Simple lightning arresters provide a spark gap for extremely high voltage.

Electronic lightning arresters are generally constructed by stacking multiple MOVs (*Figure 25–20*). These MOVs have a higher voltage and current rating than the MOVs used in common SPDs. Lightning arresters generally have a turn-on voltage of about 500 volts and can momentarily conduct currents of 50,000 to 60,000 amperes to ground.

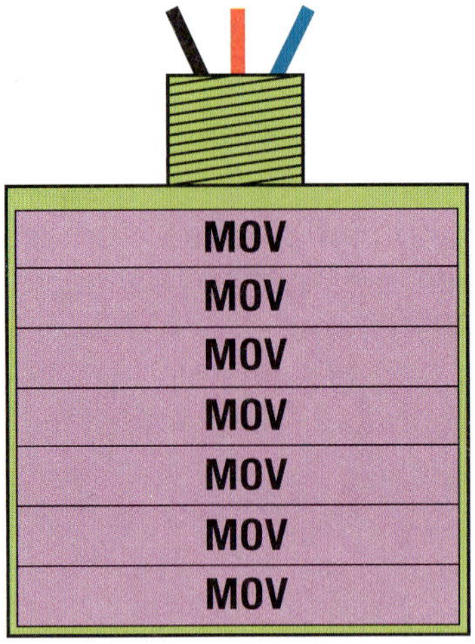

Figure 25–20 Electronic lightning arresters are generally constructed by stacking multiple MOVs.

Surge protectors and lightning arresters that protect entire home electrical circuits are available and required by *NFPA 70®: National Electrical Code®*. These protectors are connected to the main panel box (*Figure 25–21*). Surge protectors and lightning arresters of this type connect to the main panel bus through a double-pole circuit breaker (*Figure 25–22*). The surge protector and lightning arresters contain two black conductors that connect to the circuit breaker and one white or green conductor that connects to the neutral bus. It is important to note that this type of protection method is only aimed at protecting the circuitry inside the building from the harmful effects of a lightning strike and do not provide protection to the structure of the building during a lightning storm.

Figure 25–21 Lightning and surge protectors intended to protect an entire home.

Figure 25–22 Lighting and surge protectors designed to protect the entire home are connected to the service equipment.

To provide lightning protection for the structure of the building, a method must be used to encapsulate the building in a zone of protection provided by a lightning protection system. The exact details of how this system is designed and installed are beyond the scope of this text, but more information can be found in *NFPA 780: Standard for the Installation of Lightning Protection Systems*. The basic premise is that if we install a separate path for lightning to bleed off into the Earth, the impact of the lightning strike on the building itself will be minimal. These systems consist of air terminals on top of the building, down conductors to carry the energy to the ground, and a series of electrodes in the Earth to dissipate the energy from the strike. This system has proven effective over the years at mitigating the impact of a direct lightning strike to a building. However, we still must utilize the surge protection devices to minimize the impact of a lightning strike on the utility lines that can be transmitted to the building through the electrical system.

Summary

- Voltage surges generally last longer than voltage spikes.
- The frequency of a voltage surge or spike is the reciprocal of the time in seconds that the surge or spike lasts.
- One of the components used to help eliminate voltage surges and spikes is the capacitor.
- The capacitive reactance of a capacitor is determined by the frequency and capacitance of the capacitor.
- Capacitive reactance decreases with an increase in frequency.
- Capacitors oppose a change of voltage.
- Inductors oppose a change of current.

- MOVs are voltage-sensitive resistors.
- MOVs exhibit a very high resistance until their voltage rating is reached, and then they change to a very low resistance.
- MOV response times are generally 40 to 60 nanoseconds.
- Isolation transformers are used to help reduce voltage spikes and surges.
- Lightning arresters (often simple spark gap devices) are used to help conduct lightning strikes to ground.

Review Questions

1. Which of the following statements best describes the difference between a voltage surge and a voltage spike?

   A) A voltage surge is shorter duration than a spike.

   B) A voltage spike is lower magnitude than a surge.

   C) A voltage spike is shorter duration than a surge.

   D) A voltage surge is always caused by lightning.

2. True or False: A voltage surge can cause excess heat resulting in damage to semi-conductors and other electrical devices.

3. If a voltage surge lasts for 20 microseconds, what is the frequency of the surge?

   A) 50 Hz

   B) 50 kHz

   C) 50 mHz

   D) 50 MHz

4. If a voltage spike lasts for 1.5 nanoseconds, what is the frequency of the spike?

   A) 667 MHz

   B) 667 kHz

   C) 667 GHz

   D) 667 Hz

5. Which of the following methods is used to mitigate the impact of a voltage surge or voltage spike?

   A) Inserting capacitance in the system

   B) Inserting induction in the system

   C) Supplying the system through an isolation transformer

   D) Any of these

6. Which of the following is one reason why capacitors are able to help minimize the damage caused by voltage spikes and surges?
   A) Capacitors bleed off excess voltage to ground.
   B) Capacitive reactance increases exponentially during surges and spikes to limit current flow.
   C) Capacitors provide high resistance to limit the current due to the dielectric material.
   D) Capacitors provide a voltage gap.
7. What unit of measure is used to indicate the amount of protection against surges and spikes?
   A) Volts
   B) Amperes
   C) Joules
   D) Watts
8. True or False: A surge protection device meant to protect the whole house will typically have a lower joule rating than a relocatable power tap SPD.
9. Lightning protection systems protect the building from the potential damage of a lightning strike by creating which of the following?
   A) Voltage gaps in the down conductors
   B) Current limiters in the panelboard
   C) Fused protection of the conductors
   D) A zone of protection covering the building
10. True or False: A typical lightning arrester will protect only the circuits in the home from voltage spikes due to lightning but will not provide protection to the building if the lightning strikes the building.

Section 9

Three-Phase Power

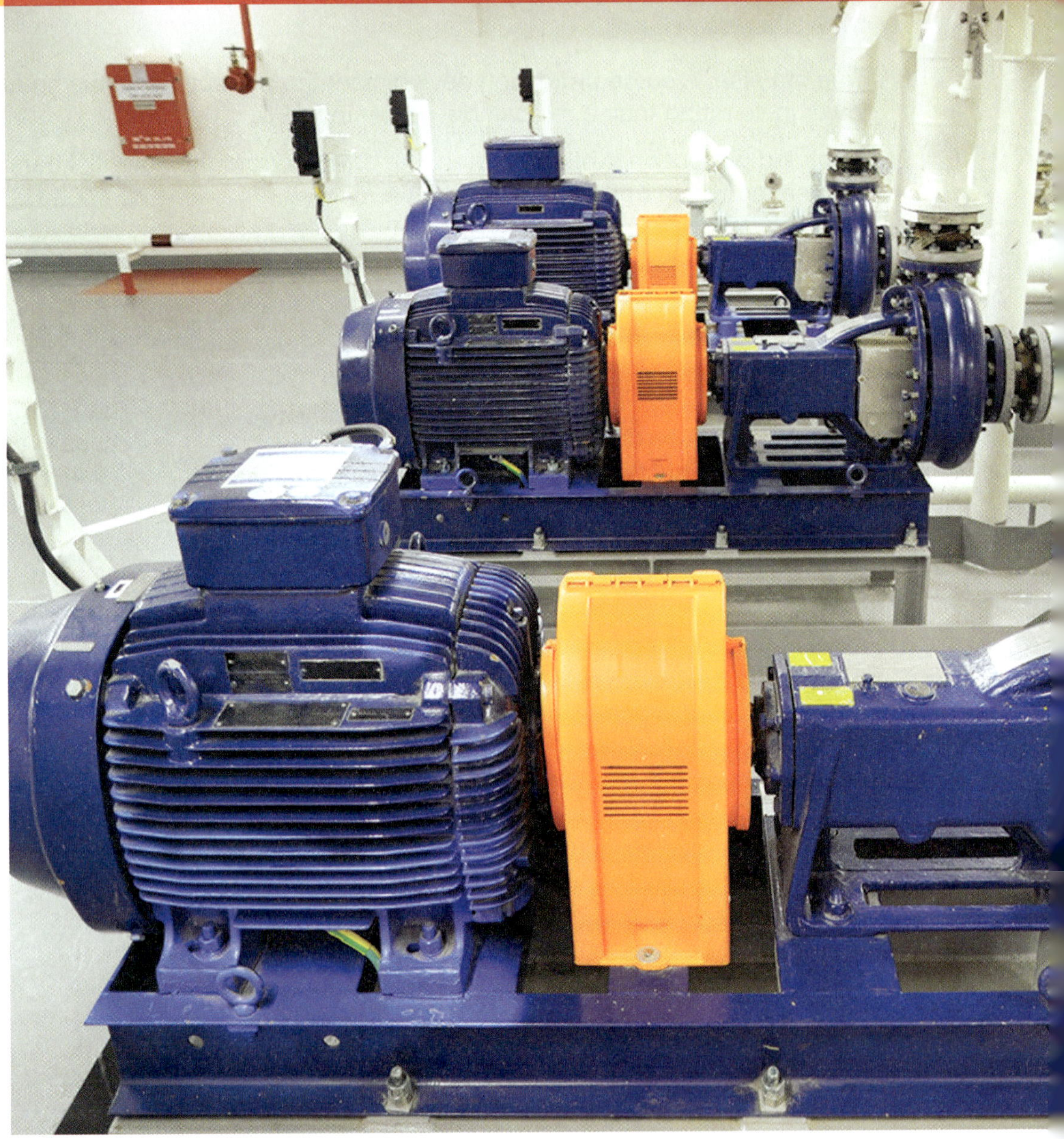

Unit 26 Three-Phase Circuits

Key Terms

Delta connection

Line current

Line voltage

Phase current

Phase voltage

Three-phase VARs

Three-phase watts

WYE connection

Why You Need to Know

Most of the power produced in the world is three-phase. The importance of understanding the principles concerning three-phase power cannot be overstated. Three-phase power is used to run almost all industry throughout the United States and Canada. Even single-phase residential electrical systems are often derived from a three-phase power system. This unit will:

- discuss the basics of three-phase power generation as the most common electric power source in the world.
- explain how three-phase power is produced and the basic connections used by devices intended to operate on three-phase power.
- illustrate how to calculate voltage, current, and power for different types of three-phase loads.
- explain why electricians must be able to understand and calculate the different voltages in a three-phase circuit.
- determine the power factor and how to correct it in a three-phase system.

Outline

Objectives

After studying this unit, you should be able to

- differentiate between three-phase and single-phase systems.
- identify defining characteristics of delta and WYE connected three-phase systems.
- explain the differences between single-phase and three-phase power.
- calculate unknown values in three-phase circuits.
- calculate three-phase power factor and correction methods.

Preview

Three-phase power systems are the most common type of power system found in the world today. The benefits of three-phase power were first demonstrated by Nikola Tesla. In the early days of electric power generation, Tesla not only led the battle concerning whether the nation should be powered with low-voltage DC or high-voltage AC, but he also proved that an alternating-current, three-phase power system was the most efficient way that electricity could be produced, transmitted, and utilized.

26–1 Three-Phase Systems

There are several reasons why three-phase power is superior to single-phase power:

1. The horsepower rating of three-phase motors and the kilovolt-ampere rating of three-phase transformers are about 150% greater than for single-phase motors or transformers with a similar frame size.
2. The power delivered by a single-phase system pulsates (*Figure 26–1*). The power falls to zero three times during each cycle. The power delivered by a three-phase circuit also pulsates, but it never falls to zero (*Figure 26–2*). In a three-phase system, the power delivered to the load is the same at any instant. This produces superior operating characteristics for three-phase motors.

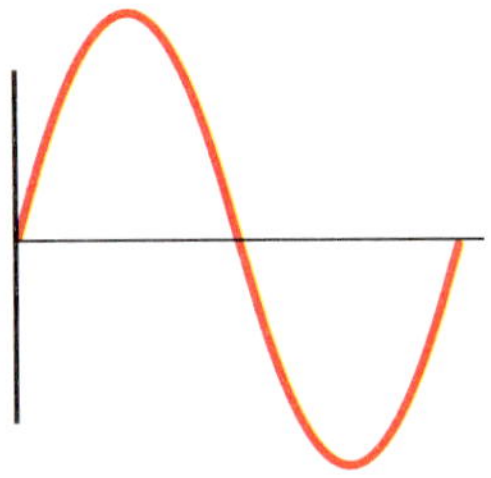

Figure 26–1 Single-phase power falls to zero three times each cycle.

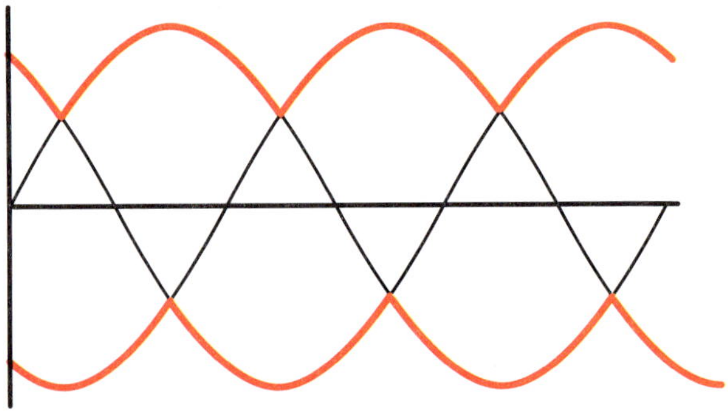

Figure 26–2 Three-phase power never falls to zero.

3. In a balanced three-phase system, the conductors only need to be about 75% of the size of conductors for a single-phase, two-wire system of the same kilovolt-ampere (kVA) rating. This savings helps offset the cost of supplying the third conductor required by three-phase systems.

A single-phase alternating voltage can be produced by rotating a magnetic field through the conductors of a stationary coil, as shown in *Figure 26–3*.

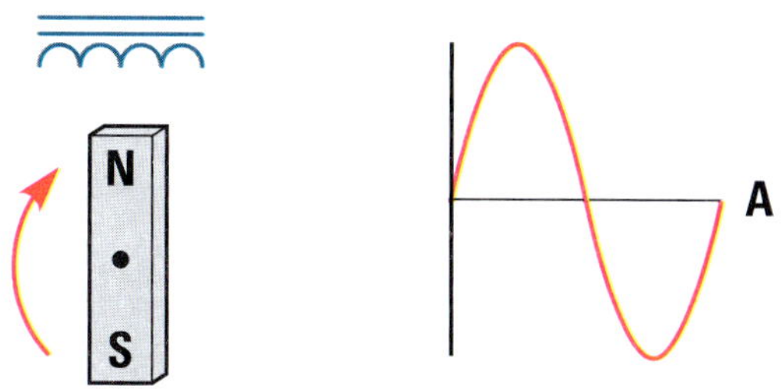

Figure 26–3 Producing a single-phase voltage.

Because alternate polarities of the magnetic field cut through the conductors of the stationary coil, the induced voltage changes polarity at the same speed as the rotation of the magnetic field. The alternator shown in *Figure 26–3* is single phase because it uses only a single coil.

If three separate coils are spaced 120° apart, as shown in *Figure 26–4*, three voltages 120° out of phase with each other are produced when the magnetic field cuts through the coils. This is how a three-phase voltage is produced. There are two basic three-phase connections utilized today: the WYE system and the delta system.

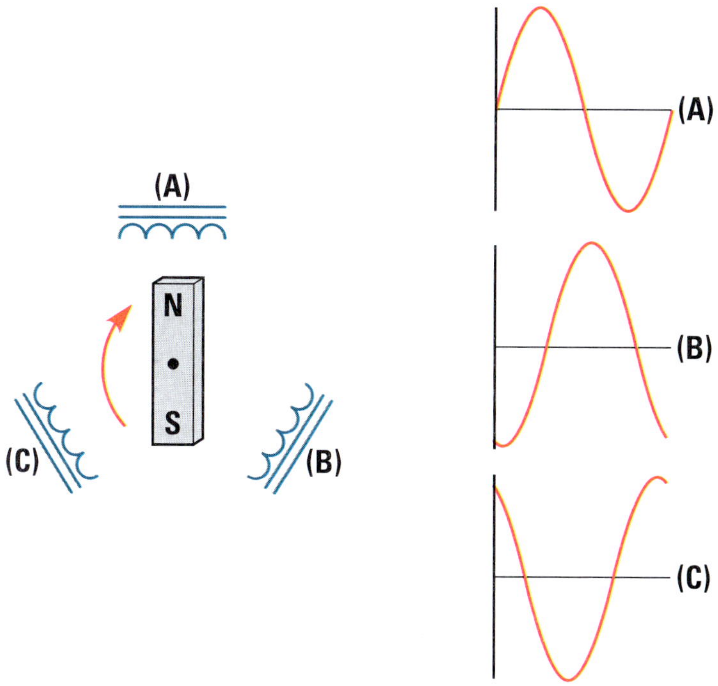

Figure 26–4 The voltages of a three-phase system are 120° out of phase with each other.

WYE Connected Systems

It is important to understand that in a three-phase power system, electricity is generated into three individual coils. These individual coils are sometimes referred to as a winding but are more accurately called a phase. The **WYE connection** is created by connecting one end of each of the phases together (*Figure 26–5*). The voltage measured across a single phase is known as the **phase voltage** (*Figure 26–6*). The voltage measured between the free ends of the coils, or lines, is known as the line-to-line voltage, or simply as the **line voltage**.

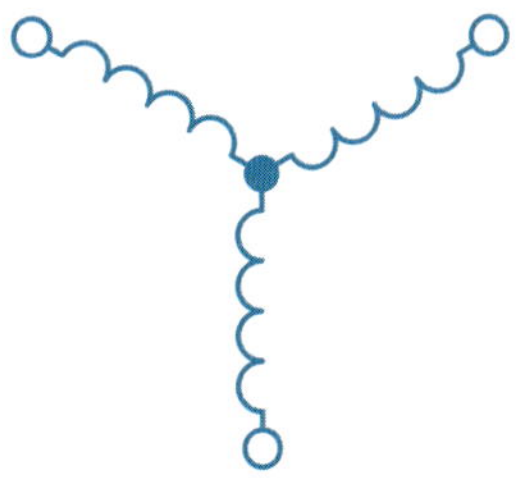

Figure 26–5 A WYE connection is formed by connecting one end of each of the windings together.

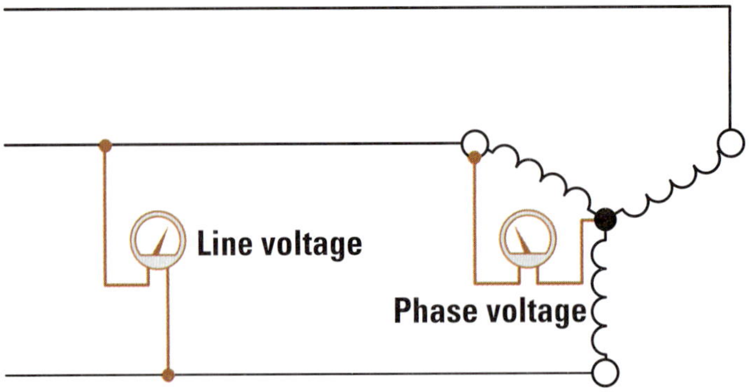

Figure 26–6 Line and phase voltages are different in a WYE connection.

In *Figure 26–7*, test instruments have been placed in various locations and connection points in a WYE-connected system. Voltmeters have been connected between the lines and across the phase. If the generator is set up so a voltage of 120 volts is induced into each of the individual phases, an interesting observation is made because of the WYE connection. The line-to-line voltage is measured at or around 208 volts, and the load can be supplied with this line-to-line voltage. However, the voltmeter connected across a single phase will indicate a value of 120 volts.

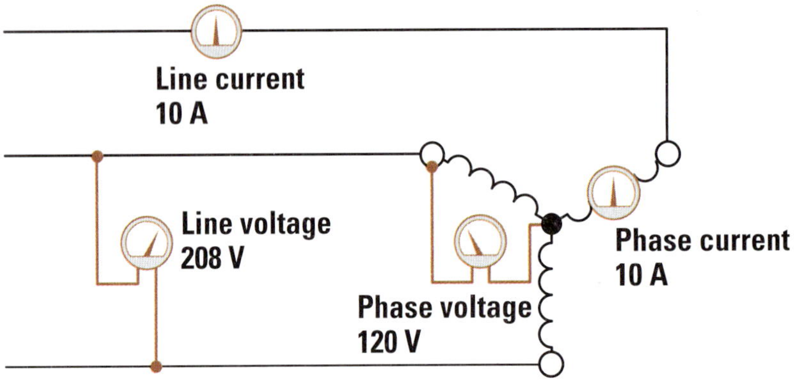

Figure 26–7 Illustration of the relationship between phase and line-to-line values for voltage and current in a WYE-connected system.

This relationship between the phase voltage and the line-to-line voltage is a foundational concept in understanding these systems and is best summed up in the following statement: ***WYE-connected systems produce a line voltage that is higher than the phase voltage by a factor of the square root of 3 (1.732)***. The formula used to calculate the line voltage in a WYE-connected system is:

$$E_{Line} = E_{Phase} \times 1.732$$

And it can be rearranged to find the phase voltage:

$$E_{Phase} = \frac{E_{Line}}{1.732}$$

Notice in *Figure 26–7* that 10 amperes of current flow in both the phase and the line. The current relationship in a WYE-connected system can best be summarized by the following statement: ***in a WYE-connected system, phase current and line current are the same***.

Voltage Relationships in a WYE-Connected System

It is often asked why the line-to-line voltage in a 120Y/208-volt system does not equal 240 volts. After all, to measure line-to-line voltage, you are measuring across two individual coils that are producing 120 volts and are connected in series, so why the difference? The reason that 120 volts plus 120 volts does not equal 240 volts here is because we need to remember that the individual phases are "out-of-phase" with each other. Specifically, we need to remember that each phase is 120° electrically apart from the next phase and can only be added using vector addition. If the three voltages are drawn 120° apart, it will be seen that the vector sum of these voltages is 208 volts (*Figure 26–8*). Another illustration of vector addition is shown in *Figure 26–9*. In this illustration, two-phase voltage vectors are added using the tip-to-tail method, and the resultant is drawn from the starting point of one vector to the end point of the other. The parallelogram method of vector addition for the voltages in a WYE-connected three-phase system is shown in *Figure 26–10.*

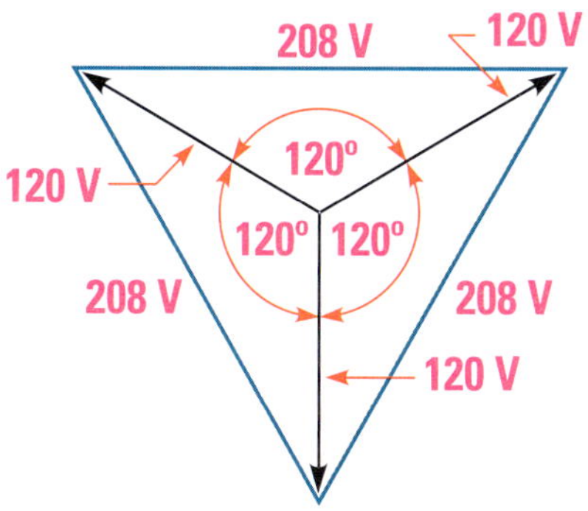

Figure 26–8 Vector sum of the voltages in a three-phase WYE connection.

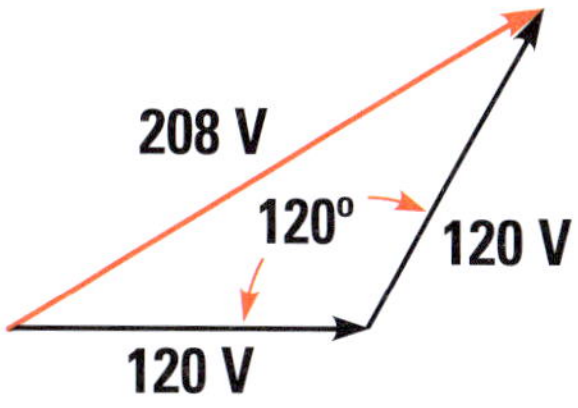

Figure 26–9 Adding voltage vectors of two-phase voltage values.

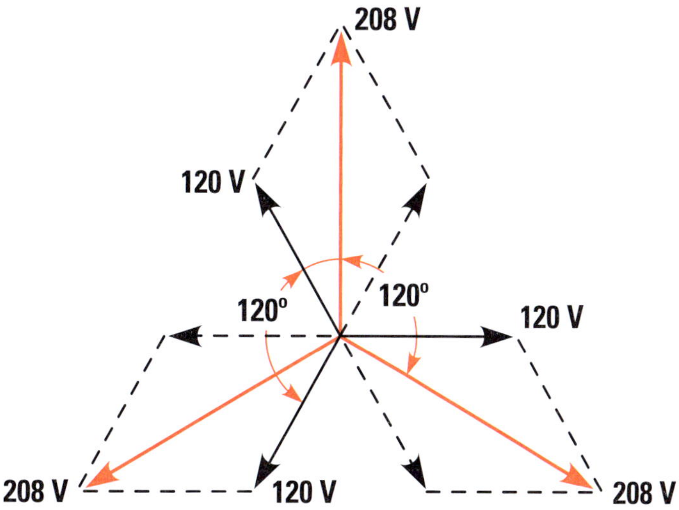

Figure 26–10 The parallelogram method of adding three-phase vectors.

Delta Connected Systems

In *Figure 26–11*, three individual coils have been connected so the end of one phase is connected to the beginning of the next phase, and so-on, to form a **delta connection.** This connection receives its name from the fact that a schematic diagram of the connection produces a triangular form and resembles the Greek letter delta (Δ). In *Figure 26–12*, test instruments have been connected to illustrate the relationship between voltage and current on both the lines and the phases. ***In a delta connection, line voltage and phase voltage are the same***. Notice that both voltmeters indicate a value of 480 volts in *Figure 26–12*. This is because when measuring voltage across a single phase winding in a delta connection, the voltmeter is measuring line-to-line as well.

$$E_{Line} = E_{Phase}$$

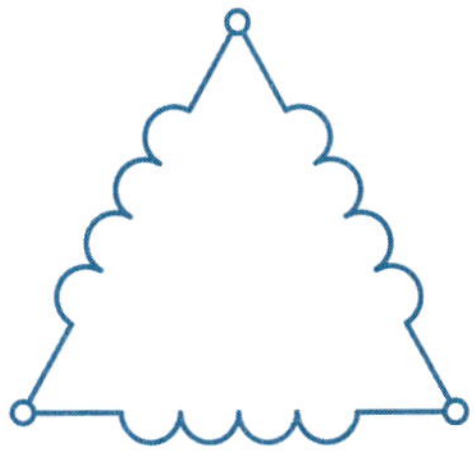

Figure 26–11 Three-phase delta connection.

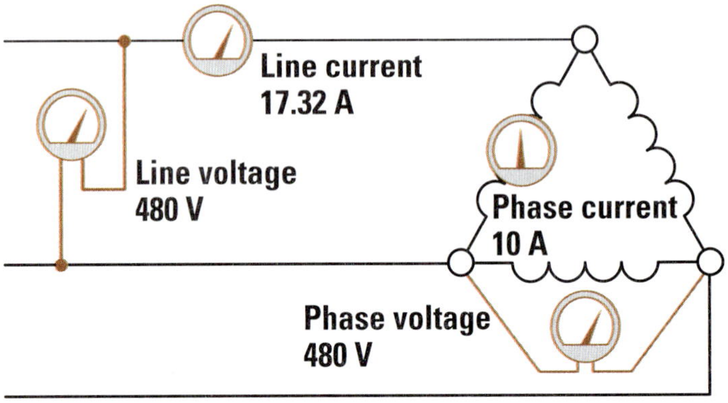

Figure 26–12 Voltage and current relationships in a delta connection.

The line current and phase current are different, however, because the line current shares current from two phase windings that are connected 120° out of phase with each other. ***The line current of a delta connection is higher than the phase current by a factor of the square root of 3 (1.732)***. In the example shown, it is assumed that each of the phase windings has a current flow of 10 amperes. The current in each of the lines, however, is 17.32 amperes. The reason for this difference in current is that current flows through different windings at different times in a three-phase circuit. During some periods of time, current will flow between two lines only. At other times, current will flow from two lines to the third (*Figure 26–13*). The delta connection is similar to a parallel connection because there is always more than one path for current flow. Because these currents are 120° out of phase with each other, vector addition must be used when finding the sum of the currents (*Figure 26–14*). Formulas for determining the current in a delta connection are:

$$I_{Line} = I_{Phase} \times 1.732$$

and

$$I_{Phase} = \frac{I_{Line}}{1.732}$$

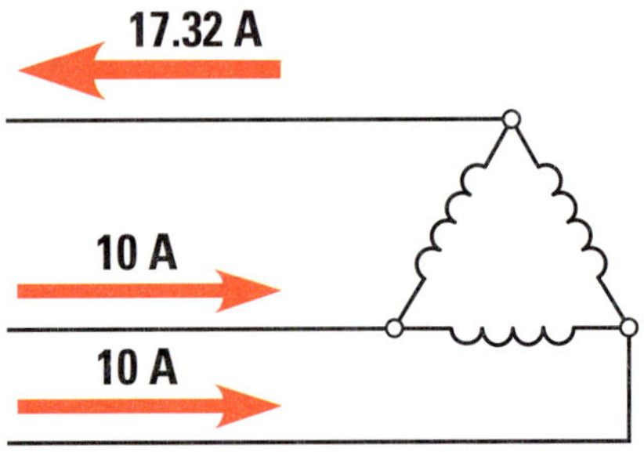

Figure 26–13 Division of currents in a delta connection.

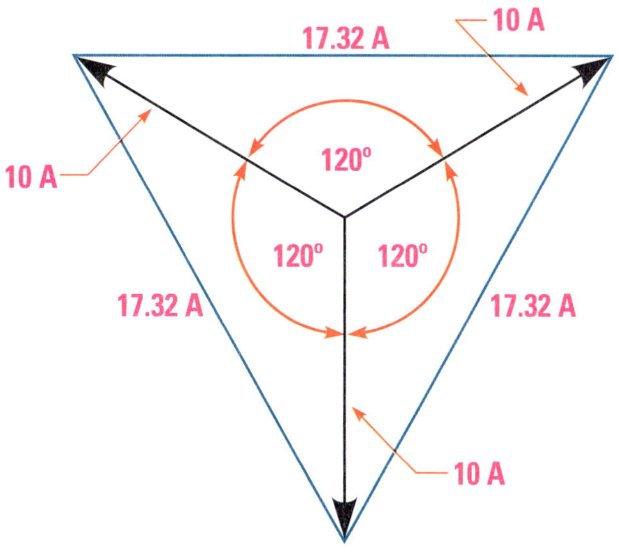

Figure 26–14 Vector addition is used to calculate the sum of the currents in a delta connection.

26–2 Three-Phase Power

Calculating power in a three-phase system is slightly different than previously discussed. In this calculation, the fact that there are three phases must be accounted for, and we must also consider that these phases are out of phase with each other. If *line* values of voltage and current are known, the apparent power of the circuit can be calculated using the following formula:

$$VA = \sqrt{3} \times E_{Line} \times I_{Line}$$

If the *phase* values of voltage and current are known, the apparent power can be calculated using the following formula:

$$VA = 3 \times E_{Phase} \times I_{Phase}$$

Notice that in the first formula, the line values of voltage and current are multiplied by the square root of 3. In the second formula, the phase values of voltage and current are multiplied by 3. The first formula is more often used because it is generally more convenient to obtain line values of voltage and current because they can be measured with a voltmeter and clamp-on ammeter.

Watts and VARs

Watts and VARs can be calculated in a similar manner. **Three-phase watts** can be calculated by multiplying the apparent power by the power factor:

$$P = \sqrt{3} \times E_{Line} \times I_{Line} \times PF$$

or

$$P = 3 \times E_{Phase} \times I_{Phase} \times PF$$

Note: When calculating the power of a purely resistive load, the voltage and current are in phase with each other, and the power factor is 1.

Three-phase VARs can be calculated in a similar manner, except that voltage and current values of a purely reactive load are used. For example, a purely capacitive load is shown in *Figure 26–15*. In this example, the line voltage is 560 volts, and the line current is 30 amperes. Capacitive VARs can be calculated using the formula:

$$VARs_C = \sqrt{3} \times E_{Line\ (Capacitive)} \times I_{Line\ (Capacitive)}$$

$$VARs_C = 1.732 \times 560\ V \times 30\ A$$

$$VARs_C = 29{,}097.6$$

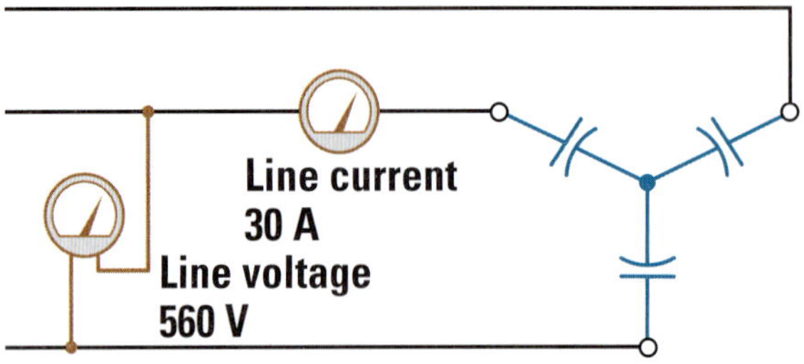

Figure 26–15 Purely capacitive three-phase load.

26–3 Calculating Unknown Values in Three-Phase Circuits

In the following examples, values of line and phase voltage, line and phase current, and power are calculated for different types of three-phase connections.

Example 26–1

A WYE-connected three-phase alternator supplies power to a delta-connected resistive load (*Figure 26–16*). The alternator has a line voltage of 480 volts. Each resistor of the delta load has 8 ohms of resistance. Find the following values:

$E_{L(Load)}$—line voltage of the load

$E_{P(Load)}$—phase voltage of the load

$I_{P(Load)}$—phase current of the load

$I_{L(Load)}$—line current to the load

$I_{L(Alt)}$—line current delivered by the alternator

$I_{P(Alt)}$—phase current of the alternator

$E_{P(Alt)}$—phase voltage of the alternator

P—true power

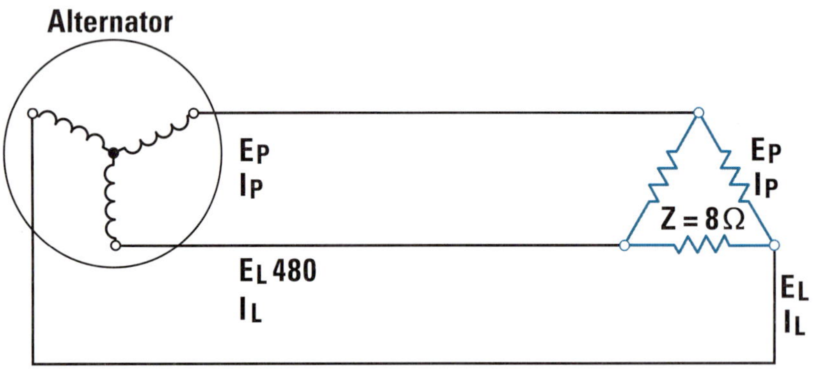

Figure 26–16 Calculating three-phase values using a WYE-connected power source and a delta-connected load (example circuit).

Example 26–1 (continued)

Solution

The load is connected directly to the alternator. Therefore, the line voltage supplied by the alternator is the line voltage of the load:

$$E_{L(Load)} = 480\ V$$

The three resistors of the load are connected in a delta connection. In a delta connection, the phase voltage is the same as the line voltage:

$$E_{P(Load)} = E_{L(Load)}$$

$$E_{P(Load)} = 480\ V$$

Each of the three resistors in the load is one phase of the load. Now that the phase voltage is known (480 V), the amount of phase current can be calculated using Ohm's law:

$$I_{P(Load)} = \frac{E_{P(Load)}}{Z}$$

$$I_{P(Load)} = \frac{480\ V}{8\ \Omega}$$

$$I_{P(Load)} = 60\ A$$

The three load resistors are connected as a delta with 60 A of current flow in each phase. The line current supplying a delta connection must be 1.732 times greater than the phase current:

$$I_{L(Load)} = I_{P(Load)} \times 1.732$$

$$I_{L(Load)} = 60\ A \times 1.732$$

$$I_{L(Load)} = 103.92\ A$$

The alternator must supply the line current to the load or loads to which it is connected. In this example, only one load is connected to the alternator. Therefore, the line current of the load is the same as the line current the alternator must supply:

$$I_{L(Alt)} = 103.92\ A$$

The phase windings of the alternator are connected in a WYE connection. In a WYE connection, the phase current and line current are equal. The phase current of the alternator is, therefore, the same as the alternator line current:

$$I_{P(Alt)} = 103.92\ A$$

The phase voltage of a WYE connection is less than the line voltage by a factor of the square root of 3. The phase voltage of the alternator is:

$$E_{P(Alt)} = \frac{E_{L(Alt)}}{1.732}$$

$$E_{P(Alt)} = \frac{480\ V}{1.732}$$

$$E_{P(Alt)} = 277.136\ V$$

Example 26–1 (continued)

In this circuit, the load is purely resistive. Therefore, the voltage and current are in phase with each other, which produces a unity power factor of 1. The true power in this circuit is calculated using the formula:

$$P = 1.732 \times E_{L(Alt)} \times I_{L(Alt)} \times PF$$

$$P = 1.732 \times 480\ V \times 103.92\ A \times 1$$

$$P = 86{,}394.931\ W$$

Example 26–2

A delta-connected alternator is connected to a WYE-connected resistive load (*Figure 26–17*). The alternator produces a line voltage of 240 V, and the resistors have a value of 6 Ω each. Find the following values:

$E_{L(Load)}$—line voltage of the load
$E_{P(Load)}$—phase voltage of the load
$I_{P(Load)}$—phase current of the load
$I_{L(Load)}$—line current to the load
$I_{L(Alt)}$—line current delivered by the alternator
$I_{P(Alt)}$—phase current of the alternator
$E_{P(Alt)}$—phase voltage of the alternator
P—true power

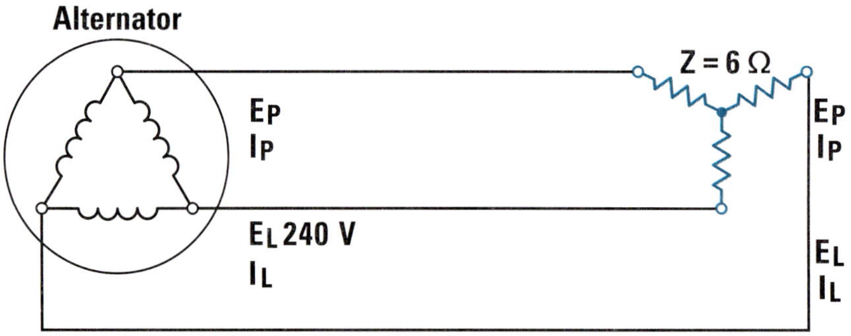

Figure 26–17 Calculating three-phase values using a delta-connected source and a WYE-connected load (example circuit).

Solution

As was the case in the previous example, the load is connected directly to the output of the alternator. The line voltage of the load must therefore be the same as the line voltage of the alternator:

$$E_{L(Load)} = 240\ V$$

The phase voltage of a WYE connection is less than the line voltage by a factor of 1.732:

$$E_{P(Load)} = \frac{240\ V}{1.732}$$

$$E_{P(Load)} = 138.568\ V$$

Example 26–2 (continued)

Each of the three 6-Ω resistors is one phase of the WYE-connected load. Because the phase voltage is 138.568 V, this voltage is applied to each of the three resistors. The amount of phase current can now be determined using Ohm's law:

$$I_{P(Load)} = \frac{E_{P(Load)}}{Z}$$

$$I_{P(Load)} = \frac{138.568\text{ V}}{6\ \Omega}$$

$$I_{P(Load)} = 23.095\text{ A}$$

The amount of line current drawn by a WYE-connected load will be the same as the phase current of the load:

$$I_{L(Load)} = 23.095\text{ A}$$

Because only a single load is connected to the alternator, the line current supplied to the load is the same as the line current of the alternator:

$$I_{L(Alt)} = 23.095\text{ A}$$

The phase windings of the alternator are connected in delta. In a delta connection, the phase current is less than the line current by a factor of 1.732:

$$I_{P(Alt)} = \frac{I_{L(Alt)}}{1.732}$$

$$I_{P(Alt)} = \frac{23.095\text{ A}}{1.732}$$

$$I_{P(Alt)} = 13.334\text{ A}$$

The phase voltage of a delta is the same as the line voltage:

$$E_{P(Alt)} = 240\text{ V}$$

Because the load in this example is purely resistive, the power factor has a value of unity, or 1. Power is calculated by using the line values of voltage and current:

$$P = 1.732 \times E_L \times I_L \times PF$$

$$P = 1.732 \times 240\text{ V} \times 23.095\text{ A} \times 1$$

$$P = 9600.13\text{ W}$$

Example 26–3

The phase windings of an alternator are connected in WYE. The alternator produces a line voltage of 440 V and supplies power to two resistive loads connected in parallel. One load contains resistors with a value of 4 Ω each that are connected in a WYE configuration. The second load contains resistors with a value of 6 Ω each that are connected in a delta configuration (*Figure 26–18*). Find the following circuit values:

$E_{L(Load\ 2)}$—line voltage of Load 2
$E_{P(Load\ 2)}$—phase voltage of Load 2

Example 26–3 (continued)

$I_{P(Load\ 2)}$—phase current of Load 2

$I_{L(Load\ 2)}$—line current to Load 2

$E_{P(Load\ 1)}$—phase voltage of Load 1

$I_{P(Load\ 1)}$—phase current of Load 1

$I_{L(Load\ 1)}$—line current to Load 1

$I_{L(Alt)}$—line current delivered by the alternator

$I_{P(Alt)}$—phase current of the alternator

$E_{P(Alt)}$—phase voltage of the alternator

P—true power

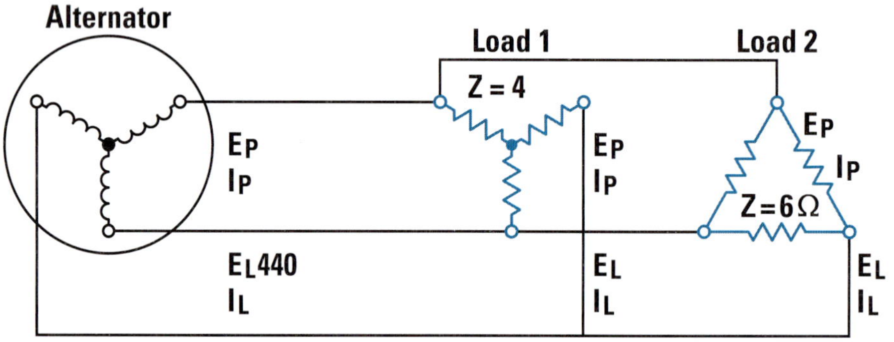

Figure 26–18 Calculating three-phase values using a WYE-connected source and two three-phase loads (example circuit).

Solution

Both loads are connected directly to the output of the alternator and in parallel with each other. Therefore, the line voltage for both Load 1 and Load 2 is the same as the line voltage of the alternator:

$$E_{L(Load\ 2)} = 440\ V$$

$$E_{L(Load\ 1)} = 440\ V$$

Load 2 is connected as a delta. The phase voltage is the same as the line voltage:

$$E_{P(Load\ 2)} = 440\ V$$

Each of the resistors that constitutes a phase of Load 2 has a value of 6 Ω. The amount of phase current can be found using Ohm's law:

$$I_{P(Load\ 2)} = \frac{E_{P(Load\ 2)}}{Z}$$

$$I_{P(Load\ 2)} = \frac{440\ V}{6\ \Omega}$$

$$I_{P(Load\ 2)} = 73.333\ A$$

The line current supplying a delta-connected load is 1.732 times greater than the phase current. The amount of line current drawn by Load 2 can be calculated by increasing the phase current value by 1.732:

$$I_{L(Load\ 2)} = I_{P(Load\ 2)} \times 1.732$$

$$I_{L(Load\ 2)} = 73.333\ A \times 1.732$$

$$I_{L(Load\ 2)} = 127.013\ A$$

Example 26–3 (continued)

The resistors of Load 1 are connected to form a WYE. The phase voltage of a WYE connection is less than the line voltage by a factor of 1.732:

$$E_{P(Load\ 1)} = \frac{E_{L(Load\ 1)}}{1.732}$$

$$E_{P(Load\ 1)} = \frac{440\ V}{1.732}$$

$$E_{P(Load\ 1)} = 254.042\ V$$

Now that the voltage applied to each of the 4-Ω resistors is known, the phase current can be calculated using Ohm's law:

$$I_{P(Load\ 1)} = \frac{E_{P(Load\ 1)}}{Z}$$

$$I_{P(Load\ 1)} = \frac{254.042\ V}{4\ \Omega}$$

$$I_{P(Load\ 1)} = 63.511\ A$$

The line current supplying a WYE-connected load is the same as the phase current. Therefore, the amount of line current drawn by Load 1 is:

$$I_{L(Load\ 1)} = 63.511\ A$$

The alternator must supply the line current needed to operate both loads. In this example, both loads are resistive. The total line current supplied by the alternator is the sum of the line currents of the two loads:

$$I_{L(Alt)} = I_{L(Load\ 1)} + I_{L(Load\ 2)}$$

$$I_{L(Alt)} = 63.511\ A + 127.013\ A$$

$$I_{L(Alt)} = 190.524\ A$$

Because the phase windings of the alternator in this example are connected in a WYE, the phase current is the same as the line current:

$$I_{P(Alt)} = 190.524\ A$$

The phase voltage of the alternator is less than the line voltage by a factor of 1.732:

$$E_{P(Alt)} = \frac{440\ V}{1.732}$$

$$E_{P(Alt)} = 254.042\ V$$

Both of the loads in this example are resistive and have a unity power factor of 1. The total power in this circuit can be found by using the line voltage and total line current supplied by the alternator:

$$P = 1.732 \times E_L \times I_L \times PF$$

$$P = 1.732 \times 440\ V \times 190.524\ A \times 1$$

$$P = 145{,}194.53\ W$$

Example 26–4

A WYE-connected three-phase alternator with a line voltage of 560 V supplies power to three different loads (*Figure 26–19*). The first load is formed by three resistors with a value of 6 Ω each, connected in a WYE; the second load comprises three inductors with an inductive reactance of 10 Ω each, connected in a delta; and the third load comprises three capacitors with a capacitive reactance of 8 Ω each, connected in a WYE. Find the following circuit values:

$E_{L(Load\ 3)}$—line voltage of Load 3 (capacitive)

$E_{P(Load\ 3)}$—phase voltage of Load 3 (capacitive)

$I_{P(Load\ 3)}$—phase current of Load 3 (capacitive)

$I_{L(Load\ 3)}$—line current to Load 3 (capacitive)

$E_{L(Load\ 2)}$—line voltage of Load 2 (inductive)

$E_{P(Load\ 2)}$—phase voltage of Load 2 (inductive)

$I_{P(Load\ 2)}$—phase current of Load 2 (inductive)

$I_{L(Load\ 2)}$—line current to Load 2 (inductive)

$E_{L(Load\ 1)}$—line voltage of Load 1 (resistive)

$E_{P(Load\ 1)}$—phase voltage of Load 1 (resistive)

$I_{P(Load\ 1)}$—phase current of Load 1 (resistive)

$I_{L(Load\ 1)}$—line current to Load 1 (resistive)

$I_{L(Alt)}$—line current delivered by the alternator

$E_{P(Alt)}$—phase voltage of the alternator

P—true power

$VARs_L$—reactive power of the inductive load

$VARs_C$—reactive power of the capacitive load

VA—apparent power

PF—power factor

$\angle\theta$—angle theta

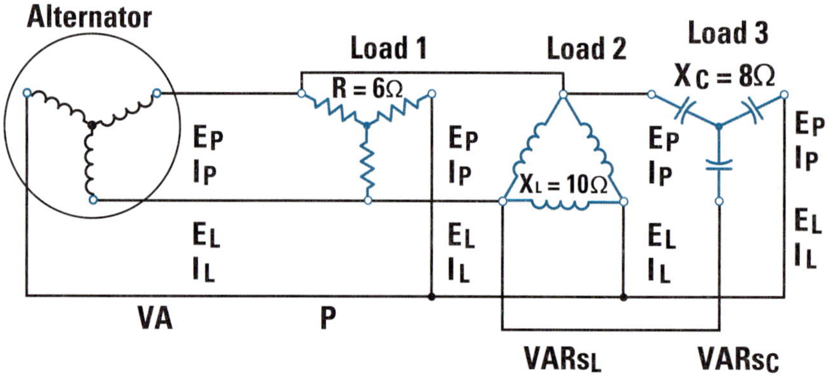

Figure 26–19 Calculating three-phase values with a WYE-connected source supplying power to a resistive, inductive, and capacitive load (example circuit).

Example 26–4 (continued)

Solution

All three loads are connected directly to the output of the alternator and are therefore in parallel with each other. The line voltage connected to each load is the same as the line voltage of the alternator:

$$E_{L(Load\ 3)} = 560\ V$$

$$E_{L(Load\ 2)} = 560\ V$$

$$E_{L(Load\ 1)} = 560\ V$$

Capacitive Load Calculations

Load 3 is formed from three capacitors with a capacitive reactance of 8 ohms each, connected in a WYE. Because this load is WYE-connected, the individual phase voltage drop is less than the line voltage by a factor of 1.732:

$$E_{P(Load\ 3)} = \frac{E_{L(Load\ 3)}}{1.732}$$

$$E_{P(Load\ 3)} = \frac{560\ V}{1.732}$$

$$E_{P(Load\ 3)} = 323.326\ V$$

Now that the voltage applied to each capacitor is known, the phase current can be calculated using Ohm's law:

$$I_{P(Load\ 3)} = \frac{E_{P(Load\ 3)}}{X_C}$$

$$I_{P(Load\ 3)} = \frac{323.326\ V}{8\ \Omega}$$

$$I_{P(Load\ 3)} = 40.416\ A$$

The line current required to supply a WYE-connected load is the same as the phase current:

$$I_{L(Load\ 3)} = 40.416\ A$$

The reactive power of Load 3 can be found using a formula similar to the formula for calculating apparent power. Because Load 3 is purely capacitive, the current and voltage are 90° out of phase with each other, and the power factor is zero:

$$VARs_C = 1.732 \times E_{L(Load\ 3)} \times I_{L(Load\ 3)}$$

$$VARs_C = 1.732 \times 560\ V \times 40.416\ A$$

$$VARs_C = 39{,}200.287$$

Inductive Load Calculations

Load 2 comprises three inductors connected in a delta with an inductive reactance of 10 ohms each. Because the load is connected in delta, the phase voltage is the same as the line voltage:

$$E_{L(Load\ 2)} = 560\ V$$

Example 26–4 (continued)

The phase current can be calculated by using Ohm's law:

$$I_{P(Load\ 2)} = \frac{E_{P(Load\ 2)}}{X_L}$$

$$I_{P(Load\ 2)} = \frac{560\ V}{10\ \Omega}$$

$$I_{P(Load\ 2)} = 56\ A$$

The amount of line current needed to supply a delta-connected load is 1.732 times greater than the phase current of the load:

$$I_{L(Load\ 2)} = I_{P(Load\ 2)} \times 1.732$$

$$I_{L(Load\ 2)} = 56\ A \times 1.732$$

$$I_{L(Load\ 2)} = 96.992\ A$$

Because Load 2 is made up of inductors, the reactive power can be calculated using the line values of voltage and current supplied to the load:

$$VARs_L = 1.732 \times E_{L(Load\ 2)} \times I_{L(Load\ 2)}$$

$$VARs_L = 1.732 \times 560\ V \times 96.992\ A$$

$$VARs_L = 94{,}074.481$$

Resistive Load Calculations

Load 1 consists of three resistors with a resistance of 6 ohms each, connected in a WYE. In a WYE connection, the phase voltage is less than the line voltage by a factor of 1.732. The phase voltage for Load 1 is the same as the phase voltage for Load 3:

$$E_{P(Load\ 1)} = 323.326\ V$$

The amount of phase current can now be calculated using the phase voltage and the resistance of each phase:

$$I_{P(Load\ 1)} = \frac{E_{P(Load\ 1)}}{R}$$

$$I_{P(Load\ 1)} = \frac{323.326\ V}{6\ \Omega}$$

$$I_{P(Load\ 1)} = 53.888\ A$$

Because the resistors of Load 1 are connected in a WYE, the line current is the same as the phase current:

$$I_{L(Load\ 1)} = 53.888\ A$$

Because Load 1 is purely resistive, true power can be calculated using the line and phase current values:

$$P = 1.732 \times E_{L(Load\ 1)} \times I_{L(Load\ 1)}$$

$$P = 1.732 \times 560\ V = 53.888\ A$$

$$P = 52{,}267.049\ W$$

26–4 Three-Phase Source Calculations

The alternator in *Example 26-4* must supply the line current for each of the three loads that are connected in parallel. However, the line currents are out of phase with each other. To find the total line current delivered by the alternator, vector addition must be used. The current flow in Load 1 is resistive and in phase with the line voltage. The current flow in Load 2 is inductive and lags the line voltage by 90°. The current flow in Load 3 is capacitive and leads the line voltage by 90°. A formula similar to the formula used to find total current flow in an RLC parallel circuit can be employed to find the total current delivered in this circuit:

$$I_{L(Alt)} = \sqrt{I_{L(Load\ 1)}^{2} + (I_{L(Load\ 2)} - I_{L(Load\ 3)})^{2}}$$

$$I_{L(Alt)} = \sqrt{(53.888\ A)^2 + (96.992\ A - 40.416\ A)^2}$$

$$I_{L(Alt)} = 78.133\ A$$

The apparent power can now be found using the line voltage and current values of the alternator:

$$VA = 1.732 \times E_{L(Alt)} \times I_{L(Alt)}$$

$$VA = 1.732 \times 560\ V \times 78.133\ A$$

$$VA = 75{,}782.759$$

The circuit power factor is the ratio of apparent power and true power:

$$PF = \frac{W}{VA}$$

$$PF - \frac{52{,}267.049\ W}{75{,}782.759\ VA}$$

$$PF = 69\%$$

Angle Theta

Remember that the power factor is equal to the cosine of angle theta and that we can find angle theta by taking the inverse cosine of the power factor:

$$PF = \text{Cos}\ \Theta$$

$$0.69 = \text{Cos}\ \Theta$$

$$\Theta = \text{Cos}^{-1}\ 0.69$$

$$\Theta = 46.39°$$

The current is 46.39° out of phase with the voltage in this circuit. Because inductive VARs are greater than capacitive VARs, the current is lagging the voltage. A vector diagram of this circuit is shown in *Figure 26–20*.

Power Factor Correction

Correcting the power factor of a three-phase circuit is similar to the procedure used to correct the power factor of a single-phase circuit. Because the example circuit is more inductive than capacitive, adding more capacitance to the circuit has the effect of adding more capacitive VARs to the circuit and thus counteracting the inductive VARs and bringing the power factor closer to unity.

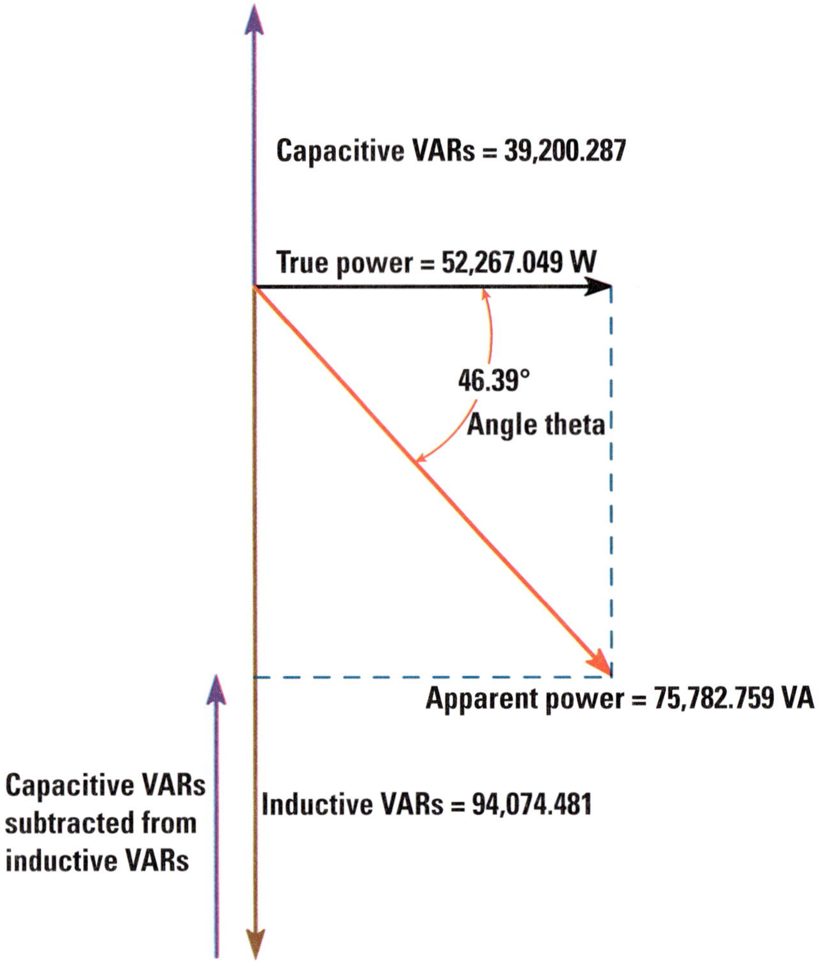

Figure 26–20 Vector diagram showing current values. Note: current lags the voltage by 46.39° because the inductive current is greater than the capacitive current.

Example 26–5

A three-phase motor is connected to a 480-V, 60-Hz line (*Figure 26–21*). A clamp-on ammeter indicates a running current of 68 A at full load, and a three-phase wattmeter indicates a true power of 40,277 W. Calculate the motor power factor first. Then, find the amount of capacitance needed to correct the power factor to 95%. Assume that the capacitors used for power factor correction are to be connected in WYE and the capacitor bank is then to be connected in parallel with the motor.

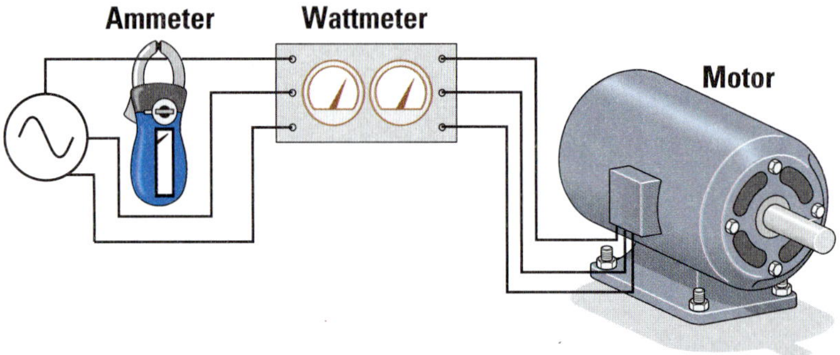

Figure 26–21 Determining apparent and true power for a three-phase motor for power factor correction.

Example 26–5 (continued)

When determining the amount of capacitance needed to correct power factor, it is helpful to follow a procedure. The procedure in this example will consist of nine steps.

Step 1: Calculate the apparent power of the circuit.

Step 2: Determine the power factor of the circuit.

Step 3: Determine the reactive power of the circuit.

Step 4: Determine the amount of apparent power that would produce the desired power factor.

Step 5: Determine the amount of reactive power that would produce the desired amount of apparent power.

Step 6: Determine the capacitive VARs necessary to produce the desired reactive power.

Step 7: Determine the amount of capacitive current necessary to produce the capacitive VARs needed.

Step 8: Determine the capacitive reactance of each capacitor.

Step 9: Determine the capacitance value of each capacitor.

Solution

Step 1: *Determine the apparent power of the circuit.*

$$VA = 1.732 \times E_L \times I_L$$

$$VA = 1.732 \times 480\ V \times 68\ A$$

$$VA = 56{,}532.48$$

Step 2: *Determine the power factor of the circuit.*

$$PF = \frac{P}{VA}$$

$$PF = \frac{40{,}277\ W}{56{,}532.48\ V}$$

$$PF = 0.7124 \textit{ or } 71.24\%$$

Step 3: *Determine the reactive power of the circuit.*

$$VARs_L = \sqrt{VA^2 - P^2}$$

$$VARs_L = \sqrt{56{,}532.46^2 - 40{,}277^2}$$

$$VARs_L = 39{,}669.693$$

Step 4: *Determine the amount of apparent power that would produce the desired power factor.*

$$VA = \frac{P}{PF}$$

$$VA = \frac{40{,}277}{0.95}$$

$$VA = 42{,}396.842$$

Example 26–5 (continued)

Step 5: *Determine the amount of reactive power that would produce the desired amount of apparent power.*

$$VARs_L = \sqrt{VA^2 - P^2}$$

$$VARs_L = \sqrt{42{,}396.842^2 - 40{,}277^2}$$

$$VARs_L = 13{,}238.409$$

Step 6: *Determine the capacitive VARs necessary to produce the desired reactive power.*

$$VARs_C = VARs_{present\ time} - VARs_{desired}$$

$$VARs_C = 39{,}669.693 - 13{,}238.409$$

$$VARs_C = 26{,}431.284$$

To correct the power factor to 95%, the inductive VARs must be reduced from 39,669.693 to 13,238.409. This can be done by connecting a bank of capacitors in the circuit that will produce a total of 26,431.284 capacitive VARs. This amount of capacitive VARs will reduce the inductive VARs to the desired amount (*Figure 26–22*).

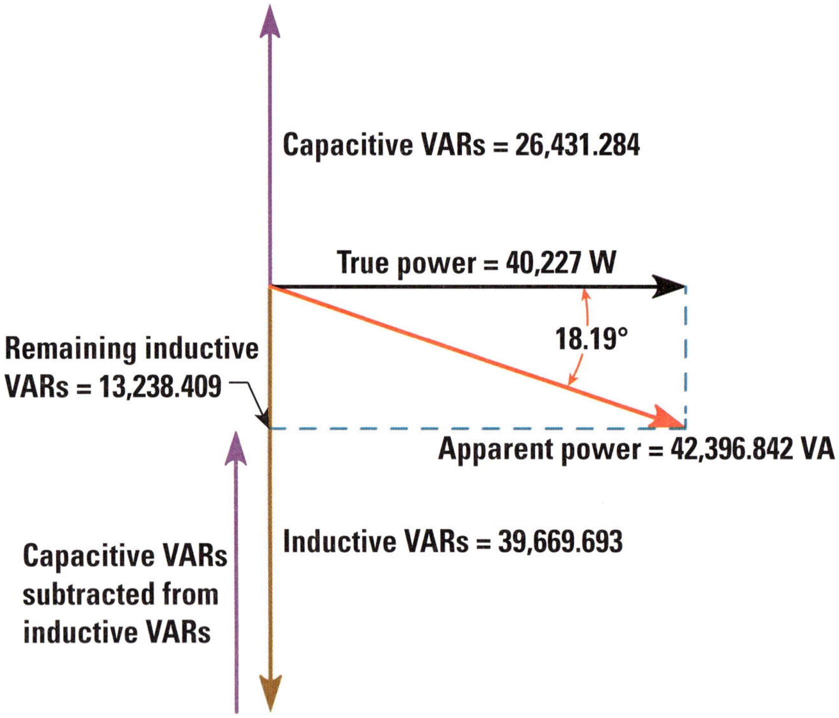

Figure 26–22 Vector diagram for power, VA, and VARs to correct motor power factor.

Step 7: *Determine the amount of capacitive current necessary to produce the capacitive VARs needed.*

$$I_C = \frac{VARs_C}{E_L \times 1.732}$$

$$I_C = \frac{26{,}431.284\ VARs_C}{480 \times 1.732}$$

$$I_C = 31.793\ A$$

Example 26–5 (continued)

The capacitive load bank is to be connected in a WYE. Therefore, the phase current is the same as the line current. The phase voltage, however, is less than the line voltage by a factor of 1.732. The phase voltage is 277.136 volts.

Step 8: *Determine the capacitive reactance of each capacitor.*

Ohm's law can be used to find the capacitive reactance needed to produce a phase current of 31.793 amperes with a voltage of 277.136 volts:

$$X_C = \frac{E_{Phase}}{I_{Phase}}$$

$$X_C = \frac{277.136\ V}{31.793\ A}$$

$$X_C = 8.717\ \Omega$$

Step 9: *Determine the capacitance value of each capacitor:*

$$C = \frac{1}{2\pi f X_C}$$

$$C = \frac{1}{377 \times 8.717}$$

$$C = 304.293\ \mu F$$

When a bank of WYE-connected capacitors with a value of 304.293 μF each is connected in parallel with the motor, the power factor will be corrected to 95% (*Figure 26–23*).

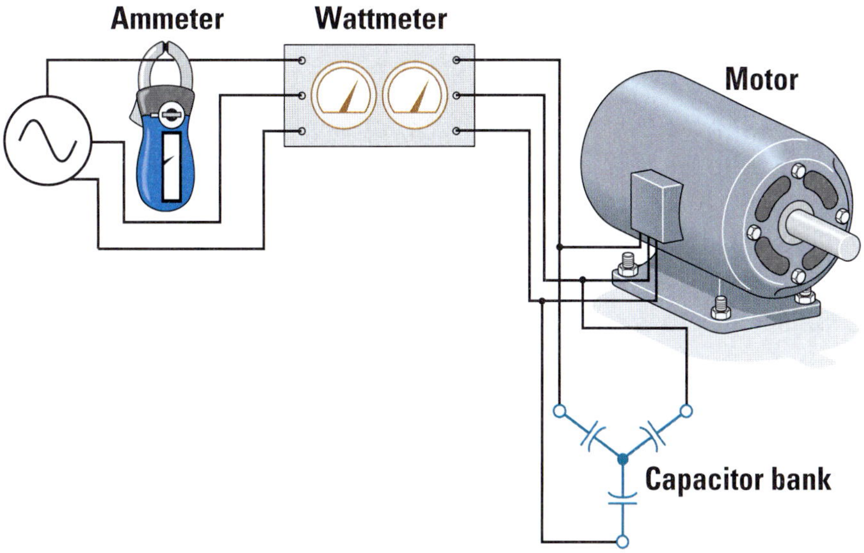

Figure 26–23 A WYE-connected bank of capacitors is used to correct motor power factor.

Summary

- The voltages of a three-phase system are 120° out of phase with each other.
- The two common types of three-phase connections are WYE and delta.
- WYE connections are characterized by the fact that one terminal of each of the devices is connected together.
- In a WYE connection, the phase voltage is less than the line voltage by a factor of 1.732. The phase current and line current are the same.
- Delta connections are characterized by the fact that the end of each phase is connected to the start of the next phase and take the shape of a triangle.
- In a delta connection, the phase voltage is the same as the line voltage. The phase current is less than the line current by a factor of 1.732.

Review Questions

1. Which of the following statements best describes one of the major advantages of utilizing three-phase power systems over single-phase systems?
   A) Three-phase systems require larger conductors.
   B) Three-phase equipment has lower kVA ratings.
   C) Three-phase systems have dual voltages.
   D) Three-phase systems provide power that never falls to zero.
2. Three-phase systems are the superior systems for the connection of equipment such as motors for which of the following reasons?
   A) The horsepower rating of motors is much greater than single-phase motors of similar size.
   B) The horsepower rating of motors is much less than single-phase motors of similar size.
   C) Voltage in a three-phase system has better rotation in a motor.
   D) Three-phase motors can pull more current than single phase motors.
3. Which of the following statements about a WYE-connected electrical system best describes the relationship of voltage and current both line-to-line and across each phase?
   A) Voltage is the same across each phase as it is line-to-line.
   B) Current on the lines will be 1.732 times greater than each individual phase.
   C) Voltage line-to-line will be 1.732 times greater than each individual phase.
   D) Current will be double on the lines than it is on each phase because there are two phases.
4. Which of the following statements about a delta-connected electrical system best describes the relationship of voltage and current both line-to-line and across each phase?
   A) Voltage line-to-line will equal 1.732 times the voltage on each phase.
   B) Current on the lines will be 1.732 times the current on each phase.
   C) Current will be the same both line-to-line and across each phase.
   D) Voltage will be two times greater line-to-line than on each phase.

5. True or False: Power in a three-phase system is calculated identically to power in a single-phase system, by multiplying voltage and current.
6. In a single-phase system, the power supplied crosses the zero point multiple times per cycle. How often does power cross the zero point in a three-phase system?
   A) Zero
   B) Once
   C) Twice
   D) Three times
7. A WYE-connected load has a voltage of 480 V applied to it line-to-line. A single-phase resistive load of 10 Ω is connected. How many amperes does the resistive load draw on this circuit?
   A) 48 A
   B) 24 A
   C) 27.7 A
   D) 20.8 A
8. A 480-V, three-phase delta system has a connected load with 16 Ω of impedance on one single phase. How much current should be expected to be measured on each of the lines supplying power to the load? (Round to the nearest ampere.)
   A) 30 A
   B) 60 A
   C) 17 A
   D) 52 A
9. A three-phase, delta-connected load has a phase voltage of 240 V and a phase current of 18 A. What is the power factor of this load if a wattmeter measures 6.235 kW on the circuit?
   A) 83%
   B) 48%
   C) 63%
   D) 74%
10. If the power factor in the previous question is a lagging power factor, what must be added to the circuit to bring the power factor closer to unity?
    A) More inductive reactance
    B) More capacitive reactance
    C) More resistance
    D) More amperes

Practical Applications

1. You are working in an industrial plant. A bank of capacitors is to be used to correct the power factor of a 480-V, three-phase motor. The capacitor bank is connected in a WYE configuration. Each of the three capacitors is rated at 25 μF and 600 VDC. You have been told to reconnect these capacitors in a delta. Can these capacitors be changed from a WYE to a delta without harm to the capacitors?

2. You are a journeyman electrician working in an industrial plant. A 480-V, three-phase, 60-Hz power panel has a current draw of 216 A. A three-phase wattmeter indicates a true power of 86 kW. You have been instructed to reduce the current draw on the panel by adding three 400-μF capacitors to the system. The capacitors are to be connected in a WYE. Each capacitor will be connected to one of the three-phase lines. What should be the current flow on the system after the bank of capacitors is added to the circuit?

3. You are an electrician working in an industrial plant. A 30-hp three-phase induction motor has a current draw of 36 amperes at full load. The motor is connected to a 480-volt line. A three-phase wattmeter indicates a true power of 22 kW. Determine the power factor of the motor and the amount of capacitance needed to correct the power factor to 95%. Also determine the minimum voltage rating of the capacitors. The capacitors are to be connected in a WYE.

Practice Problems

1. Refer to the circuit shown in *Figure 26–16* to calculate the missing values, but assume the alternator has a line voltage of 240 V and the load has a resistance of 12 Ω per phase. Find all the missing values.

$E_{P(A)}$ ______	$E_{P(L)}$ ______
$I_{P(A)}$ ______	$I_{P(L)}$ ______
$E_{L(A)}$ 240 V	$E_{L(L)}$ ______
$I_{L(A)}$ ______	$I_{L(L)}$ ______
P ______	$Z_{(PHASE)}$ 12 Ω

2. Refer to the circuit shown in *Figure 26–17* to calculate the missing values, but assume the alternator has a line voltage of 4160 V and the load has a resistance of 60 Ω per phase. Find all the missing values.

$E_{P(A)}$ ______	$E_{P(L)}$ ______
$I_{P(A)}$ ______	$I_{P(L)}$ ______
$E_{L(A)}$ 4160 V	$E_{L(L)}$ ______
$I_{L(A)}$ ______	$I_{L(L)}$ ______
P ______	$Z_{(PHASE)}$ 60 Ω

3. Refer to the circuit shown in *Figure 26–18* to calculate the missing values, but assume the alternator has a line voltage of 560 V. Load 1 has a resistance of 5 Ω per phase, and Load 2 has a resistance of 8 Ω per phase. Find all the missing values.

$E_{P(A)}$ ______	$E_{P(L1)}$ ______	$E_{P(L2)}$ ______
$I_{P(A)}$ ______	$I_{P(L1)}$ ______	$I_{P(L2)}$ ______
$E_{L(A)}$ 560 V	$E_{L(L1)}$ ______	$E_{L(L2)}$ ______
$I_{L(A)}$ ______	$I_{L(L1)}$ ______	$I_{L(L2)}$ ______
P ______	$Z_{(PHASE)}$ 5 Ω	$Z_{(PHASE)}$ 8 Ω

4. Refer to the circuit shown in *Figure 26–19* to calculate the missing values, but assume the alternator has a line voltage of 480 V. Load 1 has a resistance of 12 Ω per phase. Load 2 has an inductive reactance of 16 Ω per phase, and Load 3 has a capacitive reactance of 10 Ω per phase. Find all the missing values.

$E_{P(A)}$ ______	$E_{P(L1)}$ ______	$E_{P(L2)}$ ______	$E_{P(L3)}$ ______
$I_{P(A)}$ ______	$I_{P(L1)}$ ______	$I_{P(L2)}$ ______	$I_{P(L3)}$ ______
$E_{L(A)}$ 480 V	$E_{L(L1)}$ ______	$E_{L(L2)}$ ______	$E_{L(L3)}$ ______
$I_{L(A)}$ ______	$I_{L(L1)}$ ______	$I_{L(L2)}$ ______	$I_{L(L3)}$ ______
VA ______	$R_{(PHASE)}$ 12 Ω	$X_{L(PHASE)}$ 16 Ω	$X_{C(PHASE)}$ 10 Ω
	P ______	$VARs_L$ ______	$VARs_C$ ______

Section 10 Transformers

Unit 27 Single-Phase Transformers

Key Terms

Autotransformers

Control transformer

Distribution transformer

Excitation current

Flux leakage

Inrush current

Isolation transformers

Laminated

Neutral conductor

Primary winding

Secondary winding

Step-down transformer

Step-up transformer

Tape-wound core

Toroid core

Transformer

Turns ratio

Volts-per-turn ratio

Why You Need to Know

Understanding the basic principles of how transformers can change the voltage in a circuit is important to the information presented in later units. Many AC motors, for example, operate on these same principles. This unit

- discusses transformers and how they are divided into three major types.
- determines values of voltage and current and illustrates different methods that can be employed to determine these values.
- describes how windings determine the primary and secondary voltage and how the nameplate rating provides key information when providing protection.
- discusses installation and testing.

Outline

27–1 Introduction to Transformers

27–2 Transformer Calculations

27–3 Autotransformers

27–4 Voltage and Current Relationships in a Transformer

27–5 Transformer Nameplates

27–6 Connecting Transformer Secondaries

Objectives

After studying this unit, you should be able to

- explain the basic principles of transformer operation.
- explain how autotransformers differ from isolation transformers.
- calculate unknown circuit values on both the primary and secondary side of a transformer.
- identify key information about a transformer from the nameplate.
- explain the process for connecting the secondary windings of multiple transformers together.
- describe the impact that transformer polarity makes on the voltage and current characteristics of a transformer.

Preview

Transformers are among the most common devices found in the electrical field. They range in size from less than one cubic inch to the size of rail cars. Their ratings can range from milli-volt-amperes (mVA) to giga-volt-amperes (GVA). It is imperative that anyone working in the electrical field have an understanding of transformer types and connections. This unit presents transformers intended for use in single-phase installations.

27–1 Introduction to Transformers

A **transformer** is an electromagnetically operated device that can alter circuit characteristics such as voltage, current, and impedance. Transformers are one of the most efficient electrical devices that we will encounter. Their efficiencies commonly range from 90% to 99% at full load. Transformers can be divided into the following three classifications:

1. Isolation transformer
2. Autotransformer
3. Current transformer (current transformers were discussed in Unit 9)

Voltage, current, and power values on the primary and secondary sides of a transformer are proportional with respect to the ratio of turns between the primary windings and secondary windings. This does not mean that the exact number of turns of wire on each winding must be known to determine the different values of voltage and current for a transformer. What must be known is the *ratio* of turns. For example, assume a transformer has two windings. One winding, the primary, has 1000 turns of wire, and the other, the secondary, has 250 turns of wire (*Figure 27–1*). The **turns ratio,** often referred to as simply the transformer ratio or transformer voltage ratio, of this transformer is 4 to 1, or 4:1 (1000 turns/250 turns = 4). This indicates that there are four turns of wire on the primary for every one turn of wire on the secondary. It should be noted that the turns ratio is always stated as a ratio, such as 4 to 1. It is incorrect to say that the ratio is 4.

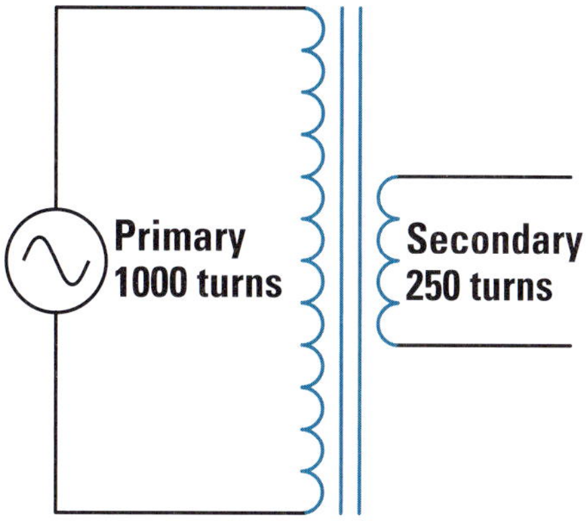

Figure 27–1 A transformer's turns ratio will determine the output of the transformer.

Transformer Formulas

Different formulas can be used to find the values of voltage and current for a transformer. The following is a list of standard formulas, where:

N_P = number of turns in the primary

N_S = number of turns in the secondary

E_P = voltage of the primary

E_S = voltage of the secondary

I_P = current in the primary

I_S = current in the secondary

$$\frac{E_P}{E_S} = \frac{N_P}{N_S}$$

$$\frac{E_P}{E_S} = \frac{I_S}{I_P}$$

$$\frac{N_P}{N_S} = \frac{I_S}{I_P}$$

or

$$E_P \times N_S = E_S \times N_P$$

$$E_P \times I_P = E_S \times I_S$$

$$N_P \times I_P = N_S \times I_S$$

The **primary winding** of a transformer is the power input winding. It is the winding that is connected to the incoming power supply. The **secondary winding** is the load winding, or output winding. It is the side of the transformer that is connected to the load and represents the desired circuit parameters that necessitated the transformer (*Figure 27–2*). For example, if the building power system is at 480 volts but receptacle voltage must be at 120 volts, a transformer that outputs a voltage of 120 volts when 480 volts is applied would be installed.

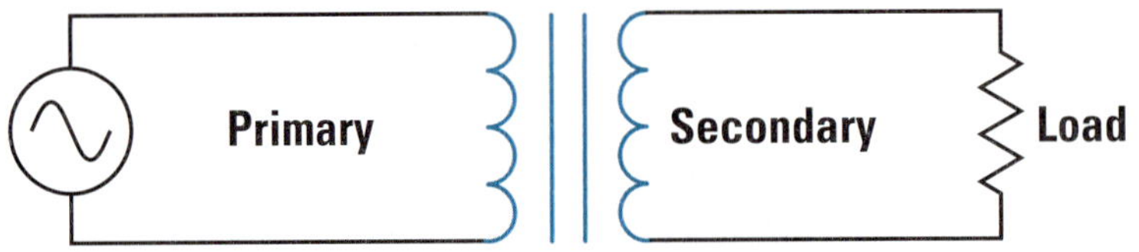

Figure 27–2 The primary and secondary windings of an isolation transformer have no direct electrical connection.

Single-Phase Transformers

The transformers shown in *Figures 27–1 and 27–2* are **isolation transformers.** This means that the secondary winding is isolated from the primary winding. There is no direct electrical connection between the primary and secondary windings. This type of transformer, however, has primary and secondary windings that are magnetically coupled. This line isolation is often a very desirable characteristic. The isolation transformer greatly reduces any voltage spikes that originate on the supply side before they are transferred to the load side. While the majority of transformers are installed to transform the characteristics of the circuit, some isolation transformers are built with a turns ratio of 1:1, have the same input and output voltage, and are used for the purpose of isolation only.

The reason that the isolation transformer can greatly reduce any voltage spikes before they reach the secondary is because of the rise time of current through an inductor. Recall from Unit 13 that DC current in an inductor rises at an exponential rate (*Figure 27–3*). As the current increases in value, the expanding magnetic field cuts through the conductors of the coil and induces a voltage that opposes the applied voltage. The amount of induced voltage is proportional to the rate of change of current. This simply means that the faster current attempts to increase, the greater the opposition to that increase is. Spike voltages and currents are generally of very short duration, which means that they increase in value very rapidly (*Figure 27–4*). This rapid change of value causes the opposition to the change to increase just as rapidly. By the time the spike has been transferred to the secondary winding of the transformer, it has been eliminated or greatly reduced (*Figure 27–5*).

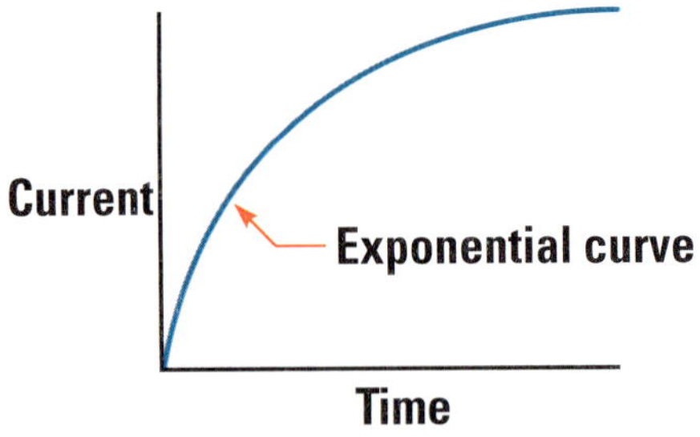

Figure 27–3 DC current through an inductor rises at an exponential rate.

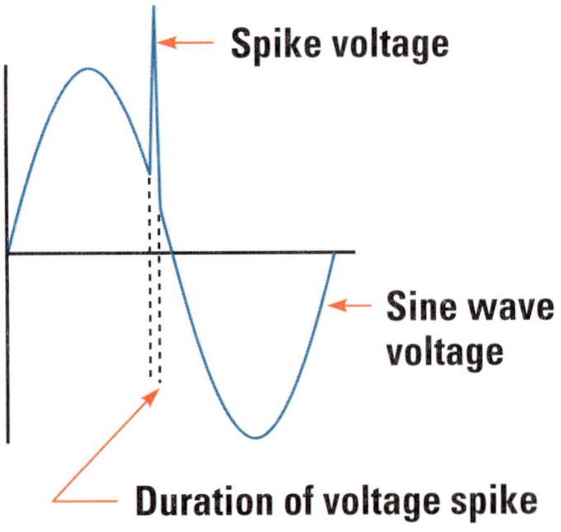

Figure 27–4 Voltage spikes are generally of very short duration.

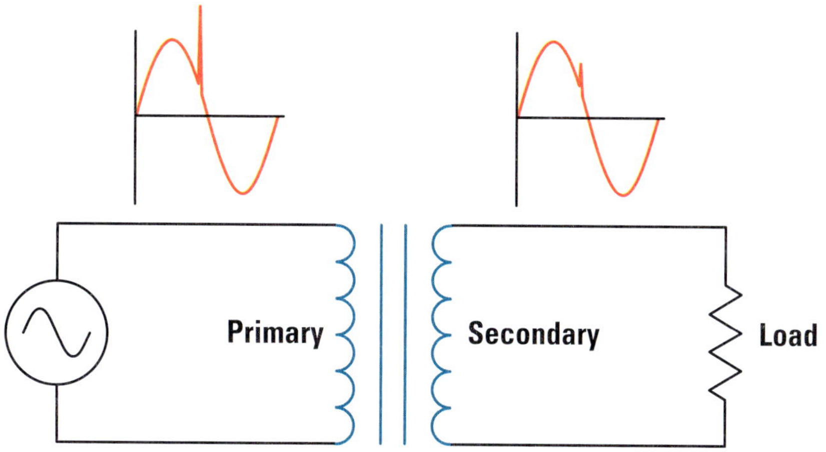

Figure 27–5 The isolation transformer greatly reduces the voltage spike on the secondary winding.

The basic construction of an isolation transformer is shown in *Figure 27–6*. A metal core is used to provide good magnetic coupling between the two windings. The core is generally made of laminations stacked together. Laminating the core helps reduce power losses caused by eddy current induction.

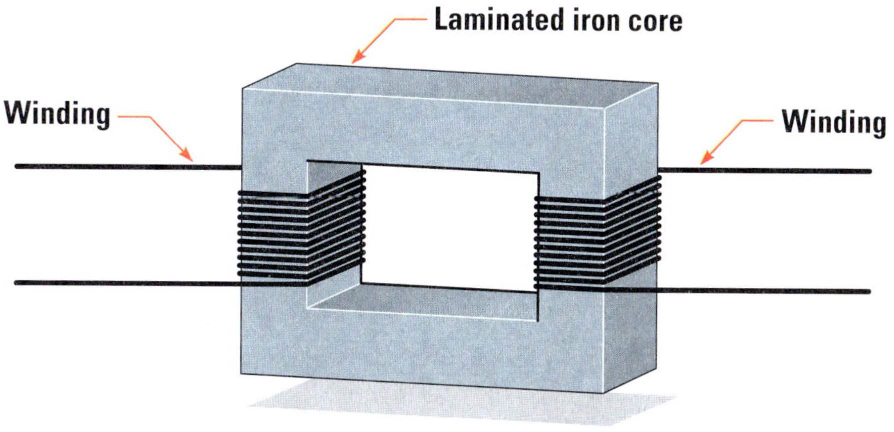

Figure 27–6 Illustration of the basic concept of an isolation transformer.

Basic Operating Principles

In *Figure 27–7,* one winding of an isolation transformer has been connected to an AC supply, and the other winding has been connected to a load. As current increases from zero to its peak positive point, a magnetic field expands outward around the winding connected to the supply. When the current decreases from its peak positive point toward zero, the magnetic field collapses. When the current increases toward its negative peak, the magnetic field again expands but with the opposite polarity. The field again collapses when the current decreases from its negative peak toward zero. This continually expanding and collapsing magnetic field cuts the windings of the primary and induces a voltage into it. This induced voltage opposes the applied voltage and limits the current flow of the primary. However, keep in mind that it takes a little bit of time to induce enough countervoltage to limit the current, and upon start up, a transformer is essentially a short-circuit. Transformers have large initial start-up current values that level out after a short period of time.

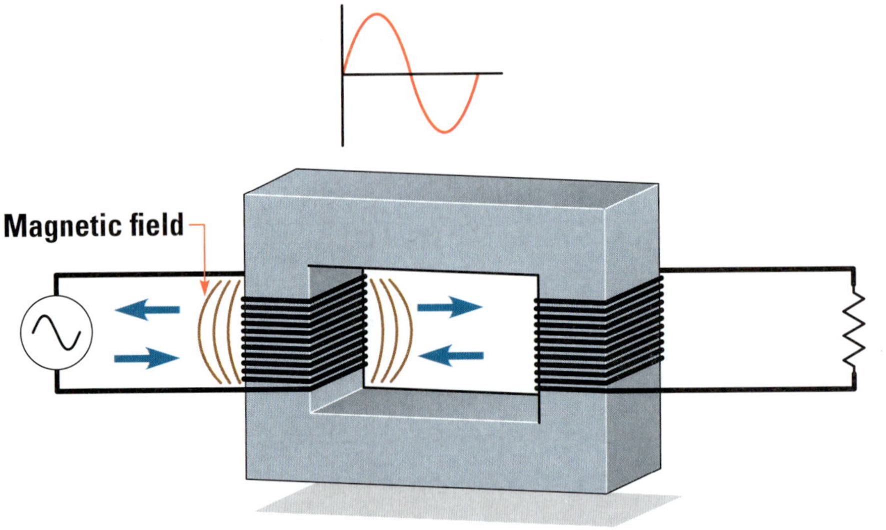

Figure 27–7 Continuously expanding and collapsing magnetic field produced by AC supply on the primary winding current in a transformer.

Excitation Current

There will always be some amount of current flow in the primary of any voltage transformer regardless of type or size even if there is no load connected to the secondary. This current flow is called the **excitation current** of the transformer. The excitation current is the amount of current required to magnetize the core of the transformer. The excitation current remains constant from no load to full load. Generally, the excitation current is such a small part of the full-load current that it is often omitted when making calculations.

Mutual Induction

Because the secondary windings of an isolation transformer are wound on the same core as the primary, the magnetic field produced by the primary winding also cuts the windings of the secondary (*Figure 27–8*). This continually changing magnetic field induces a voltage into the secondary winding. The amount of voltage induced in the secondary is determined by the ratio of the number of turns of wire in the secondary to those in the primary. For example, assume the primary has 240 turns of wire and is connected to 120 VAC. This gives the transformer a **volts-per-turn ratio** of 0.5 (120 V/240 turns = 0.5 V per turn). Now assume the secondary winding contains 100 turns of wire. Because the transformer has a volts-per-turn ratio of 0.5 volt per turn, the secondary voltage is 50 volts (100 turns × 0.5 V/turn = 50 V).

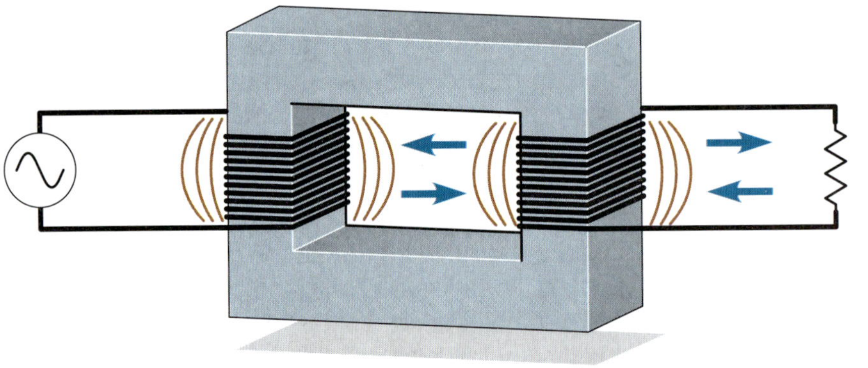

Figure 27–8 The magnetic field of the primary induces a voltage into the secondary.

27–2 Transformer Calculations

In the following examples, values of voltage, current, and turns for different transformers are calculated.

Assume that the isolation transformer shown in *Figure 27–2* has 240 turns of wire on the primary and 60 turns of wire on the secondary. This is a ratio of 4:1 (240 turns/60 turns = 4). Now assume that 120 volts are connected to the primary winding. What is the voltage of the secondary winding?

$$\frac{E_P}{E_S} = \frac{N_P}{N_S}$$

$$\frac{120\text{ V}}{E_S} = \frac{240\text{ turns}}{60\text{ turns}}$$

$$240\text{ turns } E_S = 7200\text{ V-turns}$$

$$E_S = \frac{7200\text{ V-turns}}{240\text{ turns}}$$

$$E_S = 30\text{ V}$$

This transformer would be what is known as a **step-down transformer** because its secondary voltage is lower than its primary voltage.

Now assume the load connected to the secondary winding has an impedance of 5 ohms. The next problem is to calculate the current flow in the secondary and primary windings. The current flow of the secondary can be calculated using Ohm's law because the voltage and impedance are known:

$$I = \frac{E}{Z}$$

$$I = \frac{30\text{ V}}{5\ \Omega}$$

$$I = 6\text{ A}$$

Now that the amount of current flow in the secondary is known, the primary current can be calculated using the formula:

$$\frac{E_P}{E_S} = \frac{I_S}{I_P}$$

$$\frac{120\text{ V}}{30\text{ V}} = \frac{6\text{ A}}{I_P}$$

$$120\text{ V } I_P = 180\text{ VA}$$

$$I_P = \frac{180\text{ VA}}{120\text{ V}}$$

$$I_P = 1.5\text{ A}$$

Notice that the primary voltage is higher than the secondary voltage, but the primary current is much less than the secondary current. ***A good rule for any type of transformer is that power in must equal power out.*** If the primary voltage and current are multiplied together, the product should equal the product of the voltage and current of the secondary:

Primary	Secondary
$120 \times 1.5 = 180$ VA	$30 \times 6 = 180$ VA

Let's look at another example. Assume that the primary winding contains 240 turns of wire, and the secondary contains 1200 turns of wire. This is a turns ratio of 1:5. If 120 volts are connected to the primary winding, calculate the voltage output of the secondary winding:

$$\frac{E_P}{E_S} = \frac{N_P}{N_S}$$

$$\frac{120\text{ V}}{E_S} = \frac{240\text{ turns}}{1200\text{ turns}}$$

$$240\text{ turns } E_S = 144{,}000\text{ V-turns}$$

$$E_S = \frac{144{,}000\text{ V-turns}}{240\text{ turns}}$$

$$E_S = 600\text{ V}$$

Notice that the secondary voltage of this transformer is higher than the primary voltage. This type of transformer is known as a **step-up transformer.**

Now assume that the load connected to the secondary has an impedance of 2400 ohms. Find the amount of current flow in the primary and secondary windings. The current flow in the secondary winding can be calculated using Ohm's law:

$$I = \frac{E}{Z}$$

$$I = \frac{600\text{ V}}{2400\ \Omega}$$

$$I = 0.25\text{ A}$$

Now that the amount of current flow in the secondary is known, the primary current can be calculated using the formula:

$$\frac{E_P}{E_S} = \frac{I_S}{I_P}$$

$$\frac{120\text{ V}}{600\text{ V}} = \frac{0.25\text{ A}}{I_P}$$

$$120\text{ V } I_P = 150\text{ VA}$$

$$I_P = 1.25\text{ A}$$

Notice that the amount of power input equals the amount of power output:

Primary	Secondary
120 V × 1.25 A = 150 VA	600 V × 0.25 A = 150 VA

Calculating Isolation Transformer Values Using the Primary-to-Secondary Voltage Ratio

As illustrated in the previous examples, transformer values of voltage, current, and turns can be calculated using formulas, such as Ohm's law, but it is also possible to calculate these same values using the transformer primary-to-secondary voltage ratio, commonly referred to as the "turns ratio." Assume a transformer has a primary rated at 240 volts and a secondary rated at 96 volts (*Figure 27–9*). The primary-to-secondary voltage ratio can be calculated by dividing the primary voltage by the secondary voltage:

$$\text{Ratio} = \frac{240\text{ V}}{96\text{ V}}$$

$$\text{Ratio} = 2.5{:}1$$

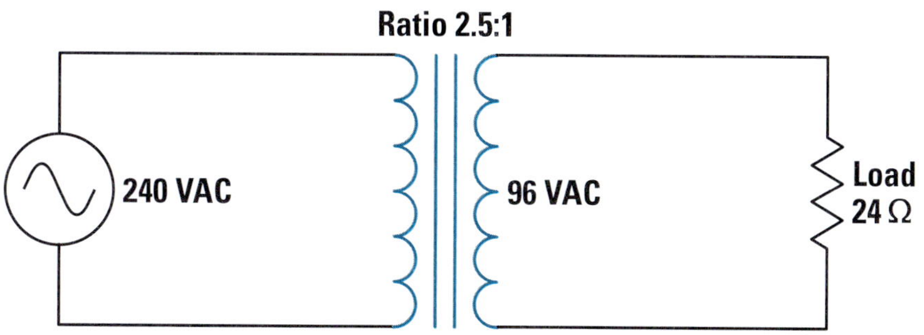

Figure 27–9 Calculating the transformer primary-to-secondary voltage ratio to determine primary and secondary values.

This primary-to-secondary voltage ratio indicates that there are 2.5 volts in the primary winding for every 1 volt in the secondary.

Now assume that a resistance of 24 ohms is connected to the secondary winding. The amount of secondary current can be found using Ohm's law:

$$I_S = \frac{96}{24}$$

$$I_S = 4\text{ A}$$

However, we cannot determine the primary current using Ohm's law because there is no impedance value on the primary side. The primary current must now be found using the ratio. Recall that the volt-amperes of the primary must equal the volt-amperes of the secondary. Because the primary voltage is greater, the primary current will have to be less than the secondary current:

$$I_P = \frac{I_S}{\text{turns ratio}}$$

$$I_P = \frac{4\text{ A}}{2.5}$$

$$I_P = 1.6\text{ A}$$

To check the answer, find the volt-amperes of the primary and secondary:

Primary	Secondary
240 V × 1.6 A = 384 VA	96 V × 4 A = 384 VA

Now let's assume an isolation transformer has a primary voltage of 120 volts and a secondary voltage of 500 volts. The secondary has a load impedance of 1200 ohms (*Figure 27–10*).

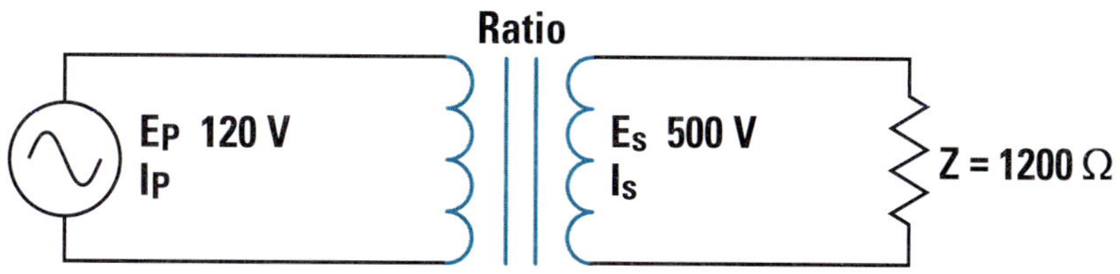

Figure 27–10 Calculating transformer values using the transformer voltage or turns ratio.

The ratio for this transformer can be found by dividing the primary voltage by the secondary voltage:

$$\text{Ratio} = \frac{120\text{ V}}{500\text{ V}}$$

$$\text{Ratio} = 1{:}4.17$$

The secondary current can be found using Ohm's law:

$$I_S = \frac{500\text{ V}}{1200\ \Omega}$$

$$I_S = 0.417\text{ A}$$

In this example, the primary voltage is lower than the secondary voltage. Therefore, the primary current must be higher:

$$I_P = I_S \times \text{turns ratio}$$

$$I_P = 0.417\text{ A} \times 4.17$$

$$I_P = 1.739\text{ A}$$

To check this answer, calculate the volt-amperes of both windings:

Primary	Secondary
120 V × 1.739 A = 208.68 VA	500 V × 0.417 A = 208.5 VA

The slight difference in answers is caused by rounding off values.

Figure 27–11 shows the transformer with all calculated values.

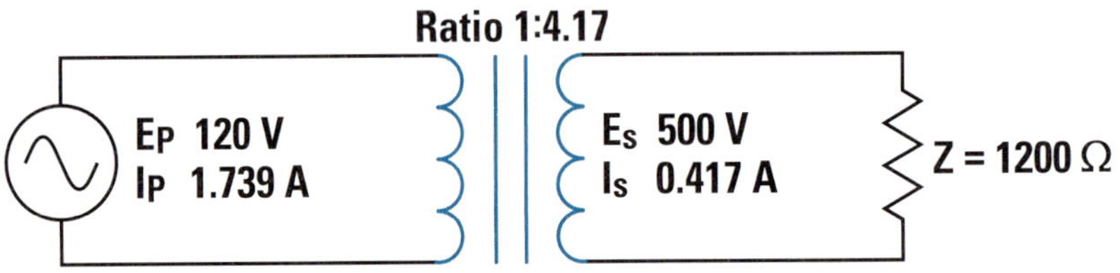

Figure 27–11 Voltage and current values show that power in equals power out.

Multiple-Tapped Windings

It is not uncommon for isolation transformers to be designed with windings that have more than one set of termination points for both the primary and secondary windings. These are called *multiple-tapped windings*. The transformer shown in *Figure 27–12* contains a secondary winding rated at 24 volts. The primary winding contains several taps, however. One of the primary lead wires is labeled C and is the common for the other leads. The other leads are labeled 120 volts, 208 volts, and 240 volts. This transformer is designed in such a manner that it can be connected to different primary voltages without changing the value of the secondary voltage. In this example, it is assumed that the secondary winding has a total of 120 turns of wire. To maintain the proper turns ratio, the primary would have 600 turns of wire between C and 120 volts, 1040 turns between C and 208 volts, and 1200 turns between C and 240 volts.

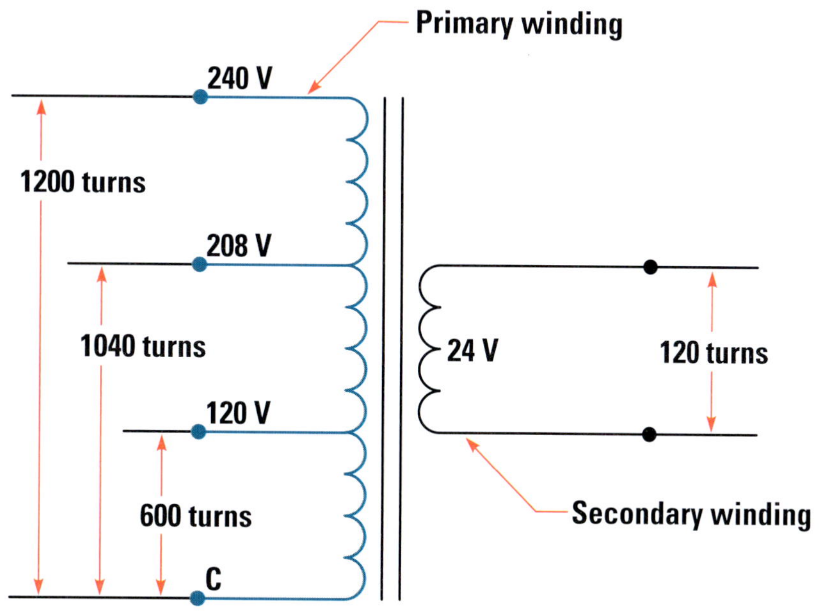

Figure 27–12 Transformer with multiple-tapped primary winding.

A transformer can also be constructed to provide several different output voltages. The isolation transformer shown in *Figure 27–13* contains a single primary winding. The secondary winding, however, has been tapped at several points. One of the secondary terminations is marked as C and is common to the other termination points. When rated voltage is applied to the primary, voltages of 12 volts, 24 volts, and 48 volts can be obtained at the secondary.

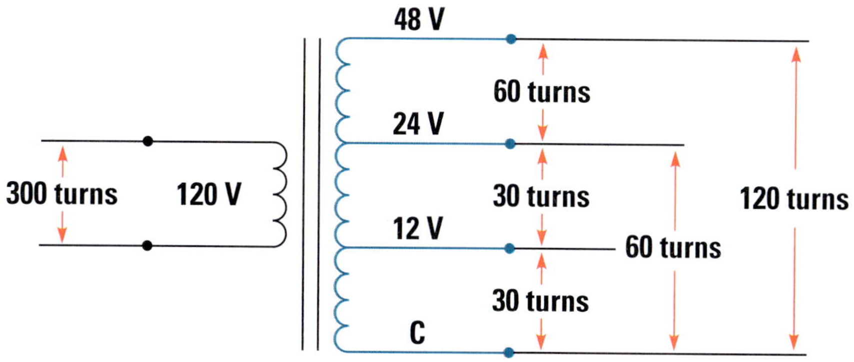

Figure 27–13 Transformer secondary with multiple taps.

The isolation transformer shown in *Figure 27–14* is similar to the transformer in *Figure 27–13*. The transformer in *Figure 27–14,* however, has multiple secondary windings instead of a single secondary winding with multiple taps. The advantage of the transformer in *Figure 27–14* is that the secondary windings are electrically isolated from each other. These secondary windings can be either step-up or step-down depending on the application of the transformer.

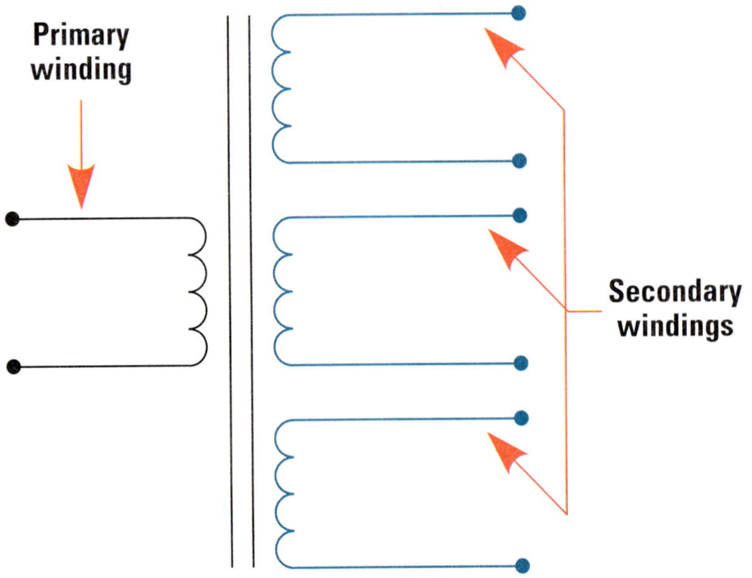

Figure 27–14 Transformer with multiple isolated secondary windings.

Calculating Values for Isolation Transformers with Multiple Secondaries

When calculating the values of an isolation transformer with multiple secondary windings, each secondary must be treated as a different transformer. For example, the transformer in *Figure 27–15* contains one primary winding and three secondary windings. The primary is connected to 120 VAC and contains 300 turns of wire. One secondary has an output voltage of 560 volts and a load impedance of 1000 ohms, the second secondary has an output voltage of 208 volts and a load impedance of 400 ohms, and the third secondary has an output voltage of 24 volts and a load impedance of 6 ohms. The current, turns of wire, and ratio for each secondary and the current of the primary will be found.

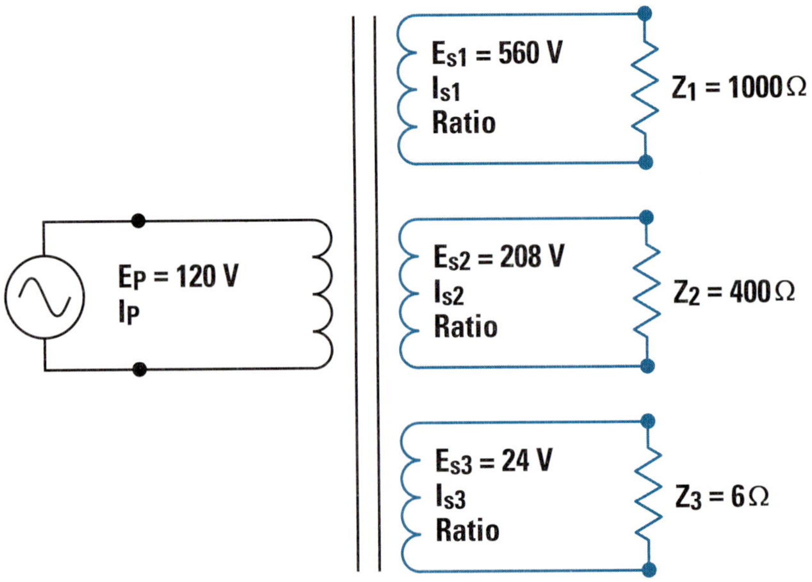

Figure 27–15 Calculating values for a transformer with multiple secondary windings.

The first step is to calculate the primary-to-secondary voltage ratio of the first secondary. The ratio can be found by dividing the primary voltage by the secondary voltage of the first winding:

$$\text{Ratio} = \frac{E_P}{E_{S1}}$$

$$\text{Ratio} = \frac{120\ \text{V}}{560\ \text{V}}$$

$$\text{Ratio} = 1{:}4.67$$

The current flow in the first secondary can be calculated using Ohm's law:

$$I_{S1} = \frac{560\ \text{V}}{1000\ \Omega}$$

$$I_{S1} = 0.56\ \text{A}$$

The amount of primary current needed to supply this secondary winding can also be found using the primary-to-secondary voltage ratio. When using this ratio to determine values on the secondary side of a transformer, remember to multiply primary values by the ratio as determined by dividing the primary voltage by the secondary voltage. If the goal is to determine primary values from secondary values, multiply the secondary values by the inverse of the ratio. For example, finding the primary current value requires multiplying the secondary current by a ratio of 4.67 to 1 instead of the 1 to 4.67 ratio determined previously:

$$I_{P(\text{FIRST SECONDARY})} = I_{S1} \times \text{voltage ratio}$$

$$I_{P(\text{FIRST SECONDARY})} = 0.56\ \text{A} \times 4.67$$

$$I_{P(\text{FIRST SECONDARY})} = 2.61\ \text{A}$$

The primary-to-secondary ratio of the second secondary winding is found by dividing the primary voltage by the secondary voltage:

$$\text{Ratio} = \frac{120\ \text{V}}{208\ \text{V}}$$

$$\text{Ratio} = 1{:}1.73$$

The amount of current flow in this secondary can be determined using Ohm's law:

$$I_{S2} = \frac{208\ \text{V}}{400\ \Omega}$$

$$I_{S2} = 0.52\ \text{A}$$

The voltage of the primary is less than this secondary. The primary, therefore, requires a greater amount of current to achieve the same amount of power, and since power in equals power out, we can find the unknown value since we know the secondary voltage and current and the primary voltage. The amount of current required to operate this secondary is calculated using the turns ratio:

$$I_{P(\text{SECOND SECONDARY})} = I_{S2} \times \text{turns ratio}$$

$$I_{P(\text{SECOND SECONDARY})} = 0.52\ \text{A} \times 1.732$$

$$I_{P(\text{SECOND SECONDARY})} = 0.9\ \text{A}$$

The turns ratio of the third secondary winding is calculated in the same way as the other two:

$$\text{Ratio} = \frac{120\ \text{V}}{24\ \text{V}}$$

$$\text{Ratio} = 5{:}1$$

The secondary current is found using Ohm's law:

$$I_{S3} = \frac{24\ \text{V}}{6\ \Omega}$$

$$I_{S3} = 4\ \text{A}$$

The primary has a higher voltage than this secondary. The primary current is, therefore, going to be less than the current on the secondary. Again, since we are using secondary values to find a primary value, we will use the inverse of the primary-to-secondary voltage ratio:

$$I_{P(\text{THIRD SECONDARY})} = I_{S3} \times \frac{1}{\text{turns ratio}}$$

$$I_{P(\text{THIRD SECONDARY})} = 4\text{A} \times \frac{1}{5}$$

$$I_{P(\text{THIRD SECONDARY})} = 0.8\ \text{A}$$

The primary must supply current to each of the three secondary windings. Therefore, the total amount of primary current is the sum of the currents required to supply each secondary:

$$I_{P(\text{TOTAL})} = I_{P1} + I_{P2} + I_{P3}$$

$$I_{P(\text{TOTAL})} = 2.61\ \text{A} + 0.9\ \text{A} + 0.8\ \text{A}$$

$$I_{P(\text{TOTAL})} = 4.31\ \text{A}$$

The transformer with all calculated values is shown in *Figure 27–16*.

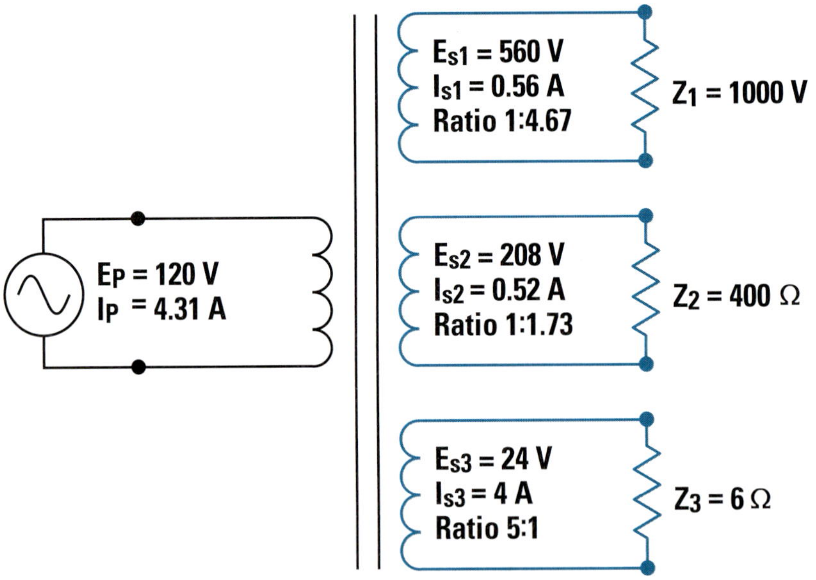

Figure 27–16 The transformer with all calculated values.

Distribution Transformers

A common type of isolation transformer is the **distribution transformer** (*Figure 27–17*). This type of transformer changes the high voltage of power company distribution lines to the utilization voltage used to supply power to the end user, such as the 240 V/120 V single-phase systems common in residential occupancies. In the *Figure 27–17* example, the primary is connected to a 7200-volt line, and the secondary produces 240 volts across the lines with a center tap at the midpoint. The center tap produces a voltage of 120 volts between itself and either of the line conductors. The center tap point is grounded and becomes the point where the **neutral conductor** will connect. If voltage is measured across the entire secondary, a voltage of 240 volts is seen. If voltage is measured from either line to the center tap, half of the secondary voltage, or 120 volts, is seen (*Figure 27–18*). If a vector diagram is drawn to illustrate this condition, you will see that the grounded neutral conductor divides the secondary winding to produce two voltages that are in phase with each other (*Figure 27–19*). Loads that are intended to operate on 240 volts, such as water heaters, electric-resistance heating units, and central air conditioners are connected directly across the lines of the secondary (*Figure 27–20*).

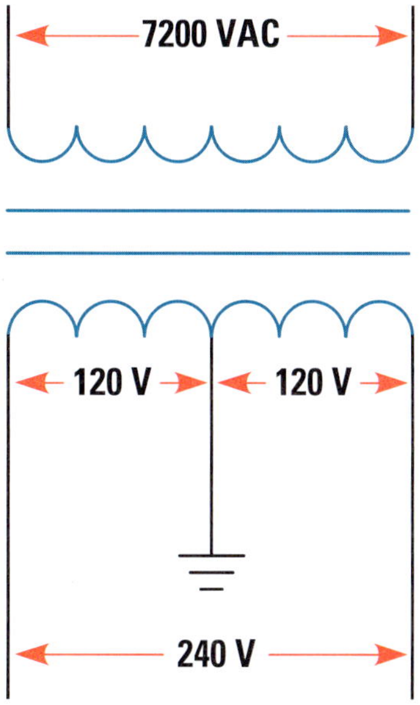

Figure 27–17 Distribution transformer.

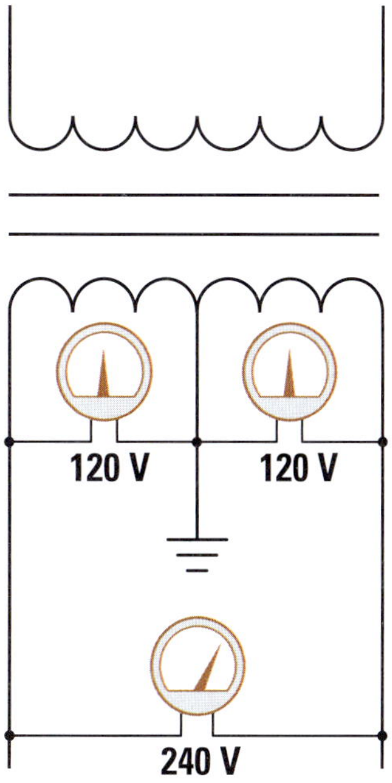

Figure 27–18 The voltage from either line to center point is 120 volts. The voltage across the entire secondary winding is 240 volts.

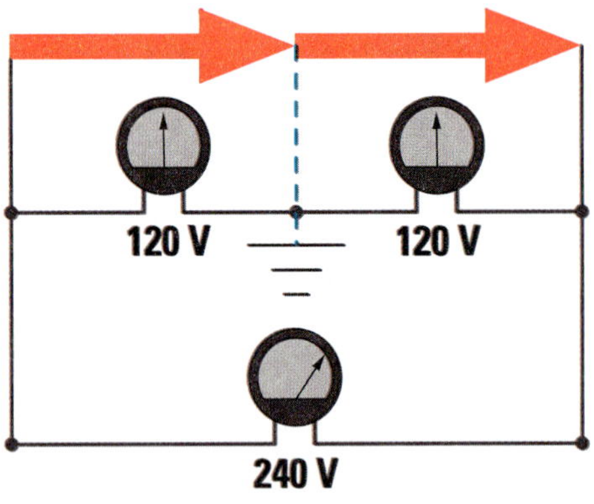

Figure 27–19 The voltages across the secondary are in phase with each other.

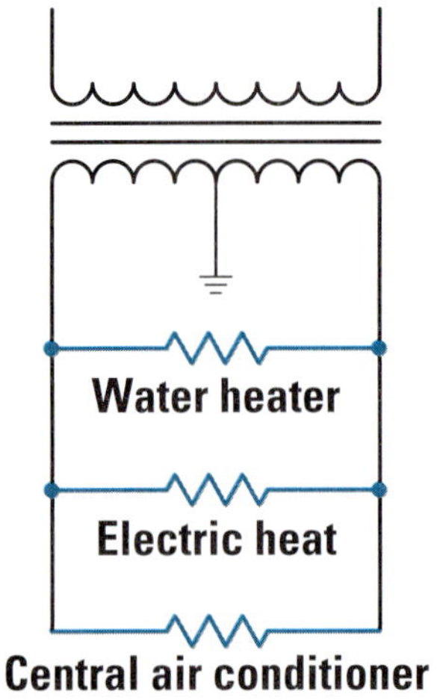

Figure 27–20 Loads of 240 volts connect directly across the secondary winding.

Loads that are intended to operate on 120 volts connect from the center tap, or neutral, to one of the secondary lines. The function of the neutral is to carry the difference in current between the two secondary lines and maintain a balanced voltage. In *Figure 27–21,* one of the secondary lines has a current flow of 30 amperes, and the other has a current flow of 24 amperes. The neutral conducts the sum of the unbalanced load. In this example, the neutral current is 6 amperes (30A − 24A = 6A).

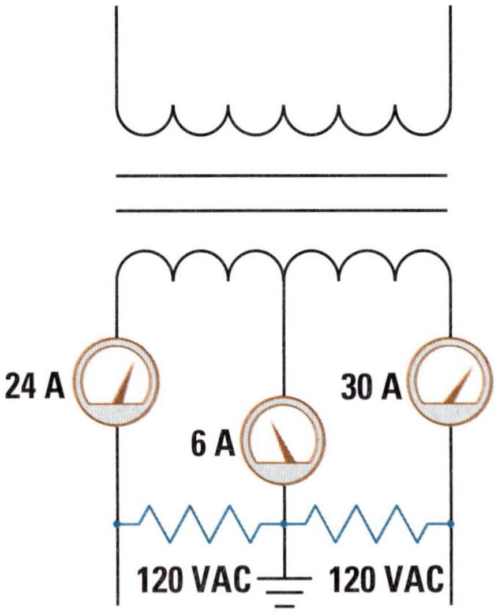

Figure 27–21 The neutral carries the sum of the unbalanced load.

Control Transformers

Another common type of isolation transformer found throughout industry is the **control transformer** (*Figure 27–22*). The control transformer is used to reduce the line voltage to the value needed to operate control circuits. A common type of control transformer contains two primary windings and one secondary. The primary windings are generally rated at 240 volts each, and the secondary is rated at 120 volts. This arrangement provides a 2:1 primary-to-secondary voltage ratio.

Figure 27–22 Control transformer with fuse protection added to the secondary winding.

One of the primary windings in *Figure 27–23* is labeled H_1 and H_2. The other is labeled H_3 and H_4. The secondary winding is labeled X_1 and X_2. If the primary of the transformer is to be connected to 240 volts, the two primary windings must be connected in parallel by connecting H_1 and H_3 together and H_2 and H_4 together. When the primary windings are connected in parallel, the same voltage is applied across both windings. This has the same effect as using one primary winding with a voltage of 240 volts and maintains the primary-to-secondary voltage ratio of 2:1 and a secondary voltage of 120 volts.

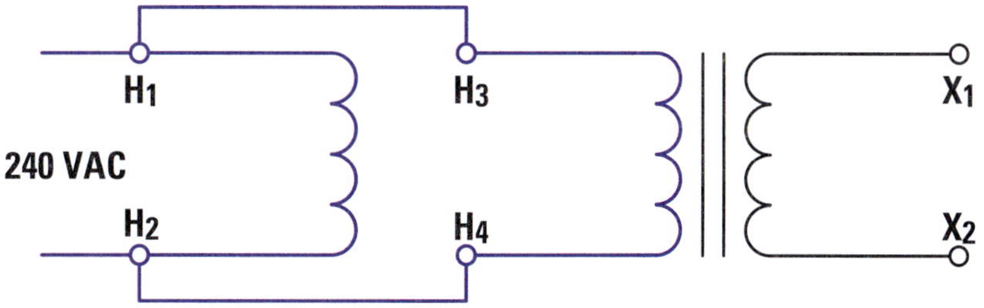

Figure 27–23 Control transformer connected for 240-volt operation.

If the transformer is to be connected to 480 volts, the two primary windings are connected in series by connecting H_2 and H_3 together (*Figure 27–24*). The incoming power is connected to H_1 and H_4. Connecting the primary windings in series in effect doubles the number of turns on the primary side and produces a turns ratio of 4:1. Therefore, when 480 volts are connected to the primary, the secondary voltage will remain at 120 volts.

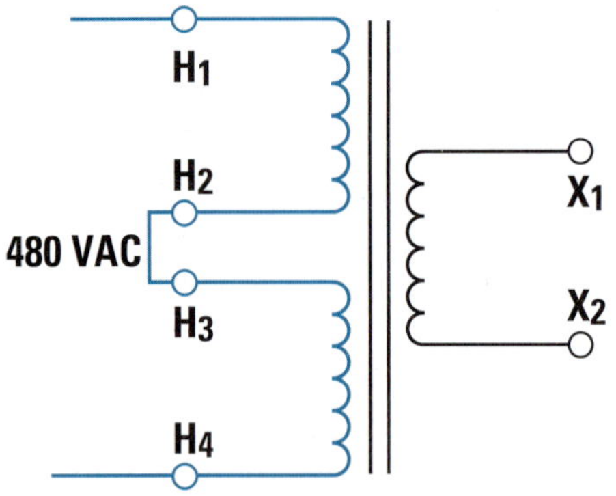

Figure 27–24 Control transformer connected for 480-volt operation.

The primary leads of a control transformer are generally cross-connected (*Figure 27–25*). This is done so metal links can be used to connect the primary for 240- or 480-volt operation. If the primary is to be connected for 240-volt operation, the metal links will be connected under screws (*Figure 27–26*). Notice that leads H_1 and H_3 are connected together, and leads H_2 and H_4 are connected together. Compare this connection with the connection shown in *Figure 27–23*.

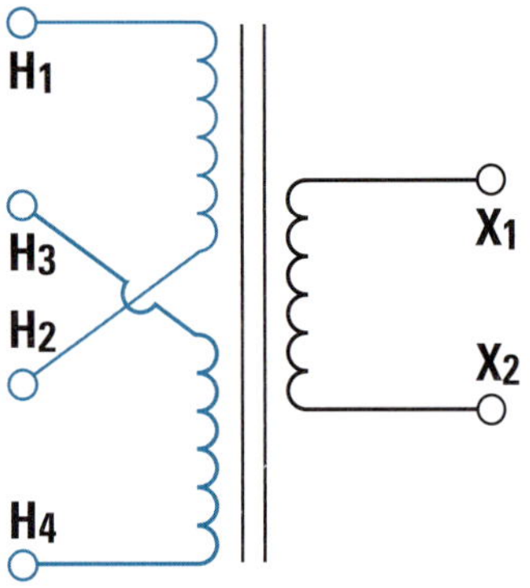

Figure 27–25 The primary windings of a control transformer are crossed.

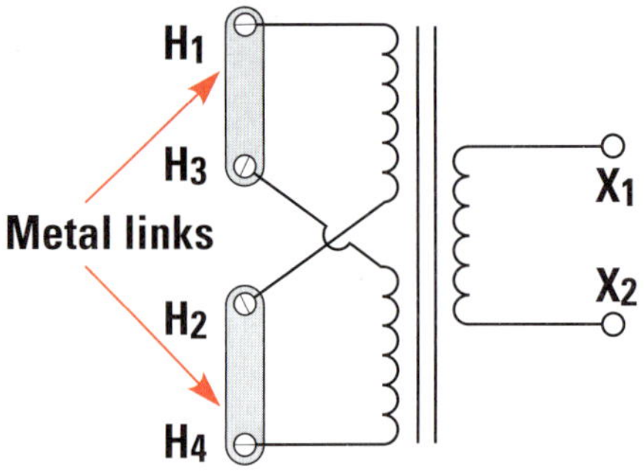

Figure 27–26 Metal links connect transformer for 240-volt operation.

If the transformer is to be connected for 480-volt operation, terminals H_2 and H_3 are connected as shown in *Figure 27–27*. Compare this connection with the connection shown in *Figure 27–24*.

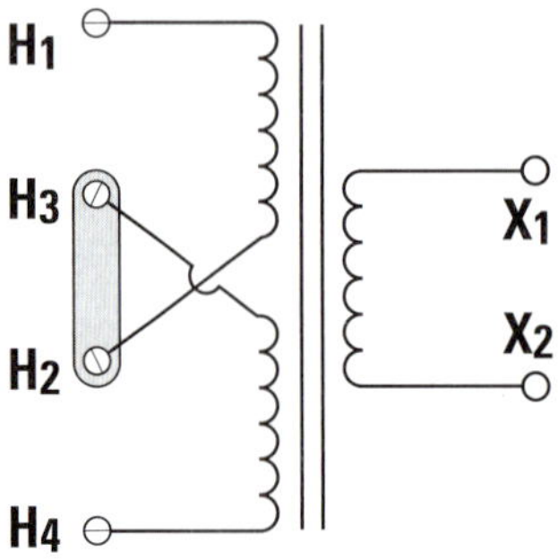

Figure 27–27 Control transformer connected for 480-volt operation.

Transformer Core Types

Several types of cores are used in the construction of transformers. Most transformer cores are made from thin steel sheets **laminated** together to form a solid metal core. The core for a 630-mega-volt-ampere (MVA) three-phase transformer is shown in *Figure 27–28*. Laminated cores are preferred because a thin layer of oxide forms on the surface of each lamination and acts as an insulator to reduce the formation of eddy currents inside the core material. The amount of core material needed for a particular transformer is determined by the power rating of the transformer. The amount of core material must be sufficient to prevent saturation at full load. The type and shape of the core generally determine the amount of magnetic coupling between the windings and, to some extent, the efficiency of the transformer.

Figure 27–28 Core of a 630-MVA three-phase transformer.
Courtesy of SPX Transformer Solutions, Inc.

The transformer illustrated in *Figure 27–29* is known as a core-type transformer. The windings are placed around each end of the core material. As a general rule, the low-voltage winding is placed closest to the core, and the high-voltage winding is placed over the low-voltage winding.

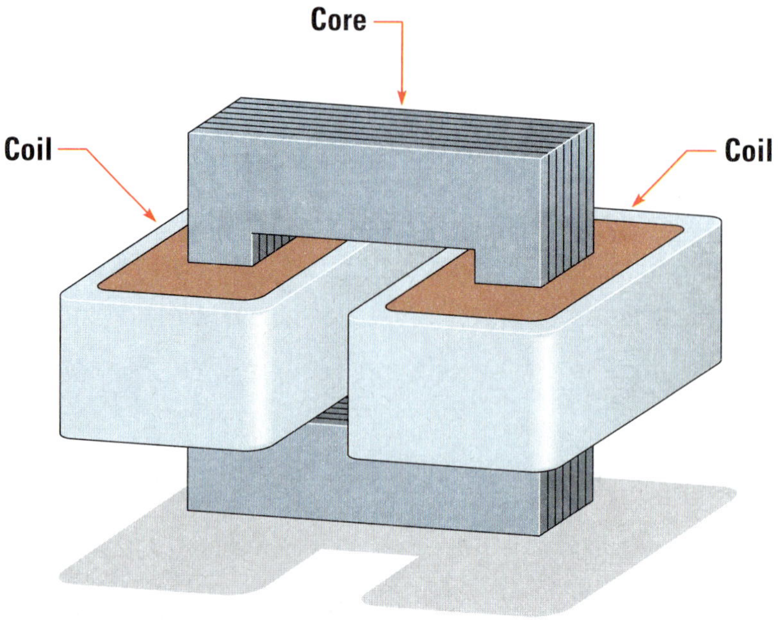

Figure 27–29 A core-type transformer.

The shell-type transformer is constructed in a similar manner to the core type, except that the shell type has a metal core piece through the middle of the window (*Figure 27–30*). The primary and secondary windings are wound around the center core piece with the low-voltage winding being closest to the metal core. This arrangement permits the transformer to be surrounded by the core and provides excellent magnetic coupling. When the transformer is in operation, all the magnetic flux must pass through the center core piece. It then divides through the two outer core pieces.

Figure 27–30 A shell-type transformer.

The H-type core shown in *Figure 27–31* is similar to the shell-type core in that it has an iron core through its center around which the primary and secondary windings are wound. The H core, however, surrounds the windings on four sides instead of two. This extra metal helps reduce stray leakage flux and improves the efficiency of the transformer. The H-type core is often found on high-voltage distribution transformers.

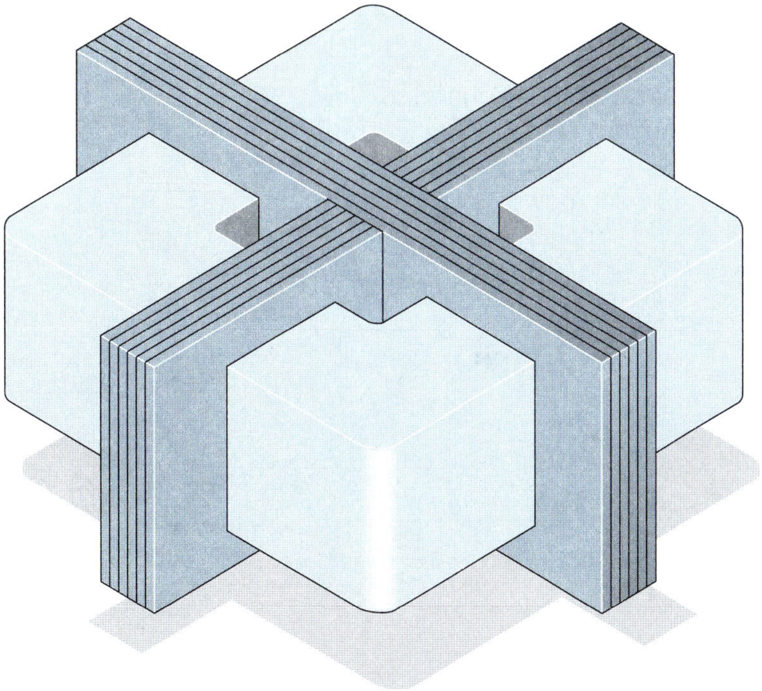

Figure 27–31 A transformer with an H-type core.

The **tape-wound core** or **toroid core** (*Figure 27–32*) is constructed by tightly winding one long continuous silicon steel tape into a spiral. The tape may or may not be housed in a plastic container, depending on the application. This type of core does not require laminating steel together. Because the core is one continuous length of metal, **flux leakage** is kept to a minimum. Flux leakage is the number of magnetic flux lines that do not follow the metal core and are lost to the surrounding air. The tape-wound core is one of the most efficient core designs available.

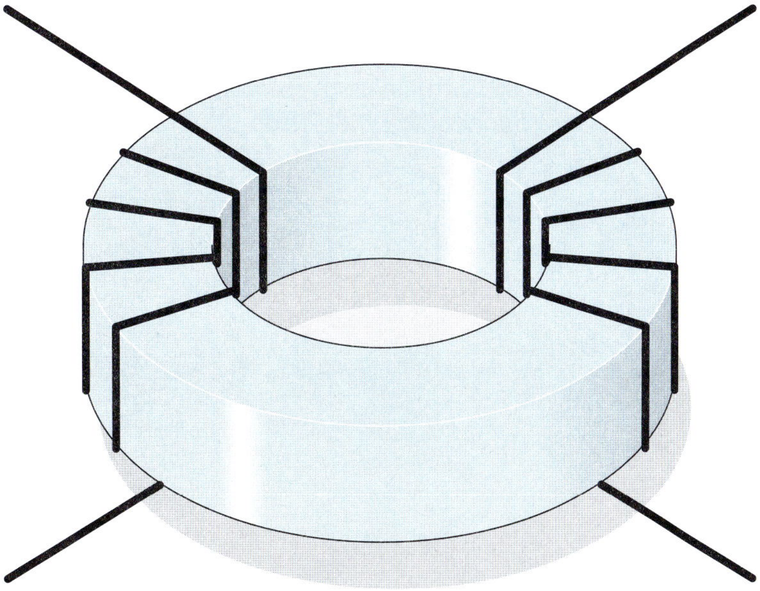

Figure 27–32 A toroid transformer.

Transformer Inrush Current

A reactor is an inductor used to add inductance to the circuit. Although transformers and reactors are both inductive devices, there is a great difference in their operating characteristics. Reactors are often connected in series with a low-impedance load to prevent **inrush current** (the amount of current that flows when power is initially applied to the circuit) from becoming excessive (*Figure 27–33*). Transformers, however, can produce extremely high inrush currents when power is first applied to the primary winding. The type of core used when constructing inductors and transformers is primarily responsible for this difference in characteristics.

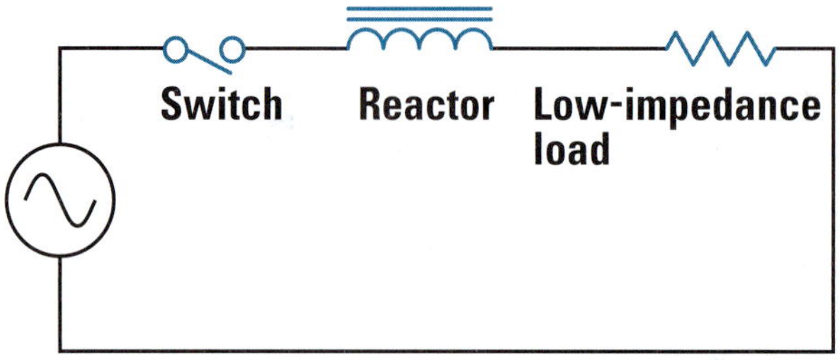

Figure 27–33 Reactors use low-impedance loads to help prevent inrush current from becoming excessive when power is first turned on.

Magnetic Domains

Magnetic materials contain tiny magnetic structures in their molecular material known as *magnetic domains* (see Unit 4). These domains can be affected by outside sources of magnetism. *Figure 27–34* illustrates a magnetic domain that has not been polarized by an outside magnetic source.

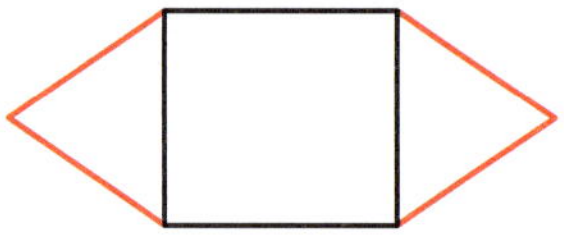

Figure 27–34 Magnetic domain in neutral position.

Now assume that the north pole of a magnet is placed toward the top of the material that contains the magnetic domains (*Figure 27–35*). Notice that the structure of the domain has changed to realign the molecules in the direction of the outside magnetic field. If the polarity of the magnetic pole is changed (*Figure 27–36*), the molecular structure of the domain changes to realign itself with the new magnetic lines of flux. This external influence can be produced by an electromagnet as well as a permanent magnet.

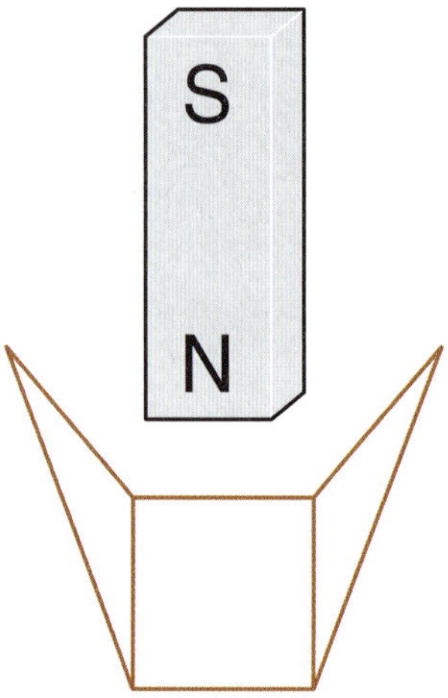

Figure 27–35 Domain influenced by a north magnetic field.

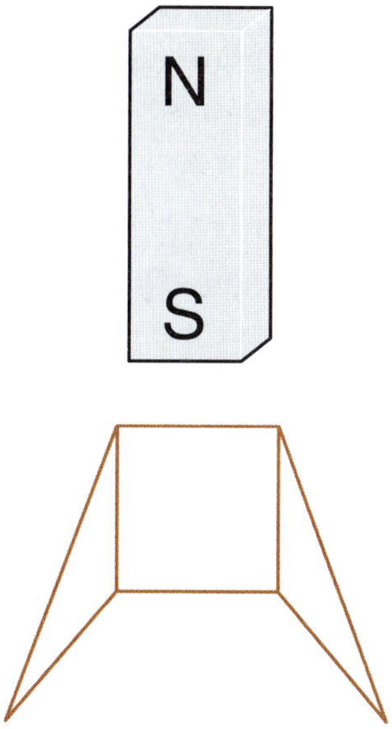

Figure 27–36 Domain influenced by a south magnetic field.

In certain types of cores, the molecular structure of the domain snaps back to its neutral position when the magnetizing force is removed. This type of core is used in the construction of reactors or chokes (*Figure 27–37*). A core of this type is constructed by separating sections of the steel laminations with an air gap. This air gap breaks the magnetic path through the core material and is responsible for the domains returning to their neutral position once the magnetizing force is removed.

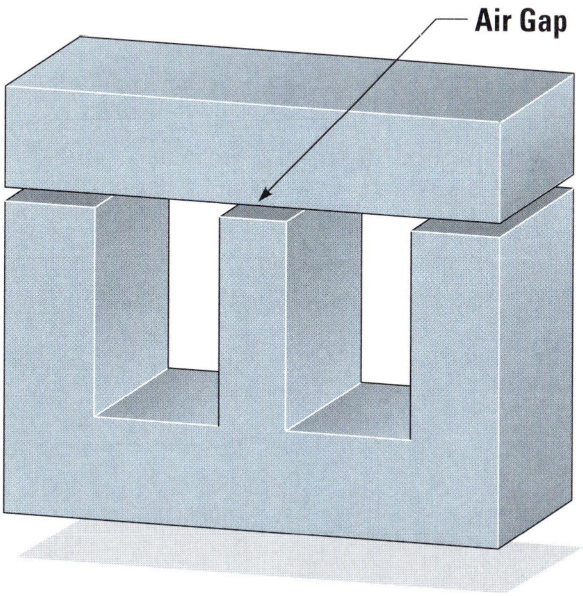

Figure 27–37 The core of an inductor contains an air gap.

The core construction of a transformer, however, does not contain an air gap. The steel laminations are connected together in such a manner as to produce a very low reluctance path for the magnetic lines of flux. In a transformer core, the domains remain in their set position once the magnetizing force has been removed. This type of core "remembers" where it was last set. This is one of the reasons why transformers can have extremely high inrush currents when they are first connected to the power line.

After all, until the inductive reactance builds up, the winding of a transformer is basically a direct connection between supply lines. In other words, a short circuit.

The amount of inrush current in the primary of a transformer is limited by three factors:

1. the amount of applied voltage,
2. the resistance of the wire in the primary winding, and
3. the flux change of the magnetic field in the core. The amount of flux change determines the amount of inductive reactance produced in the primary winding when power is applied.

Figure 27–38 illustrates a simple isolation-type transformer. The AC applied to the primary winding produces a magnetic field around the winding. As the current changes in magnitude and direction, the magnetic lines of flux change also. Because the lines of flux in the core are continually changing polarity, the magnetic domains in the core material are changing also. As stated previously, the magnetic domains in the core of a transformer remember their last set position. For this reason, the point on the waveform at which current is disconnected from the primary winding can have a great bearing on the amount of inrush current when the transformer is reconnected to power. For example, assume the power supplying the primary winding is disconnected at the zero-crossing point (*Figure 27–39*). In this instance, the magnetic domains would be set at the neutral point. When power is restored to the primary winding, the core material can be magnetized by either magnetic polarity. This permits a change of flux, which is the dominant current-limiting factor. In this instance, the amount of inrush current would be relatively low because the inductive reactance can build quickly.

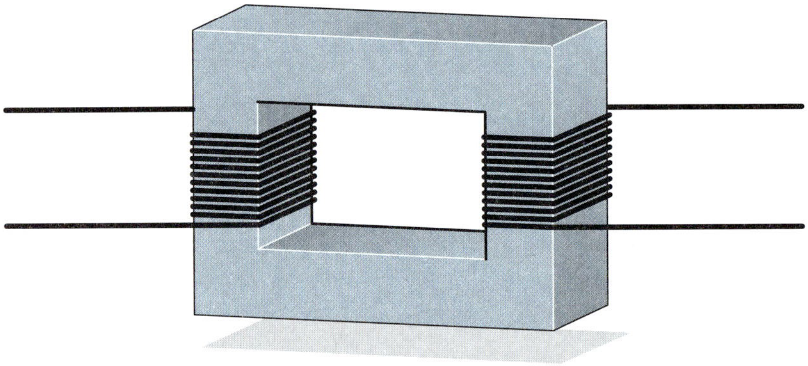

Figure 27–38 Isolation transformer.

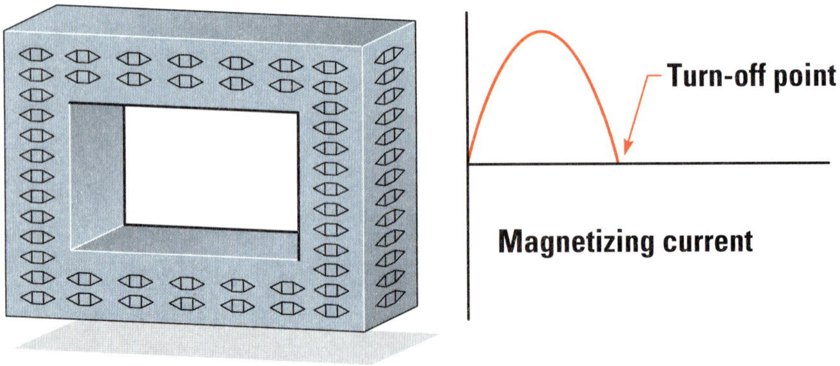

Figure 27–39 Magnetic domains are left in the neutral position when power is disconnected at the zero-crossing point.

If the power supplying current to the primary winding is interrupted at the peak point of the positive or negative half cycle, however, the domains in the core material will be set at that position. *Figure 27–40* illustrates this condition. It is assumed that the current was stopped as it reached its peak positive point. If the power is reconnected to the primary winding during the positive half cycle, only a very small amount of flux change can take place. Because the core material is saturated in the positive direction, the primary winding of the transformer is essentially an air-core inductor, which greatly decreases the inductive characteristics of the winding. The inrush current in this situation would be limited by the resistance of the winding and a very small amount of inductive reactance.

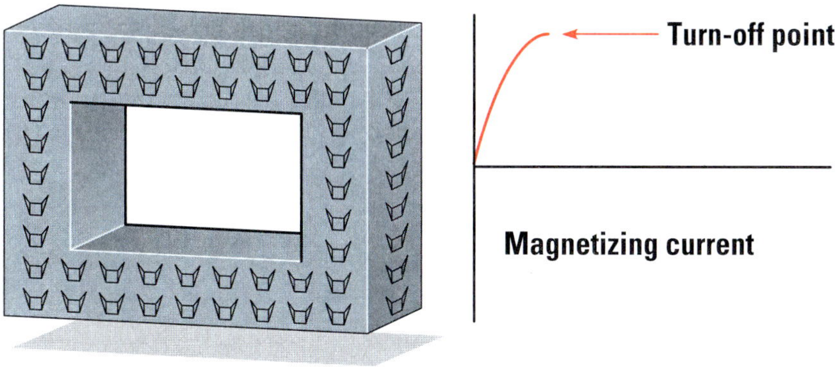

Figure 27–40 Domains are set at one end of magnetic polarity.

27–3 Autotransformers

Autotransformers are one-winding transformers. They use the same winding for both the primary and secondary. The primary winding in *Figure 27–41* is between points B and N and has a voltage of 120 volts applied to it. If the turns of wire are counted between points B and N, it can be seen that there are 120 turns of wire. Now assume that the selector switch is set to point D. The load is now connected

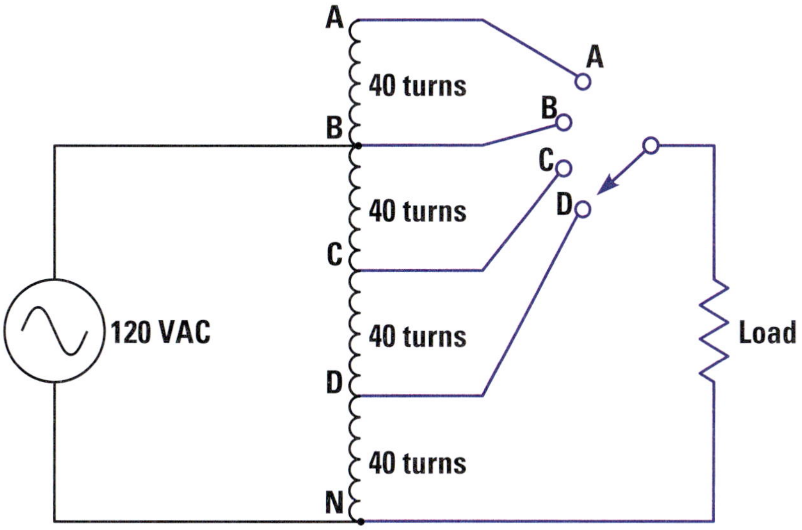

Figure 27–41 An autotransformer has only one winding used for both the primary and secondary.

between points D and N. The secondary of this transformer contains 40 turns of wire. If the amount of voltage applied to the load is to be calculated, the following formula can be used:

$$\frac{E_P}{E_S} = \frac{N_P}{N_S}$$

$$\frac{120\ \text{V}}{E_S} = \frac{120\ \text{turns}}{40\ \text{turns}}$$

$$120\ \text{turns}\ E_S = 4800\ \text{V-turns}$$

$$E_S = 40\ \text{V}$$

Assume that the load connected to the secondary has an impedance of 10 ohms. The amount of current flow in the secondary circuit can be calculated using the formula:

$$I = \frac{E}{Z}$$

$$I = \frac{40\ \text{V}}{10\ \Omega}$$

$$I = 4\ \text{A}$$

The primary current can be calculated by using the same formula that was used to calculate primary current for an isolation type of transformer:

$$\frac{E_P}{E_S} = \frac{I_S}{I_P}$$

$$\frac{120\ \text{V}}{40\ \text{V}} = \frac{4\ \text{A}}{I_P}$$

$$120\ \text{V}\ I_P = 160\ \text{VA}$$

$$I_P = 1.333\ \text{A}$$

The amount of power input and output for the autotransformer must be the same, just as they are in an isolation transformer:

Primary	**Secondary**
$120\ \text{V} \times 1.333\ \text{A} = 160\ \text{VA}$	$40\ \text{V} \times 4\ \text{A} = 160\ \text{VA}$

Now assume that the rotary switch is connected to point A. The load is now connected to 160 turns of wire. The voltage applied to the load can be calculated by:

$$\frac{E_P}{E_S} = \frac{N_P}{N_S}$$

$$\frac{120\ \text{V}}{E_S} = \frac{120\ \text{turns}}{160\ \text{turns}}$$

$$120\ \text{turns}\ E_S = 19{,}200\ \text{V-turns}$$

$$E_S = 160\ \text{V}$$

Notice that the autotransformer, like the isolation transformer, can be either a step-up or step-down transformer.

If the rotary switch shown in *Figure 27–41* were to be removed and replaced with a sliding tap that made contact directly to the transformer winding, the turns ratio could be adjusted continuously. This type of transformer is commonly referred to as a Variac or Powerstat depending on the manufacturer. A cutaway view of a variable autotransformer is shown in *Figure 27–42*. The windings are wrapped around a tape-wound toroid core inside a plastic case. The tops of the windings have been milled flat to provide a commutator. A carbon brush makes contact with the windings.

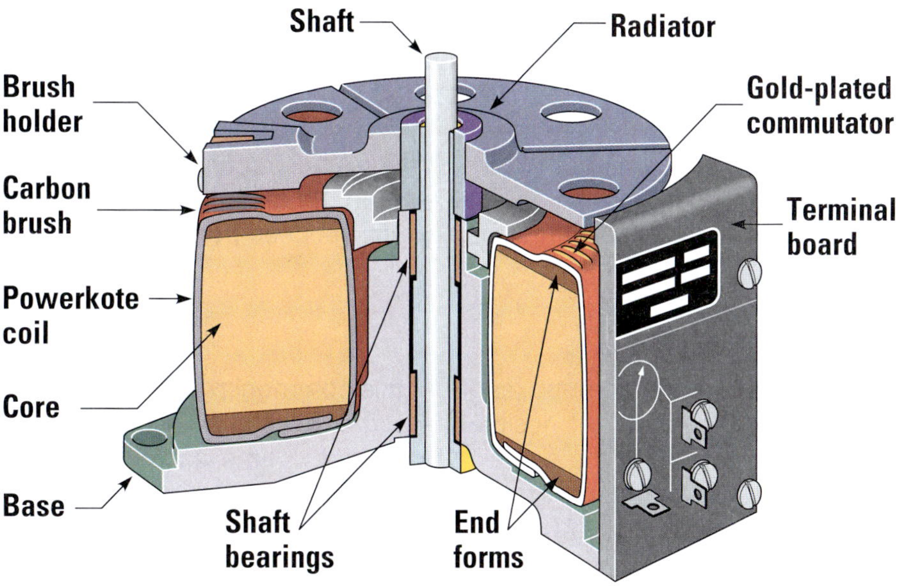

Figure 27–42 Cutaway view of a Powerstat.

Autotransformers are often used by power companies to provide a small increase or decrease to the line voltage. They help provide voltage regulation to large power lines. A three-phase autotransformer is shown in *Figure 27–43*. This transformer is contained in a housing filled with transformer oil, which acts as a coolant and prevents moisture from forming in the windings.

Figure 27–43 Three-phase autotransformer.

The autotransformer does have one disadvantage. Because the load is connected to one side of the power line, there is no line isolation between the incoming power and the load. This can cause problems with certain types of equipment and must be a consideration when designing a power system.

27–4 Voltage and Current Relationships in a Transformer

In a 60-hertz AC circuit, the voltage changes polarity 60 times per second. When discussing transformer polarity, it is necessary to consider the relationship between the different windings at the same point in time. It is therefore assumed that this point in time is when the peak positive voltage is being produced across the winding.

Polarity Markings on Schematics

When a transformer is shown on a schematic diagram, it is common practice to indicate the polarity of the transformer windings by placing a dot beside one end of each winding as shown in *Figure 27–44*. These dots signify that the polarity is the same at that point in time for each winding. For example, assume the voltage applied to the primary winding is at its peak positive value at the terminal indicated by the dot. The voltage at the dotted lead of the secondary will be at its peak positive value at the same time.

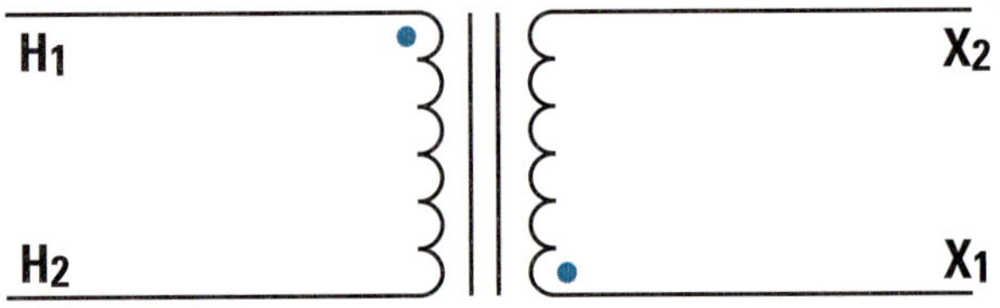

Figure 27–44 Transformer polarity dots.

This same type of polarity notation is used for transformers that have more than one primary or secondary winding. An example of a transformer with a multisecondary is shown in *Figure 27–45*.

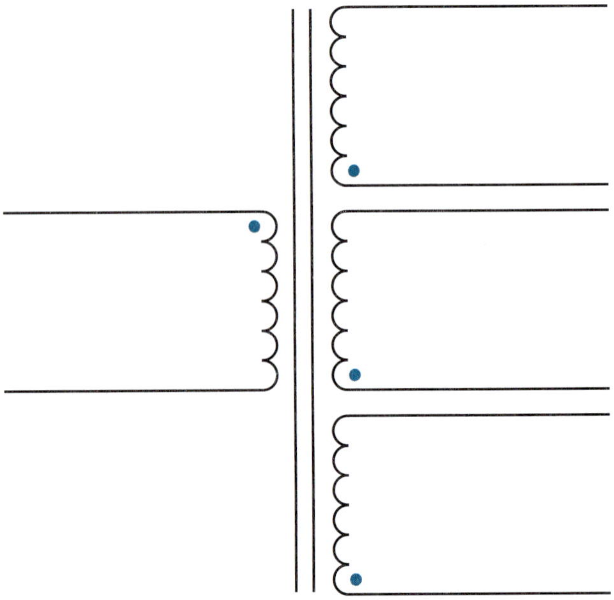

Figure 27–45 Polarity marks for multiple secondaries.

Additive and Subtractive Polarities

The polarity of transformer windings can be determined by connecting them as an autotransformer and testing for additive or subtractive polarity, often referred to as a buck-boost connection. This is done by connecting one lead of the secondary to one lead of the primary and measuring the voltage across both windings (*Figure 27–46*). The transformer shown in the example has a primary voltage rating of 120 volts and a secondary voltage rating of 24 volts. This same circuit has been redrawn in *Figure 27–47* to show the connection more clearly. Notice that the secondary winding has been connected in series with the primary winding. The transformer now contains only one winding and is, therefore, an autotransformer. When 120 volts are applied to the primary winding, the voltmeter connected across the secondary indicates either the *sum* of the two voltages or the *difference* between the two voltages. If this voltmeter indicates 144 volts (120 V + 24 V = 144 V), the windings are connected additive (boost), and polarity dots can be placed as shown in *Figure 27–48*. Notice in this connection that the secondary voltage is added to the primary voltage.

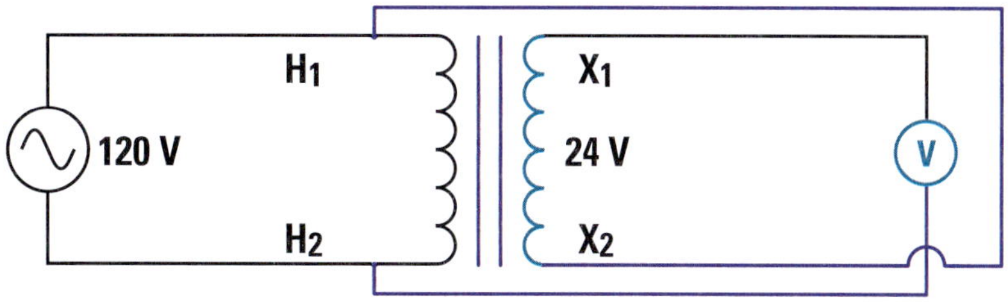

Figure 27–46 Connecting the secondary and primary windings forms an autotransformer.

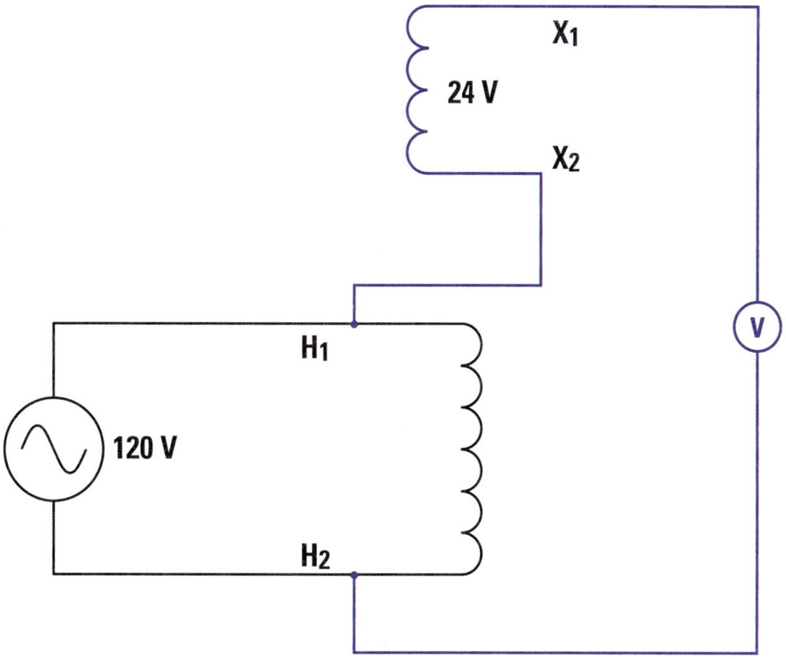

Figure 27–47 Redrawing the connection.

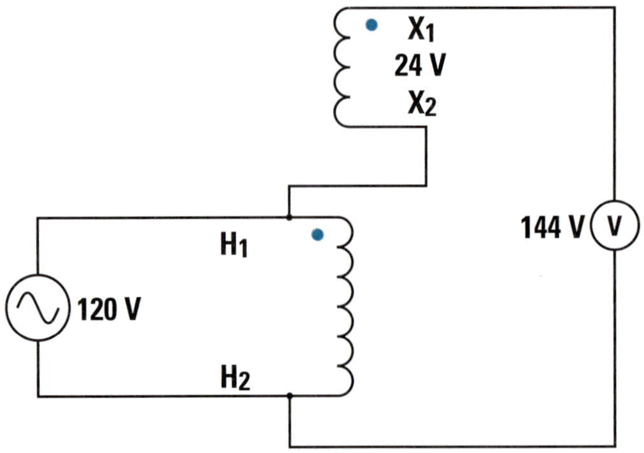

Figure 27–48 Polarity dots indicate an additive or "boost" polarity.

If the voltmeter connected to the secondary winding indicates a voltage of 96 volts (120 V − 24 V = 96 V), the windings are connected subtractive (buck), and polarity dots are placed as shown in *Figure 27–49*.

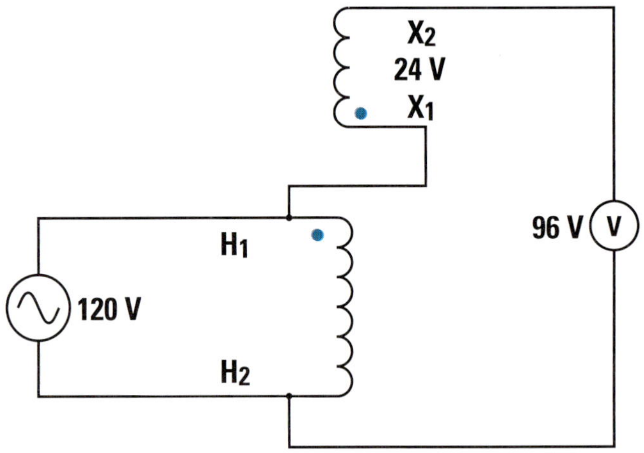

Figure 27–49 Polarity dots indicate a subtractive or "buck" polarity.

Using Arrows to Place Dots

To help in the understanding of additive and subtractive polarity, arrows can be used to indicate a direction of polarity in the windings. In *Figure 27–50,* arrows have been added to indicate the polarity of the winding. In this example, the transformer is connected additive, or boost, and both arrows point in the same direction. Notice that the arrow points to the dot. In *Figure 27–51,* it is seen that the values of the two arrows add to produce 144 volts.

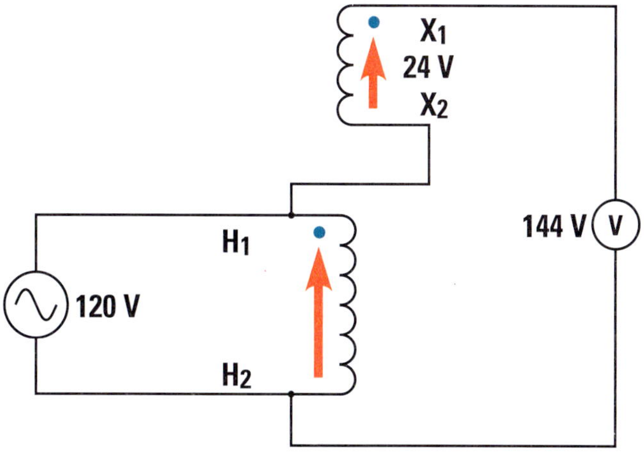

Figure 27–50 Arrows help indicate the polarity of individual windings.

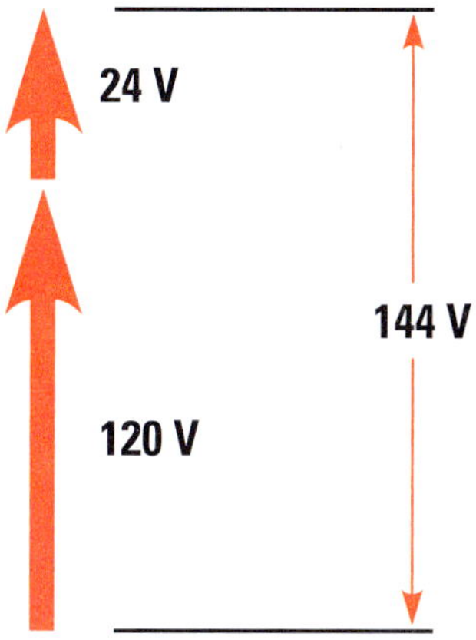

Figure 27–51 The values of the arrows add to indicate additive polarity (boost connection).

In *Figure 27–52,* the secondary winding is connected in the opposite polarity. In this instance, the arrows point in opposite directions, and the voltage of one winding opposes the voltage of the other. The result is that the smaller value "bucks" the higher voltage and reduces the output by being in the opposite polarity, as shown in *Figure 27–53*.

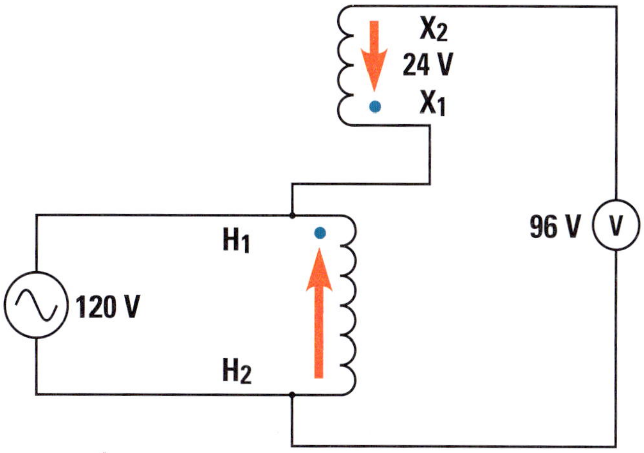

Figure 27–52 The arrows help indicate subtractive polarity.

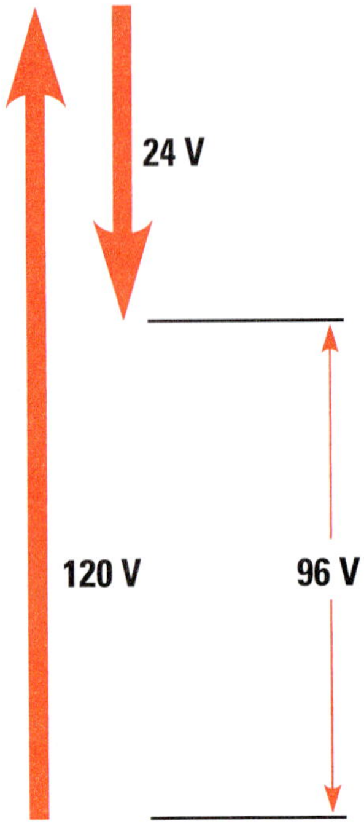

Figure 27–53 The values of the arrows subtract (buck connection).

When the primary of a transformer is connected to power but there is no load connected to the secondary, current is limited only by the inductive reactance of the primary. At this time, the transformer is essentially an inductor, and the excitation current lags the applied voltage by 90° (*Figure 27–54*). The primary current induces a voltage in the secondary. This induced voltage is proportional to the rate of change of current. The secondary voltage is at its maximum during the periods that the primary current is changing the most (0°, 180°, and 360°), and it will be zero when the primary current is not changing (90° and 270°). A plot of the primary current and secondary voltage shows that the secondary voltage lags the primary current by 90° (*Figure 27–55*). Because the secondary voltage lags the primary current by 90° and the applied voltage leads the primary current by 90°, the secondary voltage is 180° out of phase with the applied voltage and in phase with the induced voltage in the primary.

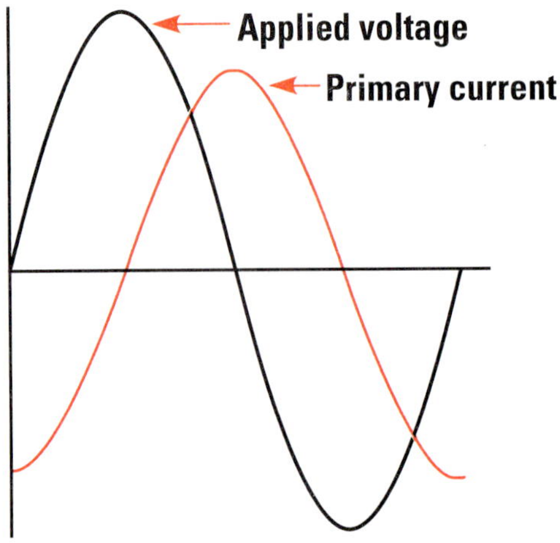

Figure 27–54 At no load, the primary current lags the voltage by 90°.

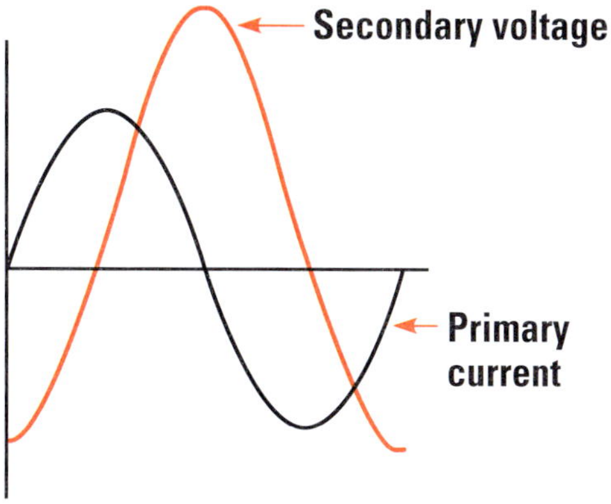

Figure 27–55 The secondary voltage lags the primary current by 90°.

Adding Load to the Secondary

When a load is connected to the secondary, current begins to flow in the secondary windings. Because the transformer is an inductive device, the secondary current lags the secondary voltage by 90°. Because the secondary voltage lags the primary current by 90°, the secondary current is 180° out of phase with the primary current (*Figure 27–56*).

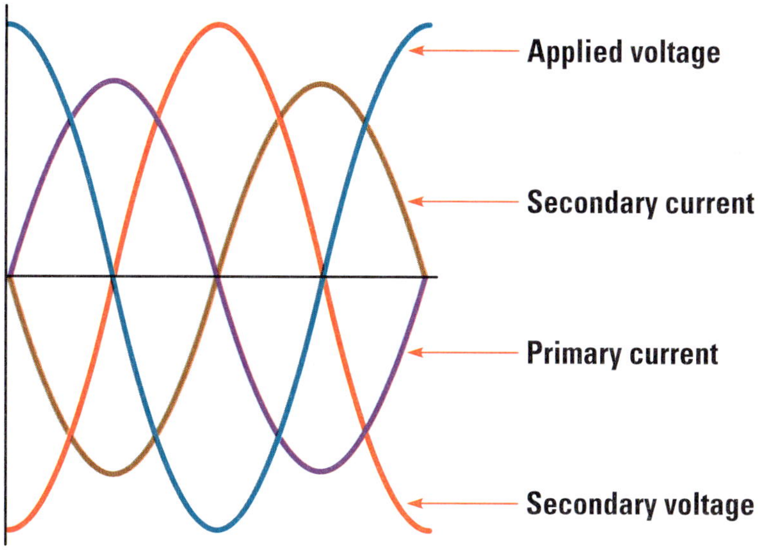

Figure 27–56 Voltage and current relationships of the primary and secondary windings.

The current of the secondary induces a countervoltage in the secondary windings that is in opposition to the countervoltage induced in the primary. The countervoltage of the secondary weakens the countervoltage of the primary and permits more primary current to flow. As secondary current increases, primary current increases proportionally.

Because the secondary current causes a decrease in the countervoltage produced in the primary, the current of the primary is limited less by inductive reactance and more by the resistance of the windings as load is added to the secondary. If a wattmeter were connected to the primary, you would see that the true power would increase as load was added to the secondary.

Testing the Transformer

Several tests can be performed to determine the condition of the transformer. A simple test for grounds, shorts, or opens can be made with an ohmmeter (*Figure 27–57*). Ohmmeter A is connected to one lead of the primary and one lead of the secondary. This test checks for shorted windings between the primary and secondary. The ohmmeter should indicate infinity for transformers with no direct connection, such as isolation transformers. If there is more than one primary or secondary winding, all isolated windings should be tested for shorts. Ohmmeter B illustrates testing the windings for grounds. One lead of the ohmmeter is connected to the case of the transformer, and the other is connected to the winding. All windings should be tested for grounds, and the ohmmeter should indicate infinity for each winding. Ohmmeter C illustrates testing the windings for continuity. The wire resistance of the winding should be indicated by the ohmmeter.

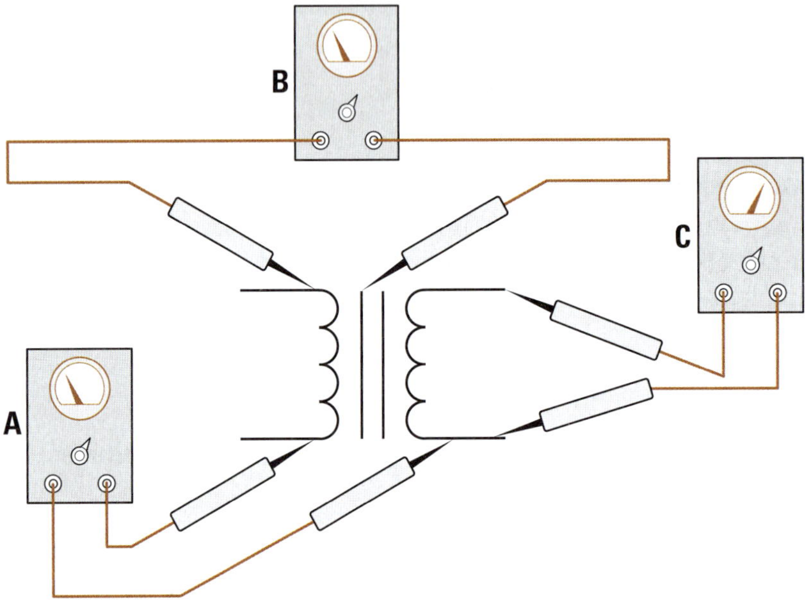

Figure 27–57 Testing a transformer with an ohmmeter.

If the transformer appears to be in good condition after these initial ohmmeter tests, it should then be tested for shorts and grounds with a megohmmeter. A MEGGER will reveal problems of insulation breakdown that an ohmmeter will not. Large oil-filled transformers should have the condition of the dielectric oil tested periodically. This test involves taking a sample of the oil and performing certain tests for dielectric strength and contamination. These tests should be incorporated into a facility's electrical equipment maintenance program and should occur at specified intervals in the program.

27–5 Transformer Nameplates

Transformers covered by the scope of *NFPA 70®: National Electrical Code®* must contain a nameplate that lists information concerning the transformer. *NEC® 450.11* requires the following information:

1. Name of manufacturer
2. Rated kilovolt-ampere
3. Frequency
4. Primary and secondary voltage
5. Impedance of transformers rated 25 kVA and larger
6. Required clearances of transformers with ventilating openings

7. Amount and kind of insulating liquid where used
8. The temperature class for the insulating system of dry-type transformers

Notice that the transformer is rated in kilovolt-amperes, not kilowatts, because the true power output of the transformer is determined by the power factor of the load. Other information that may be listed is temperature rise in °C, model number, and whether the transformer is single-phase or three-phase. Many nameplates also contain a schematic diagram of the windings to aid in connection. An example of a transformer nameplate is shown in *Figure 27–58*.

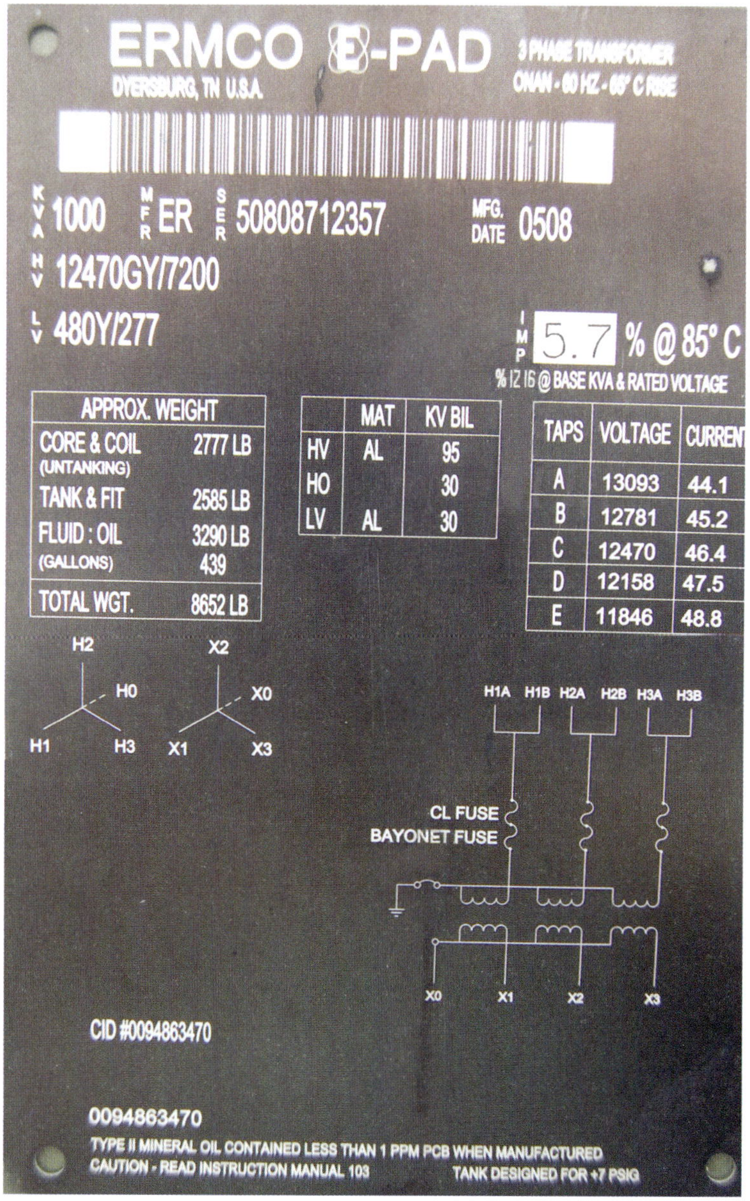

Figure 27–58 An example of a nameplate for a 1000 kVA transformer.

Determining Maximum Current

The nameplate does not list the current rating of the windings. Because power input must equal power output, the current rating for a winding can be determined by dividing the kilovolt-ampere rating by the winding voltage. For example, assume a transformer has a kilovolt-ampere rating of 0.5 kilovolt-ampere, a primary voltage of 480 volts, and a secondary voltage of 120 volts. To determine the maximum current that can be supplied by the secondary, divide the VA rating by the secondary voltage:

$$I_S = \frac{VA}{E_S}$$

$$I_S = \frac{500\ VA}{120\ V}$$

$$I_S = 4.167\ A$$

The primary current can be calculated in the same way:

$$I_P = \frac{VA}{E_P}$$

$$I_P = \frac{500\ VA}{480\ V}$$

$$I_P = 1.042\ A$$

Transformers with multiple secondary windings will generally have the current rating listed with the voltage rating.

Transformer Impedance

Transformer impedance is determined by the physical construction of the transformer. Factors such as the amount and type of core material, the wire size used to construct the windings, the number of turns, and the degree of magnetic coupling between the windings greatly affect the transformer's impedance. Impedance is expressed as a percent (%Z or %IZ) and is measured by connecting a short circuit across the low-voltage winding of the transformer and then connecting a variable-voltage source to the high-voltage winding (*Figure 27–59*). The variable voltage is then increased until rated current flows in the low-voltage winding. The transformer impedance is determined by calculating the percentage of variable voltage compared to the rated voltage of the high-voltage winding.

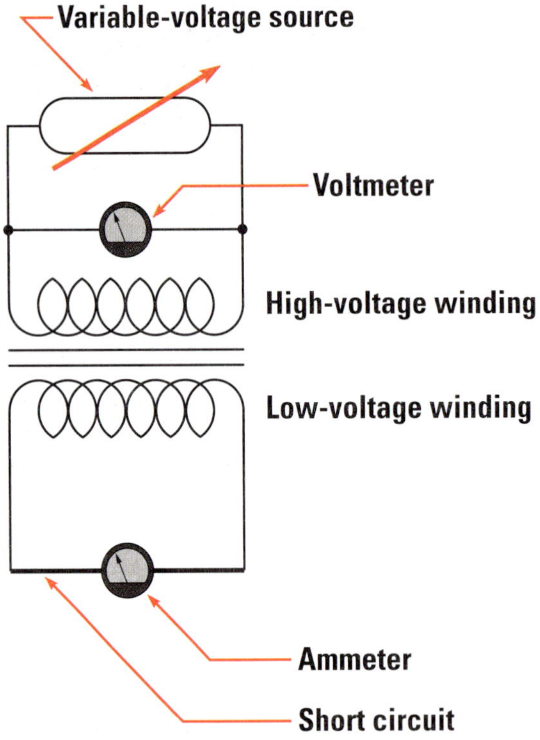

Figure 27–59 Determining transformer impedance.

Example 27–1

Assume that the transformer shown in *Figure 27–59* is a 2400/480-V, 15-kVA transformer. To determine the impedance of the transformer, first calculate the full-load current rating of the secondary winding:

$$I = \frac{VA}{E}$$

$$I = \frac{15{,}000\ VA}{480\ V}$$

$$I = 31.25\ A$$

Next, increase the source voltage connected to the high-voltage winding until a current of 31.25 A flows in the low-voltage winding. For the purpose of this example, assume that the voltage value is 138 V. Finally, determine the percentage of applied voltage compared to the rated voltage:

$$\%Z = \frac{\text{source voltage}}{\text{rated voltage}} \times 100$$

$$\%Z = \frac{138\ V}{2400\ V} \times 100$$

$$\%Z = 0.0575 \times 100$$

$$\%Z = 5.75$$

The impedance of this transformer is 5.75%.

Transformer impedance is a major factor in determining the amount of voltage drop a transformer will exhibit between no load and full load and in determining the amount of current flow in a short-circuit condition. Short-circuit current can be calculated using the formula:

$$\text{(Single phase) } I_{SC} = \frac{VA}{E \times \%Z}$$

$$\text{(Three phase) } I_{SC} = \frac{VA}{E \times \sqrt{3} \times \%Z}$$

because one of the formulas for determining current in a single-phase circuit is:

$$I = \frac{VA}{E}$$

and one of the formulas for determining current in a three-phase circuit is:

$$I = \frac{VA}{E \times \sqrt{3}}$$

The preceding formulas for determining short-circuit current can be modified to show that the short-circuit current can be calculated by dividing the rated secondary current by the %Z:

$$I_{SC} = \frac{I_{Rated}}{\%Z}$$

Example 27–2

A single-phase transformer is rated at 50 kVA and has a secondary voltage of 240 V. The nameplate reveals that the transformer has an internal impedance (%Z) of 2.5%. What is the short-circuit current for this transformer?

$$I_{secondary} = \frac{50{,}000\ VA}{240\ V}$$

$$I_{secondary} = 208.3\ A$$

$$I_{short\ circuit} = \frac{208.3\ A}{\%Z}$$

$$I_{short\ circuit} = \frac{208.3\ A}{0.025}$$

$$I_{short\ circuit} = 8{,}333.3\ A$$

It is sometimes necessary to calculate the amount of short-circuit current when determining the correct fuse rating for a circuit. The fuse must have a high enough "interrupt" rating to clear the fault in the event of a short circuit.

27–6 Connecting Transformer Secondaries

As a general rule, connecting the secondary windings of transformers in series does not present a problem. Because the current is the same in a series circuit, the problem of high circulating current does not exist in a series connection. The secondary windings can be connected in series to produce a higher output voltage. Assume that two transformers have a secondary voltage of 120 volts (*Figure 27–60*).

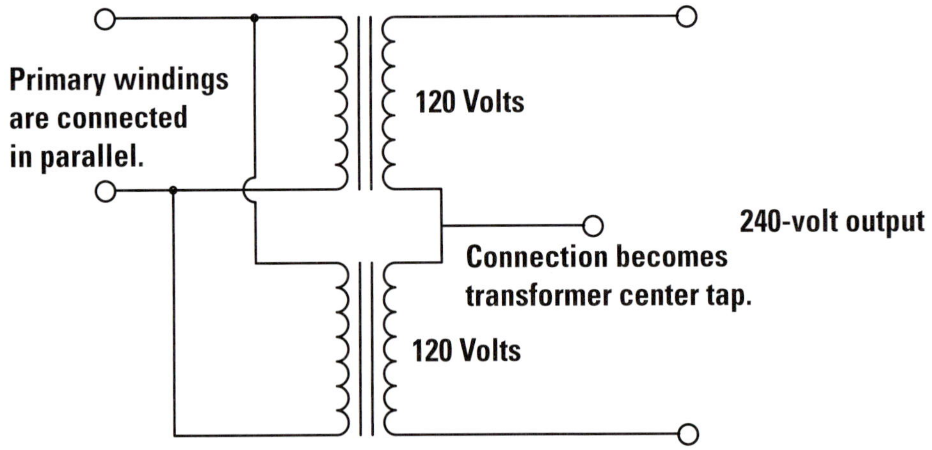

Figure 27–60 Transformer secondary windings connected in series.

The primary windings can be connected in parallel without a problem. When making this connection, the polarity of the two secondary windings must be connected in an additive polarity. The series connection of the two secondary windings will produce an output of 240 volts and can be center tapped. If the polarity is not correct, the output voltage will be zero (0).

Parallel Transformer Connections

It is sometimes necessary to connect the secondary windings of transformers in parallel to increase the current capacity, but generally, it is not done unless there is no other alternative (*Figure 27–61*). Connecting transformer primary windings in parallel is not a problem, but connecting the secondary windings in parallel can cause high circulating currents or extremely unbalanced currents that can lead to transformer failure. Transformers that are to be connected in parallel must have the following:

- the same kVA rating,
- the same turns ratio,
- the same impedance,
- the same secondary voltage, and
- the secondary windings must have the same polarity.

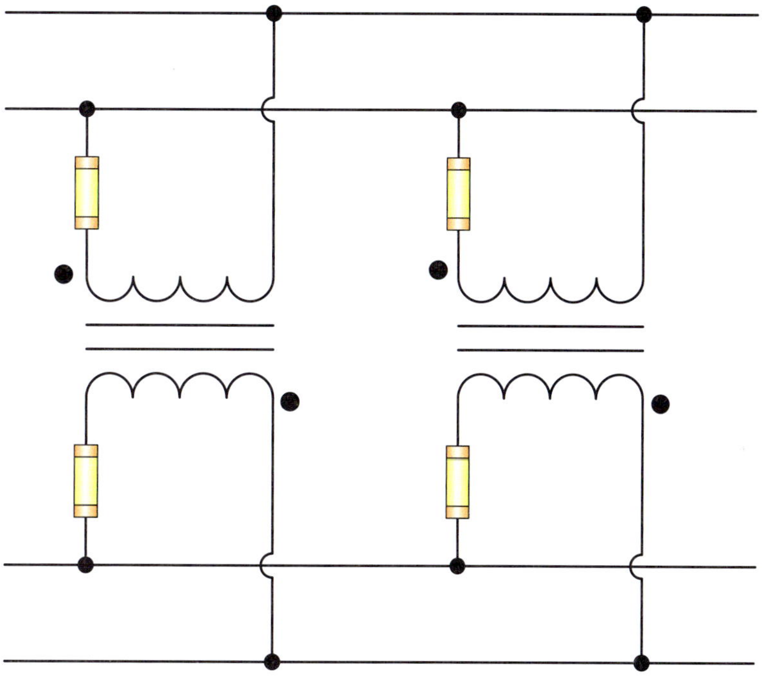

Figure 27–61 Transformers are sometimes connected in parallel to increase the current capacity of the circuit.

If any of these factors are different, it can cause the failure of one or both transformers. Assume, for example, that two transformers have the same kVA rating, same secondary voltage, and same turns ratio, but the impedance is not the same. When load is added to the parallel connection, the transformer with the higher internal impedance will exhibit a greater voltage drop, causing the other transformer to supply more of the load current. This unbalance can lead to the failure of the transformer that is supplying the majority of the load current.

In another example, assume that two transformers have the same kVA rating, same secondary voltage, and same impedance, but the turns ratio is different. When power is applied to the connection, the difference in turns ratio can cause a very high no-load circulating current. This circulating current can cause burn-out of both transformers. When load is added, the secondary current of each transformer will be a combination of both the load current and circulating current.

Connecting Parallel Transformers

Care should be exercised when paralleling transformers to ensure that the polarity is correct. If the polarity is incorrect, it produces a short circuit. A simple test can be conducted to check for the proper polarity of the windings being connected in parallel. In *Figure 27-62*, we have set up an example of the polarity test. Assume that one transformer is already connected to a line with a 4160-volt supply. Also assume that the secondary voltage is to be 120 volts. Connect one of the secondary leads to the line, and then energize the primary winding (*Figure 27–62*).

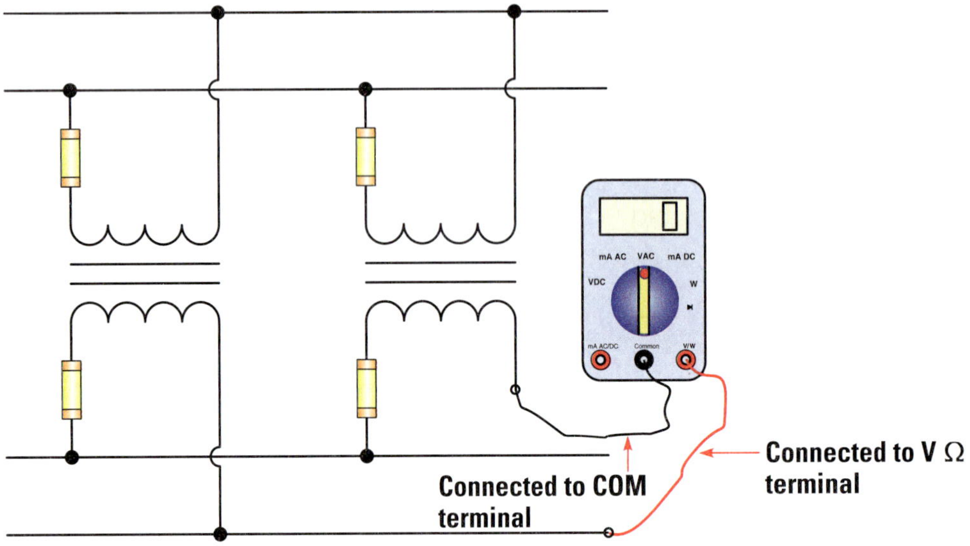

Figure 27–62 Checking the polarity of a parallel transformer connection.

Connect a voltmeter between the secondary lead that has not been connected and the line of the intended connection. If the polarity of the two transformers is correct, the voltmeter should indicate zero (0) volts. If the polarity is not correct, the voltmeter will indicate double the amount of secondary voltage. In this example, the voltmeter would indicate 240 volts if the polarity were not correct.

Precautions When Servicing Parallel Transformers

If it should become necessary to remove one of the transformers for service, the secondary should be disconnected first. This can generally be accomplished by removing the secondary fuse (*Figure 27–63*).

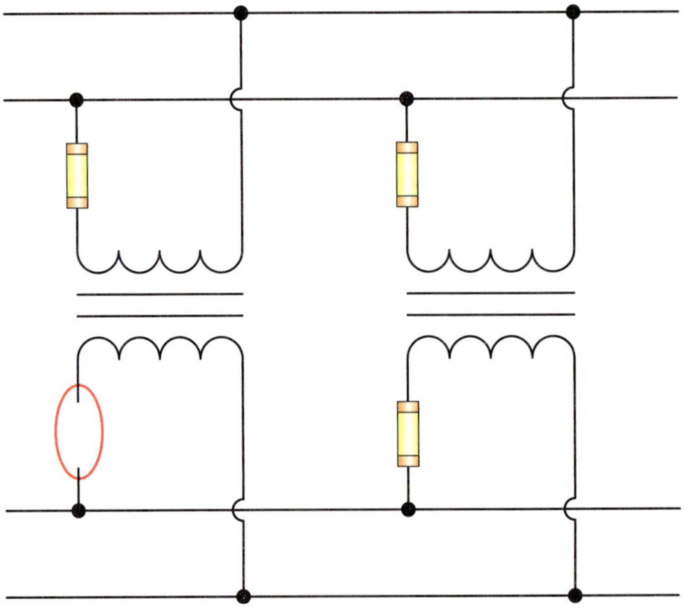

Figure 27–63 The secondary winding should be disconnected before disconnecting the primary winding.

In this example, the primary is connected to 4160 volts. If the secondary windings are not disconnected first, the parallel transformer is still energizing the secondary of the transformer to be replaced, and that transformer then becomes a step-up transformer. The primary winding will still have a voltage of 4160 even if it is disconnected from the power line. Any time parallel transformers are employed, a sign warning workers of a potential backfeed source should be located at each primary fuse.

Summary

- All values of voltage, current, and impedance in a transformer are proportional to the transformer's primary-to-secondary voltage ratio or turns ratio.
- Transformers can change values of voltage, current, and impedance but cannot change the frequency.
- The primary winding of a transformer is connected to the power line.
- The secondary winding is connected to the load.
- A transformer that has a lower secondary voltage than primary voltage is a step-down transformer.
- A transformer that has a higher secondary voltage than primary voltage is a step-up transformer.
- An isolation transformer has its primary and secondary windings electrically and mechanically separated from each other.
- When a coil induces a voltage into another coil, it is known as mutual induction.
- Transformers can have very high inrush current when first connected to the power line.
- Transformers with an air gap in their core material have lower inrush current as the core material is easily magnetized regardless of polarity.
- Autotransformers have only one winding, which is used as both the primary and secondary.
- Autotransformers have a disadvantage in that they have no line isolation between the primary and secondary windings.
- Isolation transformers help filter voltage and current spikes between the primary and secondary sides.
- Polarity dots are often added to schematic diagrams to indicate transformer polarity.
- Transformers can be connected as additive or subtractive polarity.
- As a general rule, transformer secondary windings should not be connected in parallel.

Review Questions

1. The voltage and current values on the primary side of a transformer will be proportional to the values on the secondary side of the transformer by which of the following values?

   A) The transformer kVA rating

   B) The ratio of primary turns to secondary turns

   C) The impedance of the transformer

   D) Transformer polarity

2. In a step-down transformer, the voltage will go down from the primary to the secondary. Which of the following best describes what will happen to the current from the primary to the secondary of the transformer?

   A) The current will also go down in a step-down transformer.

   B) The current will remain the same in a step-down transformer.

   C) The current will go up in a step-down transformer.

   D) It depends on how much load is connected to the step-down transformer.

3. Select the choice that best describes the difference between an isolation transformer and an autotransformer.

   A) An isolation transformer has no direct electrical connection between the primary and secondary.

   B) An autotransformer has no direct connection between the primary and secondary.

   C) An autotransformer will always be a step-down transformer.

   D) An isolation transformer will always be a step-up transformer.

4. True or False: An autotransformer is a type of single-phase transformer that has no direct electrical connection between the supply and the output of the transformer.

5. A 900 VA transformer has a primary voltage of 480 V and a secondary voltage of 24 V. If the secondary winding supplies 14.5 amperes worth of lamps and small motors, how much current will be on the primary conductors at full-load?

   A) 58 amperes

   B) 37.5 amperes

   C) 1.875 amperes

   D) 0.725 amperes

6. A 5 kVA transformer has an output of 3750 VA. The primary voltage is 277 V, and the secondary voltage is 120 V. Calculate the primary current and select the most correct answer.

   A) 13.54 amperes

   B) 31.25 amperes

   C) 18.05 amperes

   D) 41.67 amperes

7. The nameplate of a transformer lists the transformer's %Z. This refers to which of the following?

   A) The power rating of the transformer

   B) The power factor as a percentage

   C) The internal opposition to current of the transformer

   D) The amount overcurrent the windings can handle

8. True or False: The nameplate of a transformer will provide the maximum amount of current that the secondary windings are rated for.

9. When connecting transformers in parallel, which of the following statements must be true?
   A) Transformers must have the same kVA.
   B) Transformers must have the same impedance.
   C) Transformers must have the same turns ratio.
   D) All of these are true.
10. When servicing parallel transformers, what condition is often likely to be present if the secondary windings of both transformers are not disconnected before beginning work on one of the transformers?
    A) Overvoltage of the connected transformer
    B) Backfeed of the disconnected transformer
    C) Damage to windings of both transformers
    D) Not enough information
11. A transformer has a single primary winding rated at 240 V and dual secondary windings rated at 16 V each. If the transformer is connected for buck-boost operation with the secondary windings connected in series and in additive polarities, what would be the voltage output of this transformer?
    A) 208 V
    B) 224 V
    C) 256 V
    D) 272 V
12. True or False: When a buck-boost transformer is connected in a buck configuration, the output voltage will always be less than the input voltage.

Practical Applications

1. You are working in an industrial plant. You must install a single-phase transformer. The transformer has the following information on the nameplate:

Primary voltage—13,800 V

Secondary voltage—240 V

Impedance—5%

150 kVA

The secondary fuse has a current rating of 800 A and an interrupt rating of 10,000 A. Is this interrupt rating of the fuses sufficient for the short-circuit current available in this installation?

---

2. You have been given a transformer to install on a 277-V line. The transformer nameplate is shown in *Figure 27–64*. The transformer must supply a 120-V, 20-A circuit. The transformer capacity should be no less than 115% of the rated load. Does the transformer you have been given have enough kVA capacity to supply the load? To which transformer terminals would you connect the incoming power? To which transformer terminals would you connect the load?

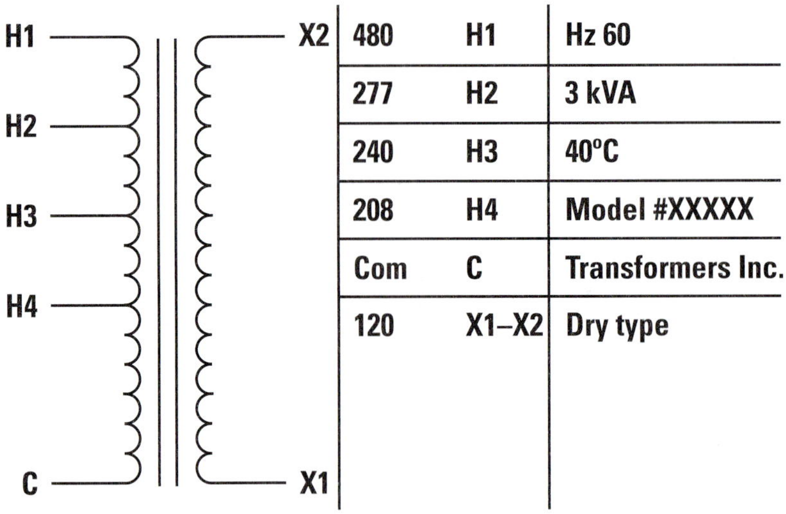

Figure 27–64 Transformer nameplate.

Practice Problems

Refer to *Figure 27–65* to answer the following questions. Find all the missing values.

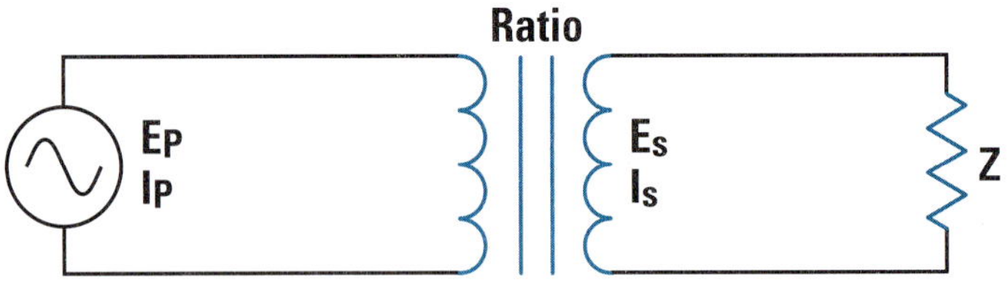

Figure 27–65 Isolation transformer practice problems.

1.

E_P 120 V | E_S 24 V
I_P ________ | I_S ________
N_P 300 turns | N_S ________
Ratio ________ | $Z = 3\ \Omega$

2.

E_P 240 V | E_S 320 V
I_P ________ | I_S ________
N_P ________ | N_S 280
Ratio ________ | $Z = 500\ \Omega$

3.

E_P ________ | E_S 160 V
I_P ________ | I_S ________
N_P ________ | N_S 80 turns
Ratio 1:2.5 | $Z = 12\ \Omega$

4.

E_P 48 | E_S 240 V
I_P ________ | I_S ________
N_P 220 turns | N_S ________
Ratio ________ | $Z = 360\ \Omega$

5.

E_P ________ | E_S ________
I_P 16.5 A | I_S 3.25 A
N_P ________ | N_S 450 turns
Ratio ________ | $Z = 56\ \Omega$

6.

E_P 480 V | E_s ________
I_P ________ | I_s ________
N_P 275 turns | N_s 525 turns
Ratio ________ | $Z = 1.2\ k\Omega$

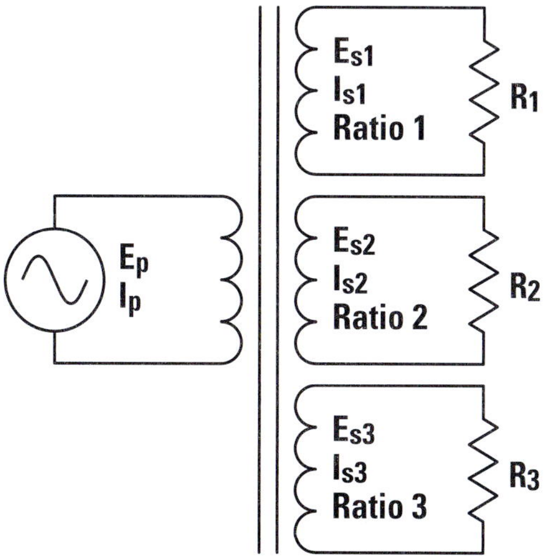

Figure 27–66 Single-phase transformer with multiple secondaries.

Refer to *Figure 27–66* to answer the following questions. Find all the missing values.

7.

E_P 208 V	E_{S1} 320 V	E_{S2} 120 V	E_{S3} 24 V
I_P ______	I_{S1} ______	I_{S2} ______	I_{S3} ______
NP 800 turns	N_{S1} ______	N_{S2} ______	N_{S3} ______
	Ratio 1:	Ratio 2:	Ratio 3:
	R_1 12 kΩ	R_2 6 Ω	R_3 8 Ω

8.

E_P 277 V	E_{S1} 480 V	E_{S2} 208 V	E_{S3} 120 V
I_P ______	I_{S1} ______	I_{S2} ______	I_{S3} ______
N_P 350 turns	N_{S1} ______	N_{S2} ______	N_{S3} ______
	Ratio 1:	Ratio 2:	Ratio 3:
	R_1 200 Ω	R_2 60 Ω	R_3 24 Ω

Unit 28 Three-Phase Transformers

Key Terms

Delta–WYE

Dielectric oil

High leg

One-line diagram

Open-delta

Orange wire

Single-phase loads

Tagging

Three-phase bank

WYE–delta

Why You Need to Know

Three-phase transformers are used throughout industry. Almost all power generated in the United States and Canada is three-phase. Transformers step up voltage for transmission and step it down again for use inside a home, industrial plant, or commercial building. This unit

- presents the difference between a three-phase transformer and a three-phase transformer bank.
- determines different voltage and current values in a three-phase transformer.
- defines phase values in calculating the values of a transformer.
- explains how harmonics are identified and overcome.
- describes different three-phase transformer connections such as delta–WYE, WYE–delta, open-delta, T connected, and Scott connected.
- discusses installation and testing.

Outline

Objectives

After studying this unit, you should be able to

- differentiate between the two main three-phase transformer configurations.
- calculate unknown voltage and current values for a three-phase transformer connection.
- explain how delta transformer connections are used to supply systems.
- explain how WYE transformer connections are used to supply systems.
- describe how less common transformer connections are used to supply loads.

Preview

Three-phase transformers are used throughout industry to change values of three-phase voltage and current. Because three-phase power is the most common way in which power is produced, transmitted, and used, an understanding of how three-phase transformer connections are made is critical to understanding these systems. This unit discusses different types of three-phase transformer connections and presents examples of how values of voltage and current for these connections are calculated.

28–1 Three-Phase Transformers

A three-phase transformer is constructed by winding three single-phase transformers on a single core (*Figure 28–1*). An illustration of a three-phase transformer is shown in *Figure 28–2*. The transformer shown will be filled with a **dielectric oil.** The dielectric oil performs several functions. Because it is a dielectric, it provides electric insulation between the windings and the case. It is also used to help provide cooling and to prevent the formation of moisture, which can deteriorate the winding insulation.

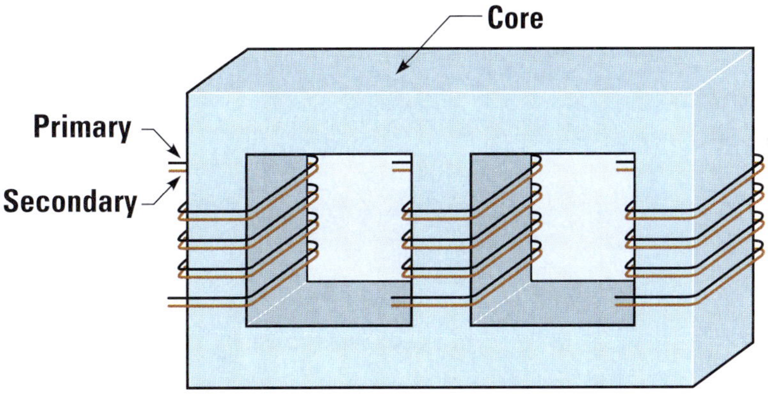

Figure 28–1 Basic construction of a three-phase transformer.

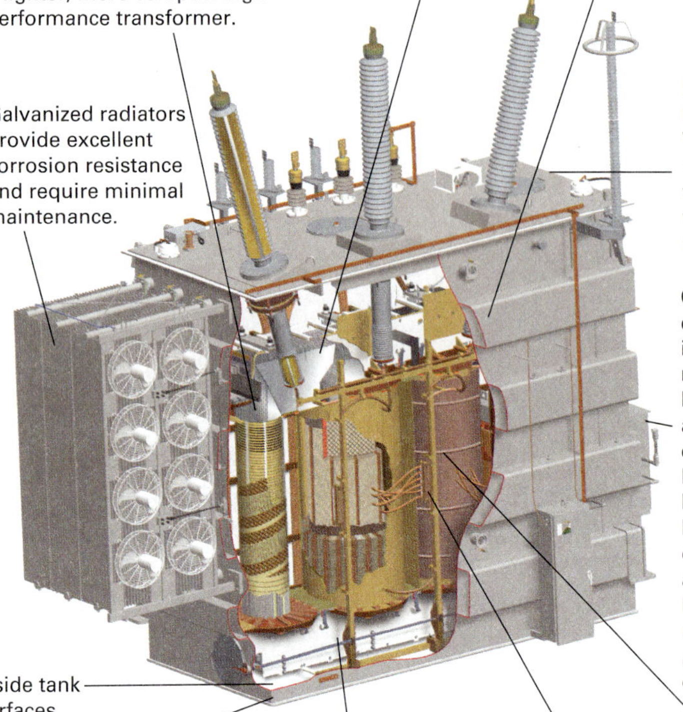

Figure 28–2 Cut-away view of an oil-filled three-phase transformer. (Courtesy of SPX Transformers Solutions, Inc.)

A pad-mounted three-phase transformer is shown in *Figure 28–3*. Transformers of this type often use a blanket of dry nitrogen gas over the dielectric oil to prevent the formation of moisture due to condensation. External gauges are generally provided to indicate the level of dielectric oil, temperature, and internal pressure (*Figure 28–4*).

Figure 28–3 Pad-mounted transformer. (Courtesy of SPX Transformers Solutions, Inc.)

Figure 28–4 External gauges used to monitor a sealed transformer.

Three-Phase Transformer Connections

Three-phase transformers are typically connected in either a delta or WYE configuration. A **WYE–delta** transformer, for example, has its primary winding connected in a WYE and its secondary winding connected in a delta (*Figure 28–5*). A **delta–WYE** transformer would have its primary winding connected in delta and its secondary connected in WYE (*Figure 28–6*).

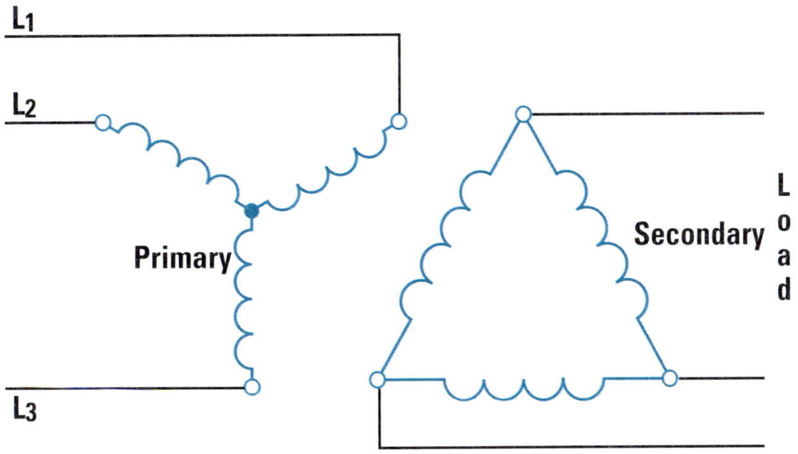

Figure 28–5 WYE–delta-connected three-phase transformer.

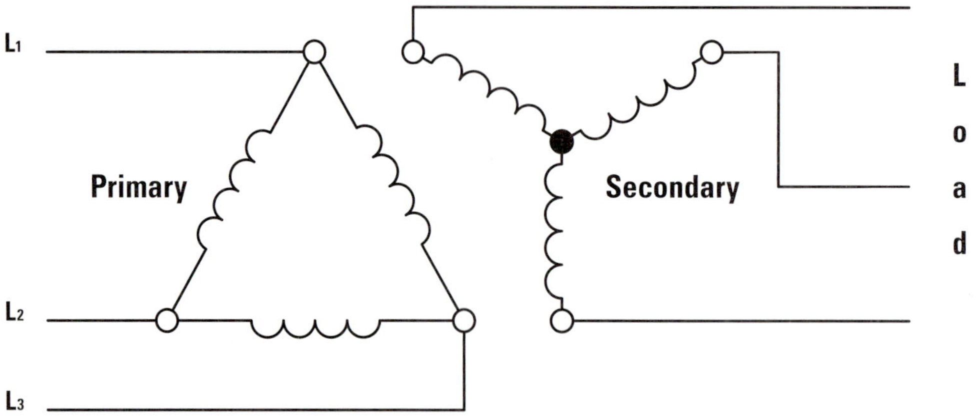

Figure 28–6 Delta–WYE-connected three-phase transformer.

Connecting Single-Phase Transformers into a Three-Phase Bank

If three-phase transformation is needed and the proper three-phase transformer is not available, three single-phase transformers can be connected to form a **three-phase bank.** These types of installations are commonly found among utility-owned pole-mounted transformers (*Figure 28–7*). When three single-phase transformers are used to make a three-phase bank, their primary and secondary windings are connected in a WYE or delta connection. The three transformer windings in *Figure 28–8* are labeled A, B, and C. One end of each primary lead is labeled H_1, and the other end is labeled H_2. One end of each secondary lead is labeled X_1, and the other end is labeled X_2.

Figure 28–7 Three single-phase transformers connected to form a three-phase bank.

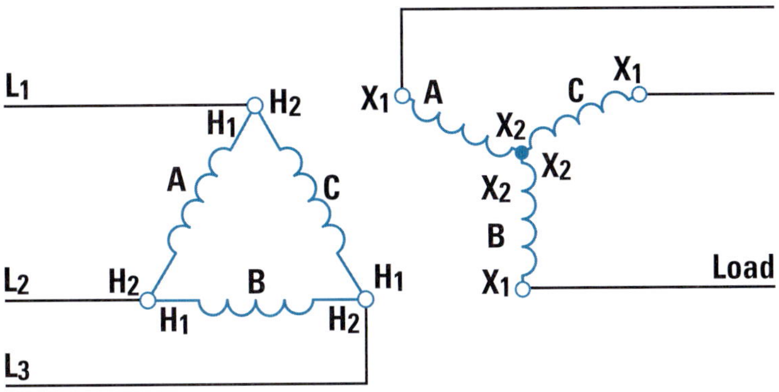

Figure 28–8 Identifying the windings.

Figure 28–9 shows three single-phase transformers labeled A, B, and C. The primary leads of each transformer are labeled H_1 and H_2, and the secondary leads are labeled X_1 and X_2. The schematic diagram of *Figure 28–8* is used to connect the three single-phase transformers into a three-phase Delta–WYE connection as shown in *Figure 28–10*.

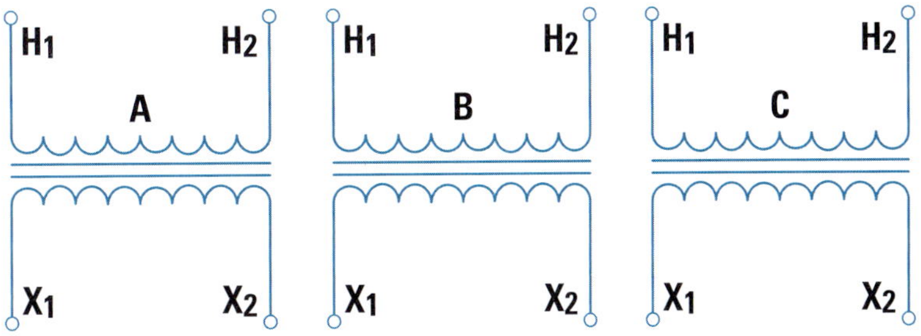

Figure 28–9 Three single-phase transformers.

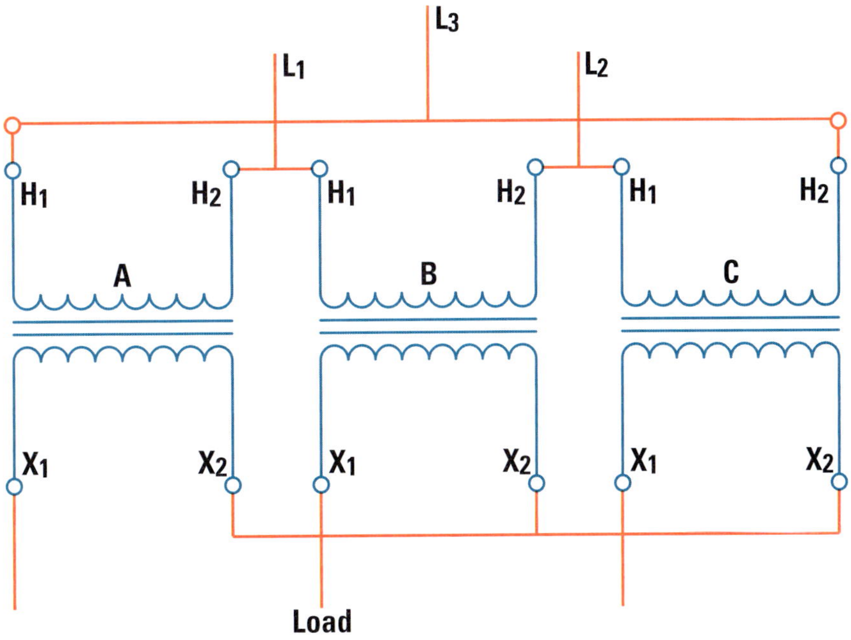

Figure 28–10 Connecting three single-phase transformers to form a Delta–WYE three-phase bank.

The primary windings are connected in a delta configuration. The schematic in *Figure 28–8* shows that the H_1 lead of one winding is tied to the H_2 lead of the next winding, and each of the power lines connect at each of the points where H_1 and H_2 connect. Notice in *Figure 28–10* that the H_1 leads of the A phase connect to the H_2 lead of the C phase, H_1 of the C phase connects to H_2 of the B phase, and H_1 of the B phase connects to H_2 of the A phase to complete the delta configuration. Also notice that each of the incoming lines connects at these connection points.

Figure 28–8 shows that the X_2 leads of all three secondaries are connected together, and the load is supplied by lines from the X_1 lead of each phase. Notice that this same connection has been made in *Figure 28–10*.

Although *Figure 28–8* illustrates the proper schematic symbology for a three-phase transformer connection, some electrical schematics and wiring diagrams do not illustrate three-phase transformer connections in this manner. One type of diagram, called the **one-line diagram,** would illustrate a delta–WYE connection as shown in *Figure 28–11*. These diagrams are generally used to show the main power distribution system of a large industrial plant. The one-line diagram in *Figure 28–12* shows the main power to the plant and the transformation of voltages to different distribution feeders. Notice that each transformer shows whether the primary and secondary are connected as a WYE or delta and the secondary voltage.

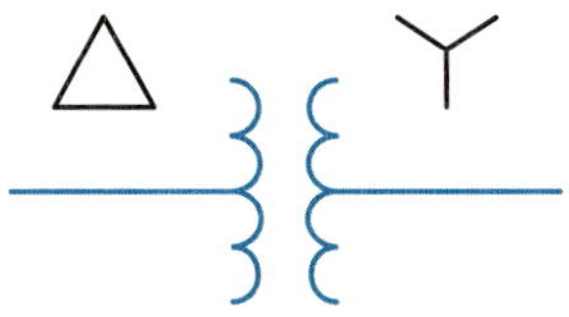

Figure 28–11 One-line diagram symbol used to represent a delta–WYE three-phase transformer connection.

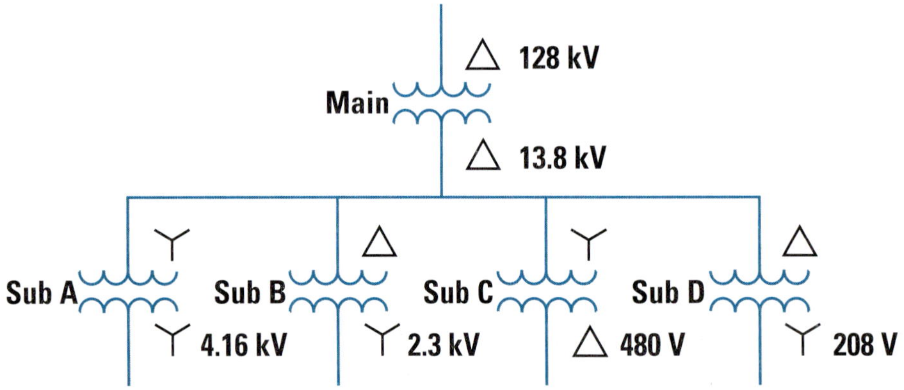

Figure 28–12 One-line diagrams are generally used to show the main power distribution of a plant.

Closing a Delta

When **closing a delta,** connections should be checked for proper polarity before making the final connection and applying power. If the phase winding of one transformer is reversed, an extremely high current will flow when power is applied. Proper phasing can be checked with a voltmeter, as shown in *Figure 28–13*. If power is applied to the transformer bank before the delta connection is closed, the voltmeter should indicate 0 volts. If one phase winding has been reversed, however, the voltmeter will indicate double the amount of voltage. For example, assume that the output voltage of a delta secondary is 240 volts. If the voltage is checked before the delta is closed, the voltmeter should indicate a voltage of 0 V if all windings have been phased properly. If one winding has been reversed, however, the voltmeter will indicate a voltage of 480 volts (240 V + 240 V). This test will confirm whether a phase winding has been reversed, but it will not indicate whether the reversed winding is located in the primary or secondary. If either the primary or the secondary winding has been reversed, the voltmeter will indicate double the output voltage.

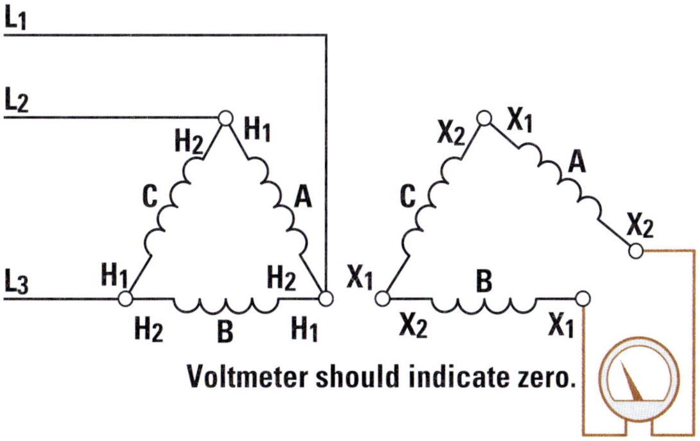

Figure 28–13 Testing for proper transformer polarity before closing the delta.

Note, however, that a voltmeter is a high-impedance device. It is not unusual for a voltmeter to indicate some amount of voltage before the delta is closed, especially if the primary has been connected as a WYE and the secondary as a delta. When this is the case, however, the voltmeter will generally indicate close to the normal output voltage if the connection is correct and double the output voltage if the connection is incorrect.

28–2 Three-Phase Transformer Calculations

To calculate the values of voltage and current for three-phase transformers, the formulas used for making transformer calculations and three-phase calculations must be followed. Another very important rule is that **only phase values of voltage and current can be used when calculating transformer values.** Refer to Transformer A in *Figure 28–9*. All transformation of voltage and current takes place between the primary and secondary windings. Because these windings form the phase values of the three-phase connection, only phase and not line values can be used when calculating transformed voltages and currents.

Example 28–1

A three-phase transformer connection is shown in *Figure 28–14.* Three single-phase transformers have been connected to form a WYE–delta bank. The primary is connected to a three-phase line of 13,800 V, and the secondary voltage is 480 V. A three-phase resistive load with an impedance of 2.77 Ω per phase is connected to the secondary of the transformer. Calculate the following values for this circuit:

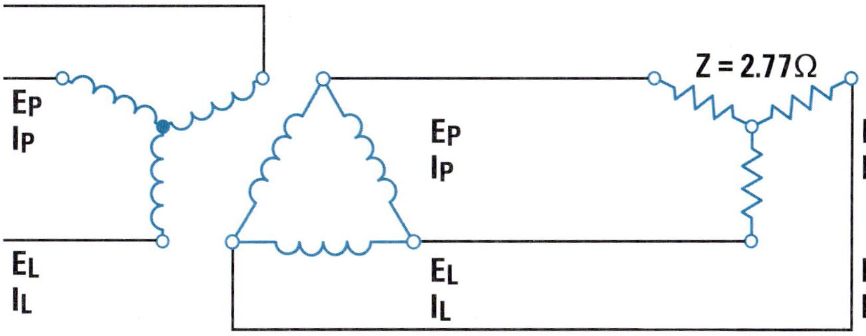

Figure 28–14 Example Circuit 1 three-phase transformer calculation.

Example 28–1 (continued)

$E_{P(PRIMARY)}$—phase voltage of the primary

$E_{P(SECONDARY)}$—phase voltage of the secondary

Ratio—turns ratio of the transformer

$E_{P(LOAD)}$—phase voltage of the load bank

$I_{P(LOAD)}$—phase current of the load bank

$I_{L(SECONDARY)}$—secondary line current

$I_{P(SECONDARY)}$—phase current of the secondary

$I_{P(PRIMARY)}$—phase current of the primary

$I_{L(PRIMARY)}$—line current of the primary

Solution

The primary windings of the three single-phase transformers have been connected to form a WYE connection. In a WYE connection, the phase voltage is less than the line voltage by a factor of 1.732 (the square root of 3). Therefore, the phase value of the primary voltage can be calculated using the formula:

$$E_{P(PRIMARY)} = \frac{E_L}{1.732}$$

$$E_{P(PRIMARY)} = \frac{13{,}800\text{ V}}{1.732}$$

$$E_{P(PRIMARY)} = 7967.667\text{ V}$$

The secondary windings are connected as a delta. In a delta connection, the phase voltage and line voltage are the same:

$$E_{P(SECONDARY)} = E_{L(SECONDARY)}$$

$$E_{P(SECONDARY)} = 480\text{ V}$$

The turns ratio can be calculated by comparing the phase voltage of the primary with the phase voltage of the secondary:

$$\text{Ratio} = \frac{\text{primary voltage}}{\text{secondary voltage}}$$

$$\text{Ratio} = \frac{7967.667\text{ V}}{480\text{ V}}$$

$$\text{Ratio} = 16.6{:}1$$

The load bank is connected in a WYE connection. The voltage across the phase of the load bank will be less than the line voltage by a factor of 1.732:

$$E_{P(LOAD)} = \frac{E_{L(LOAD)}}{1.732}$$

$$E_{P(LOAD)} = \frac{480\text{ V}}{1.732}$$

$$E_{P(LOAD)} = 277.136\text{ V}$$

Example 28–1 (continued)

Now that the voltage across each of the load resistors is known, the current flow through the phase of the load can be calculated using Ohm's law:

$$I_{P(LOAD)} = \frac{E}{R}$$

$$I_{P(LOAD)} = \frac{277.136\ V}{2.77}$$

$$I_{P(LOAD)} = 100.049\ V$$

Because the load is connected as a WYE connection, the line current is the same as the phase current:

$$I_{L(SECONDARY)} = 100.049\ A$$

The secondary of the transformer bank is connected as a delta. The phase current of the delta is less than the line current by a factor of 1.732:

$$I_{P(SECONDARY)} = \frac{I_L}{1.732}$$

$$I_{P(SECONDARY)} = \frac{100.049\ A}{1.732}$$

$$I_{P(SECONDARY)} = 57.765\ A$$

The amount of current flow through the primary can be calculated using the turns ratio. Because the primary has a higher voltage than the secondary, it will have a lower current. (Volts times amperes input must equal volts times amperes output.)

$$\text{Primary current} = \frac{\text{secondary current}}{\text{turns ratio}}$$

$$I_{P(Primary)} = \frac{57.765\ A}{16.6}$$

$$I_{P(Primary)} = 3.48\ A$$

Because all transformed values of voltage and current take place across the phases, the primary has a phase current of 3.48 A. In a WYE connection, the phase current is the same as the line current:

$$I_{L(PRIMARY)} = 3.48\ A$$

The transformer connection with all calculated values is shown in *Figure 28–15*.

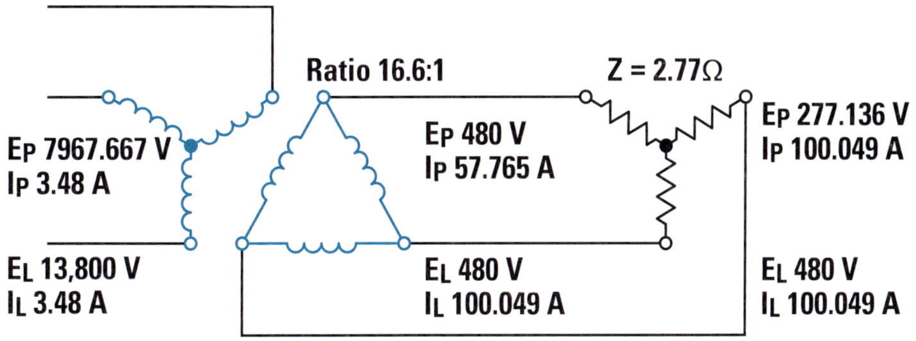

Figure 28–15 Example Circuit 1 with all missing values.

Example 28–2

A three-phase transformer is connected in a delta–delta configuration (*Figure 28–16*). The load is connected as a WYE, and each phase has an impedance of 7 Ω. The primary is connected to a line voltage of 4160 V, and the secondary line voltage is 440 V. Find the following values:

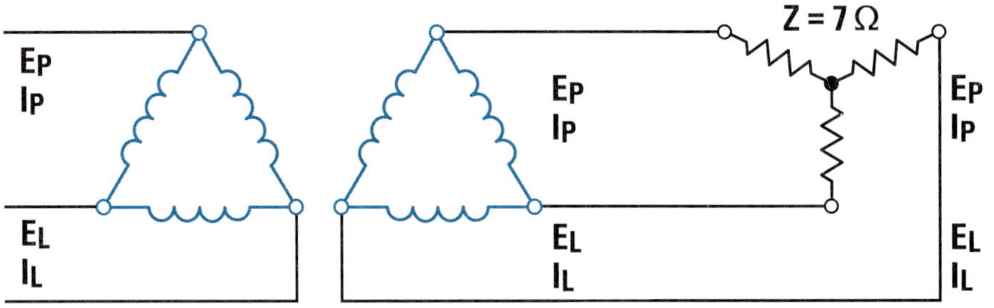

Figure 28–16 Example Circuit 2 three-phase transformer calculation.

$E_{P(PRIMARY)}$—phase voltage of the primary
$E_{P(SECONDARY)}$—phase voltage of the secondary
Ratio—turns ratio of the transformer
$E_{L(LOAD)}$—line voltage of the load
$E_{P(LOAD)}$—phase voltage of the load bank
$I_{P(LOAD)}$—phase current of the load bank
$I_{L(LOAD)}$—line current of the load
$I_{L(SECONDARY)}$—secondary line current
$I_{P(SECONDARY)}$—phase current of the secondary
$I_{P(PRIMARY)}$—phase current of the primary
$I_{L(PRIMARY)}$—line current of the primary

Solution

The primary is connected as a delta. The phase voltage will be the same as the applied line voltage:

$$E_{P(PRIMARY)} = E_{L(PRIMARY)}$$
$$E_{P(PRIMARY)} = 4160\ V$$

The secondary of the transformer is connected as a delta also. Therefore, the phase voltage of the secondary will be the same as the line voltage of the secondary:

$$E_{P(SECONDARY)} = 440\ V$$

All transformer values must be calculated using phase values of voltage and current. The turns ratio can be found by dividing the phase voltage of the primary by the phase voltage of the secondary:

$$\text{Ratio} = \frac{E_{P(PRIMARY)}}{E_{P(SECONDARY)}}$$

$$\text{Ratio} = \frac{4160\ V}{440\ V}$$

$$\text{Ratio} = 9.45{:}1$$

Example 28–2 (continued)

The load is connected directly to the output of the secondary. The line voltage applied to the load must therefore be the same as the line voltage of the secondary:

$$E_{L(LOAD)} = 440\ V$$

The load is connected in a WYE. The voltage applied across each phase will be less than the line voltage by a factor of 1.732:

$$E_{P(LOAD)} = \frac{E_{L(LOAD)}}{1.732}$$

$$E_{P(LOAD)} = \frac{440\ V}{1.732\ V}$$

$$E_{P(LOAD)} = 254.042\ V$$

The phase current of the load can be calculated using Ohm's law:

$$I_{P(LOAD)} = \frac{E_{P(LOAD)}}{Z}$$

$$I_{P(LOAD)} = \frac{254.042\ V}{7\ \Omega}$$

$$I_{P(LOAD)} = 36.292A$$

The amount of line current supplying a WYE-connected load will be the same as the phase current of the load:

$$I_{L(LOAD)} = 36.292\ A$$

Because the secondary of the transformer is supplying current to only one load, the line current of the secondary will be the same as the line current of the load:

$$I_{L(SECONDARY)} = 36.292\ A$$

The phase current in a delta connection is less than the line current by a factor of 1.732:

$$I_{P(SECONDARY)} = \frac{I_{L(SECONDARY)}}{1.732}$$

$$I_{P(SECONDARY)} = \frac{36.292\ A}{1.732}$$

$$I_{P(SECONDARY)} = 20.954\ A$$

The phase current of the transformer primary can now be calculated using the phase current of the secondary and the turns ratio:

$$I_{P(PRIMARY)} = \frac{I_{P(SECONDARY)}}{\text{turns ratio}}$$

$$I_{P(PRIMARY)} = \frac{20.954\ A}{9.45}$$

$$I_{P(PRIMARY)} = 2.217\ A$$

In this example, the primary of the transformer is connected as a delta. The line current supplying the transformer will be higher than the phase current by a factor of 1.732:

$$I_{L(PRIMARY)} = I_{P(PRIMARY)} \times 1.732$$

$$I_{L(PRIMARY)} = 2.217\ A \times 1.732$$

$$I_{L(PRIMARY)} = 3.84\ A$$

The circuit with all calculated values is shown in *Figure 28–17*.

Example 28–2 (continued)

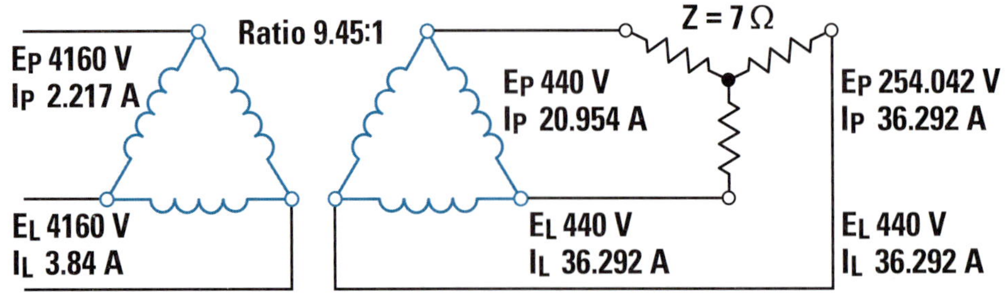

Figure 28–17 Example Circuit 2 with all missing values.

Open-Delta Connection

The **open-delta** transformer connection can be made with only two transformers instead of three (*Figure 28–18*). This connection is rarely used any more but was used when the amount of three-phase power needed is not excessive, such as in a small business. It should be noted that the output power of an open-delta connection is only 86.6% of the rated power of the two transformers. For example, assume two transformers, each having a capacity of 25 kilovolt-amperes, are connected in an open-delta connection. The total output power of this connection is 43.5 kilovolt-amperes (50 kVA × 0.866 = 43.3 kVA).

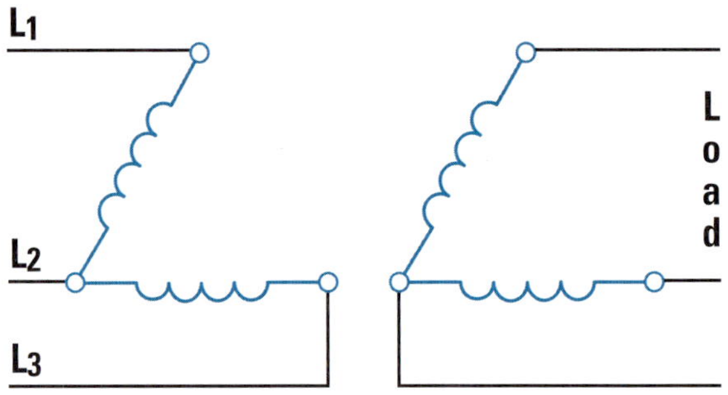

Figure 28–18 Open-delta connection.

Another figure given for this calculation is 57.7%. This percentage assumes a closed-delta bank containing three transformers. If three 25-kilovolt-amperes transformers were connected to form a closed-delta connection, the total output power would be 75 kilovolt-amperes (3 × 25 kVA = 75 kVA). If one of these transformers were removed and the transformer bank was operated as an open-delta connection, the output power would be reduced to 57.7% of its original capacity of 75 kilovolt-amperes. The output capacity of the open-delta bank is 43.3 kilovolt-amperes (75 kVA × 0.577 = 43.3 kVA).

The voltage and current values of an open-delta connection are calculated in the same manner as a standard delta–delta connection when three transformers are employed. The voltage and current rules for a delta connection must be used when determining line and phase values of voltage and current.

Single-Phase Loads

When true three-phase loads are connected to a three-phase transformer bank, there are no problems in balancing the currents and voltages of the individual phases. *Figure 28–19* illustrates this condition. In this circuit, a delta–WYE three-phase transformer bank is supplying power to a WYE-connected three-phase load in which the impedances of the three phases are the same. Notice that the amount of current flow in the phases is the same. This is the ideal condition and is certainly desired for all three-phase transformer loads. Although this is the ideal situation, it is not always possible to obtain a balanced load. Three-phase transformer connections are often used to supply a combination of three-phase and **single-phase loads,** which tends to unbalance the system.

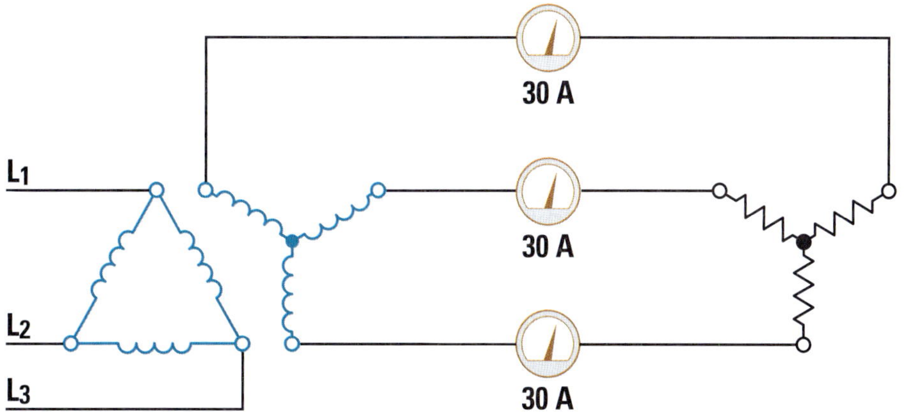

Figure 28–19 Three-phase transformer connected to a balanced three-phase load.

28–3 Three-Phase Delta Connections

One type of three-phase transformer configuration that historically has been used to supply power to single-phase and three-phase loads is shown in *Figure 28–20*. This type of transformer bank is rarely used in practice today as it creates a "high-leg" circuit and removes a third of the transformer from supplying single-phase loads. To illustrate how this type of circuit operates, we will assume that it is supplying a three-phase load with a power requirement of 75 kilovolt-amperes and single-phase loads requiring an additional 50 kilovolt-amperes. Three 25-kilovolt-ampere transformers could be used to supply the three-phase power needed (25 kVA × 3 = 75 kVA), if that were all that was connected, but the addition of the single-phase load requires one of the transformers to be larger. This transformer must supply both the three-phase and single-phase load, which requires it to have a rating of 75 kilovolt-amperes (25 kVA + 50 kVA = 75 kVA).

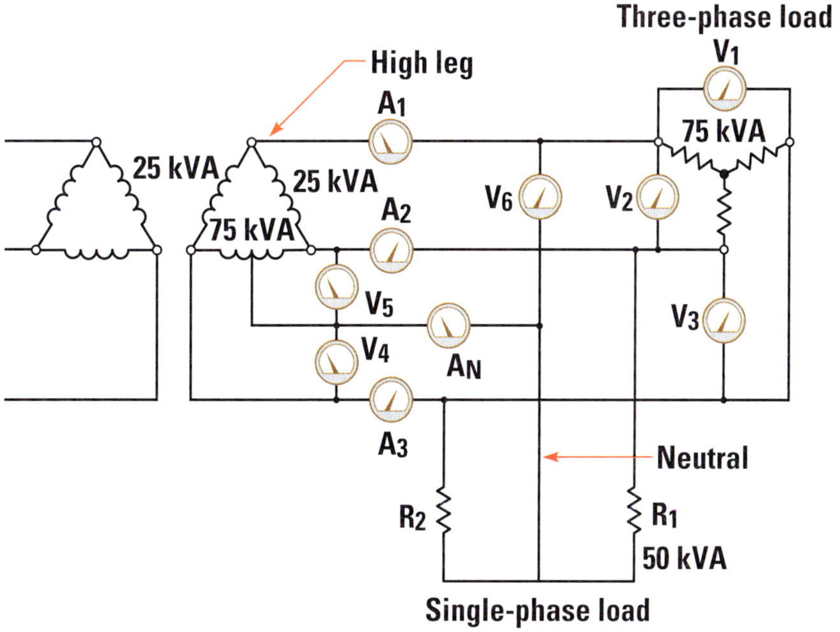

Figure 28–20 Delta connected transformer bank with center-tapped phase for single-phase loads.

Voltage Values

The connection shown in *Figure 28–20* has a line-to-line voltage of 240 volts. The three voltmeters, V_1, V_2, and V_3, have all been connected across the three-phase lines and should indicate 240 volts each. Voltmeters V_4 and V_5 have been connected between the two lines of the larger transformer and its center tap. These two voltmeters will indicate a voltage of 120 volts each. Notice that it is these two lines and the center tap that are used to supply the single-phase power needed. The center tap of the larger transformer is used as a neutral conductor for the single-phase loads. Voltmeter V_6 has been connected between the center tap of the larger transformer and the line of the smaller transformer. This line is known as a **high leg** because the voltage between this line and the neutral conductor will be higher than the voltage between the neutral and either of the other two conductors. The high-leg voltage can be calculated by increasing the single-phase center-tapped voltage value by 1.732. In this case, the high-leg voltage will be 207.84 volts (120 V $\times$ 1.732 $=$ 207.84 V). When this type of connection is employed, the *NEC* requires that the high leg be identified by connecting it to an **orange wire** or by **tagging** it at any point where a connection is made if the neutral conductor is also present. The high-leg will not be able to supply any phase-to-ground type loads and can only be used for three phase equipment.

Current Values

If the three-phase load is the only load operating and it operates at maximum capacity, ammeters A_1, A_2, and A_3 will indicate a current flow of 180.4 amperes each [75 kVA/(240 V $\times$ 1.732) $=$ 180.4 A]. Notice that only when the three-phase load is in operation is the current on each line balanced.

If the three-phase load is not in operation and only the single-phase load is operating and at maximum capacity, ammeters A_2 and A_3 will each indicate a value of 208 amperes (50 kVA/240 V $=$ 208 A). Ammeter A_1 will indicate a current flow of 0 amperes because all the load is connected between the other two lines of the transformer connection. Ammeter A_N will also indicate a value of 0 amperes provided that the single-phase load is evenly distributed between each phase conductor. The neutral conductor carries only the unbalanced load between the two-phase conductors.

Now assume that one side of the single-phase load, Resistor R_2, has been opened, and no current flows through it. If the other line maintains a current flow of 208 amperes, the neutral conductor will have a current flow of 208 amperes also (208 A $-$ 0 A $=$ 208 A).

Now assume that Resistor R_2 has a value that will permit a current flow of 58 amperes on that phase. The neutral current will now be 150 amperes (208 A $-$ 58 A $=$ 150 A).

If both the three-phase and single-phase loads are operating at the same time and the three-phase load is at maximum capacity but the single-phase load is operating in such a manner that 125 amperes flow through Resistor R_1 and 50 amperes flow through Resistor R_2, the ammeters will indicate the following values:

$$A_1 = 180.4\text{ A}$$

$$A_2 = 305.4\text{ A } (180.4\text{ A} + 125\text{ A} = 305.4\text{ A})$$

$$A_3 = 230.4\text{ A } (180.4\text{ A} + 50\text{ A} = 230.4\text{ A})$$

$$A_N = 75\text{ A } (125\text{ A} - 50\text{ A} = 75\text{ A})$$

Closed Delta without Center Tap

In the circuit shown in *Figure 28–21,* the transformer bank has been connected in a WYE–delta configuration. Notice that there is no transformer secondary with a center-tapped winding. In this circuit, there is no neutral conductor. The three loads have been connected directly across the three-phase lines. Because these three loads are connected directly across the lines, they form a delta-connected load. This is the most common use of delta transformers in industrial settings, to supply three-phase loads only.

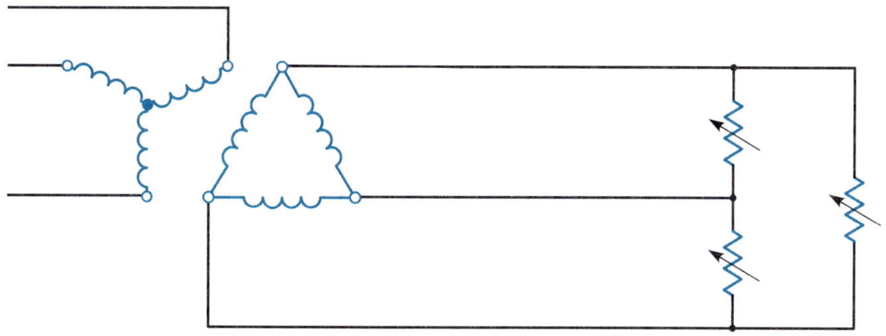

Figure 28–21 Single-phase loads supplied by a WYE–delta transformer connection.

28–4 Three-Phase WYE Connections

The circuit shown in *Figure 28–22* is a three-phase transformer connection with a delta-connected primary and WYE-connected secondary. The secondary has been center-tapped to form a neutral conductor. This is the most common connection used to provide power for single-phase loads as well as three-phase loads. Typical voltages for this type of connection are 208Y/120 and 480Y/277. The neutral conductor carries the vector sum of the unbalanced current. In this circuit, however, the sum of the unbalanced current is not the difference between two lines such as in the center-tapped delta connection. In that connection, the two lines were 180° out of phase when compared with the center tap. In the WYE connection, the lines are 120° out of phase. When all three lines are carrying the same amount of amperage, the neutral current is zero.

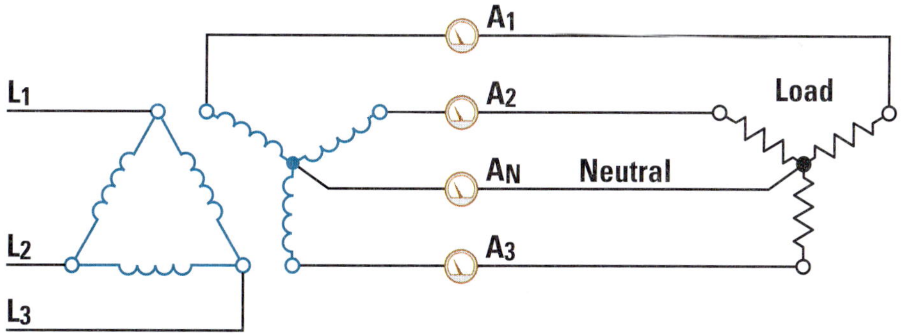

Figure 28–22 Three-phase four-wire connection.

A WYE-connected secondary with center tap can, under certain conditions, experience problems of extreme unbalance. **If this transformer connection is powered by a three-phase three-wire system, the primary winding is best connected in a delta configuration.** If the primary is connected as a WYE connection, the circuit can become exceedingly unbalanced under load. This is why most WYE-connected secondary transformers already come with the primary connected in a delta configuration.

If the incoming power is a three-phase four-wire system as shown in *Figure 28–23,* however, a WYE-connected primary can be used without problem. The neutral conductor connected to the center tap of the primary prevents the unbalance problems. It is a common practice with this type of connection to tie the neutral conductor of both primary and secondary together as shown. When this is done, however, line isolation between the primary and secondary windings is lost.

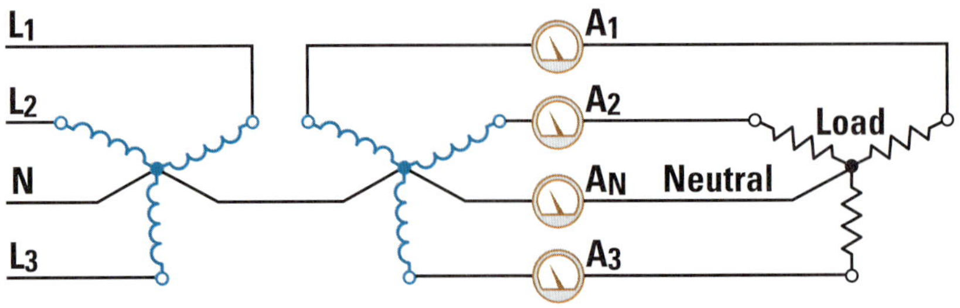

Figure 28–23 Neutral conductor is supplied by the incoming power.

28–5 Uncommon Transformer Connections

There are several transformer configurations that exist but simply are not realistic or practical in everyday use. One such connection that involves the use of two transformers to supply three-phase power is called a *T connection* (*Figure 28–24*). In this connection, one transformer is generally referred to as the main transformer and the other is called the "teaser" transformer. The main transformer must contain a center tap in both the primary and secondary windings, and the teaser transformer is typically tapped at 86.6% for both the primary and secondary windings. The advantages of the T connection over the open-delta connection is that it maintains a better phase balance.

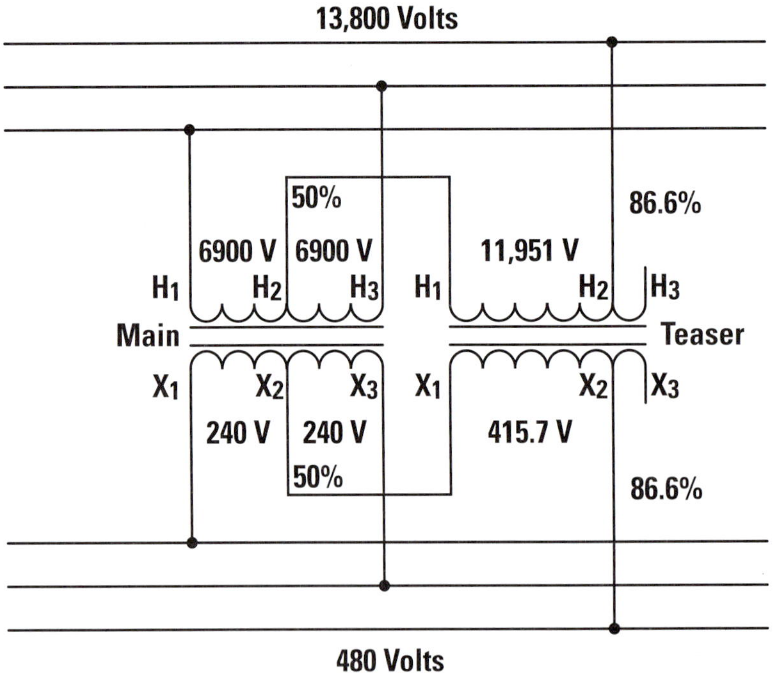

Figure 28–24 A T-connected transformer configuration.

Scott Connection

The *Scott connection* is used to convert three-phase power into two-phase power using two single-phase transformers. The Scott connection is a T-connected transformer on the primary side and separate single-phase transformers on the secondary side. The secondary windings of each transformer provide the individual phases of a two-phase system (*Figure 28–25*). The voltages of the secondary windings are 90° out of phase with each other. The Scott connection was used to provide two-phase power for the operation of two-phase motors and is extremely rare now as two-phase motors are mainly found in the history books.

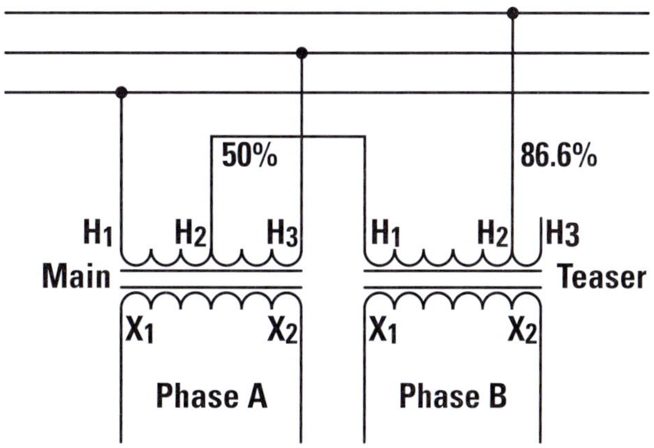

Figure 28–25 The Scott connection.

Zig-Zag Connection

The *zig-zag* or *interconnected-WYE* transformer is another uncommon connection that is primarily used for grounding purposes. It is mainly used to establish a neutral point for the grounding of fault currents. The zig-zag connection is basically a three-phase autotransformer whose windings are divided into six equal parts (*Figure 28–26*). In the event of a fault, the zig-zag connection forces the current to flow equally in the three legs of the autotransformer, offering minimum impedance to the flow of fault current. A schematic diagram of the zig-zag connection is shown in *Figure 28–26*.

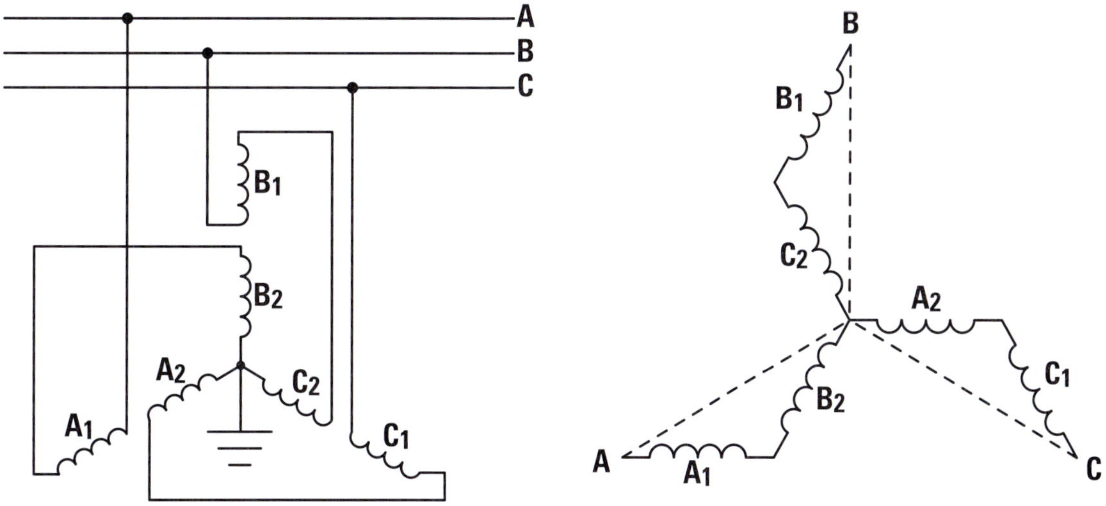

Figure 28–26 Zig-zag connection.

Six-Phase Transformer Connections

Three-phase transformers are sometimes connected in such a manner as to provide more than three phases. In some instances, it is desirable to have a power system with more than three phases. A good example of this is when it is necessary to convert or rectify alternating current into direct current with a minimum amount of ripple (pulsations of voltage). Power supplies that produce a low amount of ripple require less filtering. A schematic diagram of a six-phase connection and the phase-angle relationship of each output as compared to the center point or neutral are shown in *Figure 28–27.* Each of the six phases produces a phase angle of 60° between the output and the center point or neutral.

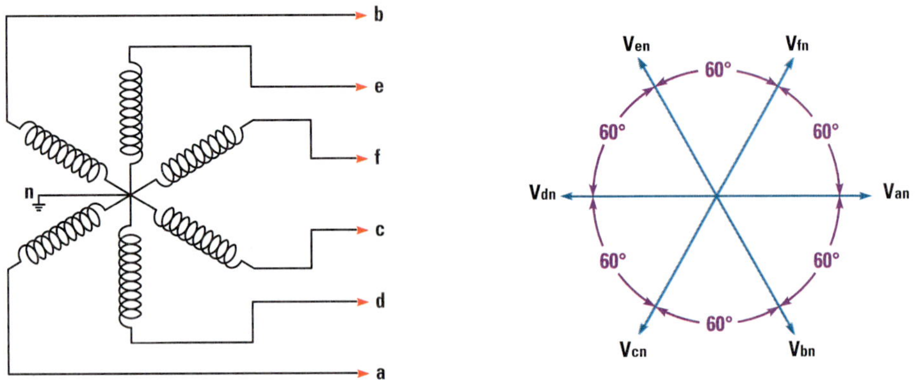

Figure 28–27 Schematic of six-phase connection and phase angles between each phase and neutral.

Six-phase connections are often used with rectifiers to produce a smoother DC output voltage. A six-phase output connected to a six-phase half-wave rectifier will produce the same ripple as a three-phase full-wave rectifier. The advantage of the half-wave rectifier is that the output voltage is higher because each phase has a voltage drop of only one diode. If the six-phase output is connected to a six-phase full-wave rectifier, the output DC voltage will have half the ripple of the half-wave rectifier (*Figure 28–28*). Another difference is that the half-wave rectifier requires a center-tap connection, and the full-wave rectifier does not.

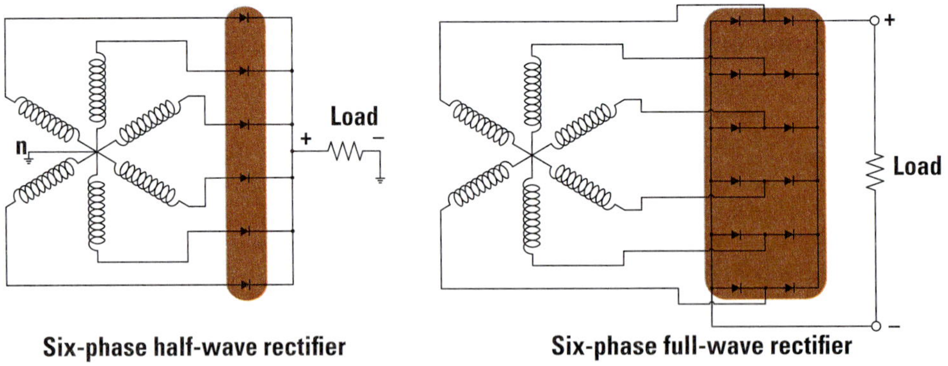

Figure 28–28 Six-phase transformer connections are often used to supply rectifiers for converting AC voltage into DC voltage.

The Double-Delta Connection

Another six-phase connection is the double delta. The transformers used in this connection must have two secondary windings that provide equal voltages. The secondary windings are connected in such a manner that two delta connections are formed. The two delta windings are connected in reverse, so they have an angular difference of 180° with respect to each other (*Figure 28–29*). A schematic diagram showing the double-delta connection is shown in *Figure 28–30.* The primary windings may be connected in WYE or delta.

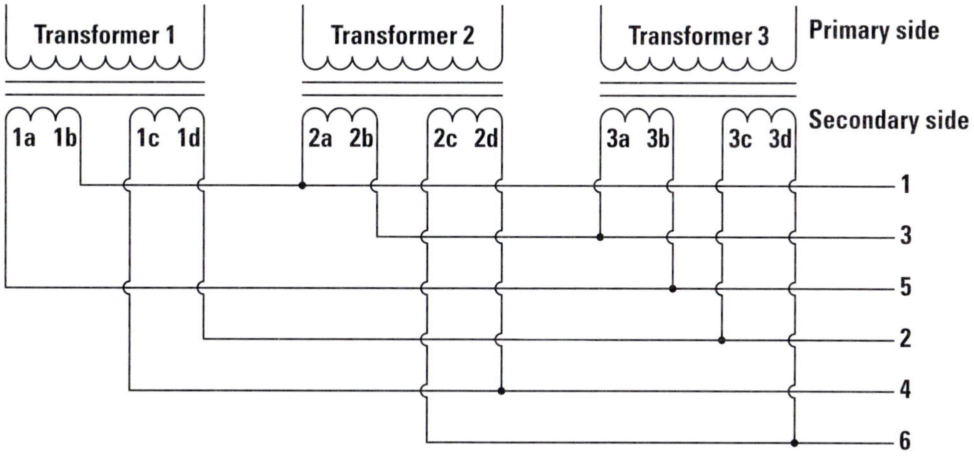

Figure 28–29 Six-phase transformer connection made with transformers that contain two separate secondary windings.

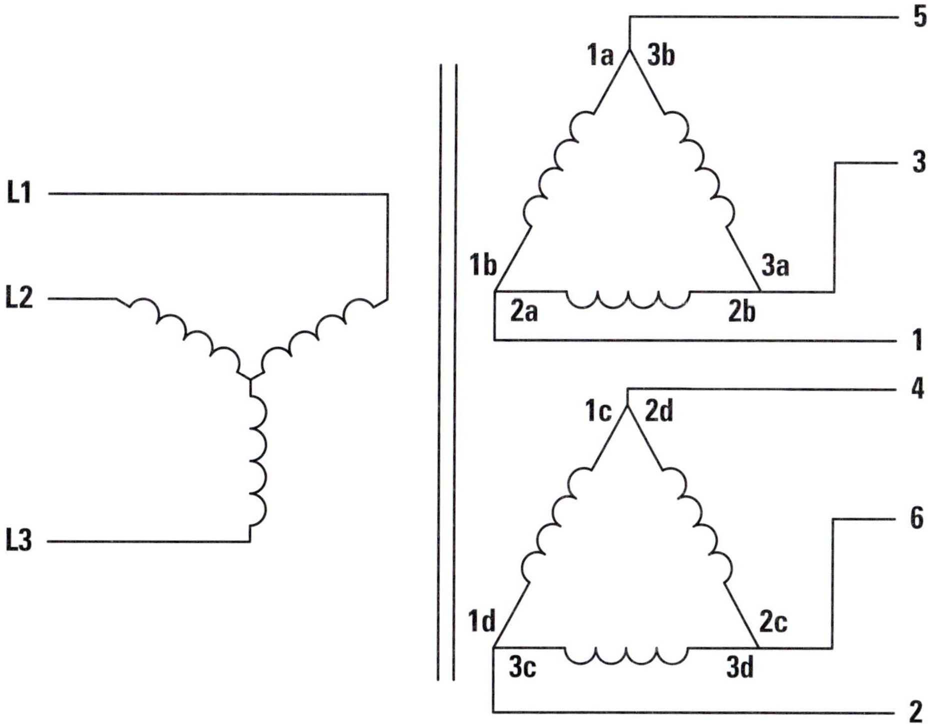

Figure 28–30 Schematic diagram of a six-phase transformer connection made with transformers that contain two separate secondary windings.

The Double-WYE Connection

The double-WYE connection is similar to the double-delta in that the transformers must have two secondary windings that produce equal voltage (*Figure 28–31*). A schematic diagram of a double-WYE connection is shown in *Figure 28–32*.

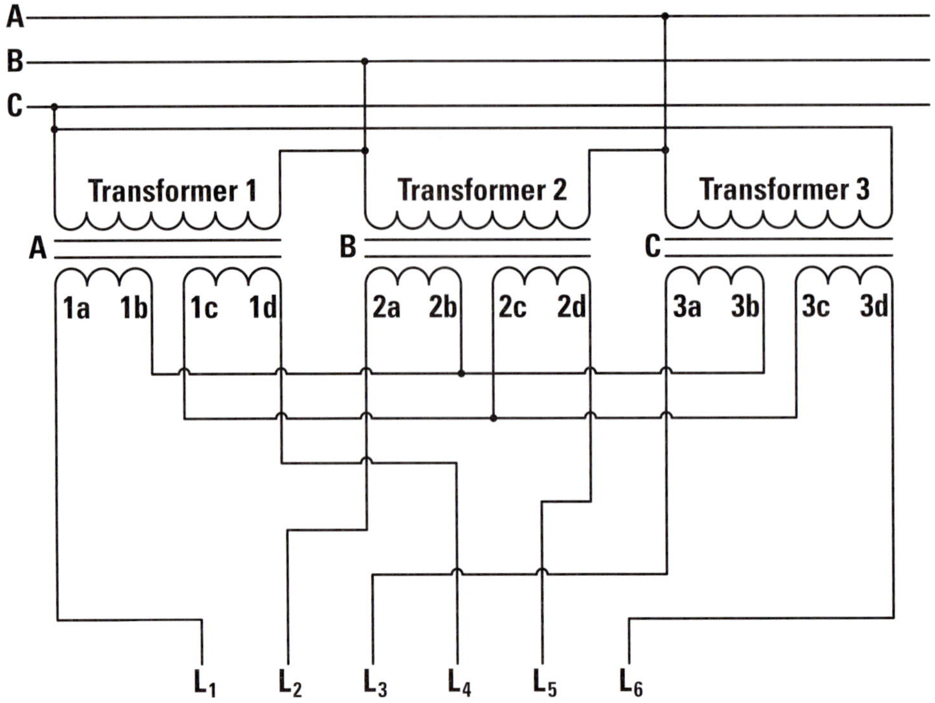

Figure 28–31 Double-WYE connection.

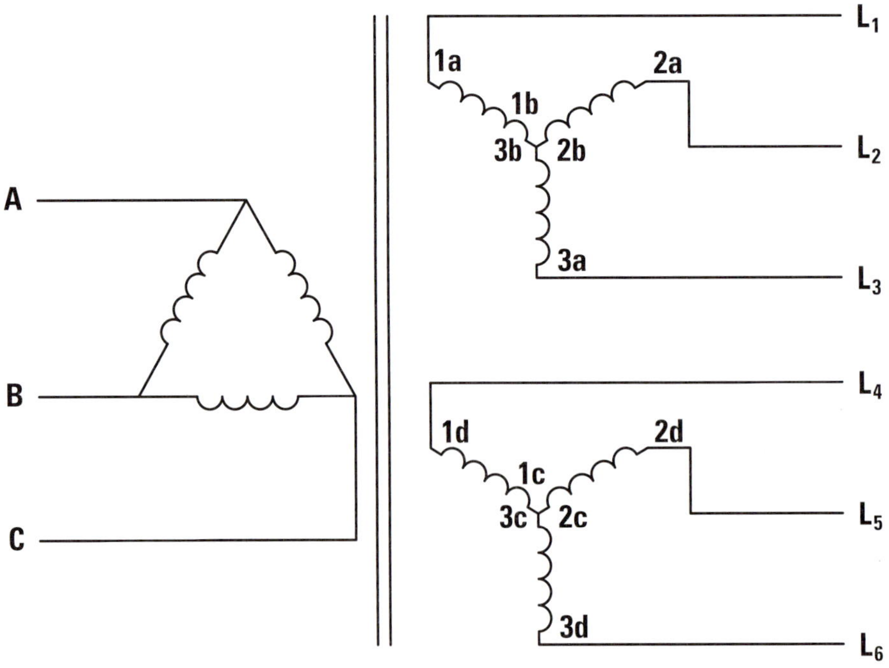

Figure 28–32 Schematic of a double-WYE connection.

Summary

- Three-phase transformers are constructed by winding three separate transformers on the same core material.
- Single-phase transformers can be used as a three-phase transformer bank by connecting their primary and secondary windings in either a WYE or delta configuration.
- When calculating three-phase transformer values, the rules for three-phase circuits must be followed as well as the rules for transformers.
- Phase values of voltage and current must be used when calculating the values associated with the transformer.
- The total power output of a three-phase transformer bank is the sum of the rating of the three individual transformers.
- An open-delta connection can be made with the use of only two transformers.
- When an open-delta connection is used, the total output power is 86.6% of the sum of the power rating of the two transformers.
- It is possible to center-tap one of the transformers in a delta connection to provide power for single-phase loads. When this is done, the phase connection with a higher voltage to ground becomes the high leg.
- The *NEC* requires that a high leg be identified by an orange wire or by tagging.
- The center connection of a WYE is used to provide a neutral conductor for single-phase loads. This produces a three-phase four-wire system. Common voltages produced by this type of connection are 208/120 and 480/277.

Review Questions

1. A transformer bank is connected so the X_2 terminals on the secondary windings are all connected together to form the neutral point. Which type of transformer was created?

   A) T-connected transformer

   B) Delta-connected transformer

   C) WYE-connected transformer

   D) Open-delta transformer

2. A transformer bank is connected so the midpoint of one winding is used as a neutral conductor, and one phase has a line-to-neutral voltage that is higher than the other two lines. Which type of transformer was created?

   A) T-connected transformer

   B) Delta-connected transformer

   C) WYE-connected transformer

   D) Zig-zag-connected transformer

3. An ammeter reads a line current of 425 amperes on a delta-connected three-phase transformer. What is the current through any one of the individual phase windings?

   A) 425 amperes

   B) 736.1 amperes

   C) 212.5 amperes

   D) 245.38 amperes

4. A WYE-connected, 225-kVA, three-phase transformer is connected to a three-phase load that draws 100 amperes on the primary at a voltage of 480 V. How much current is flowing in each of the phase windings on the secondary if the voltage is 120 V line-to-neutral?

   A) 400 amperes

   B) 230.8 amperes

   C) 624.5 amperes

   D) 1082 amperes

5. A three-phase transformer bank has the secondary connected in a WYE configuration. The center tap is used as a neutral conductor. If the voltage across any phase conductor and neutral is 277 V, how much voltage would be applied to a three-phase load connected to the secondary of this transformer bank?

   A) 480 volts

   B) 277 volts

   C) 208 volts

   D) 554 volts

6. True or False: Delta-connected transformers are typically used to supply only three-phase loads because one leg has a higher voltage to neutral than the other two legs.

7. A three-phase transformer connection has a delta-connected secondary, and one of the transformers has been center-tapped to form a neutral conductor. The phase-to-neutral value for the two lines that connect to the center-tapped secondary winding is 120 V. What is the voltage to neutral for the third line?

   A) 120 volts

   B) 240 volts

   C) 208 volts

   D) 277 volts

8. A three-phase transformer bank has the primary and secondary windings connected in a WYE configuration. Only the secondary has a center tap being used as a neutral to supply single-phase loads. Which of the following are likely outcomes of this connection?

   A) Unbalanced condition on secondary

   B) Unbalanced condition on primary

   C) Overcurrent condition on primary

   D) Overvoltage condition on secondary

9. A three-phase transformer connection has a delta-connected primary and a WYE-connected secondary. The center tap of the WYE is used as a neutral conductor. If the line-to-line voltage is 480 V, what is the voltage between all three phase conductors and the neutral conductor?
   A) 120 volts
   B) 277 volts
   C) 240 volts
   D) 480 volts
10. A WYE-connected secondary is best supplied from which type of transformer connection if the primary supply is a three-phase, three-wire feeder?
   A) WYE connection
   B) Zig-zag connection
   C) T-connection
   D) Delta connection
11. True or False: A T-connected transformer can supply three phase loads utilizing two windings instead of three.
12. Which of the following are common reasons why you might find a six-phase transformer connection?
   A) To supply six individual single-phase loads
   B) To supply a six-phase motor
   C) To supply a DC rectifier
   D) To supply two three-phase transformers

Practice Problems

Refer to the transformer shown in Figure 28–14 and find all the missing values.

1.

Primary	Secondary	Load
E_P ______	E_P ______	E_P ______
I_P ______	I_P ______	I_P ______
E_L 4160 V______	E_L 440 V______	E_L ______
I_L ______	I_L ______	I_L ______
Ratio	$Z = 3.5\ \Omega$	

2. Primary

E_P ______

I_P ______

E_L 7200 V______

I_L ______

Ratio

Secondary

E_P ______

I_P ______

E_L 240 V______

I_L ______

$Z = 4\ \Omega$

Load

E_P ______

I_P ______

E_L ______

I_L ______

Refer to the transformer connection shown in *Figure 28–33* and fill in the missing values.

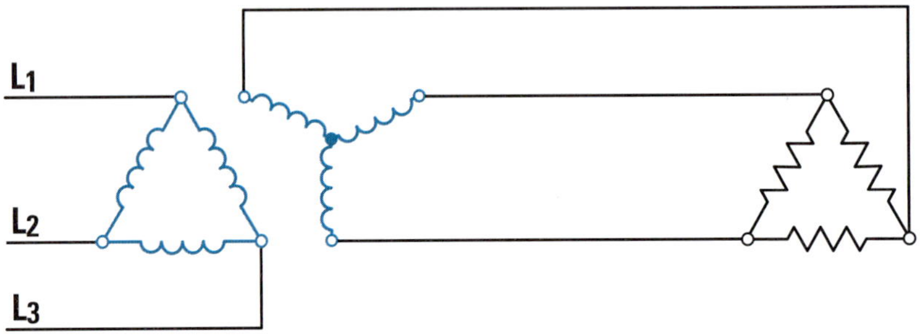

Figure 28–33 Practice problems circuit.

3. Primary

E_P ______

I_P ______

E_L 13,800 V______

I_L ______

Ratio

Secondary

E_P ______

I_P ______

E_L 480 V______

I_L ______

$Z = 2.5\ \Omega$

Load

E_P ______

I_P ______

E_L ______

I_L ______

4. Primary

E_P ______

I_P ______

E_L 23,000 V______

I_L ______

Ratio

Secondary

E_P ______

I_P ______

E_L 208 V______

I_L ______

$Z = 3\ \Omega$

Load

E_P ______

I_P ______

E_L ______

I_L ______

Section 11 DC Machines

Unit 29 DC Generators

Key Terms

Armature
Armature reaction
Brushes
Commutator
Compound generators
Compounding
Countertorque
Cumulative compound
Differential compound
Field excitation current
Frogleg-wound armatures
Generator
Lap-wound armatures
Pole pieces
Series field diverter
Series field windings
Series generator
Voltage regulator
Wave-wound armatures

Why You Need to Know

DC power is what initially started the electrification of the industrialized world. DC machines were ideal for industry due to their versatility and reliability as well as having easily controlled aspects such as speed and torque. This unit is devoted to the study of DC generators. Although AC machinery powers the vast majority of industry, the study of DC generators is very important because it is the foundation for the study of AC generators or alternators. It is also important to note that DC power is becoming more popular again with the advent of DC microgrids and battery storage systems. This unit

- describes the basic methods for generating electricity and how mechanical energy is converted to electric energy.
- explains many of the principles of induced voltage.
- describes how AC is produced in the armature of all rotating machines and how it is converted to DC before leaving the generator.
- discusses the various components of a DC generator and the principles of magnetic induction.
- explains the differences between series and shunt field windings and the characteristics of each.

Outline

29–1 What Is a Generator?
29–2 Generator Components
29–3 Generator Types
29–4 Compounding, Countertorque, and Reactivity in DC Generators
29–5 Operating DC Generators

Objectives

After studying this unit, you should be able to

- explain how DC generators convert mechanical energy to electrical energy.
- describe key factors that impact the voltage produced in a DC generator.
- discuss the functionality of major generator components.
- differentiate between the three primary types of DC generators.
- describe the compounding and countertorque properties of DC generators.
- describe methods for correcting armature reaction.
- explain the process for setting the brushes for operation of a DC generator.

Preview

Although most of the electric power generated throughout the world is AC, DC is used in several industrial applications. Many industrial plants use DC generators to produce the power needed to operate large DC motors. DC motors have characteristics that make them superior to AC motors for certain applications. Many plants also utilize DC microgrid systems and battery energy storage systems that store significant amounts of energy in DC and can supply local DC loads. DC generators and motors are also used in diesel locomotives. The diesel engine in most locomotives is used to operate a large DC generator. The generator is used to provide power to DC motors connected to the wheels.

29–1 What Is a Generator?

A **generator** is an electrical device that converts mechanical energy into electric energy. DC generators operate on the principle of magnetic induction. In Unit 13, it was shown that a voltage is induced in a conductor when it cuts magnetic lines of flux (*Figure 29–1*). In this example, the ends of the wire loop have been connected to two sliprings mounted on the shaft. Brushes are used to make contact with the sliprings and carry the current from the loop to the outside circuit.

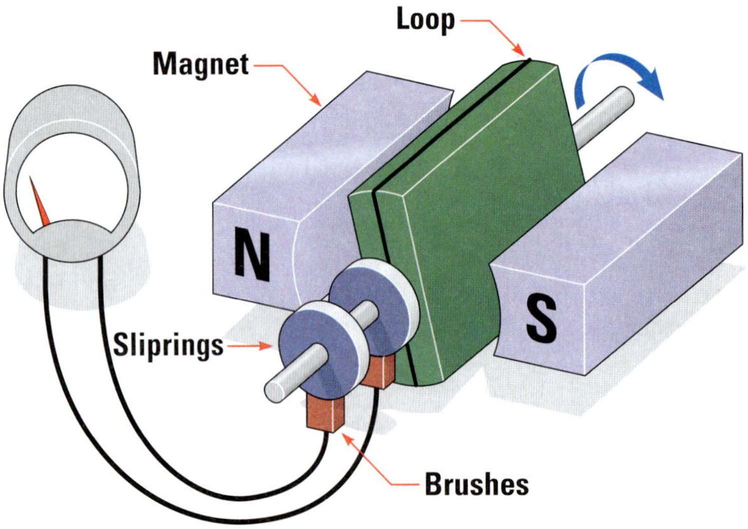

Figure 29–1 A voltage is induced in the loop conductor as it rotates and cuts magnetic lines of flux.

In *Figure 29–2*, an end view of the shaft and wire loop is shown. At this particular instant, the loop of wire is parallel to the magnetic lines of flux, and no cutting action is taking place. Because the lines of flux are not being cut by the loop, no voltage is induced in the loop.

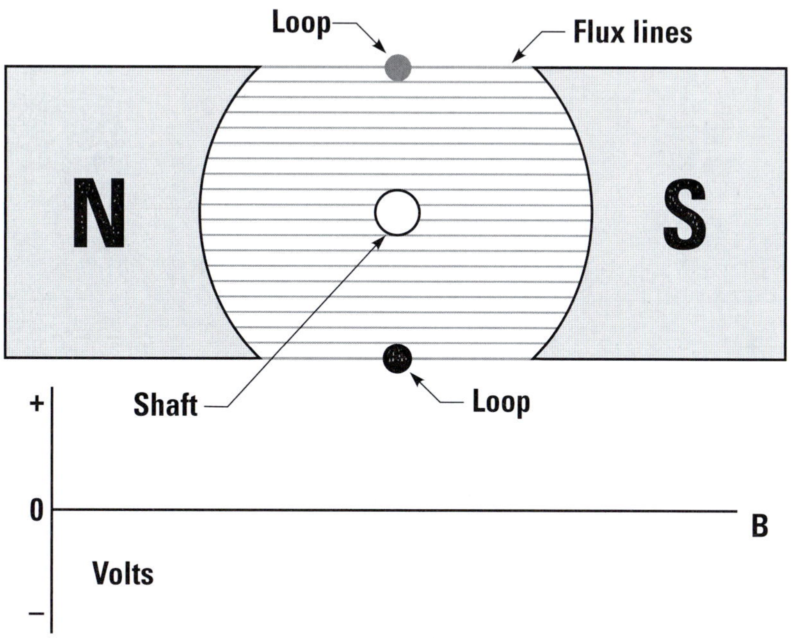

Figure 29–2 The loop conductor is traveling parallel to the lines of flux, and no cutting action is taking place.

In *Figure 29–3*, the shaft has been turned 90° clockwise. The loop of wire cuts through the magnetic lines of flux, and a voltage is induced in the loop. When the loop is rotated 90°, it is cutting the maximum number of lines of flux per second, and the voltage reaches its maximum, or peak, value.

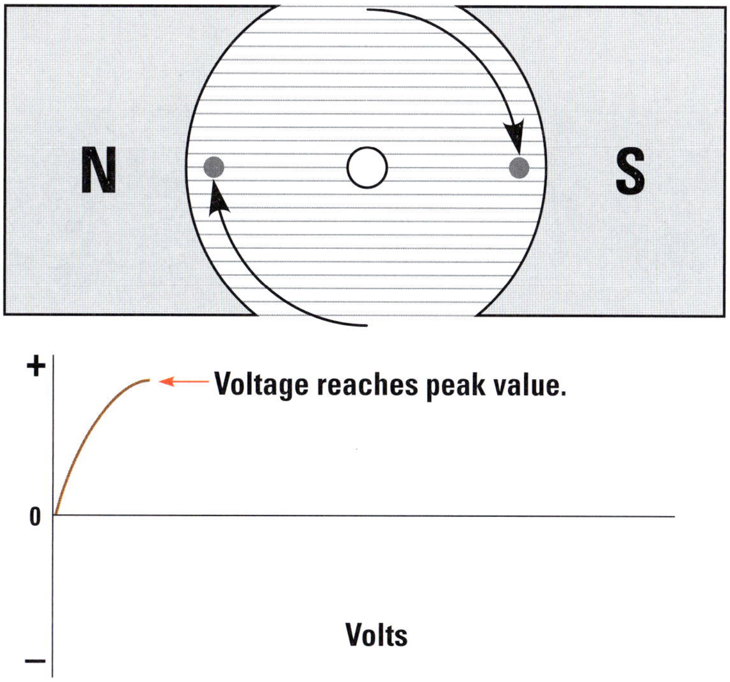

Figure 29–3 Maximum lines of flux are cut at 90° of rotation.

After another 90° of rotation (*Figure 29–4*), the loop has completed 180° of rotation and is again parallel to the lines of flux. As the loop was turned, the voltage decreased until it again reached zero.

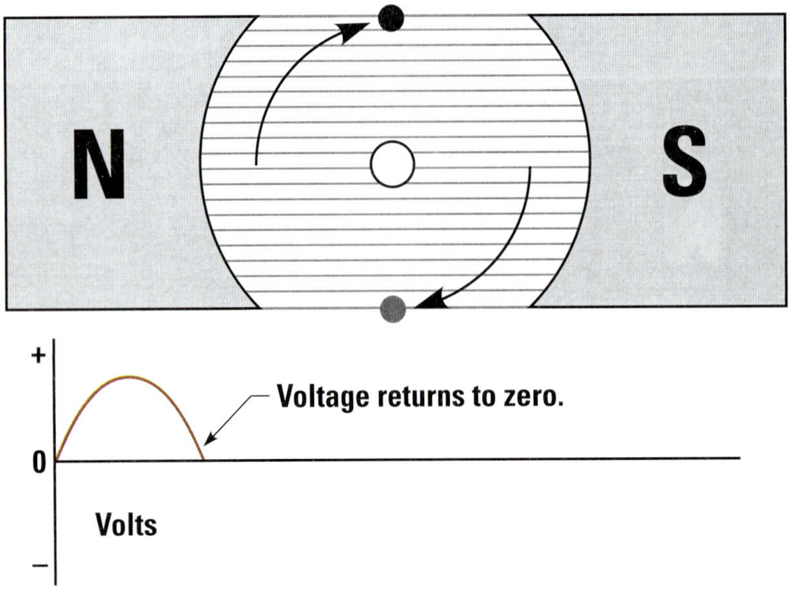

Figure 29–4 Induced voltage is back to zero after 180° of rotation.

As the loop continues to turn, the conductors again cut the lines of magnetic flux (*Figure 29–5*). This time, however, the conductor that previously cut through the flux lines of the south magnetic field is cutting the lines of the north magnetic field, and the conductor that previously cut the lines of the north magnetic field is cutting the lines of the south field. Because the conductors are cutting the flux lines of opposite magnetic polarity, the polarity of voltage reverses. After 270° of rotation, the loop has rotated to the position shown, and the maximum amount of voltage in the negative direction is being produced.

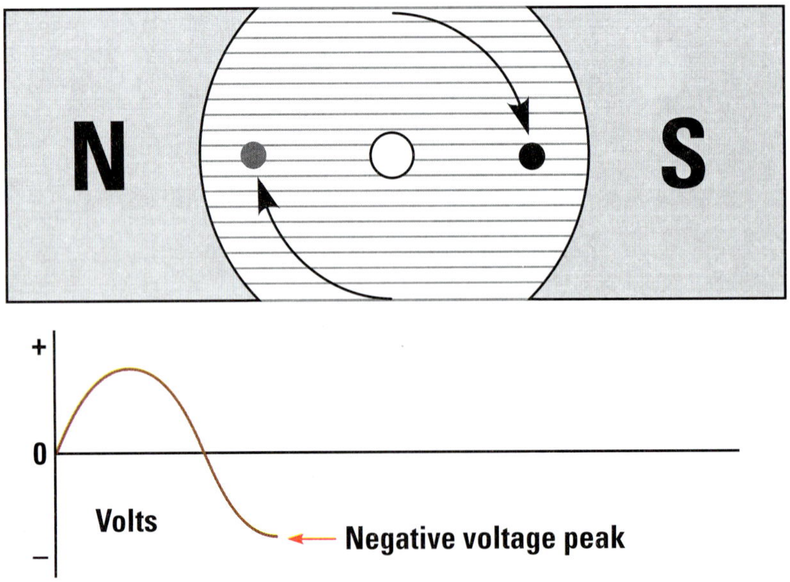

Figure 29–5 The negative voltage peak is reached after 270° of rotation.

After another 90° of rotation, the loop has completed one rotation of 360° and returned to its starting position (*Figure 29–6*), The voltage decreased from its negative peak back to zero. Notice that the voltage produced in the **armature** (the rotating member of the machine) alternates polarity. **The voltage produced in all rotating armatures is always going to be an alternating voltage.**

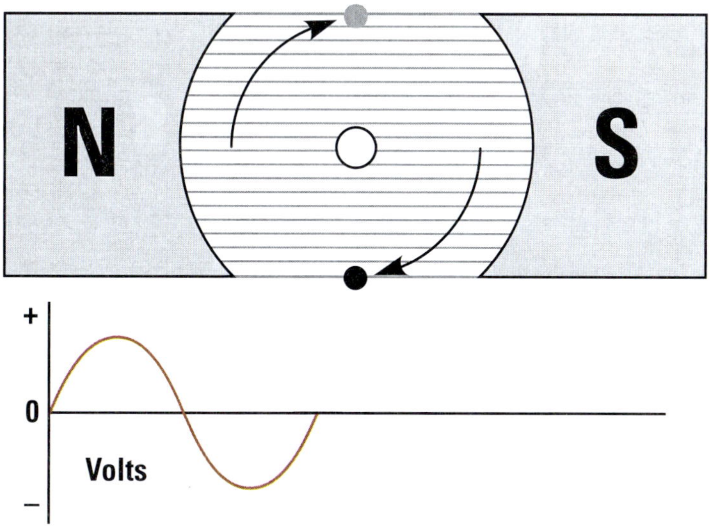

Figure 29–6 Voltage returns to the starting point after 360° of rotation.

Because DC generators must produce DC current instead of AC current, some device must be used to change the alternating voltage produced in the armature windings into direct voltage before it leaves the generator. This job is performed by the **commutator.** The commutator is constructed from a copper ring split into segments with insulating material between the segments (*Figure 29–7*). Brushes riding against the commutator segments carry the power to the outside circuit.

In *Figure 29–8*, the loop has been placed between the poles of two magnets. At this point in time, the loop is parallel to the magnetic lines of flux, and no voltage is induced in the loop. Note that the brushes make contact with both of the commutator segments at this time. This position, where there

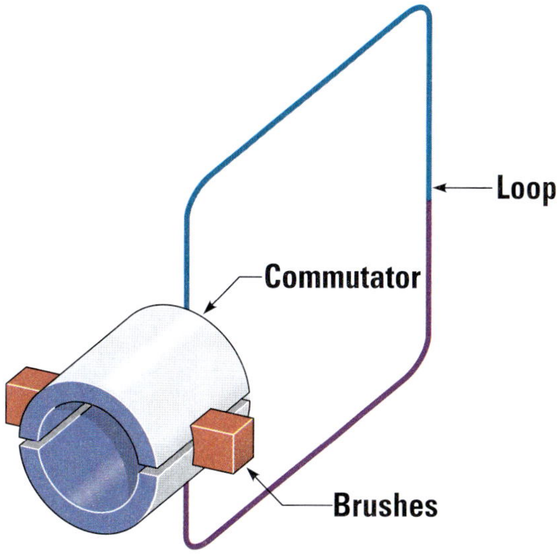

Figure 29–7 The commutator is used to convert the AC voltage produced in the armature into DC voltage.

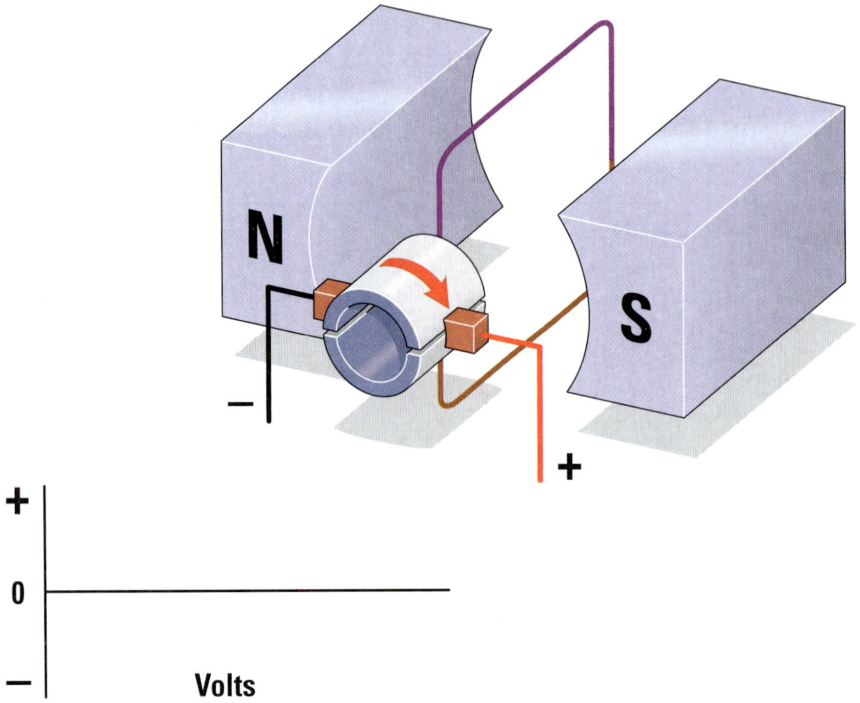

Figure 29–8 The loop is parallel to the lines of flux, and no voltage is produced.

is no induced voltage, is called the *neutral plane*. The brushes should be set to make contact between commutator segments when the armature windings are in the neutral plane position.

As the loop rotates, the conductors begin to cut through the magnetic lines of flux. The conductor cutting through the south magnetic field is connected to the commutator segment that is in contact with the positive brush, and the conductor cutting through the north magnetic field is connected to the negative brush (*Figure 29–9*). Because the loop is cutting lines of flux, a voltage is induced into the loop. After 90° of rotation, the voltage reaches its most positive point.

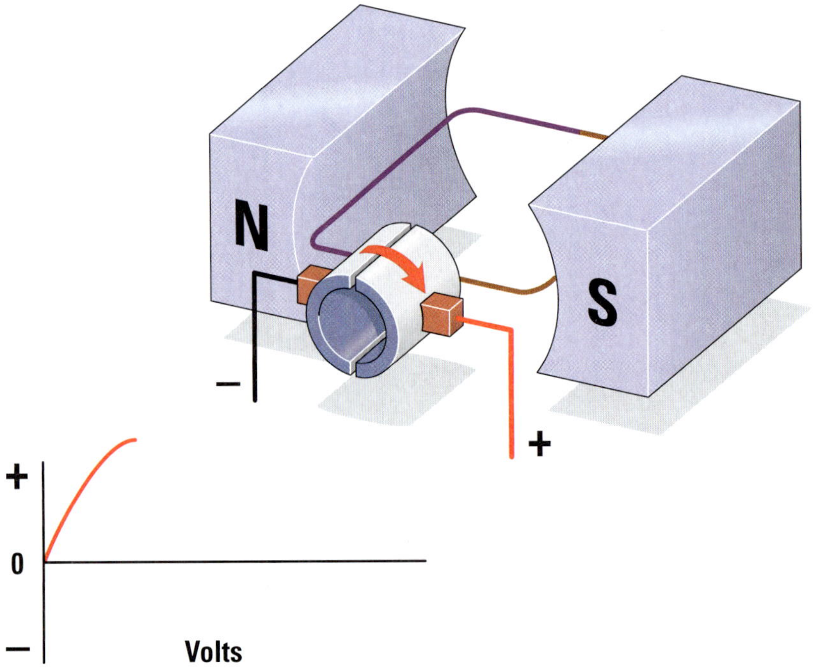

Figure 29–9 The loop has rotated 90° and produced peak positive voltage.

As the loop continues to rotate, the voltage decreases to zero. After 180° of rotation, the conductors are again parallel to the lines of flux, and no voltage is induced in the loop. Note that the brushes again make contact with both segments of the commutator at the time when there is no induced voltage in the conductors (*Figure 29–10*).

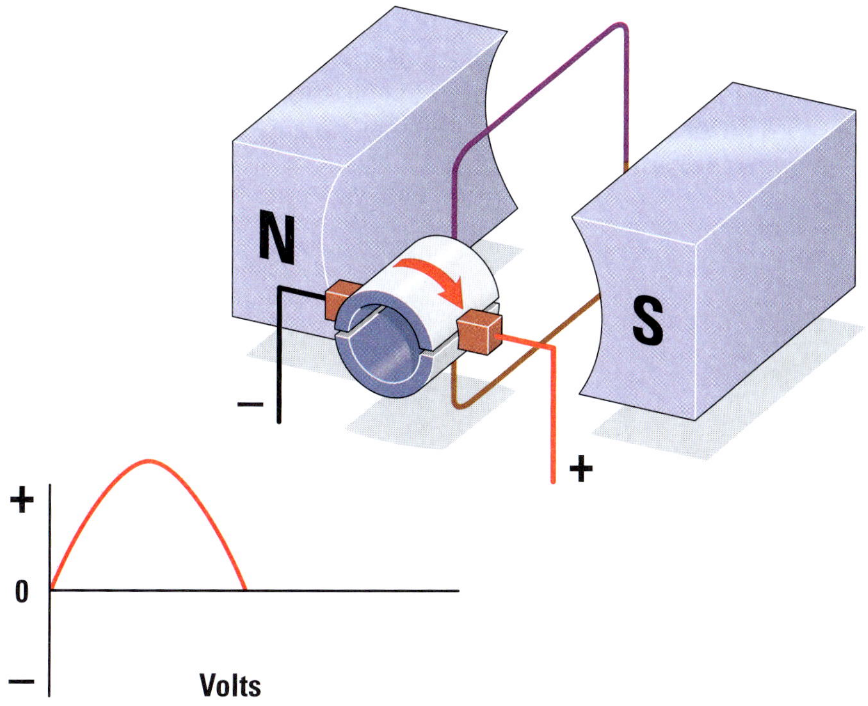

Figure 29–10 The loop has rotated 180° and again, no voltage is induced in conductors.

During the next 90° of rotation, the conductors again cut through the magnetic lines of flux. This time, however, the conductor that previously cut through the south magnetic field is now cutting the flux lines of the north magnetic field, and the conductor that previously cut the lines of flux of the north magnetic field is cutting the lines of flux of the south magnetic field (*Figure 29–11*). Because these conductors are

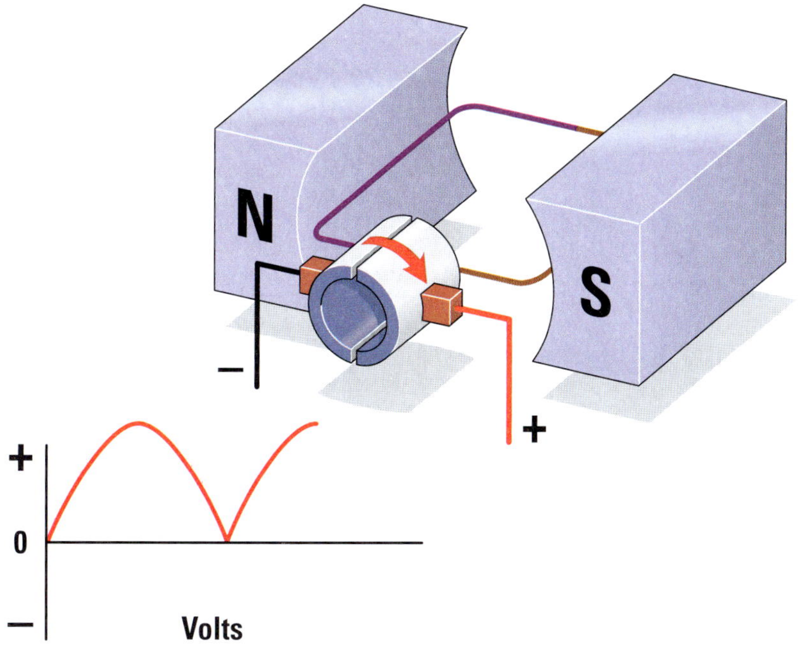

Figure 29–11 The commutator maintains the proper polarity.

cutting the lines of flux of opposite magnetic polarities, the polarity of induced voltage is different for each of the conductors. The commutator, however, maintains the correct polarity to each brush. The conductor cutting through the north magnetic field will always be connected to the negative brush, and the conductor cutting through the south field will always be connected to the positive brush. Because the polarity at the brushes has remained constant, the voltage will increase to its peak value in the same direction.

As the loop continues to rotate (*Figure 29–12*), the induced voltage again decreases to zero when the conductors become parallel to the magnetic lines of flux. Notice that, during this 360° rotation of the loop, the polarity of voltage remained the same for both halves of the waveform. The voltage is pulsating or fluctuating. It does turn on and off, but it never reverses polarity. Because the polarity for each brush remains constant, the output voltage is DC.

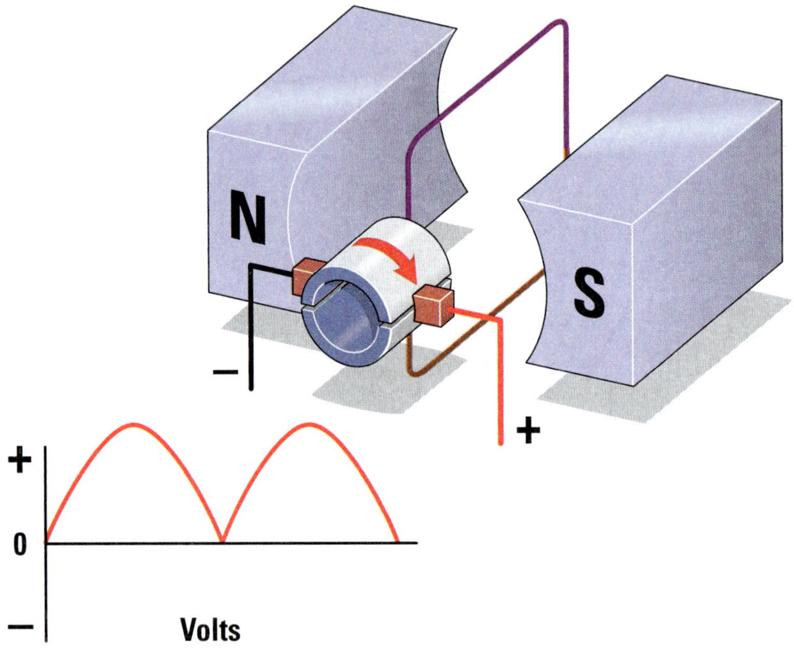

Figure 29–12 The loop completes one complete rotation, and the voltage polarity never changes.

To increase the amount of output voltage, it is common practice to increase the number of turns of wire for each loop (*Figure 29–13*). If a loop contains 20 turns of wire, the induced voltage is 20 times

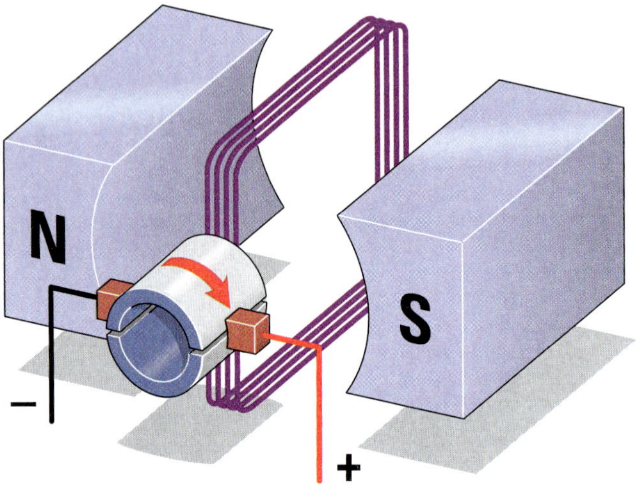

Figure 29–13 Increasing the number of turns in the armature increases the output voltage as more conductors cut lines of flux.

greater than that for a single conductor loop. The reason for this is that each loop is connected in series with the other loops. Because the loops form a series path, the voltage induced in the loops adds. In this example, if each loop has an induced voltage of 2 volts, the total voltage for this winding would be 40 volts (2 V × 20 loops = 40 V).

It is also common practice to use more than one loop of wire (*Figure 29–14*). When more than one loop is used, the average output voltage is higher and there is less pulsation or ripple in the DC voltage. Adding more loops to the armature has the effect of cutting the rotational distance between polarity changes in the armature. For example, adding loops at a 90° angle to the original loop creates four peak voltages per rotation instead of two peaks for a single loop.

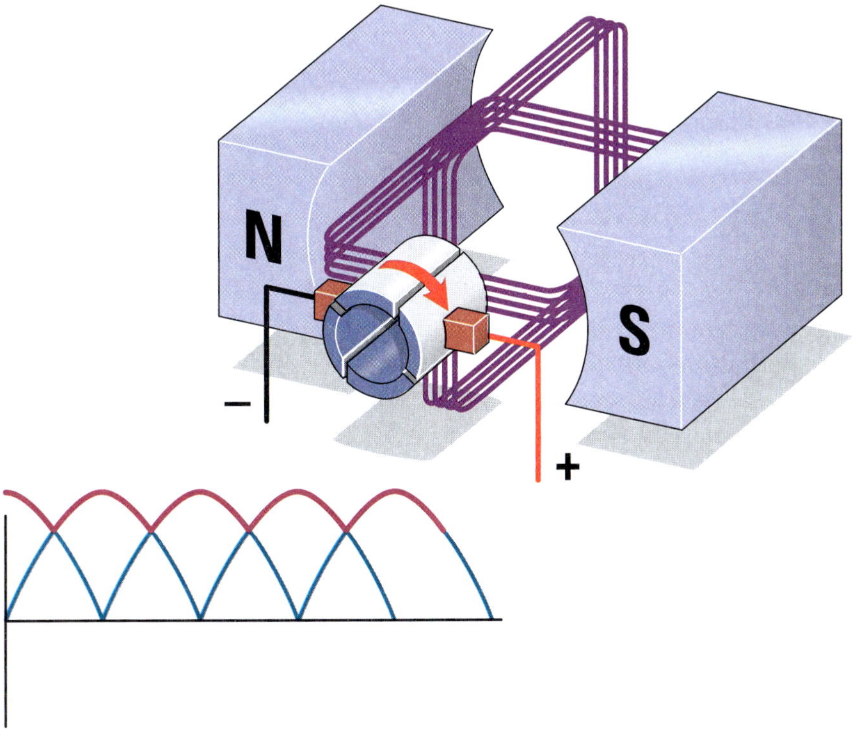

Figure 29–14 Increasing the number of loops produces a smoother output voltage.

The loops are generally placed in slots of an iron core (*Figure 29–15*). The iron acts as a magnetic conductor by providing a low-reluctance path for magnetic lines of flux to increase the inductance

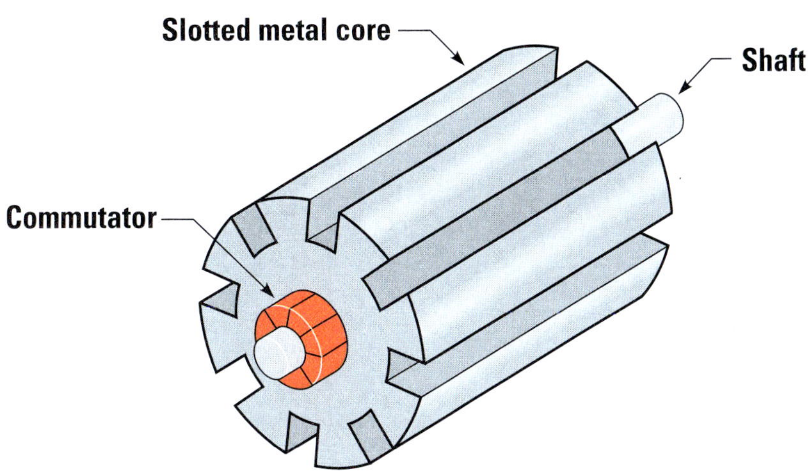

Figure 29–15 The loops of wire are wound around slots in a metal core.

of the loops and provide a higher induced voltage. The commutator is connected to the slotted metal (typically iron) core. The entire assembly of iron core, commutator, and windings is called the armature (*Figures 29–16A and 29–16B*). The windings of armatures are connected in different ways depending on the requirements of the machine. The three basic types of armature windings are the *lap, wave,* and *frogleg*.

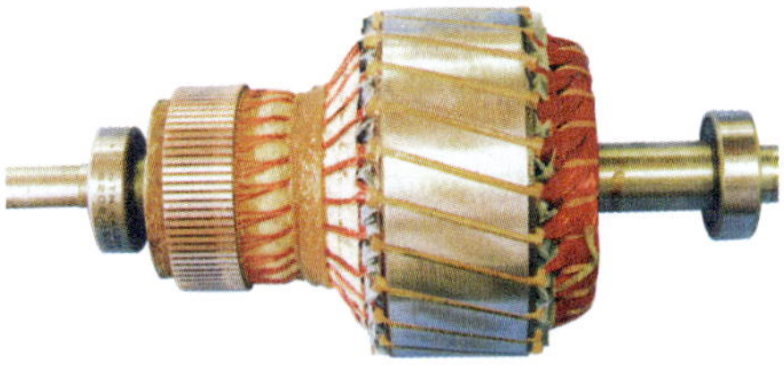

Figure 29–16A DC machine armature.

Figure 29–16B Cutaway view of an armature.

29–2 Generator Components

Lap-Wound Armatures

Lap-wound armatures are used in machines designed for high current but not necessarily high voltage levels. These armatures are generally constructed with large wire because of the high current. A good example of where lap-wound armatures are used is in the starter motor of almost all automobiles. One characteristic of machines that use a lap-wound armature is that they have as many pairs of brushes as there are pairs of poles. The windings of a lap-wound armature are connected in parallel (*Figure 29–17*). This permits the current capacity of each winding to be added and provides a higher operating current. Lap-wound armatures have as many parallel paths through the armature as there are pole pieces.

Figure 29–17 Lap-wound armatures have parallel windings and are intended for high-current operation.

Wave-Wound Armatures

Wave-wound armatures are used in machines designed for high voltage but cannot handle high amounts of current. These armatures have their windings connected in series, as shown in *Figure 29–18*. When the windings are connected in series, the voltage of each winding adds, but the current

Figure 29–18 Wave-wound armatures have series connected windings and are used in machines intended for high-voltage operation.

capacity remains the same. A good example of where wave-wound armatures are used is in the small generator in hand-cranked megohmmeters. Wave-wound armatures never contain more than two parallel paths for current flow regardless of the number of pole pieces, and they never contain more than one set of brushes (a set being one brush or group of brushes for positive and one brush or group of brushes for negative).

Frogleg-Wound Armatures

Frogleg-wound armatures are probably the most used. These armatures are used in machines designed for use with moderate levels of both voltage and current. The windings of a frogleg-wound armature are connected in series-parallel, as shown in *Figure 29–19*. Most large DC machines use frogleg-wound armatures.

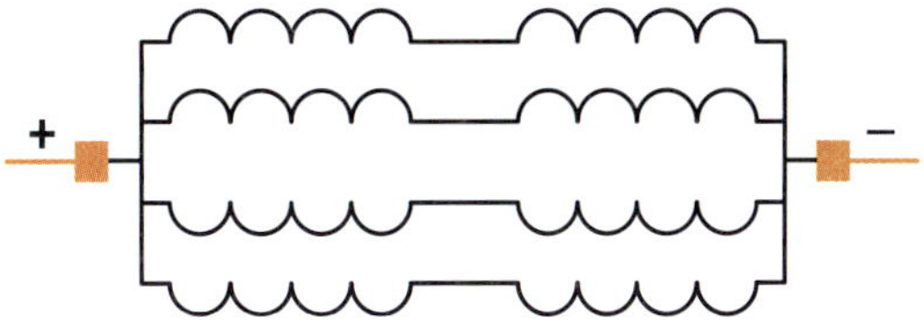

Figure 29–19 Frogleg-wound armatures are connected in a series-parallel configuration and can handle elevated current and voltage levels

Brushes

The **brushes** ride against the commutator segments and are used to connect the armature to the external circuit of the DC machine (*Figure 29–20*). Brushes are made from a material that is softer than the copper bars of the commutator. This permits the brushes, which are easy to replace, to wear instead of the commutator. The brush leads are generally marked A_1 and A_2 and are referred to as the *armature leads*.

Figure 29–20 Brushes make contact with segments of the commutator. Shutterstock, Jay Petersen.

Pole Pieces

The **pole pieces** are located inside the housing of the DC machine (*Figure 29–21*). The pole pieces provide the magnetic field necessary for the operation of the machine. They are constructed of some type of good magnetic conductive material such as soft iron or silicon steel. These easily magnetized materials help strengthen the field around the pole pieces and maximize the effectiveness of the conductors cutting through the lines of flux. Some DC generators use permanent magnets to provide the magnetic field instead of electromagnets. These machines are generally small and rated about one horsepower or less. A DC generator that uses permanent magnets as its field is referred to as a *magneto*.

Figure 29–21 Pole pieces are constructed of soft iron and placed on the inside of the housing.

Field Windings

Most DC machines use wound electromagnets to provide the magnetic field. Two types of field windings are used. One is the series field, and the other is the shunt field. **Series field windings** are made with relatively few turns of very large wire and have a very low resistance. They are so named because they are connected in series with the armature. The terminal leads of the series field are labeled S_1 and S_2. It is not uncommon to find the series field of large-horsepower machines wound with square or rectangular wire (*Figure 29–22*). The use of square wire permits the windings to be laid closer together, which increases the number of turns that can be wound in a particular space. Additionally, smaller square and rectangular wire can be used to yield the same surface area of larger round wire (*Figure 29–23*). An armature being rewound with rectangular wire is shown in *Figure 29–24*.

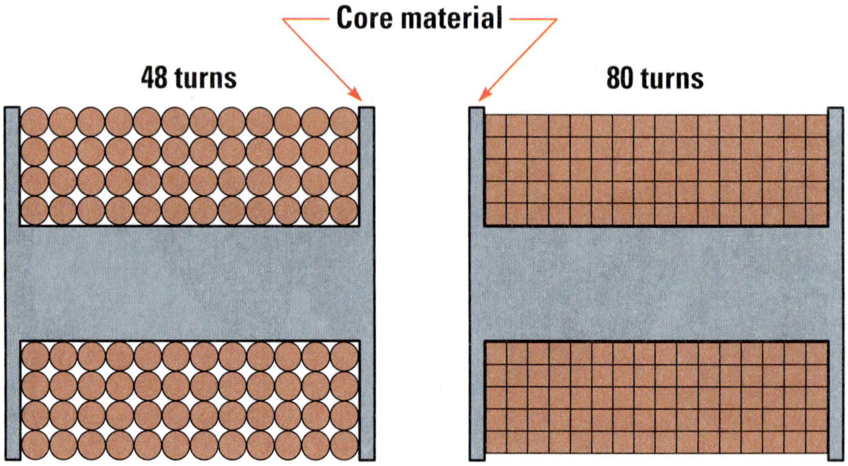

Figure 29–22 Square wire permits more turns than round wire in the same area.

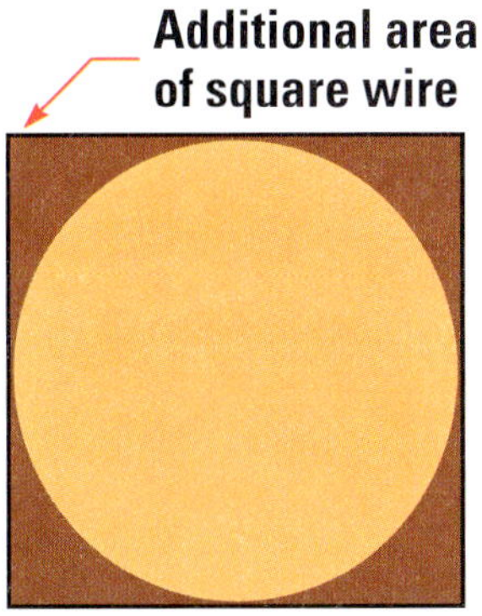

Figure 29–23 A square wire of equal size contains more surface area than round wire.

Figure 29–24 Armature wound with rectangular wire. H.L. Hailey Enterprises, Inc. - Longview, TX

Shunt field windings are made with many turns of small wire. Because the shunt field is constructed with relatively small wire, it has a much higher resistance than the series field. The shunt field is intended to be connected in parallel with, or shunt, the armature. The resistance of the shunt field must be high because its resistance is used to limit current flow through the field. The shunt field is often referred to as the "field," and its terminal leads are labeled F_1 and F_2.

When a DC machine uses both series and shunt fields, each pole piece contains both windings (*Figure 29–25*). The windings are wound on the pole pieces in such a manner that when current flows

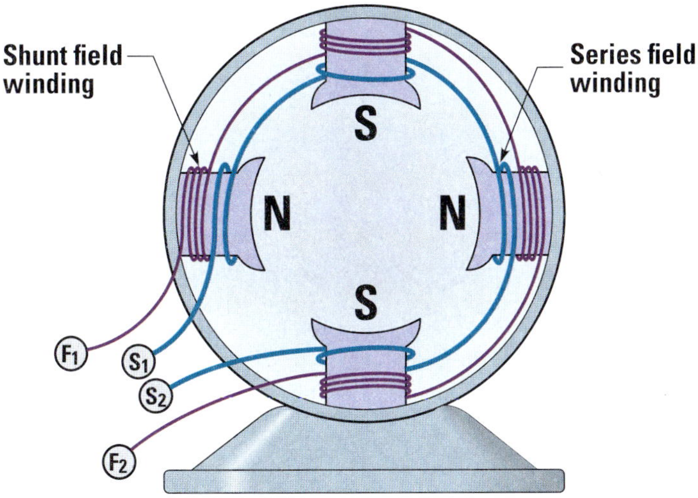

Figure 29–25 Pole pieces contain both series and shunt windings that result in alternating pole piece magnetic polarities.

through the winding, it produces alternating magnetic polarities. In *Figure 29–25*, two pole pieces form north magnetic polarities and two form south magnetic polarities. A DC machine with two field poles and one *interpole* is shown in *Figure 29–26*. Interpoles are discussed later in this unit.

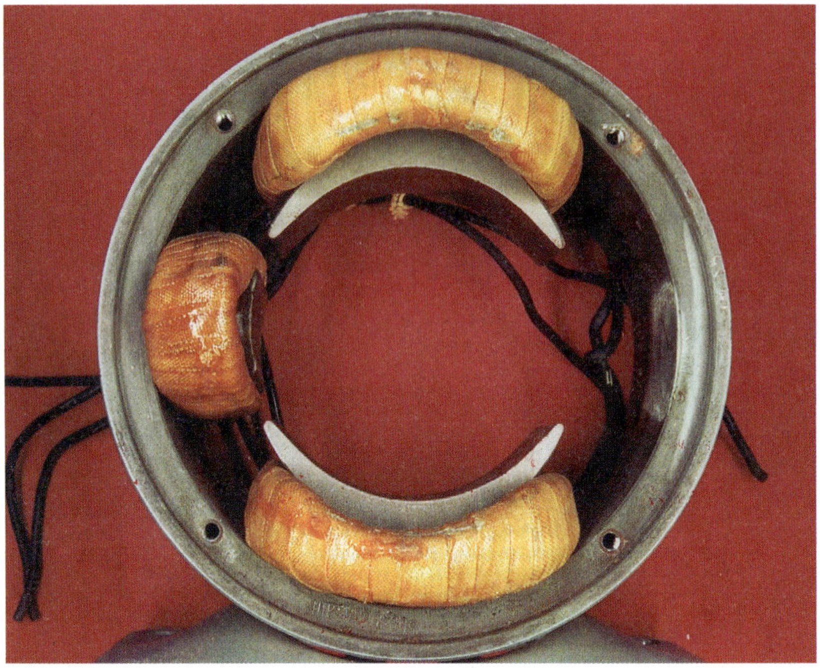

Figure 29–26 A two-pole DC machine with one interpole.

29–3 Generator Types

There are three basic types of DC generators: the series, the shunt, and the compound. The type is determined by the arrangement and connection of field coils. The **series generator** contains only a series field connected in series with the armature (*Figure 29–27*). A schematic diagram used to

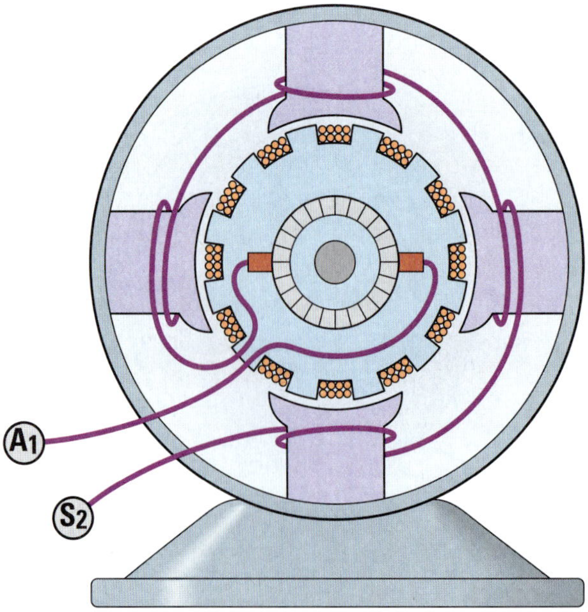

Figure 29–27 The series field is connected in series with the armature.

represent a series-connected DC machine is shown in *Figure 29–28*. The series generator must be *self-excited,* which means that the pole pieces contain some amount of residual magnetism. This residual magnetism produces an initial output voltage that permits current to flow through the field if a load is connected to the generator. The amount of output voltage produced by the generator is proportional to the following three factors:

1. the number of turns of wire in the armature,
2. the strength of the magnetic field of the pole pieces, and
3. the speed of the cutting action (speed of rotation).

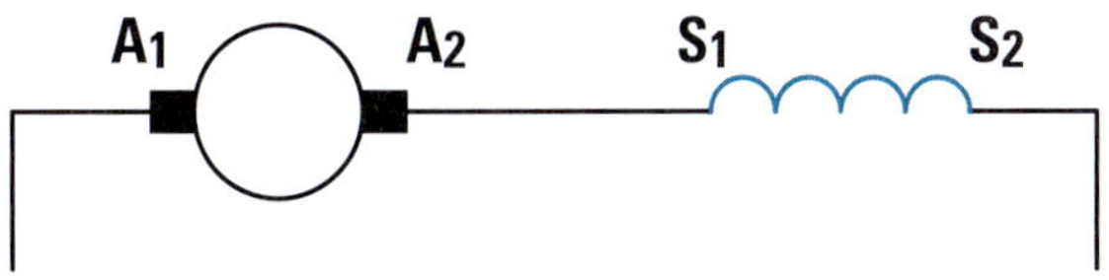

Figure 29–28 Schematic drawing of a series generator.

To understand why these three factors determine the output voltage of a generator, recall from Unit 13 that 1 volt is induced in a conductor when it cuts magnetic lines of flux at a rate of 1 weber per second (1 weber = 100,000,000 lines of flux). When conductors are wound into a loop, each turn acts like a separate conductor. Because the turns are connected in series, the voltage induced into each conductor adds. If one conductor has an induced voltage of 0.5 volt and there are 20 turns, the total induced voltage would be 10 volts.

The second factor is the strength of the magnetic field. Flux density is a measure of the strength of a magnetic field. If the number of turns of wire in the armature remains constant and the speed remains constant, the output voltage can be controlled by the number of flux lines produced by the field poles. Increasing the lines of flux increases the number of flux lines cut per second and, therefore, the output voltage. The magnetic field strength can be increased until the iron of the pole pieces reaches saturation.

Induced voltage is proportional to the number of flux lines cut per second. If the strength of the magnetic field remains constant and the number of turns of wire in the armature remains constant, the output voltage is determined by the speed at which the conductors cut the flux lines. Increasing the speed of the armature increases the speed of the cutting action, which increases the output voltage. Likewise, decreasing the speed of the armature decreases the output voltage.

Connecting Load to the Series Generator

When a load is connected to the output of a series generator, the initial voltage produced by the residual magnetism of the pole pieces produces a current flow through the load (*Figure 29–29*). Because the series field is connected in series with the armature, the current flowing through the armature and load must also flow through the series field. This causes the magnetism of the pole pieces to become

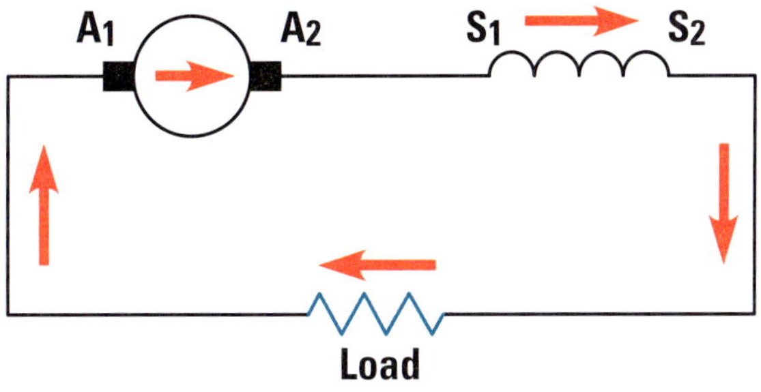

Figure 29–29 Residual magnetism produces an initial voltage used to provide a current flow through the load.

stronger and produce more magnetic lines of flux. When the strength of the magnetic pole pieces increases, the output voltage increases also.

If another load is added (*Figure 29–30*), more current flows, and the pole pieces produce more magnetic lines of flux, which again increases the output voltage. Each time a load is added to the series generator, its output voltage increases. Notice that the additional load is connected in parallel. This increase of voltage continues until the iron in the pole pieces and armature becomes saturated. At that point, an increase of load results in a decrease of output voltage (*Figure 29–31*).

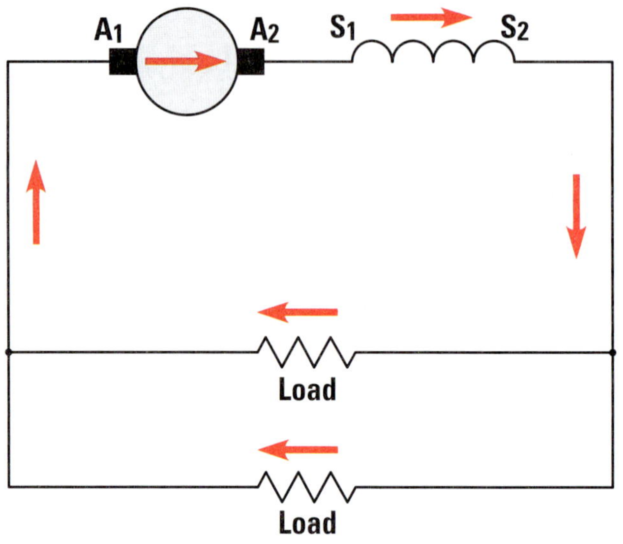

Figure 29–30 If more load is added in parallel, current flow and output voltage increase.

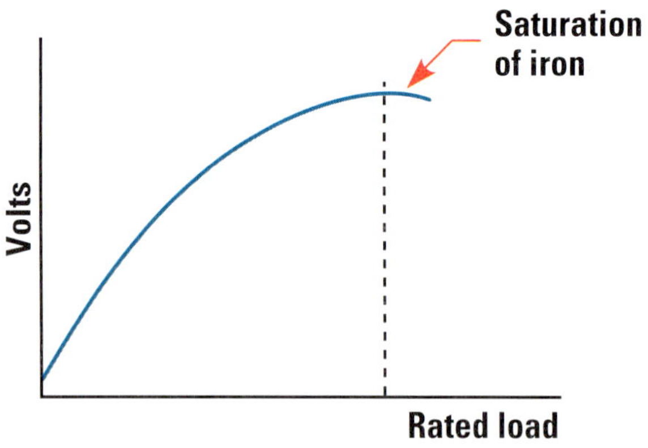

Figure 29–31 Characteristic curve of a series generator.

Shunt Generators

Shunt generators contain only a shunt field winding connected in parallel with the armature (*Figure 29–32*). A schematic diagram used to represent a shunt-connected DC machine is shown in *Figure 29–33*. Shunt generators can be either self-excited or separately excited. Self-excited shunt generators are similar to self-excited series generators in that residual magnetism in the pole pieces is used to produce an initial output voltage. In the case of a shunt generator, however, the initial voltage is used to produce a current flow through the shunt field. This current increases the magnetic field strength of the pole pieces, which produces a higher output voltage. This buildup of voltage continues until a maximum value, determined by the speed of rotation, the turns of wire in the armature, and the turns of wire on the pole pieces, is reached.

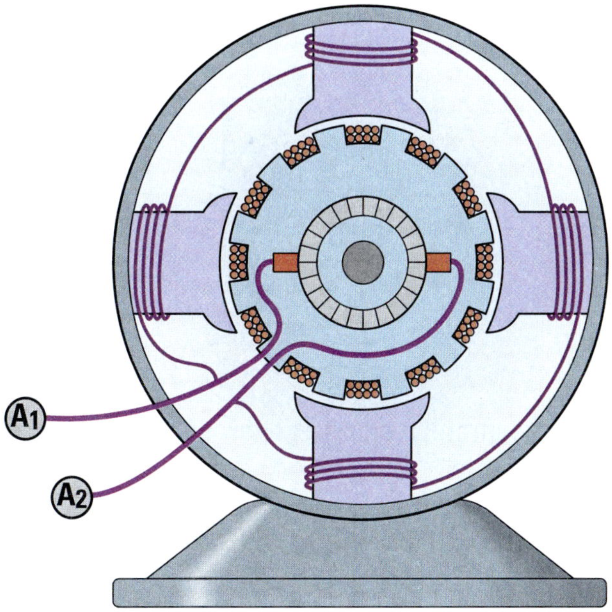

Figure 29–32 Shunt field windings are connected in parallel with the armature.

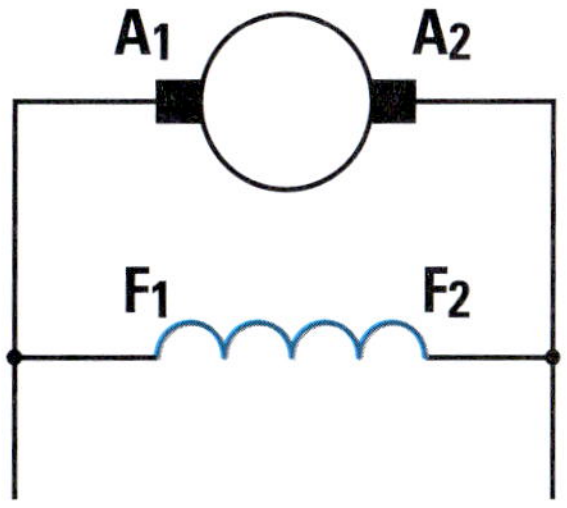

Figure 29–33 Schematic drawing of a shunt generator.

Another difference between the series generator and the self-excited shunt generator is that the series generator must be connected to a load before voltage can increase. The load is required to form a complete path for current to flow through the armature and series field (*Figure 29–29*). In a self-excited shunt generator, the shunt field winding provides a complete circuit across the armature, permitting the full output voltage to be obtained before a load is connected to the generator.

Separately excited generators have their fields connected to an external source of DC (*Figure 29–34*). The advantages of the separately excited machine are that it gives better control of the output voltage and that its voltage drop is less when load is added. The characteristic curves of both self-excited and separately excited shunt generators are shown in *Figure 29–35*.

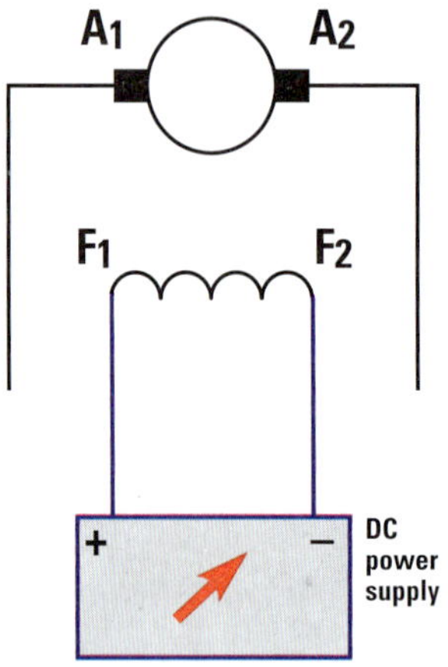

Figure 29–34 Separately excited shunt generators must have an external power source to provide excitation current for the shunt field.

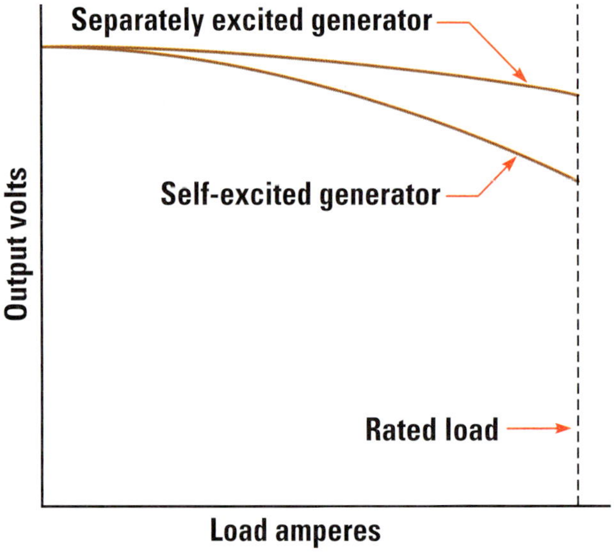

Figure 29–35 Separately excited shunt generators have less voltage drop than self-excited shunt generators as the connected load increases.

The self-excited generator exhibits a greater drop in voltage when load is added because the armature voltage is used to produce the current flow in the shunt field. Each time the voltage decreases, the current flow through the field decreases, causing a decrease in the amount of magnetic flux lines in the pole pieces. This decrease of flux in the pole pieces causes a further decrease of output voltage. The separately excited machine does not have this problem because the field flux is held constant by the external power source.

Field Excitation Current

Regardless of which type of shunt generator is used, the amount of output voltage is generally controlled by the amount of **field excitation current**. Field excitation current is the DC that flows through the shunt field winding. This current is used to turn the iron pole pieces into electromagnets. Because one of the factors that determines the output voltage of a DC generator is the strength of the magnetic field, the output voltage can be controlled by the amount of current flow through the field coils. A simple method of controlling the output voltage is by the use of a shunt field rheostat. Remember, a rheostat is a type of variable resistor. The rheostat is connected in series with the shunt field winding (*Figure 29–36*). By adding or removing resistance connected in series with the shunt field winding, the amount of current flow through the field can be controlled. This, in turn, controls the strength of the magnetic field of the pole pieces.

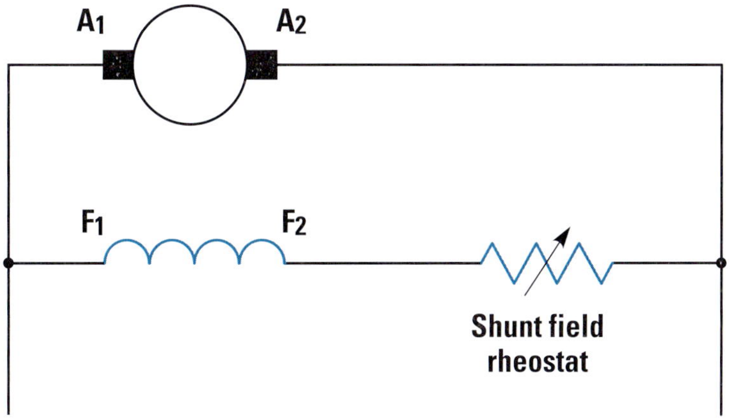

Figure 29–36 A rheostat can be placed in series with shunt field windings to control the output voltage.

When it is important that the output voltage remain constant regardless of load, an electronic **voltage regulator** can be used to adjust the shunt field current (*Figure 29–37*). The voltage regulator connects in series with the shunt field in a similar manner as a rheostat. The regulator, however, senses the amount of voltage across the load. If the output voltage drops, the regulator permits more current to flow through the shunt field. If the output voltage becomes too high, the regulator decreases the current flow through the shunt field.

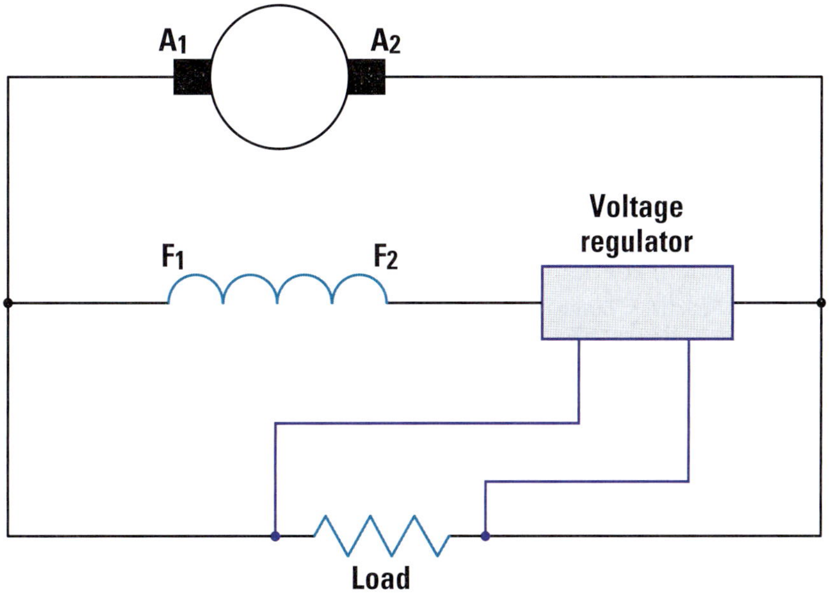

Figure 29–37 The voltage regulator controls the amount of shunt field current.

Generator Losses

When load is added to the shunt generator, the output voltage drops. This voltage drop is due to losses that are inherent to the generator. The largest of these losses is generally caused by the resistance of the armature. In *Figure 29–38*, it is assumed that the armature has a wire resistance of 10 ohms. When a load is connected to the output of the generator, current flows from the armature, through the load, and back to the armature. As current flows through the armature, the resistance of the wire causes a voltage drop. Assume that the armature has a current flow of 2 amperes. If the resistance of the armature is 10 ohms, it will require 20 volts to push the current through the resistance of the armature.

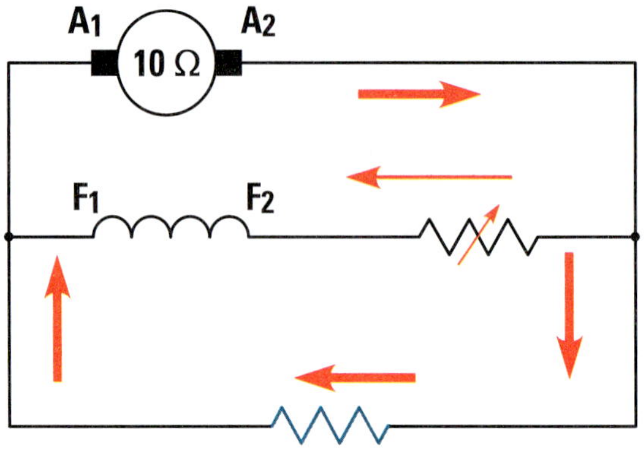

Figure 29–38 Resistance in the armature wire causes a drop in output voltage.

Now assume that the armature wire only has a resistance of 2 ohms. The same 2 amperes of current flow require only 4 volts to push the current through the armature resistance. A low-resistance armature is generally a very desirable characteristic for DC machines. **In the case of a generator, the voltage regulation is determined by the resistance of the armature.** A generator with good voltage regulation has a small amount of voltage drop as load is added.

Some other losses are I^2R losses, eddy current losses, and hysteresis losses. Recall that I^2R is one of the formulas for finding power, or watts. In the case of a DC machine, I^2R describes the power loss associated with heat due to the resistance of the wire in both the armature and field windings.

Eddy currents are currents that are induced into the metal core material by the changing magnetic field as the armature spins through the flux lines of the pole pieces. Eddy currents were discussed in previous units. These swirling currents produce heat that heats the surrounding metal and causes a power loss. Many machines are constructed with laminated pole pieces and armature cores to help reduce eddy currents. The surface of each lamination forms a layer of iron oxide, which acts as an insulator to help prevent the formation of eddy currents.

Hysteresis losses are losses due to molecular friction. As discussed previously, AC is produced inside the armature. This reversal of the direction of current flow causes the molecules of iron in the core to realign themselves each time the current changes direction. The molecules of iron are continually rubbing against each other as they realign magnetically. The friction of the molecules rubbing together causes heat, which is a power loss. Hysteresis loss is proportional to the speed of rotation of the armature. The faster the armature rotates, the more current reversals there are per second and the more heat is produced because of friction.

Compound Generators

Compound generators contain both series and shunt fields. Most large DC machines are compound-wound. The series and shunt fields can be connected in two ways. One connection is called *long shunt* (*Figure 29–39*). The long shunt connection has the shunt field connected in parallel with both the armature and series field. This is the most used of the two connections.

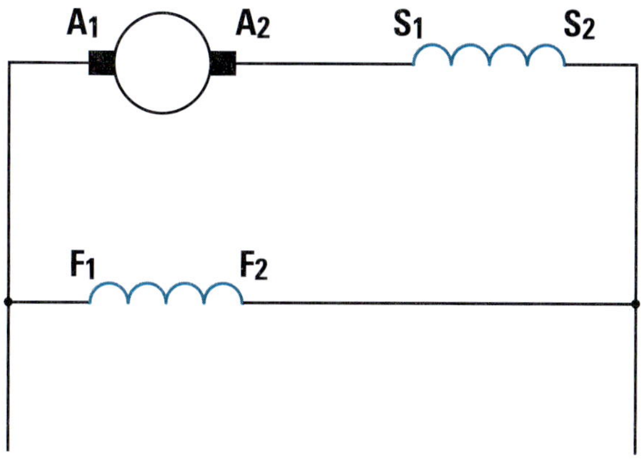

Figure 29–39 Schematic drawing of a long shunt compound generator.

The second connection is called *short shunt* (*Figure 29–40*). The short shunt connection has the series field in series with the armature; however, the shunt field is only connected in parallel with the armature and not in parallel with the series field as in the long shunt connection. This is a very common connection for DC generators that must be operated in parallel with each other.

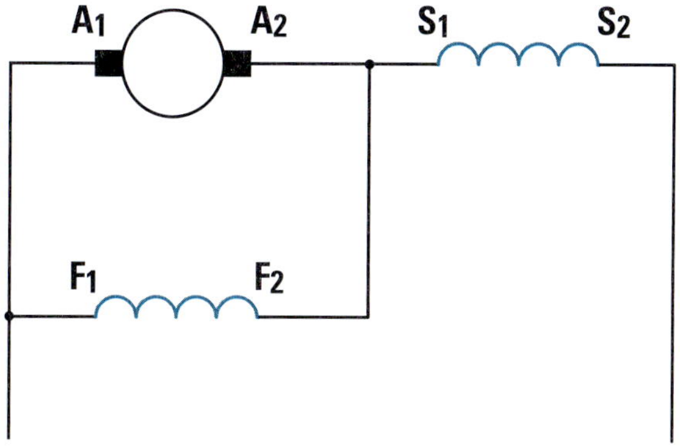

Figure 29–40 Schematic drawing of a short shunt compound generator.

29–4 Compounding, Countertorque, and Reactivity in DC Generators

The relationship of the strengths of the two fields in a generator determines the amount of **compounding** for the machine. A machine is *overcompounded* when the series field has too much control and the output voltage increases each time a load is added to the generator. Basically, the generator begins to take on the characteristics of a series generator. Overcompounding is characterized by the fact that the output voltage at full load is greater than the output voltage at no load (*Figure 29–41*).

When the generator is *flat-compounded,* the output voltage is the same at full load as it is at no load. Flat compounding is accomplished by permitting the series field to increase the output voltage by an amount that is equal to the losses of the generator.

If the series field is too weak, however, the generator becomes *undercompounded.* This condition is characterized by the fact that the output voltage is less at full load than it is at no load. When a generator is undercompounded, it has characteristics similar to those of a shunt generator.

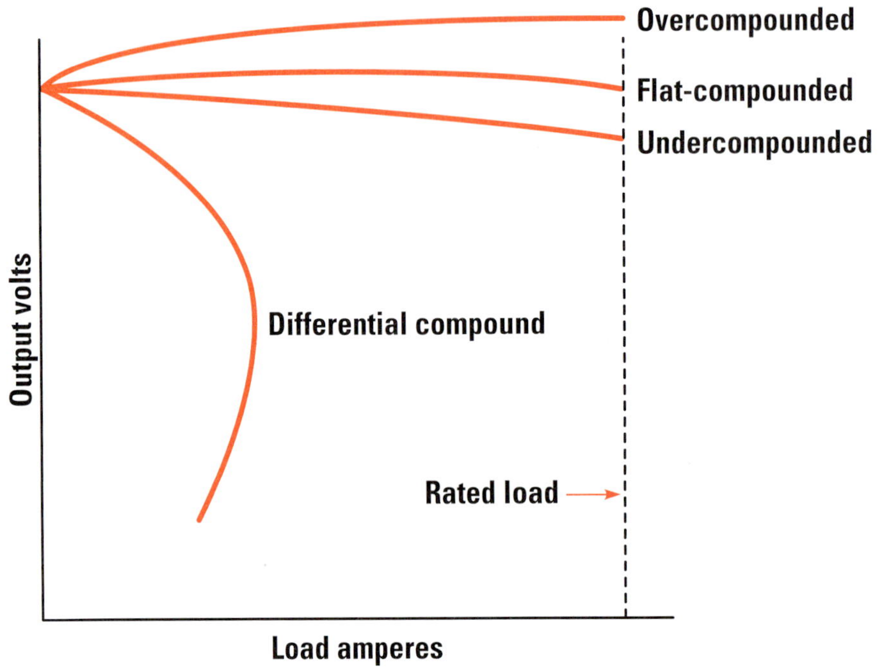

Figure 29–41 Characteristic curves of compound generators.

Controlling Compounding

Most DC machines are constructed in such a manner that they are overcompounded unless some means to control the compounding is used. The control device weakens the series field strength and thereby controls the amount of compounding. The amount of compounding is controlled by connecting a low-value variable resistor in parallel with the series field (*Figure 29–42*). This resistor is known as the series field shunt rheostat or the **series field diverter**. The rheostat permits part of the current that normally flows through the series field to flow through the resistor. This reduces the amount of magnetic flux produced by the series field, which reduces the amount of compounding.

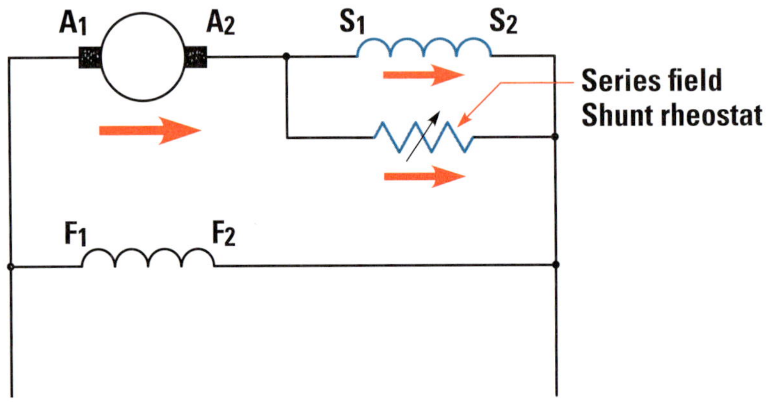

Figure 29–42 The series field shunt rheostat controls the amount of compounding.

Cumulative and Differential Compounding

DC generators are generally connected as a **cumulative compound** generator. This means the shunt and series fields are connected in such a manner that when current flows through them, they aid each other in the production of magnetism (*Figure 29–43*). In the example shown, each of the field windings would produce the same magnetic polarity for the pole piece.

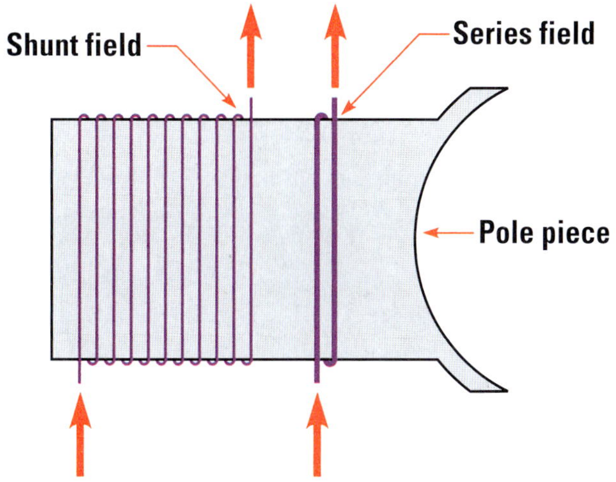

Figure 29–43 In a cumulative-compound machine, the current flows in the same direction through both the series and the shunt field.

A **differential-compound** generator has its fields connected in such a manner that they oppose each other in the production of magnetism (*Figure 29–44*). In this example, the shunt and series fields are attempting to produce opposite magnetic polarities for the same pole piece. This results in the magnetic field becoming weaker as current flow through the series field increases. Although there are some applications for a differential-compound machine, they are very limited.

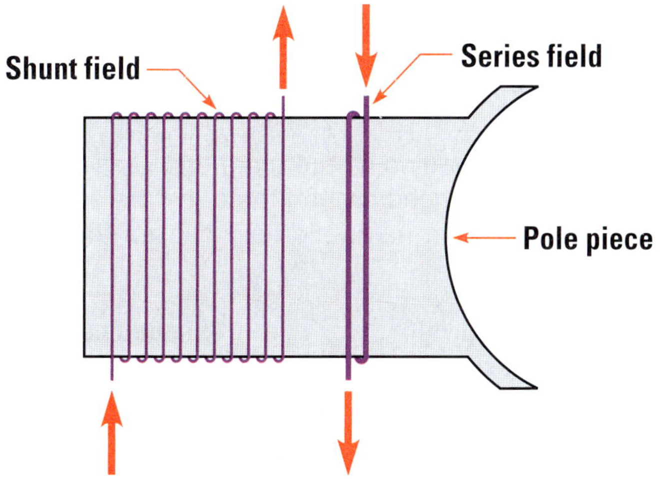

Figure 29–44 In a differential-compound machine, the current flows through the shunt field in a direction opposite that of the current flow through the series field.

Countertorque

When a load is connected to the output of a generator, current flows from the armature, through the load, and back to the armature. As current flows through the armature, a magnetic field is produced around the armature (*Figure 29–45*). However, the magnetic field of the armature is opposite in polarity to that of the pole pieces. Because these two magnetic fields are opposite in polarity, they are attracted to each other. This magnetic attraction causes the armature to become hard to turn. This turning resistance is called **countertorque**, and it must be overcome by the device used to drive the generator. This is the reason that, as load is added to the generator, more power is required to turn the armature. Because

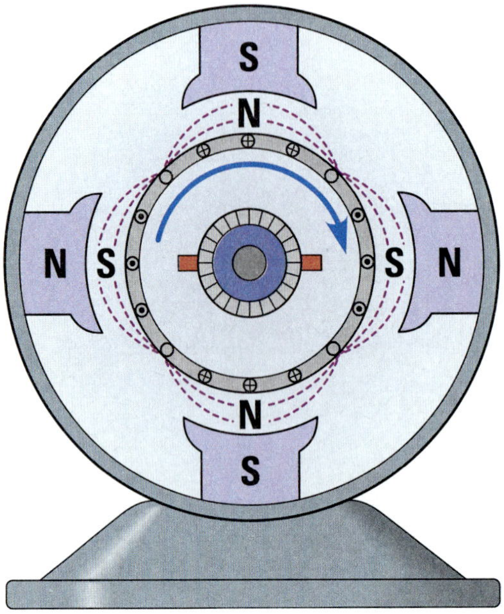

Figure 29–45 A magnetic field is produced around the armature.

countertorque is produced by the attraction of the two magnetic fields, it is proportional to the output or armature current if the field excitation current remains constant. **Countertorque is a measure of the useful electric energy produced by the generator.**

Countertorque is often used to provide a braking action in DC motors. When this type of braking action is used, it is referred to as *dynamic braking* or *regenerative braking.*

Armature Reaction

Armature reaction is the twisting or bending of the magnetic lines of flux of the pole pieces. It is caused by the magnetic field that is produced around the armature as it supplies current to the load (*Figure 29–46*). This distortion of the main magnetic field causes the position of the neutral plane

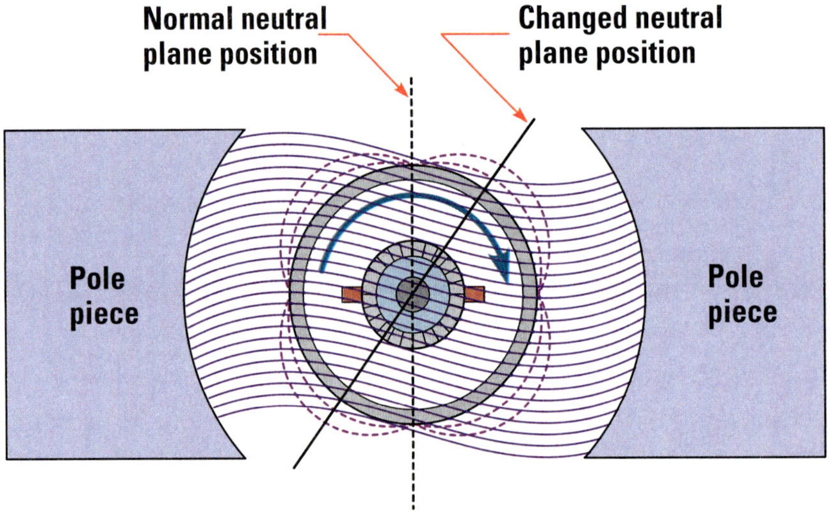

Figure 29–46 Armature reaction warps flux lines and changes the position of the neutral plane.

to change position. When the neutral plane changes, the brushes no longer make contact between commutator segments at a time when no voltage is induced in the armature. This results in power loss and arcing and sparking at the brushes, which can cause overheating and damage to both the commutator and brushes. The amount of armature reaction is proportional to armature current.

Correcting Armature Reaction

Armature reaction can be corrected in several ways. One method is to rotate the brushes an equal amount to the shift of the neutral plane (*Figure 29–47*). This method is only satisfactory, however, if the generator delivers a constant current. Because the distortion of the main magnetic field is proportional to armature current, the brushes have to be adjusted each time the load current changes. In the case of a generator, the brushes are rotated in the direction of rotation of the armature. In the case of a motor, the brushes are rotated in a direction opposite that of armature rotation.

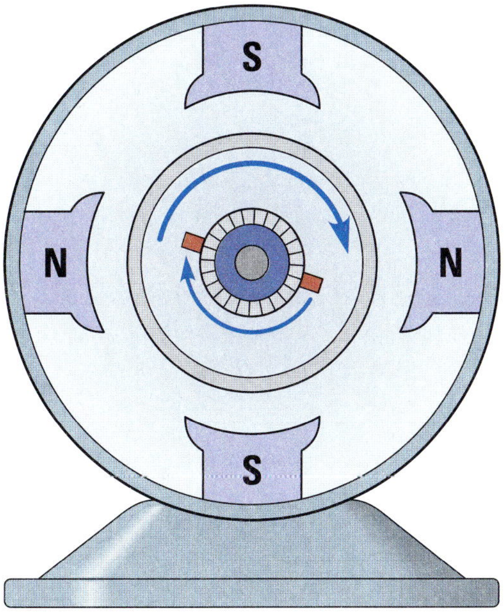

Figure 29–47 In a generator, the brushes are rotated in the direction of armature rotation to correct armature reaction.

Another method that is used often is to insert small pole pieces, called interpoles or commutating poles, between the main field poles (*Figure 29–26*). The interpoles are sometimes referred to as the *commutating winding* because they are wound with a few turns of large wire similar to the series field winding. The interpoles are connected in series with the armature, which permits their strength to increase with an increase of armature current (*Figure 29–48*). Interpole connections are often made

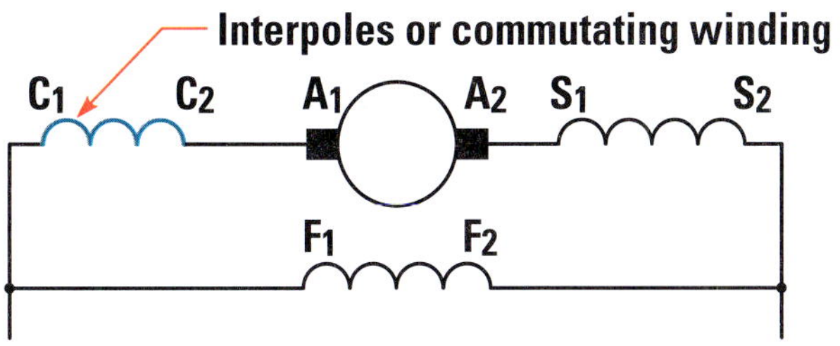

Figure 29–48 Interpoles are connected in series with the armature.

inside the housing of the machine. When the interpole connection is made internally, the A_1 lead is actually connected to one end of the interpole winding. When the interpole windings are brought out of the machine separately, they are generally labeled C_1 and C_2, which stands for commutating field. It is not unusual, however, to find them labeled S_3 and S_4.

In a generator, the magnetic field of the armature tends to bend the main magnetic field upward (*Figure 29–46*). In a motor, the armature field bends the main field downward. The function of the interpoles is to restore the field to its normal condition. Therefore, the polarity of the interpoles will correspond to the field windings depending on whether the machine is a generator or a motor (*Figure 29–49*).

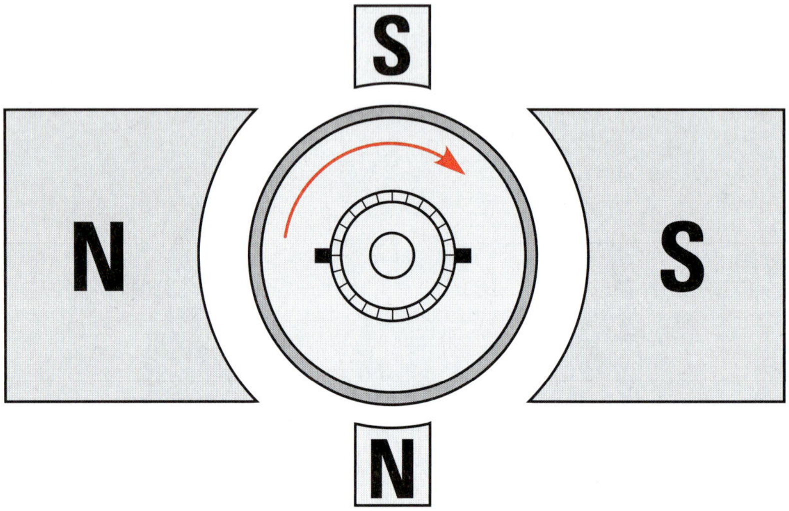

Figure 29–49 Interpoles in a generator must have the same polarity as the pole piece directly ahead of them.

Interpoles do have one disadvantage. They restore the field only in their immediate area and are not able to overcome all the field distortion. Large DC generators use another set of windings called *compensating windings* to help restore the main magnetic field. Compensating windings are made by placing a few large wires in the face of the pole piece parallel to the armature windings (*Figure 29–50*). The compensating winding is connected in series with the armature so its strength increases with an increase of output current.

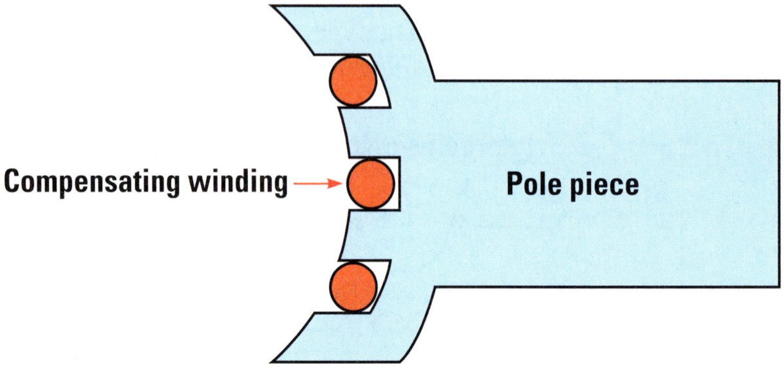

Figure 29–50 Compensating winding helps correct armature reaction in large DC generators.

29–5 Operating DC Generators

Most DC machines are designed in such a manner that the position of the brushes on the commutator can be adjusted. This is commonly referred to as setting the neutral plane of a DC machine. The simplest method of setting the brushes to the neutral plane position is to connect an AC voltmeter across the shunt field leads. Low-voltage AC is then applied to the armature (*Figure 29–51*). The armature acts like the primary of a transformer, and the shunt field acts like the secondary. If the brushes are not set at the neutral plane position, the changing magnetic field of the armature induces a voltage into the shunt field. The neutral plane position of the brushes can be found by observing the action of the AC voltmeter. The brushes are set to the neutral plane position when the voltmeter is at its lowest possible reading.

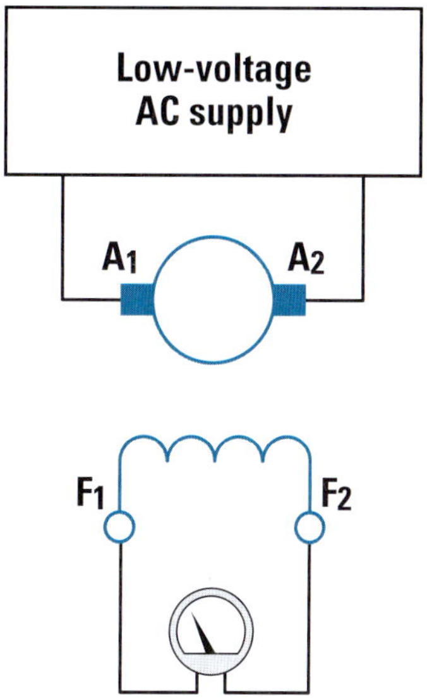

Figure 29–51 Schematic drawing of the procedure to set the neutral plane of a DC machine.

Paralleling DC Generators

There may be occasions when one DC generator cannot supply enough current to operate the connected load. In such a case, another generator is connected in parallel with the first. DC generators must never be connected in parallel without an equalizer connection (*Figure 29–52*). The equalizer connection is used to connect the series fields of the two machines in parallel with each other. This connection prevents one generator from being turned into a motor.

If the two generators are connected in parallel and the equalizing connection has not been made, the machine with the greatest excitation takes the entire load and begins operating the other machine as a motor.

The generator that begins motoring has the current flow through its series field reversed, which causes it to operate as a differential-compound motor (*Figure 29–53*). If the motoring generator is not removed from the line, the magnetic field strength of the series field will become greater than the field strength of the shunt field. This causes the polarity of the residual magnetism in the pole pieces to reverse. This is often referred to as *flashing the field*. Flashing the field results in the polarity of the output voltage being reversed when the machine is restarted as a generator. The equalizer connection prevents field reversal even if the generator becomes a motor.

The resistance of the equalizer cable should not exceed 20% of the resistance of the series field winding of the smallest paralleled generator. This ensures that the current flow provided to the series fields will divide in the approximate inverse ratio of the respective series field winding.

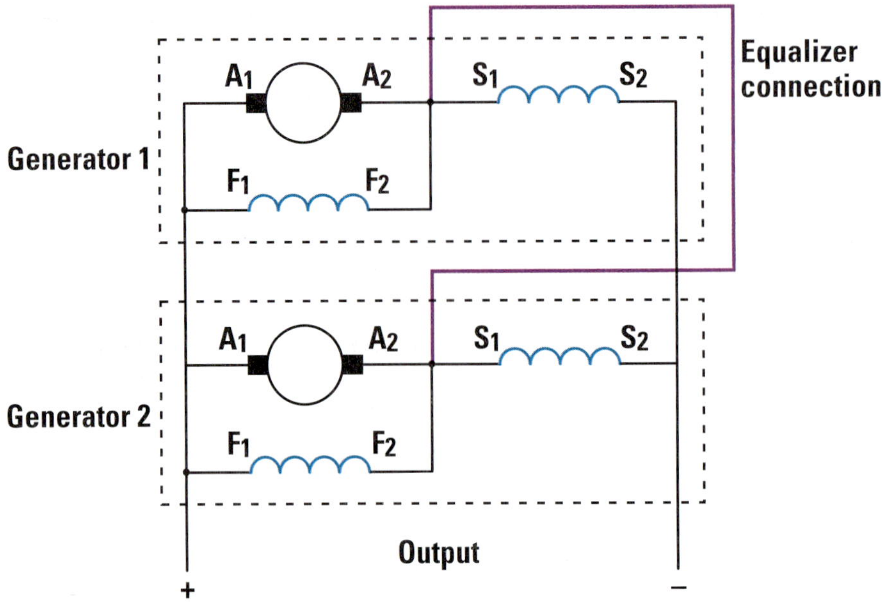

Figure 29–52 The equalizer connection is used to connect the series fields of each generator in parallel with each other.

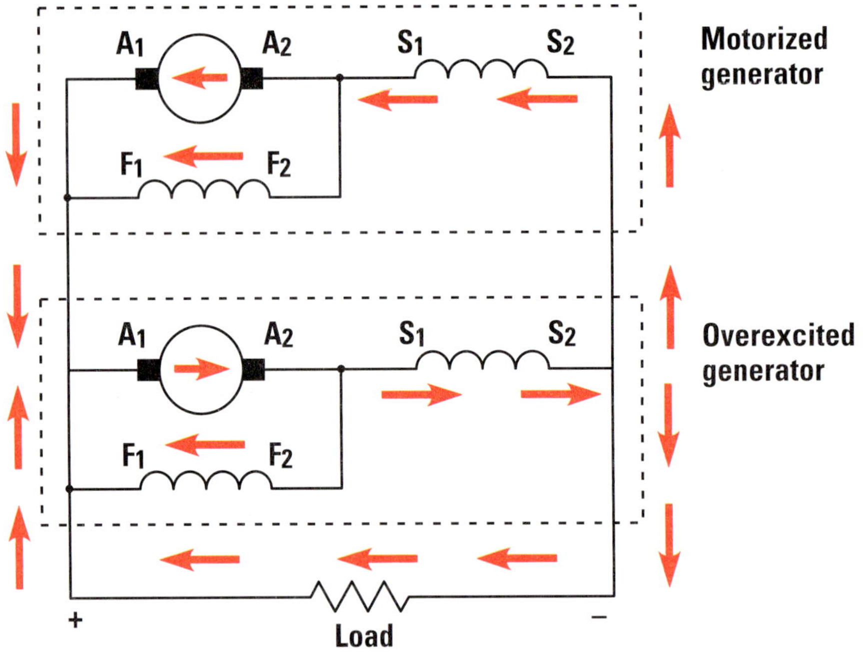

Figure 29–53 One generator supplies all the load, and the other becomes a motor.

Summary

- A generator is a machine that converts mechanical energy into electric energy.
- Generators operate on the principle of magnetic induction.
- AC is produced in all rotating armatures.

- The commutator changes the AC produced in the armature into DC.
- The brushes are used to make contact with the commutator and to carry the output current to the outside circuit.
- The position at which there is no induced voltage in the armature is called the neutral plane.
- Lap-wound armatures are used in machines designed for high-current operation.
- Wave-wound armatures are used in machines designed for low-current operation.
- Frogleg-wound armatures are the most used and are intended for machines designed for moderate levels of current.
- The loops of wire, iron core, and commutator are made as one unit and are referred to as the armature.
- The armature connection is marked A_1 and A_2.
- Series field windings are made with a few turns of large wire and have a very low resistance.
- Series field windings are connected in series with the armature.
- Series field windings are marked S_1 and S_2.
- Shunt field windings are made with many turns of small wire and have a high resistance.
- Shunt field windings are connected in parallel with the armature.
- The shunt field windings are marked F_1 and F_2.
- Three factors that determine the voltage produced by a generator are

  a. the number of turns of wire in the armature,

  b. the strength of the magnetic field of the pole pieces, and

  c. the speed of the armature.

- Series generators increase their output voltage as load is added.
- Shunt generators decrease their output voltage as load is added.
- The voltage regulation of a DC generator is proportional to the resistance of the armature.
- Compound generators contain both series and shunt field windings.
- A long shunt compound generator has the shunt field connected in parallel with both the armature and series field.
- A short shunt compound generator has the shunt field connected in parallel with only the armature and is in series with the series field.
- When a generator is overcompounded, the output voltage is higher at full load than it is at no load.
- When a generator is flat-compounded, the output voltage is the same at full load and no load.
- When a generator is undercompounded, the output voltage is less at full load than it is at no load.
- Cumulative-compound generators have their series and shunt fields connected in such a manner that they aid each other in the production of magnetism.
- Differential-compound generators have their series and shunt field windings connected in such a manner that they oppose each other in the production of magnetism.
- Armature reaction is the twisting or bending of the main magnetic field.
- Armature reaction is caused by the interaction of the main magnetic field with the field produced in the armature.
- Armature reaction is proportional to armature current.
- Interpoles are small pole pieces connected between the main field poles used to help correct armature reaction.
- Interpoles are connected in series with the armature.

- Interpole leads are sometimes marked C_1 and C_2 or S_3 and S_4.
- Interpole leads are not always brought out of the machine.
- The neutral plane can be set by connecting an AC voltmeter to the shunt field and a source of low-voltage AC to the armature. The brushes are then adjusted until the voltmeter indicates the lowest possible voltage.
- When a generator supplies current to a load, countertorque is produced, which makes the armature harder to turn.
- Countertorque is proportional to the armature current if the field excitation current remains constant.
- Countertorque is a measure of the useful electric energy produced by the generator.

Review Questions

1. What type of voltage is produced in all rotating armatures when the rotation of the armature wire cuts through magnetic lines of flux produced by the pole pieces?

   A) DC voltage

   B) AC voltage

   C) Harmonic voltage

   D) High voltage

2. In the simple generator shown in *Figure 29-1*, at what point of rotation is the voltage being induced in the loop wire at its maximum value? Zero degrees of rotation is considered top-center.

   A) 0°

   B) 180°

   C) 45°

   D) 90°

3. If the simple generator shown in *Figure 29-1* produces a voltage of 10 volts with a single wire loop, which of the following best describes the impact that will be made if the loop is increased to 4 conductors per loop? Assume all else remains the same.

   A) The voltage will increase by four times.

   B) The current will increase by four times.

   C) The voltage will decrease by four times.

   D) The current will decrease by four times.

4. Adding more loops on the armature in a DC generator will affect the voltage produced by the generator. Which of the following statements best describes the impact this has on voltage?

   A) Adding more loops in the armature increases the voltage produced.

   B) Adding more loops in the armature decreases the voltage produced.

   C) Adding more loops decreases the amount of rotation angle between maximum voltage peaks.

   D) Adding more loops decreases the frequency of the system.

5. Which of the following DC generator components would be desirable in a generator that needs to supply a high amount of current to the load?

   A) Frog-leg wound armature

   B) Wave-wound armature

   C) Split-rings

   D) Lap-wound armature

6. The commutator segments allow the AC voltage produced in the rotating armature to be converted to DC voltage. Which component of the generator interacts with the commutator to produce the useable output voltage?

   A) Armature

   B) Brushes

   C) Field windings

   D) Voltage regulator

7. True or False: The field winding in a shunt-type DC generator is always connected in series with the armature windings.

8. Which of the following types of generators will see an increase in the output voltage as more load is connected to the generator until the magnetic material in the pole pieces become saturated?

   A) Series field generator

   B) Shunt field generator

   C) Compound field generator

   D) Frog-leg field generator

9. Which of the following conditions would best characterize an issue with overcompounding in the generator?

   A) The output voltage decreases when load is added.

   B) The full-load voltage is the same as no-load.

   C) The output voltage increases as load is added.

   D) The generator becomes very difficult to turn.

10. Which term best applies to the force that is created internal to the generator as more load is added to the circuit and more and more power is required to rotate the armature?

    A) Counter-electromotive force

    B) Differential compounding

    C) Field excitation current

    D) Countertorque

11. True or False: When setting the brushes for ideal operation of a DC generator, a lower-voltage AC power supply is connected to the armature windings.

12. During the process of setting the brushes for the proper neutral plane, the brushes are in the correct orientation when which of the following statements are true?
    A) When the current in the armature is lowest
    B) When the voltage on the field windings is lowest
    C) When the voltage on the field windings is high
    D) When the current in the armature is highest
13. True or False: Interpoles are often used to correct armature reaction by connecting small pole pieces between the main pole pieces and connect extra windings in parallel with the armature windings.
14. Due to the distortion of the magnetic field in large DC generators, which of the following is used to correct the field distortion?
    A) Series field windings
    B) Parallel field windings
    C) Compensating windings
    D) Commutator windings

Practical Applications

You are working as an electrician in a large steel manufacturing plant, and you are in the process of doing preventive maintenance on a large DC generator. You have megged both the series and shunt field windings and found that each has over 10 MΩ to ground. Your ohmmeter, however, indicates a resistance of 1.5 Ω across terminals S_1 and S_2. The ohmmeter indicates a resistance of 225 Ω between terminals F_1 and F_2. Are these readings normal for this type of machine, or is there a likely problem? Explain your answer.

Unit 30 DC Motors

Key Terms

Brushless DC motors
Compound motor
Constant-speed motors
Counter-electromotive force (CEMF)
Cumulative-compound motors
Differential-compound motors
Field-loss relay
Motor
Permanent magnet motors
Printed-circuit motor
Pulse-width modulation
Series motor
ServoDisc motor
Servomotors
Shunt motor
Speed regulation
Torque

Why You Need to Know

DC motors are used in many applications that require variable speed. They are also known for their ability to produce a large amount of torque at low speed. DC motors are commonly employed in the automotive industry to start vehicles and operate such devices as electric seats, electric windows, and blowers. The air-conditioning industry employs brushless DC motors in many air handlers for central heating and cooling systems. This unit

- introduces many different types of DC motors. Some of these types include series, shunt, compound, brushless, permanent magnet, and disc servo.
- explains how torque is produced and how it is used in HVAC air handlers and explains when variable speeds are required.
- discusses the different methods of connecting DC motors based on required rotation and speed control requirements.

Outline

Objectives

After studying this unit, you should be able to

- explain how the principles of operation for DC motors convert electrical energy to kinetic energy.
- differentiate between the major types of DC motors.
- explain the principles of DC motor operation for speed control and the direction of rotation.
- describe the operating principles for DC brushless motors to describe their operation.
- explain the operation principles of permanent magnet DC motors.
- apply the right-hand motor rule to determine direction of rotation.

Preview

DC motors are used throughout industry in applications where variable speed and torque characteristics are desirable. A good example is the San Francisco cable car system. It is powered by four 550-horsepower direct current motors. The automotive industry also uses DC motors to start internal combustion engines and to operate blower fans, power seats, and other devices where a small motor is needed.

30–1 DC Motor Principles

DC motors operate on the principle of repulsion and attraction of magnetism. Motors use the same types of armature windings and field windings as the generators discussed in Unit 29. A **motor,** however, performs the opposite function of a generator. A *motor* is a device used to convert electric energy into mechanical energy. DC motors were, in fact, the first electric motors to be invented.

DC motors are the same basic machines as DC generators. However, a generator creates electricity from turning the armature, whereas in a motor, power connected to the armature causes it to turn instead. A simple one-loop armature is illustrated in *Figure 30–1*. As electrons enter the loop through the negative brush, flow around the loop, and exit through the positive brush, a current flows through the loop and creates a magnetic field. *Figure 30–2* illustrates how this field in the armature interacts with the main magnetic field created by the pole pieces. The X indicates electrons moving away from the observer like the back of an arrow moving away. The dot represents electrons moving toward the observer like the point of an approaching arrow. The left-hand rule for magnetism can be used to check the direction of the magnetic field around the conductors.

Torque

Magnetic lines of force travel from north polarity to south polarity (left to right in *Figure 30–2*). When magnetic lines of flux flow in the same direction, they repel each other. When they flow in opposite directions, they attract each other. The magnetic lines of flux around the conductors cause the loop to be pushed in the direction shown by the arrows. The force with which the magnetic fields repel, attract, and rotate the armature is called **torque.** Two factors that determine the amount of torque produced by a DC motor are

1. the strength of the magnetic field of the pole pieces.
2. the strength of the magnetic field of the armature.

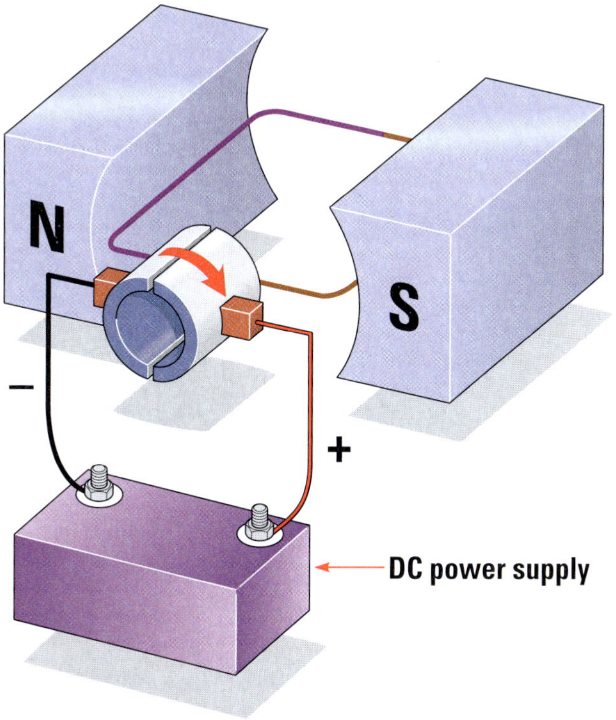

Figure 30–1 DC is supplied to the armature loop to power a DC motor.

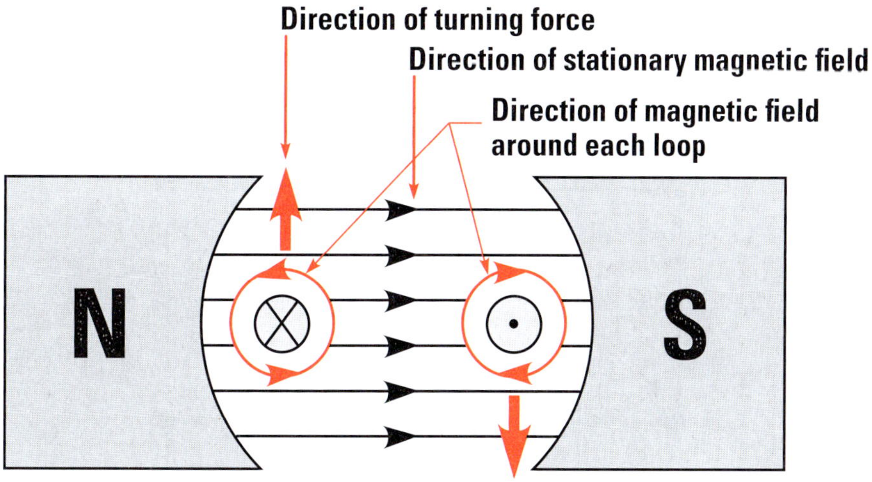

Figure 30–2 Flux lines in the same direction repel each other, and flux lines in the opposite direction attract each other.

Notice that there is no mention of speed or cutting action. One characteristic of a DC motor is that it can develop maximum torque at 0 RPM .

Increasing the Number of Loops

In the simple illustration here, a single-loop armature was used to illustrate the operating principle of a DC motor. In actual practice, armatures are constructed with many turns of wire per loop and many loops. This provides a strong continuous turning force for the armature (*Figure 30–3*).

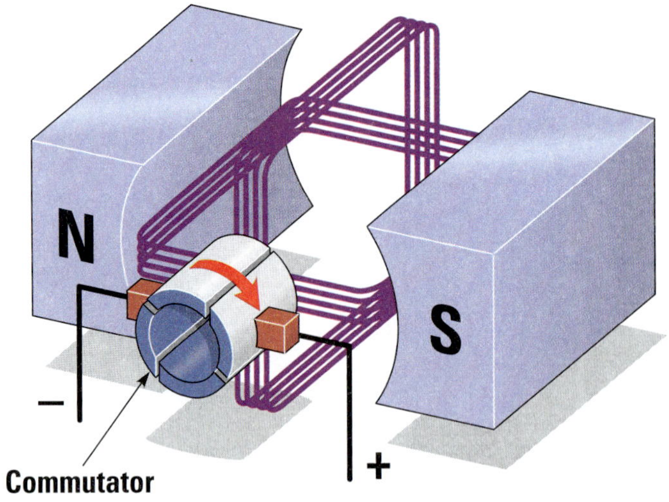

Figure 30–3 Increasing the number of loops and turns increases the torque.

The Commutator

When a DC machine is used as a generator, the commutator performs the function of changing the AC produced in the armature into DC before it exits the machine through the brushes. When a DC machine is used as a motor, the commutator performs the function of a rotary switch and maintains the correct direction of current flow through the armature windings. For the motor to develop a turning force, the magnetic field polarity of the armature must remain constant in relation to the polarity of the pole pieces. The commutator forces the direction of current flow to remain constant through certain sections of the armature as it rotates. The brushes are now used to provide power to the armature instead of delivering the power to the circuit from a DC generator (*Figure 30–4*).

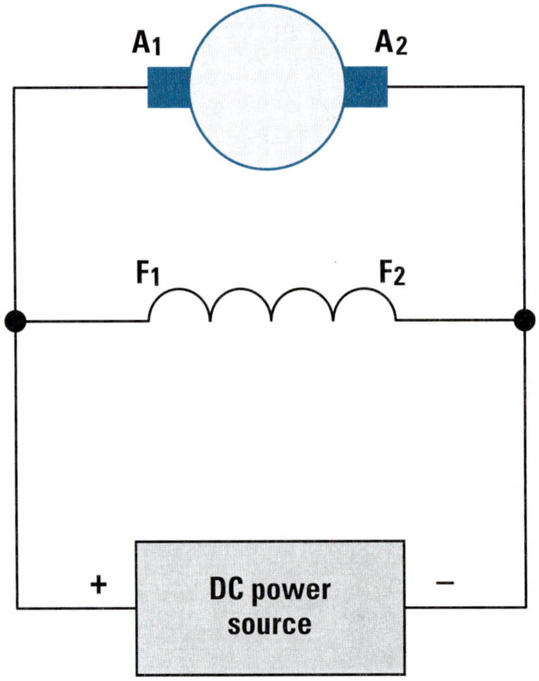

Figure 30–4 Both the field windings and the armature are powered from an external power source in a DC motor.

30–2 DC Motor Types

The DC **shunt motor** has the field windings connected in parallel with the armature (*Figure 30–4*). This permits an external power source to maintain a constant magnetic field. The shunt motor has very good speed characteristics. The full-load speed generally remains within 10% of the no-load speed. Shunt motors are often referred to as **constant-speed motors.** Characteristic curves for shunt, series, and compound motors are shown in *Figure 30–5*. Note that the shunt motor maintains the most constant speed as load is added and armature current increases.

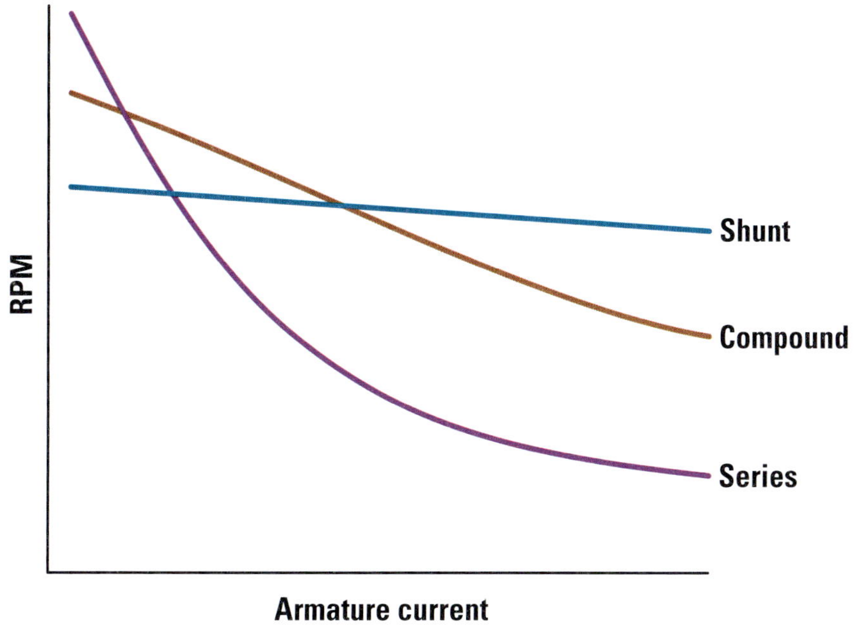

Figure 30–5 Speed vs. current curves for different DC motor types.

Counter-Electromotive Force

When the windings of the armature spin through the magnetic field produced by the pole pieces, a voltage is induced into the armature. This induced voltage is opposite in polarity to the applied voltage and is known as **counter-electromotive force (CEMF)** (*Figure 30–6*). It is the CEMF that limits the flow

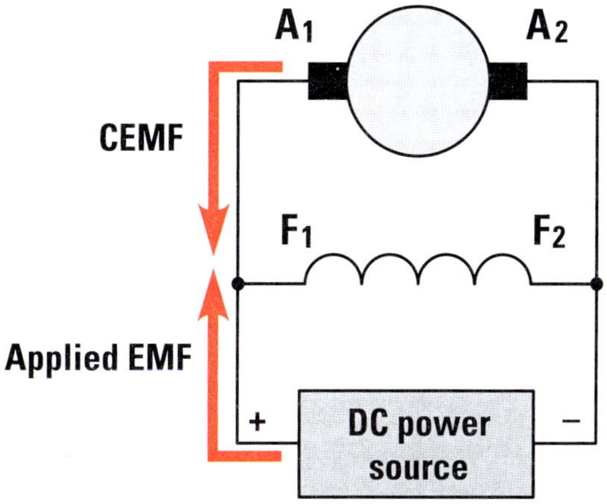

Figure 30–6 CEMF limits the flow of current through the armature.

of current through the armature when the motor is in operation. The amount of CEMF produced in the armature is proportional to the following three factors:

1. the number of turns of wire in the armature,
2. the strength of the magnetic field of the pole pieces, and
3. the speed of the armature.

Speed-Torque Characteristics

When a DC motor is first started, the inrush of current can be high because no CEMF is being produced by the armature. The only current-limiting factor is the resistance of the wire in the armature. As the armature turns and generates CEMF, it opposes the applied voltage, causing current flow to decrease. If the motor is not connected to a load, the armature continues to increase in speed until the CEMF is almost the same value as the applied voltage. At this point, the motor produces enough torque to overcome its own losses. Some of these losses are

1. I^2R loss in the armature windings,
2. windage loss,
3. bearing friction, and
4. brush friction.

When a load is added to the motor, the torque is not sufficient to support the load at the speed at which the armature is turning. The armature therefore slows down. When the armature slows down, CEMF is reduced, and more current flows through the armature windings. This produces an increase in magnetic field strength and an increase in torque. This is the reason that armature current increases when load is added to the motor.

Speed Regulation

The amount by which the speed decreases as load is added is called the **speed regulation. The speed regulation of a DC motor is proportional to the resistance of the armature.** The lower the armature resistance, the better the speed regulation. The reason for this is that armature current determines the torque produced by the motor if the field excitation current is held constant. To produce more torque, more current must flow through the armature, which increases the magnetic field strength of the armature. The amount of current flowing through the armature is determined by the CEMF and the armature resistance.

Assume an armature wire has a resistance of 6 ohms. Now assume that when load is added to the motor, an additional 3 amperes of armature current are required to produce the torque necessary to overcome the added load. In this example, a voltage of 18 volts is required to increase the armature current by 3 amperes ($3\ A \times 6\ \Omega = 18\ V$). This means that the speed of the armature must drop enough that the CEMF is 18 volts less than it was before. The reduction in CEMF permits the applied voltage to push more current through the resistance of the armature.

Now assume that the armature has a resistance of 1 ohm. If load is added that requires an additional 3 amperes of armature current, the speed of the armature must drop enough to permit a 3-volt reduction in CEMF ($3\ A \times 1\ \Omega = 3\ V$). The armature does not have to reduce speed as much to cause a 3-volt reduction in CEMF as it does for a reduction of 18 volts.

Series Motors

The operating characteristics of the DC **series motor** are very different from those of the shunt motor. The reason is that the series motor has only a series field connected in series with the armature (*Figure 30–7*). The armature current and the field current are the same. The speed of the series motor is controlled by the amount of load connected to the motor. When load is increased, the speed of the motor decreases. This causes a reduction in the amount of CEMF produced in the armature and an increase in armature and series field current. Because the current increases in both the armature and series field, the torque increases by the square of the current. In other words, if the current doubles, the torque increases four times. Curves showing the relationship of torque and armature current for the three main types of DC motors are shown in *Figure 30–8*. Notice that the series motor produces the most torque of the three motors.

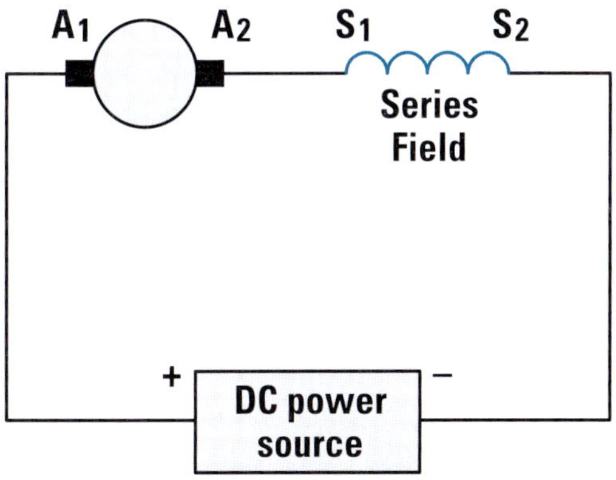

Figure 30–7 Series motor connection.

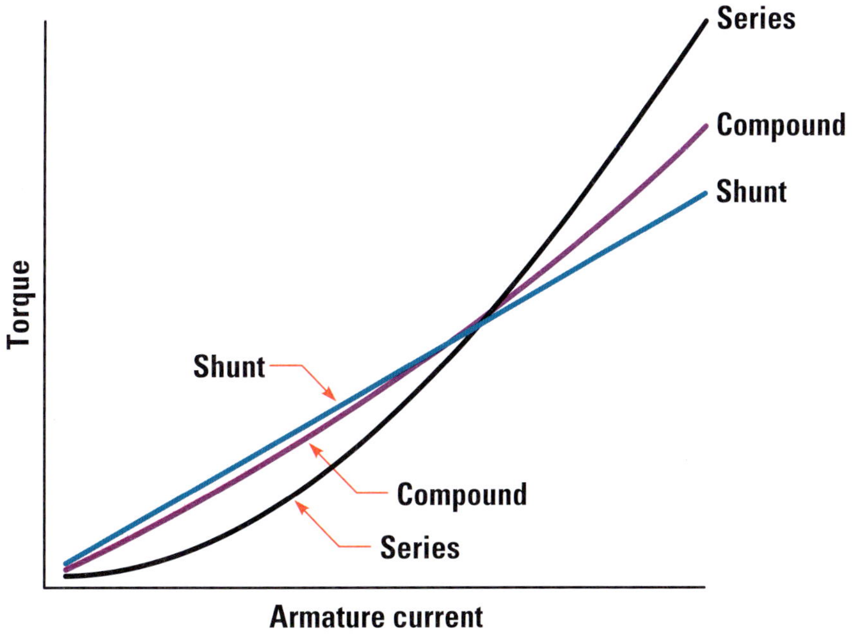

Figure 30–8 Torque vs. current curves of common types of DC motors.

Series Motor Speed Characteristics

Series motors have no natural speed limit and therefore must never be operated in a no-load condition. Large series motors that suddenly disconnect from their load can race to speeds that destroy the motor when centrifugal force slings both the windings out of the slots in the armature and the copper bars out of the commutator. For this reason, belts or chains should never be used to connect a series motor to a load.

Series motors can also develop extremely high starting torques. An average of about 450% of full torque is common. These motors are generally used for applications that require a high starting torque, such as the starter motor on an automobile, cranes and hoists, and electric buses and streetcars.

Compound Motors

The **compound motor** uses both a series field and a shunt field. This motor is used to combine the operating characteristics of both the series and the shunt motor. The series field of the compound motor permits the motor to develop high torque, and the shunt field permits speed control and regulation. The compound motor is used more than any other type of DC motor in industry. The compound motor does not develop as much torque as the series motor, but it does develop more than the shunt motor. The speed regulation of a compound motor is not as good as a shunt motor, but it is much better than a series motor (*Figure 30–5*).

Compound motors can be connected as short shunt or long shunt just as compound generators can (*Figure 30–9*). The long shunt connection is more common because it has superior speed regulation. Compound motors can also be **cumulative-compound motors** or **differential-compound motors.** Although there are some applications for differential-compound motors, it is generally a connection to be avoided. When a motor is connected differential-compound, the shunt field determines the direction of rotation of the motor at no load or light load. When load is added to the motor, the series field becomes stronger. If enough load is added, the magnetic field of the pole pieces reverses polarity and the motor suddenly stops, reverses direction, and begins operating like a series motor. This can damage the motor and the equipment to which the motor is connected.

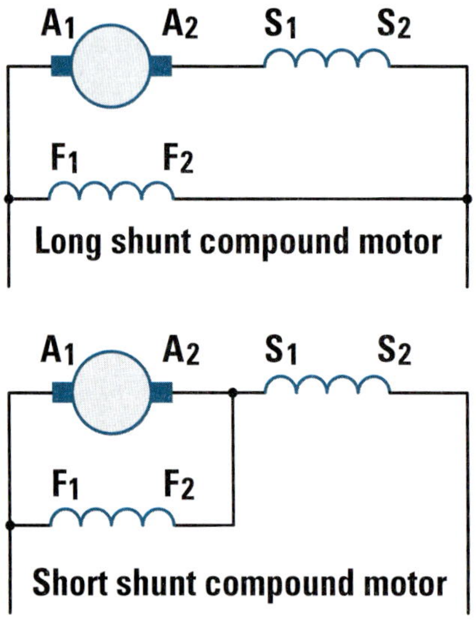

Figure 30–9 Long shunt and Short shunt compound motor connections.

If a compound motor is to be connected, the following steps can be followed to prevent the motor from being accidentally connected differential compound:

1. Disconnect the motor from the load.
2. Connect the series field and armature windings together to form a series motor connection, leaving the shunt field disconnected. Connect the motor to the power source (*Figure 30–10*).

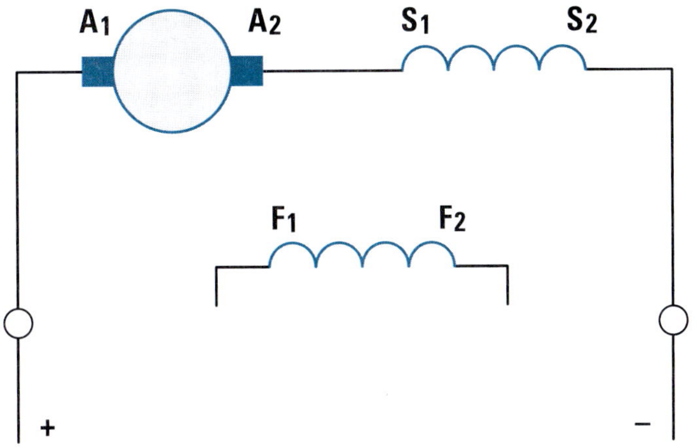

Figure 30–10 The motor is first connected as a series motor.

3. Turn the power on momentarily to determine the direction of rotation. This application of power must be of very short duration because the motor is now being operated as a series motor with no load. The idea is to "bump" the motor just to check for direction of rotation. If the motor turns in the opposite direction than is desired, reverse the connection of the armature leads. This reverses the direction of rotation of the motor.
4. Connect the shunt field leads to the incoming power (*Figure 30–11*). Again, turn on the power and check the direction of rotation. If the motor operates in the desired direction, it is connected cumulative compound. If the motor turns in the opposite direction, it is connected differential compound, and the shunt field leads should be reversed.

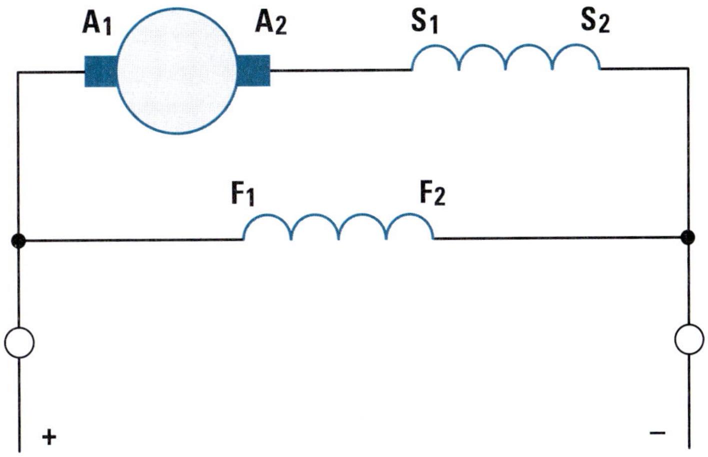

Figure 30–11 Connect the shunt field and again test for the direction of rotation.

This test can be used to check for a differential- or cumulative-compound connection because the shunt field controls the direction of rotation at no load. If the motor operates in the same direction as both a series and compound motor, the magnetic polarity of the pole pieces must be the same for both connections. This indicates that both the series and shunt field windings must be producing the same magnetic polarity and are therefore connected cumulative compound.

Terminal Identification for DC Motors

DC motors have the same terminal identification as that used for DC generators. *Figure 30–12* illustrates this standard identification. Terminals A_1 and A_2 are connected to the armature through the brushes. The ends of the series field are identified with S_1 and S_2, and the shunt field leads are marked F_1 and F_2. Some DC machines provide access to another set of windings called the commutating field or interpoles. The ends of this winding are labeled C_1 and C_2 or S_3 and S_4. It is common practice to provide access to the interpole winding on machines designed to be used as motors or generators.

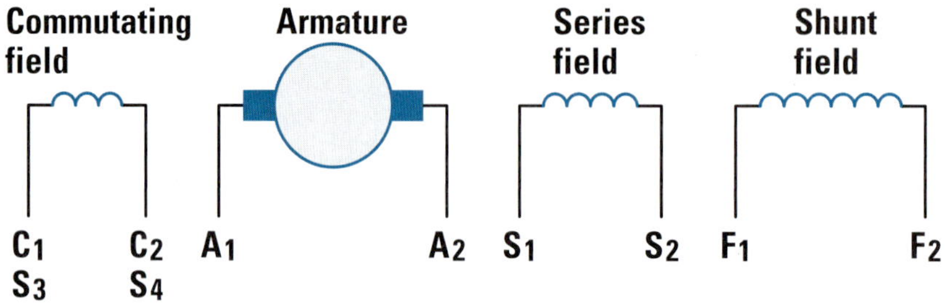

Figure 30–12 Terminal identification for DC machines.

30–3 Operating Characteristics of a DC Motor

The direction of rotation of a DC motor is determined by facing the commutator end of the motor. This is generally the back or rear of the motor. If the windings have been labeled in a standard manner, it is possible to have a good idea of which direction the motor will turn when connected. *Figure 30–13* illustrates the standard connections for a series motor. The standard connections for a shunt motor are illustrated in *Figure 30–14,* and the standard connections for a compound motor are shown in *Figure 30–15*.

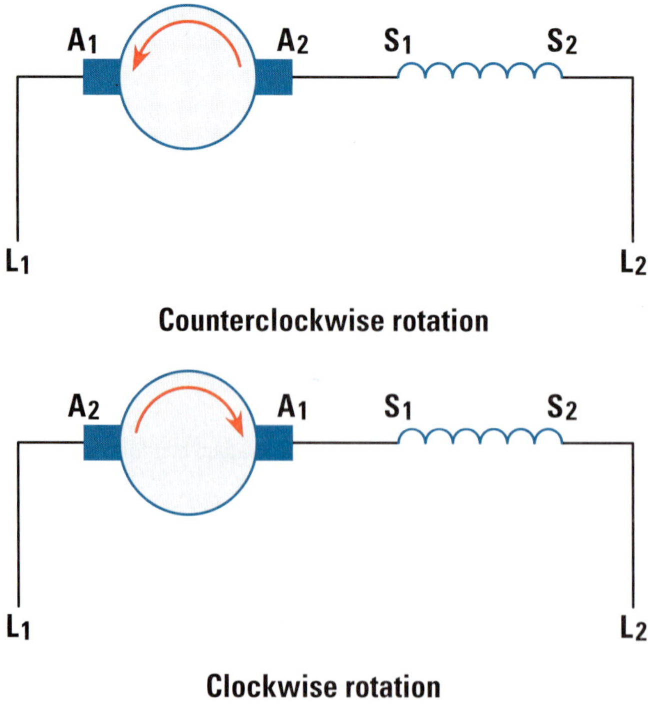

Figure 30–13 Series motor standard connections.

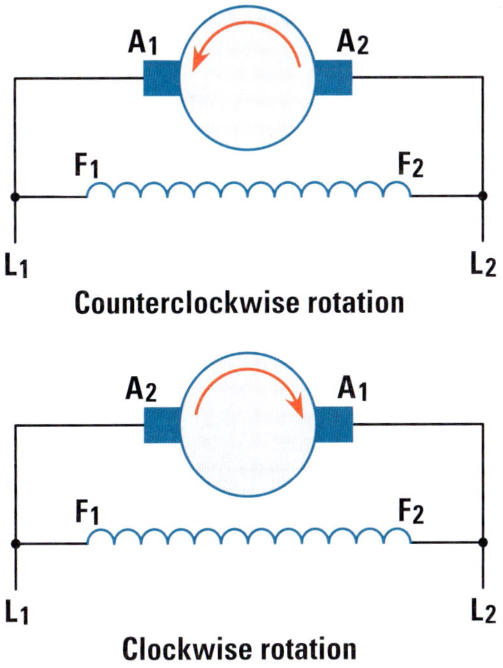

Figure 30–14 Shunt motor standard connections.

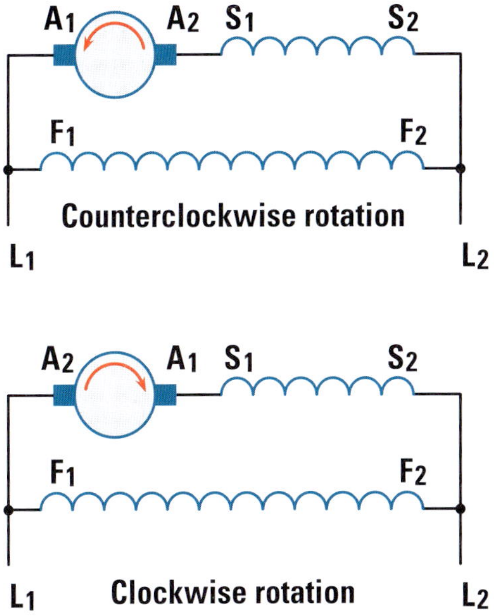

Figure 30–15 Compound motor standard long shunt connections.

When the direction of rotation of a DC motor must be reversed, it is common practice to change the connection of the armature leads only. This is done to prevent changing a cumulative-compound motor into a differential-compound motor.

Although it is standard practice to change the connection of the armature leads to reverse the direction of rotation, it is not uncommon to reverse the rotation of small shunt motors by changing the connection of the field leads. If a motor contains only a shunt field, there is no danger of changing the motor from a cumulative- to a differential-compound motor. The shunt field leads are often changed on small motors because the amount of current flow through the field is much less than the current flow through the armature. This permits a small double-pole double-throw (DPDT) switch to be used as a control for reversing the direction of rotation (*Figure 30–16*).

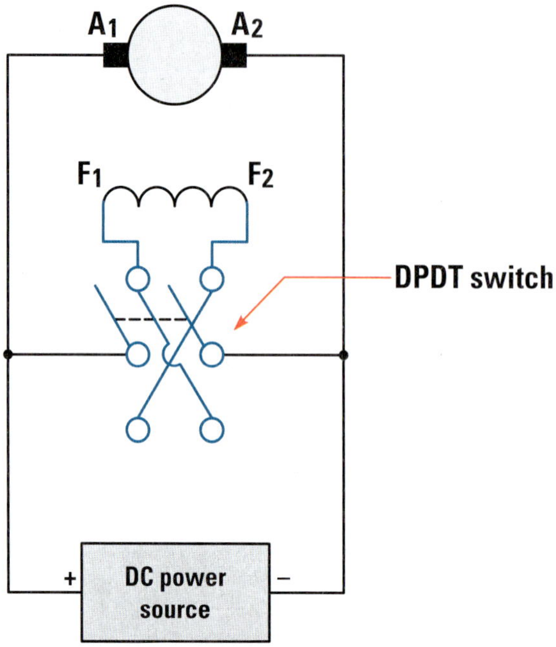

Figure 30–16 Reversing the rotation of a shunt motor by changing the shunt field leads.

Large compound motors often use a control circuit similar to the one shown in *Figure 30–17* for reversing the direction of rotation. This control circuit uses magnetic contactors to reverse the flow of current through the armature. If the circuit is traced, it will be seen that when the forward or reverse direction is chosen, only the current through the armature changes direction. The current flow through the shunt and series fields remains the same.

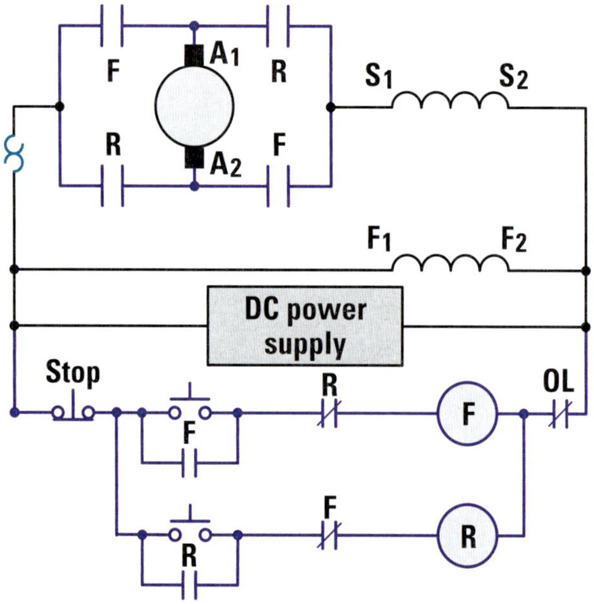

Figure 30–17 Forward-reverse control for a compound motor.

Speed Control

There are several ways of controlling the speed of DC motors. The method employed is generally dictated by the requirements of the load. When full voltage is connected to both the armature and the shunt field, the motor operates at its base speed. If the motor is to be operated below base speed, or under speed, full voltage is maintained to the shunt field, and the amount of armature current is reduced. The reduction of armature current causes the motor to produce less torque, and the speed decreases. One method of reducing armature current is to connect resistance in series with the armature (*Figure 30–18*). In the example shown, three resistors are connected in series with the armature. Contacts S_1, S_2, and S_3 can be used to bypass certain steps of resistance and permit more armature current to flow.

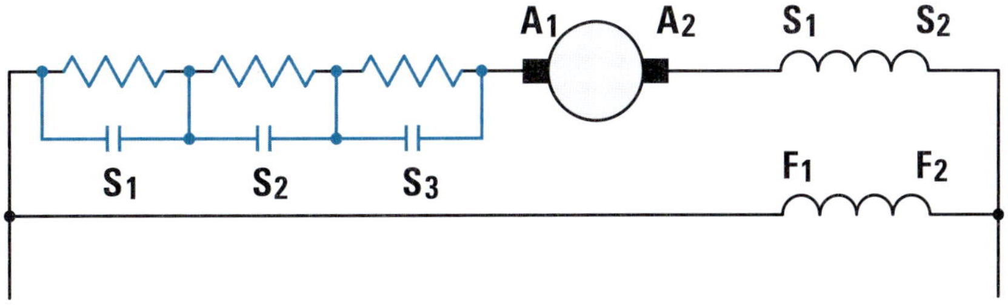

Figure 30–18 Resistors in series can limit armature current and control speed.

Although adding resistance in the armature circuit does permit the motor to be underspeeded, it has several disadvantages:

1. Resistors waste power in the form of heat.
2. Speed can only be controlled in steps.
3. Cannot increase or decrease speed smoothly.

Most large DC motors use an electronic controller to supply variable voltage to the armature circuit separately from the field (*Figure 30–19*). This permits continuous adjustment of the speed from zero to full RPM. Electronic power supplies can also sense the speed of the motor and maintain a constant speed as load is changed. Most of these power supplies can also be current-limiting, and the maximum current output can be set to a value that does not permit the motor to be harmed.

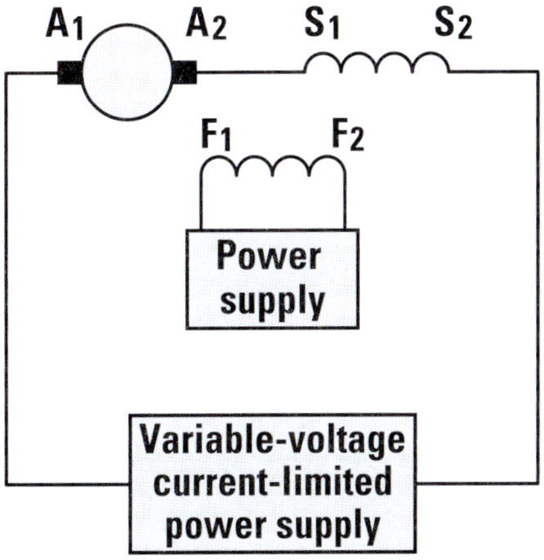

Figure 30–19 Armature current is controlled by a variable-voltage power supply.

A DC motor can also be overspeeded by connecting full voltage to the armature and reducing the current flow through the shunt field. In *Figure 30–20,* a shunt field rheostat has been connected in series with the shunt field. As resistance is added to the shunt field, field current decreases, which causes a decrease in the flux density of the pole pieces. This decrease in flux density produces less CEMF in the armature, which permits more armature current to flow. The increased armature current causes an increase in the magnetic field strength of the armature. This increased magnetic field strength of the armature produces a net gain in torque, which causes the motor speed to increase.

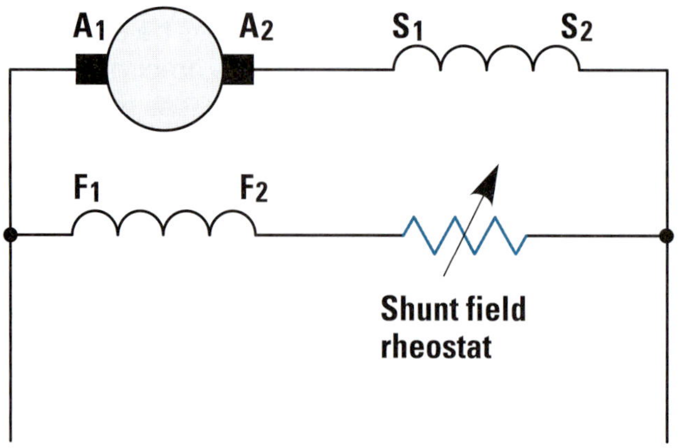

Figure 30–20 The shunt field rheostat can be used to overspeed the motor.

The Field-Loss Relay

Most large compound DC motors have a protective device connected in series with the shunt field called the **field-loss relay** (*Figure 30–21*). The function of the field-loss relay is to disconnect power to the armature if current flow through the shunt field decreases below a certain level. If the shunt field current stops completely, the compound motor becomes a series motor and increases rapidly in speed. This can cause damage to both the motor and the load.

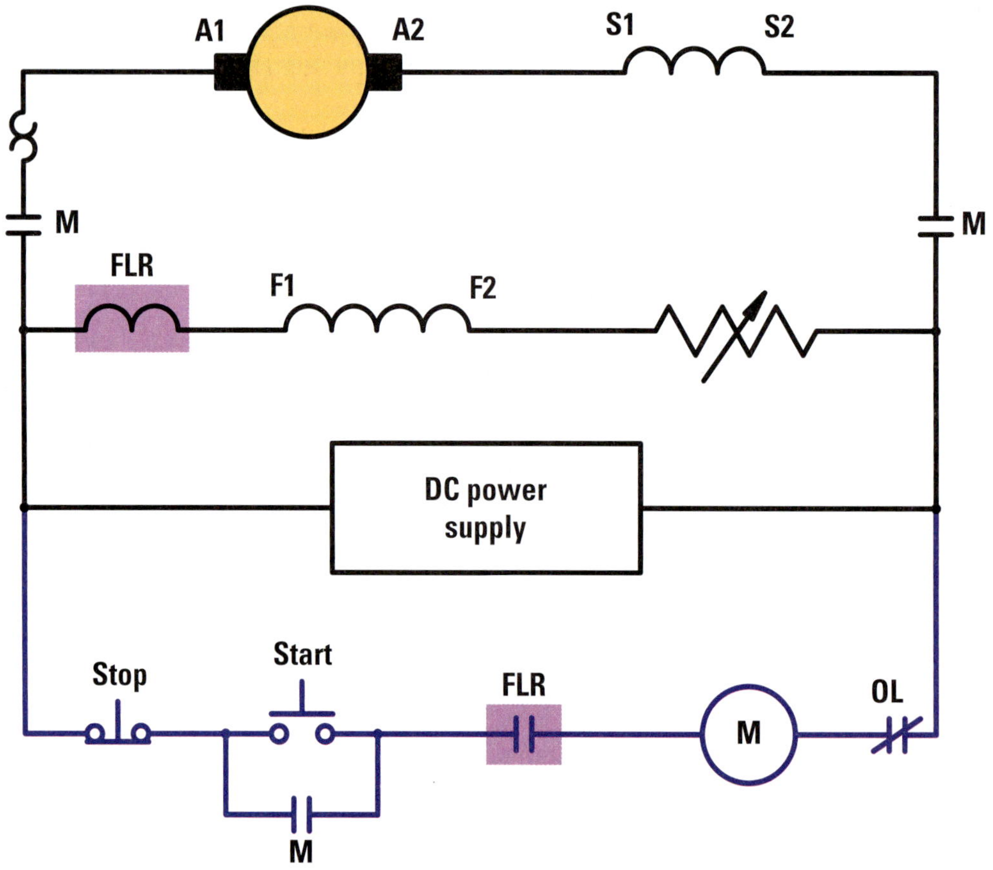

Figure 30–21 The field-loss relay disconnects power to the armature if the shunt field current stops.

Many large DC compound motors intended to operate in an overspeed condition actually contain two separate shunt fields (*Figure 30–22*). One shunt field is connected to a fixed voltage and maintains a constant field to provide an upper limit to motor speed. This shunt field is connected to the field-loss relay. The second shunt field is connected to a source of variable voltage. This shunt field is used to increase speed above the base speed. For this type of motor, base speed is achieved by applying full voltage to the armature and both shunt fields. Note that most large DC motors have voltage always applied to the shunt field, even when the motor is not in operation.

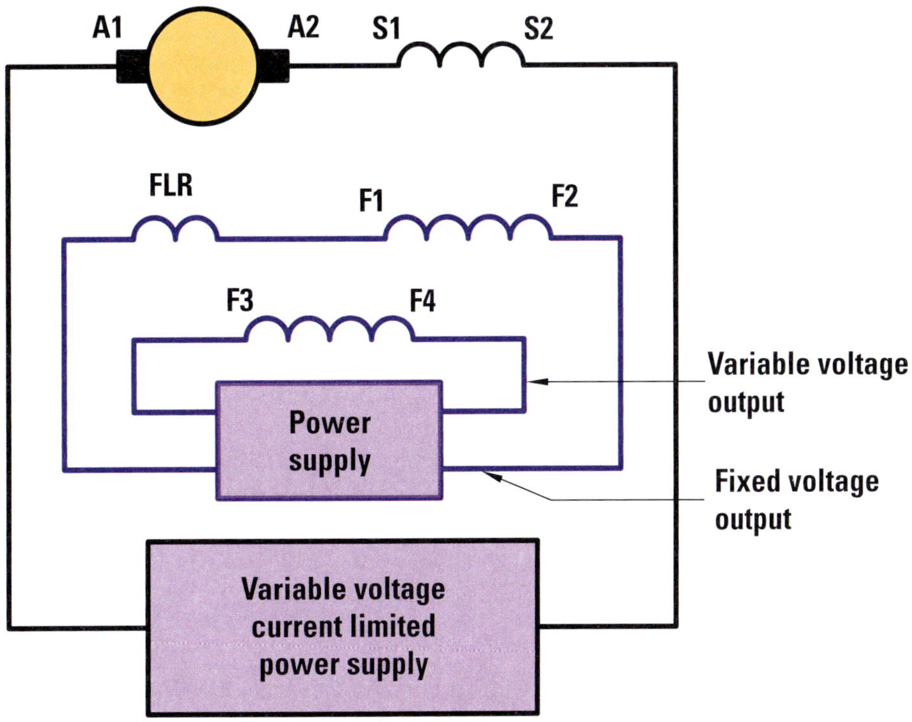

Figure 30–22 One shunt field is used to provide a stable speed; the other shunt field provides overspeed control.

Horsepower

When James Watt first began marketing steam engines, he found that he needed a way to compare them to the horses they were to replace. After conducting experiments, Watt found that the average horse could do work at a rate of 550 foot-pounds per second. This became the basic horsepower measurement. Horsepower can also be expressed in the basic electric unit for power, which is the watt:

$$1 \text{ hp} = 746 \text{ W}$$

Watts can also be converted to other power units such as:

$$1 \text{ W} = 3.412 \text{ BTUs per hour}$$

$$1055 \text{ W} = 1 \text{ BTU per second}$$

$$4.18 \text{ W} = 1 \text{ calorie per second}$$

$$1.36 \text{ W} = 1 \text{ ft-lb per second}$$

To determine the horsepower output of a motor, the rate at which the motor is doing work must be known. The following formula can be used to determine the horsepower output of a motor:

$$\text{hp} = \frac{(1.59)(\text{torque})(\text{RPM})}{100{,}000}$$

where

$$\text{hp} = \text{horsepower}$$
$$1.59 = \text{a constant}$$
$$\text{torque} = \text{torque in lb-in.}$$
$$\text{RPM} = \text{speed (angular velocity in revolutions per minute)}$$
$$100{,}000 = \text{a constant}$$

Example 30–1

How much horsepower is being produced by a motor turning a load of 350 lb-in. at a speed of 1375 RPM?

Solution

$$\text{hp} = \frac{1.59 \times 350 \text{ lb-in.} \times 1375 \text{ RPM}}{100{,}000}$$
$$\text{hp} = 7.652$$

Once the output horsepower is known, it is possible to determine the efficiency of the motor by using the formula:

$$\text{Eff.} = \frac{\text{power out}}{\text{power in}} \times 100$$

Example 30–2

A DC motor is connected to a 120-V DC line and has a current draw of 1.3 A. The motor is operating a load that requires 8 lb-in. of torque and is turning at a speed of 1250 RPM. What is the efficiency of the motor?

Solution

The first step is to determine the horsepower output of the motor:

$$\text{hp} = \frac{1.59 \times 8\text{lb-in.} \times 1250 \text{ RPM}}{100{,}000}$$
$$\text{hp} = 0.159$$

Now that the output horsepower is known, horsepower can be changed into W using the formula:

$$1 \text{ hp} = 746 \text{ W}$$
$$0.159 \times 746 = 118.614 \text{ W}$$

The amount of input power can be found by using the formula:

$$\text{W} = \text{V} \times \text{A}$$
$$\text{W} = 120 \text{ V} \times 1.3 \text{ A}$$
$$\text{W} = 156$$

The efficiency of the motor can be found by using the formula:

$$\text{Eff.} = \frac{\text{power out}}{\text{power in}} \times 100$$

Example 30–2 (continued)

The answer is multiplied by 100 to change it to a percent:

$$\text{Eff.} = \frac{118.614 \text{ W}}{156 \text{ W}} \times 100$$

$$\text{Eff.} = 76\%$$

Torque is often measured in lb-ft instead of lb-in. Another formula often used to determine horsepower when the torque is measured in lb-ft is:

$$\text{hp} = \frac{(2\pi)(\text{torque})(\text{RPM})}{33{,}000}$$

where

hp = horsepower

π = 3.1416

torque = lb-ft

RPM = speed in revolutions per minute

33,000 = a constant

30–4 Brushless DC Motors

Brushless DC motors do not contain a wound armature, a commutator, or brushes. The armature or rotor (rotating member) contains permanent magnets. The rotor is surrounded by fixed stator windings (*Figure 30–23*). The stationary armature or stator winding is generally three-phase, but some motors are designed to operate on different systems.

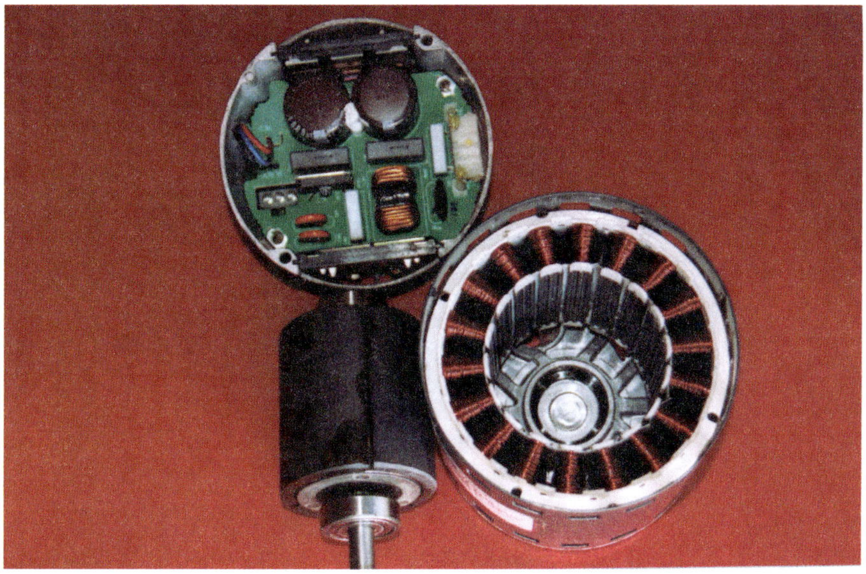

Figure 30–23 Brushless DC motor. Motor is shown with permanent magnet rotor, stator winding, and electronic components for converting the applied DC voltage into three-phase AC voltage.

The three-phase power is provided by a converter that changes the DC into AC (*Figure 30–24*). AC is used to create a rotating magnetic field inside the stator of the motor. This rotating magnetic field attracts the permanent magnets of the rotor and causes the rotor to turn in the same direction as the rotating field. Rotating magnetic fields are discussed in Unit 32.

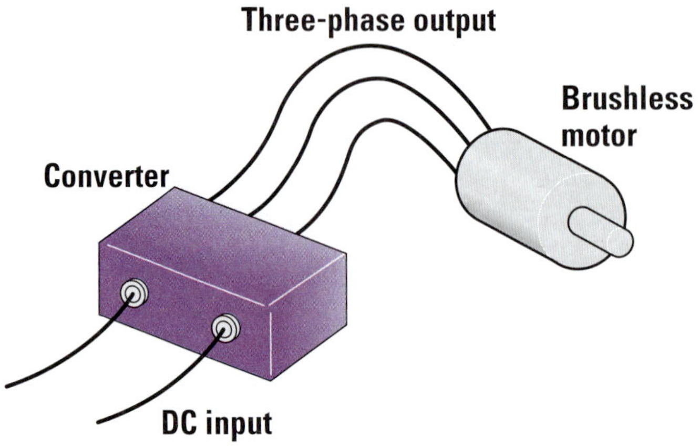

Figure 30–24 A converter changes DC into three-phase AC to power a brushless DC motor.

Some converters that supply power to brushless motors produce sine waves, but most produce a trapezoidal AC waveform (*Figure 30–25*). Motors powered by the trapezoidal waveform produce about 10% more torque than those powered by sine waves, but motors powered by sine waves operate more smoothly and with less torque ripple at low speeds. Motors that require very smooth operation such as grinders, mills, or other machines that produce a finished surface are powered by sine waves. Applications where high torque and low speed are required often employ brushless motors that have a very high number of stator poles. Some of these motors have as many as 64 stator poles per phase. The method for determining the speed of a three-phase motor is further discussed in Unit 32.

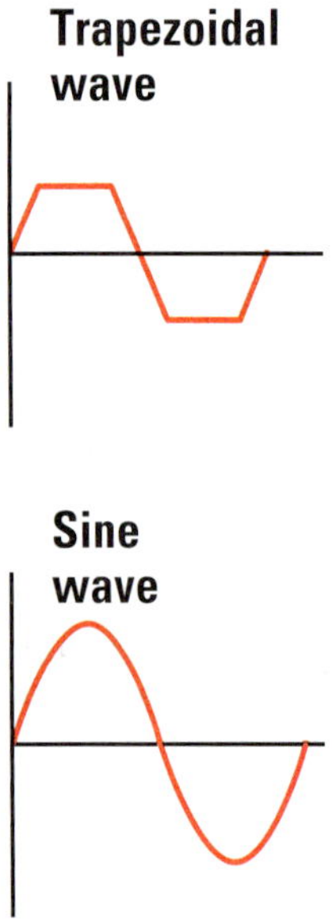

Figure 30–25 Most converters produce a trapezoidal wave, but some produce a sine wave.

The use of this many poles to provide low speed and high torque is often referred to as *magnetic gearing* because it eliminates the need for mechanical gears and other speed-reducing equipment. This, in turn, eliminates friction and backlash associated with mechanical gears. Motors that use a high pole count for low-speed high-torque applications are often called *ring motors* or *ring torquers*.

Inside-Out Motors

Another type of brushless DC motor uses a rotor shaped like a hollow cylinder or cup. The stator windings are wound inside the rotor (*Figure 30–26*). The permanent magnet rotor is mounted on the outside of the stator windings. The size and shape of the rotor cause it to act like a flywheel, which gives these motors a large amount of inertia. **Motors with a large amount of inertia exhibit superior speed-regulation characteristics.** Inside-out motors are often used to drive the hard disks in computers, to operate tape cartridge drives, to provide power for robots, and to operate fans and blowers in high-speed air-conditioning systems.

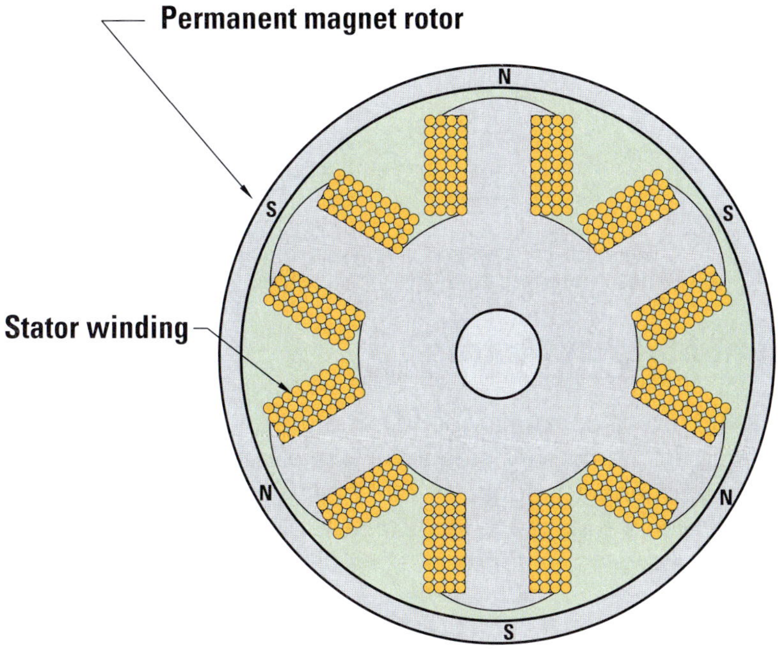

Figure 30–26 Inside-out brushless DC motor.

Differences between Brush-Type and Brushless Motors

Because brushless motors do not have a commutator or brushes, they are generally smaller and cost less than brush-type motors. The added expense of the converter, however, makes the cost about the same as a brush-type motor. Brushless motors dissipate heat more quickly because the stator windings can dissipate heat faster than a wound armature. They are more efficient than brush-type motors, require less maintenance, and as a general rule have less downtime. Brushless motors are commonly used in portable power tools because of the advantages over brush-type alternatives.

Converters

As stated previously, converters are used to change the DC into multiphase AC to produce a rotating magnetic field in the stator of the brushless DC motor. An example circuit for a DC-to-three-phase converter is shown in *Figure 30–27*. In this example, Transistor Q_1 through Transistor Q_6 are used as switches. By turning the transistors on or off in the proper sequence, current can be routed through the stator windings in such a manner that it produces three AC signals that are 120° out of phase with each other. The commutation sequence provides an action similar to that of the commutator in a brush-type motor.

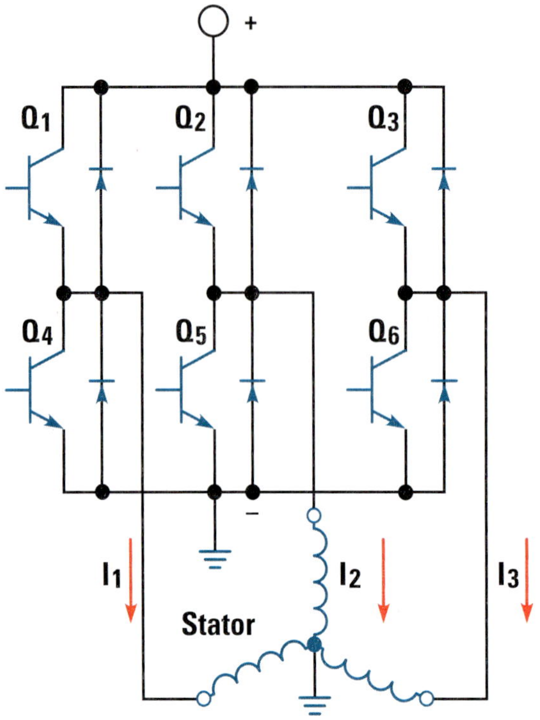

Figure 30–27 Example of a circuit to change DC voltage into three-phase AC voltage for a brushless DC motor.

30–5 Permanent Magnet Motors

Permanent magnet (PM) motors contain a wound armature and brushes like a conventional DC motor. The pole pieces, however, are permanent magnets. This eliminates the need for shunt or series field windings (*Figure 30–28*). Permanent magnet motors have a higher efficiency than conventional field-wound motors because power is supplied to the armature circuit only. These motors have been popular for many years in applications where batteries must be used to supply power to the motor.

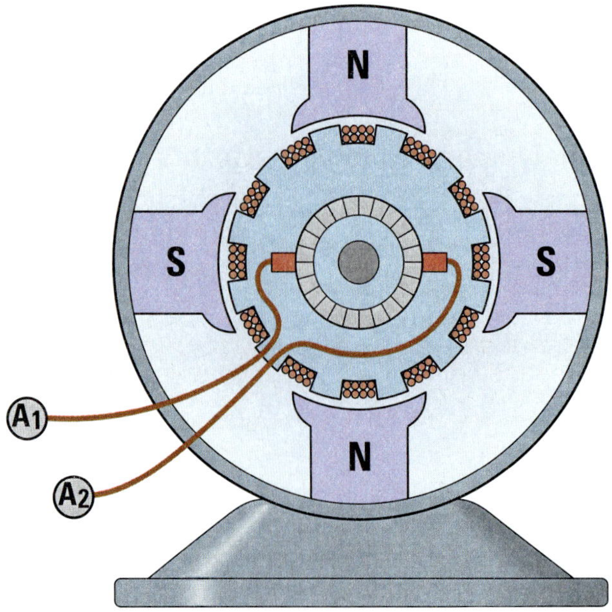

Figure 30–28 Permanent magnet motors only need power to the armature.

The horsepower rating of PM (permanent magnet) motors has increased significantly since the introduction of rare-earth magnets such as samarium-cobalt and neodymium in the mid-1970s. These materials have replaced alnico and ferrite magnets in most PM motors. The increased strength of rare-earth magnets permits motors to produce torques that range from 7.0 ounces per inch to 4500 foot-pounds. Permanent magnet motors with horsepower ratings of over 15 horsepower are now available. The torque-to-weight ratios of PM motors equipped with rare-earth magnets can exceed those of conventional field-wound motors by 40 to 90%. The power-to-weight ratios can exceed conventional motors by 50 to 200%. Permanent magnet motors of comparable horsepower ratings are smaller and lighter in weight than conventional field-wound motors.

> **Green Tip**
>
> Direct current motors that contain permanent magnet field poles are more efficient because no power is required to produce the magnetic fields of the pole pieces.

Operating Characteristics

The field flux of PM motors remains constant at all times. This gives the motor operating characteristics very similar to those of conventional separately excited shunt motors. The speed can be controlled by the amount of voltage applied to the armature, and the direction of rotation is reversed by reversing the polarity of voltage applied to the armature leads (*Figure 30–29*).

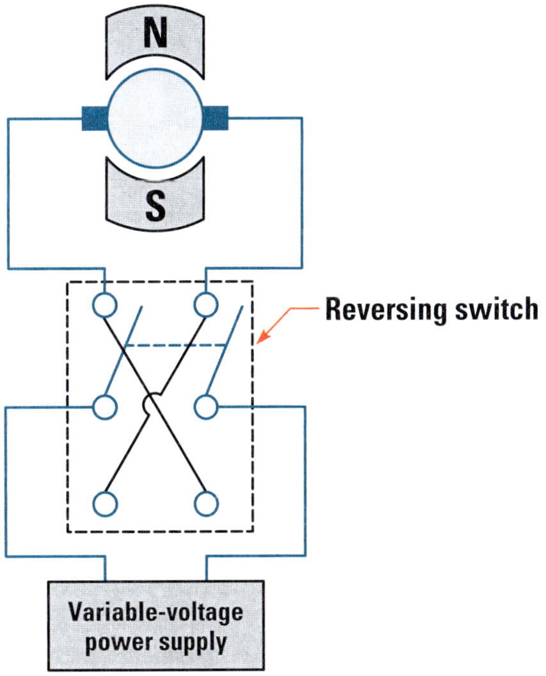

Figure 30–29 PM motors can be reversed by reversing the polarity to the armature, and the speed can be controlled by variable voltage.

DC Servomotors

Small PM motors are used as **servomotors.** These motors have small, lightweight armatures that contain very little inertia. This permits servomotors to be operated at high speed and then stopped or reversed very quickly. Servomotors generally contain from two to six poles. They are used to operate tape drives on computers and to power the spindles on numerically controlled (NC) machines such as milling machines and lathes.

DC ServoDisc Motors

One type of DC servomotor that is totally different in design is the **ServoDisc motor.** This motor uses permanent magnets to provide a constant magnetic field like conventional servomotors, but the design of the armature is completely different. In a conventional servomotor, the armature is constructed in the same way as other DC motors. The permanent magnets are mounted on the motor housing in such a manner that they create a radial magnetic field that is perpendicular to the windings of the armature (*Figure 30–30*). The armature core is made of iron because it must conduct the magnetic lines of flux between the two pole pieces.

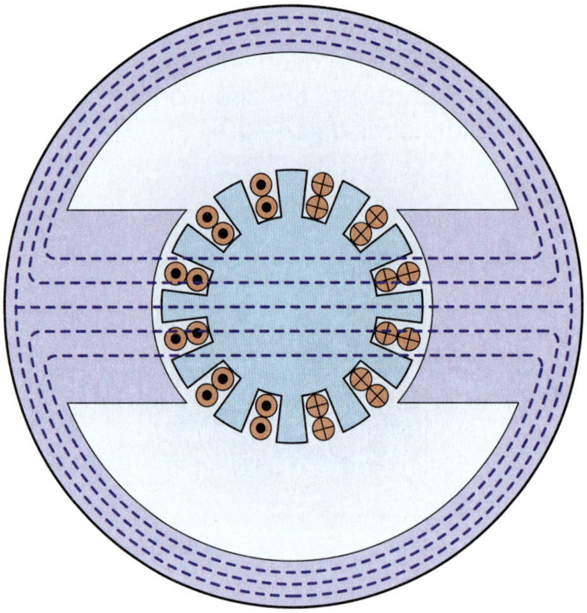

Figure 30–30 Magnetic lines of flux must travel a great distance between the poles of the permanent magnets.

The armature of the ServoDisc motor does not contain any iron. It is made of two to four layers of copper conductors formed into a thin disc. The conductors are "printed" on a fiberglass material in much the same way as a printed circuit board. For this reason, the ServoDisc motor is often called a **printed-circuit motor.** Because the disc armature is very thin, it permits the permanent magnets to be mounted on either side of the disc and parallel to the shaft of the motor (*Figure 30–31*). Because the disc is very thin, the air gap between the two magnets can be very small.

Torque is produced when current flowing though the copper conductors of the disc produces a magnetic field that reacts with the magnetic field of the permanent magnets. The permanent magnet pairs are arranged around the circumference of the motor housing in such a manner that they provide alternate magnetic fields (*Figure 30–32*). The conductors on each side of the armature produce a force that is tangent to the pole piece when current flows. The ServoDisc motor produces a relatively strong torque for its size and weight.

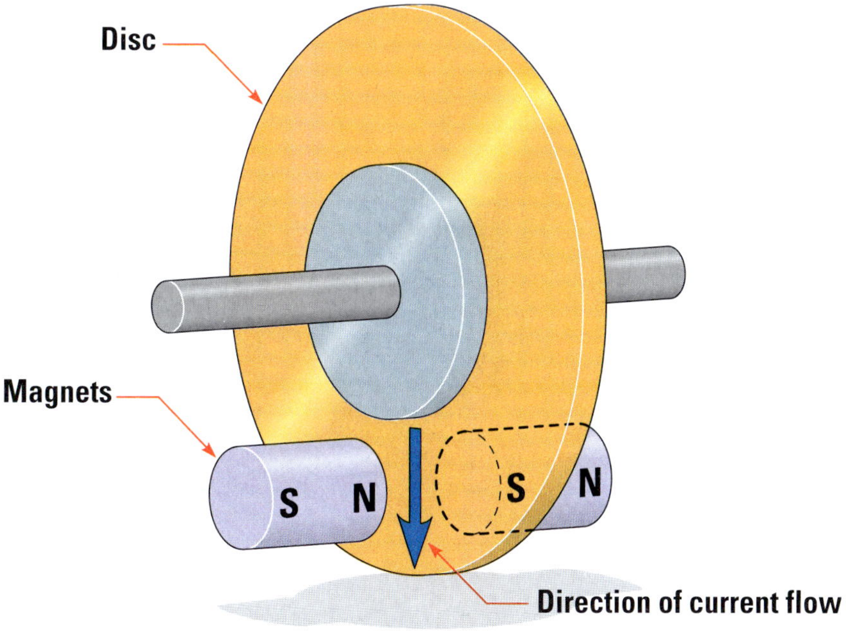

Figure 30–31 Basic construction of a ServoDisc motor.

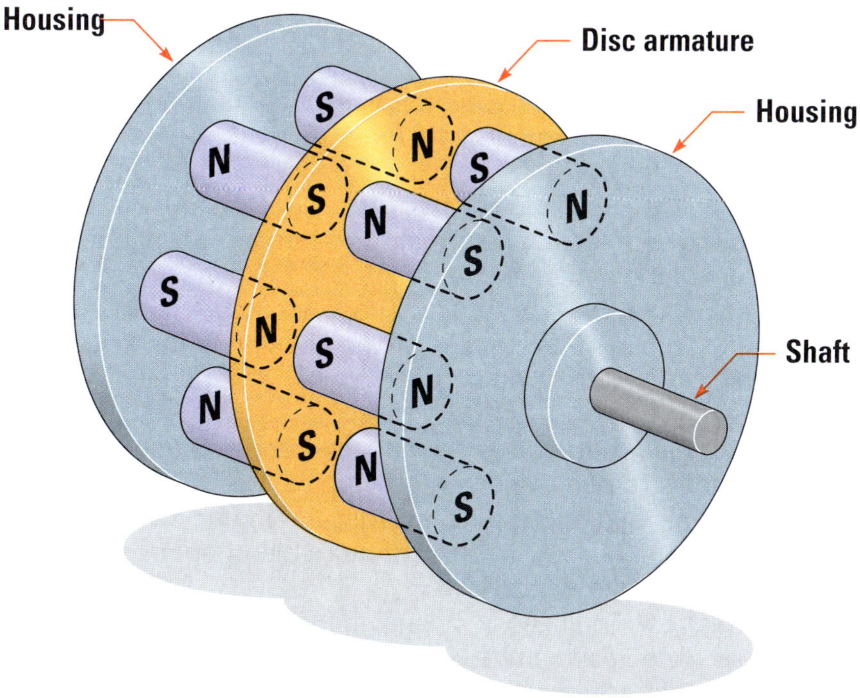

Figure 30–32 The permanent magnets are arranged to produce alternate magnetic polarities.

The conductors of the armature are so arranged on the fiberglass disc that they form a commutator on one side of the armature. Brushes riding against this commutator supply DC to the armature conductors. Because the armature contains no iron, it has almost no inductance. This greatly reduces any arcing at the brushes, which results in extremely long brush life. A typical ServoDisc motor is shown in *Figure 30–33*.

Figure 30–33 DC ServoDisc motors. Courtesy of Danaher Motion Technologies

Characteristics of ServoDisc Motors

The unique construction of the disc-type servomotor gives it some operating characteristics that are different from those of other types of permanent magnet DC motors. Permanent magnet DC motors with iron armatures have a problem with cogging at low speeds. Cogging is caused by the reaction of the iron armature with the field of the permanent magnets. If the armature is turned by hand, it jumps, or "cogs," from one position to another. As the armature is turned, the magnetic flux lines of the pole pieces pass through the core of the armature (*Figure 30–30*). Because the armature is slotted, certain areas offer a path of lower reluctance than others. These areas are more strongly attracted to the pole pieces, which causes the armature to jump from one position to the next. Because the armature of the disc motor contains no iron, there is no cogging action. This permits very smooth operation at low speeds.

Another characteristic of disc motors is extremely fast acceleration. The thin, low-inertia disc armature permits an exceptional torque–inertia ratio. A typical ServoDisc motor can accelerate from 0 to 3000 RPM in about 60° of rotation. In other words, the motor can accelerate from 0 to 3000 RPM in one-sixth of a revolution. The low-inertia armature also permits rapid stops and reversals. ServoDisc can operate at speeds over 4000 RPM.

The speed of the ServoDisc can be varied by changing the amount of voltage supplied to the armature. The voltage is generally varied using **pulse-width modulation.** Most amplifiers for the ServoDisc produce a pulsating DC voltage at a frequency of about 20 kilohertz. At a frequency of 20 kilohertz, the pulses have a width of 50 microseconds (pulse width = 1/frequency). Assume that the pulses have a peak voltage of 24 volts and are turned on for 40 microseconds and off for 10 microseconds during each pulse (*Figure 30–34*). In this example, 24 volts are supplied to the motor for 80% of the time, producing an average voltage of 19.2 volts (24 V × 0.80 = 19.2 V). Now assume that the pulse width is changed so the on time is 10 microseconds and the off time is 40 microseconds (*Figure 30–35*). A voltage of 24 volts is now supplied to the motor 20% of the time, resulting in an average value of 4.8 volts (24 V × 0.20 = 4.8 V).

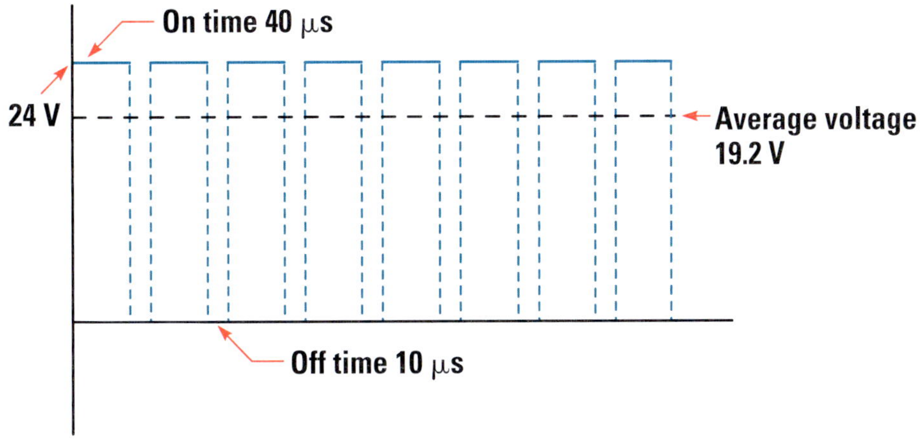

Figure 30–34 The amount of average voltage is determined by the pulse width.

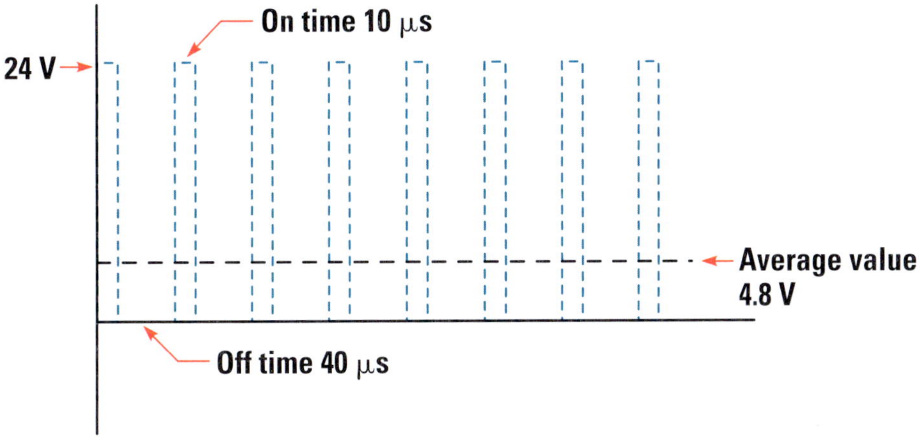

Figure 30–35 Reducing the pulse width effectively reduces the average voltage.

Because of the basic design of the motor, the ServoDisc motor can be constructed in a very small, thin case (*Figure 30–36*). This makes it very useful in some applications where other types of servomotors could not be used.

Figure 30–36 The DC ServoDisc motors can be constructed with small, thin cases. Courtesy of Danaher Motion Technologies

30–6 The Right-Hand Motor Rule

In Unit 13, it was shown that it is possible to determine the direction of current flow through the armature of a generator using the fingers of the left hand when the polarity of the field poles and the direction of rotation of the armature are known. Similarly, the fingers of the right hand can be used to determine the direction of rotation of the armature when the magnetic field polarity of the pole pieces and the direction of current flow through the armature are known (*Figure 30–37*). The thumb indicates the direction of thrust or movement of the armature. The forefinger indicates the direction of the field flux assuming that flux lines are in the direction of north to south, and the center finger indicates the direction of current flow through the armature.

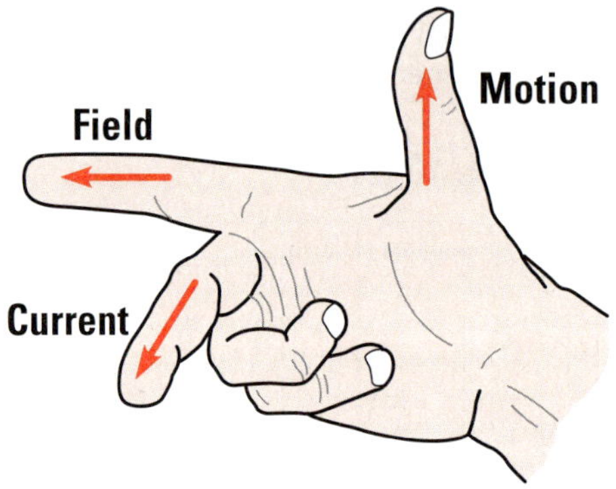

Figure 30–37 Right-hand motor rule.

Summary

- A motor is a machine that converts electric energy into mechanical energy.
- DC motors operate on the principle of attraction and repulsion of magnetism.
- Two factors that determine the torque produced by a motor are:
  a. the strength of the magnetic field of the pole pieces.
  b. the strength of the magnetic field of the armature.
- Torque is proportional to armature current if field excitation current remains constant.
- Current flow through the armature is limited by CEMF and armature wire resistance.
- Three factors that determine the amount of CEMF produced by a motor are:
  a. the strength of the magnetic field of the pole pieces.
  b. the number of turns of wire in the armature.
  c. the speed of the armature.
- Three basic types of DC motors are the series, the shunt, and the compound.
- Shunt motors are sometimes known as constant-speed motors.
- Series motors must never be operated under a no-load condition.
- Series motors can develop extremely high speed with no load.
- Compound motors contain both series and shunt field windings.
- When full voltage is applied to both the armature and shunt field, the motor operates at base speed.

- When full voltage is applied to the field and reduced voltage is applied to the armature, the motor operates below base speed.
- When full voltage is applied to the armature and reduced voltage is applied to the shunt field, the motor operates above base speed.
- The direction of rotation of a DC motor can be changed by reversing the connection of either the armature or the field leads.
- It is common practice to reverse the connection of the armature leads to prevent changing a compound motor from a cumulative- to a differential-compound connection.
- The shunt field relay is used to disconnect power to the armature if shunt field current drops below a certain level.
- Brushless DC motors do not contain a wound armature, a commutator, or brushes.
- The stator windings for brushless motors are generally three-phase AC windings.
- Brushless DC motors require a converter to change the DC into multiphase AC.
- Brushless motors generally require less maintenance and have less downtime than conventional DC motors.
- PM motors use permanent magnets as the pole pieces and do not require series or shunt field windings.
- PM motors are more efficient than conventional DC motors.
- The operating characteristics of a PM motor are similar to those of a shunt motor with external excitation.
- Servomotors have lightweight armatures with low inertia.
- ServoDisc motors contain an armature that is made of several layers of copper conductors.
- Because there is no iron in the armature of a disc servomotor, it does not have a problem with cogging.
- ServoDisc motor can be accelerated very rapidly.

Review Questions

1. True or False: A motor is an electrical device that is used to convert electrical energy into mechanical energy and perform work by using electromagnetic fields to turn the armature.
2. Which of the following statements best describes how a DC motor utilizes magnetic fields to cause the armature to rotate?
   A) Changing the polarity of the field windings to create a rotating magnetic field
   B) Magnetic fields never change polarity in a DC motor
   C) Brushes change the polarity of the field in the armature to repel and attract to the field windings
   D) Not enough information provided
3. Which of the following motors must never be operated in a no-load condition as it would lead to potential catastrophic failure?
   A) DC shunt motors
   B) DC series motors
   C) Cumulative-compound motors
   D) Differential-compound motors

4. Which of the following motor connections must be avoided as it can suddenly stop and reverse direction?
   A) DC shunt motor
   B) DC series motor
   C) Cumulative-compound motor
   D) Differential-compound motor
5. When controlling the speed of a DC motor, which of the following actions can be taken to reduce the speed of the motor?
   A) Current to the armature is modified
   B) Voltage to the field windings is modified
   C) Current to the shunt field is reduced
   D) A rheostat is connected to the field windings
6. When changing the direction of rotation in a DC motor, what must be done to accomplish this task and not risk creating a differential-compound motor?
   A) Reverse the polarity of the shunt field terminations of a compound motor
   B) A double-pole double-throw switch
   C) Reverse the polarity of the armature leads
   D) Reverse the polarity of the permanent magnets
7. True or False: Brushless DC motors operate by placing fixed permanent magnets in place of the field windings and applying excitation current to the wound armature.
8. Of the following statements, which best describes the major advantage brushless DC motors have over brush-type DC motors?
   A) Brushless motors are heavier than brush-type motors.
   B) Brushless motors have easier to replace parts.
   C) Brushless motors have higher horsepower.
   D) Brushless motors are lightweight and have fewer parts to service.
9. The speed of a permanent magnet motor can be controlled best by performing which of the following?
   A) Controlling the voltage applied to the armature of the motor
   B) Controlling the strength of the magnetic field
   C) Controlling the current into the field windings
   D) Speed cannot be controlled in a PM motor
10. Which of the following is a quality of a DC servomotor that is due to the fact that they have very little inertia and are a type of permanent magnet motor?
    A) DC servomotors move in small short steps.
    B) DC servomotors have low RPM and high-torque.
    C) DC servomotors change direction very quickly.
    D) DC servomotors are high-horsepower motors.

11. Which direction will the armature turn if the magnetic field lines travel away from you, and the current flows to the right?

    A) Up

    B) Down

    C) Left

    D) Right

12. True or False: The right-hand motor rule is opposite but similar to the left-hand generator rule.

Practical Applications

1. You are working as a plant electrician. You are told that a large DC motor was purchased from a remanufacturing vendor and was installed on the previous shift. The motor seemed to operate properly until load was added. The motor then stopped, reversed direction, and began to run at high speed. What is the probable cause of the trouble, and what would you do to correct the problem? Explain your answer.

Section
12 AC Machines

Unit

31 Three-Phase Alternators

Key Terms

Alternators

Brushless exciter

Field-discharge resistor

Parallel alternators

Phase rotation

Revolving-armature-type alternator

Revolving-field-type alternator

Rotor

Sliprings

Stator

Synchroscope

Why You Need to Know

Alternators produce most of the electric power in the world. Some applications for small single-phase alternators are portable generators for home emergencies or providing power for portable power tools on a worksite, but most alternators are three-phase and provide power to everything from concert venues to entire cities. This unit

- explains the principles of operation for almost all alternators regardless of size.
- discusses what determines output frequency and how output voltage is controlled.
- explains how alternators are connected in parallel to provide more power when needed.
- discusses the different types of alternators and the operation of each.
- explains how to interpret the NEC when determining how to connect and determine protective-devices.

Outline

31–1 Three-Phase Alternators

31–2 AC Alternator Output

31–3 Connecting AC Alternators in Parallel

31–4 Field-Discharge Protection

Objectives

After studying this unit, you should be able to

- identify components of an AC alternator by their purpose.
- differentiate between rotating field and rotating armature alternators.
- illustrate how speed of rotation and number of poles affect the output of an alternator.
- distinguish key factors in connecting AC alternators in parallel.
- describe the role of a field-discharge resistor.

Preview

Most of the electric power in the world today is produced by AC generators, or alternators. Electric power companies often use alternators producing 13,800 volts and rated in gigawatts (1 gigawatt = 1,000,000,000 W) to produce the power used throughout the United States and Canada. The entire North American continent is powered by AC generators connected together in parallel to form a massive grid of generation, transmission, and distribution equipment. These alternators are powered by turbines called prime movers, which are powered by oil, coal, natural gas, hydro energy, solar energy, wind energy, or nuclear energy. The number of energy sources for prime movers seems to be growing by the minute as engineers find new and innovative ways to tackle the growing demand for more power. The standard frequency in the United States and Canada is 60 Hz. In other parts of the world, this frequency may vary. The power produced in many of the European countries is 50 Hz.

31–1 Three-Phase Alternators

Alternators operate on the same principle of electromagnetic induction as DC generators, but they have no commutator to change the AC produced in the armature into DC. There are two basic types of alternators: the revolving-armature type and the revolving-field type.

Revolving-Armature-Type Alternators

The **revolving-armature-type alternator** is the less used of the two basic types. This alternator uses an armature similar to that of a DC machine with the exception that the loops of wire are connected to **sliprings** instead of to a commutator (*Figure 31–1*). Three separate windings are connected in either delta or wye. The armature windings are rotated inside a magnetic field (*Figure 31–2*). Power is carried to the outside circuit via brushes riding against the sliprings. This alternator is the less used because it is very limited in the amount of output voltage and kilovolt-ampere (kVA) capacity it can develop.

Revolving-Field-Type Alternators

The **revolving-field-type alternator** uses a stationary armature called the **stator** and a rotating magnetic field. This design permits higher voltage and kilovolt-ampere ratings because the outside circuit is connected directly to the stator and is not routed through sliprings and brushes. This type of alternator is constructed by placing three sets of windings 120° apart (*Figure 31–3*). In *Figure 31–3,* the winding of Phase 1 winds around the top center pole piece. It then proceeds 180° around the stator and winds around the opposite pole piece in the opposite direction. Phase 2 winds around the top pole piece directly to the left

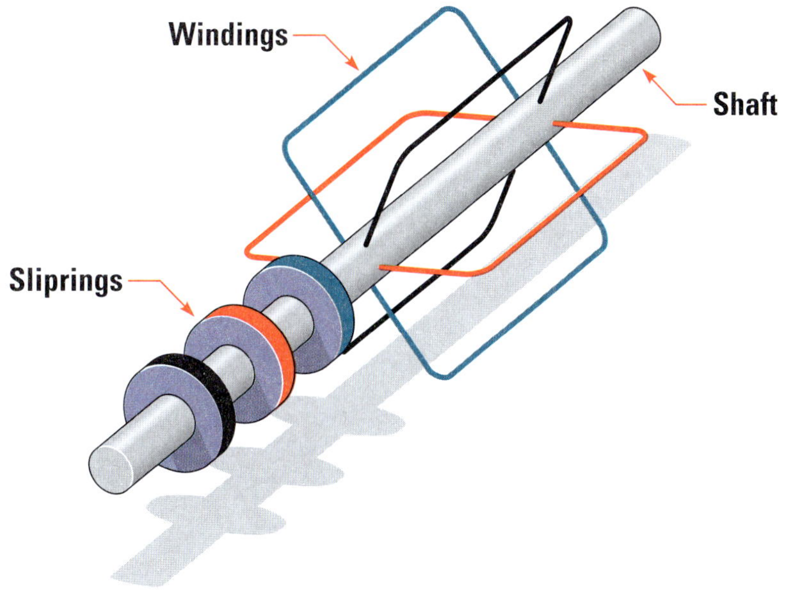

Figure 31–1 Basic design of a three-phase revolving-type armature.

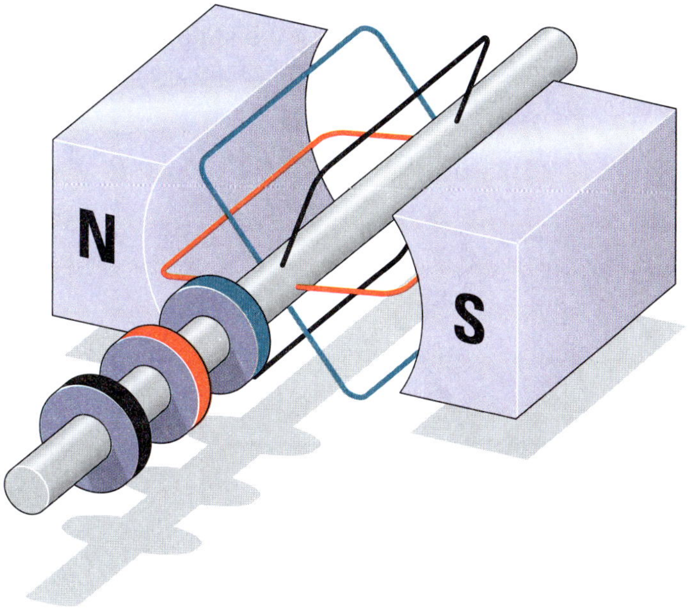

Figure 31–2 The armature conductors rotate inside a magnetic field.

of the top center pole piece. The second phase winding is wound in an opposite direction to the first. It then proceeds 180° around the stator housing and winds around the opposite pole piece in the opposite direction. The finish end of Phase 2 connects to the finish end of Phase 1. The start end of Phase 3 winds around the top pole piece to the right of the top center pole piece. This winding is wound in a direction opposite to Phase 1 also. The winding then proceeds 180° around the stator frame to its opposite pole

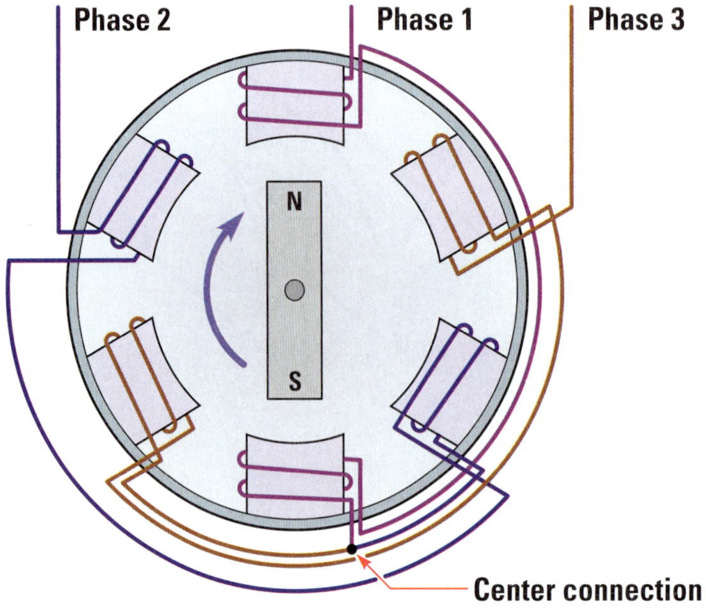

Figure 31–3 Basic design of a three-phase alternator.

piece and winds around it in an opposite direction. The finish end of Phase 3 is then connected to the finish ends of Phases 1 and 2. This forms a wye connection for the stator winding. When the magnet is rotated, voltage is induced in the three windings. Because these windings are spaced 120° apart, the induced voltages are 120° out of phase with each other (*Figure 31–4*).

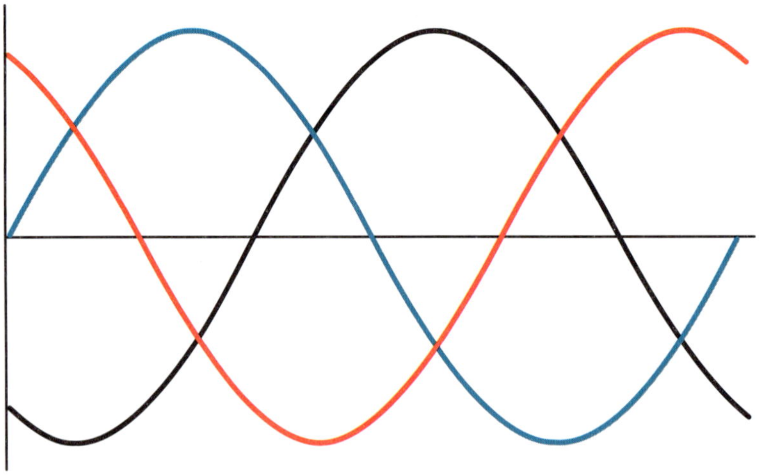

Figure 31–4 The alternator produces three sine wave voltages 120° out of phase with each other.

The stator shown in *Figure 31–3* is drawn in a manner to aid in understanding how the three phase windings are arranged and connected. In actual practice, the stator windings are placed in a smooth cylindrical core without projecting pole pieces (*Figure 31–5*). This design provides a better path for magnetic lines of flux and increases the efficiency of the alternator.

Figure 31–5 Wound stator.

The Rotor

The **rotor** is the rotating member of the machine. It provides the magnetism needed to induce voltage into the stator windings. The magnets of the rotor are electromagnets and require some source of external DC to excite the alternator. This DC source is known as excitation current. The alternator cannot produce an output voltage until the rotor has been excited. Some alternators use sliprings and brushes to provide the excitation current to the rotor (*Figure 31–6*). The DC excitation current can be varied to change the strength of the magnetic field.

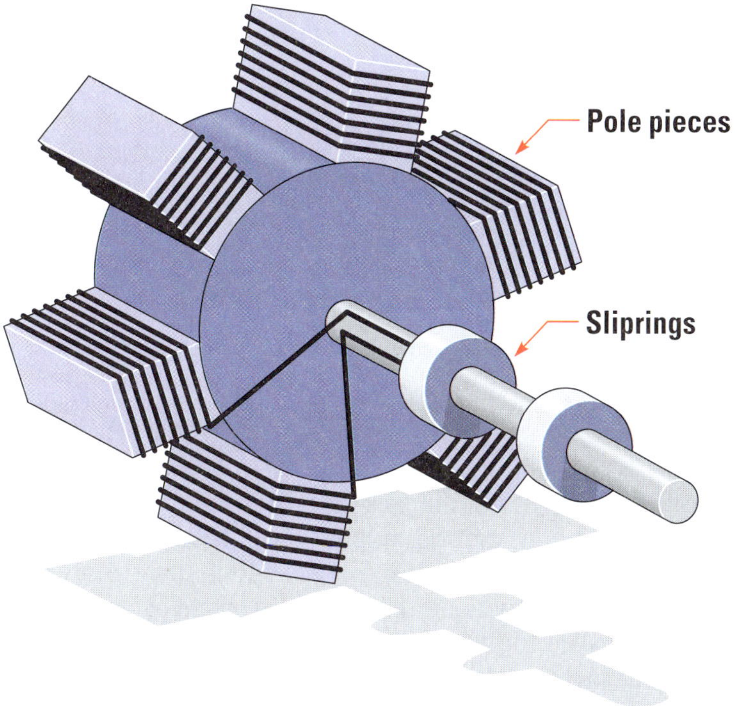

Figure 31–6 The rotor contains pole pieces that become electromagnets.

The Brushless Exciter

Most large alternators use an exciter that contains no brushes. This is accomplished by adding a separate small alternator of the armature type on the same shaft of the rotor of the larger alternator. The armature rotates between wound stationary electromagnets. The DC excitation current is connected to the wound stationary magnets (*Figure 31–7*). The amount of voltage induced in the rotor can be varied

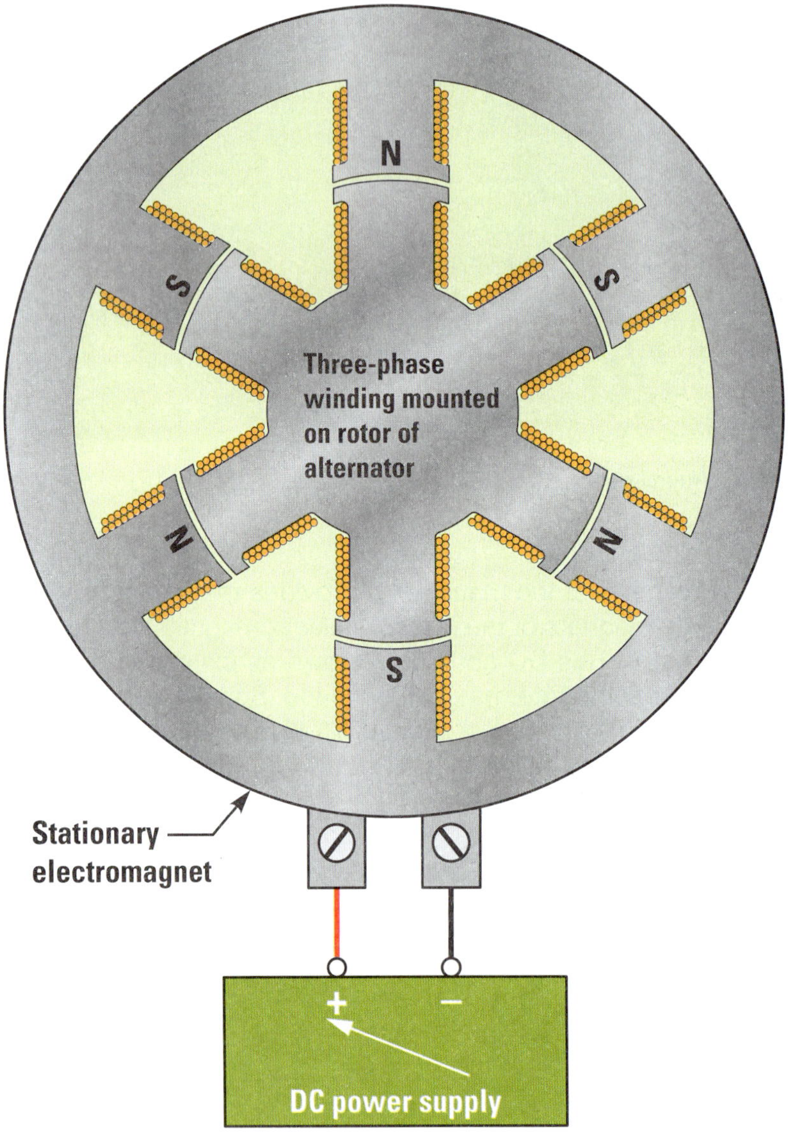

Figure 31–7 The brushless exciter uses stationary electromagnets around a rotating armature.

by changing the amount of excitation current supplied to the electromagnets. The output voltage of the armature is connected to a three-phase bridge rectifier mounted on the rotor shaft (*Figure 31–8*). The bridge rectifier converts the three-phase AC voltage produced in the armature into DC voltage before it is applied to the main rotor windings. Because the armature, rectifier, and rotor winding are connected to the main rotor shaft, they all rotate together, and no brushes or sliprings are needed to provide excitation current for the large alternator. A photograph of the **brushless exciter** assembly is shown in *Figure 31–9*. The field winding is placed in slots cut in the core material of the rotor (*Figure 31–10*).

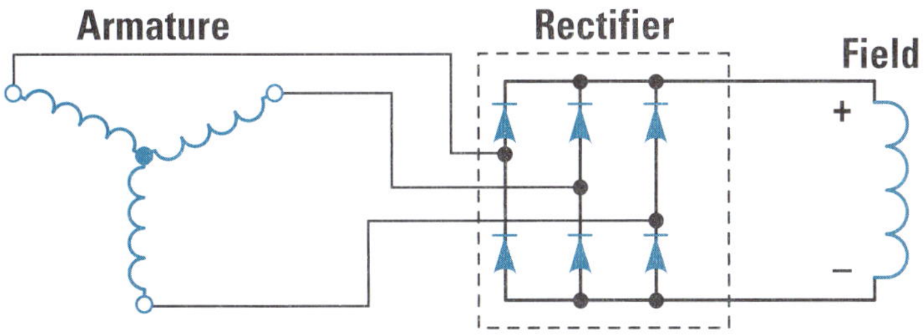

Figure 31–8 Basic brushless exciter circuit with bridge rectifier to change AC to DC for excitation in an alternator.

Figure 31–9 Brushless exciter assembly with the stationary electromagnet outside case removed. Courtesy of Electric Machinery Corporation

Figure 31–10 Rotors contain slots for placement of windings. © Smoczyslaw/Shutterstock.com

31–2 AC Alternator Output

The output frequency of an alternator is determined by the following two factors:

1. the number of stator poles, and
2. the speed of rotation of the rotor.

Because the number of poles in the stator is constant for a particular machine, the output frequency is controlled by adjusting the speed of the rotor. The following chart shows the speed of rotation needed to produce 60 hertz for alternators with different numbers of poles.

RPM	Stator Poles
3600	2
1800	4
1200	6
900	8

The following formula can also be used to determine the frequency when the poles and revolutions per minute (RPM) are known:

$$f = \frac{PS}{120}$$

where

f = frequency in hertz

P = number of poles per phase

S = speed in RPM

120 = a constant

Example 31–1

What is the output frequency of an alternator that contains six poles per phase and is turning at a speed of 1000 RPM?

Solution

$$f = \frac{6 \times 1000 \text{ RPM}}{120}$$

$$f = 50 \text{ Hz}$$

Output Voltage

Three factors determine the amount of output voltage of an alternator:

1. the length of the armature or stator conductors (number of turns),
2. the strength of the magnetic field of the rotor, and
3. the speed of rotation of the rotor.

The following formula can be used to calculate the amount of voltage induced in the stator winding:

$$E = \frac{BLV}{10^8}$$

where

10^8 = flux lines equal to 1 weber
E = induced voltage (in volts)
B = flux density in gauss
L = length of the conductor (in cm)
v = velocity (in cm/s)

One of the factors that determines the amount of induced voltage is the length of the conductor. This factor is often stated as the number of turns of wire in the stator because the voltage induced in each turn adds. Increasing the number of turns of wire has the same effect as increasing the length of one conductor. Remember that an induced voltage of 1 volt occurs when magnetic lines of flux are cut at a rate of 1 weber per second.

Controlling Output Voltage

The number of turns of wire in the stator cannot be changed in a particular machine without rewinding the stator, and the speed of rotation is generally maintained at a certain level to provide a constant output frequency. Therefore, the output voltage is controlled by increasing or decreasing the strength of the magnetic field of the rotor. The magnetic field strength is typically controlled by controlling the DC excitation current to the rotor. Increasing or decreasing this voltage has the effect of increasing or decreasing the flux line density created by the electromagnets, and, therefore, it increases or decreases the number of flux lines cut by the stator windings.

31–3 Connecting AC Alternators in Parallel

It often becomes necessary to use more than one alternator to supply the load. When more than one alternator is to be used, they are connected in parallel with each other. Several conditions must be met before **parallel alternators** can be used:

1. The phases must be connected in such a manner that the phase rotation of all the machines is the same. (A with A, B with B, C with C).
2. The phase rotation of each alternator must be the same direction. If the output voltage of Alternator 1 is rotating in a clockwise direction, the output voltage of Alternator 2 must rotate in a clockwise direction also (*Figure 31–11*).

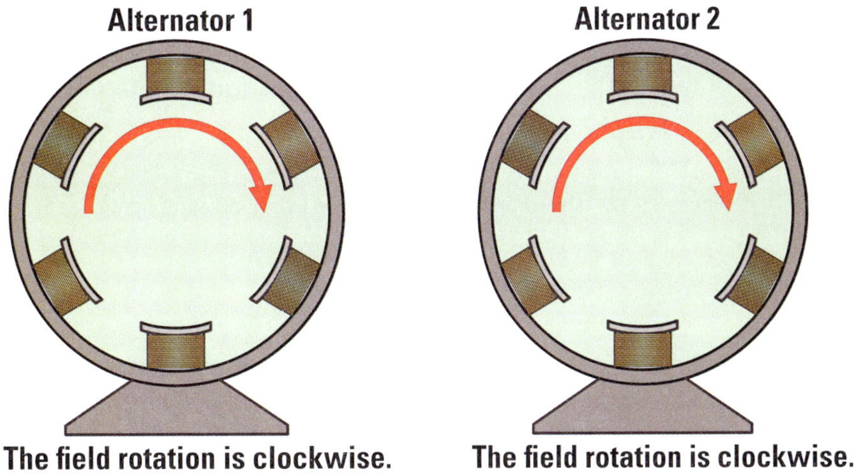

Figure 31–11 The field rotation must be the same for the two alternators.

3. The output voltage magnitude of the two alternators must be the same.
4. The frequency must be the same.

Determining Phase Rotation

One method of detecting when the **phase rotation** (the direction of magnetic field rotation) of one alternator is matched to the phase rotation of the other is with the use of a synchronizing switch (*Figure 31–12*). In *Figure 31–12,* the two alternators that are to be paralleled are connected together through a synchronizing switch. A set of three lamps acts as a resistive load between the two machines when the switch contacts are in the open position. The voltage developed across the lamps is proportional to the difference in voltage between the two alternators. The lamps are used to indicate two conditions:

1. The lamps indicate when the phase rotation of one machine is matched to the phase rotation of the other. When both alternators are operating, both are producing a voltage. The lamps blink on and off when the phase rotation of one machine is not synchronized to the phase rotation of the other machine. If all three lamps blink on and off at the same time, or in unison, the phase rotation of Alternator 1 is correctly matched to the phase rotation of Alternator 2. If the lamps blink on and off but not in unison, the phase rotation between the two machines is not correctly matched, and two lines of Alternator 2 should be switched.
2. The lamps also indicate when the phase of one machine is synchronized with the phase of the other machine. If the positive peak of Alternator 1 does not occur at the same time as the positive peak of Alternator 2, there is a potential between the two machines. This permits the lamps to glow. The brightness of the lamps indicates how far out of synchronism the two machines are. When the peak voltages of the two alternators occur at the same time, there is no potential difference between them. The lamps should be off at this time. The synchronizing switch should never be closed when the lamps are glowing.

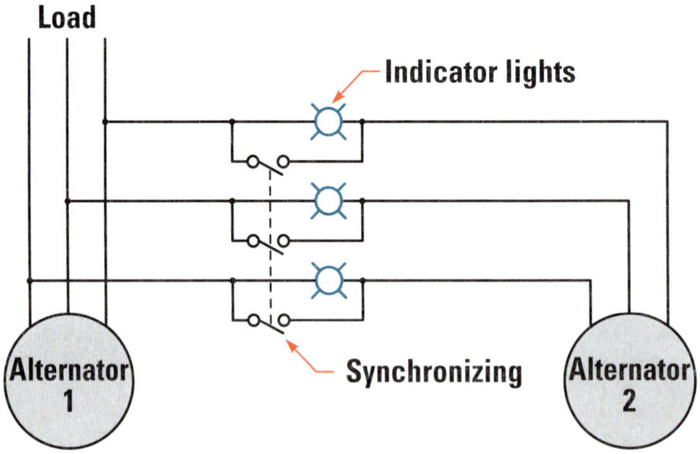

Figure 31–12 Determining phase rotation using indicator lights.

The Synchroscope

Another instrument often used for paralleling two alternators is the **synchroscope** (*Figure 31–13*). The synchroscope measures the difference in voltage and frequency of the two alternators. The pointer of the synchroscope is free to rotate in a 360° arc. The alternator already connected to the load is considered to be the base machine. The synchroscope indicates whether the frequency of the alternator to be parallel to the base machine is fast or slow. When the voltages of the two alternators are in phase, the pointer covers the shaded area on the face of the meter. When the two alternators are synchronized, the synchronizing switch can be closed.

Figure 31–13 Synchroscope.

If a synchroscope is not available, the two alternators can be paralleled using three lamps, as described earlier. If the three-lamp method is used, an AC voltmeter is connected across the same phase of each machine to measure the volt drop across the lamp. Even if the lamps blink at the same time indicating in-phase rotation, the voltage might be off. If a voltmeter indicates a difference on the same line, adjust the voltage of the second alternator accordingly. When the potential difference between the two machines is zero, that is the point at which the synchronizing switch should be closed (*Figure 31–14*).

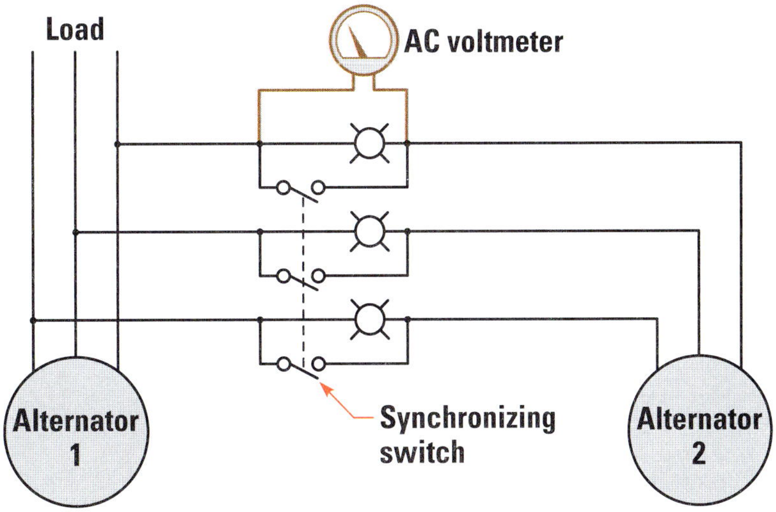

Figure 31–14 Using an AC voltmeter to indicate when the two alternators are in phase.

Sharing the Load

After the alternators have been paralleled, the power input to Alternator 2 must be increased to permit it to share part of the load. For example, if the alternator is being driven by a steam turbine, the power of the turbine would have to be increased. When this is done, the power to the load remains constant. The power output of the base alternator decreases, and the power output of the second alternator increases.

31–4 Field-Discharge Protection

When the DC excitation current is disconnected, the collapsing magnetic field can induce a high voltage in the rotor winding. This voltage can be high enough to arc contacts and damage the rotor winding or other circuit components. One method of preventing the induced voltage from becoming excessive is with the use of a **field-discharge resistor.** A special double-pole single-throw switch with a separate blade is used to connect the resistor to the field before the switch contacts open. When the switch is closed and DC is connected to the field, the circuit connecting the resistor to the field is open (*Figure 31–15*). When the switch is opened, the special blade connects the resistor to the field before the main contacts open (*Figure 31–16*).

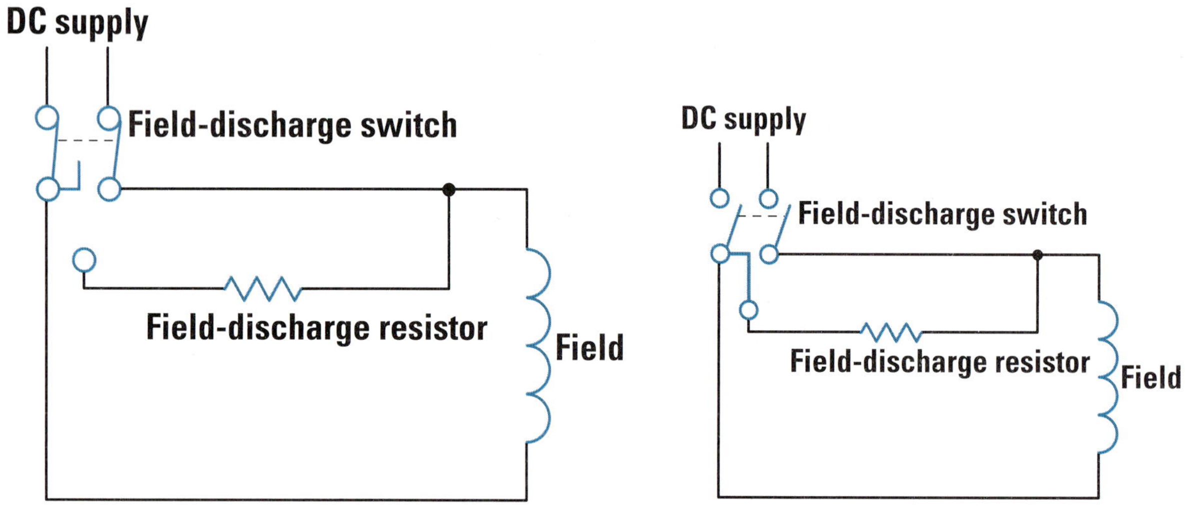

Figure 31–15 Field-discharge resistor switch used to minimize induced voltage in rotor windings.

Another method of preventing the high voltage discharge is to connect a diode in parallel with the field (*Figure 31–16*). The diode is connected in such a manner that when excitation current is flowing, the diode is reverse-biased and no current flows through the diode.

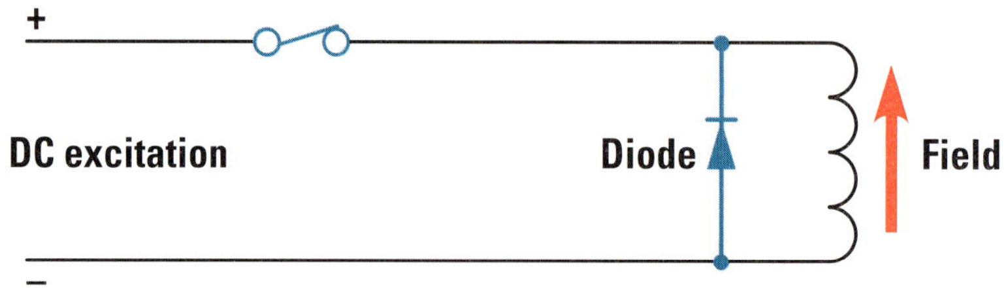

Figure 31–16 Normal current flow does not pass through a reverse-biased diode.

When the switch opens and the magnetic field collapses, the induced voltage is opposite in polarity to the applied voltage (*Figure 31–17*). The diode is now forward-biased, permitting current to flow through the diode. The energy contained in the magnetic field is dissipated in the form of heat by the diode and field winding.

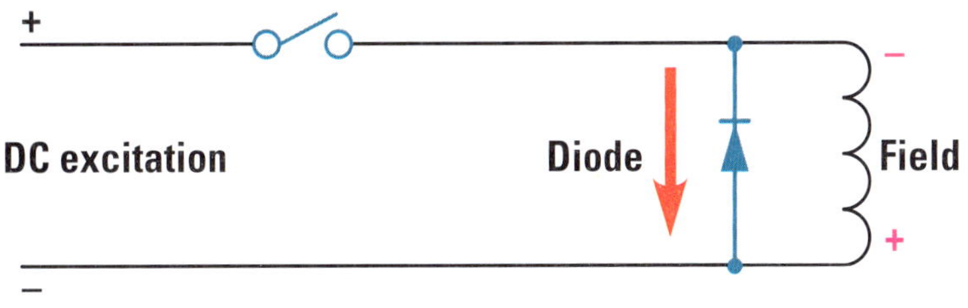

Figure 31–17 Induced current will flow through the diode.

Summary

- There are two basic types of three-phase alternators: the revolving-armature type and the revolving-field type.
- The rotating-armature type is the less used because of its limited voltage and power rating.
- The rotor of the revolving-field-type alternator contains electromagnets.
- DC must be supplied to the field before the alternator can produce an output voltage.
- The DC supplied to the field is called excitation current.
- The output frequency of an alternator is determined by the number of stator poles and the speed of rotation.
- Three factors that determine the output voltage of an alternator are
  a. the length of the conductor of the armature or stator winding,
  b. the strength of the magnetic field of the rotor, and
  c. the speed of the rotor.
- The output voltage is controlled by the amount of DC excitation current.
- Before two alternators can be connected in parallel, the output voltage of the two machines should be the same, the frequency of each machine must be the same, and the phase rotation of the machines must be the same and must be in phase with each other.
- Three lamps connected between the two alternators can be used to test for synchronized phase rotation.
- A synchroscope can be used to determine phase rotation and difference of frequency between two alternators.
- Two devices used to prevent a high voltage being induced in the rotor when the DC excitation current is stopped are a field-discharge resistor and a diode.
- Many large alternators use a brushless exciter to supply DC to the rotor winding.

Review Questions

1. Which of the following components of an AC alternator allows the voltage induced in a rotating armature to be applied to the connected circuit?

   A) Stator

   B) Commutator

   C) Sliprings

   D) Rotor

2. Which of the following terms best describes the stationary armature that is used in a revolving-field type alternator?
   A) Stator
   B) Commutator
   C) Sliprings
   D) Rotor
3. Which type of AC alternator requires the use of brushes and sliprings to deliver the electricity induced within the armature of the alternator?
   A) Rotating-field alternators
   B) Rotating-armature alternators
   C) Series-field alternator
   D) Compounding-field alternator
4. True or False: Rotating field type alternators can supply higher amounts of power and higher voltage values because the connection is made directly to the armature.
5. A six-pole alternator is rotated at 1200 RPM to produce a frequency of 60 hz. What will the frequency be if the speed is increased to 3600 rpm?
   A) 60 hz
   B) 120 hz
   C) 240 hz
   D) 180 hz
6. Which of the following describes the best practical method for controlling the output voltage of an AC alternator?
   A) Adding more turns of wire to the armature
   B) Controlling the strength of the magnetic field in the rotor
   C) Increasing how fast the rotor is spun inside the stator
   D) Adding variable resistors in series with the output
7. Why isn't speed of rotation typically used to control the output voltage of an AC alternator?
   A) Modifying the speed of rotation decreases the current.
   B) Modifying the speed of rotation increases the harmonics in the system.
   C) Modifying the speed of rotation modifies the frequency of the system.
   D) Modifying the speed of rotation increases the maintenance schedule.
8. Before connecting AC alternators in parallel with each other, which of the following statements must be verified to be true?
   A) Phases must be connected to the same phases.
   B) Phase rotation must be the same.
   C) The voltage and frequency must be the same.
   D) All must be verified as true.

9. True or False: A device called a synchroscope is often used to measure the voltage and frequency between two alternators for the purpose of paralleling the sources.
10. Which of the following describes a method for ensuring the power sources are synched when connecting two AC alternators in parallel?
    A) Set of three lamps is connected across an open switch to indicate synchronized rotation
    B) Set of three lamps is connected in series with a closed switch
    C) Set of three lamps on each alternator indicates rotation
    D) Set of three lamps at the load indicates if sources are synched
11. Rapidly disconnecting the DC excitation current in an alternator can damage windings in the rotor and other circuit components. Which statement best describes how this happens?
    A) The alternator becomes a motor and turns the prime mover in the opposite direction.
    B) The alternator builds up a static charge that dissipates through the rotor windings.
    C) The rapidly collapsing field in the rotor induces a voltage spike in the windings.
    D) The inertia of the spinning rotor causes mechanical damage when stopped abruptly.
12. True or False: The field-discharge resistor bleeds off the induced voltage created when DC excitation current is removed by connecting to ground.

Unit 32 Three-Phase Motors

Key Terms

Amortisseur winding
Code letter
Direction of rotation
Dual-voltage motors
Percent slip
Phase rotation meter
Rotating magnetic field
Rotor frequency
Single-phasing
Squirrel-cage rotor
Synchronous condenser
Synchronous speed
Wound-rotor motor

Why You Need to Know

Three-phase motors are the backbone of industry. They range in size from fractional horsepower to several thousand horsepower. It is imperative that anyone working in the electrical field have a thorough understanding of the different types of three-phase motors and their operating characteristics. This unit

- describes common types of three-phase motors. Some are designed to operate on a single voltage, and others can be made to operate on two different voltages. Some motors have their stator windings connected in wye, and others are connected in delta.
- explains how stator windings are numbered and how to connect them to operate on the voltage supplied to the motor.
- describes how to calculate protective devices, different wiring methods, and operational principles.

Outline

Objectives

After studying this unit, you should be able to

- explain the basic operating principles of three-phase motors.
- identify key factors that impact the operation of three-phase motors.
- differentiate between high and low voltage three-phase motor terminations.
- identify motor starting and running characteristics from the design of a motor's squirrel-cage rotor.
- identify key information contained on a three-phase motor nameplate.
- differentiate the operating principles of a wound-rotor motor from those of a squirrel-cage motor.
- explain the operation principle and common uses of synchronous motors.

Preview

Three-phase motors are used throughout the United States and Canada as the main workhorse for industrial manufacturing. These motors convert three-phase AC into mechanical energy to operate all types of machinery. Three-phase motors are smaller and lighter and have higher efficiencies per horsepower than single-phase motors. They are extremely rugged and require relatively little maintenance. Many of these motors are operated 24 hours a day, 7 days a week for many years with minimal problems.

32–1 Three-Phase Motors

The three most common types of three-phase motors are:

1. the squirrel-cage induction motor,
2. the wound-rotor induction motor, and
3. the synchronous motor.

All three motors operate on the same principle, and they all use the same basic design for the stator windings. The difference among them is the type of rotor used. Two of the three motors are induction motors and operate on the principle of electromagnetic induction in a manner similar to that of transformers. In fact, AC induction motors were patented as rotating transformers by Nikola Tesla. The stator winding of a motor is often referred to as the *motor primary*, and the rotor is referred to as the *motor secondary*.

The Rotating Magnetic Field

Three-phase motors operate on the principle of a **rotating magnetic field.** The magnetic field rotates due to:

1. the fact that the voltages in a three-phase system are 120° out of phase with each other.
2. the fact that the three voltages change polarity at regular intervals.
3. the arrangement of the stator windings around the inside of the motor.

Figure 32–1 shows how three-phase AC sine waves are 120° out of phase with each other and the stator winding of a three-phase motor. The stator illustrates a two-pole three-phase motor. Two pole means that there are two poles per phase. AC motors do not generally have actual pole pieces as shown in *Figure 32–1*, but they will be used here to aid in understanding how the rotating magnetic field is created in a three-phase motor. Notice that the pole pieces connected to each phase are located opposite from each other. Each phase's opposing pole pieces are wound in such a manner that when current flows through the winding, they will develop opposite magnetic polarities. The windings of poles B1 and C1 are wound in the same direction in relation to each other but in an opposite direction from the winding of pole A1. The start end of the winding for poles A1 and A2 is connected to Line 1, the start end of the winding for poles B1 and B2 is connected to Line 2, and the start end of the winding for poles C1 and C2 is connected to Line 3. The finish ends of all three windings are joined to form a wye connection for the stator.

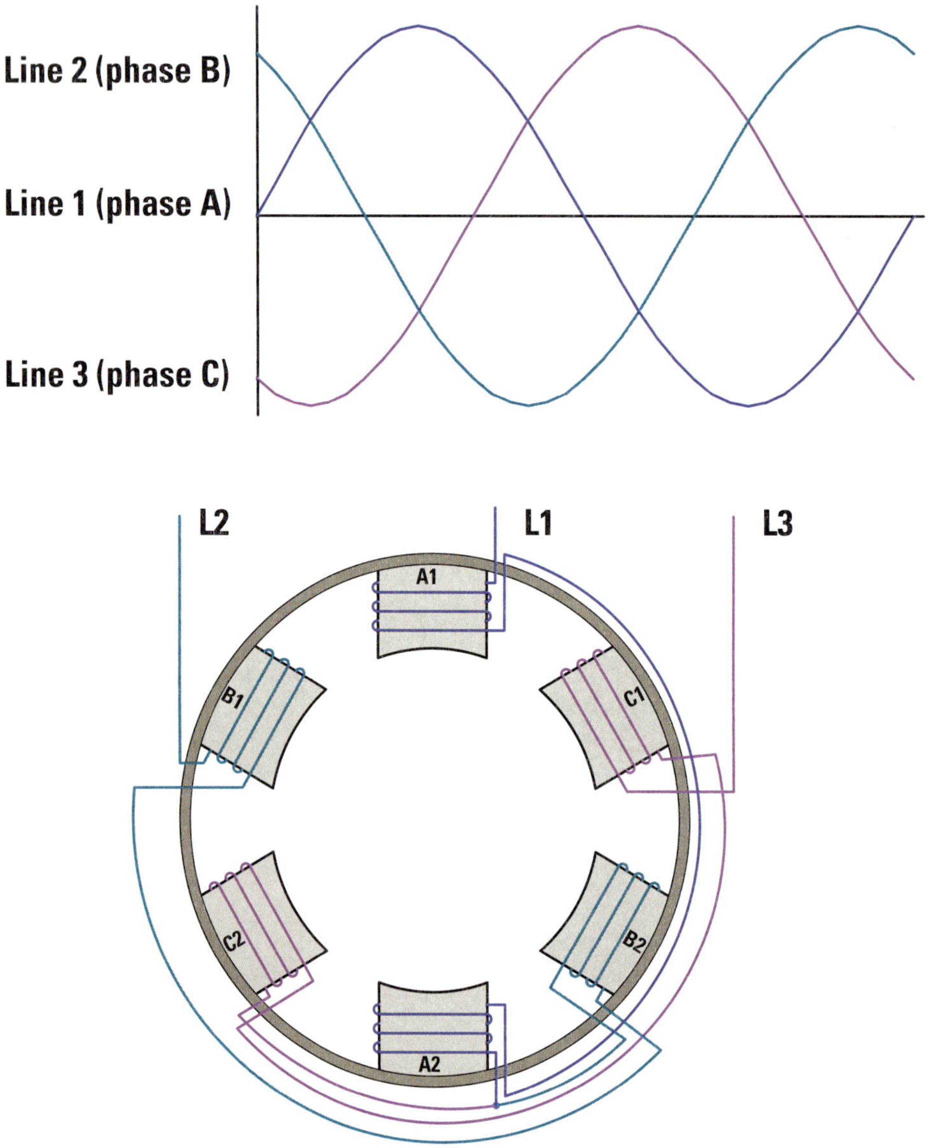

Figure 32–1 Three-phase stator and three sine wave voltages.

To understand how the magnetic field rotates around the inside of the stator, refer to *Figure 32–2*. A dashed line labeled A has been drawn to illustrate the condition of the three voltages at a point in time. The arrows drawn inside the motor indicate the greatest concentration of magnetic lines of flux; the arrows are pointing in the direction that indicates magnetic lines of flux from north to south. Line 1 has reached its maximum peak voltage in the positive direction, and Lines 2 and 3 are less than maximum and in the negative direction. Notice that the magnetic field is more concentrated between poles A1 and A2. Weaker lines of magnetic flux also exist between poles B1 and B2 and C1 and C2. Also note that poles A1, B1, and C1 all form a south magnetic polarity. Poles A2, B2, and C2 form a north magnetic polarity.

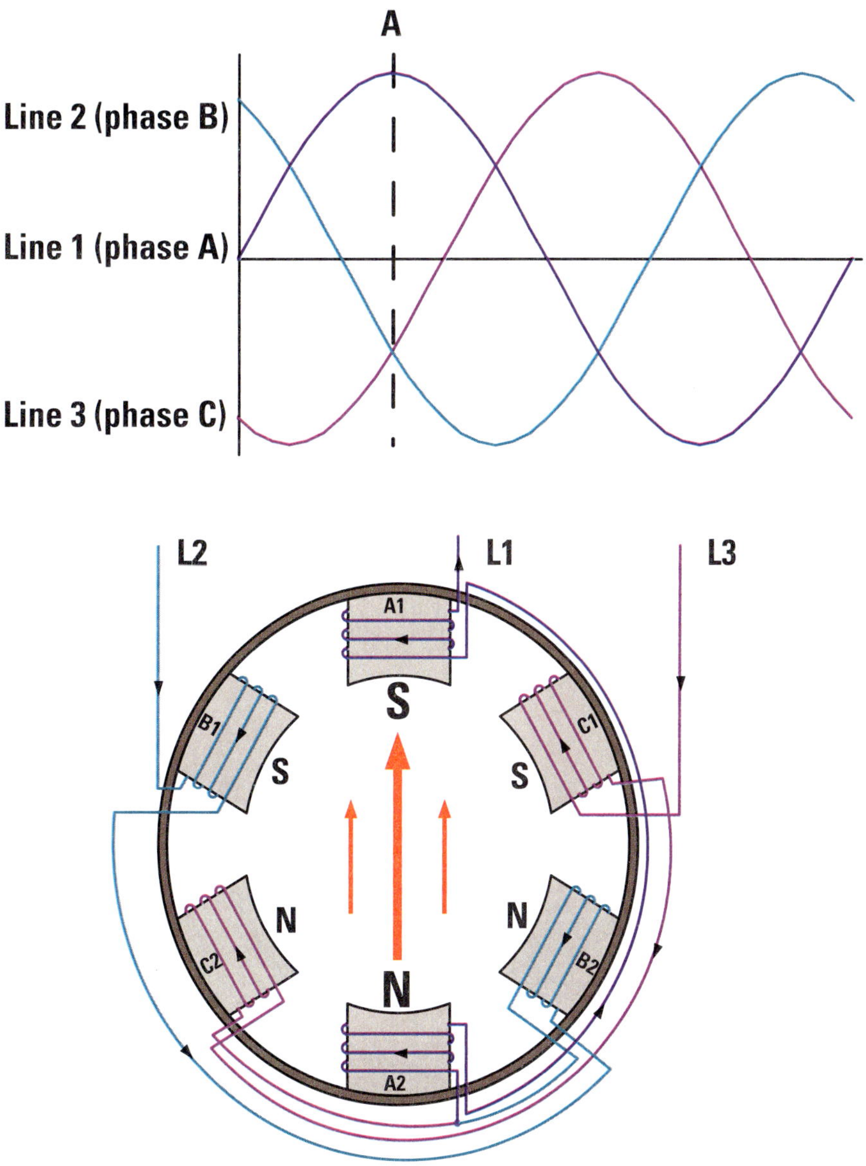

Figure 32–2 The magnetic field is concentrated between poles A1 and A2 when Line 1 is at peak voltage.

In *Figure 32–3*, line B is drawn at a point in time when the voltage of Line 3 is at the zero point, and the voltages of Lines 1 and 2 are less than maximum but opposite in polarity. The magnetic field is now concentrated between the pole pieces of phases A and B. Because Line 3 is at zero, Phase C has no current flow and therefore no magnetic field.

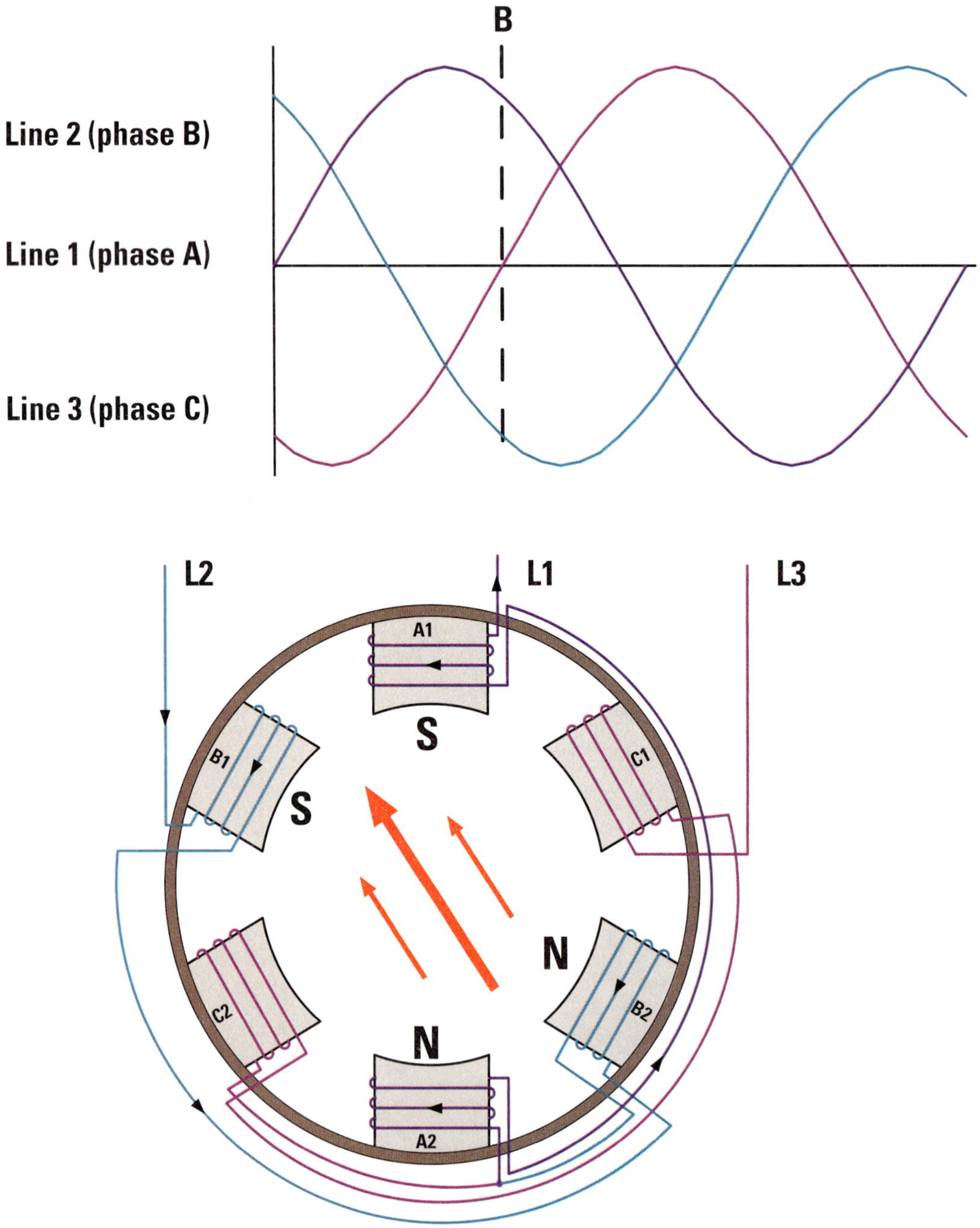

Figure 32–3 The magnetic field is concentrated between phases A and B when Line 3 is at the zero point.

In *Figure 32–4*, line C is drawn when Line 2 has reached its maximum negative value and Lines 1 and 3 are both less than maximum and have a positive polarity. The magnetic field is mainly concentrated between poles B1 and B2.

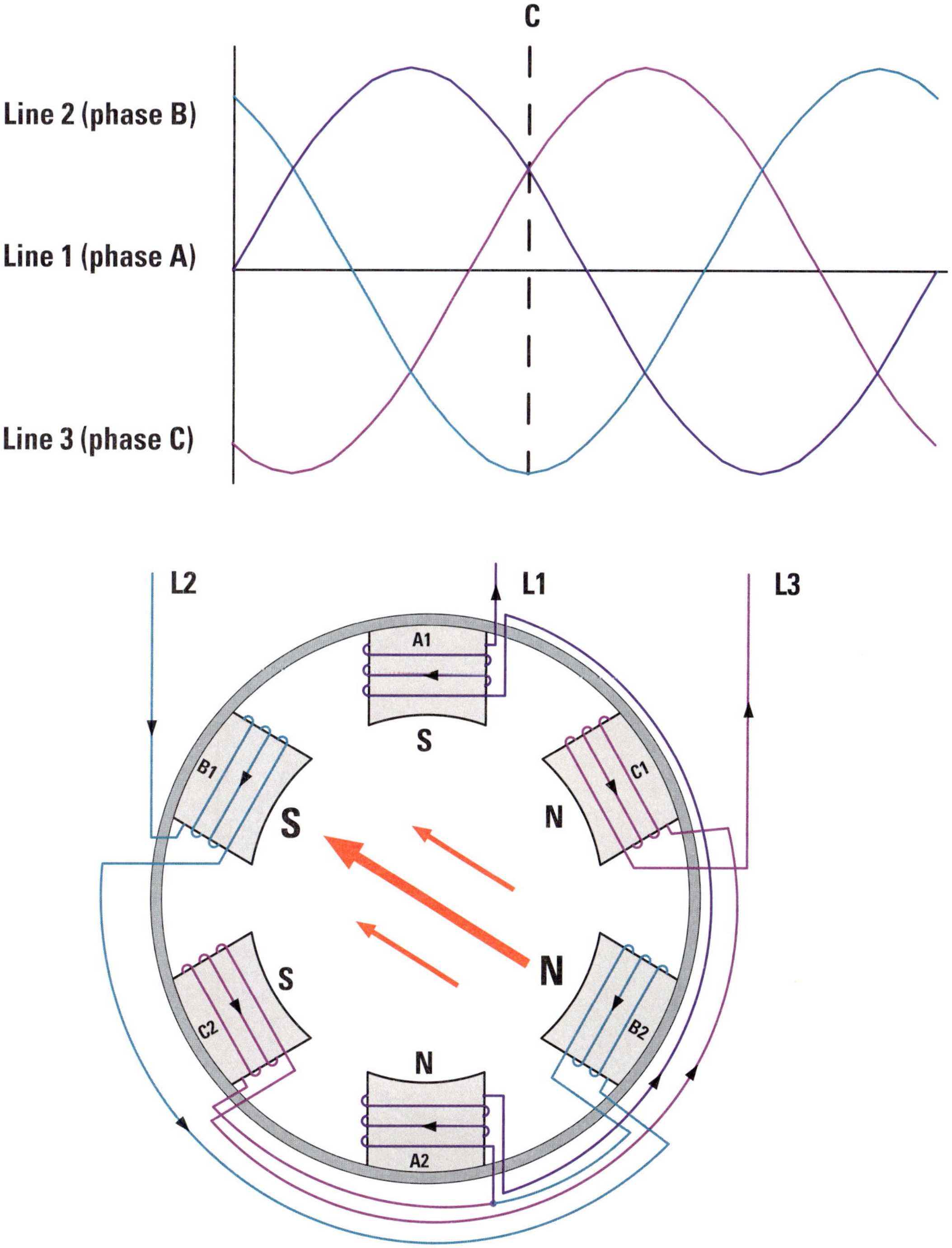

Figure 32–4 The magnetic field is concentrated between poles B1 and B2 when Line 2 is at maximum negative polarity.

Line D indicates when Line 1 is zero and Lines 2 and 3 are less than maximum and opposite in polarity (*Figure 32–5*). The magnetic field is now concentrated between the poles of phases B and C.

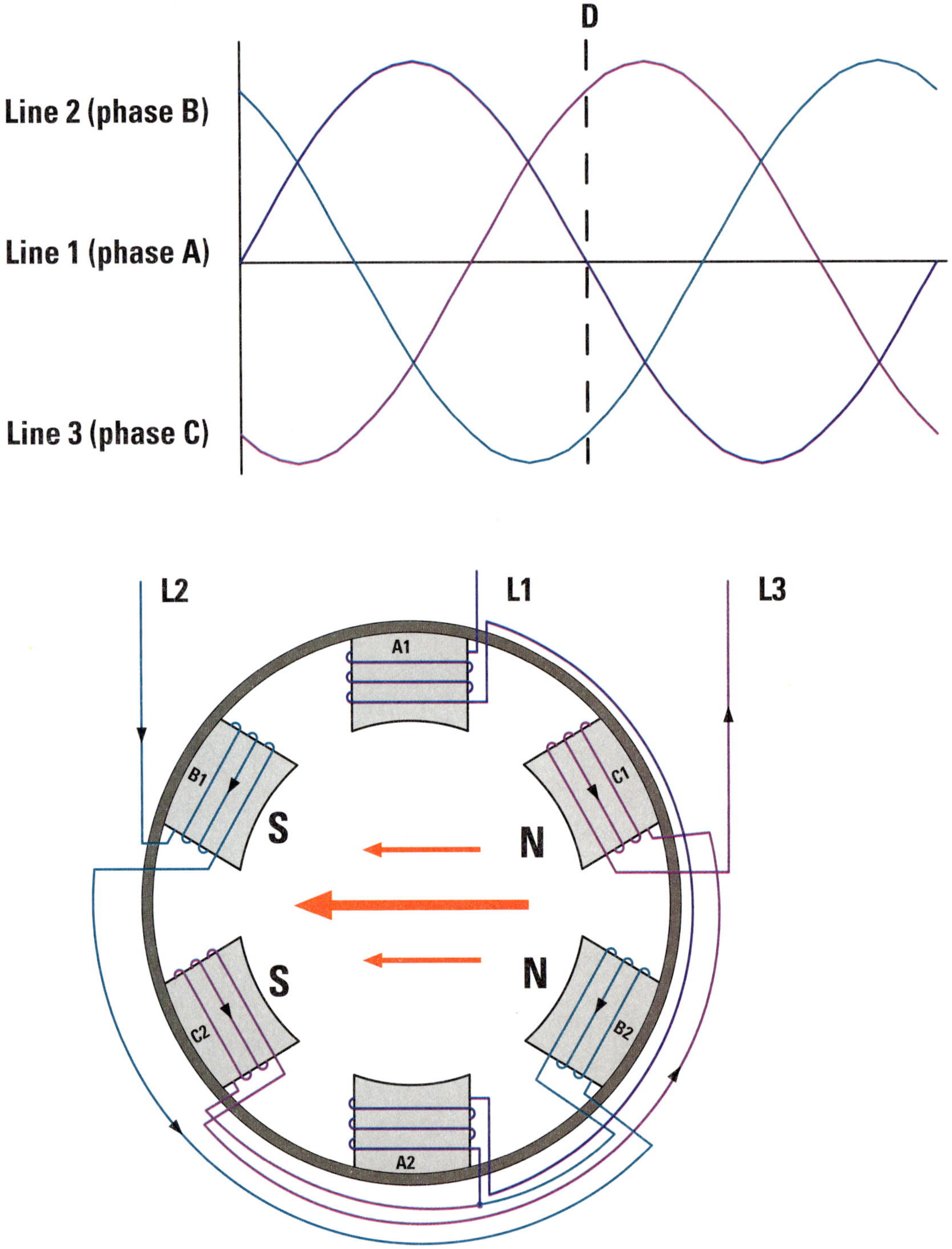

Figure 32–5 The magnetic field is concentrated between phases B and C when Line 1 is at the zero point.

In *Figure 32–6*, Line E is drawn to indicate a point in time when Line 3 has reached its peak positive point and Lines 1 and 2 are less than maximum and negative. The magnetic field is now mainly concentrated between poles C1 and C2.

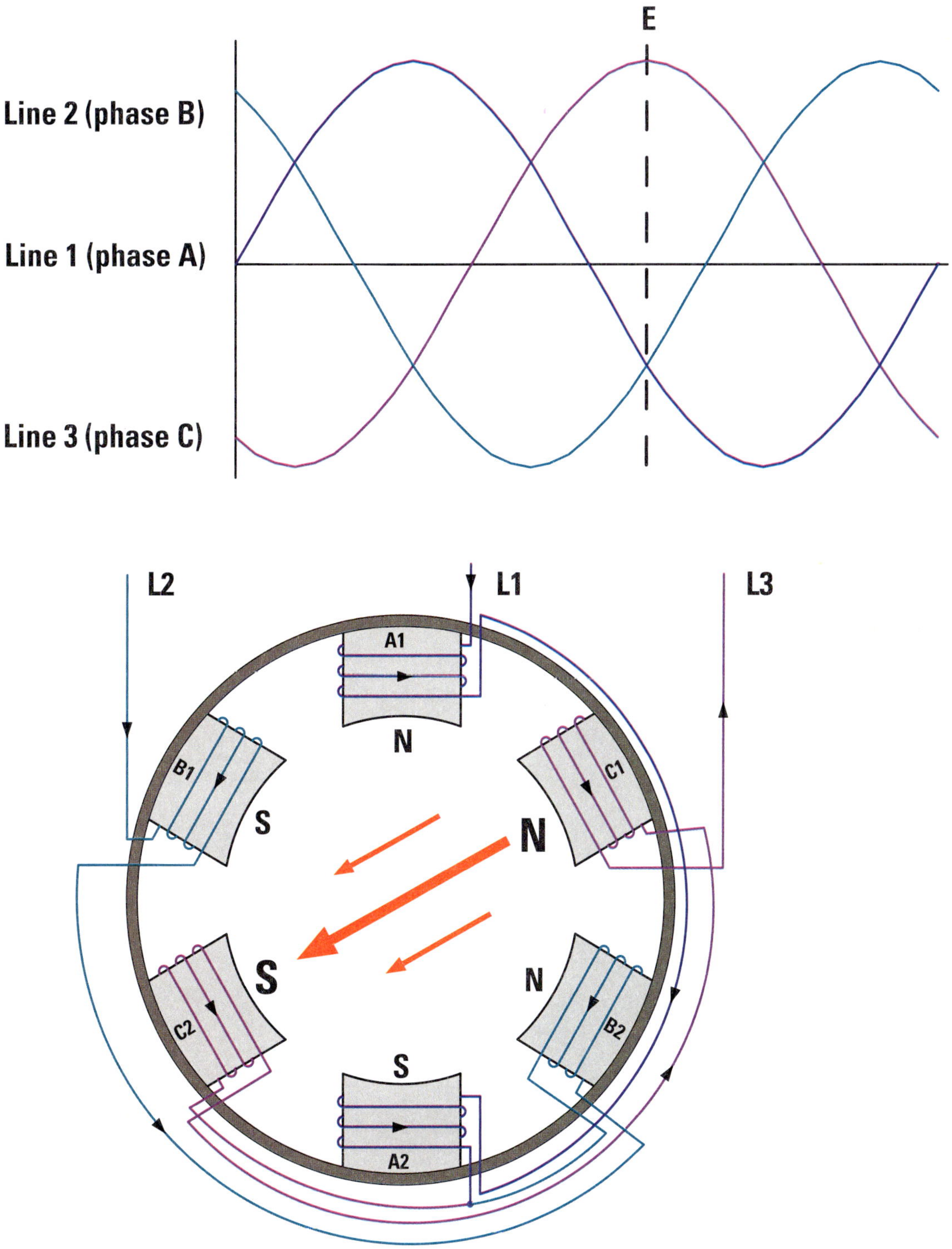

Figure 32–6 The magnetic field is concentrated between poles C1 and C2 when Line 3 reaches the maximum value.

Line F indicates when Line 2 is zero and Lines 1 and 3 are less than maximum and have opposite polarities (*Figure 32–7*). The magnetic field is now mainly concentrated between the poles of phases A and C.

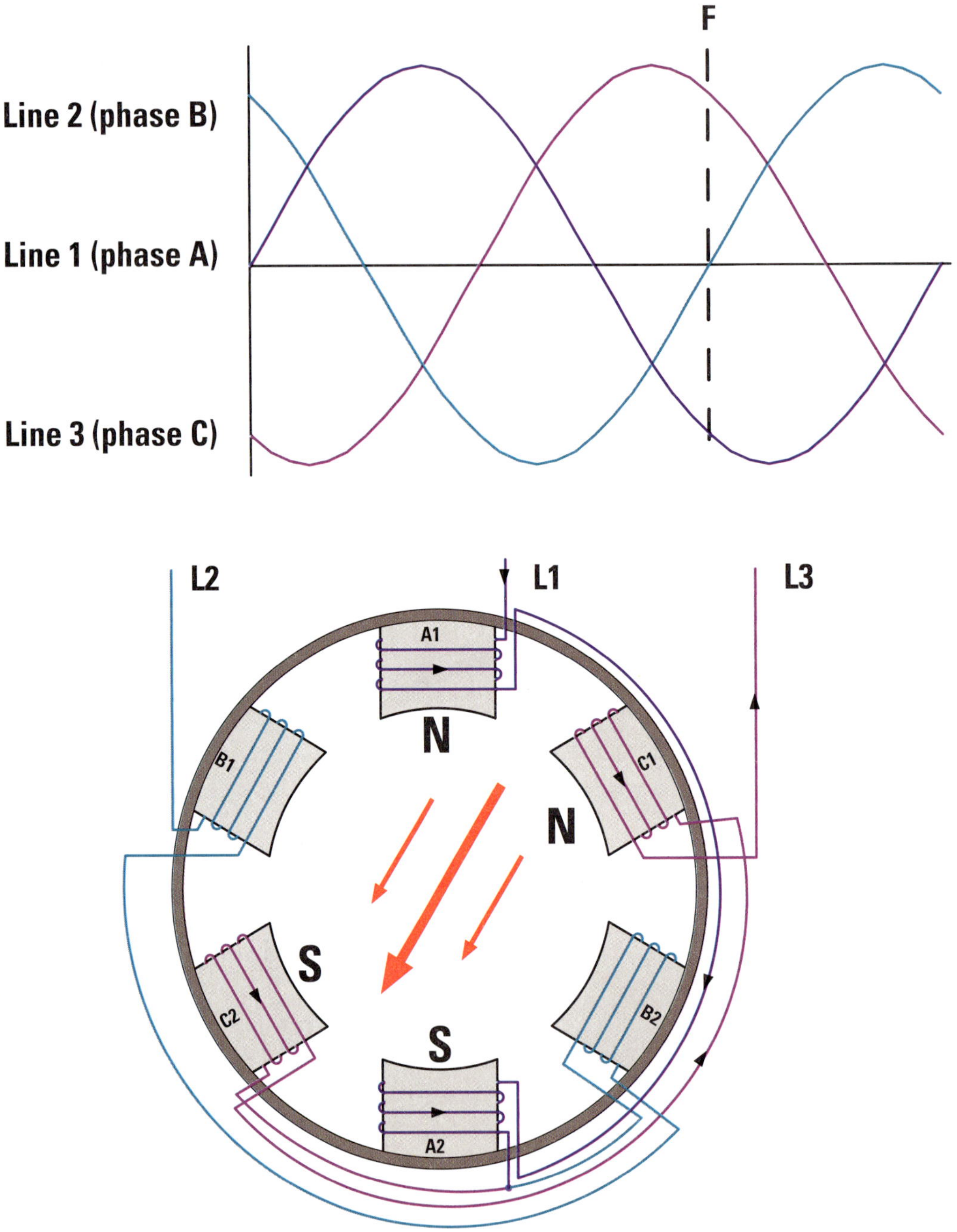

Figure 32–7 The magnetic field is concentrated between phases A and C when Line 2 crosses the zero point.

In *Figure 32–8*, Line G indicates a point in time when Line 1 has reached its maximum negative value and Lines 2 and 3 are less than maximum and have a positive polarity. The magnetic field is again concentrated between poles A1 and A2. However, pole A1 has a north magnetic polarity instead of pole A2.

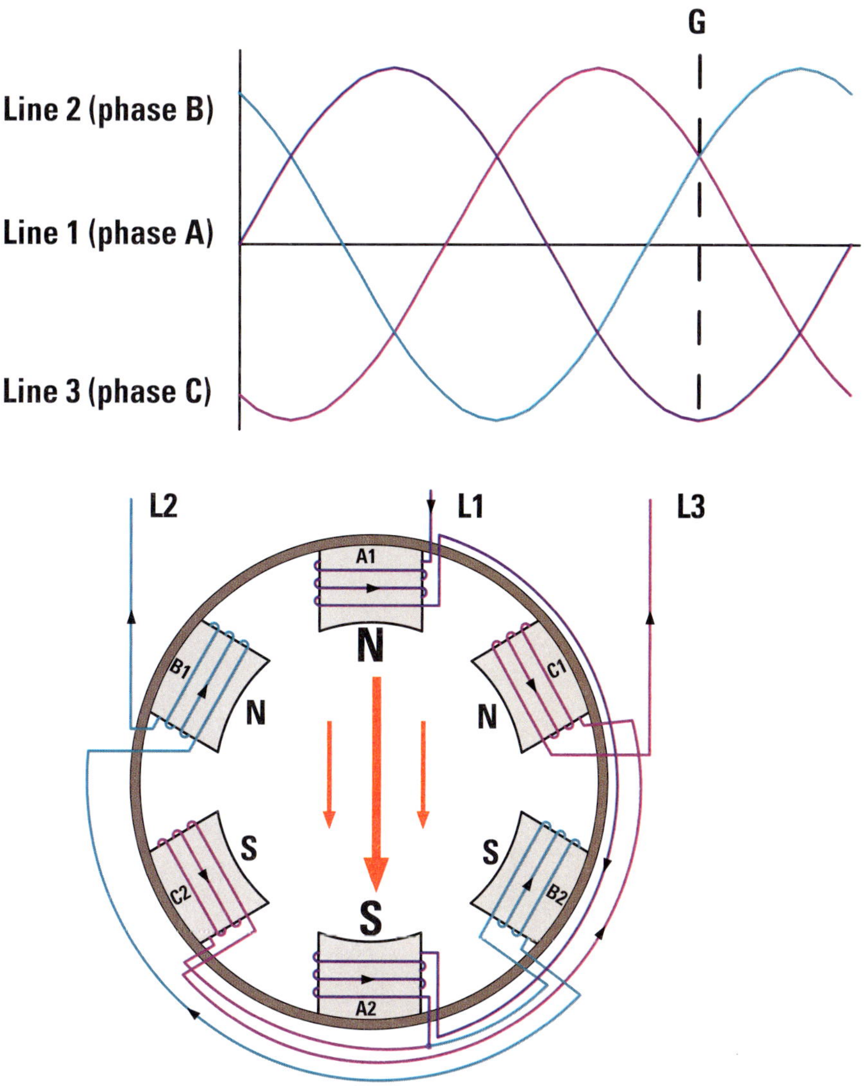

Figure 32–8 The magnetic field is again concentrated between poles A1 and A2. Note that the polarity of the magnetic fields has reversed. The magnetic field has rotated 180° during one half-cycle.

Figure 32–9 shows the position of the magnetic field at position H. The field has now rotated 270°. At the end of one complete cycle, the magnetic field completes a full 360° of rotation (*Figure 32–10*). The speed of the rotating magnetic field is 3600 RPM in a two-pole motor connected to a 60-Hz line.

Synchronous Speed

The speed at which the magnetic field rotates is called the **synchronous speed**. Two factors that determine the synchronous speed of the rotating magnetic field are:

1. the number of stator poles (per phase), and
2. the frequency of the applied voltage.

The following chart shows the synchronous speed at 60 hertz for different numbers of stator poles:

RPM	Stator Poles
3600	2
1800	4
1200	6
900	8

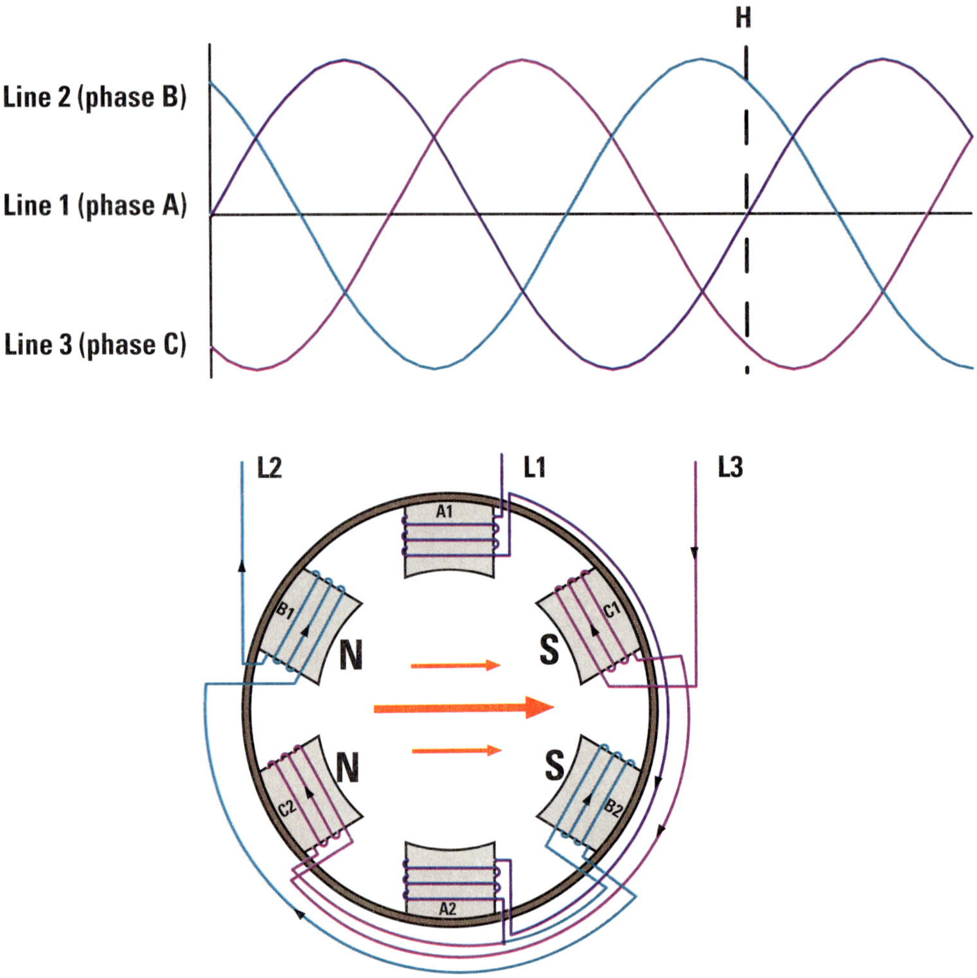

Figure 32–9 The magnetic field has rotated a total of 270° and is concentrated between phases B and C.

The stator winding of a three-phase motor is shown in *Figure 32–11*. The synchronous speed can be calculated using the formula:

$$S = \frac{120\,f}{P}$$

where

S = speed in RPM

f = frequency in Hz

P = number of stator poles (per phase)

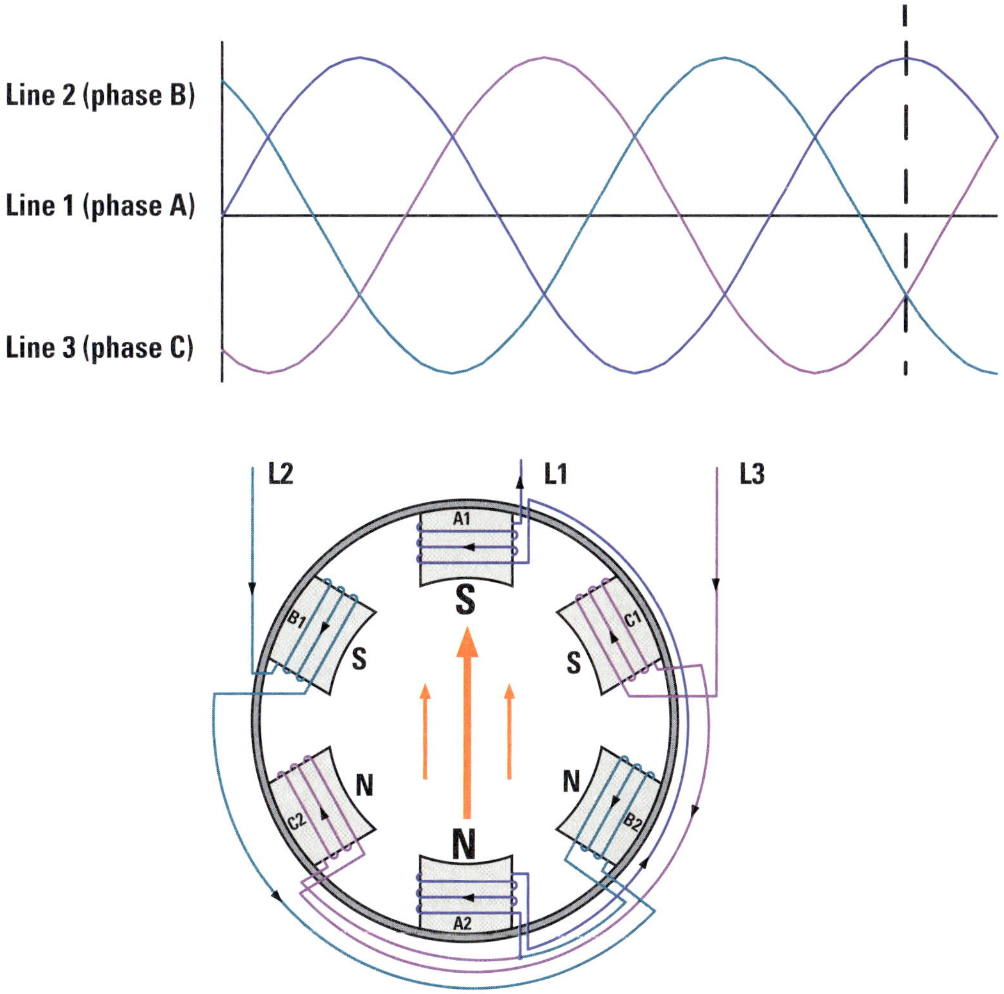

Figure 32–10 The magnetic field has rotated 360° after one complete cycle.

Figure 32–11 Stator of a three-phase motor.

Example 32–1

What is the synchronous speed of a four-pole motor connected to 50 hertz?

Solution

$$S = \frac{120 \times 50\ \text{Hz}}{4}$$

$$S = 1500\ \text{RPM}$$

Example 32–2

What frequency should be applied to a six-pole motor to produce a synchronous speed of 400 RPM?

Solution

First, change the base formula to find frequency. Once that is done, known values can be substituted in the formula:

$$f = \frac{PS}{120}$$

$$f = \frac{6 \times 400\ \text{RPM}}{120}$$

$$f = 20\ \text{Hz}$$

Determining the Direction of Rotation for Three-Phase Motors

On many types of machinery, the direction of rotation of the motor is critical. **The direction of rotation of any three-phase motor can be changed by reversing any two of its stator leads.** This causes the direction of the rotating magnetic field to reverse. When a motor is to be connected to a machine that requires a specific direction of rotation, there are two methods commonly used to check the **direction of rotation**. The first method is to perform a quick "bump" test prior to making the final connections to ensure proper rotation. In this test, power can be momentarily applied to the motor to observe its direction of rotation; however, this must be done with the motor not coupled to the load. If the rotation is incorrect, any two leads can be swapped to reverse the motor's rotation.

There may be occasions when it is not practical or is very inconvenient to apply power to the motor before it is connected to the load. In such a case, a **phase rotation meter** can be used (*Figure 32–12*). The phase rotation meter compares the phase rotation of two different three-phase connections. The meter contains six terminal leads. Three of the leads are connected to one side of the meter and are labeled MOTOR. These three motor leads are labeled A, B, and C. The LINE leads are located on the other side of the meter and are labeled A, B, and C.

To determine the direction of rotation of the motor, first zero the meter by following the instructions provided by the manufacturer. Then, set the meter selector switch to MOTOR, and connect the three MOTOR leads of the meter to the "T" leads of the motor (*Figure 32–13*). The phase rotation meter contains a zero-center voltmeter. One side of the voltmeter is labeled INCORRECT, and the other side is labeled CORRECT. While observing the zero-center voltmeter, manually turn the motor shaft in the direction of desired rotation. The zero-center voltmeter will immediately swing in the CORRECT or INCORRECT direction.

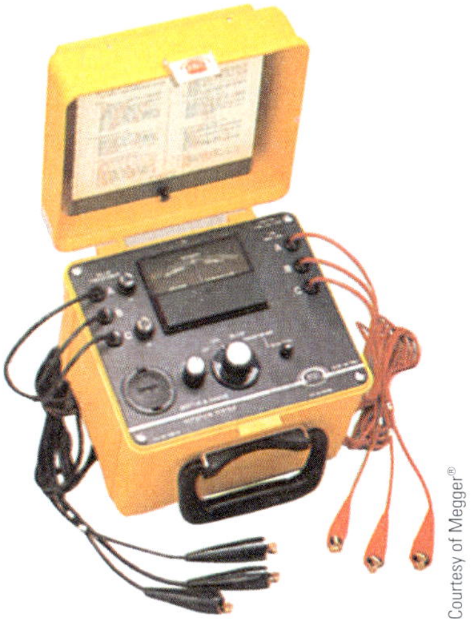

Figure 32–12 Phase rotation meter.

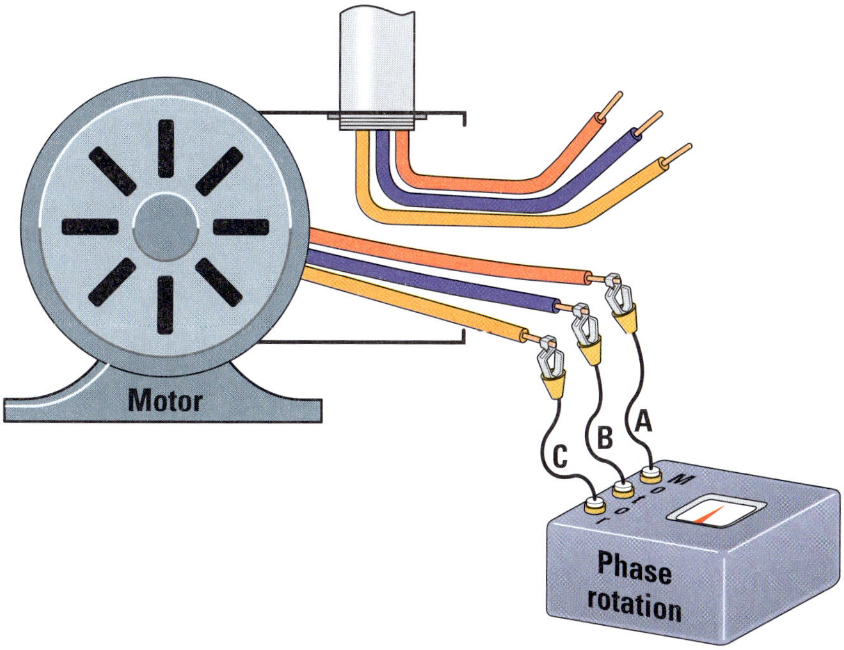

Figure 32–13 Connecting the phase rotation meter to the motor.

If the voltmeter needle indicates CORRECT, label the motor T leads A, B, and C to correspond with the MOTOR leads from the phase rotation meter. If the voltmeter needle indicates INCORRECT, change any two of the MOTOR leads from the phase rotation meter and again turn the motor shaft. The voltmeter needle should now indicate CORRECT. The motor T leads can now be labeled to correspond with the MOTOR leads from the phase rotation meter.

32–2 Three-Phase Motor Connections

Many of the three-phase motors used in industry are designed to be operated on two nominal voltage systems, such as 208-volt and 480-volt systems. Motors of this type, called **dual-voltage motors,** contain two sets of windings per phase. Most dual-voltage motors bring out nine T leads at the terminal box. A standard method used to number these leads is shown in *Figure 32–14*. Starting with Terminal 1, the leads are numbered in a decreasing clockwise spiral, as shown. Another method of determining the proper lead numbers is to add three to each terminal. For example, starting with Lead 1, add three to one. Three plus one equals four. The phase winding that begins with 1 ends with 4. Now add three to four. Three plus four equals seven. The beginning of the second winding for phase one is seven. This method will work for the windings of all phases. If in doubt, draw a diagram of the phase windings and number them in a spiral.

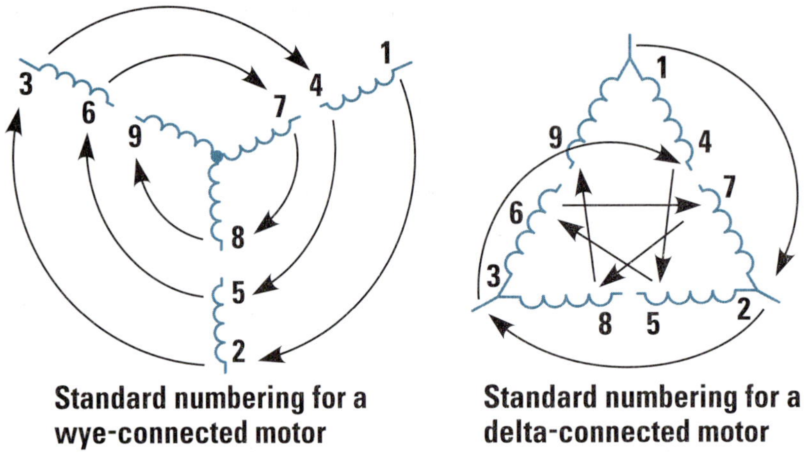

Figure 32–14 Standard lead numbering for 9-lead three-phase motors.

High-Voltage Connections

Three-phase motors can be constructed to operate in either wye or delta. If a motor is to be connected to high voltage, the phase windings are connected in series. In *Figure 32–15*, a schematic diagram and terminal connection chart for high voltage are shown for a wye-connected motor. In *Figure 32–16*, a schematic diagram and terminal connection chart for high voltage are shown for a delta-connected motor. Notice that in both cases, the windings are connected in series.

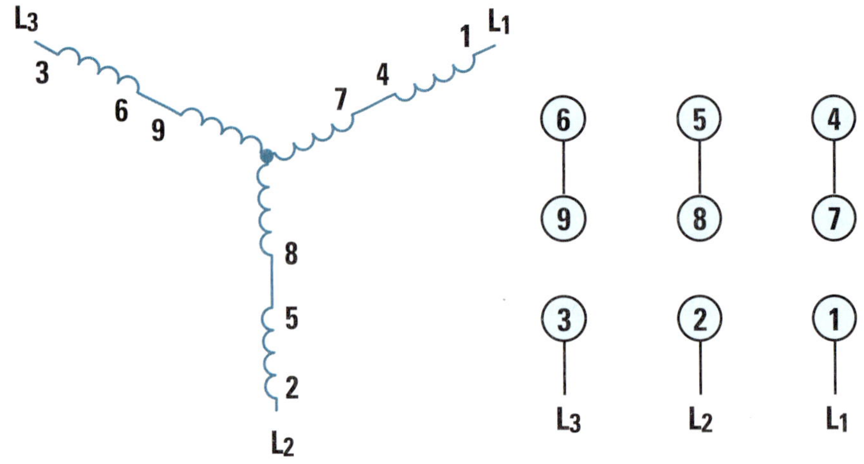

Figure 32–15 High-voltage wye connection for 9-lead motor.

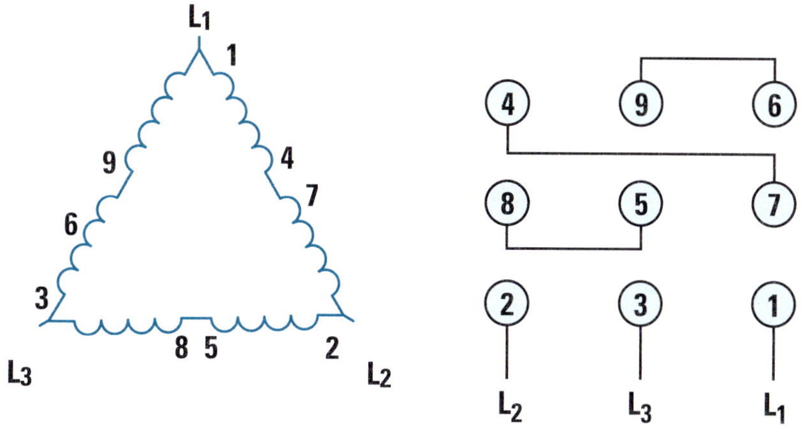

Figure 32–16 High-voltage delta connection for 9-lead motor.

Low-Voltage Connections

When a motor is to be connected for low-voltage operation, the phase windings must be connected in parallel. *Figure 32–17* shows the basic schematic diagram for a wye-connected motor with parallel phase windings. In actual practice, however, it is not possible to make this exact connection with a nine-lead motor. The schematic shows that Terminal 4 connects to the other end of the phase winding that starts with Terminal 7. Terminal 5 connects to the other end of Winding 8, and Terminal 6 connects to the other end of Winding 9. In actual motor construction, the opposite ends of Windings 7, 8, and 9 are connected together inside the motor and are not brought outside the motor case. The problem is solved, however, by forming a second wye connection by connecting Terminals 4, 5, and 6 together as shown in *Figure 32–18*.

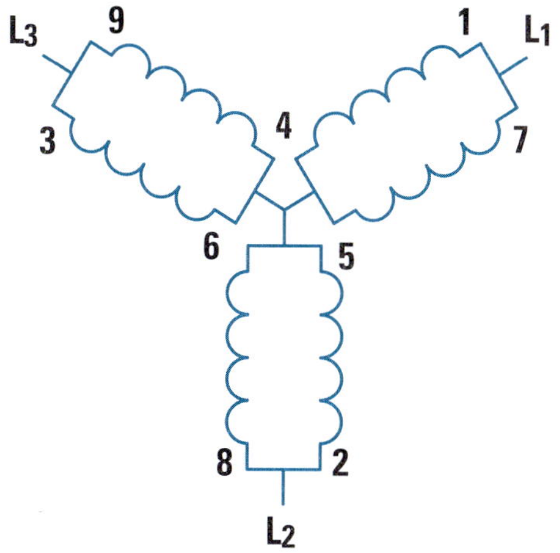

Figure 32–17 Stator windings connected in parallel for a low-voltage connection.

The phase windings of a delta-connected motor must also be connected in parallel for use on low voltage. A schematic for this connection is shown in *Figure 32–19*. A connection diagram and terminal connection chart for this hookup are shown in *Figure 32–20*.

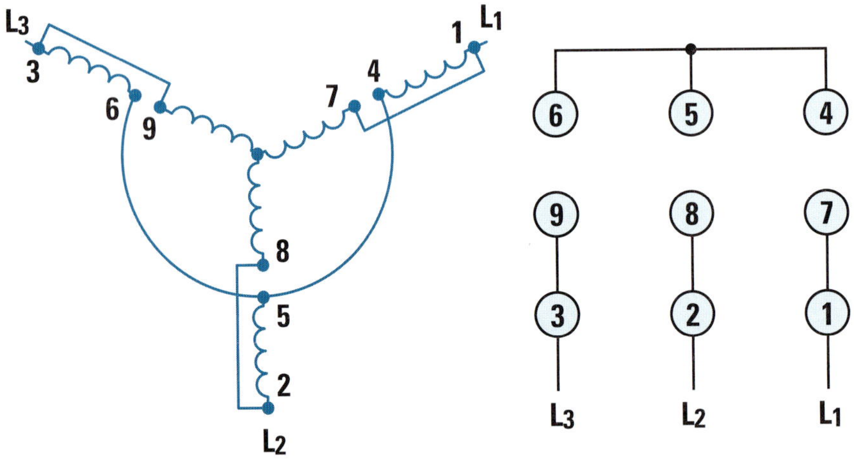

Figure 32–18 Low-voltage wye connection of a 9-lead motor.

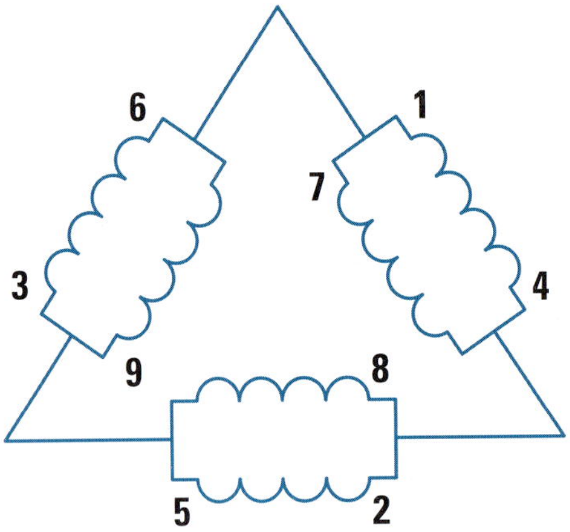

Figure 32–19 Parallel delta connection for low-voltage applications.

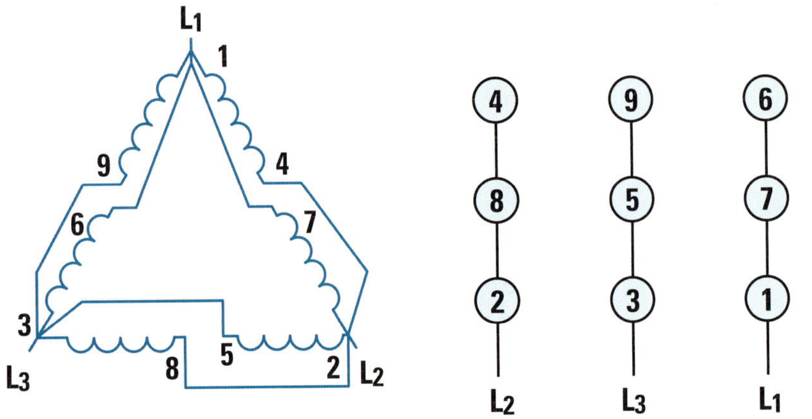

Figure 32–20 Low-voltage delta connection of a 9-lead motor.

Some dual-voltage motors contain 12 T leads instead of 9. In this instance, the opposite ends of Terminals 7, 8, and 9 are brought out for connection. *Figure 32–21* shows the standard numbering for both delta- and wye-connected motors. Twelve leads are brought out if the motor is intended to be used for wye-delta starting. When this is the case, the motor must be designed for normal operation with its windings connected in delta. If the windings are connected in wye during starting, the starting current of the motor is reduced to one-third of what it is if the motor is started as a delta.

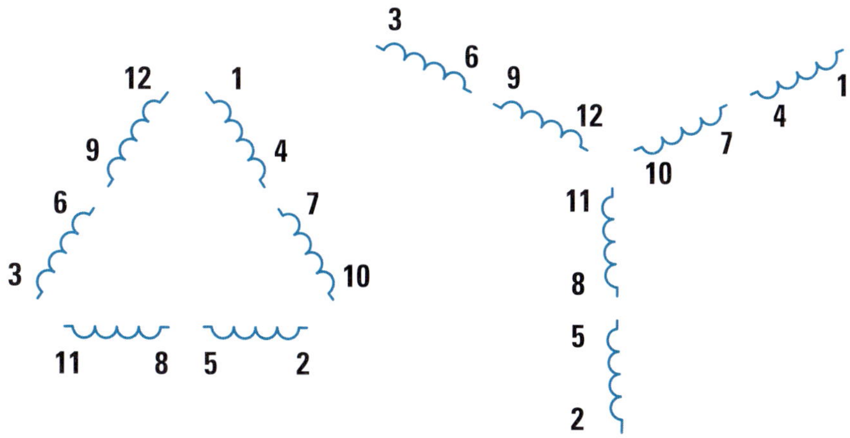

Figure 32–21 A schematic drawing of both a delta- and wye-connected 12-lead motor.

Voltage and Current Relationships for Dual-Voltage Motors

When a motor is connected to the higher voltage, the current flow will be half as much as when it is connected for low-voltage operation. The reason is that when the windings are connected in series for high-voltage operation, the impedance is four times greater than when the windings are connected for low-voltage operation. For example, assume a dual-voltage motor is intended to operate on 480 volts or 240 volts. Also assume that during full load, the motor windings exhibit an impedance of 10 ohms each. When the winding is connected in series (*Figure 32–22*), the impedance per phase is 20 ohms (10 Ω + 10 Ω = 20 Ω). If a voltage of 480 volts is connected to the motor, the phase voltage is:

$$E_{PHASE} = \frac{E_{LINE}}{1.732}$$

$$E_{PHASE} = \frac{480}{1.732}$$

$$E_{PHASE} = 277\ V$$

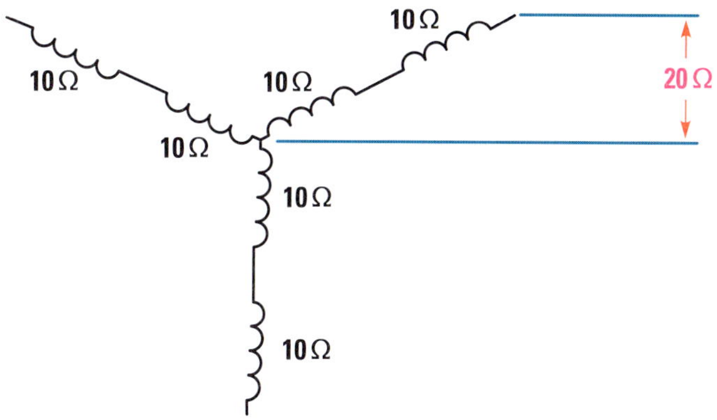

Figure 32–22 Impedance adds in series.

The amount of current flow through the phase can be calculated using Ohm's law:

$$I = \frac{E}{Z}$$

$$I = \frac{277\ V}{20\ \Omega}$$

$$I = 13.85\ A$$

If the stator windings are connected in parallel, the total impedance is found by adding the reciprocals of the impedances of the windings (*Figure 32–23*):

$$Z_T = \frac{1}{\frac{1}{Z_1} + \frac{1}{Z_2}}$$

$$Z_T = 5\ \Omega$$

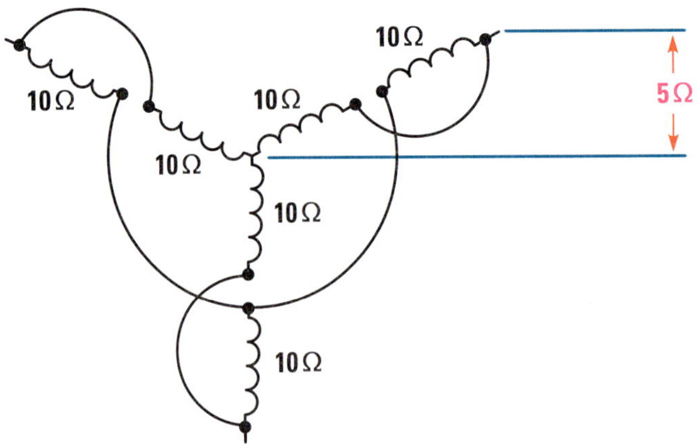

Figure 32–23 Total impedance is always less when inductors are in parallel.

If a voltage of 240 volts is connected to the motor, the voltage applied across each phase is 138.6 volts (240 V/1.732 = 138.6 V). The amount of phase current can now be calculated using Ohm's law:

$$I = \frac{E}{Z}$$

$$I = \frac{138.6\ V}{5\ \Omega}$$

$$I = 27.7\ A$$

32–3 Squirrel-Cage Induction Motors

The **squirrel-cage rotor** induction motor receives its name from the type of rotor used in the motor. A squirrel-cage rotor is made by connecting bars to two end rings (*Figure 32–24*). A shaft placed through the center of the cage permits the cage to spin around the shaft. A squirrel-cage rotor is shown in *Figure 32–25*.

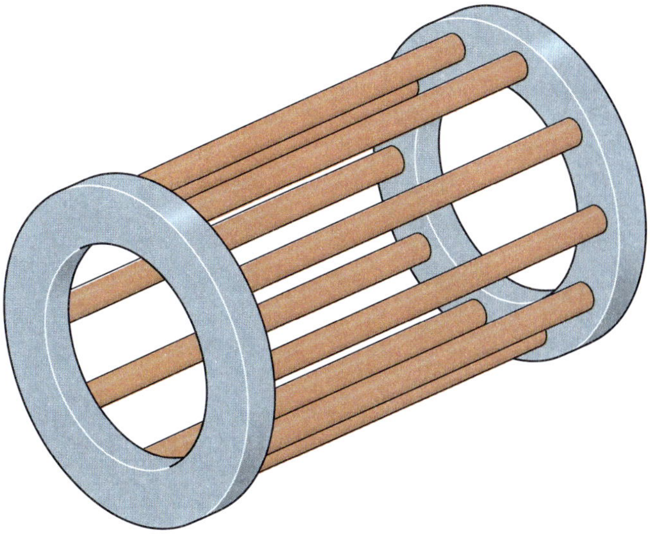

Figure 32–24 Basic squirrel-cage rotor without laminations.

Figure 32–25 Squirrel-cage rotor.

Principle of Operation

The squirrel-cage motor is an induction motor. That means the current flow in the rotor is produced by induced voltage from the rotating magnetic field of the stator. In *Figure 32–26*, a squirrel-cage rotor is

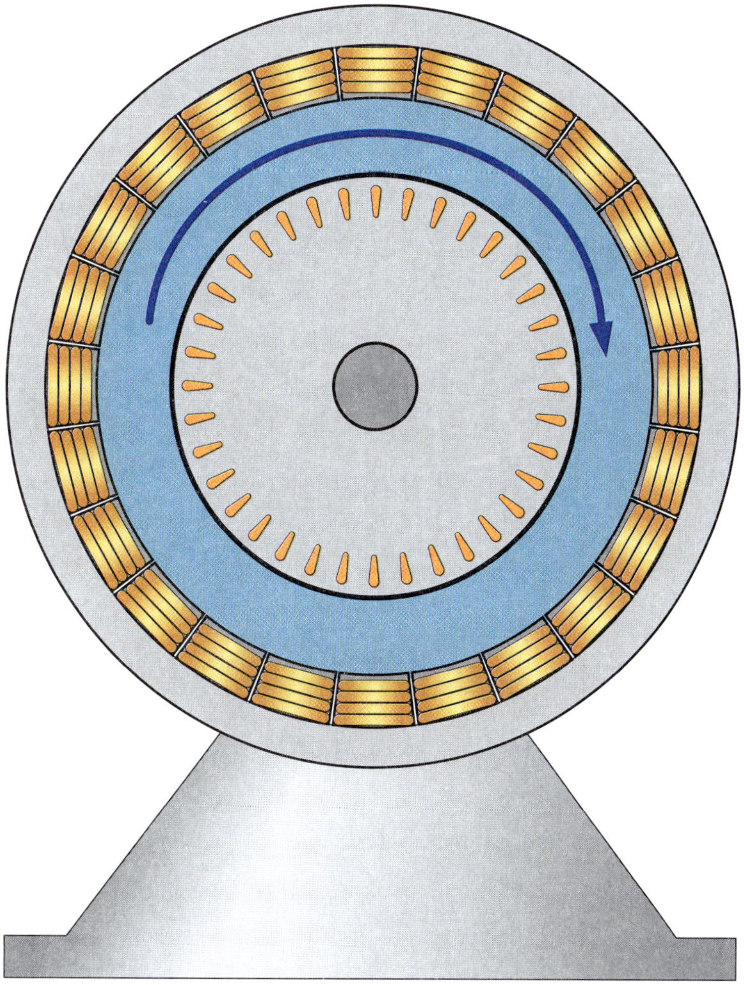

Figure 32–26 Voltage is induced into the rotor by the rotating magnetic field.

shown inside the stator of a three-phase motor. It will be assumed that the motor, shown in *Figure 32–26,* contains four poles per phase, which produces a rotating magnetic field with a synchronous speed of 1800 RPM when the stator is connected to a 60-hertz line. When power is first connected to the stator, the rotor is not turning. The magnetic field of the stator cuts the rotor bars at a rate of 1800 RPM. This cutting action induces a voltage into the rotor bars. This induced voltage will be the same frequency as the voltage applied to the stator. The amount of induced voltage is determined by three factors:

1. The strength of the magnetic field of the stator
2. The number of turns of wire cut by the magnetic field (in the case of a squirrel-cage rotor, this will be the number of bars in the rotor)
3. The speed of the cutting action

Because the rotor is stationary at this time, maximum voltage is induced into the rotor. The induced voltage causes current to flow through the rotor bars. As current flows through the rotor, a magnetic field is produced around each bar (*Figure 32–27*).

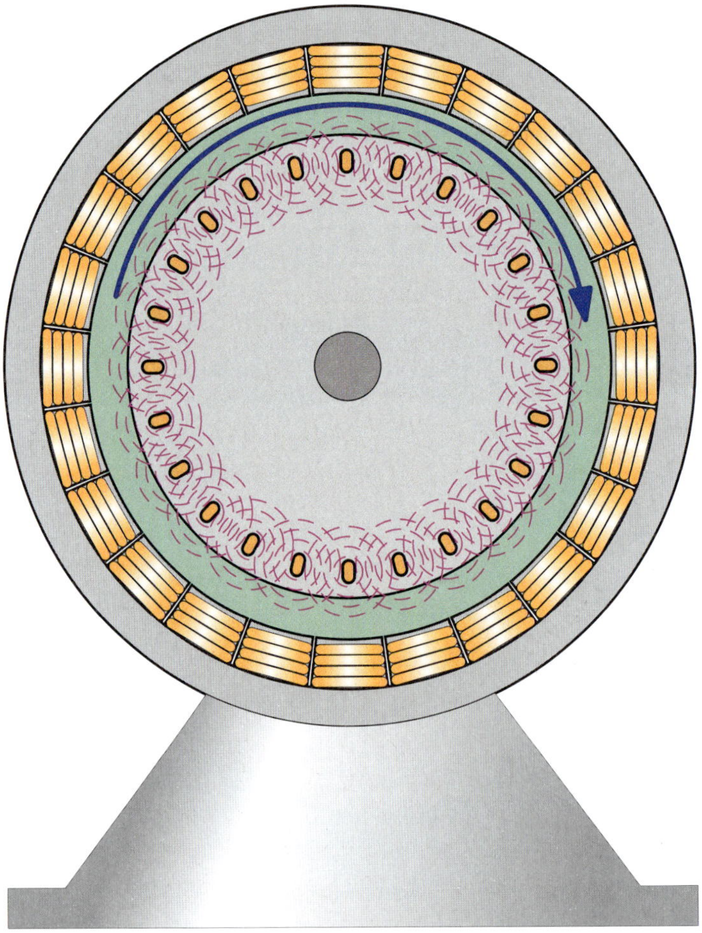

Figure 32–27 A magnetic field is produced around each rotor bar.

The magnetic field of the rotor is attracted to the magnetic field of the stator, and the rotor begins to turn in the same direction as the rotating magnetic field.

As the speed of the rotor increases, the rotating magnetic field cuts the rotor bars at a slower rate. For example, assume the rotor has accelerated to a speed of 600 RPM. The synchronous speed of the rotating magnetic field is 1800 RPM. Therefore, the rotor bars are being cut at a rate of 1200 RPM

(1800 RPM − 600 RPM = 1200 RPM). Because the rotor bars are being cut at a slower rate, less voltage is induced in the rotor, reducing rotor current. When the rotor current decreases, the stator current decreases also. If the motor is operating without a load, the rotor continues to accelerate until it reaches a speed close to that of the rotating magnetic field.

Torque

The amount of torque produced by an AC induction motor is determined by the following three factors:

1. The strength of the magnetic field of the stator
2. The strength of the magnetic field of the rotor
3. The phase angle difference between rotor and stator fields

$$T = K_T \times \varphi_S \times I_R \times \cos \theta_R$$

where

$$\begin{aligned} T &= \text{torque in lb-ft} \\ K_T &= \text{torque constant} \\ \varphi_S &= \text{stator flux (constant at all speeds)} \\ I_R &= \text{rotor current} \\ \cos \theta_R &= \text{rotor power factor} \end{aligned}$$

Notice that one of the factors that determines the amount of torque produced by an induction motor is the strength of the magnetic field of the rotor. **An induction motor can never reach synchronous speed.** If the rotor were to turn at the same speed as the rotating magnetic field, the rotor would not cut lines of flux in the rotating field, and there would be no induced voltage in the rotor and consequently no rotor current. Without rotor current, there could be no magnetic field developed by the rotor and therefore no torque or turning force. A motor operating at no load will accelerate until the torque developed is proportional to the windage and bearing friction losses.

If a load is connected to the motor, it must furnish more torque to operate the load. This causes the motor to slow down. When the motor speed decreases, the rotating magnetic field cuts the rotor bars at a faster rate. This causes more voltage to be induced in the rotor and therefore more current. The increased current flow produces a stronger magnetic field in the rotor, which causes more torque to be produced. The increased current flow in the rotor causes increased current flow in the stator. This is why motor current increases as load is added.

Another factor that determines the amount of torque developed by an induction motor is the phase angle difference between stator and rotor field flux. **Maximum torque is developed when the stator and rotor flux are in phase with each other.** Note in the preceding formula that one of the factors that determines the torque developed by an induction motor is the cosine of the rotor power factor. The cosine function reaches its maximum value of 1 when the phase angle is 0 ($\cos 0° = 1$).

Starting Characteristics

When a squirrel-cage motor is first started, it has a current draw several times greater than its normal running current. The actual amount of starting current is determined by the type of rotor bars, the horsepower rating of the motor, and the applied voltage. The type of rotor bars is indicated by the code letter found on the nameplate of a squirrel-cage motor. *Table 430.7(B)* of the *NEC* can be used to calculate the locked rotor current (initial starting current) of a squirrel-cage motor when the applied voltage, horsepower, and code letter are known.

Example 32–3

An 800-hp, three-phase squirrel-cage motor is connected to 2300 V. The motor has a locked rotor code letter of J. What is the starting (locked rotor) current of this motor?

Solution

Table 430.7(B) of the *NEC* gives a value of 7.1 to 7.99 kVA per hp as the locked rotor current of a motor with a code letter J (*Figure 32–28*). An average value of 7.5 is used for this calculation because we are looking at the starting current. However, when determining locked rotor current for motor protection, always use the maximum value of 7.99. The apparent power can be calculated by multiplying 7.5 times the hp rating of the motor:

$$\text{kVA} = 7.5\ \text{kVA/hp} \times 800\ \text{hp}$$

$$\text{kVA} = 6000$$

Table 430.7(B) Locked-Rotor Indicating Code Letters

Code Letter	Kilovolt-Amperes per Horsepower with Locked Rotor
A	0 – 3.14
B	3.15 – 3.54
C	3.55 – 3.99
D	4.0 – 4.49
E	4.5 – 4.99
F	5.0 – 5.59
G	5.6 – 6.29
H	6.3 – 7.09
J	7.1 – 7.99
K	8.0 – 8.99
L	9.0 – 9.99
M	10.0 – 11.19
N	11.2 – 12.49
P	12.5 – 13.99
R	14.0 – 15.99
S	16.0 – 17.99
T	18.0 – 19.99
U	20.0 – 22.39
V	22.4 and up

Figure 32–28 *NEC Table 430.7(B).* Reproduced with permission of NFPA from NFPA 70®, National Electrical Code®, 2023 edition. Copyright© 2022, National Fire Protection Association. For a full copy of the NFPA 70®, please go to www.nfpa.org.

The line current supplying the motor can now be calculated using the formula:

$$I_{(LINE)} = \frac{VA}{E_{(LINE)} \times 1.732}$$

$$I_{(LINE)} = \frac{6{,}000{,}000\ \text{VA}}{2300\ \text{V} \times 1.732}$$

$$I_{(LINE)} = 1506.175\ \text{A}$$

Example 32–3 (continued)

This large starting current is caused by the fact that the rotor is not turning when power is first applied to the stator. Because the rotor is not turning, the squirrel-cage bars are cut by the rotating magnetic field at the maximum rate. Remember that one of the factors that determines the amount of induced voltage is the speed of the cutting action. This high induced voltage causes a large amount of current to flow in the rotor. The large current flow in the rotor causes a large amount of current flow in the stator. Because a large amount of current flows in both the stator and rotor, a strong magnetic field is established in both.

It would first appear that the starting torque of a squirrel-cage motor is high because the magnetic fields of both the stator and rotor are strong at this point. Recall that the third factor for determining the torque developed by an induction motor is the difference in phase angle between stator flux and rotor flux. Because the rotor is being cut at a high rate of speed by the rotating stator field, the bars in the squirrel-cage rotor appear to be very inductive at this point because of the high frequency of the induced voltage. This causes the phase angle difference between the induced voltage in the rotor and rotor current to be almost 90° out of phase with each other, producing a lagging power factor for the rotor (*Figure 32–29*). This causes the rotor flux to lag the stator flux by a large amount; consequently, a relatively weak starting torque per ampere of starting current when compared with other types of three-phase motors is developed (*Figure 32–30*).

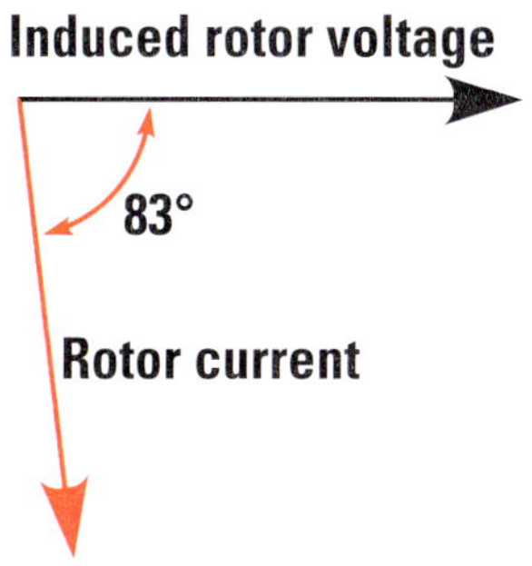

Figure 32–29 Rotor current is almost 90° out of phase with the induced voltage at the moment of starting.

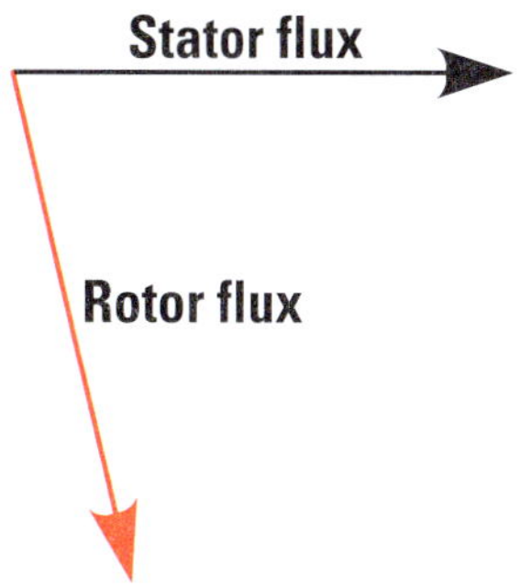

Figure 32–30 Rotor flux lags the stator flux by a large amount during starting.

Percent Slip

The speed performance of an induction motor is measured in **percent slip**. The percent slip can be determined by finding the difference in synchronous speed and rotor speed and dividing the difference by the synchronous speed. For example, assume an induction motor has a synchronous speed of 1800 RPM, and, at full load, the rotor turns at a speed of 1725 RPM. The difference between the two speeds is 75 RPM (1800 RPM − 1725 RPM = 75 RPM). The percent slip can be determined using the formula:

$$\text{Percent slip} = \frac{\text{synchronous speed} - \text{rotor speed}}{\text{synchronous speed}} \times 100$$

$$\text{Percent slip} = \frac{75\ \text{RPM}}{1800\ \text{RPM}} \times 100$$

$$\text{Percent slip} = 4.16\%$$

A rotor slip of 2 to 5% is common for most squirrel-cage induction motors. The amount of slip for a particular motor is greatly affected by the type of rotor bars used in the construction of the rotor. Squirrel-cage motors are considered to be constant-speed motors because there is a small difference between no-load speed and full-load speed.

Rotor Frequency

In the previous example, the rotor slips behind the rotating magnetic field by 75 RPM. This means that at full load, the bars of the rotor are being cut by magnetic lines of flux at a rate of 75 RPM. Therefore, the voltage being induced in the rotor at this point in time is at a much lower frequency than when the motor was started. The **rotor frequency** can be determined using the formula:

$$f = \frac{P \times S_R}{120}$$

where

$$f = \text{frequency in Hz}$$

$$P = \text{number of stator poles}$$

$$S_R = \text{rotor slip in RPM}$$

$$f = \frac{4 \times 75\ \text{RPM}}{120}$$

$$f = 2.5\ \text{Hz}$$

Because the frequency of the current in the rotor decreases as the rotor approaches synchronous speed, the rotor bars become less inductive. The current flow through the rotor becomes limited more by the resistance of the bars and less by inductive reactance. The current flow in the rotor becomes more in phase with the induced voltage, which causes less phase angle shift between stator and rotor flux (*Figure 32–31*). This is the reason that squirrel-cage motors generally have a relatively poor starting torque per ampere of starting current when compared with other types of three-phase motors but a good running torque. Although the starting torque per ampere of starting current is lower than other types of three-phase motors, the starting torque can be high because of the large amount of inrush current.

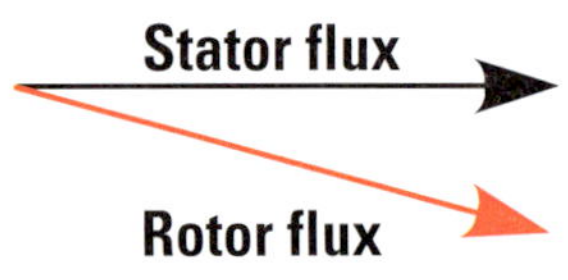

Figure 32–31 Rotor and stator flux become more in phase with each other as motor speed increases.

Reduced Voltage Starting

Because many squirrel-cage motors require a large amount of starting current, it is sometimes necessary to reduce the voltage during the starting period. When the voltage is reduced, the starting torque is reduced also. If the applied voltage is reduced to 50% of its normal value, the magnetic fields of both the stator and rotor are reduced to 50% of normal. The 50% reduction of the magnetic fields causes the starting torque to be reduced to 25% of normal. A chart showing a typical torque curve for a squirrel-cage motor is shown in *Figure 32–32*.

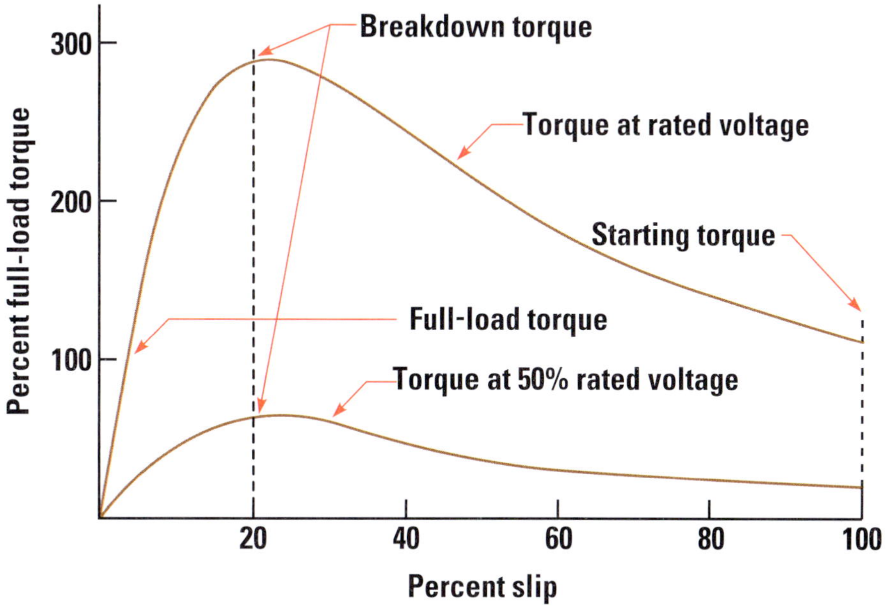

Figure 32–32 Typical torque curves for a squirrel-cage motor.

The torque formula given earlier can be used to show why this large reduction of torque occurs. Both the stator flux, φ_S, and the rotor current, I_R, are reduced to half their normal value. The product of these two values, torque, is reduced to one fourth. The torque varies as the square of the applied voltage for any given value of slip.

Code Letters

Squirrel-cage rotors are not all the same. Rotors are made with different types of bars. The type of rotor bars used in the construction of the rotor determines the operating characteristics of the motor. AC squirrel-cage motors are given a **code letter** on their nameplate. The code letter indicates the type of bars used in the rotor. *Figure 32–33* shows a rotor with type A bars. This type of rotor has the highest

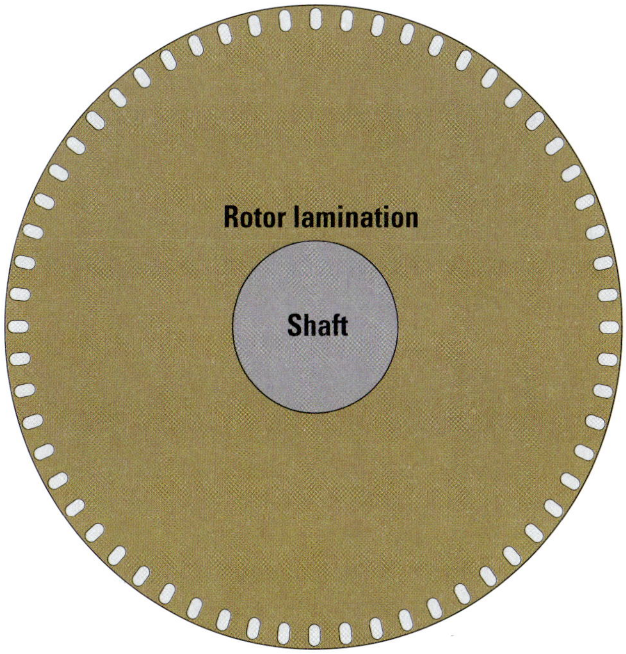

Figure 32–33 Type A rotor designs have the highest internal resistance and poor running characteristics.

internal resistance of any squirrel-cage rotor. This means that the starting torque per ampere will be high because the rotor current is closer to being in phase with the induced voltage. Also, the high resistance of the rotor bars limits the amount of current flow in the rotor when starting. This produces a low starting current for the motor. Type A rotors have very poor running characteristics, however. Because the bars are so resistive, a large amount of voltage will have to be induced into the rotor to produce an increase in rotor current and therefore an increase in the rotor magnetic field. This means that when load is added to the motor, the rotor must slow down a great amount to produce enough current in the rotor to increase the torque. Motors with type A rotors have the highest percent slip of any squirrel-cage motor. These motors are generally used in applications where starting is a problem, such as a motor that must accelerate a large flywheel from 0 RPM to its full speed. Flywheels can have a very large amount of inertia, which may require several minutes to accelerate them to their running speed when they are started.

Figure 32–34 shows a rotor like what can be found in motors with code letters B through E. These rotor bars have a lower internal resistance than Type A rotors. Rotors of this type have fair starting torque, low starting current, and fair speed regulation.

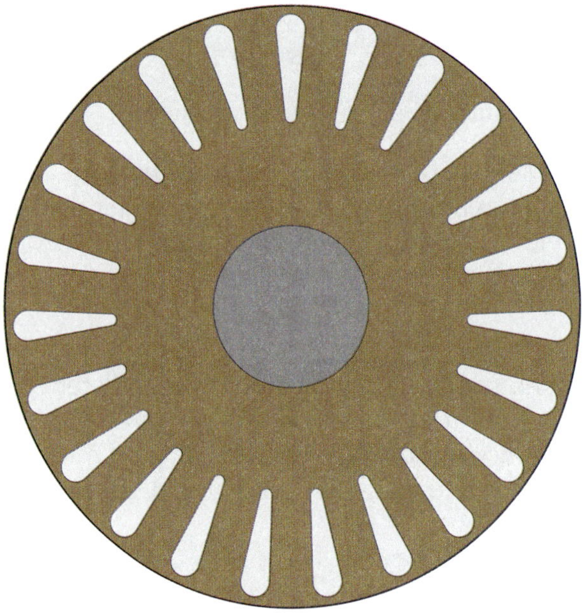

Figure 32–34 Type B–E rotors have fair starting torque, low starting current, and fair speed regulation.

Figure 32–35 shows a rotor found in motors with code letters F through V. This rotor has low starting torque per ampere, high starting current, and good running torque. Motors containing rotors of this type generally have very good speed regulation and low percent slip.

The Double-Squirrel-Cage Rotor

Some motors use a rotor that contains two sets of squirrel-cage windings (*Figure 32–36*). The outer winding consists of bars with a relatively high resistance located close to the top of the iron core. Because these bars are located close to the surface, they have a relatively low reactance. The inner winding consists of bars with a large cross-sectional area, which gives them a low resistance. The inner winding is placed deeper in the core material, which causes it to have a much higher reactance.

When the double-squirrel-cage motor is started, the rotor frequency is high. Because the inner winding is inductive, its impedance will be high compared with the resistance of the outer winding. During this period of time, most of the rotor current flows through the outer winding. The resistance of the outer

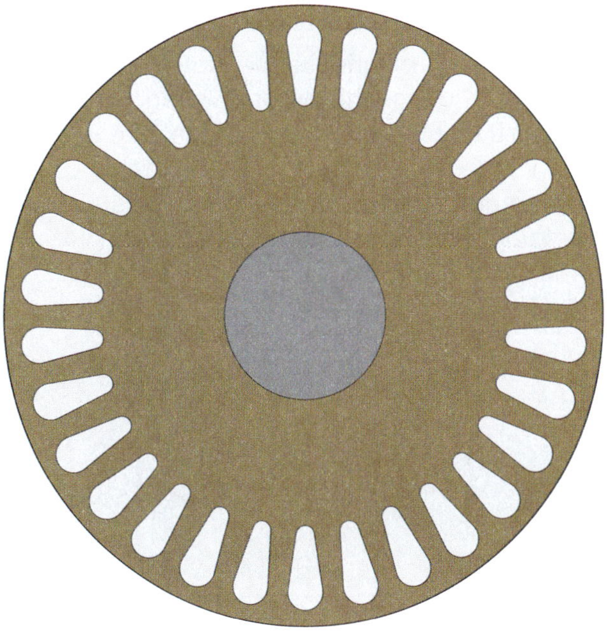

Figure 32–35 Type F–V rotors have very good speed regulation and low percent slip.

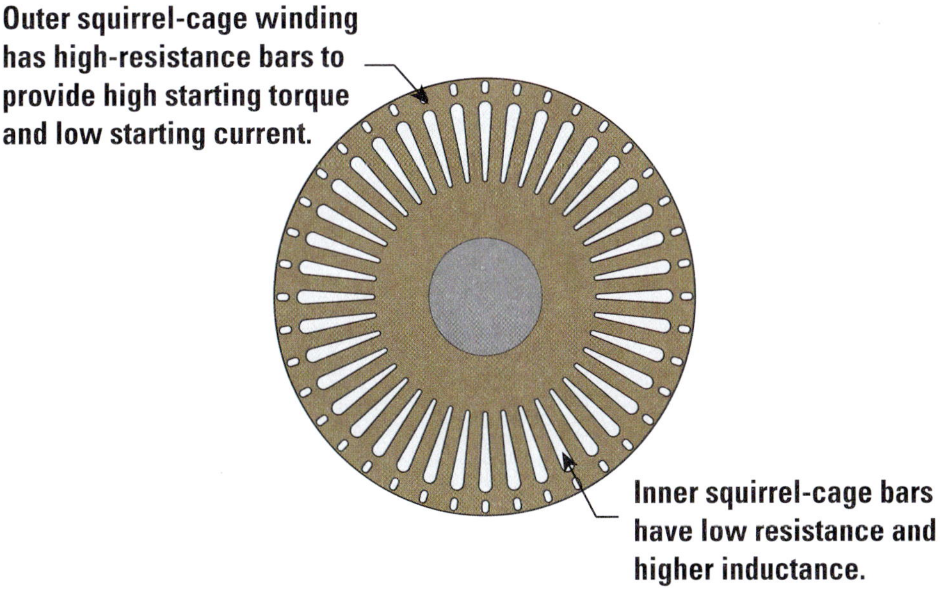

Figure 32–36 Double-squirrel-cage rotor used in NEMA design code C motors.

winding limits the current flow through the rotor, which limits the starting current to a relatively low value. Because the current is close to being in phase with the induced voltage, the rotor flux and stator flux are close to being in phase with each other and a strong starting torque is developed. The starting torque of a double-squirrel-cage motor can be as high as 250% of the rated full-load torque.

When the rotor reaches its full-load speed, rotor frequency decreases to 2 or 3 hertz. The inductive reactance of the inner winding has now decreased to a low value. Most of the rotor current now flows through the low-resistance inner winding. This type of motor has good running torque and excellent speed regulation.

Power Factor of a Squirrel-Cage Induction Motor

At no load, most of the current is used to magnetize the stator and rotor. Because most of the current is magnetizing current, it is inductive and lags the applied voltage by close to 90°. A very small resistive component is present, caused mostly by the resistance of the wire in the stator and the power needed to overcome bearing friction and windage loss. At no load, the motor appears to be a resistive-inductive series circuit with a large inductive component compared with resistance (*Figure 32–37*). A power factor of about 10% is common for a squirrel-cage motor at no load.

Figure 32–37 At no load, the motor appears to have a large amount of inductance and a very small resistance.

As load is added, electrical energy is converted into mechanical energy, and as actual work is performed, the in-phase component of current increases. The circuit now appears to contain more resistance than inductance (*Figure 32–38*). This causes the phase angle between applied voltage and motor current to decrease, causing the power factor to increase. In practice, the power factor of an induction motor at full load is about 85 to 90%. Because motors are inductive loads, the power factor is always a lagging power factor until capacitance is added to correct it.

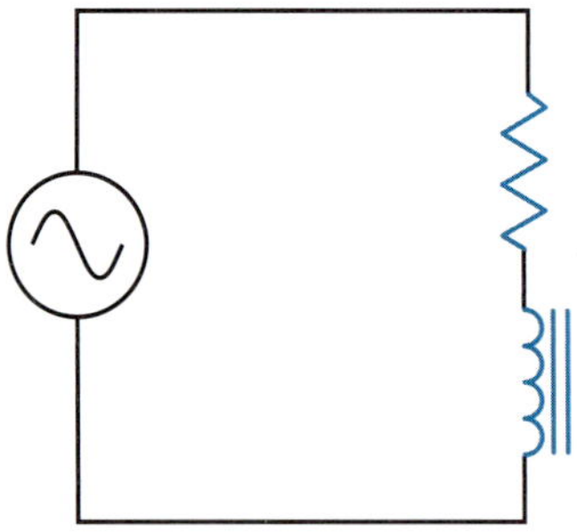

Figure 32–38 At full load, actual work is performed, and the motor appears as though the resistive properties are greater than the inductive.

Single-Phasing

Three lines supply power to a three-phase motor. If one of these lines should open, the motor will be connected to single-phase power (*Figure 32–39*). This condition is known as **single-phasing.**

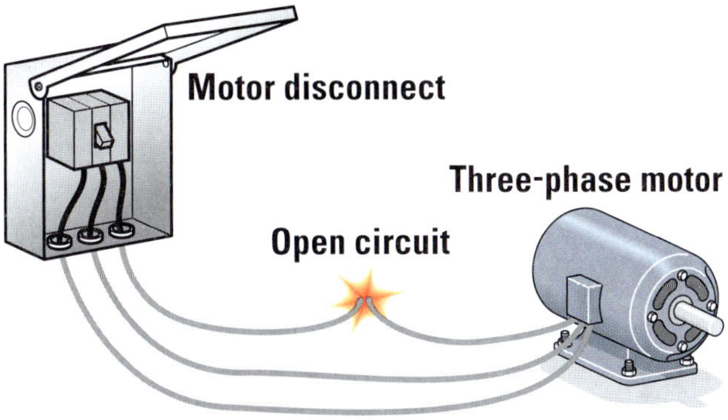

Figure 32–39 Single-phasing occurs when one line of a three-phase system is open.

If the motor is not running and single-phase power is applied to the motor, the induced voltage in the rotor sets up a magnetic field in the rotor. This magnetic field opposes the magnetic field of the stator (Lenz's law). As a result, practically no torque is developed in either the clockwise or counterclockwise direction, and the motor will not start. The current supplying the motor will be excessive, however, and damage to the stator windings can occur.

If the motor is operating under load at the time the single-phasing condition occurs, the rotor will continue to turn at a reduced speed. The moving bars of the rotor cut the stator field flux, which continues to induce voltage and current in the bars. Because of reduced speed, the rotor has high-reactive and low-resistive components, causing the rotor current to lag the induced voltage by almost 90°. This lagging current creates rotor fields midway between the stator poles, resulting in greatly reduced torque. The reduction in rotor speed causes high current flow and will most likely damage the stator winding if the motor is not disconnected from the power line.

Effects of Voltage Variation on Motors

Motors are affected when operated at other than their rated nameplate voltage. The National Electrical Manufacturers Association (NEMA) rated motors are designed to operate at ±10% of their rated voltage. *Table 32–1* shows the approximate change in full-load current and starting current for typical electric motors when operated over their rated voltage (110%) and under their rated voltage (90%).

Voltage Variation	Full-Load Current	Starting Current
110%	7% increase	10–12% increase
90%	11% decrease	10–12% decrease

Table 32–1 Effects of Voltage Variation

Motors are intended to operate on systems with balanced voltage (the voltage is the same between all phases). Unbalanced voltage is one of the leading causes of motor failure. Unbalanced voltage is generally caused when single-phase loads are supplied by three-phase systems.

The values listed in *Table 32–1* assume that the voltages across the phase conductors as measured between phases AB, BC, and AC are balanced. Greater harm is caused when the voltages are unbalanced. The greatest example of voltage unbalance occurs when one phase of a three-phase system is lost, and the motor begins single-phasing. This causes a 173% increase of current in two of the motor windings. If the normal full-load current of a motor is 20 amperes, the two windings still connected to power will have a current of 34.6 amperes, and the one winding that has lost power will have a current of 0 amperes. NEMA recommends that the unbalanced voltage not exceed ±1%. The following steps illustrate how to determine the percent of voltage unbalance in a three-phase system:

1. Take voltage measurements between all phases. In this example, assume the voltage between AB = 496 volts, BC = 460 volts, and AC = 472 volts.
2. Find the average voltage.

$$\begin{array}{r} 496 \\ 460 \\ \underline{472} \\ 1428 \end{array} \qquad 1428/3 = 476 \text{ volts}$$

3. Subtract the average voltage from the voltage reading that results in the greatest difference:

$$496 - 476 = 20 \text{ volt}$$

4. Determine the percent difference:

$$\frac{100 \times \text{Greatest voltage difference}}{\text{Average voltage}}$$

$$\frac{100 \times 20}{476} = 4.2\% \text{ voltage unbalance}$$

This difference in the voltage levels will cause the operating temperatures of the windings to increase. The percent of temperature increase in the motor caused by the voltage unbalance is equal to twice the percent squared [2 × (percent voltage unbalance)^2]:

$$2 \times 4.2 \times 4.2 = 35.28\% \text{ temperature increase in the winding with the highest current.}$$

Motor Nameplate Data

The NEMA specifies that every motor nameplate must list specific items such as:

- Manufacturer's name
- Rated voltage
- Full-load current
- Frequency
- Number of phases
- Full-load speed
- Temperature rise or insulation system class
- Duty or time rating
- Horsepower
- Locked rotor indicating code letter
- Service factor
- Frame size
- Efficiency
- NEMA design code

It should be noted that not all motor manufacturers comply with NEMA specifications, and their nameplates may or may not contain all the information specified by NEMA. In some instances, information not specified by NEMA may also be listed on a nameplate. An example of a NEMA motor nameplate is shown in *Figure 32–40*. Each item on the nameplate will be discussed.

Manufacturer's name

HP: 10	RPM: 1720
Hz: 60	Phase: 3
Type: Induction	Frame: 143T
Volts: 230/460	FLA: 28/14
Encl.: ODP	Duty: Cont.
Max. Temp.: 40°C	SF: 1.15
Code: K	NEMA Design: B
NEMA FL Eff.: .77	Ins.: B
Model No.: xxx123	SER#: 123456

Connection

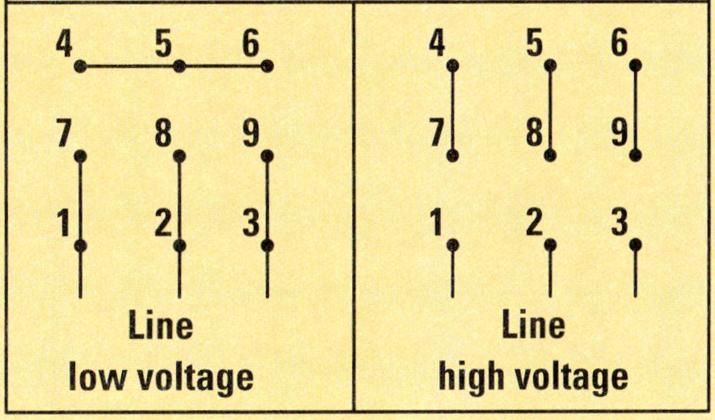

Figure 32–40 Example of a motor nameplate in compliance with NEMA specifications.

Manufacturer's Name The very top of the nameplate shown in Figure 32–40 is the manufacturer's name. This lists the manufacturer of the motor.

Horsepower Motors have a horsepower rating that is determined by the amount of torque they can produce at a specific speed under full load. The horsepower listed on the nameplate in this example is 10 HP.

RPM The RPM indicates the speed in revolutions per minute that the motor will run at rated full load. The motor will run faster at light load or no load. The example nameplate shown indicates that the motor will have a speed of 1720 RPM when the motor is under full load.

Frequency The frequency is measured in hertz. The standard frequency used throughout the United States and Canada is 60 Hz. The example nameplate indicates that the motor is designed to operate only on a frequency of 60 Hz.

Phase Phase indicates the number of phases on which the motor is designed to operate. The example nameplate indicates that the motor is to operate on three-phase power.

Motor Type Three-phase motors can be divided into three general types: squirrel-cage induction, wound-rotor induction, and synchronous. Motors listed as "induction" will generally be squirrel-cage type, which describes the type of rotor used in the motor. Wound-rotor induction motors are easily recognized by the fact that they contain three sliprings on the rotor shaft. Synchronous motors are not induction-type motors. The example nameplate lists this motor as an induction-type motor and, therefore, a squirrel-cage type motor.

Frame The frame number indicates the size and type of motor frame. A chart is generally needed to determine the exact dimensions of the frame. The example nameplate indicates that the motor has a 143T frame.

The letter that appears at the end of the frame numbers represents different frame styles. This will be useful information in ordering a replacement motor as this will specify which motor frame to purchase. For further information on NEMA frame size and mounting types, go to www.nema.org.

Voltage Volts indicates the operating voltage of the motor. The example nameplate indicates that the motor in this example is designed to operate on different voltages depending on the connection of the stator windings. These motors are generally referred to as "dual-voltage motors." If the motor is connected for low-voltage operation, it will operate on 208 or 230 volts. If it is connected for high-voltage operation, it will operate on 460 volts.

Full-Load Current The ampere rating indicates the amount of current the motor should draw at full load. It will draw less current at light load or no load. Note that there are two currents listed because the example motor is a dual voltage motor. The higher value of current corresponds to the low voltage connection, and the lower value is for a high voltage connection. The example nameplate shows the current as FLA (full-load amps). Some nameplates simply list the current as AMPS.

Enclosure The example nameplate indicates that the motor has an ODP (open drip-proof) type enclosure. Motors have different types of enclosures depending on the application. Some of the common enclosures are:

ODP *Open drip-proof*. These are very common. The case has openings to permit ventilation through the motor windings.

TEFC *Totally enclosed fan cooled*. The motor case is sealed to prevent the entrance of moisture or dirt. A fan is used to help cool the motor.

TENV *Totally enclosed non-vented*. These motors are generally used in harsh environments such as chemical plants. They are designed to be hosed down.

EXP *Explosion proof*. Totally enclosed and non-vented. Designed to be used in areas that have hazardous atmospheres.

Duty Cycle The example nameplate lists the duty cycle as continuous. The duty cycle indicates the amount of time the motor is expected to operate. A motor with a continuous duty cycle is rated to run at a substantially constant load for an indefinitely long time. Most motors are rated for continuous duty.

Intermittent-duty motors are intended to operate for shorter periods of time. An example of an intermittent-duty motor is the starter motor on an automobile. These motors develop a large amount of horsepower in a small case size. If these motors are operated for an extended period of time, they overheat and experience significant internal damage.

Temperature Rise The temperature rise of a motor is the maximum amount of rise in temperature the motor can exhibit above the ambient operating temperature and remain within the allowable operating temperature at full load. The example nameplate indicates a maximum temperature rise of 40°C for this motor. This value is generally used to help select proper overload protection.

Service Factor Motor service factor (SF) is a multiplier that, when applied to the rated horsepower of a motor, specifies the allowable overload of the motor at the specified voltage and frequency. It gives some parameters in estimating horsepower needs and actual running horsepower requirements. The example nameplate indicates that the motor has a horsepower of 10 hp and a service factor of 1.15. The service factor, however, indicates that the motor is capable of producing 11.5 hp (10 × 1.15). However, this does not mean that the motor can be used for an application requiring 11.5 hp. Selecting motors with a service factor greater than 1 allows for cooler winding temperatures at rated load, protects against intermittent heat rises, and helps to offset low or unbalanced line voltages. Service factors are established for operations at rated voltage, frequency, and ambient temperature at sea-level conditions.

Locked Rotor Code Letter The locked rotor code letter is determined by the construction of the squirrel-cage rotor. It can be used to determine the approximate amount of inrush current when the motor is started. The locked rotor code letter should not be confused with the NEMA design code on many motors. To determine the approximate starting current of a squirrel-cage induction motor using the locked rotor code letter, multiply the horsepower of the motor by the kVA per horsepower factor and divide by the applied voltage. If the motor is three-phase, be sure to include the square root of 3 factor in the calculation. A chart showing the kVA per horsepower rating for different code letters is shown in *Figure 32–41*. Refer to Table 430.7(B) in the *NEC.*

Code	kVA per horsepower	Approximate midrange value
A	0.00–3.14	1.7
B	3.15–3.54	3.3
C	3.55–3.99	3.8
D	4.00–4.49	4.3
E	4.50–4.99	4.7
F	5.00–5.59	5.3
G	5.60–6.29	5.9
H	6.30–7.09	6.7
J	7.10–7.99	7.5
K	8.00–8.99	8.5
L	9.00–9.99	9.5
M	10.00–11.19	10.6
N	11.20–12.49	11.8
P	12.50–13.99	13.2
R	14.00–15.99	15
S	16.00–17.99	17
T	18.00–19.99	19
U	20.00–22.39	21
V	22.40–and up	?

Figure 32–41 Code letters are used to determine the locked rotor current of a motor.

Example 32–4

A 15-hp three-phase motor is connected to 480 volts. The motor has a locked rotor code J. What is the approximate starting current for this motor?

Solution

The chart in *Figure 32–41* lists that the approximate midrange value for code J is 7.5.

$$I_{\text{Locked Rotor}} = \frac{15 \times 7.5 \text{ kVA}}{480 \times \sqrt{3}}$$

$$I_{\text{Locked Rotor}} = \frac{15 \times 7500}{480 \times 1.732}$$

$$I_{\text{Locked Rotor}} = \frac{112{,}500}{831.36}$$

$$I_{\text{Locked Rotor}} = 135.32 \text{ A}$$

$$\text{Amps} = \frac{\text{Hp} \times \text{kVA factor}}{\text{Volts} \times \sqrt{3}}$$

$$\text{Amps} = \frac{15 \times 7.5}{480 \times 1.732}$$

$$\text{Amps} = \frac{112.5 \text{ kVA}}{480 \times 1.732}$$

$$\text{Amps} = \frac{112{,}500 \text{ VA}}{831.36}$$

$$\text{Amps} = 135.5$$

NEMA Design Code Induction motors have different operating characteristics determined by their design. Such factors as the amount of iron used in the stator, the wire size, the number turns of wire, and the rotor design all play a part in the operating characteristics of the motor. To obtain some uniformity in motor operating characteristics, NEMA assigns design letters to general-purpose motors based on factors such as locked rotor torque, breakdown torque, slip, starting current, and other values. The NEMA design letters are A, B, C, and D.

A Design letter A motors exhibit normal starting torque, high starting current, and low slip. Design A motors must have less than 5% slip and must have less slip than Design B motors of equivalent ratings. Motors with design letter A exhibit very high starting current, often 500 to 800% of full-load current. Typical applications for these motors are driving fans, blowers, pumps, lathes, milling machines, and other machine tools.

B Design B motors are the most common. They exhibit normal starting torque, low starting current, and low slip. Design letter B motors produce about the same amount of starting torque as Design A motors but have about 25% less starting current. Rotor slip is less than 5% at full load. They have sufficient lock rotor starting torque to start most industrial loads. The example nameplate in *Figure 32–40* indicates that this motor has a NEMA design letter B.

C Motors with the design letter C have high starting torques and low starting currents. Rotor slip is less than 5% at full load. Design letter C motors are more costly than other motors because they use the double-squirrel-cage rotor. They are used to start heavy loads such as loaded pumps, compressors, and conveyors.

D Design D motors have high starting torque, about 275% of rated full-load torque, and low starting current. These motors are equipped with Type A rotors. The higher-resistance rotor bars produce a higher starting torque because the increased resistance causes the rotor flux to be more in phase with the stator flux, but the increased resistance results in a greater amount of rotor slip. Rotor slip for design letter D motors is typically 7 to 11% at full load but may go up to 17% or higher. These motors are generally used to start high-inertia loads such as large flywheels used with punch presses or shears.

Motor Efficiency The full-load efficiency indicates the overall efficiency of the motor. The efficiency is the amount of electrical energy supplied to the motor that is converted into mechanical energy. The remaining power is a loss and is mostly converted into heat. The example nameplate indicates that the motor in this example has an efficiency of 77%.

Insulation Classification The example nameplate indicates that this motor has an insulation classification B. The classification of insulation greatly affects the life span of the motor. Motors that operate in hotter climates should have a higher insulation temperature rating. The thermal capacity of different insulations is rated as A, B, F, and H. The chart shown in *Figure 32–42* lists the maximum operating temperature of the winding insulation.

Class	Maximum operating temperature
A	105°C
B	130°C
F	155°C
H	180°C

Figure 32–42 NEMA insulation temperature ratings and classification letters.

Model and Serial Numbers The model number is assigned by the manufacturer. It can be used to purchase a motor with identical characteristics. Many motors with the same model number can generally be obtained. The serial number is also assigned by the manufacturer. The serial number, however, is used to identify a particular motor. No other motor should have the same serial number.

Connection Diagrams The connection diagrams for both low- and high-voltage connections are also given on the example nameplate. Most dual-voltage motors contain nine leads in the terminal connection box. These leads are numbered T1 through T9. The connection diagram is used to make high- or low-voltage connections for the motor. Power is connected to T1, T2, and T3.

Figure 32–43 shows the schematic symbol used to represent a three-phase squirrel-cage motor.

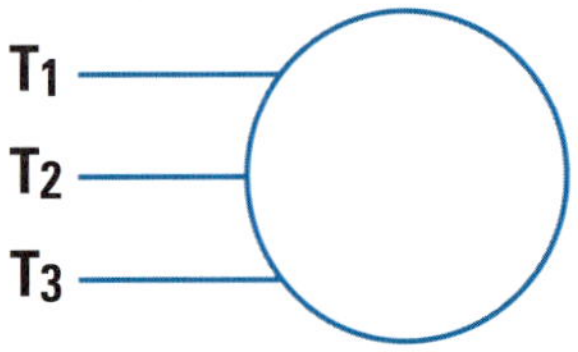

Figure 32–43 Schematic symbol of a three-phase squirrel-cage induction motor.

Consequent-Pole Squirrel-Cage Motors

Consequent-pole squirrel-cage motors permit the synchronous speed to be changed by changing the number of stator poles. If the number of poles is doubled, the synchronous speed is reduced by one-half. A two-pole motor has a synchronous speed of 3600 RPM when operated at 60 hertz. If the number of poles is doubled to four, the synchronous speed becomes 1800 RPM. The number of stator poles can be changed by changing the direction of current flow through alternate pairs of poles.

Figure 32–44 illustrates this concept. In (*A*), two coils are connected in series and current flows through them in the same direction. The two coils produce the same magnetic polarity and are a single pole. In (*B*), the coils have been reconnected in such a manner that current flows through them in opposite directions. The coils now produce the opposite magnetic polarities and produce two different poles.

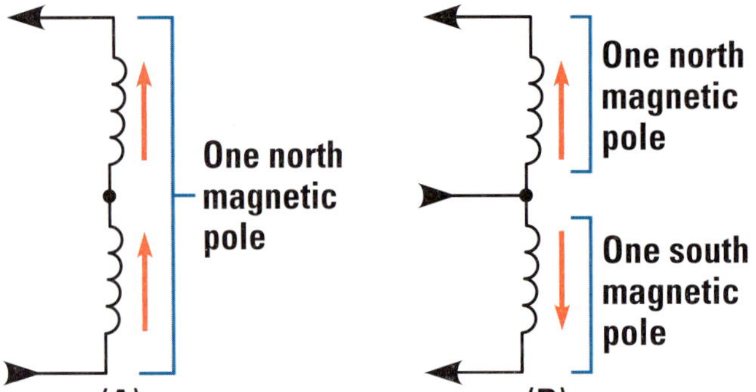

Figure 32–44 The number of poles can be changed by altering the connection to reverse the current flow through alternate poles.

Consequent-pole motors with one stator winding bring out six leads labeled T_1 through T_6. Depending on the application, the windings will be connected as a series delta or a parallel wye. If it is intended that the motor maintain the same horsepower rating for both high and low speed, the high-speed connection will be a series delta (*Figure 32–45*). The low-speed connection will be a parallel wye (*Figure 32–46*).

If it is intended that the motor maintain constant torque for both low and high speeds, the series-delta connection will provide low speed, and the parallel wye will provide high speed.

Because the speed range of a consequent-pole motor is limited to a 1:2 ratio, motors intended to operate at more than two speeds contain more than one stator winding. A consequent-pole motor with three speeds, for example, has one stator winding for one speed only and a second winding with taps. The tapped winding may provide synchronous speeds of 1800 and 900 RPM, and the separate winding may provide a speed of 1200 RPM. Consequent-pole motors with four speeds contain two separate stator windings with taps. If the second stator winding of the motor in this example were to be tapped, the motor would provide synchronous speeds of 1800, 1200, 900, and 600 RPM.

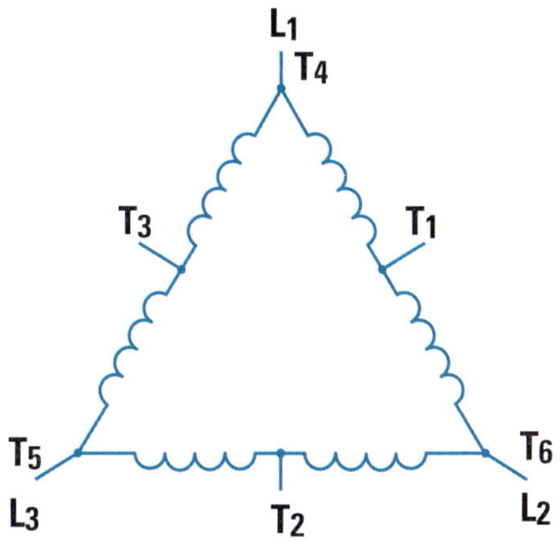

Speed	L1	L2	L3	Open	Together
Low	T1	T2	T3	———	T4 T5 T6
High	T6	T4	T5	All Others	———

Figure 32–45 High-speed series-delta connection.

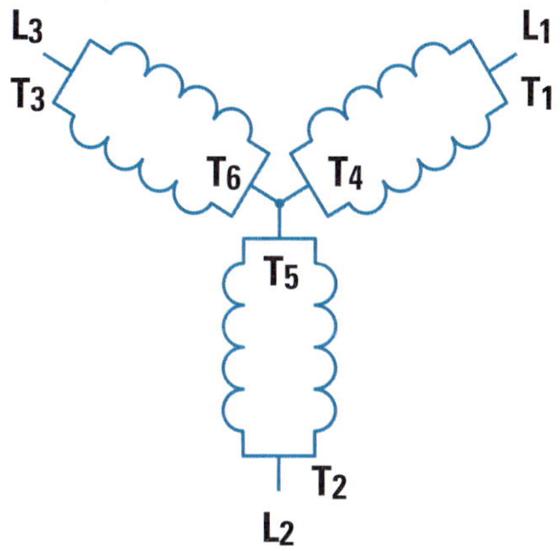

Speed	L1	L2	L3	Open	Together
Low	T1	T2	T3	———	T4 T5 T6
High	T6	T4	T5	All Others	———

Figure 32–46 Low-speed parallel-wye connection.

Motor Calculations

In the following example, output horsepower and motor efficiency are calculated. It is assumed that a 1/2-horsepower squirrel-cage motor is connected to a load. A wattmeter is connected to the motor, and the load torque measurement is calibrated in pound-inches. The motor is operating at a speed of 1725 RPM and producing a torque of 16 lb-in. The wattmeter is indicating an input power of 500 watts.

The actual amount of horsepower being produced by the motor can be calculated by using the formula:

$$hp = \frac{6.28 \times RPM \times L \times P}{33,000}$$

where

$$hp = \text{horsepower}$$
$$6.28 = \text{a constant}$$
$$RPM = \text{speed in revolutions per minute}$$
$$L = \text{distance in feet}$$
$$P = \text{pounds}$$
$$33{,}000 = \text{a constant}$$

Because the formula uses feet for the distance and the torque of the motor is rated in pound-inches, L will be changed to 1/12 of a foot, or 1 inch. To simplify the calculation, the fraction 1/12 will be changed into its decimal equivalent (0.08333). To calculate the output horsepower, substitute the known values in the formula:

$$hp = \frac{6.28 \times 1725\text{ RPM} \times 0.08333\text{ ft/in} \times 16\text{ lb-in.}}{33{,}000}$$

$$hp = \frac{14{,}444\text{ ft-lb/min}}{33{,}000}$$

$$hp = 0.438$$

One horsepower is equal to 746 watts. The output power of the motor can be calculated by multiplying the output horsepower by 746:

$$\text{Power out} = 746\text{ W/hp} \times 0.438\text{ hp}$$

$$\text{Power out} = 326.5\text{ W}$$

The efficiency of the motor can be calculated by using the formula:

$$\text{Eff.} = \frac{\text{power out}}{\text{power in}} \times 100$$

$$\text{Eff.} = \frac{326.5\text{ W}}{500\text{ W}} \times 100$$

$$\text{Eff.} = 0.653 \times 100$$

$$\text{Eff.} = 65.3\%$$

32–4 Wound-Rotor Induction Motors

The **wound-rotor motor** is very popular in industry because of its high starting torque and low starting current. The stator winding of the wound-rotor motor is the same as that of the squirrel-cage motor. The difference between the two motors lies in the construction of the rotor. Recall that the squirrel-cage rotor is constructed of bars connected together at each end by a shorting ring, as shown in *Figure 32–24*.

The rotor of a wound-rotor motor is constructed by winding three separate coils on the rotor 120° apart. The rotor will contain as many poles per phase as the stator winding. These coils are then connected to three sliprings located on the rotor shaft (*Figure 32–47*). Brushes connected to the sliprings provide an external connection to the rotor. This permits the rotor circuit to be connected to a set of resistors (*Figure 32–48*).

Figure 32–47 Rotors of wound-rotor induction motors. Courtesy of Wazee Companies, LLC, Denver, Colorado

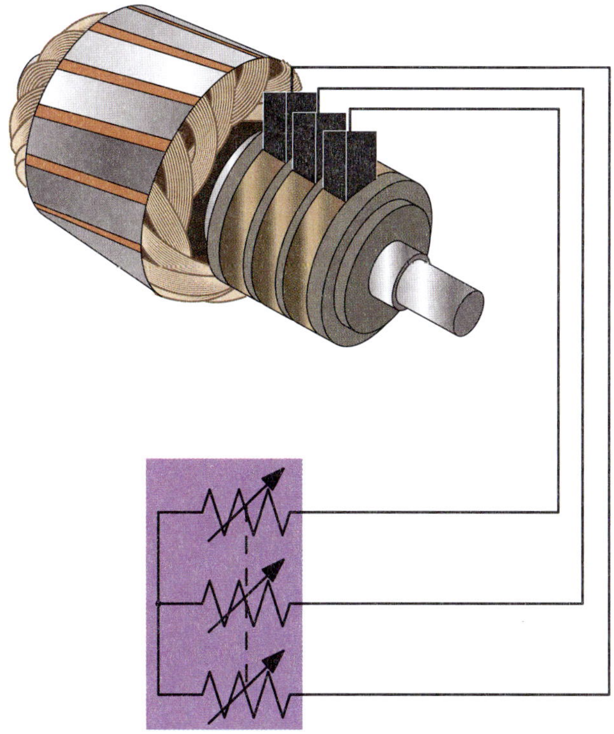

Figure 32–48 The rotor of a wound-rotor motor is connected to external variable resistors.

The stator terminal connections are generally labeled T_1, T_2, and T_3. The rotor connections are commonly labeled M_1, M_2, and M_3. The M_2 lead is generally connected to the middle slip ring, and the M_3 lead is connected close to the rotor windings. The direction of rotation for the wound-rotor motor is reversed by changing any two stator leads. Changing the M leads will have no effect on the direction of rotation. The schematic symbol for a wound-rotor motor is shown in *Figure 32–49*.

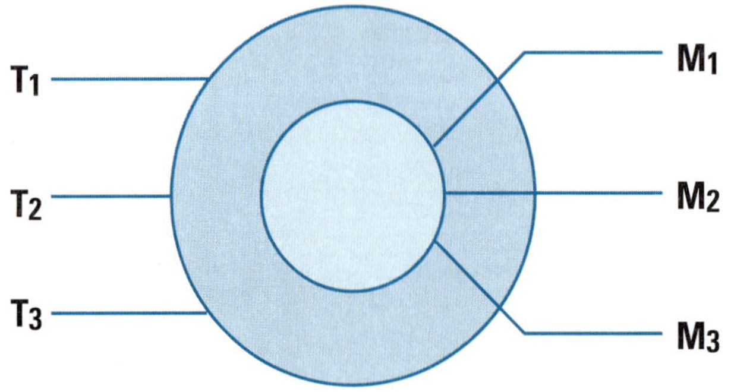

Figure 32–49 Schematic symbol for a wound-rotor induction motor.

Principles of Operation

When power is applied to the stator winding, a rotating magnetic field is created in the motor. This magnetic field cuts through the windings of the rotor and induces a voltage into them. The amount of current flow in the rotor is determined by the amount of induced voltage and the total impedance of the rotor circuit ($I = E/Z$). The rotor impedance is a combination of inductive reactance created in the rotor windings and the external resistance. The impedance could be calculated using the formula for resistance and inductive reactance connected in series:

$$Z = \sqrt{R^2 + X_L^2}$$

As the rotor speed increases, the frequency of the induced voltage decreases just as it does in the squirrel-cage motor. The reduction in frequency causes the rotor circuit to become more resistive and less inductive, decreasing the phase angle between induced voltage and rotor current.

When current flows through the rotor, a magnetic field is also produced. This magnetic field is attracted to the rotating magnetic field of the stator. As the rotor speed increases, the induced voltage decreases because of less cutting action between the rotor windings and rotating magnetic field. The decrease in induced voltage produces less current flow in the rotor and therefore less torque. If the rotor circuit resistance is reduced, more current can flow, which will increase motor torque, and the rotor will increase in speed. This action continues until all external resistance has been removed from the rotor circuit by shorting the M leads together and the motor is operating at maximum speed. At this point, the wound-rotor motor is operating in the same manner as a squirrel-cage motor.

Starting Characteristics of a Wound-Rotor Motor

Although the overall starting torque of a wound-rotor motor is less than that of an equivalent horsepower squirrel-cage motor, due to reduced current in both the rotor and stator during the starting period, the starting torque will be higher per ampere than in a squirrel-cage motor. The starting current is less because resistance is connected in the rotor circuit during starting. This resistance limits the amount of current that can flow in the rotor circuit. Because the stator current is proportional to the rotor current due to transformer action, the stator current is less also. Recall that one of the factors that determines motor torque is the phase angle difference between stator flux and rotor flux. Because resistance is connected in the rotor circuit, stator and rotor flux are closer to being in phase with each other, producing a high starting torque per ampere for the wound-rotor induction motor.

Speed Control

The speed of a wound-rotor motor can be controlled by permitting resistance to remain in the rotor circuit during operation. When this is done, the rotor and stator current is limited, which reduces the strength of both magnetic fields. The reduced magnetic field strength permits the rotor to slip behind the rotating magnetic field of the stator. The resistors of speed controllers must have higher power ratings than the resistors of starters because they operate for extended periods of time.

32–5 Synchronous Motors

The three-phase synchronous motor has several characteristics that separate it from the other types of three-phase motors. Some of these characteristics are as follows:

1. The synchronous motor is not an induction motor. It does not depend on induced current in the rotor to produce a torque.
2. It will operate at a constant speed from full load to no load.
3. The synchronous motor must have DC excitation to operate.
4. It will operate at the speed of the rotating magnetic field (synchronous speed).
5. It has the ability to correct its own power factor and the power factor of other devices connected to the same line.

Rotor Construction

The synchronous motor has the same type of stator windings as the other two three-phase motors. The rotor of a synchronous motor has windings similar to the rotor of an alternator *(see Figure 31–6)*. Wound pole pieces become electromagnets when DC is applied to them. The excitation current can be applied to the rotor through two sliprings located on the rotor shaft or by a brushless exciter. The brushless exciter for a synchronous motor is the same as that used for the alternator discussed in Unit 31.

Starting a Synchronous Motor

The rotor of a synchronous motor also contains a set of squirrel-cage bars similar to those found in a type A rotor. This set of squirrel-cage bars is used to start the motor and is known as the **amortisseur winding** (*Figure 32–50*). When power is first connected to the stator, the rotating magnetic field cuts through the squirrel-cage bars. The cutting action of the field induces a current into the squirrel-cage winding. The current flow through the amortisseur winding produces a rotor magnetic field that is attracted to the rotating magnetic field of the stator. This causes the rotor to begin turning in the direction of rotation of the stator field. When the rotor has accelerated to a speed that is close to the synchronous speed of the field, DC is connected to the rotor through the sliprings on the rotor shaft or by a brushless exciter (*Figure 32–51*).

Figure 32–50 Synchronous-motor rotor with amortisseur winding. Courtesy of Electric Machinery Corporation

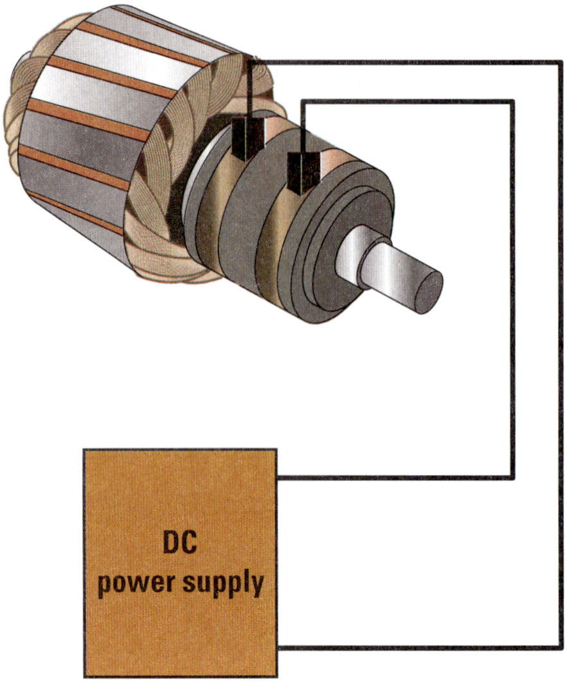

Figure 32–51 DC excitation current supplied through sliprings.

When DC is applied to the rotor, the windings of the rotor become electromagnets. The electromagnetic field of the rotor locks in step with the rotating magnetic field of the stator. The rotor will now turn at the same speed as the rotating magnetic field. When the rotor turns at the synchronous speed of the field, there is no more cutting action between the stator field and the amortisseur winding. This causes the current flow in the amortisseur winding to cease.

Notice that the synchronous motor starts as a squirrel-cage induction motor. Because the rotor uses bars that are similar to those used in a type A rotor, they have a relatively high resistance, which gives the motor good starting torque and low starting current. **A synchronous motor must never be started with DC connected to the rotor.** If DC is applied to the rotor, the field poles of the rotor become electromagnets. When the stator is energized, the rotating magnetic field begins turning at synchronous speed. The electromagnets are alternately attracted and repelled by the stator field. As a result, the rotor does not turn. The rotor and power supply most likely will be damaged by high induced voltages, however.

The Field-Discharge Resistor

When the stator winding is first energized, the rotating magnetic field cuts through the rotor winding at a fast rate of speed. This causes a large amount of voltage to be induced into the winding of the rotor. To prevent this from becoming excessive, a resistor is connected across the winding. This resistor is known as the *field-discharge resistor* (*Figure 32–52*). It also helps reduce the voltage induced into the rotor by the collapsing magnetic field when the DC is disconnected from the rotor. The field-discharge resistor is connected in parallel with the rotor winding during starting. If the motor is manually started, a field-discharge switch is used to connect the excitation current to the rotor. If the motor is automatically started, a special type of relay is used to connect excitation current to the rotor and disconnect the field-discharge resistor.

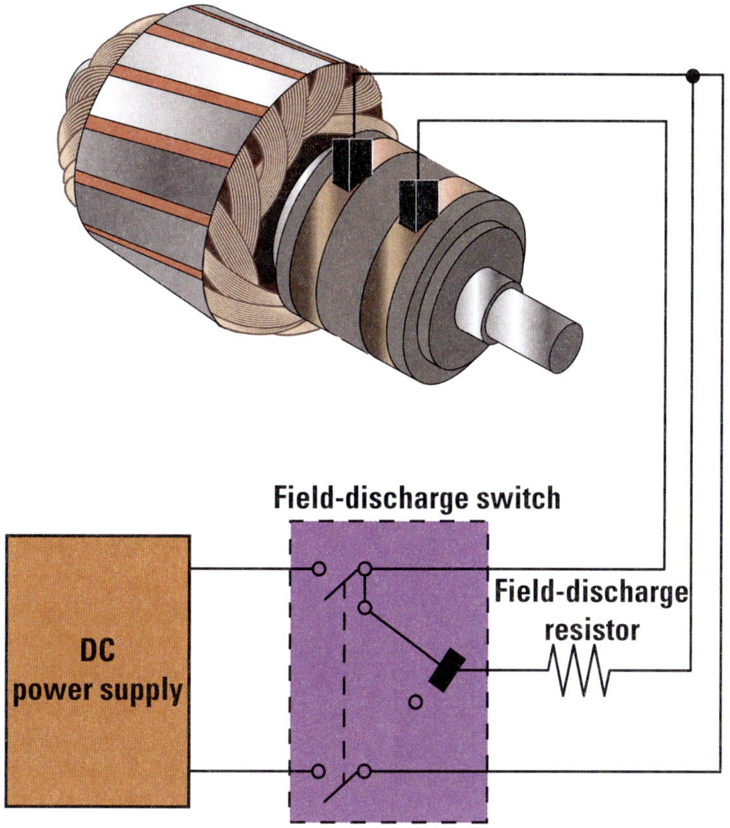

Figure 32–52 The field-discharge resistor is connected in parallel with the rotor winding during starting.

Constant-Speed Operation

Although the synchronous motor starts as an induction motor, it does not operate as one. After the amortisseur winding has been used to accelerate the rotor to about 95% of the speed of the rotating magnetic field, DC is connected to the rotor and the electromagnets lock in step with the rotating field. Notice that the synchronous motor does not depend on induced voltage from the stator field to produce a magnetic field in the rotor. The magnetic field of the rotor is produced by an external DC power source applied to the rotor.

As load is added to the motor, the magnetic field of the rotor remains locked with the rotating magnetic field of the stator, and the rotor continues to turn at the same speed. The added load, however, causes the magnetic fields of the rotor and stator to become stressed (*Figure 32–53*). If the load on the motor becomes too great, the rotor is pulled out of sync with the rotating magnetic field. The amount of torque necessary to cause this condition is called the *pullout torque*. The pullout torque for most synchronous motors ranges from 150 to 200% of rated full-load torque. If pullout torque is reached, the motor must be stopped and restarted.

Power Factor Correction

The synchronous motor has the ability to correct its own power factor and the power factor of other devices connected to the same line. The amount of power factor correction is controlled by the amount of excitation current in the rotor. If the excitation current to the rotor of a synchronous motor is too small, the motor has a lagging power factor like other induction motors. As rotor excitation current is increased, the synchronous motor appears to be more capacitive. When the excitation current reaches a point at which the power factor of the motor is at unity or 100%, it is at the normal excitation level. At this point, the current supplying the motor drops to its lowest value.

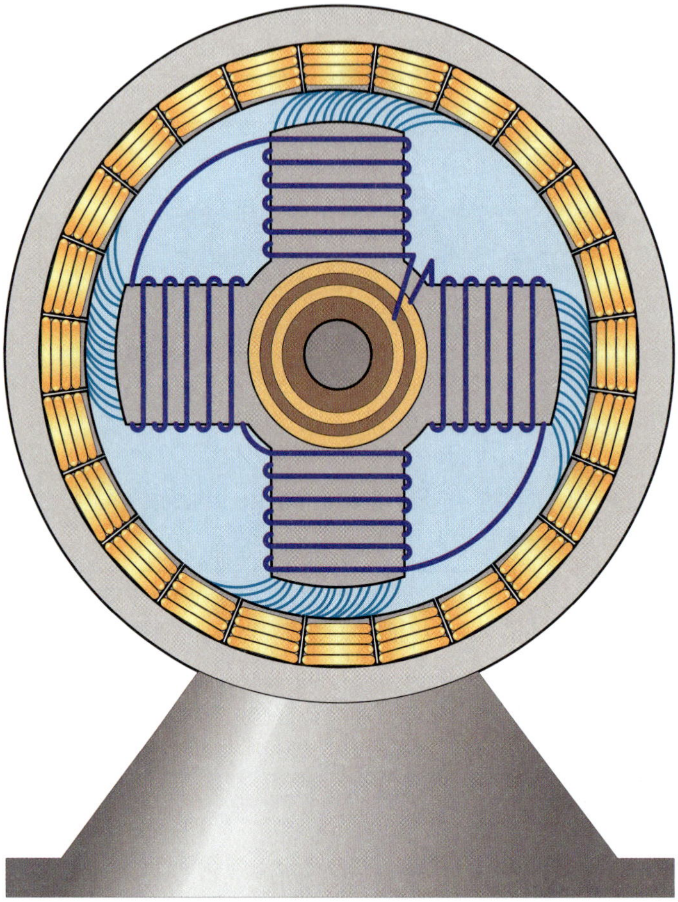

Figure 32–53 The magnetic field becomes stressed as load is added.

If the excitation current is increased above the normal level, the motor has a leading power factor and appears as more of a capacitive load. The power factor at this point, however, is leading and not lagging. Changes in the amount of excitation current do not affect the speed of the motor.

Interaction of the Stator and Rotor Magnetic Fields

Figure 32–54 illustrates how the magnetic flux of the rotor field aids or opposes the stator field. When the excitation current to the rotor is less than the amount required for normal excitation, the stator supplies some portion of the magnetizing current to aid the rotor (*Figure 32–54(A)*). The current waveform shown in *Figure 32–54(A)* depicts how the current required to provide the additional magnetism is lagging the applied voltage. The synchronous motor has a lagging power factor at this time.

In *Figure 32–54(B)*, the excitation is at the normal excitation value. Because the rotor and stator are in sync with each other, the current is in phase with the voltage and the motor power factor is at unity, or 100%. The amount of current supplied to the motor is at its lowest value during this period.

In *Figure 32–54(C)*, the excitation current is greater than that needed for normal excitation. The current supplied to the rotor will lead the applied voltage by 90°. The current waveform shown in *Figure 32–54(C)* illustrates that the motor now has a leading power factor.

Synchronous Motor Applications

Synchronous motors are very popular in industry, especially in the large horsepower ratings (motors up to 5000 hp are not uncommon). They have a low starting current per horsepower and a high starting torque.

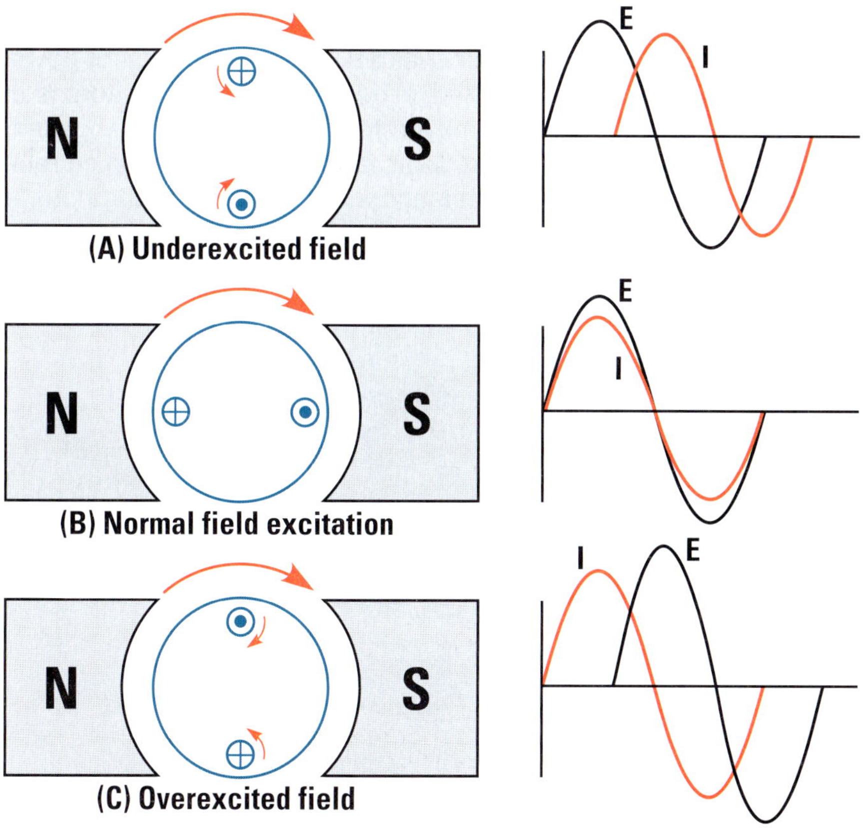

Figure 32–54 Field excitation in a synchronous motor.

They operate at a constant speed from no load to full load and maintain maximum efficiency. Synchronous motors are used to operate DC generators, fans, blowers, pumps, and centrifuges. They can correct their own power factor and can correct the power factor of other inductive loads connected to the same feeder (*Figure 32–55*). Synchronous motors are sometimes operated at no load and are used for power factor correction only. When this is done, the motor is referred to as a **synchronous condenser.**

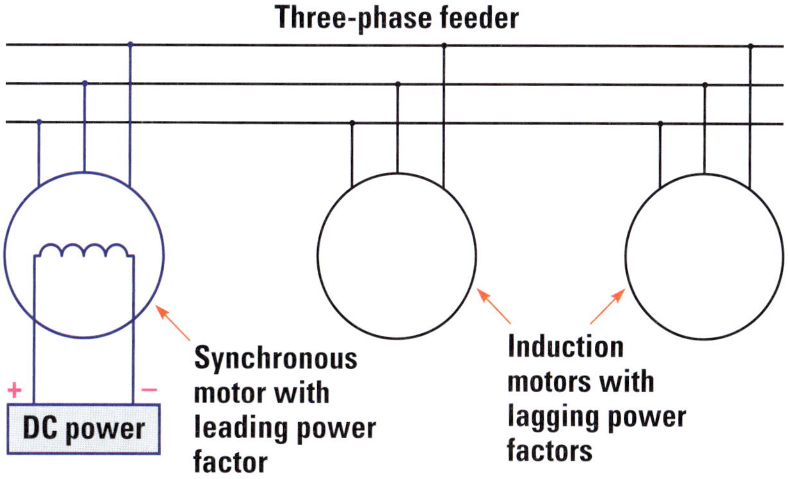

Figure 32–55 Synchronous motor can be used to correct the power factor of other motors.

Advantages of the Synchronous Condenser

The advantage of using a synchronous condenser over a bank of capacitors for power factor correction is that the amount of correction is easily controlled. When a bank of capacitors is used for correcting power factor, capacitors must be added to or removed from the bank if a change in the amount of correction is needed. When a synchronous condenser is used, only the excitation current must be changed to cause an alteration of power factor. The schematic symbol for a synchronous motor is shown in *Figure 32–56*.

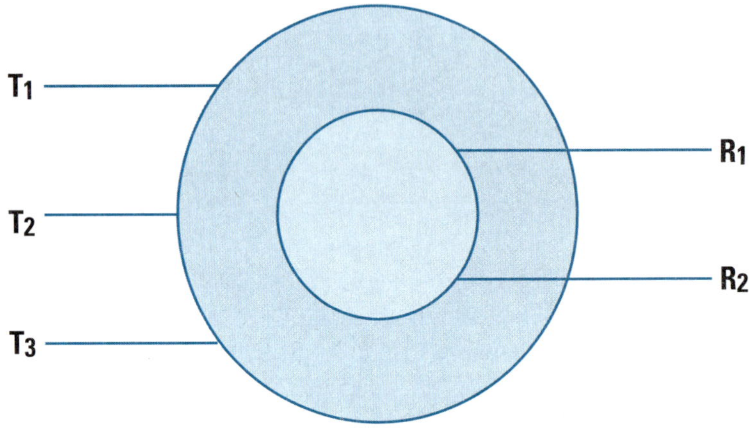

Figure 32–56 Schematic symbol for a synchronous motor.

Summary

- Three basic types of three-phase motors are
  a. squirrel-cage induction motor.
  b. wound-rotor induction motor.
  c. synchronous motor.
- All three-phase motors operate on the principle of a rotating magnetic field.
- Three factors that cause a magnetic field to rotate are
  a. the fact that the voltages of a three-phase system are 120° out of phase with each other.
  b. the fact that voltages change polarity at regular intervals.
  c. the arrangement of the stator windings.
- The speed of the rotating magnetic field is called the synchronous speed.
- Two factors that determine the synchronous speed are
  a. the number of stator poles per phase.
  b. the frequency of the applied voltage.
- The direction of rotation of any three-phase motor can be changed by reversing the connection of any two stator leads.

- The direction of rotation of a three-phase motor can be determined with a phase rotation meter before power is applied to the motor.
- Dual-voltage motors have 9 or 12 leads brought out at the terminal connection box.
- Dual-voltage motors intended for high-voltage connection have their phase windings connected in series.
- Dual-voltage motors intended for low-voltage connection have their phase windings connected in parallel.
- Motors that bring out 12 leads are generally intended for wye–delta starting.
- Three factors that determine the torque produced by an induction motor are
  a. the strength of the magnetic field of the stator.
  b. the strength of the magnetic field of the rotor.
  c. the phase angle difference between rotor and stator flux.
- Maximum torque is developed when stator and rotor flux are in phase with each other.
- The code letter on the nameplate of a squirrel-cage motor indicates the type of rotor bars used in the construction of the rotor.
- The type A rotor has the lowest starting current, highest starting torque, and poorest speed regulation of any type of squirrel-cage rotor.
- The double-squirrel-cage rotor contains two sets of squirrel-cage windings in the same rotor.
- Consequent-pole squirrel-cage motors change speed by changing the number of stator poles.
- Wound-rotor induction motors have wound rotors that contain three-phase windings.
- Wound-rotor motors have three sliprings on the rotor shaft to provide external connection to the rotor.
- Wound-rotor motors have higher starting torque per ampere and lower starting current than squirrel-cage motors of the same horsepower.
- The speed of a wound-rotor motor can be controlled by permitting resistance to remain in the rotor circuit during operation.
- Synchronous motors operate at synchronous speed.
- Synchronous motors operate at a constant speed from no load to full load.
- When load is connected to a synchronous motor, stress develops between the magnetic fields of the rotor and stator.
- Synchronous motors must have DC excitation from an external source.
- DC excitation is provided to some synchronous motors through two sliprings located on the rotor shaft, and other motors use a brushless exciter.
- Synchronous motors have the ability to produce a leading power factor by overexcitation of the DC supplied to the rotor.
- Synchronous motors have a set of type A squirrel-cage bars used for starting. This squirrel-cage winding is called the amortisseur winding.
- A field-discharge resistor is connected across the rotor winding during starting to prevent high voltage in the rotor due to induction.
- Changing the DC excitation does not affect the speed of the motor.

Review Questions

1. Three-phase motors all operate on the same basic principle, electromagnetic induction. Which of the following statements best describes the similarity among all three-phase motor types?
   A) The rotor is used to create a rotating magnetic field.
   B) Three-phase motors all produce limited power and torque.
   C) Three-phase motors all use the same basic stator design.
   D) Three-phase motors all use the same basic rotor design.
2. What is the main difference between all types of three-phase motors?
   A) The main difference is the rotor design.
   B) The main difference is the stator design.
   C) The main difference is how the field is generated.
   D) The main difference is the excitation voltage type.
3. If three-phase motors are operating based on a rotating magnetic field, which of the following is not a factor in producing this rotation?
   A) Voltage peaks are 120° out of phase with each other.
   B) Three-phase voltages change polarity regularly.
   C) The arrangement of the stator windings
   D) The rotation of the prime mover
4. True or False: The direction of rotation for any three-phase motor can be reversed by simply changing any two of the supply leads connecting to the stator.
5. When connecting a motor for a high-voltage wye connection, which of the following describes how this connection is made?
   A) Connect the motor windings in parallel
   B) Connect the motor windings in series
   C) Connect only half of the motor leads
   D) Wye connections cannot be high voltage
6. When connecting a motor for a low-voltage delta connection, which of the following best describes the connection of the motor leads?
   A) L1 to T1,T7,T6; L2 to T2,T8,T4; L3 to T3, T5, T9
   B) L1 to T1, T7; L2 to T2, T8; L3 to T3, T9; T4-T5-T6
   C) L1 to T1; L2 to T2; L3 to T3; T4-T7; T5-T8; T6-T9
   D) L1 to T1; L2 to T2; L3 to T3; T4-T5; T5-T6 T6-T7

7. For squirrel-cage motors where starting the motor is likely to be difficult, such as in flywheel or water pump applications, which of the following squirrel cage designs is most likely to be used?
   A) Type B rotor
   B) Type F rotor
   C) Type A rotor
   D) Type E rotor
8. For motors that require starting and running characteristics of varying degrees, which type of rotor is likely to be found?
   A) Type A rotor
   B) Type C rotor
   C) Type E rotor
   D) Double squirrel-cage rotor
9. True or False: The motor nameplate will specify the power factor of the individual motor.
10. Which of the following items must be found on the nameplate of the motor?
    A) Horsepower
    B) Code letter
    C) Voltage
    D) All of these
11. Which of the following is an advantage to using a wound-rotor induction motor?
    A) Lower overall starting current and higher starting torque per ampere
    B) Low starting torque
    C) Speed control
    D) Constant speed at load
12. When comparing the starting characteristics of a wound-rotor motor to those of a typical squirrel-cage motor, which statement is true with regards to the starting torque per ampere of starting current?
    A) The starting torque per ampere is lower.
    B) The starting torques per ampere is equal.
    C) The starting torque per ampere is higher.
    D) The starting torque per ampere varies.
13. What is the speed of a 6 pole synchronous motor connected to a circuit with a frequency of 60 hz?
    A) 1800 rpm
    B) 1200 rpm
    C) 3600 rpm
    D) 900 rpm

14. What is the term given to synchronous motors that operate under no load and solely in a power factor correction capacity?

    A) Power factor correction motor

    B) Synchronous correction motor

    C) Power factor synchronous correction motor

    D) Synchronous condenser

Practical Applications

1. You are working as a plant electrician. It is your job to install a 300-hp three-phase squirrel-cage induction motor. The supply voltage is 480 V. The power company has determined that the maximum amount of starting current that can be permitted by any motor in the plant is 3000 A. The motor nameplate is as follows:

Phase: 3	FLA: 352
Volts: 480	RPM: 1755
Frame: XXX	Code: L

Will it be possible to start this motor across the line, or will it be necessary to use a special process to reduce starting current?

2. You have been given the task of connecting a nine-lead three-phase dual-voltage motor to a 240-V line. You discover that the nameplate has been painted and you cannot see the connection diagram. To make the proper connections, you must know if the motor stator winding is connected wye or delta. How could you determine this using an ohmmeter? Explain your answer.

3. You are an electrician working in an industrial plant. A 30-hp three-phase squirrel-cage motor keeps tripping out on overload. The motor is connected to a 240-volt line. Voltage measurements indicate the following voltages between the different phases: A–B = 276 volts, B–C = 221 volts, and A–C = 267 volts. Determine the percentage of heat rise in the winding with the highest current.

Unit 33 Single-Phase Motors

Key Terms

Centrifugal switch
Compensating winding
Conductive compensation
Consequent-pole motor
Encoder
Holtz motor
Inductive compensation
Multispeed motors
Neutral plane
Repulsion motor
Run winding
Servo motor
Shaded-pole induction motor
Shading coil
Split-phase motors
Start winding
Stepping motors
Synchronous motors
Torque motor
Two-phase
Universal motor
Warren motor

Why You Need to Know

Single-phase motors are used mainly in residential applications and rarely in industrial and commercial locations with the exception of smaller fractional horsepower applications. Unlike three-phase motors, there are many different types of single-phase motors, and they do not all operate on the same principle. There are some that operate on the principle of a rotating magnetic field, but others do not. Some single-phase motors are designed to operate at more than one speed, while others have more constant speed applications. This unit

- presents several different types of single-phase motors and explains how they operate.
- explains how to determine the appropriate motor to be used in a given situation by evaluating the operating principles of each.

Outline

Objectives

After studying this unit, you should be able to

- describe the operating principles of single-phase motors and three-phase motors.
- explain how the different split-phase motor connections affect the starting and running characteristics of motors.
- identify key factors in the operation of shaded-pole and multispeed motors.
- describe the operation of repulsion-type motors.
- explain the operation of single-phase synchronous motors and three-phase synchronous motors.
- describe the basic operating principles of stepping and universal motors.

Preview

Although most of the large motors used in industrial settings are three-phase, sometimes, single-phase motors must be used. Single-phase motors are used almost exclusively to operate home appliances such as air conditioners, refrigerators, well pumps, and fans. They are generally designed to operate on 120 volts or 240 volts. They range in size from fractional horsepower to several horsepower, depending on the application.

33–1 Single-Phase Motors

In Unit 32, it was stated that there are three basic types of three-phase motors and that all operate on the principle of a rotating magnetic field. This is not true for single-phase motors. There are not only many different types of single-phase motors, but they also have different operating principles.

Split-Phase Motors

Split-phase motors fall into the following three general classifications:

1. The resistance-start induction-run motor
2. The capacitor-start induction-run motor
3. The capacitor-start capacitor-run, or permanent split capacitor, motor

Although all these motors have different operating characteristics, they are similar in construction and use the same operating principle. Split-phase motors receive their name from the manner in which they operate. Like three-phase motors, split-phase motors operate on the principle of a rotating magnetic field. A rotating magnetic field, however, cannot be produced with only one phase. Split-phase motors therefore split the current flow through two separate windings to simulate a **two-phase** power system. A rotating magnetic field can be produced with a two-phase system.

The voltages of a two-phase system are 90° out of phase with each other. These two out-of-phase voltages can be used to produce a rotating magnetic field in a manner similar to a three-phase system. Because two voltages or currents are out of phase with each other to produce a rotating magnetic field, split-phase motors use two separate windings to create a phase difference between the currents in each of these windings. These motors literally split one phase and produce a second phase, hence the name split-phase motor.

Stator Windings

The stator of a split-phase motor contains two separate windings, the **start winding** and the **run winding**. The start winding is made of small wire and is placed near the top of the stator core. The run winding is made of relatively large wire and is placed in the bottom of the stator core. *Figures 33–1A* and *B* are photographs of two split-phase stators. The stator in *Figure 33–1A* is used for a resistance-start induction-run motor or a capacitor-start induction-run motor. The stator in *Figure 33–1B* is used for a capacitor-start capacitor-run motor. Both stators contain four poles, and the start winding is placed at a 90° angle from the run winding.

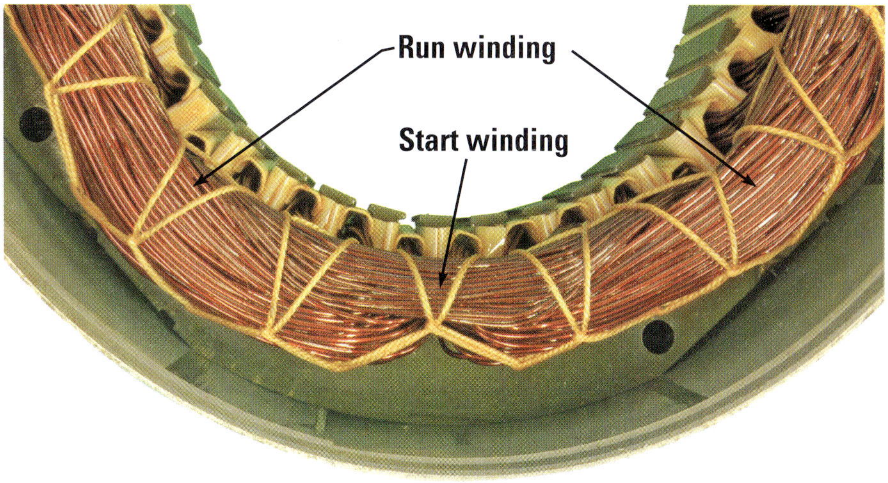

Figure 33–1A Notice that the start winding contains much smaller wire than the run winding in a resistance-start induction-run stator.

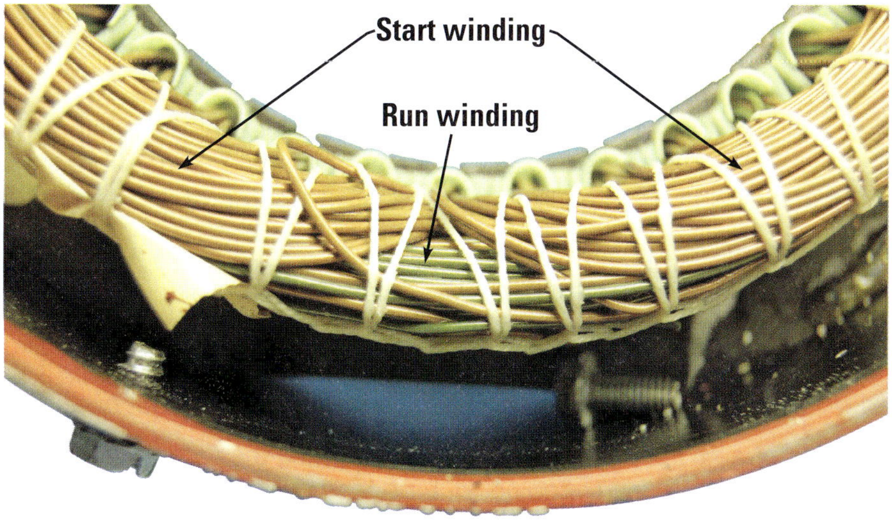

Figure 33–1B The wire size is the same for both start and run windings in a capacitor-start capacitor-run stator.

Notice the difference in size and position of the two windings of the stator shown in *Figure 33–1A*. The start winding is made from small wire and is placed near the top of the stator core. This causes it to have a higher resistance than the run winding. The start winding is located between the poles of the run winding. The run winding is made with larger wire and is placed near the bottom of the core. This gives it higher inductive reactance and less resistance than the start winding. These two windings are connected in parallel with each other *(Figure 33–2)*.

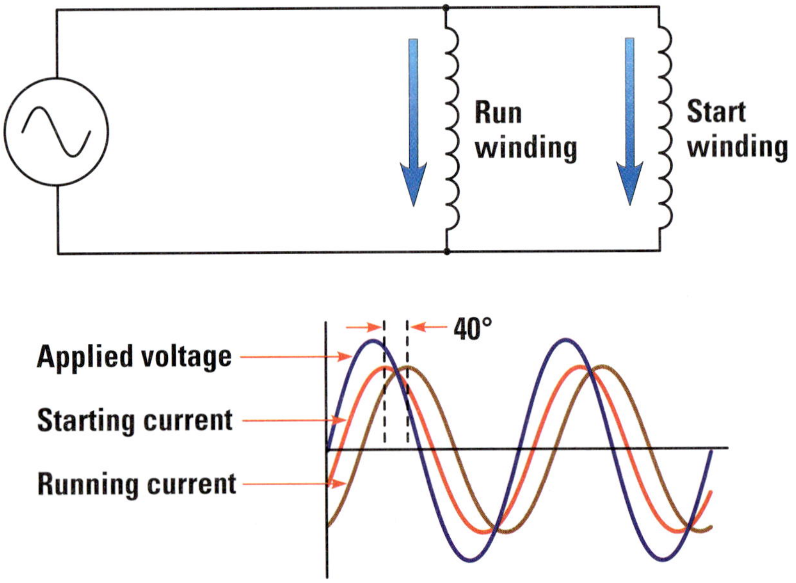

Figure 33–2 Start and run windings are connected in parallel with each other.

When power is applied to the stator, current flows through both windings. Because the start winding is more resistive, the current flow through it is more in phase with the applied voltage than the current flow through the run winding. The current flow through the run winding lags the applied voltage due to inductive reactance. These two out-of-phase currents are used to create a rotating magnetic field in the stator. The rotating field moves at synchronous speed, which can be determined in the same manner as three-phase motors.

33–2 Split-Phase Motor Connections

The resistance-start induction-run motor receives its name from the fact that the out-of-phase condition between start and run winding current is caused by the start winding being more resistive than the run winding. The amount of starting torque produced by a split-phase motor is determined by the following three factors:

1. The strength of the magnetic field of the stator
2. The strength of the magnetic field of the rotor
3. The phase angle difference between current in the start winding and current in the run winding (Greater torque is produced when these two currents are nearly 90° out of phase with each other.)

Although these two currents are out of phase with each other, they are not 90° out of phase. The run winding is more inductive than the start winding, but it does have some resistance, which prevents the current from being truly 90° out of phase with the voltage. The start winding is more resistive than the run winding, but it does have some inductive reactance, preventing the current from being in phase with the applied voltage. Therefore, a phase angle difference of 35 to 40° is produced between these two currents, resulting in a rather poor starting torque *(Figure 33–3)*.

Disconnecting the Start Winding

A stator rotating magnetic field is necessary only to start the rotor turning. Once the rotor has accelerated to approximately 75% of rated speed, the start winding can be disconnected from the circuit and the motor will continue to operate with only the run winding energized. Motors often use a **centrifugal switch** to disconnect the start windings from the circuit. The contacts of the centrifugal switch are connected in series with the start winding *(Figure 33–4)*. The centrifugal switch contains a set

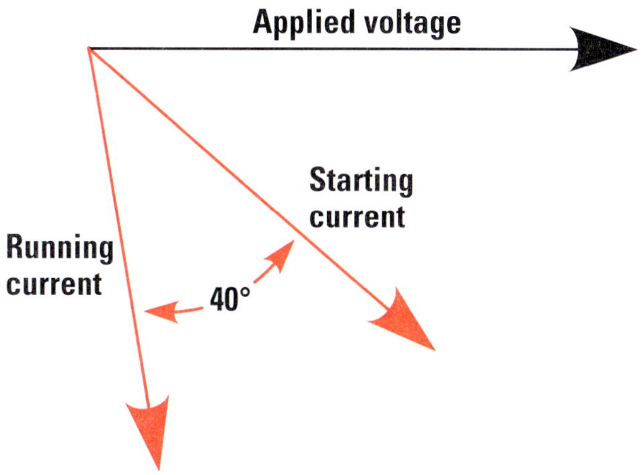

Figure 33–3 Running current and starting current are often around 35 to 40° out of phase with each other.

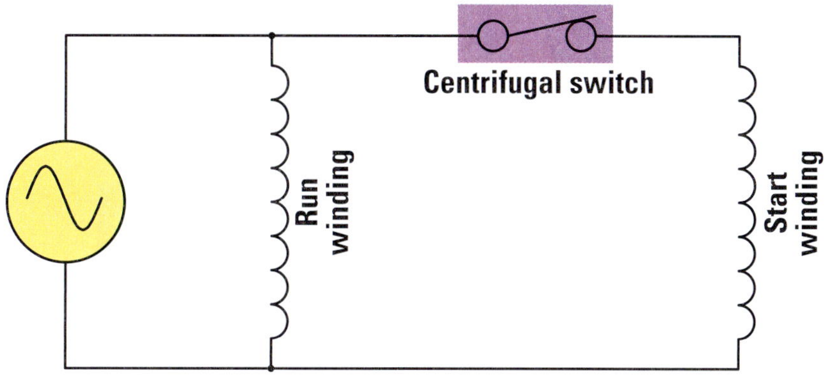

Figure 33–4 A centrifugal switch is used to disconnect the start winding from the circuit.

of spring-loaded weights. When the shaft is not turning, the springs hold a fiber washer in contact with the movable contact of the switch *(Figure 33–5)*. The fiber washer causes the movable contact to complete a circuit with a stationary contact.

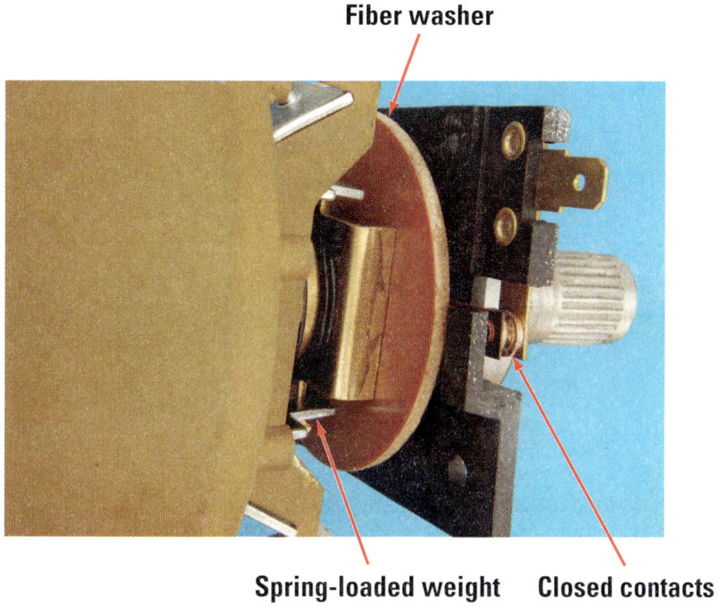

Figure 33–5 The centrifugal switch is closed when the rotor is not turning.

When the rotor accelerates to about 75% of rated speed, centrifugal force causes the weights to overcome the force of the springs. The fiber washer retracts and permits the contacts to open and disconnect the start winding from the circuit *(Figure 33–6)*. The start winding of this type of motor is intended to be energized only during the period of time that the motor is actually starting. If the start winding is not disconnected, it will be damaged by excessive current flow.

Figure 33–6 The contact opens when the rotor reaches about 75% of rated speed.

Starting Relays

Resistance-start induction-run and capacitor-start induction-run motors are sometimes hermetically sealed, such as in air-conditioning and refrigeration compressors. When these motors are hermetically sealed, a centrifugal switch cannot be used to disconnect the start winding. A starting relay that can be mounted externally must be used to disconnect the start windings from the circuit. There are three basic types of starting relays used with the resistance-start and capacitor-start motors as follows:

1. Hot-wire relay
2. Current relay
3. Solid-state starting relay

The *hot-wire relay, often referred to as a bimetallic switch,* functions as both a starting relay and an overload relay. In the circuit shown in *Figure 33–7*, a thermostat controls the operation of the motor. When the thermostat closes, current flows through a resistive wire and two normally closed contacts connected to the start and run windings of the motor. The high starting current of the motor rapidly heats the resistive wire, causing it to expand. The expansion of the wire causes the spring-loaded start winding contact to open and disconnect the start winding from the circuit, reducing motor current. If the motor should become overloaded, the resistive wire also expands enough to open the overload contact, disconnect the run winding, and turn off the motor.

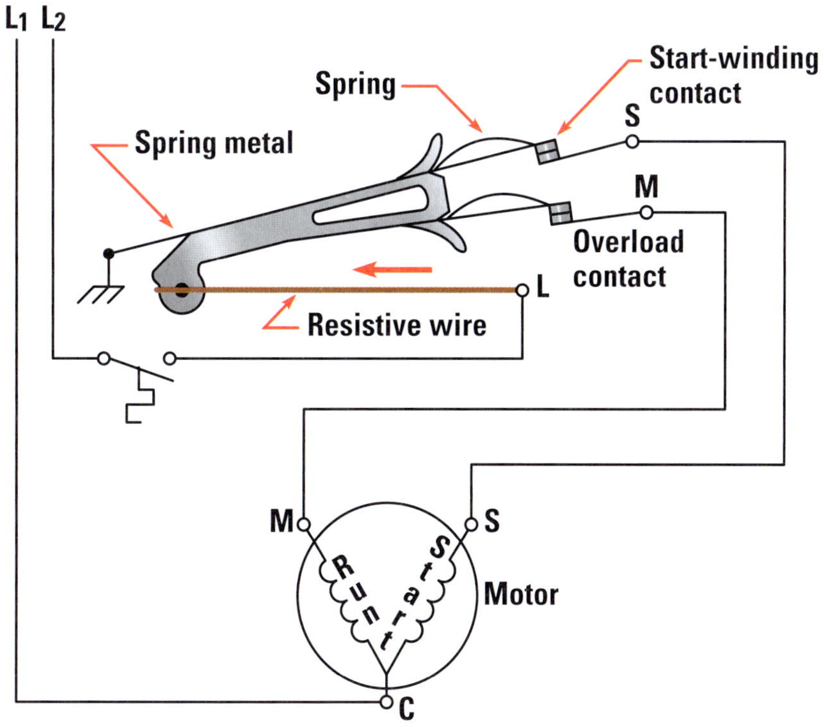

Figure 33–7 Hot-wire relay connection for control of start-winding and thermal overload protection of a motor.

A *current relay* is another type of start-winding control device that operates on the principle of a magnetic field instead of expanding metal. The current relay contains a coil with a few turns of large wire and a set of normally open contacts. The coil of the relay is connected in series with the run winding of the motor, and the contacts are connected in series with the start winding *(Figure 33–8)*. When the thermostat contact closes, power is applied to the run winding of the motor. Because the start winding is open, the motor cannot start, causing a high current to flow in the run winding circuit. This high current flow produces a strong magnetic field in the coil of the relay, causing the normally open contacts to close and start the motor. When the motor starts, the run-winding current is greatly reduced, permitting the start contacts to reopen and disconnect the start winding from the circuit.

The *solid-state starting relay* performs the same basic function as the current relay and is the most common type of control device used. The solid-state starting relay is generally more reliable and less expensive than the current relay or the centrifugal switch. The solid-state starting relay is actually an electronic component known as a *thermistor*. A thermistor is a device that exhibits a change of resistance with a change of temperature. This particular thermistor has a positive coefficient of temperature, which means that when its temperature increases, its resistance increases also. The schematic diagram in *Figure 33–9* illustrates the connection of the solid-state starting relay. The thermistor is connected in series with the start winding of the motor. When the motor is not in operation, the thermistor is at a low temperature and its resistance is low, typically 3 or 4 ohms. When power is applied, current flows to both the run and start windings of the motor. The current flowing through the thermistor causes an increase

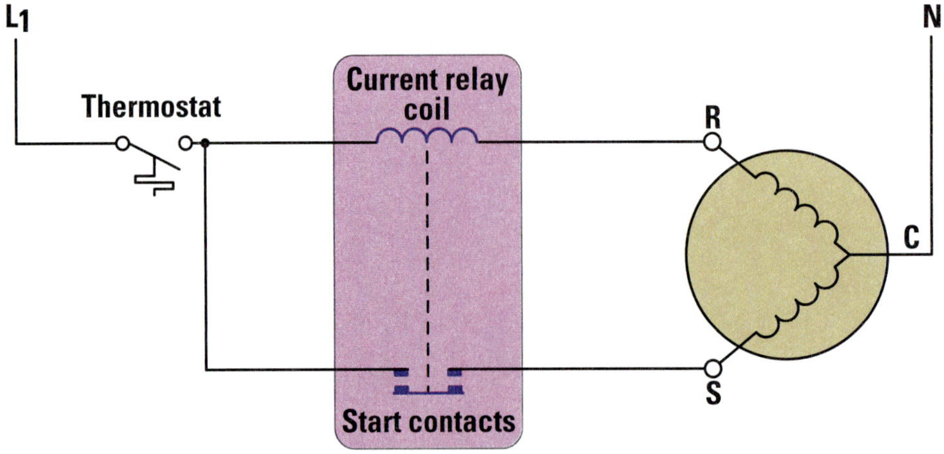

Figure 33–8 Current relay connection for control of start winding in a split-phase motor.

in temperature. This increased temperature causes the resistance of the thermistor to suddenly change to a high value of several thousand ohms. The change of temperature is so sudden that it has the effect of opening a set of contacts. Although the start winding is never completely disconnected from the power line, the amount of current flow though it is very small, typically 0.03 to 0.05 amperes, and does not affect the operation of the motor. This small amount of *leakage current* maintains the temperature of the thermistor and prevents it from returning to a low value of resistance. After the motor is disconnected from the powerline, a cooldown time of 2 to 3 minutes should be allowed to permit the thermistor to return to a low resistance before the motor is restarted.

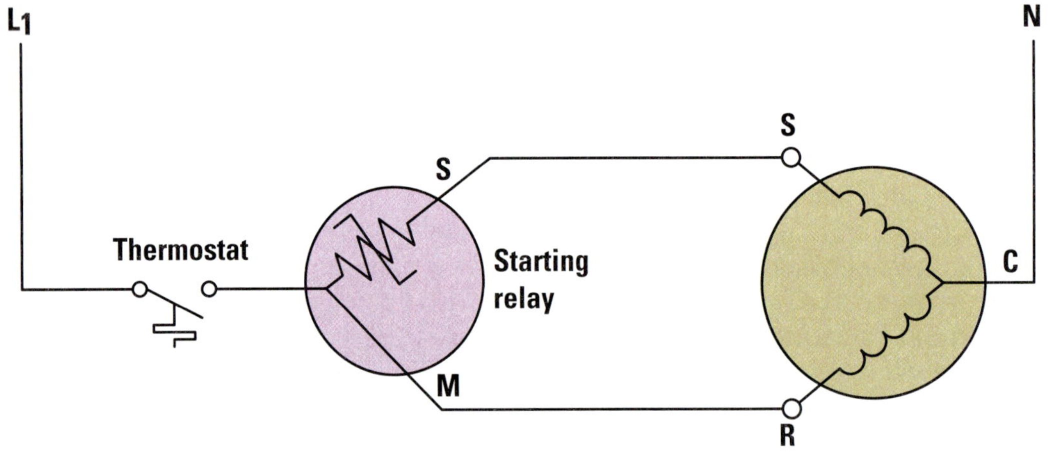

Figure 33–9 Solid-state starting relay connection for control of the start winding.

Relationship of Stator and Rotor Fields

The split-phase motor contains a squirrel-cage rotor very similar to those used with three-phase squirrel-cage motors *(Figure 33–10)*. When power is connected to the stator windings, the rotating magnetic field induces a voltage into the bars of the squirrel-cage rotor. The induced voltage causes current to flow in the rotor, and a magnetic field is produced around the rotor bars. The magnetic field of the rotor is attracted to the stator field, and the rotor begins to turn in the direction of the rotating magnetic field. After the centrifugal switch opens, only the run winding induces voltage into the rotor. This induced voltage is in phase with the stator current. The inductive reactance of the rotor is high, causing the rotor current to be almost 90° out of phase with the induced voltage. This causes the pulsating magnetic field of the rotor to lag the pulsating magnetic field of the stator by 90°. Magnetic poles, located midway between the stator poles, are created in the rotor *(Figure 33–11)*. These two pulsating magnetic fields produce a rotating magnetic field of their own, and the rotor continues to rotate.

Figure 33–10 Squirrel-cage rotor used in a split-phase motor.

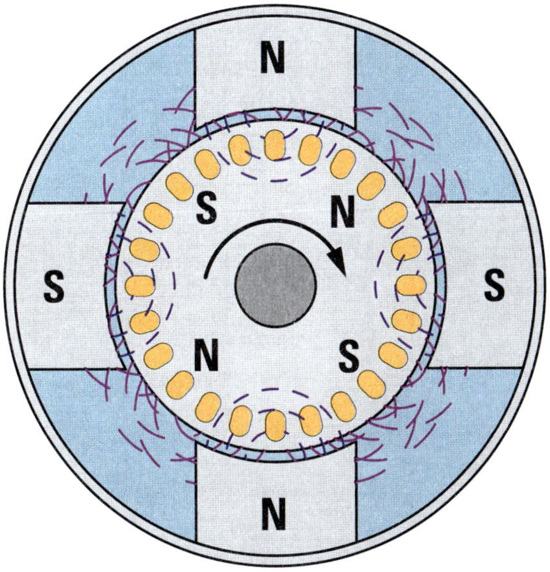

Figure 33–11 A rotating magnetic field is produced by the stator and rotor flux.

Direction of Rotation

The direction of rotation for the motor is determined by the direction of rotation of the rotating magnetic field created by the run and start windings when the motor is first started. The direction of motor rotation can be changed by reversing the connection of either the start winding or the run winding, but not both.

Capacitor-Start Induction-Run Motors

The capacitor-start induction-run motor is very similar in construction and operation to the resistance-start induction-run motor. The capacitor-start induction-run motor, however, has an AC electrolytic capacitor connected in series with the centrifugal switch and start winding *(Figure 33–12)*. The capacitor-start induction-run motor produces a starting torque that is substantially higher than that of the resistance-start induction-run motor. Recall that one of the factors that determines the starting torque for a split-phase motor is the phase angle difference between the start-winding current and run-winding current. The starting torque of a resistance-start induction-run motor is low because the phase angle difference between these two currents is only about 40° *(Figure 33–3)*.

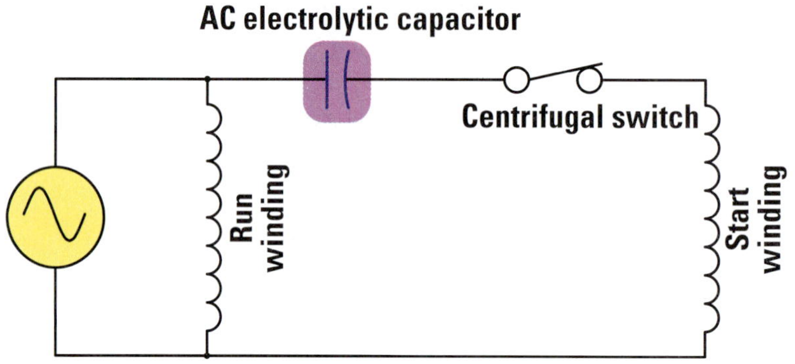

Figure 33–12 An AC electrolytic capacitor is connected in series with the start winding.

When a capacitor is connected in series with the start winding, it causes the start-winding current to lead the applied voltage. This leading current produces a 90° phase shift between the run-winding current and start-winding current *(Figure 33–13)*. Maximum starting torque is developed at this point.

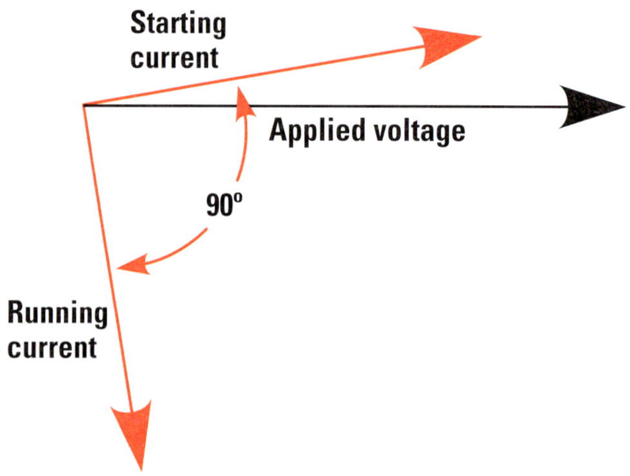

Figure 33–13 Run-winding current and start-winding current are 90° out of phase with each other.

Although the capacitor-start induction-run motor has a high starting torque, the motor should not be started more than about eight times per hour. Frequent starting can overheat the start capacitor. If the capacitor must be replaced, care should be taken to use a capacitor of the correct microfarad rating. A capacitor-start induction-run motor is shown in *Figure 33–14*.

Figure 33–14 Capacitor-start induction-run motor.

Dual-Voltage Split-Phase Motors

Many split-phase motors are designed for operation on 120 or 240 volts. *Figure 33–15* shows the schematic diagram of a split-phase motor designed for dual-voltage operation. These motors often contain two run windings and two start windings. Single-phase motor termination leads are numbered in a manner similar to three-phase motors. One of the run windings has lead numbers of T_1 and T_2. The other run winding has its leads numbered T_3 and T_4. One set of start windings is labeled T_5 and T_6, and the other set is labeled T_7 and T_8.

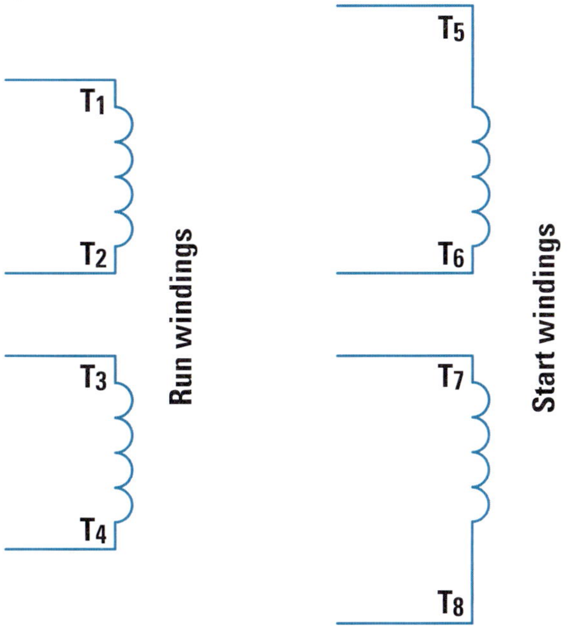

Figure 33–15 Dual-voltage windings for a split-phase motor.

If the motor is to be connected for high-voltage operation, the run windings and start windings are connected in series, as shown in *Figure 33–16*. The start windings are then connected in parallel with the run windings. If the opposite direction of rotation is desired, T_5 and T_8 are changed.

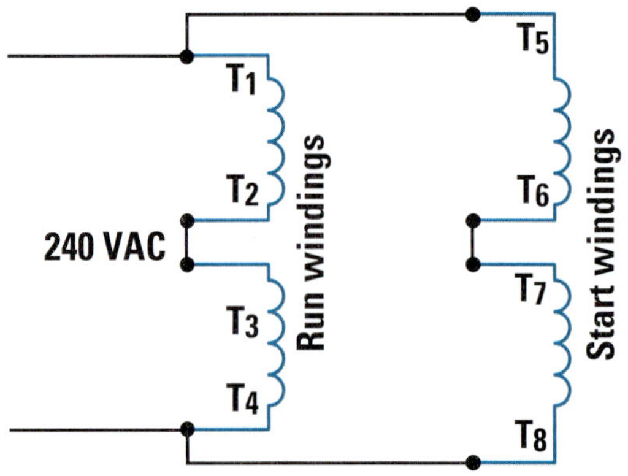

Figure 33–16 High-voltage connection for a split-phase motor with two run and two start windings.

For low-voltage operation, the windings must be connected in parallel, as shown in *Figure 33–17*. This connection is made by first connecting the run windings in parallel by hooking T_1 and T_3 together and T_2 and T_4 together. The start windings are paralleled by connecting T_5 and T_7 together and T_6 and T_8 together. The start windings are then connected in parallel with the run windings. If the opposite direction of rotation is desired, T_5 and T_7 remain connected together as do T_6 and T_8. The combined connection of T_5 and T_7 is reversed with the combined connection of T_6 and T_8.

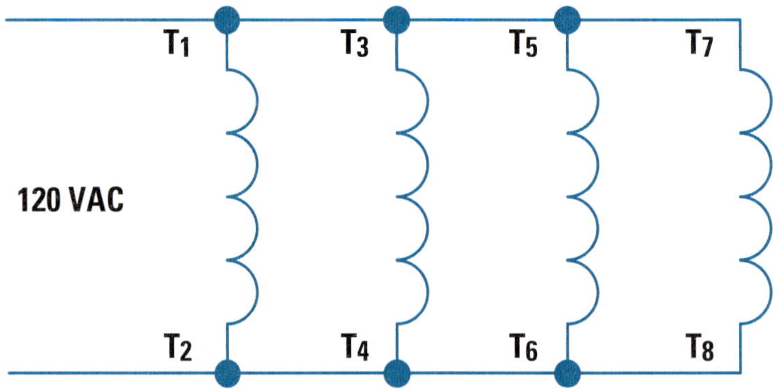

Figure 33–17 Low-voltage connection for a split-phase motor with two run and two start windings.

Not all dual-voltage single-phase motors contain two sets of start windings. *Figure 33–18* shows the schematic diagram of a motor that contains two sets of run windings and only one start winding. In this illustration, the start winding is labeled T_5 and T_6. Some motors, however, identify the start winding by labeling it T_5 and T_8.

Regardless of which method is used to label the terminal leads of the start winding, the connection is the same. If the motor is to be connected for high-voltage operation, the run windings are connected in series, and the start winding is connected in parallel with one of the run windings *(Figure 33–19)*. In this type of motor, each winding is rated at 120 volts. If the run windings are connected in series across 240 volts, each winding has a voltage drop of 120 volts. By connecting the start winding in parallel across

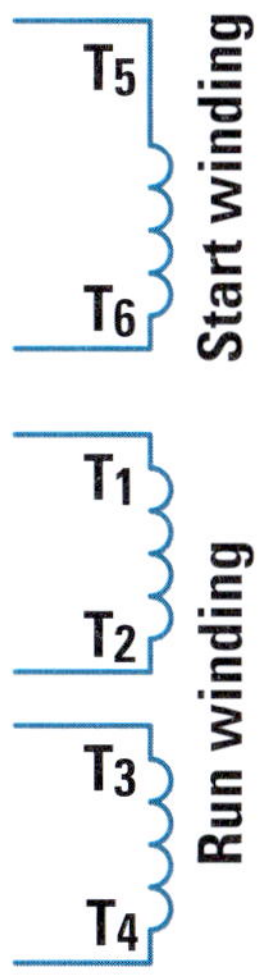

Figure 33–18 Dual-voltage motor with one start winding labeled T_5 and T_6.

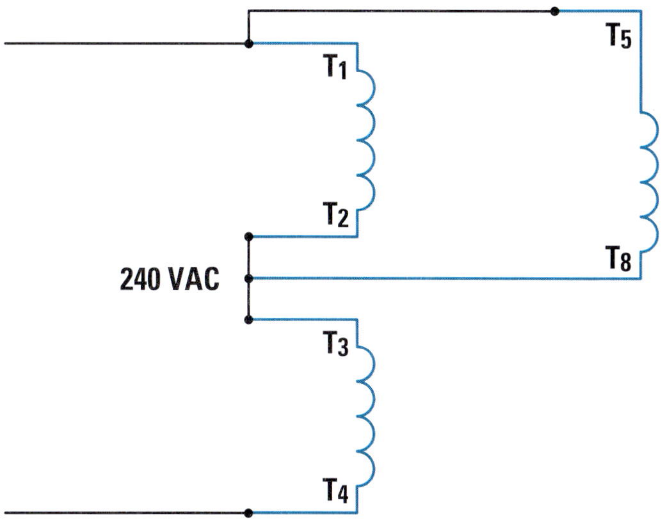

Figure 33–19 High-voltage connection with one start winding.

only one run winding, it receives only 120 volts when power is applied to the motor. If the opposite direction of rotation is desired, T_5 and T_8 should be reversed.

If the motor is to be operated on low voltage, the windings are connected in parallel as shown in *Figure 33–20*. Because all windings are connected in parallel, each receives 120 volts when power is applied to the motor.

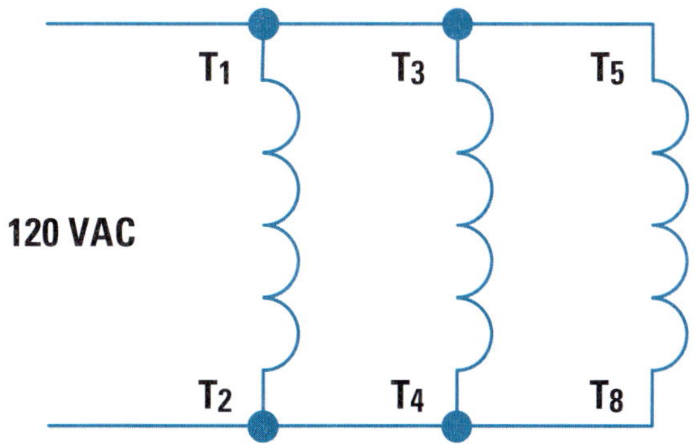

Figure 33–20 Low-voltage connection for a split-phase motor with one start winding.

Determining the Direction of Rotation for Split-Phase Motors

The direction of rotation of a single-phase motor can generally be determined when the motor is connected. The direction of rotation is determined by facing the back or rear of the motor. *Figure 33–21* shows a connection diagram for rotation. If clockwise rotation is desired, T_5 should be connected to T_1. If counterclockwise rotation is desired, T_8 (or T_6) should be connected to T_1. However, the type of motor used determines the actual connection. For example, *Figure 33–19* shows the connection of a motor with two run windings and only one start winding. If this motor were to be connected for clockwise rotation, terminal T_5 would have to be connected to T_1, and terminal T_8 would have to be connected to T_2 and T_3. If counterclockwise rotation is desired, terminal T_8 would have to be connected to T_1, and terminal T_5 would have to be connected to T_2 and T_3.

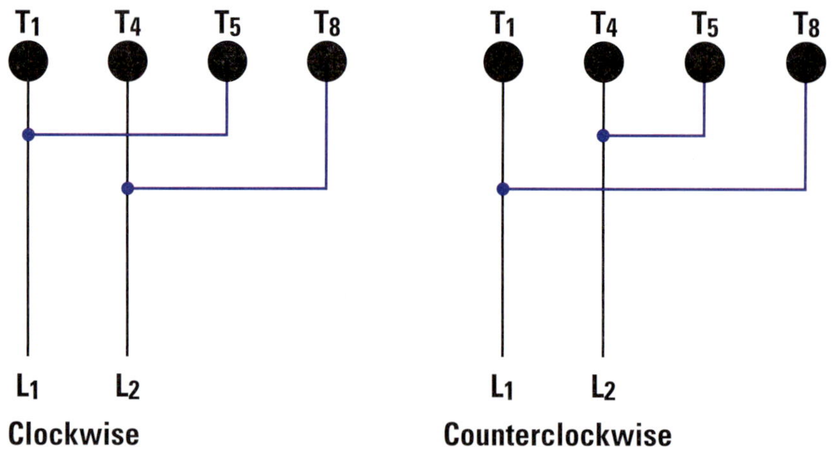

Figure 33–21 Determining direction of rotation for a split-phase motor.

Capacitor-Start Capacitor-Run, or Permanent Split Capacitor, Motors

Although the capacitor-start capacitor-run motor, also often called the permanent split capacitor motor, is a split-phase motor, it operates on a different principle than the resistance-start induction-run motor or the capacitor-start induction-run motor. The capacitor-start capacitor-run motor is designed in such a manner that its start winding remains energized at all times. A capacitor is connected in series with the winding to provide a continuous leading current in the start winding *(Figure 33–22)*. Because the start winding remains energized at all times, no centrifugal switch is needed to disconnect the start winding as the motor approaches full speed. The capacitor used in this type of motor is generally of the oil-filled type because it is intended for continuous use.

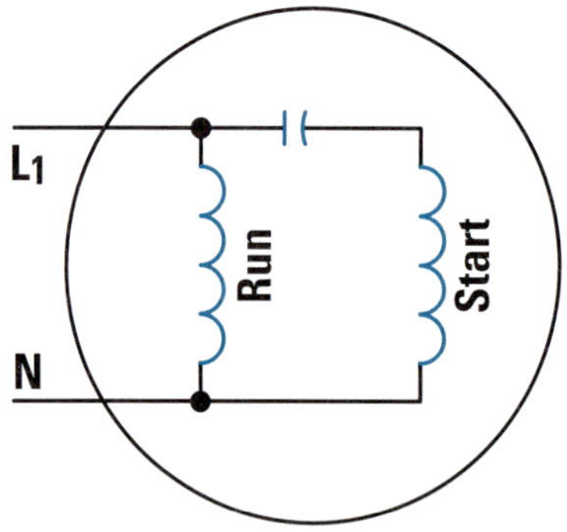

Figure 33–22 A capacitor-start capacitor-run, or permanent split capacitor, motor.

The capacitor-start capacitor-run motor actually operates on the principle of a rotating magnetic field in the stator. Because both run and start windings remain energized at all times, the stator magnetic field continues to rotate, and the motor operates as a two-phase motor. This motor has excellent starting and running torque. It is quiet in operation and has a high efficiency. Because the capacitor remains connected in the circuit at all times, the motor power factor is close to unity.

Although the capacitor-start capacitor-run motor does not require a centrifugal switch to disconnect the capacitor from the start winding, some motors use a second capacitor during the starting period to help improve starting torque *(Figure 33–23)*. A good example of this can be found on the compressor of a central air-conditioning unit designed for operation on single-phase power. If the motor is not hermetically sealed, a centrifugal switch will be used to disconnect the start capacitor from the circuit when the motor reaches approximately 75% of rated speed. Hermetically sealed motors, however, must use some type of external switch to disconnect the start capacitor from the circuit.

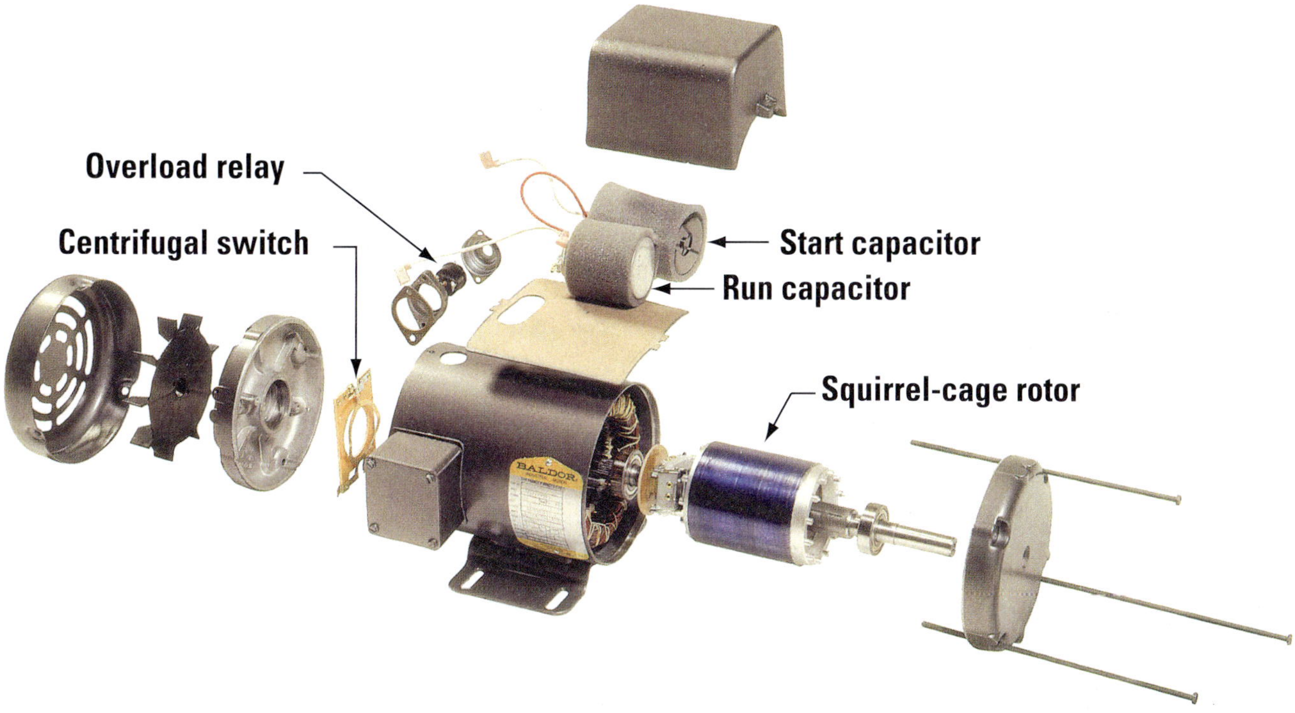

Courtesy of Baldor Electric Company

Figure 33–23 Capacitor-start capacitor-run motor, or permanent split capacitor motor, with additional starting capacitor.

The capacitor-start capacitor-run motor, or permanent split-capacitor motor, a common motor found in the air-conditioning and refrigeration industry, generally employs an external starting relay to disconnect the starting capacitor when a centrifugal switch cannot be used. The starting relay operates by sensing an increase in the voltage developed in the start winding when the motor is operating. A schematic diagram of a starting relay circuit is shown in *Figure 33–24*. In this circuit, the relay is used to disconnect the starting capacitor from the circuit when the motor reaches about 75% of its full speed. The starting-relay coil, SR, is connected in parallel with the start winding of the motor, and a normally closed SR contact is connected in series with the starting capacitor.

As the rotor begins to turn, its magnetic field induces a voltage into the start winding, producing a higher voltage across the start winding than the applied voltage. When the motor has accelerated to about 75% of its full speed, the voltage across the start winding is high enough to energize the coil. This causes the normally closed SR contact to open and disconnect the start capacitor from the circuit. Because the start winding of this motor is never disconnected from the powerline, the coil of the potential starting relay remains energized as long as the motor is in operation.

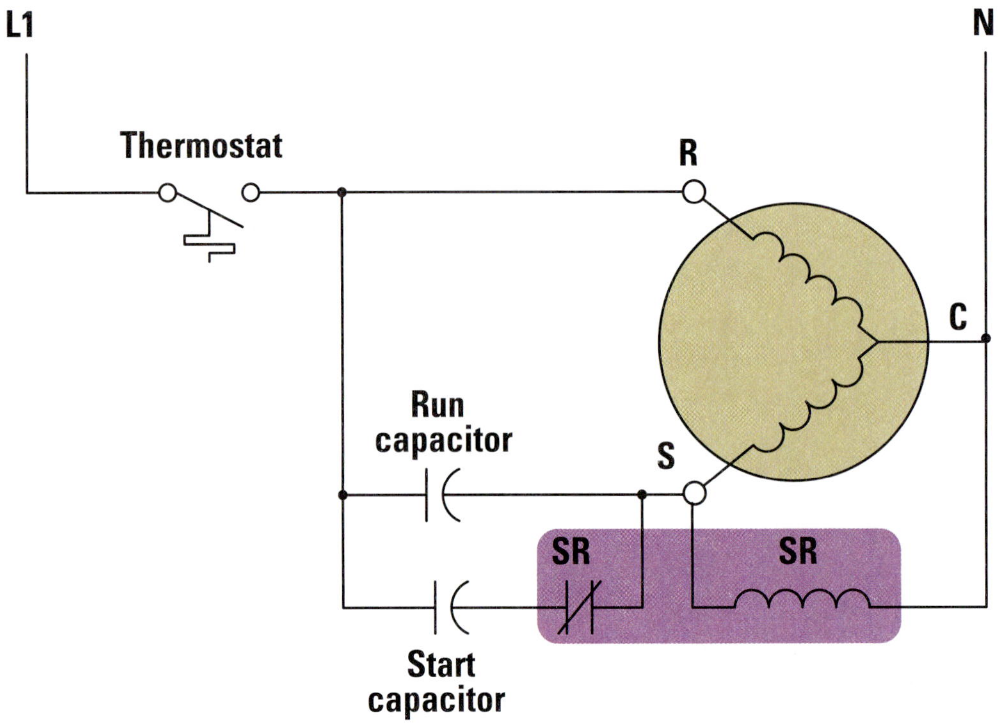

Figure 33–24 Potential relay connection for control of capacitor start winding.

33–3 Shaded-Pole Induction and Multispeed Motors

The **shaded-pole induction motor** is popular because of its simplicity and long life. This motor contains no start windings or centrifugal switch. It contains a squirrel-cage rotor and operates on the principle of a rotating magnetic field. The rotating magnetic field is created by a **shading coil** wound on one side of each pole piece. Shaded-pole motors are generally fractional-horsepower motors and are used for low-torque applications such as operating fans and blowers.

The Shading Coil

The shading coil is wound around one end of the pole piece *(Figure 33–25)* and is actually a large loop of copper wire or a copper band. The two ends are connected to form a complete circuit. The shading coil acts in the same manner as a transformer with a shorted secondary winding. When the current of the AC waveform increases from zero toward its positive peak, magnetic lines of flux cut through the shading coil and a voltage is induced in the coil. Because the coil is a low-resistance short circuit, a large amount of current flows in the loop. This current causes an opposition to the change of magnetic flux *(Figure 33–26)*.

When the AC reaches its peak value, it is no longer changing, and no voltage is being induced into the shading coil. Because there is no current flow in the shading coil, there is no opposition to the magnetic flux. The magnetic flux of the pole piece is now uniform across the pole face *(Figure 33–27)*.

When the AC begins to decrease from its peak value back toward zero, the magnetic field of the pole piece begins to collapse. A voltage is again induced into the shading coil. This induced voltage creates a current that opposes the change of magnetic flux *(Figure 33–28)*. This causes the magnetic flux to be concentrated in the shaded section of the pole piece.

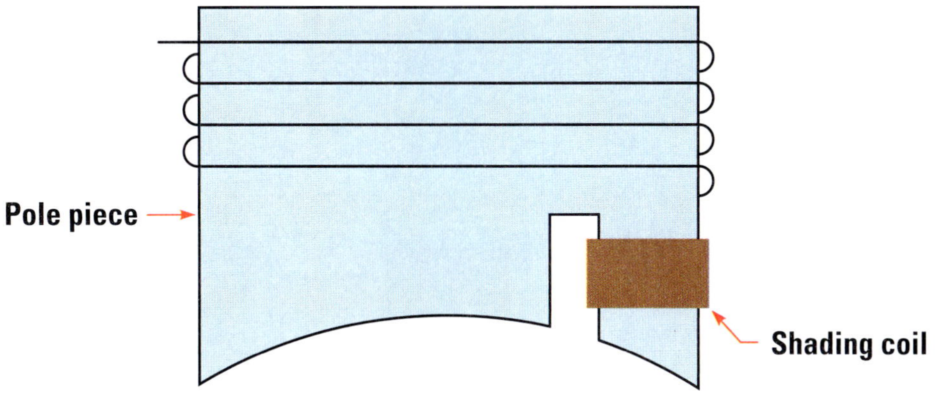

Figure 33–25 A shaded pole piece has a coil around one end of the pole.

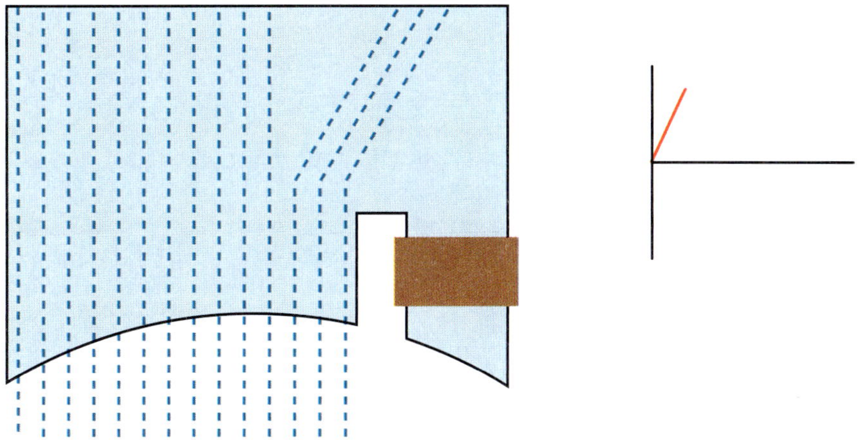

Figure 33–26 The shading coil opposes a change of flux as current increases.

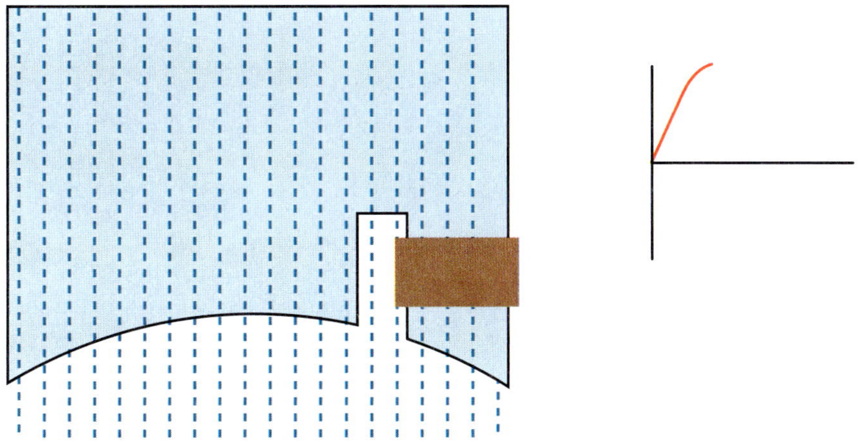

Figure 33–27 There is opposition to magnetic flux when the current is not changing.

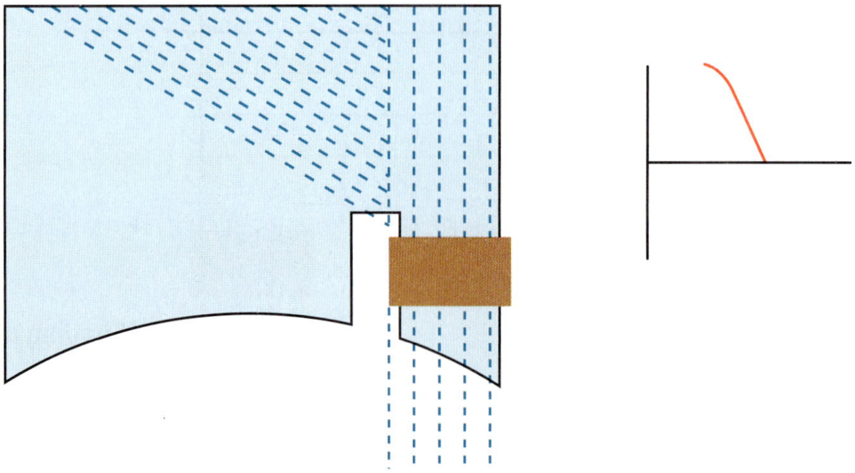

Figure 33–28 The shading coil opposes a change of flux when the current decreases.

When the AC passes through zero and begins to increase in the negative direction, the same set of events happens except that the polarity of the magnetic field is reversed. If these events were to be viewed in rapid order, the magnetic field would be seen to rotate across the face of the pole piece. A pole piece with a shading coil is shown in *Figure 33–29*.

Figure 33–29 The shading coil is a large copper conductor wound around one side of the pole piece.

Speed

The speed of the shaded-pole induction motor is determined by the same factors that determine the synchronous speed of other induction motors: frequency and number of stator poles. Shaded-pole motors are commonly wound as four- or six-pole motors. *Figure 33–30* shows a drawing of a four-pole shaded-pole induction motor.

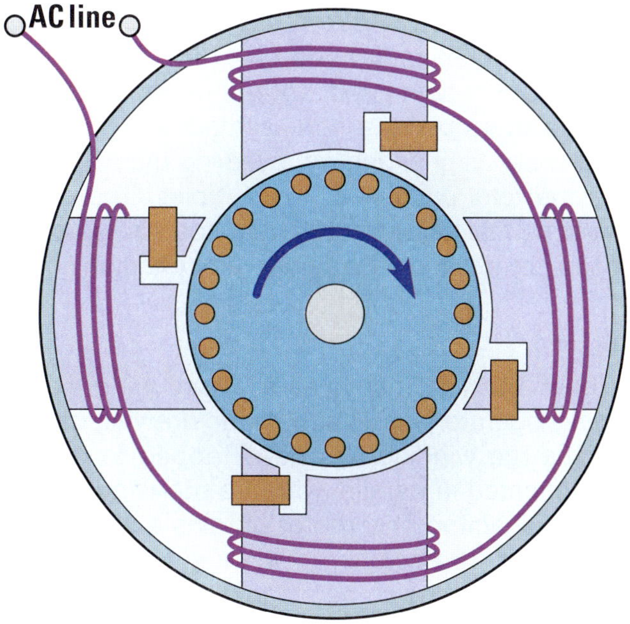

Figure 33–30 Four-pole shaded-pole induction motor.

General Operating Characteristics

The shaded-pole motor contains a standard squirrel-cage rotor. The amount of torque produced is determined by the strength of the magnetic field of the stator, the strength of the magnetic field of the rotor, and the phase angle difference between rotor and stator flux. The shaded-pole induction motor has low starting and running torque.

The direction of rotation is determined by the direction in which the rotating magnetic field moves across the pole face. The rotor turns in the direction shown by the arrow in *Figure 33–30*. It is impossible to change the direction of these motors without modifying the stator windings. Therefore, they are considered to be nonreversible. *Figure 33–31* shows a photograph of the stator winding and rotor of a shaded-pole induction motor.

Figure 33–31 Stator winding and rotor of a shaded-pole induction motor.

Multispeed Motors

There are two basic types of single-phase **multispeed motors**. One is the consequent-pole type, and the other is a specially wound *capacitor-start* capacitor-run motor or shaded-pole induction motor. The single-phase **consequent-pole motor** operates in the same basic way as the three-phase consequent pole discussed in Unit 32. The speed is changed by reversing the current flow through alternate poles and increasing or decreasing the total number of stator poles. The consequent-pole motor is used where high running torque must be maintained at different speeds. A good example of where this type of motor is used is in two-speed compressors for central air-conditioning units.

Multispeed Fan Motors

Multispeed fan motors have been used for many years. These motors are generally wound for two to five steps of speed and typically operate fans. A schematic drawing of a three-speed motor is shown in *Figure 33–32*. Notice that the run winding has been tapped to produce low, medium, and high speed. The start winding is connected in parallel with the run-winding section. The other end of the start winding is connected to a capacitor. This motor obtains a change of speed by inserting inductance in series with the run winding. The actual run winding for this motor is between the terminals marked *high* and *common*. The winding shown between *high* and *medium* is connected in series with the main run winding. When the rotary switch is connected to the medium speed position, the inductive reactance of this coil limits the amount of current flow through the run winding. When the current of the run winding is reduced, the strength of the magnetic field of the run winding is reduced and the motor produces less torque. This causes a greater amount of slip and causes the motor speed to decrease.

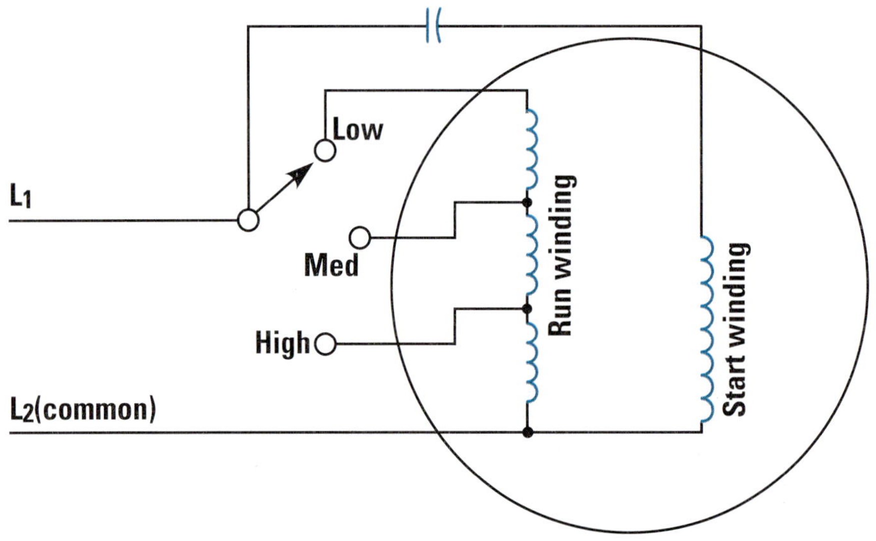

Figure 33–32 Schematic drawing of a three-speed motor.

Common speeds for a four-pole motor of this type are 1625, 1500, and 1350 RPM. Notice that this motor does not have wide ranges between speeds as would be the case with a consequent-pole motor. Most induction motors would overheat and damage the motor winding if the speed were reduced to this extent. This type of motor, however, has much higher impedance windings than most other motors. The run windings of most split-phase motors have a wire resistance of 1 to 4 ohms. This motor generally has a resistance of 10 to 15 ohms in its run winding. It is the high impedance of the windings that permits the motor to be operated in this manner without damage.

33–4 Repulsion-Type Motors

The following are three basic repulsion-type motors:

1. The repulsion motor
2. The repulsion-start induction-run motor
3. The repulsion-induction motor

Each of these three types has different operating characteristics.

A **repulsion motor** operates on the principle that like magnetic poles repel each other. The stator of a repulsion motor contains only a run winding very similar to that used in the split-phase motor. The rotor is called an armature because it contains a slotted metal core with windings placed in the slots. The windings are connected to a commutator. A set of brushes makes contact with the surface of the commutator bars. The entire assembly looks very much like a DC armature and brush assembly. One difference, however, is that the brushes of the repulsion motor are shorted together. Their function is to provide a current path through certain parts of the armature, not to provide power to the armature from an external source.

Operation

When AC power is connected to the stator winding, a magnetic field with alternating polarities is produced in the poles. This alternating field induces a voltage into the windings of the armature. When the brushes are placed in the proper position, current flows through the armature windings, producing a magnetic field of the same polarity in the armature. The armature magnetic field is repelled by the stator magnetic field, causing the armature to rotate. Repulsion motors contain the same number of brushes as there are stator poles. Repulsion motors are commonly wound for four, six, or eight poles.

Brush Position

The position of the brushes is very important. Maximum torque is developed when the brushes are placed 15° on either side of the pole pieces. *Figure 33–33* shows the effect of having the brushes placed

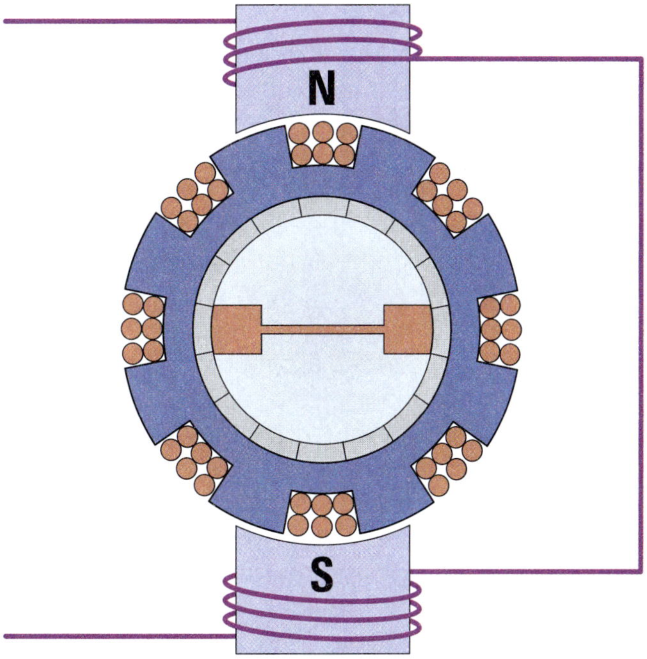

Figure 33–33 Brushes are placed at a 90° angle to the poles and no induction takes place.

at a 90° angle to the pole pieces. When the brushes are in this position, a circuit is completed between the coils located at a right angle to the poles. In this position, there is no induced voltage in the armature windings and no torque is produced by the motor.

In *Figure 33–34*, the brushes have been moved to a position so they are in line with the pole pieces. In this position, a large amount of current flows through the coils directly under the pole pieces. This current produces a magnetic field of the same polarity as the pole piece. Because the magnetic field produced in the armature is at a 0° angle to the magnetic field of the pole piece, no twisting or turning force is developed and the armature does not turn.

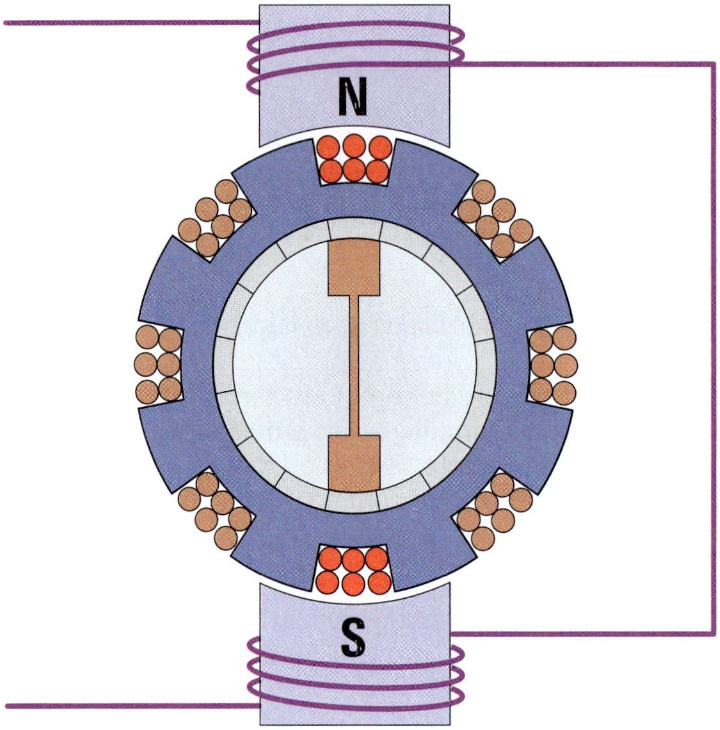

Figure 33–34 The brushes are set at a 0° angle to the pole pieces, and the armature does not turn.

In *Figure 33–35*, the brushes have been shifted in a clockwise direction so they are located 15° from the pole piece. The induced voltage in the armature winding produces a magnetic field of the same polarity as the pole piece. The magnetic field of the armature is repelled by the magnetic field of the pole piece, and the armature turns in the clockwise direction.

In *Figure 33–36*, the brushes have been shifted counterclockwise to a position 15° from the center of the pole piece. The magnetic field developed in the armature again repels the magnetic field of the pole piece, and the armature turns in the counterclockwise direction.

The direction of armature rotation is determined by the setting of the brushes. The direction of rotation for any type of repulsion motor is changed by setting the brushes 15° on either side of the pole pieces. Repulsion-type motors have the highest starting torque of any single-phase motor. The speed of a repulsion motor, not to be confused with the repulsion-start induction-run motor or the repulsion-induction motor, can be varied by changing the AC voltage supplying power for the motor. The repulsion motor has excellent starting and running torque but can exhibit unstable speed characteristics. The repulsion motor can race to very high speed if operated with no mechanical load connected to the shaft.

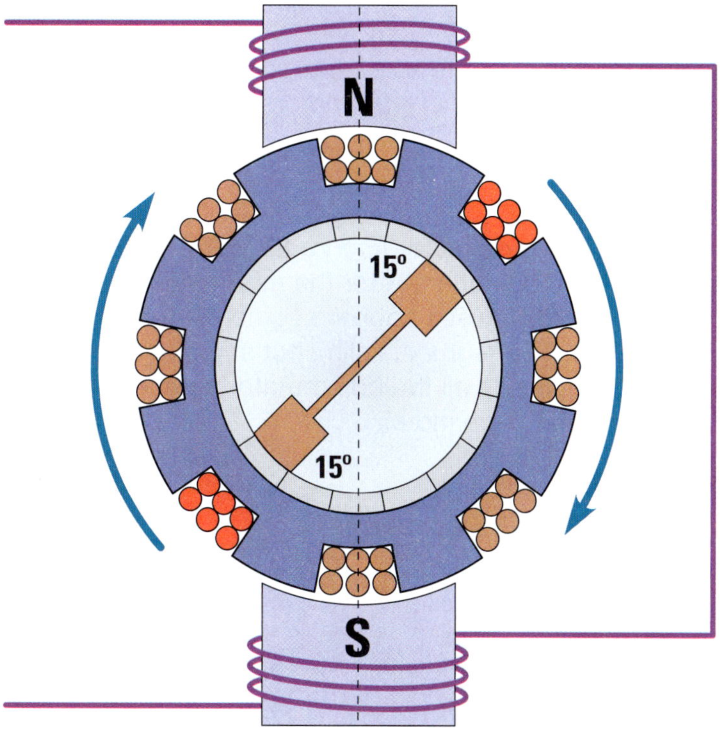

Figure 33–35 The brushes have been shifted clockwise 15°, and the poles repel the armature in a clockwise direction.

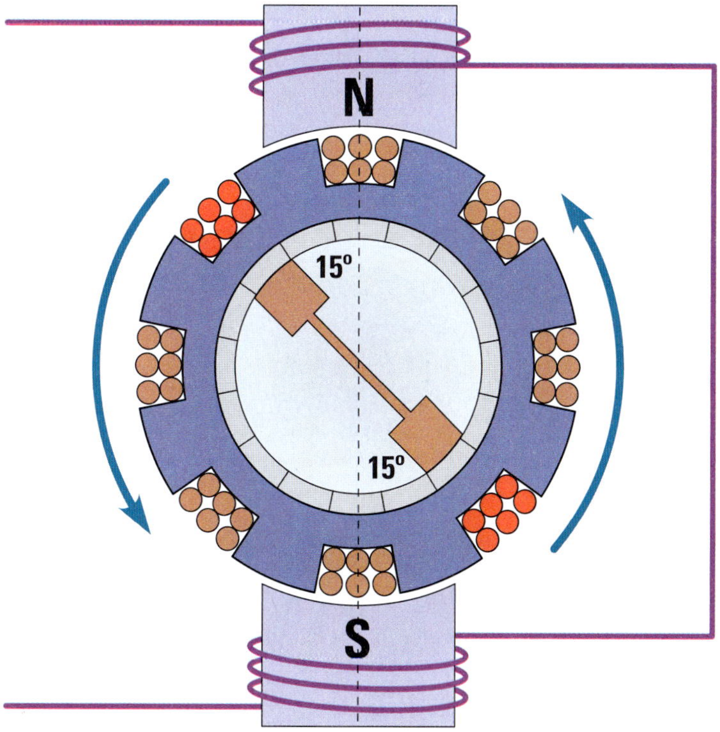

Figure 33–36 The brushes have been shifted counterclockwise 15°.

Repulsion-Start Induction-Run Motors

The repulsion-start induction-run motor starts as a repulsion motor but runs like a squirrel-cage motor. There are two types of repulsion-start induction-run motors:

1. The brush-riding type
2. The brush-lifting type

The brush-riding type uses an axial commutator *(Figure 33–37)*. The brushes ride against the commutator segments when the motor is in operation. After the motor has accelerated to approximately 75% of its full-load speed, centrifugal force causes copper segments of a short-circuiting ring to overcome the force of a spring *(Figure 33–38)*. The segments sling out and make contact with the segments of the commutator. This effectively short-circuits all the commutator segments together, and the motor operates in the same manner as a squirrel-cage motor.

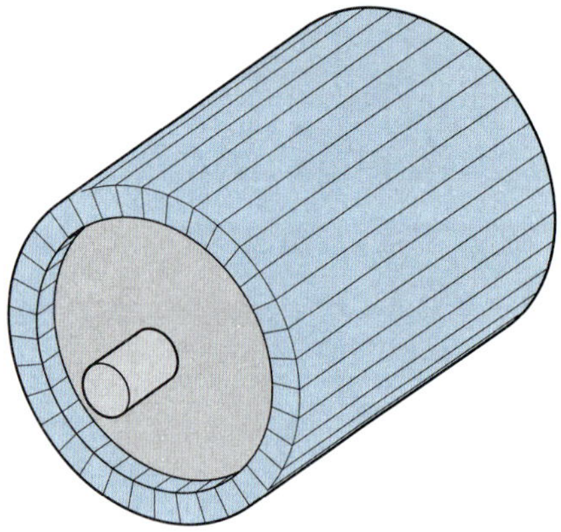

Figure 33–37 Axial commutator.

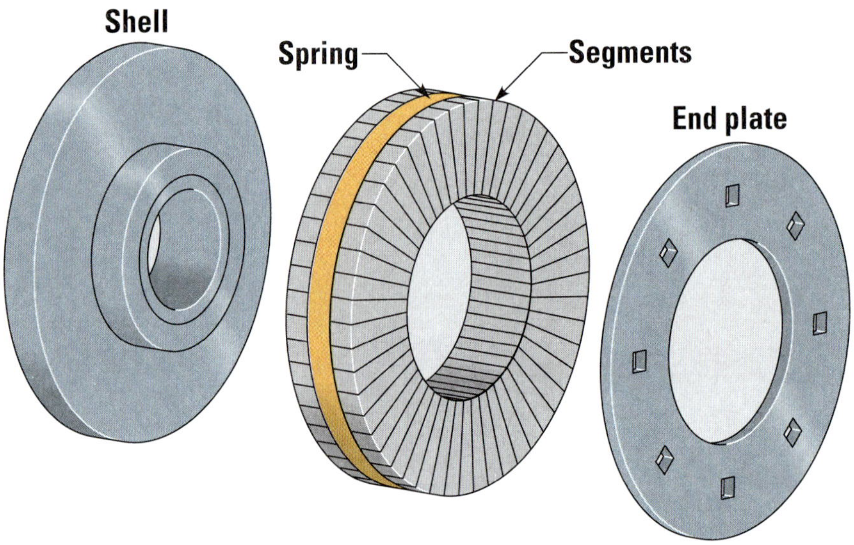

Figure 33–38 Short-circuiting ring for brush-riding-type repulsion-start induction-run motor.

The brush-lifting-type motor uses a radial commutator *(Figure 33–39)*. Weights are mounted at the front of the armature. When the motor reaches about 75% of full speed, these weights swing outward due to centrifugal force and cause two push rods to act against a spring barrel and short-circuiting necklace. The weights overcome the force of the spring and cause the entire spring barrel and brush holder assembly to move toward the back of the motor *(Figure 33–40)*. The motor is so designed that the short- circuiting necklace will short-circuit the commutator bars before the brushes lift off the surface of the radial commutator. The motor will now operate as a squirrel-cage induction motor. The armature of a brush-lifting-type repulsion-start induction-run motor with a radial commutator is shown in *Figure 33–41*.

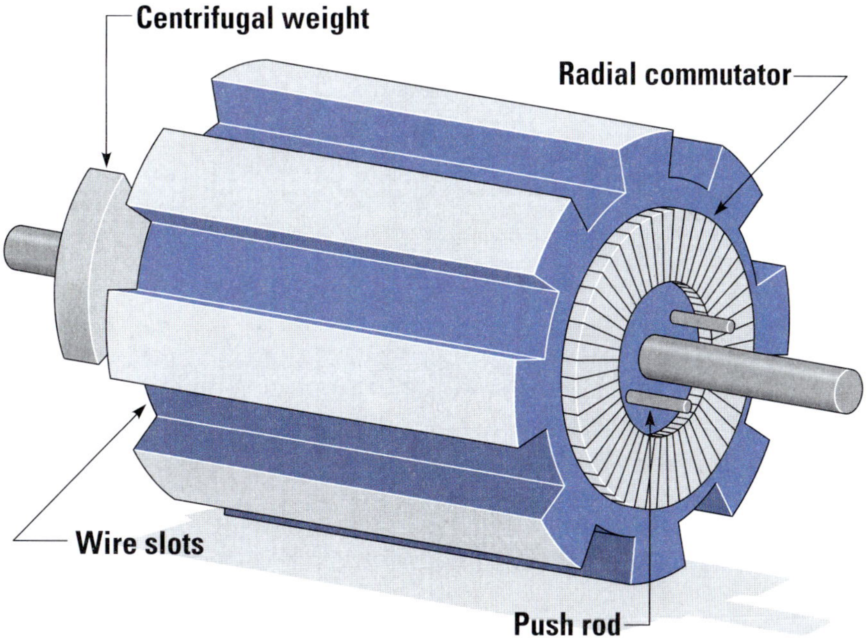

Figure 33–39 A radial commutator is used with the brush-lifting-type motor.

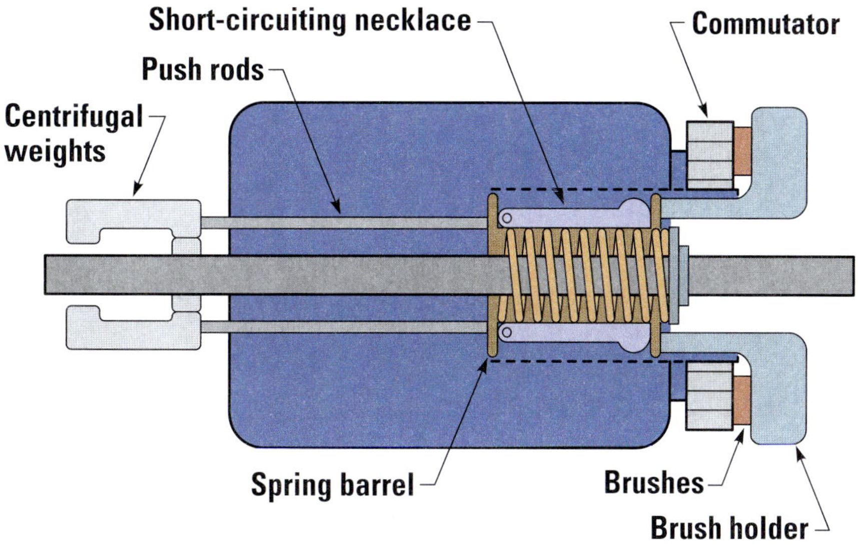

Figure 33–40 Brush-lifting-type repulsion-start induction-run motor.

Figure 33–41 Rotor of a bush-lifting-type repulsion-start induction-run motor with radial commutator.

Repulsion-Induction Motors

The repulsion-induction motor is basically the same as the repulsion motor except that a set of squirrel-cage windings are added to the armature *(Figure 33–42)*. This type of motor contains no centrifugal mechanism or short-circuiting device. The brushes ride against the commutator at all times. The repulsion-induction motor has very high starting torque because it starts as a repulsion motor. The squirrel-cage winding, however, gives it much better speed characteristics than a standard repulsion motor. This motor has very good speed regulation between no load and full load.

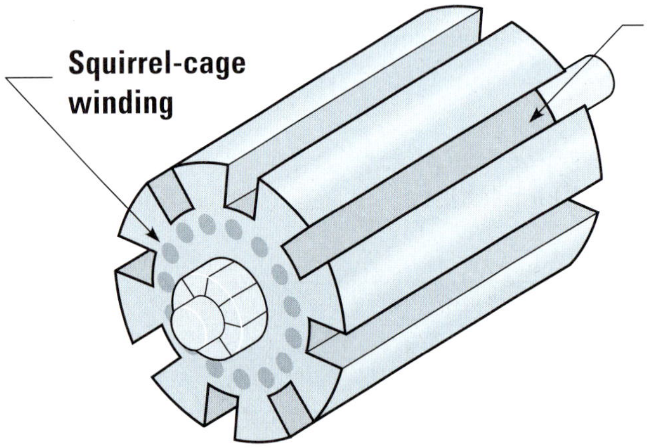

Figure 33–42 Repulsion-induction motors contain both armature and squirrel-cage windings.

33–5 Single-Phase Synchronous Motors

Single-phase **synchronous motors** are small and develop only fractional horsepower. They operate on the principle of a rotating magnetic field developed by a shaded-pole stator. Although they will operate at synchronous speed, they do not require DC excitation. They are used in applications where constant speed is required such as clock motors, timers, and recording instruments. They are also used as the driving force for small fans because they are small and inexpensive to manufacture. There are two basic types of synchronous motor: the Warren, or General Electric, motor and the Holtz motor. These motors are also referred to as hysteresis motors.

Warren Motors

The **Warren motor** is constructed with a laminated stator core and a single coil. The coil is generally wound for 120-VAC operation. The core contains two poles, which are divided into two sections each. One half of each pole piece contains a shading coil to produce a rotating magnetic field *(Figure 33–43)*. Because the stator is divided into two poles, the synchronous field speed is 3600 RPM when connected to 60 hertz.

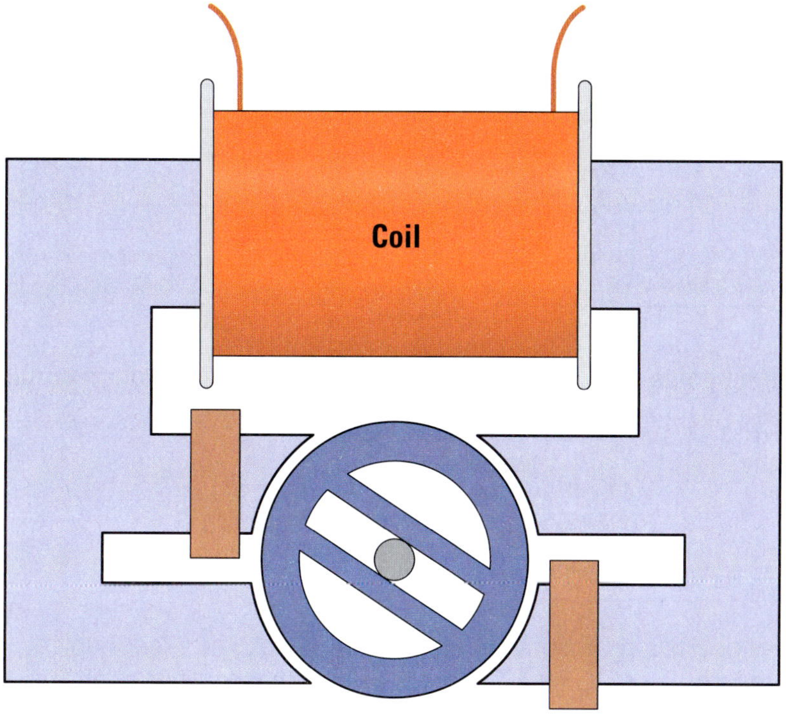

Figure 33–43 An illustration of a Warren motor.

The difference between the Warren and Holtz motor is the type of rotor used. The rotor of the Warren motor is constructed by stacking hardened steel laminations onto the rotor shaft. These disks have high hysteresis loss. The laminations form two crossbars for the rotor. When power is connected to the motor, the rotating magnetic field induces a voltage into the rotor and a strong starting torque is developed, causing the rotor to accelerate to near-synchronous speed. Once the motor has accelerated to near-synchronous speed, the rotor locks in with the rotating magnetic field. These types of motors operate at 3600 RPM.

Holtz Motors

The **Holtz motor** uses a different type of rotor *(Figure 33–44)*. This rotor is cut in such a manner that six slots are formed. These slots form six salient (projecting or jutting) poles for the rotor. A squirrel-cage winding is constructed by inserting a metal bar at the bottom of each slot. When power is connected to the motor, the squirrel-cage winding provides the torque necessary to start the rotor turning. When the rotor approaches synchronous speed, the salient poles lock in step with the field poles each half cycle. This produces a rotor speed of 1200 RPM (one-third of synchronous speed) for the motor.

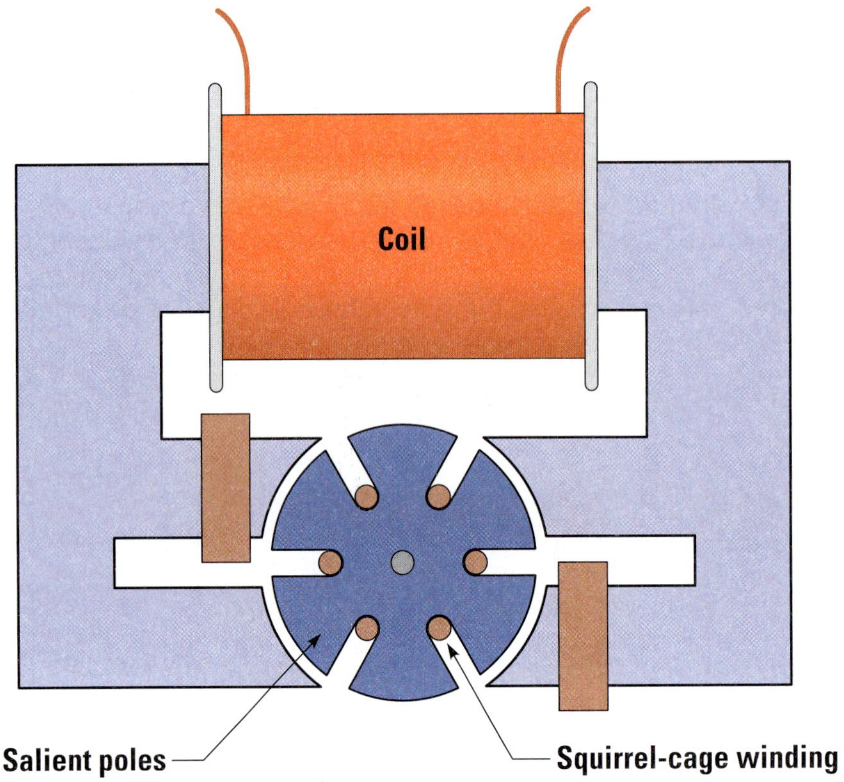

Figure 33–44 An illustration of a Holtz motor.

33–6 Stepping and Universal Motors

Stepping motors are devices that convert electric impulses into mechanical movement. Stepping motors differ from other types of DC or AC motors in that their output shaft moves through a specific angular rotation each time the motor receives a pulse. The stepping motor allows a load to be controlled as to speed, distance, or position. These motors are very accurate in their control performance. There is generally less than 5% error per angle of rotation, and this error is not cumulative regardless of the number of rotations. Stepping motors are operated on DC power but can be used as a two-phase synchronous motor when connected to AC power.

Theory of Operation

Stepping motors operate on the theory that like magnetic poles repel and unlike magnetic poles attract. Consider the circuit shown in *Figure 33–45*. In this illustration, the rotor is a permanent magnet, and the stator windings consist of two electromagnets. If current flows through the winding of Pole A creating a north magnetic pole and through pole B creating a south magnetic pole, the rotor could turn in either direction.

Now consider the circuit shown in *Figure 33–46*. In this circuit, the motor contains four stator poles instead of two. Current flows through stator Pole A and still produces a north magnetic field, and Pole B still produces a south magnetic field. The current flow through stator Pole C, however, produces a south magnetic field, and the current flow through Pole D produces a north magnetic field. In this illustration, there is no doubt as to the direction or angle of rotation. In this example, the rotor shaft turns 90° in a counterclockwise direction.

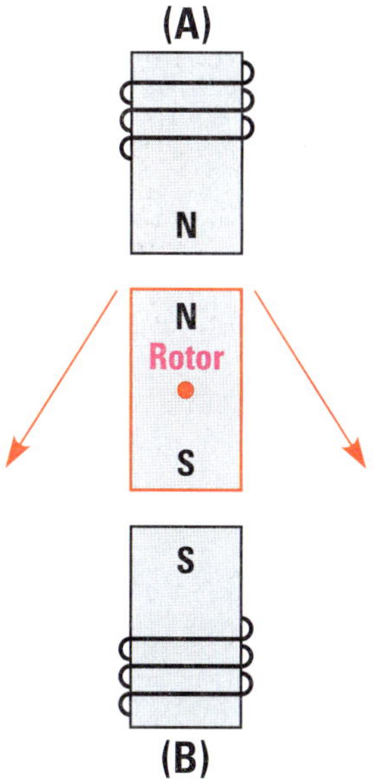

Figure 33–45 The rotor could turn in either direction.

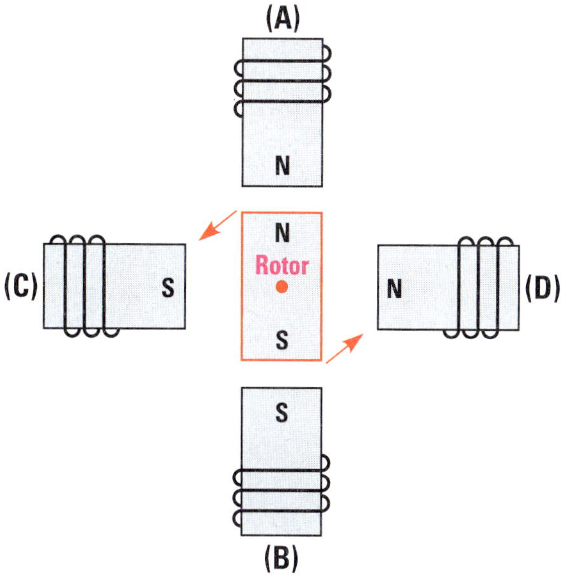

Figure 33–46 The direction of rotation is easily determined with the addition of more poles.

Figure 33–47 shows yet another condition. In this example, the current flow through Poles A and C is in such a direction as to form a north magnetic pole, and the direction of current flow through Poles B and D forms south magnetic poles. In this illustration, the permanent magnetic rotor has rotated to a position between the actual pole pieces.

To allow for better stepping resolution, most stepping motors have eight stator poles, and the pole pieces and rotor have teeth machined into them, as shown in *Figure 33–48*. In actual practice, the number of teeth machined into the stator and rotor determines the angular rotation achieved each time the motor is stepped. The stator-rotor tooth configuration shown in *Figure 33–48* produces an angular rotation of 1.8° per step.

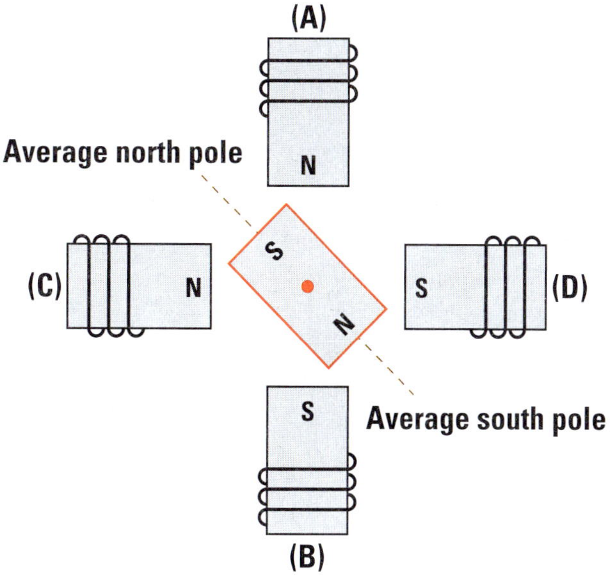

Figure 33–47 The magnet aligns with the average magnetic pole.

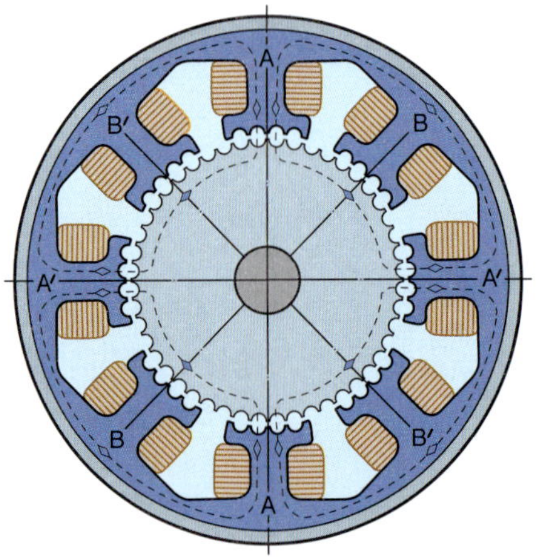

Figure 33–48 Illustration of the cross-section of a stepping motor.

AC Operation

Stepping motors can also be operated on AC voltage instead of DC voltage. In this mode of operation, they become two-phase, AC, synchronous, constant-speed motors and are classified as *permanent magnet induction motors*. A stepping motor is shown in Figure *33–49*, and a cutaway view of a stepping motor shown in *Figure 33–50*. Notice that this motor has no brushes, sliprings, commutator, gears, or belts. Bearings maintain a constant air gap between the permanent magnet rotor and the stator windings. A typical eight-stator-pole stepping motor will have a synchronous speed of 72 RPM when connected to a 60-hertz power source.

Courtesy of Danaher Motion Technologies

Figure 33–49 Stepping motor.

Figure 33–50 Cutaway of a stepping motor.

A resistive-capacitive network can be used to provide the 90° phase shift needed to change single-phase AC into two-phase AC, the type of AC required to run a stepping motor. A simple forward-off-reverse switch can be added to provide directional control. A sample circuit of this type is shown in *Figure 33–51*. The correct values of resistance and capacitance are necessary, otherwise it can result in random direction of rotation when the motor is started, change of direction when the load is varied, erratic and unstable operation, and failure to start. The manufacturer's recommendations should always be followed for the particular type of stepping motor used.

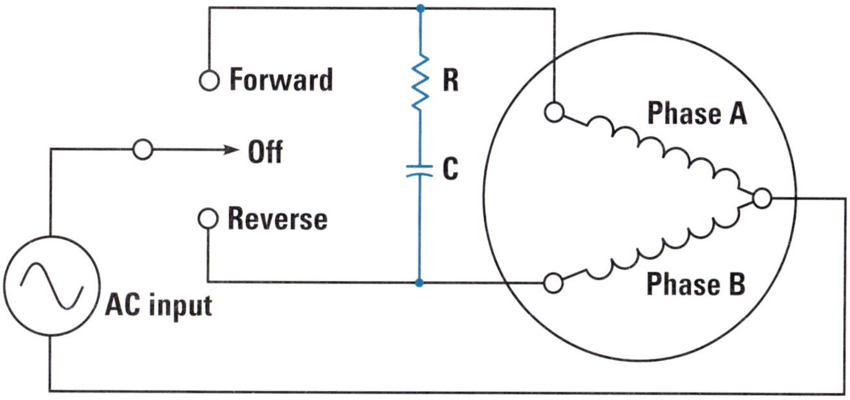

Figure 33–51 Schematic drawing of stepping motor connected to AC supply with directional control.

Stepping Motor Characteristics

When stepping motors are used as two-phase synchronous motors, they can start, stop, or reverse direction of rotation virtually instantly. The motor will start within about $1\frac{1}{2}$ cycles of the applied voltage and will stop within 5 to 25 milliseconds. Because the rotor is a permanent magnet, there is no induced current in the rotor. There is no high inrush of current when the motor is started. The starting and running currents are the same. This simplifies how we supply the motor. The permanent magnets in the rotor also provide holding torque when turned off, and if more holding torque is needed, DC voltage can be applied to one or both windings when the motor is turned off. An example circuit of this type is shown in *Figure 33–52*. If DC is applied to one winding, the holding torque will be approximately 20% greater than the rated torque of the motor. If DC is applied to both windings, the holding torque will be about $1\frac{1}{2}$ times greater than the rated torque.

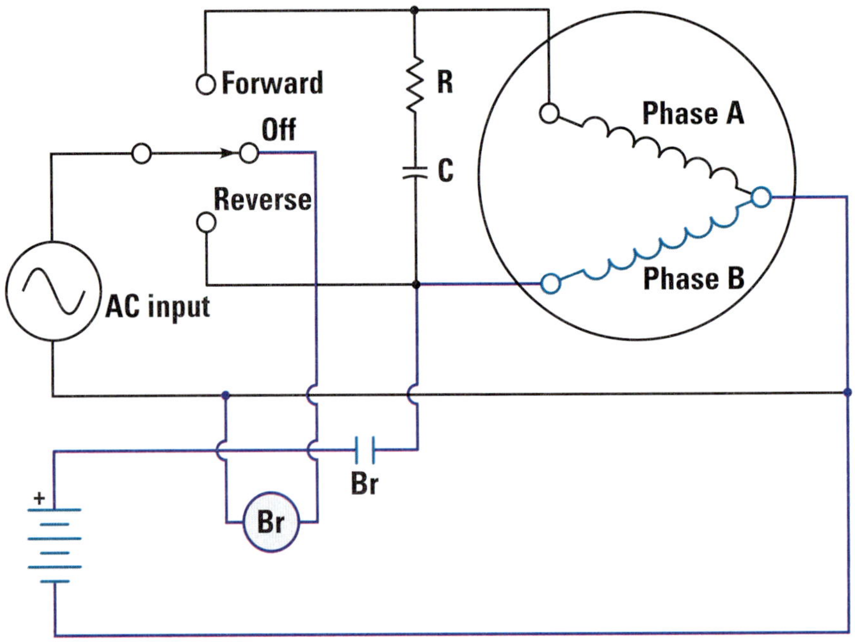

Figure 33–52 Applying DC voltage to increase holding torque.

Universal Motors

The **universal motor** is often referred to as an AC series motor. This motor is very similar to a DC series motor in its construction in that it contains a wound armature and brushes *(Figure 33–53)*. The universal motor, however, has the addition of a **compensating winding.** Universal motors also contain a laminated core to help prevent eddy currents, and the compensating winding is wound around the stator to counteract the inductive reactance in the armature winding.

The universal motor is so named because it can be operated on AC or DC voltage. When the motor is operated on DC, the compensating winding is connected in series with the series field winding *(Figure 33–54)*.

Connecting the Compensating Winding for AC

When the universal motor is operated with AC power, the compensating winding can be connected in two ways. If it is connected in series with the armature, as shown in *Figure 33–55*, it is known as **conductive compensation.**

Figure 33–53 Armature and brushes of a universal motor.

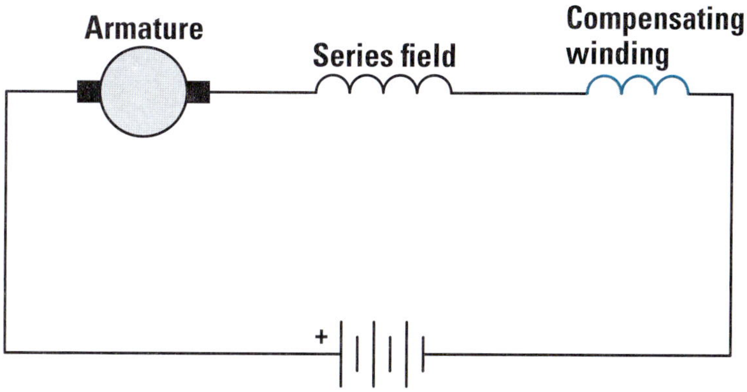

Figure 33–54 The compensating winding is connected in series with the field winding when connected to a DC supply.

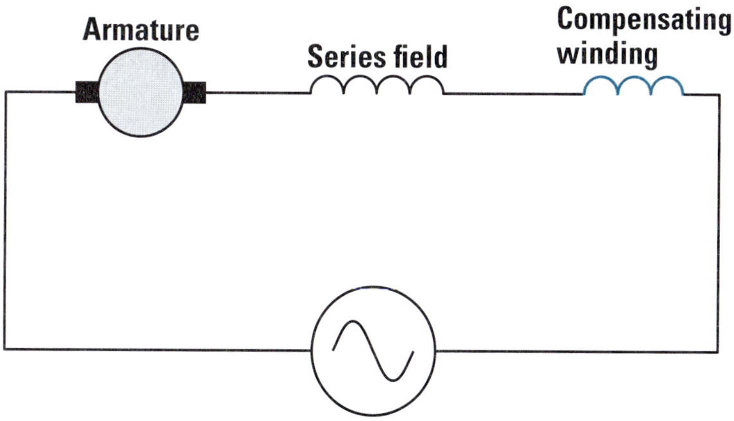

Figure 33–55 Conductive compensation resulting from series windings.

The compensating winding can also be connected by shorting its leads together, as shown in *Figure 33–56*. When connected in this manner, the winding acts like a shorted secondary winding of a transformer. Induced current permits the winding to operate when connected in this manner. This connection is known as **inductive compensation** and is only possible with AC.

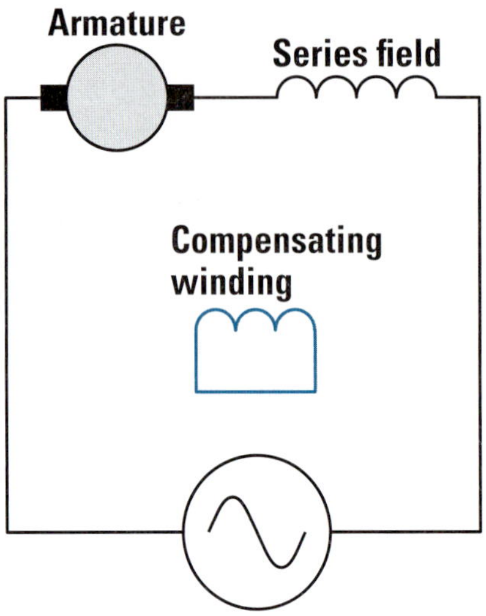

Figure 33–56 Inductive compensation caused by shorting the compensation winding in an AC-operated universal motor.

The Neutral Plane

Because the universal motor contains a wound armature, commutator, and brushes, the brushes should be set at the **neutral plane** position. This can be done in the universal motor in a manner similar to that of setting the neutral plane of a DC machine. When setting the brushes to the neutral plane position in a universal motor, either the series or the compensating winding can be used. To set the brushes to the neutral plane position using the series winding *(Figure 33–57)*, AC is connected to the armature leads. A voltmeter is connected to the series winding. Voltage is then applied to the armature. The brush position is then moved until the voltmeter reaches the lowest point.

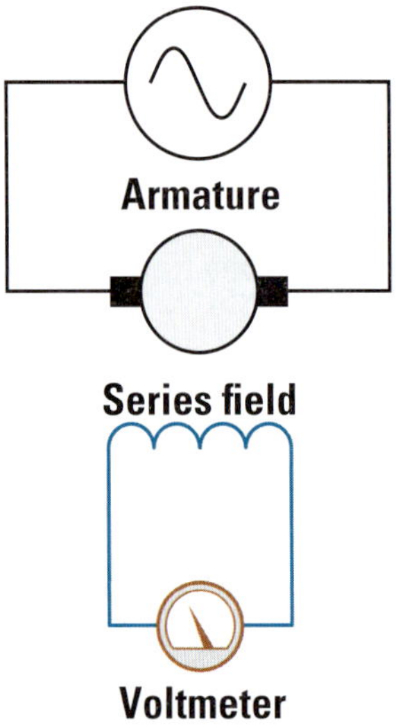

Figure 33–57 Setting the brushes to the neutral plane position.

If the compensating winding is used to set the neutral plane, AC is again connected to the armature, and a voltmeter is connected to the compensating winding. The brushes are then moved until the voltmeter indicates its highest or peak voltage, so the opposite of setting the brushes using the series field winding.

Speed Regulation

The speed regulation of the universal motor is very poor. Because this is a series motor, it has the same poor speed regulation as a DC series motor. If the universal motor is connected to a light load or no load, its speed is almost unlimited. It is not unusual for this motor to be operated at several thousand revolutions per minute. Universal motors are used in a number of portable appliances where high horsepower and light weight are needed, such as drill motors, skill saws, and vacuum cleaners. The universal motor is able to produce a high horsepower for its size and weight because of its high operating speed.

Changing the Direction of Rotation

The direction of rotation of the universal motor can be changed in the same manner as changing the direction of rotation of a DC series motor. To change the direction of rotation, change the armature leads with respect to the field leads.

Servo Motors

Servo motors intended for use with alternating current can be obtained in a variety of sizes and styles. AC servo motors are commonly used for precise position control and can be found on computer numerical control (CNC) machines. These motors generally have a stationary field or stator and a permanent magnet rotor. The permanent magnet rotor permits the starting current to remain at a very low level. In fact, the starting current is essentially the same as the running current. The stator winding of an AC servo motor is shown in *Figure 33–58*.

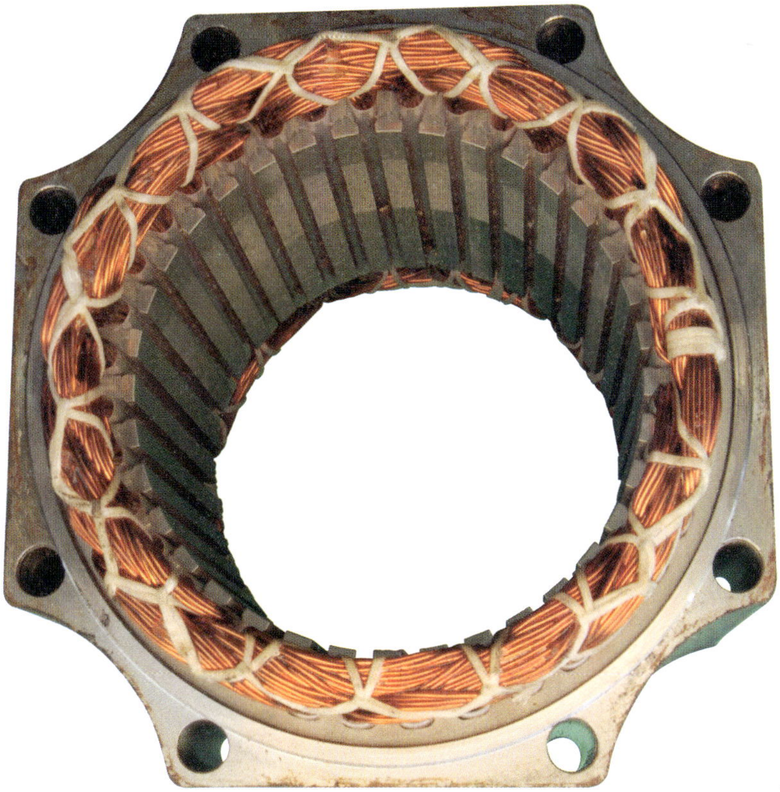

Source: Tyler Tools, Tyler, Texas

Figure 33–58 Stator winding of an AC servo motor.

AC servo motors used for position control must have some method of determining the speed, number of revolutions, and direction of rotation. The device that accomplishes this task is called an **encoder.** Encoders can vary in design depending on the complexity of the application. Many encoders employ the use of light-emitting diodes and photodiodes to indicate the position of the wheel because of their speed of operation and reliability. A simple design of an encoder is shown in *Figure 33–59*. Two light-emitting diodes (LEDs) and two photodiodes are positioned on opposite sides of the slotted wheel. The LEDs and photodiodes are positioned in such a manner that one LED and photodiode is in a location that the LED is blocked by the wheel, preventing the photodiode from receiving light. The other LED and photodiode are positioned so that the LED is in front of a slot permitting the photodiode to receive light. This arrangement permits the encoder to determine which direction the shaft turns. As the wheel turns, the encoder counts the slots to know how many revolutions the shaft turns and the speed. Encoders often contain much more complex slotted wheels depending on the application of the motor *(Figure 33–60)*. An encoder mounted on the shaft of an AC servo motor is shown in *Figure 33–61*. A permanent magnet rotor with an encoder mounted on the shaft is shown in *Figure 33–62*. An exploded diagram of an AC servo motor is shown in *Figure 33–63*.

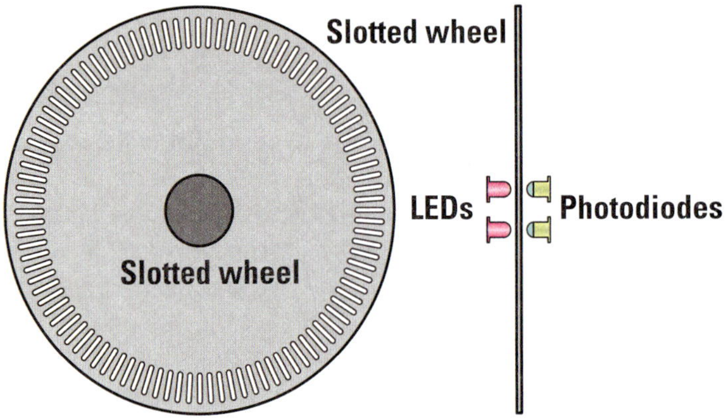

Figure 33–59 Simplified illustration of a servo motor encoder.

Figure 33–60 The slotted wheel of an encoder.

Figure 33–61 An encoder on the shaft of a servo motor.

Figure 33–62 Permanent magnet rotor with an encoder mounted on the shaft.

Torque Motors

Torque motors are a special type of motor designed to operate with the rotor stalled without damage to the motor. AC torque motors can be obtained that operate on single-phase or three-phase power. These motors contain a high-impedance stator winding that can be operated at full load without overheating. Torque motors are ideal for tension control and are often employed to power take-up reels or as a continuous brake *(Figure 33–64)*.

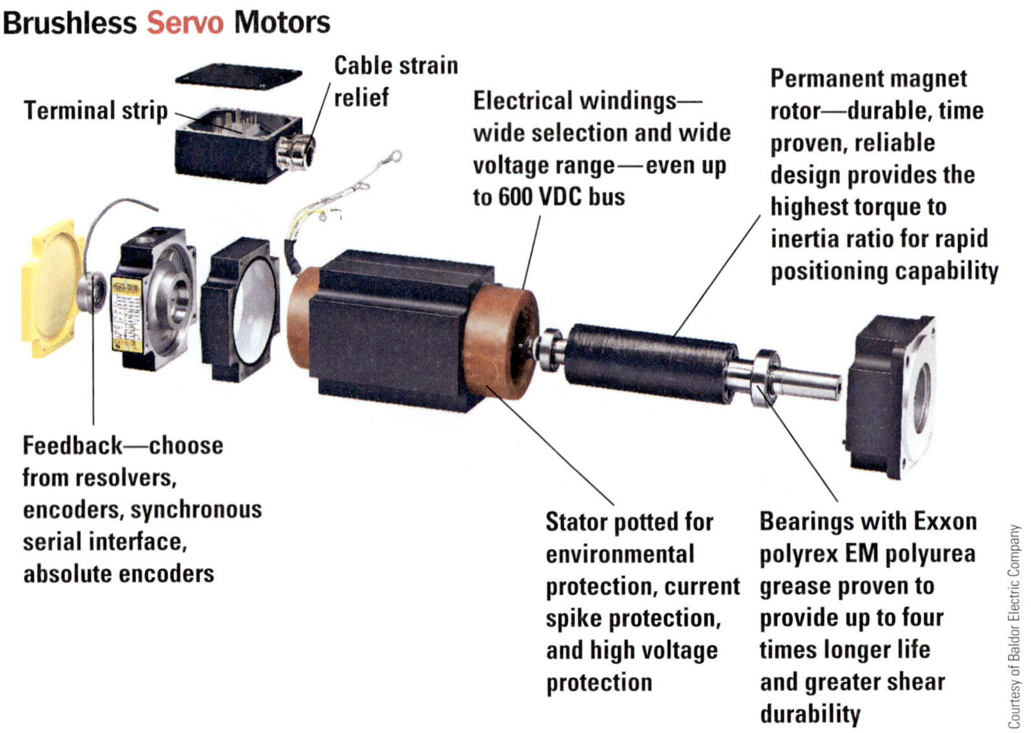

Figure 33–63 Exploded diagram of an AC servo motor.

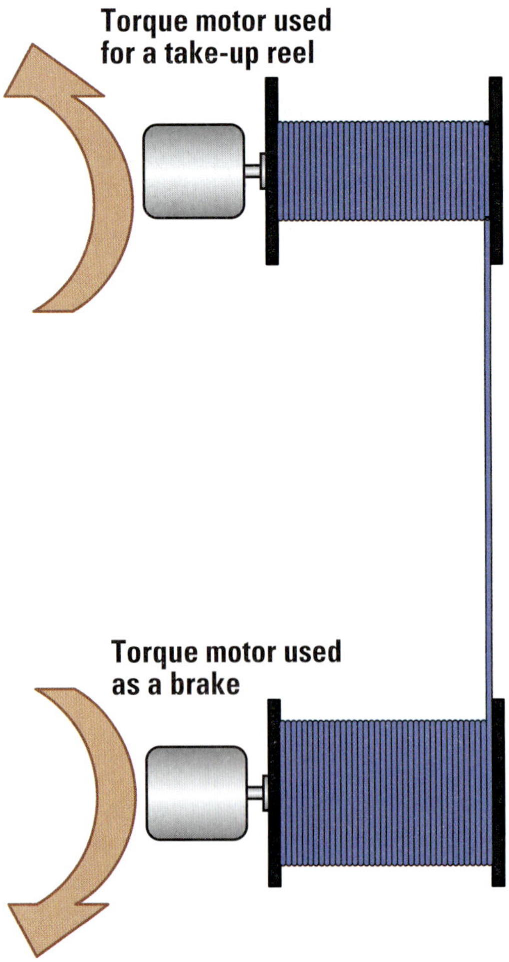

Figure 33–64 Torque motors are often used for take-up reels and as electric brakes.

AC torque motors are controlled by the amount of voltage applied to the motor. The two most common methods of supplying varying voltage to the motor are solid-state inverters, sometimes called a solid-state variac, and variable autotransformers. A single-phase variable autotransformer is shown in *Figure 33–65*. Control of small torque motors is sometimes accomplished by connecting a variable power resistor in series with the motor, as shown in *Figure 33–66*.

Figure 33–65 Example of a variable autotransformer.

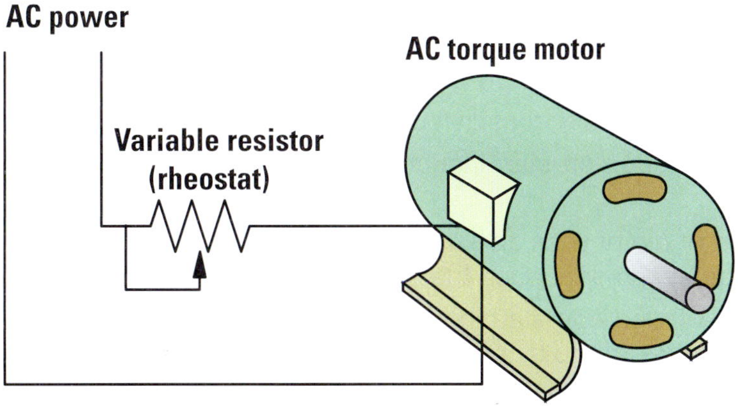

Figure 33–66 A high-wattage variable resistor is connected in series with the torque motor.

Summary

- Not all single-phase motors operate on the principle of a rotating magnetic field.
- Split-phase motors start as two-phase motors by producing an out-of-phase condition for the current in the run winding and the current in the start winding.
- The resistance of the wire in the start winding of a resistance-start induction-run motor is used to produce a phase angle difference between the current in the start winding and the current in the run winding.
- The capacitor-start induction-run motor uses an AC electrolytic capacitor to increase the phase angle difference between starting and running current. This causes an increase in starting torque.
- Maximum starting torque for a split-phase motor is developed when the start-winding current and run-winding current are 90° out of phase with each other.
- Most resistance-start induction-run motors and capacitor-start induction-run motors use a centrifugal switch to disconnect the start windings when the motor reaches approximately 75% of full-load speed.
- The capacitor-start capacitor-run motor operates like a two-phase motor because both the start and run windings remain energized during motor operation.
- Most capacitor-start capacitor-run motors use an AC oil-filled capacitor connected in series with the start winding.
- The capacitor of the capacitor-start capacitor-run motor helps to correct the power factor.
- Shaded-pole induction motors operate on the principle of a rotating magnetic field.
- The rotating magnetic field of a shaded-pole induction motor is produced by placing shading loops or coils on one side of the pole piece.
- The synchronous speed of a single-phase motor is determined by the number of stator poles and the frequency of the applied voltage.
- Consequent-pole motors are used when a change of motor speed is desired and high torque must be maintained.
- Multispeed fan motors are constructed by connecting windings in series with the main run winding.
- Multispeed fan motors have high-impedance stator windings to prevent them from overheating when their speed is reduced.
- There are three basic repulsion-type motors: the repulsion motor, the repulsion-start induction-run motor, and the repulsion-induction motor.
- Repulsion motors have the highest starting torque of any single-phase motor.
- The direction of rotation of repulsion motors is changed by setting the brushes 15° on either side of the pole pieces.
- The direction of rotation for split-phase motors is changed by reversing the start winding in relation to the run winding.
- Shaded-pole motors are generally considered to be nonreversible.
- There are two types of repulsion-start induction-run motors: the brush-riding type and the brush-lifting type.
- The brush-riding type of motor uses an axial commutator and a short-circuiting device, which short-circuits the commutator segments when the motor reaches approximately 75% of full-load speed.
- The brush-lifting type of repulsion-start induction-run motor uses a radial commutator. A centrifugal device causes the brushes to move away from the commutator and a short-circuiting necklace to short-circuit the commutator when the motor reaches about 75% of full-load speed.
- The repulsion-induction motor contains both a wound armature and squirrel-cage windings.

- There are two types of single-phase synchronous motor: the Warren and the Holtz.
- Single-phase synchronous motors are sometimes called hysteresis motors.
- The Warren motor operates at a speed of 3600 RPM.
- The Holtz motor operates at a speed of 1200 RPM.
- Stepping motors generally operate on DC but can be used on AC when capacitors are used to create a phase shift and are used to produce angular movements in steps.
- Stepping motors are generally used for position control.
- Stepping motors operate at a speed of 72 RPM when connected to 60-hertz power.
- Stepping motors can produce a holding torque when DC is connected to their windings.
- Universal motors operate on DC or AC.
- Universal motors contain a wound armature and brushes.
- Universal motors are also called AC series motors.
- Universal motors have a compensating winding that helps overcome inductive reactance.
- The direction of rotation for a universal motor can be changed by reversing the armature leads with respect to the field leads.
- AC servo motors are used for position control.
- AC servo motors generally contain an encoder that determines speed, number of shaft rotations, and direction of rotation.
- Torque motors are designed to operate with the rotor stalled without overheating.

Review Questions

1. Which of the following statements best describes the relationship between the start and run windings of a split-phase motor?

   A) The start winding is connected 90° out of phase with the run winding.

   B) The start winding is connected in series to the run winding.

   C) The start winding and run winding are the same winding.

   D) The start winding is connected at a 15° angle to the run winding.

2. To produce maximum starting torque in a split-phase motor, how many degrees out of phase should the start- and run-winding currents be with each other?

   A) 15°

   B) 30°

   C) 60°

   D) 90°

3. What type of split-phase motor develops the highest starting torque?

   A) The resistance-start induction-run motor

   B) The capacitor-start induction-run motor

   C) Stepping motors

   D) Repulsion-type motors

4. What is the main advantage of the capacitor-start capacitor-run motor over capacitor-start induction-run and resistance-start induction-run motors?
   A) The motor has excellent starting torque but unstable running characteristics.
   B) The motor uses a centrifugal switch to disconnect the capacitor.
   C) The capacitor stays connected and operates the motor as a two-phase near unity power factor motor.
   D) The motor is quiet during operation.
5. Which of the following statements best describes how a shaded pole induction motor operates on the rotating magnetic field principle when connected to a single-phase circuit?
   A) Shaded poles use different pole material to create varying lines of flux.
   B) Shaded pole motors use a shading coil to create a changing magnetic field through induction.
   C) Shaded pole motors use a start and run winding to create a phase shift of 90°.
   D) Shaded pole motors have alloy armatures to create varying magnetic properties.
6. True or False: Most multispeed fan motors contain multiple run windings connected in series where the selector switch moves between taps to add or remove the extra windings from the run winding.
7. Repulsion-type motors operate on the principle of magnetism that states like poles repel each other. Which of the following describes what must take place in a repulsion motor to see rotation of the rotor?
   A) Brushes must be at a 90° angle to pole pieces.
   B) Brushes must be at a 0° angle to pole pieces.
   C) Power must be applied to the armature windings.
   D) Brushes must be at 15° angle to produce maximum starting torque.
8. Due to the poor running characteristics of a purely repulsion-type motor, these motors often switch to an induction-type motor by utilizing a squirrel-cage in the rotor. At what percentage of full-speed do these motors switch from repulsion to induction-type motors?
   A) 75%
   B) 80%
   C) 85%
   D) 90%
9. How do single-phase synchronous motors establish a rotating field in the stator?
   A) They create synchronous two-phase voltage.
   B) They apply voltage to the rotor.
   C) They use shaded poles in the stator.
   D) They use a DC rectifier to simulate rotation.

10. A Holtz-type synchronous single-phase motor uses a rotor with salient poles and a squirrel-cage winding to start and run. What is the synchronous speed of a Holtz motor?
    A) 3600 rpm
    B) 1200 rpm
    C) 1800 rpm
    D) 900 rpm
11. How does a stepping motor create rotation of the rotor?
    A) Stepping motors rotate from magnetic fields created by electrical pulses.
    B) Stepping motors use shaded-poles.
    C) Stepping motors are split-phase motors.
    D) Stepping motors are gear driven.
12. True or False: Universal motors can operate on either AC or DC voltage because they contain a compensating winding that can either be shorted together or connected in series with the armature for AC connections or connected in series with the field windings for DC connections.

Practical Applications

1. You are an electrical contractor, and you have been called to a home to install a well pump. The homeowner has purchased the pump but does not know how to connect it. You open the connection terminal cover and discover that the motor contains eight terminal leads marked T_1 through T_8. The motor is to be connected to 240 V. At present, the T leads are connected as follows: T_1, T_3, T_5, and T_7 are connected together; and T_2, T_4, T_6, and T_8 are connected together. L_1 is connected to the group of terminals with T_1, and L_2 is connected to the group of terminals with T_2. Is it necessary to change the leads for operation on 240 V? If so, how should they be connected?

Unit 34 Motor Installation

Key Terms

Electronic overload relay

Magnetic overload relay

Overload protection

Overload relays

Termination temperature

Thermal overload relay

Why You Need to Know

Motors are the workhorses of the industrial world, and in the United States, they almost always are required to be installed in accordance with *NFPA 70®: National Electrical Code® (NEC®)*. This unit covers a high-level overview of the selection of

- conductor size.
- fuse or circuit breaker size.
- overload size,
- motor starter size.

Outline

34–1 Determining Motor Full-Load Current

34–2 Determining Conductor Size for a Single Motor

34–3 Overload Protection and Motor Starters

34–4 Short-Circuit and Ground-fault Protection

34–5 Multiple Motor Calculations

Objectives

After studying this unit, you should be able to

- differentiate between motor full-load current and nameplate full-load amps.
- apply proper *NEC* full-load current table based on motor type and characteristics.
- employ *NEC* requirements for sizing conductors that supply a single motor.
- illustrate how motors are protected from overload conditions.
- calculate the proper size branch circuit short-circuit and ground-fault protection required by the *NEC*.
- apply *NEC* requirements to select the proper size feeder short-circuit and ground-fault protective device and feeder conductors to supply multiple motors.

Preview

It is important to select the conductor size, fuse or circuit breaker size, motor starter size, and overload size associated with the installation of a new motor. Conductor size, fuse or circuit breaker size, and overload size are selected in accord with the requirements of the *NEC*. Motor starter sizes are selected with either NEMA or IEC requirements. This unit also covers multiple motors connected to a common feeder.

34–1 Determining Motor Full-Load Current

The *NEC* utilizes two different values for motor full-load current depending on what part of the circuit is being determined. One is the current listed on the motor nameplate, and the other is from tables found in Article 430 of the *National Electrical Code® (NEC®).* The nameplate current value is used to determine the **overload protection** for the motor, but the value found in *NEC* tables is used when determining everything else.

The tables in the *NEC* are based on the running current for different types of motors at different voltage levels. Article 430 in the *NEC* contains the following tables:

- Table 430.247 Full-Load Current In Amperes, Direct Current Motors
- Table 430.248 Full-Load Currents In Amperes, Single-Phase Alternating-Current Motors
- Table 430.249 Full-Load Current, Two-Phase Alternating-Current Motors (4-Wire)
- Table 430.250 Full-Load Current, Three-Phase Alternating-Current Motors

The examples in this Unit will primarily use *Table 430.248* (shown in *Figure 34–1*) and *Table 430.250* (shown in *Figure 34–2*) to determine the full-load running current values for motors. Note that the tables list the amount of current that the motor is expected to draw under a full-load condition. These tables list the ampere rating of the motors according to horsepower (HP) and connected voltage. *NEC Section 430.6(A)* states that these tables are used to determine conductor size, short-circuit–protection size, and ground fault protection size instead of the nameplate current of the motor.

Single-Phase AC Motors

The voltages listed in *Table 430.248* are 115, 200, 208, and 230. However, the voltages are meant to correspond to nominal voltage systems. For example, 115-volt values in the table shall be permitted for motors connected to voltages of 110 to 120 volts, and the 230-volt column can be used for motors

Table 430.248 full-load currents in amperes, single-phase alternating-current motors

The following values of full-load currents are for motors running at usual speeds and motors with normal torque characteristics. The voltages listed are rated motor voltages. The currents listed shall be permitted for system voltage ranges of 110 to 120 and 220 to 240 volts.

Horsepower	115 volts	200 volts	208 volts	230 volts
1/6	4.4	2.5	2.4	2.2
1/4	5.8	3.3	3.2	2.9
1/3	7.2	4.1	4.0	3.6
1/2	9.8	5.6	5.4	4.9
3/4	13.8	7.9	7.6	6.9
1	16	9.2	8.8	8.0
1 1/2	20	11.5	11.0	10
2	24	13.8	13.2	12
3	34	19.6	18.7	17
5	56	32.2	30.8	28
7 1/2	80	46.0	44.0	40
10	100	57.5	55.0	50

Figure 34–1 *NEC Table 430.248.*

connected to voltages of 220 to 240 volts. This means that if the motor is connected to a 120-volt line, it is permissible to use the currents listed in the 115-volt column. If the motor is connected to a 220- or 240-volt line, the 230-volt column can be used.

Example 34–1

A 5-HP single-phase AC motor is connected to a 240-volt line. What will be the full-load running current of this motor?

Locate 5 HP in the far-left column of *Table 430.248.* Follow across to the 230-volt column. The full-load current will be 28 amperes.

Three-Phase Motors

NEC Table 430.250 (*Figure 34–2*) is used to determine the full-load current of three-phase motors. Notice that the right side of *Table 430.250* is devoted to the full-load currents of synchronous-type motors. The currents listed are for synchronous motors that are to be operated at unity, or 100% power factor. Because synchronous motors are often made to have a leading power factor by overexciting the

Table 430.250 full-load current, three-phase alternating-current motors

The following values of full-load currents are typical for motors running at speeds usual for belted motors and motors with normal torque characteristics.

The voltages listed are rated motor voltages. The currents listed shall be permitted for system voltage ranges of 110 to 120, 220 to 240, 440 to 480, and 550 to 1000 volts.

	Induction-type squirrel cage and wound rotor (amperes)							Synchronous-type unity power factor* (amperes)			
Horsepower	115 volts	200 volts	208 volts	230 volts	460 volts	575 volts	2300 volts	230 volts	460 volts	575 volts	2300 volts
½	4.4	2.5	2.4	2.2	1.1	0.9	—	—	—	—	—
¾	6.4	3.7	3.5	3.2	1.6	1.3	—	—	—	—	—
1	8.4	4.8	4.6	4.2	2.1	1.7	—	—	—	—	—
1½	12.0	6.9	6.6	6.0	3.0	2.4	—	—	—	—	—
2	13.6	7.8	7.5	6.8	3.4	2.7	—	—	—	—	—
3	—	11.0	10.6	9.6	4.8	3.9	—	—	—	—	—
5	—	17.5	16.7	15.2	7.6	6.1		—	—	—	—
7½	—	25.3	24.2	22	11	9	—	—	—	—	—
10	—	32.2	30.8	28	14	11	—	—	—	—	—
15	—	48.3	46.2	42	21	17	—	—	—	—	—
20	—	62.1	59.4	54	27	22	—	—	—	—	—
25	—	78.2	74.8	68	34	27	—	53	26	21	—
30	—	92	88	80	40	32	—	63	32	26	—
40	—	120	114	104	52	41	—	83	41	33	—
50	—	150	143	130	65	52	—	104	52	42	—
60	—	177	169	154	77	62	16	123	61	49	12
75	—	221	211	192	96	77	20	155	78	62	15
100	—	285	273	248	124	99	26	202	101	81	20
125	—	359	343	312	156	125	31	253	126	101	25
150	—	414	396	360	180	144	37	302	151	121	30
200	—	552	528	480	240	192	49	400	201	161	40
250	—	—	—	—	302	242	60	—	—	—	—
300	—	—	—	—	361	289	72	—	—	—	—
350	—	—	—	—	414	336	83	—	—	—	—
400	—	—	—	—	477	382	95	—	—	—	—
450	—	—	—	—	515	412	103	—	—	—	—
500	—	—	—	—	590	472	118	—	—	—	—

*For 90 and 80 percent power factor, the figures shall be multiplied by 1.1 and 1.25 respectively.

Figure 34–2 *NEC Table 430.250.*

rotor current, the full-load current rating must be increased when this is done. If the motor is operated at 90% power factor, the rated full-load current listed in the table must be increased by 10%. If the motor is to be operated at 80% power factor, the full-load current is to be increased by 25%.

Example 34–2

A 60-HP squirrel-cage induction motor is connected to a 480-volt three-phase circuit. What will be the full-load current value for this motor?

Find 60 HP in the far-left column of *Table 430.250* (*Figure 34–2*) and follow across to the 460-volt column. The motor has a full-load current rating of 77 amperes.

Example 34–3

A 200-HP, 2300-volt synchronous motor is to be operated at 90% power factor. What will be the full-load current rating of this motor?

Locate 200 HP in the far-left column of *Table 430.250.* Follow across to the 2300-volt column listed under synchronous-type motors. The table indicates a value of 40 amperes. Because the motor is operated at 90% power factor, increase this value by 10%:

$$40 \times 1.10 = 44 \text{ amperes}$$

34–2 Determining Conductor Size for a Single Motor

NEC Section 430.6(A) states that the conductor for a motor connection shall be based on the full-load current values from the tables. *NEC Section 430.22* states that conductors supplying a single motor operating in a continuous-duty application shall have an ampacity of not less than 125% of the motor full-load current. A motor is considered to be continuous-duty if it operates at a substantially constant load for an indefinitely long time. *NEC Article 310* is used to select the conductor size after the required ampacity has been determined. The exact table employed will be determined by the wiring conditions. The most frequently used table is *Table 310.16,* shown in *Figure 10–6.*

Termination Temperature Limitations

Another factor that must be taken into consideration when determining the conductor size is the **termination temperature,** a temperature rating of the devices and terminals as specified in *NEC Section 110.14(C).* This section states that the conductor is to be selected and coordinated as to not exceed the lowest temperature rating of any connected termination, any connected conductor, or any connected device. The conductors listed in the first column of *Table 310.16* have a temperature rating of 60°C, the conductors in the second column 75°C, and the conductors in the third column 90°C. The temperature ratings of devices such as circuit breakers, fuses, and terminals are found in the UL (Underwriters Laboratories) product directories. When the termination temperature rating is neither listed nor known, *NEC Section 110.14(C)(1)(a)* states that for circuits rated at 100 amperes or less, or for terminations marked for 14 AWG (American Wire Gauge) through 1 AWG conductors, terminations must be used with conductors rated at 60°C. This does not mean that only the types of insulation listed in the 60°C column can be used. Conductors with higher temperature insulation can be used provided the ampacity given to these conductors does not exceed that listed in the 60°C column for the given size conductor. For example, assume that a copper conductor with type XHHW insulation is to be connected to a 50-ampere circuit breaker that does not have a listed temperature rating. According to *NEC Table 310.16,* an 8 AWG copper conductor with XHHW insulation is rated to carry 55 amperes of current. Type XHHW insulation is located in both the 75°C column and the 90°C column, but the temperature rating of the circuit breaker is not known. Therefore, the ampacity of an XHHW conductor must be selected from the ampacity ratings in the 60°C column. A 6 AWG copper conductor with type XHHW insulation would be used.

For terminations rated for over 100 amperes or marked for conductor sizes larger than 1 AWG, *Section 110.14(C)(1)(b)* states that conductors with 75°C insulation must be used, or if a conductor with a higher temperature rating is used, its ampacity cannot exceed that listed in the 75°C column.

NEC Section 110.14(C)(1)(a)(4) has a special provision for motors with marked National Electrical Manufacturers Association (NEMA) design code B, C, or D. This section states that conductors rated at 75°C or higher may be selected from the 75°C column even if the ampacity is 100 amperes or less.

Example 34–4

A 50-HP, continuous-duty, three-phase squirrel-cage induction motor is connected to a 480-volt line. The conductors are run in conduit to the motor. The nameplate states that the motor is a NEMA Design B. The termination temperature rating of the devices is not known. Copper conductors with THWN insulation are to be used for this motor connection. What is the minimum size conductors that should be used?

The first step is to determine the full-load current of the motor. This is determined from *Table 430.250* (*Figure 35–2*). The table indicates a current of 65 amperes for this motor. The current must be increased by 25%, according to *NEC Section 430.22(A):*

$$65 \times 1.25 = 81.25, \text{ or } 81 \text{ amperes}$$

Example 34–4 (continued)

Table 310.16 is used to determine what size conductor is needed to supply this load. Locate the column that contains THWN insulation in the copper section of the table. THWN is located in the 75°C column. Because this circuit is less than 100 amperes and the termination temperature is not known, the conductor size would normally be selected from the ampacities listed in the 60°C column. However, *Section 110.14(C)(1)(a)(4)* allows Design B motors to use 75°C conductors. A 4 AWG copper conductor with type THWN insulation has an ampacity of 85 amperes in the 75°C column and is permitted.

Example 34–5

A 20-HP, continuous-duty, three-phase squirrel-cage induction motor is connected to 240 volts. The conductors are to be copper with THHN insulation. The nameplate lists a NEMA Design B for this motor. What is the minimum size conductors required for this motor?

Table 430.250 lists a current of 54 amperes for this motor, and *Section 430.22(A)* requires conductors that can supply 125% of this full load value:

$$54 \times 1.25 = 67.5\text{, or } 68 \text{ amperes}$$

Because the motor has a NEMA code letter listed on the nameplate, the conductor size can be selected from the 75°C column. Three 4 AWG copper conductors with THHN insulation will be used to connect this motor.

34–3 Overload Protection and Motor Starters

Overload protection should not be confused with the fuses and circuit breakers that are installed to provide short-circuit and ground-fault protection. Overload protection is designed to protect the motor from an overload condition. For example, if a motor has a full-load current rating of 10 amperes and the motor is connected to a circuit that is protected by a 20-ampere circuit breaker (*Figure 34–3*), *a current of 15 amperes flowing would constitute an overload but would not trip the 20-ampere circuit breaker.* The motor is drawing 150% of full-load current. This much overload can overheat the motor and damage the windings. **Overload relays** are designed to open the circuit when the current becomes 115 to 125% of the motor nameplate full-load current. The setting of the overload depends on the properties of the motor that is to be protected.

Overload relays are divided into two separate sections: the current-sensing section and the contact section. The current-sensing section is connected in series with the motor and senses the amount of motor current. This section is typically connected to voltages that range from 120 to 600 volts. The contact section is part of the control circuit and operates at the control circuit voltage. Control circuit voltages generally range from 24 to 120 volts. The contact section *does not* disconnect the motor from the line. The overload contacts are much too small to control motor currents. A magnetic relay, referred to as a motor starter, with contacts large enough to connect the motor to the line is used to connect and disconnect the motor from the circuit. The contact section of the overload relay is connected in series with the coil of the motor starter. A typical start–stop push-button control circuit is shown in *Figure 34–4.* Note that the overload heaters are connected in series with the motor, allowing them to sense the amount of motor current, and the overload contact is connected in series with the coil of the motor starter. If the overload contact opens, it will de-energize the motor starter coil, causing it to disconnect the motor from power.

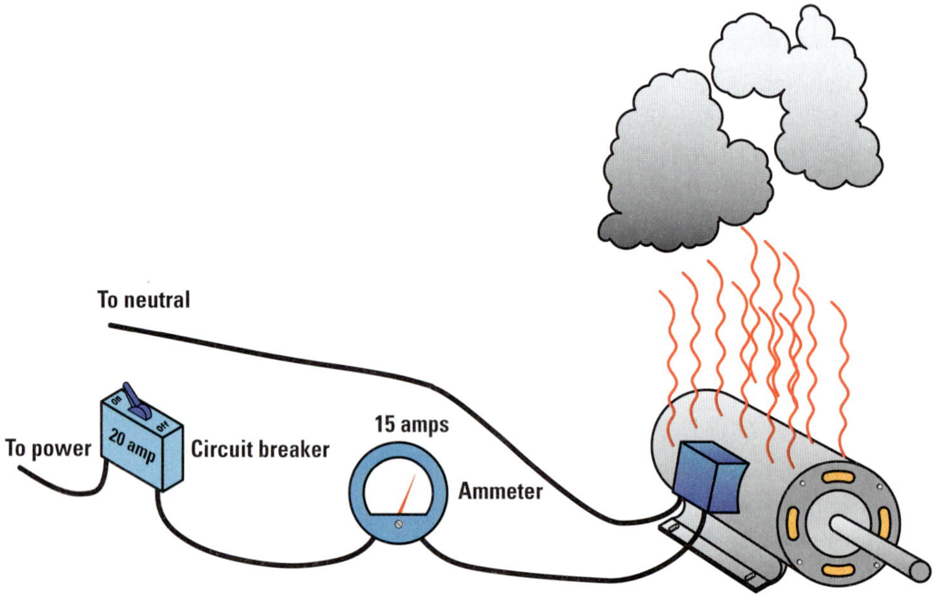

Figure 34–3 Motor circuit protected by a 20-ampere circuit breaker in overload conditions.

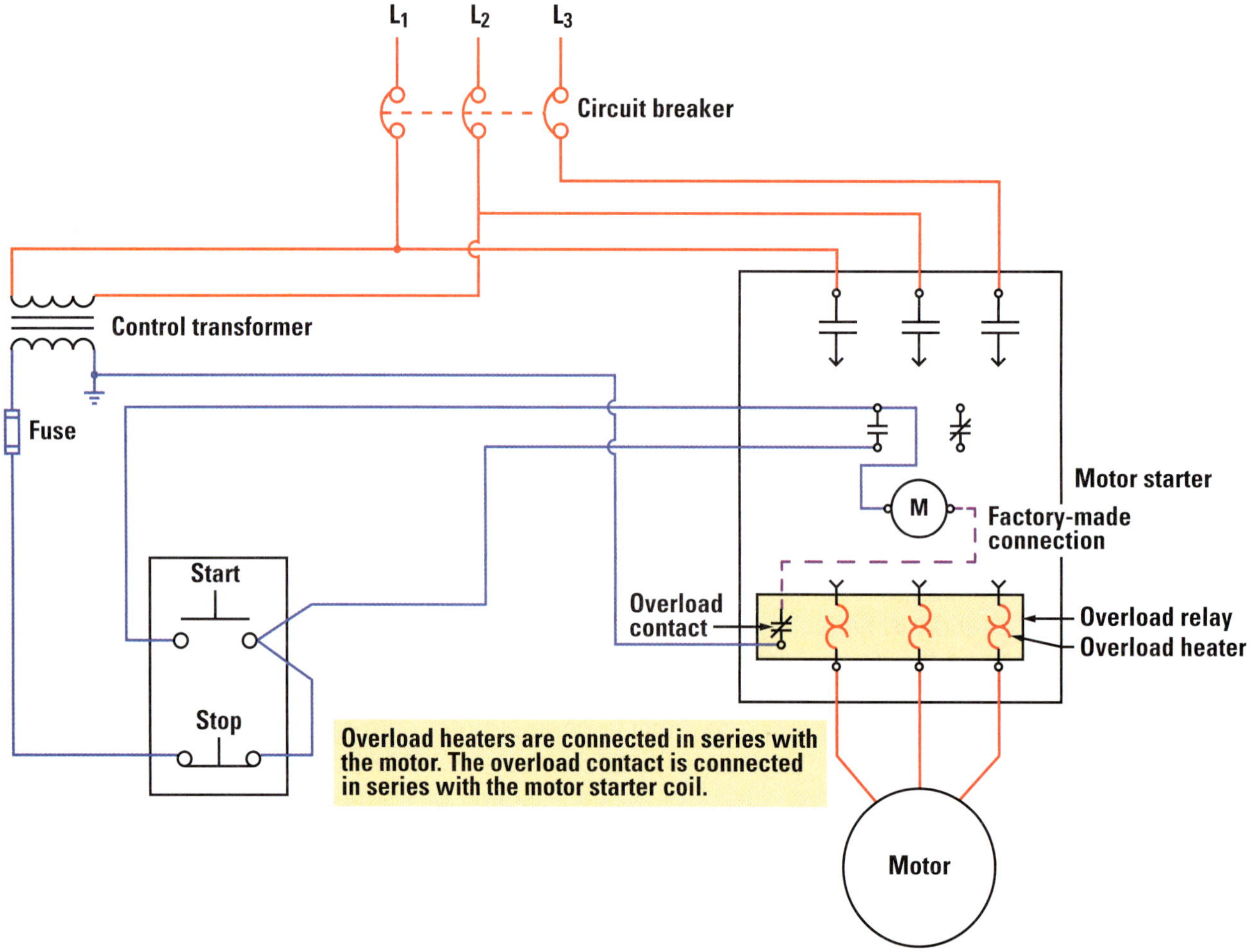

Figure 34–4 Example of a start–stop push-button control circuit and motor starter.

Overload relays are generally divided into one of three types: electronic, magnetic, and thermal. **Electronic overload relays** sense the amount of motor current by sensing the strength of the magnetic field of the conductors supplying current to the motor. These overload relays have internal timers that permit the motor time to start before opening the circuit. **Magnetic overload relays** connect a coil in series with the conductors supplying power to the motor. They are somewhat similar to electronic overload relays in that they sense motor current by magnetic field strength. Magnetic overload relays are generally used in areas that have extreme variations in ambient temperature because they are not sensitive to surrounding air temperature.

Thermal overload relays are by far the most common found in industry and will be the ones discussed in this unit. Thermal overload relays work by connecting a heating element in series with the motor. The amount of motor current determines the temperature of the heater. Overload heaters may appear different depending on the manufacturer and type of overload relay, but all perform the same function. Two different overload heaters are shown in *Figure 34–5.*

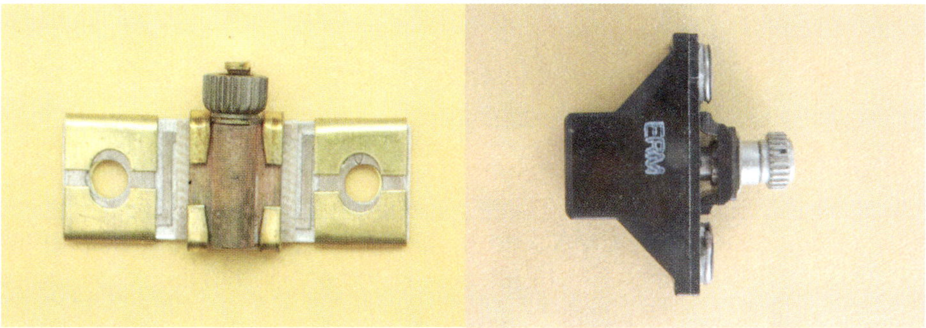

Figure 34–5 Overload heaters may appear different depending on manufacturer and type of overload relay.

NEC Section 430.32 is used to determine the overload size for motors (*Figure 34–6*). Motors with a marked service factor of 1.15 or greater or motors with a marked temperature rise of 40°C or less must have overload protection sized at no more than 125% of the nameplate current rating, and all other motors must have overload protection sized at no more than 115% of the nameplate current rating. When selecting the overload heater size for a particular motor, manufacturer charts are used to determine the proper heater size. *Figures 34–7* and *34–8* are examples of typical overload heater charts supplied by manufacturers of overload relays. The heater sizes in the chart shown in *Figure 34–8* are based on NEMA starter sizes. It may be necessary to first determine the starter size to select the proper heater.

Example 34–6

A 15-HP three-phase squirrel-cage motor is connected to 480 volts. The nameplate shows full-load amps of 18.6 amperes and a service factor of 1.00. Using the overload heater chart shown in *Figure 34–7,* what are the maximum size heaters that can be used to protect this motor?

Because the motor has a service factor (SF) of 1.00, the overload is to be sized at no more than 115% of the nameplate current. The full-load current of 18.6 amperes falls between 18.1 and 19.8 amperes. The proper overload heater size is XX46. Three XX46 heaters would be used to protect this motor.

430.32 Continuous-duty motors

(A) More than 1 horsepower. Each motor used in a continuous duty application and rated more than 1 HP shall be protected against overload by one of the following means in 430.32(A)(1) through (A)(4).

(1) Separate overload device. A separate overload device that is responsive to motor current. This device shall be selected to trip or shall be rated at no more than the following percent of the motor nameplate full-load current rating:

Motors with a marked service factor 1.15 or greater	125%
Motors with a marked temperature rise 40°C or less	125%
All other motors	115%

Figure 34–6 Overload protection requirements can be found in *Section 430.32* of the *NEC*.

Overload heater selection for NEMA starter sizes 00 – 1. Heaters are calibrated for 115% of motor full load current. For heaters that correspond to 125% of motor full load current use the next size larger heater.

Heater code	Motor full load current	Heater code	Motor full load current	Heater code	Motor full load current
XX01	.25 - .27	XX18	1.35 - 1.47	XX35	6.5 - 7.1
XX02	.28 - .31	XX19	1.48 - 1.62	XX36	7.2 - 7.8
XX03	.32 - .34	XX20	1.63 - 1.78	XX37	7.9 - 8.5
XX04	.35 - .38	XX21	1.79 - 1.95	XX38	8.6 - 9.4
XX05	.39 - .42	XX22	1.96 - 2.15	XX39	9.5 - 10.3
XX06	.43 - .46	XX23	2.16 - 2.35	XX40	10.4 - 11.3
XX07	.47 - .50	XX24	2.36 - 2.58	XX41	11.4 - 12.4
XX08	.51 - .55	XX25	2.59 - 2.83	XX42	12.5 - 13.5
XX09	.56 - .62	XX26	2.84 - 3.11	XX43	13.6 - 14.9
XX10	.63 - .68	XX27	3.12 - 3.42	XX44	15.0 - 16.3
XX11	.69 - .75	XX28	3.43 - 3.73	XX45	16,4 - 18.0
XX12	.76 - .83	XX29	3.74 - 4.07	XX46	18.1 - 19.8
XX13	.84 - .91	XX30	4.08 - 4.39	XX47	19.9 - 21.7
XX14	.92 - 1.00	XX31	4.40 - 4.87	XX48	21,8 - 23.9
XX15	1.01 - 1.11	XX32	4.88 - 5.3	XX49	24.0 - 26.2
XX16	.1.12 - 1.22	XX33	5.4 - 5.9		
XX17	1.23 - 1.34	XX34	6.0 - 6.4		

Figure 34–7 Example of a typical overload heater chart.

Overload heater selection for NEMA starter sizes 0 (00) – 4. Heaters are calibrated for 115% of motor full load current. For heaters that correspond to 125% of motor full load current use the next size larger heater.					
Heater type number	Size 0	Size 1	Size 2	Size 3	Size 4
X10	0.18	0.18	....	....	....
X11	0.20	0.20	....	....	....
X12	0.22	0.22	....	....	....
X13	0.24	0.24	....	....	....
X14	0.26	0.26	....	....	....
X15	0.29	0.29	....	....	....
X16	0.32	0.32	....	....	....
X17	0.35	0.35	....	....	....
X18	0.38	0.38	....	....	....
X19	0.42	0.42	....	....	....
X20	0.46	0.46	....	....	....
X21	0.51	0.51	....	....	....
X22	0.56	0.56	....	....	....
X23	0.62	0.62	....	....	....
X24	0.68	0.68	....	....	....
X25	0.75	0.75	....	....	....
X26	0.82	0.82	....	....	....
X27	0.90	0.90	....	....	....
X28	0.99	0.99	....	....	....
X29	1.09	1.09	....	....	....
X30	1.20	1.20	....	....	....
X31	1.32	1.32	....	....	....
X32	1.45	1.45	....	....	....
X33	1.59	1.59	....	....	....
X34	1.75	1.75	....	....	....
X35	1.93	1.93	....	....	....
X36	2.12	2.12	....	....	....
X37	2.33	2.33	....	....	....
X38	2.56	2.56	....	....	....
X39	2.81	2.81	....	....	....
X40	3.09	3.09	....	....	....
X41	3.40	3.40	....	....	....
X42	3.74	3.74	....	....	....
X43	4.11	4.11	....	....	....
X44	4.52	4.52	....	....	....
X45	4.97	4.97	....	....	....
X46	5.46	5.46	5.60	....	....
X47	6.01	6.01	6.15	....	....
X48	6.60	6.60	6.76	....	....
X49	7.26	7.26	7.43	....	....

Heater chart continued					
Heater type number	Size 0	Size 1	Size 2	Size 3	Size 4
X50	7.98	7.98	8.17	....	....
X51	8.78	8.78	8.98	....	....
X52	9.65	9.65	9.87	....	....
X53	10.6	10.6	10.8	....	....
X54	11.7	11.7	11.9	....	....
X55	12.8	12.8	13.1	....	....
X56	14.1	14.1	14.4	....	....
X57	15.4	15.4	15.7	....	....
X58	16.8	16.8	17.1	....	....
X59	18.3	18.3	18.6	....	....
X60	....	19.8	20.1	....	....
X61	....	21.3	21.7	25.5	....
X62	....	22.7	23.1	28.1	....
X63	....	24.4	24.8	31.0	32.0
X64	....	26.2	28.6	34.0	35.0
X65	....	28.2	30.5	37.0	38.5
X66	....	....	33.0	40.0	42.5
X67	....	....	35.5	43.5	46.5
X68	....	....	38.0	47.0	51.0
X69	....	....	40.5	51.0	55.0
X70	....	....	43.5	55.0	59.0
X71	....	....	47.0	59.0	64.0
X72	....	....	....	63.0	69.0
X73	....	....	....	67.0	74.0
X74	....	....	....	71.0	79.0
X75	....	....	....	76.0	84.0
X76	....	....	....	80.0	90.0
X77	....	....	....	85.0	96.0
X78	....	....	....	90.0	102
X79	....	....	....	....	107
X80	....	....	....	....	113
X81	....	....	....	....	118
X82	....	....	....	....	124
X83	....	....	....	....	130
X84	....	....	....	....	135

Figure 34–8 Example of an overload heater chart supplied by a manufacturer.

Example 34–7

A 60-HP, three-phase, squirrel-cage motor is connected to 480 volts. The nameplate shows a full-load current of 74.8 amperes and an SF of 1.20 for this motor. Using the chart shown in *Figure 34–8,* determine the overload heater size necessary to protect this motor.

Because the nameplate indicates a service factor greater than 1.15, the overload is to be sized for 125% of the nameplate current. The note at the top of the heater chart states that for heaters that correspond to 125% of full-load current, use the next larger size. The NEMA Starter chart in Figure 34–9 indicates that this motor requires a size 4 motor starter. The nameplate current of 74.8 amperes falls between 74 and 79 amperes in the size 4 column of the chart. When selecting overload heater sizes and the current falls between two values on the chart, always use the higher value. The heater number that is closest to but not less than 74.8 in the size 4 column is X74. However, the heater is to be sized for 125% of motor current, which would permit the next-larger heater. Keep in mind that the required overload protection device is a maximum rating and as long as there are no issues with the starting current of the motor, a smaller heater provides better protection of the motor and can save it from damage. For this installation, either the X74 or X73 heater can be used.

Example 34–7 (continued)

Maximum horsepower rating nonplugging and nonjogging duty				Maximum horsepower rating nonplugging and nonjogging duty			
NEMA size	Load volts	Single phase	Poly phase	NEMA size	Load volts	Single phase	Poly phase
00	115	½	....	3	115	7½	....
	200	....	1½		200	....	25
	230	1	1½		230	15	30
	380	....	1½		380	....	50
	460	....	2		460	....	50
	575	....	2		575	....	50
0	115	1	....	4	115	....	....
	200	....	3		200	....	40
	230	2	3		230	....	50
	380	....	5		380	....	75
	460	....	5		460	....	100
	575	....	5		575	....	100
1	115	2	....	5	115	....	....
	200	....	7½		200	....	75
	230	3	7½		230	....	100
	380	....	10		380	....	150
	460	....	10		460	....	200
	575	....	10		575	....	200
*1P	115	3	....	6	115	....	....
	200	....	....		200	....	150
	230	5	....		230	....	200
	380	....	....		380	....	300
	460	....	....		460	....	400
	575	....	....		575	....	400
2	115	3	....	7	230	....	300
	200	....	10		460	....	600
	230	7½	15		575	....	600
	380	....	25	8	230	....	450
	460	....	25		460	....	900
	575	....	25		575	....	900

Figure 34–9 NEMA size chart for motor starters.

Overload Protection for Large Motors

Large motors often have full-load currents that exceed the rating of the largest overload heaters. *NEC Table 430.250* lists a current of 590 amperes for a 500-HP squirrel-cage motor connected to 480 volts. The largest overload heater listed on the chart in *Figure 34–8* is 135 amperes. When large motors are supplied, current transformers (CTs) are sometimes used to reduce the current supplied to the overload heaters (*Figure 34–10*). Current transformers have a standard output of 5 amperes under a short-circuit condition when rated current flows through the primary. The outputs of the current transformers are connected to the overload heaters (*Figure 34–11*). The large number of heater sizes less than 5 amperes

Figure 34–10 Current transformers are used to reduce the current to overload heaters.

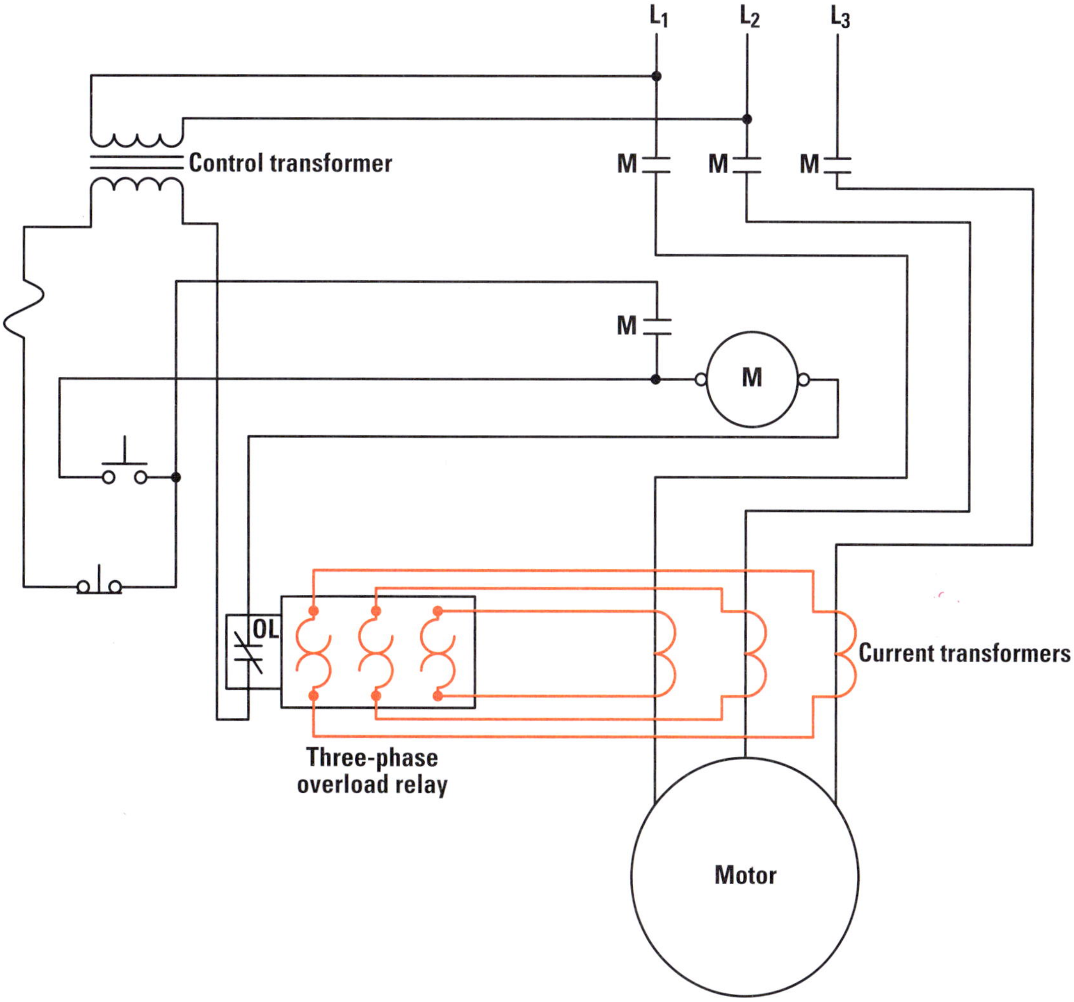

Figure 34–11 The output of the current transformers are connected to the overload heaters.

permits more precise sizing of heaters from current transformer outputs. It should be noted that when using current transformers to reduce the current supplied to the overload heater, it is no longer necessary to consider the NEMA starter size when selecting the correct heater size.

Example 34–8

A 150-HP motor is connected to 480 volts and has a nameplate current of 148 amperes. The motor has an SF of 1.10. Current transformers with a ratio of 250:5 are used to reduce the current supplying the overload heaters. Using the chart shown in *Figure 34–8,* what size overload heaters should be used to protect this motor?

Determine the output current of the current transformers when 148 amperes flow through the primary of the current transformer:

$$\frac{250}{5} = \frac{148}{X}$$

$$250\ X = 740\ (5 \times 148)$$

$$X = 2.96 \text{ amperes}$$

Because the motor has a service factor of 1.10, the overload heaters will be sized at 115% of the motor full-load current. X40 heaters will be used to protect the motor. Large motor starters are often equipped with current transformers. A size 5 motor starter with current transformers is shown in *Figure 34–12.*

Figure 34–12 Size 5 motor starter with current transformers.

Motor Starter Size

The size of the motor starter must also be considered when installing a motor. Starter sizes are rated by motor type, horsepower, and connected voltage. The two most common ratings are NEMA® (National Electrical Manufacturers Association) and IEC (International Electrotechnical Commission). A chart showing common NEMA size starters for alternating current motors is shown in *Figure 34–9,* and one for IEC starters is shown in *Figure 34–13.*

Example 34–9

A 50-HP three-phase motor is connected to 480 volts. What size NEMA starter should be used to connect this motor to the line?

The chart in *Figure 34–9* indicates a size 3 starter for this motor.

Example 34–10

What size IEC starter should be used to connect the motor in the above example?

The chart in *Figure 34–13* indicates a size K starter to connect the motor to the line.

34–4 Short-Circuit and Ground-fault Protection

The rating of the short-circuit/ground-fault protective device is determined by *NEC Table 430.52(C)(1)* (*Figure 34–14*). This table specifies the percentage of full-load current based on the type of motor and what type of short-circuit/ground-fault protective device being used. Most inductive motors are protected by dual-element time delay fuses or inverse time circuit breakers due to the in-rush current during starting. Dual-element time delay fuses and inverse time circuit breakers are designed to permit some amount of excessive current for a short period of time to permit motors to start.

Each of these columns lists the percentage of full-load current (table value, not nameplate) that is to be used in determining the ampere rating of the protective device. *NEC Section 430.52(C)(1)* states that the protective device is to have a rating or setting not to exceed the value calculated in accordance with *Table 430.52(C)(1).* However, if the calculated value does not correspond to a standard size or rating of a fuse or circuit breaker, then it shall be permissible to use the next higher standard size. The standard sizes of fuses and circuit breakers are listed in *NEC Table 240.6(A)* (*Figure 34–15*).

Table 430.52(C)(1) lists squirrel-cage motor types by NEMA design letters. *Section 430.7(A)(9)* requires that motor nameplates be marked with design letters A, B, C, or D. Motors manufactured before this requirement, however, do not list design letters on the nameplate. Most common squirrel-cage motors used are Design B motors and for purposes of selecting protective devices are considered to be Design B unless otherwise listed. Design B motors exhibit a little higher efficiency than other types of motors, but they also have a higher starting current.

Size	Max amps	Motor voltage	Maximum horsepower 1Φ	Maximum horsepower 3Φ
A	7	115	¼	
		200		1½
		230	½	1½
		460		3
		575		5
B	10	115	½	
		200		2
		230	1	2
		460		5
		575		7½
C	12	115	½	
		200		3
		230	2	3
		460		7½
		575		10
D	18	115	1	
		200		5
		230	3	5
		460		10
		575		15
E	25	115	2	
		200		5
		230	3	7½
		460		15
		575		20
F	32	115	2	
		200		7½
		230	5	10
		460		20
		575		25
G	37	115	3	
		200		7½
		230	5	10
		460		25
		575		30
H	44	115	3	
		200		10
		230	7½	15
		460		30
		575		40
J	60	115	5	
		200		15
		230	10	20
		460		40
		575		40
K	73	115	5	
		200		20
		230	10	25
		460		50
		575		50
L	85	115	7½	
		200		25
		230	10	30
		460		60
		575		75

Size	Max amps	Motor voltage	Maximum horsepower 1Φ	Maximum horsepower 3Φ
M	105	115	10	
		200		30
		230	10	40
		460		75
		575		100
N	140	115	10	
		200		40
		230	10	50
		460		100
		575		125
P	170	115		
		200		50
		230		60
		460		125
		575		125
R	200	115		
		200		60
		230		75
		460		150
		575		150
S	300	115		
		200		75
		230		100
		460		200
		575		200
T	420	115		
		200		125
		230		125
		460		250
		575		250
U	520	115		
		200		150
		230		150
		460		350
		575		250
V	550	115		
		200		150
		230		200
		460		400
		575		400
W	700	115		
		200		200
		230		250
		460		500
		575		500
X	810	115		
		200		250
		230		300
		460		600
		575		600
Z	1215	115		
		200		450
		230		450
		460		900
		575		900

Figure 34–13 IEC size motor starters (60 Hz).

Table 430.52(C)(1) maximum rating or setting of motor branch-circuit short-circuit and ground-fault protective devices

	Percentage of full-load current			
Type of motor	**Nontime delay fuse**	**Dual element (time-delay) fuse**	**Instantaneous trip breaker**	**Inverse time breaker**
Single-phase motors	300	175	800	250
AC polyphase motors other than wound-rotor	300	175	800	250
Squirrel cage other than Design B energy-efficient and Design B premium efficiency	300	175	800	250
Design B energy-efficient and Design B premium efficiency	300	175	1100	250
Synchronous	300	175	800	250
Wound-rotor	150	150	800	150
DC (constant voltage)	150	150	250	150

Note: For certain exceptions to the values specified, see 430.54.
The values in the nontime Delay Fuse column apply to time delay Class CC fuses.
The values given in the last column also cover the ratings of nonadjustable inverse time types of circuit breakers that may be modified as in 430.42(C)(1). Exceptions No. 1 and No. 2.
Synchronous motors of the low-torque, low-speed type (usually 450 RPM or lower) such as used to drive reciprocating compressors, pumps, and so forth that start unloaded do not require a fuse rating or circuit breaker setting in excess of 200 percent of full-load current.

Figure 34–14 *NEC Table 430.52(C)(1)* is used to determine the rating of the branch-circuit short-circuit and ground-fault protective device.

Standard ampere ratings				
10	15	20	25	30
35	40	45	50	60
70	80	90	100	110
125	150	175	200	225
250	300	350	400	450
500	600	700	800	1000
1200	1600	2000	2500	3000
4000	5000	6000		

Figure 34–15 *NEC Table 240.6(A)* lists standard sizes of fuses and circuit breakers. (Reproduced with permission of NFPA from NFPA 70®, National Electrical Code®, 2023 edition. Copyright © 2022, National Fire Protection Association. For a full copy of the NFPA 70®, please go to www.nfpa.org.)

Example 34–11

A 100-HP, three-phase squirrel-cage induction motor is connected to a 240-volt line. The motor does not contain a NEMA design code. A dual-element time delay fuse is to be used as the short-circuit and ground-fault protective device. Determine the maximum permitted rating of fuses for this circuit.

Table 430.250 lists a full-load current of 248 amperes for this motor. *Table 430.52(C)(1)* indicates that a dual-element time delay fuse is to be calculated at 175% of the full-load current rating for an AC poly-phase (more than one phase) squirrel cage motor. Because the motor does not list a NEMA design code on the nameplate, it will be assumed that the motor is Design B:

$$248 \times 1.75 = 434 \text{ amperes}$$

However, this does not correspond to a standard size fuse. The nearest standard fuse size above the computed value listed in *NEC Table 240.6(A)* (*Figure 34–15*) is 450 amperes, so 450 ampere fuses will be used to protect this motor. However, this is again the maximum value for the fuse and any fuse less than 450 amperes can be used as long as there are no issues starting the motor.

If for some reason the selected size fuse will not permit the motor to start due to the fuse blowing, *NEC Section 430.52(C)(1)(b)(2)* states that the rating of a dual-element time delay fuse may be increased to a maximum of 225% of the full load motor current. However, if 225% does not correspond to a standard size fuse, the next higher size is not permitted to be used.

Example Problems

In the following examples, the conductor size, overload protective device size, and fuse or circuit-breaker size will be determined for different motors.

Example Problem 34–1

A 7.5-HP, single-phase squirrel-cage induction motor is connected to a 230-volt line. The motor is continuous-duty. The conductors are to be copper with type XHHW-2 insulation. The motor is protected by an inverse time circuit breaker. The motor nameplate lists the following information:

HP 7½, Amps 36, Phase 1, Volts 230, SF 1.00, NEMA Design B Code V

NEC Table 430.248 (*Figure 34–1*) lists a running current of 40 amperes for a 7.5-HP single-phase motor connected to 230 volts. To determine the minimum required conductor size, the running current must be increased by 25%:

$$40 \times 1.25 = 50 \text{ amperes}$$

Table 310.16 is used to find the conductor that can supply this level of ampacity. *Section 110.14(C)* states that for currents of 100 amperes or less, the ampacity rating of the conductor may be determined from the 75°C column if a motor lists a NEMA Design Code on the nameplate. In this instance, the insulation type is located in the 90°C column. Therefore, the ampacity of the conductor must not exceed that in the 75°C column instead of the 90°C column. An 8 AWG copper conductor with type XHHW-2 insulation will be used.

The NEMA and IEC starter sizes are chosen from the charts shown in *Figures 34–9* and *34–13*. The chart in *Figure 34–9* indicates a NEMA size 2 starter, and the chart in *Figure 34–13* an IEC size H starter.

The overload heater size is determined from *NEC Section 430.32(A)(1)*. Because the service factor is less than 1.15, the heading ALL OTHER MOTORS will be used. The heater size will be selected on the basis of 115% of motor full-load current. The heater should be selected from the column listing size 2 starters. A size X67 heater is selected from the chart shown in *Figure 34–8*.

The inverse time circuit breaker size is selected using *NEC Table 430.52(C)(1)* (*Figure 34–14*). The table indicates that the full-load current value from *NEC Table 430.248* is to be increased by 250%:

$$40 \times 2.50 = 100 \text{ amperes}$$

NEC Table 240.6(A) (*Figure 34–15*) lists 100 amperes as one of the standard-size circuit breakers. A 100-ampere three-phase circuit breaker will be used to protect the motor circuit from short-circuit and ground-fault. *Figure 34–16* shows all values for this motor installation.

Example Problem 34–1 (continued)

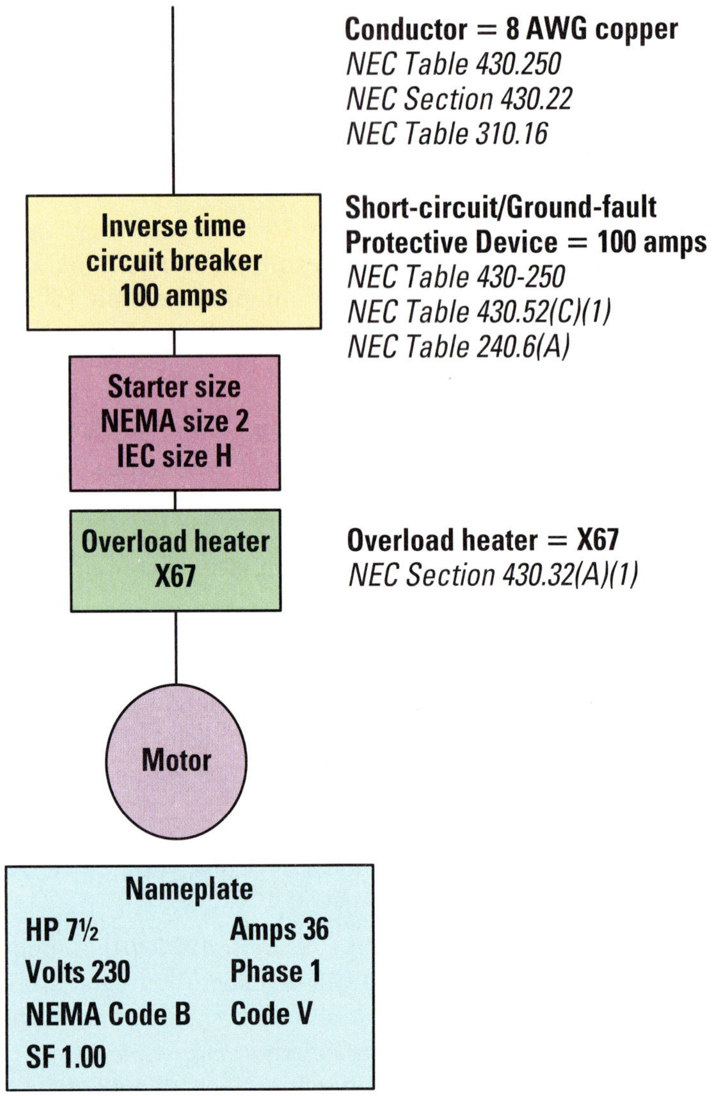

Figure 34–16 Calculations for Example Problem 34–1.

Example Problem 34–2

A 60-HP, three-phase squirrel cage induction motor is connected to a 480-volt line. The motor is continuous duty. The conductors are to be copper with type THWN insulation. The motor is protected by dual element time delay fuses. The motor nameplate lists the following information:

HP 60, Amps 71, Phase 3, Volts 480, SF 1.10, NEMA Code B Code U

NEC Table 430.250 (*Figure 34–2*) lists a running current of 77 amperes for a 60-HP, three-phase, induction motor connected to 460 volts. To determine the minimum conductor size, the running current must be increased by 25%:

$$77 \times 1.25 = 96.25, \text{ or } 96 \text{ amperes}$$

Table 310.16 is used to find the conductor size. *Section 110.14(C)* states that for currents of 100 amperes or less, the ampacity rating of the conductor may be determined from the 75°C column if a motor lists

Example Problem 34–2 (continued)

a NEMA Design Code on the nameplate. The conductor size is determined using the 75°C column. A 3 AWG copper conductor with type THWN insulation will be used.

The NEMA and IEC starter sizes are chosen from the charts shown in *Figures 34–9* and *34–13.* The chart in *Figure 34–9* indicates a NEMA size 4 starter, and the chart in *Figure 34–13* an IEC size L starter.

The overload heater size is determined from *NEC Section 430.32(A)(1).* Because the service factor is less than 1.15, the heading ALL OTHER MOTORS will be used. The heater size will be selected on the basis of 115% of motor full-load current. This motor requires a NEMA size 4 starter; therefore, the heater should be selected from the size 4 column. A size X74 heater is selected from the chart shown in *Figure 34–8.*

The dual-element time delay fuse size is selected using *NEC Table 430.52(C)(1)* (*Figure 34–14*). The table indicates that the full-load current value from *NEC Table 430.250* is to be increased by 175%:

$$77 \times 1.75 = 134.75, \text{ or } 135 \text{ amperes}$$

NEC Table 240.6(A) (*Figure 34–15*) lists 150 amperes as the nearest standard-size fuse. Three 150-ampere fuses will be used to protect the motor circuit. *Figure 34–17* shows all values for this motor installation.

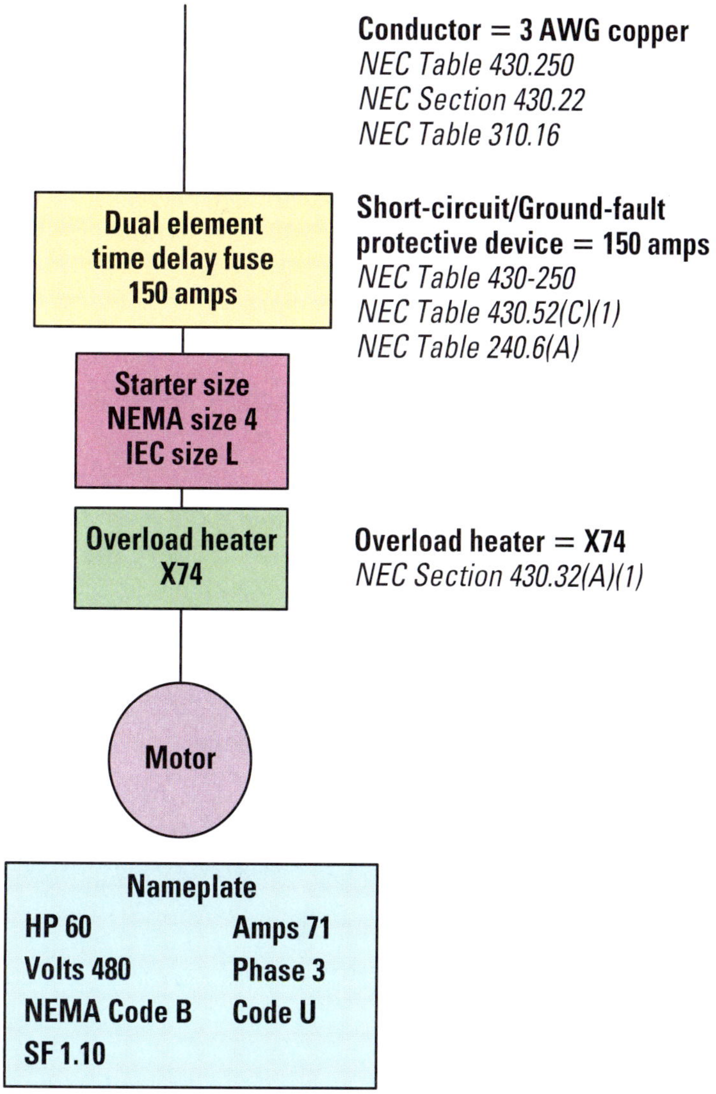

Figure 34–17 Calculations for Example Problem 34–2.

Example Problem 34–3

A 100-HP, three-phase, squirrel-cage induction motor is connected to a 480-volt line. The motor is continuous-duty. The conductors are to be copper with type THWN insulation. The motor is protected by dual-element time delay fuses. The motor nameplate lists the following information:

HP 100, Amps 117, Phase 3, Volts 480, SF 1.25, NEMA Code B Code J

Because of the high running current, current transformers with a ratio of 200:5 will be used to reduce the current to the overload heaters.

NEC Table 430.250 (*Figure 34–2*) list a running current of 124 amperes for a 100-HP, three-phase induction motor connected to 460 volts. To determine the minimum conductor size, the running current must be increased by 25%:

$$124 \times 1.25 = 155 \text{ amperes}$$

Table 310.16 is used to find the conductor size with sufficient ampacity to supply this motor. *NEC Section 110.14(C)* states that for currents greater than 100 amperes, the ampacity rating of the conductor may be determined from the 75°C column; 2/0 AWG copper conductors with type THWN insulation will be used.

The NEMA and IEC starter sizes are chosen from the charts shown in *Figures 34–9* and *34–13*. The chart in *Figure 34–9* indicates a NEMA size 4 starter, and the chart in *Figure 34–13* indicates an IEC size N starter.

The maximum overload heater size is determined from *NEC Section 430.32(A)(1)*. Because the service factor is greater than 1.15, the heater size will be selected on the basis of 125% of the motor full-load nameplate current. Current transformers with a ratio of 200:5 are to be employed to reduce the amount of current supplied to the overload heaters:

$$\frac{200}{5} = \frac{117}{X}$$

$$200\,X = 585\ (5 \times 117)$$

$$X = 2.925$$

The overload heater chart in *Figure 34–7* shows that heater XX26 is rated for currents between 2.84 and 3.11. This overload heater, however, must be sized at 125% of the nameplate current instead of 115%. Therefore, the next size heater must be used. XX27 heaters will be used to protect this motor.

The dual-element time delay fuse size is selected using *NEC Table 430.52(C)(1)* (*Figure 34–14*). The table indicates that the full-load current value from *NEC Table 430.250* is to be increased by 175%.

$$124 \times 1.75 = 217 \text{ amperes}$$

NEC Table 240.6(A) (*Figure 34–15*) lists 225 amperes as the nearest standard-size fuse. Three 225-ampere fuses will be used to protect the motor circuit. *Figure 34–18* shows all values for this motor installation.

Example Problem 34–3 (continued)

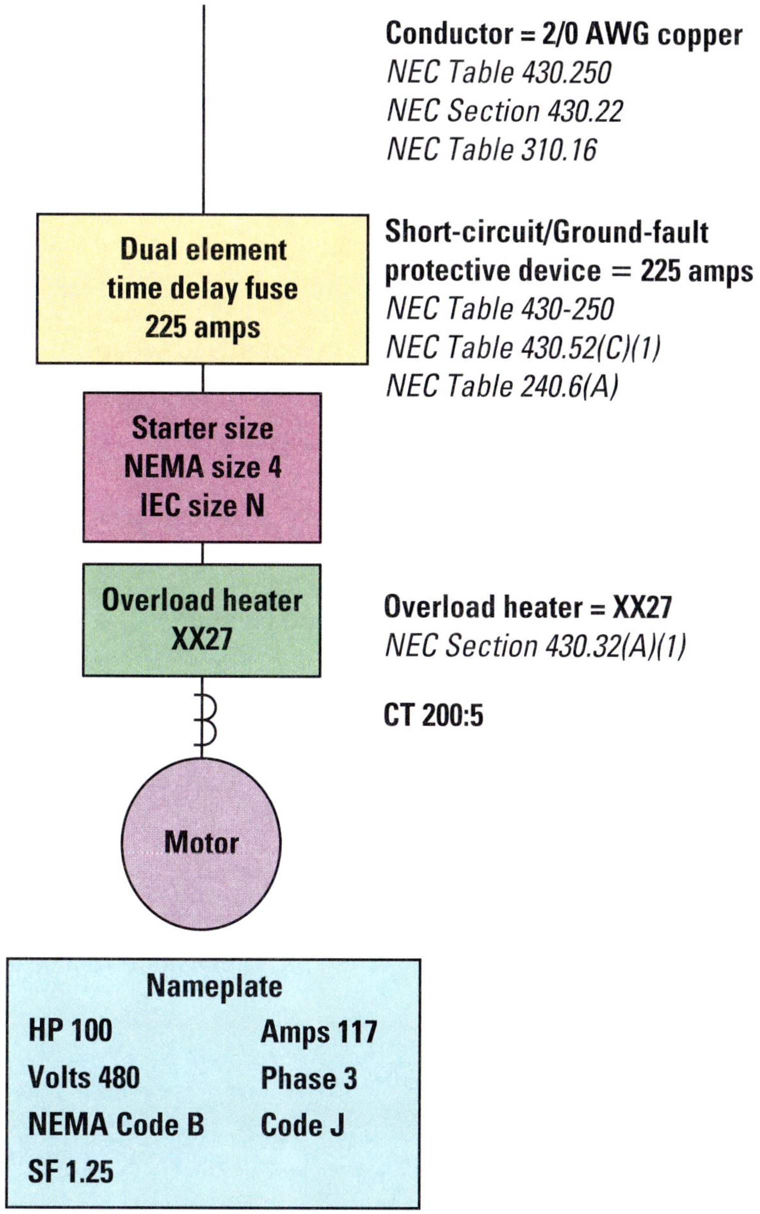

Figure 34–18 Calculations for Example Problem 34–3.

Example Problem 34–4

A 150-HP, three-phase synchronous motor is connected to a 480-volt line. The motor is continuous-duty and is to be operated with a 90% power factor. The conductors are to be copper with type THWN insulation. The motor is protected by dual element time delay fuses. The motor nameplate lists the following information:

HP 150, Amps 142, Phase 3, Volts 480, SF 1.25, NEMA Code A Code C, PF 90% Type Synchronous

Because of the high running current, current transformers with a ratio of 250:5 will be used to reduce the current to the overload heaters.

NEC Table 430.250 (*Figure 34–2*) lists a running current of 151 amperes for a 150-HP, three-phase synchronous motor connected to 460 volts. Because the motor is to be operated at 90% power factor, the table current must be increased by 10%:

$$151 \times 1.10 = 166.1, \text{ or } 166 \text{ amperes}$$

This current will be the table current value for this motor in every application throughout this example.

To determine the minimum conductor size, the running current must be increased by 25%:

$$166 \times 1.25 = 207.5, \text{ or } 208 \text{ amperes}$$

Table 310.16 is used to find the conductor size. *NEC Section 110.14(C)* states that for currents greater than 100 amperes, the ampacity rating of the conductor may be determined from the 75°C column; 4/0 AWG copper conductors with type THWN insulation will be used.

The NEMA and IEC starter sizes are chosen from the charts shown in *Figures 34–9* and *34–13*. The chart in *Figure 34–9* indicates a NEMA size 5 starter, and the chart in *Figure 34–13* indicates an IEC size R starter.

The overload heater size is determined from *NEC Section 430.32(A)(1)*. Because the service factor is greater than 1.15, the heater size will be selected on the basis of 125% of the motor full-load nameplate current. Current transformers with a ratio of 250:5 are to be employed to reduce the amount of current supplied to the overload heaters:

$$\frac{250}{5} = \frac{141}{X}$$

$$250\ X = 710\ (5 \times 142)$$

$$X = 2.84$$

The overload heater chart in *Figure 34–7* shows that heater XX26 is rated for 3.09 amperes. This overload heater, however, must be sized at 125% of the nameplate current instead of 115%. Therefore, the next size heater must be used. XX27 heaters will be used to protect this motor.

The dual-element time delay fuse size is selected using *NEC Table 430.52(C)(1)* (*Figure 34–14*). The table indicates that the full-load current value from *NEC Table 430.250* is to be increased by 175%:

$$166 \times 1.75 = 290.6, \text{ or } 291 \text{ amperes}$$

NEC Table 240.6(A) (*Figure 35–15*) lists 300 amperes as the nearest standard size fuse. Three 300-ampere fuses will be used to protect the motor circuit. *Figure 34–19* shows all values for this motor installation.

Example Problem 34–4 (continued)

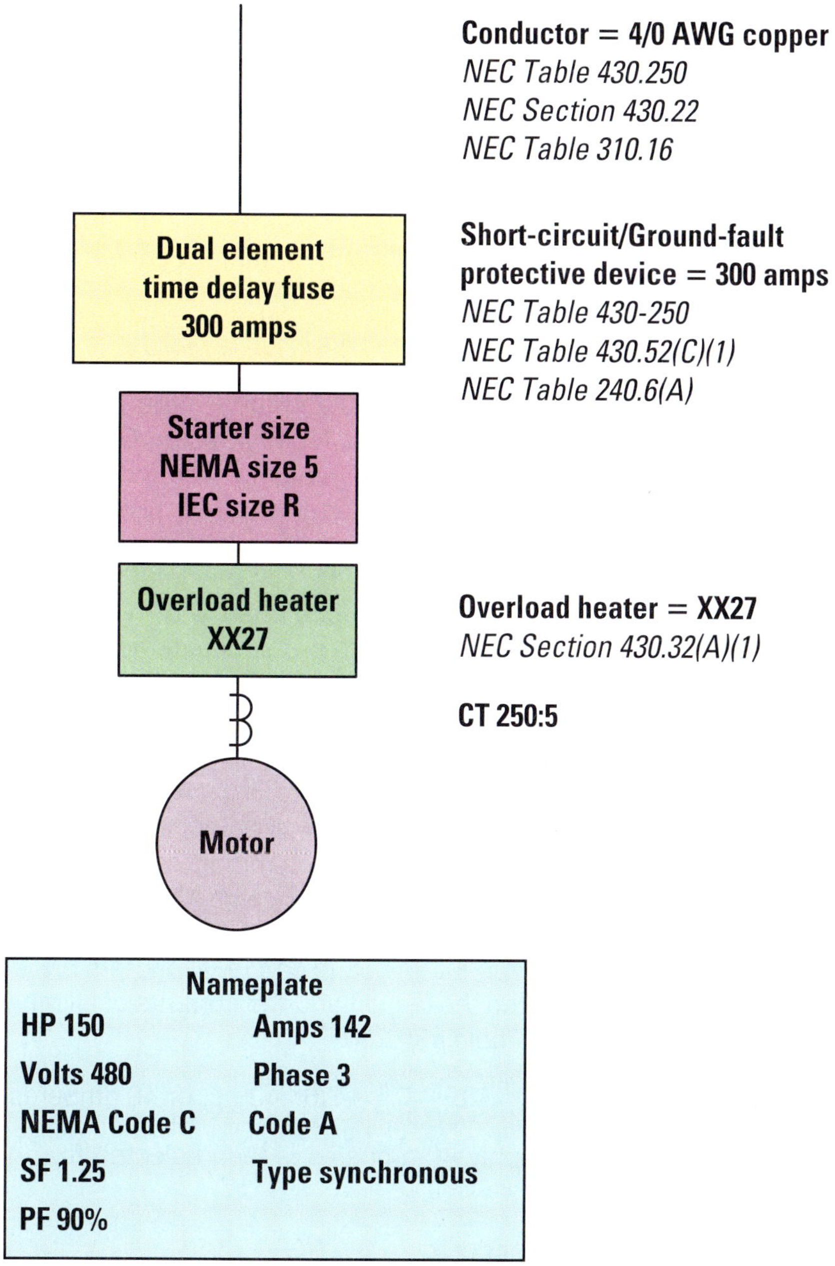

Figure 34–19 Calculations for Example Problem 34–4.

34–5 Multiple Motor Calculations

Motor feeder short-circuit and ground-fault protective devices for multiple motor connections can be found in *NEC Section 430.62(A)*, and conductor sizes can be found in *NEC Section 430.24*. The requirements for a feeder are very similar to a single motor, there are just more motors. Let's look at an example of five motors that are connected to a common feeder. The feeder is 480 volts and three-phase, and the conductors are to be copper with type THWN insulation. Each motor is to be protected with an inverse time circuit breaker and a separate overload device. The main feeder is also protected by an inverse time circuit breaker. The termination temperature rating of the connected devices is not known. All the motors are the same, and the nameplates state the following:

HP 5, Bolts 480, Amps 6.5, SF 1.00, NEMA Code B Code J, Phase 3

Determine the conductor size, circuit breaker size, NEMA and IEC starter size for each motor, and overload heater size. Determine the conductor size and fuse size for the feeder supplying all the connected motors.

NEC Table 430.250 lists a current of 7.6 amperes for a 5-HP motor connected to 460 volts. To determine the conductor size, increase the table current by 25%:

$$7.6 \times 1.25 = 9.5$$

NEC Table 310.16 lists a 14 AWG with an ampacity of 20 amperes in the 75°C column as the nearest standard-size conductor. Because the motor contains a NEMA Design Code, the conductor can be installed in accordance with its ampacity from the 75°C column. The individual motors will be connected with 14 AWG copper conductors. The NEMA starter size is 0 selected from the chart shown in *Figure 34–9,* and the IEC starter size B is selected from the chart shown in *Figure 34–13.*

The motor nameplate lists a service factor less than 1.15. The overload heater will be sized at 115% of the motor nameplate current of 6.5 amperes. The overload chart shown in *Figure 34–7* lists XX35 as the correct heater size. Three XX35 heaters will be used to protect each motor.

NEC Table 430.52(C)(1) (*Figure 34–14*) indicates that an inverse time circuit breaker is to be sized at 250% of the current listed in *NEC Table 430.250:*

$$7.6 \times 2.50 = 19 \text{ amperes}$$

The nearest standard-size circuit breaker listed in *NEC Table 240.6(A)* (*Figure 34–15*) is 20 amperes. A 20-ampere inverse time circuit breaker is used to protect each motor.

The main feeder conductor size is computed in accordance with *NEC Section 430.24* by increasing the largest amperage rating of the motors connected to the feeder by 25% and then adding the full-load current ratings of the other motors to this amount. In this example, all motors are the same. Therefore, the running current of one motor is considered the largest and the currents of all other motors are added to the calculation as determined from *NEC Table 430.250:*

$$7.6 \times 1.25 = 9.5 + 7.6 + 7.6 + 7.6 + 7.6 = 39.9, \text{ or } 40 \text{ amperes}$$

A conductor with sufficient ampacity to supply 40 amperes is then selected from the 75°C column of *NEC Table 310.16:*

$$\text{Conductor} = 8 \text{ AWG}$$

The size of the short-circuit and ground-fault protective device is determined by *Section 430.62(A).* The *NEC* states that the rating or setting of the feeder short-circuit and ground-fault protective device ***shall not be greater than*** the rating or setting of the largest branch circuit short-circuit and ground fault protective device for any single motor supplied by the feeder plus the sum of all the full-load currents of the other motors connected to the feeder. The largest protective device is a 20-ampere circuit breaker. The running currents of all other motors as per the *NEC* tables will be added to this value.

$$20 + 7.6 + 7.6 + 7.6 + 7.6 = 50.4 \text{ or } 50 \text{ amperes}$$

NEC Table 240.6(A) (*Figure 34–14*) indicates that a 50-ampere inverse time circuit breaker is a standard size circuit breaker. *Figure 34–20* shows all calculations for this installation.

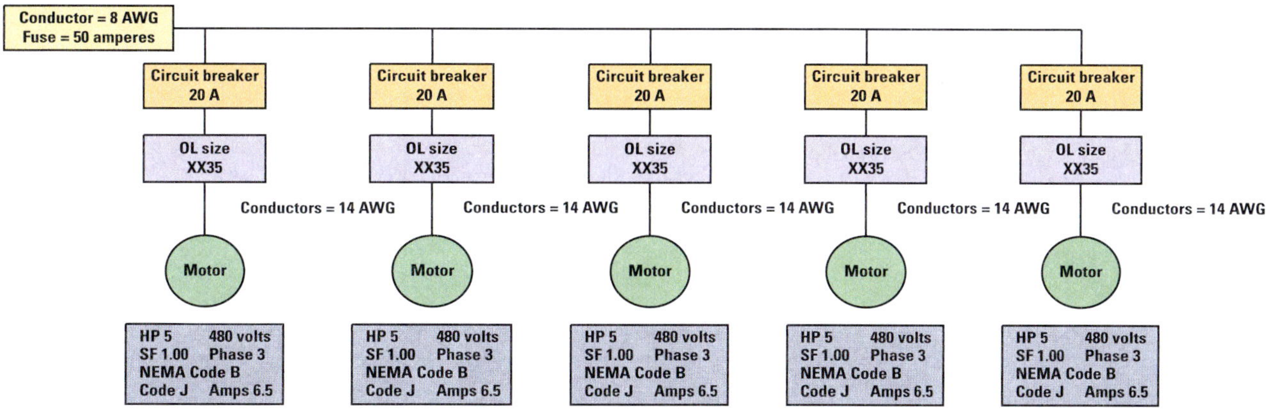

Figure 34–20 Calculations for multiple motor calculation.

Summary

- When installing motors, there are two different current values to consider—full-load current from the tables in the *NEC* and the motor nameplate current.
- Full-load currents from the tables in *NEC Article 430* are used when determining conductor, short-circuit and ground-fault protection, and the disconnect and controller sizes.
- Nameplate full-load amps current is used when determining overload sizes.
- Fuses and circuit breakers are used to protect the circuit from short-circuit and ground-fault conditions.
- Overload devices or fuses are used to protect the motor.
- Overload relays contain two sections—the current-sensing section and the contact section.
- Overload relays do not disconnect the motor from the power source.
- Motor controllers called starters are used to connect and disconnect the motor from the power source.
- Thermal overload relays are the most commonly used in industry.
- Thermal overload relays sense motor current by connecting a heater in series with the motor.
- The current rating of an overload relay is determined by the overload heater used.
- Overload heater charts are supplied by the manufacturer of the overload relay.
- Dual-element time delay fuses and inverse time circuit breakers are generally used to protect inductive loads such as transformers and motors.
- The two most common ratings for motor starters are NEMA and IEC.
- Motor starter sizes are chosen by the horsepower rating of the motor and connected voltage.

Review Questions

1. Which NEC table is used to determine the full-load running current for single-phase motors for the purposes of sizing conductors and fuses?

   A) Table 430.248

   B) Table 430.249

   C) Table 430.250

   D) Table 430.251(A)

2. A 5-HP single-phase motor is connected to 208 volts. What is the full-load running current for this motor?
   A) 56 amperes
   B) 28 amperes
   C) 32.2 amperes
   D) 30.8 amperes
3. A 100-HP three-phase squirrel-cage induction motor is connected to 480 volts. Which table in Article 430 of the NEC should be used to determine motor current for conductor sizing?
   A) Table 430.247
   B) Table 430.250
   C) Table 430.248
   D) Table 430.52(C)(1)
4. A 75-HP synchronous motor connected to a 480-volt three-phase feeder is running a water pump. The motor is operating at an 80% power factor. How many full-load amperes must be used to determine conductor size?
   A) 78 amperes
   B) 155 amperes
   C) 97.5 amperes
   D) 101 amperes
5. True or False: The NEC requires motor branch circuit conductors to be sized at 125% of the full-load current value on the nameplate.
6. What minimum size THHN copper conductor is required to supply the branch circuit conductors to a 100-HP, 480-volt, three-phase Design B motor?
   A) 1/0 AWG
   B) 2/0 AWG
   C) 1 AWG
   D) 3/0 AWG
7. A squirrel-cage induction motor has a nameplate full-load current of 79 amperes and a marked temperature rise of 30°C. The motor requires a NEMA size 4 starter. Using the overload heater chart in *Figure 34–8*, what size overload heaters should be used to protect this motor?
   A) X50
   B) X76
   C) X71
   D) X74

8. Continuous-duty motors marked with service factors greater than 1.15 are permitted to be protected from overload by a separate overload device rated no greater than which of the following?

   A) 125%

   B) 115%

   C) 140%

   D) 130%

9. A 480-volt, three-phase, 10-HP motor is supplied by a set of dual-element time-delay fuses. What is the maximum size permitted to supply this branch circuit if there are no issues with starting current?

   A) 25 amperes

   B) 30 amperes

   C) 40 amperes

   D) 20 amperes

10. True or False: If the maximum size short-circuit and ground-fault protective device required from Table 430.52(C)(1) does not correspond to a standard size fuse or circuit breaker rating, the next higher standard size is permitted to be used.

11. A motor feeder circuit supplies 10 three-phase 480-volt motors rated at 25 HP each. What is the maximum size inverse-time circuit breaker that can be used to supply this feeder? Starting current is not an issue.

    A) 400 amperes

    B) 375 amperes

    C) 350 amperes

    D) 325 amperes

12. True or False: The NEC requires a feeder to be protected from short-circuit and ground-fault issues by a device that is sized at the sum of the ratings for all individual motors.

Practical Applications

1. A 30-HP squirrel-cage induction motor is connected to a 240-volt three-phase feeder. The motor is to be replaced with a 50-HP squirrel cage induction motor connected to 480 volts. The motor has a NEMA Code B, a full-load current of 49.8 amps, and an SF of 1.10. The present conductors are 3 AWG copper with type THWN insulation. Are these conductors sufficient to connect the 50-HP motor, or will they have to be replaced?

---

2. Refer to the preceding Practical Application question. When the 30-HP motor is replaced with the 50-HP, will the existing motor starter have to be replaced? If so, what size starter should be selected? A NEMA size 3 starter was used for the 30-HP motor.

---

3. Refer to the first Practical Application question in this set. When the 50-HP motor replaces the 30-HP motor, the overload heaters will have to be changed. Using the chart in *Figure 34–8*, what size overload heaters should be used to protect the 50-HP motor?

Practice Problems

1. A 10-HP single-phase motor is connected to 240 volts. The motor nameplate indicates a NEMA Code B, a full-load running current of 47.5 amperes, and a temperature rise of 40°C. What is the minimum size copper conductors with THWN insulation allowed to supply this motor?

2. A 30-HP three-phase squirrel-cage induction motor is connected to 240 volts. The motor is to be protected by an inverse time circuit breaker. What is the maximum size device permitted to protect the branch circuit from short-circuit and ground-fault protection for this motor?

3. A 75-HP synchronous motor is connected to a 480-volt three-phase feeder. The conductors are to be copper with type XHHW-2 insulation. The nameplate lists a running current of 74 amperes, a NEMA code letter B, and a service factor of 1.00. The motor is to operate at unity, or 100% power factor. Determine the minimum conductor size, maximum overload heater size using the chart shown in *Figure 34–8,* NEMA starter size, and the maximum dual-element time-delay fuse that can be used for branch-circuit short-circuit and ground-fault protection for this motor installation.

4. A 40-HP dual-voltage squirrel-cage motor is connected to a 240-volt three-phase feeder. The nameplate lists: HP 40, Volts 240/480, Full-load amps 98/49, Service Factor 1.25, NEMA Code B. The conductors are to be copper with THWN insulation. The circuit is protected by dual-element time delay fuses. Determine the minimum conductor size, the maximum overload heater size using the chart shown in *Figure 34–8,* and branch-circuit short-circuit and ground-fault protection for this installation.

5. A 350-HP squirrel-cage induction motor is connected to a 480-volt three-phase feeder. The nameplate lists the following: HP 350, Service factor 1.10, Full-load amps 408, NEMA Code B. The conductors are to be copper with THWN insulation, and 600:5 current transformers are used to reduce the current supplied to the overload heaters. The motor circuit is protected with dual-element time delay fuses. Determine the minimum conductor size and the branch-circuit short-circuit and ground-fault protection for this installation.

6. Three squirrel-cage motors are connected to a 480-volt three-phase feeder. The nameplates list the following:

   Motor 1: HP 10, SF 1.25, Full-load amps 13.1, NEMA Code B

   Motor 2: HP 25, SF 1.10, Full-load amps 32.8, NEMA Code B

   Motor 3: HP 15, SF 1.00, Full-load amps 19.6, NEMA Code B

   All conductors are to be copper, with THWN insulation. The main feeder and each of the motors are protected by dual-element time delay fuses. Determine the minimum conductor size for each motor and the feeder supplying the motors, the maximum overload heater size for each motor using the chart in *Figure 34–8*, the branch-circuit short-circuit and ground-fault protection for each motor, and the short-circuit and ground-fault protection the main feeder.

Unit 35 Harmonics

Key Terms

Negative-sequence harmonic

Nonlinear loads

Phasor rotation

Positive-sequence harmonic

Pulse-width modulation

Triplen harmonics

Why You Need to Know

Modern electrical equipment is utilizing more and more electronic devices to control voltage, current, and frequency. These electronic devices often operate by turning on and off at different rates, causing pulsations on the power line. These pulsations can cause the normal sine wave to become distorted. Harmonics are a result of this distortion. This unit

- discusses the frequency of different harmonics.
- discusses the name and sequence of different harmonics and the problems associated with them.
- describes the methods of detecting harmonics.
- describes the methods of dealing with harmonic problems.

Outline

Objectives

After studying this unit, you should be able to

- differentiate harmonic currents from normal currents.
- analyze a circuit's load characteristics to determine if harmonic currents are likely.
- explain the impact of harmonic currents on single-phase circuits.
- explain the impact of harmonic currents on three-phase circuits.
- describe how the transformer harmonic derating factor is used to determine the impact of harmonic currents on transformers.

Preview

Harmonics are a source of problems in many industries. These problems range from overheating conductors and transformers to circuit breakers unexpectedly tripping under what seem like normal conditions, to motors burning out for no apparent reason. Understanding the conditions that cause harmonics—and how to correct the problems—is essential for anyone working in the electrical field.

35–1 What Are Harmonics?

Harmonics are voltages or currents that operate at a frequency that is a multiple of the fundamental power frequency. If the fundamental power frequency is 60 hertz, for example, the second harmonic would be 120 hertz, the third harmonic would be 180 hertz, and so on (*Figure 35–1*). Harmonics are produced by **nonlinear loads** that draw current in pulses rather than in a continuous manner. Harmonics on single-phase circuits are generally caused by devices such as computer power supplies, electronic dimmers, LED drivers, and so on. Three-phase harmonics are generally produced by variable-frequency drives for AC motors and electronic drives for DC motors. A good example of a pulsating load is one that converts AC into DC and then regulates the DC voltage by **pulse-width modulation** (*Figure 35–2*).

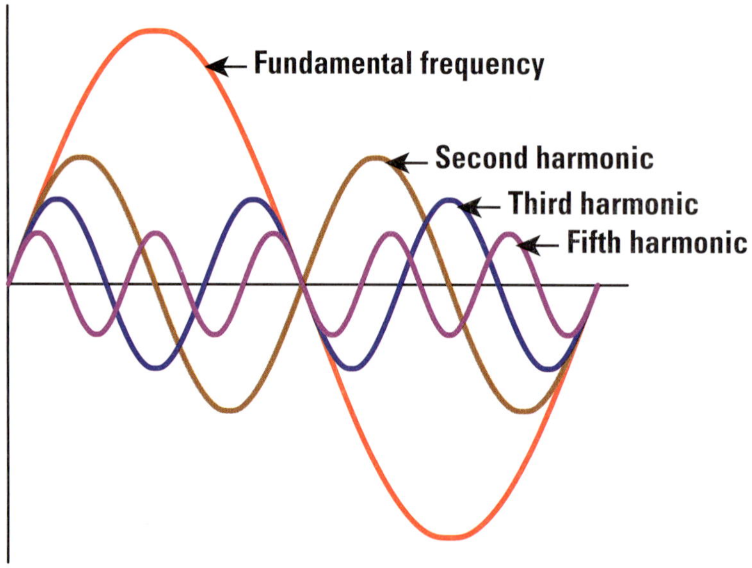

Figure 35–1 Harmonic frequencies are a multiple of a base or fundamental frequency.

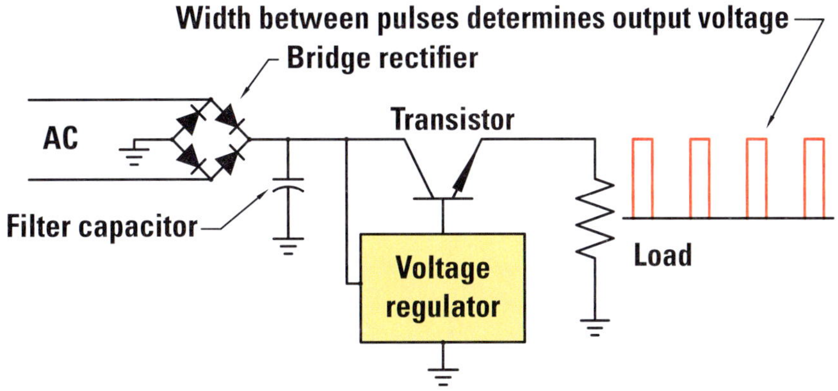

Figure 35–2 Pulse-width modulation regulates the output voltage by varying the time the transistor conducts compared to the time it is turned off.

Many regulated power supplies operate in this manner. The bridge rectifier in *Figure 35–2* changes the AC into pulsating DC. A filter capacitor is used to smooth the pulsations. The transistor turns on and off to supply power to the load. The amount of time the transistor is turned on compared to the time it is turned off determines the output DC voltage. Each time the transistor turns on, it causes the capacitor to begin discharging. When the transistor turns off, the capacitor will begin to charge again. Current is drawn from the AC line each time the capacitor charges. These pulsations of current produced by the charging capacitor can cause the AC sine wave to become distorted. These distorted current and voltage waveforms flow back into the other parts of the power system (*Figure 35–3*).

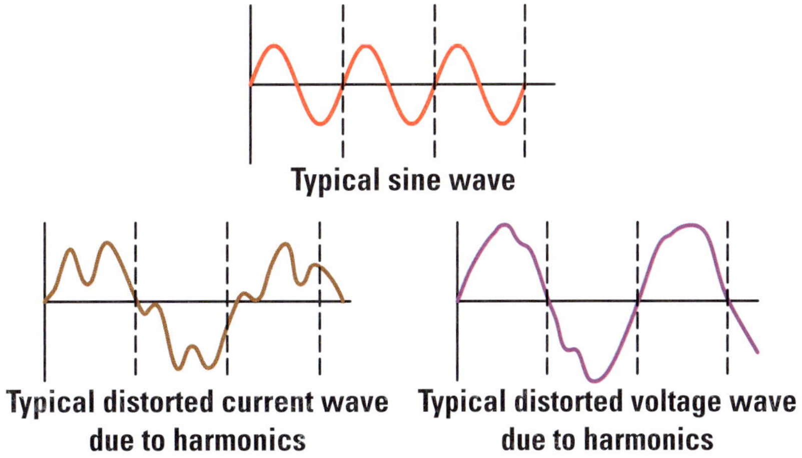

Figure 35–3 Harmonics cause a distortion of the typical AC sine wave.

Harmonic Effects

Harmonics can have very detrimental effects on electric equipment. Some common symptoms of harmonics problems are overheated equipment and circuit breakers that trip when no fault or overload exists. Harmonics are classified by name, frequency, and sequence. The name refers to whether the harmonic is the second, third, fourth, or so on of the fundamental or base frequency. The frequency of the harmonic is always a multiple of the base frequency. The second harmonic of a 60-hertz fundamental frequency operates at 120 hertz, the third at 180 hertz, the fourth at 240 hertz, and so on. The sequence refers to the **phasor or phasor rotation** with respect to the fundamental waveform. The operating principle of a three-phase induction motor is a rotating magnetic field. A **positive-sequence harmonic** would cause the magnetic field to rotate in the same direction as the fundamental frequency. A **negative-sequence harmonic** would cause the magnetic field to rotate in the opposite direction of the fundamental frequency. A particular set of harmonics called "triplens" has a zero sequence. **Triplen harmonics** are the odd multiples of the third harmonic (third, ninth, fifteenth, twenty-first, etc.). A chart showing the sequence of the first nine harmonics is shown in *Table 35–1*.

Name	Fund.	2nd	3rd	4th	5th	6th	7th	8th	9th
Frequency	60	120	180	240	300	360	420	480	540
Sequence	+	±	0	+	±	0	+	±	0

Table 35–1 Name, Frequency, and Sequence of the First Nine Harmonics

Harmonics with a positive sequence generally cause overheating issues in electrical equipment. Negative-sequence harmonics cause the same heating problems as positive harmonics. However, there are additional concerns with motors. Because the phasor rotation of a negative harmonic is opposite that of the fundamental frequency, it will tend to weaken the rotating magnetic field of an induction motor, causing it to produce less torque. The reduction of torque causes the motor to operate below normal speed. The reduction in speed results in excessive motor current and overheating.

Although triplens do not have a phasor rotation, they can cause a great deal of trouble from overheating in a three-phase four-wire system, such as a 208/120-volt or 480/277-volt system. In a common 208/120-volt WYE-connected system, the primary is generally connected in delta, and the secondary is connected in WYE (*Figure 35–4*).

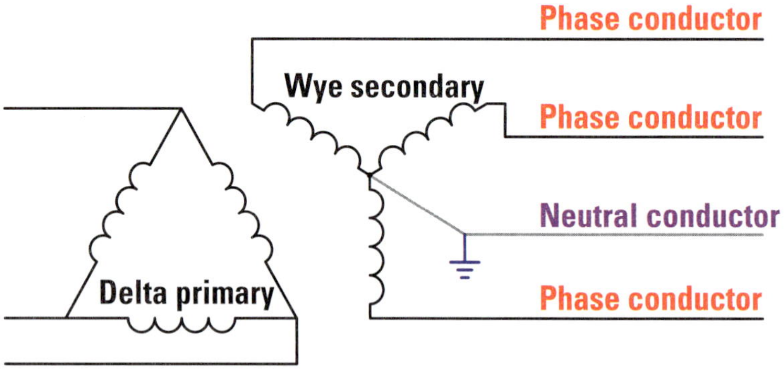

Figure 35–4 In a three-phase four-wire WYE-connected system, the center of the WYE-connected secondary is tapped to form a neutral conductor.

Single-phase loads that operate on 120 volts are connected between any phase conductor and the neutral conductor. The neutral current is the vector sum of the phase currents. In a balanced three-phase circuit (all phases having equal current), the neutral current is zero. Although single-phase loads tend to cause an unbalanced condition, the vector sum of the currents causes the neutral conductor to carry less current than any of the phase conductors. This is true for loads that are linear and draw a continuous sine wave current. When pulsating (nonlinear) currents are connected to a three-phase four-wire system, triplen harmonic frequencies disrupt the normal phasor relationship of the phase currents and can cause the phase currents to add in the neutral conductor instead of cancel. For this reason, *NFPA 70®: National Electrical Code®* requires the neutral conductor of three-phase, four-wire systems to be counted as a current carrying conductor when a major portion of the load being supplied is nonlinear. Because the neutral conductor is not protected by a fuse or circuit breaker, there is real danger of excessive heating in the neutral conductor. There are several cable manufacturers that produce cables with what is commonly referred to as a "super-neutral" to prevent this overheating.

Harmonic currents are also reflected in the delta primary winding, where they circulate and cause overheating. Other heating problems are caused by eddy current and hysteresis losses. Transformers are typically designed for 60-hertz operation. Higher harmonic frequencies produce greater core losses than the transformer is designed to handle. Transformers that are connected to circuits that produce harmonics must sometimes be derated or replaced with transformers that are specially designed to operate with harmonic frequencies.

Transformers are not the only point in the system to be affected by harmonic currents. Generators can also be affected in the same way as transformers. This is especially true for standby generators used to power data-processing equipment in the event of a power failure. Some harmonic frequencies can even distort the zero crossing of the waveform produced by the generator.

Thermal-magnetic circuit breakers use a bimetallic trip mechanism that is sensitive to the heat produced by the circuit current. These circuit breakers are designed to respond to the heating effect of the true-RMS current value. If the current becomes too great, the bimetallic mechanism trips the breaker open. Harmonic currents cause a distortion of the RMS value, which can cause the breaker to trip when it should not or not to trip when it should. Thermal-magnetic circuit breakers, however, are generally better protection against harmonic currents than electronic circuit breakers. Electronic breakers sense the peak value of current. The peaks of harmonic currents can be higher than the fundamental sine wave peak values (*Figure 35–5*). Although the peaks of harmonic currents are often higher than the fundamental frequency, they can be lower. In some cases, electronic breakers may trip at low currents, and in other cases, they may not trip at all.

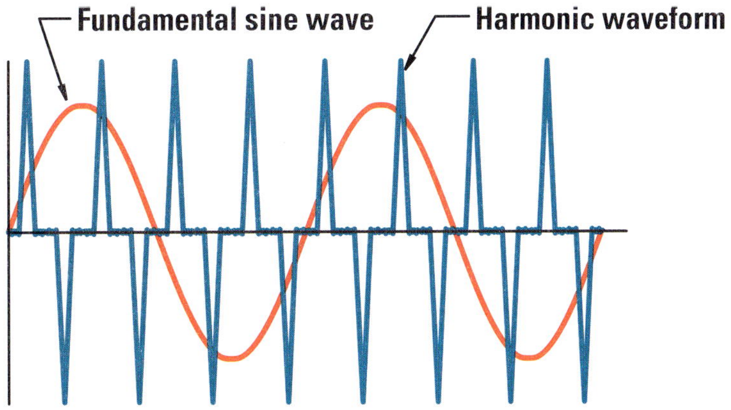

Figure 35–5 Harmonic waveforms can have higher peak values than the fundamental waveform.

Triplen harmonic currents can also cause problems with neutral buss ducts and connecting lugs. A neutral buss is sized to carry the rated phase current. Because triplen harmonics can cause the neutral current to be higher than anticipated, it is possible for the neutral buss to become overloaded.

Panelboards and buss ducts are designed to carry currents that operate at 60 hertz or whatever frequency they have been designed for. Harmonic currents produce magnetic fields that operate at higher frequencies. If these fields should become mechanically resonant with the panel or buss duct enclosures, the panels and buss ducts can vibrate and produce buzzing sounds at the harmonic frequency.

Telecommunications equipment is also often affected by interference from harmonic currents. Telecommunication cables are often run on the same support structures as distribution power lines. To minimize interference, communication cables are run as far from phase conductors as possible but are relatively close to the neutral conductor. Harmonic currents in the neutral conductor can induce high-frequency currents into the communication cable. These high-frequency currents can be heard as a high-pitch buzzing sound on telephone lines.

35–2 Harmonic Concerns in Single-Phase Systems

There are several steps to follow to determine if there is a problem with harmonics. One step is to do a survey of the equipment. This is especially important in determining if there is a problem with harmonics in a single-phase system.

1. Make an equipment check. Equipment such as computers, printers, and lighting equipment with electronic ballasts/LED drivers are known to produce harmonics. Any piece of equipment that draws current in pulses can produce harmonics.
2. Review maintenance records to see whether there have been problems with circuit breakers tripping for no apparent reason.
3. Check transformers for overheating. If the cooling vents are unobstructed and the transformer is operating excessively hot, harmonics could be the problem. Check transformer currents with an ammeter capable of indicating a true-RMS current value. Make sure that the voltage and current ratings of the transformer have not been exceeded.

It is necessary to use an ammeter that responds to true RMS current when making this check. Some ammeters respond to the average value, not the RMS value. Meters that respond to the true-RMS value generally state this on the meter. Meters that respond to the average value are generally less expensive and do not state that they are true-RMS meters.

Meters that respond to the average value use a rectifier to convert the AC into DC. This value must be increased by a factor of 1.111 to change the average reading into the RMS value for a sine wave current. True-RMS responding meters calculate the heating effect of the current. The chart in *Figure 35–6* shows some of the differences between average-indicating meters and true-RMS meters. In a distorted waveform, the true-RMS value of current will no longer be *Average* × 1.111 (*Figure 35–7*). The distorted waveform can cause the average value to be as much as 50% less than the RMS value.

Ammeter type	Sine wave response	Square wave response	Distorted wave response
Average responding	Correct	Approximately 10% high	As much as 50% low
True-RMS responding	Correct	Correct	Correct

Figure 35–6 Comparison of average responding and true-RMS responding ammeters.

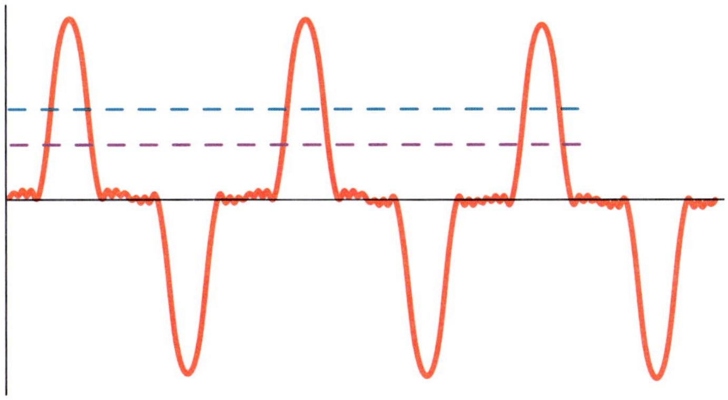

Figure 35–7 Average current values are generally less than the true-RMS value in a distorted waveform.

Another method of determining whether a harmonic problem exists in a single-phase system is to make two separate current checks. One check is made using an ammeter that indicates the true-RMS value, and the other is made using a meter that indicates the average value (*Figure 35–8*). In this example, it is assumed that the true-RMS ammeter indicates a value of 36.8 amperes and the average ammeter indicates a value of 24.8 amperes. Determine the ratio of the two measurements by dividing the average value by the true-RMS value:

$$\text{Ratio} = \frac{\text{Average}}{\text{RMS}}$$

$$\text{Ratio} = \frac{24.8\text{ A}}{36.8\text{ A}}$$

$$\text{Ratio} = 0.674$$

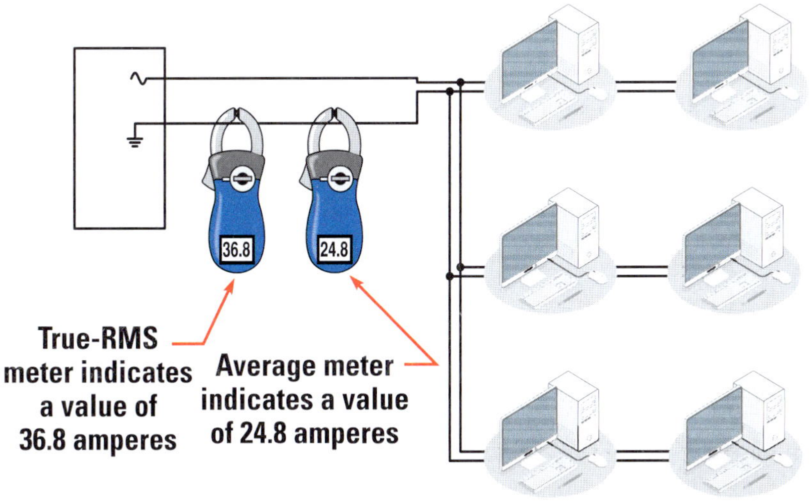

Figure 35–8 Determining harmonic problems using two ammeters.

A ratio of 1 would indicate no harmonic distortion. A ratio of 0.5 would indicate extreme harmonic distortion. This method does not reveal the name or sequence of the harmonic distortion, but it does give an indication that there is a problem with harmonics. To determine the name, sequence, and amount of harmonic distortion present, a harmonic analyzer should be employed (*Figure 35–9*).

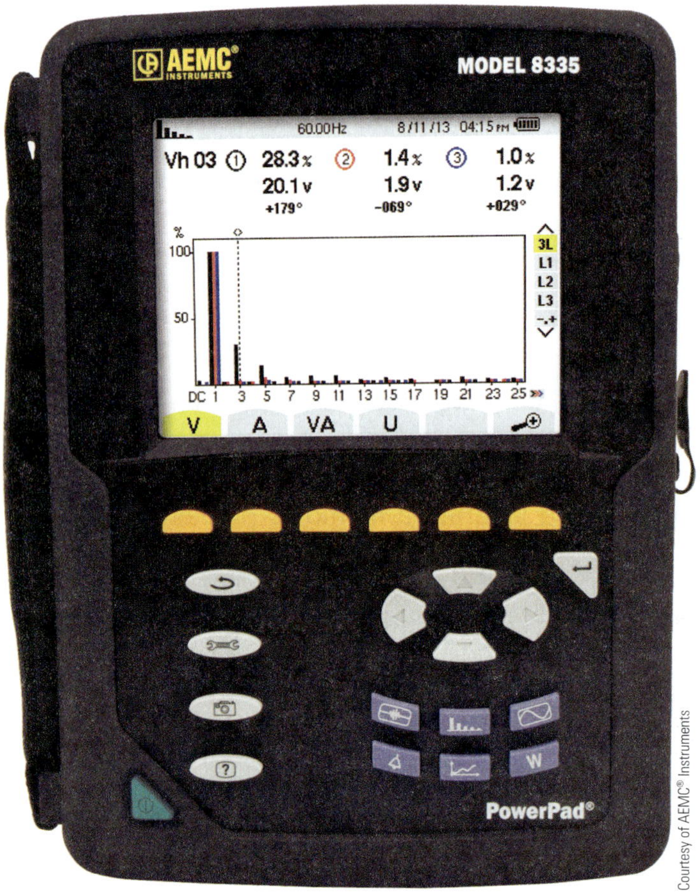

Courtesy of AEMC® Instruments

Figure 35–9 Power meter with harmonic display.

35–3 Harmonic Concerns in Three-Phase Systems

Determining whether a problem with harmonics exists in a three-phase system follows a similar process. Because harmonic problems in a three-phase system generally occur in a WYE-connected four-wire system, this example assumes a delta-connected primary and WYE-connected secondary with a center-tapped neutral, as shown in *Figure 35–4*. To test for harmonic distortion in a three-phase, four-wire system, measure all phase currents and the neutral current with both a true-RMS indicating ammeter and an average-indicating ammeter. It is assumed that the three-phase system being tested is supplied by a 200-kilovolt-ampere transformer, and the current values shown in *Table 35–2* were recorded. The current values indicate that a problem with harmonics does exist in the system. Note the higher current measurements made with the true-RMS indicating ammeter and also the fact that the neutral current is higher than any phase current.

Dealing with Harmonic Problems

After it has been determined that harmonic problems exist, something must be done to deal with the problem. It is generally not practical to remove the equipment causing the harmonic distortion, so other methods must be employed. It is a good idea to consult a power-quality expert to determine the exact nature and amount of harmonic distortion present. Some general procedures for dealing with harmonics follow.

Conductor	True-RMS responding ammeter	Average indicating ammeter
Phase 1	365	292
Phase 2	396	308
Phase 3	387	316
Neutral	488	478

Table 35–2 Measured Current Values

1. In a three-phase, four-wire system, the 60-hertz part of the neutral current can be reduced by balancing the current on the phase conductors. If all phases have equal current flow, the neutral current would be zero.
2. If triplen harmonics are present on the neutral conductor, harmonic filters can be added at the load. These filters can help reduce the amount of harmonics on the line. Harmonic filters are constructed with a combination of inductors, capacitor, and resistors. These filters are tuned to the frequency of a specific harmonic and are designed to conduct the energy of the harmonic to ground. The design of the filter is dictated by the frequency of the harmonic to be eliminated (*Figure 35–10*).

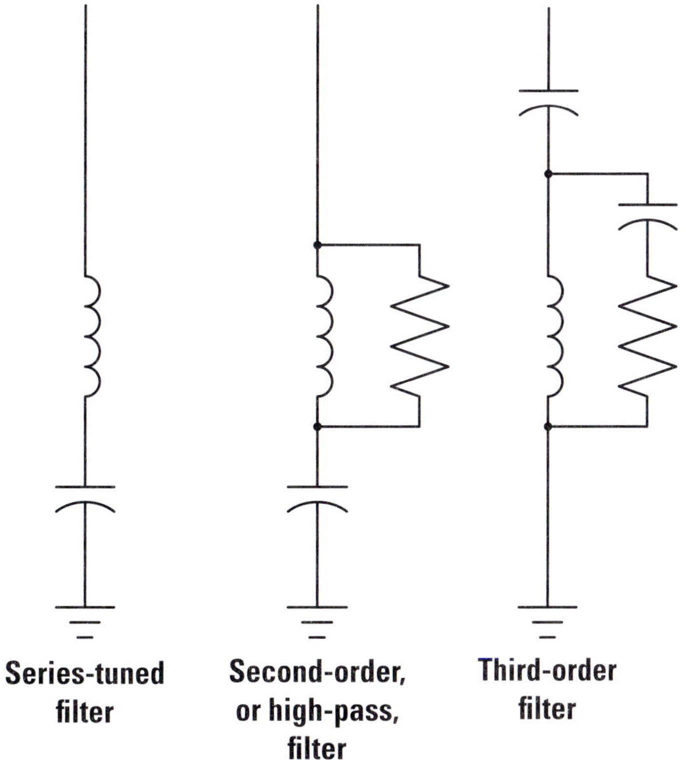

Figure 35–10 The construction of a harmonic filter is dictated by the frequency of the harmonic to be eliminated.

3. Pull extra neutral conductors. The more common practice is to use a separate neutral for each phase instead of using a shared neutral.
4. Install a larger ("super") neutral conductor. If it is impractical to supply a separate neutral conductor for each phase, increase the size of the common neutral.
5. Derate or reduce the amount of load on the transformer. Harmonic problems generally involve overheating of the transformer. In many instances, it is necessary to derate the transformer to a point at which it can handle the extra current caused by the harmonic distortion. When this is done, it is generally necessary to add a second transformer and divide the load between the two.

35–4 Transformer Harmonic Derating Factor

Probably the most practical and straightforward method for determining the derating factor for a transformer is that recommended by the Computer and Business Equipment Manufacturers Association. In this method, we will again make two ampere measurements. One is the true-RMS current of the phases, and the second is the instantaneous peak phase current. The instantaneous peak current can be determined with an oscilloscope connected to a current probe or with an ammeter capable of indicating the peak value of current. Most digital clamp-on ammeters have the ability to indicate average, true-RMS, and peak values of current. For this example, it is assumed that peak current values are measured for the 200-kilovolt-ampere transformer discussed previously. These values are added to the previous data obtained with the true-RMS and average-indicating ammeters (*Table 35–3*).

Conductor	True-RMS responding ammeter	Average indicating ammeter	Instantaneous peak current
Phase 1	365 A	292 A	716 A
Phase 2	396 A	308 A	794 A
Phase 3	387 A	316 A	737 A
Neutral	488 A	478 A	957 A

Table 35–3 Peak Currents Are Added to the Chart

The formula for determining the transformer harmonic derating factor is:

$$\text{THDF} = \frac{(1.414)(\text{RMS phase current})}{\text{instantaneous peak current}}$$

This formula produces a derating factor somewhere between 0 and 1.0. Because instantaneous peak value of current is equal to the RMS value × 1.414, if the current waveforms are sinusoidal (no harmonic distortion), the formula produces a derating factor of 1.0. If harmonic currents are present and adding to the system, the measured instantaneous current will be greater than the RMS value × 1.414 and will give a result less than 1.0; this will be the derating factor. Once the derating factor is determined, multiply the derating factor by the kilovolt-ampere capacity of the transformer. The product is the maximum load that should be placed on the transformer.

If the phase currents are unequal, find an average value by adding the currents together and dividing by three:

$$\text{Phase (RMS)} = \frac{365\text{ A} + 396\text{ A} + 387\text{ A}}{3}$$

$$\text{Phase (RMS)} = 382.7\text{ A}$$

$$\text{Phase (Peak)} = \frac{716\text{ A} + 794\text{ A} + 737\text{ A}}{3}$$

$$\text{Phase (Peak)} = 749$$

$$\text{THDF} = \frac{(1.414)(382.7\text{ A})}{749\text{ A}}$$

$$\text{THDF} = 0.722$$

The 200-kilovolt-ampere transformer in this example should be derated to 144.4 kilovolt-amperes (200 kVA $\times$ 0.722).

Summary

- Harmonics are generally caused by loads that pulse the power line.
- Harmonic distortion on single-phase lines is often caused by computer power supplies, copy machines, fax machines, and LED drivers.
- Harmonic distortion on three-phase power lines is generally caused by variable-frequency drives and electronic DC drives.
- Harmonics can have a positive rotation, negative rotation, or no rotation.
- Positive-rotating harmonics rotate in the same direction as the fundamental frequency.
- Negative-rotating harmonics rotate in the opposite direction of the fundamental frequency.
- Triplen harmonics are the odd multiples of the third harmonic.
- Harmonic problems can generally be determined by using a true-RMS ammeter and an average-indicating ammeter or by using a true-RMS ammeter and an ammeter that indicates the peak value.
- Triplen harmonics generally cause overheating of the neutral conductor on three-phase, four-wire systems.

Review Questions

1. True or False: A harmonic current is any current that exists on the line at a frequency other than the applied frequency of the source. Harmonics are typically in multiples of the fundamental frequency.
2. Which of the following frequencies would indicate the presence of a triplen harmonic on a 60-hz system?

   A) 360 hz

   B) 180 hz

   C) 240 hz

   D) 120 hz

3. Which of the following load types is an indication that harmonic currents might be present in the circuit?
   A) A computer server room
   B) LED lighting
   C) Electronic lighting control units
   D) All of these
4. The NEC requires neutral conductors that serve line-to-neutral nonlinear loads to be counted as which of the following due to the presence of harmonic currents?
   A) Current-carrying conductors
   B) Non-current carrying conductors
   C) Ungrounded conductors
   D) Super conductors
5. What method can be used to determine if harmonic currents exist in a single-phase system?
   A) Multiply the true-RMS current by the average current
   B) Divide the applied voltage by the true-RMS current
   C) Divide the true-RMS current by the average value
   D) Multiply the instantaneous peak by the true-RMS current value
6. Which of the following are concerns when single-phase systems experience harmonic currents?
   A) Excessive current on conductors
   B) Overheating of terminations
   C) Overheating of equipment
   D) All of these
7. True or False: Single-phase circuits that supply several LED light fixtures have no concerns about harmonics being present.
8. Harmonic currents in a three-phase system are commonly found in which of the following systems?
   A) Three-phase, three-wire, WYE-connected
   B) Three-phase, four-wire, WYE-connected
   C) Three-phase, three-wire, delta-connected
   D) Three-phase, four-wire, delta-connected
9. In a three-phase WYE-connected system that supplies a system where a major portion of its load is nonlinear loads, which of the following must be done when sizing conductors?
   A) Adding harmonic filters at the load
   B) Install separate neutral conductors for each circuit
   C) Install a super neutral for multi-wire circuits
   D) All of these

10. True or False: Harmonic currents can cause a transformer to overheat due to additional current on the windings. One way to mitigate this issue is to derate the rating of the transformer.
11. When determining the transformer harmonic derating factor by measuring RMS current and peak current, what does it mean when the formula equals 1.0?
    A) No harmonic currents present
    B) Significant harmonic currents
    C) Transformer must be derated
    D) Not enough information
12. Upon taking current readings of a large utility-scale distribution transformer, it was determined that the peak instantaneous current was 1732 amperes, and the RMS current value was measured at 948 amperes. What is the THDF?
    A) 0.62
    B) 0.77
    C) 0.81
    D) 0.93

Appendix A

Identifying the Leads of a Three-Phase, WYE-Connected, and Dual-Voltage Motor

The terminal markings of a three-phase motor are standardized and used to connect the motor for operation on 240 or 480 volts. *Figure A–1* shows these terminal markings and their relationship to the other motor windings. If the motor is connected to a 240-volt line, the motor windings are connected parallel to each other, as shown in *Figure A–2*. If the motor is to be operated on a 480-volt line, the motor windings are connected in series, as shown in *Figure A–3*.

As long as these motor windings remain marked with proper numbers, connecting the motor for operation on a 240- or 480-volt power line is relatively simple. If these numbers are removed or damaged, however, the lead must be reidentified before the motor can be connected. The following procedure can be used to identify the proper relationship of the motor windings using a 6V lantern battery, an ohmmeter, and an analog voltmeter:

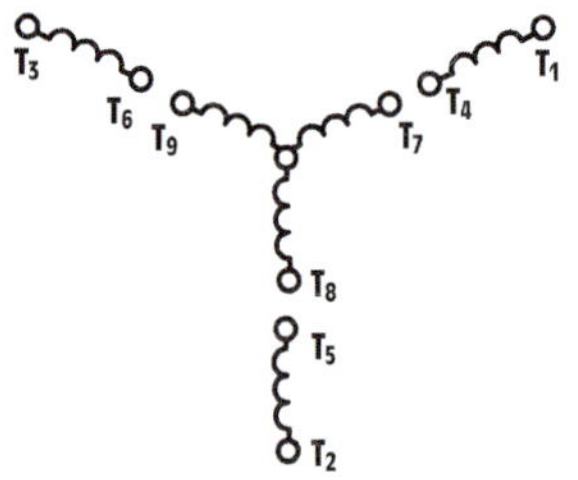

Figure A–1 Standard terminal markings for a three-phase motor.

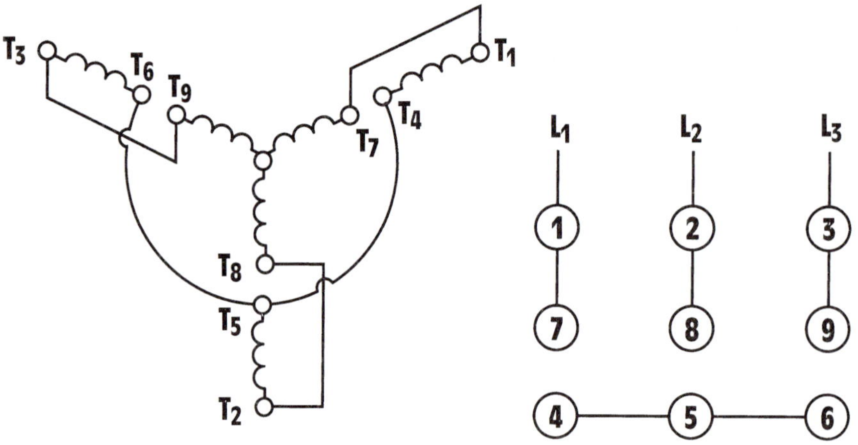

Figure A–2 Low-voltage connection.

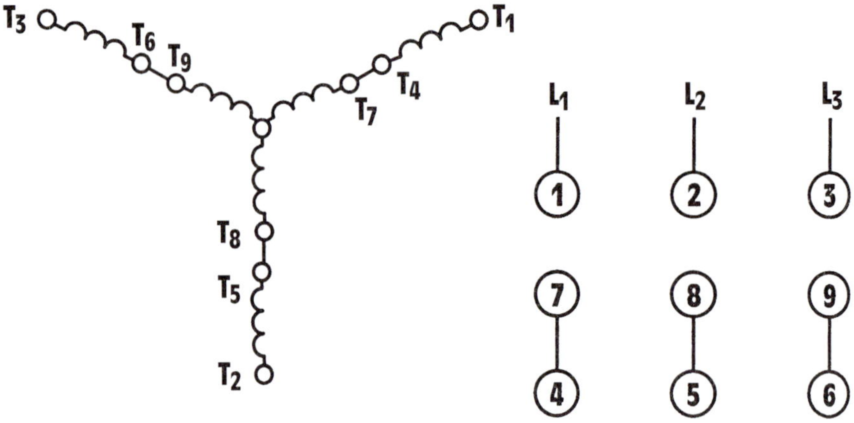

Figure A–3 High-voltage connection.

1. Using the ohmmeter, divide the motor windings into four separate groups. One group will have continuity to three leads, and the other three groups will have continuity between only two leads (refer to *Figure A–1*). These pairs will eventually be labeled as T_1-T_4, T_2-T_5, and T_3-T_6.

Caution

Make sure to use alligator clips to take measurements as holding the probes to the wires with your fingers will introduce your body resistance to the measurements.

Mark the three leads that have continuity with each other as T_7, T_8, and T_9. You can mark these in any order as this process will identify the other leads associated with T_7, T_8, and T_9. Connect one terminal of the battery to the lead labeled T_9, polarity is not important yet, and connect the voltmeter to any of the three two-wire pairs. Make sure to select the mV setting on the voltmeter. With the other terminal of the battery, briefly connect to T_8 and release. Repeat this momentary contact process, called flashing, while watching the voltmeter. If the voltmeter shows little to no deflection, the pair connected to the voltmeter will be leads T_1 and T_4 (*Figure A–4*). If the voltmeter shows a significant deflection, check another pair until you find the pair with no deflection.

2. To determine which lead of the pair is T_1 and which is T_4, reconnect the battery to leads marked T_7 and T_8 with the positive (+) terminal on T_7 and flash the battery negative (−) on T_8. Connect the voltmeter to the T_1-T_4 pair so that the meter shows a positive deflection (*Figure A-5*). The lead connected to the positive voltmeter lead will be T_1 and the other lead will be T_4.

3. Next reconnect the battery to the leads marked T_7 and T_9 with the positive to T_9 and flashing the negative on T_7. Connect the voltmeter to the other pairs until you find a positive meter deflection. Once this is found, mark the lead connected to the positive lead of the voltmeter as T_3 (*Figure A-6*). The other lead in the pair is T_6.

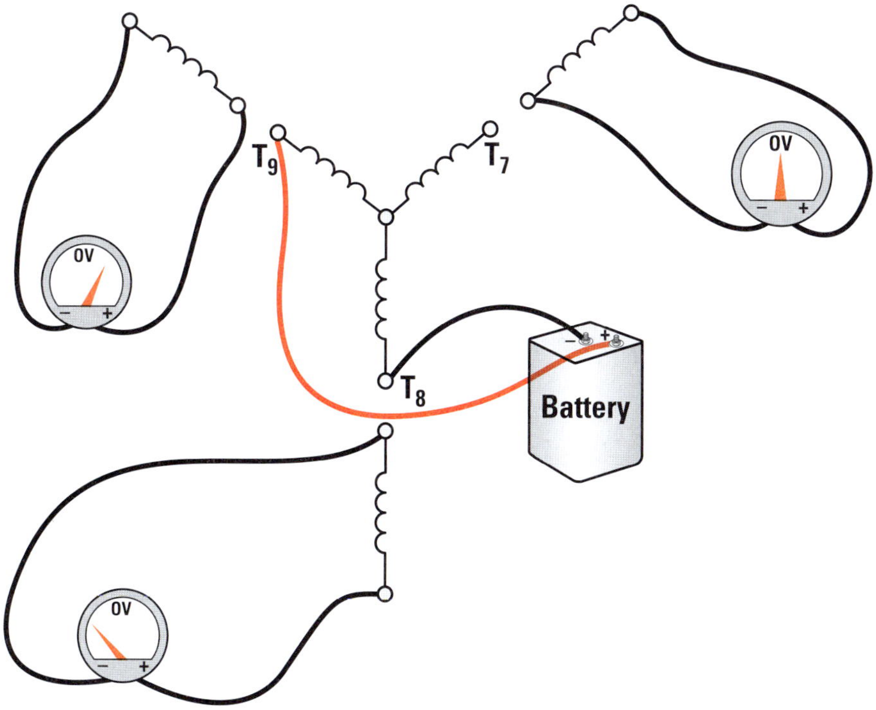

Figure A–4 T_7 and T_8 connected to a lantern battery to find group with no voltmeter deflection.

4. The last step is to reconnect the battery to the leads marked T_8 and T_9, with the battery positive connected to T8 and flashing the negative to T_9. The voltmeter will indicate T_2 is connected to the positive lead when there is a positive deflection (*Figure A-7*). The other lead is T_5. At this point, all 9 leads are accurately identified.

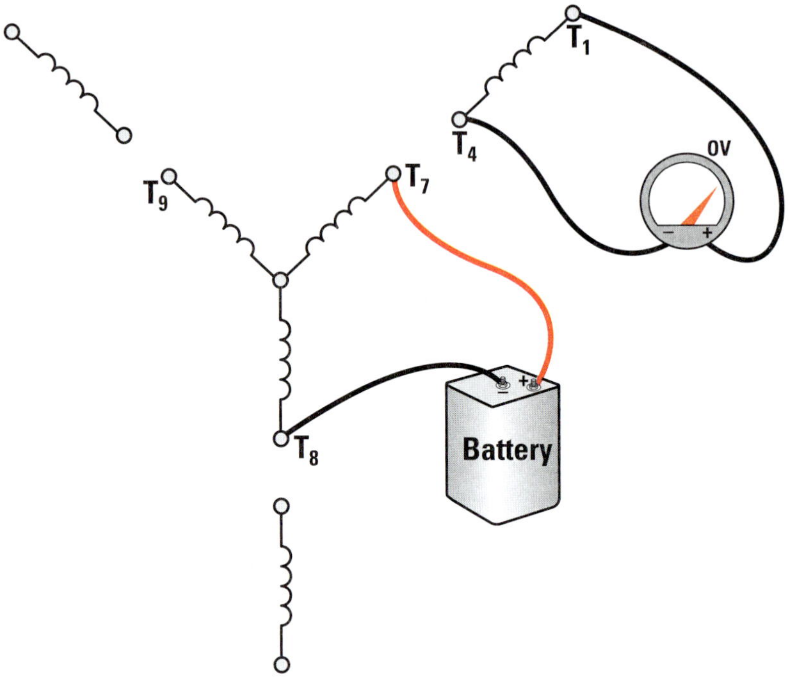

Figure A–5 Battery connected to T_7 and T_8 to find the proper polarity of T_1 and T_4.

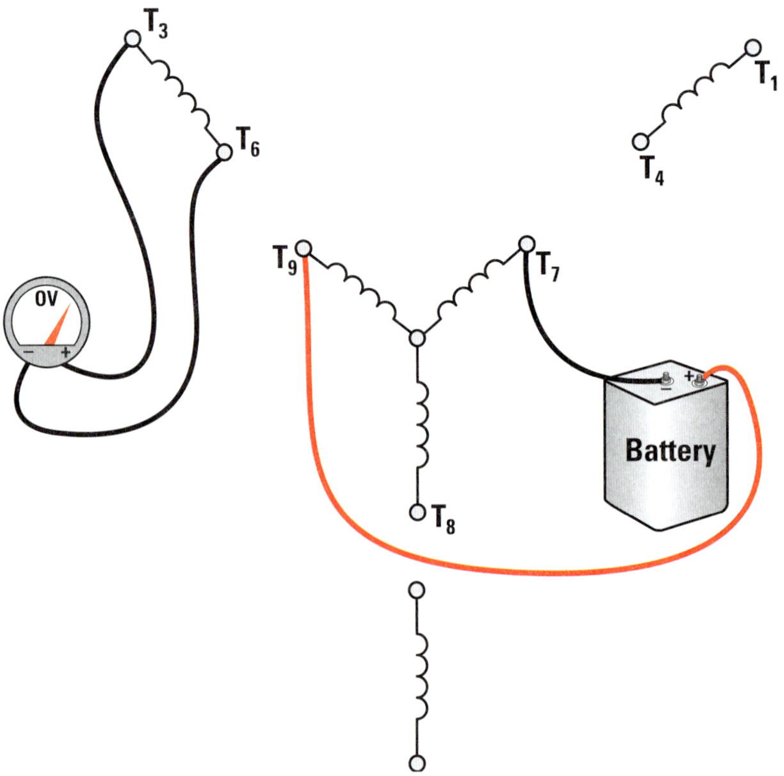

Figure A–6 Battery is connected to T_7 and T_9 to determine the proper polarity of T_3 and T_6.

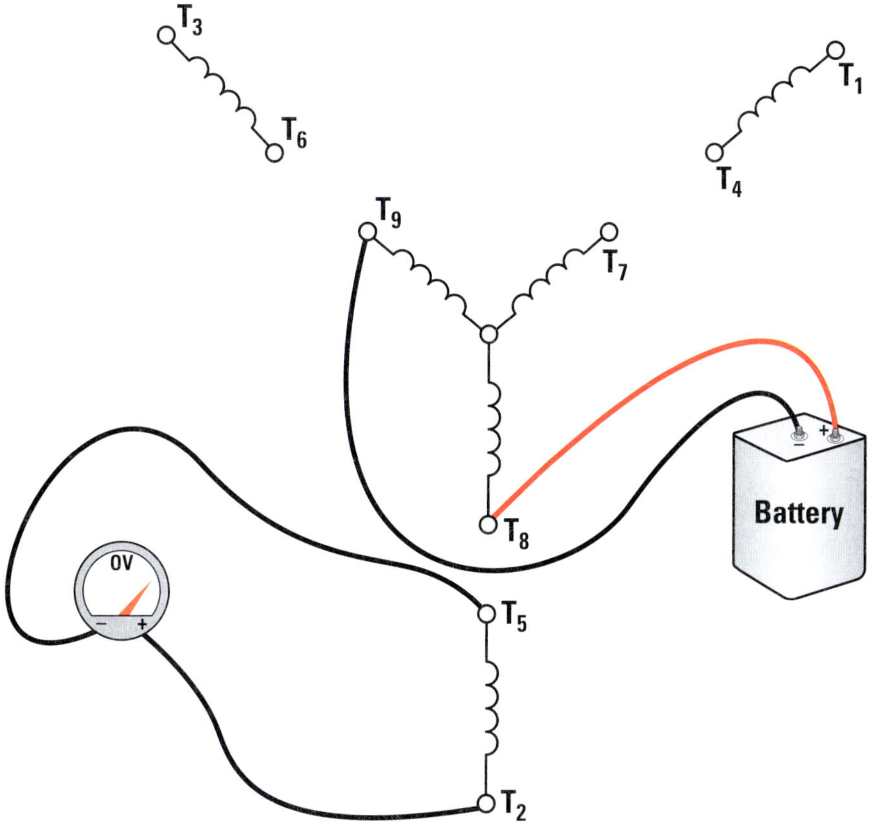

Figure A–7 Battery is connected to T_8 and T_9 to determine the proper polarity of T_2 and T_5.

Appendix B

AC Formulas

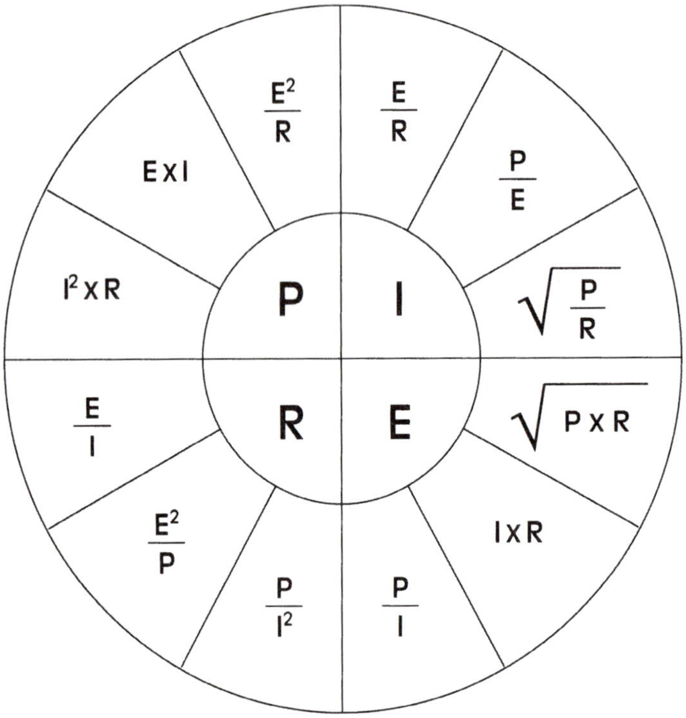

Instantaneous and Maximum Values

The instantaneous value of voltage and current for a sine wave is equal to the peak, or maximum, value of the waveform times the sine of the angle. For example, a waveform has a peak value of 300 volts. What is the voltage at an angle of 22°?

$E_{(INST)} = E_{(MAX)} \times \sin \angle$

$E_{(INST)} = 300 \times 0.3746$ (sin of 22°)

$E_{(INST)} = 112.328\,V$

$$E_{MAX} = \frac{E_{INST}}{\sin \angle}$$

$$\sin \angle = \frac{E_{INST}}{E_{max}}$$

Changing Peak, RMS, and Average Values

To change	To	Multiply by
Peak	RMS	0.707
Peak	Average	0.637
Peak	Peak-to-peak	2
RMS	Peak	1.414
Average	Peak	1.567
RMS	Average	0.9
Average	RMS	1.111

Pure Resistive Circuit

$E = I \times R$ $\quad I = \frac{E}{R}$ $\quad R = \frac{E}{I}$ $\quad P = E \times I$

$E = \frac{P}{I}$ $\quad I = \frac{P}{E}$ $\quad R = \frac{E^2}{P}$ $\quad P = I^2 \times R$

$E = \sqrt{P \times R}$ $\quad I = \sqrt{\frac{P}{R}}$ $\quad R = \frac{P}{I^2}$ $\quad P = \frac{E^2}{R}$

Series Resistive Circuits

$R_T = R_1 + R_2 + R_3$

$I_T = I_1 = I_2 = I_3$

$E_T = E_1 + E_2 + E_3$

$P_T = P_1 + P_2 + P_3$

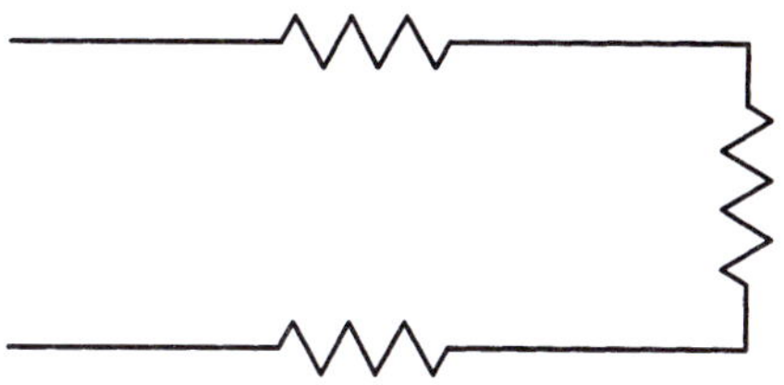

Parallel Resistive Circuits

$$R_T = \frac{1}{\frac{1}{R_1} + \frac{1}{R_2} + \frac{1}{R_3}}$$

$$R_T - \frac{R_1 \times R_2}{R_1 + R_2}$$

$I_T = I_1 + I_2 + I_3$

$E_T = E_1 = E_2 = E_3$

$P_T = P_1 + P_2 + P_3$

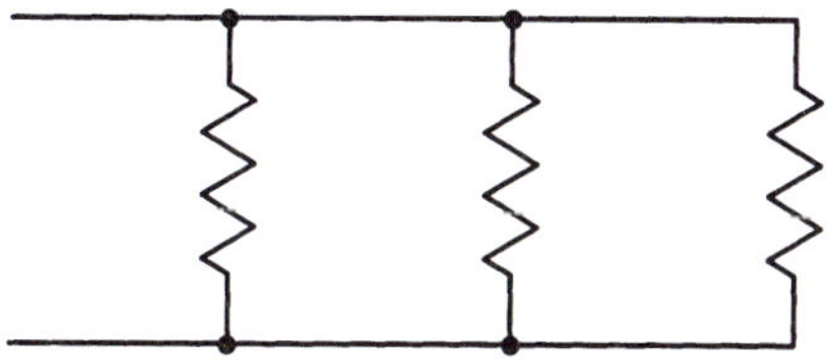

Pure Inductive Circuits

In a pure inductive circuit, the current lags the voltage by 90°. Therefore, there is no true power, or watts, and the power factor is 0. VARs is the inductive equivalent of watts.

$L = \frac{X_L}{2\pi f}$ $\quad X_L = 2\pi fL$

$E_L = I_L \times X_L$ $\quad I_L = \frac{E_L}{X_L}$ $\quad X_L = \frac{E_L}{I_L}$ $\quad VARs_L = E_L \times I_L$

$E_L = \sqrt{VARs_L \times X_L}$ $\quad I_L = \frac{VARs_L}{E_L}$ $\quad X_L = \frac{{E_L}^2}{VARs_L}$ $\quad VARs_L = {I_L}^2 \times X_L$

$E_L = \frac{VARs_L}{I_L}$ $\quad I_L = \sqrt{\frac{VARs_L}{X_L}}$ $\quad X_L = \frac{VARs_L}{{I_L}^2}$ $\quad VARs_L = \frac{{E_L}^2}{X_L}$

Series Inductive Circuits

$E_{LT} = E_{L1} + E_{L2} + E_{L3}$

$X_{LT} = X_{L1} + X_{L2} + X_{L3}$

$I_{LT} = I_{L1} = I_{L2} = I_{L3}$

$VARs_{LT} = VARs_{L1} + VARs_{L2} + VARs_{L3}$

$L_T = L_1 + L_2 + L_3$

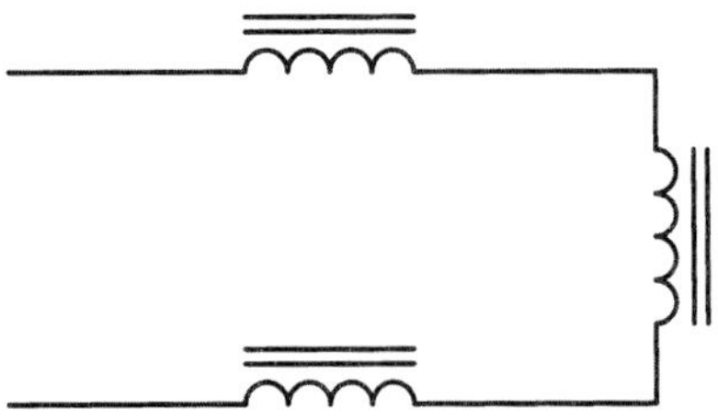

Parallel Inductive Circuits

$E_{LT} = E_{L1} = E_{L2} = E_{L3}$

$I_{LT} = I_{L1} + I_{L2} + I_{L3}$

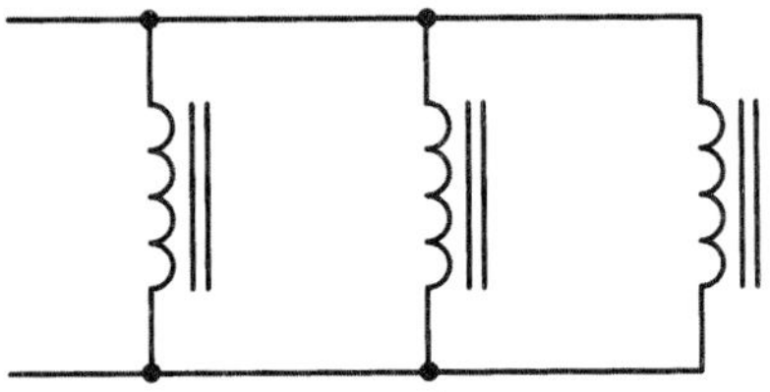

$$X_{LT} = \frac{X_{L1} \times X_{L2}}{X_{L1} + X_{L2}}$$

$$X_{LT} = \frac{1}{\frac{1}{X_{L1}} + \frac{1}{X_{L2}} + \frac{1}{X_{L3}}} \qquad L_T = \frac{1}{\frac{1}{L_1} + \frac{1}{L_2} + \frac{1}{L_3}} \qquad L_T = \frac{L_1 \times L_2}{L_1 + L_2}$$

$VARs_{LT} = VARs_{L1} + VARs_{L2} + VARs_{L3}$

Pure Capacitive Circuits

In a pure capacitive circuit, the current leads the voltage by 90°. For this reason there is no true power, or watts, and no power factor. VARs is the equivalent of watts in a pure capacitive circuit.

The value of C in the formula for finding capacitance is in farads and must be changed to the capacitive units being used.

$$C = \frac{1}{2\pi f X_C} \qquad X_C = \frac{1}{2\pi f C}$$

$$E_C = I_C \times X_C \qquad I_C = \frac{E_C}{X_C} \qquad X_C = \frac{E_C}{I_C} \qquad VARs_C = E_C \times I_C$$

$$E_C = \sqrt{VARs_C \times X_C} \qquad I_C = \frac{VARs_C}{E_C} \qquad X_C = \frac{E_C^2}{VARs_C} \qquad VARs_C = I_C^2 \times X_C$$

$$E_C = \frac{VARs_C}{I_C} \qquad I_C = \sqrt{\frac{VARs_C}{X_C}} \qquad X_C = \frac{VARs_C}{I_C^2} \qquad VARs_C = \frac{E_C^2}{X_C}$$

Series Capacitive Circuits

$$E_{CT} = E_{C1} + E_{C2} + E_{C3}$$

$$I_{CT} = I_{C1} = I_{C2} = I_{C3}$$

$$X_{CT} = X_{C1} + X_{C2} + X_{C3}$$

$$VARs_{CT} = VARs_{C1} + VARs_{C2} + VARs_{C3}$$

$$C_T = \frac{1}{\frac{1}{C_1} + \frac{1}{C_2} + \frac{1}{C_3}} \qquad C_T = \frac{C_1 \times C_2}{C_1 + C_2}$$

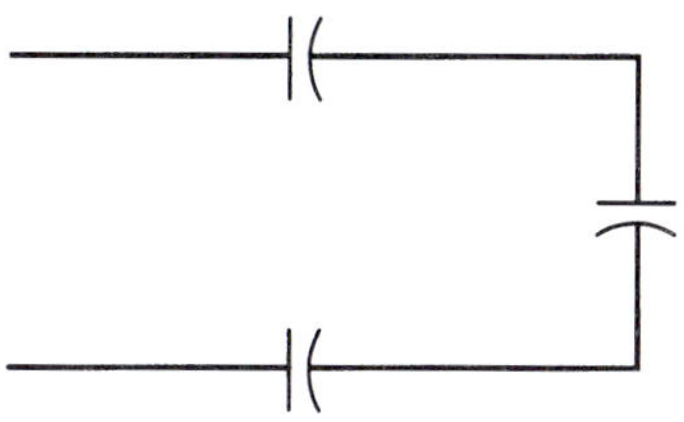

Parallel Capacitive Circuits

$$E_{CT} = E_{C1} = E_{C2} = E_{C3} \qquad I_{CT} = I_{C1} + I_{C2} + I_{C3}$$

$$X_{CT} = \frac{1}{\frac{1}{X_{C1}} + \frac{1}{X_{C2}} + \frac{1}{X_{C3}}} \qquad X_{CT} = \frac{X_{C1} \times X_{C2}}{X_{C1} + X_{C2}}$$

$$C_T = C_1 + C_2 + C_3 \qquad VARs_{CT} = VARs_{C1} + VARs_{C2} + VARs_{C3}$$

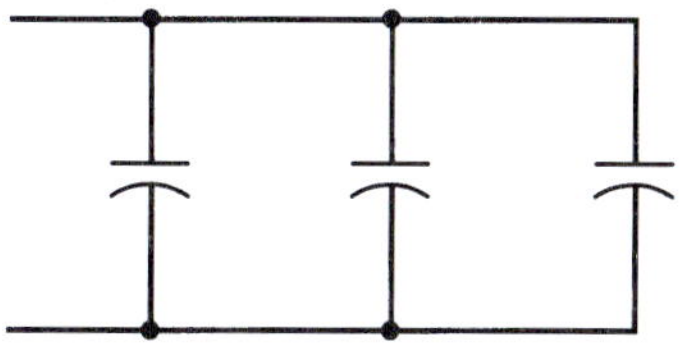

Resistive-Inductive Series Circuits

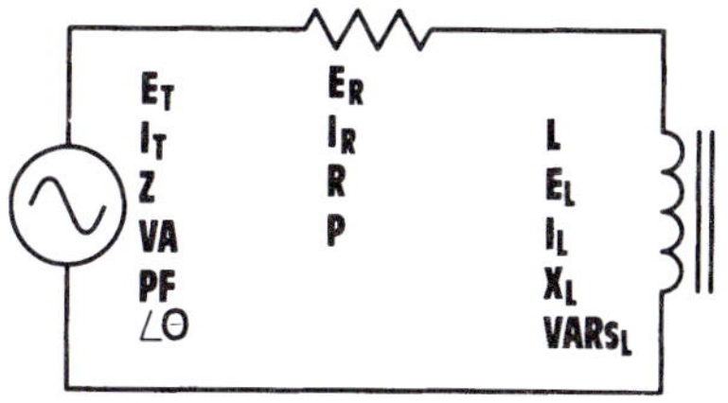

To find values for the resistor, use the formulas in the Pure Resistive section.

To find values for the inductor, use the formulas in the Pure Inductive section.

$$E_T = \sqrt{E_R^2 + E_L^2} \qquad PF = \frac{P}{VA} \qquad R = \frac{P}{I_R^2} \qquad Z = \frac{R}{PF}$$

$$E_T = I_T \times Z \qquad PF = \frac{E_R}{E_T} \qquad R = Z \times PF \qquad Z = \frac{E_T^2}{VA}$$

$$E_T = \frac{VA}{I_T} \qquad PF = \cos \angle\theta \qquad VARs_L = \sqrt{VA^2 - P^2} \qquad P = E_R \times I_R$$

$$E_T = \frac{E_R}{PF} \qquad R = \sqrt{Z^2 - X_L^2} \qquad Z = \sqrt{R^2 + X_L^2} \qquad P = \sqrt{VA^2 - VARs_L^2}$$

$$E_T = \sqrt{VA \times Z} \qquad R = \frac{E_R}{I_R} \qquad Z = \frac{E_T}{I_T} \qquad P = \frac{E_R^2}{R}$$

$$PF = \frac{R}{Z} \qquad R = \frac{E_R^2}{P} \qquad Z = \frac{VA}{I_T^2} \qquad P = I_R^2 \times R$$

Resistive-Inductive Series Circuits (continued)

$P = VA \times PF$

$E_L = I_L \times X_L$

$E_L = \sqrt{E_T^2 - E_R^2}$

$E_L = \sqrt{VARs_L \times X_L}$

$E_L = \dfrac{VARs_L}{I_L}$

$VARs_L = E_L \times I_L$

$VA = E_T \times I_T$

$VA = I_T^2 \times Z$

$VA = \dfrac{E_T^2}{Z}$

$VA = \sqrt{P^2 + VARS_L^2}$

$VA = \dfrac{P}{PF}$

$E_R = I_R \times R$

$E_R = \sqrt{P \times R}$

$E_R = \dfrac{P}{I_R}$

$E_R = \sqrt{E_T^2 - E_L^2}$

$E_R = E_T \times PF$

$I_L = I_R = I_T$

$I_L = \dfrac{E_L}{X_L}$

$I_L = \dfrac{VARs_L}{E_L}$

$I_L = \sqrt{\dfrac{VARs_L}{X_L}}$

$VARs_L = \dfrac{E_L^2}{X_L}$

$L = \dfrac{X_L}{2\pi f}$

$I_T = I_R = I_L$

$I_T = \dfrac{E_T}{Z}$

$I_T = \dfrac{VA}{E_T}$

$I_T = \sqrt{\dfrac{VA}{Z}}$

$I_R = I_T = I_L$

$I_R = \dfrac{E_R}{R}$

$I_R = \dfrac{P}{E_R}$

$I_R = \sqrt{\dfrac{P}{R}}$

$X_L = \sqrt{Z^2 - R^2}$

$X_L = \dfrac{E_L}{I_L}$

$X_L = \dfrac{E_L^2}{VARs_L}$

$X_L = \dfrac{VARs_L}{I_L^2}$

$X_L = 2\pi fL$

$VARs_L = I_L^2 \times X_L$

Resistive-Inductive Parallel Circuits

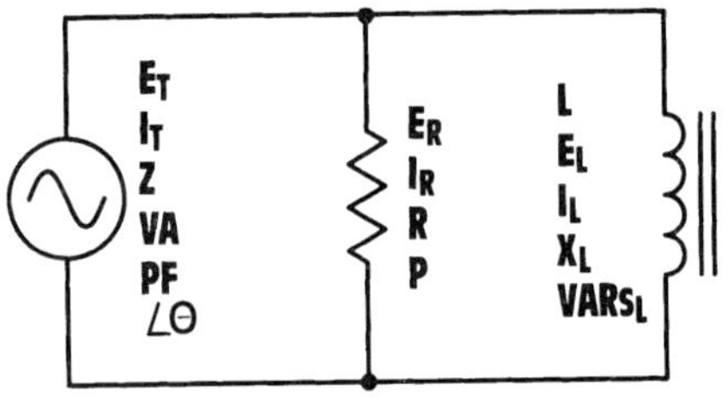

$Z = \dfrac{1}{\sqrt{\left(\dfrac{1}{R}\right)^2 + \left(\dfrac{1}{X_L}\right)^2}}$

$Z = \dfrac{E_T}{I_T}$ $Z = \dfrac{E_T^2}{VA}$

$VA = E_T \times I_T$

$VA = I_T^2 \times Z$

$VA = \dfrac{E_T^2}{Z}$

$VA = \sqrt{P^2 + VARs_L^2}$

$VA = \dfrac{P}{PF}$

$VARs_L = I_L^2 \times X_L$

$X_L = \dfrac{E_L}{I_L}$

$X_L = \dfrac{E_L^2}{VARs_L}$

$X_L = \dfrac{VARs_L}{I_L^2}$

$Z = \dfrac{VA}{I_T^2}$

$Z = R \times PF$

$PF = \dfrac{Z}{R}$

$PF = \dfrac{1}{\left(\dfrac{R}{Z}\right)}$

$PF = \dfrac{P}{VA}$

$PF = \dfrac{I_R}{I_T}$

$PF = \cos \angle\theta$

$VARs_L = \sqrt{VA^2 - P^2}$

$X_L = \dfrac{1}{\sqrt{\left(\dfrac{1}{Z}\right)^2 - \left(\dfrac{1}{R}\right)^2}}$

$X_L = 2\pi fL$

$R = \dfrac{E_R}{I_R}$

$E_T = E_R = E_L$

$E_T = I_T \times Z$

$L = \dfrac{X_L}{2\pi f}$

$I_T = \sqrt{I_R^2 + I_L^2}$

$I_T = \dfrac{E_T}{Z}$

$E_L = I_L \times X_L$

$E_L = E_T = E_R$

Resistive-Inductive Parallel Circuits (continued)

$E_L = \sqrt{VARs_L \times X_L}$

$E_L = \dfrac{VARs_L}{I_L}$

$VARs_L = E_L \times I_L$

$E_L = I_R \times R$

$E_R = \sqrt{P \times R}$

$E_R = \dfrac{P}{I_R}$

$E_R = E_T = E_L$

$R = \dfrac{P}{{I_R}^2}$

$E_T = \dfrac{VA}{I_T}$

$E_T = \sqrt{VA \times Z}$

$I_T = \dfrac{VA}{E_T}$ $\quad I_T = \sqrt{\dfrac{VA}{Z}}$

$I_T = \dfrac{I_R}{PF}$

$I_L = \sqrt{{I_T}^2 - {I_R}^2}$

$I_L = \dfrac{E_L}{X_L}$

$I_L = \dfrac{VARs_L}{E_L}$

$I_L = \sqrt{\dfrac{VARs_L}{X_L}}$

$VARs_L = \dfrac{{E_L}^2}{X_L}$

$I_R = \sqrt{{I_T}^2 - {I_L}^2}$

$I_R = \dfrac{E_R}{R}$

$I_R = \dfrac{P}{E_R}$

$I_R = \sqrt{\dfrac{P}{R}}$

$I_R = I_T \times PF$

$R = \dfrac{1}{\sqrt{\left(\dfrac{1}{Z}\right)^2 - \left(\dfrac{1}{X_L}\right)^2}}$

$P = E_R \times I_R$

$R = \dfrac{{E_R}^2}{P}$

$P = \dfrac{{E_R}^2}{R}$

$R = \dfrac{Z}{PF}$

$P = {I_R}^2 \times R$

$P = \sqrt{VA^2 - {VARs_L}^2}$

$P = VA \times PF$

Resistive-Capacitive Series Circuits

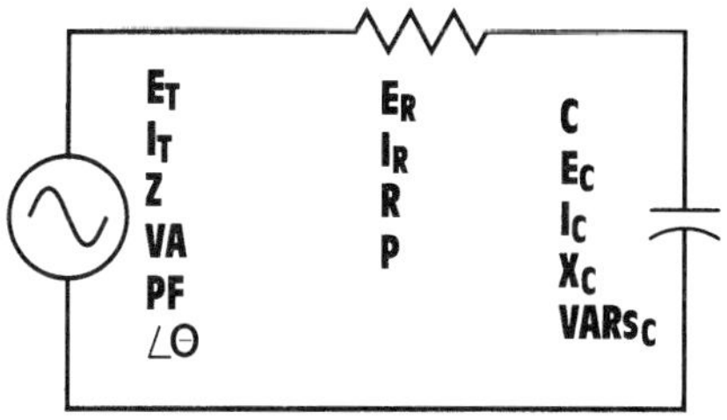

$Z = \dfrac{{E_T}^2}{VA}$

$Z = \sqrt{R^2 + {X_C}^2}$

$Z = \dfrac{E_T}{I_T}$

$VA = E_T \times I_T$

$VA = {I_T}^2 \times Z$

$VA = \dfrac{{E_T}^2}{Z}$

$VA = \sqrt{P^2 + {VARs_C}^2}$

$VA = \dfrac{P}{PF}$

$I_R = I_T = I_C$

$I_R = \dfrac{E_R}{R}$

$Z = \dfrac{VA}{{I_T}^2}$

$Z = \dfrac{R}{PF}$

$PF = \dfrac{R}{Z}$

$PF = \dfrac{P}{VA}$

$PF = \dfrac{E_R}{E_T}$

$PF = \cos \angle\theta$

$R = \sqrt{Z^2 - {X_C}^2}$

$R = \dfrac{E_R}{I_R}$

$R = \dfrac{{E_R}^2}{P}$

$E_T = \sqrt{{E_R}^2 + {E_C}^2}$

$E_T = I_T \times Z$

$E_T = \dfrac{VA}{I_T}$

$E_T = \dfrac{E_R}{PF}$

$P = E_R \times I_R$

$P = \sqrt{VA^2 - {VARs_C}^2}$

$P = \dfrac{{E_R}^2}{R}$

$P = {I_R}^2 \times R$

$P = VA \times PF$

$E_C = I_C \times X_C$

$E_C = \sqrt{{E_T}^2 - {E_R}^2}$

$I_T = I_R = I_C$

$I_T = \dfrac{E_T}{Z}$

$I_T = \dfrac{VA}{E_T}$

$C = \dfrac{1}{2\pi f X_C}$

Resistive-Capacitive Series Circuits (continued)

$$E_R = I_R \times R$$

$$E_R = \sqrt{P \times R}$$

$$E_R = \frac{P}{I_R}$$

$$E_R = \sqrt{E_T^2 - E_C^2}$$

$$E_R = E_T \times PF$$

$$I_C = I_R = I_T$$

$$I_C = \frac{E_C}{X_C}$$

$$I_R = \frac{P}{E_R}$$

$$I_R = \sqrt{\frac{P}{R}}$$

$$X_C = \sqrt{Z^2 - R^2}$$

$$X_C = \frac{E_C}{I_C}$$

$$VARs_C = \sqrt{VA^2 - P^2}$$

$$R = \frac{P}{I_R^2}$$

$$R = Z \times PF$$

$$X_C = \frac{E_C^2}{VARs_C}$$

$$X_C = \frac{VARs_C}{I_C^2}$$

$$E_C = \sqrt{VARs_C \times X_C}$$

$$E_C = \frac{VARs_C}{I_C}$$

$$X_C = \frac{1}{2\pi fC}$$

$$VARs_C = E_C \times I_C$$

$$I_C = \frac{VARs_C}{E_C}$$

$$I_C = \sqrt{\frac{VARs_C}{X_C}}$$

$$VARs_C = I_C^2 \times X_C$$

$$VARs_C = \frac{E_C^2}{X_C}$$

Resistive-Capacitive Parallel Circuits

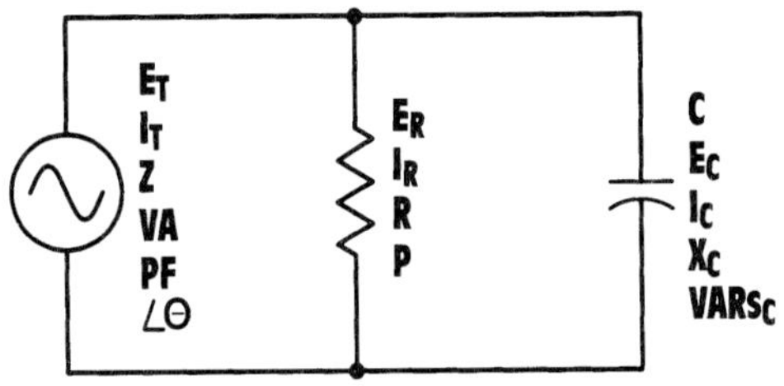

$$Z = \frac{1}{\sqrt{\left(\frac{1}{R}\right)^2 + \left(\frac{1}{X_C}\right)^2}}$$

$$I_T = \sqrt{I_R^2 + I_C^2}$$

$$I_T = \frac{E_T}{Z}$$

$$I_T = \frac{VA}{E_T}$$

$$I_T = \frac{I_R}{PF}$$

$$I_T = \sqrt{\frac{VA}{Z}}$$

$$Z = \frac{VA}{I_T^2}$$

$$VA = E_T \times I_T$$

$$VA = I_T^2 \times Z$$

$$VA = \frac{E_T^2}{Z}$$

$$VA = \sqrt{P^2 + VARs_C^2}$$

$$VA = \frac{P}{PF}$$

$$E_T = E_R = E_C$$

$$E_T = I_T \times Z$$

$$E_T = \frac{VA}{I_T}$$

$$C = \frac{1}{2\pi fX_C}$$

$$Z = \frac{E_T}{I_T} \qquad Z = \frac{E_T^2}{VA}$$

$$PF = \frac{Z}{R}$$

$$PF = \frac{1}{\left(\frac{R}{Z}\right)}$$

$$PF = \frac{P}{VA}$$

$$PF = \frac{I_R}{I_T}$$

$$PF = \cos \angle\theta$$

$$E_T = \sqrt{VA \times Z}$$

$$Z = R \times PF$$

$$P = E_R \times I_R$$

$$P = \sqrt{VA^2 - VARs_C^2}$$

$$P = \frac{E_R^2}{R}$$

$$P = I_R^2 \times R$$

$$P = VA \times PF$$

$$E_R = I_R \times R$$

$$E_R = \sqrt{P \times R}$$

$$E_R = \frac{P}{I_R}$$

$$E_R = E_T = E_c$$

$$I_C = \sqrt{I_T^2 - I_R^2}$$

$$I_C = \frac{E_C}{X_C}$$

$$I_C = \frac{VARs_C}{E_C}$$

$$I_C = \sqrt{\frac{VARs_C}{X_C}}$$

$$I_R = \sqrt{I_T^2 - I_C^2}$$

$$I_R = \frac{E_R}{R}$$

$$I_R = \frac{P}{E_R}$$

$$I_R = \sqrt{\frac{P}{R}}$$

$$X_c = \frac{1}{\sqrt{\left(\frac{1}{Z}\right)^2 - \left(\frac{1}{R}\right)^2}}$$

$$X_c = \frac{E_C}{I_C}$$

Resistive-Capacitive Parallel Circuits (continued)

$$R = \frac{1}{\sqrt{\left(\frac{1}{Z}\right)^2 - \left(\frac{1}{X_C}\right)^2}}$$

$$R = \frac{E_R}{I_R}$$

$$R = \frac{E_R^2}{P}$$

$$R = \frac{P}{I_R^2}$$

$$R = \frac{Z}{PF}$$

$$X_c = \frac{E_c^2}{VARs_C}$$

$$X_c = \frac{VARs_C}{I_c^2}$$

$$X_c = \frac{1}{2\pi fC}$$

$$E_C = I_C \times X_C$$

$$E_C = E_T = E_R$$

$$E_C = \sqrt{VARs_C \times X_C}$$

$$E_C = \frac{VARs_C}{I_C}$$

$$VARs_C = I_C^2 \times X_C$$

$$VARs_C = \frac{E_C^2}{X_C}$$

$$VARs_C = E_C \times I_C$$

$$VARs_C = \sqrt{VA^2 - P^2}$$

Resistive-Inductive-Capacitive Series Circuits

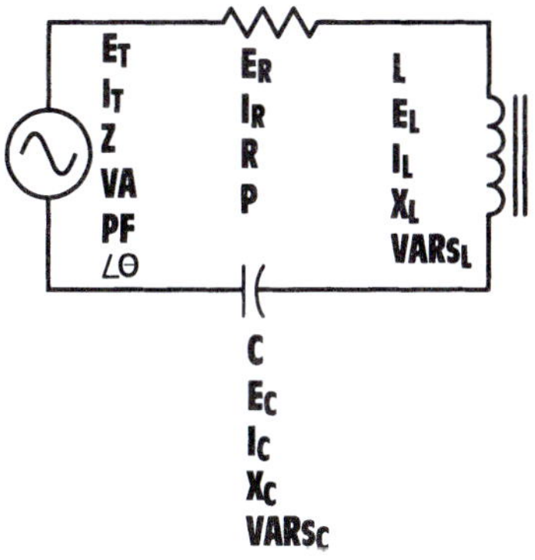

$$E_T = \sqrt{E_R^2 + (E_L - E_c)^2}$$

$$E_T = I_T \times Z$$

$$E_T = \frac{VA}{I_T}$$

$$E_T = \frac{E_R}{PF}$$

$$Z = \sqrt{R^2 + (X_L - X_C)^2}$$

$$Z = \frac{E_T}{I_T}$$

$$Z = \frac{VA}{I_T^2}$$

$$Z = \frac{R}{PF}$$

$$PF = \frac{R}{Z}$$

$$PF = \frac{P}{VA}$$

$$PF = \frac{E_R}{E_T}$$

$$PF = \cos \angle\theta$$

$$I_R = I_T = I_C = I_L$$

$$I_R = \frac{E_R}{R}$$

$$I_R = \frac{P}{E_R}$$

$$I_R = \sqrt{\frac{P}{R}}$$

$$I_c = I_R = I_T = I_L$$

$$I_c = \frac{E_C}{X_C}$$

$$I_C = \frac{VARs_C}{E_C}$$

$$I_C = \sqrt{\frac{VARs_C}{X_C}}$$

$$P = E_R \times I_R$$

$$P = \sqrt{VA^2 - (VARs_L - VARs_C)^2}$$

$$P = \frac{E_R^2}{R}$$

$$P = I_R^2 \times R$$

$$R = \sqrt{Z^2 - (X_L - X_C)^2}$$

$$R = \frac{E_R}{I_R}$$

$$R = \frac{E_R^2}{P}$$

$$X_C = \frac{1}{2\pi fC}$$

$$X_C = \frac{E_c}{VARs_c}$$

$$X_C = \frac{E_C}{I_C}$$

$$VARs_C = I_C^2 \times X_C$$

$$VARs_C = \frac{E_C^2}{X_C}$$

$$VARs_C = E_C \times I_C$$

$$I_T = I_R = I_L = I_C$$

$$I_T = \frac{E_T}{Z}$$

$$VA = E_T \times I_T$$

$$VA = I_T^2 \times Z$$

$$VA = \sqrt{P^2 - (VARs_L - VARs_C)^2}$$

$$P = VA \times PF$$

$$E_R = \sqrt{P \times R}$$

$$R = Z \times PF$$

$$R = \frac{P}{I_R^2}$$

$$X_C = \frac{VARs_C}{I_C^2}$$

$$L = \frac{X_L}{2\pi f}$$

$$X_L = \frac{E_L}{I_L}$$

$$X_L = \frac{E_L^2}{VARs_L}$$

Resistive-Inductive-Capacitive Series Circuits (continued)

$$I_T = \frac{VA}{E_T}$$

$$I_T = \sqrt{\frac{VA}{Z}}$$

$$VA = \frac{P}{PF}$$

$$VA = \frac{E_T^2}{Z}$$

$$E_R = I_R \times R$$

$$E_R = \frac{P}{I_R}$$

$$E_R = \sqrt{E_T^2 - (E_L - E_C)^2}$$

$$E_R = E_T \times PF$$

$$E_C = I_C \times X_c$$

$$E_C = \sqrt{VARs_C \times X_C}$$

$$E_C = \frac{VARs_C}{I_C}$$

$$C = \frac{1}{2\pi f X_C}$$

$$X_L = 2\pi f L$$

$$X_L = \frac{VARs_L}{I_L^2}$$

$$E_L = I_L \times X_L$$

$$E_L = \sqrt{VARs_L \times X_L}$$

$$E_L = \frac{VARs_L}{I_L}$$

$$I_L = I_R = I_T = I_c$$

$$I_L = \frac{E_L}{X_L}$$

$$I_L = \frac{VARs_L}{E_L}$$

$$I_L = \sqrt{\frac{VARs_L}{X_L}}$$

$$VARs_L = E_L \times I_L$$

$$VARs_L = \frac{E_L^2}{X_L}$$

$$VARs_L = I_L^2 \times X_L$$

Resistive-Inductive-Capacitive Parallel Circuits

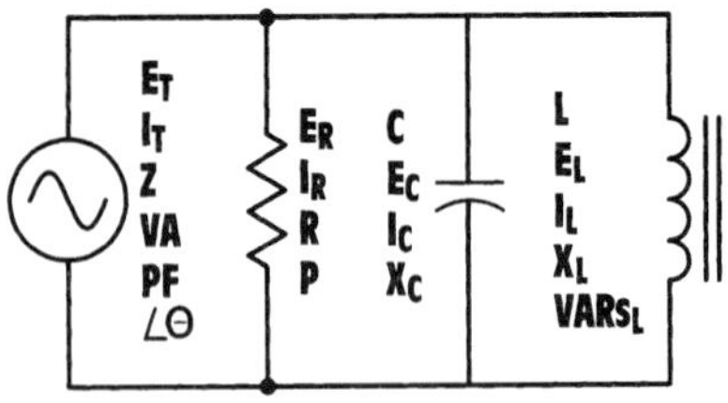

$$Z = \frac{1}{\sqrt{\left(\frac{1}{R}\right)^2 + \left(\frac{1}{X_L} - \frac{1}{X_C}\right)^2}}$$

$$Z = \frac{E_T}{I_T}$$

$$Z = \frac{E_T^2}{VA}$$

$$I_L = \frac{E_L}{X_L}$$

$$VA = E_T \times I_T$$

$$Z = \frac{VA}{I_T^2}$$

$$VA = I_T^2 \times Z$$

$$Z = R \times PF$$

$$VA = \frac{E_T^2}{Z}$$

$$VA = \sqrt{P^2 + (VARs_L - VARs_c)^2}$$

$$VA = \frac{P}{PF}$$

$$PF = \frac{Z}{R}$$

$$PF = \frac{P}{VA}$$

$$PF = \frac{1}{\left(\frac{R}{Z}\right)}$$

$$PF = \frac{I_R}{I_T}$$

$$PF = \cos \angle\theta$$

$$X_L = \frac{E_L}{I_L}$$

$$I_L = \frac{VARs_L}{E_L}$$

$$I_L = \sqrt{\frac{VARs_L}{X_L}}$$

$$E_T = E_R = E_L = E_C$$

$$E_T = I_T \times Z$$

$$E_T = \frac{VA}{I_T}$$

$$E_T = \sqrt{VA \times Z}$$

$$I_T = \sqrt{I_R^2 + (I_L - I_C)^2}$$

$$I_T = \frac{E_T}{Z}$$

$$E_L = I_L \times X_L$$

$$E_L = \frac{VARs_L}{I_L}$$

$$X_L = 2\pi f L$$

$$X_L = \frac{E_L^2}{VARs_L}$$

$$X_L = \frac{VARs_L}{I_L^2}$$

$$I_T = \frac{VA}{E_T} \qquad I_T = \sqrt{\frac{VA}{Z}}$$

$$I_T = \frac{I_R}{PF}$$

$$E_L = E_T = E_R = E_c$$

$$E_L = \sqrt{VARs_L \times X_L}$$

$$L = \frac{X_L}{2\pi f}$$

$$VARs_L = \frac{E_L^2}{X_L}$$

$$VARs_L = I_L^2 \times X_L$$

$$VARs_L = E_L \times I_L$$

$$E_R = I_R \times R$$

$$E_R = \sqrt{P \times R}$$

$$E_R = \frac{P}{I_R}$$

Resistive-Inductive-Capacitive Parallel Circuits (continued)

$$E_R = E_T = E_L = E_c$$

$$P = \sqrt{VA^2 - (VARs_L - VARs_C)^2}$$

$$E_C = \frac{VARs_C}{I_C}$$

$$E_C = I_C \times X_C$$

$$E_C = E_T = E_R = E_L$$

$$E_C = \sqrt{VARs_C \times X_C}$$

$$I_R = \sqrt{I_T^2 - (I_L - I_C)^2}$$

$$I_R = \frac{E_R}{R}$$

$$I_R = \frac{P}{E_R}$$

$$I_R = \sqrt{\frac{P}{R}}$$

$$P = VA \times PF$$

$$I_C = \frac{E_C}{X_C}$$

$$I_C = \frac{VARs_C}{E_C}$$

$$I_C = \sqrt{\frac{VARs_C}{X_C}}$$

$$I_R = I_T \times PF$$

$$R = \frac{E_R}{I_R}$$

$$R = \frac{P}{I_R^2} \qquad R = \frac{Z}{PF}$$

$$P = E_R \times I_R$$

$$P = \frac{E_R^2}{R}$$

$$X_C = \frac{E_C}{I_C}$$

$$X_C = \frac{E_C^2}{VARs_C}$$

$$X_C = \frac{VARs_C}{I_C^2}$$

$$R = \frac{1}{\sqrt{\left(\frac{1}{Z}\right)^2 - \left(\frac{1}{X_L} - \frac{1}{X_C}\right)^2}}$$

$$R = \frac{E_R^2}{P}$$

$$P = I_R^2 \times R$$

$$X_C = \frac{1}{2\pi fC}$$

$$VARs_C = E_C \times I_C$$

$$VARs_C = I_C^2 \times X_C$$

$$VARs_C = \frac{E_C^2}{X_C}$$

Transformers

$$\frac{E_P}{E_S} = \frac{N_P}{N_S} \qquad \frac{E_P}{E_S} = \frac{I_S}{I_P} \qquad \frac{N_P}{N_S} = \frac{I_S}{I_P} \qquad \frac{Z_P}{Z_s} = \left(\frac{N_P}{N_s}\right)^2 \qquad Z_P = Z_s\left(\frac{N_P}{N_S}\right)^2 \qquad Z_s = Z_P\left(\frac{N_S}{N_P}\right)^2$$

E_P = Voltage of the primary

E_S = Voltage of the secondary

I_P = Current of the primary

I_S = Current of the secondary

N_P = Number of turns of the primary

N_S = Number of turns of the secondary

Z_P = Impedance of the primary

Z_S = Impedance of the secondary

Three-Phase Connections

WYE Connection

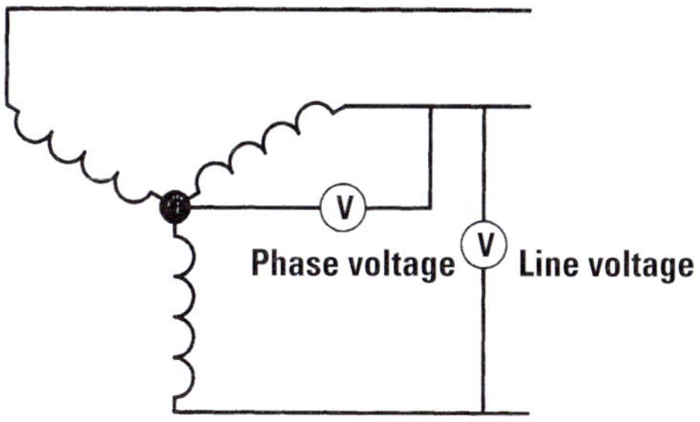

In a **WYE connection**, the line current and phase current are the same:

$$I_{LINE} = I_{PHASE}$$

In a WYE connection, the line voltage is higher than the phase voltage by a factor of the square root of 3:

$$E_{LINE} = E_{PHASE} \times \sqrt{3}$$

$$E_{PHASE} = \frac{E_{LINE}}{\sqrt{3}}$$

Delta Connection

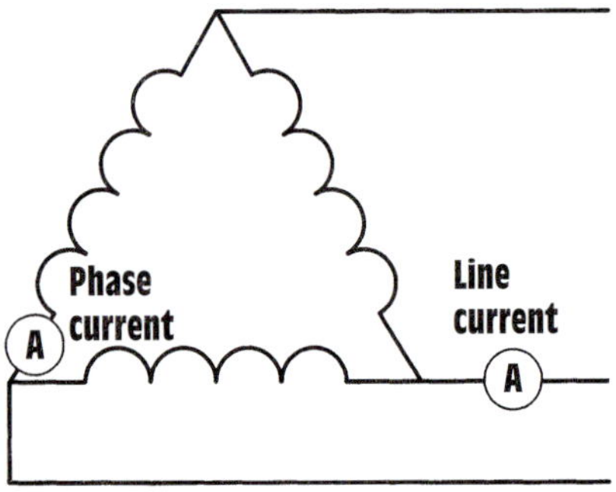

In a **delta connection**, the line voltage and phase voltage are the same:

$E_{LINE} = I_{PHASE}$

In a delta-connected system, the line current is higher than the phase current by a factor of the square root of 3:

$I_{LINE} = I_{PHASE} \times \sqrt{3}$

$I_{PHASE} = \frac{I_{LINE}}{\sqrt{3}}$

Open Delta Connection

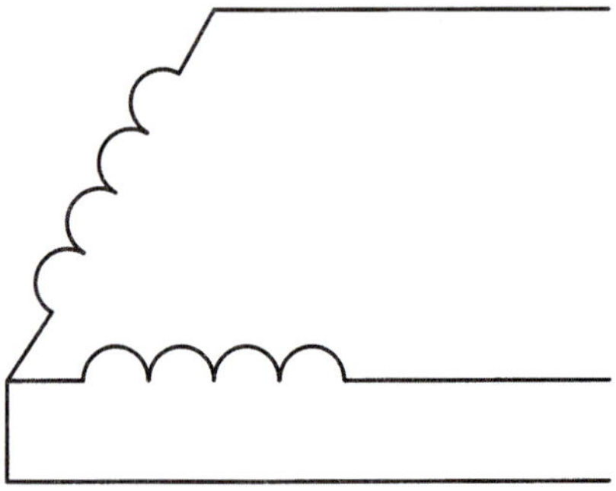

Open-delta connections provide 86.6% of the sum of the power rating of the two transformers. Example: The two transformers shown are rated at 60 kilovolt-amperes each. The total power rating for this connection would be:

$60 + 60 = 120$ kVA

120 kVA $\times 0.866 = 103.92$ kVA

All values of voltage and current are computed in the same manner as for a closed-delta connection.

$P = \sqrt{3} \times E_{LINE} \times I_{LINE} \times PF$

$P = 3 \times E_{PHASE} \times I_{PHASE} \times PF$

Apparent and True Power

$VA = \sqrt{3} \times E_{LINE} \times I_{LINE}$

$VA = 3 \times E_{PHASE} \times I_{PHASE}$

Appendix C

Greek Alphabet

Name of letter	Uppercase	Lowercase	Designates
Alpha	A	α	Angles
Beta	B	β	Angles; flux density
Gamma	Γ	γ	Conductivity; photon
Delta	Δ	δ	Change of a quantity
Epsilon	E	ε	Permittivity (dielectric constant)
Zeta	Z	ζ	Impedance; coefficients; coordinates
Eta	H	η	Hysteresis coefficient; efficiency
Theta	Θ	θ	Phase angle
Iota	I	ι	
Kappa	K	κ	Dielectric constant; coefficient of coupling; susceptibility
Lambda	Λ	λ	Wavelength
Mu	M	μ	Permeability; the prefix micro-; amplification factor
Nu	N	ν	Change of quantity
Xi	Ξ	ξ	
Omicron	O	ο	
Pi	Π	π	3.1416
Rho	P	ρ	Resistivity
Sigma	Σ	σ	Conductivity
Tau	T	τ	Time constant; time phase displacement
Upsilon	Υ	υ	Reluctivity
Phi	Φ	φ	Angles; magnetic flow
Chi	X	χ	Susceptibility
Psi	Ψ	ψ	Dielectric flux; phase difference
Omega	Ω	ω	Ohm; unit of resistance angular velocity

Appendix D

Metals

Metal	Sym.	Spec. grav.	Melt Point °C	Melt Point °F	Elec. Cond. % copper	Lb/in.^{3}
Aluminum	Al	2.710	660	1220	64.9	0.0978
Antimony	Sb	6.620	631	1167	4.42	0.2390
Arsenic	As	5.730			4.90	0.2070
Beryllium	Be	1.830	1287	2336	9.32	0.0660
Bismuth	Bi	9.800	271	512	1.50	0.3540
Brass (70–30)		8.510	900	1652	28.0	0.3070
Bronze (5% Sn)		8.870	1000	1382	18.0	0.3200
Cadmium	Cd	8.650	321	610	22.7	0.3120
Calcium	Ca	1.550	842	1562	50.1	0.0560
Cobalt	Co	8.900	1495	2723	17.8	0.3210
Copper	Cu	8.900	1085	1981	100.0	0.3210
Gold	Au	19.30	1064	1945	71.2	0.6970
Graphite		2.250	3500	6332	0.001	0.0812
Indium	In	7.300	156	311	20.6	0.2640
Iridium	Ir	22.40	2446	4442	32.5	0.8090
Iron	Fe	7.200	1400 to 1500	2552 to 2732	17.6	0.2600
Malleable		7.200	1600 to 1500	2912 to 2732	10.0	0.2600
Wrought		7.700	1600	2912	10.0	0.2780
Lead	Pb	11.40	327	621	8.35	0.4120
Magnesium	Mg	1.740	650	1204	38.7	0.0628
Manganese	Mn	7.200	1246	2273	0.90	0.2600
Mercury	Hg	13.65	−38.8	−37.7	1.80	0.4930
Molybdenum	Mo	10.20	2623	4748	36.1	0.3680
Monel (63–37)		8.870	1300	2372	3.00	0.3200
Nickel	Ni	8.90	1455	2646	25.0	0.3210
Phosphorus	P	1.82	44.2	111.6	10^{-17}	0.0657
Platinum	Pt	21.46	1768	3214	17.5	0.0657

Metal	Sym.	Spec. grav.	Melt Point °C	°F	Elec. Cond. % copper	Lb/in.^{3}
Potassium	K	0.860	63.0	146.1	28.0	0.0310
Selenium	Se	4.81	221	430	14.4	0.1740
Silicon	Si	2.40	1414	2577	10^{-5}	0.0866
Silver	Ag	10.50	962	1764	106	0.3790
Steel (carbon)		7.84	1330 to 1380	2436 to 2516	10.0	0.2830
Stainless						
(18–8)		7.92	1500	2732	2.50	0.2860
(13–Cr)		7.78	1520	2768	3.50	0.2810
(18–Cr)		7.73	1500	2732	3.00	0.2790
Tantalum	Ta	16.6	3017	5462	13.9	0.5990
Tellurium	Te	6.20	450	846	10^{-5}	0.2240
Thorium	Th	11.7	1750	3182	9.10	0.4220
Tin	Sn	7.30	232	449	15.0	0.2640
Titanium	Ti	4.50	1668	3034	2.10	0.1620
Tungsten	W	19.3	3422	6181	31.5	0.6970
Uranium	U	18.7	1135	2075	2.80	0.6750
Vanadium	V	5.96	1910	3470	6.63	0.2150
Zinc	Zn	7.14	420	788	29.1	0.2580
Zirconium	Zr	6.40	1855	3371	4.20	0.2310

Appendix E

Scientific Notation

In scientific notation, the coefficient of the power of 10 is always expressed with one digit to the left of the decimal point and the required power of 10.

Example:

$$350{,}000 = 3.5^{\times 10^{5}}$$

$$875 = 8.75^{\times 10^{2}}$$

$$0.002 = 2^{\times 10^{-2}}$$

$$0.000{,}045 = 4.5^{\times 10^{-5}}$$

Notice that in each example, there is one digit to the left of the decimal place. $9500 = 95^{\times 10^{-2}}$ and $0.0056 = 0.56^{\times 10^{-2}}$ are not examples of scientific notation.

Scientific notation is used in almost all scientific calculations. It was first used for making calculations with a slide rule. A slide rule is a tool that can perform mathematical operations such as multiplication and division; find square roots; find logs of numbers; and find sines, cosines, and tangents of angles. When finding a number on a slide rule, only the actual digits are used, not decimal points or zeros (unless they come between two other digits such as 102). To a slide rule, the numbers 0.000012, 0.0012, 0.12, 1.2, 12, 120, 12000, or 12000000 are all the same number: 12. Since the slide rule recognizes only the basic digits of any number, imagine the problem of determining where to place a decimal point in an answer. As long as only simple calculations are done, there is no problem in determining where the decimal point should be placed.

Example 1

Multiply 12×20. It is obvious where the decimal point should be placed in this problem:

$$240.00$$

Example 2

Now assume that the following numbers are to be multiplied together:

$$0.000041 \times 380{,}000 \times 0.19 \times 720 \times 0.0032$$

In this problem, it is not obvious where the decimal point should be placed in the answer. Scientific notation can be used to simplify the numbers so an estimated answer can be obtained. Scientific notation is used to change any number to a simple whole number by dividing or multiplying the number by a power of 10. Any number can be multiplied by 10 by moving the decimal point one place to the right. Any number can be divided by 10 by moving the decimal point one place to the left. For example, the number 0.000041 can be changed to the whole number 4.1 by multiplying it by 10 five times. Therefore, if the number 4.1 is divided by 10 five times it will be the same as the original number, 0.000041. In this problem, the number 0.000041 will be changed to 4.1 by multiplying it by 10 five times. The new number is 4.1 times 10 to the negative fifth:

$$4.1 \times 10^{-5}$$

Example 2 (continued)

Because superscripts and subscripts are often hard to produce in printed material, it is a common practice to use the letter *E* to indicate the exponent in scientific notation. The preceding notation can also be written

$$4.1\text{E} - 05$$

The number 380,000 can be reduced to a simple whole number of 3.8 by dividing it by 10 five times. The number 3.8 must, therefore, be notated to indicate that the original number is actually 3.8 multiplied by 10 five times:

$$3.8 \times 10^{+5}$$

or

$$3.8\text{E}05$$

The other numbers in the problem can also be changed to simple whole numbers using scientific notation:

$$0.19 \text{ becomes } 1.9\text{E} - 01$$

$$720 \text{ becomes } 7.2\text{E}02$$

$$0.0032 \text{ becomes } 3.2\text{E} - 03$$

Now that the numbers have been simplified using scientific notation, an estimate can be obtained by multiplying these simple numbers together and adding the exponents: 4.1 is about 4; 3.8 is about 4; 1.9 is about 2; 7.2 is about 7; and 3.2 is about 3. Therefore:

$$4 \times 4 \times 2 \times 7 \times 3 = 672$$

When the exponents are added:

$$(\text{E} - 05) + (\text{E}05) + (\text{E} - 01) + (\text{E}02) + (\text{E} - 03) = \text{E} - 02$$

The estimated answer becomes 672E − 02. When the calculation is completed, the actual answer becomes:

$$682.03\text{E} - 02$$

or

$$6.8203$$

Using Scientific Notation with Calculators

In the early 1970s, scientific calculators, often referred to as *slide-rule calculators,* became commonplace. Most of these calculators have the ability to display from 8 to 10 digits, depending on the manufacturer. Scientific calculations, however, often involve numbers that contain more than 8 or 10 digits. To overcome the limitation of an 8- or 10-digit display, slide-rule calculators depend on scientific notation. When a number becomes too large for the calculator to display, scientific notation is used automatically. Imagine, for example, that it became necessary to display the distance (in kilometers) that light travels in one year (approx. 9,460,800,000,000 km). This number contains 13 digits. The calculator would display this number as shown:

9.4608 12

The number 12 shown to the right of 9.4608 is the scientific notation exponent. This number could be written 9.4608E12, indicating that the decimal point should be moved to the right 12 places.

If a minus sign should appear ahead of the scientific notation exponent, it indicates that the decimal point should be moved to the left. The number on the following display contains a negative scientific notation exponent:

7.5698 − 06

This number could be rewritten as:

0.0000075698

Entering Numbers in Scientific Notation

Slide-rule calculators also have the ability to enter numbers in scientific notation. To do this, the exponent key must be used. There are several ways in which manufacturers mark the exponent key. Some are marked **EXP**, and others are marked **EE**. Some manufacturers use **E** and **EXX**. Regardless of how the manufacturer chooses to indicate the functions, they work the same.

EXP   EE   E   EXX

Assume that the number 549E08 is to be entered. The following keystrokes are used:

5 4 9 EE 8

There are also different ways in which scientific notation may appear on the display. Several are shown:

5.49 10   5.49 $10^{\times 10^{10}}$ 5.49 E10

Notice that although the number is entered as 549 EE 8, the calculator automatically changes the number so there is only one digit to the left of the decimal place.

If a number with a negative exponent is to be entered, the change sign (+/−) key should be used. Assume that the number 1.276E − 4 is to be entered. The following key strokes would be used:

1 • 2 7 6 EXP +/− 4

The display would show the following:

1.276 − 04

Setting the Display

Some calculators permit the answer to be displayed in any of three different ways. One of these ways is with *floating decimals (FD)*. When the calculator is set for this mode of operation, the answers are displayed with the decimal point appearing in the normal position. The only exception to this is if the number to be displayed is too large. In this case, the calculator automatically displays the number in scientific notation. In the *scientific mode (Sci)*, the calculator displays all entries and answers in scientific notation.

When set in the *engineering mode (Eng)*, the calculator displays all entries and answers in *engineering notation*. Engineering notation is very similar to scientific notation except that all answers are displayed in steps of 1000 instead of 10. This simply means that the scientific notation units are in steps of three instead of one. When in engineering notation, the calculator does not display exponents of 1, 2, 4, 5, 7, 8, and so on. When displayed in steps of three, the notation corresponds to standard engineering notation units such as kilo, mega, giga, milli, micro, and so on. For example, assume a calculator is set in the scientific mode and displays the number shown:

5.69836 05

Now assume that the calculator is reset to the engineering mode. The number would now be displayed as shown:

569.836 03

The number could now be read as "569.836 kilo," because kilo means 1000, or 1E03.

Answers to Practice Problems

Unit 2 Electrical Quantities and Ohm's Law

Volts (E)	Amperes (I)	Ohms (R)	Watt (P)
153 V	0.056 A	2732.143 Ω	8.568 W
305.5 V	0.65 A	470 Ω	198.575 W
24 V	5.167 A	4.645 Ω	124 W
3.59 V	0.00975 A	368.179 Ω	0.035 W
76.472 V	0.0112 A	6.8 k Ω	0.86 W
460 V	6.389 A	72 Ω	2938.889 W
48 V	1.2 A	40 Ω	57.6 W
123.2 V	154 A	0.8 Ω	18,972.8 W
277 V	2.744 A	100.959 Ω	760 W
14.535 V	0.0043 A	3380.206 Ω	0.0625 W
54.083 V	0.000416 A	130 k Ω	0.0225 W
96 V	0.0436 A	2.2 k Ω	4.189 W

Unit 5 Resistors

1st Band	2nd Band	3rd Band	4th Band	Value	% Tolerance
Red	Yellow	Brown	Silver	240 Ω	10
Blue	Gray	Red	Gold	6800 Ω	5
Orange	Orange	Orange	Gold	33 k Ω	5
Brown	Red	Black	Red	12 Ω	2
Brown	Green	Silver	Silver	0.15 Ω	10
Brown	Gray	Green	Silver	1.8 M Ω	10
Brown	Black	Yellow	None	100 k Ω	20
Brown	Black	Orange	Gold	10 k Ω	5
Violet	Green	Black	Red	75 Ω	2
Yellow	Violet	Red	None	4.7 k Ω	20
Gray	Red	Green	Red	8.2 M Ω	2
Green	Blue	Gold	Red	5.6 Ω	2

Unit 6 Series Circuits

1.	E_T 120 V	E_1 16.34 V	E_2 13.68 V	E_3 28.5 V	E_4 38 V	E_5 23.56 V
	I_T 0.038 A	I_1 0.038 A	I_2 0.038 A	I_3 0.038 A	I_4 0.038 A	I_5 0.038 A
	R_T 3160 Ω	R_1 430 Ω	R_2 360 Ω	R_3 750 Ω	R_4 1000 Ω	R_5 620 Ω
	P_T 4.56 W	P_1 0.621 W	P_2 0.52 W	P_3 1.083 W	P_4 1.444 W	P_5 0.894 W
	E_T 50 V	E_1 6 V	E_2 16.5 V	E_3 11 V	E_4 9 V	E_5 7.5 V
	I_T 0.005 A	I_1 0.005 A	I_2 0.005 A	I_3 0.005A	I_4 0.005 A	I_5 0.005 A
	R_T 10 kΩ	R_1 1.2 kΩ	R_2 3.3 kΩ	R_3 2.2 kΩ	R_4 1.8 kΩ	R_5 1.5 kΩ
	P_T 0.25 W	P_1 0.03 W	P_2 0.0825 W	P_3 0.055 W	P_4 0.045 W	P_5 0.0375 W
	E_T 340 V	E_1 44 V	E_2 94 V	E_3 60 V	E_4 40 V	E_5 102 V
	I_T 0.002 A	I_1 0.002 A	I_2 0.002 A	I_3 0.002 A	I_4 0.002 A	I_5 0.002 A
	R_T 170 kΩ	R_1 22 kΩ	R_2 47 kΩ	R_3 30 kΩ	R_4 20 kΩ	R_5 51 kΩ
	P_T 0.68 W	P_1 0.088 W	P_2 0.188 W	P_3 0.12 W	P_4 0.08 W	P_5 0.204 W
2.	E_1 = 18.46 V	E_2 = 40.62 V	E_3 = 33.23 V	E_4 = 27.69 V		
3.	E_T 36V	E_1 8.4 V	E_2 9.3 V	E_3 11.25 V	E_4 7.05 V	
	I_T 0.015	I_1 0.015	I_2 0.015	I_3 0.015	I_4 0.015	
	R_T 2400 Ω	R_1 560 Ω	R_2 620 Ω	R_3 750 Ω	R_4 470 Ω	
	P_T 0.54 W	P_1 0.126 W	P_2 0.140 W	P_3 0.169 W	P_4 0.106 W	
4.	E_T 48 V	E_1 24.6 V	E_2 9.9 V	E_3 8.1 V	E_4 5.4 V	
	I_T 0.0003	I_1 0.0003	I_2 0.0003	I_3 0.0003	I_4 0.0003	
	R_T 160 kΩ	R_1 82 kΩ	R_2 33 kΩ	R_3 27 kΩ	R_4 18 kΩ	
	P_T 0.0144 W	P_1 0.00738 W	P_2 0.00297 W	P_3 0.00243 W	P_4 0.00162 W	

Unit 7 Parallel Circuits

1.	E_T 120.03 V	E_1 120.03 V	E_2 120.03 V	E_3 120.03 V	E_4 120.03 V
	I_T 0.942 A	I_1 0.177 A	I_2 0.146 A	I_3 0.255 A	I_4 0.364 A
	R_T 127.42 Ω	R_1 680 Ω	R_2 820 Ω	R_3 470 Ω	R_4 330 Ω
	P_T 113.068 W	P_1 21.425 W	P_2 17.542 W	P_3 27.007 W	P_4 43.619 W
2.	E_T 277.056 V	E_1 277.056 V	E_2 277.056 V	E_3 277.056 V	E_4 277.056 V
	I_T 0.00639 A	I_1 0.00231 A	I_2 0.00139 A	I_3 0.00154 A	I_4 0.00115 A
	R_T 43,357.746 Ω	R_1 119,937.662 Ω	R_2 199,320.863 Ω	R_3 179,906.494 Ω	R_4 240,918.261 Ω
	P_T 1.771 W	P_1 0.640 W	P_2 0.384 W	P_3 0.427 W	P_4 0.319 W
3.	E_T 47.994 V	E_1 47.994 V	E_2 47.994 V	E_3 47.994 V	E_4 47.994 V
	I_T 13.399 A	I_1 3 A	I_2 4.799 A	I_3 3.2 A	I_4 2.4 A
	R_T 3.582 Ω	R_1 16 Ω	R_2 10 Ω	R_3 15 Ω	R_4 20 Ω
	P_T 643.072 W	P_1 143.982 W	P_2 230.323 W	P_3 153.581 W	P_4 115.186 W

4.	E_T 240.78 V	E_1 240.78 V	E_2 240.78 V	E_3 240.78 V	E_4 240.78 V
	I_T 0.0143 A	I_1 0.00293 A	I_2 0.0032 A	I_3 0.00429 A	I_4 0.00387 A
	R_T 16,802.591 Ω	R_1 82 Ω	R_2 75 k Ω	R_3 56 k Ω	R_4 62 k Ω
	P_T 3.436 W	P_1 0.704 W	P_2 0.769 W	P_3 1.031 W	P_4 0.932 W
5.	R_T = 83.735 W	I_1 = 0.0116 A	I_2 = 0.00891 A		
	I_3 = 0.0140 A	I_4 = 0.0155 A			
6.	R_T = 53,171.513W	I_1 = 1.18 mA	I_2 = 0.886 mA		
	I_3 = 0.742 mA	I_4 = 3.19 mA			
7.	E_T 24 V	E_1 24 V	E_2 24 V	E_3 24 V	E_4 24 V
	I_T 0.5 A	I_1 0.08 A	I_2 0.1 A	I_3 0.12 A	I_4 0.2 A
	R_T 48 Ω	R_1 300 Ω	R_2 240 Ω	R_3 200 Ω	R_4 120 Ω
	P_T 12 W	P_1 1.92 W	P_2 2.4 W	P_3 2.88 W	P_4 4.8 W
8.	E_T 12 V	E_1 12 V	E_2 12 V	E_3 12 V	E_4 12 V
	I_T 0.24 A	I_1 0.0667 A	I_2 0.0333 A	I_3 0.06 A	I_4 0.08 A
	R_T 50 W	R_1 180 Ω	R_2 360 Ω	R_3 200 Ω	R_4 150 Ω
	P_T 2.88 W	P_1 0.8 W	P_2 0.4 W	P_3 0.72 W	P_4 0.96 W

Unit 8 Combination Circuits

1.	I_T 0.0241 A	E_1 36.15 V	E_2 12.103 V	E_3 26.6 V	E_4 38.782 V
	R_T 3109.216 Ω	I_1 0.0241 A	I_2 0.0133 A	I_3 0.0133 A	I_4 0.0107 A
2.	I_T 0.00946 A	E_1 208.12 V	E_2 85.14 V	E_3 56.76 V	E_4 141.9 V
	R_T 37 k Ω	I_1 0.00946 A	I_2 0.00473 A	I_3 0.00473 A	I_4 0.00473 A
3.	I_T 0.0616 A	E_1 5.051 V	E_2 5.456 V	E_3 7.502 V	E_4 12.943 V
	R_T 292.118 Ω	I_1 0.0616 A	I_2 0.0341 A	I_3 0.0341 A	I_4 0.0275 A
4.	E_T 712.95 V	E_1 712.95 V	E_2 295.478 V	E_3 417.472 V	E_4 417.472 V
	R_T 891.187 Ω	I_1 0.475 A	I_2 0.325 A	I_3 0.209 A	I_4 0.116 A
5.	E_T 8033.068 V	E_1 8033.068 V	E_2 4275 V	E_3 3762 V	E_4 3762 V
	R_T 12,358.566 Ω	I_1 0.365 A	I_2 0.285 A	I_3 0.171 A	I_4 0.144 A
6.	E_T 58.55 V	E_1 58.55 V	E_2 19.693 V	E_3 38.813 V	E_4 38.813 V
	R_T 48.792 Ω	I_1 0.781 A	I_2 0.419 A	I_3 0.176 A	I_4 0.243 A
7.	E_T 250 V	E_1 55 V	E_2 50 V	E_3 18.8 V	E_4 11.2 V
	I_T 0.25 A	I_1 0.25 A	I_2 0.1 A	I_3 0.04 A	I_4 0.04 A
	R_T 1000 Ω	R_1 220 Ω	R_2 500 Ω	R_3 470 Ω	R_4 280 Ω
	P_T 62.5 W	P_1 13.75 W	P_2 5 W	P_3 0.752 W	P_4 0.448 W
	E_5 40 V	E_6 30 V	E_7 52.5 V	E_8 67.5 V	E_9 75 V
	I_5 0.1 A	I_6 0.06 A	I_7 0.15 A	I_8 0.15 A	I_9 0.25 A

	R_5 400 Ω	R_6 500 Ω	R_7 350 Ω	R_8 450 Ω	R_9 300 Ω
	P_5 4 W	P_6 1.8 W	P_7 7.875 W	P_8 10.125 W	P_9 18.75 W
8.	E_T 24 V	E_1 3.358 V	E_2 4.797	E_3 0.478 V	E_4 1.248 V
	I_T 0.024 A	I_1 0.024 A	I_2 0.00961 A	I_3 0.00385 A	I_4 0.00385 A
	R_T 1000 Ω	R_1 139.917 Ω	R_2 499.167 Ω	R_3 124.156 Ω	R_4 324.156 Ω
	P_T 0.576 W	P_1 0.0806 W	P_2 0.0461 W	P_3 0.00184 W	P_4 0.00481 W
	E_5 2.112 V	E_6 1.726 V	E_7 3.597 V	E_8 5.042 V	E_9 12 V
	I_5 0.0096 A	I_6 0.00576 A	I_7 0.0144 A	I_8 0.0144 A	I_9 0.024 A
	R_5 219.771 Ω	R_6 299.653 Ω	R_7 249.792 Ω	R_8 350.139 Ω	R_9 500 Ω
	P_5 0.0203 W	P_6 0.00995 W	P_7 0.0518 W	P_8 0.0726 W	P_9 0.288 W

Unit 10 Determining Conductor Sizes

1. 10.4 Ω cm/ft × 450 ft = 4680 Ω cm

$$\frac{4680\ \Omega\ \text{cm}}{66.370\ \text{cm}} = .0705\ \Omega$$

2. 1.817 Ω × 16,510 cm = 29,998.67 Ω cm

$$\frac{29{,}998.67\ \Omega\ \text{cm}}{500\ \text{ft}} = 60\ \frac{\Omega\ \text{cm}}{\text{ft}}\text{(Iron)}$$

3. 430 A (from *Table 310.16*) × 0.71 (correction factor [*Table 310.15(B)(1)(1)*]) = 305.3 A

4. 30 A × 0.70 = 21 A [*Table 310.15(C)(1)*]

5. 4/0 AWG (746 × 50 = 37,300 watts) (37,300/250 = 149.2 amps) (149.2 × 125% = 186.5 amps) (Correct for ambient temperature: 186.5/0.82 = 227.4 amps.) (Conductor is chosen from the 75°C column of Table 310.16.)

6. 2/0 AWG (25,000/480 = 52.1 amperes) (52.1 × 125% = 65.1 amperes) (Because the initial current demand is 65.1 amperes, the conductors must be chosen from the 60°C column of Table 310.16.) (Adjust for ambient temperature: 65.1/0.58 = 112.2 amps.) (Adjust for number of conductors in the raceway: 112.2/0.80 = 140.2 amps.)

7. R = 1.27 Ω

$$R = R_{ref}[1 + \alpha(T - T_{ref})]$$

$$R = 10.4[1 + 0.0039(87 - 20)]$$

$$R = 13.12$$

$$R = \frac{K \times L}{CM}$$

$$R = \frac{13.12 \times 250}{2583}$$

$$R = 1.27\ \Omega$$

8. R = 18.36 V

$$R - R_{ret}[1 + \alpha(T - T_{ret})]$$

$$R = 17[1 + 0.004(40 - 20)]$$

$$R = 18.36\ \Omega$$

Use the *NEC* to determine the ampacity of the following conductors:

Size	Material	Insulation	Amb. Temp.	Conductors in Raceway	Amps
#10 AWG	Copper	RHW	44°C	3	28.7 A
350 kcmil	Copper	XHH	128°F	6	212.8 A
#2 AWG	Aluminum	TW	86°F	2	75 A
3/0 AWG	Aluminum	XHHW-2	38°C	9	111.48 A
500 kcmil	Copper	THWN	48°C	6	228 A
#6 AWG	Copper	THW-2	150°F	3	24.75 A
2/0 AWG	Aluminum	UF	86°F	12	57.5 A
750 kcmil	Aluminum	RHW-2	34°C	6	334.08 A

Unit 14 Basic Trigonometry and Vectors

∠X	∠Y	Side A	Side B	Hyp.
40°	50°	5.142	6.128	8
57°	33°	72	46.759	85.85
37.1°	52.9°	38	50.249	63
52°	38°	17.919	14	22.739
48°	42°	173.255	156	233.138

Unit 15 Alternating Current

Sine Wave Values

Peak volts	Inst. volts	Degrees
347	208	36.829
780	536.64	43.5
80.464	24.3	17.6
224	5.65	1.445
48.7	43.976	64.6
339.411	240	45

Peak volts	Inst. volts	Degrees
87.2	23.7	15.771
156.9	155.488	82.3
110.387	62.7	34.6
1256	400	18.571
15,720	3269.76	12
126.795	72.4	34.8

Peak, RMS, and Average Values

Peak volts	RMS	Average
12.7	8.979	8.09
76.073	53.8	48.42
257.794	182.426	164.2
1235	873.145	786.695
339.36	240	216
26.06	18.443	16.6

Peak volts	RMS	Average
339.7	240.168	216.389
17.816	12.6	11.34
14.13	9.999	9
123.7	87.456	78.797
105.767	74.8	67.32
169.56	119.988	108

Unit 16 Inductance in AC Circuits

Inductive Circuits

1.

Inductance (H)	Frequency (Hz)	Induct. Rct. (W)
1.2	60	452.4
0.085	400.052	213.628
0.75	1000	4712.389
0.65	600	2450.448
3.6	29.999	678.584
2.619	25	411.459

Inductance (H)	Frequency (Hz)	Induct. Rct. (W)
0.5	60	188.5
0.85	1199.934	6408.849
1.6	20	201.062
0.45	400	1130.976
4.8	80.001	2412.743
0.0065	1000	40.841

2. 480 Hz $\left(X_L = \dfrac{99.526}{2 \times 3.1416 \times 0.033}\right)$
3. 0.0729 henrys $(Z = 120/4)$ $(Z = 30\ \Omega)$ $(X_L = \sqrt{30^2 - 12^2})$
4. 296.722 Ω $(X_L = 2 \times 3.1416 \times 60)$ $(X_L = 282.744\ \Omega)$ $(Z = \sqrt{90^2 - 282.722^2})$
5. $Q = 5.239$ $(Z = 240/3)$ $(Z = 80\ \Omega)$ $(X_L = \sqrt{80^2 - 15^2})$ $(X_L = 78.581\ \Omega)$ $(Q = 78.581/15)$

Unit 17 Resistive-Inductive Series Circuits

1.

E_T 480 V	E_R 247.128 V	E_L 411.489 V
I_T 20.594 A	I_R 20.594 A	I_L 20.594 A
Z 23.308 Ω	R 12 Ω	X_L 19.981 Ω
VA 9885.12	P 5089.354 W	$VARs_L$ 8474.204
PF 51.485%	∠Θ 59.003°	L 0.053 H

2.

E_T 130 V	E_R 78 V	E_L 104 V
I_T 6.5 A	I_R 6.5 A	I_L 6.5 A
Z 20 Ω	R 12 Ω	X_L 16 Ω
VA 845	P 507 W	$VARs_L$ 676
PF 60%	∠Θ 53.13°	L 0.0424 H

3.

E_T 120 V	E_R 96 V	E_L 72 V
I_T 1.2 A	I_R 1.2 A	I_L 1.2 A
Z 20 Ω	R 80 Ω	X_L 60 Ω
VA 144	P 115.2 W	$VARs_L$ 86.4
PF 80%	∠Θ 36.87°	L 0.15915 H

4.

E_T 239.806 V	E_R 187.049 V	E_L 150.059 V
I_T 1.563 A	I_R 1.563 A	I_L 1.563 A
Z 153.427 Ω	R 119.673 Ω	X_L 96.007 Ω
VA 374.817	P 292.357 W	$VARs_L$ 234.553
PF 78%	∠Θ 38.739°	L 0.0382 H

5. 187.4 volts
6. 84.8 PF
7. 195.86 VARs
8. 439.38 VARs
9. 0.03 W
10. 81.98 W
11. 390.3 watts
12. 37.98 VARs
13. 807.77 W
14. 77.9% PF
15. 90.05 volts
16. 154.89 volts

Unit 18 Resistive-Inductive Parallel Circuits

1.

E_T 240.005 V	E_R 240.005 V	E_L 240.005 V
I_T 34.553 A	I_R 17.143 A	I_L 30 A
Z 6.946 Ω	R 14 Ω	X_L 8 Ω
VA 8292.893	P 4114.406 W	$VARs_L$ 7200.15
PF 49.61%	$\angle\Theta$ 60.25°	L 0.02122 H

2.

E_T 277 V	E_R 277 V	E_L 277 V
I_T 39 A	I_R 15 A	I_L 36 A
Z 7.103 Ω	R 18.467 Ω	X_L 7.694 Ω
VA 10,803	P 4155 W	$VARs_L$ 9972
PF 38.46%	$\angle\Theta$ 67.36°	L 0.0204 H

3.

E_T 72 V	E_R 72 V	E_L 72 V
I_T 2 A	I_R 1.6 A	I_L 1.2 A
Z 36 Ω	R 45 Ω	X_L 60 Ω
VA 144	P 115.2 W	$VARs_L$ 86.4
PF 80%	$\angle\Theta$ 36.87°	L 0.15915 H

4.

E_T 150.062 V	E_R 150.062 V	E_L 150.062 V
I_T 2.498 A	I_R 1.948 A	I_L 1.563 A
Z 60.073 Ω	R 77.034 Ω	X_L 96.007 Ω
VA 374.817	P 292.357 W	$VARs_L$ 234.553
PF 78%	$\angle\Theta$ 38.74°	L 0.0382 H

5. 199.69 Ω
6. 346.4 volts
7. 0.8 amps
8. 240 Ω
9. 174.5 VARs
10. 202.6 VA
11. 0.144 amps
12. 83.2% PF
13. 0.8 amps
14. 7.09 VA
15. 0.127 henrys
16. 1.632 amps

Unit 19 Capacitors

RC Time Constants

1.

Resistance	Capacitance	Time constant	Total time
150 kΩ	100 μF	15 s	75 s
350 kΩ	20 μF	7 s	35 s
142,857,142.9	350 pF	0.05 s	0.25 s

40 MΩ	0.05 μF	2 s	10 s
1.2 MΩ	0.47 μF	0.564 s	2.82 s
4166.67 Ω	12 μF	0.05 s	0.25 s
86 kΩ	3.488 μF	0.3 s	1.5 s
120 kΩ	470 pF	56.4 μs	282 μs
80 kΩ	250 nF	20 ms	100 ms
3.75 Ω	8 μF	30 μs	150 μs
100 kΩ	1.5 μF	150 ms	750 ms
33 kΩ	4 μF	132 ms	660 ms

2. 34.28 μF **3.** 200 μF **4.** 2.48047 μF **5.** 84 seconds

Unit 20 Capacitance in AC Circuits

Capacitive Circuits

1.

Capacitance	X_C	Frequency
38 μF	69.803 Ω	60 Hz
5.049 μF	78.8 Ω	400 Hz
250 pF	4.5 k Ω	141.471 kHz
234 μF	0.068 Ω	10 kHz
13.263 μF	240 Ω	50 Hz
10 μF	36.8 Ω	432.486 Hz
560 nF	0.142 Ω	2 MHz
176.835 nF	15 k Ω	60 Hz
75 μF	560 Ω	3.789 Hz
470 pF	1.693 k Ω	200 kHz
58.513 nF	6.8 k Ω	400 Hz
34 μF	450 Ω	10.402 Hz

2. 564.37 Ω **3.** 3.57 μF **4.** 282.19 Hz

5. 3.127 amps (C = 20 + 40 + 60) (C = 110 μF) (X_C = 24.114 Ω) (E = 24.114 × 8.6) (E = 207.38 volts) (X_C for 40 μF capacitor = 66.313 Ω) (I_C = 207.38/66.313)

6. 180 Ω (Find C at 60 Hz.) (Use the value of C to find X_C at 100 Hz.)

7. The existing capacitor has a value of 132.6 μF. The needed capacitance is 262.2 μF. (480/24 = 20 Ω) (Use the value of X_C to determine existing capacitor value.) (Determine the value of X_C needed to limit the current to 16 amperes.) (480/16 = 30 Ω) (Determine the total capacitance needed to produce 30 Ω of X_C.) (C = 88.4 μF) Convert the formula for total capacitance of series-connected capacitors to determine the value of the second capacitor:

$$\left(C_{Needed} = \frac{1}{\frac{1}{88.4} - \frac{1}{132.6}}\right) (C_{Needed} = 265.2\ \mu F)$$

Unit 21 Resistive-Capacitive Series Circuits

1.	E_T 480 V	E_R 288 V	E_C 384 V
	I_T 24 A	I_R 24 A	I_C 24 A
	Z 20 Ω	R 12 Ω	X_C 16 Ω
	Va 11,520	P 6912 W	$VARs_C$ 9216
	PF 60%	$\angle\theta$ 53.13°	C 165.782 μF
2.	E_T 130 V	E_R 78 V	E_C 104 V
	I_T 6.5 A	I_R 6.5 A	I_C 6.5 A
	Z 20 Ω	R 12 Ω	X_C 16 Ω
	VA 845	P 507 W	$VARs_C$ 676
	PF 60%	$\angle\theta$ 53.13°	C 165.782 μF
3.	E_T 346.709 V	E_R 277.368 V	E_C 207.975 V
	I_T 1.246 A	I_R 1.246 A	I_C 1.246 A
	Z 278.258 Ω	R 222.607 Ω	X_C 166.914 Ω
	VA 432	P 345.6 W	$VARs_C$ 259.2
	PF 80%	$\angle\theta$ 36.87°	C 15.8915 μF
4.	E_T 186.335 V	E_R 126.708 V	E_C 135.936 V
	I_T 1.61 A	I_R 1.61 A	I_C 1.61 A
	Z 115.736 Ω	R 78.701 Ω	X_C 84.432 Ω
	VA 300	P 204 W	$VARs_C$ 219.964
	PF 68%	$\angle\theta$ 47.156°	C 4.7125 μF

5. 196.07 volts **7.** 51.9% PF **9.** 9.2 μF

6. 47.3 VARs **8.** 330.9 volts **10.** 3,680.6 watts

Unit 22 Resistive-Capacitive Parallel Circuits

1.	E_T 122.689 V	E_R 122.689 V	E_C 122.689 V
	I_T 10.463 A	I_R 8.764 A	I_C 5.718 A
	Z 11.725 Ω	R 14 Ω	X_C 21.456 Ω
	VA 1283.695	P 1075.246 W	$VARs_C$ 701.536
	PF 83.76%	$\angle\theta$ 33.07°	C 132.626 μF
2.	E_T 54 V	E_R 54 V	E_C 54 V
	I_T 2.5 A	I_R 1.5 A	I_C 2 A
	Z 21.6 Ω	R 36 Ω	X_C 27 Ω
	VA 135	P 81 W	$VARs_C$ 108
	PF 60%	$\angle\theta$ 53.13°	C 14.736 μF

3.	E_T 307.202 V	E_R 307.202 V	E_C 307.202 V
	I_T 140.991 A	I_R 65.6 A	I_C 124.8 A
	Z 2.17888 Ω	R 4.683 Ω	X_C 2.462 Ω
	VA 43,312.717 VA	P 20,152.451	$VARs_C$ 38,338.81 VARs
	PF 46.53%	∠θ 62.29°	C 107.653 μF
4.	E_T 69.549 V	E_R 69.549 V	E_C 69.549 V
	I_T 7.5 A	I_R 6.999 A	I_C 2.696 A
	Z 9.273 Ω	R 9.937 Ω	X_C 25.797 Ω
	VA 521.618	P 486.75 W	$VARs_C$ 187.5
	PF 93.31%	∠θ 21.09°	C 6.169 μF

5. 2,857.96 Ω **7.** 110 Ω **9.** 81.2% PF

6. 0.436 amps **8.** 2,883.4 VARs **10.** 24.99 μF

Unit 23 Resistive-Inductive-Capacitive Series Circuits

1.	E_T 120 V	E_R 72 V	E_L 200 V	E_C 104 V
	I_T 2 A	I_R 2 A	I_L 2 A	I_C 2 A
	Z 60 Ω	R 36 Ω	X_L 100 Ω	X_C 52 Ω
	VA 240	P 144 W	$VARs_L$ 400	$VARs_C$ 208
	PF 60%	∠θ 53.13°	L 0.265	C 51.01 μF
2.	E_T 35.678 V	E_R 24 V	E_L 21.6 V	E_C 48 V
	I_T 0.6 A	I_R 0.6 A	I_L 0.6 A	I_C 0.6 A
	Z 59.464 Ω	R 40 Ω	X_L 36 Ω	X_C 80 Ω
	VA 21.407	P 14.4 W	$VARs_L$ 12.96	$VARs_C$ 28.8
	PF 67.27%	∠θ 47.7°	L 0.0143	C 4.974 μF
3.	E_T 23.988 V	E_R 14.993 V	E_L 75 V	E_C 56.25 V
	I_T 1.25 A	I_R 1.25 A	I_L 1.25 A	I_C 1.25 A
	Z 19.19	R 11.994 Ω	X_L 60 Ω	X_C 45 Ω
	VA 29.985	P 18.741 W	$VARs_L$ 93.75	$VARs_C$ 70.313
	PF 62.5%	∠θ 51.32°	L 0.159 H	C 58.945 μF
4.	E_T 349.057 V	E_R 185 V	E_L 740 V	E_C 444 V
	I_T 0.148 A	I_R 0.148 A	I_L 0.148 A	I_C 0.148 A
	Z 2358.493 Ω	R 1250 Ω	X_L 5000 Ω	X_C 3000 Ω
	VA 51.8	P 27.38 W	$VARs_L$ 109.52	$VARs_C$ 65.712
	PF 52.86%	∠θ 58.06°	L 0.796 H	C 0.053 μF

5. 101.24 volts

6. 87.4 % PF

7. The power factor is leading because there are more capacitive VARs than inductive VARs.

8. 35.6 volts **9.** 59.52 Ω **10.** 283.15 Ω

Unit 24 Resistive-Inductive-Capacitive Parallel Circuits

1.	E_T 120 V	E_R 120 V	E_L 120 V	E_C 120 V
	I_T 3.387 A	I_R 3.333 A	I_L 3 A	I_C 2.4 A
	Z 35.431 Ω	R 36 Ω	X_L 40 Ω	X_C 50 Ω
	VA 406.44	P 400 W	$VARs_L$ 360	$VARs_C$ 288
	PF 98.41%	$\angle\Theta$ 10.2°	L 0.106 H	C 53.05 μF
2.	E_T 243.885 V	E_R 243.885 V	E_L 243.885 V	E_C 243.885 V
	I_T 22.267 A	I_R 15.745 A	I_L 7.873 A	I_C 23.618 A
	Z 10.953 Ω	R 15.49 Ω	X_L 30.977 Ω	X_C 10.326 Ω
	VA 5430.58	P 3840 W	$VARs_L$ 1920	$VARs_C$ 5760
	PF 70.71%	$\angle\Theta$ 45.01°	L 0.0123 H	C 38.532 μF
3.	E_T 23.999 V	E_R 23.999 V	E_L 23.999 V	E_C 23.999 V
	I_T 2.004 A	I_R 2 A	I_L 0.4 A	I_C 0.533 A
	Z 11.973 Ω	R 12 Ω	X_L 60 Ω	X_C 45 Ω
	VA 48.106	P 48 W	$VARs_L$ 9.6	$VARs_C$ 12.791
	PF 99.78%	$\angle\Theta$ 3.62°	L 0.159	C 58.94 μF
4.	E_T 480 V	E_R 480 V	E_L 480 V	E_C 480 V
	I_T 100 A	I_R 60 A	I_L 150 A	I_C 70 A
	Z 4.8 Ω	R 8 Ω	X_L 3.2 Ω	X_C 6.857 Ω
	VA 48 kVA	P 28.8 kW	$VARs_L$ 72 kVARs	$VARs_C$ 33.6 VARs
	PF 60%	$\angle\Theta$ 53.13°	L 0.509 mH	C 23.211 μF

5. 16,571.4 W
6. 92.3% PF
7. $\angle\Theta = 56.98°$
8. 312.2 watts
9. 480 volts
10. 3.33 amperes

Unit 26 Three-Phase Circuits

1.	$E_{P(A)}$ 138.568 V	$E_{P(L)}$ 240 V
	$I_{P(A)}$ 34.64 A	$I_{P(L)}$ 20 A
	$E_{L(A)}$ 240 V	$E_{L(L)}$ 240 V
	$I_{L(A)}$ 34.64 A	$I_{L(L)}$ 34.64 A
	P 14,399.155 W	Z_{PHASE} 12 Ω
2.	$E_{P(A)}$ 4160 V	$E_{P(L)}$ 2401.849 V
	$E_{P(a)}$ 23.112 A	$E_{P(L)}$ 40.031 A
	$E_{L(A)}$ 4160 V	$E_{L(L)}$ 4160 V
	$I_{L(A)}$ 40.031 A	$I_{L(L)}$ 40.031 A
	P 288,428.159 W	Z_{PHASE} 60 Ω

3.

$E_{P(A)}$ 323.326 V	$E_{P(L1)}$ 323.326 V	$E_{P(L2)}$ 560 V
$I_{P(A)}$ 185.91 A	$I_{P(L1)}$ 64.665 A	$I_{P(L2)}$ 70 A
$E_{L(A)}$ 560 V	$E_{L(L1)}$ 560 V	$E_{L(L2)}$ 560 V
$I_{L(A)}$ 185.905 A	$I_{L(L1)}$ 64.665 A	$I_{L(L2)}$ 121.24 A
P 180,317.83 W	Z_{PHASE} 5 Ω	Z_{PHASE} 8 Ω

4.

$E_{P(A)}$ 277.136 V	$E_{P(L1)}$ 277.136 V	$E_{P(L2)}$ 480 V	$E_{P(L3)}$ 277.136 V
$I_{P(A)}$ 34.485 A	$I_{P(L1)}$ 23.095 A	$I_{P(L2)}$ 30 A	$I_{P(L3)}$ 27.714 A
$E_{L(A)}$ 480 V	$E_{L(L1)}$ 480 V	$E_{L(L2)}$ 480 V	$E_{L(L3)}$ 480 V
$I_{L(A)}$ 34.049 A	$I_{L(L1)}$ 23.095 A	$I_{L(L2)}$ 51.96 A	$I_{L(L3)}$ 27.714 A
VA 27,838.09	Z_{PHASE} 12 Ω	$Z_{L(PHASE)}$ 16 Ω	$X_{C(PHASE)}$ 10 Ω
	P 19,200.259 W	$VARs_L$ 43,197.466	$VARs_C$ 23,040.311

Unit 27 Single-Phase Transformers

1.

E_P 120 V	E_S 24 V
I_P 1.6 A	I_S 8 A
Ratio 5:1	$Z = 3\ \Omega$

2.

E_P 240 V	E_S 320 V
I_P 0.853 A	I_S 0.643 A
Ratio 1:1.333	$Z = 500\ \Omega$

3.

E_P 64 V	E_S 160 V
I_P 33.333 A	I_S 13.333 A
Ratio 1:2.5	$Z = 12\ \Omega$

4.

E_P 48 V	E_S 240 V
I_P 3.333 A	I_S 0.667 A
Ratio 1:5	$Z = 360\ \Omega$

5.

E_P 35.848 V	E_S 182 V
I_P 16.5 A	I_S 3.25 A
Ratio 1:5.077	$Z = 56\ \Omega$

6. E_P 480 V E_S 916.346 V
I_P 1.458 A I_S 0.764 A

Ratio 1:1.909 Z = 1.2 k Ω

7. E_P 208 V E_S 1 320 V E_{S2} 120 V E_{S3} 24 V
I_P 11.93 A I_{S1} 0.0267 A I_{S2} 20 A I_{S3} 3 A

Ratio 1 1:1.54 Ratio 2 1.73:1 Ratio 3 1:8.67
R_1 12 k Ω R_2 6 Ω R_3 8 Ω

8. E_P 277 V E_{S1} 480 V E_{S2} 208 V E_{S3} 120 V
I_P 8.93 A I_{S1} 2.4 A I_{S2} 3.47 A I_{S3} 5 A

Ratio 1 1:1.73 Ratio 2 1.33:1 Ratio 3 2.31:1
R_1 200 Ω R_2 60 Ω R_3 24 Ω

Unit 28 Three-Phase Transformers

1. E_P 2401.8 V E_P 440 V E_P 254.04 V
I_P 7.67 A I_P 41.9 A I_P 72.58 A
E_L 4160 V E_L 440 V E_L 440 V
I_L 7.67 A I_L 72.58 A I_L 72.58 A
Ratio 5.46:1 Z 3.5 Ω

2. E_P 4157.04 V E_P 240 V E_P 138.57 V
I_P 1.15 A I_P 20 A I_P 34.64 A
E_L 7200 V E_L 240 V E_L 240 V
I_L 1.15 A I_L 34.64 A I_L 34.64 A
Ratio 17.32:1 Z 4 Ω

3. E_P 13,800 V E_P 277 V E_P 480 V
I_P 6.68 A I_P 332.54 A I_P 192 A
E_L 13,800 V E_L 480 V E_L 480 V
I_L 11.57 A I_L 332.54 A I_L 332.54 A
Ratio 49.76:1 Z 2.5 Ω

4. E_P 23,000 V E_P 120 V E_P 208 V
I_P 0.626 A I_P 120.08 A I_P 69.33 A
E_L 23,000 V E_L 208 V E_L 208 V
I_L 1.08 A I_L 120.08 A I_L 120.08 A
Ratio 191.66:1 Z 3 Ω

Unit 34 Motor Installation

1. Conductors = #6 AWG (50 × 1.25 = 62.5 or 63 amps) (Conductor selected from 75°C column of Table 310.16)
2. 200-amp circuit breaker (80 × 2.50 = 200)
3. Conductor size = #3 AWG (78 × 1.25 = 97.5 or 98 amps. NOTE: Table current was not increased because motor is to operate at unity power factor)

   NEMA starter = #4, IEC starter = M

   (OL heaters = X73. Sized for 115% of nameplate current.) (Selected from size 4 column)

   Fuse size = 150 amp (78 × 1.75 = 136.5 or 137 amps)

4. Conductor = #1 AWG (104 × 1.25 = 130 amps) (Answer taken from 75°C column)

   NEMA starter = 4, IEC starter = M

   OL heater = X79 (Chart indicates X78 but increase one size to size OL at 125%)

   Fuse = 200 Amps (104 × 1.75 = 182 amps)

5. Conductor = 900 kcmil (414 × 1.25 = 517.5 or 518 amps) (Conductor selected from 75°C column)

   Fuse = 800 amperes (414 × 1.75 = 724.5 or 725 A)

6. Motor #1

   Conductor = #14 AWG (14 × 1.25 = 17.5 or 18 amps) (Selected from 75°C column)

   OL heater = X57 (the current value indicates a size X56. The next higher size is used because the heater is sized for 125% of FLA.)

   Fuse = 25 amps (14 × 1.75 = 24.5 or 25 amps)

   Motor #2

   Conductor = #8 AWG (34 × 1.25 = 42.5 or 43 amps) (Selected from the 75°C column)

   OL heater = X66 (Selected from the size 2 starter column)

   Fuse = 60 amps (34 × 1.75 = 59.5 or 60 amps) (Heater sized for 115% of motor FLA)

   Motor #3

   Conductor = #10 AWG (21 × 1.25 = 26.25 or 26) (Selected from 75°C column)

   OL heater = X60 (Selected from size 2 column)

   Fuse = 40 amps (21 × 1.75 = 36.75 or 37 amps)

   Main Feeder

   Conductor = #4 AWG (34 × 1.25 = 42.5 + 14 + 21 = 77.5 or 78 amps)

   Fuse = 90 amps (60 + 14 + 21 = 95) (Closest without going over is 90 amps)

Glossary

A

Acids any of a large group of substances having a pH less than 7.

Adjacent next to or beside.

Alkalis substances having a pH greater than 7.

Alternating current (AC) current that reverses its direction of flow periodically. Reversals generally occur at regular intervals.

Alternator a machine used to generate AC by rotating conductors through a magnetic field.

Ambient air temperature the temperature surrounding a device.

American Wire Gauge (AWG) a measurement of the diameter of a wire. The gauge scale was formerly known as the Brown and Sharp scale. The scale has a fixed constant of 1.123 between gauge sizes.

Ammeter an instrument used to measure the flow of current.

Ammeter shunt a device used to change the full-scale value of an ammeter by providing another path for current flow. Most ammeter shunts produce a voltage drop of 50 millivolts when rated current flows through them.

Amortisseur winding a squirrel-cage winding on the rotor of a synchronous motor used for starting purposes only.

Ampacity the maximum current-carrying capacity of a conductor without exceeding the temperature limitations of its insulation.

Ampere (A) a unit of measure for the rate of current flow. One ampere equals 1 coulomb per second.

Ampere-turns a basic unit for measurement of magnetism. The product of the number of turns of wire times current flow.

Amplitude the highest value reached by a signal, voltage, or current.

Analog meters meters that rely on a mechanical pointer and employ a scale to measure indicated quantities.

Angle theta ($\angle\theta$) generally used to express the phase angle difference between voltage and current in a reactive circuit.

Anode the positive terminal of an electric device.

Apparent power (VA) the value found by multiplying the applied voltage by the total current of an AC circuit. Apparent power is measured in volt-amperes (VA) and should not be confused with true power, measured in watts.

Arc a luminous discharge of electricity through gas.

Arc-rated clothing has been tested to standardized arc rating test methods to determine the arc rating of the garment and that has been assigned an arc rating in cal/cm^2 that describes the performance of the garment when exposed to an electric arc discharge.

Armature the rotating member of a motor or generator. The armature generally contains windings and a commutator.

Armature reaction the twisting or bending of the main magnetic field of a motor or generator. Armature reaction is proportional to armature current.

Artificial respiration providing a person with an external method of breathing. Generally employing the mouth-to-mouth method.

Atom the smallest part of an element that contains all the properties of that element.

Atomic number the atomic number is equal to the number of protons (positively charged particles) in the nucleus of an atom.

Attraction the force that causes one object to pull another object to itself.

Autotransformers transformers that use only one winding for both primary and secondary.

Average value the value of DC voltage when it is rectified from AC voltage. The ratio is dependent on the type of AC waveform that has been rectified.

Ayrton shunt an ammeter shunt that permits multiple range values.

B

Bandwidth a range of frequencies.

Battery a device used to convert chemical energy into electric energy. A group of voltaic cells connected together in a series or parallel connection.

Bidirectional an object that moves in two directions.

British thermal unit (BTU) the amount of heat necessary to raise or lower 1 pound of water 1° Fahrenheit.

Brushes sliding contacts, generally made of carbon, used to provide connection to rotating parts of machines.

Brushless DC motors DC motors that do not depend on a commutator and brushes to operate.

Brushless exciter a device used to excite the rotor of an alternator or synchronous motor that does not use brushes or sliprings.

C

Capacitance (C) the electrical size of a capacitor.

Capacitive reactance (X_C) the current-limiting property of a capacitor in an AC circuit.

Capacitor a device made with two conductive plates separated by an insulator or dielectric.

Capacitor charge rate T = RC

T = time in seconds

R = resistance in ohms

C = capacitance in farads

Carbon film resistor a resistor that is produced by coating a nonconductive material with a film of carbon.

Cardiopulmonary resuscitation (CPR) a method of alternating closed heart massage and artificial respiration for a person who is not breathing and has no pulse.

Cathode the negative terminal of an electric device.

Cell a single unit that converts chemical energy into electrical energy.

Centrifugal switch a switch sensitive to speed of rotation, generally used to disconnect the start windings of a split-phase motor.

Chassis ground ground connection to the case of a piece of equipment.

Circuit branch a point in an electric circuit where current can flow in more than one direction.

Circuit breakers devices designed to open circuits both manually and automatically under an abnormal amount of current flow. These devices are rated by voltage, current, and horsepower.

Circuit impedance the total current-limiting property of an AC circuit.

Circular mil an area measure of wire determined by squaring the diameter of the wire in mils (1 mil = 0.001 inch). Example: A wire with a diameter of 0.020 inch would have a circular mil area of 400 cmil ($20 \times 20 = 400$).

Clamp-on ammeter an ammeter that has a movable jaw that can be clamped around a conductor.

Code letter a letter found on the nameplate of a motor that describes the types of bars used in the squirrel-cage rotor.

Color code a system where colors are used to represent numbers.

Combination circuit a circuit that contains both series and parallel paths.

Commutator strips or bars of metal insulated from each other and arranged around one end of an armature. They provide connection between the armature windings and the brushes. The commutator is used to ensure proper direction of current flow through the armature windings.

Compensating winding the winding embedded in the main field poles of a DC machine. The compensating winding is used to help overcome armature reaction.

Complete path the path that exists in a circuit when it can be traced from one terminal of a power source back to the other terminal without a break.

Composition carbon resistors resistors that are made by mixing carbon with other compounds to produce a substance that has a known resistance.

Compound generators generators that contain both series and shunt field windings.

Compound motor a motor that contains both series and shunt field windings.

Compounding generally indicates the voltage increase caused by the series field as compared to the voltage drop of the generator.

Conductive compensation a method of connecting the compensating winding in a universal motor.

Conductor a device or material that permits current to flow through it easily.

Confined spaces spaces that have very limited means of entrance and exit and that are not intended for prolonged occupancy.

Consequent-pole motor a type of motor that changes speeds by changing the number of stator poles.

Constant-speed motors motors that run at the same approximate speed from no load to full load.

Control transformer a common type of isolation transformer generally used to change line voltage of 240 or 480 volts into 120 volts to operate motor-control equipment.

Conventional current flow theory a theory that considers current to flow from the most positive source to the most negative source.

Copper sulfate a chemical compound formed by combining copper, sulfur, and oxygen ($CuSO_4$).

Correction factor pertaining to conductors, the de-rating of the current capacity of conductors due to ambient temperature.

Cosine in trigonometry, it is the ratio of the adjacent side of the angle and the hypotenuse.

Coulomb a quantity of electrons equal to 6.25×10^{18}.

Counter-electromotive force (CEMF) the voltage induced in the armature of a DC motor that opposes the applied voltage and limits armature current; also called back-EMF.

Countertorque the magnetic force developed in the armature of a generator that makes the shaft hard to turn. Countertorque is proportional to armature current and is a measure of the electric energy produced by the generator.

Cumulative compound when the series and shunt fields are connected in such a manner as to aid each other in the production of magnetism.

Cumulative-compound motors motors that are so connected that their series and shunt fields aid each other in the production of magnetism.

Cuprous cyanide a chemical compound used to electroplate copper to iron.

Current capacity the amount of current a power source is capable of supplying.

Current dividers a circuit that is constructed to produce a specific amount of current flow through a branch.

Current flow through the capacitor the amount of current flowing through a capacitor.

Current flow through the inductor (I_L) the amount of current flowing through an inductor.

Current flow through the resistor (I_R) amount of current flowing through a resistor.

Current lags voltage the condition that occurs in an AC circuit containing inductance.

Current transformer a transformer used for metering alternating current. Current transformers generally have a standard output current value of 5 A at rated primary current.

Cycle one complete AC waveform.

D

d'Arsonval movement a meter movement using a permanent magnet and a coil of wire. The basic meter movement used in many analog type voltmeters, ammeters, and ohmmeters.

Damp locations locations that are protected from the weather and are not subject to saturation with water or other liquids but are subject to moderate degrees of moisture.

De-energized circuit a circuit that has no power applied to it.

Delta connection a circuit formed by connecting three electric devices in series to form a closed loop. It is used most often in three-phase connections.

Delta–WYE a three-phase transformer connection with a delta-connected primary and a WYE-connected secondary.

Demagnetize to remove magnetism from an object.

Dielectric an electric insulator.

Dielectric constant a method of determining the insulation properties of a dielectric using air as a reference.

Dielectric oil a type of insulating oil often used to cool transformers.

Dielectric stress the molecular stress placed on a dielectric material proportional to the level of applied voltage.

Differential-compound motors motors that are so connected that their series and shunt fields oppose each other in the production of magnetism.

Differential-compound when the series and shunt field are connected in such a manner as to oppose each other in the production of magnetism.

Digital multimeter (DMM) is an electronic measuring device used to measure different electrical properties, such as voltage, current, and resistance that displays the value in a digital readout.

Direct current (DC) current that does not reverse its direction of flow.

Direction of rotation the direction, clockwise or counterclockwise, that the rotor of a motor turns.

Disconnect to physically remove an object from a circuit or a switch that is employed to remove power for a circuit.

Distribution transformer a common transformer used by utility companies to change their line voltage into the voltage necessary to supply the premises wiring system.

Dry locations locations not normally subject to dampness or wetness.

Dual-voltage motors motors that can be connected to operate on either of two voltages.

E

Earth ground a ground established by driving a metal rod into the earth.

Eddy current circular induced current contrary to the main currents. Eddy currents are a source of heat and power loss in magnetically operated devices.

Effective value often referred to as RMS voltage or current. The amount of AC voltage or current that will produce the same heating effects as a like amount of DC voltage or current.

Electrical noise Unwanted electrical frequencies that can cause noise on a radio.

Electrically safe work condition is a state in which an electrical conductor or circuit part has been disconnected from energized parts, locked/tagged in accordance with established standards, tested for the absence of voltage, and, if necessary, temporarily grounded for personnel protection.

Electrolysis the decomposition of a chemical compound or metal caused by an electric current.

Electrolyte a chemical compound capable of conducting electric current by being broken down into ions. Electrolytes can be acids or alkalis.

Electrolytic a type of capacitor that is polarity sensitive.

Electromagnetic induction voltage produced by lines of magnetic flux cutting a conductor.

Electromagnets magnets that rely on electricity to become magnetized.

Electromotive force (EMF) electrical pressure, the force that pushes electrons through a wire; the force that causes electrons to flow through a circuit. The most common definition of voltage.

Electromotive series of metals a list of metals that can be employed for constructing voltaic cells.

Electron one of the three major parts of an atom. The electron carries a negative charge.

Electron flow theory the most popular theory of electric current flow that states electricity flows from the more negative part of the power source to the more positive.

Electron impact the collision that occurs when an electron strikes another electron or object.

Electron orbit the shell in which an electron moves around the nucleus of an atom.

Electron spin patterns the pattern or direction in which electrons spin as they orbit the nucleus of an atom.

Electronic overload relay An overload relay that uses electronic components to detect motor current.

Electroplating the act of depositing the atoms of one type of metal onto another.

Electroscope a device used to determine whether an object contains a positive or negative charge.
Electrostatic charge the static charge that is developed by adding or removing electrons from an object.
Electrostatic charges voltage developed by an excess or lack of electrons on an object.
Element (1) one of the basic building blocks of nature. An atom is the smallest part of an element; and (2) one part of a group of devices.
Encoder a device that determines the speed, number of revolutions, and direction of a servo motor.
Energized circuit a circuit that has power applied to it.
Equipment grounding conductor the conductor that is used to ensure equipment exists at ground potential. This conductor also ensures that all normally non-current carrying metal surfaces are bonded together, and it provides an effective ground-fault current path. It is generally green in color or bare.
Excitation current (1) the DC used to produce electromagnetism in the fields of a DC motor or generator or in the rotor of an alternator or synchronous motor; and (2) the AC that flows in the primary of a transformer with no load connected.
Exponential the rate at which current rises and falls in an inductor or the rate at which capacitors charge and discharge.
Exponential curve a natural curve that is the rate at which current can rise or fall in an inductor or the rate of charge and discharge for a capacitor.

F

Farad (F) the basic unit of capacitance.
Fibrillation the state in which the heart quivers and does not pump blood through the body.
Field excitation current the current used to produce a magnetic field in a pole piece.
Field-discharge resistor a resistor used to reduce the voltage spike caused by a collapsing magnetic field when the DC power supplied to the rotor of a synchronous motor or alternator is disconnected.
Field-loss relay (FLR) a current relay connected in series with the shunt field of a DC motor. The relay causes power to be disconnected from the armature in the event that field current drops below a certain level.
Fixed resistors resistors that have a value that cannot be changed.
Flux magnetic lines of force.
Flux density the number of magnetic lines contained in a certain area. The area measurement depends on the system of measurement.
Flux leakage the number of magnetic flux lines that radiate into space and do not cut through the conductors of an inductor or secondary winding of a transformer.
Frequency the number of complete cycles of AC voltage that occur in 1 s.
Frogleg-wound armatures armatures that have their windings connected in such a way as to form a series-parallel connection.
Fuses devices used to protect a circuit or electric device from excessive current. Fuses operate by melting a metal link when current becomes excessive.

G

Galvanometer a meter movement requiring microamperes to cause a full-scale deflection. Many galvanometers have a zero center, which permits them to measure both positive and negative values.
Generator a device used to convert mechanical energy into electric energy.
Ground point the point at which a grounding conductor is attached.

H

Henry (H) the basic unit of inductance.
Hertz (Hz) the international unit of frequency.
High-frequency occurrences High frequencies that are generally produced by voltage spikes.
High leg a certain type of three-phase power connection that exhibits a higher voltage on one leg when measured to ground than on the other legs.
HIPOT a high-voltage tester often used to test the dielectric of a capacitor.
Holding contacts contacts used for the purpose of maintaining current flow to the coil of a relay.
Holtz motor a type of single-phase synchronous motor.
Horseplay the act of playing jokes on others, generally through some physical act.
Horsepower a measure of power for electric and mechanical devices.
Hydrogen a gas.
Hydrometer a device used to measure the specific gravity of a fluid, such as the electrolyte used in a battery.
Hypotenuse the longest side of a right triangle.
Hysteresis loss power loss in a conductive material caused by molecular friction. Hysteresis loss is proportional to frequency.

I

Idiot proofing designing a piece of equipment so that it cannot be misused or connected in an incorrect manner.
Impedance (Z) the total opposition to current flow in an electric circuit.
In phase a condition that occurs when voltage and current reach all peak values at the same time.
Induced voltage voltage that is produced in a conductor by magnetic lines of flux cutting the conductor.
Inductance (L) property of a circuit that tends to oppose any change of current.
Inductive compensation a method of connecting the compensating windings in a universal motor.

Inductive reactance (X_L) the current-limiting property of an inductor in an AC circuit.

Inrush current the amount of current that is drawn when an inductive device is first connected to the power line.

Insulation a resistive coating that prevents the flow of electrons.

Insulator a material used to electrically isolate two conductive surfaces.

Internal resistance the resistance inherent to a power source.

Ion a charged atom.

Ionization potential the amount of voltage or potential that must be applied to cause ionization of a gas.

Isolation transformers transformers in which the secondary winding is electrically isolated from the primary winding.

J

JAN (Joint Army-Navy) standard Joint Army Navy standard.

Joule (J) a basic unit of electric energy. A joule is the amount of work done when 1 A flows through 1 W for 1 s. A joule is equal to 1 W-s.

L

Lagging power factor power factor caused by current lagging the voltage in an inductive circuit.

Laminated something made of thin sheets stacked together.

Lap-wound armatures armatures that have their windings connected in such a way as to form a parallel connection.

Leading power factor power factor caused by current leading the voltage in a capacitive circuit.

Leakage current the amount of current that flows through the dielectric separating the plates of a capacitor.

Left-hand generator rule a rule for determining the direction of current flow through an armature when the direction of motion and polarity of the magnetic field are known.

Left-hand rule a rule used for determining the polarity of an electromagnet when the direction of current flow is known.

Lenz's law a law that explains the relationships of movement, voltage polarity, and magnetic field polarity that pertain to magnetic induction.

Lightning a natural static charge that can produce extremely high voltage arcs.

Lightning arrester a device designed to help prevent damage to equipment by transferring the energy of a lightning strike safely to ground.

Lightning bolts high-voltage arcs caused by lightning discharge. Lightning bolts can flow from cloud to cloud, cloud to ground, or ground to cloud.

Lightning rods metal poles that are generally pointed on one end. They are mounted above the structure that is to be protected from lightning.

Line voltage the voltage as measured between the lines of a three-phase system.

Linear wave a waveform that is produced when voltage or current rises or falls in direct proportion to time.

Lines of flux lines of magnetic force.

Load the power consumed by a piece of equipment or a circuit while performing its function.

Load test a test for determining the condition of charge for a battery.

Lockout a mechanical device used to prevent the operation of some other component.

Lockout and tagout the act of placing a lock and warning tag on the disconnect of a piece of equipment that is not to be energized.

Lodestones natural magnets.

M

Magnetic domains a section within a magnetic material where atoms are aligned to produce a north or south polarity.

Magnetic molecules refer to magnetic domains.

Magnetic overload relay Overload relays that sense motor current by inserting a coil in series with the motor.

Magnetomotive force (mmf) the magnetic force produced by current flowing through a conductor or coil.

Matter anything that occupies space and possesses mass.

Maximum operating temperature the highest operating temperature of a conductor. Generally determined by the type of insulation around the conductor.

MEGGER a registered name by Biddle Industries. A high-voltage tester used for testing conductor insulation.

Metal film resistors resistors that are produced by coating a nonconductive material with a film of metal.

Metal glaze resistors resistors that are produced by applying a mixture of metal and glass to a nonconductive material.

Metal oxide varistor (MOV) a semiconductor device that changes its resistance in accordance with the voltage applied across it.

Metallic salt a chemical compound that is generally in a crystal state and exhibits the properties of a metal.

Meter a device for measuring some quantity such as voltage, current, resistance, etc.

Mil-foot a standard for measuring the resistivity of wire. A mil-foot is the resistance of a piece of wire 1 mil in diameter and 1 foot in length.

Milliampere 1/1000 of an ampere (0.001 ampere).

Molecules the smallest part of a compound that is made of individual atoms.

Motor a device used to convert electric energy into mechanical energy.

Moving-coil meter a meter that is constructed by connecting a point to a coil that is permitted to move when current flows through it.

Multirange ammeters ammeters that can be set for more than one full-range value.
Multirange voltmeters voltmeters that can be set for more than one full-range value.
Multispeed motor a motor that can be operated at more than one speed.
Multiturn variable resistors variable resistors that require multiple turns to change their resistance value. They are generally employed where precise control of the resistance value is critical.

N

NFPA 70®: National Electrical Code® (NEC®) the ANSI approved national code that governs the installation of electrical equipment in industrial, residential, and commercial facilities. Published by the NFPA, the National Fire Protection Association.
Negative one polarity of a voltage, current, or charge.
Negative-sequence harmonic causes the rotating magnetic field of a polyphase motor to rotate in a direction opposite that of the fundamental frequency.
Neutral conductor The circuit conductor that is connected to the neutral point in the system, and if the system is grounded, it is also the grounded conductor. The neutral is generally white or gray in color.
Neutral plane the point at which there is no voltage induced in the armature of a motor or generator.
Neutron one of the principal parts of an atom. The neutron has no charge and is part of the nucleus.
Nickel-cadmium (nicad) cell a voltaic cell using nickel and cadmium as cell plates with an electrolyte of potassium hydroxide producing a voltage of approximately 1.2 volts.
Node a joining point where electric connections are made.
Noise-to-signal ratio The ratio of electrical noise compared to the desired signal.
Nonlinear load loads that draw current in pulses.
Nonpolarized capacitors capacitors that can be connected to either polarity of voltage. Generally used in alternating current circuits.
Nucleus the center or middle.
Nuisance static charges static electric charges that develop on objects when it is not desirable for the object to become charged.

O

Occupational Safety and Health Administration (OSHA) a government organization responsible for enforcement of the Occupational Safety and Health Act of 1970.
Ohm (Ω) the unit of measure for electrical resistance.
Ohm's law a set of electrical laws developed by Georg S. Ohm that permits electrical quantities to be calculated mathematically.
Ohmmeter a device used to measure resistance.
One-line diagram a type of wiring diagram that shows basic power runs and does not show actual connections.
Open-delta a type of three-phase transformer connection that employs only two transformers.
Opposite across from.
Orange wire one method of color coding the high leg connections of some three-phase power systems.
Oscillator a device used to change DC voltage into AC voltage.
Oscilloscope a voltmeter that displays a waveform of voltage in proportion to its amplitude with respect to time.
Out of phase the condition in which two components do not reach their positive or negative peaks at the same time.
Overload protection is installed to protect equipment from operation in excess of the normal, full-load current rating. This can be in the form of a fuse, relay, heater, or electronic device.
Overload relay a relay used to protect a motor from damage caused by overloads. The overload relay senses motor current and disconnects the motor from the line if the current is excessive for a certain length of time.

P

Parallel alternators alternators connected in parallel.
Parallel block a group of components connected in parallel.
Parallel circuits circuits that contain more than one path for current flow.
Parallel conductors conductors connected in parallel.
Parallelogram method a method of determining vector angles.
Peak value the highest value of voltage or current reached by a waveform.
Percent slip the percent of speed difference between the speed of the rotor and the speed of the rotation magnetic field of a motor.
Permanent magnet (PM) motors motors that rely on permanent magnets to supply the magnetism in the field poles instead of series or shunt field windings.
Permanent magnets magnets that do not require an external force to maintain their magnetic properties.
Permeability a measurement of a material's ability to conduct magnetic lines of flux. The standard is air, which has a permeability of 1.
Phase angle shift see phase shift.
Phase rotation the direction of rotation of a magnetic field produced by the voltages of a poly-phase power system.
Phase rotation meter a meter used to determine the rotation of a motor.
Phase shift a change in the phase relationship between two quantities of voltage or current.
Phase voltage the amount of voltage measured across a single phase of a three-phase load or power supply.
Phasor rotation see phase rotation.

Piezoelectricity the production of electricity by applying pressure to a crystal.
Plates metal foil used to store electrons in a capacitor.
Polarized capacitors capacitors that are sensitive to polarity.
Pole pieces the parts of a DC machine that form magnetic poles.
Positive an electric charge developed when an object contains more protons than electrons.
Positive-sequence harmonic causes the magnetic field of a polyphase motor to rotate in the same direction as the fundamental frequency.
Pot common short term for potentiometer.
Potential difference the difference between the positive and negative charges developed by a power source. Voltage is often described as potential difference.
Potentiometer a variable resistor with a sliding contact that is used as a voltage divider.
Power measured in watts, it describes the amount of electrical energy inverted to some other form.
Power factor (PF) a comparison of the true power (watts) to the apparent power (volt-amperes) in an AC circuit.
Power factor correction adding capacitance or inductance to a circuit to cause the current and voltage to become more in phase with each other.
Precipitators devices that employ static electric charges to help capture and remove small particles from the air.
Primary cell a voltaic cell that cannot be recharged.
Primary winding the winding of a transformer to which the power line is connected.
Printed-circuit motor a common name for a ServoDisc motor.
Proton one of the three principal parts of an atom. The proton has a positive charge.
Pulse-width modulation a method of controlling the output voltage of DC power supply.
Pythagoras a Greek mathematician.
Pythagorean theorem a theorem that states that the square of the sides of a right triangle are equal to the square of the hypotenuse.

Q

Quadrature power same as VARs or wattless power.
Quality (Q) of a component. Can be calculated by comparing the resistance to the reactance.

R

R-L time constant a formula that can be employed to determine the rate at which current will rise or fall in an inductive circuit.
RC time constant the time necessary to charge or discharge a capacitor when it is connected in series with resistance.
Reactance the opposition to current flow in an AC circuit offered by pure inductance or pure capacitance.
Reactive power (VARs) also called wattless power. Calculated by multiplying reactive voltage by reactive current. Can be produced by inductive or capacitive loads.
Reduce to make smaller or more compact.
Reluctance resistance to magnetism.
Repulsion motor a type of single-phase motor that uses repulsion of magnetism to operate.
Repulsion the force that causes one object to push another object away from itself.
Residual magnetism the amount of magnetism left in an object after the magnetizing force has been removed.
Resistance the opposition to current flow in an AC or DC circuit.
Resistive loads loads that contain pure resistance.
Resonance the point at which circuit inductance and capacitance become equal.
Revolving-armature-type alternator an alternator that uses stationary magnetic fields and the armature containing the field winding turns. Power is carried to the load by sliprings and brushes.
Revolving-field-type alternator an alternator that rotates a magnetic field inside stationary windings.
Rheostat a variable resistor.
Right angle an angle of 90 degrees.
Right triangle a triangle that contains a right angle.
Ripple the pulsations of current or voltage that occur when an AC voltage is rectified to DC.
RMS (root-mean-square) value the value of AC voltage that will produce as much power when connected across a resistor as a like amount of DC voltage.
Rotating magnetic field the revolving magnetic field inside a polyphase motor.
Rotor the rotating member of an AC machine.
Rotor frequency the frequency of the voltage induced in the rotor of an induction motor.
Run winding the winding of a split-phase motor that remains energized during operation.

S

Safety data sheet (SDS) written material supplied by various manufacturers that lists safety information about a particular product.
Saturation the maximum amount of magnetic flux a material can hold.
Scaffolds portable structures that can be assembled to provide a high working platform.
Secondary cell a voltaic cell that can be recharged.
Secondary winding the winding of a transformer to which a load is connected.
Selenium an element that exhibits photosensitive properties. Its resistance changes inversely with illumination.
Semiconductor a material that contains four valence electrons and is used in the production of solid-state devices.

Series circuit a circuit that contains only one path for current flow.

Series field diverter a resistor connected in parallel with the series field to divert some of the current away from the field and cause it to produce less magnetic field strength.

Series field windings low-resistance windings connected in series with the armature.

Series generator a generator that contains only a series field winding.

Series motor a motor that contains only a series field winding.

ServoDisc motor a motor that employs a printed circuit board as the armature.

Servomotor motor used for precise position control.

Shaded-pole induction motor a type of single-phase motor that uses shading coils to produce a rotating magnetic field.

Shaded-pole motor an AC induction motor that develops a rotating magnetic field by shading part of the stator windings with a shading coil.

Shading coil a large copper wire or band connected around part of a magnetic pole piece to oppose a change of magnetic flux.

Short circuit an electric circuit that contains very low resistance to limit the flow of current due to a connection that bypasses the circuit load.

Shunt motor a motor that contains only a shunt field winding.

Simple parallel circuit a circuit that contains only elements connected in parallel.

Sine the angular relationship of the opposite side of a right triangle and the hypotenuse.

Sine wave a waveform that is proportional to the sine of the angle of ratio and the peak value of voltage or current.

Single-phase loads loads that operate on only one phase of an AC power system.

Single-phasing a condition that occurs when one line of a three-phase power system is lost.

Skin effect the tendency of electrons to move toward the outside of a conductor in an AC circuit.

Sliprings circular bands of metal placed on the rotating part of a machine. Carbon brushes riding in contact with the sliprings provide connection to the external circuit.

SOH—CAH—TOA is a pneumonic device used to remember right-triangle trigonometry functions.

Specific gravity the ratio of the volume and weight of a substance to an equal volume and weight of water. Water has a specific gravity of 1.

Speed rate of change.

Speed regulation a comparison of a motor's speed between no load and full load.

Split-phase motors a type of single-phase motor that uses resistance or capacitance to cause a shift in the phase of the current in the run winding and the current in the start winding. The three primary types of split-phase motors are resistance-start induction-run, capacitor-start induction-run, and capacitor-start capacitor-run motors.

Squirrel-cage rotor the rotor of an AC induction motor constructed by connecting metal bars together at each end.

Star connection another name for WYE connection.

Start winding the winding of a split-phase motor that is connected out of phase with the run winding to produce a rotating magnetic field.

Static stationary or not moving.

Stator the stationary winding of an AC motor.

Step-down transformer a transformer that produces a lower voltage at its secondary than is applied to its primary.

Step-up transformer a transformer that produces a higher voltage at its secondary than is applied to its primary.

Stepping motors a type of motor that moves a predetermined amount of angular rotation each time it receives a pulse of power.

Strength of magnetic field flux density.

Sulfuric acid a highly corrosive compound used in many processes (H_2SO_4).

Surface area the amount of surface of an object, generally expressed in square inches, square feet, etc.

Synchronous condenser a term used to describe a synchronous motor that is used to provide power factor correction.

Synchronous motors motors that operate at the speed of the rotating magnetic field.

Synchronous speed the speed of the rotating magnetic field of an AC induction motor.

Synchroscope an instrument used to determine the phase angle difference between the voltages of two alternators.

T

Tagging a method of identifying conductors or components.

Tangent the angular relationship of the opposite side of a right triangle and the adjacent side.

Tank circuit an inductor and capacitor connected in parallel.

Tape-wound core a core made by winding a continuous length of magnetic metal into a spiral.

Termination temperature The temperature of an electrical termination.

Thermal fuse A fuse designed to open due to excessive heat.

Thermal overload relay An overload relay that senses motor current by connecting a heater in series with the motor.

Thermal runaway is a phenomenon in which the lithium-ion cell enters an uncontrollable, self-heating state, often resulting in a fire or violent ejection of cell gas.

Thermocouple a device for producing a voltage due to a difference of temperature.

Three-phase bank single-phase transformers connected in such a way as to operate on a three-phase system.

Three-phase VARs reactive power in a three-phase circuit.

Three-phase watts the true power in a three-phase system. The formula used is dependent on whether line or phase values of voltage and current are to be employed.

Thundercloud a cloud that contains charged particles.

Tolerance the amount of variation from a standard value that is permissible.

Toroid core a doughnut-shaped electromagnet.

Torque the turning force developed by a motor.

Torque motor motor designed to operate with the rotor stalled without damage to the motor.

Total current (I_T) total amount of current flow in a circuit.

Total voltage the voltage applied to the circuit.

Trace the current path tracing the path of current through a circuit.

Transformer an electric device that changes one value of AC voltage into another value of AC voltage.

Triangle wave a linear wave form.

Triplen harmonics the odd multiples of the third harmonic.

True power generally expressed as watts. Describes the amount of electrical energy converted into some other form.

Turns of wire the number of turns of wire in a coil.

Turns ratio the ratio of the number of turns in the primary winding of a transformer compared to the turns of wire in the secondary winding.

Two-phase a power system that contains two phases 90° out-of-phase with each other.

U

Unidirectional moving in only one direction.

Unity 100%, or 1.

Universal motor a type of motor that can be operated on AC or DC power.

Useful static charges static electric charges that are intentionally placed on objects.

V

Valence electrons electrons located in the outer orbit of an atom.

Variable capacitors capacitors whose value can be changed.

Variable resistor a resistor whose resistance value can be varied between its minimum and maximum values.

Vector a line having a specific length and direction.

Vector addition a method of adding angles.

Volt (V) a measure of electromotive force. The potential necessary to cause 1 coulomb to produce 1 joule of work.

Voltage divider a circuit that generally is constructed by connecting resistors in series to produce desired voltage drops across the resistors.

Voltage drop the amount of voltage required to cause an amount of current to flow through a certain resistance.

Voltage drop across the capacitor the amount of voltage measured across a capacitor.

Voltage drop across the resistor the amount of voltage measured across a resistor.

Voltage polarity the polarity of the voltage at a particular point in a circuit. Generally of concern in direct current circuits.

Voltage rating a rating that indicates the amount of voltage that can be safely connected to a device.

Voltage regulator a device or circuit that maintains a constant value of voltage.

Voltage spike a momentary increase of voltage lasting 2 nanoseconds or less.

Voltage surge A momentary increase of voltage lasting 50 microseconds or less.

Voltaic cell a device that converts chemical energy into electric energy.

Voltaic pile a group of voltaic cells connected together to form a battery.

Voltmeter an instrument used to measure a level of voltage.

Volts-per-turn ratio determined by dividing the voltage applied to the primary of a transformer by the number of turns of wire in the primary.

W

Warren motor a type of single-phase synchronous motor.

Watt (W) a measure of true power.

Wattmeter a meter used to measure true power or watts.

Wave-wound armatures armatures that have their windings connected in such a way as to form a series connection.

Weber (Wb) a measure of magnetic lines of flux in the MKS (Meter-Kilogram-Second) system. 1 Wb = 100 million lines of flux.

Wet locations areas that are subject to saturation with water or other liquids.

Wire-wound resistors resistors that are constructed by winding resistive wire around a nonconductive material.

Wound-rotor motor a three-phase motor containing a rotor with windings and sliprings. This rotor permits control of rotor current by connecting external resistance in series with the rotor winding.

WYE connection a connection of three components made in such a manner that one end of each component is connected. This connection is generally used to connect devices to a three-phase power system.

WYE–delta a three-phase transformer connection that has its primary connected in WYE and a secondary connected in delta.

X

X-rays beams of high-energy protons.

Index

A

B

C

J

K

L

P

Q

R

T

X

Y

Z